THE
MATHEMATICAL PAPERS OF
ISAAC NEWTON
VOLUME IV
1674-1684

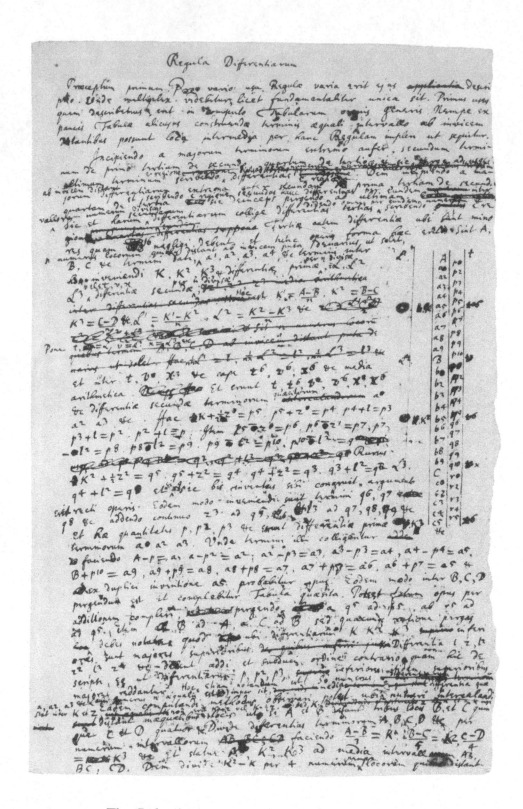

The 'Rule of differences' in interpolation (**1**, 1, §3).

THE
MATHEMATICAL PAPERS OF
ISAAC NEWTON

VOLUME IV

1674-1684

EDITED BY
D. T. WHITESIDE

WITH THE ASSISTANCE IN PUBLICATION OF
M. A. HOSKIN AND A. PRAG

CAMBRIDGE
AT THE UNIVERSITY PRESS
1971

CAMBRIDGE UNIVERSITY PRESS
Cambridge, New York, Melbourne, Madrid, Cape Town, Singapore, São Paulo

Cambridge University Press
The Edinburgh Building, Cambridge CB2 8RU, UK

Published in the United States of America by Cambridge University Press, New York

www.cambridge.org
Information on this title: www.cambridge.org/9780521077408

First published 1971
This digitally printed version 2008

A catalogue record for this publication is available from the British Library

Library of Congress Catalogue Card Number: 65–11203

ISBN 978-0-521-07740-8 hardback
ISBN 978-0-521-04583-4 paperback
ISBN 978-0-521-72054-0 paperback set (8 volumes)

PREFACE

The present volume, fourth in sequence in the present edition, contains a number of important minor papers which display the wide spectrum of Newton's mathematical interests during the period 1674–84: interpolation by finite differences, number theory, trigonometry, pure and analytical geometry of conics and cubics, geometrical calculus and infinite series. A few have been previously published at various times during the last century (all in more or less incomplete form and for the most part inadequately transcribed) but the greater part here make their first appearance in print. While not holding out the wide, panoramic prospects of Newton's more sustained mathematical treatises, the present pieces have each their own delights and prominences, and they unquestionably fill a not inconsiderable gap in our understanding of the totality of development of his mathematical expertise. More generally, they serve—dimly and fitfully—to illuminate a period in Newton's life (particularly the eight years from 1676) of which very little is known. The reader will set his own estimate on their individual value and significance.

For permitting reproduction of manuscripts in their custody I am uniquely indebted to the Librarian and Syndics of the University Library, Cambridge, and to a private owner, source of many items in previous volumes, who continues to desire to remain anonymous. In the case of the several institutions who have, over the years, promoted my researches in various ways I would again make special mention of the courtesy and helpfulness of the staff of Cambridge University Library, notably that of the Anderson Room where I have spent many long hours gathering and collating material. For financial assistance during the preparation of this volume I continue to be indebted to the Sloan Foundation, the Leverhulme Trust and the Master and Fellows of Trinity College, Cambridge. Freedom from financial worry is a very necessary concomitant of the undistracted scholarship my present undertaking demands, and I am grateful to be so generously supported. Once more, too, a very special word for my oldest patron, Sir Harold Hartley: applied to him, the adage of being ninety years young remains unhackneyed and I express my appreciation of the warmth of his continuing interest.

To my assistants—or rather mentors—an unblunted acknowledgement of their help. Mr A. Prag has again undertaken the tedium of preparing a name-index but has otherwise played throughout an active rôle, one which by its very nature can leave no obvious trace in the finished book, in eliminating a multitude of errors and false emphases in the text. To Dr M. A. Hoskin, particularly involved in ensuring the absolute accuracy of the reproduced text-figures, my continuing thanks for his omnipresent help.

To the Syndics of the Cambridge University Press, finally, my deep appreciation yet again of the untiring efforts of all the uncredited subeditors, draughtsmen, printers and production assistants who have cooperated in producing so magnificent an example of the printer's art. In a very real sense this is their volume.

D.T.W.
1 May 1970

EDITORIAL NOTE

Little needs to be said editorially of a volume which closely follows its predecessors in style. For explanation of the principal conventions introduced we again refer the reader to pages x–xiv of the first volume. As ever, we aim at the ideal of being as strictly faithful to the autograph manuscripts reproduced as the confines of the printed page allow, taking as few liberties as possible (silent or otherwise) with their idiosyncracies and contracted forms. All insertion in Newton's text is made within square brackets, and is introduced only to smooth out minor illogicalities and eliminate trivial slips of his pen. Within the demands of readability and the restrictions of modern idiom we have kept the English translations (set, in the case of principal texts, on right-hand pages facing the Latin original) deliberately literal, conceiving our purpose there to be to help the reader understand Newton's own words rather than to prepare a polished paraphrase existing in its own right. In both Latin and English texts two thick vertical bars in the margin alongside denote that the passage so singled out is cancelled in the manuscript: once more we urge the reader not to confuse this cancelling symbol with the thin vertical strokes '‖' we have elsewhere used to mark a Newtonian page-division. Certain special, non-standard notations are explained as they occur. In **3**, **1**, §§1/2 we employ the convention 'fl (A)' to denote the English equivalent of Newton's 'fl: A' or simply 'fl A' (see III: 330, note (8)); and in **3**, **2**, §1 again (compare III: 72, note (86)) introduce the anachronism of 'translating' Newton's literal fluxions p, q and r (of fluent variables x, y and z) by their dotted equals $\dot{x}$, $\dot{y}$ and $\dot{z}$. For backwards reference we use, as convenience dictates, either the abbreviated form 'I, **1**, 3, §3.2' (by which understand '[Volume] I, [Part] **1**, [Section] 3, [Subsection] 3, [Division] 2') or, more shortly, 'I: 104–11' (meaning the same passage in '[Volume] I: [pages] 104–11'). Once more this convention should not be confused with the special usage employed in the concluding appendix, where for instance '[4: 186]' (see page 672, note (54)) refers to page 186 of the fourth volume (Cambridge, 1967) of the Royal Society's edition of *The Correspondence of Isaac Newton*.

GENERAL INTRODUCTION

In this fourth instalment of our edition of Newton's extant mathematical papers we gather together a number of minor pieces and related fragments, calculations and predrafts, which in our judgement were composed during the decade from mid-1674. (The only extended treatise—the two states of his self-styled 'lectures' on algebraic techniques and their detailed application to various problems of arithmetic and geometry—which came from his ever-fertile pen during this long and superficially unproductive middle period of his Cambridge years we reserve for the next volume.) The largely jejune, unforthcoming character of this somewhat ill-matched miscellany of papers (on finite differences, indeterminate analysis, spherical trigonometry and the pure and Cartesian geometry of conics and cubics) is doubtless symptomatic of Newton's reluctance at this time to devote more than a minimal amount of his creative effort—otherwise given over to his still inadequately explored contemporary physico-chemical, alchemical and religious studies—to mathematical research. Only at the very end of the decade, just a few weeks before Edmond Halley's famous visit to Cambridge in the summer of 1684 radically altered the even tenor of his placid, routine existence, do his writings again show excitement in the theory of fluxions and infinite series. Except, perhaps, for the opening papers on interpolation by means of finite differences[1] none but the concluding 'Matheseos Universalis Specimina' (which embodies Newton's rejuvenated interest, provoked by his receiving in June 1684 a presentation copy of David Gregory's *Exercitatio Geometrica*[2]) would seem to have been externally motivated unless maybe by a chance reading of some book in the college, university and private libraries accessible to him in Cambridge. There is, in inevitable sequel, a near universal dearth of background information regarding the history and sequence of their composition: the rare insights which contemporary documentation does allow we have heartlessly squeezed bone-dry in moulding the numerous, often highly tentative conjectures which we have dared, in lieu of firmer evidence, to present at appropriate places in our introductions and footnotes. The reader will be the best judge of the relevance and success of our many guesses and assessments.

To round out this somewhat arid commentary we may here be permitted to review the little that is known—that mostly trivial and unconnected—concerning Newton's personal activities and enthusiasms at this time. Certain

---

(1) Initially inspired, it would seem, by a plea early in 1675 from the obscure London 'philo-accomptant' John Smith and subsequently reinforced the next year when he came to draft his *epistola posterior* for Leibniz; see pages 6–7, 8 note (1) and 22 note (1) below.

(2) See pages 413–16; and compare note (34) below.

details of the rare participations in university affairs allowed him by the statutes of his Lucasian professorship[3] have been rescued by Edleston from the oblivion of the Cambridge archives: his voting slips backing candidates—usually successfully so—for various college and university posts and for membership of Parliament reveal a modicum of dutiful interest on his part in the conduct of public affairs, while his several gifts of books and money to the new library at Trinity are evidence at once of his devotion to the betterment of the academic facilities of his college and of the continuing bookishness of his mind.[4] As Aston, Wickins and others of the small circle of his youthful friends at Cambridge continued to depart, some reluctantly, some to obtain preferment,[5] Newton clearly grew increasingly introverted in his outlook and he acquired few new intimates to replace his old-standing acquaintances. In the middle 1680's, if we may believe his amanuensis, his social circle had shrunk to embrace only John Ellis, Fellow—and later Master—of Caius College; John Laughton, afterwards to become University Librarian; John Vigani, the Professor of Chemistry; and his fellow Lincolnshire countryman and college senior, Humphrey Babington.[6] Here, no doubt, the traditional anecdotes of his withdrawal from social life, his lack of concern for his appearance and his forgetfulness of meal-times[7] begin to approximate the historical reality, but

(3) See III: xxvii, note (17).

(4) Joseph Edleston, *Correspondence of Sir Isaac Newton and Professor Cotes* (London, 1850): xxvi–xxix, *passim*. Presentations by Newton to the Wren library of such works as Nehemiah Grew's newly appeared *Musæum Regalis Societatis* are recorded in 1675, 1679 and 1680; in 1676 he subscribed £40 towards the new library and loaned a further £100 four years later.

(5) Francis Aston, refused a royal dispensation to continue his Fellowship at Trinity without taking holy orders (see III: xxiv, note (10)), left early in 1675 to enjoy the pleasurable refinements of London society; John Wickins accepted the comfortable living of Stoke Edith, near Monmouth, early in 1684. Newton evidently keenly missed Wickins' service as part-time assistant, for he lost no time in writing to Dr Walker, master of his old grammar school at Grantham, in quest of a permanent amanuensis educated enough to be able to turn Newton's heavily cancelled and overwritten drafts into an accurate unblemished transcript. Walker's choice, his young namesake Humphrey Newton, arrived in Cambridge about the spring of 1684 ('In yᵉ last year of K. Charles 2ᵈ' he told Conduitt in January 1728 [King's College. Keynes MS 135.1]) and was at once set to work copying several mathematical pieces, soon abandoned for the finished states of the 'De Motu Corporum' and *Principia Mathematica* ('wᶜʰ stupendous Work, by his Order, I copied out, before it went to yᵉ Press' [*ibid.*]).

(6) In his letter to Conduitt on 17 January 1727/8 (see the previous note) Humphrey Newton included among 'others of his Acquaintance' at this time 'Dʳ Babington of Trinity... Mʳ Ellis of Keys, Mʳ Loughan of Trinity, & Mʳ Vigani, a Chymist', adding that 'Mʳ Laughton, who was then yᵉ Library keeper of Trin. Coll. resorted much to his Chamber'.

(7) Many of these, in fact, derive from Humphrey Newton's reminiscences to Conduitt on 17 January and 14 February 1727/8 (King's College. Keynes MS 135.1/2, published in a much modernized transcript in David Brewster's *Memoirs of the Life, Writings, and Discoveries of Sir Isaac Newton*, 2, Edinburgh, 1855: 91–8). Wickins' equivalent account, reported by his son Nicholas to Conduitt on 16 January of the same year (Keynes MS 137, reproduced in Brewster's *Memoirs*, 2: 88–90), tells of 'Sʳ Isaac's forgetfullness of his food, when intent upon

they should not be overstressed. Newton was never the conventional cardboard figure of a sour, dour Puritan, but retained throughout an enjoyment, however rarely manifested, of social contact and conversation, taking in the company of his 'few Visitors...much Delight and Pleasure at an Evening, when they come to wait upon Him'[8] and nonchalantly closing a letter to William Briggs in September 1682 with the remark that 'You have now ye summ of what I can think of set down in a tumultuary way as I could get time from my Sturbridge Fair friends'.[9] In a revealing phrase his elder Lincolnshire compatriot, the Platonist philosopher Henry More (with whom Newton many times discussed his 'Apocalyptical Notions'), wrote in 1680 of 'his countenance which is ordinarily melancholy and thoughtfull' but on occasion, when his interest was roused, 'mighty lightsome and chearfull, and...in a maner transported'.[10]

It remains nonetheless true that Newton came gradually during the middle 1670's to sever the many contacts he had earlier made with the outside world. The list of his college exits and redits[11] during the ensuing decade of muted activity records brief excursions from Cambridge, mostly of only a few days at a time, in each of the years from 1674 to 1683, but we have little documented knowledge of the purpose or destination of his journeys: we would guess that many of these were annual visits home to Lincolnshire to stay with his mother and (after 1679) her family rather than to enjoy the various advantages and delights which London had to offer. Between early February and the middle of March 1675 Newton was, we know, in the capital successfully stage-managing the royal dispensation which confirmed his statutory right, as Lucasian

his Studies; And of his Rising in a pleasant manner wth ye Satisfaction of having found out some Proposition; without any concern for, or seeming Want of his Nights Sleep, wch he was sensible he had lost thereby'.

(8) Humphrey Newton to Conduitt, 17 January 1727/8 (see notes (5) and (7)).

(9) Newton to Briggs, 12 September 1682 (*Correspondence of Isaac Newton*, **2**, Cambridge, 1960: 385).

(10) Henry More to John Sharp, 16 August 1680 (British Museum. Sloane MS 4276: 41, reproduced in M. H. Nicolson, *Conway Letters. The Correspondence of Anne, Viscountess Conway, Henry More, and their Friends, 1642–84* (London, 1930): 478–9). More went on to state that 'Mr Newton has a singular Genius to Mathematicks, and I take him to be a good serious man'. F. E. Manuel in his *A Portrait of Isaac Newton* (Cambridge, Massachusetts, 1968: 108) avers, largely on the evidence of a funeral ring bequeathed by More to Newton, that '[t]heir pleasant relations continued until More's death in 1687' but this is perhaps overdone. At the end of his will (Nicolson, *Conway Letters*: 482–3) More ordained: 'Item I give funerall Rings to the Master and all the Fellows of Christ Colledge as alsoe to Mr Newton Professor of Mathematicks to my Cousin Pigots...Doctor Sharpe...Doctor Fowler Doctor John More Doctor Davies.... And if I have left out any whom it may seem fit to have bequeathed funerall Rings to I leave it to the discretion of my Executor to make a Supply not caring to burthen my mind with such Curiosities...'. The clear inference is that those to whom More bequeathed rings were acquaintances not close to his heart.

(11) Edleston, *Correspondence* (note (4)): lxxxv.

professor, to retain his college fellowship without taking holy orders;[12] he then took the opportunity to be formally admitted to the Royal Society (on 18 February) and a month later (on 18 March) to repeat, with Hooke's scarcely willing assistance, his 'spectrum experiment' before the assembled Fellows.[13] Four years later his extended absences in 1679 from Cambridge (between 15 and 24 May, again between early June and 19 July, and a last long leave between 28 July and 27 November) were to attend his mother on her deathbed and subsequently no doubt to settle her estate, of which he was both executor and principal beneficiary.[14] Of the depth and extent of any emotional trauma which the loss of a mother for whom he had at the least an open affection we have no exact knowledge and it seems a little vain to guess at its future impact on the nature and quality of his researches and in moderating his puritanical morality.[15] At one point only, replying to Hooke the day after his last return in late 1679 to excuse himself from a 'Philosophical correspondence [in which] I cannot but acknowledge my self every way... tempted to concurr', did he hint at the sorrow which must have possessed his mind, justifying himself with the remark that

---

(12) See III: xxiv, note (10). We ought there to have said that on becoming Master of Trinity in March 1673 Isaac Barrow sought vigorously to enforce the college statute (confirmed in June 1650) which decreed that 'socii qui Magistri artium sunt' should, 'post septem annos in eo gradu plene confectos', be ordained 'Presbyteri' upon penalty of automatically forfeiting their fellowship. (Compare William Whewell, 'Barrow and his Academical Times' [*The Theological Works of Isaac Barrow, D.D.*, **9**, Cambridge, 1859: i–lv]: xi.) As we have seen (II: xi) Newton had been elected a major Fellow of Trinity in March 1668.

(13) Edleston, *Correspondence* (note (4)): xlix, note (42).

(14) Hannah Smith's will, proved at Lincoln on 11 June 1679, is reproduced by C. W. Foster in his 'Sir Isaac Newton's Family' (*Reports and Papers of the Archaeological and Architectural Societies* [*The Architectural & Archaeological Society of the County of Lincoln*], **39**, 1928: 1–62): 50–3. Apart from receiving various parcels of land at Buckminster and the 'Ling house' and its estate at Woolsthorpe, Newton was given 'the whole use and profit of all legacies... until such time as they shall become due'. Since by the terms of Hannah's marriage settlement with Barnabas Smith more than thirty years before Newton already possessed the manor house in Woolsthorpe and certain dues, he was henceforth—over and above his comfortable professorial stipend and fellow's dividends—a rich man.

(15) Several recent writers—notably F. E. Manuel—have made Newton's (wholly undocumented) lonely love as a child for his mother the key to an understanding of the idiosyncracies and abnormalities of his adult psyche, and not least his repression and hypothesized latent homosexuality. In the dearth of exact information regarding Newton's early development it is all too easy to pass to the risible extreme: in one not too untypically purple passage of his attempt at a psychological *Portrait* (see note (10)) Manuel queries, for instance, whether in silently condoning 'the [unproven] act of fornication between his friend Halifax and his niece [Catherine Barton]' in the early 1700's Newton was in fact 'vicariously having carnal intercourse with his [dead] mother', so consummating a 'child-fantasy of desire' (page 262). We prefer to be bound more closely, if less provocatively, by substantiated historical fact.

I am at present unfurnished wth matter answerable to your expectations. For I have been this last half year in Lincolnshire cumbred wth concerns amongst my relations till yesterday when I returned hither; so y^t I have had no time to entertein Philosophicall meditations or so much as to study or mind any thing els but Countrey affairs. And before that, I had for some years past been endeavouring to bend my self from Philosophy to other studies in so much y^t I have long grutched the time spent in y^t study unless it be perhaps at idle hours sometimes for a diversion: wch makes me almost wholy unacquainted wth what Philosophers at London or abroad have of late been imployed about....And having thus shook hands wth Philosophy, & being also at present taken of wth other business, I hope it will not be interpreted out of any unkindness to you or ye R. Society that I am backward in engaging my self in these matters, though formerly I must acknowledge I was moved by other reasons, to decline as much as Mr Oldenburg's importunity & ways to engage me in disputes would permit, all correspondence with him about them.[16]

Through the Royal Society's secretary, Henry Oldenburg, Newton had in fact continued during the middle 1670's to conduct a vigorous correspondence on optics, culminating in his deposit in the Society's archives in December 1675 of a 'Hypothesis explaining the Properties of Light discussed of in my severall Papers [in 1672]' and a complementary 'Discourse of Observations',[17] but dwindling rapidly thereafter in the face of hostile criticism on the part of Hooke and Lucas, and ending in an impatient note to Aubrey in mid-1678 requesting not to be pestered further in these matters.[18] Earlier, after a nine months' gap following a somewhat curt letter to the London 'philomath' John Collins which in September 1673 terminated further discussion on his part of the merits and defects of his 'Catadioptricall' telescope,[19] Newton had warmly responded in June 1674 to a request from Collins to review Robert Anderson's recently appeared *The Genuine Use and Effects of the Gunne* by passing

---

(16) Newton to Robert Hooke, 28 November 1679 (*Correspondence*, **2**: 300–1).

(17) These were communicated to Oldenburg as enclosures in Newton's letter of 7 December as a counterblast to Hooke's severe critique of the optical papers he had sent to be read at the Royal Society in 1672; see R. S. Westfall, 'Newton's Reply to Hooke and the Theory of Colors' (*Isis*, **54**, 1963: 82–96, especially 88–91). The former paper is reproduced in Newton's *Correspondence*, **1**, 1959: 362–86; the latter (whose autograph draft is ULC. Add. 3970.3: 518^r–528^v, revised on 501^r–517^r) was first published by Thomas Birch in his *History of the Royal Society of London*, **3** (London, 1757): 272–305 [=(ed. I. B. Cohen) *Isaac Newton's Papers & Letters on Natural Philosophy*, Cambridge, 1958: 202–35; compare also *Correspondence*, **1**: 390–2].

(18) 'I understand you have a letter from Mr Lucas for me. Pray forbear to send me anything more of that nature' (*Correspondence*, **2**: 269); compare R. S. Westfall, 'Newton Defends his First Publication: The Newton–Lucas Correspondence' (*Isis*, **57**, 1966: 299–314). Aubrey, who sought briefly to replace Oldenburg as Newton's intermediary in 1678, may not have received the exact phrasing of this autograph draft but doubtless the rebuff was communicated in an equally forceful form.

(19) 'I shall not trouble you any further wth discourses about ye Perspective' (*Correspondence*, **1**: 307).

on his own thoughts on resisted projectile motion, though he warily added that 'If you should have occasion to speak of this to yᵉ Author, I desire you would not mention me becaus I have no mind to concern my self further about it'.[20] Over the next two years, while never ceasing to insist on his essential lack of interest in speculations which for him had grown dry, and ever unwilling to publish his early papers on the subject,[21] Newton did allow himself to be drawn out of his shell several times on mathematical queries (speedily transmitted which were by Collins and Oldenburg) raised by Michael Dary and John Smith in London, and above all in answer to Leibniz' pleas from Paris to be given details of his current mathematical research.[22] But the continued hostility evinced by Hooke and Lucas towards his optical theories gradually eroded even this hesitant willingness to share and discuss his mathematical ideas. Already on 5 September 1676 he could write to Collins in London that 'I doubt I shal put you to too much trouble to transcribe Mʳ Leibnitz's whole letter [of 17/27 August] if it be so long, & therefore I shall desire you to send me only a general account of it, wᵗʰ such passages as you think may concern me...'.[23] Two months later this reluctance to be concerned with details had hardened into a resolution, communicated to Collins on 8 November, to veto all future proposed schemes for publishing his mathematical and scientific papers during his lifetime;[24] for, as he wrote to Oldenburg ten days afterward,

I see I have made my self a slave to Philosophy, but if I get free of [this] buisiness I will resolutely bid adew to it eternally, excepting what I do for my privat satisfaction or

---

(20) *Correspondence*, 1: 309. The technical aspects of Newton's 1674 projectile curve will be discussed in the sixth volume.

(21) Writing to James Gregory on 1 May and 29 June 1675 Collins observed that 'Mʳ Newton being lately here [in February and March preceding] did not seeme to have any intent to publish anything as yet' but 'gives in his Lectures yearly to the publick Library [see III: xviii], and prosecutes his Chimicall studies and Experiments' (*Correspondence*, 1: 341, 345), adding on 19 October that 'Mʳ Newton...I have not writt to or seene this 11 or 12 Months [!]...not troubling him as being intent upon Chimicall studies and practises, and both he and Dʳ Barrow beginning to thinke Mathˡˡ Speculations to grow at least dry if not somewhat barren' (from Collins' draft in private possession; compare *Correspondence*, 1: 356). A year later, doubtless having his 'De Analysi' in mind (compare II: 168), in a postscript to his letter of 26 October 1676 he firmly requested Oldenburg 'Pray let none of my mathematical papers be printed wᵗʰout my special licence' (*Correspondence*, 2: 163).

(22) See the concluding appendix to this volume (pages 657–74 below).

(23) *Correspondence*, 2: 96; see page 671, note (52).

(24) 'You seem to desire yᵗ I would publish my method [of infinite series and fluxions] & I look upon your advice as an act of singular friendship, being I beleive censured by divers for my scattered letters in yᵉ *Transactions* about such things as no body els would have let come out wᵗʰout a substantial discours. I could wish I could retract what has been done, but by that, I have learnt what's to my convenience, wᶜʰ is to let what I write ly by till I am out of yᵉ way' (*Correspondence*, 2: 179).

leave to come out after me. For I see a man must either resolve to put out nothing new or to become a slave to defend it.[25]

Newton's transmission to Oldenburg on 24 October of his *epistola posterior* for Leibniz marked the temporary end of his willingness to participate in external scientific activity of any kind.

To be sure, the near-simultaneous deaths the following year, not only of Oldenburg (in September 1677) but of his college senior—and, since 1673, Trinity's Master—Isaac Barrow (the previous May) cut away most of the footholds Newton had succeeded in fashioning for himself in London scientific society. Even his contact with his London intelligencer John Collins, now in increasingly straitened financial circumstances, was seemingly severed after October 1678 when he addressed a last request to Newton: 'Vouchsafe to consider this most usefull Probleme [of drawing a touch line to the rumb Spirall] and impart the same as an Appendage to your Letters to Leibnitz which we [Collins and John Wallis?] hope you will consent to have printed in English.'[26] Oldenburg's sudden demise left an uncomfortable vacuum at the

(25) *Correspondence*, 2: 182–3. Probably it was at this time (late autumn 1676) that he abandoned the collected edition—already partially set in type—of his 1672 optical correspondence. See I. B. Cohen, 'Versions of Isaac Newton's first published paper', *Archives Internationales d'Histoire des Sciences*, 11, 1958: 357–75; and A. R. Hall, 'Newton's First Book (I)', *ibid.*, 13, 1960: 39–54, especially 49–53. A corrected original of the figure on page 14 of the single sheet C[1ʳ–4ᵛ = pages 9–16] of this abortive printing is preserved in the Cambridge Portsmouth papers, now pasted in at the end of the earlier, autograph text of Newton's Lucasian 'lectures' on optics (ULC. Add. 4002: 128ᵃ; the figures on 128ᵇ and 128ᶜ are no doubt equivalent drawings for corresponding diagrams on later sheets in the abortive printing).

(26) *Correspondence*, 2: 286; see page 658, note (6) below. As Collins wrote to Thomas Baker on 23 May 1677, 'The truth of it is, it hath been my misfortune to be concerned in public employments...wherein I have not been paid, and have great arrears due to me, for want whereof I am almost ruined; and having a numerous family to maintain, to wit a wife and seven small children, I am forced to undertake such occasional business as offers, and by consequence to neglect a correspondence with the learned which though unworthy I much covet' (S. P. Rigaud, *Correspondence of Scientific Men of the Seventeenth Century*, 2 (Oxford, 1841): 23). The Council minutes of the Royal Society record that on 21 January 1679/80 'Dr. *Croune* proposing from Mr. *Collins*, that the latter was ready to print two volumes of algebra, written by Dr. *Wallis*, Mr. *Baker*, Mr. *Newton* &c provided the society would engage to take off 60 copies after the rate of 1ᵈ.½ a sheet; it was ordered, that Mr. *Collins* should be desired to make his proposal in writing, and that the society would farther consider of encouraging the proposal' (Birch, *History of the Royal Society* (note (17)), 4: 4). Insofar as Newton came into it, this was probably a last dying ember of Collins' intended publication ten years before of Mercator's Latin translation of Kinckhuysen's Dutch *Algebra* with Newton's appended 'improvements' (see II: 277–91). No mention of Newton's name occurs in a further proposal of Collins on 12 July 1682 'for printing a book of algebra in such manner, as by the said proposal appeared' (Birch, *History*, 4: 155) and nothing more is heard of Collins' venture, though an edition of Baker's *Geometrical Key; or The Gate of Equations Unlocked* was published two years afterwards (London, 1684) under the Society's *imprimatur*. The last contact of any kind between Newton and Collins occurred in April 1682 when the latter, as he announced on

heart of scientific activity in England which it was not easy to fill.[27] But though Newton was not alone in suffering a chilly exclusion from the gossip of his scientific contemporaries in London and abroad and the circle of regular correspondents which Oldenburg had artfully and, when he saw need, provocatively coordinated, in the seclusion of a Fenland university, without anyone to share his mathematical excitements and where, as he told Hooke on 28 November 1679, 'I know no body...addicted to making Astron. Observations: & my own shortsightedness & tenderness of health makes me something unfit',[28] Newton's sense of intellectual desolation must have been extreme. In February 1679 we find him taking the unprecedented step of seeking to open a correspondence with Robert Boyle by presenting his 'apprehensions' regarding the physical properties of a conjectured 'æthereal substance capable of contraction & dilatation, strongly elastick, & in a word much like air in all respects, but far more subtile', but what reply (if any) he drew we do not know.[29] The following November, after his mother's death, Newton responded, slowly at first but with a growing interest, to a feeler from Hooke (who had recently taken over the duties of the Royal Society's corresponding secretary) to consider anew the allied topics of free fall at the earth's surface and the motions of the solar planets in their elliptical orbits: a fruitful thrust and parry during the next few weeks provoked also a growing irritation on Newton's part at what he conceived to be Hooke's schoolmasterish tone, and the correspondence terminated in mid-January 1680 after Newton refused

---

16 May following to a governor of Christ's Hospital (probably Flamsteed), received out of the blue 'a Letter from Mr Isaac Newton publick Professor of ye Mathematicks in ye University of Cambridge, desiring me to recommend Mr Edwd Pagett...as a Person to bee admitted Lecturer of Navigation to ye Hosp^{ll} Boys that are to be instructed therein' (*Correspondence*, **2**: 376; compare 373); with Collins' endorsement that Newton's glowing 'account' of Paget's character and attainments 'I must needs judge, according to my mean abilities, to be altogether true' (*ibid.*: 376–7) backing him, Paget was appointed to the post. Collins died the next year on 10 November 1683.

(27) More than two years after Oldenburg's death Flamsteed wrote to Richard Towneley on 13 February 1679/80 that 'Our meetings at the Royal Society want Mr Oldenburg's correspondences and on that account are not so well furnished nor frequented as formerly, but I hope a little time will put us into order and produce some of those curiosities and experiments which lie imprisoned ageless by us' (Royal Astronomical Society, London. MS 243 (Fl). 45).

(28) *Correspondence*, **2**: 301. In alleviation of this dismal picture of scientific activity at Cambridge he added 'Yet it's likely I may sometime this winter when I have more leisure then at present attempt what you propound...'. The unexpected appearance the next winter in the skies over East Anglia of the 1680–1 comet did arouse considerable amateur interest in observing the heavens: Newton himself called upon the willing services of Bainbridge and John Ellis as collaborators in his own sightings (ULC. Add. 4004: 99^v).

(29) *Correspondence*, **2**: 288–95, especially 288–9. The next letter between the two which is preserved, that from Boyle on 19 August 1682 (*ibid.*: 379), refers to a preceding 'obligeing Letter' from Newton accompanied by a 'Role of Papers', now both lost.

to respond further to Hooke's letters.[30] A more qualitative dialogue on the nature of comets and the shape of their paths, which took place with John Flamsteed (the recently appointed Astronomer Royal) during the winter and spring of 1680–1 was equally shortlived.[31] During the next four years Newton's rare letters are largely brisk and non-committal, and even in his untypically elaborate response to William Briggs on his favourite topic of visual perception there is no mistaking the tone of withdrawn diffidence which underlies his opening observation that

> I am of all men grown y[e] most shy of setting pen to paper about any thing that may lead into disputes, yet...I shall set down my suspicions about your Theory,...on this condition, that if I can write but plain enough to make you understand me, I may leave all to your use w[th]out pressing it further on. For I designe not to confute or convince you but only to present & submit my thoughts to your consideration & judgment.[32]

Though Newton at once proceeded to be more than a little critical of Briggs' theory of vision, the shy self-distrust latent in his words is not wholly a polite contrivance.

So continued, we imagine, the even tenor of Newton's middle years at Cambridge, to be seriously disrupted by external event only with the arrival in June 1684 of David Gregory's newly published 'geometrical exercise'. Despite the humble modesty of the letter which accompanied Gregory's *Exercitatio Geometrica* and its plea to be allowed his 'free thoughts and character of this exercitation, which I assure you I will justly value more then that of all the rest of y[e] world',[33] Newton was considerably put out to have priority of publication of a particular case of his binomial theorem and certain allied expansions into infinite series so suddenly snatched away from him and from so unexpected a quarter, and in a spasm of energy he produced in quick succession two short tracts, neither completed, elaborating his own claims to priority of discovery and expounding the derivation and application of more sophisticated infinite series developments. On these, the 'Matheseos Univer-

---

(30) *Correspondence*, **2**: 297–313. We will return to consider certain technical aspects of this exchange of letters in the sixth volume. In a celebrated passage of his later letter to Halley on 14 July 1686 Newton affirmed: 'This is true, that his [Hooke's] Letters occasioned my finding the method of determining [planetary] Figures, w[ch] when I had tried in y[e] Ellipsis, I threw the calculation by being upon other studies & so it rested for about 5 years till upon your request I sought for y[t] paper, & not finding it did it again' (*ibid.*: 444).

(31) *Correspondence*, **2**: 315–20, 336–67. It is well known that Newton was encouraged by Flamsteed to abandon his first thought that comets travel, in the vicinity of the sun at least, in straight lines with a (nearly) uniform speed. Some unpublished researches into cometary orbits based upon this Keplerian supposition will be reproduced in the next volume (v, **1**, **2**, Appendix 3).

(32) *Correspondence*, **2**: 381.

(33) *Correspondence*, **2**: 396.

b

salis Specimina' and the 'De computo serierum',[34] he was—or so, lacking any confirming evidence, we choose to believe—still working the following August when Edmond Halley paid his famous visit to Cambridge to ask him 'what he thought the Curve would be that would be described by the Planets supposing the force of attraction towards the Sun to be reciprocal to the square of their distance from it'.[35] With the all but dead spark of his interest in celestial dynamics thus rekindled and the bitterness of his optical disputes of earlier years temporarily forgotten (if not forgiven) Newton laid aside his mathematical researches and began to draft the first of several versions of a little treatise 'De Motu Corporum' which over the next year and a half was to be developed into the multi-faceted work on mathematical physics which he published to the world in 1687 as his *Philosophiæ Naturalis Principia Mathematica*.

With this magnificent codification of his investigations over many years into the motion (free and resisted) of bodies his mathematical researches during the decade from 1674 cannot, *en masse*, be compared; on the other hand they reveal in large part the continuing originality and fertility of his mind as it ranged over a wide spectrum of topics in number theory, algebra, trigonometry, analytical geometry and calculus, while the papers on finite differences and the pure geometry of conics (partially, indeed, subsumed into his *Principia* in 1685–6) are an index of how far in advance of his contemporaries he could on occasion travel. But let the texts of these researches now speak directly for themselves.

---

(34) Reproduced as **3**, 2, §§1/2 below; compare *Correspondence*, **2**: 400–1.

(35) In the words of the memorandum dictated by Abraham de Moivre to John Conduitt in November 1727 (see I: 5, note (9)). The passage continues: 'Sr I. replied immediately that it would be an Ellipsis. The Doctor [Halley] struck with joy & amazement asked him how he knew it. Why saith he I have calculated it. Whereupon Dr Halley asked him for his calculation without any further delay. Sr Isaac looked among his papers but could not find it, but he promised him to renew it, & then to send it him. Sr Isaac in order to make good his promise fell to work again, but he could not come to that conclusion wch he thought he had before examined with care. However he attempted a new way [that of ULC. Add. 3965.2?] which tho^u longer than the first, brought him again to his former conclusion, then he examined carefully what might be the reason why the calculation he had undertaken before did not prove right, &...That being perceived, he made both his calculations agree together.'

ANALYTICAL TABLE
OF CONTENTS

PART 1

RESEARCHES IN ALGEBRA, NUMBER THEORY
AND TRIGONOMETRY

(*c.* 1675–*c.* 1684)

INTRODUCTION   3

Earlier researches into finite differences: Newton's ignorance of Briggs and Harriot, 4. Possible influence of Mercator (1668) and Leibniz (1673), 5. Newton's primitive difference-scheme for systematic root-extractions (May 1675, for John Smith), 6. Interpolation by means of a general parabola: derived rules for 'subcenturying' tabulations, rejected from the *epistola posterior* to Leibniz (October 1676), 6. The 'Regula differentiarum': general divided differences (late 1676) and subsequent publication of the advancing-differences formulas (in *Principia*, 1687), 7. 'Numerical problems' (in elementary Diophantine analysis, late 1670's): Newton's debt to Viète, Schooten, Wallis and (?) Fermat, 8. Construction of rational points on conics and cubics, 10. Newton's early knowledge of plane and spherical trigonometry (derived from Oughtred, Norwood, Gunter and Ward): his 'Epitome Trigonometriæ' and the revised, augmented 'Compendium' (1683), 11. His intended posthumous edition of St John Hare's 'Trigonometria' with complements of his own (early 1684?), 12. Computation of annuities, 12. A near-anticipation of Cotes' quadratic factorization of the general binomial, 13.

1. APPROACHES TO A GENERAL THEORY OF FINITE DIFFERENCES   14

§1 (ULC. Res. 1894. 3 (3)). Subtabulation of a given numerical list of square, cube and fourth roots. Newton's *ad hoc* difference schemes for subtabulating by tenths and hundredths, 15. His observations on their use, 17. Editorial explanation and justification of Newton's underlying assumptions, 18.

§2 (ULC. Add. 3964. 5: 3ʳ–4ᵛ). A discarded sheet from Newton's *epistola posterior* to Leibniz. His geometrical model for constructing the general formulas of angular section, 22. Their use in tabulating sines and cosines, 24. Construction of a table of logarithms of integers (to base 10) and their ensuing subtabulation, 26. Particular rules for 'subcenturying' equal intervals of argument (third/fourth finite differences supposed constant), 28. The general rule for subtabulating at unequal intervals (stated for constant third differences), 32.

PART 2
RESEARCHES IN PURE AND ANALYTICAL GEOMETRY
(*c.* 1676–*c.* 1680)

PART 3
THE 'GEOMETRIA CURVILINEA' AND 'MATHESEOS UNIVERSALIS SPECIMINA'

are examined most simply and clearly by Euclidean geometry augmented by fluxional postulates.

Book 1. Definitions of fluxional increment and decrement, 424. Axioms and postulates, 426. The basic theorems: evaluation of the fluxion of a product (in the model of an incremented right triangle), 428. The model abandoned and the theorems restated in abstract form, 432. The fluxion of a power, 434. Theorems relating the fluxions of sides and base-segments of a right triangle whose hypotenuse or side is given, 436. And those of the sides of a general right triangle, 440. The fluxions of the basic trigonometrical functions of an angle (in the model of a circle arc), 440. Of an angle increased or decreased by a constant, 442. Of the sum or difference of two angles, 444. Of multiples of an angle, 446. Theorems on the fluxions of the sides, base-segments and perpendiculars of a general (scalene) triangle: given the base and vertex angle, 448. Given a side and vertex angle, 450. The general relationship of the fluxions of the sides and base-segments, 452. A lemma establishing the existence of the orthocentre in a triangle, 454. Further theorems relating the fluxions of the sides, angles, base- and perpendicular-segments of a general triangle: much tedious near-duplication, 456. Newton says as much in a marginal note on Proposition 24, 462. A scholium: for 'acute-angled' triangle read 'obtuse-angled' with suitable changes of sign, 464. The fluxion of the segment intercepted in a line by a moving line related to the fluxion of the angle of rotation, 466. Computation of the instantaneous (tangent) direction of a curve defined in generalized Cartesian coordinates, 468. The instantaneous normal in a bipolar coordinate system: the case where the included angle is fixed, 470. The general case, 472. Fluxional increase in a compound Cartesian/polar system (abandoned), 472. The fluxion of a curve's arc related to that of its abscissa in 'Newtonian' coordinates, 474.

Book 2. Determination of maxima and minima (by equating the fluxion to zero), 474. Construction of the normal at a point on an ellipse, 476. Examples of tangents in bipolars: the ellipse, Cartesian oval and trident, 478. The tangent to a general conchoid or cissoid (defined as the locus of a fixed point in the plane of a moving angle, one side passing through a fixed pole, the end-point of the other sliding on a fixed line): the centre of 'motion' (curvature) is an immediate corollary, 480. Tangent constructions for other loci: the curve of constant total distance from given fixed points, the 3/4-line locus, spirals, quadratrix, 482.

§2. Two schemes for revising the 'Geometria'. [1] (ULC. Add. 3963.7: 61ᵛ). A first sketch, changing little: a lemma is added justifying the last ratio of an arc and its chord to be one of equality, 484. [2] (ULC. Add. 3960.5: 53–4/66/57/60/59/64). A much amplified revise of the first book. The new opening section 'on proportionals': its proofs are based on the fluxional increase of a triangle whose vertex angle is fixed, 486. Variant proofs appealing to the semicircle as geometrical model, 492. The revised section on the fluxions of right triangles, 494. Of general triangles with one angle fixed, 496. Of general (unrestricted) triangles, 498. The 'fluxions of lines' (defined in various rectilinear/polar coordinate systems), 502. The 'fluxions of surfaces' (title and proposition numbers only), 504.

Appendix 1. Variant drafts of the 'Geometria'. [1] (ULC. Add. 3963.7: 47ʳ–49ʳ). A first extended version of Book 1: definitions, axioms, postulates and Propositions 1–13, 506. Propositions 4–10 are taken over *verbatim* from Theorems 1 and 2 of the addendum to the 1671 tract, 509. [2] (*ibid.*: 52ʳ). Proposition 12 in [1] is split into two separate theorems, 512. [3] (*ibid.*: 52ʳ). A first draft of Proposition 14 in the main text, 513. [4] (*ibid.*: 52ᵛ). A first version of Proposition 18, 514. [5] (*ibid.*: 53ᵛ). Two trivial variants on Proposition 20 omitted from the main text, 515. [6] (*ibid.*: 55ʳ). A variant proof of Proposition 22, 518.

APPENDIX 2 (ULC. Add. 3960.5:63). Preliminary calculations and proofs for the revised section 'on proportionals', 518.

APPENDIX 3. Newton's first public announcement of fluxions (in *Principia*, ₁1687: **2**, Lemma II). Fluxions are proportional to nascent moments of fluents, 522. Newton's (inevitably unsuccessful) attempt to evade an appeal to vanishing increments in his proof of the fundamental product theorem, 523. The sequel essentially repeats Propositions 4–10 of the 'Geometria', 524.

2. SPECIMENS OF A UNIVERSAL SYSTEM OF MATHEMATICS 526

§1. The 'Specimina': initial state. [1] (private). Prefatory remarks: David Gregory's *Exercitatio* (including results in infinite-series expansions obtained by his uncle James 'from a single series of Newton's') has decided Newton to publish a commented edition of his 1676 correspondence with Leibniz on the topic, 526. The 'Specimina' is intended to supplement Gregory's introduction, not to replace it, 528. Selections from Newton's *epistola prior* to Leibniz and from Leibniz' reply, 530. Newton's edited version of his ensuing *epistola posterior* to Leibniz: the sections not directly related to infinite series are mostly omitted, 532. Leibniz' two delayed replies (summer 1677, unanswered), 536. The titles of six chapters and a conclusion (unimplemented) 'on tangents and other Leibnizian matters' are announced, 538.

[2] (private/ULC. Add. 3964.3: 7^r–10^r/15^r–20^r). Chapter 1, 'on the roots of affected equations'. To Leibniz' query about the expression of 'impossible' roots by infinite series (with real coefficients) Newton replies blandly (and erroneously) that such roots are all denoted by divergent series, 540. Leibniz' query on the expression of multiple roots is answered by finding their separate infinite-series expansions from the several roots of the primary 'fictitious' equation, 540. Examples of this, 542. Chapter 2, 'on the properties of series'. The 'progression' (sequential pattern) of a series $\{s_i\}$ is discovered by identifying the multiplying factor s_{i+1}/s_i, 544. The 'termination' (aggregate) of a series found by inverse-differencing, 546. The geometrical model in which the term f_{-1} is interpolated in a line 'of parabolic kind' whose ordinates f_i, $i = 0, 1, 2, 3, \ldots$ are the first term of the series and the initial first, second, third, ... differences of its terms, 548. The use of finite differences in transforming a series to a (hopefully) more quickly convergent 'transmuted' form, 550. An attempted example (but the transmuted series converges less swiftly), 552. Chapter 3, 'revealing the broad expanse of analysis by infinite equations'. The types of problem reducible by division, root-extraction and other methods to an equation involving an infinite series, 554. Square-root extraction instanced in finding the Cartesian equation of a curve, the sum of whose linear distances at each point from four fixed points is constant, 556. A 'mechanical' example involving expansion of the inverse-sine as a series, 558. Rough 'freehand' interpolation used to determine an approximation to the first term of a series, 560. Greater accuracy is obtained by using finite-difference interpolation of a 'regular' curve, 562. Use of series in problems of maxima and minima, areas, lengths of arc, tangents, ... is sketched, 564. Chapter 4, on resolving problems by fluxions. The basic rule (taken over from the 1671 tract) for evaluating the fluxion of a fluent quantity of two or more variables, 566. Its application to constructing the subtangent at a point on a Cartesian curve, 570. Other applications: maxima and minima, areas, arc-lengths, curvature, 572. Deriving series-expansions of quantities determined by more general relationships between the sides of the differential triangle (and an abscissa), 574. Chapter 5, on Newton's 'more general method of reducing problems to infinite series'. The form of

A first verbal redraft of [1], 637. [3] (*ibid.*: 9^v–10^v). A fuller revise, augmented by a discussion of the operation of inverse-differencing. [4] (*ibid.*: 17^v). A first opening to Chapter 3: an intended solution of the Apollonian 'inclination' problem (abandoned when seen to be more complicated than a straightforward solution of the resulting quartic?), 641. [5] (*ibid.*: 17^v). An attempt to derive the defining equation of an Archimedean spiral in a special Newtonian coordinate system (abandoned when multiple roots are seen to complicate matters?), 642.

APPENDIX 3 (ULC. Add. 3964.3: 11^r–12^v). Preliminary versions of Chapter 2 of the 'De computo serierum'. [1] Series transformation by dividing through by $(1 \pm x)^k$, 644. A possible inspiration from James Gregory, 645. More sophisticated approximations of this type: in instance, the convergence of the series expansion of $\log 2$ is much quickened (the first five terms now yield a value true to six places), 649. [2] Successively refined difference-approximations to general series expansions, 650. [3] The 'transmutation' $x/(1+x) \to z$: its effect computed for several simple logarithmic and trigonometrical series, 651. [4] The draft revise of Chapter 2: Newton's modified difference-transmutation is now incorporated, 652. As an example the inverse-tangent series is transmuted into one for an inverse-sine, 653.

APPENDIX
MATHEMATICAL TOPICS IN NEWTON'S CORRESPONDENCE
(1674–1676)

June 1674 and November 1676 mark the opening and close of Newton's mathematical correspondence at this time: except for an exchange with Hooke (late 1679, on fall under gravity) he remains silent till June 1684, 657. At Collins' request (June 1674) Newton sets out the series solution of $y^3 + a^2y - b^3 = 0$ as 'a form for all cubic equations', 659. The binomial cube root as a particular case, 660. Correspondence with Michael Dary (October 1674) on the trinomial $z^n + bz^{n-1} - R = 0$, 660. As an alternative Newton proposes iteration of $z_{i+1} = {}^{n-1}\sqrt{[R/(z_i + b)]}$, 661. To Dary (January 1675) he communicates a Huygenian approximation to the general ellipse arc (founded on his 1671 series rectification), 662. For John Smith (May 1675) he frames his difference-scheme for easy computation of square and cube roots, given every hundredth one, 663. To compute the latter he suggests the rule $\sqrt[n]{A} \approx B + n^{-1}(A - B^n)/B^{n-1}$, 664. Autumn 1675: in Newton's eyes 'mathematical speculations grow somewhat barren' and his correspondence lapses briefly, 664. Collins is set right about the solution of equations by intersecting curves (September 1676), 665. Leibniz asks for Newton's series method: Newton responds with the binomial expansion in his *epistola prior* (June 1676), 666. His series solution of 'affected' equations, numerical and algebraic, 667. Particular series expansions: inverse-sine, natural logarithm and exponential function; generalized Viète formula for sines of multiple angles; solution of Kepler's problem, 668. Series for ellipse arc; quadratrix' subtangent, area and arc-length; 'second segments' of a spheroid: two Huygenian approximations to the general arc of a circle and conic, 669. An unwieldly 'mechanical' construction of conic area, 670. To Leibniz' request for further information Newton communicates his lengthy *epistola posterior* (October 1676; see 3, Appendix 1), 670. Its opening biographical remarks are an accurate report of his extant early manuscripts and his debt therein to Wallis, 672. The central passages in the *epistola* borrow heavily from his 1671 tract: its fluxional Problems 1 and 2 are enunciated in anagrammatic form (impossible to penetrate), but its general tone is of friendly helpfulness, 673. Leibniz' delay in receiving the letter (not till June 1677) and Oldenburg's death (September 1677) terminate the interchange of insights and criticisms, 674.

INDEX OF NAMES   **675**

LIST OF PLATES

RESEARCHES IN ALGEBRA, NUMBER THEORY AND TRIGONOMETRY

(*c.* 1675 - *c.* 1684)

INTRODUCTION

This opening part of the present volume reproduces a gathering of miscellaneous autograph manuscripts composed by Newton at different times during the years 1675–84 and dealing with such varied topics as finite differences, theory of numbers, plane and spherical trigonometry, the computation of annuities and the quadratic factorization of the general algebraic binomial. Within each main subdivision the papers now printed are set out in roughly chronological sequence, but editorial convenience alone has dictated that the groups themselves should be bedded together and no claim to an underlying homogeneity of theme or treatment is presupposed in this mating. To be sure, the essential separateness of the several mathematical strands here arbitrarily interwoven mirrors the fundamental lack of continuity in both topic and output which characterizes so much of Newton's mathematical research at this period. Within the confines of each group of papers it will none the less be relevant to make certain general observations.

The extent to which Newton's advances in the field of finite differences depended on his prior knowledge of the work of his predecessors is difficult to assess. More than forty years ago D. C. Fraser made a determined attempt[1] to locate the source of Newton's early interest in interpolation in his reading of Henry Briggs' *Arithmetica Logarithmica*, but in the light of our present improved knowledge of his introduction to higher mathematics this seems highly

---

(1) Duncan Fraser, 'Newton's Interpolation Formulas. An unpublished Manuscript by Sir Isaac Newton' (*Journal of the Institute of Actuaries*, **58**, 1927: 53–95 [= *Newton's Interpolation Formulas* (London, 1927): 53–95], especially 57–60). Fraser was encouraged by the apparent parallels between his 'unpublished Manuscript' (the 'Regula Differentiarum' reproduced as 1, §3 below) and the seventh and twelfth chapters of Briggs' book (*Arithmetica Logarithmica sive Logarithmorum Chiliades Triginta* (London, 1624): 12–1[5], 24–7) to assert: 'Newton never mentions Briggs [but it] is a natural supposition, that when Newton went into the country in 1665 he took Briggs' work with him....The raw material was there in profusion. Newton saw its value, divined the underlying principles, and evolved a new mathematical instrument. The arithmetical rules of Briggs, which had to be separately determined for every individual case, were replaced by a single general law, which included every case. ...it is highly probable that [Newton's results] were obtained by meditation on the facts previously assembled by the labours of Briggs' (pp. 58–9). For the reader who wishes to check the implications of this wholly unsupported conjecture the best secondary account of Briggs' contributions to the theory of interpolation by finite differences is Charles Hutton's introduction to his revised fifth edition of Henry Sherwin's *Mathematical Tables*; ... *To which is prefixed, A Large and Original History of the Discoveries and Writings relating to those Subjects* (London, ₁1785): 1–124, especially 61–84 [reprinted in F. Maseres, *Scriptores Logarithmici*; *or a Collection of Several Curious Tracts on the Nature and Construction of Logarithms mentioned in Dr. Hutton's Historical Introduction*, **1** (London, 1791): i–cxxi, especially lxiii–lxxvi]. See also J.-B. Delambre, *Histoire de l'Astronomie Moderne*, **1** (Paris, 1821): 535–44; and compare D. T. Whiteside, 'Patterns of Mathematical Thought in the later Seventeenth Century' (*Archive for History of Exact Sciences*, **1**, 1961: 179–388): 233–6.

improbable: indeed, if the young Newton ever glanced at the *Arithmetica* (or equivalent passages in Gellibrand's posthumous edition of Briggs' *Trigonometria Britannica*), his impressions are nowhere recorded and he could not have appreciated several of the subtleties contained therein.[2] More recently J. A. Lohne has outlined the riches of an unpublished tract 'De Numeris Triangularibus et inde De Progressionibus Arithmeticis Magisteria magna',[3] composed by Thomas Harriot around 1600, in which the general advancing differences interpolation formula $f_n = f_0 + \sum_{1 \leqslant i \leqslant k} \binom{n}{i} \delta_0^i$ is established for functions whose k-th unit differences δ^k are constant, and is then used to interpolate values corresponding to fractional values of the argument x: Newton, however, was no more likely to have suspected the existence of this manuscript (then locked away in Petworth House) than any other of his contemporaries to whom Harriot was known only as an algebraist. By his own account[4] Newton was already familiar in 1665 with the technique of subtabulating a function f_{jn}, given for $k+1$ (successive) n-intervals of argument, $\alpha \leqslant j \leqslant \alpha + k$, by identifying each given instance f_{jn} with a corresponding 'progression' $\sum_{0 \leqslant i \leqslant k} \binom{jn}{i} a_i$ and, by inversion, evaluating the constants a_i.[5] Three years later Nicolaus Mercator

---

(2) We saw earlier (II: 222, note (59)) that Newton was unaware in 1669 of Briggs' anticipation in his *Trigonometria Britannica* (Gouda, 1633) of his own improvement on Viète's method for extracting the (real) roots of numerical equations. Likewise, it will be remarked in the sequel (1, §3: note (25) below) that in his 'Regula Differentiarum' he errs in sign in stating a subtabulation rule of which two particular cases were correctly cited by Briggs in both his *Arithmetica* and *Trigonometria*.

(3) British Museum MS Add. 6782: 107–46, described by Lohne in his 'Thomas Harriot als Mathematiker' (*Centaurus*, 11, 1965: 19–45), §3: 31–4.

(4) See I: 130–3, where Newton employs this technique to subtabulate unit instances of the binomial expansion $f_m \equiv a^2(b+x)^m$, $m = jn = 0, 1, 2, 3, \ldots$. Subsequently (I: 534–9) he made an analogous application of this scheme to the problem of isolating quadratic and cubic factors ($k = 2, 3$) of a given 'numerall equation'.

(5) These are, in fact, the i-th unit differences δ_0^i. This subtabulation method was independently discovered shortly afterwards by François Regnaud and published in Gabriel Mouton's *Observationes Diametrorum Solis et Lunæ Apparentium.... Pro cujus, & aliarum tabularum constructione seu perfectione, quædam numerorum proprietates non inutiliter deteguntur* (Lyons, 1670): Liber III, Caput III. *De nonnullis numerorum proprietatibus*: 384–96: 'Problema Quartum. Datis in eadem serie aliquot numeris, qui unâ eadémque ratione continuè progredientes, habeant quotlibet differentias harumque ultimas æquales; intermedios quotlibet numeros, qui eamdem progressionis legem retineant, invenire.' Following Regnaud's scheme Mouton tabulates (pp. 385–6) the 51 progressions

$$f_i = f + \binom{i}{1}e + \binom{i}{2}d + \binom{i}{3}c + \binom{i}{4}b + \binom{i}{5}a, \quad 0 \leqslant i \leqslant 50,$$

and their first five orders of differences ($f = f_0$, $e = \delta_0^1$, $d = \delta_0^2$, ..., $a = \delta_0^5$); then gives (pp. 387–90 and 390–6 respectively) two examples of the application of this array to n-section of the

publicized his independent discovery of Harriot's advancing differences inter-polation theorem (the so-called Newton–Gauss formula) in his *Logarithmotechnia*, making explicit an approach which Briggs had earlier made use of in a restricted way in his *Arithmetica*.[6] We have previously argued[7] that Mercator's discussion in that work of the series expansion of $\log(1+x)$ provoked Newton to draft his own 'De Analysi per æquationes numero terminorum infinitas' as a counter-blast in 1669, but Mercator's publication of the fundamental finite-differences formula did not encourage him in his 1671 tract to improve upon the *ad hoc* method there given for intercalating mean values in a canon of logarithms listed at equal intervals of the argument.[8] We do not know Newton's reaction when Leibniz' further variant on the advancing differences theorem was communi-cated to him by Oldenburg soon after receiving it in February 1673[9] but it could only have been mild and short-lived at best. For the catalyst which began

---

argument interval, $n = 3, 10$. In the trisection case Mouton's subtabulation rule yields in effect, to $O(\delta^6)$, $\delta_0^i = (\Delta_0^1 - \Delta_0^2 + \frac{5}{3}\Delta_0^3 - \frac{10}{3}\Delta_0^4 + \frac{22}{3}\Delta_0^5)^i$, $i = 1, 2, 3, 4, 5$, in which Δ_0^j is put for the adjusted j-th difference $(\frac{1}{3})^j\Delta_0^j$; that is, on introducing fractional binomial coefficients,

$$\delta_0^i = \left(\sum_{1 \leqslant j \leqslant 5} \binom{\frac{1}{3}}{j} \Delta_0^j\right)^i \quad \text{to} \quad O(\Delta^6).$$

(Compare 1, §3: note (25) below where Briggs' equivalent rule yields

$$\delta_i^i = (\Delta_0^1 - \tfrac{1}{3}\Delta_{-1}^3 + \tfrac{1}{3}\Delta_{-2}^5 \ldots)^i.)$$

The best secondary account of the Mouton-Regnaud subtabulation rules is Frédéric Maurice's 'Mémoire sur les Interpolations, contenant surtout...la démonstration générale et complète de la méthode de quintisection de Briggs et de celle de Mouton, quand les indices sont équidifférents....' (*Connaissance des Temps ou des Mouvements Célestes...pour l'an 1847* (Paris, 1844): 181–222 [= (English translation by T. B. Sprague and J. H. Williams), *Journal of the Institute of Actuaries* 14, 1867: 1–36].)

(6) See Mercator's *Logarithmotechnia: sive Methodus construendi Logarithmos Nova, accurata, & facilis; Scripto Antehàc Communicata, Anno Sc. 1667* (London, 1668): Propositio III: 11–14. In Chapter 12 of his *Arithmetica Logarithmica* (note (1) above): 24–7 Briggs had given a subtabula-tion algorithm, true to the order of the third differences, which appeals implicitly to the formula $f_{\frac{1}{10}n} = f_0 + \frac{1}{10}n\Delta_0^1 - \frac{1}{200}n(10-n)\Delta_0^2$, $-5 \leqslant n \leqslant +5$.

(7) See II: 165–7.

(8) See III: 232–4.

(9) Namely, in Leibniz' letter to Oldenburg of 3 February 1672/3 (Royal Society MS LXXXI, No. 30), first published by Newton in his *Commercium Epistolicum D. Johannis Collins, et Aliorum de Analysi promota* (London, 1712): 32–6 as documentation of John Pell's charge against Leibniz of plagiarizing the theorem from Mouton's *Observationes* (*sic*!). The auto-graph transcript of this letter which Oldenburg subsequently passed on to Newton is now ULC. Add. 3971.1: 27^r–28^v. Leibniz does not extend his method of differences ('quas voco Generatrices') to treat of the problem of subtabulation at fractional values of the argument, giving as his unique example the generation of cubes $f_n \equiv n^3$, for which $\Delta_0^1 = 1$, $\Delta_0^2 = 6$, $\Delta_i^3 = 6$ and so generally $n^3 \equiv 0 + \binom{n}{1} + 6\binom{n}{2} + 6\binom{n}{3}$.

the explosive series of researches into finite-difference interpolation here reproduced we must rather look to the London 'Philo-Accomptant' John Smith who in the spring of 1675 approached Newton, doubtless through their mutual acquaintance John Collins, for help with his project of compiling a table of the square, cube and fourth roots of the natural numbers from 1 to 10000.

Newton's reply to Smith on 8 May following is reproduced as the opening text below and there is no need to dwell here upon its detail: in brief, he there suggests that each set of hundred roots $(100p)^{1/k}$, $1 \leqslant p \leqslant 100$ should (for $k = 2, 3, 4$) be computed outright—later communicating an iterative rule to lighten the calculations involved—but that it is then simplest to subtabulate the ninety-nine values between each of these by means of their adjusted first, second and third differences. Though it would appear that Smith began at once to recompute his roots according to Newton's improved method,[10] he could, unfortunately, persuade no stationer to underwrite the cost of publishing his tables: whether these were ever completed is not known. At first glance all the effort which Newton had put into the preparation of his ingenious difference scheme—one 'excellently well suited to those particular purposes for which he design'd it & ... very curious'[11] but not seemingly more generally applicable— was wasted. However, the manuscripts reproduced in sequel (all probably composed during the next two years) indicate that Newton clung tenaciously to the general problem of interpolation,[12] slowly worrying out rules for subtabulating and extending given tabular instances of a function of one variable. By the time he came, in the early autumn of 1676, to draft his *epistola posterior* to Leibniz he had achieved the general insight that if the n-th differences of a set of values of a function, f_x say, given at equal intervals of the argument x are constant then the function may adequately be approximated by the general 'parabola' $a + bx + cx^2 + \ldots + hx^n$, where the constants $a, b, c, \ldots, h$ are readily evaluated from the given instances $f_\alpha, f_{\alpha+1}, f_{\alpha+2}, \ldots, f_{\alpha+n}$ and so in terms of the various i-th order differences Δ_k^i constructible from them in a variety of ways.

---

(10) According to Collins, who reported to Gregory on 29 June 1675 that 'one Mr Smith is a making [tables of square roots and cube roots] for all numbers from 0 to 10000, Mr Newton having imparted a ready method for maintaining the last difference true (according to the determined number of places the table is to consist of) and consequently the carrying on of such tables is performed by addition of differences, upon supposition that the roots of every 100th number are first in store' (*James Gregory Tercentenary Memorial Volume* (London, 1939): 309). Collins, however, was ever optimistic in such matters.

(11) Cotes' assessment for William Jones in January 1712 upon receiving a transcript of Newton's letter (J. Edleston, *Correspondence of Sir Isaac Newton and Professor Cotes* (London, 1850): 220; compare 1, §1: note (14) below).

(12) That of finding a 'Regula quæ ad innumera æqualia intervalla recte se habet, [quia tum] recte se habebit in locis intermedijs' as he phrased it in 1712 in a draft (ULC. Add. 3960.6: 91) of Propositio XI of his *Analysis per quantitates fluentes et earum momenta*.

In a sheet[13] discarded from the letter finally posted to Oldenburg on 24 October for onward transmission to Leibniz, Newton originally intended to communicate seven rules, 'wholly unworthy to be passed on',[14] for 'subcenturying' the entries in a given logarithmic or trigonometric canon which it might be useful for intending compilers of such tables to know. (No doubt Newton again had John Smith in mind.) After a last minute change of heart he replaced these rules in the *epistola* itself by the merest hint that such terms might be intercalated 'by a certain method which I could almost have described here for the use of calculators',[15] adding apologetically in his covering letter to Oldenburg that he should have omitted more of the surrounding context 'since to avoid greater tediousness I left out something else on w^ch [these things] have some dependance'.[16] Unknown to Oldenburg or even Collins, however, Newton went on to draft two short papers, the unfinished 'Regula Differentiarum' and an untitled account of his general method which was later to serve as the source for his definitive 'Methodus Differentialis'.[17] In these he brilliantly extended his previous interpolation techniques to the case where the differences of the argument of the tabulated function need no longer be equal, discovering not only the so-called Newton–Stirling and Newton–Bessel finite difference formulas but also their generalizations, and deriving—if not quite correctly[18]—general Briggsian precepts for subtabulation. During the two years 1675–76, in effect, Newton laid down the elements of the modern elementary theory of interpolation by finite differences but, except for a short lemma included in his *Principia*[19] a decade afterwards where he stated (without proof) his generalization of the

(13) Reproduced as 1, §2 below.

(14) See 1, §2: note (23).

(15) '...per Methodum quandam quam in usum Calculatorum ferè hic descripsissem' (*Correspondence of Isaac Newton*, **2**, 1960: 126). We should not anticipate the rancour of the later calculus priority dispute by regarding Newton's suppression of these rules as deliberately malicious. Courtesy apart, he had no need to reply to a letter from a stranger several years his junior, let alone compose the lengthy answer in which he made an honest attempt to lay bare to Leibniz the essence of his discoveries in algebra and infinite series. It is much more likely that Newton grew increasingly dissatisfied with his proposed 'Regulæ' but had insufficient time or inclination to recast them to his satisfaction before sending the completed *epistola* off to Oldenburg. It is true that he never responded to Leibniz' further query in June 1677 (*Correspondence*, **2**: 216) about the generality of his interpolation method, but his failure to reply was symptomatic of a growing reluctance on Newton's part to enter into scientific correspondence with anyone.

(16) *Correspondence*, **2**: 110.

(17) Published, shortly after its final revision, by William Jones in his 1711 compendium of Newton's short mathematical pieces, the *Analysis Per Quantitatum Series, Fluxiones, ac Differentias*; compare 1, §4: note (1).

(18) See especially 1, §3: notes (23) and (25).

(19) *Philosophiæ Naturalis Principia Mathematica* (London, ₁1687): Liber III, Lemma V: 481–3, reproduced in 1, Appendix below.

advancing differences formula, diffidently kept back his insights and discoveries therein for nearly forty years more.[20]

In sequel we reproduce the two extant states of Newton's solution of certain 'Numerical problems' in indeterminate analysis, together with an allied paper on generating rational solutions to such Diophantine problems when one or more particular instances are given or evident by inspection. As a young man he had shown a fitful curiosity in the general quadratic expression

$$ax^2 + bx + c + dxy + ey + fy^2 = 0,$$

seeking a 'rule ... to make x & y rationall',[21] but it would seem that his curiosity in such matters soon abated. How and why Newton came to renew his interest in elementary number theory in the later 1670's and what incited him to compose the present manuscripts are questions to which the known facts allow no answer, though we may perhaps guess that he intended his researches to form the basis of a university lecture course on the topic (probably never given). A close examination of Newton's text suggests that he was not unaware of previous publications on the subject by Viète, Bachet, Schooten and Wallis[22] but his principal source of background information was inevitably the classical

---

(20) In the decade before Newton published his general interpolation method (see note (17)) at least two men were driven to rediscover it on the basis of the *Principia* lemma alone. Introducing his own variant method 'De Curvis Algebraicis per quotlibet data puncta ducendis', appended to his *Phoronomia, sive de Viribus et Motibus Corporum Solidorum et Fluidorum Libri Duo* (Amsterdam, 1716: 389–93), Jakob Hermann remarked that 'solutionem Summus Newtonus primus invenit:...tamen...mihi videre non contigit, excepta ea, quam in Lemmate v. Lib. III. *Princ. Phil. Natur.* sine omni analysi & demonstratione tradita, ubi...ego solutionem annis 1704. & 1705. aggressus sum & obtinui' (p. 389). More surprisingly, in the preface to his tract 'De Methodo Differentiali Newtoniana' (*Opera Miscellanea* [appended to Smith's edition of his *Harmonia Mensurarum*, Cambridge, 1722]: 22–33) Roger Cotes, who was soon to edit the second edition of Newton's *Principia* for him, observed that 'Anno 1707 [ubi propositiones præcedentes conscripsi] Nesciebam...a *Newtono* compositum fuisse Tractatum de eodem argumento' and that he remained ignorant 'usque dum ejusdem Exemplar typis impressum Anno 1711 dono accepissem a doctissimo Editore Domino *Jones*' (*Postscriptum*: 32). Even eight years after Jones' publication the young James Stirling felt called upon to remind the world just how far Newton had 'promoted' the theory of interpolation by finite differences: 'Per hasce artes *Newtonianas* universa doctrina Approximationum reducitur ad solutionem Problematis, *Invenire Lineam Geometricam quæ per data quotcunque puncta transibit.* [compare 1, §4.2 below]...Existimo igitur *Newtonum* perduxisse methodum Approximandi [per quantitatum differentias] ad summum perfectionis fastigium; dum ex unico simplicissimo principio totam hanc doctrinam longe lateque patentem deducit. Quapropter credendum est animum *Newtoni* non satis perspectum fuisse iis, qui ejus methodos appellant particulares, & alias tanquam suas & solas genuinas atque generales venditant, quæ aliæ non erant quam Corollaria facillima à *Newtonianis*' ('Methodus Differentialis *Newtoniana* Illustrata' [= *Philosophical Transactions*, **30**, No. 362 (for September/October 1719): §III: 1050–70]: 1051). Despite his concluding prod at Leibniz and Johann Bernoulli, Stirling's assertion of Newton's preeminence in the field of finite differences came soon to be widely admitted and in 1761

account of the analysis of indeterminate equations, Diophantus' *Arithmetica*, read by him in Bachet's 1621 *princeps* edition. Like his mentor, Newton is here principally concerned to find particular rational solutions to the Diophantine equations of second, third and fourth degrees which he poses, bothering little with the tighter demands set by the restriction to integer solutions which the seventeenth-century master of such analysis, Pierre de Fermat, systematically ordained in his own researches—though Newton is happy to provide these when they arise naturally from his approach—and worrying hardly at all over the possibility of their non-existence.[23] As we would expect, his handling of such equations is never less than competent, revealing a firm grasp of essentials and at times an elegance in exposition never attained by those of his English contemporaries, notably John Wallis, who interested themselves in problems of number theory. Yet we are here a long way away from Whiston's universal genius who 'in Mathematicks, could sometimes see almost by Intuition, even without Demonstration':[24] indeed, it is patent that in dealing with *Problemata numeralia* Newton has little of Fermat's profound insight into the hidden properties of integers and none of his flair for establishing novel methods of attack on the more recalcitrant problems of elementary number theory. On the

---

J. J. Lalande could begin his influential 'Mémoire sur les Interpolations' with the observation that 'La méthode générale des interpolations a été traitée dans toute la généralité possible par M. *Newton*' (*Mémoires de Mathématique et de Physique tirées des Registres de l'Académie Royale des Sciences de l'Année M.DCCLXI* (Paris, 1763): 125).

(21) See 1: 542–3.

(22) For details see 2, §1: note (2) below.

(23) In a classical paper (*Novi Commentarii Academiæ Scientiarum Petropolitanæ*, 6, 1756/7: 155–84 [= *Leonhardi Euleri Opera Omnia* (1) **2**. *Commentationes Arithmeticæ* 2 (Leipzig/Berlin, 1915): 428–45]) Euler later drew a careful line between 'Alia...problemata [analysis Diophanteæ] ita...comparata, ut solutiones generales exhiberi queant, quæ omnes plane numeros satisfacientes in se complectuntur' and 'alia [quæ] nonnisi solutiones particulares admittunt, vel saltem per methodos cognitas nonnisi tales solutiones eruere licet', observing that 'in genere...convenit prioris ordinis problemata multo facilius resolvi quam ea, quæ ad alterum ordinem referuntur, quippe quæ plerumque singularem sagacitatem cum eximiis artificiis coniunctam requirunt, in quibus maxima vis huius Analysis cernitur' (pp. 155–6). If Newton appreciated Euler's distinction, he was by preference drawn to attack the former, more generally solvable type of problem.

(24) *Memoirs of the Life and Writings of Mr. William Whiston.... Written by himself* [*in the 79th, 80th, 81st and 82d Years of his Age*] (London, ₁1749): 39. But Whiston was, we may suspect, a poor judge of mathematical genius and originality and, indeed, here continued '...as was the Case in that famous Proposition in his *Principia* [Liber 1, Lemma XII = ₁1687: 47], that *All Parallelograms circumscribed about the Conjugate Diameters of an Ellipsis are equal*; which he told Mr. *Cotes* he used before it had ever been demonstrated by any one, as it was afterward'. Newton was evidently pulling Cotes' leg since, as he accurately remarked in justification of the lemma, 'Constat ex Conicis' (understand 'Apollonij VII, 31', which also proves the complementary theorem for the hyperbola) and the property is all but self-evident when the ellipse is viewed as the orthogonal projection of a circle.

contrary, in his impatience with equations which do not have a readily determinable particular rational solution Newton is much too quick to dismiss as uninteresting such 'propositions négatives' as the Eulerian equation $x^2 + 1 = y^3$ and the cubic case $x^3 + y^3 = z^3$ of Fermat's 'great' theorem,[25] apparently failing to appreciate their subtleties, and is undoubtedly at his best in the final piece where, for the first time, he develops the powerful technique of constructing rational points on a Cartesian curve (by drawing straight lines, in particular chords and tangents, of rational slope through given rational points) which is the geometrical equivalent of a favourite arithmetical method of Diophantus for generating rational solutions to the indeterminate equation in two variables which defines it.[26]

In the three following papers on plane and spherical trigonometry, compiled by him about the summer of 1683, Newton is clearly more at home, though once again we know disappointingly little regarding their composition. In the autumn of 1665, as we saw, he had moulded his annotations on Richard Norwood's *Trigonometrie*, Gunter's *Sector* and above all Oughtred's *Trigonometria* into a short tract 'Of ye resolution of streight lined [and] sphæricall triangles', ostensibly to use it in the solution of 'Questions in Navigation'.[27] The lack of

---

(25) See 2, §1: notes (16) and (70).

(26) Compare 2, §2: note (1).

(27) The two extant versions of this are reproduced in 1: 466–70/470–5.

(28) In sketching the contents of Bernhard Varenius' *Geographia Generalis, In qua affectiones generales Telluris explicantur* (Amsterdam, ₁1650), whose second edition (Cambridge, 1672) carried the subtitle *Summâ curâ quam plurimis in locis emendata, & XXXIII Schematibus novis, ære incisis, unâ cum Tabb. aliquot quæ desiderabantur aucta & illustrata. Ab Isaaco Newton, Math. Prof. Lucasiano Apud Cantabrigienses* (compare 11: 288, note (43)), A. R. Hall has remarked that its '*Pars Respectiva*...contains...some propositions in spherical trigonometry' ('Newton's First Book (11)', *Archives Internationales d'Histoire des Sciences*, **13**, 1960: 55–61, especially 58). The twelve propositions 'De distantia locorum [in globo]' referred to (Liber 111, Caput xxxiii = (₂1672), pp. 471–80) reduce, in equivalent 'abstract' trigonometrical form, mostly to determining the sides, hypotenuse and one angle of a right spherical triangle, given any three of them, but these are usually constructed manually 'per globum...omissis methodis per calculum [trigonometricum] & planisphærium [catholicum] agentibus, etsi accuratioribus' (p. 480) though in one instance (Propositio IX: 475–8: 'Datâ latitudine & longitudine duorum locorum, invenire eorum distantiam') Varenius does outline the use of the universal stereographic projection 'Quoniam planisphærium aptius vel commodius est ad usum, præsertim nautis, & multi ejus beneficio problemata solvere amant'. In 1672 Newton evidently did not think it worthwhile to replace Varenius' chapter with an equivalent summary of the trigonometry of the right spherical triangle. (On the equivalent application of the universal 'planisphere' to the problem—propounded by Johannes Werner in the third book of his long unpublished work 'De Meteoroscopijs' (*c.* 1500) and subsequently popularized by Apianus (Bienewitz) in his *Astronomicum Cæsareum* (Ingolstadt, 1540)—see J. D. North, 'Werner, Apian, Blagrave and the Meteoroscope', *British Journal for the History of Science*, **3**, 1966: 57–65, especially 59–61.)

(29) See 3, §1: note (2).

any documentary evidence to the contrary suggests that over the next decade and a half he felt no real need to improve upon that exposition of the elements of trigonometry, and no doubt on the rare occasions when he touched upon the subject in his researches or in his professorial activities the existing published treatments filled all his requirements.[28] Why in 1683 Newton began to compile a fresh summary of existing trigonometrical knowledge is not known with certainty but his reason may be accurately guessed. His first text on this topic now reproduced, a brief 'Epitome of Trigonometry', is manifestly based on a standard undergraduate précis of the period, Seth Ward's *Idea Trigonometriæ Demonstratæ*,[29] and we may reasonably conclude that Newton's manuscript was likewise intended for the instruction of Cambridge students: a conjecture strongly reinforced by the existence of a slightly variant transcript of the piece in the hand of Henry Wharton, who in his title asserts that it was 'given' him in 1683—probably at one of the gatherings of the 'select' group of students who received private instruction on mathematics in Newton's college rooms rather than in a formal public lecture.[30] In its original state the piece is wholly derivative in content, but in an abortive revise Newton added to it a group of Napierian half-angle theorems which he proved in a wholly novel way by projecting the given spherical triangle stereographically into a corresponding plane (non-rectilinear) triangle, making good use of both the circle- and angle-preserving properties of the projection.[31] The following paper, a more expansive 'Compendium of Trigonometry', is manifestly an augmented revise of the parent 'Epitome' which contains little that is new. In line with his opening dictum that here 'regard is paid not so much to brevity as to the learner's capacity' Newton goes into basic principles at some length, expatiating upon the definitions of the main trigonometrical functions (with some interesting if invalid etymological derivations)[32] and elaborating twelve introductory

(30) Wharton's transcript is reproduced in **3**, Appendix 1 below. His biographer later stated that 'From his first admission into [*Caius*] College [in February 1680]...under the Care and Tuition of Mr. *John Ellys*, one of the Senior Fellows of the same...he pursued his Studies with an indefatigable Industry.....He attained likewise...no mean Skill in *Mathematicks*. Which last was much encreased by the kindness of Mr. *Isaac Newton*, Fellow of *Trinity College*, the incomparable *Lucas-Professor* of *Mathematicks* in the University, who was pleased to give him further instructions in that noble Science, amongst a select Company in his own private Chamber' (see the anonymous 'Life of Mr. *Hen. Wharton*' prefixed to the second edition of *Fourteen Sermons Preach'd in Lambeth Chapel...In the Years MDCLXXXVIII. MDCLXXXIX. By the Learned Henry Wharton M.A.* (London, ₂1700): A3ʳ–A4ʳ). It is possibly not irrelevant that Wharton's tutor Ellis was one of Newton's closest friends at this time.

(31) See **3**, §1: notes (24) and (25). The former property had been established long before by Ptolemy in his *Planispherium*: the latter, though discovered by Harriot three-quarters of a century earlier, was still unpublished but known in the later 1670's to the circle of mathematicians—including Newton—with whom John Collins corresponded.

(32) See **3**, §2: notes (3) and (4).

lemmas which will later prove to be 'of use'. In the body of his treatise, however, he tends to forget that his intended audience is one of beginners and his arguments become increasingly taut and compressed: any Cambridge student for whom this was his introduction to the trigonometry of the sphere, in particular, would have had a decidedly difficult time with it. In time, no doubt, Newton came to realize this as well as anybody. Certainly, he thought much about the artificialities and obscurities of the conventional presentation of spherical trigonometry, and in the anonymous commentary which he wrote shortly afterwards on St John Hare's manuscript 'Trigonometria' he introduces—anticipating later equivalent developments of the topic by Francis Blake and Euler[33]—a simplified geometrical derivation of the sine, cosine and tangent formulas for the general spherical triangle by relating the edges and angles of polyhedra circumscribing the spherical 'solid' angle which it forms. For further simplicity the circumscribing solids are collapsed into the nets formed by their plane sides: an elegant refinement (unknown to Blake or Euler) which allows the main half-angle theorems as immediate corollaries. Who the mysterious St John Hare was, why he should have inspired Newton to the unprecedented action of editing his mathematical papers for posthumous publication, and why no contemporary record of his presence at Cambridge seems to exist, all are questions difficult to determine. We ourselves hesitate between regarding him as a favourite (if unofficial) pupil and an intimate friend.[34] His researches in trigonometry, sold at auction thirty years ago with the mass of Newton's other miscellaneous papers, have temporarily disappeared from public view, but to give some taste of the quality of his mind we have appended an excerpt dealing in outline with the spherical case, taken from an unpublished, essentially derivative student 'Mathematico-algebraical compendium' (dated 1675) which is in his hand.[35] One day, doubtless, more will come to be known about this shadowy, insubstantially documented corner of Newton's personal life.

The last section reproduces two minor worksheets, devoted respectively to the computation of annuities when the deposited capital 'discompts' compound interest annually, and to determining a quadratic factor of the general binomial $a^n \pm x^n$, n integral. The former topic was of great practical, if of minor theoretical, interest to a man as interested as Newton in making wise investment of his money: a draft letter in which he commissioned annuities for the three surviving children of his recently widowed half-sister Hannah in the 1690's reinforces the

---

(33) For details see 3, §3: note (9). Euler, in particular, was very proud to have discovered a unified geometrical approach which 'Doctrinam sphæricam complectitur' in this way.

(34) The little we have been able to find out, positively and negatively, about Hare is given in 3, §3: note (1), along with our own undocumented conjecture as to his identity.

(35) See 3, Appendix 4 below. The trigonometrical portions are based narrowly on Seth Ward's *Idea Trigonometriæ Demonstratæ* (Oxford, 1654).

point.[36] The latter computations, in which Newton comes near to anticipating Roger Cotes' general quadratic factorization of the binomial, were evidently of main interest to him because of their relevance to the evaluation of complex algebraic integrals: above all, treading in their path he was able in his *epistola posterior* to respond to Leibniz' communication to him in August 1676 of the 'simplest' series $1-\frac{1}{3}+\frac{1}{5}-\frac{1}{7}+\dots$ for evaluating $\frac{1}{4}\pi$ by returning the elegant variant expression $1+\frac{1}{3}-\frac{1}{5}-\frac{1}{7}+\dots$ for $\pi/2\sqrt{2}$, one whose subtlety Leibniz himself was seemingly unable to penetrate.[37]

(36) The relevant portions of this autograph draft are given in 4, §1: note (5) below.
(37) See 4, §2: notes (28) and (29) for details.

1

APPROACHES TO A GENERAL THEORY OF FINITE DIFFERENCES

[1675–6]

§1. COMPLETION OF A NUMERICAL TABLE OF ROOTS BY A PRIMITIVE SUBTABULATION (MAY 1675).[1]

From a contemporary copy in the University Library, Cambridge[2]

[3]If you would compleat[4] a table [of Square, Cube & Squaresquare roots] to 8 decimal places let the root of every hundredth number be extracted to 10 decimal places and then compute every tenth number and afterwards every number by the following methods.

---

(1) Extracted from a letter sent on 8 May 1675 to John Smith, an accountant and compiler of numerical tables then resident in London. As a 'Student of Mathematicks' six years before Smith had, according to John Collins in his 1676 *Historiola* of James Gregory's mathematical achievement (Royal Society MS LXXXI, No. 46: 8, quoted in H. W. Turnbull's *Gregory Tercentenary Memorial Volume* (London, 1939): 345, note 1), collaborated with John Newton 'in making a table of Areas of Segm^ts of a Circle, to facilitate guaging', and we may accept Edleston's conjecture (see next note) that he was the 'Philo-Accomptant' who had published a *Stereometrie* at London in 1673. (Collins had first met Smith some time before 1670 when, it would appear, he procured for him the post of 'clerk' under the Commissioner of Accompts upon his own retirement from it.) In the autumn of 1674 Smith began to contemplate the computation of an extended table of square, cube and fourth roots—an intention which Collins was not slow to advertise. On 23 November he wrote to James Gregory that 'We have one Mr Smith here [in London] taking pains to afford us tables of the square and cube roots of all numbers from unit to 10000, which will much facilitate Cardan's rules [for solving cubics]: and he would go on with alacrity if, beginning at 1000, and designing his table to be tried to 10 or 12 places of figures and preparing some in store to begin with, and to raise differences from; he could have a series to maintain the last difference always true, so as to carry on the rest by additional work' (*Gregory Volume*: 291–2). The next spring, similarly, he remarked for Leibniz' benefit that 'when we have a table in print of Square and Cube rootes as now we have of the Squares and Cubes of all numbers from 0 to 10000, these rules of Cardan will not be such a Scarecrow as they have hitherto been thought and we hope ere long such tables will be published here by Mr John Smith' (Collins to Oldenburg, 10 April 1675 = Royal Society MS LXXXI, No. 36; compare Oldenburg to Leibniz, 12 April 1675 [in C. I. Gerhardt, *Der Briefwechsel von G. W. Leibniz mit Mathematikern*, 1 (Berlin, 1899): 118]). The circumstances in which Newton was persuaded to help Smith by suggesting a suitable 'series for the maintaining

[In the first Table.][5]

Let n signify every 100th number and F its root whither Square Cube or Squaresquare. $n-50$ $n-40$ $n-30$ &c every tenth number and A B C D [&c] their Roots[,] o p q r &c the differences of these roots[,] op pq qr &c their 2d differences, (that is op the difference of o and p. pq the difference of p and q &c) and m their 3d difference[,] that is the common difference of $*o$ and op[&c]. Further let α β γ δ &c signify the difference of these roots from the next less[,] namely α the difference of A the root of $n-50$ and the like root of $n-51$, β the difference of [the roots of] $n-40$ &

[Tab 1]

$n-50$	A	$*o$	α	
	o	m	o	
$n-40$	B	op	β	
	p	m	π	
$n-30$	C	pq	γ	
	q	m	χ	
$n-20$	D	qr	δ	
	r	m	ρ	
$n-10$	E	rs	ϵ	
	s	m	σ	
n	F	st	ζ	$\dfrac{m}{10}$
	t	m	τ	
$n+10$	G	tv	$[\eta]$	
	v	m	υ	
$n+20$	H	vx	θ	
	x	m	ϕ	
$n+30$	I	xy	ι	
	y	m	ψ	
$n+40$	K	yz	κ	
	z	m	ω	
$n+50$	L	$z*$	λ	

Tab 2

$n-6$	^4E	
	$^5\epsilon$	
$n-5$	$^5E\,F^5$	
	ζ^4	
$n-4$	F^4	
	ζ^3	
$n-3$	F^3	
	ζ^2	
$n-2$	F^2	
	ζ^1	
$n-1$	F^1	
	ζ	
n	F	$\dfrac{st}{100}$
	$^1\zeta$	
$n+1$	^1F	
	$^2\zeta$	
$n+2$	^2F	
	$^3\zeta$	
$n+3$	^3F	
	$^4\zeta$	
$n+4$	^4F	
	$^5\zeta$	
$n+5$	$^5F\,G^5$	
	η^4	
$n+6$	G^4	
	η^3	
$n+7$	G^3	
	η^2	
$n+8$	G^2	
	η^1	
$n+9$	G^1	
	η	
$n+10$	G	$\dfrac{tv}{100}$
	$^1\eta$	
$n+11$	^1G	
	$^2\eta$	
$n+12$	^2G	&c

of the last differences in his tables true supposing he makes them but to ten places of figures' (Collins to Gregory, 1 May 1675 = *Gregory Volume*: 298) are not known but there can be no doubt that he did so at Collins' request, either by letter or through Barrow.

(2) ULC. Res 1894.9 (third manuscript), a transcript made by John Keill about 1710 from a 'Coppy' of Newton's (already lost?) autograph, newly found by William Jones 'among Mr Collins's papers', which had then just come into his possession. (Presumably, as was the case with his copies of Newton's following letters to Smith on 24 July and 27 August 1675, this primary *copia vera* was a transcript made by Smith for Collins' private use.) A variant secondary copy in Jones' own hand, somewhat fuller than Keill's version but lacking Newton's third concluding note, is now in Roger Cotes' papers in Trinity College, Cambridge (MS R. 16. 38: 308r–9r), having been enclosed by Jones in a letter to Cotes on 1 January 1711/12 (*ibid.*: 307r): both the letter and its enclosure were first published by Edleston in his *Correspondence of*

$n-41_{[,]}$ ζ the difference of the roots of n and $n-1_{[,]}$ $\eta^{(6)}$ the difference of the roots of $n+10$ and $n+9$ &c$_{[,]}$ and o π χ ρ &c signify the differences of α β γ [&c] and $\frac{m}{10}$ the common difference of o π χ ρ [&c].$^{(7)}$

In the 2^d Table.

Let $n-6$ $n-5$ $n-4$ $n-3$ &c signify the single numbers$_{[,]}$ ^4E ^5E or F^5 F^4 F^3 &c their roots$_{[,]}$ $^5\epsilon$ ζ^4 ζ^3 ζ^2 &c the differences of those roots$_{[,]}$ and $\frac{st}{100}$ the common differences of those differences for the ten numbers between $n-5$ and $n+5$[;] and so for the ten numbers between $n+5$ and $n+15$ let G^5 G^4 G^3 &c signify the roots$_{[,]}$ η^4 η^3 η^2 [&c] their first differences and $\frac{tv}{100}$ their 2^d differences, and the like for every denary between $n-50$ and $n+50$.$^{(8)}$

This Explication of the Table being premised you may compute them thus.

Out of n extract the $\begin{cases}\text{Square}\\\text{Cube}\\\text{Squaresquare}\end{cases}$ Root F. make

$$\begin{cases}\dfrac{10F}{2n}=\omega.^{(9)} & \dfrac{10\omega}{2n}=st. & \dfrac{30st}{2n}=m.\\[2mm]\dfrac{10F}{3n}=\omega. & \dfrac{20\omega}{3n}=st. & \dfrac{50st}{3n}=m.\\[2mm]\dfrac{10F}{4n}=\omega. & \dfrac{30\omega}{4n}=st. & \dfrac{70st}{5n}=m.\end{cases} \quad \omega+\tfrac{1}{2}st+\tfrac{1}{6}m=s. \quad \dfrac{\omega}{10}+\dfrac{\frac{1}{2}st}{100}=\zeta.^{(10)}$$

Sir Isaac Newton and Professor Cotes (London, 1850): 214–15 and 215–19 respectively. See also *The Correspondence of Isaac Newton*, **1**, 1959: 342–4; and D. C. Fraser, 'Newton's Interpolation Formulas', *Journal of the Institute of Actuaries* **51**, 1918: 77–106, especially 101–5 [= *Newton's Interpolation Formulas* (London, 1927): 25–9].

(3) In Jones' copy (note (2) above) the letter begins: 'I have consider'd y^e buisiness of computing Tables of Square, Cube, & Sq. Sq^r Roots; and y^e best way of p^rforming it, y^t I can think of is y^t which follows.'

(4) Jones' version reads 'compute'.

(5) This subheading, evidently omitted by Keill, is taken from Jones' transcript (note (2) above), and a 'Tab 1' is inserted correspondingly in the accompanying scheme.

(6) As in Jones' copy Keill's transcript omits 'η' in the fourth column of Table 1, while raising the five entries vertically beneath it (v, θ, ϕ, ι, ψ) up to the next higher row, thus creating a blank entry in line with 'y'. The error was evidently present in (Smith's?) primary copy of Newton's letter.

(7) Observe in each case that the first differences (o, p, q, ..., z; α, β, γ, ..., λ) are formed, in modern style, by taking upper from lower preceding quantities ($o = B-A$, $p = C-B$, and so on; but that the remaining second and third differences are computed, in Newton's usual way, by taking lower from upper preceding terms ($op = o-p$, $m = op-pq$, $\pi = \beta-\gamma$, $\frac{1}{10}m = \pi-\chi$, and so on). The effect of this inconsistency is to make all the differences (of

& $\dfrac{st}{10} + \dfrac{55m}{1000} = \sigma$. And these quantities F st m s ζ & σ being thus found the rest are given by addition and substraction.[11] For $st+m=rs$. $rs+m=qr$ &c. $st-m=tv$. $tv-m=vx$ &c. Again $s+rs=r$. $r+qr=q$ &c. $s-st=t$. $t-tv=v$ &c. Again $F-s=E$. $E-r=D$ &c. $F+t=G$. $G+v=H$ &c. Further $\sigma + \dfrac{m}{[10]} = \rho$. $\rho + \dfrac{m}{10} = \chi$ [&c]. $\sigma - \dfrac{m}{[10]} = \tau$. $\tau - \dfrac{m}{10} = \upsilon$ &c. Lastly $\zeta + \sigma = \epsilon$. $\epsilon + \rho = \delta$ [&c]. $\zeta - \tau = \eta$. $\eta - \upsilon = \theta$ &c.

These quantities being thus computed in the first table to every tenth number[,] the roots may be computed in the 2^d table to every number by addition and substraction only. for $\zeta + \dfrac{st}{100} = \zeta^1$. $\zeta^1 + \dfrac{st}{100} = \zeta^2$ [&c]. $\zeta - \dfrac{st}{[100]} = {}^1\zeta$. ${}^1\zeta - \dfrac{st}{[100]} = {}^2\zeta$ &c. again $F - \zeta = F^1$. $F^1 - \zeta^1 = F^2$ &c. $F + {}^1\zeta = {}^1F$. ${}^1F + {}^2\zeta = {}^2F$ &c. Thus you must proceed to five figures on either hand; and then doe the like in the next ten figures saying $\eta + \dfrac{tv}{100} = \eta^1$. $\eta^1 + \dfrac{tv}{100} = \eta^2$ [&c] and the like for every denary between $n-50$ and $n+50$.

In these computations Note[,] first that they must be done every where to 10 or 11 places if you will have a table of roots exact to 8 places.[12]

2dly if ${}^5F$ & G^5 the roots of $n+5$ found two waycs agree to 8 places it argues the whole work from wch they were derived to be true[,] and so of the Roots of $n+15$ $n+25$ $n-5$ &c and also [of] the terms $A *o$ & α[,] L $z*$ & λ where two works meet. [Let this therefore be y^e Proof of y^e work.][13]

$(n+i)^k$, $k = \frac{1}{2}, \frac{1}{3}, \frac{1}{4}$) positive. Since the contracted layout of Newton's array of differences may not be immediately comprehensible, a more modern equivalent scheme is given in the following 'Explanation'.

(8) Compare the preceding note. Here likewise Newton forms his first differences by taking, in modern style, upper from lower preceding quantities (${}^5\epsilon = {}^5E/F^5 - {}^4E$, $\zeta = F - F^1$ and so on) but constructs his second differences in his more usual converse way by taking lower from upper preceding first differences ($\frac{1}{100}st = \zeta^4 - \zeta^3$ and so on).

(9) This duplicate use of ω should be distinguished from the entry 'ω' $[= -\Delta^1 \delta^1_{39}]$ in the fourth column of Table 1. In the following 'Explanation' we denote it by $\omega^{[r]}$.

(10) Strictly, a term '$\dfrac{+\frac{1}{6}m}{1000}$' has been omitted from ζ's value (compare Edleston's *Correspondence* (note (2) above): 218, note *), but it is clear, as noticed by Fraser ('Newton's Interpolation Formulas' (note (2) above): 103, note †), that within Newton's order of approximation (10^{-9}) it is negligible and hence allowably—and designedly—omitted by him. Compare note (18).

(11) Jones' transcript (note (2) above) has the more Newtonian variant 'subduction'.

(12) See the final paragraph in the following 'Explanation'.

(13) This sentence is inserted from Jones' transcript (note (2) above), which in turn omits Newton's third observation following.

3[dly] the Square roots of all numbers between 2500 and 10000 and of all Cube numbers between 1250 and 10000 and of all Squaresquare numbers between

625 and 10000 are only to be thus computed. For the $\begin{Bmatrix} \text{Square} \\ \text{Cube} \\ \text{Squaresquare} \end{Bmatrix}$ roots of

numbers less than $\begin{Bmatrix} 2500 \\ 1250 \\ 625 \end{Bmatrix}$ may be found by halfing the $\begin{Bmatrix} \text{Square} \\ \text{Cube} \\ \text{Squaresquare} \end{Bmatrix}$ roots

of every $\begin{Bmatrix} 4 \\ 8 \\ 16 \end{Bmatrix}$ number below 10000. For $\frac{1}{2}\sqrt{9996}=\sqrt{2499}$ and $\frac{1}{2}\sqrt{9992}=\sqrt{2498}$.

And $\frac{1}{2}\sqrt{c\,9992}=\sqrt{c\,1249}$ and $\frac{1}{2}\sqrt{c\,9984}=\sqrt{c\,1248}$. And $\frac{1}{2}\sqrt{qq\,9984}=\sqrt{qq\,624}$ and $\frac{1}{2}\sqrt{qq\,9968}=\sqrt{qq\,623}$.[14]

Explanation. Between every hundredth root $(i=100p)$ in the sequence $[f_i =]\ (n+i)^k$, $k=\frac{1}{2}, \frac{1}{3}, \frac{1}{4}$, it is Newton's intention to subtabulate the 99 intervening ones by means of an appropriate scheme of differences. In outlining his procedure we may conveniently make use of the difference operators defined recursively by

$$\begin{Bmatrix} \Delta_i^1 = f_{i+10} - f_i \\ \Delta_i^r = \Delta_{i+10}^{r-1} - \Delta_i^{r-1} \end{Bmatrix} \quad \text{and} \quad \begin{Bmatrix} \delta_i^1 = f_{i+1} - f_i \\ \delta_i^r = \delta_{i+1}^{r-1} - \delta_i^{r-1} \end{Bmatrix}.$$

---

(14) In Jones' version the letter concludes: 'This S[r] is w[t] has occur'd to me about your design, which I hope will do your business, the whole work being p[r]form'd by Addit. & Subduct: excepting y[t] in y[e] computation of every 100[th] number, there is required y[e] Extraction of one root, & three divisions to find F, ω, st, & m' (see Edleston's *Correspondence* (note (2) above): 219). We may fittingly quote Cotes' reply to Jones in January 1712, giving him his 'hearty thanks' for sending along this transcript of Newton's letter: '[S[r] Isaac's] method seems to be excellently well suited to those particular purposes for which he design'd it, & I do not doubt I shall find it very curious when I have leasure to examine it to the bottom' (*ibid.*: 220).

(15) See his following letters to Smith of 24 July and 27 August 1675 (*Correspondence of Isaac Newton*, **1**, 1959: 348–9 and 350–1 respectively), enclosing 'what occurrs to mee...about facilitating y[e] Extrac[t]ion of [roots]'. As he wrote, his 'former Method might be applyed to determin all by every 1000[th]...[root], but not with advantage, for it will require the Extrac[t]ion of [roots] to 14 or 15 places, besides a greater number of Additions, Subduc[t]ions & Divisions in those greater numbers' (*ibid.*: 348). In preference to this bleak alternative he proposed that the value of each hundredth root A^k, $k = \frac{1}{2}, \frac{1}{3}, \frac{1}{4}$, should be extracted, either by 'common Arithmetick' or logarithmically, to five decimal places as B and that 'the other 6 places' should be obtained from this 'Quotient' by means of the approximate equality, accurate to $O(10^{-12})$, $A \approx k[(1/k-1)B + A/B^{1/k-1}]$. See the concluding appendix (pages 664–5) for the justification of this approximation and further details.

(16) We may be sure that Newton noticed the intriguing progression of the coefficients

As a first step Newton subtabulates every tenth root, interpolating on either side of $F=f_0$ five terms

m				
	*0	A	α	
m	0		o	
	op	B	β	$\tfrac{1}{10}m$
m	p		π	
	pq	C	γ	$\tfrac{1}{10}m$
m	q		χ	
	qr	D	δ	$\tfrac{1}{10}m$
m	r		ρ	
	rs	E	ε	$\tfrac{1}{10}m$
m	s		σ	
	st	F	ζ	$\tfrac{1}{10}m$
m	t		τ	
	tv	G	η	$\tfrac{1}{10}m$
m	v		υ	
	vx	H	θ	$\tfrac{1}{10}m$
m	x		φ	
	xy	I	ι	$\tfrac{1}{10}m$
m	y		ψ	
	yz	K	κ	$\tfrac{1}{10}m$
m	z		ω	
	z*	L	λ	

$$f_i = F\left[1 + \frac{i}{n}\binom{k}{1} + \frac{i^2}{n^2}\binom{k}{2} + \frac{i^3}{n^3}\binom{k}{3}\cdots\right]$$

($A=f_{-50}$, $B=f_{-40}$, ..., $E=f_{-10}$, $G=f_{10}$, ..., $L=f_{50}$) by constructing the first differences

$$\Delta_i^1 = F\left[\frac{10}{n}\binom{k}{1} + \frac{20(i+5)}{n^2}\binom{k}{2}\right.$$
$$\left. + \frac{30i(i+10)+1000}{n^3}\binom{k}{3}\cdots\right]$$

($o=\Delta_{-50}^1$, $p=\Delta_{-40}^1$, ..., $t=\Delta_0^1$, ..., $z=\Delta_{40}^1$) and from these the second differences

$$-\Delta_i^2 = F\left[-\frac{200}{n^2}\binom{k}{2} - \frac{600(i+10)}{n^3}\binom{k}{3}\cdots\right]$$

($*o = -\Delta_{-60}^2$, $op = -\Delta_{-50}^2$, ..., $tv = -\Delta_0^2$, ..., $z* = -\Delta_{40}^2$) and from these in turn, finally, the third differences (assumed equal)

$$m = +\Delta_i^3 = F\left[\frac{6000}{n^3}\binom{k}{3}\cdots\right].$$

The root $F=n^k$ is to be extracted outright by 'common Arithmetick' or, as Newton will later explain,[15] by an iteration procedure, while the entries s, st and m are to be constructed, on taking $\omega^{[\prime]}=(10k/n)F$, by computing

$$st = -\Delta_{-10}^2 = F\left[-\frac{200}{n^2}\binom{k}{2}\cdots\right] \approx -\frac{10(k-1)}{n}\times\omega^{[\prime]},$$

$$m = \Delta_i^3 = F\left[\frac{6000}{n^3}\binom{3}{k}\cdots\right] \approx -\frac{10(k-2)}{n}\times st,$$

and
$$s = \Delta_{-10}^1 = F\left[\frac{10}{n}\binom{k}{1} - \frac{100}{n^2}\binom{k}{2} + \frac{1000}{n^3}\binom{k}{3}\cdots\right] \approx \omega^{[\prime]} + \tfrac{1}{2}st + \tfrac{1}{6}m.^{(16)}$$

1, $\tfrac{1}{2}$, $\tfrac{1}{6}$, ... here, but at this time he could hardly have attained the insight that this expansion of $s = \Delta_{-10}^1$ is the opening of the Taylor series

$$f_0 - f_{-10} = -[(-10)f_0' + \tfrac{1}{2}(-10)^2 f_0'' + \tfrac{1}{6}(-10)^3 f_0''' \cdots],$$

where the p-th derivative $\qquad f_0^{[p]} = p!\binom{k}{p}n^{k-p}$

and hence $\omega^{[\prime]} = 10f_0'$ while, to Newton's order of accuracy,

$$\left(-10^{-4}\binom{k}{4}n^{k-4} \approx \tfrac{1}{24}10^4 f_0''''\right), \quad st \approx -100f_0'' \quad \text{and} \quad m \approx 1000f_0'''.$$

It is evident that, with F, s, st and m given, the difference table may be completed simply by addition and subtraction.

To subtabulate further (namely, by interpolating nine roots between each root f_i now found) Newton ingeniously introduces an allied difference scheme, in which the first column contains every tenth difference δ^1_{-10h-1}, where

$$\delta^1_i = F\left[\frac{1}{n}\binom{k}{1} + \frac{2i+1}{n^2}\binom{k}{2} + \frac{3i(i+1)}{n^3}\binom{k}{3}\cdots\right]$$

$(\alpha = \delta^1_{-51}, \ \beta = \delta^1_{-41}, \ \ldots, \ \epsilon = \delta^1_{-11}, \ \zeta = \delta^1_{-1}, \ \ldots, \ \lambda = \delta^1_{49})$, and then differences these, obtaining a second column of entries

$$-\Delta^1\delta^1_i = \delta^1_i - \delta^1_{i+10} = F\left[-\frac{20}{n^2}\binom{k}{2} - \frac{30(2i+11)}{n^3}\binom{k}{3}\cdots\right]$$

$(o = -\Delta^1\delta^1_{-51}, \ \pi = -\Delta^1\delta^1_{-41}, \ \ldots, \ \sigma = -\Delta^1\delta^1_{-11}, \ \tau = -\Delta^1\delta^1_{-1}, \ \ldots, \ \omega = -\Delta^1\delta^1_{39})$ and in turn from these a final column of 'third' differences

$$\Delta^2\delta^1_i = F\left[\frac{600}{n^3}\binom{k}{3}\cdots\right] \approx \tfrac{1}{10}m.^{(17)}$$

By calculation it follows at once that

$$\zeta = \delta^1_{-1} = F\left[\frac{1}{n}\binom{k}{1} - \frac{1}{n^2}\binom{k}{2} + \frac{1}{n^3}\binom{k}{3}\cdots\right] \approx \tfrac{1}{10}\omega^{[']} + \tfrac{1}{200}st + \tfrac{1}{6000}m^{(18)}$$

and $\sigma = -\Delta^1\delta^1_{-11} = F\left[-\frac{20}{n^2}\binom{k}{2} + \frac{330}{n^3}\binom{k}{3}\cdots\right] \approx \tfrac{1}{10}st + \tfrac{55}{1000}m$, and from these $\alpha = \delta^1_{-51}, \ \ldots, \ \lambda = \delta^1_{49}$ may all be obtained by addition and subtraction.

---

(17) This approximation is independent of Newton's present particular choice of generating function f_i. In a general scheme of differences, to $O(\delta^6)$, it is true that

$$\tfrac{1}{1000}\Delta^3_i - \tfrac{1}{100}\Delta^2\delta^1_{i+4\frac{1}{2}} = \tfrac{33}{8}\delta^5_0.$$

(18) Equivalently, as a Taylor expansion

$$\zeta = f_0 - f_{-1} = -[(-1)f'_0 + \tfrac{1}{2}(-1)^2 f''_0 + \tfrac{1}{6}(-1)^3 f'''_0 \cdots],$$

but see note (16). The last term $\tfrac{1}{6000}m \approx \tfrac{1}{6}f'''_0$ is negligible within Newton's order of approximation (10^{-9}).

(19) It is likewise a general property of any difference scheme (compare note (17)) that, to $O(\delta^5)$, $\tfrac{1}{100}\Delta^2_i - \delta^2_{i+9} = \tfrac{33}{4}\delta^4_0$.

(20) In an aside to Collins, entered below his transcript of Newton's letter to him of 27 August 1675, Smith observed in confirmation that 'I finde all the Roots (found by the longest Radius [*sc.* 10^{14}] of [Briggs'] logarithms [in his *Arithmetica Logarithmica* (London, 1624), presumably, or in Vlacq's 1629 amplification of Briggs' canon]) false & uncertaine from the 8th place of the Decimal onwards; though the logarithm it selfe & the work upon it be duly proved' (*Correspondence of Isaac Newton*, 1: 351).

(21) See notes (17) and (19).

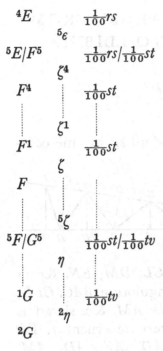

^4E $\frac{1}{100}rs$

$^5\epsilon$

$^5E/F^5$ $\frac{1}{100}rs/\frac{1}{100}st$

ζ^4

F^4 $\frac{1}{100}st$

ζ^1

F^1 $\frac{1}{100}st$

ζ

F

$^5\zeta$

$^5F/G^5$ $\frac{1}{100}st/\frac{1}{100}tv$

η

^1G $\frac{1}{100}tv$

$^2\eta$

^2G

It remains only to complete the column of first differences δ^1_i by interpolating nine mean differences between each in the second table ($^5\epsilon=\delta^1_{-6}$, $\zeta^4=\delta^1_{-5}$, ..., $\zeta^1=\delta^1_{-2}$, $\zeta=\delta^1_0$, ..., $^5\zeta=\delta^1_4$, $\eta^4=\delta^1_5$, ..., $^2\eta=\delta^1_{11}$). This Newton accomplishes by setting the 'common [second] difference' of the roots f_{i+10j}, $-5\leqslant i,j\leqslant 5$, as

$$-\tfrac{1}{100}\Delta^2_{10(j-1)},$$

that is, by choosing $\delta^2_{10j-1}=\frac{1}{100}\Delta^2_{10j-10}$,[19] and making $\delta^1_{i+10j-1}=\delta^2_{10j-1}$, $-5\leqslant i\leqslant 5$. Where 'two works meet', in the entries $\delta^2_{10j-1\pm5}$, the duplicate terms may suitably be harmonized; for instance, the entry $-\delta^2_{-6}$ can be a combination of the alternatives $\frac{1}{100}rs$ and $\frac{1}{100}st$ suggested by Newton's scheme.

An essential limitation on the accuracy of the roots constructed by this approach is set by the working assumption that the differences Δ^3_i, $\Delta^2\delta^1_i$ and δ^2_i are constant. Within Newton's chosen range of values of the argument ($k=\frac{1}{2}$, $n>2500$; $k=\frac{1}{3}$, $n>1250$; $k=\frac{1}{4}$, $n>625$) the implicit neglect of $-10^{-4}\Delta^4_i\approx -10^{-4}\binom{k}{4}n^{k-4}$ and $\delta^3_i\approx\binom{k}{3}n^{k-3}$ is of terms of $O(10^{-9})$, and so, as Newton himself realizes, the resulting table of roots will be exact only to eight places.[20] In comparison the errors in setting $\frac{1}{1000}\Delta^3_i=\frac{1}{100}\Delta^2\delta^1_{i+4\frac{1}{2}}$ and $\frac{1}{100}\Delta^2_i=\delta^2_{i+9}$, of respective orders δ^5_0 and δ^4_0,[21] are wholly negligible.

§2. RULES FOR SUBTABULATION DISCARDED FROM NEWTON'S 'EPISTOLA POSTERIOR' TO LEIBNIZ (OCTOBER 1676).[1]

From the original in the University Library, Cambridge[2]

— [in] Tabulam referenda.[3]

Tabula verò sinuum ex quâ hæc et alia multa dependent sic optimè computabitur. In angulo *HAI* inscribe æquales rectas *AB, BC, CD, DE, EF* &c et anguli *BAC, CBD, DCE, EDF* &[c] erunt in arithmetica progressione numerorum 1, 2,

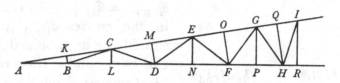

3, 4, 5 &c. Ad *AI* et *AH* demitte perpendicula *BK, CL, DM, EN*, &c et posito *AB* radio, erunt *AK, BL, CM, DN* &c cosinus angulorum *BAC, CBD, DCE, EDF*, &c. Quare cum sit *AB.AK :: AC.AL :: AD.AM.* &c, si radius *AB* statuatur unitas & detur primus cosinus *AK*, cæteri sic eruentur. Est $2AK = AC$. $AK \times AC = AL$. $AL - AB = BL$. $AL + BL = AD$. $AK \times AD = AM$. $AM - AC = CM$. $AM + CM = AE$. $AK \times AE = AN$. $AN - AD = DN$ &c.[4] Et eodem modo si una cum *AK* detur *FP* cosinus majoris alicujus anguli *GFP* multiplicis vel non multiplicis ipsius *BAK*, potest angulus *BAK* repetitis vicibus huic addi vel auferri. Scilicet est $AB^q - AK^q (BK^q) . AB^q :: FG^q - FP^q (GP^q) . AG^q$. $AK \times AG = AP$. $AP + FP = AH$. $AK \times AH = AQ$. $AQ - AG = GQ$. $AQ + GQ = AI$.

---

(1) The discarded draft sheet for Newton's second letter to Leibniz which contains these subtabulation rules is here reproduced in its entirety to show its salient points to best effect, although its opening paragraphs are—except for being inverted—little different from their equivalents in the final version posted to Oldenburg on 24 October (*Correspondence of Isaac Newton*, **2**, 1960: 110–29, especially 124–6). The reason for his deciding ultimately not to send these 'Regulæ' on to Leibniz is discussed in the preceding introduction, but it will be clear that they are merely offshoots of the more general interpolation methods codified in the two following manuscripts (§§3 and 4 below).

(2) ULC. Add. 3964.5: 3ʳ–4ᵛ, a single folded folio sheet recorded by Samuel Horsley on 23 October 1777 (on a slip now loose in ULC. Add. 4005) as 'From Nº 4 of selected papers.... A fragment in Latin concerning Interpolations which seems to have been part of a letter to Oldenburg'. Packaged with it, he noted, were 'A fragment in English concerning quadratures by equidistant ordinates' (now ULC. Add. 3964.4: 21ʳ, to be reproduced in the seventh volume) and 'Two Problems', but he added his judgement that 'None of this to be published unless the entire pieces should be found'. A century later, when the cataloguers of Lord Portsmouth's Newton collection came upon the allied manuscript 'Regula Differentiarum' (§3 following) they set it without explanation or ado in the same folder (ULC. Add. 3964.5) as the present draft sheet. Subsequently, D. C. Fraser was encouraged to conclude that the two pieces now thus placed together composed two parts of a single manuscript on interpola-

Translation

...to be entered into a table.[3]

But in fact a table of sines, on which these and many other things depend, will be best calculated in this way. Within the angle $H\hat{A}I$ inscribe equal line segments AB, BC, CD, DE, EF, ... and the angles $B\hat{A}C$, $C\hat{B}D$, $D\hat{C}E$, $E\hat{D}F$, ... will then be in arithmetical proportion as the numbers 1, 2, 3, 4, 5, ... To AI and AH let fall the perpendiculars BK, CL, DM, EN, ... and, on taking AB to be the radius, AK, BL, CM, DN, ... will be the cosines of the angles $B\hat{A}C$, $C\hat{B}D$, $D\hat{C}E$, $E\hat{D}F$, ... Hence, since $AB:AK = AC:AL = AD:AM = ...$, if the radius AB be set as unity and the first cosine AK be given, the remainder will be derived as follows: $2AK = AC$, $AK \times AC = AL$, $AL - AB = BL$; $AL + BL = AD$, $AK \times AD = AM$, $AM - AC = CM$; $AM + CM = AE$, $AK \times AE = AN$,

$$AN - AD = DN; \quad ....[4]$$

And in the same manner if together with AK there be given FP, the cosine of some larger angle $G\hat{F}P$ (a multiple of $B\hat{A}K$ or not), the angle $B\hat{A}K$ can repeatedly be added to or taken from it. For plainly

$$AB^2 - AK^2 \text{ (or } BK^2) : AB^2 = FG^2 - FP^2 \text{ (or } GP^2) : AG^2, \quad AK \times AG = AP,$$

$$AP + FP = AH, \quad AK \times AH = AQ, \quad AQ - AG = GQ; \quad AG + GQ = AI,$$

---

tion and this he duly published as such (*Journal of the Institute of Actuaries*, **58**, 1927: 53–95 [= *Newton's Interpolation Formulas* (London, 1927): 53–95], especially 75–84), carefully omitting the first three paragraphs of the present piece from his reproduced text since they 'appear in reversed order in the letter to Oldenburg of 24 October 1676...already published': correspondingly, on the inserted photocopy the lines preceding 'Succenturiare terminos...' have been painted out.

(3) Newton quotes the phrase (*Correspondence*, **2**: 114, line 11) immediately preceding the point in his *Epistola* at which the present sheet is to be inserted.

(4) More compactly, $2AK = AC$, $AK \times AC - AB = BL$, $AB + 2BL = AD$,

$$AK \times AD - AC = CM, \quad AC + 2CM = AE, \quad ....$$

In general terms, on setting $H\hat{A}I = \alpha$, $B = B_1$, $C = B_2$, $D = B_3$, ..., $K = b_1$, $L = b_2$, $M = b_3$, ..., it follows that $AB_1 = 1$, $Ab_1 = \cos\alpha$, $AB_n = \sin n\alpha/\sin\alpha$ and $B_{n-1}b_n = \cos n\alpha$. Newton's equations are then successive instances of the general scheme of recursion

$$AB_{n+1} = AB_{n-1} + 2B_{n-1}b_n, \quad B_n b_{n+1} = Ab_1 \times AB_{n+1} - AB_n.$$

It will be evident that this technique for constructing the cosines of multiple angles is essentially that expounded in Problem 8 of the geometrical appendix to his 'Observations' on Kinckhuysen's *Algebra* (see II: 444).

$AK \times AI = AR$. $AR - AH = HR$, &c. Et retro $AP - FP = AF$. $AK \times AF = AO$. $AG - AO = EO$. $AO - EO = AE$. $AK \times AE = AN$. $AF - AN = DN$. &c.[5] Sic igitur per unicam tantum multiplicationem, novus cosinus perpetim colligitur: quæ quidem multiplicatio ubi angulus A valde parvus est multum abreviabitur adhibendo sinum versum $AB - AK$ vice cosinus AK. Sit enim ille sinus versus v, et erit $v \times AC = x$. $AC - x = AL$. $v \times AD = y$. $AD - y = AM$. &c vel $v \times AG = x$. $AG - x = AP$, &c.[6]

Cognito itacʒ ex superioribus[7] Quadrantalis arcus longitudine (viz: 1.5707963267949 existente radio 1) sume partem ejus correspondentem angulo A.[8] Hujus partis quadratum dic z, & duo tresve primi termini hujus supra[9] expositæ seriei $\dfrac{z}{2} - \dfrac{zz}{24} + \dfrac{z^3}{720} - \dfrac{z^4}{40320}$ &c dabunt sinum versum anguli A; quo habito cosinus multiplicium ejus eruendi erunt per methodum jam explicatam. Sic primo cosinus senorum vel quinorum graduum inter 0^{gr} et 60^{gr} vel inter 30^{gr} et 90^{gr} computari possunt, & postea cosinus graduum vel dimidiorum graduum.[10] Nam non convenit progredi per nimios saltus. Postea cosinus reliquorum 30 graduum prodeunt ex his per solam additionem vel subductionem siquidem cosinus anguli alicujus N sit summa cosinuum angulorum $60^{gr} + N$ & $60^{gr} - N$, ut notum est.[11] Denicʒ cosinus intermedij ad decimas et centesimas partes graduum intercalandi erunt per methodum post descriptam:

---

(5) Since by supposition $AB = FG = 1$, when $FP = \cos \widehat{GFP} = \cos k\alpha$, say, is given, so is $AG = \sin k\alpha / \sin \alpha$ and hence we may apply the scheme of the previous note, finding in turn $P = b_k$, $H = B_{k+1}$, $Q = b_{k+1}$, ... and again, in inverse sequence $F = B_{k-1}$, $O = b_{k-1}$, ...

(6) At once, where $v = 1 - \cos \alpha$, there follows

$$1 - v = AL/AC = AM/AD = \ldots = AP/AG = \ldots.$$

In the revised version of this paragraph included in the letter as sent (*Correspondence*, **2**: 125) Newton amplified his scheme to include the step-by-step determination of $B_n b_n = \sin n\alpha$ as well, successively developing both

$$B_{n+1} b_{n+1} = 2 \sin \alpha B_{n-1} b_n + B_{n-1} b_{n-1} \quad \text{and} \quad B_n b_{n+1} = 2 \cos \alpha B_{n-1} b_n - B_{n-2} b_{n-1}.$$

(This is essentially Viète's recursive scheme, though derived by a different geometrical model; see I: 83, note (51).)

(7) Understand the text on *Correspondence*, **2**: 121–2, repeated virtually word for word by Newton from his 1671 tract (III: 224–6).

(8) In other words, Newton requires his reader to compute the radian measure of $\hat{A}$.

(9) Newton had sent the series for versin $z = 1 - \cos z$ to Leibniz in his *epistola prior* of 13 June 1676 (*Correspondence of Isaac Newton*, **2**: 25), an expansion overlooked by Leibniz in his reply of 17 August since he himself announced (*ibid.*: 61) the series for $\cos a$ as a 'direct' deduction by his own method. Earlier in his present letter Newton had called Leibniz' notice to the point with the words 'Credo Cl. Leibnitium, dum posuit seriem pro determinatione cosinus ex arcu dato, vix anim[ū] advertisse [ad] seriem meam pro determinatione sinus versi ex eodem arcu, siquidem hæc idem sunt' (*ibid.*: 122, corrected by ULC. Add. 3977.4).

$AK \times AI = AR$, $AR - AH = HR$; .... And backwards $AP - FP = AF$,

$AK \times AF = AO$, $AG - AO = EO$; $AO - EO = AE$, $AK \times AE = AN$,

$AF - AN = DN$; ....[5] So therefore, merely by a single multiplication, a new cosine is perpetually acquired—and this multiplication indeed, where the angle at A is exceedingly small, will be much shortened by employing the versed sine $AB - AK$ in place of the cosine AK. For let that versed sine be v and then $v \times AC = x$, $AC - x = AL$; $v \times AD = y$, $AD - y = AM$; ..., or alternatively

$$v \times AG = x, \quad AG - x = AP; \quad ....^{(6)}$$

Accordingly, when from the preceding discussion[7] the length of a quadrantal arc (namely 1.57079 63267 949 where the radius is 1) has been ascertained, take the part of it which corresponds to the angle $\hat{A}$.[8] Call the square of this part z and the first two or three terms of this previously[9] expounded series

$$\tfrac{1}{2}z - \tfrac{1}{24}z^2 + \tfrac{1}{720}z^3 - \tfrac{1}{40320}z^4 \cdots$$

will yield the versed sine of the angle $\hat{A}$; when you have it, the cosines of its multiples will be derived by the method now explained. Thus, in the first instance, the cosines of every fifth or sixth degree between 0° and 60° or between 30° and 90° can be computed, and subsequently the cosines of every degree or half degree.[10] For it is not advantageous to proceed by leaps which are too broad. Afterwards the cosines of the remaining thirty degrees are producible from these by addition or subtraction alone, seeing that the cosine of any angle N is the sum of the cosines of the angles $60° + N$ and $60° - N$, as is well known.[11] Intermediate cosines, finally, at every tenth and hundredth part of a degree will need to be intercalated by a method described below, but with the cosines

---

(10) Compare Alexander Anderson's scheme for an 'angular canon', noted by Newton early in 1665 (I, 3, 2, §3). Here, similarly, it is suggested that, when the cosines of angles of $5k°$ or $6k°$, $k = 1, 2, \ldots$, have been computed outright by the preceding cosine series (as $\cos\frac{1}{36}k\pi$ and $\cos\frac{1}{30}k\pi$ respectively), then the theorems

$\cos 2\alpha = 2\cos^2\alpha - 1$, $\cos 3\alpha = 4\cos^3\alpha - 3\cos\alpha$ and $\cos 5\alpha = 16\cos^5\alpha - 20\cos^3 + 5\cos\alpha$

derived from the above scheme should be used to subtabulate these to every degree or half degree.

(11) See, for example, Henry Briggs, *Trigonometrica Britannica: sive De Doctrina Triangulorum Libri Duo* (Gouda, 1633): 42, where it is introduced as an aid to the computation of sine tables in the form 'Si duorum Arcuum summa æquetur Gradibus 60:0': summa Sinuum datorum Arcuum æquabitur Sinui Arcus comp[o]siti è Gradibus 60, & alterutro datorum Arcuum'. Notice that Newton follows Briggs not only in suggesting that the cosines of angles be computed at intervals of every degree and half degree by angle bisection, trisection and quintusection, but ordaining (contrary to accepted contemporary practice) that further subdivision be made to hundredths (rather than sixtieths) of a degree.

substitutis tamen prius logarithmis inventorum cosinuum si Tabula artificialium sinuum[12] desideretur.

Logarithmorum etiam (nam lubet hic excurrere) constructio hæc videtur
optima. Per methodum supra traditam[13] quære Logarithmos Hyperbolicos
numerorum 10, 0.98, 0.99, 1.01, 1.02, id quod fit spatio unius & alterius horæ.
Dein divide Logarithmos quatuor posteriorum per Logarithmum numeri 10,
& addito indice[14] 2, habebis veros[15] Logarithmos numerorum 98, 99, 100, 101,
102 in Tabulam inserendos. Hos interpolabis[16] per dena intervalla, & habebis
Logarithmos omnium numerorum inter 980 & 1020; dein omnes inter 980 &
1000 iterum interpolabis per dena intervalla et habebis Tabulam eatenus
constructam. Dein ex his eruendi erunt Logarithmi omnium primorum
numerorum & eorum multiplicium minorum quam 100: Ad quod nihil
requiritur præter[17] additionem et substractionem siquidem sit

$$\sqrt{⑩ \frac{9984 \times 1020}{9945}} = 2. \quad \sqrt{④ \frac{8 \times 9963}{984}} = 3. \quad \frac{10}{2} = 5. \quad \sqrt{\frac{98}{2}} = 7. \quad \frac{99}{9} = 11.$$

$$\frac{1001}{7 \times 11} = 13. \quad \frac{102}{6} = 17. \quad \frac{988}{4 \times 13} = 19. \quad \frac{9936}{16 \times 27} = 23. \quad \frac{986}{2 \times 17} = 29. \quad \frac{992}{32} = 31.$$

$$\frac{999}{27} = 37. \quad \frac{984}{24} = 41. \quad \frac{989}{23} = 43. \quad \frac{987^{[18]}}{27} = 47. \quad \frac{9911}{11 \times 17} = 53. \quad \frac{9971}{13 \times 13} = 59.$$

$$\frac{9882}{2 \times 81} = 61. \quad \frac{9849}{3 \times 49} = 67. \quad \frac{994}{14} = 71. \quad \frac{9928}{8 \times 17} = 73. \quad \frac{9954}{7 \times 18} = 79. \quad \frac{996}{12} = 83.$$

$$\frac{9968}{7 \times 16} = 89. \quad \frac{9894}{6 \times 17} = 97.$$ Et habitis sic Logarithmis omnium numerorum
minorum quam 100, restat[19] tantùm hos etiam semel atqɜ iterum interpolare
per dena intervalla.[20]

---

(12) Read 'cosinuum' (cosines), strictly. With the following subtabulation rules omitted,
the letter as sent to Leibniz orders rather coyly at this point 'Tunc ad decimas et centesimas
partes graduum pergendum est per aliam methodum, substitutis tamen priùs Logarithmis
sinuum inventorum si ejus generis Tabula desideretur' (*Correspondence*, **2**: 126).

(13) Understand the text reproduced on *Correspondence*, **2**: 113–14, which is based closely
on an equivalent passage in Problem 9 of the 1671 tract (III: 226–30). Likewise the present
paragraph (repeated with few variants on *Correspondence*, **2**: 124) represents an augmented
version of III: 230–2.

(14) The equivalent 'præfixis indicibus' is cancelled.

(15) That is, 'common' (to base 10).

(16) Observe Newton's (now standard) distinction between 'interpolating' known values
and 'intercalating' new ones between these; compare III: 233, note (519).

(17) The equivalent 'Quod fit per solam' is cancelled.

(18) H. W. Turnbull's note at this point in the letter as sent to Leibniz (*Correspondence*, **2**:
158, note (61)) is in error: the numerical slip is present not only in this rejected sheet but also
in the transcript (Wickins', now ULC. Add. 3977.4) which Newton retained for his private use.

found replaced beforehand by their logarithms should a table of logarithmic sines[12] be wanted.

In the case of logarithms also (for that here affords a pleasant digression) this construction is seemingly the best. By the method recounted above[13] seek out the hyperbolic logarithms of the numbers 10, 0·98, 0·99, 1·01 and 1·02: this will take but an hour or two's time. Then divide the logarithms of the four latter ones by the logarithm of the number 10 and when the index 2 is added you will have the true[15] logarithms of the numbers 98, 99, 100, 101 and 102 ready to be inserted in the table. These you shall interpolate[16] by ten intervals at a time, so obtaining the logarithms of all numbers between 980 and 1020, and then all between 980 and 1000 you shall again interpolate by ten intervals at a time and so obtain the table thus far constructed. Next, from these will have to be ascertained the logarithms of all prime numbers and their multiples less than 100, to accomplish which nothing is required beyond[17] addition and subtraction, seeing that

$$\sqrt[10]{\frac{9984 \times 1020}{9945}} = 2, \quad \sqrt[4]{\frac{8 \times 9963}{984}} = 3, \quad \frac{10}{2} = 5, \quad \sqrt{\frac{98}{2}} = 7, \quad \frac{99}{9} = 11,$$

$$\frac{1001}{7 \times 11} = 13, \quad \frac{102}{6} = 17, \quad \frac{988}{4 \times 13} = 19, \quad \frac{9936}{16 \times 27} = 23, \quad \frac{986}{2 \times 17} = 29,$$

$$\frac{992}{32} = 31, \quad \frac{999}{27} = 37, \quad \frac{984}{24} = 41, \quad \frac{989}{23} = 43, \quad \frac{987}{2[1]} = 47,$$

$$\frac{9911}{11 \times 17} = 53, \quad \frac{9971}{13 \times 13} = 59, \quad \frac{9882}{2 \times 81} = 61, \quad \frac{9849}{3 \times 49} = 67, \quad \frac{994}{14} = 71,$$

$$\frac{9928}{8 \times 17} = 73, \quad \frac{9954}{7 \times 18} = 79, \quad \frac{996}{12} = 83, \quad \frac{9968}{7 \times 16} = 89, \quad \frac{9894}{6 \times 17} = 97.$$

And when the logarithms of all numbers less than 100 have been obtained in this way, it remains[19] merely to interpolate these also once and then again by ten intervals at a time.[20]

(19) When, namely, a table of the logarithms of the numbers 1 to 10000 is required.

(20) In place of the remainder of this sheet, omitted *in toto* from the letter finally sent, Newton wrote: 'Quæ de hoc genere Tabularum dicuntur, ad alias transferri possunt ubi ratiocinia geometrica locum non obtinent. Sufficit autem per has series computare triginta vel viginti aut fortè pauciores terminos Tabulæ in debitis distantijs, siquidem termini intermedij facilè interseruntur per Methodum quandam quam in usum Calculatorum ferè hic descripsissem. Sed pergo ad alia' (*Correspondence*, 2: 126). He evidently appreciated that this omission left the preceding sketch of the construction of logarithmic and cosine tables hanging somewhat in the air, for in his covering letter to Oldenburg on 24 October (*ibid.*: 110) he spoke of his 'answer' as 'too tedious. I could wish I had left out some things since to avoid greater tediousness I left out something else on wch they have some dependance. But I had rather you should have it any way, then write it over again being at present otherwise incumbred'.

Succenturiare[21] terminos Tabulæ alicujus ubi primæ differentiæ sufficiunt, ut ferè fit in Logarithmis, res obvia est et nota. Sed ubi secundæ, tertiæ, aliæqȝ differentiæ interveniunt, desiderari videtur Regula generalis eliciendi has differentias & more optimo componendi. Formulam[22] quæ mihi maximè placet, licet vix sit hujus loci, & minimè digna quæ perscribatur ad Leibnitium,[23] cum tamen applicabilis sit ad computum Tabularum cujuscunqȝ generis, non gravabor hic fusiùs describere in usum Calculatorum, quibus siquos noveris[24] et res tanti videatur communicare possis.

Reg. 1.

In Tab 1 &c.

Reg. 2.

Quinetiam &c.

Reg 3.

Aliquando &c.[25]

---

(21) The rare verb 'succenturiare' occurs in classical Latin only in the military phrase 'to admit to [a vacant place in] a century' (that is, a squad of 100 men) and in the derived sense 'to substitute', one still found occasionally in contemporary speech. Only three months before, for example, Oldenburg had written to Leibniz (on 26 July): 'en tibi varia et accumulata Collinii nostri communicata...donec...alia a D^{no} Newtono succenturientur' (C. I. Gerhardt, *Der Briefwechsel von G. W. Leibniz mit Mathematikern*, **1** (Berlin, 1899): 169). In context the intended significance of Newton's 'subcentury' (framed on the analogy of 'subdivide', perhaps?) is precisely defined and we cannot accept D. C. Fraser's more general rendering of it into English as 'The process of supplying the missing [terms]' (*Newton's Interpolation Formulas* (note (2) above): 85).

(22) From the last paragraph of this sheet (see note (47) below) it seems clear that Newton's general formula is the general interpolation theorem—for equal differences of the argument?—developed in §§3/4 following (with adjusted and divided differences respectively).

(23) Newton was already at this time aware of Leibniz' considerable interest in methods of interpolating by means of finite differences, for Oldenburg had earlier sent him a transcript of Leibniz' letter to him of 3 February 1672/3: there, in apparent ignorance of its previous publication by Briggs and Mercator, Leibniz had presented his variant on the formula (first discovered by Thomas Harriot in still unpublished researches) for expressing the general value

$$f_x = f_0 + \binom{x}{1}\Delta_0^1 + \binom{x}{2}\Delta_0^2 \ \ldots$$

in terms of the (advancing) differences of given instances tabulated at equal intervals of the argument x. (See the preceding introduction, particularly note (15).) Though in hindsight we may know it to have been false, it was a reasonable assumption on Newton's part that Leibniz in the three years intervening had probed deeper into the theory of finite-difference interpolation.

(24) It is, of course, Oldenburg who is addressed.

To subtabulate the terms of any table by hundredths[21] when, as usually happens in logarithms, first differences suffice is an obvious matter and one well known. But when second, third and other differences come into it, a general rule for eliciting these differences and compounding them to best effect would seem desirable. The formula[22] which I find most agreeable, though barely relevant to my present purpose and wholly unworthy to be passed on in a letter to Leibniz,[23] I shall none the less—since it is applicable to the computation of tables of any kind whatsoever—not hesitate to describe here somewhat lavishly for the use of calculators: to these, if you[24] know any and the matter seems important enough, you might communicate it.

Rule 1. In Table 1....

Rule 2. Indeed....

Rule 3. Sometimes....[25]

---

(25) The accompanying sheet on which Newton amplified these first three rules (and the following observation) is lost. From what Newton has to say of them in the sequel it is clear that all three rules are concerned with placing a general term $A+p$ 'corresponding' to some intermediate term $S+n$ (by $A+p = f_n$, say) within the sequence of 'cardinal' terms $A = f_0$, $B = f_1$, $C = f_2$ and $D = f_3$. Further, the maximum error of Rule 1 is stated to be roughly $\frac{1}{16} = \binom{\frac{1}{2}}{3}$ times the ignored third difference, that of Rule 3 to be about

S	A
$S+n$	$A+p$
$S+1$	B
$S+2$	C
$S+3$	D

$\frac{3}{128} = \binom{\frac{3}{2}}{4}$ times the ignored fourth difference. From this we may conjecture that Rule 1 and Rule 3 are instances of the simple advancing difference formula

$$p = f_n - f_0 = ne + \binom{n}{2}f + \binom{n}{3}g + \binom{n}{4}h \dots,$$

where

$$e = \Delta_0^1 = B - A, \quad f = \Delta_0^2 = C - 2B + A, \quad g = \Delta_0^3 = D - 3C + 3B - A, \quad h = \Delta_0^4 = \dots,$$

and so on: namely,

(Rule 1) $p \approx n(B-A) + \frac{1}{2}(n^2 - n)(C - 2B + A)$

and (Rule 3) $p \approx n(B-A) + \frac{1}{2}(n^2 - n)(C - 2B + A) + \frac{1}{6}(n^3 - 3n^2 + 2n)(D - 3C + 3B - A)$.

Since for $0 \leqslant n \leqslant 2$ the maximum value of $\pm\binom{n}{3}$ (when $n = 1 \pm 1/\sqrt{3}$) is $1/9\sqrt{3} \approx \frac{1}{16}$, while similarly (when $n = \frac{3}{2} \pm \frac{1}{2}\sqrt{5}$ and $\frac{3}{2}$) for $0 \leqslant n \leqslant 3$ it follows that $-\frac{1}{24} \leqslant \binom{n}{4} \leqslant \frac{3}{128}$, this restoration explains Newton's assessment of the 'maximum' errors in his Rules, $|\frac{1}{16}g|$ and $|\frac{3}{128}h|$ respectively. No further reference is made to Rule 2. Fraser states that here 'Values are supposed to be given at unequal intervals and the formula is true to second differences' (*Newton's Interpolation Formulas* (note (2) above): 55), but this is only his guess. Perhaps, in line with Rule 7 below, Newton inverted his Rule 1

$$\left(p = ne + \binom{n}{2}f + O(g)\right) \quad \text{as} \quad n = \frac{p}{e} - \binom{p/e}{2}\frac{f}{e} + O\left(\frac{g}{e}\right),$$

thus deriving the argument $S+n$ which corresponds to the general term $A+p$.

In hisce computationibus ------- & e contra.[26]

Hoc etiam observandum est quod errores maximi in Regula prima sunt decima sexta pars tertiarum differentiarum terminorum cardinalium A, B, C, circiter; & quod in tertia Regula sunt tantùm $\frac{3}{128}$ sive $\frac{1}{43}$ pars istarum[27] differentiarum circiter. Atcɜ adeo prima Regula tutò adhiberi potest ubi tertiæ differentiæ non sunt majores quam 8, & tertia Regula ubi quartæ differentiæ non sunt majores quam 21.[28] Sic enim error non excedet dimidium unitatis in ultimis figuris Tabulæ. Et proinde si Cardinales Termini non erraverint dimidio unitatis in ultimis locis, Tabula vix errabit integra unitate in istis ultimis locis. Sed interim notandum est, quod intercalando dena loca, in Reg 3 additiones & subductiones millesimarum partium differentiarum tertiarum (quæ partes sunt differentiæ tertiæ terminorum intercalandorum) fieri debent ad tria figurarum loca ultra ultimas figuras cardinalium Terminorum, & eædem operationes in secundis differentijs in Reg. 1 & 3 fieri debent ad duo loca, & in primis differentijs ad unum locum ultra istas ultimas figuras,[29] si Tabulam ad uscɜ istas ultimas figuras desideras accuratam.

Reg 4.

Siquando unicus terminus intercalandus est. Sume Quatuor continuos terminos A, B, C, D in quorum medio novus terminus desideratur, & summam duorum mediorum $B+C$ dic R ac summam extremorum $A+[D]$ dic S: eritcɜ $\frac{9R-S}{16}$ terminus intercalandus, errore existente minore quam quadragesima pars differentiæ quartæ terminorum A, B, C, D, E: nempe $\frac{3}{128}$ pars istius differentiæ circiter, ut in Reg 3.[30] Nam hæ duæ Regulæ habent idem fundamentum, ut et ea quæ sequitur.

Reg. 5.

Si seriem[31] A, B, C, D velis interpolare in duobus continuis locis inter B et C: intercalandi[32] termini erunt $\dfrac{60B+30C-5A-4D}{81}$ & $\dfrac{30B+60C-4A-5D}{81}$, errore existente minore quam $\frac{5}{243}$ pars præfatæ differentiæ quartæ, hoc est vix quinquagesima pars ejus.[33]

---

(26) These opening and closing words of Newton's observation are too vague to allow its restoration. We may perhaps conjecture that it had to do with the signs of the various differences: thus, $e = B-A$ is to be taken positive when $B > A$, and 'conversely so' when $B < A$.

(27) Understand 'quartarum' (fourth).

(28) $8 = \left[\dfrac{16}{2}\right]$; $21 = \left[\dfrac{128}{2 \times 3}\right]$.

(29) 'cardinalium terminorum' (of the cardinal terms) is cancelled.

In computations of this sort...and conversely.[26]

This also should be observed: the greatest errors in the first Rule are, roughly, one-sixteenth of the third differences of the cardinal terms A, B and C; but in the third Rule are only a $\frac{3}{128}$ (that is, $\frac{1}{43}$) part of those[27] differences, roughly. In consequence the first Rule may safely be employed when the third differences are not greater than 8, and the third Rule when the fourth differences are not greater than 21.[28] For in this way the error will not exceed half a unit in the last figures of the table, and as a result, if the cardinal terms were not in error by half a unit in the final places, the table will scarcely err by a whole unit in those final places. But at the same time it should be noted that, in intercalating ten places at a time in Rule 3, the additions and subtractions of thousandths of the third differences (which are the third differences of the terms to be intercalated) ought to be performed to three places of figures beyond the last ones of the cardinal terms, and the same operations ought in second differences, in Rules 1 and 3, to be performed to two places, and in first differences to one place, beyond those final figures[29] if you want the table to be accurate up to those final figures.

Rule 4. On occasion a single term has to be intercalated. Take four successive terms A, B, C and D, in whose middle a new term is desired, call the sum, $B+C$, of the two middles R and that, $A+D$, of the two extremes S: then $\frac{1}{16}(9R-S)$ will be the term to be intercalated, with an error less than one-fortieth of the fourth difference of the terms A, B, C, D, E—to be more precise, a $\frac{3}{128}$th part of that difference or thereabout, as in Rule 3.[30] For these two Rules, like that which follows, have the same foundation.

Rule 5. Should you wish to interpolate the series[31] A, B, C, D in two successive places between B and C: the[32] terms to be intercalated will be

$$\tfrac{1}{81}(60B+30C-5A-4D) \quad \text{and} \quad \tfrac{1}{81}(30B+60C-4A-5D),$$

with an error less than a $\frac{5}{243}$rd part of the aforesaid fourth difference—in other words, barely a fiftieth of it.[33]

---

(30) In the terms of note (25) the 'terminus intercalandus' is, by Rule 3,

$$f_{\frac{3}{2}} = A+\tfrac{3}{2}e+\tfrac{3}{8}f-\tfrac{1}{16}g+\tfrac{3}{128}h \ldots,$$

that is, to $O(\tfrac{3}{128}h)$, $f_{\frac{3}{2}} = \tfrac{1}{16}[9(B+C)-(A+D)]$.

(31) 'terminorum continuam' (of terms in sequence) is cancelled.

(32) Newton first added 'desiderati duo' (two desired). See also note (16).

(33) Here, to order $\binom{\frac{4}{3}}{4}h = \binom{\frac{5}{3}}{4}h = \dfrac{5}{243}h$, the terms to be intercalated are (by Rule

$$f_{\frac{4}{3}} = A+\tfrac{4}{3}e+\tfrac{3}{9}f-\tfrac{5}{81}g = \tfrac{1}{81}(-5A+60B+30C-4D)$$

and $$f_{\frac{5}{3}} = A+\tfrac{5}{3}e+\tfrac{5}{9}f-\tfrac{5}{81}g = \tfrac{1}{81}(-4A+30B+60C-5D).$$

Reg. 6.

Sed ubi differentiæ quartæ tantæ sunt ut hæ Regulæ nequeant adhiberi, sume sex continuos terminos A, B, C, D, E, F in quorum medio terminus intercalandus est. Et fac $C+D=R$. $B+E=S$ et $A+F=T$. Et erit $\dfrac{150R-25S+3T}{256}$ terminus ille intercalandus, errore existente minore quàm ducentesima pars differentiæ sextæ terminorum A, B, C, D, E, F, G. nempe $\frac{5}{1024}$ istius differentiæ circiter.[34] Hac Regula itaqჳ interpolabitur Tabula usqჳ dum Regula quarta vel Tertia adhiberi possit. Atqჳ ita omnis difficultas Tabularum eo ferè redigitur ut sciamus invenire paucos cardinales Terminos ab invicem uniformiter distantes: ad quod Series nostræ[35] plerumqჳ conducent.[36]

Reg. 7.

Hisce demum[37] Regulam adjiciam in usum constructorum Tabularum. Norunt Calculatores[38] eruere Terminum e Tabula aliqua correspondentem numero dato, & vice versa numerum correspondentem Termino dato per proportionales partes differentiarum quando differentiæ primæ ad id sufficiunt.[39] Sed ubi differentiæ secundæ aliæqჳ interveniunt hæretur. Sint igitur A, B, C, D termini in Tabula[40] numeris $S-1$. S. $S+1$. $S+2$ correspondentes: et sit $B+p$ terminus intermedius correspondens numero intermedio $S+n$. Sit etiam e prima differentia terminorum A & C, f secunda differentia terminorum A, B, C et g tertia differentia terminorum A, B, C, D, et si ex dato $S+n$[41] quæratur $B+p$ erit

$S-1$	A	G
S	B	H
$S+n$	$B+p$	$H+q$
$S+1$	C	I
$S+2$	D	K

$$\tfrac{1}{2}ne+\tfrac{1}{2}nnf-\frac{n-n^3}{6}g=p:$$

Errore existente exigua parte quartæ differentiæ terminorum A, B, C, D, E.[42]

---

(34) To order $\binom{\frac{5}{2}}{4}\Delta_0^6 = -\dfrac{5}{1024}\Delta_0^6$, on taking $i = \Delta_0^5 = F-5E+10D-10C+5B-A$, the term here to be intercalated is

$$f_{\frac{5}{2}} = A+\tfrac{5}{2}e+\tfrac{15}{8}f+\tfrac{5}{16}g-\tfrac{5}{128}h+\tfrac{3}{256}i = \tfrac{1}{256}(150(C+D)-25(B+E)+3(A+F)).$$

The corresponding results for $f_{\frac{7}{2}}$ and $f_{\frac{9}{2}}$, no doubt established by the 'Newton–Stirling' central difference formula (see note (42)), were first published by James Stirling in his 'Methodus Differentialis *Newtoniana* Illustrata' (*Philosophical Transactions* **30**, No. 362 [for September/October 1719]: 1050–70, especially 1062; see also his *Methodus Differentialis: sive Tractatus de Summatione et Interpolatione Serierum Infinitarum* (London, 1730): Pars Secunda, Propositio XXXIII: 152–3).

(35) The logarithmic, exponential and trigonometric series listed by Newton for Leibniz both in his *Epistola prior* of 13 June 1676 (*Correspondence*, **2**: 25–9) and earlier in the present letter (*ibid*.: 121–5). Compare note (10) above.

Rule 6. But when fourth differences are so great that these rules cannot be employed, take six successive terms A, B, C, D, E and F, in whose middle a term is to be intercalated, and make $C+D = R$, $B+E = S$, $A+F = T$: then

$$\tfrac{1}{256}(150R - 25S + 3T)$$

will be the term to be intercalated, with an error less than a two-hundredth of the sixth difference of the terms A, B, C, D, E, F, G, *viz.* $\tfrac{5}{1024}$ of that difference or thereabouts.[34] By this rule, consequently, the table will be interpolated until the point where Rule 4 or 3 may be applied. And so all the difficulty of tabulation is virtually reduced to our knowing how to find a few cardinal terms uniformly distant one from another: for this our series[35] will for the most part serve.[36]

Rule 7. To these, lastly, let me add a[37] rule for the benefit of constructors of tables. Calculators[38] are familiar with how to derive a term which, in some tabulation, corresponds to a given number and, conversely, how to elicit the number corresponding to a given term by proportional parts of the differences when first differences are sufficient for it.[39] But when second and other differences come into play, they are at a loss. Let therefore A, B, C, D be the terms in a table[40] which correspond to the numbers $S-1$, S, $S+1$, $S+2$ and suppose $B+p$ the intermediate term corresponding to the intermediate number $S+n$; further, let e be the first difference of the terms A and C, f the second difference of the terms A, B and C, and g the third difference of the terms A, B, C and D: then if $B+p$ in line with the given number $S+n$[41] be sought, $p = \tfrac{1}{2}ne + \tfrac{1}{2}n^2 f - \tfrac{1}{6}(n-n^3)\,g$ with an error which is a slight fraction of the fourth difference of the terms A, B, C, D, E.[42] But in practice it is sufficient to put

---

(36) This last sentence replaces 'Cæterùm his regulis non tantùm Tabulæ construi, sed et series...componantur' (For the rest, by these rules not only [can] tables be constructed but also series...may be composed), evidently copied in part from a lost prior draft. Understand in the gap something like 'erui possunt ex quibus termini omnes intermedij' (can be derived, from which all intermediate terms).

(37) 'unam' (single) is cancelled.

(38) Newton first wrote 'Mathematici' (arithmeticians).

(39) Namely, in the case of those simple (logarithmic and trigonometric) functions whose values are in 1, 1 correspondence with their arguments and whose rate of growth changes smoothly.

(40) In the accompanying figure Newton first inserted a last entry 'E' in the middle column (corresponding to a next term $S+3$ in the left-hand column) and then cancelled it. The right-hand column is not referred to in the text, but evidently it was intended that the intermediate term $H+q = \phi_n \equiv \phi(S+n)$ is to be inserted in the sequence $G = \phi_{-1}$, $H = \phi_0$, $I = \phi_1$ and $K = \phi_2$.

(41) Since $B+p = f_n \equiv f(S+n)$ is constructed by Newton from the sequence $A = f_{-1}$, $B = f_0$, $C = f_1$ and $D = f_2$, he perhaps intended to write 'e regione dati $S+n$' and so we translate it.

(42) Here (compare note (25)) $e = C-A = (B-A)+(C-B)$, $f = (C-2B+A)$ and $g = (D-3C+3B-A)$ and so the formula for $p = f_n - f_0$ is an instance of the 'Newton-

Sed usibus sufficiat ponere $\dfrac{ne+nnf}{2}=[p]$ quia tunc error $\dfrac{n-n^3}{6}\,g$ semper erit minor decima quinta parte tertiæ differentiæ g.[43] Sic igitur ex dato $S+n$ invenitur $B+p$. Et vice versa si ex dato $B+p$ quæratur $S+n$ erit

$$n=\frac{2p}{e}-\frac{4ppf}{e^3}+\frac{2ppf-8p^3}{3e^4}\,g$$

errore existente vix majore quam quarta differentia terminorum divisa per primam.[44] Sed sufficit ponere $n=\dfrac{2p}{e}-\dfrac{4ppf}{e^3}$. Hic suppono differentias e, f, g collectas esse auferendo superiores terminos de inferioribus (A d[e] B, B de C &c) & superiorum differentias de inferiorum differentijs,[45] & proinde quando e vel f vel g sic prodit negativa quantitas, signum ejus in Regulis mutandum erit. Et idem observandum est de signo quantitatum n et p.

Has Regulas derivo a consideratione nostrarum[46] infinitarum serierum concipiendo terminos Tabulæ exprimi per talem seriem puta

$$a+bx+cxx+dx^3 \quad \&c$$

ubi indefinita quantitas x denotat differentiam inter numerum termino correspondentem et alium quemvis proximum numerum datum, & bx, cxx, dx^3 &c quantitates sunt a quibus differentiæ primæ, secundæ, tertiæ, cæteræcʒ originem trahunt.[47]

Sed de his plusquam satis.

Stirling' expansion $p = \dfrac{1}{2}\dbinom{n}{1}\,[\Delta^1_{-1}+\Delta^1_0]+\dfrac{n^2}{2!}\,\Delta^2_{-1}+\dfrac{1}{2}\dbinom{n+1}{3}\,[\Delta^3_{-2}+\Delta^3_{-1}]\ldots$ (see §4: note (19) below), in which third differences are assumed constant. The implicit assumption is made that $-1 \leqslant n \leqslant 1$, when the neglected fourth difference $\left|\dfrac{1}{4}n\dbinom{n+1}{3}\Delta^4_{-2}\right|$ will be at most (when $n^2=\dfrac{1}{2}$) $\dfrac{1}{96}\Delta^4$. See also note (47).

(43) The maximum value $1/9\sqrt{3}$ (slightly less than $\frac{1}{15}$) of $\pm\dbinom{n+1}{3}$, $-1 \leqslant n \leqslant 1$, is attained when $n^2=\frac{1}{3}$.

(44) An error has crept in here. When the root n of

$$n-2(p/e)+n^2(f/e)+\tfrac{1}{3}(n^3-n)\,(g/e)+\tfrac{1}{12}(n^4-n^2)(h/e)\ldots=0$$

(in which $h=\Delta^4_{-2}$) is extracted, we will find successively

$$n = 2(p/e)+\alpha(f/e) = 2(p/e)-4(p^2/e^2)\,(f/e)+\beta(g/e)$$

$$= 2(p/e)-4(p^2/e^2)\,(f/e)-\tfrac{1}{3}(8p^3/e^3-2p/e)\,(g/e)+8(p^3/e^3)\,(f^2/e^2)+\gamma(h/e).$$

Even when terms in f^2 are assumed negligible in comparison with those in eg, the last term in Newton's value for n should read ' $+\dfrac{2pee-8p^3}{3e^4}\,g$ '.

$p = \frac{1}{2}(ne + n^2 f)$ since then the error $\frac{1}{6}(n - n^3) g$ will always be less than a fifteenth of the third difference g.[43] In this way, then, in line with given $S + n$ is found $B + p$. And conversely if in line with given $B + p$ is required $S + n$, there will be $n = 2p/e - 4p^2 f/e^3 + \frac{1}{3}(2p^2 f - 8p^3) g/e^4$ with an error scarcely greater than the fourth difference of the terms divided by the first.[44] But it is enough to put $n = 2p/e - 4p^2 f/e^3$. Here I suppose the differences e, f and g to be gathered by taking away upper terms from lower ones (A from B, B from C, and so on) and differences of upper ones from those of lower ones;[45] accordingly when e, f or g turns out on this assumption to be negative, its sign will have to be changed in the rules. The same observation holds for the signs of the quantities n and p.

These rules I derive from an attentive inspection of our[46] infinite series by conceiving the terms of a table to be expressed by some such series as

$$a + bx + cx^2 + dx^3 \ldots,$$

where the variable quantity x denotes the difference between the number corresponding to a term and any other given number nearest to it, while bx, cx^2, dx^3, ... are quantities from which the first, second, third and other differences take their origin.[47]

But of these things more than enough.

────────────

(45) In agreement with modern convention, that is.

(46) Newton's and Leibniz'? Or does he mean just his own?

(47) This fundamental assumption is developed at length in §4 below. As an instance, the expansion cited in the previous paragraph (see note (42)) is derived at once by setting $f_x \equiv f(S + x) = a + bx + cx^2 + dx^3$, for then

$$\begin{cases} f_{-1} = A = a - b + c - d \\ f_0 = B = a \\ f_1 = C = a + b + c + d \\ f_2 = D = a + 2b + 4c + 8d \end{cases} \quad \text{and so} \quad \begin{cases} B = a \\ \frac{1}{2}e = \frac{1}{2}(C - A) = b + d \\ \frac{1}{2}f = \frac{1}{2}(C - 2B + A) = c \\ \frac{1}{6}g = \frac{1}{6}(D - 3C + 3B - A) = d \end{cases};$$

hence $p = f_n - f_0 = bn + cn^2 + dn^3 = (\frac{1}{2}e - \frac{1}{6}g)n + \frac{1}{2}fn^2 + \frac{1}{6}gn^3$.

§3. THE 'REGULA DIFFERENTIARUM' (OCTOBER 1676?).[1]

From the original in the University Library, Cambridge[2]

REGULA DIFFERENTIARUM.[3]

Præceptum primum. Pro vario usu Regulæ varia erit ejus descriptio.[4] Unde multiplex videbitur, licet fundamentaliter unica sit.[5] Primus usus quem describemus erit in computo Tabularum omnis generis. Nempe ex paucis Tabulæ alicujus construendæ terminis æquali intervallo ab invicem distantibus possunt loca intermedia per hanc Regulam impleri ut sequitur.

Incipiendo a majorum[6] terminorum extremo aufer, secundum terminum de primo[,] tertium de secundo, quartum de tertio & sic perge ad usqꝫ ultimum terminum scribendo e regione differentias divisas per numerum intervallorum quibus termini ab invicem distant. Dein incipiendo a majorum[6] differentiarum extrema aufer secundam de prima, tertiam de secunda, quartam de tertia[,] sic deinceps pergendo ad ultimam et scribendo e regione secundas hasce differentias

---

(1) These three precepts for subtabulation by finite differences and a following general rule for interpolation by adjusted differences are entered on a hastily composed, extensively cancelled and much overwritten worksheet which was apparently never revised by Newton into a more finished form. (Its roughness and disarray are well illustrated in the photocopy of its first page which is reproduced in frontispiece to the present volume.) In part at least, Newton's text is closely related to the preceding rejected sheet from his *Epistola posterior* to Leibniz in 1676 (§2 above), but at the same time its greater depth and generality clearly marks an advance upon the relatively crude difference schemes and particularized interpolation rules there expounded. We may perhaps guess that it represents a further abortive attempt to communicate to Leibniz the essence of Newton's new-found insight into finite differences: certainly, a dating of late 1676 is strongly suggested by its handwriting.

(2) ULC. Add. 3964.5: 1^r–2^r, published in full—with some considerable errors in transcription and interpretation—by D. C. Fraser in his 'Newton's Interpolation Formulas. An unpublished Manuscript by Sir Isaac Newton', *Journal of the Institute of Actuaries*, **58**, 1927: 78–84 [= *Newton's Interpolation Formulas* (London, 1927): 78–84]; compare Fraser's 'Newton and Interpolation' in (ed. W. J. Greenstreet) *Isaac Newton, 1642–1727* (London, 1927): 45–69, especially 54–8. We should, in particular, firmly resist his assertion ('Newton and Interpolation': 58) that 'The propositions of the *Methodus* [essentially §4 following] must have been before him when [Newton] was compiling the *Regula*, but when he came to review them, it may be supposed that the idea came to him that the laborious algebra of the *Methodus* was quite unnecessary, and that the fundamental proposition could be presented in a new form as a self-contained scheme which verified itself'. This conjecture is based on Fraser's prior assumption (*ibid.*: 46) that 'it appears to be probable that [the *Methodus Differentialis*] was written quite early in the period 1665–1671'. Granted that he did not know of Newton's autograph original (ULC. Add. 4004: 82^r–84^r) of the *Methodus*, firmly set by its handwriting in the mid-1670's, it should none the less have been evident to him that the clean-cut, logically ordered text of a work printed in 1711 (see §4: note (1) below) was not a prior state of a rough manuscript left unfinished by Newton thirty-five years before.

Translation

THE RULE OF DIFFERENCES.[3]

First Precept. Corresponding to its varying applications there will be a variety of written forms[4] of the Rule. Consequently, though it is fundamentally unique[5] it will appear to be manifold. The first application which we shall describe will be to the computation of tables of every sort. To be exact, from foreknowledge of a few terms set equally distant from one another in some table to be constructed, intermediate positions may be filled in as follows by this Rule.

Starting with the extreme of the greater[6] terms, take away successively the second term from the first, the third from the second, the fourth from the third and so on up to the last term, writing in line the differences divided by the number of intervals by which the terms differ one from another. Next, starting with the extreme of the greater[6] differences, take away successively the second one from the first, the third from the second, the fourth from the third and so on up to the last, writing in line these second differences divided by the same

---

(3) By this 'Rule of Differences' we should understand not so much the 'una regula generalis' of advancing adjusted differences which concludes the present piece as the concept which underlies all interpolation by finite differences. As Newton wrote to Leibniz on 24 October 1676, 'Ejus fundamentum est commoda expedita et generalis solutio hujus Problematis[:] *Curvam geometricam describere quæ per data quotcunꝗ puncta transibit....* quamvis prima fronte intractabile videatur; tamen res aliter se habet. Est enim fere ex pulcherrimis quæ solvere desiderem' (*Correspondence of Isaac Newton*, **2**, 1960: 119). In the present group of manuscripts it is Newton's basic assumption that all functions to be interpolated may be expressed as the infinite parabola $f_x = a + bx + cx^2 + ...$ (see especially §2: note (47) and §4 following), but he might equally as well have preceded James Stirling in his preference for the 'Harriot–Briggs' form $f_x = A + B\binom{x}{1} + C\binom{x}{2} ...$ $(A = f_0, B = \delta_0^1, C = \delta_0^2, ...)$, which is finite for integral $x \geqslant 0$. (Compare James Stirling's remark in his *Methodus Differentialis: sive Tractatus de Summatione et Interpolatione Serierum Infinitarum* (London, 1730: 'Pars Secunda de Interpolatione': 108) that 'notandum est formam Ordinatæ...quam assumit *Newtonus* in demonstrando suæ methodi fundamento, male huic negotio destinatam esse. Nam valor cujusque Coefficientis prodit in Serie infinita'.) Equally, we may be sure that Newton would not have spurned the use in suitable cases of Stirling's 'hyperbola Logarithmica' (*Methodus*: 143) $f_x' = \alpha + \beta r^x + \gamma r^{2x} + \delta r^{3x} ...$: indeed, as we shall see in the final volume, in a concluding scholium to the printed version of §4 below he foreshadowed the employment of the general curve $F_x = af(0) + bf(x) + cf(x^2) + df(x^3) ...$ (William Jones, *Analysis Per Quantitatum Series, Fluxiones, ac Differentias* (London, 1711): 101).

(4) 'applicatio' (application) is cancelled.

(5) All variants of the rule, when expressed in explicit polynomial form, reduce to the same unique function $f_x = a + bx + cx^2 + dx^3 ...$ which is fundamental to Newton's scheme.

(6) Newton assumes that all values of the function f_x to be interpolated are laid out in an arithmetical progression from less to greater x. Observe, in consequence, that his (adjusted) differences are throughout taken contrary to modern convention; as a result the odd (first, third, ...) differences will be opposite in sign to their present-day equivalents.

per eundem intervallorum numerum divisas. Sic et harum secundarum differentiarum collige differentias tertias dividendo omnes per eundem numerum & scribendo e regione. [7]Tertiæ autem differentiæ ubi sunt minores quam 16[8] negligi debent et tunc operis forma hæc erit.[9] Sint A, B, C &c termini dati, n numerus locorum quibus distant ab invicem puta denarius ut solet, a, a^2, a^3, a^4 &c termini inter hos inveniendi[,] K, K^2, K^3 &[c] differentiæ primæ per n divisæ, L, L^2, L^3 [&c] differentiæ secundæ per n divisæ. Hoc est

$$K = \frac{A-B}{n}, \quad K^2 = \frac{B-C}{n}, \quad K^3 = \frac{C-D}{n} \quad \&c.$$

$$L = \frac{K-K^2}{n}, \quad L^2 = \frac{K^2-K^3}{n} \quad \&c.^{[10]} \quad \text{Pone } t=L, \ v=L^2.$$

$x = L^3$ &c et inter t, v, x &c cape t^6, v^6, x^6 &c media arithmetica. Et erunt t, t^6, v, v^6, x, x^6 &c differentiæ secundæ terminorum quæsitorum[11] a, a^2[,] a^3 &c. Fac $K + \frac{1}{2}t^6 = p^5$. $p^5 + t^6 = p^4$. $p^4 + t = p^3$. $p^3 + t = p^2$. $p^2 + t = p$. Item $p^5 - t^6 = p^6$. $p^6 - t^6 = p^7$. $p^7 - v = p^8$. $p^8 - v = p^9$. $p^9 - v = p^{10}$. $p^{10} - v = q$. Rursus

$$K^2 + \frac{1}{2}v^6 = q^5. \quad q^5 + v^6 = q^4. \quad q^4 + v^6 = q^3.$$

$q^3 + v = q^2$. $q^{[2]} + v = q$. et si q sic bis inventus sibi congruit, argumento erit recti operis. Eodem modo inveniendi sunt termini q^6, q^7, q^8 &c addendo continuo v^6 ad q^5, q^6 & x ad q^7, q^8, q^9 &c. Et hæ quantitates p, p^2, p^3 &c erunt differentiæ primæ terminorum a, a^2, a^3 [&c]. Unde termini illi colligentur faciendo $A - p = a$, $a - p^2 = a^2$,

$$a^2 - p^3 = a^3, \quad a^3 - p^{[4]} = a^4, \quad a^4 - p^{[5]} = a^5,$$

$$B + p^{10} = a^9, \quad a^9 + p^9 = a^8, \quad a^8 + p^8 = a^7,$$

$a^7 + p^7 = a^6$, $a^6 + p^{[6]} = a^5$ & ex duplici inventione a^5 probabitur opus. Eodem modo inter B, C, D

L	A		t
	a	p	
	a^2	p^2	
	a^3	p^3	
	a^4	p^4	
K	a^5	p^5	t^6
	a^6	p^6	
	a^7	p^7	
	a^8	p^8	
	a^9	p^9	
L^2	B	p^{10}	v
	b	q	
	b^2	q^2	
	b^3	q^3	
	b^4	q^4	
K^2	b^5	q^5	v^6
	b^6	q^6	
	b^7	q^7	
	b^8	q^8	
	b^9	q^9	
L^3	C	q^{10}	x
	c	r	
	c^2	r^2	
	c^3	r^3	
	c^4	r^4	
K^3	c^5	r^5	x^6
	&c		

---

(7) Newton first continued 'Quartas differentias suppono [negligendas?]' (the fourth differences I take [to be negligible?]).

(8) Strictly, $9\sqrt{3}$; that is, $1/\mathrm{Max}\left|\binom{n}{3}\right|$, $0 \leqslant n \leqslant 2$. See §2: note (25).

(9) In sequel (and in his accompanying difference table) Newton originally wrote l, z, l^2, z^2, l^3, z^3 for t, t^6, v, v^6, x, x^6 but failed to make the change-over in several places of his text. It is clear (see the frontispiece plate) that he also intended to cancel the superscript '1' in all occurrences of a^1, b^1, c^1, p^1, q^1, r^1, K^1 and L^1. In the text here reproduced all these intended transpositions and cancellations have silently been made. Note, too, that consistently in our

number of intervals. In the same manner, too, from the second differences gather third differences, dividing them all by the same number and writing them in line. [7]But third differences should be neglected when they are less than 16[8] and in that event the work will have this form.[9] Let A, B, C, ... be the given terms; n the number of places they are distant from one another (say 10, as is usually the case); a_1, a_2, a_3, a_4, ... the terms to be found between these; K_1, K_2, K_3, ... the first differences divided by n; L_1, L_2, L_3, ... the second differences divided by n: that is, $K_1 = (A - B)/n$, $K_2 = (B - C)/n$, $K_3 = (C - D)/n$, ...; $L_1 = (K_1 - K_2)/n$, $L_2 = (K_2 - K_3)/n$, ...[10] Put $t_1 = L$, $v_1 = L_2$, $x_1 = L_3$, ... and between t_1, v_1, x_1, ... take the arithmetic means t_6, v_6, x_6, ... Then will t_1, t_6, v_1, v_6, x_1, x_6, ... be the second differences of the terms a_1, a_2, a_3, ... required.[11] Make $K_1 + \frac{1}{2}t_6 = p_5$, $p_5 + t_6 = p_4$, $p_4 + t_1 = p_3$, $p_3 + t_1 = p_2$, $p_2 + t_1 = p$; and likewise $p_5 - t_6 = p_6$, $p_6 - t_6 = p_7$, $p_7 - v_1 = p_8$, $p_8 - v_1 = p_9$, $p_9 - v_1 = p_{10}$, $p_{10} - v_1 = q_1$. And again $K_2 + \frac{1}{2}v_6 = q_5$, $q_5 + v_6 = q_4$, $q_4 + v_6 = q_3$, $q_3 + v_1 = q_2$, $q_2 + v_1 = q_1$. It will then be a check on the accuracy of the work if the two values of q_1 found in this way agree. In the same manner the terms q_6, q_7, q_8, ... are to be found by continually adding v_6 to q_5, q_6 and x_1 to q_7, q_8, q_9, ... and so on. These quantities p_1, p_2, p_3, ... will then be the first differences of the terms a_1, a_2, a_3, ... Hence those terms are gathered by making $A - p_1 = a_1$, $a_1 - p_2 = a_2$, $a_2 - p_3 = a_3$, $a_3 - p_4 = a_4$, $a_4 - p_5 = a_5$, $B + p_{10} = a_9$, $a_9 + p_9 = a_8$, $a_8 + p_8 = a_7$, $a_7 + p_7 = a_6$, $a_6 + p_6 = a_5$ and the operation will be tested from agreement of the doubly found value of a_5. You must proceed in the same manner between B, C and D, and

English version we have, in accord with modern practice, lowered all numerical suffixes to subscript position.

(10) Newton's original sentence ran equivalently 'Sint A, B, C &c termini dati, a^1, a^2, a^3, a^4 &c termini inter hos inveniendi, K, K^2, K^3, &[c] differentiæ primæ, L, L^2, L^3 [&c] differentiæ secundæ; & z^1, z^2, z^3 [*sc.* t^6, v^6, x^6] media arithmetica inter differentias secundas. Hoc est $z^1 = \dfrac{L^1 + L^2}{2}$. $z^2 = \dfrac{L^2 + L^3}{2}$ &c. Sit n numerus locorū quibus termini A, B, C, D ab invicem distant, puta denarius ut solet. Fac $\dfrac{1}{nn}L^1 = l^1$, $\dfrac{1}{nn}L^2 = l^2$, $\dfrac{1}{nn}L^3 = l^3$ &c.' In other words, in Newton's first scheme the first and second order differences K^i, L^i were not adjusted.

(11) This replaces 'intercalandorum' (to be intercalated).

pergendum est et complebitur Tabula quæsita. Potest totum opus per additionem compleri pergendo a q^5 ad p^5, ab r^5 ad q^5, item ab B ad A, a C ad B, sed quacunqꝫ ratione pergas debes notare quod ubi differentiarum K, K^2, K^3 inferiores sunt majores superioribus[12] Differenti[æ] t, t^6, v, v^6, x, x^6 &c debent addi et subduci ordine contrario quam hic descripsi, eo ut differentiarum p, p^2, p^3, [&c] inferiores itidem superioribus majores reddantur. Hoc etiam notandum est. Si numerus differentiarum [terminorū] a, a^2, a^3 &c (cui numerus n æqualis est) impar sit, media differentia[13] quæ stat inter K et t^6 non erit $K+\frac{1}{2}t^6$ vel $K-\frac{1}{2}t^6$ sed K.[14]

[*Præceptum 2.*] Eadem computandi methodus observari potest ubi numeri intercalandi distant inæqualibus locis[:] ut si A et B distant tribus locis[,] B et C quinque[,] C et D quatuor &c. Divide differentias terminorum A, B, C, D &c per numerum intervallorum[15] faciendo $\dfrac{A-B}{3}=K$.

$\dfrac{B-C}{5}=K^2$, $\dfrac{C-D}{4}=K^3$ &c et statue K, K^2, K^3 ad media intervallorum AB, BC, CD. Dein divide K^2-K[16] per 4[17] numerum nempe locorum quibus distant, et K^2-K^3 per $4\frac{1}{2}$[17] numerum locorum quibus illa distant &c et habebis differentias secundas L^2, L^3 &c.[18] Tunc quia K stat e regione intervalli inter a et a^2 fac $K=p^2$ nimirum differentiæ primæ quam respicit, [&] $p^2-L^2=p^3$. Et quia K^2 stat e regione intervalli inter b^2 et b^3 fac $K^2=q^3$. $q^3+L^2=q^2$. $q^2+L=q$. $q^3-L^3=q^4$. $q^4-L^3=q^5$. Sed quia K^3 stat e regione c^2 sive inter r^2 et r^3 fac $K^3+\frac{1}{2}L^3=r^3$. $K^3-\frac{1}{2}L^3=r^2$. $r^2-L^3=r$. $r^3+L^4=r^4$ &c.[19]

L	K	A	p	[t]
	K —	a	p	
		a^2	p^2	
L^2		B	p^3	v
		b	q	
	K^2 —	b^2	q^2	
		b^3	q^3	
		b^4	q^4	
L^3		C	q^5	x
		c	r	
	K^3 —	c^2	r^2	
		c^3	r^3	
L^4		D	r^4	y

---

(12) The clarification 'de quibus auferri jussi' (from which I ordered them to be taken) is cancelled.

(13) Newton first wrote 'medius terminus' (mean term) in error.

(14) The particular case, for equal differences ($n = 10$) of argument and when third differences are neglected, of Precepts 2 and 3 below. See notes (19) and (24).

(15) The amplification 'AB, BC, CD [&c]' is cancelled.   (16) Read '$K-K^2$'.

(17) That is, $\frac{1}{2}(3+5)$ and $\frac{1}{2}(5+4)$ respectively, since A, B, C and D differ by 3, 5 and 4 places.

(18) The text has an uncancelled following phrase 'una cum earum medijs arithmeticis z^2, z^3 &c' (together with their arithmetic means z_2, z_3, ...) which is superfluous in Newton's revise. Compare note (10).

(19) The generalization of Precept 1 to arguments listed at unequal intervals. Since Newton neglects third differences, his general interpolation function is $f_x = a+bx+cx^2$ and we may set $A=f_\alpha$, $B=f_\beta$, $C=f_\gamma$, $D=f_\delta$ (and hence $p^2 = -\delta^1_{\frac{1}{2}(\alpha+\beta-1)} = -b-c(\alpha+\beta)$,

$$q^3 = -\delta^1_{\frac{1}{2}(\beta+\gamma-1)} = -b-c(\beta+\gamma), \quad r^2 = -\delta^1_{\frac{1}{2}(\gamma+\delta)-1} = -b-c(\gamma+\delta-1),$$

$$r^3 = -\delta^1_{\frac{1}{2}(\gamma+\delta)} = -b-c(\gamma+\delta+1), \quad t = v = x = y = +\delta^2 = 2c.$$

the required table will be filled up. The whole operation can be completed by addition on proceeding from q_5 to p_5 and from r_5 to q_5, and likewise from B to A and from C to B, but whatever procedure you use you should note that, when the lower of the differences K_1, K_2, K_3, ... are greater than the upper ones,[12] then the differences t_1, t_6, v_1, v_6, x_1, x_6, ... should be added and subtracted in the order contrary to that I have here described, so as to make the lower of the differences p, p_2, p_3, ... correspondingly greater than the upper ones. This too must be noted: if the number of differences of the terms a_1, a_2, a_3, ... (equal to the number n) be odd, the mean difference[13] standing between K_1 and t_6 will not be $K_1 + \frac{1}{2}t_6$ or $K_1 - \frac{1}{2}t_6$ but K_1.[14]

Precept 2. The same method of computation can be kept when the intervals between the numbers to be intercalated are unequal in length. So if A and B are three places distant, B and C five, C and D four, and so on, divide the differences of the terms A, B, C, D, ... by the corresponding number of the intervals,[15] making $\frac{1}{3}(A-B) = K_1$, $\frac{1}{5}(B-C) = K_2$, $\frac{1}{4}(C-D) = K_3$, ..., and place K_1, K_2, K_3 at the mid-points of the intervals AB, BC, CD. Next, divide $K_1 - K_2$ by 4,[17] namely the number of places by which they are distant, and $K_2 - K_3$ by $4\frac{1}{2}$,[17] the number of places by which those are distant, and so on, and you will have the second differences L_2, L_3, ...[18] Then, because K_1 stands in line with the interval between a_1 and a_2, make K_1 equal to p_2 (that is, to the first difference which it faces) and $p_2 - L_2 = p_3$; and, since K_2 stands in line with the interval between b_2 and b_3, make $K_2 = q_3$, $q_3 + L_2 = q_2$, $q_2 + L_1 = q_1$, $q_3 - L_3 = q_4$, $q_4 - L_3 = q_5$; but, because K_3 stands in line with c_2 or the interval between r_2 and r_3, make $K_3 + \frac{1}{2}L_3 = r_3$, $K_3 - \frac{1}{2}L_3 = r_2$, $r_2 - L_3 = r_1$, $r_3 + L_4 = r_4$, and so on.[19]

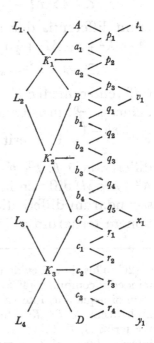

Then, in the case of the interval AB (since $\beta - \alpha$ is odd) first- and second-order adjusted differences are constructed as

$$K = \frac{A-B}{\beta-\alpha} = -(\alpha, \beta) = -b - c(\alpha+\beta) = p^2, \quad K^2 = \frac{B-C}{\gamma-\beta} = -(\beta, \gamma) = -b - c(\beta+\gamma) = q^3;$$

and $L^2 = \dfrac{K-K^2}{\frac{1}{2}(\gamma-\alpha)} = \dfrac{-(\alpha, \beta)+(\beta, \gamma)}{\frac{1}{2}(\gamma-\alpha)} = 2(\alpha, \beta, \gamma) = 2c = v$. So for the interval BC, since $\gamma - \beta$ is likewise odd. But in the case of the interval CD (since $\delta - \gamma$ is even) the adjusted differences are constructed as $K^3 = \dfrac{C-D}{\delta-\gamma} = -(\gamma, \delta) = -b - c(\gamma+\delta) = \frac{1}{2}(r^2+r^3)$ and

$$L^3 = \frac{K^2-K^3}{\frac{1}{2}(\delta-\beta)} = 2(\beta, \gamma, \delta) = 2c = x.$$

[*Præceptum 3.*] Hæc ita se habebunt ubi differentiæ tertiæ sunt minores quam 8. At ubi sunt majores quam 8, quartæ vero differentiæ non majores quam 32,[20] sint M, M^2, M^3 vel x, y, z [&c] tertiæ differentiæ[21] per n divisæ. Et si n sit numerus par, puta[22] 10 Fac $\dfrac{nn-4}{24}=\lambda$, & $K+\lambda\times M+\frac{1}{2}[s]^6$ atcB

$$K+\lambda\times M-\tfrac{1}{2}[s]^6$$

erunt differentiæ duæ mediæ p^5 & p^6 et $K^2+\lambda\times M^2+\frac{1}{2}[t]^6$ &

$$K^2+\lambda\times M^2-\tfrac{1}{2}[t]^6$$

differentiæ mediæ q^5 & q^6 et sic deinceps.[23] Sin n sit numerᵘˢ impar Fac $\dfrac{nn-1}{24}=\lambda$ et erit $K+\lambda\times M$

		A		s	
[L]			p		w^6
		a	p^2	s^2	
		a^2	p^3	s^3	
		a^3	p^4	s^4	
[M]	[K]	a^4	p^5	s^5	
		a^5	p^6	s^6	
		a^6	p^7	s^7	x
		a^7	p^8	s^8	
		a^8	p^9	s^9	
[L²]		a^9	p^{10}	s^{10}	
		B	q	t	
		b	q^2	t^2	x^6
[M²]	[K²]	b^2	q^3	t^3	
		b^3			[y]

differentiarū p, p^2, p^3 [&c] media[,][24] illa nempe quæ stat e regione K, et $K^2+\lambda\times M^2$ differentiarum q, q^2, q^3 &c media.[25] Postea complenda est columna secundarum differentiarum s, s^2, s^3 &[c per] continuā additionem & subductionem tertiarum differentiarum x y z et arithmeticè mediarum w^6 x^6 y^6 &c.

---

(20) This should evidently be '21', that is, [43/2]; see §2: note (28). (Perhaps he absentmindedly computed [$4^3/2$] in error?) Newton first continued: 'sint ut supra A, B, C, D, &c termini æqualibus intervallis ab invicem distantes, n numerus intervallorum quibus distant ab invicem, K, K^2, K^3 differentiæ primæ per n divisæ, L, L^2, L^3 secundæ, &c' (let, as before, the terms A, B, C, D, ... be distant at equal intervals from one another, n the number of intervals by which they are distant from one another, K_1, K_2, K_3 the first differences divided by n, L_1, L_2, L_3 the second ones [divided by n^2], ...).

(21) 'pariter' (likewise) is cancelled.

(22) Newton first wrote '6 vel 10' (6 or 10). The accompanying diagram illustrates this case of subtabulation by tenths. For clarity's sake we have eliminated from it a superfluous entry 'a^{10}' ($=B$) but have inserted, in line with the previous schemes, columns of first-, second- and third-order adjusted differences K, K^2; L, L^2; M, M^2.

(23) The value of λ should be '$-\dfrac{nn-4}{24}$' in all cases. In this extension of Precept 1 the third differences $M\approx x$, $M^2\approx y$, $M^3\approx z$, ... are no longer regarded as negligible. If the general function to be interpolated, $f_x = a+bx+cx^2+dx^3+O(x^4)$, is given at the even n-intervals $A=f_{-\frac{1}{2}n}$, $B=f_{\frac{1}{2}n}$, $C=f_{\frac{3}{2}n}$, ..., then Newton's adjusted differences are

$$K = \frac{A-B}{n} = -(-\tfrac{1}{2}n,\ \tfrac{1}{2}n) = -n^{-1}\Delta^1_{-\frac{1}{2}n} = -\underline{\Delta}^1_{-\frac{1}{2}n} = -b-\tfrac{1}{4}dn^2,$$

Precept 3. These rules hold where the third differences are less than 8. Where, however, they are greater than 8 but the fourth differences are not greater than 32,[20] let M_1, M_2, M_3, ... or x_1, y_1, z_1, ... be the third differences[21] divided by n. Then, if n be an even number, say[22] 10, make $\frac{1}{24}(n^2-4) = \lambda$ and

$$K_1 + \lambda M_1 + \tfrac{1}{2}s_6, \quad K_1 + \lambda M_1 - \tfrac{1}{2}s_6$$

will be the two middle differences p_5, p_6, while $K_2 + \lambda M_2 + \tfrac{1}{2}t_6$, $K_2 + \lambda M_2 - \tfrac{1}{2}t_6$ will be the two middle differences q_5, q_6, and so forth.[23] But if n be an odd number, make $\frac{1}{24}(n^2-1) = \lambda$ and $K_1 + \lambda M_1$ will be the middle one of the differences p_1, p_2, p_3, ...,[24] namely that which stands in line with K_1 and $K_2 + \lambda M_2$ the middle one of the differences q_1, q_2, q_3, ...[25] Subsequently the column of second differences s_1, s_2, s_3, ... must be completed by continual addition and subtraction of the third differences x_1, y_1, z_1, ... and their arithmetic means w_6, x_6, y_6, ... Next, by

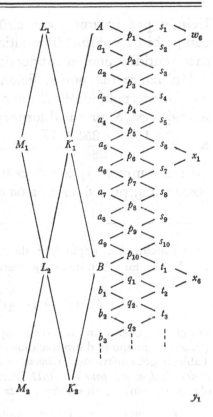

$$K^2 = -\underline{\Delta}^1_{\frac{1}{2}n} = -b + 2cn - \tfrac{13}{4}dn^2,$$

$$L = +2(-\tfrac{3}{2}n, -\tfrac{1}{2}n, \tfrac{1}{2}n) = n^{-2}\Delta^2_{-\frac{3}{2}n} = \underline{\Delta}^2_{-\frac{3}{2}n} = 2c - 3dn,$$

$$L^2 = \frac{K - K^2}{\frac{1}{2}(2n)} = +\underline{\Delta}^2_{-\frac{1}{2}n} = 2c + 3dn$$

and $$M = \frac{L - L^2}{\frac{1}{3}(3n)} = -6(-\tfrac{3}{2}n, -\tfrac{1}{2}n, \tfrac{1}{2}n, \tfrac{3}{2}n) = -n^{-3}\Delta^3_{-\frac{3}{2}n} = -\underline{\Delta}^3_{-\frac{3}{2}n} = -6d,$$

with $$p^5 = -\delta^1_{-1} = -b + c - d, \quad p^6 = -\delta^1_0 = -b - c - d,$$

$$s^6 = \tfrac{1}{2}(p^5 - p^6) = +\delta^2_{-1} = 2c \quad \text{and} \quad x = -\delta^3_{-1} = -6d = M,$$

so that finally $\frac{1}{2}(p^5 + p^6) = -\mu\delta^1_{-\frac{1}{2}} = -b - d = K - \frac{1}{24}(n^2-4)M$. A generalization of this sub-tabulation rule was later developed by Roger Cotes in his *Canonotechnia sive Constructio Tabularum per Differentias* [= *Opera Miscellanea* (appended to Robert Smith's edition of his *Harmonia Mensurarum*) (Cambridge, 1722): 35–71]: Propositio IX: 60–3, especially 63. Cotes there showed, by successive reduction between Newton–Bessel and Newton–Stirling expansions of $\frac{1}{2}(f_x - f_{-x})$ in terms of the various differences $\underline{\Delta}^{2k+1}_{-\frac{1}{2}(2k+1)n}$ (his $(2k+1)$-th 'ordines rotundi') and $\mu\delta^{2k+1}_{-\frac{1}{2}(2k+1)}$ (his 'ordines quadrati Mediales') respectively that

$$\underline{\Delta}^1_{-\frac{1}{2}n} = \mu\delta^1_{-\frac{1}{2}} + \frac{\frac{1}{4}n^2 - 1}{3!}\mu\delta^3_{-\frac{3}{2}} + \frac{(\frac{1}{4}n^2-1)(\frac{1}{4}n^2-4)}{5!}\mu\delta^5_{-\frac{5}{2}} \cdots;$$

Deinde per harum secundarū differentiarum additionem et subductionem complebitur colūna differentiarum primarum $p\ p^2\ p^3$ [&c]. Ac tandem per earum additionem et subductionem colligentur[26] numeri quæsiti $a\ a^2\ a^3$ &c.

Ubi hæ regulæ non sufficiunt sint A, B, C, D, E, F sex continui termini in medio interpolandi,[27] R summa duorum $C+D$[28] inter quos terminus interserendus est, S summa duorum proximorum $B+E$, T summa extimorum $A+F$ & erit[29] $\dfrac{150R-25S+3T}{256}$ terminus in medio loco inter C ac D consistens, errore existente minore quam ducentesima pars differentiæ sextæ terminorum propositorum[,] puta si seriei ea est natura ut differentiæ semper crescant regulariter

---

from which, upon expanding the 'powers' $\underline{\Delta}^{2k+1}_{-\frac{1}{2}(2k+1)n}$ in terms of the various $\mu\delta^{2l+1}_{-\frac{1}{2}(2l+1)}$, $k \leqslant l \leqslant kn$, and then inverting, there results

$$\mu\delta^1_{-\frac{1}{2}} = \underline{\Delta}^1_{-\frac{1}{2}n} - \frac{n^2-4}{2^2.3!}\,\underline{\Delta}^3_{-\frac{3}{2}n} + \frac{(n^2-4)(9n^2-24)}{2^4.5!}\,\underline{\Delta}^5_{-\frac{5}{2}n}\cdots.$$

Using the general unit-advancing operator E suggested by A.-M. Legendre a century later ('Sur une méthode d'interpolation employée par Briggs dans la construction de ses grandes Tables trigonométriques', *Connaissance des Tems, ou des Mouvemens Célestes, à l'usage des Astronomes et des Navigateurs, pour l'an 1817* (Paris, 1815): 219–22) a general proof of Cotes' formula is obtained: namely, on setting $n = 2k$ we may put

$$\underline{\Delta}^p_{rn} = (2k)^{-p}(E^{2k}-1)^p E^{2kr} \quad \text{and} \quad \delta^q_s = (E-1)^q E^s \quad (\mu\delta^q_{i+\frac{1}{2}} = \tfrac{1}{2}(E+1)(E-1)^q E^i),$$

and thence derive, by evaluating $(E^{2k}-1)/E^{k-1}(E^2-1)$ in powers of $(E-1)^2/E$,

$$\underline{\Delta}^p_{-\frac{1}{2}np} = \left(\mu\delta^1_{-\frac{1}{2}} + \sum_{1\leqslant j\leqslant k-1}\ \prod_{1\leqslant i\leqslant j}\left[\frac{k^2-i^2}{2i(2i+1)}\right]\mu\delta^{2j+1}_{-j-\frac{1}{2}}\right)^p.$$

(24) Newton first continued more simply 'e regione in columna' (in line column-wise).

(25) Here the value of λ should be taken throughout as $-\frac{1}{24}(n^2-1)$. Much as in note (23), but where n is now odd,

$$K = -\underline{\Delta}^1_{-\frac{1}{2}n} = -b-\tfrac{1}{4}dn^2, \quad L = +\underline{\Delta}^2_{-\frac{3}{2}n} = 2c-3dn, \quad M = \underline{\Delta}^3_{-\frac{3}{2}n} = -6d,$$

so that the 'media differentia' $p^{\frac{1}{2}(n+1)}$ is $-\delta^1_{-\frac{1}{2}} = -b-\tfrac{1}{4}d = K-\tfrac{1}{24}(n^2-1)M$. The general rule for this complementary case was likewise determined by Cotes in his *Canonotechnia* (note (24) above): Propositio VIII: 55–9, especially 58. There, by successive elimination between Newton–Bessel expansions of $\frac{1}{2}(f_x-f_{-x})$ in terms both of $\underline{\Delta}^{2k+1}_{-\frac{1}{2}(2k+1)n}$ (his $(2k+1)$-th 'ordines rotundi') and $\delta^{2k+1}_{-\frac{1}{2}(2k+1)}$ (his corresponding 'ordines quadrati'), it is determined that

$$\underline{\Delta}^1_{-\frac{1}{2}n} = \delta^1_{-\frac{1}{2}} + \frac{n^2-1}{2^2.3!}\,\delta^3_{-\frac{3}{2}} + \frac{(n^2-1)(n^2-9)}{2^4.5!}\,\delta^5_{-\frac{5}{2}}\cdots;$$

from which, on evaluating the 'powers' $\underline{\Delta}^{2k+1}_{-\frac{1}{2}(2k+1)n}$ in terms of $\delta^{2l+1}_{-\frac{1}{2}(2l+1)}$, $k \leqslant l \leqslant kn$, and inverting the resulting equations, there comes

$$\delta^1_{-\frac{1}{2}} = \underline{\Delta}^1_{-\frac{1}{2}n} - \frac{n^2-1}{2^2.3!}\,\underline{\Delta}^3_{-\frac{3}{2}n} + \frac{(n^2-1)(9n^2-1)}{2^4.5!}\,\underline{\Delta}^5_{-\frac{5}{2}n}\cdots.$$

addition and subtraction of these second differences the column of first differences $p_1, p_2, p_3, \ldots$ will be completed. And at length by their addition and subtraction the numbers $a_1, a_2, a_3, \ldots$ sought will be gathered.[26]

When these rules are insufficient, let A, B, C, D, E, F be six successive terms which have to be interpolated in their middle,[27] R the sum, $C+D$, of the two between which a term is to be inserted, S the sum, $B+E$, of the two next and T the sum, $A+F$, of the outermost: then[29] $\frac{1}{256}(150R - 25S + 3T)$ will be the term situated at the mid-point between C and D, with an error less than a two-hundredth of the sixth difference of the proposed terms, on the assumption that it is the property of the series that the differences should always increase or

---

As Newton could well have known (though there is nothing to show that he did so), Henry Briggs had half a century earlier computed the first twenty 'powers' of $\underline{\Delta}^1_{-\frac{1}{2}n}$ in the particular cases $n = 3$ and $n = 5$—probably making equivalent use of the 'Stirling' interpolation function $f_x = A + B\binom{x}{1} + C\binom{x}{2} + D\binom{x}{3} \ldots$ rather than Newton's $f_x = a + bx + cx^2 + dx^3 \ldots$ —, listing the resulting triangular arrays of 'differentiæ correctæ'

$$(n = 3) \quad \underline{\Delta}^p_{-\frac{1}{2}p} = (\delta^1_{-\frac{1}{2}} + \tfrac{1}{3}\delta^3_{-\frac{3}{2}})^p \quad (p = 1, 2, 3, \ldots, 20),$$

and

$$(n = 5) \quad \underline{\Delta}^q_{-\frac{5}{2}q} = (\delta^1_{-\frac{1}{2}} + \delta^3_{-\frac{3}{2}} + \tfrac{1}{5}\delta^5_{-\frac{5}{2}})^q \quad (q = 1, 2, 3, \ldots, 20),$$

without explicitly justifying their validity but showing their power and facility in subtabulating both logarithmic and trigonometric tables (*Arithmetica Logarithmica, sive Logarithmorum Chiliades Triginta* (London, 1624): Caput XIII: 27–32; *Trigonometria Britannica: sive De Doctrina Triangulorum Libri Duo* (Gouda, 1633): Liber I, Caput XII: 35–41). A century after Cotes derived the general subtabulation rule Legendre ('Sur une méthode d'interpolation employée par Briggs...' (note (24)); compare F. Maurice, 'Mémoire sur les Interpolations', *Connaissance des Temps...pour l'an 1847* (Paris, 1844): 181–222, especially §III: 193–8) verified it by use of the symbolic operator E: essentially, if we put $n = 2k+1$ and set

$$\underline{\Delta}^p_{rn} = (2k+1)^{-p}(E^{2k+1} - 1)^p E^{(2k+1)r} \quad \text{and} \quad \delta^q_s = (E-1)^q E^s,$$

by expressing $(E^{2k+1} - 1)/E^k(E-1)$ in terms of powers of $(E-1)^2/E$ there results

$$\underline{\Delta}^p_{-\frac{1}{2}np} = \left(\delta^1_{-\frac{1}{2}} + \sum_{1 \leqslant j \leqslant k} \prod_{1 \leqslant i \leqslant j} \left[\frac{(2k+1)^2 - (2i-1)^2}{2^2 \times 2i(2i+1)}\right] \delta^{2j+1}_{-j-\frac{1}{2}}\right)^p.$$

(26) 'habebuntur' (will be had) is cancelled. It is, of course, only a convenient assumption—which may not always be true—that the third differences $w, x, y, z, \ldots$ are roughly in arithmetic progression and hence that their middles $w^6, x^6, y^6, \ldots$ are their arithmetic means.

(27) Newton has cancelled 'ut supra' (as above).

(28) 'quorumvis sibi proximorum' (any [two] next to each other) is deleted.

(29) Newton first continued '$\frac{9R - S}{16}$ terminus ille interserendus [inter C et D] errore existente minore quam quadragesima pars differentiæ quartæ' ($\frac{1}{16}(9R - S)$ will be the term to be inserted [between C and D], with an error of less than a fortieth part of the fourth difference). This is copied word for word from Rule 4 of the preceding section (§2).

vel semper decrescant. Hoc modo per bisectionem procedi potest usɋ dum[30] differentiæ quartæ minores sint quam 32.[31]

Possent aliæ hujusmodi regulæ tradi sed mallem rem omnem una regula generali complecti et ostendere quomodo series quævis in loco imperato intercalari[32] possit. Exponatur series per lineas *Ap, Bq, Cr, Ds, Et, Fv, Gw* &c ad lineam *AG* perpendiculariter erectas & intervalla terminorum per partes lineæ illius *AB, BC, CD, DE* &c.[33] Fac $\frac{A-B}{AB}=b$,

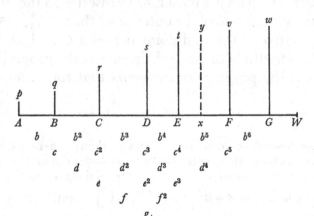

$\frac{B-C}{BC}=b^2$, $\frac{C-D}{CD}=b^3$ &c. Item $\frac{b-b^2}{\frac{1}{2}AC}=c$, $\frac{b^2-b^3}{\frac{1}{2}BD}=c^2$, $\frac{b^3-b^4}{\frac{1}{2}CE}=c^3$ &c. Dein $\frac{c-c^2}{\frac{1}{3}AD}=d$, $\frac{c^2-c^3}{\frac{1}{3}BE}=d^2$, $\frac{c^3-c^4}{\frac{1}{3}CF}=d^3$ &[c]. Porro $\frac{d-d^2}{\frac{1}{4}AE}=e$. $\frac{d^2-d^3}{\frac{1}{4}BF}=e^2$ &c. Tunc $\frac{e-e^2}{\frac{1}{5}AF}=f$ [&c] et sic in sequentibus usɋ ad ad finem operis, dividendo semper differentias primas per intervalla terminorum quorum sunt differentiæ, secundas per dimidium duorum intervallorum quibus respondent, tertias per tertiam partem trium & sic porrò pergendo usɋ dum in ultimo loco differentia satis exigua sit.[34] Hoc peracto capiantur tum terminorum tum differentiarum primæ *A, b, c, d, e, f, g* &c. Sit differentiarum illarum numerus *n*,[35] locus quem intercalare oportet *x*, terminus intercalaris *xy*, et regrediendo ab ultima differentia puta *g* et ab ultimo terminorū ex quibus differentia illa colligebatur puta *G*, fac $f+\frac{g\times Gx}{n}=p$. $e+\frac{p\times Fx}{n-1}=q$. $d-\frac{q\times Ex}{n-2}=r$. $c-\frac{r\times Dx}{n-3}=s$. $b-\frac{s\times Cx}{n-4}=t$. $A-\frac{t\times Bx}{n-5}=v$,[36] pergendo semper juxta tenorem progressionis

---

(30) An unfinished first continuation reads 'præcedentes reg[ulæ applicari possint?]' (the preceding rules [can be applied?]).

(31) Read '21' (compare note (20) above). This paragraph essentially repeats Rule 6 of the preceding piece (see §2: note (34)).

(32) As we would expect (see §2: note (16)) Newton first wrote 'interpolari' (interpolated). The subsequent alteration is bewildering.

(33) Newton has cancelled a following passage, inserting its equivalent below: 'sintɋ terminorum differentiæ per intervalla terminorum quibus respondent divisæ[,] primæ quidem *b, b², b³,* &c, secundæ *c, c², c³,* &c, tertiæ *d, d², d³* &c[,] quartæ *e, e², e³* &c & sic ad ultimas. Hoc est.' Observe in sequel that the end-points *A, B, C, D,* ... of the lines *Ap, Bq, Cr, Ds,* ... are used to denote their magnitude. The end-point *W* of the line *ABC* ... apparently denotes

always decrease in a regular way. In this manner a bisection procedure may be employed until[30] the fourth differences prove to be less than 32.[31]

Other rules of this kind might be presented, but I would prefer to embrace everything in one single general rule and show how any series you wish may be intercalated[32] in any place commanded. Let the series be exhibited by the lines Ap, Bq, Cr, Ds, Et, Fv, Gw, ... raised at right angles to the line AG, and the intervals of the terms by the parts AB, BC, CD, DE ... of that line.[33] Make

$$\frac{A-B}{AB}=b_1, \quad \frac{B-C}{BC}=b_2, \quad \frac{C-D}{CD}=b_3,$$

...; likewise $\dfrac{b_1-b_2}{\frac{1}{2}AC}=c_1$, $\dfrac{b_2-b_3}{\frac{1}{2}BD}=c_2$,

$\dfrac{b_3-b_4}{\frac{1}{2}CE}=c_3$, ...; next $\dfrac{c_1-c_2}{\frac{1}{3}AD}=d_1$, $\dfrac{c_2-c_3}{\frac{1}{3}BE}=d_2$, $\dfrac{c_3-c_4}{\frac{1}{3}CF}=d_3$, ...; further $\dfrac{d_1-d_2}{\frac{1}{4}AE}=e_1$,

$\dfrac{d_2-d_3}{\frac{1}{4}BF}=e_2$, ...; then $\dfrac{e_1-e_2}{\frac{1}{5}AF}=f_1$, ..., and so on in sequel till the work is finished, dividing always first differences by the intervals of the terms whose differences they are, second ones by half of the two corresponding intervals, third ones by a third of the three corresponding and so forth until the difference in the final place be slight enough.[34] When this is accomplished, take the leading quantities both of the terms and the differences, A, b_1, c_1, d_1, e_1, f_1, g_1, ..., and let those differences be n in number,[35] the place it is required to intercalate call x, the term to be intercalated xy; then, going backwards from the last difference, say g_1, and from the last of the terms, say G, from which that difference was gathered, make $f_1+g_1\times\dfrac{Gx}{n}=p$, $e_1+p\times\dfrac{Fx}{n-1}=q$, $d_1-q\times\dfrac{Ex}{n-2}=r$, $c_1-r\times\dfrac{Dx}{n-3}=s$,

$b_1-s\times\dfrac{Cx}{n-4}=t$, $A-t\times\dfrac{Bx}{n-5}=v$,[36] proceeding always following the sense of

---

the last 'known' quantity to be interpolated (X, Y and Z being inappropriate for the purpose, in context at least).

(34) In continuation Newton first wrote 'puta non major unitate' (suppose not greater than a unit), then replacing it by the unfinished phrase 'puta minor sit quam' (suppose it less than...). A following cancelled amplification at this point reads 'Nam cum b, b^2, b^3 &c respondeant medijs[,] interstare supponantur e regione mediorum punctorum inter A, B, C, D, &c et c, c^2, c^3 &c e regione mediorum punctorum inter b, b^2, b^3, &c. Distantia terminorum b & b^2[,] existens summa distantiarum hinc inde a B, erit $\frac{1}{2}AB+\frac{1}{2}BC$. [&c]' (For since b_1, b_2, b_3, ... correspond to means, let them be supposed to stand between in line with the mid-points between A, B, C, D, ..., and c_1, c_2, c_3, ... in line with the mid-points between b_1, b_2, b_3, ... The distance of the terms b_1 and b_2, being the sum of their distances either way from B, will then be $\frac{1}{2}(AB+BC)$, [and so on].)

(35) Below (and in Newton's diagram) n is taken equal to 6.

hujus usqᵬ ad primum $A_{[,]}$ et erit quantitatum sic collectarum novissima puta v terminus intercalaris xy quem invenire oportuit. Ubi nota quod terminos omnes $\dfrac{Gx \times g}{n}$, $\dfrac{Fx \times p}{n-1}$ multiplicatos per Gx, Fx quæ jacent ultra x, addo: cæteros $\dfrac{Ex \times q}{n-2}$, $\dfrac{Dx \times r}{n-3}$ &c ubi Ex, Dx &c jacent citra x subduco.[37] Nota etiam quod additio quæ hic præcipitur, in subductionem mutari debet et subductio in additionem quoties termini addendi vel subducendi prodeunt negativi. Ut si foret $\dfrac{Dx \times r^{(38)}}{n-3}$ majus quam c, atqᵬ adeo s negativum, ad b adderem terminum $\dfrac{Cx \times s}{n-4}^{(38)}$. Et sic si foret b^2 major quam b atqᵬ adeo c negativum, ponerem $-c - \dfrac{Dx \times r^{(38)}}{n-3} = s$.

Regula hæc non alia eget demonstratione quam quod eadem methodo regreditur a differentijs ad terminum quæsitum xy, qua prius progrediebatur a terminis datis ad differentias: quodqᵬ terminus ille quæsitus xy idem evadit cum[39] terminis datis Ap, Bq, Cr &c ubi punctum x incidit in puncta A, B, C &c adeoqᵬ[40] terminatur ad curvam quandam lineam pw quæ transit per omnia puncta p, q, r, s, t, v, w.[41] Sed exemplo res illustranda est. Proponantur ergo

---

(36) The fine detail here needs considerable correction. Read '...et a penultimo terminorum ex quibus differentia illa colligebatur, puta F, fac

$$f + \frac{g \times Fx}{n} = p. \quad e - \frac{p \times Ex}{n-1} = q. \quad d - \frac{q \times Dx}{n-2} = r. \quad c - \frac{r \times Cx}{n-3} = s. \quad b - \frac{s \times Bx}{n-4} = t. \quad A - \frac{t \times Ax}{n-5} = v',$$

and correspondingly so in the English version.

(37) Here (compare note (36)) read '...nota quod terminos omnes $\dfrac{Fx \times g}{n}$, &c multiplicatos per Fx & quæ jacent ultra x addo: cæteros $\dfrac{Ex \times p}{n-1}$, $\dfrac{Dx \times q}{n-2}$ &c ubi Ex, Dx &c jacent citra x subduco'.

(38) Read, in line with the two previous notes, '$\dfrac{Cx \times r}{n-3}$', '$\dfrac{Bx \times s}{n-4}$' and '$-c - \dfrac{Cx \times r}{n-3}$' respectively.

(39) This replaces 'quodqᵬ linea intercalaris xy idem est ac' (and that the intercalary line xy is the same as).

(40) 'linea illa' (that [intercalary] line) is cancelled. Compare note (39).

(41) On using square brackets to denote adjusted differences and setting the corresponding divided differences in their customary round ones, we may (with Newton) settle on some arbitrary point, O say, as origin and define $OA = A$, $OB = B$, $OC = C$, ..., $Ox = x$: the problem is then to interpolate $f_x = xy$ in the given sequence $Ap = f_A$, $Bq = f_B$, $Cr = f_C$, ... in terms of the differences $\dfrac{f_x - f_A}{x - A} = t = [x, A] = (x, A)$;

this progression up to the leading term A, and the most recent, say v, of the quantities gathered in this way will be the intercaland term xy which it was required to find. Here note that all terms $\dfrac{Gx}{n} \times g$, $\dfrac{Fx}{n-1} \times p$ multiplied by any Gx, Fx lying beyond x, I add: the rest $\dfrac{Ex}{n-2} \times q$, $\dfrac{Dx}{n-3} \times r$, ... in which Ex, Dx, ... lie on the near side of x, I subtract.[37] Note also that the addition here laid down should be altered to subtraction (and subtraction to addition) each time that the terms to be added or subtracted prove to be negative. For instance, if $\dfrac{Dx}{n-3} \times r$[38] were greater than c_1, and s consequently negative, to b_1 I should add the term $\dfrac{Cx}{n-4} \times s$.[38] And so if b_2 were greater than b_1, and c_1 therefore negative, I should put $-c_1 - \dfrac{Dx}{n-3} \times r$[38] $= s$.

This Rule needs no other demonstration except to observe that regress is made from the differences to the required term xy by the same method as previously we advanced from the given terms to the differences; and that the term xy sought comes to be identical with[39] the given terms Ap, Bq, Cr, ... when the point x coincides with the points A, B, C, ..., and consequently[40] terminates in some curve pw which passes through all the points p, q, r, s, t, v, w.[41] But the

$$\frac{t-b}{\frac{1}{2}(x-B)} = s = [x, A, B] = 2!(x, A, B), \text{ where } b = [A, B];$$

$$\frac{s-c}{\frac{1}{3}(x-C)} = r = [x, A, B, C] = 3!(x, A, B, C), \text{ where } c = [A, B, C];$$

$$\frac{r-d}{\frac{1}{4}(x-D)} = q = [x, A, B, C, D] = 4!(x, A, B, C, D), \text{ where } d = [A, B, C, D];$$

$$\frac{q-e}{\frac{1}{5}(x-E)} = p = [x, A, B, C, D, E] = 5!(x, A, B, C, D, E), \text{ where } e = [A, B, C, D, E];$$

and
$$\frac{p-f}{\frac{1}{6}(x-F)} = [x, A, B, C, D, E, F] = 6!(x, A, B, C, D, E, F) \approx g,$$

where $f = [A, B, C, D, E, F]$ and $g = [A, B, C, D, E, F, G]$.

It follows at once that by inverting these equations one by one we may successively 'unwrap' the general adjusted differences formula (correct to the order of $h = [A, B, C, D, E, F, G]$):

$$f_x = f_A + (x-A)\,t = f_A + (x-A)\,b + \tfrac{1}{2}(x-A)(x-B)\,s = \ldots$$

$$= f_A + (x-A)\,b + \frac{1}{2!}(x-A)(x-B)\,c + \frac{1}{3!}(x-A)(x-B)(x-C)\,d + \ldots$$

$$+ \frac{1}{6!}(x-A)(x-B)(x-C)(x-D)(x-E)(x-F)\,g.$$

It will be evident that the 'Harriot–Briggs' advancing differences formula is the particular case of this where the intervals AB, BC, CD, ... are all equal: namely, on taking the origin at

numerorum 24, 25, 26, 27, 28, 30 logarithmi et requiratur logarithmus numeri cujusvis intermedij, et operis hæc erit forma.[42]

A and putting $AB = BC = CD = \dots = EF = \dots = k$, so that $Ap = f_0$, $b = k\Delta_0^1$, $c = k^2\Delta_0^2$, ..., $g = k^6\Delta_0^6$, ..., there results

$$f_x = f_0 + \binom{x/k}{1}\Delta_0^1 + \binom{x/k}{2}\Delta_0^2 + \binom{x/k}{3}\Delta_0^3 \dots.$$

This apart, there is little to commend adjusted differences over their divided equivalents—computationally, indeed, it rapidly becomes frustrating to have to divide each difference of n-th order not only by the difference of the bounding arguments but also by the factor n. It is significant that in all his future expositions of the general finite-difference formula Newton made unique use of divided differences.

(42) The text breaks off at this point and there is no evidence that Newton ever worked through his example. Since a logarithm intermediate between $\log_{10}24 = 1 \cdot 38021\,124$ and $\log_{10}30 = 1 \cdot 47712\,125$ has to be interpolated, it will clearly be easiest to set

$$f_x = \log_{10}(27+x) = \log_{10}27 + (\tfrac{1}{27}x - \tfrac{1}{2}(\tfrac{1}{27}x)^2 + \dots)\log_{10}e,$$

and, given f_{-3}, $f_{-2} = 1 \cdot 39794\,001$, $f_{-1} = 1 \cdot 41497\,335$, $f_0 = 1 \cdot 43136\,376$, $f_1 = 1 \cdot 44715\,803$ and f_3, seek any intervening logarithm f_x, $-3 \leqslant x \leqslant 3$, by an appropriate adjusted differences formula: say by this $\left(\text{correct to order } \dfrac{x(x-1)(x+1)(x-3)(x+2)(x+3)}{6!} g \approx 10^{-11}\right)$,

$$f_x = f_0 + x\left(b + \frac{x-1}{2}\left\{c + \frac{x+1}{3}\left[d + \frac{x-3}{4}\left(e + \frac{x+2}{5}f\right)\right]\right\}\right),$$

in which $b = [0, 1] = 0 \cdot 01579\,427$, $c = [-1, 0, 1] = -0 \cdot 00059\,615$,

$$d = [-1, 0, 1, 3] = 0 \cdot 00004\,078, \quad e = [-2, -1, 0, 1, 3] = 0 \cdot 00000\,479,$$

$$f = [-3, -2, -1, 0, 1, 3] = 0 \cdot 00000\,078$$

and $g = [x, -3, -2, -1, 0, 1, 3] \approx 0 \cdot 0000001$.

(Using this expression D. C. Fraser gives the complete computation of

$$f_{\frac{1}{5}} = \log_{10}27 \cdot 2 = 1 \cdot 43456\,8905$$

in his 'Newton's Interpolation Formulas' (note (2) above): 72–3.)

point should be illustrated by an example. So let the logarithms of the numbers 24, 25, 26, 27, 28, 30 be propounded and the logarithm of any intermediate number be required: the work will then have this form.[42]

§4. SCHEMES FOR INTERPOLATION BY CENTRAL DIFFERENCES (LATE 1676?).[1]

[1][3]

Series Arithmetica

Series correspondens.

From the original in Newton's Waste Book[a]

A + x	a	+ bx	+ cxx	+ dx³	+ ex⁴	+ fx⁵	+ gx⁶	+ hx⁷	+ ix⁸	+ kx⁹	+ lx¹⁰	[+ mx¹¹ &c]
A + 6[4]	a	+6b	+36c	+216d							+60466176l	+362797056m
A + 5	a	+5b	+25c	+125d	+625e	+3125f	+15625g	+78125h	+390625i	+1953125k	+9765625l	+48828125m
A + 4	a	+4b	+16c	+64d	+256e	+1024f	+4096g	+16384h	+65536i	+262144k	+1048576l	+4194304m
A + 3	a	+3b	+9c	+27d	+81e	+243f	+729g	+2187h	+6561i	+19683k	+59049l	+177147m
A + 2	a	+2b	+4c	+8d	+16e	+32f	+64g	+128h	+256i	+512k	+1024l	+2048m
A + 1	a	+b	+c	+d	+e	+f	+g	+h	+i	+k	+l	+m
A + 0	a											
A − 1	a	−b	+c	−d	+e	−f	+g	−h	+i	−k	+l	−m
A − 2	a	−2b	+4c	−8d	+16e	−32f	+64g	−128h	+256i	−512k	+1024l	−2048m
&c												

The first Difference of these termes.[5]

a	+ bx	+ cxx	+ dx³	+ ex⁴	+ fx⁵	+ gx⁶	+ hx⁷	+ ix⁸	+ kx⁹	+ lx¹⁰	+ mx¹¹
&c											
	b	+3c	+7d	+15e	+31f	+63g	+127h	+255i	+511k	+1023l	+2047m
	b	+c	+d	+e	+f	+g	+h	+i	+k	+l	+m
	b	−c	+d	−e	+f	−g	+h	−i	+k	−l	+m
	b	−3c	+7d	−15e	+31f	−63g	+127h	−255i	+511k	−1023l	+2047m
&c											

The second Difference.[6]

	+ cxx	+ dx³	+ ex⁴	+ fx⁵	+ gx⁶	+ hx⁷	+ ix⁸	+ kx⁹	+ lx¹⁰	+ mx¹¹
&c										
	2c	+6d	+14e	+30f	+62g	+126h	+254i	+510k	+1022l	+2046m
	2c	0	+2e	0	+2g	0	+2i	0	+2l	0
	2c	−6d	+14e	−30f	+62g	−126h	+254i	−510k	+1022l	−2046m
&c										

The 3ᵈ diff.[5]

	+ dx³	+ ex⁴	+ fx⁵	+ gx⁶	+ hx⁷	+ ix⁸	+ kx⁹	+ lx¹⁰	+ mx¹¹
&c									
	6d	+36e	+150f	+540g	+1806h	+5796i	+18150k		+171006m
	6d	+12e	+30f	+60g	+126h	+252i	+510k	+1020l	+2046m
	6d	−12e	+30f	−60g	+126h	−252i	+510k	−1020l	+2046m
	6d	−36e	+150f	−540g	+1806h	−5796i	+18150k		+171006m
							+204630k		+3669006m
							+1225230k		+36774606m
									+228718446m
&c									

	e	f	g	h	i	k	l	m
The 4th diff.(6)	&c							
	24e	+120f	+480g	+1680h	+5544i	+17640k	+2040l	+168960m
	24e	+0	+120g	+0	+504i	+0	+	+0
	24e	−120f	+480g	−1680h	+5544i	−17640k	+	−168960m
						−186480[k]		−3498000[m]
						−1020600[k]		−33105600[m]
								−191943840[m]
The 5th diff.(5)		&c						
		120f	+1080g	+6720h	+35280i	+168840k	+	+3329040[m]
		120f	+360g	+1680h	+5040i	+17640k	+	+168960m
		120f	−360g	−1680h	−5040i	−17640k	−	+168960m
		120f	−1080g	−6720h	−35280i	+[168840]k	−	+3329040[m]
								+29607600[m]
								+15883824[0m]
The 6th Diff.(6)			&c					
			720g	+5040h	+30240i	+151200k	+	+3160080m
			720g	+0	+10080i	+0	+	+0
			720g	−5040h	+30240i	−151200k	+	−3160080m
								−26278560[m]
								−129230640[m]
The 7th diff.(5)				&c				
				5040h	+20160i	+514080k	+	+2318480m
				5040h	+20160i	+151200k	+	+3160080m
				5040h	−20160i	−151200k	−	+3160080m
				5040h	−20160i	+514080k	−	+2318480m
								+10295208[0m]
The 8th diff.(6)					&c			
					40320i	+362880k	+	+19958400m
					40320i	+0	+	+0
					40320i	−362880k	+	−19958400m
								−79833600m
The 9th diff.(7)								
						362880k	+	+19958400m
						362880k	−	+19958400m
								+59875200m
The 10th diff.(8)							+	
							+	
The 11th diff.(9)								39916800m
								[&c]

[The notes for this table appear on pp. 54–55.]

The use of these differences is for composing rules to find the differences of y^e terms of a table wch is to be interpoled by y^e continuall addition of those differences & also for drawing a geometric line through as many given points as you please.

Prob. [1]

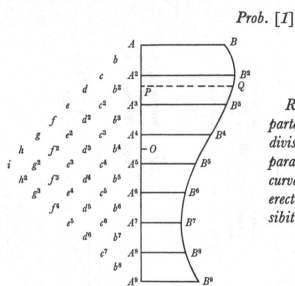

Recta aliqua AA^9 in æquales quotcunqꝫ partes AA^2, A^2A^3, A^3A^4, A^4A^5,[10] &c divisa et ad puncta divisionum erectis parallelis AB, A^2B^2, A^3B^3 &c: invenire curvam geometricam[11] quæ per omnium erectarum extremitates B, B^2, B^3 &c transibit.

(1) A systematic exposition, from first principles, of interpolation by means of central differences, in which Newton derives the now usually named Newton–Stirling and Newton–Bessel formulas (when the arguments are given at equal intervals) and their generalizations. In the former case—following contemporary practice and computational good sense—straightforward arithmetical differences are employed, but in general Newton abandons his previous preference for adjusted differences (compare §3: note (41)) in favour of their divided equivalents. Throughout, all differences are formed in their modern way. This manuscript is, in a narrow sense, unpublished but its content (except for the opening tabulation of differences to eleventh order) eventually appeared in print, lightly revised and partly transposed, as Propositions 1, 3 and 4 of the *Methodus Differentialis* in William Jones' compendium, *Analysis Per Quantitatum Series, Fluxiones ac Differentias: Cum Enumeratione Linearum Tertii Ordinis* ((London, 1711): 93–101, especially 93–4, 95–7). Our assigned date of composition must, for lack of any firm knowledge regarding its background, be tentative: in settling on late 1676 we have been guided primarily by our assessment of its handwriting and by its position in the Waste Book (see note (2) below). This conjecture agrees well with our judgement, based on the relative maturity and completeness of the present paper, that it represents a later stage in the development of Newton's researches into finite differences than the 'Regula Differentiarum' (§3 preceding).

(2) ULC. Add. 4004: 82^r–84^r. This entry is placed immediately after the 'Octob. 1676. Memorandum' (reproduced in II: 191, note (25)) on his *Epistola posterior* to Leibniz, in which Newton recorded the solution of the fluxional anagrams he then communicated. Comparison of these two autographs reveals no significant variation in handwriting style and they were evidently penned at about the same time.

(3) In the following scheme the 'corresponding series'

$$f_x \equiv f(A+x) = \sum_{0 \leqslant i \leqslant \ldots} a_i x^i \quad (a_0 = a, \; a_1 = b, \; a_2 = c, \; \ldots, \; a_{11} = m, \; \ldots)$$

Translation

[The use of these differences is for composing rules to find the differences of y^e terms of a table wch is to be interpoled by y^e continuall addition of those differences & also for drawing a geometric line through as many given points as you please.]

Problem [1]

When some straight line $A_1 A_9$ is divided into any number of equal parts $A_1 A_2$, $A_2 A_3$, $A_3 A_4$, $A_4 A_5$, ... and at the points of division there are raised the parallels $A_1 B_1$, $A_2 B_2$, $A_3 B_3$, ...: to find a geometrical curve[11] *which shall pass through the end-points B_1, B_2, B_3, ... of each of the lines erected.*

is set out line by line (at unit intervals of argument) in descending 'arithmetical sequence' $..., f_6, f_5, ..., f_1, f_0, f_{-1}, ...$; and from these tabulated instances Newton successively computes the central differences $\Delta^{2k-1}_{-(k-\frac{1}{2})\pm\frac{1}{2}}$, $\Delta^{2k-1}_{-(k-\frac{1}{2})\pm\frac{3}{2}}$ and Δ^{2k}_{-k}, $\Delta^{2k}_{-k\pm1}$, $k = 1, 2, 3, ...$, as far as the eleventh order ($\Delta^{11}_{-5\frac{1}{2}\pm\frac{1}{2}} \approx 11! \, a_{11}$). After an ambitious opening in which the coefficients of the various a_i in the expansions of the f_x are evaluated as far as $6^{11} = 362797056$ (that of $m = a_{11}$ in f_6), his ardour subsequently cools and the latter portion of the tabulation is, in fact, built up in reverse by skilful use of properties of the differences, which may be stated generally in the form

$$\Delta^{2k-1}_{-(k-\frac{1}{2})\pm\frac{1}{2}} = (2k-1)! \, a_{2k-1} \pm \tfrac{1}{2}(2k)! \, a_{2k} + \tfrac{1}{12}(k+1)(2k+1)! \, a_{2k+1} \cdots,$$

$$\Delta^{2k-1}_{-(k-\frac{1}{2})\pm\frac{3}{2}} = (2k-1)! \, a_{2k-1} \pm \tfrac{3}{2}(2k)! \, a_{2k} + \tfrac{1}{12}(k+13)(2k+1)! \, a_{2k+1} \cdots,$$

$$\Delta^{2k}_{-k} = (2k)! \, a_{2k} + \tfrac{1}{12}k(2k+2)! \, a_{2k+2} \cdots$$

(where all terms in $a_{2(k+i)+1}$, $i = 0, 1, 2, ...$, are zero)

and $\Delta^{2k}_{-k\pm1} = (2k)! \, a_{2k} \pm (2k+1)! \, a_{2k+1} + \tfrac{1}{12}(k+6)(2k+2)! \, a_{2k+2} \pm \tfrac{1}{12}(k+2)(2k+3)! \, a_{2k+3} \cdots.$

(4) The omitted terms in the following expansion of f_6, namely

'$+ 1296e + 7776f + 46656g + 279936h + 1679616i + 10077696k$',

are not needed to form the following differences.

(5) The differences Δ^{2k-1}_{-k+2}, Δ^{2k-1}_{-k+1}, Δ^{2k-1}_{-k} and Δ^{2k-1}_{-k-1}, $k = 1, 2, 3, 4$, respectively. Observe (compare note (3)) how certain of the later terms in $k = a_9$ and $m = a_{11}$ are built up, in reverse sequence, from corresponding ones in following differences, and are in turn used to construct terms in k and m in certain differences Δ^{2k-1}_{-k-2}, Δ^{2k-1}_{-k-3}, ...

(6) The differences Δ^{2k}_{-k+1}, Δ^{2k}_{k} and Δ^{2k}_{-k+1}, $k = 1, 2, 3, 4$, respectively. Again (compare notes (3) and (5)) observe how certain terms in $k = a_9$ and $m = a_{11}$ are constructed from the directly computed expansions of subsequent differences and are themselves, in turn, used to build up terms in k and m in certain following Δ^{2k}_{-k-2}, Δ^{2k}_{-k-3}, ...

(7) The differences Δ^9_{-4} and Δ^9_{-5} (lacking the terms $\pm 1814400l$ respectively) and the term in m (namely, $\tfrac{3}{2}11!$) from Δ^9_{-6}.

(8) Evidently Newton here meant to enter, to $O(a_{12})$, $\Delta^{10}_{-5} = +3628800l$ and

$$\Delta^{10}_{-6} = +362880l - 39916800m,$$

the two differences which result from the preceding '9th diff' (see note (7)).

(9) That is, to $O(a_{12})$, $\Delta^{11}_{-6} = 11! \, m$.

(10) The continuation '$A^5 A^6$, $A^6 A^7$, $A^7 A^8$, $A^8 A^9$' is cancelled.

(11) In the revised version of this problem which appeared in 1711 as Propositio III of the *Methodus Differentialis* (Jones' *Analysis* (note (1) above): 95) Newton added the accurate qualifying phrase 'generis Parabolici' (of parabolic kind). Compare §3: note (3).

Erectarum AB, A^2B^2, A^3B^3 &c quære differentias primas b, b^2, b^3, &c$_{[,]}$ secundas c, c^2, c^3 &c$_{[,]}$ tertias d, d^2, d^3 &c & sic deinceps usq̃ dum veneris ad ultimam differentiam i. Tunc incipiendo ab ultima differentia excerpe medias differentias in alternis $\begin{cases} \text{columnis} \\ \text{seriebus}^{(12)} \\ \text{ordinibus} \end{cases}$ differentiarum et arithmetica media inter duas medias reliquorum ordinum$_{[,]}$ pergendo usq̃ ad seriem primorum terminorum AB, A^2B^2, A^3B^3, &c.[13] Sint hæc k, l, m, n, o, p, q, r, s &c quorum ultimum significet ultimam differentiam, penultimum medium arithmeticum inter duas penultimas differentias, antepenultimum mediam trium antepenultimarum differentiarum, & sic deinceps usq̃ ad primum quod erit vel medius terminorum A, A^2, A^3 [&c],[14] vel arithmeticum medium inter duos medios. Prius accidit ubi numerus terminorum A, A^2, A^3 &c[14] est impar, posterius ubi par.[15]

Cas 1.

In casu priori[16] sit A^5B^5 iste medius terminus, hoc est $A^5B^5 = k$, $\dfrac{b^4+b^5}{2} = l$,

$c^4 = m$, $\dfrac{d^3+d^4}{2} = n$, $e^3 = o$, $\dfrac{f^2+f^3}{2} = p$, $g^2 = q$, $\dfrac{h+h^2}{2} = r$, $i = s$. et erecta ordinatim applicata PQ dic $A^5P = x$,[17] duc terminos hujus progressionis

$$1 \times \frac{x}{1} \times \frac{x}{2} \times \frac{xx-1}{3x} \times \frac{x}{4} \times \frac{xx-4}{5x} \times \frac{x}{6} \times \frac{xx-9}{7x} \times \frac{x}{8} \times \frac{xx-16}{9x} \times \frac{x}{10} \times \frac{xx-25}{11x} \times \frac{x}{12} \times \frac{xx-36}{13x} \text{ \&c}$$

---

(12) The 1711 version (see note (11) above) omits the second of these alternatives, reading 'Columnis vel Ordinibus' (columns or ranks).

(13) Newton first wrote 'A, A^2, A^3, &c' simply, understanding (compare §3: note (33)) that the lines AB, A^2B^2, A^3B^3, ... are denoted in magnitude by their end-points A, A^2, A^3, ... All subsequent occurrences of the contracted form were evidently intended to be amplified in a similar way, but one or two instances escaped him in revision.

A first, cancelled version of the following text carried straight on, without any subdivision into 'Cas. 1', with the words: 'Sint hæc k, l, m, n, o, p, q, r, s, t, nempe $k = i$ differentiæ ultimæ, $l = \dfrac{h+h^2}{2}$, $m = g^2$, $n = \dfrac{f^2+f^3}{2}$, $o = e^3$, $p = \dfrac{d^3+d^4}{2}$, $q = c^4$, $r = \dfrac{b^4+b^5}{2}$, $s = A^5$ [*sc.* A^5B^5]; nam series terminorum A, A^2, A^3 [*sc.* AB, A^2B^2, A^3B^3], &c hic est instar seriei differentiarum, ultimo termino seriei k, l, m, &c existente vel medio termino seriei hujus A, A^2, A^3 [*sc.* AB, A^2B^2, A^3B^3] &c si constat impari numero terminorum ut in hoc casu, vel arithmetico medio. Erige ordinatim applicatam PQ et bis[e]cta AA^9 in A^5 dic $A^5P = x$, $PQ = y$' (Let these be k, l, m, n, o, p, q, r, s, t; namely, $k = $ last difference i_1, $l = \frac{1}{2}(h_1+h_2)$, $m = g_2$, $n = \frac{1}{2}(f_2+f_3)$, $o = e_3$, $p = \frac{1}{2}(d_3+d_4)$, $q = c_4$, $r = \frac{1}{2}(b_4+b_5)$ and $s = A_5[B_5]$. For the series of terms $A_1[B_1]$, $A_2[B_2]$, $A_3[B_3]$, ... here mirrors the series of differences, the last term of the series k, l, m, ... being either the middle one of this series $A_1[B_1]$, $A_2[B_2]$, $A_3[B_3]$, ... if (as in the present case) it consists of an odd number of terms, or an arithmetic mean [*i.e.* of the two middlemost ones]. Erect the ordinate PQ and, having bisected A_1A_9 in A_5, call $A_5P = x$, $PQ = y$.) A first revision of this passage was made in the margin alongside before its slightly variant equivalent (that now reproduced) was copied into the main text.

(14) Read (see note (13) above) 'AB, A^2B^2, A^3B^3 &c'.

Of the erected lines $A_1 B_1$, $A_2 B_2$, $A_3 B_3$, ... seek the first differences b_1, b_2, b_3, ...; their second ones c_1, c_2, c_3, ...; their third ones d_1, d_2, d_3, ...; and so on successively until you reach the last difference i_1. Then, beginning with the last difference, take out the middle differences in alternate columns/series/ranks[12] of differences and the arithmetic means of the two middle-most ones in the remaining ranks, proceeding as far as the series $A_1 B_1$, $A_2 B_2$, $A_3 B_3$, ... of first terms.[13] Let these be k, l, m, n, o, p, q, r, s, ..., the last of which is to denote the last difference, the last but one the arithmetic mean between the two last but one differences, the last but two the middle one of the three last but two differences, and so on in turn up to the first, which will be either the middle one of the terms $A_1[B_1]$, $A_2[B_2]$, $A_3[B_3]$, ... or the arithmetic mean between the two middle-most. The former happens when the number of terms $A_1[B_1]$, $A_2[B_2]$, $A_3[B_3]$, ... is odd, the latter when it is even.[15]

Case 1

In the former case let[16] $A_5 B_5$ be that middle term, that is,

$$A_5 B_5 = k, \quad \tfrac{1}{2}(b_4 + b_5) = l, \quad c_4 = m, \quad \tfrac{1}{2}(d_3 + d_4) = n, \quad e_3 = o, \quad \tfrac{1}{2}(f_2 + f_3) = p,$$

$$g_2 = q, \quad \tfrac{1}{2}(h_1 + h_2) = r, \quad i_1 = s.$$ Then, having erected the ordinate PQ call $A_5 P = x$[17] and multiply the terms of this progression

$$1 \times \frac{x}{1} \times \frac{x}{2} \times \frac{x^2 - 1}{3x} \times \frac{x}{4} \times \frac{x^2 - 4}{5x} \times \frac{x}{6} \times \frac{x^2 - 9}{7x} \times \frac{x}{8} \times \frac{x^2 - 16}{9x} \times \frac{x}{10} \times \frac{x^2 - 25}{11x} \times \frac{x}{12} \times \frac{x^2 - 36}{13x} \cdots$$

---

(15) Corresponding to some origin $O \equiv A^k$ (either one of the equal unit-divisions A, A^2, A^3, ... of the base line or a point midway between two such divisions) and the general point $P \equiv A^x$, say, of the line AA^{2k-1} of arguments, Newton supposes that the general ordinate of the 'geometrical' curve drawn through B, B^2, ..., B^{2k-1} is $PQ = f(OA^x) \equiv f_{k-x}$, so that $AB = f_{k-1}$, $A^2 B^2 = f_{k-2}$, ..., $A^{2k-1} B^{2k-1} = f_{1-k}$. He then constructs the array of differences $b, b^2, b^3, ...; c, c^2, ...; d, d^2, ...; ...; h, h^2, ...; i, ...$, and from these selects the differences central round O. Two cases arise (Newton's 'Cas. 1' and 'Cas. 2') according as the number of quantities to be interpolated is odd or even. In the first, k is integral and hence the differences central round O are $\tfrac{1}{2}(b^{k-1} + b^k) = \mu\Delta^1_{-\frac{1}{2}}$, $c^{k-1} = \Delta^2_{-1}$, $\tfrac{1}{2}(d^{k-2} + d^{k-1}) = \mu\Delta^3_{-\frac{3}{2}}$, $e^{k-2} = \Delta^4_{-2}$, ..., $\tfrac{1}{2}(h^{k-4} + h^{k-3}) = \mu\Delta^7_{-\frac{7}{2}}$, $i^{k-4} = \Delta^8_{-4}$, ... In the latter instance $k + \tfrac{1}{2}$ is integral and the central differences correspondingly selected are $b^{k-\frac{1}{2}} = \Delta^1_{-\frac{1}{2}}$, $\tfrac{1}{2}(c^{k-\frac{3}{2}} + c^{k-\frac{1}{2}}) = \mu\Delta^2_{-1}$, $d^{k-\frac{3}{2}} = \Delta^3_{-\frac{3}{2}}$, ..., $h^{k-\frac{7}{2}} = \Delta^7_{-\frac{7}{2}}$, $\tfrac{1}{2}(i^{k-\frac{9}{2}} + i^{k-\frac{7}{2}}) = \mu\Delta^8_{-4}$, ... (Newton's figure illustrates Case 2, in which the eight quantities $AB = f_{\frac{7}{2}}$, $A^2 B^2 = f_{\frac{5}{2}}$, ..., $A^4 B^4 = f_{\frac{1}{2}}$, $A^5 B^5 = f_{-\frac{1}{2}}$, ..., $A^8 B^8 = f_{-\frac{7}{2}}$ are to be interpolated, so that $k = 4\frac{1}{2}$ and $O \equiv A^{4\frac{1}{2}}$ bisects $A^4 A^5$.) In either case Newton effects the interpolation with the help of the general parabola $PQ = f(OP) \equiv f_{k-x}$, where

$$f_x = A + Bx + Cx^2 + Dx^3 \cdots,$$

the various orders of whose unit-differences he has constructed above.

(16) In first draft (see note (13) above) Newton at this point inserted the phrase 'numerus primorum terminorum 9 et erit' (let the number of primary terms be 9 and then will). The effect of the omission is to allow the following formula to embrace pairs of ordinates ($A^{10+n} B^{10+n}$, $n = 0, 1, 2, ...$, and their corresponding 'negative' mates $A^{-n} B^{-n}$) central round $A^5 B^5$.

(17) Since the origin $O \equiv A^k$ is assumed to coincide with A^5 in this first case; compare note (15) above.

in se continuò.[18] Et orientur termini

$$1 \cdot x \cdot \frac{xx}{2} \cdot \frac{x^3-x}{6}, \frac{x^4-xx}{24} \cdot \frac{x^5-5x^3+4x}{120} \cdot \frac{x^6-5x^4+4xx}{720} \cdot \frac{x^7-14x^5+49x^3-36x}{5040} \,\&c$$

per quos si termini seriei k, l, m, n, o, p &c respectivè multiplicentur, aggregatum factorum $k + xl + \dfrac{xx}{2} m + \dfrac{x^3-x}{6} n + \dfrac{x^4-xx}{24} o + \dfrac{x^5-5x^3+4x}{120} p$ &c erit longitudo ordinatim applicatæ PQ.[19]

Cas 2.

In casu posteriori sint A^4B^4, A^5B^5 duo medij termini, hoc est $\dfrac{A^4B^4+A^5B^5}{2}=k$, $b^4=l, \dfrac{c^3+c^4}{2}=m, d^3=n, \dfrac{e^2+e^3}{[2]}=o,$[20] $f^2=p, \dfrac{g+g^2}{2}=q,$ & $h=r$, et erecta ordinatim applicata PQ,[21] biseca A^4A^5 in O et dicto $OP=x$ duc terminos hujus progressionis $1 \times \dfrac{x}{1} \times \dfrac{xx-\frac14}{2x} \times \dfrac{x}{3} \times \dfrac{xx-\frac94}{4x} \times \dfrac{x}{5} \times \dfrac{xx-\frac{25}{4}}{6x} \times \dfrac{x}{7} \times \dfrac{xx-\frac{49}{4}}{8x}$ &c in se continuo. Et orientur termini $1. \; x. \; \dfrac{4xx-1}{8} \cdot \dfrac{4x^3-x}{24} \cdot \dfrac{16x^4-40xx+9}{384}$ &c per quos si termini

---

(18) Newton first wrote down the progression erroneously as

$$`1 \times \frac{x^2}{2} \times \frac{xx-1}{3,4} \times \frac{xx-4}{5,6} \times \frac{xx-9}{7,8} \times \frac{xx-16}{9,10} \times \frac{xx-25}{11\times12} \times \frac{xx-36}{13\times14} \times \&c\,`,$$

from which he drew in sequel the terms

$$`1 \cdot \frac{xx}{2} \cdot \frac{x^4-xx}{24} \cdot \frac{x^6-5x^4+4xx}{720} \cdot \frac{x^8-14x^6+49x^4-36xx}{40320} \cdot \&c\,`.$$

(19) The not inaptly named 'Newton–Stirling' central difference formula which Newton earlier intended to communicate to Leibniz (see §2: note (42)). James Stirling first published his variant of the formula in his 'Methodus Differentialis *Newtoniana* Illustrata' (*Philosophical Transactions*, **30**, No. 362 [for September/October 1719]: 1050–70, especially 1055) as the 'Casus Secundus' of its basic 'Propositio'. (See also his *Methodus Differentialis: sive Tractatus de Summatione et Interpolatione Serierum Infinitarum* (London, 1730): Pars Secunda, 'De Interpolatione Serierum': Propositio XX, Casus primus: 104–5. As we have already seen (§3: notes (23) and (25)) Roger Cotes also made wide use of the formula in his *Canonotechnia sive Constructio Tabularum per Differentias* [= *Opera Miscellanea* (Cambridge, 1722): 36–71].) Here on taking $f_x = A + Bx + Cx^2 + Dx^3 + Ex^4 + Fx^5 \ldots$, we may suppose the general ordinate PQ to be $f(A^5A^x) \equiv f_{5-x}$, so that $AB = f_4$, $A^2B^2 = f_3$, ..., $A^5B^5 = f_0$, ..., $A^9B^9 = f_{-4}$, ... Newton's differences are then

$$\tfrac12(b^{i+4}+b^{i+5}) = \tfrac12(f_{-i+1}-f_{-i-1})$$
$$= B + 2iC + (3i^2+1)D + (4i^2+4)iE + (5i^4+10i^2+1)F \ldots,$$
$$c^{i+4} = b^{i+4}-b^{i+5} = 2[C+3iD+(6i^2+1)E+(10i^2+5)iF \ldots],$$
$$\tfrac12(d^{i+3}+d^{i+4}) = \tfrac12(c^{i+3}-c^{i+5}) = 6[D+4iE+(10i^2+5)F \ldots],$$
$$e^{i+3} = d^{i+3}-d^{i+4} = 24[E+5iF \ldots],$$
$$\tfrac12(f^{i+2}+f^{i+3}) = \tfrac12(e^{i+2}-e^{i+4}) = 120[F \ldots],$$

one into another continually.[18] There will arise the terms 1, x, $\frac{1}{2}x^2$, $\frac{1}{6}(x^3-x)$, $\frac{1}{24}(x^4-x^2)$, $\frac{1}{120}(x^5-5x^3+4x)$, $\frac{1}{720}(x^6-5x^4+4x^2)$,

$$\frac{1}{5040}(x^7-14x^5+49x^3-36x), \quad \dots,$$

and if the terms of the series $k, l, m, n, o, p, \dots$ be respectively multiplied by these, the aggregate $k+lx+\frac{1}{2}mx^2+\frac{1}{6}n(x^3-x)+\frac{1}{24}o(x^4-x^2)+\frac{1}{120}p(x^5-5x^3+4x) \dots$ of the products will be the length of the ordinate PQ.[19]

Case 2

In the latter case let A_4B_4, A_5B_5 be the two middle terms, that is,

$$\tfrac{1}{2}(A_4B_4+A_5B_5) = k, \quad b_4 = l, \quad \tfrac{1}{2}(c_3+c_4) = m, \quad d_3 = n, \quad [\tfrac{1}{2}](e_2+e_3) = o,^{[20]}$$

$f_2 = p$, $\frac{1}{2}(g_1+g_2) = q$ and $h_1 = r$; then, having raised the ordinate PQ,[21] bisect A_4A_5 at O and, calling $OP = x$, multiply the terms of this progression

$$1 \times \frac{x}{1} \times \frac{x^2-\frac{1}{4}}{2x} \times \frac{x}{3} \times \frac{x^2-\frac{9}{4}}{4x} \times \frac{x}{5} \times \frac{x^2-\frac{25}{4}}{6x} \times \frac{x}{7} \times \frac{x^2-\frac{49}{4}}{8x} \dots$$

one into another continually. There will arise the terms

$$1, \; x, \; \tfrac{1}{8}(4x^2-1), \; \tfrac{1}{24}(4x^3-x), \; \tfrac{1}{384}(16x^4-40x^2+9), \; \dots,$$

and if the terms of the series $k, l, m, n, o, p, \dots$ be respectively multiplied by these,

---

and so on. Hence

$$k = f_0 = A, \quad l = \mu\Delta^1_{-\frac{1}{2}} = B+D+F \dots, \quad m = \Delta^2_{-1} = 2C+2E \dots,$$

$$n = \mu\Delta^3_{-\frac{3}{2}} = 6D+30F \dots, \quad o = \Delta^4_{-2} = 24E \dots, \quad p = \mu\Delta^5_{-\frac{5}{2}} = 120F \dots;$$

from which, by inversion,

$$A = k, \quad B = l-\tfrac{1}{6}n+\tfrac{1}{30}p \dots, \quad C = \tfrac{1}{2}n-\tfrac{1}{24}o \dots, \quad D = \tfrac{1}{6}n-\tfrac{1}{24}p \dots, \quad E = \tfrac{1}{24}o \dots, \quad F = \tfrac{1}{120}p \dots;$$

so that, when the latter values are entered in place of $A, B, C, \dots$ and the terms in $k, l, m, \dots$ are gathered, there results

$$f_x = f_0 + \binom{x}{1}\mu\Delta^1_{-\frac{1}{2}} + \tfrac{1}{2}x\binom{x}{1}\Delta^2_{-1} + \binom{x+1}{3}\mu\Delta^3_{-\frac{3}{2}} + \tfrac{1}{4}x\binom{x+1}{3}\Delta^4_{-2} + \binom{x+2}{5}\mu\Delta^5_{-\frac{5}{2}} \dots.$$

(20) Newton's text reads '$e^2+e^3=o$' simply, an error which reappeared uncorrected in print in his *Methodus Differentialis* (note (1) above): 96 in 1711.

(21) That is, in the notation of note (15), $f(A^{4\frac{1}{2}}A^x) \equiv f_{4\frac{1}{2}-x}$.

seriei k, l, m, n, o, p &c respectivè multiplicentur, aggregatum factorum

$$k + xl + \frac{4xx-1}{8}m + \frac{4x^3-x}{24}n + \frac{16x^4-40xx+9}{384}o \text{ \&c}$$

erit longitudo ordinatim applicatæ PQ.[22]

Sed hic notandum est 1 Quod intervalla AA^2, A^2A^3, A^3A^4 &c hic supponuntur esse unitates, Et quod differentiæ colligi debent auferendo inferiores quantitates de superioribus A^2B^2 de AB, A^3B^3 de A^2B^2, b^2 de b &c [hoc est] faciendo $AB - A^2B^2 = b$, $A^2B^2 - A^3B^3 = b^2$. $b - b^2 = c$ &c[23] adeoꝗ quando differentiæ illæ hoc modo prodeunt negativæ signa earum ubiꝗ mutanda sunt.

---

[2]

For solving this Problem generally *Datis quotcunꝗ punctis Curvam describere quæ per omnia transibit:* Note these differences[24]

---

(22) Much as before (compare note (19) above) we may here set

$$f_x = A + Bx + Cx^2 + Dx^3 + Ex^4 + Fx^5 \ldots$$

(where now $AB = f_{\frac{1}{2}}$, $A^2B^2 = f_{\frac{3}{2}}$, ..., $A^4B^4 = f_{\frac{1}{2}}$, $A^5B^5 = f_{-\frac{1}{2}}$, ..., $A^8B^8 = f_{-\frac{7}{2}}$, ...) and hence determine

$$b^{i+4} = f_{-i+\frac{1}{2}} - f_{-i-\frac{1}{2}} = B + 2iC + (3i^2 + \tfrac{1}{4})D + (4i^2 + 1)iE + (5i^4 + \tfrac{5}{2}i^2 + \tfrac{1}{16})F \ldots,$$

$$\tfrac{1}{2}(c^{i+3} + c^{i+4}) = \tfrac{1}{2}(b^{i+3} - b^{i+5}) = 2[C + 3iD + (6i^2 + \tfrac{5}{2})E + (5i^2 + \tfrac{25}{4})iF \ldots],$$

$$d^{i+3} = c^{i+3} - c^{i+4} = 6[D + 4iE + (10i^2 + \tfrac{5}{2})F \ldots],$$

$$\tfrac{1}{2}(e^{i+2} + e^{i+3}) = \tfrac{1}{2}(d^{i+2} - d^{i+4}) = 24[E + 5iF \ldots],$$

$$f^{i+2} = e^{i+2} - e^{i+3} = 120[F \ldots],$$

and so on. In particular

$$k = \tfrac{1}{2}(A^4B^4 + A^5B^5) = \mu f_0 = A + \tfrac{1}{4}C + \tfrac{1}{16}E \ldots, \quad l = b^4 = \Delta^1_{-\frac{1}{2}} = B + \tfrac{1}{4}D + \tfrac{1}{16}F \ldots,$$

$$m = \tfrac{1}{2}(c^3 + c^4) = \mu\Delta^2_{-1} = 2C + 5E \ldots, \quad n = d^3 = \Delta^3_{-\frac{3}{2}} = 6D + 15F \ldots,$$

$$o = \tfrac{1}{2}(e^2 + e^3) = \mu\Delta^4_{-2} = 24E \ldots, \quad p = f^2 = \Delta^5_{-\frac{5}{2}} = 120F \ldots, \ldots;$$

from which, by inversion and successive reduction, there comes

$$A = k - \tfrac{1}{8}m + \tfrac{9}{384}o \ldots, \quad B = l - \tfrac{1}{24}n + \tfrac{9}{1920}p \ldots, \quad C = \tfrac{1}{2}m - \tfrac{5}{48}o \ldots,$$

$$D = \tfrac{1}{6}n - \tfrac{1}{48}p, \quad E = \tfrac{1}{24}o \ldots, \quad F = \tfrac{1}{120}p \ldots;$$

and finally, after substituting these values of A, B, C, ... in the series for f_x and gathering the various k, l, m, ... together, there results

$$f_x = \mu f_0 + x\Delta^1_{-\frac{1}{2}} + \binom{x+\frac{1}{2}}{2}\mu\Delta^2_{-1} + \tfrac{1}{3}x\binom{x+\frac{1}{2}}{2}\Delta^3_{-\frac{3}{2}} + \binom{x+\frac{3}{2}}{4}\mu\Delta^4_{-2} + \tfrac{1}{5}x\binom{x+\frac{3}{2}}{4}\Delta^5_{-\frac{5}{2}} \ldots.$$

the aggregate $k+lx+\frac{1}{8}m(4x^2-1)+\frac{1}{24}n(4x^3-x)+\frac{1}{384}o(16x^4-40x^2+9)$ ... of the products will be the length of the ordinate PQ.[22]

But it should here be noted, first, that the intervals $A_1A_2, A_2A_3, A_3A_4, \ldots$ are in this instance taken to be units; and that the differences ought to be gathered by taking away lower quantities from upper ones (A_2B_2 from A_1B_1, A_3B_3 from A_2B_2, b_2 from b_1, and so on: in other words by making

$$A_1B_1-A_2B_2 = b_1, \quad A_2B_2-A_3B_3 = b_2, \quad b_1-b_2 = c_1, \ldots)\text{[23]}$$

and consequently, when those differences prove in this way to be negative, their signs are to be changed throughout.

---

[2]

[For solving this Problem generally]: *Given any number of points, to describe a curve which shall pass through all of them*, [Note these differences][24]

This so-called 'Newton–Bessel' formula was independently discovered about 1708 by Roger Cotes and called by him 'harum omnium correctionum pro termino intermedio synopsi[s] non inelegan[s]' (*Canonotechnia* (note (19) above): 66) and reappeared, with due acknowledgement of Newton's priority, as the 'Casus Tertius' of the fundamental 'Propositio' of Stirling's 'Methodus Differentialis *Newtoniana* Illustrata' (note (19) above) and, in a slightly revised form, in his later *Methodus Differentialis* (London, 1730): Propositio XX, Casus secundus: 105–6. The attribution to F. W. Bessel (who was but one of several to employ it in his researches into numerical analysis in the early nineteenth century) seems distinctly inappropriate. Why not, in honour of its two co-inventors, call it the 'Newton–Cotes' formula?

(23) This is necessary (and in agreement with modern practice, though not apparently so) since Newton has chosen the direction of $OP = x$ in the converse sense from O to that of the sequence of the points $A, A^2, A^3, \ldots$. In both Newton's present central-difference formulas, to change the direction of $OP = x$ (that is, to substitute $x \to -x$) it is necessary only to alter the sense in which the individual differences are taken (or, equivalently, to change the sign of the differences, $l, n, o, \ldots$ of odd order), the formulas themselves holding true without variation.

(24) Newton generalizes his preceding scheme, in which (adjusted) differences of a function set out for equal intervals of its argument were computed. Here now (compare note (3) above), given $f(A+x) \equiv f_x = a+bx+cx^2+dx^3 \ldots$, he successively constructs the first, second, third and fourth order divided differences of $\alpha = f_p$, $\beta = f_q$, $\gamma = f_r$, $\delta = f_s$ and $\epsilon = f_t$: namely,

$$\zeta = (f_p-f_q)/(p-q) = (p,q), \quad \eta = (f_q-f_r)/(q-r) = (q,r), \quad \theta = (r,s), \quad \kappa = (s,t);$$

$$\lambda = [(p,q)-(q,r)]/(p-r) = (p,q,r), \quad \mu = (q,r,s), \quad \nu = (r,s,t);$$

$$\xi = [(p,q,r)-(q,r,s)]/(p-s) = (p,q,r,s), \quad \pi = (q,r,s,t);$$

and
$$\sigma = [(p,q,r,s)-(q,r,s,t)]/(p-t) = (p,q,r,s,t).$$

A	a
$A+x$	$a+bx+cxx+dx^3+ex^4+fx^5$ &c
$A+p$	$a+bp+cpp+dp^3+ep^4+fp^5=\alpha$
$A+q$	$a+bq+cqq+dq^3+eq^4+fq^5=\beta$
$A+r$	$a+br+crr+dr^3+er^4+fr^5=\gamma$
$A+s$	$a+bs+css+ds^3+es^4+fs^5=\delta$
$A+t$	$a+bt+ctt+dt^3+et^4+ft^5=\epsilon$

$$p-q)\alpha-\beta=b+c\times\overline{p+q}+d\times\overline{pp+pq+qq}+e\times\overline{p^3+ppq+pqq+q^3}+f\times p^4 \ \&c=\zeta$$

$$q-r)\beta-\gamma=b+c\times\overline{q+r}+d\times\overline{qq+qr+rr}+e\times\overline{q^3+qqr+qrr+r^3}+f\times q^4 \ \&c=\eta$$

$$r-s)\gamma-\delta=b+c\times\overline{r+s}+d\times\overline{rr+rs+ss}+e\times\overline{r^3+rrs+rss+s^3}+f\times r^4 \ \&c=\theta$$

$$s-t)\delta-\epsilon=b+c\times\overline{s+t}+d\times\overline{ss+st+tt}+e\times\overline{s^3+sst+stt+t^3}+f\times s^4 \ \&c=\kappa$$

$$p-r)\zeta-\eta=c+d\times\overline{p+q+r}+e\times\overline{pp+pq+qq+pr+qr+rr}+f\times p^3 \ \&c=\lambda$$

$$q-s)\eta-\theta=c+d\times\overline{q+r+s}+e\times\overline{qq+qr+rr+qs+rs+ss}+f[\times]q^3 \ \&c=\mu$$

$$r-t)\theta-\kappa=c+d\times\overline{r+s+t}+e\times\overline{rr+rs+ss+rt+st+tt}+f[\times]r^3 \ \&c=\nu$$

$$p-s)\lambda-\mu=d+e\times\overline{p+q+r+s}+f\times\overline{pp+pq+pr+ps+qq+qr+qs+rr+rs+ss}=\xi$$

$$q-t)\mu-\nu=d+e\times\overline{q+r+s+t}+f\times\overline{qq+qr+qs+qt+rr+rs+rt+ss+st+tt}=\pi$$

$$p-t)\xi-\pi=e+f\times\overline{p+q+r+s}=\sigma.^{(25)}$$

Prob [2]

Curvam Geometricam describere quæ per data quotcunq̃ puncta transibit.

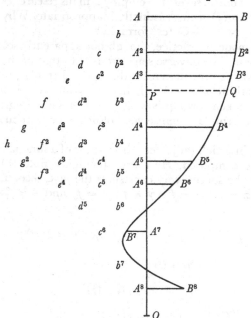

Sint ista puncta B, B^2, B^3, B^4, B^5, B^6, B^7 &c. Et ad rectam quamvis AA[8] demitte perpendicula BA, B^2A^2, B^3A^3 &c et fac $\dfrac{AB-A^2B^2}{AA^2}=b$

$$\frac{A^2B^2-A^3B^3}{A^2A^3}=b^2, \quad \frac{A^3B^3-A^4B^4}{A^3A^4}=b^3,$$

$$\frac{A^4B^4-A^5B^5}{A^4A^5}=b^4, \quad \frac{A^5B^5-A^6B^6}{A^5A^6}=b^5,$$

$$\frac{A^6B^6+A^7B^7}{A^6A^7}=b^6.$$

$$\frac{-A^7B^7-A^8B^8}{A^7A^8}=b^7.^{(26)} \text{ Deinde}$$

$$\frac{b-b^2}{AA^3}=c. \quad \frac{b^2-b^3}{A^2A^4}=c^2. \quad \frac{b^3-b^4}{A^3A^5}=c^3 \ \&c.$$

A	a
$A+x$	$a+bx+cx^2+dx^3+ex^4+fx^5\ldots$

$A+p$	$a+bp+cp^2+dp^3+ep^4+fp^5=\alpha$
$A+q$	$a+bq+cq^2+dq^3+eq^4+fq^5=\beta$
$A+r$	$a+br+cr^2+dr^3+er^4+fr^5=\gamma$
$A+s$	$a+bs+cs^2+ds^3+es^4+fs^5=\delta$
$A+t$	$a+bt+ct^2+dt^3+et^4+ft^5=\epsilon$

$$p-q)\alpha-\beta=b+c(p+q)+d(p^2+pq+q^2)+e(p^3+p^2q+pq^2+q^3)+f(p^4\ldots)=\zeta$$

$$q-r)\beta-\gamma=b+c(q+r)+d(q^2+qr+r^2)+e(q^3+q^2r+qr^2+r^3)+f(q^4\ldots)=\eta$$

$$r-s)\gamma-\delta=b+c(r+s)+d(r^2+rs+s^2)+e(r^3+r^2s+rs^2+s^3)+f(r^4\ldots)=\theta$$

$$s-t)\delta-\epsilon=b+c(s+t)+d(s^2+st+t^2)+e(s^3+s^2t+st^2+t^3)+f(s^4\ldots)=\kappa$$

$$p-r)\zeta-\eta=c+d(p+q+r)+e(p^2+pq+q^2+pr+qr+r^2)+f(p^3\ldots)=\lambda$$

$$q-s)\eta-\theta=c+d(q+r+s)+e(q^2+qr+r^2+qs+rs+s^2)+f(q^3\ldots)=\mu$$

$$r-t)\theta-\kappa=c+d(r+s+t)+e(r^2+rs+s^2+rt+st+t^2)+f(r^3\ldots)=\nu$$

$$p-s)\lambda-\mu=d+e(p+q+r+s)+f(p^2+pq+pr+ps+q^2+qr+qs+r^2+rs+s^2)=\xi$$

$$q-t)\mu-\nu=d+e(q+r+s+t)+f(q^2+qr+qs+qt+r^2+rs+rt+s^2+st+t^2)=\pi$$

$$p-t)\xi-\pi=e+f(p+q+r+s+t)=\sigma.^{(25)}$$

Problem [2]

To describe a geometrical curve which shall pass through any number of given points.
Let those points be B_1, B_2, B_3, B_4, B_5, B_6, B_7, ... To any straight line A_1A_7 let
fall the perpendiculars B_1A_1, B_2A_2, B_3A_3, ... and make $\dfrac{A_1B_1-A_2B_2}{A_1A_2}=b_1$,

$\dfrac{A_2B_2-A_3B_3}{A_2A_3}=b_2$, $\dfrac{A_3B_3-A_4B_4}{A_3A_4}=b_3$, $\dfrac{A_4B_4-A_5B_5}{A_4A_5}=b_4$, $\dfrac{A_5B_5-A_6B_6}{A_5A_6}=b_5$,

$\dfrac{A_6B_6+A_7B_7}{A_6A_7}=b_6$, $\dfrac{-A_7B_7-A_8B_8}{A_7A_8}=b_7$;[26]   next $\dfrac{b_1-b_2}{A_1A_3}=c_1$, $\dfrac{b_2-b_3}{A_2A_4}=c_2$,

---

(25) Assuming now that $f_x=a+bx+cx^2+dx^3+ex^4+fx^5$, exactly. In this case
$$(p,q,r,s,x)=e+f(p+q+r+s+x),$$
for all x, and hence $(p,q,r,s,t,x)=[(p,q,r,s,x)-\sigma]/(x-t)=f$, constant. The general
divided differences interpolation formula ('Cas. 2' of the following Appendix; compare §3:
note (41)) follows immediately by successively inverting the construction of the various
differences: in turn $f_x=f_p+(x-p)(p,x)$, $(p,x)=\zeta+(x-q)(p,q,x)$,
$$(p,q,x)=\lambda+(x-r)(p,q,r,x),\quad (p,q,r,x)=\xi+(x-s)(p,q,r,s,x)$$

Tunc $\dfrac{c-c^2}{AA^4}=d.\ \dfrac{c^3-c^3}{A^2A^5}=d^2.\ \dfrac{c^3-c^4}{A^3A^6}=d^3$ &c. Tunc $\dfrac{d-d^2}{AA^5}=e.\ \dfrac{d^2-d^3}{A^2A^6}=e^2.\ \dfrac{d^3-d^4}{A^3A^7}=e^3$

&c. Sic pergendum est ad ultimam differentiam. Differentijs sic collectis & divisis per intervalla ordinatim applicatarum; in alternis earum columnis sive seriebus vel ordinibus[27] excerpe medias incipiendo ab ultima et in reliquis columnis excerpe media arithmetica inter duas medias, pergendo usq ad seriem primorum terminorum AB, A^2B^2, &c. Sunto hæc k, l, m, n, o, p, q, r &c quorum ultimus terminus significet ultimam diff.[3] penultimus medium arithmeticum inter duas penultimas, antepenultimus mediam trium antepenultimarum &c. Et primus k erit media ordinatim applicata si numerus datorum punctorum est impar, vel medium arithmeticum inter duas medias si numerus e[o]rum est par.[28]

Cas 1.

In casu priori sit A^4B^4 ista media ordinatim applicata, hoc est $A^4B^4=k$, $\dfrac{b^3+b^4}{2}=l.\ c^3=m,\ \dfrac{d^2+d^3}{2}=n,\ e^2=o.\ \dfrac{f+f^2}{[2]}=p.$[29] $g=q$. Et erecta ordinatim applicata PQ et in basi AA^5 sumpto quovis puncto O dic $OP=x$, et duc in se gradatim terminos hujus progressionis

$$1\times\overline{x-OA^4}\times x-\dfrac{\overline{+OA^3+OA^5}}{2}\times\dfrac{\overline{x-OA^3}\times\overline{x-OA^5}}{x-\dfrac{OA^3+OA^5}{2}}\times x-\dfrac{\overline{+OA^2+OA^6}}{2}\times\,\&c$$

et ortam progressionem asserva. Vel quod perinde est, duc terminos progressionis

$$1\times\overline{x-OA^4}\times\overline{x-OA^3}\times\overline{x-OA^5}\times\overline{x-OA^2}\times\overline{x-OA^6}\times\overline{x-OA}\times\overline{x-OA^7}\times\,\&c$$

in se gradatim et terminos exortos duc respective in terminos hujus progressionis

$1.\ x-\dfrac{+OA^3+OA^5}{2}.\ x-\dfrac{+OA^2+OA^6}{2}.\ x-\dfrac{+OA+OA^7}{2}$. &c et orientur termini intermedij: tota progressione existente

$$1.\ x-OA^4.\ xx-\dfrac{+OA^3+2OA^4+OA^5}{2}x+\dfrac{OA^3\times OA^4+OA^5\times OA^4}{[2]}.\ \&c.$$

and $(p, q, r, s, x) = \sigma+(x-t)f$, so that

$$f_x = f_p+(x-p)[\zeta+(x-q)\{\lambda+(x-r)(\xi+(x-s)[\sigma+(x-t)f])\}].$$

In the following Problem, however, Newton prefers to develop only the generalizations of the Newton–Stirling and Newton–Bessel central-difference formulas which he has constructed in the preceding Problem 1 (Cases 1 and 2 respectively).

(26) Observe that the ordinate A^7B^7 (lying to the left of the base line AA^8 in Newton's figure) has correctly been taken negative in sign.

(27) Compare note (12) above. Here the 1711 printed version (*Analysis* (note (1)): 98) accurately transcribes 'Columnis sive Seriebus vel Ordinibus'.

(28) Much as in note (15), corresponding to some origin O—either (Case 1) one of the divisions $A, A^2, A^3, \ldots$ of the base line, or (Case 2) a point midway between two such divisions—

$\dfrac{b_3 - b_4}{A_3 A_5} = c_3, \dots$; then $\dfrac{c_1 - c_2}{A_1 A_4} = d_1, \dfrac{c_2 - c_3}{A_2 A_5} = d_2, \dfrac{c_3 - c_4}{A_3 A_6} = d_3, \dots$; then $\dfrac{d_1 - d_2}{A_1 A_5} = e_1$,

$\dfrac{d_2 - d_3}{A_2 A_6} = e_2, \dfrac{d_3 - d_4}{A_3 A_7} = e_3, \dots$. You must continue in this fashion to the last difference. When the differences have been thus collected and divided by the differences of the ordinates, in their alternate columns or series or ranks[27] take out the middle ones starting from the last, and in the remaining columns take out the arithmetic means between the two middle-most, continuing up to the series $A_1 B_1, A_2 B_2, \dots$ of first terms. Let these be $k, l, m, n, o, p, q, r, \dots$, the last term of which shall denote the last difference, the last but one the arithmetic mean between the two last but one differences, the last but two the middle one of the three last but two, and so on. The first term k will then be the middle ordinate if the number of given points is odd, or alternatively the arithmetic mean between the two middle-most if their number is even.[28]

Case 1

In the former case let $A_4 B_4$ be that middle ordinate, that is, $A_4 B_4 = k$, $\frac{1}{2}(b_3 + b_4) = l$, $c_3 = m$, $\frac{1}{2}(d_2 + d_3) = n$, $e_2 = o$, $\frac{1}{2}(f_1 + f_2) = p$,[29] $g_1 = q$. Then, on erecting the ordinate PQ and taking an arbitrary point O in the base $A_1 A_5$, call $OP = x$ and multiply step by step one into another the terms of this progression

$$1 \times (x - OA_4) \times (x - \tfrac{1}{2}[OA_3 + OA_5]) \times \frac{(x - OA_3)\,(x - OA_5)}{x - \tfrac{1}{2}[OA_3 + OA_5]}$$
$$\times (x - \tfrac{1}{2}[OA_2 + OA_6]) \times \dots,$$

retaining the progression which arises. Or what amounts to the same, multiply the terms of the progression

$$1 \times (x - OA_4) \times (x - OA_3)\,(x - OA_5) \times (x - OA_2)\,(x - OA_6)$$
$$\times (x - OA_1)\,(x - OA_7) \times \dots$$

step by step into each other and the resulting terms multiply respectively into the terms of this progression $1, x - \frac{1}{2}(OA_3 + OA_5), x - \frac{1}{2}(OA_2 + OA_6)$, $x - \frac{1}{2}(OA_1 + OA_7), \dots$, and the intermediate terms will arise, the full progression being $1, x - OA_4, x^2 - \frac{1}{2}(OA_3 + 2OA_4 + OA_5)\,x + \frac{1}{2}(OA_3 + OA_5)\,OA_4, \dots$.

---

and the general point $P \equiv A^x$ in the base line, Newton supposes that the general ordinate of the parabola drawn through $B, B^2, B^3, \dots$ is, on setting $OA^i = a_i$, $PQ \equiv A^i B^i = f(OA^i) \equiv f_{a_i}$, where now $AB = f_{a_1}$, $A^2 B^2 = f_{a_2}$, $A^3 B^3 = f_{a_3}, \dots$, $A^8 B^8 = f_{a_8}, \dots$. He next constructs the array $b, b^2, b^3, \dots; c, c^2, \dots; d, d^2, \dots; h, \dots$ of divided differences and from these (as in Problem 1) selects the differences central round O, finally accomplishing his interpolation by assuming $f_x \equiv A + Bx + Cx^2 + Dx^3 \dots$ to as many terms as numerical accuracy requires.

(29) The text reads '$f + f^2 = p$'; compare note (20) above. The 1711 printed text has the present corrected reading (*Analysis* (note (1)): 98).

Vel dic $OA = \alpha$, $OA^2 = \beta$, $OA^3 = \gamma$, $OA^4 = \delta$, $OA^5 = \epsilon$, $OA^6 = \zeta$, $OA^7 = \eta$,

$$\frac{OA^3 + OA^5}{2} = \theta, \quad \frac{OA^2 + OA^6}{2} = \kappa. \quad \frac{OA + OA^7}{2} = \lambda.^{(30)}$$

Et ex progressione $1 \times \overline{x - \delta} \times \overline{\overline{x - \gamma} \times \overline{x - \epsilon}} \times \overline{\overline{\overline{x - \beta} \times \overline{x - \zeta}} \times \overline{x - \alpha} \times \overline{x - \eta}}$ collige terminos a quibus multiplicatis per $1. x - \theta. x - \kappa. x - \lambda$ &c collige alios terminos intermedios, tota serie prodeunte

$$1. \quad x - \delta. \quad xx \genfrac{}{}{0pt}{}{-\delta}{-\theta} x + \delta\theta. \quad x^3 \genfrac{}{}{0pt}{}{-\delta}{-2\theta} xx + \genfrac{}{}{0pt}{}{+\gamma\epsilon}{[2]\delta\theta} x - \gamma\delta\epsilon.^{(31)} \quad \&c.$$

Per cujus terminos multiplica terminos seriei $k. l. m. n. o$ &c et aggregatum productorum $k + \overline{x - \delta} \times l + \overline{xx \genfrac{}{}{0pt}{}{-\delta}{-\theta} x + \delta\theta} \times m$ &c erit longitudo ordinatim applicatæ $PQ.^{(32)}$

Cas. 2.

In casu posteriori sint $A^4 B^4$ et $A^5 B^5$ duæ mediæ ordinatim applicatæ, hoc est $\frac{A^4 B^{[4]} + A^5 B^5}{2} = k. \ b^4 = l. \ \frac{c^3 + c^4}{2} = m. \ d^3 = n. \ \frac{e^2 + e^3}{2} = o. \ f^2 = p.$ &c. Et alternorum k, m, o, q &c coefficientes orientur ex multiplicatione terminorum hujus progressionis in se

$$1 \times \overline{\overline{x - OA^4} \times \overline{x - OA^5}} \times \overline{\overline{x - OA^3} \times \overline{x - OA^6}} \times \overline{\overline{x - OA^2} \times \overline{x - OA^7}}$$

$$\times \overline{\overline{x - OA} \times \overline{x - OA^8}} \quad \&c$$

---

(30) Otherwise, $2\theta = \gamma + \epsilon$, $2\kappa = \beta + \zeta$ and $2\lambda = \alpha + \eta$.

(31) That is, $x^3 - [\delta + (\gamma + \epsilon)]x^2 + [\gamma\epsilon + \delta(\gamma + \epsilon)]x - \gamma\delta\epsilon = (x - \gamma)(x - \delta)(x - \epsilon)$.

(32) The generalization of the Newton–Stirling formula (Case 1 of the preceding Problem 1); see note (19). In Newtonian terms (compare note (28)), on taking $OA^i = a_i$ and so

$$A^i B^i = f_{a_i} = A + Ba_i + Ca_i^2 + Da_i^3 + Ea_i^4 \ldots,$$

the various differences prove to be

$$b^i = (a_i, a_{i+1}) = B + C(a_i + a_{i+1}) + (D(a_i^2 + a_i a_{i+1} + a_{i+1}^2) + E(a_i^3 + a_i^2 a_{i+1} + a_i a_{i+1}^2 + a_{i+1}^3) \ldots,$$

$$c^i = (a_i, a_{i+1}, a_{i+2}) = C + D(a_i + a_{i+1} + a_{i+2}) + E(a_i^2 + a_i a_{i+1} + a_{i+1}^2 + a_i a_{i+2} + a_{i+1} a_{i+2} + a_{i+2}^2) \ldots,$$

$d^i = D + E(a_i + a_{i+1} + a_{i+2} + a_{i+3}) \ldots, e^i = E \ldots$, and so on, much as in the preceding tabulation of divided differences. Hence, on setting

$$a_3 = \gamma, \quad a_4 = \delta, \quad a_5 = \epsilon, \quad \tfrac{1}{2}(a_3 + a_5) = \tfrac{1}{2}(\gamma + \epsilon) = \theta, \quad \ldots,$$

Or call $OA_1 = \alpha$, $OA_2 = \beta$, $OA_3 = \gamma$, $OA_4 = \delta$, $OA_5 = \epsilon$, $OA_6 = \zeta$, $OA_7 = \eta$, $\frac{1}{2}(OA_3 + OA_5) = \theta$, $\frac{1}{2}(OA_2 + OA_6) = \kappa$, $\frac{1}{2}(OA_1 + OA_7) = \lambda$;[30] then from the progression $1 \times (x - \delta) \times (x - \gamma)\,(x - \epsilon) \times (x - \beta)\,(x - \zeta) \times (x - \alpha)\,(x - \eta)$ gather the terms, and from these multiplied by 1, $x - \theta$, $x - \kappa$, $x - \lambda$, ... gather other intermediate terms, the full series proving to be 1, $x - \delta$, $x^2 - (\delta + \theta)\,x + \delta\theta$,

$$x^3 - (\delta + 2\theta)\,x^2 + (\gamma\epsilon + 2\delta\theta)\,x - \gamma\delta\epsilon,^{(31)} \quad \ldots$$

By the terms of this multiply the terms of the series k, l, m, n, o, ... and the aggregate of the products $k + l(x - \delta) + m(x^2 - (\delta + \theta)\,x + \delta\theta)$ ... will be the length of the ordinate PQ.[32]

Case 2

In the latter case let $A_4 B_4$ and $A_5 B_5$ be the two central ordinates, that is, $\frac{1}{2}(A_4 B_4 + A_5 B_5) = k$, $b_4 = l$, $\frac{1}{2}(c_3 + c_4) = m$, $d_3 = n$, $\frac{1}{2}(e_2 + e_3) = o$, $f_2 = p$, ... Then the coefficients of alternate terms k, m, o, q, ... arise from multiplication of those of this series

$$1 \times (x - OA_4)\,(x - OA_5) \times (x - OA_3)\,(x - OA_6) \times (x - OA_2)\,(x - OA_7)$$
$$\times (x - OA_1)\,(x - OA_8) \ldots$$

Newton's selected central differences are respectively

$$k = A^4 B^4 = f_{a_4} = A + B\delta + C\delta^2 + D\delta^3 \ldots,$$

$$l = \tfrac{1}{2}(b^3 + b^4) = B + C(\theta + \delta) + D(\tfrac{1}{2}(\gamma^2 + \epsilon^2) + \delta\theta + \delta^2) \ldots,$$

$$m = c^3 = C + D(2\theta + \delta) \ldots,$$

$$n = \tfrac{1}{2}(d^2 + d^3) = D \ldots,$$

and so on similarly; so that, inversely,

$$D = n, \quad C = m - n(2\theta + \delta) \ldots, \quad B = l - m(\theta + \delta) + n\gamma\epsilon \ldots$$

and $A = k - l\delta + m\delta\theta - n\gamma\delta\epsilon$ .... Finally, on putting $a_i = x$, there follows

$$PQ \equiv f_x = k + l(x - \delta) + m(x - \delta)\,(x - \theta) + m(x - \gamma)\,(x - \delta)\,(x - \epsilon) \ldots$$

This formula follows readily—as Newton no doubt soon came to realize—as the mean of the two series defined recursively by the 'advancing difference' sequences (compare note (25) above) $f_x = f_{a_i} + (x - a_i)\,(x, a_i)$ and

$$(x, a_{i-p}, a_{i-p+1}, \ldots, a_{i+p}) = \begin{cases} (a_{i-p-1}, a_{i-p}, \ldots, a_{i+p}) + (x - a_{i-p-1}) \\ (a_{i-p}, a_{i-p+1}, \ldots, a_{i+p+1}) + (x - a_{i+p+1}) \end{cases} (a_{i-p-1}, \ldots, a_{i+p+1})$$
$$+ (x - a_{i-p-1})\,(x - a_{i+p+1})\,(x, a_{i-p-1}, \ldots, a_{i+p+1})$$

in the equivalent form

$$f_x = f_{a_i} + (x - a_i)\,[\tfrac{1}{2}\{(a_{i-1}, a_i) + (a_i, a_{i+1})\} + (x - \tfrac{1}{2}[a_{i-1} + a_{i+1}])\,(a_{i-1}, a_i, a_{i+1}) \ldots].$$

et reliquorum coefficientes ex multiplicatione horū per terminos hujus progressionis

$$x - \frac{+OA^4 + OA^5}{2} \cdot x - \frac{+OA^3 + OA^6}{2} \cdot x - \frac{+OA^2 + OA^7}{2} \cdot x - \frac{+OA + OA^8}{2} \cdot \text{\&c.}$$

Hoc est erit

$$k + x - \frac{+OA^4 + OA^5}{2} \times l + \overline{xx - \overline{OA^4[+]OA^5} \times x + OA^4 \times OA^5} \times m \text{ \&c}$$

=ordinatim applicatæ *PQ*

$$= k \begin{matrix} +x \\ -\frac12 OA^4 \\ -\frac12 OA^5 \end{matrix} \times\!\!\!\times\; l \begin{matrix} +x \\ -OA^4 \end{matrix} \times\!\!\!\times \begin{matrix} x \\ -OA^5 \end{matrix} \times\!\!\!\times\; m \begin{matrix} +x \\ -OA^4 \end{matrix} \times\!\!\!\times \begin{matrix} x \\ -OA^5 \end{matrix} \times\!\!\!\times \begin{matrix} x \\ -\frac12 OA^3 \\ -\frac12 OA^6 \end{matrix} \times\!\!\!\times\; n$$

$$\begin{matrix} +x \\ -OA^3 \end{matrix} \times\!\!\!\times \begin{matrix} x \\ -OA^4 \end{matrix} \times\!\!\!\times \begin{matrix} x \\ -OA^5 \end{matrix} \times\!\!\!\times \begin{matrix} x \\ -OA^6 \end{matrix} \times\!\!\!\times\; o \; [\text{\&c}].$$

Sive dic $x - \dfrac{+OA^4 + OA^5}{2} = \pi.\ \overline{x - OA^4} \times \overline{x - OA^5} = \rho.\ \rho \times x - \dfrac{+OA^3 + OA^6}{2} = \sigma.$

$\rho \times \overline{x - OA^3} \times \overline{x - OA^6} = \tau.\quad \tau \times x - \dfrac{+OA^2 + OA^7}{2} = \upsilon.\quad \tau \times \overline{x - OA^2} \times \overline{x - OA^7} = \phi.$

$\phi \times x - \dfrac{+OA + OA^8}{2} = \chi.\ \phi \times \overline{x - OA} \times \overline{x - OA^8} = \psi.\ [\text{\&c}]$ et erit

$$k + \pi l + \rho m + \sigma n + \tau o + \upsilon p + \phi q + \chi r + \psi s \quad [\text{\&c}] = PQ.^{(33)}$$

---

(33) The corresponding generalization of the Newton–Bessel formula (Case 2 of Problem 1); see note (22). Much as in the preceding Case 1 (compare note (32)) Newton selects the central divided differences (where $a_3 = \gamma$, $a_4 = \delta$, $a_5 = \epsilon$, $a_6 = \zeta$, $\frac12(a_4 + a_5) = \theta'$, $\frac12(a_3 + a_6) = \chi'$, ...)

$$k = \tfrac12(a_4 + a_5) = A + B\theta' + C(2\theta'^2 - \delta\epsilon) + D(4\theta'^3 - 3\theta'\delta\epsilon) \,...,$$

$$l = b^4 = (a_4, a_5) = B + 2C\theta' + D(4\theta'^2 - \delta\epsilon) \,...,$$

$$m = \tfrac12(c^3 + c^4) = \tfrac12[(a_3, a_4, a_5) + (a_4, a_5, a_6)] = C + D(\chi' + 2\theta') \,...,$$

$$n = d^3 = (a_3, a_4, a_5, a_6) = D \,...,$$

into one another, and the coefficients of the remainder from multiplication of these by the terms of this progression $x - \frac{1}{2}[OA_4 + OA_5]$, $x - \frac{1}{2}[OA_3 + OA_6]$, $x - \frac{1}{2}[OA_2 + OA_7]$, $x - \frac{1}{2}[OA_1 + OA_8]$, ... That is, the ordinate PQ will equal

$$k + l(x - \tfrac{1}{2}[OA_4 + OA_5]) + m(x^2 - [OA_4 + OA_5]\,x + OA_4 \times OA_5) \ldots$$
$$= k + l(x - \tfrac{1}{2}[OA_4 + OA_5]) + m(x - OA_4)\,(x - OA_5)$$
$$+ n(x - OA_4)\,(x - OA_5)\,(x - \tfrac{1}{2}[OA_3 + OA_6])$$
$$+ o(x - OA_3)\,(x - OA_4)\,(x - OA_5)\,(x - OA_6) \ldots.$$

Otherwise, call $x - \frac{1}{2}[OA_4 + OA_5] = \pi$, $(x - OA_4)\,(x - OA_5) = \rho$,

$$\rho(x - \tfrac{1}{2}[OA_3 + OA_6]) = \sigma, \quad \rho(x - OA_3)\,(x - OA_6) = \tau,$$
$$\tau(x - \tfrac{1}{2}[OA_2 + OA_7]) = \upsilon, \quad \tau(x - OA_2)\,(x - OA_7) = \phi,$$
$$\phi(x - \tfrac{1}{2}[OA_1 + OA_7]) = \chi, \quad \phi(x - OA_1)\,(x - OA_8) = \psi, \quad \ldots$$

and then $PQ = k + l\pi + m\rho + n\sigma + o\tau + p\upsilon + q\phi + r\chi + s\psi \ldots.$[33]

---

and so on; so that, on reversing $D = n \ldots$, $C = m - n(\delta + \epsilon + \chi') \ldots$,

$$B = l - m(\delta + \epsilon) + n([\delta + \epsilon]\chi' + \delta\epsilon) \ldots \quad \text{and} \quad A = k - l\theta' + m\delta\epsilon - n\delta\epsilon\chi' \ldots;$$

from which there results

$$PQ \equiv f_x = k + l(x - \theta') + m(x - \delta)\,(x - \epsilon) + n(x - \delta)\,(x - \epsilon)\,(x - \chi') \ldots.$$

This follows more readily, as Newton must soon have come to realize, as the mean of the two series defined recursively by

$$f_x = \begin{Bmatrix} f_{a_{i-\frac{1}{2}}} + (x - a_{i-\frac{1}{2}}) \\ f_{a_{i+\frac{1}{2}}} + (x - a_{i+\frac{1}{2}}) \end{Bmatrix} (a_{i-\frac{1}{2}}, a_{i+\frac{1}{2}}) + (x - a_{i-\frac{1}{2}})\,(x - a_{i+\frac{1}{2}})\,(x, a_{i-\frac{1}{2}}, a_{i+\frac{1}{2}})$$

and

$$(x, a_{i-p+\frac{1}{2}}, \ldots, a_{i+p-\frac{1}{2}}) = \begin{Bmatrix} (a_{i-p-\frac{1}{2}}, \ldots, a_{i+p-\frac{1}{2}}) + (x - a_{i-p-\frac{1}{2}}) \\ (a_{i-p+\frac{1}{2}}, \ldots, a_{i+p+\frac{1}{2}}) + (x - a_{i+p+\frac{1}{2}}) \end{Bmatrix} (a_{i-p-\frac{1}{2}}, \ldots, a_{i+p+\frac{1}{2}})$$
$$+ (x - a_{i-p-\frac{1}{2}})\,(x - a_{i+p+\frac{1}{2}})\,(x, a_{i-p-\frac{1}{2}}, \ldots, a_{i+p+\frac{1}{2}})$$

in the equivalent form

$$f_x = \tfrac{1}{2}[f_{a_{i-\frac{1}{2}}} + f_{a_{i+\frac{1}{2}}}] + (x - \tfrac{1}{2}[a_{i-\frac{1}{2}} + a_{i+\frac{1}{2}}])\,(a_{i-\frac{1}{2}}, a_{i+\frac{1}{2}})$$
$$+ (x - a_{i-\frac{1}{2}})\,(x - a_{i+\frac{1}{2}})\,[\tfrac{1}{2}\{(a_{i-\frac{3}{2}}, a_{i-\frac{1}{2}}, a_{i+\frac{1}{2}}) + (a_{i-\frac{1}{2}}, a_{i+\frac{1}{2}}, a_{i+\frac{3}{2}})\} \ldots].$$

APPENDIX
INTERPOLATION BY ADVANCING
DIFFERENCES (1686)

Extracted from Newton's *Principia Mathematica*[1]

Invenire lineam curvam generis Parabolici, quæ per data quotcunque puncta transibit.

Sunto puncta illa *A, B, C, D, E, F*, &c. & ab iisdem ad rectam quamvis positione datam *HN* demitte perpendicula quotcunque *AH, BI, CK, DL, EM, FN*.[2]

---

(1) *Philosophiæ Naturalis Principia Mathematica* (London, ₁1687): 481–3, implemented by the insertion of two omitted 'Errata Sensum turbantia' listed on signature [Ooo 4ʳ]. The explicit function of this Lemma v of Newton's 'De Mundi Systemate Liber Tertius' is to introduce the immediately following 'Lemma vi. *Ex observatis aliquot locis Cometæ invenire locum ejus ad tempus quodvis intermedium datum*' (see note (6) below), but its importance in a wider context lies in its being his first known statement of the 'Harriot–Briggs' advancing differences formula for interpolating over equal intervals of argument and its (divided differences) generalization, here made known to the world for the first time. Apart from an insignificantly variant prior transcript in Humphrey Newton's hand at the equivalent place in the press copy (now Royal Society MS LXIX) of the whole volume—corrected by Newton himself at a few places—no previous manuscript of the present piece is known, but a composition date of mid-1686 will not be far out. The autumn before Newton had experienced difficulty in computing cometary orbits to sufficient accuracy 'but am now going about it' as he told Flamsteed on 19 September (*Correspondence of Isaac Newton*, **2**, 1960: 419). The printer's manuscript for Book 3 of the *Principia* arrived in London on 28 March 1687, 'yesterday...sennight' in Halley's acknowledgement eight days later (*ibid.*: 473).

(2) Much as before, in his (suppressed) analysis Newton will suppose that with respect to some origin in the base line *HN* (which is conveniently taken to coincide with *H*) the general ordinate of the curve *ABC*... 'of parabolic kind' may be defined by $SR = f_{HS}$, where $f_x = \alpha + \beta x + \gamma x^2$ .... His present insistence on choosing perpendicular (rather than general oblique) ordinates is dictated by the cometary context in which the Lemma is set: it is not, of course, a necessary restriction.

(3) The ordinates *EM* and *FN*, lying below the base line *HN* in Newton's figure, are accordingly set as negative magnitudes. Observe that the differences are (contrary to modern practice) taken in the sense opposite to the sequence of points *H, I, K, L*, ...: the effect is to make Newton's odd-order differences b, ^2b, ^3b, ...; d, ^2d, ...; f ... opposite in sign to their present-day equivalents. The otherwise trivial change in notation for the various differences ('^ib' instead of the previous 'b^i' and so on) has the obvious advantage that they will no longer be misunderstood to be powers of the initial differences, b, c, d, ..., of each order.

(4) The 'Harriot–Briggs' advancing (adjusted) differences formula for interpolation over unit intervals of argument. If, more generally, we set the equal intervals as

$$HI = IK = ... = MN = h$$

and (with respect to origin *H* in *HN*) take

$$HS = x, \quad SR = f_{HS} \equiv f_x = a + Bx + Cx^2 + Dx^3 + Ex^4 ...,$$

Cas. 1. Si punctorum *H, I, K, L, M, N* æqualia sunt intervalla *HI, IK, KL,* &c. collige perpendiculorum *AH, BI, CK* &c differentias primas b, ^2b, ^3b, ^4b, ^5b, &c. secundas c, ^2c, ^3c, ^4c, &c. tertias d, ^2d, ^3d, &c. id est, ita ut sit $HA - BI$ $= b$, $\quad BI - CK = {}^2b$, $\quad CK - DL = {}^3b$, $DL + EM = {}^4b$, $\quad -EM + FN = {}^5b$,[3]

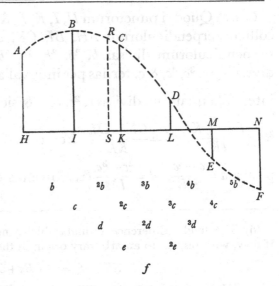

&c. dein $b - {}^2b = c$ &c. & sic pergatur ad differentiam ultimam, quæ hic est f. Deinde erecta quacunque perpendiculari *RS*, quæ fuerit ordinatim applicata ad curvam quæsitam: ut inveniatur hujus longitudo, pone intervalla *HI, IK, KL, LM,* &c unitates esse, & dic $AH = a$, $-HS = p$, $\frac{1}{2}p$ in $-IS = q$, $\frac{1}{3}q$ in $+SK = r$, $\frac{1}{4}r$ in $+SL = s$, $\frac{1}{5}s$ in $+SM = t$; pergendo videlicet ad usque penultimum perpendiculum *ME*, & præponendo signa negativa terminis *HS, IS,* &c qui jacent ad partes puncti *S* versus *A*, & signa affirmativa terminis *SK, SL,* &c qui jacent ad alteras partes puncti *S*. Et signis probe observatis erit $RS = a + bp + cq + dr + es + ft$ &c.[4]

---

then $AH = f_0 = a$, $BI = f_h$, $CK = f_{2h}$, ..., $FN = f_{5h}$, ..., and so Newton's differences are

$$^ib = f_{(i-1)h} - f_{ih} = -\Delta^1_{(i-1)h} = -[B + C(2i-1) + D(3i^2 - 3i + 1) + E(4i^3 - 6i^2 + 4i - 1) \ldots],$$

$$^ic = {}^ib - {}^{i+1}b = +\Delta^2_{(i-1)h} = 2[C + D(3i) + E(6i^2 + 1) \ldots],$$

$$^id = {}^ic - {}^{i+1}c = -\Delta^3_{(i-1)h} = -6[D + E(4i + 2) \ldots],$$

$$^ie = {}^id - {}^{i+1}d = +\Delta^4_{(i-1)h} = 24[E \ldots],$$

and so on. In particular $-b = \Delta^1_0 = B + C + D + E \ldots$, $c = \Delta^2_0 = 2C + 6D + 14E \ldots$,

$$-d = \Delta^3_0 = 6D + 36E \ldots, \quad e = \Delta^4_0 = 24E \ldots;$$

and so, inversely,

$$B = -b - \tfrac{1}{2}c - \tfrac{1}{3}d - \tfrac{1}{4}e \ldots, \quad C = \tfrac{1}{2}c + \tfrac{1}{2}d + \tfrac{11}{24}e \ldots, \quad D = -\tfrac{1}{6}d - \tfrac{1}{4}e \ldots, \quad E = \tfrac{1}{24}e \ldots.$$

Hence $SR \equiv f_x = a + bp + cq + dr + es \ldots$, where

$$p = -x/h = -\binom{x/h}{1}, \quad q = \tfrac{1}{2}p(1 - x/h) = \binom{x/h}{2}, \quad r = \tfrac{1}{3}q(2 - x/h) = -\binom{x/h}{3},$$

$$s = \tfrac{1}{4}r(3 - x/h) = \binom{x/h}{4}, \ldots.$$

Newton's result is the particular case for which $h = 1$.

Cas. 2. Quod si punctorum H, I, K, L, &c inæqualia sint intervalla HI, IK, &c. collige perpendiculorum AH, BI, CK, &c differentias primas per intervalla perpendiculorum divisas $b, {}^2b, {}^3b, {}^4b, {}^5b$ [&c]; secundas per intervalla bina divisas $c, {}^2c, {}^3c, {}^4c$, &c. tertias per intervalla terna divisas $d, {}^2d, {}^3d$, &c. quartas per intervalla quaterna divisas $e, {}^2e$, &c. & sic deinceps; id est ita ut sit $b = \dfrac{AH - BI}{HI}$,

$^2b = \dfrac{BI - CK}{IK}$, $\quad ^3b = \dfrac{CK - DL}{KL}$ &c. dein $c = \dfrac{b - {}^2b}{HK}$, $\quad ^2c = \dfrac{{}^2b - {}^3b}{IL}$, $\quad ^3c = \dfrac{{}^3b - {}^4b}{KM}$ &c.

postea $d = \dfrac{c - {}^2c}{HL}$, $\quad ^2d = \dfrac{{}^2c - {}^3c}{IM}$ &c. Inventis differentiis, dic $AH = a$, $-HS = p$, p in

---

(5) The divided differences formula which generalizes Case 1 above; compare §4: note (25). If now, with respect to an arbitrary origin in the base line HN, we set

$$SR \equiv f_x = A + Bx + Cx^2 + Dx^3 + Ex^4 \ldots$$

with $AH = f_{a_1} = a$, $BI = f_{a_2}$, $CK = f_{a_3}$, ..., $FN = f_{a_5}$, ..., then Newton's differences are respectively (compare §4: note (32))

$${}^ib = -(a_i, a_{i+1}) = -B - C(a_i + a_{i+1}) - D(a_i^2 + a_i a_{i+1} + a_{i+1}^2)$$
$$- E(a_i^3 + a_i^2 a_{i+1} + a_i a_{i+1}^2 + a_{i+1}^3) \ldots,$$

$${}^ic = +(a_i, a_{i+1}, a_{i+2}) = C + D(a_i + a_{i+1} + a_{i+2})$$
$$+ E(a_i^2 + a_i a_{i+1} + a_{i+1}^2 + a_i a_{i+2} + a_{i+1} a_{i+2} + a_{i+2}^2) \ldots,$$

$${}^id = -(a_i, a_{i+1}, a_{i+2}, a_{i+3}) = -D - E(a_i + a_{i+1} + a_{i+2} + a_{i+3}) \ldots,$$

$${}^ie = +(a_i, a_{i+1}, a_{i+2}, a_{i+3}, a_{i+4}) = E \ldots,$$

and so forth. In particular

$$b = -(a_1, a_2), \quad c = +(a_1, a_2, a_3), \quad d = -(a_1, a_2, a_3, a_4), \quad e = +(a_1, a_2, a_3, a_4, a_5), \quad \ldots,$$

so that, on making $p = a_1 - x$, $q/p = a_2 - x$, $r/q = a_3 - x$, $s/r = a_4 - x$, $t/s = a_5 - x$, ..., there results $f_x = a - p(x, a_1)$, where successively

$$(x, a_1) = B + C(x + a_1) + D(x^2 + a_1 x + a_1^2) + E(x^3 + a_1 x^2 + a_1^2 x + a_1^3) \ldots$$
$$= -b + (q/p)(x, a_1, a_2),$$

$$(x, a_1, a_2) = C + D(x + a_1 + a_2) + E(x^2 + (a_1 + a_2)x + a_1^2 + a_1 a_2 + a_2^2) \ldots$$
$$= -c + (r/q)(x, a_1, a_2, a_3),$$

$$(x, a_1, a_2, a_3) = D + E(x + a_1 + a_2 + a_3) \ldots = -d + (s/r)(x, a_1, a_2, a_3, a_4),$$

$$(x, a_1, a_2, a_3, a_4) = E \ldots = -e + (t/s)(x, a_1, a_2, a_3, a_4, a_5), \ldots.$$

Newton's result follows at once, upon eliminating each of the left-hand divided differences in turn, in the form

$$f_x = a + p[b + (q/p)\{c + (r/q)(d + (s/r)[e + (t/s)\{f + \ldots\}])\}].$$

Case 1 follows on taking $a_i = (i-1)h$, $i = 1, 2, 3, \ldots$, when

$$b = -\frac{1}{h}\Delta_0^1, \quad c = \frac{1}{2h^2}\Delta_0^2, \quad d = -\frac{1}{3!\,h^3}\Delta_0^3, \quad e = \frac{1}{4!\,h^4}\Delta_0^4, \ldots$$

$-IS = q$, q in $+SK = r$, r in $+SL = s$, s in $+SM = t$; pergendo scilicet ad usque perpendiculum penultimum ME, & erit ordinatim applicata

$$RS = a + bp + cq + dr + es + ft \quad \&c.^{(5)}$$

Corol. Hinc areæ curvarum omnium inveniri possunt quamproximè. Nam si curvæ cujusvis quadrandæ inveniantur puncta aliquot, & Parabola per eadem duci intelligatur: erit area Parabolæ hujus eadem quam proximè cum area curvæ illius quadrandæ. Potest autem Parabola per Methodos notissimas semper quadrari.[6]

---

and $p = -x$, $q = x(x-h)$, $r = -x(x-h)(x-2h)$, $s = x(x-h)(x-2h)(x-3h)$, ..., yielding at length the advancing differences formula for interpolation over equal h-intervals

$$f_x = f_0 + \binom{x/h}{1}\Delta_0^1 + \binom{x/h}{2}\Delta_0^2 + \binom{x/h}{3}\Delta_0^3 + \binom{x/h}{4}\Delta_0^4 \dots.$$

The following Lemma VI in Book 3 of the *Principia* ($_1$1687: 483) outlines the application of this general case to the problem of interpolating cometary longitudes and latitudes, given five observed places: 'Designent *HI, IK, KL, LM* tempora inter observationes, *HA, IB, KC, LD, ME*, observatas quinque longitudines Cometæ, *HS* tempus datum inter observationem primam & longitudinem quæsitam. Et si per puncta *A, B, C, D, E* duci intelligatur curva regularis *ABCDE*; & per Lemma superius inveniatur ejus ordinatim applicata *RS*, erit *RS* longitudo quæsita. Eadem methodo ex observatis quinque latitudinibus invenitur latitudo ad tempus datum.' A final paragraph adds the practical remark that 'Si longitudinum observatarum parvæ sint differentiæ, puta graduum tantum 4 vel 5; suffecerint observationes tres vel quatuor ad inveniendam longitudinem & latitudinem novam. Sin majores sint differentiæ, puta graduum 10 vel 20, debebunt observationes quinque adhiberi'. Though Newton did not, in any of the three editions of the *Principia* published in his lifetime, give a numerical example to illustrate this application, he did in fact make some use of his general divided differences formula in the autumn of 1695 to check, probably at Halley's instigation, the elements of orbit of the comet which appeared during the winter of 1680–1: the most detailed of these computations (ULC. Add. 3965.14: 586^r) will be reproduced in the seventh volume.

(6) In other words, if $SR = f_{HS}$ is the defining equation of the curve through the points $A \equiv (a_1, f_{a_1})$, $B \equiv (a_2, f_{a_2})$, ..., $F \equiv (a_n, f_{a_n})$, then the area $\int_{a_1}^{a_n} f_x \, . \, dx$ beneath it may be approximated as $\left[ax + \frac{1}{2}bx^2 + \frac{1}{3}cx^3 \dots + \frac{1}{n}ex^n \right]_{a_1}^{a_n}$ to the same order of accuracy that

$$a + bx + cx^2 \dots + ex^{n-1}$$

approximates f_x. About 1710 Newton drafted a concluding Scholium (ULC. Add. 4005.15: 71^r/3965.14: 236^v) to the text reproduced as §4 above, echoing his present Corollary with the phrase: 'Utilis est hæc methodus ad Tabulas construendas per interpolationem serierum [quibus Problema expeditius] reducetur. Ut si quatuor sint Ordinatæ *A, B, C, D* ad æqualia intervalla disposita... Area inter Ordinatam primam et quartam erit $\left[\dfrac{A + 3B + 3C + D}{8}\right]$.'

(The version which subsequently appeared in print in Newton's *Methodus Differentialis* [= *Analysis Per Quantitatum Series, Fluxiones, ac Differentias* (London, 1711): 93–101, especially

2

PROBLEMS IN ELEMENTARY NUMBER THEORY

[?late 1670's][1]

§1. THE SOLUTION OF SIMPLE DIOPHANTINE EQUATIONS.[2]

From the originals in the University Library, Cambridge

[1][3]

Prob. 1. Invenire 3 continue proportionales numeros quorum summa erit quadratum.

$$xx + xy + yy = zz =^{[4]} xx + 2ex + ee. \quad \frac{yy - ee}{2e - y} = x. \quad [\text{Pone}] \quad y = 3. \quad e = 2. \quad [\text{erit}] \quad x = 5.$$

25. 15. 9 $\div$ [quorum summa] = 49.[5]

100] reads somewhat differently.) In proof, if we set $A = f_{-\frac{1}{2}}$, $B = f_{-\frac{1}{6}}$, $C = f_{\frac{1}{6}}$ and $D = f_{\frac{1}{2}}$ where $f_x \approx a + bx + cx^2 + dx^3$, then $A + D \approx 2a + \frac{1}{2}c$, $B + C \approx 2a + \frac{1}{18}c$ and so

$$\int_{-\frac{1}{2}}^{\frac{1}{2}} f_x \, . \, dx \approx a + \frac{1}{12}c \approx \frac{1}{8}(A + 3B + 3C + D).$$

In the case of the three equidistant ordinates $A = f_{-\frac{1}{2}}$, $B = f_0$ and $C = f_{\frac{1}{2}}$ the corresponding approximation of f_x as $a + bx + cx^2$ yields 'Simpson's' rule, namely

$$\int_{-\frac{1}{2}}^{\frac{1}{2}} f_x \, . \, dx \approx \frac{1}{6}(A + 4B + C).$$

When Roger Cotes in February 1711 read the presentation copy of the *Analysis* given him by its editor, William Jones, he seized on this 'Regula qua ex datis quatuor Ordinatis ad æqualia intervalla sitis metitur aream inter primam & ultimam' as 'pulcherrima in primis & utilissima' and extended it to cover all cases of approximation by parallel, equidistant ordinates ranging in number from 3 to 11, setting these 'Cotes' formulas' as a 'Postscriptum' (pp. 32–3) to his tract *De Methodo Differentiali Newtoniana* [= *Opera Miscellanea* (Cambridge, 1722): 23–32; compare Joseph Edleston's *Correspondence of Sir Isaac Newton with Professor Cotes* (London, 1850): 206]. The accuracy of Cotes' results is attested by their exact agreement with an equivalent table of areas approximated by odd numbers (3, 5, 7, 9, 11) of equidistant ordinates computed independently by James Stirling some eight years later and published in his 'Methodus Differentialis Newtoniana Illustrata' (*Philosophical Transactions*, **30**, No. 362 [for September/October 1719]: 1050–70, especially 1063; revised as Proposition XXXI of his *Methodus Differentialis* (London, 1730): 146–7).

(1) This tentative date is assigned uniquely from our assessment of Newton's handwriting style in the two autograph manuscripts (evidently composed at roughly the same period) which are here reproduced. No documentary confirmation of this suggested chronology appears to exist and we can only guess at Newton's intention in composing them: perhaps they represent his preliminary researches for a projected university lecture course on number theory or fuel for discussion at the open hours in his rooms which, by statute, he was required to hold twice weekly during term-time (see III: xxii–xxiii). At his death Newton's library con-

Translation

[1][3]

Problem 1. To find three numbers in continued proportion whose sum shall be a square.
Take $x^2 + xy + y^2 = z^2 = $[4] $(x+e)^2$, then $x = (y^2 - e^2)/(2e - y)$. Put $y = 3$, $e = 2$ and so $x = 5$, when the continued proportionals will be 25, 15, 9 and their sum 49.[5]

---

tained Bachet's lavish folio *princeps* edition of Diophantus' Ἀριθμητική (*Diophanti Alexandrini Arithmeticorvm Libri Sex. Et de Nvmeris Mvltangvlis Liber Vnvs. Nunc primùm Græcè & Latinè editi, atque absolutissimis Commentariis illustrati* (Paris, 1621); compare T. L. Heath, *Diophantus of Alexandria. A Study in the History of Greek Algebra* (Cambridge, ₂1910): 26–8) but its present location is not known. (It would seem to have been sold at auction as part of the Musgrave library in the early 1920's; see R. de Villamil, *Newton: the Man* (London, 1931): 74.) This, rather than Samuel Fermat's re-edition (Toulouse, 1670—inferior in its text but with the 'observations' which his father, Pierre, crowded into the margins of his copy of Bachet's edition, and Jacques de Billy's valuable *Doctrinæ Analyticæ Inventum Novum. Collectum...ex varijs Epistolis quas ad eum diversis temporibus misit D. P. de Fermat...*), appears to have been the source text for Newton's present mild incursions into elementary Diophantine analysis. François Viète's five books of *Zetetica* (Tours, 1593), a brief but fruitful commentary on Books 1–5 of the *Arithmetica*, were conveniently available to him in Frans van Schooten's *Francisci Vietæ Opera Mathematica* (Leyden, 1646): 42–81, but whether Newton consulted them in composing his present 'Problemata' is doubtful. The evidence of his debt to the fifth book of Schooten's *Exercitationum Mathematicarum Libri Quinque* (Leyden, 1657: Liber v. 'Sectiones Miscellaneæ', especially *Sectiones* VI–XIII: 404–36) and to pertinent passages in John Wallis' *Commercium Epistolicum. De Quæstionibus quibusdam Mathematicis, nuper habitum* (Oxford, 1658) is a little more tangible: both these—unlike the *Zetetica*—he had annotated as an undergraduate. (See I: 50–6 and 116–19 respectively.) The techniques employed by Newton are, in their essential structure, Diophantus' but—like Viète, Bachet, Fermat, Descartes and Schooten before him—he makes good use of the new-found freedom and precision of the algebraic free variable in applying them. In our following footnotes reference to Diophantus' *Arithmetica* is made according to Bachet's text. We would warn the reader that the numbering of the problems is slightly different in Paul Tannery's definitive text in the Teubner edition of *Diophanti Alexandrini Opera Omnia cum Græcis Commentariis* (Leipzig, 1893).

(2) Two versions of a paper on 'Numerical problems' involving indeterminate analysis of second, third and fourth degrees. As we shall emphasize in following footnotes, these for the most part complement and generalize equivalent problems in Diophantus' *Arithmetica*.

(3) ULC. Add. 3964.1: 3ʳ–4ᵛ, a rough, heavily cancelled and rewritten draft of the 'Problemata'. Newton's text is here slightly transposed to accord with his revised numbering of the 'Problems'.

(4) On taking $z = x + e$, that is.

(5) Namely, the square of $z = (y^2 - ey + e^2)/(2e - y)$. On multiplying through by $2e - y$ there arises the more convenient parametrization $x = a^2 - e^2$, $y = a(2e - a)$, $z = a^2 - ae + e^2$. The problem—there given its present numerical solution—is set by Diophantus in his *Arithmetica* as a lemma (v, 7 [Bachet]) to a problem (v, 9 = v, 7 [Tannery]) which requires the construction of two right triangles equal in area. On parametrizing the sides of these by $z^2 + x^2$, $z^2 - x^2$, $2xz$ and $z^2 + y^2$, $z^2 - y^2$, $2yz$ (see Problem 3), the condition is that $x(z^2 - x^2) = y(z^2 - y^2)$ or, $x \neq y$, $x^2 + xy + y^2 = z^2$. (See Viète's *Zetetica* (note (1) above): Liber v, Zeteticum x/Zeteticum XI; compare also J. E. Hofmann's article 'Sur un problème de Roberval', *Revue d'Histoire des Sciences et de leurs Applications* **3**, 1950: 312–18.)

Prob 2. Invenire 3 ÷÷ quorum summa erit cubus. $xx + xy + yy = z^3$.[6] $x = ey$. $z = fy$.

[erit] $ee + e + 1 = f^3 y$. $\dfrac{ee + e + 1}{f^3} = y$. [Pone] $e = 2$. $f = 1$. [erit] $y = 7$. $x = 14$. 196.

98. 49 ÷÷ [quorum summa] $= 343 = 7^{\text{cub}}$.

Prob 3. Invenire 2 quadrata quorum summa vel diff. erit quadr.

$$xx + yy = zz = xx + 2ex + ee. \qquad \frac{yy - ee}{2e} = x. \qquad x + e = \frac{yy + ee}{2e} = z.\text{[7]}$$

Vel sic. $x = f - e$. $ff - 2fe + ee + yy = zz$.[8]

Prob 4. Invenire 2 quadratos quorū sum[m]a vel diff est cubus. $xx + yy = z^3$. $ey = x$.

$fy = z$. $ee + 1 = f^3 y$. $\dfrac{ee + 1}{f^3} = y$. [Pone] $e = 2$. $f = 1$. [erit] $y = 5$. $x = 10$. $xx + yy = 5$,

5, 5.[9]

Prob 5. Invenire 2 quadrata quorum summa vel diff quadrato quadratum est.[10]

Prob 6. Invenire 2 cubos quorum summa vel diff quadratum est. $x^3 + y^3 = zz$. $x = ey$.

$z = fy$. $e^3 y + y = ff$. $y = \dfrac{ff}{e^3 + 1}$. [Pone] $e = 2$. $f = 3$. [erit] $y = 1$. $x = 2$. $1 + 8 = 9$.[11]

Prob 7. Invenire 2 cubos quorum summa quadrato-quadratum est. $x^3 + y^3 = z^4$. $ey = x$.

$fy = z$. $e^3 + 1 = f^4 y$. $\dfrac{e^3 + 1}{f^4} = y$. [Pone] $e = 1$. $f = 1$. [erit] $y = 2$. $x = 2$.

$$x^3 + y^3 = 16 = 2^{\text{qq}}.$$

Et sic in infinitum.

Prob 8. Invenire quadratum quod additum lateri efficit quadratū.

$$x^2 + x = y^2 = xx - 2ex + ee.$$

$\dfrac{ee}{1 + 2e} = x$. [Ubi] $e = 1$. [erit] $x = \tfrac{1}{3}$. $\tfrac{1}{9} + \tfrac{1}{3} = \tfrac{4}{9}$.

---

(6) Newton first continued abortively with

$$`x = -\tfrac{1}{2}y + \sqrt{z^3 - \tfrac{3}{4}yy} = \frac{-y + \sqrt{4z^3 - 3yy}}{2} \cdot \sqrt{4z^3 - 3yy} = [2x + y]`.$$

(7) This solution to the Babylonian problem of constructing Pythagorean triples (compare O. Neugebauer, *The Exact Sciences in Antiquity* (Copenhagen, 1951): 35–41) essentially follows Diophantus' *Arithmetica* II, 8; for its many subsequent solutions see L. E. Dickson's *History of the Theory of Numbers.* II: *Diophantine Analysis* (Washington, 1920): 165–70. The general parametrization $(y^2 - e^2)^2 + (2ey)^2 = (y^2 + e^2)^2$ is immediate.

(8) Newton first continued 'Pone $\dfrac{yy}{4e} = f$' (Put $y^2/4e = f$). When this is done, there results $z = f + e$: the equivalent parametrization $(f - e)^2 + (2e^{\frac{1}{2}}f^{\frac{1}{2}})^2 = (f + e)^2$ is an immediate corollary but, since $e^{\frac{1}{2}}f^{\frac{1}{2}}$ will not in general be rational, it is not of great use.

(9) As Newton must have known, this problem becomes a good deal more subtle when y is restricted. Thus, when $y = 3$ no solution is possible in rationals (see Dickson, *History*, II (note (7) above): 534). Twenty years earlier Wallis had published Fermat's letter to Kenelm

Problem 2. To find three continued proportionals whose sum shall be a cube.

Take $x^2+xy+y^2=z^3$.[6] With $x=ey$, $z=fy$ there will be $e^2+e+1=f^3y$ and $y=(e^2+e+1)/f^3$. Put $e=2$, $f=1$ and hence $y=7$, $x=14$, when the continued proportionals will be 196, 98, 49 and their sum $343=7^3$.

Problem 3. To find two squares whose sum or difference shall be a square.

Take $x^2+y^2=z^2=(x+e)^2$, whereupon $x=(y^2-e^2)/2e$,

$$z=x+e=(y^2+e^2)/2e.^{[7]}$$

Or thus. Take $x=f-e$, whence $f^2-2fe+e^2+y^2=z^2$.[8]

Problem 4. To find two squares whose sum or difference is a cube.

Take $x^2+y^2=z^3$. With $x=ey$, $z=fy$ there will be $e^2+1=f^3y$ and

$$y=(e^2+1)/f^3.$$

Put $e=2$, $f=1$, whence $y=5$, $x=10$ and $x^2+y^2=5\times5\times5$.[9]

Problem 5. To find two squares whose sum or difference is a fourth power.[10]

Problem 6. To find two cubes whose sum or difference is a square.

Take $x^3+y^3=z^2$. With $x=ey$, $z=fy$ there follows $e^3y+y=f^2$ or

$$y=f^2/(e^3+1).$$

Put $e=2$, $f=3$ and there will be $y=1$, $x=2$ and $1+8=9$.[11]

Problem 7. To find two cubes whose sum is a fourth power.

Take $x^3+y^3=z^4$. With $x=ey$, $z=fy$ there will be $e^3+1=f^4y$ or

$$y=(e^3+1)/f^4.$$

Put $e=1$, $f=1$, whence $y=2$, $x=2$ and $x^3+y^3=16=2^4$. And so on indefinitely.

Problem 8. To find a square which added to its side makes up a square.

Take $x^2+x=y^2=(x-e)^2$, whereupon $x=e^2/(1+2e)$. When $e=1$ there will be $x=\frac{1}{3}$ and $\frac{1}{9}+\frac{1}{3}=\frac{4}{9}$.

Digby on 15 August 1657 (N.S.) in which he asserted (correctly) that only two integral solutions ($x^2=4$, $z=2$; $x^2=121$, $z=5$) were possible when $y=2$: 'je dis que si on cherche un quarré, qui adjousté à 4 face un cube, il n'en trouvera jamais que deux en nõbres entiers, sçavoir 4 & 121: ...apres cela toute l'infinité des nombres n'en sçauroit furnir un troisiesme, qui ait la mesme proprieté' (*Commercium Epistolicum* (note (1) above): 22–3). See also note (18).

(10) A late addition in the manuscript: for its solution see Problem 9 in the revise below.

(11) The least integer solution. The 'simpler' substitution $e=1$, $f=2$ yields the equality $2^3+2^3=4^2$, an answer evidently reserved for the next problem. For the subsequent history of this Diophantine equation see Dickson's *History*, II (note (7) above): 578–81. Observe that Newton in the sequel avoids the unsolvable Fermatian problem: 'Invenire 2 cubos quorum summa vel differentia cubus est'.

Prob 9. Invenire numerum cujus quadratum et cubus simul efficiunt cubū. $x^3 + xx = y^3$.

$ex = y$. $x + 1 = e^3 x. \dfrac{1}{e^3 - 1} = x$. [Pone] $e = 2$. [fit] $\frac{1}{7} = x$. $\frac{1}{343} + \frac{7}{343}^{(12)} = \frac{8}{343} = \text{cub } \frac{2}{7}$.

Prob 10. Invenire cubū qui additus radici fit cubus.

$x^3 + x = y^3 = x^3 + 3ex^2 + 3eex + e^3$. $de = x$. $d = 3ddee + 3dee + ee$.

$$ee = \frac{d}{3dd + 3d + 1}. \quad d = ff. \quad 3f^4 + 3ff + 1 = gg^{(13)} = 1 + 2h + hh.$$

$kff = h$. $3ff + 3 = 2k + kkff$.$^{(14)}$

[*Prob*].$^{(15)}$ *Invenire cubum qui additus lateri cubum efficit.* $x^3 + x = y^3$. $ex = y$.

$$xx + 1 = e^3 xx. \quad xx = \frac{1}{e^3 - 1}.$$

$\left[\text{Pone } z = \dfrac{1}{x}.\right]$ $e^3 - 1 = zz$. Quære duos quadratos æquales cubo per Prob [sequens].$^{(16)}$ vizt $f^3 = gg + hh$.

Prob [10]. Invenire quadratum quod dato quadrato additum efficit cubum. $aa + xx = y^3$.

Sit $a = d^3 + e$. $x = f + g^3$. [erit] $d^6 \begin{matrix} + 2d^3 e + ee \\ + ff \end{matrix} + 2fg^3 + g^6 = y^3$. [Pone] $3d^4 gg = 2d^3 e + ff$.

$3ddg^4 = ee + 2fg^3$.$^{(17)}$ [erit] $e = ggd$. $f = gdd$. [adeoꝗ] $a = d^3 + ggd$.$^{(18)}$

Prob 11. Duos quadratos numeros invenire qui per datos numeros multiplicati et additi datum numerum componunt.$^{(19)}$ $axx + byy = c$. Ad Problematis hujus solutionem

---

(12) That is, $\frac{1}{49}$.

(13) On putting $g = f/e$.

(14) Newton leaves his reader the (impossible) problem of choosing k so that

$$(3 - 2k)/(k^2 - 3)$$

is a perfect square. See note (16).

(15) This and the following (uncancelled) problem are placed after Problem 14 in the manuscript, but their present resiting seems more logical.

(16) The text reads (following Newton's first numbering) 'Prob 5', that is 'Prob 4' in the revised numbering which we here follow, but we retain the more logical reading of his cancelled first phrase 'Quære e et z per sequ: prob.' Neither Problem 4 nor Problem 10 afford any help in finding a rational solution of $a^2 + x^2 = y^3$ when, as here (where $a = 1$), the integer can be put into neither of the forms $(e^2 + 1)/f^3$ or $d(d^2 + g^2)$. In fact, as Euler stated sixty years later ('Theorematum quorundam Arithmeticorum Demonstrationes', *Commentarii Academiæ Scientiarum Petropolitanæ* **10** (1738) [1747]: 125–46, especially 146: Theorema 10, Corollarium 1 [= *Leonhardi Euleri Opera Omnia* (1), *Commentationes Arithmeticæ*, **1** (Leipzig/Berlin, 1915): 58]), the equation $e^3 - 1 = z^2$ has no rational solution of any kind apart from the trivial instance $e = 1$, $z = 0$. (For set $e = 1 + b/a$, where the positive integers a, b are coprime: then immediately $ab(3a^2 + 3ab + b^2)$ must be square. If $b = 3c$, then $ca(3c^2 + 3ca + a^2)$ must be square, with $ca < ab$. If $b \neq 3c$, then a, b and $3a^2 + 3ab + b^2$ are coprime, and so each is square; so put $3a^2 + 3ab + b^2 = (am/n - b)^2$ where m, n are positive integers, that is,

$$a/b = (2m + 3n)n/(m^2 - 3n^2),$$

Problem 9. To find a number whose square and cube together make up a cube.

Take $x^3+x^2=y^3$. With $y=ex$ there will be $x+1=e^3x$ or $x=1/(e^3-1)$. Put $e=2$ and there comes $x=\frac{1}{7}$, yielding $\frac{1}{343}+\frac{7}{343}$[12] $=\frac{8}{343}=\left(\frac{2}{7}\right)^3$.

Problem 10. To find a cube which, when added to its root, becomes a cube.

Take $x^3+x=y^3=(x+e)^3$. With $x=de$ there results $d=3d^2e^2+3de^2+e^2$ or $e^2=d/(3d^2+3d+1)$. Put $d=f^2$ and $3f^4+3f^2+1=g^2$[13] $=(1+h)^2$ or, with $h=kf^2$, $3f^2+3=2k+k^2f^2$.[14]

Problem.[15] *To find a cube which, when added to its side, makes up a cube.*

Take $x^3+x=y^3$. With $y=ex$ there will be $x^2+1=e^3x^2$ or $x^2=1/(e^3-1)$. Put $z=1/x$ and $e^3-1=z^2$. Seek two squares which are equal to a cube by the following[16] problem, viz: $f^3=g^2+h^2$.

Problem 10. To find a square which, when added to a given square, makes up a cube.

Suppose $a^2+x^2=y^3$. Let $a=d^3+e$, $x=f+g^3$ and then

$$d^6+(2d^3e+f^2)+(e^2+2fg^3)+g^6=y^3.$$

Put $2d^3e+f^2=3d^4g^2$, $e^2+2fg^3=3d^2g^4$[17] and there will be $e=g^2d$, $f=gd^2$ and consequently $a=d^3+g^2d$.[18]

Problem 11. To find two square numbers which, when multiplied by given numbers and added, together make a given number.[19]

---

and consequently, since $m \neq 3\mu$ (for $3\mu^2-n^2$ cannot be a square), $a=(2m+3n)n$ and $b=m^2-3n^2$ are each square: on setting $b=(m-np/q)^2$, there follows $m/n=(3p^2+q^2)/2pq$ and so $a(pq/n)^2=pq(3p^2+3pq+q^2)$ is square, with $pq<ab$. Whence the existence of any pair of integers a, b yielding a solution $e=1+b/a$ impossibly implies the existence of an infinity of lower positive pairs.) Having come, no doubt, to suspect this impossibility of solution, in his revise below Newton later quietly discarded the present problem. We may well be disappointed that he did not attempt to test the solvability of his equation, other than empirically at least.

(17) Namely, so as to make the left-hand side of the equation the perfect cube $(d^2+g^2)^3$.

(18) This parametrization is not generally valid, failing in particular (see notes (16) and (9)) when $a^2=1$ and $a^2=9$. The problem of solving in rationals the general Diophantine equation $y^3=x^2+k$, k a general integer, is still unsolved except in certain cases and for most low values of k; see Dickson's *History*, II (note (7) above): 533–9, and compare L. J. Mordell, *A Chapter in the Theory of Numbers* (Cambridge, 1947).

(19) The first discussion of this generalized Pythagorean equation known to Dickson (*History*, II (note (7) above): 407) is cited from Euler's posthumous papers (*Opera Postuma, Mathematica et Physica*, 1 (St Petersburg, 1862): 490, printing a manuscript of about 1769). Newton's present approach is manifestly an extension of Viète's generalization of Diophantus' *Arithmetica*, II, 10, which poses the problem of resolving $2^2+3^2=13$ into two other rational squares: on taking these to be $(x+2)^2+(mx-3)^2$ and setting $m=2$, there results $x=\frac{8}{5}$. Pursuing this approach, Viète found in Zeteticum II of Book 4 of his *Zetetica* (Tours, 1593) [=*Opera Mathematica* (Leyden, 1646): 62–3] the parametrization

$$B^2+D^2=(A+B)^2+([S/R]A-D)^2, \quad \text{where} \quad A=2R(-RB+SD)/(R^2+S^2),$$

so that $(B^2+D^2)(\alpha^2+\beta^2)=(\alpha\beta\pm\beta D)^2+(\alpha D\mp\beta B)^2$ on setting $S^2-R^2=\alpha$ and $2RS=\pm\beta$. In geometrical equivalent Newton's generalization, given that the point (d,e) is on the ellipse $ax^2+by^2=c$, constructs the straight line $y=e-(q/p)(x-d)$ of slope $-q/p=-f$ through it

requiritur casus aliquis quo possibilitas solutionis constet.[20] Sit iste $add + bee = c$. Pone $x = d + p$. $y = e - q$. ergo $add + 2adp + app + bee - 2beq + bqq = c$. Dele æquales et restabit $2adp + app - 2beq + bqq = 0$. Pone $q = fp$ et erit

$$2ad + ap - 2bef + bffp = 0. \quad \text{Vel} \quad \frac{2bef - 2ad}{bff + a} = p.$$

Prob 12. Invenire omnes integros quadratos numeros vel omnes cubos vel omnes quadrato-quadratos &c quorum[21] *diff*[erentia] *æquatur dato numero.* Si $xx - yy = n$ ergo n divisibilis est per $x - y$ et $x - y$ est aliquis divisor ipsius n. Sit iste d et erit $\frac{n}{d} = x + y$

& $\dfrac{\frac{n}{d} \pm n}{2} = \dfrac{x}{y}$. Si $x^3 - y^3 = n$ erit $xx + xy + yy = \dfrac{n}{d}$. Aufer dd et restabit $3xy = \dfrac{n}{d} - [dd]$.

Unde $\dfrac{\frac{n}{d} - dd}{3}$ divisibilis est per x restante y. Nota ergo omnes rectangulos divisores

ipsius $\dfrac{\frac{n}{d} - dd}{3}$ quorum differentia est d. Vel si mavis adde $\dfrac{\frac{n}{d} - dd}{3}$ ad $\dfrac{n}{d}$[ɔ] summa

$\dfrac{\frac{4n}{d} - dd}{3}$ erit quadratum $xx + 2xy + yy$. Sit radix ejus r, $\dfrac{r \pm d}{2} = \dfrac{x}{y}$. Rursus si

$x^5 - y^5 = n$ erit $x^4 + x^3y + xxyy + xy^3 + y^4 = \dfrac{n}{d}$. Aufer d^4 et restabit

$$5x^3y - 5xxyy + 5xy^3 = \frac{n}{d} - [d]^4.$$

Ergo $\dfrac{\frac{n}{d} - d^4}{5}$ divisibilis est per xy. Quare nota[22] ejus divisores quorum se[c]undi divisores rectanguli differunt numero d.[23] Eodem modo cubi omnes vel quadrato-cubi &c inveniri possunt quorum summa æquatur dato numero.[24]

Prob 13. Invenire 3[25] *cubos quorum summa quadrato-cubus est.* $v^3 + x^3 + y^3 = z^5$. $dy = v$. $ey = x$. $fy = z$. $d^3 + e^3 + 1 = f^5yy$. [puta] $d^3 + e^3 + 1 = tt$.[26] [Cape] $d = 3 - e$ vel $= 4 \times 3 - e$ vel $= 9 \times 3 - e$ &c vel $= 3gg - e$. puta $= 3 - e$. [erit]

---

and thence determines its second (of necessity rational) meet $(d + p, e - q)$ with the ellipse, so finding $p = 2(bef - ad)/(a + bf^2)$ where $f = q/p$. In consequence Newton's implied parametrization is

$$ad^2 + be^2 = a\left(\frac{d(-ap^2 + bq^2) \pm e(2bpq)}{ap^2 + bq^2}\right)^2 + b\left(\frac{e(-ap^2 + bq^2) \mp d(2bpq)}{ap^2 + bq^2}\right)^2.$$

Compare Dickson's *History*, II: 225–6.

(20) Strictly, not so. 'If m, n are any integers, the equation $mx^2 + ny^2 = 1$ is solvable for rational numbers x, y, if the congruence $mx^2 + ny^2 \equiv 1 \pmod{p^e}$ is solvable in integers x, y for every prime p and positive integer e' (Dickson, *History*, II (note (7) above): 408). Given one

Take $ax^2 + by^2 = c$. In order to solve this problem some instance of it is needed to establish the possibility of its solution.[20] Let this be $ad^2 + be^2 = c$. Put $x = d+p$, $y = e-q$, whence $ad^2 + 2adp + ap^2 + be^2 - 2beq + bq^2 = c$. Delete equals and there will remain $2adp + ap^2 - 2beq + bq^2 = 0$. Put $q = fp$ and then $2ad + ap - 2bef + bf^2 p = 0$ or $p = (2bef - 2ad)/(bf^2 + a)$.

Problem 12. To find all integral squares or cubes or fourth powers, and so on, whose[21] *difference equals a given number.*

If $x^2 - y^2 = n$, therefore n is divisible by $x-y$ and $x-y$ is some divisor of n. Let that be d and then $x+y = n/d$ and $x, y = \frac{1}{2}(n/d \pm n)$. If $x^3 - y^3 = n$, then

$$x^2 + xy + y^2 = n/d.$$

Take away d^2 and there will remain $3xy = n/d - d^2$. Hence $\frac{1}{3}(n/d - d^2)$ is divisible by x, leaving y. Note therefore all factor-pairs of $\frac{1}{3}(n/d - d^2)$, whose difference is d. Or if you prefer it, add $\frac{1}{3}(n/d - d^2)$ to n/d, and the sum $\frac{1}{3}(4n/d - d^2)$ will be the square $(x+y)^2$. Let its root be r and $x, y = \frac{1}{2}(r \pm d)$. Again, if $x^5 - y^5 = n$, then $x^4 + x^3 y + x^2 y^2 + xy^3 + y^4 = n/d$. Take away d^4 and there will remain $5x^3 y - 5x^2 y^2 + 5xy^3 = n/d - d^4$. Therefore $\frac{1}{5}(n/d - d^4)$ is divisible by xy. Consequently, note[22] those of its divisors which have secondary factor-pairs differing by the number d.[23] In the same manner all cubes or fifth powers, and so on, may be found whose sum is equal to a given number.[24]

Problem 13. To find three[25] *cubes whose sum is a fifth power.*

Suppose $v^3 + x^3 + y^3 = z^5$, that is, with $v = dy$, $x = ey$, $z = fy$,

$$d^3 + e^3 + 1 = f^5 y^2;$$

say, $d^3 + e^3 + 1 = t^2$.[26] Take $d = 3g^2 - e$, $g = 1, 2, 3, \ldots$; say $d = 3 - e$. Then

---

solution (d, e) of $ax^2 + by^2 = c$, Newton proceeds to construct the second solution $(d+p, e-fp)$: in geometrical equivalent (compare Case 1 of §2 following) the second solution is found by determining the second meet of $y - e + f(x-d) = 0$ with the conic $ax^2 + by^2 = c$.

(21) 'summa vel' (sum or) is cancelled. Correspondingly in the next sentence Newton has deleted 'Si $xx + yy = n$'.

(22) Newton first wrote 'tenta omnes' (test all).

(23) The continuation 'Vel sic de $\frac{n}{d} - d^4$ aufer $5d$' (Or thus: from $n/d - d^4$ take away $5d$) is cancelled.

(24) It seems curious that Newton should think it worthwhile to insert this refined technique for gaining solutions by inspection as a problem in its own right.

(25) '2' is cancelled. Correspondingly Newton began his argument '$x^3 + y^3 = z^5$. $ey = x$. $fy = z$. $e^3 + 1 = f^5 yy$. $e^3 + 1 = zz$.[!] $z = 1+v$' and then broke off. Evidently in this case $(x/z)^3 + (y/z)^3 = z^2$, an equation already treated in Problem 7. Likewise, on putting

$$(v/z)^3 + (x/z)^3 + (y/z)^3 = z^2$$

the present problem might more simply have been stated: 'Invenire 3 cubos quorum summa quadratum est.'

(26) That is, on setting $f^5 = t^2/y^2$. In sequel Newton sets $f = 1$, so that $t = y = z$, and in effect solves the restricted problem $v^3 + x^3 + y^3 = y^5$ only.

$$28-27e+9ee=tt=9ee+6eh+hh.$$

[adeoœ] $\dfrac{28-hh}{27+6h}=e.$ [27] [Posito] $h=-2.$ [fit] $e=\frac{8}{5}.$ $t=\frac{14}{5}.$ $d=\frac{7}{5}.$ $\dfrac{t}{\sqrt{f^5}}=y.$ [hoc est

capiendo] $f=1.$ $z=y=\frac{14}{5}.$ $\frac{98}{25}=v.$ $\frac{112}{25}=x.$ $v^3+x^3+y^3=$

$$\frac{941192+1404928+343000}{25\times25\times25}=\frac{2689120}{25,25,25}=\frac{537824}{5,25,25}.$$

Cujus radix Quadr-cubica est $\frac{14}{5}.$

[Vel] solve sic. $d^3+e^3+1=f^5yy=tt.$ [Cape] $d=k-e.$

$$k^3-3kke+3kee[+1]=[tt].$$

[Ergo si] $3k=gg.$ $[t]=ge+h.$ [erit] $k^3-3kke[+1]=2geh+hh.$ [28]

Prob 10.[29] *Invenire 3 numeros quorum summa est quadratum et summa quadratorū cub⁵.*
$x+y+z=e^2.$ $xx+yy+zz=f^3.$ [Pone] $gx=y,$ $hx=z.$ $kx=f.$ [fit] $1+gg+hh=k^3x.$
$\dfrac{1+gg+hh}{k^3}=x.$ [ac] $e^2=x+gx+hx.$ [adeoœ] $\dfrac{e^2}{1+g+h}=x=\dfrac{1+gg+hh}{k^3}.$ [30] [Sit]

$ek=1+g+n.$ [erit]

$$e^3k^3=1+3h+3hh+h^3+3n+3nn+n^3+3hhn+3hnn+6hn$$
$$=1+h+hh+h^3+g+gg+g^3+ggh+ghh.\ ^{(31)}$$

$2h+2hh+3n+3nn+n^3+3hnn+3hhn+6hn=g+gg+g^3+ggh+ghh.$ [seu]

$$h=\frac{2+3nn+6n-gg}{2g-4-6n}+\sqrt{\qquad}^{(32)}.$$

[Cape $1+g+h=cc.$ fit] $e^{[2]}k^3c=2+2g+2gg-2cc-2gcc+c^4.$ [33]

---

(27) Newton first continued '$h=-4.$ [erit] $e=4.$ $t=3e+h=8$', rightly cancelled since it yields the trivial solution $d=-1,$ and so $v=-y$ or $x^3=y^5=2^{15}.$

(28) And so $e=(k^3-h^2+1)/(3k^2+2gh),$ which reduces to Newton's previous solution—effectively (see note (26)) of $(v/y)^3+(x/y)^3+1=y^2$—when $k=3.$ As stated by him, the present problem is too general to have an interesting rational solution and further restriction (say, by adding the condition that $v,$ x and y be in arithmetical progression; compare Dickson's *History*, II (note (7) above): 585 ff.) is needed to make it worth pursuing.

(29) This cancelled first version of the following problem is set in the manuscript immediately before Problem 11 above. If we interpret Newton's heavily cancelled text correctly it originally required the reader 'Invenire 3 numeros quorum summa est cubus et summa quadratorū est cub⁵', beginning its argument

'$x+y+z=e^3.$ $xx+yy+zz=f^3.$ $e^3-x-y=z.$ $2xx+2yy+2xy-2xe^3-2ye^3+e^6=f^3.$

Quoniam maximæ potestates $2xx$ & $2yy$ destruendæ sunt & tamen non sunt quadratæ (Since the greatest powers $2x^2$ and $2y^2$ have to be destroyed, and yet are not square). $gx=y.$ $hx=z.$

$$kx=f.\quad \frac{1+gg+hh}{k^3}=x.\quad e^3=x+gx+hx.\quad \frac{e^3}{1+g+h}=\frac{1+gg+hh}{k^3}.$$

$e^3k^3=1+g+h+gg+hh+g^3+ggh+h^3.$' He then tried a new tack, putting '$1+g+h=l.$ $ek=lm$' and hence deriving—but without at first comprehending that it holds only for $m=1$ (and so $(1+g+h)^2=1+g^2+h^2$)—the further argument

'$llm^3=1+gg+hh=gg+2gn+nn.$ $\dfrac{hh-nn+1}{2n}=g=l-h-1.$

$28-27e+9e^2 = t^2 = (3e+h)^2$ and consequently $e = (28-h^2)/(27+6h)$.[27] On setting $h = -2$ there comes $e = \frac{8}{5}$, $t = \frac{14}{5}$, $d = \frac{7}{5}$ and $t/f^{\frac{5}{2}} = y$; that is, on taking $f = 1$, there is $z = y = \frac{14}{5}$, $v = \frac{98}{25}$, $x = \frac{112}{25}$ and $v^3 + x^3 + y^3 =$

$$(941192 + 1404928 + 34300)/25^3 = 2689120/25^3 = 537824/5 \times 25^2,$$

whose fifth root is $\frac{14}{5}$.

Or solve it this way. [...] $d^3 + e^3 + 1 = f^5 y^2 = t^2$. Take $d = k - e$, and so $k^3 - 3k^2 e + 3ke^2 + 1 = t^2$. Hence if $3k = g^2$ and $t = ge + h$, then

$$k^3 - 3k^2 e + 1 = 2geh + h^2.\text{[28]}$$

Problem 10.[29] *To find three numbers whose sum is a square and the sum of whose squares a cube.*

Say $x + y + z = e^2$, $x^2 + y^2 + z^2 = f^3$. Put $y = gx$, $z = hx$, $f = kx$ and there comes $1 + g^2 + h^2 = k^3 x$ (or $x = (1 + g^2 + h^2)/k^3$) and $x + gx + hx = e^2$, so that

$$e^2/(1+g+h) = x = (1+g^2+h^2)/k^3.\text{[30]}$$

Let $ek = 1 + g + n$ and there will be

$$e^3 k^3 = 1 + 3h + 3h^2 + h^3 + 3n + 3n^2 + n^3 + 3h^2 n + 3hn^2 + 6hn$$

$$= 1 + h + h^2 + h^3 + g + g^2 + g^3 + g^2 h + gh^2,\text{[31]}$$

so that $2h + 2h^2 + 3n + 3n^2 + n^3 + 3hn^2 + 3h^2 n + 6hn = g + g^2 + g^3 + g^2 h + gh^2$, or

$$h = (2 + 3n^2 + 6n - g^2)/(2g - 4 - 6n) + \sqrt{\text{- - -}}.\text{[32]}$$

Take $1 + g + h = c^2$ and there comes $e^2 k^3 c = 2 + 2g + 2g^2 - 2c^2 - 2gc^2 + c^4.$[33]

$lm^3 = g + n$. $2n + hh - nn + 1 + 2nh$ in $m^3 = hh + nn + 1$. $nn = \dfrac{2nhm^3 + 2nm^3 + hhm^3 + m^3 - 1 - h^2}{1 + m^3}$.

$$n = \frac{hm^3 + m^3}{1+m^3} + \sqrt{\frac{2hhm^6 + 2hm^6 + 2m^6 - hh - 1}{[1 + 2m^3 + m^3]}}.$$

Then he saw his error and put '$m = 1$. [adeoq] $h + 1 = n$. $g = \dfrac{-h}{h+1}$. $l = \dfrac{hh+h+1}{h+1}$', from which on taking '$h = -2$. $g = -2$. $l = -3$. $e = +1$' he drew the particular solution

$$\text{'}x = -\tfrac{1}{3}.\ y = +\tfrac{2}{3}.\ z = \tfrac{2}{3}\text{'}.$$

(30) Newton first continued '[Sit] $ek = \dfrac{1+g+h}{3}$', but quickly saw that this led to troublesome powers of 3 in the resulting denominators.

(31) That is, $(1+g+h)(1+g^2+h^2) = e^2 k^3$, whence $e = 1$. We omit a following line in which Newton substituted '$pn = g$. $qn = h$' in this value.

(32) Realizing that extracting the root of the preceding quadratic in h is pointless Newton omits to compute the quantity under the radical sign.

(33) That is, $1 + g^2 + (-[1+g] + c^2)^2 = e^2 k^3/c^2$. The right-hand side of this 'equation' is written slightly above the line and we may suppose that Newton intended to divide it by a term $(c^3 = (1+g+h)^{\frac{3}{2}})$ which balances the equality before he abandoned his computation.

Prob 14. Invenire tres numeros quorum summa quadratum est et quadratorum summa cubus sit. $xx+yy+zz=f^3$. [Pone] $gx=y$. $hx=z$. $kx=f$. [erit] $1+gg+hh=k^3x$.

[seu] $\dfrac{1+gg+hh}{k^3}=x$. Sit insuper $x+y+z=ee$ sive $x+gx+hx=ee$ & erit

$\dfrac{ee}{1+g+h}=x=\dfrac{1+gg+hh}{k^3}$. Pone $1+g+h$ in $l=e$ & erit $llk^3\times\overline{1+g+h}=1+gg+hh$.

Pone $llk^3=p$. et extracta radice $g=\frac{1}{2}p\pm\sqrt{\frac{1}{4}pp+ph+p-hh-1}$. Pone radicem

hanc $=\frac{1}{2}p+n$ & erit $ph+p-hh-1=pn+nn$, & $p=\dfrac{nn+hh+1}{h-n+1}$. Unde habetur

radix et inde $g=p+n$ vel $=-n$.[34] Assume $lk=m$ & erit $\dfrac{p}{mm}=k$. $\dfrac{m}{k}=l$. $1+g+h$

in $l=e$. $el=x$. $gx=y$, $hx=z$.[35] Ut si sumas $h=1$. $n=-1$, erit $p=1=g$. Et assumendo insuper $m=1$ erit $k=1=l$, & $e=x=3=y=z$. Jam $3+3+3=9$ & $9+9+9=27$. Ru[r]sus si assumas $h=2$ & $n=1=m$: erit $p=3$. $g=4$. $k=3$. $l=\frac{1}{3}$. $\frac{7}{3}=e$. $x=\frac{7}{9}$. $y=\frac{28}{9}$, $z=\frac{14}{9}$ quorum summa est quad $\frac{49}{9}$ & summa quadratorum cubus $\frac{343}{27}$. Vel quoniam est $\dfrac{1+g+h\text{ in }m^6}{pp}=x=\dfrac{7m^6}{9}$, assume $m=3$ et numeros integros habebis viz $567=x$. $5268=y$.[36] $1134=z$.

[*Prob*] 15.[37] *Invenire tres numeros quorum singuli additi cubo summæ componunt cubos.* $a+b+c=x$. $x^3+a=r^3$. $x^3+b=s^3$. $x^3+c=t^3$. [adeoq] $3x^3+x=r^3+s^3+t^3$. [Cape]

$r=ex$. $s=fx$, $t=gx$. $3x^2+1=e^3x^2+f^3xx+g^3xx$. [sive] $xx=\dfrac{1}{e^3+f^3+g^3-3}$. [Cape]

$e^3+f^3+g^3-3=yy$.[38] $e=k+m$. $f=l-m$. [erit]

$$k^3+3kkm+3kmm+l^3-3llm+3lmm+g^3-3=yy.$$

---

(34) That is, $g = \frac{1}{2}p \pm (\frac{1}{2}p+n)$. In his first example below Newton takes $g = -n$, in the second he sets $g = p+n$.

(35) Newton's solution to his problem (usually attributed to Euler) is not always a convenient one. What is probably the simplest solution in integers, namely $x = 5$, $y = 9$ and $z = 35$ ($e = 7$, $f = 11$), here follows somewhat arbitrarily by putting $h = 7$, $g = -n = \frac{9}{5}$ and $m = \frac{11}{5}$. (Compare L. Euler, *Opera Postuma* (note (19) above), 1: 256; also Dickson's *History*, II (note (7) above): 599.)

(36) Read '$2268=y$'.

(37) This, together with the considerably more difficult complementary problem attempted abortively by Newton in his next paragraph (see notes (42), (43) and (44)) is Diophantus' celebrated problem of three cubes (*Arithmetica*, v, 18/19 in Bachet's edition of 1621 = v, 15/16 [Tannery]). Diophantus himself merely derived particular solutions to the two cases of his problem without invoking any obviously generalizable method. (In the terms of Newton's current solution it will be obvious that the core of the present problem is to choose e, f and g such that $\pm(e^3+f^3+g^3-3) = y^2$. In his v, 18 [Bachet] Diophantus set $e = m+1$, $f = 2-m$, $g = 2$ and computed $m = \frac{2}{15}$ by equating $e^3+f^3+g^3-3$ to $(3m-4)^2$; but in v, 19 he arbitrarily chose $y=\frac{3}{2}$ and then stated that $3-[(\frac{5}{6})^3+(\frac{4}{6})^3-(\frac{2}{6})^3] = (\frac{3}{2})^2$, adding that he had elsewhere (in his lost treatise on *Porisms*?) shown how to express the difference of two cubes as the sum of two other cubes, so that, by implication, a positive solution could be derived from this.) Ludolph van Ceulen in Problem 68 of his *Van den Circkel* (Delft, 1596) drew attention to

Problem 14. To find three numbers whose sum is a square and the sum of whose squares shall be a cube.

Suppose $x^2 + y^2 + z^2 = f^3$. Put $y = gx$, $z = hx$, $f = kx$ and there will be

$$1 + g^2 + h^2 = k^3 x \quad \text{or} \quad x = (1 + g^2 + h^2)/k^3.$$

Take in addition $x + y + z = e^2$, that is, $x + gx + hx = e^2$, and there will be $e^2/(1 + g + h) = x = (1 + g^2 + h^2)/k^3$. Put $e = (1 + g + h) l$ and then

$$l^2 k^3 (1 + g + h) = 1 + g^2 + h^2.$$

Set $l^2 k^3 = p$ and, on extracting the root, $g = \frac{1}{2}p \pm \sqrt{[\frac{1}{4}p^2 + ph + p - h^2 - 1]}$. Put this root equal to $\frac{1}{2}p + n$ and then $ph + p - h^2 - 1 = pn + n^2$, and

$$p = (n^2 + h^2 + 1)/(h - n + 1).$$

From this the root is ascertained and thence $g = p + n$ or $-n$.[34] Assume $lk = m$ and there will be $p/m^2 = k$, $m/k = l$, $(1 + g + h) l = e$, $el = x$, $gx = y$, $hx = z$.[35] Should you, for instance, take $h = 1$, $n = -1$, then $p = g = 1$ and, on assuming in addition $m = 1$, there will then be $k = l = 1$ and $e = x = y = z = 3$. At once $3 + 3 + 3 = 9$ and $9 + 9 + 9 = 27$. Again, if you assume $h = 2$ and $n = m = 1$, then $p = 3$, $g = 4$, $k = 3$, $l = \frac{1}{3}$, $e = \frac{7}{3}$, and so, $x = \frac{7}{9}$, $y = \frac{28}{9}$, $z = \frac{14}{9}$, whose sum is the square $\frac{49}{9}$ and the sum of whose squares is the cube $\frac{343}{27}$. Alternatively, since $x = p^{-2}(1 + g + h) m^6 = \frac{7}{9}m^6$, assume $m = 3$ and you will have integers, namely, $x = 567$, $y = [2]268$, $z = 1134$.

Problem 15.[37] *To find three numbers which, when added singly to the cube of their sum, form cubes.*

Suppose $a + b + c = x$ with $x^3 + a = r^3$, $x^3 + b = s^3$, $x^3 + c = t^3$ and consequently $3x^3 + x = r^3 + s^3 + t^3$. Take $r = ex$, $s = fx$, $t = gx$, so that

$$3x^2 + 1 = e^3 x^2 + f^3 x^2 + g^3 x^2 \quad \text{or} \quad x^2 = 1/(e^3 + f^3 + g^3 - 3).$$

Take $e^3 + f^3 + g^3 - 3 = y^2$,[38] $e = k + m$, $f = l - m$ and then there will be

$$k^3 + 3k^2 m + 3km^2 + l^3 - 3l^2 m + 3lm^2 + g^3 - 3 = y^2.$$

---

the inadequacies of Diophantus' method of solution in the latter case (v, 19), and following his lead Bachet, Girard, Fermat and Nicolaus Huberti worked hard to elaborate more general approaches to the problem during the next half century. (See Dickson's *History*, II (note (7) above): 607–8.) Huberti's rehash of van Ceulen's solution, published by Frans van Schooten in 1657 in his *Exercitationes Mathematicæ* (note (1) above): Liber v, Sectio XIII. 'Quæstio XIX, libri v Arithmeticorum Diophanti. Invenire tres numeros, ut cubus summæ eorum, quovis ipsorum detracto, faciat cubum', was perhaps the means of attracting Newton's attention to what Schooten described as a question '[c]ujus solutio à Diophanto perplexè tradita [est]' (*ibid.*: 434). However, his following discussion of the problem is, in its detail, wholly original.

(38) On taking $y = 1/x$, that is. Newton first continued

$$'f = m + n. \ g = m - n. \ e^3 + 2m^3 + 6mnn - 3 = yy.$$

[Posito] $3k+3l=pp$. [sit] $y=pm+n$. [adeoq̃] $yy=ppmm+2pmn+nn$. Pone valores yy æquales et deletis æqualib^s erit

$$k^3+3kkm+l^3-3llm+g^3-3=2pmn+nn.$$

sive $\dfrac{k^3+l^3+g^3-3-nn}{2pn-3kk+3ll}=m$. Assume ergo quemlibet numerum pro p, puta 3 vel -3,$^{(39)}$ ut et alium numerum pro k, puta 1 vel 0 vel $\frac{1}{2}$ vel $\frac{2}{3}$ vel $\frac{3}{2}$ &c et fac $\dfrac{pp}{3}-k=l$. Dein pro g assume numerum aliquem majorem unita[t]e et alium quemvis pro n ita ut invento m sit uterq̃ $k+m$ & $l-m$ major 1. Dein pone $\dfrac{1}{pm+n}$ $^{(40)}=x$. $k+m$ in $x=r$. & $l-m$ [in x]$=s$. $gx=t$. $r^3-x^3=a$. $s^3-x^3=b$. $t^3-x^3=c$. Q.E.F. Ut si assumas $p=3$. $k=1$. $g=2$ & $n=1$ erit $l=2$. $m=\dfrac{13}{15}$.

$x=\dfrac{5}{18}$. $r=\dfrac{5}{9}$. $s=\dfrac{14}{27}$. $t=\dfrac{17}{54}$. [!] $a=\dfrac{875}{18,\,18,\,18}$ &c.$^{(41)}$

Si tres numeri desiderantur qui de cubo summæ detracti relinquunt cubos, quoniam in hoc casu$^{(42)}$ r^3, s^3, ac t^3 minores sunt quam x^3 adeoq̃ e, f ac g minores unitate, assume g & k minores unitate et p ac n tales ut sint $k+m$ & $l-m$ minores etiam unitate sed non negativi et prodibunt numeri qui quæstioni satisfacient.$^{(43)}$ Ut si assumas $p=\frac{3}{2}$. & $k=0$ erit $l=\frac{3}{4}$. Vel si $p=\frac{3}{2}$. & $k=-\frac{1}{4}$ erit $l=1$. Dein tenta $g=\frac{1}{4}$ vel $\frac{3}{4}$ vel $\frac{1}{2}$ [vel] $\frac{5}{4}$ &c et numerum quemvis pro n. Ut si assumas $k=0$. $p=\frac{3}{2}$. $l=\frac{3}{4}$. $g=\frac{5}{4}$ & $n=\frac{1}{2}$ erit $m=-\frac{7}{66}$. $\frac{44}{15}=x$. $\frac{14}{45}=r$ & $\frac{85}{45}=s$ & $\frac{17}{9}=t$.$^{(44)}$

[Posito] $6m=pp$. [erit] $e^3+\dfrac{p^6}{108}-3+ppnn=yy=ppnn+2pnq+qq$ [ubi $y=pn+q$]. Ergo

$$\dfrac{e^3+\dfrac{p^6}{108}-3-qq}{2pq}=n.$$

[Fac] $p=2$. [fit] $m=\frac{2}{3}$. $n=\dfrac{e^3-\frac{19}{9}[!]-qq}{4q}$. [hoc est ubi] $e=3$. $q=5$. [erit] $n=-\frac{1}{45}[!]$. $f=\frac{29}{45}$. $g=\frac{31}{45}$. $pn+q=y=\frac{223}{45}$. $\frac{45}{223}=x$'. An abortive first revise breaks off after '$p=3$. $m=\frac{3}{2}$'.

(39) Choice of $p=3\pi$ allows us to set the denominator of m's value as $3(2n\pi-k^2+l^2)$ with $k+l=3\pi^2$, but there is no real benefit in so doing since the denominator, on substitution of $3(k+l)=p^2$, becomes $p(2n-p[k-l])$.

(40) That is, $1/y$; compare note (38). Newton has cancelled an unnecessary introductory restriction '*sit pm+n affirmativum*' (let $pm+n$ be positive) at the end of the previous sentence.

(41) Understand '$b=\dfrac{18577}{54,\,54,\,54}$. $c=\dfrac{1538}{54,\,54,\,54}$'. A cancelled second set of examples at this point reads 'Vel si assumas (Or should you assume) $p=3$. $k=1\frac{1}{2}=g$. & $n=\frac{1}{2}$ erit $l=1\frac{1}{2}$. $m=\frac{1}{24}$. $x=\frac{8}{5}$. $r=\frac{37}{15}$. $s=\frac{36}{15}$. $t=\frac{35}{15}$. Vel si assumas $p=3$. $k=g=n=\frac{3}{2}$ erit $l=\frac{3}{2}$. $m=\frac{5}{24}$. $x=\frac{8}{7}$. $r=\frac{41}{21}$. $s=\frac{36}{21}$. $t=\frac{31}{21}$'. The several inconsistencies here will be evident.

Setting $3k+3l = p^2$, let $y = pm+n$ and so $y^2 = p^2m^2+2pmn+n^2$. Set the two values of y^2 equal and, with equals deleted, there will result

$$k^3+3k^2m+l^3-3l^2m+g^3-3 = 2pmn+n^2, \quad \text{or} \quad m = \frac{k^3+l^3+g^3-3-n^2}{2pn-3k^2+3l^2}.$$

Assume any number you please, therefore, in place of p, say 3 or -3,[39] and also another number for k, say 1 or 0 or $\frac{1}{2}$ or $\frac{2}{3}$ or $\frac{3}{2}$ ..., and make $\frac{1}{3}p^2-k = l$. Then in place of g assume some number greater than unity and any other you wish for n, provided that, when m is found, both $k+m$ and $l-m$ be greater than 1. Then put $1/(pm+n)$[40] $= x$, $(k+m)\,x = r$, $(l-m)\,x = s$, $gx = t$, $r^3-x^3 = a$, $s^3-x^3 = b$, $t^3-x^3 = c$. As was to be done. If, for instance, you assume $p = 3$, $k = 1$, $g = 2$ and $n = 1$, then $l = 2$, $m = \dfrac{13}{15}$, $x = \dfrac{5}{18}$, $t = \dfrac{5}{9}$, $r = \dfrac{14}{27}$, $= s\,\dfrac{17}{54}$, $a = \dfrac{875}{18^3}$, and so on.[41]

If three numbers be desired which, when removed from the cube of their sum, leave cubes, since in this case[42] r^3, s^3 and t^3 are less than x^3 and consequently e, f and g less than unity, assume g and k less than unity, p and n such that $k+m$ and $l-m$ are likewise less than unity but not negative: numbers which shall satisfy the question will then result.[43] For instance, should you assume $p = \frac{3}{2}$ and $k = 0$, then $l = \frac{3}{4}$; or if $p = \frac{3}{2}$ and $k = -\frac{1}{4}$, then $l = -1$. Next, try $g = \frac{1}{4}$ or $\frac{3}{4}$ or $\frac{1}{2}$ or $\frac{5}{4}$ ... and any number you wish in place of n. So, if you assume $k = 0$, $p = \frac{3}{2}$, $l = \frac{3}{4}$, $g = \frac{5}{4}$ and $n = \frac{1}{2}$, then $m = -\frac{7}{66}$, $x = \frac{44}{15}$, $r = \frac{14}{45}$, $s = \frac{85}{45}$ and $t = \frac{17}{9}$.[44]

---

(42) Namely, Diophantus' *Arithmetica*, v, 19 (in Bachet's 1621 numbering): in Newton's terms it is required to find rationals a, b and c such that, on making $a+b+c = x$, there results $x^3-a = r^3$, $x^3-b = s^3$ and $x^3-c = t^3$, where r, s and t are rational. Here, on setting $r = ex$, $s = fx$ and $t = gx$ with $1/x = y$, it follows that $3-(e^3+f^3+g^3) = y^2$.

(43) This is all but incomprehensible. For $3-(e^3+f^3+g^3)$ to be an exact square it is an unnecessary restriction to assume that *each* of e, f and g is less than unity: indeed Newton himself assumes $g = \frac{5}{4}$ in his following example. If, with him, we make the further assumption that $k = 0$, so that $e = m$, $f = l-m$ and $3l = p^2$, then there results

$$3-g^3-\tfrac{1}{27}p^6+\tfrac{1}{3}p^4m-p^2m^2 = y^2.$$

Here the term in m^2 can no longer be destroyed—as Newton apparently continues to think—by assuming $y = pm+n$ and his further argument on this basis is fundamentally mistaken.

(44) This example is intolerably vitiated and does not repay too close a study. Substitution of $k = 0$, $p = \frac{3}{2}$ (and so $l = \frac{3}{4}$), $g = \frac{5}{4}$ and $n = \frac{1}{2}$ in Newton's previous formula yields $m = -\frac{14}{51}$, while it is certainly untrue that (as Newton would have it)

$$3-(\tfrac{7}{66})^3-2(\tfrac{85}{132})^3 = (\tfrac{15}{44})^2.$$

It is clear that he forthwith abandoned his attempt to construct $3-(e^3+f^3+g^3) = y^2$ in rationals, for in the following revise of his 'Problemata' he omits all discussion of this knotty equation and the problem (v, 19) of Diophantus' *Arithmetica* which it was to resolve.

Rursus si velis duos numeros qui cubo summæ additi vel subducti efficiunt cubos, res peragetur ponendo $g=1$ eo ut t^3 et x^3 æquales adeoꝗ $c=0$. Ut si fuerit $g=1=k=n$. [$p=3$] & $l=2$ erit $m=\frac{2}{5}$. $\frac{5}{11}=x$.[45]

Proponatur hoc Prob primo sic. Invenire tres numeros qui cubo summæ sigillatim additi component totidem cubos quorum latera sunt in arithmetica progressione.[46] Et ubi hoc modo solvisti solve generalius.

Prob 16.[47] *Invenire duos cubos quorum summa vel differentia æquatur dato numero.* $x^3+y^3=a$. Puta casum aliquem datum quo rei possibilitas constet vizt $b^3+c^3=a$. Pone $b+m=x$. $c-n=y$. erit $3bbm+3bmm+m^3-3ccn+3cnn-n^3=0$. [48]Pone

$3bbm-3ccn=0$. erit et $3bmm+m^3+3cnn-n^3=0$. Scribe $\dfrac{bbm}{cc}$ pro n. erit

$3b+m+\dfrac{3b^4}{c^3}-\dfrac{b^6}{c^6}m=0$. Et $\dfrac{3bc^6+3b^4c^3}{b^6-c^6}=m=\dfrac{3bc^3}{b^3-c^3}$. [49]

Prob 17. Invenire 4 cubos quorum summa duarum æquabitur summæ vel diff[erentiæ] *aliarum duarum.*[50] $v^3+x^3=y^3+z^3$. Sit $v+x$. $y+z::1$. n. Erit dividendo

$$vv-vx+xx=yyn-yzn+zzn.$$

Scribe $nv+nx-y$ pro z. erit

$$vv-vx+xx=3yyn-3nnvy-3nnxy+n^3vv+2n^3xv+n^3xx.[51]$$

---

(45) Hence $r=\frac{7}{11}$, $s=\frac{8}{11}$ and so $a=\frac{218}{1331}$, $b=\frac{387}{1331}$. This uninteresting special case of Diophantus' problem appears to be original with Newton.

(46) This restriction is developed in detail as Problem 23 of the following revise; see note (92) below.

(47) In his *Arithmetica*, v, 19 [Bachet's edition] Diophantus remarked that he had solved this problem in his *Porisms* (see note (37)), but his solution is lost. Late in the sixteenth century François Viète took up the problem in his *Zetetica* (Tours, 1593): Liber IV, Zeteticum XVIII: 20^r [= *Opera Mathematica* (Leyden, 1646): 42–81, especially 74–5], setting

$$B^3-D^3 = (B-A)^3+([B^2/D^2]A-D)^3$$

where A is unknown, and thence determining $A = 3BD^3/(B^3+D^3)$. Similarly, in following *Zetetica*, XIX and XX (ff. 20^r/20^v) he equated

$$B^3+D^3 = (B+A)^3-([B^2/D^2]A-D)^3 \quad \text{and} \quad B^3-D^3 = (A-D)^3-([D^2/B^2]A-B)^3$$

and determined corresponding values for A. Viète's results were inserted by Bachet in his commentary on Diophantus, IV, 2 (see his edition of the *Arithmetica* (note (1) above): 179–82) and this, rather than Schooten's edition of Viète's *Opera*, was probably the source of Newton's knowledge of the problem. (In his undergraduate annotations on the *Opera* (see 1: 63–88) he had seemingly ignored Viète's *Zetetica* altogether.) Subsequently the problem was much studied by Girard, Fermat, Frénicle, Brouncker, Wallis and others. (See, for example, Wallis' *Commercium Epistolicum* (note (1) above): 22, 41, 99 ff., 118 ff., 123 ff.; and compare Dickson's *History*, II (note (7) above): 550–2.)

(48) Newton first continued 'Sit $m=p+n$. Erit

$$p^3+3ppn+3pnn+36pp+6bpn+3bnn+3bbp+3bbn-3ccn+3cnn=0.$$

Pone $pr=n$. Erit

$$pp+3ppr+3pprr+3bp+6bpr+3prr+3bb-3bbr-3ccr+3crr=0.'$$

Again, should you want two numbers which, when added to or taken from the cube of their sum, make up cubes, the demand will be met by setting $g = 1$, with the effect that t^3 and x^3 will be equal and hence $c = 0$. So if there were $g = k = n = 1$, $p = 3$ and $l = 2$, then $m = \frac{2}{5}$ and $x = \frac{5}{11}$.[45]

But let this problem be proposed in the first instance as follows: To find three numbers which, when added singly to the cube of their sum, shall make up an equal number of cubes whose sides are in arithmetical progression.[46] And when you have solved it in this form, then solve it more generally.

Problem 16.[47] *To find two cubes whose sum or difference is equal to a given number.*

Suppose $x^3 + y^3 = a$. Think of some given case which will establish the possibility of the demand, say $b^3 + c^3 = a$. Put $x = b + m$, $y = c - n$ and there will be $3b^2m + 3bm^2 + m^3 - 3c^2n + 3cn^2 - n^3 = 0$. [48] Put $3b^2m - 3c^2n = 0$ and then $3bm^2 + m^3 + 3cn^2 - n^3 = 0$. Write b^2m/c^2 in place of n. There will result

$$3b + m + 3b^4/c^3 - b^6m/c^6 = 0 \quad \text{and} \quad (3bc^6 + 3b^4c^3)/(b^6 - c^6) = m = 3bc^3/(b^3 - c^3).\text{[49]}$$

Problem 17. To find four cubes, the sum of two of which shall be equal to the sum or difference of the other two.[50]

Suppose $v^3 + x^3 = y^3 + z^3$. Let $y + z = n(v + x)$ and after division there will be $v^2 - vx + x^2 = ny^2 - nyz + nz^2$. Write $nv + nx - y$ in place of z, and there will result $v^2 - vx + x^2 = 3ny^2 - 3n^2vy - 3n^2xy + n^3v^2 + 2n^3xv + n^3x^2$.[51]

This first mistaken 'improvement' on Viète's approach may be an indication that Newton did not have the text of *Zetetica*, IV, 18–20, before him as he wrote. In any event—and despite any reluctance on his part to follow another's method slavishly—in the sequel he is forced to make the simplifying supposition that $m/n = c^2/b^2$.

(49) The parametrization of Viète's Zeteticum XIX (see note (47) above) follows at once in the form

$$b^3 + c^3 = \left(\frac{b(b^3 + 2c^3)}{b^3 - c^3}\right)^3 - \left(\frac{c(2b^3 + c^3)}{b^3 - c^3}\right)^3.$$

In geometrical equivalent (compare Case 2 of §2 below) Newton constructs an arbitrary line of slope $-n/m$ through the point (b, c) on the conchoidal cubic $x^3 + y^3 = b^3 + c^3$; subsequently, his stipulation that $n/m = b^2/c^2$ determines this line to be the tangent to the cubic at (b, c) and it will therefore meet the cubic again in the rational point $(b + m, c - [b^2/c^2]m)$, where $m = 3bc^3/(b^3 - c^3)$.

(50) Fermat's variant on the preceding problem, stated by him in the form 'Datum numerum ex duobus numeris cubis compositum dividere in duos alios numeros cubos' in his letter to Kenelm Digby on 15 August 1657 (N.S.); see Wallis' *Commercium Epistolicum* (note (1) above): 22 [= *Œuvres de Pierre de Fermat*, 2 (Paris, 1894): 344]. To Fermat's particular challenge to decompose $1^3 + 2^3$ into two other cubes Wallis in November (*ibid.*: 41) evasively returned the result $1^3 + 8^3 = (4\frac{1}{2})^3 + (7\frac{1}{2})^3$, adding twenty-one further rational solutions of the equation $v^3 + x^3 = y^3 + z^3$; compare Dickson's *History*, II (note (7) above): 551–2.

(51) There Newton abandons the problem—unnecessarily, since Binet's (most general rational) parametrization follows easily on taking $v + x = 3p$ and $y = qn + r$, when there comes

$$3p^2 - 3pv + v^2 = y^2n - 3pyn^2 + 3p^2n^3 = r^2n + (2q - 3p)rn^2 + (q^2 - 3pq + 3p^2)n^3$$

$$= r^2n + rsn^2 + \tfrac{1}{4}(s^2 + 3p^2)$$

Prob. [17.] Triangula duo æquicrura in numeris invenire quæ perimetro et area æquales erunt.[52]

Prob. 18. Triangula datum ang. habentia, perimetroʒ et area æqualibus invenire.[53]

Prob. 20. Numerū invenire quē dat[i] numer[i] dividentes, datos relinquent residuos.[54]

Prob 21. Numeros invenire qui per datos num[eros] multipl. datum numerum produ-[c]ent.[55] [Puta] $ax + by = d$. aufer a de [d] quoties licebit. Adde b semel vel sæpius si opus est ut a iterum auferri possit. Et sic perge addendo b et auferendo a donec d vel evanuerit vel ad primum numerum redierit. Si ad primum numerum redierit opus in circulum etiam rediens ostendit d nunquam tolli possi per multiplices a et b adeoʒ quæstio impossibilis erit.[56] Sin d evanuerit pone n numerum omnium additionum ipsius b[57] et erit $-n$ vel $a - n$ vel $2a - n$ vel $3a - n$ &c $= y$ et $\dfrac{d - by}{a} = x$.[58] [Sit] $7x + 5y = 62$. [erit] $62 - 8 \times 7 = 6$. $6 + 5 = 11$. $11 - 7 = 4$. $4 + 5 = 9$. $9 - 7 = 2$. $2 + 5 = 7$. $7 - 7 = 0$. Ergo

$$1 + 1 + 1 = n = 3. \qquad 7 - 3 = 4 = y. \qquad \frac{62 - 20}{7} = \frac{d - by}{a} = x = 6.$$

---

on making $2q - 3p = s$; thence, on setting $s = 2a$, $p = 2b$ and $v = n^2 + t$, there results

$$12b^2 - 6bt + t^2 + 2(t - 3b)\,n^2 + n^4 = r^2 n + 2arn^2 + (a^2 + 3b^2)\,n^3;$$

finally, take $n = a^2 + 3b^2$ and $t = ar + 3b$, and so $r^2 = 1$, and on setting $r = -1$ there follows $v = (a^2 + 3b^2)^2 - a + 3b$, $x = -(a^2 + 3b^2)^2 + a + 3b$, $y = (a + 3b)(a^2 + 3b^2) - 1$ and

$$z = (-a + 3b)(a^2 + 3b^2) + 1.$$

See Dickson's *History*, II (note (7) above): 554–5; and compare G. H. Hardy and E. M. Wright, *An Introduction to the Theory of Numbers* (Oxford, ₄1960): 199–200.

(52) This 'problème de Roberval' is taken up in detail in the following revise (Problem 25; see note (97) below).

(53) A somewhat jejune variant on the preceding problem, one justly omitted by Newton from his revised 'Problemata'. If two rational triangles of sides a, b, c and p, q, r have equal perimeters $a + b + c = p + q + r$, equal area A and the angles (say α) included between a, b and between p, q equal, then $ab = pq = 2A/\sin\alpha$ and $(a^2 + b^2 - c^2)/ab = (p^2 + q^2 - r^2)/pq = 2\cos\alpha$, so that

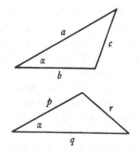

$$(a + b)^2 - c^2 = (p + q)^2 - r^2$$

and hence also $a + b - c = p + q - r$, or $a + b = p + q$ and $c = r$. The two solutions possible ($a = p$, $b = q$, $c = r$; $a = q$, $b = p$, $c = r$) are the trivial cases in which the two triangles are congruent.

(54) The classical 'Chinese problem of remainders' (see Dickson's *History*, II (note (7) above): 57–63), solved at least as early as the first century A.D. in the *Suan-ching* of Sun-Tsŭ, apparently rediscovered independently by Brahmagupta six centuries later and introduced into Europe by Leonardo of Pisa in his *Liber Abbaci* (1202), later receiving a generalized treatment in Book 1 of Michael Stifel's *Arithmetica Integra* (Nuremberg, 1544): 38ᵛ. In his scholium to the following Problem 21 Newton remarks that it is 'solved by differences', clearly

Problem 17. *To find two rational isosceles triangles which are equal in perimeter and area.*[52]

Problem 18. *To find [rational] triangles which have a given angle and are equal in perimeter and area.*[53]

Problem 20. *To find a number which shall, after division by given numbers, leave given remainders.*[54]

Problem 21. *To find numbers which, when multiplied by given numbers, shall produce a given number.*[55]

Say $ax + by = d$. Take a from d as many times as permissible, then add b once or more often, as needed, so that a may again be taken away. And so continue, adding b and taking away a, until d either vanishes or returns to a prime number. If it returns to a prime the corresponding circularity in the working reveals that d can never be eliminated by multiples of a and b, and the question will consequently be impossible.[56] If, however, d vanishes, put n for the total number of additions of b,[57] and then $y = -n$ or $a-n$ or $2a-n$ or $3a-n$..., with

$$x = (d - by)/a.^{[58]}$$

Take $7x + 5y = 62$ and there will be $62 - 8 \times 7 = 6$, $6 + 5 = 11$, $11 - 7 = 4$, $4 + 5 = 9$, $9 - 7 = 2$, $2 + 5 = 7$, $7 - 7 = 0$. Hence $n = 1 + 1 + 1 = 3$,

$$y = 7 - 3 = 4, \quad x = (d - by)/a - (62 - 20)/7 = 6.$$

---

having in view the method of Nicolaus Huberti which Schooten published in 1657 in his *Exercitationes Mathematicæ* (note (1) above): Liber v, Sectio vii: 407–10 and which, as we have seen (i: 52), Newton noted as an undergraduate.

(55) When they are added together, that is.

(56) This can, of course, never happen when a and b are coprime or, more generally, when d is a multiple of the highest common factor of a and b.

(57) Newton first wrote 'pone n numerum per quem multiplicasti a' (put n for the number by which you multiplied a).

(58) In general, if $ax + by = d$ is satisfied by $y = -n$ and so $x = (d + bn)/a$, then it is also satisfied by $y = \lambda a - n$, $x = (d + bn)/a - \lambda b = (d - [\lambda a - n] b)/a$. At the beginning of the century Bachet had pioneered the method of using Euclid's algorithm (*Elements*, x, 3) to find the highest common factor of a and b (which must also be a factor of d for an integral solution) and then multiplying through by the quotient of d and that factor: thus in Newton's present example Euclid's algorithm yields $7 \times -2 + 5 \times 3 = 1$ and hence

$$7 \times -124 + 5 \times 186 = 62 = 7(5k - 124) + 5(186 - 7k)$$

as the general solution, with Newton's lowest integer solution resulting when $k = 26$. (Compare Problem 5 of Bachet's *Problemes Plaisans et Delectables, Qui se Font par les Nombres* (Lyons, $_1$1612); see also Dickson's *History*, ii (note (7) above): 44.5). Here Newton prefers a refinement of the brute-force method—of small theoretical interest—published by John Kersey in 1673 (*The Elements of that Mathematical Art commonly called Algebra, expounded in Four Books*, **1** (London, 1673): 301), first computing $d - a\alpha = \beta < a$ and then determining $\beta = b\gamma - a\delta$ by successively putting $\gamma = 1, 2, 3, \ldots$.

Rursus $ax+by+cz=d$. Take a as often as you can from d. Add b & take away a by turns as before untill d vanish or return[59] or be a multiplex of c &c. [Sit] $6x+10y+15z=133$. [erit] $133-22$, $6=1$. $1+10=11$. $11-6=5$. $5+10=15$. Ergo $1=z$. $1+1=2=n$. $6-2=4=y$. $\dfrac{133-15-40}{6}=\dfrac{78}{6}=13$ [$=x$].

NB   This Problem is better solved by differences like [y^e] 20th. Propound it more generally thus. *To make* $\dfrac{a+bx}{c+dx}$ *integer.*[60]

Prob 21[7]. *To find perfect numbers.* i.e. wch are equal to y^e summ of their divisors.[61]

---

[2][62]   PROBLEMATA NUMERALIA

1. Invenire tres continuè proportionales numeros quorum extremi quadrata sint et summa quadratum est. $xx+xy+yy=zz$. $z=y+e$. $xx+xy=2ey+ee$. $\dfrac{xx-ee}{2e-x}=y$. Ut si $x=3$. $e=2$. erit $y=5$. et 9. 15. 25 tres continue prop[or]t^s quorum summa 49 est quad.[63]

2. Invenire tres $\div\div$ quorum extremi quadrata sint et omnium summa cubus.

$$xx+xy+yy=z^3.$$

Sit $ez=x$. $fz=y$. Erit $eezz+efzz+ffzz=z^3$. $ee+ef+ff=z$. Ut si fuerit $e=1, f=2$ erit $z=7=x$. $y=14$. et 49, 98, 196 tres $\div\div$ quorum summa 343 est 7^{cub}.

3. Invenire tres $\div\div$ quorum extremi quadrata sint et omniū summa quadrato-quadratum est. $xx+xy+yy=z^4$. Sit $ez=x, fz=y$, erit $eezz+efzz+ffzz=z^4$ vel $ee+ef+ff=zz$. Quære ergo $e_{[,]} f$ et z per Prob. 1. Ut si fuerit in exemplo ibi allato $e=3, f=5$, erit $z=7$. $x=21$. $y=35$. Et 441, 735, 1225 tres $\div\div$ quarum summa 2401 est 7^{qq}.[64]

---

(59) Newton first continued 'if y^e latter z may not be nothing. Add c once then take away a & add b as before untill d return or vanish. If y^e former add c once more &c'. In this insignificant generalization of the preceding problem it will be evident that d 'returns' and 'may not be nothing' when it is not a multiple of the highest common factor of a, b and c.

(60) Clearly, $(a+bx)/(c+dx) = k$, integral, if x and k can be chosen such that

$$-bx+ck+dkx = a.$$

(61) No doubt Newton here intended merely to repeat Euclid's construction of (even) perfect numbers (*Elements*, IX, 36): namely, if 2^n-1 is prime, then $2^{n-1}(2^n-1)$ is perfect. The subsequent history of Euclid's theorem (which, as Euler later showed, generates all even perfects, while no odd perfects are known) is sketched in L. E. Dickson, *History of the Theory of Numbers*, I (Washington, 1919): 3–33; see also T. L. Heath, *The Thirteen Books of Euclid's Elements*, 2 (Cambridge, ₂1926): 421–6. As C. J. Scriba has recently shown in his *Studien zur Mathematik des John Wallis (1616–1703)* [= *Boethius*, 6 (Wiesbaden, 1966)]: 95–103, Wallis had almost succeeded in anticipating Euler's proof that all even perfects are covered by Euclid's formula, probably inspired by Schooten's communication in February 1658 of Fermat's

Again, suppose $ax+by+cz = d$. [Take a as often as you can from d. Add b & take away a by turns as before untill d vanish or return[59] or be a multiplex of c &c.] Let $6x+10y+15z = 133$. There will be $133-22\times6 = 1$, $1+10 = 11$, $11-6 = 5$, $5+10 = 15$. Hence $z = 1$, $n = 1+1 = 2$, $y = 6-2 = 4$,

$$x = (133-15-40)/6 = 78/6 = 13.$$

N.B. [This Problem is better solved by differences like y^e 20th. Propound it more generally thus. *To make* $\dfrac{a+bx}{c+dx}$ *integer*.]

Problem 21[′]. [*To find perfect numbers*. i.e. wch are equal to y^e summ of their divisors.][61]

---

[2][62]   NUMERICAL PROBLEMS

   1. To find three numbers in continued proportion, whose extremes are to be squares and whose sum is a square.
   Take $x^2+xy+y^2 = z^2$ and $z = y+e$. Then $x^2+xy = 2ey+e^2$ and

$$y = (x^2-e^2)/(2e-x).$$

So if $x = 3$, $e = 2$, then $y = 5$ and 9, 15, 25 are three continued proportionals whose sum 49 is a square.[63]
   2. To find three continued proportionals, whose extremes are to be squares and total sum is a cube.
   Say $x^2+xy+y^2 = z^3$. Let $x = ez$, $y = fz$ and then $e^2z^2+efz^2+f^2z^2 = z^3$, or $z = e^2+ef+f^2$. If, for instance, there were $e = 1$, $f = 2$, then $z = x = 7$, $y = 14$ and 49, 98, 196 are three continued proportionals whose sum 343 is 7 cubed.
   3. To find three continued proportionals, whose extremes are to be squares and total sum is a fourth power.
   Say $x^2+xy+y^2 = z^4$. Let $x = ez$, $y = fz$ and then $e^2z^2+efz^2+f^2z^2 = z^4$ or $e^2+ef+f^2 = z^2$. Therefore seek e, f and z by Problem 1. Thus if, in the example there adduced, there were $e = 3$, $f = 5$, then $z = 7$, $x = 21$, $y = 35$ and 441, 735, 1225 are three continued proportionals whose sum 2401 is 7^4.[64]

---

challenge 'Ostendere, utrum *perfecti numeri aliâ ratione quàm ab Euclide traditur* prop. ult. lib. 9. Elem. hoc est, absque progressione dupla, *sint inveniendi necne*' (see Wallis' *Commercium Epistolicum* (note (1) above): 141 and 150). Such existence problems were never to Newton's taste and it is hardly likely that he here intended to include an examination of Fermat's *défi*.
   (62) ULC. Add. 3964.1: 1^r-2^v, a revised version of the preceding which is for the most part much more carefully composed though occasional heavily cancelled and roughly rewritten passages indicate that this is still only a working draft.
   (63) See note (5) above. This problem, like that following, is insignificantly variant from its preceding draft version.
   (64) This and the following Problem 4 are little more than corollaries to Problems 1 and 2. Their essential lack of originality scarcely befits their separate status.

4. Invenire tres ÷ quorum extremi quadrata sint et omniū summa quadrato-cubus est.
$xx+xy+yy=z^5$. Sit $ez=x$. $fz=y$ et erit $ee+ef+ff=z^3$. Quære ergo e, f et z ut in Prob 2. Atqᵷ ita pergitur in infinitum.

5. Invenire quatuor ÷ quorum summa quadratum est et extremi sunt cubi.

$$x^3+x^2y+xyy+y^3=zz.$$

[Sit] $ez=x$. $fz=y$. [erit] $e^3z+eefz+effz+f^3z=1$. [adeoqᵷ] $\dfrac{1}{e^3+eef+eff+f^3}=z$.

6. Invenire quatuor ÷ quorum extremi sunt cubi et summa cubus est.

$$x^3+xxy+xyy+y^3=z^3.$$

[Sit] $y=e-x$. $z=e-f$. [erit] $2exx-2eex=-3eef+3eff-f^3$. Pone $2eex=3eef$ sive $2x=3f$ et erit $\frac{9}{2}eff=3eff-f^3$. et $3e=-2f$. Unde [si $e=4$ erit $f=-6$.] $x=-9$. $y=13$ & $z=10$.[65]

7. Invenire quadrata duo quorum summa vel differentia quadratum est. $xx+yy=zz$.

[Sit] $z=y+e$. [erit] $xx=2ey+ee$. $y=\dfrac{xx-ee}{2e}$.[66]

8. Invenire quadrata duo quorum summa vel differentia cubus est. $xx+yy=z^3$. sit $ez=x$, $fz=y$ erit $ee+ff=z$.[67]

9. Invenire quadrata duo quorum summa vel differentia quadrato-quadratum est. $xx+yy=z^4$. $ez=x$. $fz=y$. [fit] $ee+ff=zz$. Quære e, f, et z per Prob [7]. Et sic in infinitum.

10. Invenire cubos duos quorum summa vel differentia quadratum est. $x^3+y^3=zz$. Sit $ez=x$. $fz=y$ et erit $e^3z+f^3z=1$. sive $z=\dfrac{1}{e^3+f^3}$. Vel sit $y=mx$ & $z=nx$. erit $x+m^3x=nn$, et $x=\dfrac{nn}{1+m^3}$.[68]

11. Invenire cubos duos quorum summa vel differentia[69] *quadratoquadratum est vel quadrato-cubus, et sic in infinitum.* $x^3+y^3=z^4$. sit $ez=x$. $fz=y$. erit $e^3+f^3=z$. Rursus si $x^3+y^3=z^5$. erit $e^3+f^3=zz$ unde e f et z determinandi sunt per Prob 10. Iterum si $x^3+y^3=z^6$. erit $e^3+f^3=z^3$ unde Problema solvetur si modo duo cubi inveniri possunt quorum summa vel differentia cubus est.[70]

---

(65) The generalization of Problem 1 to four continued proportionals. Since Newton's choice of $y = e-x$, $z = e-f$ determines $2ex(x-e)+f(3e^2-3ef+f^2) = 0$, the most general reduction is to set $x = kf$, yielding the equation (by hypothesis, rationally factorizable) $(3-2k)e^2+(2k^2-3)ef+f^2 = 0$. Evidently, it is simplest (with Newton) to choose $k = \frac{3}{2}$.

(66) See notes (7) and (8) above.

(67) Much as in his draft 'Prob 6' (see page 76) Newton could have completed his solution by deducing the parametrization $x = e(e^2+f^2)$, $y = f(e^2+f^2)$, $z = e^2+f^2$. He might have noticed that when z is of the form $4k+3$ no integer solution is possible.

(68) For this latter parametrization see Newton's draft Problem 6 (note (11) above).

(69) Newton first continued 'dato numero sive cubico sive non cubico æqualis est' (is equal to a given number, whether a cube or not). See next note.

4. To find three continued proportionals whose extremes are to be squares and total sum is a fifth power.

Say $x^2 + xy + y^2 = z^5$. Let $x = ez$, $y = fz$ and then $e^2 + ef + f^2 = z^3$. Therefore seek e, f and z as in Problem 2. And so it is continued indefinitely.

5. To find four continued proportionals whose sum is a square and extremes are cubes.

Say $x^3 + x^2y + xy^2 + y^3 = z^2$. Let $x = ez$, $y = fz$ and then

$$e^3z + e^2fz + ef^2z + f^3z = 1$$

and hence $z = 1/(e^3 + e^2f + ef^2 + f^3)$.

6. To find four continued proportionals whose extremes are cubes and sum is a cube.

Suppose $x^3 + x^2y + xy^2 + y^3 = z^3$. Let $y = e - x$, $z = e - f$ and then

$$2ex^2 - 2e^2x = -3e^2f + 3ef^2 - f^3.$$

Put $2e^2x = 3e^2f$, that is, $2x = 3f$, and there follows $\frac{9}{2}ef^2 = 3ef^2 - f^3$, and $3e = -2f$. Hence if $e = 4$, then $f = -6$, $x = 9$, $y = 13$ and $z = 10$.[65]

7. To find two squares whose sum or difference is a square.

Say $x^2 + y^2 = z^2$. Let $z = y + e$ and then $x^2 = 2ey + e^2$ or $y = (x^2 - e^2)/2e$.[66]

8. To find two squares whose sum or difference is a cube.

Say $x^2 + y^2 = z^3$. Let $x = ez$, $y = fz$ and then $z = e^2 + f^2$.[67]

9. To find two squares whose sum or difference is a fourth power.

Say $x^2 + y^2 = z^4$. Let $x = ez$, $y = fz$ and there results $e^2 + f^2 = z^2$. Seek e, f and z by Problem 7. And so on indefinitely.

10. To find two cubes whose sum or difference is a square.

Suppose $x^3 + y^3 = z^2$. Let $x = ez$, $y = fz$ and then $e^3z + f^3z = 1$ or

$$z = 1/(e^3 + f^3).$$

Or let $y = mx$ and $z = nx$, when $x + m^3x = n^2$ and $x = n^2/(1 + m^3)$.[68]

11. To find two cubes whose sum or difference[69] *is a fourth power or a fifth one, and so on indefinitely.*

Say $x^3 + y^3 = z^4$. Let $x = ez$, $y = fz$ and there will be $e^3 + f^3 = z$. Again, if $x^3 + y^3 = z^5$, then $e^3 + f^3 = z^2$ and hence e, f and z are to be determined by Problem 10. Further, if $x^3 + y^3 = z^6$, then $e^3 + f^3 = z^3$ and hence the problem will be resolved provided two cubes can be found whose sum or difference is a cube.[70]

---

(70) In perhaps the most celebrated annotation in the history of mathematics Fermat had recorded earlier, alongside Problem 8 of Book 2 (see note (7) above) in the margin of his copy of Bachet's 1621 edition of Diophantus' *Arithmetica*, that 'Cubum...in duos cubos, aut quadratoquadratum in duos quadratoquadratos & generaliter nullam in infinitum vltra quadratum potestatem in duos eiusdem nominis fas est diuidere[,] cuius rei demonstrationem mirabilem sane detexi. Hanc marginis exiguitas non caperet' (Samuel Fermat, *Diophanti...* *Arithmeticorum Libri Sex* (note (1) above): ₂61: 'Observatio Domini Petri de Fermat [in II, 8]'). What Fermat's 'marvellous demonstration' might have been has puzzled mathematicians and

Sunt alij modi solvendi hæc Problemata. Ut si assumas $a^3 + abb = x$ et $aab + b^3 = y$ erit $xx + yy$ cubus de $aa + bb$. Vel si assumas $[a^3] - abb = x$ et $aab - b^3 = y$ erit $xx - yy$ cubus de $aa - bb$. Sic posito $a^4 + ab^3 = x$ et $a^3 b + b^4 = y$ erit $x^3 + y^3$ quadrato-quadratus de $a^3 + b^3$. atcɟ ita in infinitum.[71]

12. *Invenire quadrata duo qui postquam in datos numeros ducti sunt, summā vel differentiam habebunt dato numero æqualem.* $axx + byy = c$. Quo quæstionis possibilitas constet, detur casus aliquis terminorū. Sit iste $app + bqq = c$. Pone $x = p + s$. $y = q - t$ et erit $app + 2aps + ass + bqq - 2bqt + btt = c = app + bqq$. seu

$$2aps + ass - 2bqt + btt = 0.$$

Pone $t = es$ et erit $2ap + as - 2bqe + bees = 0$. Et $s = \dfrac{2bqe - 2ap}{a + bee}$.[72]

13. *Invenire cubos duos qui postquam per datos numeros multiplicati fuerint summam vel differentiam habebunt dato numero æqualem.* $ax^3 + by^3 = c$. Detur casus aliquis quo Problematis possibilitas constet. Sit iste $ap^3 + bq^3 = c$. Pone $x = p + s$, $y = q - t$ et erit (deletis æqualibus) $3apps + 3apss + as^3 - 3bqqt + 3bqtt - bt^3 = 0$. Pone $apps = bqqt$ & scribendo $\dfrac{apps}{bqq}$ pro t emerget $3apss + as^3 + \dfrac{3aap^4 ss}{bq^3} - \dfrac{a^3 p^6 s^3}{bbq^6} = 0$. sive[73]

$\dfrac{+3pbbq^6 + 3abp^4 q^3}{aap^6 - bbq^6} = s = \dfrac{3pbq^3}{ap^3 - bq^3}$. Sed nota quod ubi hoc modo invenisti x et y si fuerit summa $ap^3 + bq^3 = c$, tunc differentia $ax^3 - by^3$ vel $by^3 - ax^3$ erit $= c$. Et vice versa. Sed si rursus subst[it]uis inventos numeros x et y pro p et q prodibunt duo numeri desiderati.[74]

---

historians for three centuries. In our opinion it was almost certainly the proof of non-existence which establishes that any given integral solution implies a *descente infinie* of lower (positive) integral solutions: certainly, Fermat in August 1659 in his 'Relation des nouvelles Decouvertes en la Science des Nombres' asserted that he had proved it impossible to divide a cube into two other (rational) cubes by framing such a *descente* (see his *Œuvres*, **2** (Paris, 1894): 431–6, especially 433), while in his annotation on Problem 20 of Bachet's additions to Book 6 of the *Arithmetica* (to find a rational right triangle of rational area) Fermat, here explicitly stating that his method of proof was by establishing an infinite descent of solutions, again observed that Bachet's broad margin was too small for him to set it out in detail. (Compare Heath's *Diophantus* (note (1) above): 145, note 3, and 293–5.) Whether Newton knew of Fermat's assertion of his 'great theorem' by the late 1670's is doubtful, but he must have been aware that in August 1657 Fermat had issued jointly to Brouncker and Wallis his challenge 'datum numerum cubum in duos cubos rationales dividere', adding the following April that 'c'est une de mes propositions negatives que ni luy [Wallis] ni le Seigneur Brouncker ne demontreront peut estre pas si aysement; car ie soutiens, *Qu'il n'y a aucun cube en nombres qui puisse estre divisé en deux cubes rationaux*' (Fermat to Kenelm Digby, 15 August 1657 and 7 April 1658 (N.S.), first published a few months later in Wallis' *Commercium Epistolicum* (note (1) above): 23, 160 respectively). (Wallis' still unpublished, incomplete proof of the non-existence of rational solutions of $x^3 + y^3 = z^3$ is discussed by C. J. Scriba in his *Studien zur Mathematik des John Wallis* (note (61) above): 103–7. In 1753 Euler stated that he had proved the problem impossible but his published proof (*Vollständige Anleitung zur Algebra*, **2** (St Petersburg, 1770): Chapter 15,

There are other ways of solving these problems. If, for instance, you should assume $x = a^3 + ab^2$ and $y = a^2b + b^3$, then $x^2 + y^2$ will be the cube of $a^2 + b^2$. Alternatively, should you assume $x = a^3 - ab^2$ and $y = a^2b - b^3$, then $x^2 - y^2$ is the cube of $a^2 - b^2$. In this manner, on putting $x = a^4 + ab^3$ and $y = a^3b + b^4$, $x^3 + y^3$ will be the fourth power of $a^3 + b^3$. And so indefinitely.[71]

12. To find two squares which, after multiplication by given numbers, have a sum or difference equal to a given number.

Suppose $ax^2 + by^2 = c$. To establish that the requirement is possible, let some particular case of the formula be given, say $ap^2 + bq^2 = c$. Put $x = p + s, y = q - t$ and there will be $ap^2 + 2aps + as^2 + bq^2 - 2bqt + bt^2 = c = ap^2 + bq^2$ or

$$2aps + as^2 - 2bqt + bt^2 = 0.$$

Put $t = es$ and then $2ap + as - 2bqe + be^2s = 0$, and so

$$s = (2bqe - 2ap)/(a + be^2).\text{[72]}$$

13. To find two cubes which, after multiplication by given numbers, have a sum or difference equal to a given number.

Suppose $ax^3 + by^3 = c$. Let some instance be given which shall establish the problem's possibility, say $ap^3 + bq^3 = c$. Put $x = p + s, y = q - t$ and, on deletion of equals, there will be $3ap^2s + 3aps^2 + as^3 - 3bq^2t + 3bqt^2 - bt^3 = 0$. Set

$$ap^2s = bq^2t$$

and on writing ap^2s/bq^2 in place of t there will emerge

$$3aps^2 + as^3 + 3a^2p^4s^2/bq^3 - a^3p^6s^3/b^2q^6 = 0,$$

that is,[73] $s = (3pb^2q^6 + 3abp^4q^3)/(a^2p^6 - b^2q^6) = 3pbq^3/(ap^3 - bq^3)$. But note that when you have found x and y in this way, if it was $ap^3 + bq^3 = c$, then the difference $ax^3 - by^3$ or $by^3 - ax^3$ will equal c, and conversely so. But if you again substitute the numbers x and y found in place of p and q, a pair of numbers desired will result.[74]

---

§243: 509–16 [= *Opera Omnia* (1), **1** (Leipzig/Berlin, 1911): 484–9]) is defective; see Dickson's *History*, II (note (7) above): 545–6. A watertight proof by a Fermatian *descente infinie* was finally given by C. F. Gauss (*Werke*, **2** (Leipzig/Berlin, 1863): 387–90) about 1820.)

(71) In general, the equation $x^n \pm y^n = z^{n+1}$ is parametrized by setting $x = a(a^n \pm b^n)$, $y = b(a^n \pm b^n)$ and so $z = a^n \pm b^n$.

(72) An insignificantly variant revise of Newton's earlier Problem 11 (see note (19)).

(73) On eliminating the factor as^2 throughout.

(74) Newton generalizes Problem 16 of his first draft (see notes (47) and (49) above). Much as before, in geometrical equivalent the procedure constructs an arbitrary line of slope $-t/s$ through the given point (p, q) on the conchoidal cubic $ax^3 + by^3 = c$, while the further stipulation $t/s = ap^2/bq^2$ determines the line to be tangent to the cubic at that point; it will consequently meet the cubic again in the third rational point $(p + s, q - [ap^2/bq^2]s)$, where

$$s = 3bpq^3/(ap^3 - bq^3).$$

Compare Case 2 of §2 following.

Si cubi p^3 et q^3 indeterminati supponantur æque ac x et y possit problema sic generalius solvi. Ubi ad æquationem $3apps + 3apss + as^3 - 3bqqt + 3bqtt - bt^3 = 0$ [perveneris], Pone $\frac{as}{bt} = nn$. & $q = np - v$, et emerget[75]

$$3apss + as^3 + 6btnpv - 3btvv + 3bttnp - 3bttv - bt^3 = 0.$$

Unde $p = \frac{bt^3 + 3bttv + 3btvv - as^3}{3ass + 6btnv + 3bttn}$. Assume ergo n ac t et $\frac{btnn}{a}$ erit s. postea assumendo etiam v inveniuntur p et $q_{[,]}$ unde dantur x et y. Potest et valor p sic concinnari $p = \frac{tt + 3tv + 3vv - nnss}{3nns + 6nv + 3tn}$. [76]

Cor 1. Hinc tres cubi inveniri possunt quorum summa æquatur dato cubo.[77]

Cor 2. Hinc etiam si duo cubi inventi fuerint tertio aliquo æquales[78] possunt infiniti inveniri quorum summa vel differentia æquatur dato cuivis cubo.

Prob 14. *Invenire quadratos omnes integros quorū differentia æquatur dato numero.*

$xx - yy = a$. [vel] $x + y = \frac{a}{x - y}$. Est ergo a divisibilis per $x - y$. Pone ergo pro $x - y$ divisorem aliquem ipsius a. Sit iste e. erit $\frac{a}{e} = x + y$ & $\frac{a}{2e} + \frac{e}{2} = x$. $\frac{a}{2e} - \frac{1}{2}e = y$. Tenta ergo omnes divisores a minores $\sqrt{a}$ et si nullus succedit nulli erunt x et y.[79]

Prob 15. *Invenire cubos omnes integros quorum summa vel differentia æquatur dato numero.* $x^3 + y^3 = a$. [adeoq̃] $xx - xy + yy = \frac{a}{x + y}$. Sit e (divisor aliquis ipsius a) $= x + y$. erit $\frac{ee}{3} - \frac{a}{3e} = xy$. et $\frac{4a}{3e} - \frac{ee}{3} = xx - 2xy + yy$. Unde $\frac{1}{2}e \pm \frac{1}{2}\sqrt{\frac{4a}{3e} - \frac{ee}{3}} = x$.[80] y.

Prob 16. *Invenire quadrato-cubos omnes integros quorum summa vel differentia æquatur dato numero.* $x^5 - y^5 = a$. $x^4 + x^3y + x^2y^2 + xy^3 + y^4 = \frac{a}{x - y}$. Sit e (divisor aliquis

---

(75) On replacing q in the preceding equation by its equal $np - v$.

(76) On substituting $a = bn^2t/s$ and dividing through by bt, that is. In geometrical equivalent, the conic $y^2 + qy + q^2 = n^2(x^2 + px + p^2)$ will pass through the two further meets (x, y), $x = p + s$, $y = q - t$, of the cubic $ax^3 + by^3 = ap^3 + bq^3$ with the line of slope $-a/bn^2 = -t/s$ through the point (p, q). Evidently, a sufficient condition for the point (x, y) to be rational is that it should lie on one of the family of lines $y = nx + k$, k rational, whence also $v = np - q$ is rational when s and t are rational. The preceding particular solution (in whose equivalent Cartesian model the line through (p, q) is tangent there to the cubic) follows trivially by making $v = 0$ and hence $n = q/p$. It is not clear to us why Newton has cancelled this present, more comprehensive solution. Did it perhaps appear obvious to him that the value

$$p = (-n^2s^2 + t^2 + 3tv + 3v^2)/3n(ns + t + 2v)$$

renders the equation $\quad ap^3 + b(np - v)^3 = a(p + s)^3 + b(np - v - t)^3$

an identity on setting $n^2 = as/bt$? Or did he discard it as being unnecessarily complex?

If as well as x and y the cubes p^3 and q^3 be supposed indeterminate, the problem might be solved more generally this way. When you have attained the equation $3ap^2s + 3aps^2 + as^3 - 3bq^2t + 3bqt^2 - bt^3 = 0$, put

$$as/bt = n^2 \quad \text{and} \quad q = np - v,$$

and there will emerge[75] $3aps^2 + as^3 + 6bnptv - 3btv^2 + 3bnp^2 - 3bt^2v - bt^3 = 0$. From this $p = (bt^3 + 3bt^2v + 3btv^2 - as^3)/(3as^2 + 6bntv + 3bnt^2)$. Take any n and t, therefore, and s will be bn^2t/a; subsequently, on assuming v also, there are found p and q, and from these x and y are given. The value of p may also be obtained in this equivalent form $p = (t^2 + 3tv + 3v^2 - n^2s^2)/(3n^2s + 6nv + 3nt)$.[76]

Corollary 1. Hence three cubes can be found whose sum is equal to a given cube.[77]

Corollary 2. Hence, also, if two cubes have been found equal to some third one,[78] then an infinite number can be found whose sum or difference is equal to any cube you wish.

Problem 14. To find all integral squares whose difference is equal to a given number.

Suppose $x^2 - y^2 = a$, or $x + y = a/(x - y)$. Therefore a is divisible by $x - y$. So in place of $x - y$ put some divisor of a. Let that be e, whence $x + y = a/e$ and so $x = \frac{1}{2}a/e + \frac{1}{2}e$, $y = \frac{1}{2}a/e - \frac{1}{2}e$. So attempt all divisors of a which are less than $\sqrt{a}$, and if none succeeds no x and y will exist.[79]

Problem 15. To find all integral cubes whose sum or difference is equal to a given number.

Take $x^3 + y^3 = a$, and consequently $x^2 - xy + y^2 = a/(x + y)$. Let $x + y$ be equal to e (some divisor of a) and then $xy = \frac{1}{3}e^2 - \frac{1}{3}a/e$ with $(x - y)^2 = \frac{4}{3}a/e - \frac{1}{3}e^2$. Hence $x, y = \frac{1}{2}e \pm \frac{1}{2}\sqrt{[\frac{4}{3}a/e - \frac{1}{3}e^2]}$.[80]

Problem 16. To find all integral fifth powers whose sum or difference is equal to a given number.

Suppose $x^5 - y^5 = a$ and so $x^4 + x^3y + x^2y^2 + xy^3 + y^4 = a/(x - y)$. Let $x - y = e$

---

(77) At once, on setting $a = b = 1$ in Newton's main solution, there results

$$p^3 + q^3 + (p(2p^3 + q^3)/(p^3 - q^3))^3 = k^3, \quad \text{where} \quad k = (p^3 + 2q^3)q/(p^3 - q^3).$$

Compare note (49).

(78) Which it is, of course, impossible to do! See note (70) above.

(79) This and the two following problems are not essentially variant from their draft versions in [1] (Problem 12).

(80) To complete the argument we obviously require some equivalent to 'Tenta ergo omnes divisores a et si nullus efficiat $\dfrac{4a}{3e} - \dfrac{ee}{3}$ quadratum, puta bb, necnon $\frac{1}{2}e \pm \frac{1}{2}b$ integer, nulli erunt x et y'.

ipsius $a) = x - y$. erit $\dfrac{a}{5e} - \dfrac{e^4}{5} = x^3y - xxyy + xy^3$. Unde divisor aliquis ipsius $\dfrac{a}{5e} - \dfrac{e^4}{5}$ erit[81] y et $e+y$ sive x dividet quotum. Tenta igitur siqui reperiantur his conditionibus[82] divisores. Si nulli frustra tentabis ulterius solutionem Problematis. Eadem methodus extendit ad superiores potestates in infinitum.

Prob 17. Invenire quadratum et cubum quorum summa vel differentia quadrato-quadratum est. $xx + y^3 = z^4$. Sit $ez = x$. $fz = y$. erit $ee + f^3z = zz$. Sit insuper $gz = e$. erit $ggz + f^3 = z$. & $\dfrac{f^3}{1-gg} = z$.[83]

[*Prob. 18*] *Invenire quadratum et cubum quorum summa vel differentia quadrato-cubus est.* $xx + y^3 = z^5$. Sit $ez = x$. $fz = y$. erit $ee + f^3z = z^3$. Sit $gz = e$. erit $ggz + f^3 = zz$. Sit $mg = f$ & $ng = z$. erit $ng + m^3g = nn$ & $\dfrac{nn}{n+m^3} = g$.[83]

[*Prob. 19.*] *Invenire duos cubos quorum summa æquatur summæ laterum vel differentia differentiæ.* $x^3 + y^3 = x + y$. Divide omnia per $x+y$. erit $xx - xy + yy = 1$. Pone $x = z+1$. Erit $zz + 2z - zy - y + yy = 0$. Pone $ez = y$. Erit $z + 2 - ez - e + eez = 0$. Et $z = \dfrac{e-2}{1-e+ee}$.[84]

[*Prob. 20.*] *Invenire numeros duos quorum summa quadratum est et quadratorum summa cubus.* $x + y = rr$. $xx + yy = s^3$. Sit $es = x$. $fs = y$. erit $es + fs = rr$ & $ee + ff = s$. et inde $e^3 + eef + eff + f^3 = rr$. Rursus (ut in Prob [5]) pone $pr = e$. $qr = f$ et erit $p^3r + ppqr + pqqr + q^3r = 1$. et $r = \dfrac{1}{p^3 + ppq + pqq + q^3}$.[85]

[*Prob. 21.*][86] *Invenire cubum qui subductus de latere suo relinquit cubum.* $x - x^3 = y^3$. Sit $y = e - x$. erit $x = e^3 - 3eex + 3exx$. Pone $x = 3exx$. sive $\dfrac{1}{3x} = e$. Erit etiam $e^3 - 3eex = 0$ sive $e = 3x$. Quare $\dfrac{1}{3x} = 3x$. $x = \frac{1}{3}$.[87]

[*Prob. 22.*] *Invenire duos numeros qui additi cubo summæ efficiunt alios duos cubos.*[88]

---

(81) Newton first continued 'xy et ejus rectanguli divisores x et y' (xy [will be some divisor of $\frac{1}{5}a/e - \frac{1}{5}e^4$] and x, y the divisors of that product).

(82) Understand 'congruentes' or some equivalent.

(83) These two problems, too loosely general to offer intellectual excitement, have no forerunners in Diophantus' *Arithmetica* and are not listed in Dickson's *History*.

(84) The problem is taken from Diophantus' *Arithmetica*, IV, 10, but the present method is different. In effect Diophantus set $x = (2-z)/v$, $y = z/v$ and hence reduced the problem to solving $4 - 6z + 3z^2 = v^2$; he then set $v = |2-4z|$, hence deriving $z = \frac{10}{13}$ and so $v = \frac{14}{13}$, $x = \frac{8}{7}$ and $y = \frac{5}{7}$. This solution is obtained by setting $e = 5$ in Newton's parametrization.

(85) More simply, since $(e^2 - f^2)s = (e-f)r^2$ where $s = e^2 + f^2$, or $e^4 - f^4 = (e-f)r^2$, at once $r(p^4 - q^4) = p - q$. A century later Euler examined this problem by a different approach (see his *Opera Postuma* (note (19) above), 1: 255–6; compare Dickson's *History*, II (note (7) above): 599): his solution $x = 29601$, $y = 25624$ ($r = 235$, $s = 1153$) does not follow easily

(some divisor of a) and there will be $x^3y - x^2y^2 + xy^3 = \frac{1}{5}a/e - \frac{1}{5}e^4$. Hence[81] y will be some divisor of $\frac{1}{5}a/e - \frac{1}{5}e^4$ and x, that is, $e+y$, will divide the quotient. Test therefore if there are any divisors to be found [which agree] with these conditions. If there are none, you will further attempt solution of the problem to no purpose. The same technique extends indefinitely to higher powers.

Problem 17. To find a square and a cube whose sum or difference is a fourth power.

Suppose $x^2 + y^3 = z^4$. Let $x = ez$, $y = fz$ and then $e^2 + f^3z = z^2$. Furthermore let $e = gz$, when $g^2z + f^3 = z$ and $z = f^3/(1-g^2)$.[83]

Problem 18. To find a square and a cube whose sum or difference is a fifth power.

Suppose $x^2 + y^3 = z^5$. Let $x = ez$, $y = fz$ and then $e^2 + f^3z = z^3$. Let $e = gz$ and then $g^2z + f^3 = z^2$. Let $f = mg$ and $z = ng$, when $ng + m^3g = n^2$ and

$$g = n^2/(n+m^3).^{(83)}$$

Problem 19. To find two cubes whose sum equals the sum of their sides or difference the difference.

Say $x^3 + y^3 = x + y$. Divide through by $x+y$ and there will be $x^2 - xy + y^2 = 1$. Put $x = z+1$ and then $z^2 + 2z - yz - y + y^2 = 0$. Put $y = ez$, when

$$z + 2 - ez - e + e^2z = 0 \quad \text{and} \quad z = (e-2)/(1-e+e^2).^{(84)}$$

Problem 20. To find two numbers, whose sum is a square and the sum of whose squares is a cube.

Take $x+y = r^2$, $x^2 + y^2 = s^3$. Let $x = es$, $y = fs$, so that $es + fs = r^2$ and $e^2 + f^2 = s$, and thence $e^3 + e^2f + ef^2 + f^3 = r^2$. Again (as in Problem 5) put $e = pr$, $f = qr$ and then $p^3r + p^2qr + pq^2r + q^3r = 1$, or

$$r = 1/(p^3 + p^2q + pq^2 + q^3).^{(85)}$$

Problem 21.[86] To find a cube which, when taken from its side, leaves a cube.

Say $x - x^3 = y^3$. Let $y = e - x$ and then $x = e^3 - 3e^2x + 3ex^2$. Put $x = 3ex^2$, that is, $e = 1/3x$, and there will also be $e^3 - 3e^2x = 0$ or $e = 3x$. Consequently $1/3x = 3x$ and so $x = \frac{1}{3}$.[87]

Problem 22. To find two numbers which, when added to the cube of their sum, make up two other cubes.[88]

---

from Newton's parametrization. The trivial solution $x = y = r = s = 2$ arises from setting $p = q = \frac{1}{2}$ in the latter.

(86) Newton first enunciated this problem in the complementary form 'Invenire numerum qui subductus de cubo suo relinquit cubum. $x^3 - x = y^3$' (To find a number which, when taken from its cube, leaves a cube. [Say] $x^3 - x = y^3$) but at once cancelled.

(87) Yielding $y = \frac{2}{3}$. Effectively, Newton here sets $y = 1/3x - x$. The solution is unique since, as Euler subsequently showed in his 'Theorematum quorundam Arithmeticorum Demonstrationes' (note (16) above): 145, the equation $(y/x)^3 + 1 = (1/x)^2$ is satisfied in rationals only by $y/x = 2$, $1/x = 3$.

(88) Newton would seem to have been the first to consider this particular case of Diophantus' three-cube problem worthy of special attention.

Sint x et y numeri, z summa eorum. Requiritur ut sit $z^3+x=r^3$ & $z^3+y=s^3$. Adde. erit $2z^3+x+y=r^3+s^3$. hoc est $2z^3+z=r^3+s^3$. Pone $ez=r$, $fz=s$. erit $2zz+1=e^3zz+f^3zz$. sive $zz=\dfrac{1}{e^3+f^3-2}$. Sit $e^3+f^3-2=vv$.[89] Pone $e=m+p$. $f=n-p$. erit $m^3+3mmp+3mpp+n^3-3nnp+3npp-2=vv$. Pone $3m+3n=qq$. et $v=pq+t$. Erit $m^3+3mmp+n^3-3nnp-2=2pqt+tt$. Et $p=\dfrac{m^3+n^3-2-tt}{2qt+3nn-3mm}$.

Assume ergo q et m & fac $\dfrac{qq}{3}-m=n$. Dein assume etiam t et habebis p, unde cætera omnia determinantur.

Proponi solet hocce Diophanti[90] Problema de tribus numeris et fastigium hujus generis Problematum habetur. [91]Ab alijs solvitur per fortuitam assumptionem numerorum, antequam ad conclusionem pervenitur: quod minus probamus. Potest autem de tribus vel pluribus numeris sic proponi.

[*Prob. 23.*] *Invenire tres vel plures numeros qui additi cubo summæ omnium, efficient alios totidem cubos quorum latera sunt in arithmetica progressione.*[92] Sint numeri $x.\ y.\ v.$ summa eorum z. Requiritur ut sit $z^3+x=r^3$. $z^3+y=s^3$. $z^3+v=t^3$. et r, s, t in Arithm. progressione. Ergo $3z^3+z=r^3+s^3+t^3$. Sit $ez=r$. $fz=s$, $gz=t$. Erit $3zz+1=e^3zz+f^3zz+g^3zz$. & $zz=\dfrac{1}{e^3+f^3+g^3-3}$. Sit $e^3+f^3+g^3-3=kk$. et cum numeri r, s, t, adeoꝗ et e,f,g sint in arithmetica progressione[,] ut hoc fiat pone $f-p=e$ & $f+p=g$ et erit $3f^3+6fpp-3=kk$. Pone $6f=cc$ et $k=cp+h$ et erit $3f^3-3=2cph+hh$ et $p=\dfrac{3f^3-3-hh}{2ch}$. Assume ergo c et habebis f, tunc assume etiam h et habebis p. Cætera patunt. Ut si assumas $c=3$, erit $f=\dfrac{3}{2}$. Si assumas præterea $h=\dfrac{3}{2}$ erit $p=\dfrac{13}{24}$. $k=\dfrac{25}{8}$. $z\left(=\dfrac{1}{k}\right)=\dfrac{8}{25}$. $r=\dfrac{23}{75}$. $s=\dfrac{36}{75}$. $t=\dfrac{49}{25}$. [93]

NB   *Generalius autem sic solvitur.* Sint numeri quotcunꝗ inveniendi a, b, c, d &c, eorum summa z, & numerus n. Sitꝗ $z^3+a=t^3$. $z^3+b=v^3$. $z^3+c=w^3$. $z^3+d=x$[3] &c et erit $nz^3+z=t^3+v^3+w^3+x^3$ &c. Sit $ez=t$, $fz=v$, $gz=w$, $hz=x$ &c et erit $nzz+1=e^3zz+f^3zz+g^3zz+h^3zz$ &c & $zz=\dfrac{1}{e^3+f^3+g^3+h^3\ \&c-n}$. Pone

(89) On putting $z = 1/v$, that is.

(90) See note (37) above.

(91) Newton first continued 'Nec tamen satis artificiosa solutio' (But its solution has not been sufficiently skilful).

(92) A Newtonian refinement of Diophantus' *Arithmetica*, v, 18 [Bachet]: the restriction of the problem in this way seems to have no precedent, other than its enunciation in Newton's draft Problem 15 (see note (46)).

(93) Whence $x = -\dfrac{1657}{421875}$, $y = \dfrac{1216}{15625}$ and $v = \dfrac{4153}{16875}$.

Let the numbers be x and y, with z their sum. It is required that there be $z^3+x = r^3$ and $z^3+y = s^3$. Add these and there will be $2z^3+x+y = r^3+s^3$, that is, $2z^3+z = r^3+s^3$. Put $r = ez$, $s = fz$ and then

$$2z^2+1 = e^3z^2+f^3z^2 \quad \text{or} \quad z^2 = 1/(e^3+f^3-2).$$

Let e^3+f^3-2 equal $v^{2\,(89)}$ and put $e = m+p, f = n-p$, giving

$$m^3+3m^2p+3mp^2+n^3-3n^2p+3np^2-2 = v^2.$$

Set $3m+3n = q^2$ with $v = pq+t$. There will be

$$m^3+3m^2p+n^3-3n^2p-2 = 2pqt+t^2$$

and $p = (m^3+n^3-2-t^2)/(2qt+3n^2-3m^2)$. So take any q and m and make $\frac{1}{3}q^2-m = n$. Next, also assume t and you will have p, and from these all the remainder is determined.

This problem of Diophantus[90] is usually proposed for three numbers and is considered to be the pinnacle of achievement in this type of problem.[91] By others it is solved by a fortuitous choice of numbers as a preliminary to attaining the result: this we less approve of. It can be proposed for three or more numbers in this way.

Problem 23. To find three or more numbers which, when added to the cube of their total sum, shall make an equal number of other cubes whose sides are in arithmetical progression.[92]

Let the numbers be x, y and v, their sum z. It is required that there be $z^3+x = r^3, z^3+y = s^3$ and $z^3+v = t^3$, where r, s, t are in arithmetical progression, and consequently $3z^3+z = r^3+s^3+t^3$. Let $r = ez$, $s = fz$, $t = gz$, and then $3z^2+1 = e^3z^2+f^3z^2+g^3z^2$, so that $z^2 = 1/(e^3+f^3+g^3-3)$. Let

$$e^3+f^3+g^3-3 = k^2.$$

Then, since the numbers r, s, t and so also e, f, g are in arithmetical progression, that this be so put $e = f-p$, $g = f+p$ and there will be $3f^3+6fp^2-3 = k^2$. Put $6f = c^2$ and $k = cp+h$, and there will follow $3f^3-3 = 2cph+h^2$ and so

$$p = (3f^3-3-h^2)/2ch.$$

Therefore take any c and you will have f, then assume h also and you will have p. The rest is evident. Should you, for instance, take $c = 3$, there will be $f = \frac{3}{2}$. Moreover, if you assume $h = \frac{3}{2}$, then $p = \frac{13}{24}$, $k = \frac{25}{8}$, z (or $1/k$) $= \frac{8}{25}$, $r = \frac{23}{75}$, $s = \frac{36}{75}$ and $t = \frac{49}{25}$.[93]

N.B. This is solved still more generally as follows. Let there be however many numbers $a, b, c, d, \ldots$ to be found, their sum z and their number n. Let $z^3+a = t^3$, $z^3+b = v^3$, $z^3+c = w^3$, $z^3+d = x^3$, $\ldots$, and there will be

$$nz^3+z = t^3+v^3+w^3+x^3 \ldots.$$

Let $t = ez$, $v = fz$, $w = gz$, $x = hz$, $\ldots$, and then

$$nz^2+1 = e^3z^2+f^3z^2+g^3z^2+h^3z^2 \ldots,$$

$e^3 + f^3 + g^3 + h^3$ &c $-n = kk$, $e = p - m$, $f = q + m$ et erit

$$p^3 - 3ppm + 3pmm + q^3 + 3qqm + 3qmm + g^3 + h^3 \text{ \&c } -n = kk.$$

Pone $3p + 3q = rr$ & $k = rm + s$ et erit

$$p^3 - 3ppm + q^3 + 3qqm + g^3 + h^3 \text{ \&c } -n = 2rms + ss.$$

Et $m = \dfrac{p^3 + q^3 + g^3 + h^3 \text{ \&c } -n - ss}{2rs - 3pp + 3qq}$. Assume ergo r et p. Fac $q = \dfrac{rr}{3} - p$. Dein

assume etiam $g\,h$ & s et habebis m. Tunc pone $rm + s = k.\dfrac{1}{k} = z$ &c.[94]

Potest etiam Problema ita solvi ut termini t, v, w, x &c sint in Geometrica vel alia quavis progressione. Assumatur enim progressio quævis in numeris. Sit m semisumma terminorum et ipsa progressio $m - \alpha$. $m - \beta$. $m + \gamma$. $m + \delta$ &c. Ubi æquationem $e^3 + f^3 + g^3 + h^3$ &c $-n = kk$ obtinuisti pone $e = ml - \alpha l$, $f = ml - \beta l$, $g = ml + \gamma l$, $h = ml + \delta l$ &c. $[-]3\alpha l[-]3\beta l + 3\gamma l + 3\delta l$ &c $= rr$, $k = rml + s$ et emerget (deletis æqualibus) $nm^3 l^3 - 3mm\alpha l^3 - 3mm\beta l^3 + 3mm\gamma l^3 + 3mm\delta l^3$ &c $-n = 2rmls + ss$. Et

Prob [24]. *Idem facere ea lege ut quantitates t, v, w, x &c sint in Arithmetica progressione.*[96]

Prob [25]. *Triangula duo*[97] *æquicrura invenire quorum areæ et perimetri æquales erunt.* Sint triangula *ABC, DEF.* bases $AB = 2p$, $DE = 2x$. crura[98] $AC = q$, $DF = y$. Et erit $p + q = x + y$ &

$$p\sqrt{qq - pp} = x\sqrt{yy - xx}$$

sive $ppqq - p^4 = xxyy - x^4$. Divide per

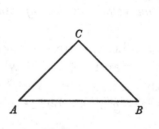

(94) An insignificant generalization of Diophantus' three-cube problem, resolved by the method employed in Problem 15 of [1].

(95) This gloomy conclusion was inserted by Newton when he broke off the (cancelled) accompanying text in despair, nor is it altogether accurate. Thus in the case of Diophantus' problem (*Arithmetica*, v, 18 [Bachet]), given that $x + y + v = z$, the equations $z^3 + x = r^3$, $z^3 + y = s^3$ and $z^3 + v = t^3$ (so that $e^3 + f^3 + g^3 - 3 = 1/z^2$ where $e = r/z, f = s/z$ and $g = t/z$) may trivially be satisfied by r, s, t in geometrical progression by setting $e = u^2$ and $f = 1$ (or $g = 1/u^2$), when $z = 1/(u^3 - u^{-3})$, $r = u^2/(u^3 - u^{-3})$, $s = 1/(u^3 - u^{-3})$, $t = u^{-2}/(u^3 - u^{-3})$ and so $\quad x = u^3/(u^3 - u^{-3})^2$, $\quad y = 0$, $\quad v = -u^{-3}/(u^3 - u^{-3})^2$.

(96) This unimplemented generalization of Diophantus' problem, in its restricted Newtonian form, was cautiously inserted when the ambitious preceding attempt proved fruitless.

(97) This replaces 'quotcunqȝ' (any number of). As we have seen (note (5)) Viète had established in his *Zetetica* in 1593 that the two right triangles of bases $z^2 - x^2$, $z^2 - y^2$ and altitudes $2xz$, $2yz$ (and so of respective hypotenuses $z^2 + x^2$, $z^2 + y^2$) are equal in area when $x^2 + xy + y^2 = z^2$. Evidently, if the two right triangles are rotated round their altitudes to become the pair of isosceles triangles of bases $2(z^2 - x^2)$ and $2(z^2 - y^2)$ respectively, these will not only have equal area $2xz(z^2 - x^2) = 2yz(z^2 - y^2)$ but also equal perimeter $4z^2$. According to Mariotte, in a report made to the Paris *Académie des Sciences* in July 1668 on John Pell's newly published English version (*An Introduction to Algebra, translated out of the High-Dutch...*

so that $z^2 = 1/(e^3 + f^3 + g^3 + h^3 \ldots - n)$. Put $e^3 + f^3 + g^3 + h^3 \ldots - n = k^2$ with $e = p - m$, $f = q + m$, and there will result

$$p^3 - 3p^2m + 3pm^2 + q^3 + 3q^2m + 3qm^2 + g^3 + h^3 \ldots - n = k^2.$$

Put $3p + 3q = r^2$ and $k = rm + s$, and there will be

$$p^3 - 3p^2m + q^3 + 3q^2m + g^3 + h^3 \ldots - n = 2rms + s^2,$$

so that $m = (p^3 + q^3 + g^3 + h^3 \ldots - n - s^2)/(2rs - 3p^2 + 3q^2)$. Take therefore any r and p, and make $q = \frac{1}{3}r^2 - p$. Next, assume also g, h and s, and you will have m. Then put $rm + s = k$, $1/k = z$, and so on.[94]

The problem can also be solved under the limitation that the terms $t, v, w, x, \ldots$ are in geometrical or any other progression. For let any arbitrary rational progression be assumed. Suppose m is the half-sum of its terms and that the progression itself is $m - \alpha$, $m - \beta$, $m + \gamma$, $m + \delta$, …. When you have obtained the equation $e^3 + f^3 + g^3 + h^3 \ldots - n = k^2$, put $e = l(m - \alpha)$, $f = l(m - \beta)$, $g = l(m + \gamma)$, $h = l(m + \delta)$, …, $(-3\alpha - 3\beta + 3\gamma + 3\delta \ldots) l = r^2$ and $k = lmr + s$. There will then emerge (after equals are deleted)

There is no solution unless they were in arithmetical progression.[95]

$$l^3 m^3 n - 3\alpha l^3 m^2 - 3\beta l^3 m^2 + 3\gamma l^3 m^2 + 3\delta l^3 m^2 \ldots - n = 2lmrs + s^2.$$

And…

Problem 24. To accomplish the same with the restriction that the quantities $t, v, w, x, \ldots$ shall be in arithmetical progression.[96]

Problem 25. To find two[97] isosceles triangles which shall be equal in area and perimeter.

Let the triangles be ABC, DEF; their bases $AB = 2p$, $DE = 2x$; their slanting sides[98] $AC = q$, $DF = y$. Then $p + q = x + y$ and $p\sqrt{[q^2 - p^2]} = x\sqrt{[y^2 - x^2]}$, that

(London, 1668)) of J. H. Rahn's *Teutsche Algebra*, this extension of Viète's *Zeteticum*, XI, was first proposed publicly by Roberval in 1633 in the form 'trouver en nombres deux triangles isoscèles, isopérimètres et égaux'. Roberval's solution to the problem is not known, but presumably it did not differ significantly from Viète's: Mariotte himself, to be sure, gave the latter's parametrization $x = A^2 - B^2$, $y = 2AB + B^2$, $z = A^2 + AB + B^2$ in his commentary on the problem. (See Pierre Costabel, 'Sur un problème de Roberval et un cas particulier d'Analyse Diophantienne', *Revue d'Histoire des Sciences*, **2**, 1948: 80–8; and J. E. Hofmann, 'À propos d'un problème de Roberval…. Les sources du problème', *Revue d'Histoire des Sciences*, **3**, 1950: 312–18. Compare also Costabel's 'Réflexions sur la méthode de Viète', *ibid.*: 319–26; and Hofmann's 'Note complémentaire', *ibid.*: 326–33.) Soon after, Descartes gave a new solution of the problem and this, published in 1657 by Schooten in his *Exercitationes Mathematicæ* (note (1) above): Liber V, Sectio XII. 'Solutio Problematis, quod anno 1633 Parisiis palàm fuit propositum, qualem adinvenit…R. des Cartes': 432–4, was annotated by Newton in 1664 (see I: 56). Rahn's attention was, no doubt, likewise drawn to the problem by reading Schooten's version of Descartes' solution, but his own discussion (*Teutsche Algebra* (Zurich, 1659): 116–19) deals, in a numerical instance only, with the allied problem of constructing an isosceles triangle equal in area and perimeter to a given one (see note (99)). The detail of Newton's present approach to the problem is essentially his own but it is interesting to observe that, in this and Problem 26 following, he embraces the three variant solutions then known.

(98) Literally 'legs', the two sides which straddle the vertex angle. The manuscript lacks any accompanying figure and that reproduced is our restoration.

æquales $p+q$ & $x+y$ et erit $ppq-p^3=xxy-x^3$. Pro y scribe $p+q-x$ et erit $ppq-p^3=pxx+qxx-2x^3$. et $q=\dfrac{p^3+pxx-2x^3}{pp-xx}$. Divide per $p-x$. fit

$$q=\frac{pp+px+2xx}{p+x}. \quad ^{(99)}$$

Prob [26]. *Triangula ijsdem conditionibus invenire ut perpendicula sint etiam rationalia.*

[Sit] $\sqrt{qq-pp}=q-e$. [erit] $-pp=-2qe+ee$. $\dfrac{ee+pp}{2e}=q^{(100)}=\dfrac{pp+px+2xx}{p+x}$.

$$p^3 \begin{array}{l} +ppx+pee+xee \\ -2epp-2epx-4exx \end{array}=0.$$

Et $p^3-2epp+eep+xpp-2exp+eex=4exx$. Divide omnia per $p+x$ et extracta utrobiᵩ radice erit $-p+e=2x\sqrt{\dfrac{e}{p+x}}$. Pone $\sqrt{\dfrac{e}{p+x}}=v$. Erit $-p+e=2vx$. pro e scribe $pvv+xvv$. erit $pvv+xvv-p=2xv$ et $p=\dfrac{2xv-xvv}{vv-1}$. $^{(101)}$

Idem aliter. Sit $[\frac{1}{2}]AB=aa-bb$. $AC=aa+bb$. et perpendic $=2ab$. Item $[\frac{1}{2}]DE=rr-ss$. $DF=rr+ss$ et perpendic $=2rs$. Erit $[\frac{1}{2}]AB+AC=[\frac{1}{2}]DE+DF_{[,]}$ id est $2aa=2rr$ et $a=r$. Item$^{(102)}$ $aa-bb\times2ab=rr-ss\times2rs=aa-ss\times2as$ et $aab-aas=b^3-s^3$. Divide per $b-s$. erit $aa=bb+bs+ss$. Pone $a=s+t$. erit $2st+tt=bb+bs$. et $s=\dfrac{bb-tt}{2t-b}$. $^{(103)}$

His addantur duo vel tria exempla$^{(104)}$ in quæstionibus de loco plano solidove & similibus ad ostendendum qua ratione quantitates ubi quæstio aliqua ad exitum perducitur in numeris illustrari poteri[n]t.$^{(105)}$

---

(99) This last sentence was added only when, halfway through penning the next problem, Newton saw the factor $p-x$ common to numerator and denominator; compare note (100). The solution $x=p$ is, of course, trivial. In his *Algebra* (see note (97)) Rahn gave a slightly variant reduction: in effect, on setting $p+q=x+y=k$, so that $q^2-p^2=k(k-2p)$ and $y^2-x^2=k(k-2x)$, there results $p^2(k-2p)-x^2(k-2x)=0$ or, on eliminating the factor $x-p$, $2(x^2+px+p^2)-k(x+p)=0$ with $q=k-p$. In Rahn's example $p=12$, $q=37$ and so $k=49$, yielding $2x^2-25x-300=0$ or $x=20$, $y=29$.

(100) Newton first continued $\Big{'}=\dfrac{p^3+pxx-2x^3}{pp-xx}\cdot p^4-2ep^3\begin{array}{l}+eepp\\-xxpp\end{array}-2exxp\begin{array}{l}-eexx\\+4ex^3\end{array}=0\,\Big{'}$ and then, evidently noticing the common factor $p-x$ for the first time, hurriedly amended his argument. Compare note (99). The altitude $\sqrt{[q^2-p^2]}=(p^2-e^2)/2e$ and hypotenuse $q=(p^2+e^2)/2e$ could have been evaluated at once by Problem 7 [= Problem 3 of Newton's preceding draft].

(101) This is basically the Cartesian solution (see note (97)) but Newton completes it somewhat differently. Descartes (compare 1: 56) chooses the equivalent parametrization $p=2ab$, $q=a^2+b^2$ (or $e=2b^2$) and then sets $x=2kd$, $y=k^2+d^2$ and so

$$\sqrt{[y^2-x^2]}=k^2-d^2:$$

is, $p^2q^2 - p^4 = x^2y^2 - x^4$. Divide by the equals $p+q$ and $x+y$ and there will be $p^2q - p^3 = x^2y - x^3$. In place of y write $p+q-x$ and there follows

$$p^2q - p^3 = px^2 + qx^2 - 2x^3 \quad \text{and so} \quad q = (p^3 + px^2 - 2x^3)/(p^2 - x^2).$$

Divide through by $p-x$ and there comes $q = (p^2 + px + 2x^2)/(p+x)$.[99]

Problem 26. To find triangles subject to the same conditions so that their perpendiculars also are rational.

Let $\sqrt{[q^2 - p^2]} = q - e$ and there will be $-p^2 = -2qe + e^2$, so that

$$(e^2 + p^2)/2e = q^{(100)} = (p^2 + px + 2x^2)/(p+x).$$

Hence $p^3 + (x - 2e)\,p^2 + (e^2 - 2ex)\,p + e^2x - 4ex^2 = 0$ and so

$$(p+x)\,(p^2 - 2ep + e^2) = 4ex^2.$$

Divide throughout by $p+x$ and, on extracting the root of either side, there will be $-p+e = 2x\sqrt{[e/(p+x)]}$. Put $\sqrt{[e/(p+x)]} = v$ and there will be $-p+e = 2vx$. In place of e write $pv^2 + xv^2$ and there will result $pv^2 + xv^2 - p = 2xv$ and so $p = (2xv - xv^2)/(v^2 - 1)$.[101]

The same another way. Let $\tfrac{1}{2}AB = a^2 - b^2$, $AC = a^2 + b^2$ and the perpendicular will then be equal to $2ab$. Likewise let $\tfrac{1}{2}DE = r^2 - s^2$, $DF = r^2 + s^2$ and the perpendicular will equal $2rs$. Then $\tfrac{1}{2}AB + AC = \tfrac{1}{2}DE + DF$, that is, $2a^2 = 2r^2$ and so $a = r$. Similarly[102] $(a^2 - b^2) \times 2ab = (r^2 - s^2) \times 2rs = (a^2 - s^2) \times 2as$ and so $a^2b - a^2s = b^3 - s^3$. Divide through by $b - s$ and there will be $a^2 = b^2 + bs + s^2$. Put $a = s + t$ and there will result $2st + t^2 = b^2 + bs$ and so

$$s = (b^2 - t^2)/(2t - b).$$[103]

To these let there be added two or three examples[104] in questions having to do with plane or solid loci and similar topics—this to show by what procedure, after some question is brought to a conclusion, its quantities may be illustrated in rationals.[105]

---

from the equality of the perimeters $2(p+q) = 2(x+y)$ there results $a+b = k+d$, so Descartes puts $k = a+z$, $d = b-z$, when the areal equality $2ab(a^2 - b^2) = 2kd(k^2 - d^2)$ yields (on eliminating the factor $4(a+b)z$) the condition $z^2 + \tfrac{3}{2}(a-b)\,z + \tfrac{1}{2}(a^2 - 4ab + b^2) = 0$ or

$$4z + 3(a-b) = \sqrt{[a^2 + 14ab + b^2]} = a+b+c$$

where $a = (2b+c)\,c/(12b - 2c)$.

   (102) Because the areas are equal.

   (103) The Viète–Roberval solution (see notes (5) and (97)). The equation $a^2 = b^2 + bs + s^2$ is resolved by the technique (Diophantus') used in Problem 1.

   (104) 'Problemata' (problems) is cancelled.

   (105) A private note by Newton to himself which, it would appear, he never implemented. We may guess that the 'quæstiones' referred to would concern the inscription of rational polygons in 'plane loci' (circles)—more generally, in 'solid loci' (conics) or other curves of higher degree—of rational diameter, subject to a given further restriction (perhaps the

Nota autem quod Prob 6 sic solvi debet. *Inveni[r]e 4 ÷÷ quorum extremi sunt cubi et summa omnium cubus.* $x^3 + xxy + xyy + y^3 = z^3$. $z = q + y$.

$$x^3 + xxy + xyy = q^3 + 3qqy + 3qyy.$$

Pone $x = 3q$. Erit $27q + 9y = q + 3y$. et $y = -\dfrac{13q}{3}$. Ut si ponatur $q = -3$ erit $y = 13$, $x = -9$ & $z = 10$. Sed cum progressionis hic inventæ termini[106] sint alternatim negativi, quomodo ad affirmativos perveniatur etiam docebo. In æquatione prima pro x scribe $r + s$ et $p + q$ pro z. Et orietur

$$\begin{aligned} r^3 + 3rrs + 3rss \quad + s^3 &= p^3 + 3ppq + 3pqq + q^3. \\ + rry + 2rsy + ssy & \\ + ryy + syy & \\ + y^3 & \end{aligned}$$

Jam cum termini $s^3 + ssy + syy + y^3$ sint in geometrica progressione pro s scribe $-\frac{13}{9}y$ et summa terminorum istorum cubus erit. Pro q scribe $[-]\frac{10}{9}[y]$ et q^3 æqualis erit huic cubo. Quare termini isti se mutuo destruent. Pone insuper terminos $3rss + 2rsy + ryy = 3pqq$ et hi etiam destru[e]ntur et ex his et cæteris quæ restant colligentur p r et y rationales.[107] Unde simul datur x atꝗ adeo progressio quæsita.

---

stipulation that their area or their circum-radius be rational). In the case of the rational cyclic quadrilateral of sides a, b, c, d with diagonals e, f, area A (and so circum-diameter D) each rational, the conditions to fulfil are that

$$e = \sqrt{\frac{(ac + bd)(ad + bc)}{ab + cd}}, \quad f = \sqrt{\frac{(ac + bd)(ab + cd)}{ad + bc}}$$

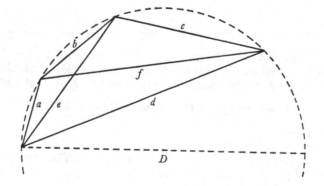

and $A = \frac{1}{2}(ab + cd)e/D = \frac{1}{4}\sqrt{[(-a + b + c + d)(a - b + c + d)(a + b - c + d)(a + b + c - d)]}$ shall all be rational when, say, a, b, c and d are integers. It will evidently be simplest to choose D as the common hypotenuse of a trio of rational right (Heronian) triangles, one side of each of which is (say) a, e and d. Here Newton could conceivably have known that Simon Jacob had already,

Note, too, that Problem 6 ought to be solved this way: *To find four continued proportionals whose extremes are cubes and whose sum total is a cube.* Say

$$x^3 + x^2y + xy^2 + y^3 = z^3$$

or, with $z = q+y$, $x^3 + x^2y + xy^2 = q^3 + 3q^2y + 3qy^2$. Put $x = 3q$, then

$$27q + 9y = q + 3y \quad \text{and so} \quad y = -\tfrac{13}{3}q.$$

If, for instance, q is set equal to -3, then $y = 13$, $x = -9$ and $z = 10$. But since the terms of the progression here found[106] are alternately negative, I will explain a method of attaining positive ones also. In the first equation write $r+s$ in place of x and $p+q$ for z, and there will arise

$$r^3 + r^2(3s+y) + r(3s^2 + 2sy + y^2) + s^3 + s^2y + sy^2 + y^3 = p^3 + 3p^2q + 3pq^2 + q^3.$$

Now since the terms $s^3 + s^2y + sy^2 + y^3$ are in geometrical progression, in place of s write $-\tfrac{13}{9}y$ and the sum of those terms will be a cube. In place of q write $-\tfrac{10}{9}y$ and q^3 will be equal to this cube. These terms will accordingly cancel each other. In addition, set the terms $r(3s^2 + 2sy + y^2)$ and $3pq^2$ equal to each other, and these too will be eliminated. Then from these and the others remaining will be gathered rational values of p, r and y.[107] As an immediate consequence x, and hence the progression required, is given.

---

a century before, set it as a problem in his posthumously published *Rechenbuch* that the cyclic quadrilateral of sides 16, 25, 33 and 60 has diagonals of length 39 and 52, its circum-diameter 65 and hence its area 714: the foundation for this question was, no doubt, Jacob's prior derivation of the trio of Pythagorean triples 16, 63, 65; 25, 60, 65; and 39, 52, 65. (See his *Ein New und Wolgegrund Rechenbuch auff den Linien uñ Ziffern* (Frankfurt, ₁1560) [= ₃1600: 309ʳ]; also G. Eneström, 'Über die ältere Geschichte des Sehnenvierecks', *Bibliotheca Mathematica*, ₃**12** (1911/12): 84.) Newton's extant papers contain only one example of this sort: specifically, the particular case in which the side d of the cyclic quadrilateral coincides with the circum-diameter D, when $e = \sqrt{[d^2 - c^2]}$ and $f = \sqrt{[d^2 - a^2]}$, so that $(d^2 - a^2)(d^2 - c^2) = (ac + bd)^2$ and hence $d^3 - (a^2 + b^2 + c^2)d - 2abc = 0$. Thus, on noticing that $7^2 + 24^2 = 25^2 = 15^2 + 20^2$, it will be enough to set $a = 7$, $b = c = 15$ and $d = 25$ to obtain a rational solution. As we shall see in the next volume, Newton, at about the time he wrote the present manuscript, made this quadrilateral a main worked example of 'Quomodo Quæstiones Geometricæ ad æquationem redigantur' in his Lucasian lectures on algebra of 'Octob. 1675' (ULC. Dd. 9.67: 55–65 [= *Arithmetica Universalis*; *sive De Compositione et Resolutione Arithmetica* (Cambridge, ₁1707): 100–13]). For the subsequent history of this type of problem compare Dickson's *History*, II, (note (7) above): 220–1.

(106) Namely, -729, $+1053$, -1521 and $+2197$.

(107) Newton uses his previous result to equate $s^3 + s^2y + sy^2 + y^3 = q^3$ by setting $s/y = -\tfrac{13}{9}$, $q/y = -\tfrac{10}{9}$: with this reduction his further equation $r(3s^2 + 2sy + y^2) = 3pq^2$ yields $p/r = \tfrac{59}{50}$, whence the remaining equation $r^3 + (3s+y)r^2 = p^3 + 3p^2q$ or $r^3 - \tfrac{10}{3}r^2y = p^3 - \tfrac{10}{3}p^2y$, that is, (since $p \neq r$) $p^2 + pr + r^2 = \tfrac{10}{3}(p+r)y$, gives (on replacing p by its equal $\tfrac{59}{50}r$ and rejecting the trivial solution $r = 0$) $r/y = \tfrac{59500}{26793}$ and consequently $p/y = \tfrac{70210}{26793}$, $x/y = \tfrac{2311}{2977}$, $z/y = \tfrac{13480}{8931}$. Obviously any positive rational value of y will yield corresponding positive rational values of x and z which solve the problem.

§2. THE GENERATION OF RATIONAL SOLUTIONS
FROM GIVEN INSTANCES.

From the original in the University Library, Cambridge[1]

DE RESOLUTIONE QUÆSTIONUM CIRCA NUMEROS.

Primo numeri quæsiti redigendi sunt ad æquationem secundum conditiones quæstionis, Deinde exponendi sunt per basem et ordinatam curvæ lineæ quam

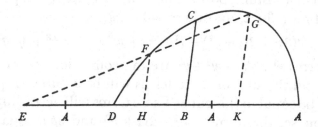

æquatio illa designabit.[2] Sit curva ista *DC* et numeri *AB*, *BC*, curva existente tali ut numerus *BC* ordinatim applicatus ad numerum *AB* in dato angulo *ABC* semper terminetur ad eam. Deinde inquirenda erunt puncta curvæ quæ efficient numeros *AB*, *BC* rationales. Casus autem in quibus hoc fieri deprehendo sunt sequentes.

1. Si numeri in æquatione non ultra gradum quadraticū ascendant ita ut curva sit Conica sectio: et detur aliquod punctum *F* in curva, quo efficitur ut

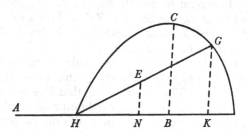

numeri *AH*, *HF* sint rationales[,] ex hoc unico exemplo regula generalis sic elicietur. In *AH* cape *HE* cujusvis rationalis longitudinis, age *EF*, secet hæc curvam in *G* et demissa *G*[*K*] parallela[3] *CB* numeri *AK*, *KG* erunt rationales.[4]

---

(1) ULC. Add. 4004: 91ᵛ. In this page of his Waste Book, evidently composed at the same period as the preceding 'Numerical problems', Newton introduces for the first time—in its geometrical guise, at least—a general technique for finding rational solutions to non-homogeneous Diophantine equations in two variables (or, equivalently, to corresponding homogeneous ones in three variables). His general principle is all but self-evident: a straight line passing through *n* − 1 rational points of an arbitrary *n*-th degree curve defined by the Cartesian

Translation

THE RESOLUTION OF QUESTIONS REGARDING NUMBERS

First, the numbers required are to be reduced to an equation in agreement with the defining conditions of the question, and then they must be displayed as the base and ordinate of the curve which that equation shall denote.[2] Let that curve be *DC* and the numbers *AB*, *BC*, the curve being such that the number *BC* ordinately applied at a given angle *ABC* to the number *AB* is ever terminated at it. Then search must be made for the points of the curve which will yield rational numbers *AB*, *BC*. The cases in which this happens are, I detect, the following.

1. If the numbers in the equation do not rise above the quadratic degree, so that the curve is a conic: Let there be given some point *F* on the curve which renders the numbers *AH*, *HF* rational, and then from this single example a general rule will be thus evoked. In *AH* take *HE* of any rational length, draw *EF* and let it meet the curve in *G*: then, when *GK* is let fall parallel to *CB*, the numbers *AK*, *KG* will be rational.[4] If the point *F* is located in the line *AH*, *FH*

---

equation $f(x, y) = 0$ (or by the equivalent homogeneous equation $f(x, y, z) = 0$) will meet it in a further rational point. The technique is, as Newton well realizes, particularly effective for conics and cubics, since in the former case any chord through a rational point on a conic which has rational slope will intersect it again in a rational point, while in the cubic case a line through any two rational points upon it—which it is often convenient to choose to be coincident (namely, when the given line is tangent or through a double point)—will likewise have its third point of intersection with the cubic rational. We need not be surprised to find that this 'chord and tangent' method was already, in analytical equivalent, widely employed by Diophantus in his *Arithmetica*: this is forcefully underlined by Isabelle Bachmakova in her comparison of 'Diophante et Fermat' (*Revue d'Histoire des Sciences*, **19** (1966): 289–306, especially 200–8) and we ourselves have also noticed above Viète's use of the technique in his *Zetetica* of 1593 (compare §1: notes (19) and (49)). The procedure was subsequently employed with profit by Euler and, in the cubic case, was discussed systematically by Lagrange in 1777 (compare L. E. Dickson, *History of the Theory of Numbers. II*: *Diophantine Analysis* (Washington, 1920): 595, 660–1; see also L. J. Mordell, 'On the Rational Solutions of the Indeterminate Equations of Third and Fourth Degrees', *Proceedings of the Cambridge Philosophical Society*, **21** (1922): 179–92). In the present paper Newton gives no examples of the method, but we have indicated instances of it in the preceding section (see §1: notes (49), (74) and (76)) and further illustrate it in following footnotes with problems taken from Diophantus' *Arithmetica*.

(2) In a general Cartesian coordinate system, that is.

(3) Newton's text has the dittograph 'pararallela'.

(4) In proof, take *A* to be the origin and the conic *DCA* to be defined by

$$ax^2 + bxy + cy^2 + dx + ey + f = 0.$$

It will then readily be proved that the general line $y = m(x-r) + s$, of rational slope m and through the rational point $F(r, s)$ on the conic, intersects it again in the point $G(X, Y)$ where $a(X+r) + b(mX+s) + cm(m(X-r) + 2s) + d + em = 0$ and $Y = m(X-r) + s$.

Si punctum *F* reperitur in linea *AH*, *FH* existente nulla, tunc cape *HN* cujusvis rationalis longitudinis, erige *NE* parallelam *BC*, & cujusvis etiam rationalis longitudinis. Age *HE* occurrentem curvæ in *G* et erunt *AK*, *KG* rationales.[5]

Quoniam hic unicum saltem exemplum requiritur, primo inquirendum est ejusmodi exemplum, dein regula generalis inde elicietur ut supra. [5]Exempli autem inveniendi[6] hæc erit methodus ni melior occurrat. Sit æquatio

$$axx + bxy + cyy + dx + ey + f = 0.$$

ubi *x* et *y* designant numeros. Et si terminus *f* desit, punctum istud[7] erit *A*. Si *x* in æquatione *axx + dx + f* = 0 sit rationale puncta duo erunt in *AB*.[8] Si [*y*] in *cyy + ey + f*[= 0] sit rationale puncta duo erunt in recta[9] quæ ducitur ab *A* parallela *BC*. Si *bb − 4ac* sit quadratus numerus affirmativus tunc curva erit Hyperbola[10] et recta ducta a puncto *A* vel *B* parallela Asymptoto secans curvam in *G* exhibebit numeros *AK*[,] *GK* rationales.[11] Sunt alij innumeri casus quibus numerandis non immoror.

2. Si æquatio ascendit ad tres dimensiones, et tria habentur exempla rationalia non in arithmetica progressione[,][12] possunt inde innumera alia reperiri. Sint enim *P*, *Q*, *R* puncta Curv[æ] ad ista exempla.[13] Junge *PR*, *RQ*, *PQ* et punct[a] *S*, *T*, *Y*[14] ubi *PR*, *RQ*, *PQ* secant curvam dabunt alios tres

---

(5) As an illustration, Diophantus' problem of constructing Pythagorean triples (*Arithmetica*, II, 8; compare §1: note (7)) may be posed in the form: to construct the rational point *G(X, Y)* on the semicircle (*C*): $x^2 + y^2 = k^2$. Evidently the line $y = kx - a$ of rational slope *EA/HA = k* through the rational point *H*(0, −*a*) will determine the general parametrization

$$X/a = 2k/(k^2 + 1), \quad Y/a = (k^2 - 1)/(k^2 + 1).$$

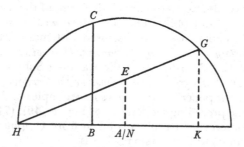

Correspondingly, the homogeneous equation $x^2 + xy + y^2 = z^2$ (*Arithmetica*, v, 7 [Bachet]; compare §1: note (5)) may be solved in rationals by finding rational points either on the ellipse $x^2 + xy + y^2 = 1$ (through the points (±1, 0) and (0, ±1)) or on the hyperbola $x^2 + x + 1 = y^2$ (through the point (0, ±1)).

(6) Newton first began 'De exemplo autem inveniendo notandum est [quod]' (For finding an example, however, it should be noted that).

being zero, then take *HN* of any rational length, erect *NE* parallel to *BC* and also of any rational length, draw *HE* meeting the curve in *G*, and *AK*, *KG* will then be rational.[5]

Seeing that a minimum of one example is here required, a pertinent example must then be sought for, and then a general rule will be elicited from it as above. For finding an example, however,[6] the procedure will be this, unless a better one happens along. Let the equation be $ax^2 + bxy + cy^2 + dx + ey + f = 0$, in which x and y denote the numbers. Then if the term f should be missing, *A* will be that point.[7] If x in the equation $ax^2 + dx + f = 0$ be rational, there will be two points in *AB*;[8] and if y in $cy^2 + ey + f = 0$ be rational, there will be two points in the straight line[9] which is drawn from *A* parallel to *BC*. Should $b^2 - 4ac$ be a positive square, then the curve will be a hyperbola[10] and the straight line drawn from the point *A* or *B* parallel to an asymptote, intersecting the curve in *G*, will furnish rational numbers *AK*, *GK*.[11] There are other countless cases, but I do not linger to enumerate them.

2. If the equation rises to three dimensions, and three rational instances are had which are not in arithmetical progression,[12] innumerable further ones may be ascertained from them. For let *P*, *Q* and *R* be the points on the curve [corresponding] to those instances. Join *PR*, *RQ*, *PQ* and the points *S*, *T*, *Y*[14]

---

(7) The point *E* in the preceding figure.

(8) Namely, the line $y = 0$ whose intersections with the conic are determined by

$$ax^2 + dx + f = 0.$$

(9) Of Cartesian equation $x = 0$, the intersections being given by $cy^2 + ey + f = 0$.

(10) Evidently if, to use anachronistic terminology, the discriminant $b^2 - 4ac$ is positive, we may then put $k = \sqrt{[b^2 - 4ac]}$, $l = (bd - 2ae)/k$ and hence write the conic's defining equation in the hyperbolic form

$$(2ax + (b+k)y + d + l)(2ax + (b-k)y + d - l) = d^2 - l^2 - 4af.$$

(11) Here the hyperbola's asymptote has rational slope $-2a/(b \pm \sqrt{[b^2 - 4ac]})$. This particular approach fails, for instance, in the case of the 'Fermatian' equation $ax^2 - y^2 = b$, a non-square (Diophantus' *Arithmetica*, VI, 15), where we must require the given point, (c, d) say, to be on the hyperbola: at once the line through (c, d) of rational slope k will meet the hyperbola again in the second rational point $(c + \lambda, d + k\lambda)$, $\lambda = 2(ac - kd)/(k^2 - a)$ if only a rational point (c, d) exists, that is, if, given the rationals a and b, a rational pair x, y can be found which satisfies $ax^2 - y^2 = b$. (In Diophantus' problem $a = 3$, $b = 11$ and the rational solution $x = 5$, $y = 8$ yields, on putting $k = -2$, the second rational solution $x = 67$, $y = 116$.)

(12) Understand the rational pairs x_1, y_1; x_2, y_2; x_3, y_3 to be in 'arithmetical progression' when $(x_1 - x_2):(x_2 - x_3) = (y_1 - y_2):(y_2 - y_3)$, that is, when the Cartesian points (x_1, y_1), (x_2, y_2), (x_3, y_3) are in the same straight line.

(13) 'correspondentes' or some equivalent is to be understood.

(14) In Newton's figure the point *Y* is located, impossibly so, on the serpentine central portion of the cubic. For the purist an accurate figure is given in the English version.

numeros. Dein junge QS et punctum X quo QS secat curvam dabit alium numerum. Et sic in infinitum.[15]

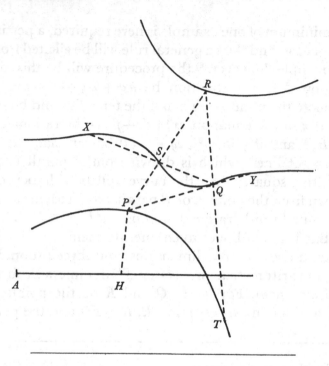

(15) We may illustrate Newton's bare paragraph with two companion problems from Diophantus' *Arithmetica*. In IV, 25, Diophantus requires rationals a, b such that $a(b+1)$ and $(a+1)b$ are each perfect cubes: on putting $a = x^2-1$, $b = k^3x$ the problem reduces to finding rational solutions x, y of $(x^2-1)(k^3x+1) = y^3$, k free but rational. This is the Cartesian defining equation of a family of unicursal cubics, all through the rational points $(0, -1)$ and $(1, 0)$,

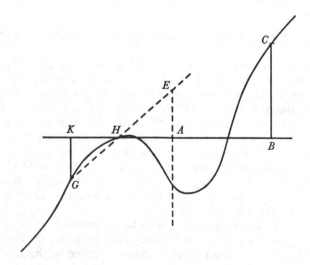

in which PR, RQ, PQ cut the curve will yield three other numbers. Next, join QS and the point X in which QS cuts the curve will give another number. And so on indefinitely.[15]

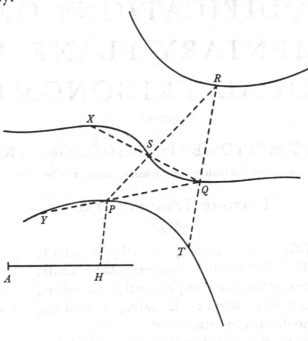

which snake round their corresponding asymptotes $y = kx + \frac{1}{3}k^{-2}$. Diophantus, in analytical equivalent, chooses the general line $y = px - 1$ through $(0, -1)$ and then puts $p = k$, so making it parallel to the asymptote (of rational slope k) and so through the second rational point at infinity in that direction; accordingly, it meets the cubic in the second (finite) rational point (X, Y), where $(X^2 - 1)(k^3X + 1) = (kX - 1)^3$ or $X = k(k^2 + 3)/(3k^2 + 1)$ and hence $Y = (k^4 - 1)/(3k^2 + 1)$. Further sets of rational points on the cubic may then be found by constructing lines through $(0, -1)$, $(1, 0)$, (X, Y) and their further intersections with the cubic since the three base points are not in 'arithmetical progression'. In Diophantus' preceding problem (IV, 24) is illustrated the case in which the two base points P, Q are coincident and so Y is the third meet of the cubic with the tangent to it at P. Here the problem is, after reduction, to satisfy $x(a - x) = y^3 - y$ in rationals: in the geometrical equivalent to Diophantus' argument the tangent $y = \frac{1}{2}ax - 1$ to the cubic at the rational point $(0, -1)$ is drawn, so determining the further rational point $(2a^{-3}(3a^2 - 4), 2a^{-2}(a^2 - 2))$. (A single given rational point is not, in general, sufficient to generate all rational points on an arbitrary cubic by this method but L. J. Mordell has shown that a finite number of such generators are adequate for this purpose; see the article by him quoted in note (1).)

3

CODIFICATIONS OF ELEMENTARY PLANE AND SPHERICAL TRIGONOMETRY

[*c.* 1683–4][1]

§1. THE 'EPITOME OF TRIGONOMETRY'.[2]

From the original draft in the University Library, Cambridge[3]

EPITOME TRIGONOMETRIÆ.

Notæ.

In triangulo ABC; A, B, C anguli; AB, BC, AC latera; A', B', C', AB', AC', BC' eorum complementa; s sinus; t tangens[;] s', t' sinus et tangens complementi;[4] R Radius; Ba Basis; Cr Crura;[5] ∠ angulus. L Latus; Z summa; X differentia; μ medium proportionale.[6]

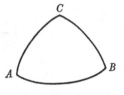

R Radius. s sinus. t tangens. ' Complementum. s', t' sinus et tangens complementi.[4] ∠ angulus. L latus. Cr crura.[5] Ba basis.[7] ∞ medium proportionale. = æqualitas. :: proportionalitas. ∥ parallelum. ⊥ perpendic.[8]

---

(1) Of the three papers here reproduced the first, the unfinished 'Epitome Trigonometriæ', is firmly dated as 'given in the year 1683' in Henry Wharton's transcript (Appendix 1 below), while the 'Compendium' which follows is undoubtedly its immediate revision: the third piece, Newton's additions to St John Hare's manuscript 'Trigonometria', is less easy to fix narrowly in time, but the existence of Humphrey Newton's fair copy of it gives a terminal date of 1689 for its composition (see §3: note (2)) and the style of Newton's handwriting in his autograph draft agrees well with a dating of *c.* 1684.

(2) A *précis* of the basic theorems of plane and spherical trigonometry, whose only essential originality lies in its use in Proposition 15 (see note (24)) of the conformal property of stereographic projection to derive 'Lambert's' half-angle formula for the oblique spherical triangle. In the general structure of his exposition and for much of his detail Newton leans heavily on Seth Ward's long popular but extremely condensed undergraduate primer (*Idea Trigonometriæ Demonstratæ* (*In usum Juventutis Oxoniensis*), Oxford, 1654), an unmarked copy of which (now Trinity College, Cambridge, NQ. 8.71) was in his library at his death: at the same time, however, Newton introduces several modifications both of notation and in the form of his proofs. Like Ward before him, his purpose in making the present compilation was no doubt to introduce the students who frequented his professorial lectures as painlessly as possible to the elements of trigonometry—a conjecture fully confirmed by the fact that Henry Wharton transcribed a somewhat augmented version of its first state among other undergraduate notes

Translation

AN EPITOME OF TRIGONOMETRY
Symbols

In the triangle ABC: A, B, C are angles, AB, BC, CA the sides, and A', B', C', AB', BC', CA' their complements; s denotes sine, t tangent, s' and t' the complementary sine and complementary tangent;[4] R is the radius, Ba base, Cr *crura*;[5] $\angle$ denotes an angle, S a side, Z sum, X difference and μ mean proportional.[6]

R is the radius; s sine, t tangent, ' complement, s' and t' the complementary sine and the complementary tangent;[4] $\angle$ angle, S side, Cr *crura*,[5] Ba base;[7] ∞ the mean proportional, = equality, :: proportionality, $\parallel$ parallel, $\perp$ perpendicular.[8]

---

on mathematics made during the summer of 1683 (see note (3)). As such, this 'Epitome' will offer little of interest to the student of Newton's advanced researches.

(3) ULC. Add. 3959.5: 47^r/47^v. This single folio sheet was manifestly penned in two stages. The original composition (on f. 47^r) comprehended the first fourteen propositions of the 'Epitome' only, though several more may perhaps have been added on a second sheet which is now lost. A slightly amplified Newtonian fair copy (no longer in existence) of this first state was transcribed soon afterwards by Henry Wharton, then a third-year undergraduate at Caius College, among other notes on elementary geometry and trigonometry compiled in summer 1683 (now in Lambeth Palace, Codex H. Wharton 592): comparison of this transcript (reproduced *in toto* as Appendix 1 below) with the original text given here will make clear the limits of Newton's first summary. When he came subsequently to draft his more comprehensive 'Compendium' of trigonometry (§2 following) many of his first revisions for the piece were entered on the verso of his manuscript of the 'Epitome' but this attempt to improve the content of the latter without substantially altering its structure was soon abandoned, leaving its revise incomplete.

(4) By 'sine', 'tangent', 'cosine' and 'cotangent' here understand, in modern equivalent, the product of these and the radius (namely 'R sin', 'R tan', 'R cos' and 'R cot' respectively, which we 'translate' as 'Sin', 'Tan', 'Cos' and 'Cot').

(5) The two upright sides which stand 'legs apart' on the base: thus where AB is base, AC and CB are the *crura*.

(6) These symbols are copied from Ward's *Idea* (note (2) above): 1 with the exception of this last notation for 'medium proportionale': Newton's 'μ' (which will at once be replaced by '∞' in the following revise) would appear to have no exact antecedent, but it is evidently a variant of Oughtred's symbol 'm' or 'M' (for 'medium'). Compare Florian Cajori's *A History of Mathematical Notations* (La Salle, Illinois, 1928): 1, 193, 194; 2, 161–2.

(7) Newton first continued 'V angulus verticalis. Z summa. X differentia. μ medium proportionale' (V 'vertical angle', Z 'sum', X 'difference' and μ 'mean proportional'). The choice of 'V' for 'vertex angle' (in line with Oughtred's following prenex capitals for + and −) is probably a Newtonian idiosyncracy which, mercifully, was discarded as quickly as it was invented.

(8) In our English version we retain Newton's symbols for the most part, but collapse ' = ' and ' :: ' into ' = ' and discard 'S', 'Cr' and 'Ba' completely. We have also replaced instances of 'μ' by Newton's preferred symbol '∞' in our modern rendering.

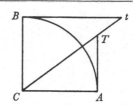

Lemma.[9]

t. R::R. t'. Nam $AT.AC$::$BC.Bt$.

Propositiones.

1. In triangulo rectilineo ABC rectangulo

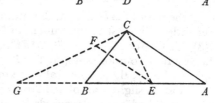

$$BC.CA::R.sB. \quad \text{Prop 1.}$$

$$BA.CA::R.tB. \quad \text{Prop 2.}^{[10]}$$

2. In triangulo quovis rectilineo ABC
Sit $CD \perp BA$, & erit $BC.CD$::$R.sB$ (Pr. 1) &

$$CD.CA::sA.R \text{ (Pr 1)}.$$

Ergo ex æquo $BC.CA$::$sA.sB$. Prop 3.

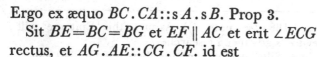

Sit $BE=BC=BG$ et $EF \parallel AC$ et erit $\angle ECG$
rectus, et $AG.AE$::$CG.CF$. id est

$$AB+BC.AB-BC::t\tfrac{1}{2}CBG^{[11]}.t\frac{C-A}{2}.^{[12]}$$

Prop 4.

Sit $BE = BC$. $KE.CE$::1.2::$EF.EA$::$KF.CA$. MLN, EPA semicirculi
centro F radijs FK, FE descripti. [Erit]

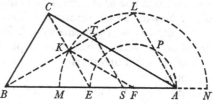

$$APL \parallel CE \perp BKL.^{[13]} \quad BN=\frac{BA+BC+AC}{2}.$$

$$BM=\frac{BA+BC-AC}{2}.$$

[Sit] $BM\mu BN=BK\mu BL=BT$.

$$BA\mu BC=BA\mu BE=BS.$$

[Erit] $BS.BT$::$R.s'\dfrac{B}{2}.^{[14]}$ Prop 5.

---

(9) Understand 'Lemma 1'. In Propositions 17 and 19 in his revised version of the 'Epitome' (see notes (28) and (29) below) Newton in fact appeals to two further lemmas which he does not explicitly cite. Here he might well have added the Pythagorean lemma

$$\text{'}s^2+s\text{'}^2=R^2\text{'}.$$

(10) It is surprising to find these two basic definitions set as 'propositions' (*sc.* to be proved) but Newton, it would appear, has simply copied Ward's two 'Fundamenta pro Solutione Rectangulorum [Triangulorum]' (*Idea* (note (2) above): 2) without appreciating their logical status.

(11) That is, '$\dfrac{C+A}{2}$'.

Lemma[9]

Tan: $R = R$: Cot. For $AT:AC = BC:Bt$.

Propositions

1. In the rectilinear right-angled triangle ABC:

$$BC:CA = R:\mathrm{Sin}\,B. \quad \textit{Proposition 1.}$$

$$BA:CA = R:\mathrm{Tan}\,B. \quad \textit{Proposition 2.}[10]$$

2. In any rectilinear triangle ABC:
Let there be $CD \perp BA$, and then (by Proposition 1)

$$BC:CD = R:\mathrm{Sin}\,B \quad \text{and} \quad CD:CA = \mathrm{Sin}\,A:R.$$

Therefore, *ex æquo* $BC:CA = \mathrm{Sin}\,A:\mathrm{Sin}\,B$. *Proposition 3.*
Let $BE = BC = BG$ and $EF \parallel AC$ and $\widehat{ECG}$ will be right. Then

$$AG:AE = CG:CF,$$

that is,

$$AB + BC:AB - BC = \mathrm{Tan}\,\tfrac{1}{2}\widehat{CBG}[11]:\mathrm{Tan}\,\tfrac{1}{2}(C - A).[12] \quad \textit{Proposition 4.}$$

Let $BE = BC$ and $KE:CE = 1:2 = EF:EA = KF:CA$, then, on drawing the semicircles MLN, EPA with centre F and radii FK, FE, there will be

$$APL \parallel CE \perp BKL,[13]$$

and so $BN = \tfrac{1}{2}(BA + BC + AC)$, $BM = \tfrac{1}{2}(BA + BC - AC)$. Let

$$BT = BM \infty BN = BK \infty BL \quad \text{and} \quad BS = BA \infty BC = BA \infty BE,$$

then $BS:BT = R:\mathrm{Cos}\,\tfrac{1}{2}B$.[14] *Proposition 5.*

---

(12) The half-tangent rule for plane triangles. Newton's proof is not, as we might expect, taken from Ward's *Idea*, but is the 'necessary Demonstration of our friend Mr *Robert Anderson*, not hitherto published by any' which Thomas Street set out in great detail in his *Astronomia Carolina. A New Theory of the Cœlestial Motions* (London, 1661): 62–4.

(13) Evidently, since BKL is (by construction) drawn perpendicular to CKE, the line KF will meet the outer circle in a second point Q such that LQ is parallel to KE and so passes through the point A: whence AL and EK are each perpendicular to BKL.

(14) Since $BS^2 = BE \times BA$ and $BT^2 = BK \times BL$, while

$$BE^2:BK^2 = BA^2:BL^2 = BE \times BA:BK \times BL$$

(because EK and AL are parallel), at once $BE:BK = BA:BL = BS:BT$ and so ST is parallel to EK, AL, that is, perpendicular to BKL. In modern terms, on taking $AB = c$, $BC = a$, $CA = b$ with $s = \tfrac{1}{2}(a+b+c)$, $BT/BS = \cos\tfrac{1}{2}B = \sqrt{[s(s-b)/ca]}$.

Item $NA = \dfrac{AC+BC-BA}{2}$. $NE = \dfrac{AC-BC+BA}{2}$. [adeoqȝ quia]

$$NA \,\mu\, NE = LA \,\mu\, LP = LA \,\mu\, KE = ST.$$

[fit] $BS \, . \, ST :: R \, . \, s\dfrac{B}{2}.$ [15]   Prop 6.

3. In triangulis sphæricis ABC, DBE rectangulis ad A et E,

Designent DF, DA, DC, CF, CE quadrantes in octava ꝑte sphæræ cujus centrum C, radij CD, CE &c.[16] Et erit $CE \, . \, EH :: CB \, . \, BL$. et

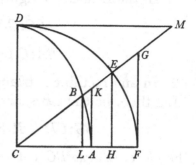

$$CF \, . \, FG :: CA \, . \, AK,$$

seu $CD \, . \, DM :: AK \, . \, CA$. [Id est]

    In tri ABC; $R \, . \, s \, C :: s \, CB \, . \, s \, AB$.   Prop 7.

      $R \, . \, t \, C :: s \, CA \, . \, t \, AB$ seu $R \, . \, t' \, C :: t \, AB \, . \, s \, CA$.   Prop 8.

    Et in tri DEB $R \, . \, s' \, DE :: s' \, BE \, . \, s' \, DB$.   Prop 9.

      $R \, . \, t' \, DE :: s' \, D \, . \, t' \, DB$ seu $R \, . \, t \, DE :: t' \, DB \, . \, s' \, D$.   Prop 10.

    Quare in tri DEB, $R \, . \, s \, B :: s \, DB \, . \, s \, DE$ (per Prop 7) &

$$R \, . \, t' \, B :: t \, DE \, . \, s \, BE.$$

Hoc est in tri ABC,

    $R \, . \, s \, B :: s' \, AB \, . \, s' \, C$.   Prop 11.

    $R \, . \, t' \, B :: t' \, C \, . \, s' \, CB$.   Prop 12.

---

(15) Evidently, if (as in note (13)) Q is the second meet of KF with the outer circle, then at once $LP = AQ = KR$, while $NA \times AM$ (or $NA \times NE$) $= LA \times AQ$. In the terminology of the previous note, Newton's result is $ST/BS = \sin\frac{1}{2}B = \sqrt{[(s-c)(s-a)/ca]}$. As an immediate corollary '$BT \, . \, ST :: R \, . \, t\dfrac{B}{2}$' (that is, in modern equivalent,

$$ST/BT = \tan\tfrac{1}{2}B = \sqrt{[(s-c)(s-a)/s(s-b)]}).$$

(16) The orthogonal projection by which the spherical triangles ABC, DBE are 'designated' in the diametral plane $DEFG$ is clarified in the oblique perspective added in our English version. For this approach Newton is narrowly indebted to Seth Ward's *Idea* (note (2) above): 8, but Ward in turn merely summarizes an equivalent passage in Henry Briggs' posthumous *Trigonometria Britannica*: *Sive De Doctrina Triangulorum Libri Duo* (Gouda, 1633): Liber II, Pars II, Caput II: 78–9. The employment of orthogonal projection in spherical trigonometry is, no doubt, anticipated on the back plates of a wide variety of medieval astrolabes, while its application to the theoretical solution of right spherical triangles was already made in the mid-fifteenth century by Regiomontanus (Johann Müller); see his *De Triangulis Omnimodis Libri Quinque* (Nuremberg, 1533): Liber IV, §XXXIIII: 122–5 [= B. Hughes, *Regiomontanus On Triangles* (Madison, Wisconsin, 1967): 260–7].

Likewise $NA = \frac{1}{2}(AC+BC-BA)$ and $NE = \frac{1}{2}(AC-BC+BA)$, and so, since $ST = NA \backsim NE = LA \backsim LP = LA \backsim KE$, there results $BS:ST = R:\operatorname{Sin}\frac{1}{2}B.$[15] *Proposition 6.*

3. In the spherical triangles ABC, DBE right-angled at A and E:

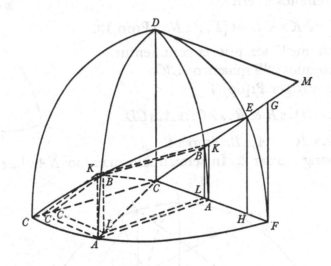

Let DF, DA, DC, CF, CE denote quadrants in the eighth part (octant) of a sphere whose centre is C and radii CD, CE, ...[16] There will then be

$$CE:EH = CB:BL, \quad \text{and} \quad CF:FG = CA:AK,$$

or $CD:DM = AK:CA$. That is, in the triangle ABC

$$R:\operatorname{Sin} C = \operatorname{Sin} CB:\operatorname{Sin} AB, \quad \textit{Proposition 7.}$$

$$R:\operatorname{Tan} C = \operatorname{Sin} CA:\operatorname{Tan} AB, \quad \text{or equivalently}$$

$$R:\operatorname{Cot} C = \operatorname{Tan} AB:\operatorname{Sin} CA; \quad \textit{Proposition 8.}$$

while in the triangle DEB

$$R:\operatorname{Cos} DE = \operatorname{Cos} BE:\operatorname{Cos} DB, \quad \textit{Proposition 9.}$$

$$R:\operatorname{Cot} DE = \operatorname{Cos} D:\operatorname{Cot} DB, \quad \text{or equivalently}$$

$$R:\operatorname{Tan} DE = \operatorname{Cot} DB:\operatorname{Cos} D. \quad \textit{Proposition 10.}$$

Consequently, in the triangle DEB (by Proposition 7)

$$R:\operatorname{Sin} B = \operatorname{Sin} DB:\operatorname{Sin} DE \quad \text{and} \quad R:\operatorname{Cot} B = \operatorname{Tan} DE:\operatorname{sin} BE;$$

that is, in the triangle ABC

$$R:\operatorname{Sin} B = \operatorname{Cos} AB:\operatorname{Cos} C, \quad \textit{Proposition 11.}$$

$$R:\operatorname{Cot} B = \operatorname{Cot} C:\operatorname{Cos} CB. \quad \textit{Proposition 12.}$$

Ergo ex his quinqȝ terminis, nimirum cruribus duobus anguli recti, et complementis tum hypotenusæ tū angulorum duorum ad hypotenusam designet M terminum quemlibet[,] L et N terminos duos utrinqȝ proximos, K et O terminos duos reliquos utrinqȝ remotos et erit

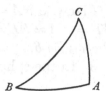

$$R \times s\, M = s'\, K \times s'\, O = t\, [L] \times t\, N. \quad \text{Prop 13.}$$

Qua una propositione[17] sex priores continentur.

4. In triangulo quovis[18] sphærico ABC

Sit $CD \perp AB$ et erit per Pr[op] 7

$$s\, BC \,.\, R :: s\, CD \,.\, s\, B \ \& \ R \,.\, s\, AC :: s\, A \,.\, s\, CD.$$

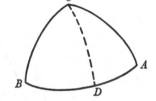

Et ex æquo $s\, BC \,.\, s\, AC :: s\, A \,.\, s\, B.$ Prop 14.

[19]Sit ABC triang. sphær & In semicirculo maximo $NA\nu$ bisecto in A, sit

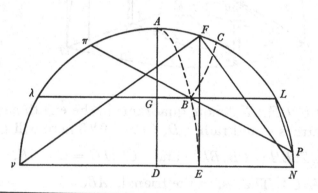

---

(17) John Napier's rule for circular parts, here presented in a slightly modified form based manifestly on Ward's *Idea* (note (2) above): 10–12: 'Quinqȝ circularium partium, tres semper in quæstionem cadunt; quarum due dantur, tertia quæritur. Harum una est media, due extremæ $\begin{cases} \text{conjunctæ.} \\ \text{disjunctæ.} \end{cases}$

'Axioma I. ubi extremæ sunt à medio disjunctæ. R. s' [extrem]æ unius :: s' [extrem]æ alterius. s, partis mediæ....

'Axioma II. ubi extremæ sunt medio conjunctæ. R. t' [extrem]æ unius :: t, [extrem]æ alterius. s, partis mediæ....' These two 'axioms' are narrowly based in turn on the two equivalent ones previously published in Briggs' *Trigonometria Britannica* (note (16) above): Liber II, Pars II, Caput II: 78. Napier himself in his original 'Porisma generale' (*Logarithmorum Canonis descriptio, Ejusque usus, in utraque Trigonometria* (Edinburgh, 1614): Liber II, Caput IV 'De simplicibus Quadrantalibus': 30–5, especially 35) had stated his rule in less immediately comprehensible form as 'Logarithmus intermediæ æquatur differentialibus vicinarum (circumpositarū), seu antilogarithmis remotarum (oppositarū) extremarū': here, on putting $L_N x$ for Napier's logarithm of x (that is, in modern equivalent, $R \log[R/x]$), we should understand the 'logarithmus' to be $L_N \text{Sin}$, the 'antilogarithmus' to be $L_N \text{Cos}$ and their 'differentialis' to be $L_N \text{Tan}$. In the case of the spherical right triangle ABC of hypotenuse $\widehat{BC} = a$ Newton's present 'quinque partes circulares' K, L, M, N, O (in which M is Napier's 'pars intermedia', L and N are his 'partes extremæ quæ circumponuntur', K and O his 'partes extremæ quæ

Therefore of these five terms, namely: the two sides around the right angle, and the complements both of the hypotenuse and the two angles adjoining it, let M denote any one at random, L and N the two nearest terms on its either side, with K and O the two remaining terms farthest from it on either hand, and there will be

$$R \times \operatorname{Sin} M = \operatorname{Cos} K \times \operatorname{Cos} O = \operatorname{Tan} L \times \operatorname{Tan} N. \quad \textit{Proposition 13.}$$

In this single proposition[17] the previous six are contained.

4. In any[18] spherical triangle ABC:

Let there be $CD \perp AB$ and then (by Proposition 7)

$$\operatorname{Sin} BC : R = \operatorname{Sin} CD : \operatorname{Sin} B, \quad \text{while} \quad R : \operatorname{Sin} AC = \operatorname{Sin} A : \operatorname{Sin} CD,$$

so that *ex æquo* $\operatorname{Sin} BC : \operatorname{Sin} AC = \operatorname{Sin} A : \operatorname{Sin} B.$ *Proposition 14.*

[19]Let ABC be a spherical triangle, then in the great semicircle $NA\nu$, bisected

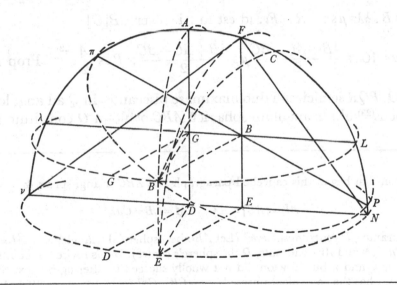

opponuntur') are, on employing his notation ' for 'complement', any cyclic permutation of b, 'C, 'a, 'B, c (namely, the *crus AC*, the complement of the angle at C, the complement of the hypotenuse, the complement of the angle at B, and the second *crus AB* respectively). The asymmetry here is only apparent for, as Napier remarked (*Descriptio*: 31–2) 'circularium partium uniformitas manifestissimè patet in rectangulis factis in superficie globi ex quinque circulis magnis' (those which, in the accompanying figure, form the self-polar spherical pentagon *ARSTU* generated by the triangle *ABC*, whose sides are, in order, the circular parts increased by a quadrant).

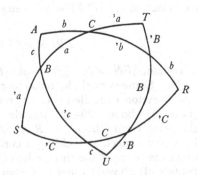

(18) 'obliquangulo' (oblique-angled) is cancelled. This unnecessary, conventional adjective was no doubt copied from Ward's *Idea* (note (2) above): 14.

$AL = AB = A\lambda$. $C[P] = CB = C[\pi]$.[20] $NF = NE = \angle BAC$. Et reliquis peractis ut in figura.

In tri BLP, s B . s $P :: \frac{1}{2}LP$. $\frac{1}{2}BL$. Item

$$GL \times s\,B \cdot \tfrac{1}{2}LP \times s\,P (= \tfrac{1}{2}BL \times s\,B) :: GL \cdot \tfrac{1}{2}BL :: R \cdot \tfrac{1}{2}EN$$

$$:: R^q \cdot R \times \tfrac{1}{2}EN = \tfrac{1}{2}FN \times \tfrac{1}{2}FN.$$

Ergo $GL\,\mu\,s\,B \cdot \frac{1}{2}LP\,\mu\,s\,P :: R \cdot \frac{1}{2}FN$. id est in tri sphær ABC

$$s\,AB\,\mu\,s\,AC \cdot s\,\frac{AC + CB - AB}{2}\,\mu\,s\,\frac{CB - AC + AB}{2} :: R \cdot s\,\frac{A}{2}. \quad ^{(21)} \quad \text{Prop 15.}$$

In tri $B\lambda\pi$, s B . s $\pi :: \frac{1}{2}\lambda\pi$. $\frac{1}{2}B\lambda$. Item

$$G\lambda \times s\,B \cdot \tfrac{1}{2}\lambda\pi \times s\,\pi(\tfrac{1}{2}B\lambda \times s\,B) :: G\lambda \cdot \tfrac{1}{2}B\lambda :: R \cdot \tfrac{1}{2}E\nu :: R^q \cdot R \times \tfrac{1}{2}E\nu = \frac{F\nu}{2} \times \frac{F\nu}{2}.$$

Ergo $G\lambda\,\mu\,s\,B \cdot \frac{1}{2}\lambda\pi\,\mu\,s\,P :: R \cdot \frac{1}{2}F\nu$. id est in tri sphær $AB[C]$

$$s\,AB\,\mu\,s\,AC \cdot s\,\frac{AB + AC - BC}{2}\,\mu\,s\,\frac{AB + AC + BC}{2} :: R \cdot s,\frac{A}{2}. \quad ^{(22)} \quad \text{Prop 16.}$$

Sint AQO, PQR semicirculi duo maximi se bisecantes in Q ad angulos rectos. Sitcჳ $\alpha\beta\gamma$ locus[23] ubi triangulum sphæricū ABC oculo in O constituto apparet.

---

(19) Newton first began this cancelled paragraph 'Sit ABC triang. sphær &

$$AL = [AB] = A\lambda. \quad CM = CB = C\mu.$$

AN, $A\nu$ Quadrantes, D centrum sphæræ' (Let ABC be a spherical triangle, with $AL = AB = A\lambda$, $CM = CB = C\mu$, AN and $A\nu$ quadrants, D the sphere's centre). In his revise the points M and μ are denoted by P and π but Newton did not wholly succeed in altering his text to suit this.

(20) The text has the equivalent phrase '$CM = CB = C\mu$' (compare note (19)). A following cancellation reads '$NF = NE = $ ang BAC. $\frac{1}{2}FN = $ s $BAC[!]$. $\frac{1}{2}F\nu = $ s $BAC[!]$.

$$\text{s } BLM = \text{s } \tfrac{1}{2}M\lambda = \text{s }\frac{AC + AB + BC}{2} \cdot \text{s }\frac{\mu L}{2} = \text{s } LMB'\text{:}$$

of course, $\frac{1}{2}FN = R - \frac{1}{2}F\nu = \mathrm{Sin}\frac{1}{2}\widehat{BAC}$.

(21) Some very shaky reasoning in deriving this Napierean result was no doubt the reason why Newton cancelled this attempted improvement on Ward's proof of this 'Axioma' (*Idea* (note (2) above): 20–1). Here the implicit equation s $B = $ s AC (that is, $\mathrm{Sin}\,B = \mathrm{Sin}\,AC$) and the proportion $GL:BL = (R$ or$)\,DN:EN$ are each a (complementary) *non-sequitur*, the errors in which cancel, so yielding a correct theorem. (This standard result, determining the half-angles in terms of the three sides of a spherical triangle, was deduced by Napier in his *Descriptio* (note (16) above): Liber II, Caput v: 48 as a corollary of a theorem by Regiomontanus (*De Triangulis Omnimodis* (note (16) above): Liber v, §II: 127–9), but was stated by him in the somewhat cumbrous form 'summa ex Logarithmis crurum subducta à summa ex Logarithmis

at A, let $AL = AB = A\lambda$, $CP = CB = C\pi$,[20] and $NF = NE = \widehat{BAC}$, with the rest completed as it is in the figure. Then, in the triangle BLP

$$\mathrm{Sin}\,B : \mathrm{Sin}\,P = \tfrac{1}{2}LP : \tfrac{1}{2}BL.$$

Likewise

$$GL \times \mathrm{Sin}\,B : \tfrac{1}{2}LP \times \mathrm{Sin}\,P \;(\text{or } \tfrac{1}{2}BL \times \mathrm{Sin}\,B) = GL : \tfrac{1}{2}BL = R : \tfrac{1}{2}EN,$$

that is, $R^2 : (R \times \tfrac{1}{2}EN \text{ or}) \; \tfrac{1}{2}FN \times \tfrac{1}{2}FN$. Therefore

$$GL \backsim \mathrm{Sin}\,B : \tfrac{1}{2}LP \backsim \mathrm{Sin}\,P = R : \tfrac{1}{2}FN,$$

and so, in the spherical triangle ABC,

$$\mathrm{Sin}\,AB \backsim \mathrm{Sin}\,AC : \mathrm{Sin}\,\tfrac{1}{2}(AC + CB - AB) \backsim \mathrm{Sin}\,\tfrac{1}{2}(CB - AC + AB)$$

$$= R : \mathrm{Sin}\,\tfrac{1}{2}A.\text{[21]} \quad \textit{Proposition 15.}$$

In the triangle $B\lambda\pi$

$$\mathrm{Sin}\,B : \mathrm{Sin}\,\pi = \tfrac{1}{2}\lambda\pi : \tfrac{1}{2}B\lambda.$$

Likewise

$$G\lambda \times \mathrm{Sin}\,B : \tfrac{1}{2}\lambda\pi \times \mathrm{Sin}\,\pi \;(\text{or } \tfrac{1}{2}B\lambda \times \mathrm{Sin}\,B) = G\lambda : \tfrac{1}{2}B\lambda = R : \tfrac{1}{2}E\nu,$$

that is, $R^2 : (R \times \tfrac{1}{2}E\nu) = \tfrac{1}{2}F\nu \times \tfrac{1}{2}F\nu$. Therefore

$$G\lambda \backsim \mathrm{Sin}\,B : \tfrac{1}{2}\lambda\pi \backsim \mathrm{Sin}\,P = R : \tfrac{1}{2}F\nu,$$

and so, in the spherical triangle ABC,

$$\mathrm{Sin}\,AB \backsim \mathrm{Sin}\,AC : \mathrm{Sin}\,\tfrac{1}{2}(AB + AC - BC) \backsim \mathrm{Sin}\,\tfrac{1}{2}(AB + AC + BC)$$

$$= R : \mathrm{Cos}\,\tfrac{1}{2}A.\text{[22]} \quad \textit{Proposition 16.}$$

Let AQO, PQR be two great semicircles bisecting each other at right angles in Q. Also let $\alpha\beta\gamma$ be the apparent position[23] of the spherical triangle ABC to an

---

aggregati & differentiæ semibasis & semidifferentiæ crurum, relinquit duplum Logarithmi dimidii anguli verticalis': that is, in the terminology of note (17),

$$L_N\mathrm{Sin}\,(\tfrac{1}{2}a + \tfrac{1}{2}(b-c)) + L_N\mathrm{Sin}\,(\tfrac{1}{2}a - \tfrac{1}{2}(b-c)) - [L_N\mathrm{Sin}\,b + L_N\mathrm{Sin}\,c] = 2L_N\mathrm{Sin}\,\tfrac{1}{2}A.$$

Ward in his *Idea* followed the more straightforward deduction presented by Henry Briggs in his *Trigonometria Britannica* (note (16) above): 82.)

(22) Much as before the errors in the implicit equation $sB = sAC$ and the proportion $G\lambda : B\lambda = (R \text{ or}) \; D\nu : E\nu$ cancel to yield a correct Napierean result. (See Napier's *Descriptio* (note (17) above): 48: 'Summa ex Logarithmis crurum subducta à summa ex Logarithmis aggregati & differentiæ semibasis & semiaggregati crurum, relinquit duplum antilogarithmi dimidii anguli verticalis'; that is, in the terminology introduced in note (17),

$$L_N\mathrm{Sin}\,(\tfrac{1}{2}(b+c) + \tfrac{1}{2}a) + L_N\mathrm{Sin}\,(\tfrac{1}{2}(b+c) - \tfrac{1}{2}a) - [L_N\mathrm{Sin}\,b + L_N\mathrm{Sin}\,c] = 2L_N\mathrm{Cos}\,\tfrac{1}{2}A.$$

Napier did not prove this immediate corollary of his preceding expression for $L_N\mathrm{Sin}\,\tfrac{1}{2}A$, contenting himself with the remark that '[hoc] alterius loci est demonstrare'.)

(23) Understand 'in plano PQR' (in the plane PQR).

Erit $\alpha\beta = t\dfrac{AB}{2}$; $\alpha\gamma = t\dfrac{AC}{2}$; $\angle\beta = \angle B$.

$\angle\gamma = \angle C,$[24] eorum differentia = differentiæ $\angle\angle\beta$, γ in tri rectilineo $\alpha\beta\gamma$.[25] Sed

$\alpha\beta + \alpha\gamma \, . \, \alpha\beta - \alpha\gamma :: t'\dfrac{\beta\alpha\gamma}{2}$[26] $. \, t\dfrac{\gamma - \beta}{2}$. Ergo

$t\dfrac{AB}{2} + t\dfrac{AC}{2} \, . \, t\dfrac{AB}{2} - t\dfrac{AC}{2} :: t'\dfrac{A}{2} \, . \, t\dfrac{C-B}{2}.$

Prop 15.

Id est[27] in tri DCO,

$t\dfrac{DO}{2} + t'\dfrac{OC}{2} \, . \, t\dfrac{DO}{2} - t'\dfrac{OC}{2} ::$

$t\dfrac{O}{2} \, . \, t\dfrac{C+D-2\,\mathrm{Rect}}{2} = t'\dfrac{C+D}{2} :: (\text{per Lem [1]}) \, t\dfrac{C+D}{2} \, . \, t'\dfrac{O}{2}.$ Prop 16.

(24) Since the projection of the spherical surface stereographically from the pole O onto the equatorial plane PQR, by which the spherical triangle ABC is mapped into the plane triangle $\alpha\beta\gamma$, is angle-preserving. Thomas Harriot had discovered (and demonstrated) this property some time after 1594 (see J. A. Lohne, 'Thomas Harriot als Mathematiker', *Centaurus*, **11** (1965): 19–45, especially 25–7; and J. V. Pepper, 'Harriot's Calculation of the Meridional Parts as Logarithmic Tangents', *Archive for History of Exact Sciences*, **4** (1968): 359–413, especially 411–12; compare also I: 475) but for another seventy years the writers who published upon the 'planisphere' projection (even François d'Aiguillon, who in his *Opticorum Libri Sex* (Antwerp, 1613): 572–637 both named it 'stereographic' and discussed its circle-preserving property at great length) remained ignorant of its conformality—a property easily proved when once suspected. In the middle 1670's, at last, word of the angle-preserving nature of the projection reached John Collins and in quick time was passed on to his acquaintances. (It may not be coincidence that at this time Collins was making great efforts to locate Harriot's unpublished mathematical papers, but the suggestive connection between the two is hard to validate.) By May 1675, in particular, Collins knew that the 'Log^{me} Curve or Spirall line' was 'no other than the Projection of the rumbe Spirall [loxodrome] on the Surface of the Earth, the Eye being at one Pole and projecting the same on a Plaine touching the Sphere at the other Pole' (Royal Society MS LXXXI, No. 39: 3) and later communicated the insight both to Leibniz (who in London in October 1676 noted that 'Rumbi in plano polū oppositum tangente projecti oculo in opposito polo existente dant lineā spiralem nauticam' [Hanover, Leibniz–Handschriften, **35**, VIII, 23: 1^v]) and to Newton (in his letter of 12 October 1678 [*Correspondence of Isaac Newton*, **2**, 1960: 286]; see page 658 below). It remained for Edmond Halley to publish a rigorous proof— essentially Harriot's—that 'In the *Stereographick Projection*, the Angles, under which the Circles intersect each other, are in all cases equal to the Spherical Angles they represent: ...this not being vulgarly known' ('An Easie Demonstration of the Analogy of the Logarithmick Tangents to the Meridian Line or sum of the Secants', *Philosophical Transactions*, **19**, No. 219 (for January/ February 1695/6): §IV: 202–14, especially 204), when he was careful to add that 'This *Lemma* I lately received from Mr. *Ab. de Moivre*, though I since understand from Dr. *Hook* that he long ago [in 1682?] produced the same thing before the *Society*' (*ibid.*: 205; compare Hooke's 1682 diary note (R. T. Gunther, *Early Science in Oxford*, **7** (Oxford, 1930): 601) deriving 'Descartes line' from the 'Rhum line on a cylinder projection').

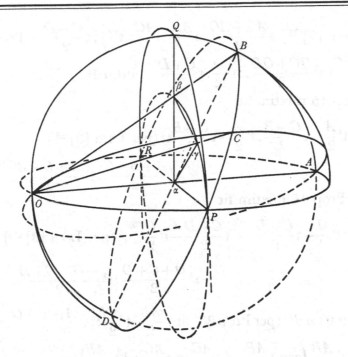

eye stationed at O. Then $\alpha\beta = \operatorname{Tan} \frac{1}{2}AB$, $\alpha\gamma = \operatorname{Tan} \frac{1}{2}AC$, $\hat{\beta} = \hat{B}$, $\gamma = \hat{C}$,[24] and the difference of the latter equals the difference of $\hat{\beta}$, $\hat{\gamma}$ in the rectilinear triangle $\alpha\beta\gamma$.[25] But $\alpha\beta + \alpha\gamma : \alpha\beta - \alpha\gamma = \operatorname{Cot} \frac{1}{2}\hat{\beta}\alpha\gamma$[26]$: \operatorname{Tan} \frac{1}{2}(\gamma - \beta)$ and therefore

$$\operatorname{Tan} \tfrac{1}{2}AB + \operatorname{Tan} \tfrac{1}{2}AC : \operatorname{Tan} \tfrac{1}{2}AB - \operatorname{Tan} \tfrac{1}{2}AC = \operatorname{Cot} \tfrac{1}{2}A : \operatorname{Tan} \tfrac{1}{2}(C - B).$$

Proposition 15.

That is,[27] in the triangle DCO,

$$\operatorname{Tan} \tfrac{1}{2}DO + \operatorname{Cot} \tfrac{1}{2}OC : \operatorname{Tan} \tfrac{1}{2}DO - \operatorname{Cot} \tfrac{1}{2}OC = \operatorname{Tan} \tfrac{1}{2}O : (\operatorname{Tan} \tfrac{1}{2}(C + D - 180^\circ) \text{ or})$$

$$\operatorname{Cot} \tfrac{1}{2}(C + D) = \text{(by Lemma [1])} \operatorname{Tan} \tfrac{1}{2}(C + D) : \operatorname{Cot} \tfrac{1}{2}O. \quad \textit{Proposition 16.}$$

---

(25) For since, as Ptolemy effectively showed in his *Planispherium*, all small circles on the sphere are mapped by stereographical projection into corresponding circles in the equatorial plane, 'arcus $\beta\gamma$ [est] circularis' (the arc $\beta\gamma$ is circular)—to rescue a phrase suppressed by Newton in his manuscript—and so at its end-points β, γ it is equally inclined to its chord $\beta\gamma$. (As we have it the text of Ptolemy's work [= (ed. J. L. Heiberg) *Opera quæ exstant omnia,* **2** (Leipzig, 1907): 225–70] is corrupt and the most general case of the circle-preserving property—an immediate extension of the particular instances dealt with by him—is not in fact considered in the extant version.)

(26) That is, ' $\operatorname{t} \dfrac{\gamma + \beta}{2}$ ' ($\operatorname{Tan} \frac{1}{2}(\gamma + \beta)$).

(27) In a cancelled preceding phrase Newton ordered: 'Produc AC, BC, donec occurrant QO producto in O et D' (Produce AC, BC till they meet QO produced in O and D).

Ergo (per Lem [2])[(28)] $s\dfrac{AB+AC}{2} \cdot s\dfrac{AB-AC}{2} :: t'\dfrac{A}{2} \cdot t\dfrac{C-B}{2}$.   Prop 17.

Et $s'\dfrac{OD-OC}{2} \cdot s'\dfrac{OC+OD}{2} :: t'\dfrac{O}{2} \cdot t\dfrac{C+D}{2}$.   Prop 18.

Porro ex Prop 15 mixtim fit

$$t\dfrac{AB}{2} \cdot t\dfrac{AC}{2} :: t'\dfrac{A}{2} + t\dfrac{C-B}{2} \cdot t'\dfrac{A}{2} - t\dfrac{C-B}{2} :: (\text{per Lem [3]}^{(29)})$$

$$s'\dfrac{A+B-C}{2} \cdot s'\dfrac{A+C-B}{2}. \quad \text{Prop 19.}$$

Similiter ex Prop 16 mixtim fit

$$t\dfrac{DO}{2} \cdot t'\dfrac{OC}{2} :: t'\dfrac{O}{2} - t\dfrac{C+D}{2} \cdot - t\dfrac{C+D}{2} - t'\dfrac{O}{2}^{(30)} :: (\text{per Lem [3]}^{(29)})$$

$$s'\dfrac{O+C+D}{2} \cdot s'\dfrac{-O+C+D}{2}. \quad \text{Prop 20.}$$

Ergo rursus in tri ABC (per Prop 20) fit $t'\dfrac{AB}{2} \cdot t\dfrac{AC}{2} :: s'\dfrac{A+B+C}{2} \cdot s'\dfrac{B+C-A}{2}$.

Quare $t\dfrac{AB}{2}$ ad $t'\dfrac{AB}{2}$ $\left(\text{seu } t\dfrac{AB}{2} \text{ ad } t\dfrac{AC}{2} \text{ et } t\dfrac{AC}{2} \text{ ad } t'\dfrac{AB}{2}\right)$ est

$$s'\dfrac{A+B-C}{2} \times s'\dfrac{-A+C+B}{2} \text{ad } s'\dfrac{-B+C+A}{2} \times s'\dfrac{A+B+C}{2}.$$

et in dimidiata ratione $t\dfrac{AB}{2}$ ad R vel R ad $t'\dfrac{AB}{2}$ ut

$$s'\dfrac{A+B-C}{2} \backsim s'\dfrac{B+C-A}{2} \text{ ad } s'\dfrac{A+C-B}{2} \backsim s'\dfrac{A+B+C}{2}. \quad \text{Prop 21.}^{(31)}$$

---

(28) Newton's text lacks a corresponding lemma, but we should evidently understand some equivalent to 'Lemma 2. $s\dfrac{AB+AC}{2} \cdot s\dfrac{AB-AC}{2} :: t\dfrac{C+B}{2} \left(= t'\dfrac{A}{2}\right) \cdot t\dfrac{C-B}{2}$' (that is,

$$\mathrm{Sin}\tfrac{1}{2}(AB+AC) : \mathrm{Sin}\tfrac{1}{2}(AB-AC) = \mathrm{Tan}\tfrac{1}{2}(C+B) \text{ (or } \mathrm{Cot}\tfrac{1}{2}A) : \mathrm{Tan}\tfrac{1}{2}(C-B)).$$

(29) This, too, is lacking in Newton's text, but understand

$$\text{'Lemma 3. } t'\dfrac{C}{2} + t\dfrac{A}{2} \cdot t'\dfrac{C}{2} - t\dfrac{A}{2} :: s'\dfrac{C-A}{2} \cdot s'\dfrac{C+A}{2}\text{'}$$

(that is, $\mathrm{Cot}\tfrac{1}{2}C + \mathrm{Tan}\tfrac{1}{2}A : \mathrm{Cot}\tfrac{1}{2}C - \mathrm{Tan}\tfrac{1}{2}A = \mathrm{Cos}\tfrac{1}{2}(C-A) : \mathrm{Cos}\tfrac{1}{2}(C+A))$ or some equivalent. No doubt Newton would prove this on some geometrical model (as in his following 'Compendium') rather than by adducing the addition formula for cosines.

(30) Read '$t'\dfrac{O}{2} + t\dfrac{C+D}{2} \cdot t\dfrac{C+D}{2} - t'\dfrac{O}{2}$' (that is,

$$\mathrm{Cot}\tfrac{1}{2}O + \mathrm{Tan}\tfrac{1}{2}(C+D) : \mathrm{Tan}\tfrac{1}{2}(C+D) - \mathrm{Cot}\tfrac{1}{2}O).$$

The mistake is carried through into Proposition 21 following.

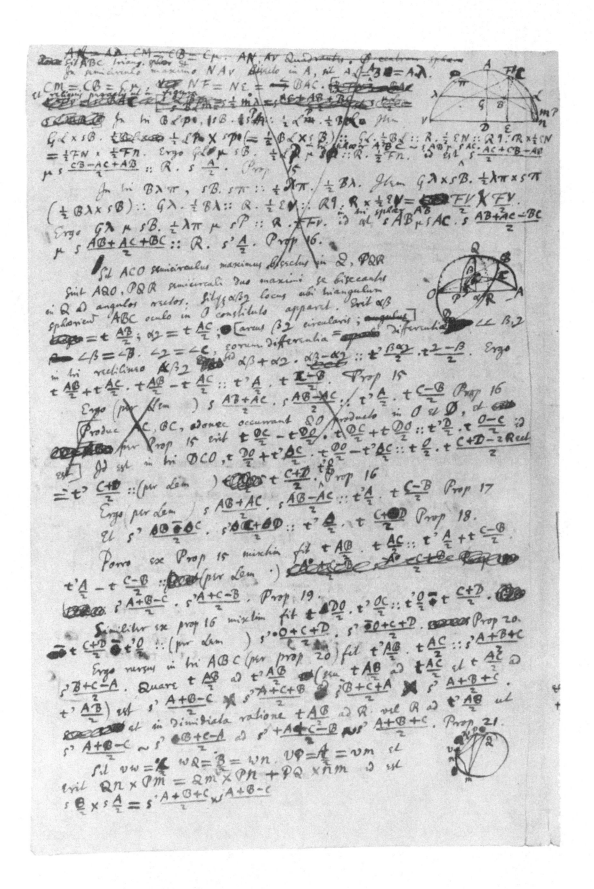

Plate I. Napier's analogies for the spherical triangle derived by stereographic projection (1, 3, §1).

Therefore (by Lemma [2]$^{(28)}$)

$$\operatorname{Sin}\tfrac{1}{2}(AB+AC):\operatorname{Sin}\tfrac{1}{2}(AB-AC)=\operatorname{Cot}\tfrac{1}{2}A:\operatorname{Tan}\tfrac{1}{2}(C-B). \qquad \textit{Proposition 17.}$$

And $\quad\operatorname{Cos}\tfrac{1}{2}(OD-OC):\operatorname{Cos}\tfrac{1}{2}(OC+OD)=\operatorname{Cot}\tfrac{1}{2}O:\operatorname{Tan}\tfrac{1}{2}(C+D).$

Proposition 18.

Further, from Proposition 15 there comes *mixtim*

$$\operatorname{Tan}\tfrac{1}{2}AB:\operatorname{Tan}\tfrac{1}{2}AC=\operatorname{Cot}\tfrac{1}{2}A+\operatorname{Tan}\tfrac{1}{2}(C-B):\operatorname{Cot}\tfrac{1}{2}A-\operatorname{Tan}\tfrac{1}{2}(C-B)$$

$$=\text{(by Lemma [3]}^{(29)})\operatorname{Cos}\tfrac{1}{2}(A+B-C):\operatorname{Cos}\tfrac{1}{2}(A+C-B). \qquad \textit{Proposition 19.}$$

Similarly, from Proposition 16 there comes *mixtim*

$$\operatorname{Tan}\tfrac{1}{2}DO:\operatorname{Cot}\tfrac{1}{2}OC=\operatorname{Cot}\tfrac{1}{2}O-\operatorname{Tan}\tfrac{1}{2}(C+D):-\operatorname{Tan}\tfrac{1}{2}(C+D)-\operatorname{Cot}\tfrac{1}{2}O^{(30)}$$

$$=\text{(by Lemma [3]}^{(29)})\operatorname{Cos}\tfrac{1}{2}(O+C+D):\operatorname{Cos}\tfrac{1}{2}(-O+C+D). \qquad \textit{Proposition 20.}$$

So back in the triangle *ABC* there comes (by Proposition 20)

$$\operatorname{Cot}\tfrac{1}{2}AB:\operatorname{Tan}\tfrac{1}{2}AC=\operatorname{Cos}\tfrac{1}{2}(A+B+C):\operatorname{Cos}\tfrac{1}{2}(B+C-A).$$

Consequently $\operatorname{Tan}\tfrac{1}{2}AB:\operatorname{Cot}\tfrac{1}{2}AB$ (that is, the compound of $\operatorname{Tan}\tfrac{1}{2}AB$ to $\operatorname{Tan}\tfrac{1}{2}AC$ and of $\operatorname{Tan}\tfrac{1}{2}AC$ to $\operatorname{Cot}\tfrac{1}{2}AB$) is $\operatorname{Cos}\tfrac{1}{2}(A+B-C)\times\operatorname{Cos}\tfrac{1}{2}(-A+C+B)$ to $\operatorname{Cos}\tfrac{1}{2}(-B+C+A)\times\operatorname{Cos}\tfrac{1}{2}(A+B+C)$; and so, on halving the ratios, $\operatorname{Tan}\tfrac{1}{2}AB$ to R (or R to $\operatorname{Cot}\tfrac{1}{2}AB$) is as $\operatorname{Cos}\tfrac{1}{2}(A+B-C)\infty\operatorname{Cos}\tfrac{1}{2}(B+C-A)$ to

$$\operatorname{Cos}\tfrac{1}{2}(A+C-B)\infty\operatorname{Cos}\tfrac{1}{2}(A+B+C). \qquad \textit{Proposition 21.}^{(31)}$$

---

(31) On correcting the signs Propositions 20 and 21 should read

$$`\dots\text{t}\frac{DO}{2}.\text{t'}\frac{OC}{2}::\dots\text{s'}\frac{-O+C+D}{2}.-\text{s'}\frac{-O-C-D}{2}. \quad \text{Prop 20.}$$

Ergo rursus in tri[angulo] *ABC* fit $\text{t'}\dfrac{AB}{2}.\text{t}\dfrac{AC}{2}::\text{s'}\dfrac{-A+B+C}{2}.\text{s'}\dfrac{A-B-C}{2}$. Quare $\text{t}\dfrac{AB}{2}$ ad $\text{t'}\dfrac{AB}{2}\dots$ est $\text{s'}\dfrac{A+B+C}{2}\times-\text{s'}\dfrac{A+B+C}{2}$ ad $\text{s'}\dfrac{A-B+C}{2}\times\text{s'}\dfrac{-A+B+C}{2}$. et in dimidiata ratione $\text{t}\dfrac{AB}{2}$ ad R vel R ad $\text{t'}\dfrac{AB}{2}$ ut

$$\text{s'}\frac{A+B-C}{2}\infty-\text{s'}\frac{A+B+C}{2} \quad \text{ad} \quad \text{s'}\frac{A-B+C}{2}\infty\text{s'}\frac{-A+B+C}{2}. \quad \text{Prop 21'.}$$

On putting $\tfrac{1}{2}(A+B+C)=S$ we see that Newton's corrected result is Lambert's formula expressing a side in terms of the three angles of a spherical triangle, namely

$$\operatorname{Tan}^2\tfrac{1}{2}(AB)/R^2=-\operatorname{Cos}S.\operatorname{Cos}(S-C)/\operatorname{Cos}(S-A).\operatorname{Cos}(S-B).$$

(See J. H. Lambert's *Beyträge zum Gebrauch der Mathematik,* **1** (Berlin, 1765): Cap. 3, §61: 406.) As we have noted (1: 469/473) Newton was already familiar in 1665 with the equivalent formula $\operatorname{Sin}^2\tfrac{1}{2}(AB)/R^2=-\operatorname{Cos}S.\operatorname{Cos}(S-C)/\operatorname{Cos}A.\operatorname{Cos}B$, the straightforward polar equivalent of Napier's formula for $\operatorname{Sin}^2\tfrac{1}{2}C/R^2$ (see note (21)).

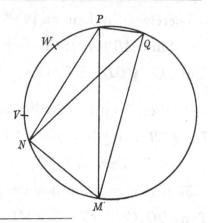

[32]Sit $VW = \dfrac{C}{2}$. $WQ = \dfrac{B}{2} = WN$. $VP = \dfrac{A}{2} = VM$

et erit $QN \times PM = QM \times PN + PQ \times NM$. id

est $s\dfrac{B}{2} \times s\dfrac{A}{2} = s\dfrac{A+B+C}{2} \times s\dfrac{A+B-C}{2}$[33]

§2. THE 'COMPENDIUM OF TRIGONOMETRY'.[1]

From the original in the University Library, Cambridge[2]

TRIGONOMETRIA.

In hoc Compendio non tam brevitati quam discentium captui consulitur. Qua de causa Principiorum explicatio fusior est.

Sect: 1.

Prænoscenda.

1. Si Circuli Quadrans lineis rectis a centro ad circumferentiam ductis dividatur in nonaginta æquales partes et pars unaquæcȝ dividatur in sexaginta æquales partes, & harum unaquæcȝ in alias sexaginta et sic continuo: partes tum anguli recti ad centrum tum arcus quadrantalis ad circumferentiam, quæ prima divisione fiunt dicuntur Gradus, eæ quæ secunda divisione Minuta prima, vel absolutè Minuta, quæ tertia divisione Minuta secunda, quæ quarta Minuta tertia, et sic in infinitum. Lineæ vero a centro ad circumferentiam ductæ, quasi a puncto lucente manantes, dicuntur Radij.[3]

---

(32) Suddenly, a lemma on plane trigonometry! It is manifest that at this point Newton gave up any hope of completing his revise of the 'Epitome'.

(33) Newton breaks off, leaving his trigonometrical identity (the analytical equivalent of the preceding Ptolemaic theorem connecting the diagonals and sides of a cyclic quadrilateral) both faulty and incomplete. In the accompanying figure the points V and W are the midpoints of $\widehat{MP} = R\hat{A}$ and $\widehat{NQ} = R\hat{B}$, so that, on taking $\widehat{VW} = \frac{1}{2}R\hat{C}$, there results

$QN = 2\mathrm{Sin}\frac{1}{2}B$, $\quad PM = 2\mathrm{Sin}\frac{1}{2}A$, $\quad QM = 2\mathrm{Sin}\frac{1}{4}(A+B+C)$, $\quad PN = 2\mathrm{Sin}\frac{1}{4}(A+B-C)$,

$PQ = 2\mathrm{Sin}\frac{1}{4}(-A+B+C)$ and $NM = 2\mathrm{Sin}\frac{1}{4}(A-B+C)$; whence correctly

$$'s\frac{B}{2} \times s\frac{A}{2} = s\frac{A+B+C}{4} \times s\frac{A+B-C}{4} + s\frac{-A+B+C}{4} \times s\frac{A-B+C}{4}'.$$

(32)Let $VW = \tfrac{1}{2}C$, $WQ = WN = \tfrac{1}{2}B$, $VP = VM = \tfrac{1}{2}A$ and there will be $QN \times PM = QM \times PN + PQ \times NM$: that is,

$$\text{Sin}\ \tfrac{1}{2}B \times \text{Sin}\ \tfrac{1}{2}A = \text{Cos}\ [\tfrac{1}{2}]\ (A+B+C) \times \text{Cos}\ [\tfrac{1}{2}?]\ (A+B-C)\ \dots^{(33)}$$

Translation

TRIGONOMETRY

In this Compendium regard is paid not so much to brevity as to the learner's capacity. For this reason explanation of principles is somewhat elaborate.

SECTION 1. PRELIMINARIES

1. If a quadrant of a circle be divided by straight lines drawn from its centre to the circumference into ninety equal parts, and each of these parts be divided into sixty equal portions, and each one of these into sixty further ones, and so on continually; then the parts, both of the right angle at the centre and of the quadrantal arc along the circumference, which are created at the first division are called DEGREES, those formed by the second division FIRST MINUTES, or just simply MINUTES, those by the third SECOND MINUTES, those by the fourth THIRD MINUTES, and so on indefinitely. The lines, indeed, which are drawn from the centre to the circumference, much like emanations from a luminous point, are called RADII (RAYS).(3)

---

(1) An elaborate revise of the preceding 'Epitome Trigonometriæ', manifestly composed soon after he abandoned the prior text in its unfinished second state. As Newton observes in his introductory paragraph, this is a compendium for the student beginner—no doubt he had his own undergraduate audiences squarely in mind—where brevity is sacrificed to the over-riding aim of clarity in exposition. Otherwise, apart from an augmentation of the sections on plane trigonometry (in which a number of newly introduced identities are given freshly concocted geometrical proof), the content of the present 'Compendium' is essentially the same as that of the 'Epitome' which is its model.

(2) ULC. Add. 3959.5: 33^r–44^v (with predrafts at 48^r and 49^r). At the top left-hand corner of the first of these autograph folio sheets Thomas Pellet in 1727 added his listing 'N° 4' (*sc.* in the rough catalogue of Newton's papers he then made; see 1: xxi). In a typical progression Newton penned the opening sheets of the draft sequence with great care (though these were later overwritten much more roughly), making subtle use of both large and small capitals in his text and drawing his figures accurately with ruler and compass; increasingly, however, he lost his patience and stylistic flow, and the later sheets are much more hastily composed, with multiple cancellations and freehand diagrams.

(3) It is impossible to render the double (geometrical and optical) sense of the Latin word by a single English equivalent. The optical sense is scarcely primary since 'radius' would seem to share the same (etymological) root as 'radix' and 'ramus' and so signify a 'rod' or 'staff': no doubt, however, the geometrical and optical senses derive basically from the same metaphor of the 'spokes' of a wheel.

2. Si a singulis circumferentiæ divisionibus ad alterutrum Quadrantis latus rectilineum demittantur perpendicula: dicuntur hæc tum angulorum ad centrum quos subtendunt tum arcuum angulis oppositorum Sɪɴᴜs ʀᴇᴄᴛɪ, vel absolute Sɪɴᴜs. Jacent enim in sinu seu cavitate arcuum, et arcubus a tergo sitis quasi pectora sunt.[4] Et sinus Quadrantis totius nonaginta graduũ qui Radius est, dicitur Sɪɴᴜs ᴛᴏᴛᴜs, eo quod Quadrantis totius sinus est et sinuum maximus. Pars autem radij qui a termino sinus recti ad arcum producitur dicitur Sɪɴᴜs ᴠᴇʀsᴜs ejusdem arcus.

3. Si ad terminum lateris alterutrius Quadrantis juxta circumferentiam erigatur perpendiculum, et lineæ a centro ad singulas circumferentiæ divisiones ductæ producantur donec occurrant perpendiculo: partes perpendiculi ad lineas illas productas terminatæ dicuntur tam angulorum ad centrum quos subtendunt quam arcuũ his oppositorum Tᴀɴɢᴇɴᴛᴇs, & lineæ a centro ad usᴄꝫ perpendiculum productæ dicuntur eorundem angulorum et arcuum Sᴇᴄᴀɴᴛᴇs. Imponuntur verò hæc nomina quod perpendiculum tangit circumferentiam ad terminum lateris Quadrantis, lineæ verò productæ ipsam secant.

In illustrationem præcedentium sit *ACB* circuli quadrans; sintᴄꝫ *Ad, Ae, Af* quælibet arcus[5] quadrantalis *AB* partes; *ACd, ACe, ACf* similes partes anguli recti *ACB*; *dg, eh, fi* perpendicula a punctis divisionum *d, e, f* ad Quadrantis latus *CA* demissa; *AP* perpendiculum ad lateris *CA* terminum *A* erectum, et *Ck, Cl, Cm* lineæ per puncta divisionum *d, e, f* ad usᴄꝫ perpendiculum illud ductæ: et erit *dg* tam anguli oppositi *AC[d]* quam subtendentis arcus *Ad* sinus rectus, *Ag* ejusdem anguli et arcûs sinus versus, *Ak* tangens ejus et *Ck* secans. Et sic anguli *ACe* et arcus *Ae* erit *eh* sinus, *Ah* sinus versus[,] *Al* tangens et *Cl* secans. Idem

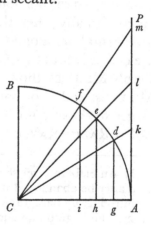

intellige de alio quovis angulo *ACf* et arcu subtendente *Af* cum eorum sinu recto *fi*, verso *Ai*, tangente *Am* et secante *Cm*. Et quot sunt nonagesimæ partes arcus quadrantalis *AB* et unius partis partes sexagesimæ in arcu *Ad*, tot gradibus constare dicitur & minutis tum arcus ille *Ad* tum angulus *ACd*, totᴄꝫ graduum et minutorum dicitur esse *dg* sinus, *Ag* sinus versus[,] *Ak* tangens et *Ck* secans. Radius vero *CB* qui totius arcus *AB* et anguli recti *ACB* sinus est, dicitur sinus totus.

---

(4) An ingenious but erroneous explanation. In historical fact the word 'sinus' was introduced in its technical sense by Gerard of Cremona as his twelfth-century Latin rendering of the Arabic 'jaib', which normally does indeed mean 'breast' or 'bosom' but which in trigonometrical context is a consonantally equivalent corruption of 'jîba', invented earlier as a phonetic 'translation' of Aryabhaṭa's 'jīva' (half-chord). See A. von Braunmühl, *Vorlesungen*

2. If from each individual point of division in the circumference perpendiculars are let fall to one or other of the quadrant's straight sides, these are called the RIGHT SINES both of the angles which they subtend at the centre and of the arcs opposite these angles, or just simply their SINES. For they lie in the *sinus* or hollow of the arcs and, with these backing arcs, are like breasts.[4] Further, the sine of the whole quadrant of ninety degrees—namely, the radius—is called the WHOLE SINE, for the reason that the sine of the whole quadrant is also the greatest of the sines. The portion, however, of the radius which extends from the end of the right Sine to the arc is called the VERSINE of the same arc.

3. If at the end of either of the quadrant's sides which adjoins the circumference a perpendicular be raised, and lines drawn from the centre to the separate divisions of the circumference be extended till they intersect this perpendicular, then the portions of the perpendicular which terminate at those extended lines are called TANGENTS equally of the angles subtended by them at the centre and of the arcs opposite to these, while the lines extended from the centre as far as the perpendicular are called SECANTS of the same angles and arcs. These names are imposed, to be sure, because the perpendicular touches the circumference at the end of the quadrant's side, while the extended lines intersect it.

In illustration of the preceding, let ABC be a quadrant of a circle; Ad, Ae, Af any parts whatsoever of the quadrantal arc[5] $\widehat{AB}$; $A\widehat{C}d$, $A\widehat{C}e$, $A\widehat{C}f$ corresponding portions of the right angle $A\widehat{C}B$; dg, eh, fi perpendiculars let fall from the points d, e, f of division to the quadrant's side CA; AP the perpendicular raised at the end A of the side CA; and Ck, Cl, Cm lines extended through the points d, e, f of division up to that perpendicular: then will dg be the right sine both of the opposite angle $A\widehat{C}d$ and the subtending arc $\widehat{Ad}$, Ag the versine of the same angle and arc, Ak their tangent and Ck their secant. And in this way eh will be the sine of the angle $A\widehat{C}e$ and of the arc $\widehat{Ae}$, Ah the versine, Al the tangent and Cl the secant. Understand the same for any other angle $A\widehat{C}f$ and subtending arc $\widehat{Af}$, with fi their right sine, Ai their versine, Am their tangent and Cm their secant. Further, that arc $\widehat{Ad}$ (and equally the angle $A\widehat{C}d$) is said to consist of as many degrees and minutes as it contains ninetieth parts of the quadrantal arc $\widehat{AB}$ and sixtieths of one of those parts, and its sine dg, versine Ag, tangent Ak and secant Ck are each said to be of an equal number of degrees and minutes. The radius CB, to be sure,—the sine of the whole arc $\widehat{AB}$ and of the right angle $A\widehat{C}B$—is called the whole sine.

über Geschichte der Trigonometrie, **1** (Leipzig, 1903): 49–50; J. Tropfke, *Geschichte der Elementar-Mathematik*, **5** (Berlin/Leipzig, 1923): 31–3; and D. E. Smith, *History of Mathematics*, **2** (Boston, 1925): 615–16.

(5) 'circumferentiæ' (circumference) is cancelled.

4.[6] Porro si angulus vel arcus quadrantalis dividatur in duas partes, dicitur una pars COMPLEMENTUM alterius AD QUADRANTEM[7] vel absolutè COMPLEMENTUM ejus. Et si semicirculus recta a centro ducta dividatur in duas partes, dicitur angulus vel arcus in una parte COMPLEMENTUM ejus in altera AD SEMICIRCULUM.[7] Sic angulus *BCd* complementum est anguli *ACd* & arcus *Bd* complementum arcus *Ad*. Et hæc complementa Quadrante aucta fiunt complementa ad semicirculum.

5. Complementorum ad semicirculum sinus tangentes et secantes eædem sunt atcq arcuum et angulorum quorum sunt complementa. Complementorum vero ad 90^{gr} [8] sinus tangentes et secantes nominibus CO-SINUUM, CO-TANGENTIUM et CO-SECANTIUM designari solent. Nos hac in re brevitati consulentes, Radium, Sinum, Tangentem, Sinum complementi et Tangentem complementi designabimus literis *R*, s, t, sc, tc respectivè: scribendo verbi gratia s*A* pro sinu anguli *A*, sc*A* pro sinu complementi ejus, & sic in cæteris.[9]

6. Si in Circulo aliquo cujus Radius assumitur numerus decimalis, puta 100000, computentur sinus ad singulos gradus et minuta, et sinus illi e regione graduum et minutorum quorum sunt sinus referantur in Tabulam, habebitur tabula illa vulgaris quæ CANON SINUUM dicitur.[10] Et in eodem circulo ad singulos gradus et minuta, computando tangentes et secantes condentur tabulæ vulgares tangentium et secantium. Hunc autem circulum appellabimus CANONICUM.

7. Tabulis vero ad unum aliquem circulum conditis, possunt sinus tangentes et secantes eorundem angulorum in alio quovis circulo per auream regulam[11] inveniri. Ut enim Radius est ad sinus tangentes et secantes in circulo canonico, ita Radius est ad sinus tangentes et secantes in circulo quovis alio. Et hinc via ad Trigonometriam patuit.

8. Hic autem triangula solum consideramus quorum latera sunt linearum in superficie sphæræ brevissima[12] inter puncta sua, id est, in plano triangula rectilinea, in superficie sphæræ triangula arcubus maximorum circulorum confecta. Circulus MAXIMUS est qui Radio Sphæræ descriptus sphæram

---

(6) Newton first began this paragraph 'Porro Angulus qui una cum altero angulo rectū angulu[m facit, dicitur istius complementum ad 90 gradus]'. (Furthermore, an angle which together with a second one forms a right angle [is called the latter's complement to 90°].)

(7) Newton has cancelled 'ad 90^{gra}' (to 90°) and 'ad 180^{gr}' (to 180°) here, respectively.

(8) Read 'ad quadrantem' (to a quadrant); see note (7).

(9) The notation is Richard Norwood's. (See his *Trigonometrie, Or, The Doctrine of Triangles* (London, 1631 [= $_2$1645]): Book 1: 20. At his death Newton's library contained a copy of the second edition (now Trinity College, Cambridge. NQ. 9.33).) Except in Lemma 4 following, however, in the sequel Newton reverts—as in his 'Epitome' (§1 above)—to Ward's notation s', t' for cosine and cotangent. In our English version we maintain our previous practice of translating 'sinus', 'cosinus', 'tangens' and 'cotangens' (that is, *R*sin, *R*cos, *R*tan and *R*cot respectively) as 'Sin', 'Cos', 'Tan' and 'Cot'.

4.[6] Furthermore, if the quadrantal angle or arc be divided into two parts, one part is called the other's COMPLEMENT TO A QUADRANT[7] or just simply its COMPLEMENT. And if a semicircle be divided into two parts by a straight line drawn from its centre, the angle or arc in the one part is called the COMPLEMENT TO A SEMICIRCLE[7] of that in the other. So the angle $B\widehat{C}d$ is the complement of the angle $A\widehat{C}d$ and the arc $\widehat{Bd}$ the complement of the arc $\widehat{Ad}$, while these complements when increased by a quadrant become complements to a semicircle.

5. The sines, tangents and secants of complements to a semicircle are the same as those of the arcs and angles whose complements they are. However, the sines, tangents and secants of complements to 90°[8] are usually designated by the names of COSINES, COTANGENTS and COSECANTS. Aiming at brevity in this matter, we shall designate the radius, sine, tangent, complementary sine and complementary tangent by the letters R, s, t, sc, tc respectively, writing s A in place of 'sine of angle A', for example, and sc A in place of its complementary sine, and so in the other cases.[9]

6. If in some circle whose radius is taken to be a power of ten, say $100\,000$, the sines shall be computed for each separate degree and minute, and those sines are then entered up in a table in line with the degrees and minutes whose sines they are, there will be had that widely used table called a CANON OF SINES.[10] And by computing tangents and secants in the same circle for each individual degree and minute the common tables of tangents and secants will be built up. This circle we shall dub CANONICAL.

7. When, to be sure, tables have been constructed for some one circle, the sines, tangents and secants of the same angles in any other circle may be found by the golden rule.[11] For as the radius is to the sines, tangents and secants in the canonical circle, so is the radius to the [corresponding] sines, tangents and secants in any other circle. From here on the way is open to trigonometry.

8. Here we take into our compass only triangles whose sides are the shortest[12] of the lines on a spherical surface between its vertices: to wit, in a plane, rectilinear triangles and, on the surface of a sphere, triangles made up of arcs of great circles. A GREAT circle is one which is described by the sphere's radius, bisecting

---

(10) An autograph 'Tabula sinuum ad semigradus' (listing $10^5\sin\theta$ to five places at 30' intervals, $0° \leqslant \theta \leqslant 90°$), found in the same packet as the present manuscript and doubtless intended to form part of its finished version, is reproduced as Appendix 2 below.

(11) The 'Rule of Three' which relates four quantities in proportion. For the terminology, current in English arithmetics of Newton's period, see D. E. Smith's *History of Mathematics* (note (5) above), **2**: 486.

(12) Or, equivalently, 'longissima' (longest)! Newton a little thoughtlessly generalizes the 'Euclidean' postulate (in fact Proclus'; see T. L. Heath, *The Thirteen Books of Euclid's Elements*, **1** (Cambridge, 1926): 168) that the straight line is the shortest distance between its end-points.

bisecat.[13] POLI ejus sunt puncta duo opposita in superficie sphæræ nonaginta gradibus hinc inde ab eo distantia. Et angulus quem circuli duo maximi continent idem est cum angulo quem plana illorum circulorum continent, et mensuratur per subtendentem tertij circuli maximi cujus polus est in concursu priorum. Omnes igitur termini trianguli sphærici arcubus circulorum Radio sphæræ descriptorum definiuntur, et inde si Radius ille idem statuatur cum Radio circuli canonici, sinus tangentes & secantes terminorum omniũ trianguli eædem erunt cum canonicis.

9. Triangula tam rectilinea quàm sphærica dicuntur RECTANGULA quæ rectum angulum habent, OBLIQUANGULA quæ rectum non habent, et in rectangulis latus quod angulo recto opponitur dicitur HYPOTENUSA, reliqua duo latera nominamus CRURA.[14] Sic et ubi laterum aliquod pro BASI trianguli habetur, reliqua latera dicuntur CRURA,[14] et ANGULUS VERTICALIS est qui basi opponitur, CATHETUS[15] quæ ab angulo illo ad basem perpendicularis est, et SEGMENTA tam anguli verticalis quam basis quæ sunt inter cathetum et crura ubi cathetus cadit intra triangulum; quod nomen etiam retineri potest ubi cadit extra.

10. Quoniam arcuum et angulorum qui majores sunt Quadrante, sinus tangentes et secantes eædem sunt ac complementorum ad semicirculum: hinc ubi ex sinu tangente aut secante colligendus est e Tabulis arcus aliquis vel angulus qui non multum differt a quadrantali, incertum esse potest utrum sumendus sit arcus angulusve Quadrantali minor aut arcûs illius vel anguli complementum ad semicirculum. Cui ambiguitati tollendæ[16] inservire possunt hæ[17] regulæ.

[1.] In omni triangulo tam sphærico quam rectilineo, major angulus majori lateri opponitur, ut et majus latus majori angulo.[18] Latus item quodvis minus est summa, majus differentia reliquorum laterum.[19]

[2.] In omnibus triangulis sphæricis tres anguli simul sumpti majores sunt duobus rectis[20] et eorum excessus supra duos rectos est ad duos rectos ut area trianguli sphærici ad quartam partem totius superficiei sphæræ, seu ad aream circuli maximi.[21] Et si angulus aliquis propior sit Quadranti quam latus

---

(13) By implication the radius is constrained to move in a unique plane.

(14) See §1: note (5).

(15) κάθετος (the '[line] let fall'); compare Euclid's *Elements* I, Definition 10.

(16) Newton first wrote 'Cui scrupulo tollendo' (In removing this doubt).

(17) 'cautiones et' (precautions and) is cancelled.

(18) The case of the plane triangle is Euclid, *Elements* I, 18/19, and the spherical case is easily proved by an equivalent argument.

(19) This is patently true when each side of the triangle is the *shortest* distance between the two vertices which are its end-points, but will not hold (compare note (12)) when, in the spherical case, any side is longer than half a great circle.

the sphere.[13] Its POLES are the two opposite points in the sphere's surface distant ninety degrees from it on either side. The angle which two great circles contain is then identical with the angle contained by the planes of those circles, and is measured by the subtended arc of the third great circle whose pole is at the former circles' meeting point. All boundaries, therefore, of a spherical triangle are delimited by arcs of circles described by the sphere's radius, and accordingly, if that radius should be decreed to be identical with the radius of the canonical circle, then the sines, tangents and secants of all the boundaries of a triangle will be identical with those in the canon.

9. Those triangles, both rectilinear and spherical, possessing a right angle are called RIGHT-ANGLED, those not having a right angle OBLIQUE-ANGLED; and in right-angled ones the side opposite the right angle is called the HYPOTENUSE, while we name the remaining two sides CRURA (LEGS).[14] So also when some one of the sides is considered as BASE, the remaining sides are called CRURA, while the VERTICAL ANGLE is that opposite the base, the CATHETUS[15] the perpendicular from that angle to the base, and, when the cathetus falls within the triangle, the SEGMENTS both of the vertical angle and of the base are those between the *cathetus* and the *crura*, but this name may also be kept when it falls outside.

10. Seeing that in the case of arcs and angles which are greater than a quadrant sines, tangents and secants are the same as those of their complements to a semicircle, accordingly when some arc or angle little different from a quadrantal one has to be ascertained from its sine, tangent or secant by means of tables, it may not be established whether the arc or angle less than a quadrant must be taken or whether the complement of that arc or angle to a semicircle. In removing such ambiguity[16] these[17] rules may be of service:

(1) In every triangle, both spherical and rectilinear, the greater angle is opposite the greater side, and likewise the greater side opposite the greater angle.[18] Similarly any side is less than the sum but greater than the difference of the remaining sides.[19]

(2) In all spherical triangles the three angles taken together are more than two right angles[20] and their excess over two right angles is to two right angles as the area of the spherical triangle to a quarter of the total spherical surface, that is, to the area of a great circle.[21] And if some angle be closer to a quadrant

---

(20) Newton first continued somewhat mysteriously '& latera duo minora una cum complemento maximi lateris ad semici[rculum]' (and the two lesser sides together with the complement of the greatest side to a semicircle...).

(21) The fundamental theorem that the area of a spherical triangle is proportional to its 'spherical excess'. Newton would doubtless have proved this by expressing the area in terms of spherical lunules: namely, since the triangle pairs $A'BC'$, $AB'C$ and $AB'C'$, $A'BC$ are congruent, the sum of the lunules $(ABA'C) + (BCB'A) + (CAC'B) = 2\Delta ABC + \text{hemisphere } (ACA'C')$ and

oppositum, duo e lateribus sunt speciei ejusdem[,] tertium Quadrante minus: sin a Quadrante remotior sit, duo ex angulis sunt speciei ejusdem, tertius quadrante major.[22] Ejusdem vero speciei esse dicuntur quæ aut quadrantalia sunt aut utracȝ Quadrante majora aut utracȝ minora.

[3.] In triangulis sphæricis rectangulis, crus et angulus ei oppositus ejusdem sunt speciei.[23] Crura etiam duo ejusdem sunt speciei ubi Hypotenusa minor est Quadrante, at diversæ ubi major.[24]

11. Usui sunt etiam Lemmata sequentia.

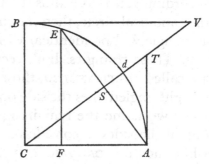

Lem. 1. Radius ad sinum, secans ad tangentem et co-secans ad Radium; duplus item sinus ad versum sinum dupli & tangens dimidij ad versum sinum totius, sunt in eadem ratione.[25] Sit arcus cujusvis *Ad* sinus *AS*, tangens *AT*, secans *CT*, co-tangens *BV*, co-secans *CV*, duplus sinus *AE*, dupli sinus *EF* ac dupli sinus versus *AF*. Et ob similia triangula *CAS*, *CTA*, *VCB*, *AEF* erit $CA.AS::CT.AT::CV.CB::AE.AF$. Item $CA.EF::AT.AF$.

Lem. 2. Radius ad Tangentem, co-sinus ad sinum, & co-tangens ad Radium sunt in eadem ratione. $CA.CT::CS.SA::BV.BC$.

Lem. 3. Radius ad secantem, co-tangens ad co-secantem et co-sinus ad Radium sunt in eadem ratione. $CA.CT::BV.CV::CS.CA$.

---

hence, where R is the sphere's radius,

$$2R^2(\hat{A}+\hat{B}+\hat{C}) = 2\Delta\,ABC+2\pi R^2$$

or, in Newton's form,

$$(\hat{A}+\hat{B}+\hat{C}-\pi):\pi = \Delta\,ABC:\pi R^2.$$

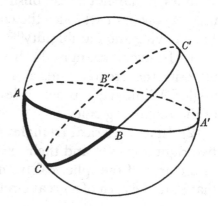

As we have seen (III: 419, note (51)), the theorem was known to Thomas Harriot about 1603 but was first published by Albert Girard in his *Inuention Nouuelle en l'Algebre* (Amsterdam, 1629): G1v–H4v and—independently?—by Cavalieri in his *Directorium Generale Uranometricum* (Bologna, 1632): 320. See also J. Tropfke, *Geschichte* (note (4) above), 5: 129–30; and J. V. Pepper, 'Harriot's Calculation of the Meridional Parts as Logarithmic Tangents', *Archive for History of Exact Sciences*, 4 (1968): 359–413, especially 367, notes 43 and 44.

(22) This is taken, doubtless by way of Seth Ward's *Idea Trigonometriæ Demonstratæ* (Oxford, 1654): 15: Cautio 3, from John Napier's *Logarithmorum Canonis descriptio, Ejusque usus, in utraque Trigonometria* (Edinburgh, 1614): Liber II, Caput III 'De Triangulis Sphæricis', Sententiæ 3/4: 29. Neither Napier nor Ward give any proof but, to be valid, the rule requires the chosen angle to be acute. (With this restriction we may easily show that, in the triangle ABC, if $\sin A > \sin a$,

than the opposite side, two of the sides are of the same species, the third less than a quadrant: but if it be more distant from a quadrant, then two of the angles are of the same species, the third greater than a quadrant.[22] Of course, elements are said to be of the same species when they are quadrantal or alternatively both greater than a quadrant or both less.

(3) In right-angled spherical triangles a *crus* and the angle opposite to it are of the same species.[23] Also, the two *crura* are of same species when the hypotenuse is less than a quadrant, but of different ones when it is greater.[24]

11. The following Lemmas are also of use.

Lemma 1. The ratios of radius to sine, secant to tangent and cosecant to radius, and likewise of twice the sine to the versine of twice the angle, and the tangent of half the angle to the versine of the whole, all are the same.[25] Of any arc Ad let the sine be AS, tangent AT, secant CT, cotangent BV, cosecant CV, twice the sine AE, the sine of the double angle EF and the versine of the double angle AF. Then because of the similar triangles CAS, CTA, VCB, AEF there will be $CA:AS = CT:AT = CV:CB = AE:AF$, and likewise $CA:EF = AT:AF$.

Lemma 2. The ratios of radius to tangent, cosine to sine and cotangent to radius are the same. For $CA:CT = CS:SA = BV:BC$.

Lemma 3. The ratios of radius to secant, contangent to cosecant and cosine to radius are the same. For $CA:CT = BV:CV = CS:CA$.

---

then $\cos b \cos c > 0$ and so b and c must be of the same 'species' (both either greater or less than a quadrant); while if $\sin A < \sin a$, then $\cos B \cos C < 0$ and so $\hat{B}$ and $\hat{C}$ are of different 'species' (one less and one greater than $\frac{1}{2}\pi$). For, in the first case, if $\sin A > \sin a$ and $\cos b \cos c < 0$, at once $\cos a = \cos b \cos c + \sin b \sin c \cos A < \sin b \sin c \cos A$, so that

$$1 < \sin^2 a + \sin^2 b \sin^2 c \cos^2 A < \sin^2 A + \sin^2 b \sin^2 c \cos^2 A$$

and therefore $1 < \sin^2 b \sin^2 c$; while, in the second, if $\sin A < \sin a$ and $\cos B \cos C > 0$, then

$$\cos A = -\cos B \cos C + \sin B \sin C \cos a < \sin B \sin C \cos a,$$

or $1 < \sin^2 A + \sin^2 B \sin^2 C \cos^2 a < \sin^2 a + \sin^2 B \sin^2 C \cos^2 a$ whence $1 < \sin^2 B \sin^2 C$. Both conclusions are manifestly contradictory.) This general rule appears to be original with Napier, though it comprehends, for instance, several theorems presented by Regiomontanus in the fourth book of his *De Triangulis Omnimodis* (Nuremberg, 1533).

(23) Evidently so since, in the triangle ABC right-angled at C (so that the *crura* are $\widehat{BC} = a$ and $\widehat{AC} = b$), $\cos A = \cos a \sin B$.

(24) For, in the triangle ABC right-angled at C, $\cos c = \cos a \cos b$ and so the *crura* $\widehat{BC} = a$ and $\widehat{AC} = b$ are of opposite 'species' only when the hypotenuse c is greater than a quadrant.

(25) In fact, where $CA = R$ and $A\hat{C}d = d\hat{C}E = \theta$, Newton shows that

$$R:\mathrm{Sin}\,\theta = \mathrm{Sec}\,\theta:\mathrm{Tan}\,\theta = \mathrm{Cosec}\,\theta:R = 2\mathrm{Sin}\,\theta:\mathrm{Versin}\,2\theta$$

and $R:\mathrm{Sin}\,2\theta = \mathrm{Tan}\,\theta:\mathrm{Versin}\,2\theta$ (or $R - \mathrm{Cos}\,2\theta$).

Lem. 4. Sinus arcuum duorum co-secantibus, tangentes co-tangentibus, & secantes co-sinubus reciprocè sunt proportionales.[26] Sint M et N arcus duo, et erit per Lem 1, $R . s M :: co\text{-}sec M . R$, & $R . s N :: co\text{-}sec N . R$. Ergo ex æquo perturbate $s M . s N :: co\text{-}sec N . co\text{-}sec M$. Simili argumento colligetur ex Lem. 2 esse $t M . t N :: tc N . tc M$. et ex Lem. 3, esse $sec M . sec N :: sc N . sc M$.

His analogijs fit ut Theoremata de sinubus & tangentibus converti possint in Theoremata de secantibus & sinubus versis, & e contra. Qua de causa in sequentibus Theoremata de sinubus et tangentibus solummodo tradentur, ne aliorum tabulis opus sit.

[27]*Lem 5.* Ut summa tangentium ad differentiam tangentium ita sinus summæ ad sinum differentiæ.

$$IL . DM :: IC . DC :: KC . FC :: KL . FM.$$

Et vicissim $IL . KL :: DM . FM :: DG . FH$.
Id est

$$t AE + t ED . t AE - t ED ::$$
$$s \overline{AE + ED} . s \overline{AE - ED}.$$

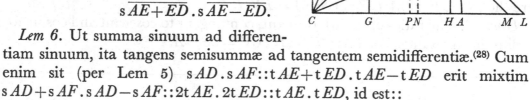

Lem 6. Ut summa sinuum ad differentiam sinuum, ita tangens semisummæ ad tangentem semidifferentiæ.[28] Cum enim sit (per Lem 5) $s AD . s AF :: t AE + t ED . t AE - t ED$ erit mixtim $s AD + s AF . s AD - s AF :: 2t AE . 2t ED :: t AE . t ED$, id est ::

$$t \frac{AD + AF}{2} . t \frac{AD - AF}{2}.$$

Lem 7. Ut summa secantium ad differentiam secantium ita tangens complementi semisummæ ad tangentem semidifferentiæ. Nam

$$\text{secans } AD . sec AF :: (\text{Lem 4}) s FB . s DB.$$
Et mixtim

$$sec AD + sec AF . sec AD - sec AF :: s FB + s DB . s FB - s DB :: (\text{Lem 6}) t EB . t ED$$
$$\text{id est} :: t' \frac{AD + AF}{2} . t \frac{AD - AF}{2}.$$

---

(26) No doubt because it states the obvious, Newton has cancelled an immediately following passage: 'Proportionales enim sunt quatuor quantitates ubi rectangulum sub extremis æquale est rectangulo sub medijs. Et hic rectangula illa semper æqualia sunt, quia æqualia quadrato radij, ut e tribus præcedentibus Lemmatis manifestum est. Verbi gratia sinus arcus M est ad sinū arcus N ut co-secans arcus N ad co-secantem arcus M: quia per Lem 1 Radius est [ad] sinum M ut cosecans M ad Radium & ad sinum N ut co-secans N ad Radium' (For four quantities are in proportion when the product of the extremes is equal to the product of the middles. Here those products are ever equal, being equal, namely, to the square of the

Lemma 4. The sines of two arcs are reciprocally proportional to their cosecants, the tangents to their cotangents, and the secants to their cosines.[26] Let M and N be the two arcs and then, by Lemma 1,

$$R : \text{Sin } M = \text{Cosec } M : R \quad \text{and} \quad R : \text{Sin } N = \text{Cosec } N : R.$$

Therefore *ex æquo* invertedly $\text{Sin } M : \text{Sin } N = \text{Cosec } N : \text{Cosec } M$. By a similar argument it will be gathered from Lemma 2 that

$$\text{Tan } M : \text{Tan } N = \text{Cot } N : \text{Cot } M;$$

and from Lemma 3 that $\text{Sec } M : \text{Sec } N = \text{Cos } N : \text{Cos } M$.

These proportions make it possible to convert theorems on sines and tangents into ones on secants and versines, and the contrary. For this reason in the sequel theorems on sines and tangents alone are presented to avoid the need for tables of other functions.

[27]*Lemma 5.* As the sum of tangents to their difference, so is the sine of the [angles'] sum to the sine of their difference. For

$$IL : DM = IC : DC = KC : FC = KL : FM$$

and inversely $IL : KL = DM : FM = DG : FH$, that is,

$$\text{Tan } \widehat{AE} + \text{Tan } \widehat{ED} : \text{Tan } \widehat{AE} - \text{Tan } \widehat{ED} = \text{Sin } (\widehat{AE} + \widehat{ED}) : \text{Sin } (\widehat{AE} - \widehat{ED}).$$

Lemma 6. As the sum of sines to their difference, so is the tangent of the [angles'] half sum to the tangent of their half difference.[28] For since (by Lemma 5) $\text{Sin } \widehat{AD} : \text{Sin } \widehat{AF} = \text{Tan } \widehat{AE} + \text{Tan } \widehat{ED} : \text{Tan } \widehat{AE} - \text{Tan } \widehat{ED}$, there will be *mixtim* $\text{Sin } \widehat{AD} + \text{Sin } \widehat{AF} : \text{Sin } \widehat{AD} - \text{Sin } \widehat{AF} = 2 \text{ Tan } \widehat{AE} : 2 \text{ Tan } \widehat{ED}$, that is,

$$\text{Tan } \widehat{AE} : \text{Tan } \widehat{ED} = \text{Tan } \tfrac{1}{2}(\widehat{AD} + \widehat{AF}) : \text{Tan } \tfrac{1}{2}(\widehat{AD} - \widehat{AF}).$$

Lemma 7. As the sum of secants to their difference, so the cotangent of the [angles'] half sum to the tangent of their half difference. For (by Lemma 4) $\text{Sec } \widehat{AD} : \text{Sec } \widehat{AF} = \text{Sin } \widehat{FB} : \text{Sin } \widehat{DB}$, and *mixtim*

$$\text{Sec } \widehat{AD} + \text{Sec } \widehat{AF} : \text{Sec } \widehat{AD} - \text{Sec } \widehat{AF} = \text{Sin } \widehat{FB} + \text{Sin } \widehat{DB} : \text{Sin } \widehat{FB} - \text{Sin } \widehat{DB}$$

$$= (\text{by Lemma 6}) \text{ Tan } \widehat{EB} : \text{Tan } \widehat{ED},$$

$$\text{that is, Cot } \tfrac{1}{2}(\widehat{AD} + \widehat{AF}) : \text{Tan } \tfrac{1}{2}(\widehat{AD} - \widehat{AF}).$$

---

radius, as is manifest from the three preceding lemmas. For instance, the sine of arc M is to the sine of arc N as the cosecant of arc N to the cosecant of arc M; since, by Lemma 1, the radius is to the sine of M as the cosecant of M to the radius, and to the sine of N as the cosecant of N to the radius).

(27) Newton's more concise first draft of Lemmas 5 and 6 following is now ULC. Add. 3959.5: 49$^\text{r}$ (top). In his present diagram he somewhat inefficiently denoted the upper endpoint of the quadrant by a second 'G': in reproduction we retain Newton's first choice of 'B' in his draft figure and have silently corrected the text to correspond.

(28) In his first draft (see note (27)) Newton concluded succinctly

$$\text{'}DG + FH . DG - FH :: IL + KL (= 2EL). IK :: EL . EK\text{'}.$$

Lem 8. Ut Radius ad sinum semisummæ ita sinus complementi semidifferentiæ ad semisummam sinuum et sinus semidifferentiæ ad semidifferentiam cosinuum.

$$CE . EN :: CQ . QP = \frac{FH+DG}{2} :: DQ . RQ.^{(29)}$$

Lem 9. Ut Radius ad sinum complementi semisummæ ita sinus semidifferentiæ ad semidifferentiam sinuum et sinus complementi semidifferentiæ ad semi-summam co-sinuum. $CE . CN :: DQ . DR = \dfrac{DG-FH}{2} :: CQ . CP = \dfrac{CG+CH}{2}.^{(30)}$

Lem. 10. Ut sinus summæ ad summā sinuum ita differentia sinuum ad sinū differentiæ. $CQ^q = CD^q (CE^q) - DQ^q$. Ipsis autem CQ, CE, et DQ proportionales sunt CP, CN, DR ideoǫ

$$CP^q (CQ^q - QP^q) = CN^q - DR^q \quad \text{et} \quad QP^q - DR^q = CQ^q - CN^q = EN^q - DQ^q$$

et $QP + DR (DG) . EN + DQ :: EN - DQ . QP - DR (FH).^{(31)}$ Q.E.D.

Lem. 11. Illud etiam scias, ex datis trianguli rectilinei duobus angulis dari tertium, & ex uno dato dari summā reliquorum, eo quod omnes tres conficiunt duos rectos.

Lem 12. Et circulis per polos laterum trianguli sphærici transeuntibus, aliud triangulum constitui cujus latera sunt complementa angulorum & anguli laterum prioris ad semicirculum.^{(32)}

Et his prælibatis, accedimus jam ad solutionem triangulorum.

SECT: 2

DE TRIANGULIS PLANIS RECTANGULIS.

THEOREMATA.

1. Ut Radius ad sinum anguli alterutrius acuti, ita Hypotenusa ad crus angulo illi oppositum.

---

(29) In modern equivalent,

$$R : \operatorname{Sin} \tfrac{1}{2}(\widehat{AD} + \widehat{AF}) = \operatorname{Cos} \tfrac{1}{2}(\widehat{AD} - \widehat{AF}) : \tfrac{1}{2}(\operatorname{Sin} \widehat{AD} + \operatorname{Sin} \widehat{AF})$$
$$= \operatorname{Sin} \tfrac{1}{2}(\widehat{AD} - \widehat{AF}) : \tfrac{1}{2}(\operatorname{Cos} \widehat{AF} - \operatorname{Cos} \widehat{AD}).$$

(30) Correspondingly,

$$R : \operatorname{Cos} \tfrac{1}{2}(\widehat{AD} + \widehat{AF}) = \operatorname{Sin} \tfrac{1}{2}(\widehat{AD} - \widehat{AF}) : \tfrac{1}{2}(\operatorname{Sin} \widehat{AD} - \operatorname{Sin} \widehat{AF})$$
$$= \operatorname{Sin} \tfrac{1}{2}(\widehat{AD} - \widehat{AF}) : \tfrac{1}{2}(\operatorname{Cos} \widehat{AD} + \operatorname{Cos} \widehat{AF}).$$

(31) That is,

$$\operatorname{Sin}(\widehat{AD} + \widehat{AF}) : (\operatorname{Sin} \widehat{AD} + \operatorname{Sin} \widehat{AF}) = (\operatorname{Sin} \widehat{AD} - \operatorname{Sin} \widehat{AF}) : \operatorname{Sin}(\widehat{AD} - \widehat{AF}).$$

(32) It is surprising that Newton does not think this constructional definition of the polar triangle of a given spherical triangle worthy of further explanation: a century before, though such mathematicians as Regiomontanus and Pitiscus made use of polar properties in particular theorems, the subtleties of this fundamental spherical duality were far from being appreciated. Only after François Viète in Caput XIX, Scholion V of his *Variorum de Rebus Mathematicis*

Lemma 8. As the radius to the sine of the half sum, so is the cosine of the half difference to the half sum of the sines, and also the sine of the half difference to the half difference of the cosines. For

$$CE : EN = CQ : (QP \text{ or}) \tfrac{1}{2}(FH + DG) = DQ : RQ.^{(29)}$$

Lemma 9. As the radius to the cosine of the half sum, so the sine of the half difference to the half sum of the cosines. For

$$CE : CN = DQ : (DR \text{ or}) \tfrac{1}{2}(DG - FH) = CQ : (CP \text{ or}) \tfrac{1}{2}(CG + CH).^{(30)}$$

Lemma 10. As the sine of the sum to the sum of sines, so the difference of the sines to the sine of the difference. For $CQ^2 = CD^2$ (or CE^2) $- DQ^2$. But to CQ, CE and DQ are proportional CP, CN, DR and consequently

$$CP^2 \text{ (or } CQ^2 - QP^2) = CN^2 - DR^2,$$

so that $QP^2 - DR^2 = CQ^2 - CN^2 = EN^2 - DQ^2$ and hence

$$QP + DR \text{ (or } DG) : EN + DQ = EN - DQ : QP - DR \text{ (or } FH).^{(31)}$$

As was to be shown.

Lemma 11. You should know also that from two angles of a rectilinear triangle given the third is given, and that from one given the sum of those remaining is given, for the reason that all three together make up two right angles.

Lemma 12. And further that, from circle arcs passing through the poles of the sides of a spherical triangle a second triangle is constructed whose sides are the complements to a semicircle of the angles, and angles of the sides, of the former.^(32)

And with this foretaste let us now pass on to the solution of triangles.

SECTION 2. ON RIGHT-ANGLED PLANE TRIANGLES

Theorems

1. As the radius to the sine of either acute angle, so the hypotenuse to the *crus* opposite that angle.

Responsorum Liber VIII (Tours, 1593 [= *Opera Mathematica* (Leyden, 1646): 422 ff.]) indicated the riches of his Ἐναλλαγὴ πλευρογωνική did the theory of the polar (supplementary) triangle achieve its due place in the standard trigonometrical expositions and even then received its popular accolade in Napier's *Descriptio* (note (22) above): Liber II, Caput VI, §11: 55: 'In omni triangulo sphærico mutari possunt latera in angulos, & anguli in latera: assumptis tamen prius pro unico quovis angulo, & suo subtendente latere suis ad semicirculum reliquis'. (See A. von Braunmühl, 'Zur Geschichte des sphärischen Polardreieckes', *Bibliotheca Mathematica*, ₃1 (1898): 65–72; and compare Tropfke's *Geschichte* (note (4) above), 5: 125–6.) Newton doubtless borrows his present exposition of polar duality from Ward's *Idea* (note (22) above): 22: 'Anguli cujuscunꝙ Trianguli [sphærici] in Latera, & Latera in Angulos commutari possunt, sumpto, pro maximo sive Latere, sive Angulo, ejus Complemento ad semicirculum.'

2. Ut Radius ad tangentem anguli alterutrius acuti, ita crus angulo illo conterminum ad crus alterum.[33]

3. Crus alterutrum medium proportionale est inter summam et differentiam hypotenusæ et cruris alterius.[34]

DEMONSTRATIO.

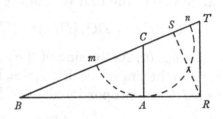

1 & 2.[35] Sit *ABC* triangulum rectangulum ad *A*. Produc *BA* ad *R* ut sit *BR* radius canonicus. Ad *BR* erige normalem *RT* occurrentem *BC* productæ in *T*, et ad *BT* demitte normalem *RS*. Erit *RS* sinus canonicus & *R*[*T*] tangens canonica anguli *B*. Sunt autem triangula rectangula *BAC*, *BSR*, *BRT* ob angulum *B* communem similia: et inde *BR*.*RS*::*BC*.*CA*. et *BR*.*RT*::*BA*.*AC*.

3. Centro *C* intervallo *CA* describe circulum secantem *BT* in *m* et *n*, et (per 36. [III] Elem) tangens *BA* erit medium proportionale inter linearum *BC*, *AC* summam *Bn* et differentiam *Bm*.[36]

CASUS PROBLEMATUM.

His paucis solvuntur quæstiones de triangulis planis rectangulis. Nam 1mo, si dantur anguli et Hypotenusa, invenientur crura[37] per Theorema 1.

2. Si dantur anguli et crus invenietur Hypotenusa per Th. 1, & alterum crus per Th. 2.

3. Si datur Hypotenusa et crus, invenientur anguli per Th: 1 et Lem: 11, et alterum crus per Th: 3.

4. Si dantur crura invenientur anguli per Th: 2 & Lem 11, Dein ex crure et angulo invenietur Hypotenusa per Th. 1.

USUS CANONIS.

Ut si detur in triangulo quovis *BAC*, angulus *B* et Hypotenusa *BC*, et quæratur crus *AC*: juxta Theor. 1 erit *R*.*sB*::*BC*.*AC*, adeoqͻ quærendo *sB* in canone & multiplicando *BC* per *sB* ac dividendo per *R* invenietur *AC*. Quod si daretur crus *AC* et Hypotenusa *BC* et quæreretur angulus *C*, rursus dirigimur (per Casum 3) ad hanc anlogiam *R*.*sB*::*BC*.*AC*. Ubi

---

(33) These first two 'theorems' are little more than restatements of the definitions of 'sine' and 'tangent' in Section 1, 'Prænoscenda' 2 and 3.

(34) The fundamental Pythagorean theorem (Euclid, *Elements* I, 47) expounded in the form $BA = \sqrt{[(BC+CA)(BC-CA)]}$.

2. As the radius to the tangent of either acute angle, so the *crus* bounding that angle to the other *crus*.[33]

3. Either *crus* is a mean proportional between the sum and the difference of the hypotenuse and other *crus*.[34]

Demonstration

1 and 2.[35] Let *ABC* be a triangle right-angled at *A*. Produce *BA* to *R* so that *BR* is the canonical radius. To *BR* erect the normal *RT* meeting *BC* produced in *T*, and to *BT* let fall the normal *RS*. The latter will then be the canonical sine and *RT* the canonical tangent of the angle *B*. But the right-angled triangles *BAC*, *BSR* and *BST* are, because of the common angle *B*, similar: in consequence $BR:RS = BC:CA$ and $BR:RT = BA:AC$.

3. With centre *C* and radius *CA* describe a circle cutting *BT* in *m* and *n*, and then (by *Elements*, III, 36) the tangent *BA* will be a mean proportional between the sum *Bn* of the lines *BC*, *AC* and their difference *Bm*.[36]

Cases of Problems

By these few propositions questions regarding plane right-angled triangles are solved. For

1. If the angles and the hypotenuse are given, the *crura*[37] will be found by Theorem 1.

2. If the angles and a *crus* are given, the hypotenuse will be found by Theorem 1 and the other *crus* by Theorem 2.

3. If the hypotenuse and a *crus* are given, the angles will be found by Theorem 1 and Lemma 11 and the other *crus* by Theorem 3.

4. If the *crura* are given, the angles will be found by Theorem 2 and Lemma 11, and then from a *crus* and an angle the hypotenuse will be found by Theorem 1.

The Use of a Canon

If, for instance, in any triangle *BAC* there be given the angle *B* and the hypotenuse *BC*, and the *crus AC* be sought, according to Theorem 1 there will be $R:\mathrm{Sin}\,B = BC:AC$; accordingly, by searching out Sin *B* in the canon and multiplying *BC* by Sin *B* and then dividing by *R*, *AC* will be found. But should the *crus AC* and hypotenuse *BC* be given and the angle *C* sought, we are again (by Case 3) directed to this proportion $R:\mathrm{Sin}\,B = BC:AC$. Here since the term

---

(35) Compare Ward's *Idea* (note (22) above): 2–3.

(36) This demonstration is effectively circular since *Elements*, III, 36 depends on I, 47, for its proof.

(37) This replaces 'latera' (sides). Newton has made an equivalent change of 'latus' into 'crus' many times in the sequel.

cum terminus quæsitus s *B* in medio loco sit, divido factum extremorum *R* × *AC* per datum medium terminum *BC* ut ille alter medius s *B* prodeat.

In numeris, sit *BC* hypotenusa constans partibus 794 & *B* angulus graduum 35 & 48^{min} et erit juxta Th: 1 Rad. sin 35^{gr} 48′ :: 794 . *AC*. Quæro igitur in tabula sinuum$^{(38)}$ e regione anguli [3]5^{gr} 48′ sinum anguli illius, qui est 58496,$^{(39)}$ ut et Radium qui stat e regione 90^{gr} et his substitutis proportio superior fiet 100000 . 58496 :: 794 . *AC*. Quare multiplico 794 per 58496, et factum divido per 100000, et quotus 464,458 erit longitudo quæsiti *AC*.

Rursus si dato crure *AC* = 464,458, et hypotenusa *BC* = 794 quæratur angulus *B* : juxta proportionem præcedentem, nempe *R* . s *B* :: *BC* . *AC*, seu 100000 . s *B* :: 794 . 464,458, multiplicandi erunt termini 100000 & 464,458 & factum 46445800 dividendum erit per 794, et quotus 58496 erit s *B*. Quare in tabula sinuum quæro sinum 58496 et e regione invenio angulum 35^{gr} 48′ cujus iste sinus est, hoc est angulum desideratum *B*.

Cæterum compendij gratia pro numeris utimur utplurimum logarithmis numerorum: quorum beneficio multiplicationes et divisiones vitamus. Ubi enim in numeris multiplicandum esset in logarithmis addimus tantum & ubi in illis dividendum esset, in his subducimus, et pro media proportionali in illis, sumimus in his tantum semisummam. Hac de causa vice sinuum et tangentium, logarithmos sinuum et tangentium in tabulas jamdudum retulerunt Mathematici: ipsis imponentes nomen sinuū et tangentiū artificialium ut a naturalibus distinguant.$^{(40)}$ In allatis igitur exemplis, per Logarithmos talis erit operatio.$^{(41)}$

1. Datis hypotenusa *BC* = 794 & angulo *B* = 35^{gr} 48′, ad inveniendum *AC* dic,

Ut Radius 100000 - - - - - - - - 10.000000
ad sinū anguli 35^{gr} 48′ - - - - - - 9.767124
ita hypotenusa 794 - - - - - - - - 3.899820
ad crus *AC* 464,45 - - - - - - - - 3.666944.

2 Datis crure *AC* = 464,45 & hypotenusa *BC* = 794, ad inveniendum angulum *B* dic,

Ut hypotenusa 794 - - - - - - - - 3.899820
ad crus 464,45 - - - - - - - - - 3.666944
ita Radius 100000 - - - - - - - 10.000000
ad sinum *B* 35^{gr} 48′ - - - - - - - 9.767124.

---

(38) Say, by extrapolation from the manuscript 'Tabula sinuum ad semigradus' reproduced in Appendix 2.

(39) That is, Sin 35° 48′ where the *sinus totus* is 10^5.

(40) The terminology is John Napier's in his 'pre-logarithmic' *Constructio* [*Tabulæ Artificialis*], published two years after his death by his son Robert with the anachronistic title *Mirifici Logarithmorum Canonis Constructio*; *Et eorum ad naturales ipsorum numeros habitudines* (Edinburgh, 1619): there the name 'Numerus artificialis sinus dati' is introduced (*ibid.*: 15) for what Napier later came to call 'Logarithmus sinus' (*Descriptio* (note (22) above): 3).

Sin B required occupies a middle position, I divide the product $R \times AC$ of the extremes by the given middle term BC to get the other middle one Sin B.

Numerically, let the hypotenuse BC consist of 794 parts and the angle B of 35 degrees 48 minutes: then by Theorem 1 Radius: Sin $35° 48' = 794 : AC$. So in a sine table[38] in line with the angle $35° 48'$ I look up the sine of that angle, which is 58 496,[39] and also the radius (which stands in line with $90°$), and when these are substituted the above proportion will become

$$100\,000 : 58\,496 = 794 : AC.$$

Hence I multiply 794 by 58 496 and divide the product by 100 000, and the quotient 464·458 will be the length of AC required.

Again, if, given the *crus* $AC = 464·458$ and hypotenuse $BC = 794$, the angle B be sought, according to the preceding proportion (viz. $R : \mathrm{Sin}\,B = BC : AC$, or $100\,000 : \mathrm{Sin}\,B = 794 : 464·458$) the terms 100 000 and 464·458 must be multiplied and their product divided by 794, and the quotient 58 496 will be Sin B. Hence in the table of sines I look for the sine 58 496 and in line with it I find the angle $35° 48'$ of which it is the sine: this is the desired angle B.

However, for brevity's sake in the place of numbers we employ for the most part logarithms of the numbers: by their benefit we avoid multiplications and divisions. For when in numbers we had to multiply, in logarithms we merely add; and when in the former we had to divide, in the latter we subtract; and in place of the mean proportional in the former, we take in the latter merely the half sum. For this reason instead of simple sines and tangents computers have for a long while now listed logarithms of sines and tangents in their tables, imposing on these the name of 'artificial' sines and tangents to distinguish them from natural ones.[40] In the examples presented, therefore, the procedure by means of logarithms will be as follows.[41]

1. Given the hypotenuse $BC = 794$ and the angle $B = 35° 48'$, to find AC say:

As the radius 100 000 - - - - - - - - - -	[−]10·000000
to the sine of the angle $35° 48'$, - - - - -	9·767124
so the hypotenuse 794 - - - - - - - - -	3·899820
to the *crus* $AC = 464·45$ - - - - - - - -	3·666944.

2. Given the *crus* $AC = 464·45$ and hypotenuse $BC = 794$, to find the angle B say:

As the hypotenuse 794 - - - - - - - - -	[−]3·899820
to the *crus* 464·45, - - - - - - - - - - -	3·666944
so the radius 100 000 - - - - - - - - - -	10·000000
to the sine of $B = 35° 48'$ - - - - - - - -	9·767124.

---

(41) Observe that Newton follows contemporary practice in assuming a conventional radius (*sinus totus*) of 10^{10} for his logarithmic sines.

His operationibus proportio ita ordinatur ut terminus quæsitus semper occupet locum ultimum: dein terminorū datorum logarithmi quæruntur in tabulis, deq̃ summa secundi ac tertij subducendo primum invenitur logarithmus quæsiti. Deniq̃ in tabulis e regione hujus logarithmi invenitur quæsitum.

Sect 3.
De Triangulis planis obliquangulis.(42)
Theoremata.

1. Triangulorum latera sunt sinubus oppositorum angulorum proportionalia.

2. Ut summa duorum laterum ad eorum differentiam ita tangens semisummæ angulorum oppositorum ad tangentem semidifferentiæ eorundem.

3. Ut semisumma trium laterum ad excessum illius semisum̄æ supra latus aliquod, ita tangens complementi semissis anguli alterutrius lateri illi contermini ad tangentem semissis anguli alterius contermini.

4. Ut excessus semisummæ trium laterum supra unum latus ad excessum ejus supra alterum latus, ita tangens semissis anguli posteriori lateri opposti ad tangentem semissis anguli priori lateri contermini.

5. Ut medium proportionale inter semisummam trium laterum et excessum illius semisummæ supra unum e lateribus ad medium proportionale inter excessum illius semisummæ supra alterum latus et excessum ejusdem supra tertium latus, ita Radius ad tangentem semissis anguli primo lateri opposti.(43)

6. Et ita tangens complementi semissis anguli illius ad Radium.

7. Ut medium proportionale inter latera duo ad medium proportionale inter semisummam trium laterum et excessum illius semisummæ supra tertium latus, ita Radius ad sinum complementi semissis anguli tertio lateri opposti.

8. Ut medium proportionale inter latera duo ad medium proportionale inter excessum semisummæ trium laterum supra unum e duobus lateribus & excessum ejus alterum, ita Radius ad sinum semissis anguli tertio lateri opposti.

Demonstratio.

1. In triangulo *ABC* demisso perpendiculo *CD*, est (per Th: 1, Sect: 2) *AC.DC*::*R.s A*, et *DC.BC*::s*B.R*. Quare ex æquo, *AC.BC*::s*B.s A*. Q.E.D.

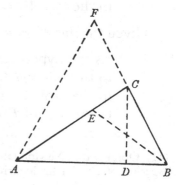

In these schemes of operation the proportion is so arranged that the required term shall always occupy the final position. Next, the logarithms of the given terms are looked up in tables, and by taking the first from the sum of the second and third the logarithm of that sought is found. Finally, in line with this logarithm in the tables the term required is found.

SECTION 3. ON OBLIQUE-ANGLED PLANE TRIANGLES[42]

Theorems

1. The sides of triangles are proportional to the sines of the angles opposite.

2. As the sum of two sides to their difference, so the tangent of the half sum of the opposite angles to the tangent of their half difference.

3. As the half sum of the three sides to the excess of that half sum over some side, so the cotangent of half either angle adjacent to that side to the tangent of half the other adjacent angle.

4. As the excess of the half sum of the three sides over one of the sides to its excess over a second side, so the tangent of half the angle opposite the latter side to the tangent of half the angle adjacent to the first one.

5. As the mean proportional between the half sum of the three sides and the excess of that half sum over one of the sides to the mean proportional between the excess of that half sum over a second side and its excess over the third one, so is the radius to the tangent of half the angle opposite to the first side.[43]

6. And so also the cotangent of half that angle is to the radius.

7. As the mean proportional between two sides to the mean proportional between the half sum of the three sides and the excess of that half sum over the third side, so the radius to the cosine of half the angle opposite the third side.

8. As the mean proportional between two sides to the mean proportional between the excess of the half sum of the three sides over one of the two sides and its excess over the other, so is the radius to the sine of half the angle opposite the third side.

Demonstration

1. In the triangle ABC, after the perpendicular CD is let fall, there is (by Theorem 1 of Section 2) $AC:DC = R:\mathrm{Sin}\,A$ and $DC:BC = \mathrm{Sin}\,B:R$. Hence *ex æquo* $AC:BC = \mathrm{Sin}\,B:\mathrm{Sin}\,A$. As was to be demonstrated.

---

(42) A first version, headed 'DE TRIANGULIS QUIBUSVIS PLANIS', of this third section is reproduced as Appendix 3.1 below, together with some preliminary drafts for the present text (Appendix 3.2). It will be evident that the former is heavily indebted for its proofs to Propositions 3, 4 and 5 of the preceding 'Epitome' but the parentage of the present revise is much less evident.

(43) A cancelled final phrase 'et tangens complementi semissis ejus ad Radium' (and the cotangent of that half [angle] to the radius) reveals that Newton first intended to subsume the following proposition under the present 'Prop. 5'.

2. Et mixtim

$$AC+BC\,.\,AC-BC::\mathrm{s}\,B+\mathrm{s}\,A\,.\,\mathrm{s}\,B-\mathrm{s}\,A::(\text{per Lem 6})\,\mathrm{t}\frac{B+A}{2}\,.\,\mathrm{t}\frac{B-A}{2}.$$

3. Id est si sumantur $AE=EB$ et $BF=AF$, in triangulo CBE,

$$EB+EC+BC\,.\,EB+EC-BC::\mathrm{t'}\frac{C}{2}\,.\,\mathrm{t}\frac{EBC}{2}.\ ^{(44)}$$

4. Et in triangulo CAF, $AC+AF-CF\,.\,AC-AF+CF::\mathrm{t}\dfrac{FCA}{2}\,.\,\mathrm{t}\dfrac{CAF}{2}.$

5, 6. Et rursus per prop 3 est $AF+CF+AC\,.\,AF+CF-AC::\mathrm{t'}\dfrac{CAF}{2}\,.\,\mathrm{t}\dfrac{FCA}{2}$,

et scribendo m pro $AF+CF+AC$. n pro $AF+CF-AC$, p pro $AC+AF-CF$

et q pro $A[C]+CF-AF$, et addendo rationes, fit $mp\,.\,nq::\mathrm{t'}\dfrac{A}{2}\,.\,\mathrm{t}\dfrac{A}{2}.$ id est

$::RR\,.\,\mathrm{tt}\dfrac{A}{2}::\mathrm{t't'}\dfrac{A}{2}\,.\,R^q.$

7, 8. Et componendo

$$mp+nq\left(=4AF\times CA\right)\,.\,mp\,.\,nq::RR+\mathrm{tt}\frac{A}{2}$$

$$\left(=\text{secanti}\,\frac{A^{(45)}}{2}\right)\,.\,RR\,.\,\mathrm{tt}\frac{A}{2}::RR\,.\,\mathrm{s's'}\frac{A}{2}\,.\,\mathrm{ss}\frac{A}{2}.$$

Esse autem $mp+nq=4AF\times AC$ manifestum est. Nam

$$mp=AC+AF+CF\times AC+AF-CF=\overline{AC+AF}^q-CF^q$$

$\&\ nq=CF+AF-AC\times CF-AF+AC=CF^q-\overline{AC-AF}^q$ $\&$ summa [erit]

$$\overline{AC+AF}^q-\overline{AC-AF}^q=(\text{per 5 \& 6 \textsc{ii} Elem})\ 4AC\times AF.$$

Casus Problematum.

1. Datis igitur angulis et uno latere invenientur reliqua per Th. 1.

2. Datis duobus lateribus et angulo uni eorum opposito invenietur angulus alteri oppositus per Th 1, Dein ex angulis et alterutro latere invenietur tertium latus itidem per Th. 1.[46]

3. Datis duobus lateribus et angulo interjecto, cum hujus dimidij complementum sit semisumma angulorum oppositorum, invenietur semidifferentia

---

(44) That is, '$AC+BC\,.\,AC-BC::t\dfrac{B+A}{2}\,.\,t\dfrac{B-A}{2}$' or, on putting $BC=a$ and $AC=b$,

$b+a:b-a=\operatorname{Tan}\tfrac{1}{2}(\hat{B}+\hat{A}):\operatorname{Tan}\tfrac{1}{2}(\hat{B}-\hat{A})$. Observe that from this point onwards Newton reverts to Ward's notation for cosine and cotangent (see note (9) above), further introducing

the contractions '$\mathrm{ss}\dfrac{A}{2}$', '$\mathrm{s's}\dfrac{A}{2}$', '$\mathrm{tt}\dfrac{A}{2}$' and '$\mathrm{t't'}\dfrac{A}{2}$' for the respective squares of $\mathrm{s}\dfrac{A}{2}$, $\mathrm{s'}\dfrac{A}{2}$, $\mathrm{t}\dfrac{A}{2}$

and $\mathrm{t'}\dfrac{A}{2}$ (on the analogy, doubtless, of 'RR' for the square of R).

2. And *mixtim*

$$AC+BC:AC-BC = \operatorname{Sin} B + \operatorname{Sin} A : \operatorname{Sin} B - \operatorname{Sin} A$$

$$= (\text{by Lemma 6}) \operatorname{Tan} \tfrac{1}{2}(B+A):\operatorname{Tan} \tfrac{1}{2}(B-A).$$

3. That is, if in the triangle CBE there be taken $AE = EB$ and $BF = AF$, then $EB+EC+BC:EB+EC-BC = \operatorname{Cot} \tfrac{1}{2}C:\operatorname{Tan} \tfrac{1}{2}\widehat{EBC}.$[44]

4. And in the triangle CAF

$$AC+AF-CF:AC-AF+CF = \operatorname{Tan} \tfrac{1}{2}\widehat{FCA}:\operatorname{Tan} \tfrac{1}{2}\widehat{CAF}.$$

5/6. Again, by Proposition 3,

$$AF+CF+AC:AF+CF-AC = \operatorname{Cot} \tfrac{1}{2}\widehat{CAF}:\operatorname{Tan} \tfrac{1}{2}\widehat{FCA}$$

and so, on writing m for $AF+CF+AC$, n for $AF+CF-AC$, p for $AC+AF-CF$ and q for $AC+CF-AF$, and then compounding the ratios, there results

$$mp:nq = \operatorname{Cot} \tfrac{1}{2}A:\operatorname{Tan} \tfrac{1}{2}A,$$

that is, $R^2:\operatorname{Tan}^2 \tfrac{1}{2}A$ or $\operatorname{Cot}^2 \tfrac{1}{2}A:R^2.$

7/8. And *componendo*

$$mp+nq \text{ (or } 4AF \times CA):mp:nq = R^2 + \operatorname{Tan}^2 \tfrac{1}{2}A$$

$$(\text{or } \operatorname{Sec}^{[2]} \tfrac{1}{2}A^{[45]}):R^2:\operatorname{Tan}^2 \tfrac{1}{2}A = R^2:\operatorname{Cos}^2 \tfrac{1}{2}A:\operatorname{Sin}^2 \tfrac{1}{2}A.$$

That, however, $mp+nq = 4AF \times AC$ is obvious. For

$$mp = (AC+AF+CF)(AC+AF-CF) = (AC+AF)^2 - CF^2$$

while $nq = (CF+AF-AC)(CF-AF+AC) = CF^2 - (AC-AF)^2$ and so their sum will be $(AC+AF)^2 - (AC-AF)^2 = (\text{by *Elements*, II, 5/6}) \; 4AC \times AF.$

Cases of Problems

1. Given, therefore, the angles and one side, the remainder will be found by Theorem 1.

2. Given two sides and the angle opposite to one of them, the angle opposite to the other will be found by Theorem 1, then from the angles and one or other side the third side will be found, likewise by Theorem 1.[46]

3. Given two sides and their included angle, since the complement of half the last is the half sum of the angles opposite, the half difference of the angles

(45) Read 'quadrato secantis $\dfrac{A}{2}$'.

(46) Newton does not record that two solutions are possible in this 'ambiguous' case. (Ward in his *Idea Trigonometriæ* does not mention it at all.)

angulorum oppositorū per Th. [2]; quæ addita semisummæ efficit angulum majorem ex oppositis, subducta efficit minorem.[47] Dein ex angulis et alterutro latere invenietur tertium latus per Theor 1.

4. Datis tribus lateribus, invenietur angulus quilibet per Th 2.[48] Dein ex angulo illo et lateribus invenientur alij anguli per Th. 1 & Lem. 5.

SECT. 4.
DE TRIANGULIS SPHÆRICIS RECTANGULIS,
THEOREMATA.

In triangulo sphærico rectangulo considerabimus tantū hos quinꝗ terminos, crura duo, & complementa tum Hypotenusæ tum duorum angulorum ad hypotenusam.[49] Et ubi horum aliquis ut terminus impar vel medius spectatur, duos utrinꝗ proximos vocabimus extremos proximos, reliquos duos extremos remotos. Dicimus igitur

1. Ut Radius ad sinum complementi extremi remoti, ita sinus complementi alterius extremi remoti ad sinum medij.

2. Ut Radius ad tangentem extremi proximi, ita tangens alterius extremi proximi ad sinum medij.

Vel ut uno Theoremate hæc duo complectamur dicimus Rectangula sub Radio et sinu medij, sub co-sinubus extremorum remotorum et sub tangentibus extremorum proximorum æqualia esse.[50]

DEMONSTRATIO.

[51]Medius terminus aut crus est, aut complementum hypotenusæ aut complementum anguli. Et in his tribus casibus ubi extremi remoti sunt Theoremata dabunt has analogias.

1. Ut Radius ad sinum hypotenusæ ita sinus anguli alterutrius ad sinum cruris oppositi.

2. Ut Radius ad sinum complementi cruris alterutrius ita sinus complementi cruris alterius ad sinum complementi hypotenusæ.

---

(47) The greater angle lies opposite the greater side, of course.

(48) Read '3, 4, 5, 6, 7 vel 8' (3, 4, ..., 7 or 8). Newton has neglected to amend his reference to 'Th 2' in the cancelled earlier version of this section (reproduced as Appendix 3.1 below).

(49) Newton first continued 'Nam angulum rectum inter terminos de quibus jam agitur non numeramus' (for we do not number the right angle among the terms now at issue). A cancelled version of the following text, indebted to Ward's *Idea* for its variant 'conjoint/disjoint' terminology (see §1: note (17)), reads equivalently: 'Et ex his quinꝗ conjunctos vocabimus qui intermedium terminum non habent, disjunctos qui habent intermedium. His præstratis dicimus

'1. In tribus terminis quorum unus a reliquis duobus disjunctus est. Ut Radius ad sinum...'.

(50) Napier's rules of circular parts, taken over with some slight alteration from the 'Epitome' (§1 preceding).

opposite will be found by Theorem 2. This, when added to their half sum, yields the greater of the opposite angles, and, when taken away, the lesser.[47] Then from the angles and one or other side the third side will be found by Theorem 1.

4. Given the three sides, any angle you desire will be found by Theorem 2.[48] Then from that angle and the sides the other angles will be found by Theorem 1 and Lemma 5.

SECTION 4. ON SPHERICAL RIGHT-ANGLED TRIANGLES

Theorems

In a spherical right-angled triangle we shall consider merely these five elements: the two *crura* and the complements both of the hypotenuse and the two angles adjacent to the hypotenuse.[49] And when some one of these is regarded as odd element out or middle one, the two nearest on either side we shall call the nearest outer ones, and the remaining two the farthest outer ones. We accordingly assert:

1. As the radius to the cosine of a farthest outer element, so the cosine of the other farthest outer one to the sine of the middle.

2. As the radius to the tangent of a nearest outer element, so the tangent of the other nearest outer one to the sine of the middle.

Or to embrace these two in a single theorem, we assert that the products of the radius and the sine of the middle element, of the cosines of the farthest outer ones, and of the tangents of the nearest outer ones are equal.[50]

Demonstration

[51]The middle element is either a *crus* or the complement of the hypotenuse or the complement of an angle. And in these three cases the theorems will, when the outer elements are the farthest ones, yield these three proportions:

1. As the radius to the sine of the hypotenuse, so the sine of either angle to the sine of the opposite *crus*.

2. As the radius to the cosine of either *crus*, so the cosine of the other *crus* to the cosine of the hypotenuse.

---

(51) Newton has cancelled a first opening: 'Si medius terminus est crus, extremi remoti erunt complementum anguli oppositi et complementum Hypotenusæ, extremi proximi complementum anguli contermini et crus alterum. Si medius terminus est complementum hypotenusæ, extremi remoti erunt crura, proximi complementa anguloru. Si medius terminus est complementum anguli, extremi remoti erunt crus oppositum et complementum anguli alterius, proximi crus conterminum.' (If the middle term is a *crus*, the outer extremes will be the complement of the angle opposite and the complement of the hypotenuse, the near extremes the complement of the adjacent angle and the second *crus*. If the middle term is the complement of the hypotenuse, the outer extremes will be the *crura*, the inner ones the complements of the angles. If the middle term is the complement of an angle, the outer extremes will be the opposite *crus* and the complement of the second angle, the near ones the adjacent *crus* [and the complement of the hypotenuse].)

3. Ut Radius ad sinum complementi cruris alterutrius, ita sinus anguli huic cruri contermini ad sinum complementi anguli alterius.

Ubi vero extremi sunt proximi, Theoremata dabunt has.

4. Ut Radius ad tangentem complementi anguli alterutrius, ita tangens cruris oppositi ad sinum cruris alterius.

5. Ut Radius ad tangentem complementi unius anguli ita tangens complementi alterius anguli ad sinum complementi hypotenusæ.

6. Ut Radius ad tangentem cruris alterutrius ita tangens complementi hypotenusæ ad sinum complementi anguli intermedij.

In tot analogias resolvantur Theoremata et non plures. Unde si has sex demonstravero, Theoremata constabunt.

Sint AB, AC, BC tres arcus quadrantales circulorum maximorum in superficie sphæræ descripti, puncto C referente tam polos circuli AB quam centrum sphæræ, et superficie CAB referente tam superficiem trianguli sphærici quod tres angulos ad A, B, C rectos habet & superficiei sphæræ totius octava pars est quam superficiem planam Quadrantis ABC[52]. Sint BD, CE alij duo arcus quadrantales cum prioribus constituentes triangula sphærica $CD[F]$, $BE[F]$[53] rectangula ad D et E. Deniꝗ sint $ER_{[,]}$ FS sinus et $AT_{[,]}$ DV tangentes arcuum

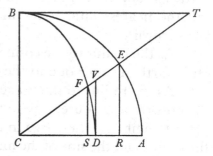

AE, DF: et triangula CFS, CVD, CER in parallelis planis constituta erunt tum sibi ipsis tum triangulo TCB similia. Quare $CE \cdot ER :: CF \cdot FS$, et $CB \cdot BT :: VD \cdot CD$. hoc est in triangulo sphærico CFD, Ut Radius ad sinum anguli C ita sinus hypotenusæ CF ad sinum cruris FD, quæ Analogia 1ma est; et Ut Radius ad tangentem complementi anguli $[C]$ ita tangens cruris DF ad sinum cruris CD, Analogia 4ta: Est in triangulo sphærico BEF, $R \cdot sc\,BE :: sc\,EF \cdot sc\,BF$, Analogia 2da; et $R \cdot t\,BE :: tc\,BF \cdot sc\,DA$ seu sc B, Analogia 6ta. Et in triangulo sphærico BEF per Analogiam primam et quartam erit $R \cdot s\,BF :: s\,F \cdot s\,BE$, et $R \cdot tc\,F :: t\,BE \cdot s\,EF$, id est in triangulo sphærico CDF, $R \cdot sc\,DF :: s\,F \cdot sc\,C$, Analogia 3^{a}; & $R \cdot tc\,F :: tc\,C \cdot sc\,CF$, Analogia 5ta.

---

(52) As with the equivalent orthogonal projection in the 'Epitome' (see §1: note (16)) C, D, F, S and V represent all points in the lines drawn perpendicular to the plane of the figure through them. Compare the oblique perspective which we have introduced into our English text.

(53) No doubt misreading his figure, Newton carelessly wrote 'CDV, BEV' in the manuscript.

3. As the radius to the cosine of either *crus*, so the sine of the angle adjacent to this *crus* to the cosine of the other angle.

When, to be sure, the outer elements are nearest ones, the theorems will yield these:

4. As the radius to the cotangent of either angle, so the tangent of the opposite *crus* to the sine of the other *crus*.

5. As the radius to the cotangent of one angle, so the cotangent of the other one to the cosine of the hypotenuse.

6. As the radius to the tangent of either *crus*, so the cotangent of the hypotenuse to the cosine of the intervening angle.

Into so many proportions may the theorems be resolved and no more. Hence, once I have demonstrated these six, the theorems will be established.

Let $\widehat{AB}$, $\widehat{AC}$, $\widehat{BC}$ be three quadrantal arcs of great circles described in the surface of a sphere, with the point C denoting both the poles of the circle AB and the sphere's centre, and the surface CAB indicating both the surface of the spherical triangle, having right angles at A, B, C, which is an eighth of the total spherical surface and the plane surface of the quadrant ABC.[52] Let $\widehat{BD}$, $\widehat{CE}$ be two other quadrantal arcs constituting with the previous ones the spherical triangles CDF, BEF[53]

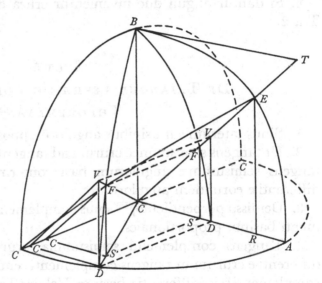

right-angled at D and E. Finally, let ER, FS be sines and AT, DV tangents of the arcs $\widehat{AE}$, $\widehat{DF}$: the triangles CFS, CVD and CER, situated in parallel planes, will then be similar both one to another and to the triangle TCB. Hence $CE:ER = CF:FS$ and $CB:BT = VD:CD$. That is, in the spherical triangle CFD, As the radius to the sine of the angle C, so the sine of the hypotenuse $\widehat{CF}$ to the sine of the *crus* $\widehat{FD}$ (which is Proportion 1); and, As the radius to the cotangent of the angle C, so the tangent of the *crus* $\widehat{DF}$ to the sine of the *crus* $\widehat{CD}$ (Proportion 4). Further, in the spherical triangle BEF, $R:\mathrm{Cos}\,\widehat{BE} = \mathrm{Cos}\,\widehat{EF}:\mathrm{Cos}\,\widehat{BF}$ (Proportion 2); and $R:\mathrm{Tan}\,\widehat{BE} = \mathrm{Cot}\,\widehat{BF}:(\mathrm{Cos}\,\widehat{DA}$ or$)\,\mathrm{Cos}\,B$ (Proportion 6). Also in the triangle BEF, by Proportions 1 and 4, there is

$$R:\mathrm{Sin}\,\widehat{BF} = \mathrm{Sin}\,F:\mathrm{Sin}\,\widehat{BE} \quad \text{and} \quad R:\mathrm{Cot}\,F = \mathrm{Tan}\,\widehat{BE}:\mathrm{Sin}\,\widehat{EF},$$

and consequently, in the triangle CDF, $R:\mathrm{Cos}\,\widehat{DF} = \mathrm{Sin}\,F:\mathrm{Cos}\,C$ (Proportion 3); and $R:\mathrm{Cot}\,F = \mathrm{Cot}\,C:\mathrm{Cos}\,\widehat{CF}$ (Proportion 5).

Casus Problematum.

1. Si datur Hypotenusa et crus invenietur angulus interjectus per Th 2, alter angulus et alterum crus per Th 1.

2. Si datur Hypotenusa & angulus invenietur crus adjacens & alter angulus per Th 2, crus alterum per Th 1.

3. Si datur crus et angulus oppositus invenietur crus alterum per Th 2, hypotenusa et angulus alter per Th 1.

4. Si datur crus et angulus adjacens, invenietur angulus alter per Th 2, crus alterum et hypotenusa per Th 1.

5. Si dantur crura duo invenientur anguli duo per Th 2 & hypotenusa per Th. 1.

6. Si dantur anguli duo invenientur crura duo per Th 1, hypotenusa per Th. 2.

Sect 5
De Triangulis sphæricis obliquangulis.
Theoremata.[54]

1. Sinus laterum sunt sinibus angulorū oppositorum proportionales.

2. Ut tangens semisummæ crurum ad tangentem semidifferentiæ crurum ita tangens semisummæ angulorum basi conterminorum ad tangentem semidifferentiæ eorundem angulorum.

3. Demisso perpendiculo,[55] sinus complementi crurum sunt sinibus complementi basium proportionales.

4. Tangens complementi semisummæ crurum est ad tangentem semidifferentiæ crurum ut tangens complementi semisummæ basium ad tangentem complementi semidifferentiæ basium. Vel sic: Tangens semisummæ basiū est ad tangentem semisummæ crurum ut tangens semidifferentiæ crurum ad tangentem semidiff[er]entiæ basium.

5. Sinus complementi angulorum ad basem sunt sinibus angulorū ad cathetum proportionales.

6. Tangens complementi semisummæ angulorum ad basem est ad tangentem semidifferentiæ eorundem ut tangens semisummæ angulorum ad cathetum sit[56] ad tangentem semidifferentiæ ipsorum.

7. Tangentes crurum sunt sinibus complementi angulorum ad cathetum reciproce proportionales.

---

(54) Originally there were only four theorems, to which Newton subsequently added seventeen more, keeping his 'Th 1' unchanged but reordering his Theorems 2, 3 and 4 as 19, 14 and 15 respectively in the present list.

Cases of Problems

1. If the hypotenuse and a *crus* are given, the intervening angle will be found by Theorem 2, the other angle and second *crus* by Theorem 1.

2. If the hypotenuse and an angle are given, the adjacent *crus* and the other angle will be found by Theorem 2, the other *crus* by Theorem 1.

3. If a *crus* and the opposite angle are given, the second *crus* will be found by Theorem 2, the hypotenuse and other angle by Theorem 1.

4. If a *crus* and the adjacent angle are given, the second angle will be found by Theorem 2, the other *crus* and the hypotenuse by Theorem 1.

5. If the two *crura* are given, the two angles will be found by Theorem 2 and the hypotenuse by Theorem 1.

6. If the two angles are given, the two *crura* will be found by Theorem 1, the hypotenuse by Theorem 2.

SECTION 5. ON SPHERICAL OBLIQUE-ANGLED TRIANGLES

Theorems[54]

1. The sines of the sides are proportional to the sines of the opposite angles.

2. As the tangent of the half sum of the *crura* to the tangent of the half difference of the *crura*, so the tangent of the half sum of the angles adjoining the base to the tangent of the half difference of those angles.

3. On letting fall the perpendicular,[55] the cosines of the *crura* are proportional to the cosines of the base (-segment)s.

4. The cotangent of the half sum of the *crura* is to the tangent of their half difference as the cotangent of the half sum of the bases to the cotangent of the bases' half difference. Or equivalently: The tangent of the half sum of the bases is to that of the half sum of the *crura* as the tangent of the half difference of the *crura* to that of the half difference of the bases.

5. The cosines of the angles at the base are proportional to the sines of the angles at the *cathetus*.

6. The cotangent of the half sum of the angles at the base is to the tangent of their half difference as the tangent of the half sum of the angles at the *cathetus* to that of the latter's half difference.

7. The tangents of the *crura* are inversely proportional to the cosines of the angles at the cathetus.

---

(55) Newton has previously called this the 'cathetus' (compare note (15 above) and he returns to this terminology below.

(56) Read 'est'. Newton may have introduced this careless subjunctive under the spurious influence of the preceding 'ut'.

8. Sinus summæ crurum est ad sinum differentiæ ipsorum ut tangens complementi semisummæ angulorum ad cathetum ad tangentem $\frac{1}{2}$ differentiæ eorundem.

9. Tangentes anguloru ad basem sunt sinibus basiu reciproce proportionales.

10. Sinus summæ angulorum ad basem est ad sinū differentiæ eorundem ut tangens semisummæ basium ad tangentem semidifferentiæ ipsarum.

12.[57] Ut summa tangentium crurum dimidiorum ad differentiam tangentium eorundem, ita tangens complementi semissis anguli inter crura ad tangentem semidifferentiæ reliquorum angulorum.

13. Ut summa tangentis semissis cruris et tangentis complementi semissis alterius cruris ad differentiam illarum tangentium ita tangens dimidij anguli inclusi ad tangentem complementi semisummæ angulorum reliquorum. et tangens semisummæ angulorum reliquorum ad tangentem complementi dimidij anguli inclusi.

14. Ut sinus semisummæ crurum ad sinum semidifferentiæ eorundem ita tangens complementi semissis anguli inter crura ad tangentem semidifferentiæ reliquorum angulorum.

15. Ut sinus complementi semisummæ crurū ad sinum complementi semidifferentiæ crurum, ita tangens complementi semissis anguli inter crura ad tangentem semisummæ reliquorum angulorum.

16. Ut tangens semissis cruris unius ad tangentem semissis cruris alterius ita sinus excessus semisummæ trium angulorum supra angulum priori cruri oppositum ad sinum excessus semisummæ ejusdem supra angulum posteriori lateri oppositum.

17. Ut tangens complementi semissis cruris alterutrius ad tangentem semissis cruris alterius ita sinus semisummæ trium angulorum ad sinum excessus illius semisummæ supra angulum intra crura.

18. Ut medium proportionale inter sinum complementi semisummæ trium laterum et sinum excessus illius semisummæ supra latus aliquod ad medium proportionale inte[r] sinum excessus illius semisummæ supra latus alterum et sinum excessus ejus supra latus tertium, ita Radius ad tangentem semissis anguli tertio[58] lateri oppositi et tangens complementi semissis ejus ad Radium.

19. Ut medium proportionale inter sinus laterum duorum ad medium proportionale inter sinū semisummæ trium laterum & sinum excessus illius semisummæ supra latus tertium ita Radius ad sinum complementi semissis anguli tertio lateri oppositi.

20. Ut mediū proportionale inter sinus laterum duorum ad medium proportionale inter sinum excessus semisummæ trium laterū supra latus unum e

---

(57) A little variant cancelled draft of this theorem is numbered '11' in correct sequence: in making the present revise (which follows immediately after the draft, but at the top of a new page) Newton forgetfully increased its number by one as though it were a new proposition.

8. The sine of the sum of the *crura* is to that of their difference as the cotangent of the half sum of the angles at the *cathetus* to the tangent of their half difference.

9. The tangents of the angles at the base are inversely proportional to the sines of the bases.

10. The sine of the sum of the angles at the base is to that of their difference as the tangent of the half sum of the bases to that of their half difference.

12.[57] As the sum of the tangents of the half *crura* to the difference of those tangents, so the cotangent of half the angle between the *crura* to the tangent of the half difference of the remaining angles.

13. As the sum of the tangent of half a *crus* and the cotangent of half the other *crus* to the difference of this tangent and cotangent, so the tangent of half the included angle to the cosine of half the sum of the remaining angles (or as the tangent of half the sum of the remaining angles to the cotangent of half the included angle).

14. As the sine of the half sum of the *crura* to that of their half difference, so the cotangent of half the angle between the *crura* to the tangent of the half sum of the remaining angles.

15. As the cosine of the half sum of the *crura* to that of their half difference, so the cotangent of half the angle between the *crura* to the tangent of the half sum of the remaining angles.

16. As the tangent of half one *crus* to that of half the other one, so the sine of the excess of the half sum of the three angles over the angle opposite the first *crus* to that of the excess of the same half sum over the angle opposite the latter side.

17. As the cotangent of half either *crus* to the tangent of half the other, so the sine of the half sum of the three angles to the sine of the excess of that half sum over the angle between the *crura*.

18. As the mean proportional between the cosine of the half sum of the three sides and the sine of the excess of that half sum over some side to the mean proportional between the sine of the excess of that half sum over a second side and that of its excess over the third side, so the radius to the tangent of half the angle opposite the [first] side, and so the cotangent of that half angle to the radius.

19. As the mean proportional between the sines of two sides to the mean proportional between the sine of the half sum of the three sides and the sine of the excess of that half sum over the third side, so the radius to the cosine of half the angle opposite the third side.

20. As the mean proportional between the sines of two sides to the mean proportional between the sine of the excess of the half sum of the three sides over

---

(58) Read 'primo'.

duobus illis et sinum excessus ejusdem semisummæ supra latus alterum, ita Radius ad sinum semissis anguli tertio lateri oppositi.

[59][21. Si arcus bisecet angulum inter duo latera, erunt sinus illorum laterum ut sinus segmentorum lateris oppositi.

22. Quod si arcus bisecet latus oppositum, erunt sinus duorum aliorum laterum ut sinus angulorum inter hæc latera et arcum inclusorum.]

DEMONSTRATIO PRÆCEDENTIŪ.[60]

1. In triangulo sphærico ABC demisso perpendiculo CD est (per Th 1. Sect 4) $s\,AC\,.\,R::s\,CD\,.\,s\,A$, & $R\,.\,s\,BC::s\,B\,.\,s\,CD$. Et ex æquo $s\,AC\,.\,s\,BC::s\,B\,.\,s\,A$, Theorema 1um.

2. Et mixtim

$$s\,AC+s\,BC\,.\,s\,AC-s\,BC::s\,B+s\,A\,.\,s\,B-s\,A,$$

id est per Lem [6] $t\dfrac{AC+BC}{2}\,.\,t\dfrac{AC-BC}{2}::t\dfrac{B+A}{2}\,.\,t\dfrac{B-A}{2}$ [>] Th: 2.

3. $s'\,AC\,.\,s'\,AD(::s'\,CD\,.\,R)::s'\,BC\,.\,s'\,BD$, Th: 3.

4. Et mixtim

$$s'\,AC+s'\,BC\,.\,s'\,BC-s'\,AC::s'\,AD+s'\,BD\,.\,s'\,BD-s'\,AD,$$

id est per Lem [6] $t'\dfrac{AC+BC}{2}\,.\,t\dfrac{AC-BC}{2}::t'\dfrac{AD+BD}{2}\,.\,t\dfrac{AD-BD}{2}$. & per Lem 4

$$t\dfrac{AD+BD}{2}\,.\,t\dfrac{AC+BC}{2}::t\dfrac{AC-BC}{2}\,.\,t\dfrac{AD-BD}{2}.$$

5. $s'\,A\,.\,s\,ACD(::s'\,CD\,.\,R)::s'\,B\,.\,s\,BCD$.

6. Et mixtim $s'\,A+s'\,B\,.-s'\,B+s'\,A::s\,ACD+s\,BCD\,.-s\,BCD+s\,ACD$, id est

per Lem [6] $t'\dfrac{A+B}{2}\,.\,t\dfrac{B-A}{2}::t\dfrac{ACD+BCD}{2}\,.\,t\dfrac{'ACD-BCD}{2}$.

7. $t'\,AC\,.\,s'\,ACD(::R\,.\,t\,CD)::t'\,BC\,.\,s'\,BCD$. Ergo

$$t\,BC\,.\,t\,AC::s'\,ACD\,.\,s'\,BCD.$$

8. Et mixtim $t\,AC+t\,BC\,.\,t\,AC-t\,BC::s'\,ACD+s'\,BCD\,.\,s'\,ACD-s'\,BCD$, id

est per Lemma[ta] 5 & 6, $s\,\overline{AC+BC}\,.\,s\,\overline{AC-BC}::t'\dfrac{ACD+BCD}{2}\,.\,t\dfrac{ACD-BCD}{2}.$

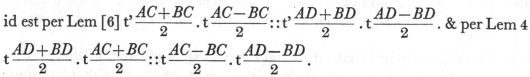

---

(59) Theorems 21 and 22 were added by Newton at a late stage when he was completing the 'Demonstratio' which follows and he never bothered to add their corresponding enunciations. This insertion is our attempt to fill this gap but, of course, any equivalent restoration will suffice.

(60) A preliminary draft (ULC. Add. 3959.5: 48ʳ) of Theorems 1–15 accompanies the manuscript text, but is both little variant and incomplete.

one of those two and the sine of the excess of the same half sum over the second side, so the radius to the sine of half the angle opposite the third side.

[59][21. If an arc should bisect the angle between two sides, then the sines of those sides will be as the sines of the segments of the opposite side.

22. But should the arc bisect the opposite side, then the sines of the two other sides will be as the sines of the angles included between these sides and the arc.]

Demonstration of the preceding[60]

1. In the spherical triangle ABC, on letting fall the perpendicular CD, there is (by Section 4, Theorem 1)

$$\operatorname{Sin} \widehat{AC} : R = \operatorname{Sin} \widehat{CD} : \operatorname{Sin} A \quad \text{and again} \quad R : \operatorname{Sin} \widehat{BC} = \operatorname{Sin} B : \operatorname{Sin} \widehat{CD},$$

so that *ex æquo* $\operatorname{Sin} \widehat{AC} : \operatorname{Sin} \widehat{BC} = \operatorname{Sin} B : \operatorname{Sin} A$. *Theorem 1.*

2. And *mixtim*

$$\operatorname{Sin} \widehat{AC} + \operatorname{Sin} \widehat{BC} : \operatorname{Sin} \widehat{AC} - \operatorname{Sin} \widehat{BC} = \operatorname{Sin} B + \operatorname{Sin} A : \operatorname{Sin} B - \operatorname{Sin} A,$$

that is, by Lemma 6

$$\operatorname{Tan} \tfrac{1}{2}(\widehat{AC} + \widehat{BC}) : \operatorname{Tan} \tfrac{1}{2}(\widehat{AC} - \widehat{BC}) = \operatorname{Tan} \tfrac{1}{2}(B + A) : \operatorname{Tan} \tfrac{1}{2}(B - A). \text{ *Theorem 2.*}$$

3. $\operatorname{Cos} \widehat{AC} : \operatorname{Cos} \widehat{AD}$ (or $\operatorname{Cos} \widehat{CD} : R$) $= \operatorname{Cos} \widehat{BC} : \operatorname{Cos} \widehat{BD}$. *Theorem 3.*

4. And *mixtim*

$$\operatorname{Cos} \widehat{AC} + \operatorname{Cos} \widehat{BC} : \operatorname{Cos} \widehat{BC} - \operatorname{Cos} \widehat{AC} = \operatorname{Cos} \widehat{AD} + \operatorname{Cos} \widehat{BD} : \operatorname{Cos} \widehat{BD} - \operatorname{Cos} \widehat{AD},$$

that is, by Lemma 6

$$\operatorname{Cot} \tfrac{1}{2}(\widehat{AC} + \widehat{BC}) : \operatorname{Tan} \tfrac{1}{2}(\widehat{AC} - \widehat{BC}) = \operatorname{Cot} \tfrac{1}{2}(\widehat{AD} + \widehat{BD}) : \operatorname{Tan} \tfrac{1}{2}(\widehat{AD} - \widehat{BD}),$$

and so by Lemma 4

$$\operatorname{Tan} \tfrac{1}{2}(\widehat{AD} + \widehat{BD}) : \operatorname{Tan} \tfrac{1}{2}(\widehat{AC} + \widehat{BC}) = \operatorname{Tan} \tfrac{1}{2}(\widehat{AC} - \widehat{BC}) : \operatorname{Tan} \tfrac{1}{2}(\widehat{AD} - \widehat{BD}).$$

5. $\operatorname{Cos} A : \operatorname{Sin} \widehat{ACD}$ (or $\operatorname{Cos} \widehat{CD} : R$) $= \operatorname{Cos} B : \operatorname{Sin} \widehat{BCD}$.

6. And *mixtim*

$$\operatorname{Cos} A + \operatorname{Cos} B : \operatorname{Cos} A - \operatorname{Cos} B = \operatorname{Sin} \widehat{ACD} + \operatorname{Sin} \widehat{BCD} : \operatorname{Sin} \widehat{ACD} - \operatorname{Sin} \widehat{BCD},$$

that is, by Lemma 6

$$\operatorname{Cot} \tfrac{1}{2}(A + B) : \operatorname{Tan} \tfrac{1}{2}(B - A) = \operatorname{Tan} \tfrac{1}{2}(\widehat{ACD} + \widehat{BCD}) : \operatorname{Tan} \tfrac{1}{2}(\widehat{ACD} - \widehat{BCD}).$$

7. $\operatorname{Cot} \widehat{AC} : \operatorname{Cos} \widehat{ACD}$ (or $R : \operatorname{Tan} \widehat{CD}$) $= \operatorname{Cot} \widehat{BC} : \operatorname{Cos} \widehat{BCD}$. Therefore

$$\operatorname{Tan} \widehat{BC} : \operatorname{Tan} \widehat{AC} = \operatorname{Cos} \widehat{ACD} : \operatorname{Cos} \widehat{BCD}.$$

8. And *mixtim*

$$\operatorname{Tan} \widehat{AC} + \operatorname{Tan} \widehat{BC} : \operatorname{Tan} \widehat{AC} - \operatorname{Tan} \widehat{BC}$$
$$= \operatorname{Cos} \widehat{ACD} + \operatorname{Cos} \widehat{BCD} : \operatorname{Cos} \widehat{ACD} - \operatorname{Cos} \widehat{BCD},$$

that is, by Lemmas 5 and 6

$$\operatorname{Sin} (\widehat{AC} + \widehat{BC}) : \operatorname{Sin} (\widehat{AC} - \widehat{BC}) = \operatorname{Cot} \tfrac{1}{2}(\widehat{ACD} + \widehat{BCD}) : \operatorname{Tan} \tfrac{1}{2}(\widehat{ACD} - \widehat{BCD}).$$

9. t' A.s AD(::R.t CD)::t' B.s BD. Ergo t B.t A::s AD.s BD.

10. Et mixtim t B+t A.t B−t A::s AD+s BD.s AD−s BD, id est (per Lem 5 & 6) s $\overline{B+A}$.s $\overline{B-A}$::t $\dfrac{AD+BD}{2}$.t $\dfrac{AD-BD}{2}$.

12.[61] Sint AZO, MZN semicirculi duo maximi in superficie hemisphærij se bisecantes[62] in Z, BAC triangulum sphæricũ, AO, BO, CO rectæ secantes planum MZN in punctis a, b, c; abc trianguli ABC projectio seu locus visus in plano MZN ubi spectatoris oculus situs est in O: et existente Oa radio erunt ab, ac tangentes dimidiorũ arcuũ AB, AC; et ceb arcus circularis[63] continens cum rectis ab, ac angulos b, c angulis B, C æquales, et angulo A æqualis erit angulus a, ut ex natura projectionis notum est. Duc rectam cfb, et angulorum fba, bca differentia æqualis erit differentiæ angulorum eba, eca, id est differentiæ angu-

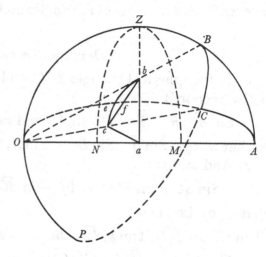

lorum B, C. At (per Th [2. Sect 3]) ab+ac.ab−ac::t' $\dfrac{bac}{2}$.t $\dfrac{fca-fba}{2}$. Hoc est

$$t\frac{AB}{2}+t\frac{AC}{2}.t\frac{AB}{2}-t\frac{AC}{2}::t'\frac{BAC}{2}.t\frac{C-B}{2}.\quad{}^{(64)}$$

13. Id est in triangulo complementali COP,

$$t\frac{OP}{2}+t'\frac{OC}{2}.t\frac{OP}{2}-t'\frac{OC}{2}::t\frac{POC}{2}.t\frac{C+P-180^{\mathrm{gr}}}{2}$$

$$=t'\frac{-C-P}{2}::t\frac{C+P}{2}.t'\frac{-POC}{2}.$$

---

(61) Newton first wrote '11'; compare note (57). Much as in the first (cancelled) version of Proposition 15 of the preceding 'Epitome' (see §1: note (21)) an incomplete preliminary 'Demonstratio' of the following group of theorems began as follows: 'Sit H centrum sphæræ, $GA\gamma$ semicirculus maximus bisectus in A, et BAE triangulum sphæricum. Cape AL, $A\lambda$

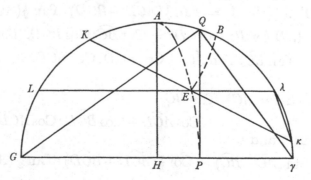

9. $\operatorname{Cot} A : \operatorname{Sin} \widehat{AD}$ (or $R : \operatorname{Tan} \widehat{CD}$) $= \operatorname{Cot} B : \operatorname{Sin} \widehat{BD}$. Therefore

$$\operatorname{Tan} B : \operatorname{Tan} A = \operatorname{Sin} \widehat{AD} : \operatorname{Sin} \widehat{BD}.$$

10. And *mixtim*

$$\operatorname{Tan} B + \operatorname{Tan} A : \operatorname{Tan} B - \operatorname{Tan} A = \operatorname{Sin} \widehat{AD} + \operatorname{Sin} \widehat{BD} : \operatorname{Sin} \widehat{AD} - \operatorname{Sin} \widehat{BD},$$

that is (by Lemmas 5 and 6)

$$\operatorname{Sin}(B + A) : \operatorname{Sin}(B - A) = \operatorname{Tan} \tfrac{1}{2}(\widehat{AD} + \widehat{BD}) : \operatorname{Tan} \tfrac{1}{2}(\widehat{AD} - \widehat{BD}).$$

12.[61] Let AZO, MZN be two great semicircles in the surface of a hemisphere which bisect each other[62] at Z; BAC a spherical triangle; AO, BO, CO straight lines cutting the plane MZN in the points a, b, c; and abc the projection of the triangle ABC, in other words, its apparent position in the plane MZN when the spectator's eye is located at O. Then, where Oa is the radius, ab, ac will be the tangents of half the arcs $\widehat{AB}$, $\widehat{AC}$; while $\widehat{ceb}$ will be a circle arc[63] containing with the straight lines ab, ac angles b, c equal to the angles B, C, and the angle a will also be equal to the angle A, as is known from the nature of the projection. Draw the straight line cfb and the difference of the angles $\hat{fba}$, $\hat{bca}$ will equal that of the angles $\hat{eba}$, $\hat{eca}$—that is, of the angles B, C. But (by Section 3, Theorem 2) $ab + ac : ab - ac = \operatorname{Cot} \tfrac{1}{2}\hat{bac} : \operatorname{Tan} \tfrac{1}{2}(\hat{fca} - \hat{fba})$, and so

$$\operatorname{Tan} \tfrac{1}{2}\widehat{AB} + \operatorname{Tan} \tfrac{1}{2}\widehat{AC} : \operatorname{Tan} \tfrac{1}{2}\widehat{AB} - \operatorname{Tan} \tfrac{1}{2}\widehat{AC} = \operatorname{Cot} \tfrac{1}{2}\widehat{BAC} : \operatorname{Tan} \tfrac{1}{2}(C - B).^{[64]}$$

13. That is, in the complementary triangle COP

$$\operatorname{Tan} \tfrac{1}{2}\widehat{OP} + \operatorname{Cot} \tfrac{1}{2}\widehat{OC} : \operatorname{Tan} \tfrac{1}{2}\widehat{OP} - \operatorname{Cot} \tfrac{1}{2}\widehat{OC}$$

$$= \operatorname{Tan} \tfrac{1}{2}\widehat{POC} : (\operatorname{Tan} \tfrac{1}{2}(C + P - 180°) \text{ or}) \operatorname{Cot} -\tfrac{1}{2}(C + P)$$

$$= \operatorname{Tan} \tfrac{1}{2}(C + P) : \operatorname{Cot} -\tfrac{1}{2}\widehat{POC}.$$

æquales lateri AE & BK, $B\kappa$ æquales lateri BE, et PG vel QG complementum anguli BAE' (Let H be the sphere's centre, $GA\gamma$ a great semicircle bisected at A, and BAE a spherical triangle. Take $\widehat{AL}$, $\widehat{A\lambda}$ equal to the side $\widehat{AE}$ and $\widehat{BK}$, $\widehat{B\kappa}$ equal to the side $\widehat{BE}$, also $\widehat{PG}$ or $\widehat{QG}$ the complement of the angle $B\hat{A}E$ ...). The figure depicted is our restoration in line with that which illustrates the corresponding passage in the 'Epitome': both Newton's diagram and the continuation of his text were evidently set on a following page which he destroyed when the more sophisticated proof by stereographic projection was introduced.

(62) Understand 'ad rectos angulos' (at right angles).

(63) See §1: note (25).

(64) In his first draft (see note (60) above) Newton noted more succinctly 'O oculus, Oa Rad, ac tang $\frac{AC}{2}$, ab tang $\frac{AB}{2}$. ang cab, $ac[e]$, $ab[e] =$ ang A, C, $[B]$. Duc rectā $b[f]c$, erit ang $ac[f] - ab[f] = ac[e] - ab[e] = C - B$. Sed $ab + ac$. $ab - ac :: \text{t}' \frac{a}{2}$. $\text{t} \frac{ac[f] - ab[f]}{2}$. Id est

$$\text{t}\frac{AB}{2} + \text{t}\frac{AC}{2} \cdot \text{t}\frac{AB}{2} - \text{t}\frac{AC}{2} :: \text{t}'\frac{A}{2} \cdot \text{t}\frac{C-B}{2} :: \text{t}'\frac{C-B}{2} \cdot \text{t}\frac{A}{2}.'$$

14. Quare per Lem 6 in triangulo *ABC* est

$$\text{s}\frac{AB+AC}{2}\cdot\text{s}\frac{AB-AC}{2}::\text{t'}\frac{A}{2}\cdot\text{t}\frac{C-B}{2}.$$

15. Et in triang. *COP*; $\text{s'}\dfrac{OC-OP}{2}\cdot\text{s'}\dfrac{OC+OP}{2}::\text{t}\dfrac{C+P}{2}\cdot\text{t'}\dfrac{-POC}{2}.$

16. Et mixtim in triang *ABC* est

$$\text{t}\frac{AB}{2}\cdot\text{t}\frac{AC}{2}::\text{t'}\frac{A}{2}+\text{t}\frac{C-B}{2}\cdot\text{t'}\frac{A}{2}-\text{t}\frac{C-B}{2}::(\text{per Lem 6})\ \text{s'}\frac{A+B-C}{2}\cdot\text{s'}\frac{A-B+C}{2}.$$

17 Et in triang *COP*; $\text{t}\dfrac{OP}{2}\cdot\text{t'}\dfrac{OC}{2}::\text{t}\dfrac{C+P}{2}+\text{t'}\dfrac{-POC}{2}\cdot\text{t}\dfrac{C+P}{2}-\text{t'}\dfrac{-POC}{2}::$

$$(\text{per Lem 6})\ \text{s'}\frac{-O-C-P}{2}\cdot\text{s'}\frac{-O+C+P-180^{[\text{gr}]}}{2}::\text{s'}\frac{C+P+O}{2}\cdot\text{s'}\frac{C+P-O}{2}.$$

18. Quare si[65] convertantur latera in angulos et anguli in latera sumendo eorum complementa ad 180^gr : Fiet (per Th 17)

$$\text{s}\frac{AB+AC+BC}{2}\cdot\text{s}\frac{AB+AC-BC}{2}::\text{t'}\frac{B}{2}\cdot\text{t}\frac{C}{2};$$

et (per Th 16) $\text{s}\dfrac{BC+AB-AC}{2}\cdot\text{s}\dfrac{BC-AB+AC}{2}::\text{t'}\dfrac{B}{2}\cdot\text{t'}\dfrac{C}{2}::\text{t}\dfrac{C}{2}\cdot\text{t}\dfrac{B}{2};$

et addendo rationes

$$\text{s}\frac{AB+AC+BC}{2}\times\text{s}\frac{AB-AC+BC}{2}\cdot\text{s}\frac{AB+AC-BC}{2}\times\text{s}\frac{BC-AB+AC}{2}::$$

$$\text{t'}\frac{B}{2}\cdot\text{t}\frac{B}{2}::\text{t't'}\frac{B}{2}\cdot RR::RR\cdot\text{tt}\frac{B}{2}.$$

19 & 20.[66] Cape $MH=MK=AB.$ $MN=AC\cdot NL=NI=BC.$ et erit

$$\tfrac{1}{2}HL=\text{s}\frac{AB+AC+BC}{2}.$$

$$\tfrac{1}{2}IK=\text{s}\frac{AB-AC+BC}{2}.$$

$$\tfrac{1}{2}HI=\text{s}\frac{AB+AC-BC}{2}\ \text{et}$$

$$\tfrac{1}{2}KL=\text{s}\frac{BC-AB+AC}{2}.$$

et $\tfrac{1}{2}HL\times\tfrac{1}{2}IK+\tfrac{1}{2}HI\times\tfrac{1}{2}KL=^{[67]}\tfrac{1}{2}HK\times\tfrac{1}{2}IL=\text{s}\,AB\times\text{s}\,BC.$ Quare per Th. 18[68] componendo erit

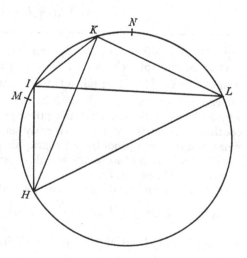

---

(65) Understand 'per Lem 12' (by Lemma 12).

14. Hence, by Lemma 6, in the triangle ABC

$$\text{Sin} \tfrac{1}{2}(\widehat{AB}+\widehat{AC}):\text{Sin}\tfrac{1}{2}(\widehat{AB}-\widehat{AC}) = \text{Cot}\tfrac{1}{2}A:\text{Tan}\tfrac{1}{2}(C-B).$$

15. And in the triangle COP

$$\text{Cos}\tfrac{1}{2}(\widehat{OC}-\widehat{OP}):\text{Cos}\tfrac{1}{2}(\widehat{OC}+\widehat{OP}) = \text{Tan}\tfrac{1}{2}(C+P):\text{Cot}-\tfrac{1}{2}\widehat{POC}.$$

16. And *mixtim*, in the triangle ABC,

$$\text{Tan}\tfrac{1}{2}\widehat{AB}:\text{Tan}\tfrac{1}{2}\widehat{AC} = \text{Cot}\tfrac{1}{2}A+\text{Tan}\tfrac{1}{2}(C-B):\text{Cot}\tfrac{1}{2}A-\text{Tan}\tfrac{1}{2}(C-B)$$
$$= \text{(by Lemma 6) } \text{Cos}\tfrac{1}{2}(A+B-C):\text{Cos}\tfrac{1}{2}(A-B+C).$$

17. And in the triangle COP,

$$\text{Tan}\tfrac{1}{2}\widehat{OP}:\text{Cot}\tfrac{1}{2}\widehat{OC} = \text{Tan}\tfrac{1}{2}(C+P)+\text{Cot}-\tfrac{1}{2}\widehat{POC}:\text{Tan}\tfrac{1}{2}(C+P)-\text{Cot}-\tfrac{1}{2}\widehat{POC},$$

that is, by Lemma 6,

$$\text{Cos}-\tfrac{1}{2}(C+P+O):\text{Cos}\tfrac{1}{2}(C+P-O-180°)$$
$$= \text{Cos}\tfrac{1}{2}(C+P+O):\text{Cos}\tfrac{1}{2}(C+P-O).$$

18. Hence, if[65] sides be converted to angles and angles to sides by taking their complements to 180°, there will come to be (by Theorem 17)

$$\text{Sin}\tfrac{1}{2}(\widehat{AB}+\widehat{AC}+\widehat{BC}):\text{Sin}\tfrac{1}{2}(\widehat{AB}+\widehat{AC}-\widehat{BC}) = \text{Cot}\tfrac{1}{2}B:\text{Tan}\tfrac{1}{2}C;$$

also (by Theorem 16)

$$\text{Sin}\tfrac{1}{2}(\widehat{BC}+\widehat{AB}-\widehat{AC}):\text{Sin}\tfrac{1}{2}(\widehat{BC}-\widehat{AB}+\widehat{AC}) = \text{Cot}\tfrac{1}{2}B:\text{Cot}\tfrac{1}{2}C$$

or $\text{Tan}\tfrac{1}{2}C:\text{Tan}\tfrac{1}{2}B$; and so by combining these ratios

$$\text{Sin}\tfrac{1}{2}(\widehat{AB}+\widehat{AC}+\widehat{BC}) \times \text{Sin}\tfrac{1}{2}(\widehat{AB}-\widehat{AC}+\widehat{BC}):\text{Sin}\tfrac{1}{2}(\widehat{AB}+\widehat{AC}-\widehat{BC})$$
$$\times \text{Sin}\tfrac{1}{2}(-\widehat{AB}+\widehat{AC}+\widehat{BC}) = \text{Cot}\tfrac{1}{2}B:\text{Tan}\tfrac{1}{2}B = \text{Cot}^2\tfrac{1}{2}B:R^2 = R^2:\text{Tan}^2\tfrac{1}{2}B.$$

19 and 20.[66] Take $MH = MK = AB$, $MN = AC$, $NL = NI = BC$ and there will be $\tfrac{1}{2}HL = \text{Sin}\tfrac{1}{2}(\widehat{AB}+\widehat{AC}+\widehat{BC})$, $\tfrac{1}{2}IK = \text{Sin}\tfrac{1}{2}(\widehat{AB}-\widehat{AC}+\widehat{BC})$,

$$\tfrac{1}{2}HI = \text{Sin}\tfrac{1}{2}(\widehat{AB}+\widehat{AC}-\widehat{BC}) \quad \text{and} \quad \tfrac{1}{2}KL = \text{Sin}\tfrac{1}{2}(-\widehat{AB}+\widehat{AC}+\widehat{BC}),$$

with $\quad \tfrac{1}{2}HL \times \tfrac{1}{2}IK + \tfrac{1}{2}HI \times \tfrac{1}{2}KL = {}^{[67]}\tfrac{1}{2}HK \times \tfrac{1}{2}IL = \text{Sin}\,\widehat{AB} \times \text{Sin}\,\widehat{BC}.$

Hence by Theorem 18[68] there will be *componendo*

---

(66) Compare the final paragraph of the preceding 'Epitome' (§1: note (33)). Newton evidently decided to cancel this (corrected) proof because of its unnecessary appeal to a theorem (Ptolemy's) which was not yet standard in elementary texts of geometry, rather than because he remained dissatisfied with its form.

(67) The theorem is Ptolemy's (*Syntaxis*, I, 10 [= (ed. J. L. Heiberg) *Opera omnia quæ exstant*, **1** (Leipzig, 1899): 31–2]; compare T. L. Heath, *A History of Greek Mathematics*, **2** (Oxford, 1921): 279). Newton would doubtless prove it in the traditional way by finding P in HK such that $\widehat{HIP} = \widehat{LIK}$ and making use of the pairs of similar triangles HIP, LIK and PIK, HIL to show that $HP = HI \times KL/IL$ and $PK = HL \times IK/IL$.

(68) '$\tfrac{1}{2}HL \times \tfrac{1}{2}IK . \tfrac{1}{2}HI \times \tfrac{1}{2}KL :: RR . \text{tt}\dfrac{B}{2}$' when appropriate substitution is made for

$$\widehat{AB} = \tfrac{1}{2}\widehat{HK}, \quad \widehat{BC} = \tfrac{1}{2}\widehat{HL} \quad \text{and} \quad \widehat{AC} = \widehat{MN}.$$

$$\mathrm{s}\,AB\times\mathrm{s}\,BC\,.\,\mathrm{s}\frac{AB+AC+BC}{2}\times\mathrm{s}\frac{AB-AC+BC}{2}\,.\,\mathrm{s}\frac{AB+AC-BC}{2}\times\mathrm{s}\frac{BC-AB+AC}{2}$$

$$::RR+\mathrm{tt}\,\frac{B}{2}\,.\,RR\,.\,\mathrm{tt}\,\frac{B}{2}::(\text{per Lem 1 \& 2})\,RR\,.\,\mathrm{s's'}\,\frac{B}{2}\,.\,\mathrm{ss}\,\frac{B}{2}\,.$$

19 & 20. Est (per Lem 10)

$$\mathrm{s}\frac{AB+AC+BC}{2}\times\mathrm{s}\frac{AB-AC+BC}{2}\left(=\mathrm{s}\frac{AB+BC}{2}+\mathrm{s}\frac{AC}{2}\times\mathrm{s}\frac{AB+BC}{2}-\mathrm{s}\frac{AC}{2}\right)$$

$$=\mathrm{ss}\frac{AB+BC}{2}-\mathrm{ss}\frac{AC}{2},\ \&\ \mathrm{s}\frac{AB+AC-BC}{2}\times\mathrm{s}\frac{BC-AB+AC}{2}$$

$$\left(=\mathrm{s}\frac{AC}{2}+\mathrm{s}\frac{AB-BC}{2}\times\mathrm{s}\frac{AC}{2}-\mathrm{s}\frac{AB-BC}{2}\right)=\mathrm{ss}\frac{AC}{2}-\mathrm{ss}\frac{AB-BC}{2},\ \text{et eorum summa}$$

$$\mathrm{ss}\frac{AB+BC}{2}-\mathrm{ss}\frac{AB-BC}{2}\left(=\mathrm{s}\frac{AB+BC}{2}+\mathrm{s}\frac{AB-BC}{2}\times\mathrm{s}\frac{AB+BC}{2}-\mathrm{s}\frac{AB-BC}{2}\right)$$

$$=(\text{per lem 10})\,\mathrm{s}\,AB\times\mathrm{s}\,BC.$$

Quare per Th 18 componendo erit

$$\mathrm{s}\,AB\times\mathrm{s}\,BC\,.\,\mathrm{s}\frac{AB+AC+BC}{2}\times\mathrm{s}\frac{AB-AC+BC}{2}\,.\,\mathrm{s}\frac{AB+AC-BC}{2}\times\mathrm{s}\frac{BC-AB+AC}{2}$$

$$::RR+\mathrm{tt}\,\frac{B}{2}\,.\,RR\,.\,\mathrm{tt}\,\frac{B}{2}::(\text{per Lem 1 \& 2})\,RR\,.\,\mathrm{s's'}\,\frac{B}{2}\,.\,\mathrm{ss}\frac{B}{2}\,.$$

21.[69] His addi possunt Theoremata de bisectione anguli vel lateris. Ut si linea *CE* bisecet angulum *C*, erit

$$\mathrm{s}\,AC\,.\,\mathrm{s}\,AE(::\mathrm{s}\,E\,.\,\mathrm{s}\,ACE::\mathrm{s}\,E\,.\,\mathrm{s}\,BCE)::\mathrm{s}\,CB\,.\,\mathrm{s}\,EB.$$

Et mixtim

$$\mathrm{s}\,AC+\mathrm{s}\,CB\,.\,\mathrm{s}\,AC-\mathrm{s}\,CB::\mathrm{s}\,AE+\mathrm{s}\,EB\,.\,\mathrm{s}\,AE-\mathrm{s}\,EB.$$

& per Lem 5

$$\mathrm{t}\frac{AC+CB}{2}\,.\,\mathrm{t}\frac{AC-CB}{2}::\mathrm{t}\frac{AE+EB}{2}\,.\,\mathrm{t}\frac{AE-EB}{2}::(\text{per Th. 2})\,\mathrm{t}\frac{B+A}{2}\,.\,\mathrm{t}\frac{B-A}{2}.$$

22. Quod si linea *CE* bisecet latus *AB*, erit

$$\mathrm{s}\,AC\,.\,\mathrm{s}\,ACE::\mathrm{s}\,AE\,.\,\mathrm{s}\,E::\mathrm{s}\,EB\,.\,\mathrm{s}\,E::\mathrm{s}\,BC\,.\,\mathrm{s}\,BCE.$$

Et mixtim $\mathrm{s}\,AC+\mathrm{s}\,BC\,.\,\mathrm{s}\,AC-\mathrm{s}\,BC::\mathrm{s}\,ACE+\mathrm{s}\,BCE\,.\,\mathrm{s}\,ACE-\mathrm{s}\,BCE.$ Id est (per Lem 5) $\mathrm{t}\frac{AC+BC}{2}\,.\,\mathrm{t}\frac{AC-BC}{2}::\mathrm{t}\frac{ACB}{2}\,.\,\mathrm{t}\frac{ACE-BCE}{2}::(\text{per Th 2})\,\mathrm{t}\frac{B+A}{2}\,.\,\mathrm{t}\frac{B-A}{2}.$

---

(69) See note (59).

$$\text{Sin } \widehat{AB} \times \text{Sin } \widehat{BC} : \text{Sin } \tfrac{1}{2}(\widehat{AB}+\widehat{AC}+\widehat{BC}) \times \text{Sin } \tfrac{1}{2}(\widehat{AB}-\widehat{AC}+\widehat{BC})$$
$$: \text{Sin } \tfrac{1}{2}(\widehat{AB}+\widehat{AC}-\widehat{BC}) \times \text{Sin } \tfrac{1}{2}(-\widehat{AB}+\widehat{AC}+\widehat{BC})$$
$$= R^2 + \text{Tan}^2 \tfrac{1}{2}B : R^2 : \text{Tan}^2 \tfrac{1}{2}B = \text{(by Lemmas 1 and 2) } R^2 : \text{Cos}^2 \tfrac{1}{2}B : \text{Sin}^2 \tfrac{1}{2}B.$$

19 and 20. There is (by Lemma 10)

$$\text{Sin } \tfrac{1}{2}(\widehat{AB}+\widehat{AC}+\widehat{BC}) \times \text{Sin } \tfrac{1}{2}(\widehat{AB}-\widehat{AC}+\widehat{BC})$$

or
$$[\text{Sin } \tfrac{1}{2}(\widehat{AB}+\widehat{BC}) + \text{Sin } \tfrac{1}{2}\widehat{AC}] \times [\text{Sin } \tfrac{1}{2}(\widehat{AB}+\widehat{BC}) - \text{Sin } \tfrac{1}{2}\widehat{AC}]$$
$$= \text{Sin}^2 \tfrac{1}{2}(\widehat{AB}+\widehat{BC}) - \text{Sin}^2 \tfrac{1}{2}\widehat{AC},$$

and
$$\text{Sin } \tfrac{1}{2}(\widehat{AC}+\widehat{AB}-\widehat{BC}) \times \text{Sin } \tfrac{1}{2}(\widehat{AC}-\widehat{AB}+\widehat{BC})$$

or
$$[\text{Sin } \tfrac{1}{2}\widehat{AC} + \text{Sin } \tfrac{1}{2}(\widehat{AB}-\widehat{BC})] \times [\text{Sin } \tfrac{1}{2}\widehat{AC} - \text{Sin } \tfrac{1}{2}(\widehat{AB}-\widehat{BC})]$$
$$= \text{Sin}^2 \tfrac{1}{2}\widehat{AC} - \text{Sin}^2 \tfrac{1}{2}(\widehat{AB}-\widehat{BC}),$$

and so their sum is $\text{Sin}^2 \tfrac{1}{2}(\widehat{AB}+\widehat{BC}) - \text{Sin}^2 \tfrac{1}{2}(\widehat{AB}-\widehat{AC})$, that is,

$$[\text{Sin } \tfrac{1}{2}(\widehat{AB}+\widehat{BC}) + \text{Sin } \tfrac{1}{2}(\widehat{AB}-\widehat{BC})] \times [\text{Sin } \tfrac{1}{2}(\widehat{AB}+\widehat{BC}) - \text{Sin } \tfrac{1}{2}(\widehat{AB}-\widehat{BC})]$$
$$= \text{(by Lemma 10) Sin } \widehat{AB} \times \text{Sin } \widehat{BC}.$$

Hence by Theorem 18 *componendo*

$$\text{Sin } \widehat{AB} \times \text{Sin } \widehat{BC} : \text{Sin } \tfrac{1}{2}(\widehat{AB}+\widehat{AC}+\widehat{BC}) \times \text{Sin } \tfrac{1}{2}(\widehat{AB}-\widehat{AC}+\widehat{BC})$$
$$: \text{Sin } \tfrac{1}{2}(\widehat{AB}+\widehat{AC}-\widehat{BC}) \times \text{Sin } \tfrac{1}{2}(-\widehat{AB}+\widehat{AC}+\widehat{BC})$$
$$= R^2 + \text{Tan}^2 \tfrac{1}{2}B : R^2 : \text{Tan}^2 \tfrac{1}{2}B = \text{(by Lemmas 1 and 2) } R^2 : \text{Cos}^2 \tfrac{1}{2}B : \text{Sin}^2 \tfrac{1}{2}B.$$

21.[69] To these may be added theorems regarding the bisection of an angle or a side. Should, for instance, the line CE bisect the angle C, then

$$\text{Sin } \widehat{AC} : \text{Sin } \widehat{AE} (= \text{Sin } E : \text{Sin } \widehat{ACE} \text{ or Sin } \widehat{BCE}) = \text{Sin } \widehat{CB} : \text{Sin } \widehat{EB},$$

and *mixtim*

$$\text{Sin } \widehat{AC} + \text{Sin } \widehat{CB} : \text{Sin } \widehat{AC} - \text{Sin } \widehat{CB} = \text{Sin } \widehat{AE} + \text{Sin } \widehat{EB} : \text{Sin } \widehat{AE} - \text{Sin } \widehat{EB},$$

so that by Lemma 5

$$\text{Tan } \tfrac{1}{2}(\widehat{AC}+\widehat{CB}) : \text{Tan } \tfrac{1}{2}(\widehat{AC}-\widehat{CB}) = \text{Tan } \tfrac{1}{2}(\widehat{AE}+\widehat{EB}) : \text{Tan } \tfrac{1}{2}(\widehat{AE}-\widehat{EB}),$$

that is, (by Theorem 2) $\text{Tan } \tfrac{1}{2}(B+A) : \text{Tan } \tfrac{1}{2}(B-A)$.

22. But should the line CE bisect the side AB, then

$$\text{Sin } \widehat{AC} : \text{Sin } \widehat{ACE} = \text{Sin } \widehat{AE} : \text{Sin } E = \text{Sin } \widehat{EB} : \text{Sin } E = \text{Sin } \widehat{BC} : \text{Sin } \widehat{BCE}.$$

And *mixtim*

$$\text{Sin } \widehat{AC} + \text{Sin } \widehat{BC} : \text{Sin } \widehat{AC} - \text{Sin } \widehat{BC} = \text{Sin } \widehat{ACE} + \text{Sin } \widehat{BCE} : \text{Sin } \widehat{ACE} - \text{Sin } \widehat{BCE},$$

that is, (by Lemma 5)

$$\text{Tan } \tfrac{1}{2}(\widehat{AC}+\widehat{BC}) : \text{Tan } \tfrac{1}{2}(\widehat{AC}-\widehat{BC}) = \text{Tan } \tfrac{1}{2}\widehat{ACB} : \text{Tan } \tfrac{1}{2}(\widehat{ACE}-\widehat{BCE})$$
$$= \text{(by Theorem 2) Tan } \tfrac{1}{2}(B+A) : \text{Tan } \tfrac{1}{2}(B-A).$$

§3. NEWTON'S COMMENTARY ON ST. JOHN HARE'S 'TRIGONOMETRY'.[1]

[Early 1684?]

From the original drafts in the University Library, Cambridge[2]

[1][3]

TRIGONOMETRIA
SUCCINCTÈ PROPOSITA
ET
NOVA METHODO DEMONSTRATA
A S^{TO} JOANNE HAREO ARM$^{[RO]}$.

In circulo quocunꝗ cujus Radij duo quivis angulum quemvis ad centrum continent, perpendiculum quod ad Radij alterutrius terminum erigitur, et ad usꝗ radium alterum producitur, quia circumferentiam tangit, dicitur Tangens arcus inter Radios, et Radius alter ad usꝗ Tangentem productus, quia secat circumferentiam, dicitur Secans ejusdem arcus, et perpendiculum quod a radij alterutrius termino in radium alterum demittitur, & in arcus illius cavitate vel

---

(1) Newton's introduction and completion of an unlocated manuscript treatise of trigonometry by a 'St John Hare, Esqre' we are unable accurately to identify. The preface is evidently modelled on the introduction to his own 'Compendium' (§2 preceding), nor is the content of the latter any more original; however, its elegant derivation of the fundamental theorems of spherical trigonometry (by collapsing a general solid angle into the net formed by its component plane surfaces), hitherto unpublished, has subsequently no exact parallel that we can trace, though in its essential structure it is equivalent to later 'solid' discussions by Francis Blake (1752) and Euler (1779). The 'Trigonometria' cited in Newton's title doubtless formed part of the 'Collection of Elementary Calculations and Figures relating to Spherical Trigonometry [probably by St John Hare], on 70 *ll. folio, and* 37 *smaller pieces of paper*' which was sold as Lot No. 323 at the 1936 Sotheby's sale of Newton papers (see 1: xxxv and John Taylor's *Catalogue of the Newton Papers Sold by Order of The Viscount Lymington* (London, 1936): 85) and has subsequently vanished. Two small pocket-books written by 'St Johan: Hare Anno Domini 1675', now to be found among the *miscellanea* in the Portsmouth papers in Cambridge University Library (Add. 3997 and 3998: the latter is essentially a rough, augmented transcript of the former) contain two versions of a 'Compendium Mathematicoalgebraicum Ex Oughtredo, Wallisio, Metio, Schooteno, Cartesioꝗ in proprios usus collectum conscriptumꝗ' which gathers detailed, workmanlike student annotations of such introductory texts as Oughtred's *Clavis Mathematicæ* (in one of the augmented Latin editions of 1648, 1652 and 1667), Wallis' *Mathesis Universalis* (Oxford, 1657) and Adrian Metius' *Arithmetica et Geometrica Nova* (Leyden, 1625) together with the elementary portions of Schooten's *Exercitationum Mathematicarum Libri V* (Leyden, 1657) and Descartes' *Geometrie* (in the second Latin edition of 1659/61). To suggest the possible flavour of Hare's missing 'Trigonometria' we reproduce in Appendix 4 below extracts from this 'Compendium' (heavily indebted, in fact, to Seth Ward's *Idea Trigonometriæ Demonstratæ*) which deal with the fundamentals of spherical trigonometry. Ten years later (see note (8) below) Newton wrote of his additions as completing an already posthumous work, so that St John Hare died some time before late 1684. With these clues it is tempting to think of Hare as an undergraduate at Cambridge in the middle 1670's, who took

Translation

[1]⁽³⁾

T R I G O N O M E T R Y

SUCCINCTLY PROPOUNDED

AND

DEMONSTRATED A NEW WAY

BY ST JOHN HARE, ESQ.

In any circle whatsoever, any two radii of which contain any angle at its centre, the perpendicular raised at the end of either radius and extended as far as the other is, because it touches the circumference, named the 'tangent' of the arc between the radii; while the second radius extended as far as this tangent is, because it intersects the circumference, named the 'secant' of the same arc; and the perpendicular let fall from the end of either radius to the other, which is located in the hollow or *sinus* (fold) of that arc, is named the 'sine' of the arc.

---

liberty of statutorial provision (compare III: xviii) to seek Newton out in his rooms at Trinity, but no one with a name approximating to his is known to have matriculated at any of the Cambridge colleges at any time during the later seventeenth century. It may be, alternatively, that Hare was an older man who came to mathematics in his middle life: in this case it is just possible that he was the St John Hare from Abbotsley in Huntingdonshire (a day's ride from Cambridge) who was taught at Westminster School, graduated at Christ Church, Oxford, in 1648, was called to the bar in 1658 and was still alive in 1681 (when he married for the second time). It is baffling that a man who achieved, after his death, the rare distinction of having Newton edit his papers for intended publication should now be so completely unknown. Perhaps, if and when his manuscript 'Trigonometria' resurfaces from oblivion, we may one day gain more insight into why Newton laid his 'friendly hand' upon it.

(2) Add. 3959.4: 24^r–27^v, basically a transcript, in the hand of his newly appointed amanuensis, Humphrey Newton, of an earlier autograph draft (ff. 30^r–32^r) which lacks the last four paragraphs of the revised text here reproduced: these were added, evidently at a moment's notice, when Newton came to correct Humphrey's secretary copy in his own hand. A variant preliminary draft (ff. 28^r/29^r) of the first part of the main text (also in Humphrey's transcript but with a few autograph alterations) is set as Appendix 5 below. The conjectural dating is based on our assessment of Newton's handwriting style, but the ante-date of spring 1684 can be put on Humphrey Newton's transcripts by his firm assertion to John Conduitt in January 1728 that 'In ye last year of K. Charles 2^d [who died in February 1685] Sr Isaac was pleas'd...to send for me up to Cambridge, of Whom I had the Opportunity as well Hon^r to wait of, for about 5 years' (King's College, Cambridge. Keynes MS 135^A, published in David Brewster's *Memoirs of the Life, Writings and Discoveries of Sir Isaac Newton*, **2** (Edinburgh, 1855): 91). The manuscript of the main text lacks all figures except those here identified as Figures 13 and 14: the other diagrams reproduced are restored, apart from Figure 15 which is based on an autograph sketch on f. 32^r, on the basis of the (not always uniquely explanatory) context. No doubt the section on plane trigonometry, here lacking, was to be taken over *verbatim* from St John Hare's 'Trigonometria'.

(3) These opening paragraphs effectively summarize Sections 1 and 2 of Newton's 'Compendium' (§2 preceding). As before, his reader is supposed to be familiar with the *regula aurea* (rule of three) for computing a fourth proportional to three given quantities and to know how to use logarithmic and trigonometric tables, but little more.

sinu locatur dicitur arcus illius sinus. Sic in circulo *BCF* cujus Radij *AB*, *AC* angulum *A* ad centrum continent, perpendicul[um] *BE* quod ad radij *AB* terminum *B* erigitur, et cum radio altero *AC* producto concurrit in *E*, Tangens est arcus *BC* qui radijs duobus interjacet; et radius alter ad usq tangentem productus *AE*, Secans est ejusdem arcus; et perpendiculum *BD* quod a Radij alterutrius termino *B* in radium alterum *AC* demittitur, est arcus ejusdem Sinus.

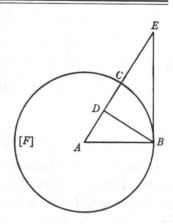

Quoniam arcus quadrantalis in partes æquales nonaginta dividitur quæ gradus dicuntur, et angulus rectus ad centrum, quem arcus ille subtendit, dividitur in partes seu gradus totidem; fit ut arcus omnes et anguli ad centrum quos subtendunt, constent semper ex æquali graduum numero, et inde ut in operationibus Trigonometricis pro se mutuò promiscuè usurpentur, utq sinus tangentes et secantes arcuum dicuntur etiam sinus tangentes et secantes angulorum ad centrum. Sic *BD* sinus dicitur, et *BE* tangens atq *AE* secans tam angul[i] *BAC* ad centrum qu[a]m arcus *BC* angulum illum ad circumferentiam subtendentis.

Trigonometriæ fundamentum est ut in assumpto circulo quocunq, cujus Radius in partes 1000 vel 10000 vel 100000 &c divisus sit, Tabula in Canon Sinuum Tangentium et Secantium ad gradus singulos singulasq graduum partes sexagesimales computetur, cujus ope in alijs quibuscunq circulis propositis sinus, tangentes et secantes ex angulis datis et vicissim anguli ex sinibus tangentibus et secantibus datis per Regulam auream inveniri possint. In circulis enim diversis, si anguli ad centrum æquantur inter se; Sinus, Tangentes et Secantes (ob similitudinem triangulorum) sunt inter se ut Radij.

Quinetiam in circulis universis, si Radij in partes 1000 vel 10000 vel 100000 &c dividi intelligantur, sinus tangentes et secantes illi ipsi erunt qui in Canone.

Usus autem Canonis in mensurandis Triangulis est ut (cum computationes per proportionem angulorum et arcuum non procedant) ex datis angulis et arcubus quærantur eorum sinus tangentes vel secantes in Canone, dein ex his et lateribus datis rectilineis per proportiones linearum quærantur vel ignota latera rectilinea triangulorum vel sinus tangentes et secantes angulorum et arcuum ignotorum, et tum demum ut ex his sinibus tangentibus vel secantibus in Canone inveniantur anguli et arcus ignoti.

Ut si proponatur triangulum *ABC* rectangulum ad *A*, et datis latere *AB* partium 112400 et angulo *B* 28^{gr} 20′ quæratur latus *AC*; quoniam *AC* tangens est anguli *B* in circulo qui centro *B* et radio *BA* describitur, si *Ba* ponatur æqualis radio circuli in

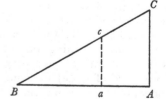

Thus in the circle BCF whose radii AB, AC contain the angle A at its centre, the perpendicular BE erected at the end-point B of the radius AB and meeting the extension of the second radius AC in E is the tangent of the arc $\overset{\frown}{BC}$ which lies between the two radii; and the extension AE of the second radius as far as the tangent is the secant of the same arc; and the perpendicular BD let fall from the end-point B of one radius to the other, AC, is the sine of this same arc.

Seeing that the quadrantal arc is divided into ninety equal parts, called degrees, while the right angle at its centre subtended by that arc is divided into the same number of parts or degrees, all arcs together with the central angles subtended by them will accordingly always consist of an equal number of degrees and hence in trigonometrical operations may indiscriminately be employed in each other's place. So too the sines, tangents and secants of arcs are called the sines, tangents and secants of their central angles. In this fashion BD is called the sine, BE the tangent and AE the secant both of the central angle $B\widehat{A}C$ and of the arc $\overset{\frown}{BC}$ subtending that angle at the circumference.

The fundamental basis of trigonometry is that in any circle taken at random, the radius of which shall have been divided into 1000, 10000, 100000, ... parts, a table of sines, tangents and secants shall be computed for each individual degree and each of its sexagesimal parts to serve as a canon, so that with its help the sines, tangents and secants in any circles proposed may be determined from their angles given by means of the golden rule [of proportions], and conversely the angles from their sines, tangents or secants given. For in different circles, if the angles at the centre are equal to one another, the sines, tangents and secants are (because of similarity of triangles) to one another as the radii.

Indeed in circles universally, if the radii are understood to be divided into 1000, 10000, 100000, ... parts, the sines, tangents and secants will be the very ones in the canon.

The use, however, of the canon in the mensuration of triangles (since computations by the proportion of angles and arcs do not there arise) is this: that from angles and arcs given their sines, tangents and secants may be looked up in the canon, that then from these and given rectilinear sides there may be ascertained through ratios of lines either the unknown rectilinear sides of triangles or the sines, tangents and secants of unknown angles and arcs, and that finally from these sines, tangents or secants may then be found the unknown angles and arcs in the canon.

For instance if in the proposed triangle ABC, right-angled at A, the side $AB = 112400$ parts and the angle $B = 28° 20'$ should be given and the side AC sought: since AC is the tangent of the angle B in the circle described with centre B and radius BA, if Ba be set equal to the radius of the circle in the canon, say

Canone seu partibus 10000,[4] et erigatur perpendiculum *ac* æqualis tangenti anguli *B* in canone, seu partibus 5392, triangula *BAC, Bac* erunt similia: Dico igitur[5] Ut Radius *Ba* 10000 ad anguli *B* tangentem *ac* 5392 ita crus *AB* 112400 ad crus *AC*: quod propterea (per Regulam auream) erit 60606. Nam si termini duo medij 5392 et 112400 se mutuo multiplicent, et factus 60606080[0] dividatur per terminum primum 10000, prodibit terminus ultimus 60606. [6]Vel (quod perinde est) si terminorum mediorum Logarithmi addantur et a summa auferatur logarithmus termini primi, manebit logarithmus termini ultimi: Ut fit in operatione sequente

	[Numeri]	[Logarithmi]
Ut Radius	10000	10.000000
ad tang 28^{gr}.20′	5392	9.731746
ita crus *AB*	112400	5.050766
		14.782512
ad crus *AC*	60606.	4.782512.

Et eodem modo intelligendæ sunt proportiones omnes quæ ad solutionem Triangulorum in Libris Trigonometricis proponi solent.

---

[2][7] DE TRIANGULIS SPHÆRICIS RECTANGULIS.

[8]In figura 1, 2 et 3 sunt *gm, mr, rg* latera trianguli sphærici *gmr* quibus anguli *gLm, mLr, rLg* ad centrum sphæræ subtenduntur. Suntᶿ *GLM, MLR, RLG* (fig 2 & 3) plana laterum illorum extra sphæram producta, & plano quarto *GMR* transversim secta. Concipe plana *GLM, RLG, GMR* super basibus *ML, RL, MR* erigi & ita ad invicem applicari ut lineæ æquales *GL* & *GL, GM* & *GM, GR* & *GR* coincidant; & habebitur figura in situ suo naturali. In hoc situ si trianguli angulus ad *r* ponatur rectus & sectionis planum *GMR* planis *GLM*,

---

(4) Newton's (autograph) accompanying figure, here accurately reproduced, is not drawn to scale.

(5) A first, cancelled continuation reads 'quod crus *AC* sit ad crus *AB* seu 112400 ut anguli *B* tangens *ac* seu 5392 ad Radium *aB* seu 10000 adeoᴄᴈ multiplicando terminos medios 112400 et 5392 in se et factum 60606080[0] dividendo per terminum extremum datum 10000 invenio latus quæsitum *AC* partium 60606. Eodem recidit si dicamus crus *AC* esse ad crus *AB* partium 112400 ut anguli *B* tangens *ac* partium 5392 ad radium *aB* partium [10000]' (that the *crus AC* shall be to the *crus AB* or 112400 as Tan *B* = *ac* or 5392 to the radius *aB* or 10000, and consequently, on multiplying the middle terms 112400 and 5392 into each other and dividing the product 606060800 by the given extreme term, I find the side *AC* sought to have 60606 parts. It comes to the same were we to say that the *crus AC* is to the *crus AB* of 112400 parts as *ac* = Tan *B* of 5392 parts to the radius *aB* of 10000 parts).

(6) The remainder of this paragraph is a late autograph addition to Humphrey Newton's transcript.

of 10 000 parts,[4] and the perpendicular *ac* equal to the tangent of the angle *B* in the canon, namely 5392 parts, be erected, then the triangles *BAC*, *Bac* will be similar. I therefore say:[5] As the radius *Ba* = 10 000 to the tangent of angle *B*, *ac* = 5392, so the *crus AB* = 112 400 to the *crus AC*—which will accordingly (by the golden rule) be 60 606. For if the two middle terms 5392 and 112 400 be multiplied into each other and the product 606 060 800 divided by the first term 10 000, there will result the last term 60 606. [6]Or (what is equivalent) if the logarithms of the middle terms be added and from their sum be taken away that of the first term, there will remain the logarithm of the last term—as is done in the following procedure:

	[Numbers]	[Logarithms]
As the radius	10 000	10.000000
to the tangent of 28° 20′,	5 392	9.731746
so the *crus AB*	112 400	5.050766
		14.782512
to the *crus AC*.	60 606	4.782512.

And in the same manner are to be understood all proportions which are usually proposed in trigonometrical books for the solution of triangles.

[2][7]   ON SPHERICAL RIGHT-ANGLED TRIANGLES

[8]In figures 1, 2 and 3 $\widehat{gm}$, $\widehat{mr}$, $\widehat{rg}$ are the sides of the spherical triangle *gmr*, and $g\widehat{L}m$, $m\widehat{L}r$, $r\widehat{L}g$ the angles subtended by them at the centre of the sphere. Also *GLM*, *MLR*, *RLG* (figures 2 and 3) are the planes of those sides extended beyond the sphere and cut crosswise by a fourth plane *GMR*. Imagine the planes *GLM*, *RLG*, *GMR* to be raised upon their bases *ML*, *RL*, *MR* and applied to one another in such a way that the equal lines *GL* and *GL*, *GM* and *GM*, *GR* and *GR* coincide: the figure will then be obtained in its natural situation. In this situation, if the angle at *r* be set a right angle and the plane of the section

---

(7) We omit a trivially variant repetition of the main title. 'Trigonometria / succinctè proposita ac nova methodo / demonstrata a S^to Johanne Hareo Arm^ro.' It would appear (see next note) that Newton's present variant treatment of spherical trigonometry was intended to complement the corresponding parts of St John Hare's 'Trigonometria' (left unfinished at his death), rather than to replace them entirely, in the published work.

(8) In his autograph draft (ULC. Add. 3959.4: 30^r) Newton has cancelled the opening sentence 'Operi posthumo & incompleto demonstrationes casuum ultimorum 5 vizt secundi tertij quarti quinti et sexti, cum explicatione sequente adjecit amica manus' (To this incomplete posthumous work the demonstrations of the last five cases, namely, the second, third, fourth, fifth and sixth, have been added by a friendly hand). Observe Newton's typical wish to remain anonymous in editing another's work. (Compare II: 283, note (16).)

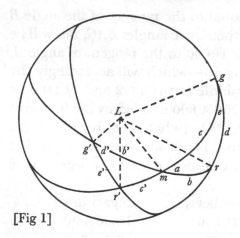

[Fig 2]

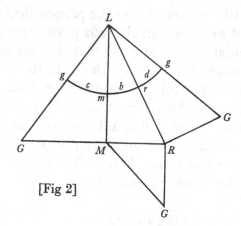

[Fig 3]

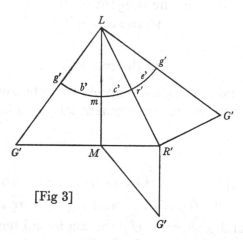

[Fig 1]

MLR perpendiculariter insistat; anguli *GML, RML, GRL, MRG* evadent recti & sectionis angulus *GMR* æqualis erit angulo trianguli ad *m*. Igitur in circulo quovis qui radio *GL* describitur erit *GM* sinus & *ML* cosinus anguli *GLM* seu lateris *gm*, atcʒ *GR* sinus & *RL* cosinus anguli *GLR* seu lateris *rg*. Et in circulo quovis qui radio *GM* describitur erit *ML* cotangens anguli *GLM* seu arcus *gm* atcʒ *GR* sinus & *MR* cosinus anguli *GMR* id est anguli trianguli ad *m*. Et sic in alijs.[9]

Trianguli propositi notentur jam anguli vocalibus & latera consonantibus; nimirum hypotenusa litera *c*, latera duo reliqua literis *b*, *d*, angulus ad latus *b* litera *a* & angulus alter litera *e*, & sint horum complementa ad angulum rectum *c'*, *b'*, *d'*, *a'*, *e'*: & casus quatuor primi ubi anguli duo non sunt in quæstione solventur per figuram secundam. In quinto & sexto transmutatur triangulum

---

(9) There is no loss of generality—indeed, the argument is slightly simplified—on supposing the points *m* and *M* to coincide, when the plane *GMR* comes to be tangent to the spherical surface *gmr* at that point. The possibilities of this 'tangential' model in elegantly deriving the basic theorems of spherical trigonometry (and also of its more general version where *gm̂* is no longer a right angle) were not again noticed, it would appear, till seventy years later when Francis Blake outlined its 'perspicuity' in a short paper, bereft of detailed proofs, read to the Royal Society in May 1752 ('Spherical Trigonometry reduced to Plane', *Philosophical Transactions*, 47, 1752: §LXXIV: 441–4). Blake was manifestly ignorant of Newton's prior researches, for he asserted (*ibid.*: 441) that 'I take it to be new, that from the axioms of only plane trigonometry, and almost independent of solids, and the doctrine of the sphere, the spherical cases are...to be solved'. Some time later, just possibly forewarned by Blake's paper, Euler gave a

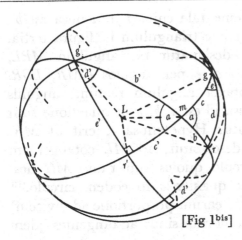

[Fig 1bis]

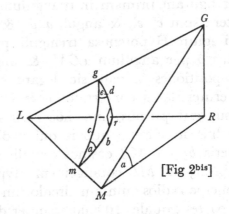

[Fig 2bis]

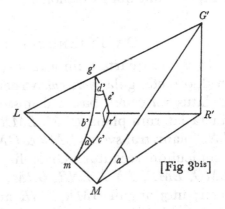

[Fig 3bis]

GMR stand perpendicularly to the planes *GLM* and *MLR*, the angles $G\widehat{M}L$, $R\widehat{M}L$, $G\widehat{R}L$, $M\widehat{R}G$ will prove to be right angles and the angle $G\widehat{M}R$ of the section will be equal to the angle of the triangle at *m*. Therefore, in any circle described with radius *GL*, *GM* will be the sine and *ML* the cosine of the angle $G\widehat{L}M$ or of the side $\widehat{gm}$, while *GR* will be the sine and *RL* the cosine of the angle $G\widehat{L}R$ or of the side $\widehat{gr}$. And in any circle described with the radius *GM*, *ML* will be the co-tangent of the angle $G\widehat{L}M$ or of the arc $\widehat{gm}$, while *GR* will be the sine and *MR* the cosine of the angle $G\widehat{M}R$, that is, of the angle of the triangle at *m*. And so in other cases.[9]

Now in the triangle proposed let the angles be denoted by vowels and the sides by consonants: that is to say, the hypotenuse by the letter *c*, the two remaining sides by the letters *b* and *d*, the angle adjoining the side *b* by the letter *a* and the other angle by the letter *e*, and let the complements of these to a right angle be *c'*, *b'*, *d'*, *a'* and *e'*. The first four cases, where the two angles are not in question, will be decided by the second figure. In the fifth and sixth the triangle is trans-

---

somewhat clumsy derivation of the cosine formula for spherical triangles on the basis of a solid figure equivalent to Newton's present 'Fig 5' in the case where the points *m*, *M* and also *n*, *N* are coincident ('Trigonometria Sphærica Universa ex primis Principiis dilucide derivata', *Acta Academiæ Scientiarum Petropolitanæ*, **3**, 1778: 72–86, especially 73–4 [= *Opera Omnia* (1), **16**. *Commentationes Geometricæ*, **1** (Zurich, 1953): 224–5]; compare A. von Braunmühl, *Vorlesungen über Geschichte der Trigonometrie*, **2** (Leipzig, 1903): 122–3). Neither Blake nor Euler thought of collapsing their solid figures into the net of their plane faces and each, in consequence, missed the refinements of the method which Newton here develops.

per figuram primam in triangulum complementale cujus hypotenusa est *b'*,
latera sunt *c'*, *e'*, & anguli *a*, *d'*, & ponitur hoc triangulum in figura tertia.
Ibi igitur Hypotenusa trianguli propositi designatur per angulum *MRL*,
latus *b* per angulum *LGM*, & anguli duo *a*, *e* per angulos *GMR*, *LGR*.
Propositiones autem sic legantur. Si (præter angulum rectum) angulus
alteruter, latus conterminum & Hypotenusa id est *a*, *b*, *c* in quæstione sunt
casus erit primus: inq̃ hoc casu si quæritur Hypotenusa *c*, erit ut *MR*,
Radius in circulo quovis qui radio illo describitur, ad *ML* cotangentem
lateris *b*; ita *MR* cosinus anguli *a* in circulo quovis qui radio *MG* describitur, ad *ML* cotangentem Hypotenusæ quæsitæ *c* in eodem circulo.[10]
Concipe radios omnium circulorum (neglecta earum proportione ad invicem)
in partes æquales 1000000 semper dividi, & horum sinus ac tangentes ijdem
semper erunt qui in Canone.

DE TRIANGULIS SPHÆRICIS OBLIQUANGULIS.

In figura quarta quinta & sexta resolvitur triangulum obliquangulum *gmn*
in duo rectangula *gmr*, *gnr* perpendiculo *gr*
in latus *mn* ceu[11] basim demisso. Et producta laterum plana *LGM*, *LMR*, & *LGN*,
LNR planis transversis *GMR* & *GNR* ad perpendiculum secantur. Unde fit ut anguli
LMG, *LMR*, *LNG*, *LNR*, *GRM*, *GRN* sint
recti; utq̃ anguli *GMR*, *GNR* angulis trianguli sphærici ad basem, angulíq̃ *GLM*,
GLN, *MLN*, *MLR*, *NLR* lateribus basi &
segmentis basis respondeant. Concipe triangula *GLM*, *GLN*, *GRM*, *GRN* super basibus *LM*, *LN*, *RM*, *RN* erecta ita applicari
ad invicem ut eorum latera æqualia *GL* &

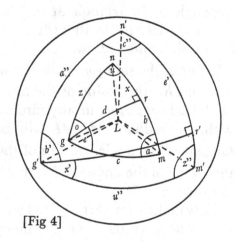

[Fig 4]

GL, *GM* & *GM*, *GN* & *GN*, *GR* & *GR* coincidant & habebitur figura in situ
naturali.

---

(10) In other words, from $\dfrac{MR}{LM} = \dfrac{MR}{GM} \times \dfrac{GM}{LM}$ Newton deduces that $R/\mathrm{Cot}\,b = \mathrm{Cos}\,a/\mathrm{Cot}\,c$
or, equivalently, $\tan b = \cos a \tan c$. Similarly, the equalities

$$\frac{GR}{LR} = \frac{GR}{MR} \times \frac{MR}{LR}, \quad \frac{GR}{GL} = \frac{GR}{GM} \times \frac{GM}{GL} \quad \text{and} \quad \frac{LM}{GL} = \frac{LR}{GL} \times \frac{LM}{LR}$$

yield the three further Cases 2, 3 and 4 where likewise 'anguli duo non sunt in quæstione' (but
only *d̂*, it is understood): namely, $\tan d = \tan a \sin b$, $\sin d = \sin a \sin c$ and $\cos c = \cos b \cos d$.
The 'complemental' Cases 5 and 6 are evidently to be derived from $\dfrac{GR}{GM} \times \dfrac{GM}{GL} = \dfrac{GR}{GL}$ (or

formed by means of the first figure into the complementary triangle whose hypotenuse is b', sides c' and e', and angles a and d'. This triangle is represented in the third figure. In the former, then, the hypotenuse of the triangle propounded is designated by the angle $M\widehat{R}L$, its side b by the angle $L\widehat{G}M$ and its two angles a, e by the angles $G\widehat{M}R$, $L\widehat{G}R$. The propositions, however, are to be chosen in this order. If (apart from the right angle) one or other angle, the adjoining side and the hypotenuse—that is, a, b and c—are at question, this will be the first case. And if, in this case, the hypotenuse c is sought, there will be: as MR, the radius in any circle described with that radius, to ML, the cotangent of the side b, so MR, the cosine of angle a in any circle described with radius MG, to ML, the cotangent in the same circle of the required hypotenuse c.[10] Conceive the radii of all circles (neglecting their proportions individually to each other) always to be divided into 1 000 000 equal parts, and their sines and tangents will always be those in the canon.

ON SPHERICAL OBLIQUE-ANGLED TRIANGLES

In the fourth, fifth and sixth figures the oblique-angled triangle *gmn* is resolved into two right-angled ones *gmr*, *gnr* by the perpendicular $\widehat{gr}$ let fall perpendicularly onto the side $\widehat{mn}$ functioning as[11] the base. Also the planes LGM, LMR and LGN, LNR are, after extension, cut perpendicularly by the crosswise planes GMR and GNR. As a result, accordingly, the angles $L\widehat{M}G$, $L\widehat{M}R$, $L\widehat{N}G$, $L\widehat{N}R$, $G\widehat{R}M$ and $G\widehat{R}N$ are right, while the angles $G\widehat{M}R$, $G\widehat{N}R$ correspond to the angles of the spherical triangle at the base, and the angles $G\widehat{L}M$, $G\widehat{L}N$, $M\widehat{L}N$, $M\widehat{L}R$, $N\widehat{L}R$ to the sides, the base and the segments of the base respectively. Imagine the triangles GLM, GLN, GRM, GRN to be raised upon their bases LM, LN, RM, RN and so applied to one another that their equal sides GL and GL, GM and GM, GN and GN coincide, and the figure will be obtained in its natural situation.

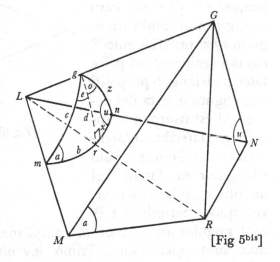

[Fig 5bis]

$\sin a \cos b = \cos e$) and $\dfrac{MR}{GR} \times \dfrac{GR}{LR} = \dfrac{MR}{LR}$ ($\cot a \cot e = \cos c$). No doubt Newton would have written these six cases in the form of proportions and using the abbreviated notation, borrowed from John Newton, which he uses below (see note (21)).

(11) The same reading is found in Newton's autograph draft (ULC. Add. 3959.4: 30^{r}), but did Newton intend merely to write 'seu' (or)?

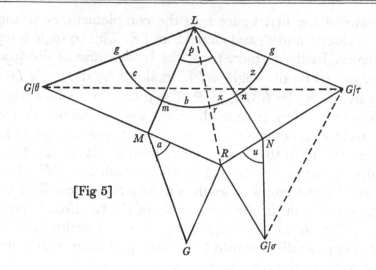

[Fig 5]

Trianguli propositi notentur jam latera literis *c*, *z*, basis litera *p*, se[g]menta basis literis *b*, *x*, anguli ad basem literis *e*, *o*. Et propositiones in casu primo resolventur per figuram quintam usurpando triangulum *gmn* pro triangulo proposito: illæ in casu secundo usurpando triangulum *gmn* pro triangulo[(12)] cujus anguli sunt ad polos laterum trianguli propositi ut in figura quarta delineatur, id est usurpando angulos pro lateribus & latera pro angulis. Namcҕ horum triangulorum latera vel angulis mutuis æquantur, vel quod perinde est,[(13)]

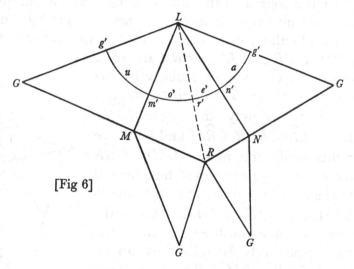

[Fig 6]

sunt angulorum mutuorum complementa ad duos rectos.[(14)] In casu utrocҕ duæ sunt operationes. Primò invenitur vel basis segmentum *b* in casu primo vel angulus ad cathetum *e* in casu secundo: deinde in casu utrocҕ invenitur latus vel angulus quæsitus. Operatio prima in columna prima, secunda in columnis quatuor sequentibus docetur.[(15)] Cum autem perpendiculum nunc intra triangulum, nunc extra vel ad partes *u z* vel ad partes *a c* cadat, modus primus figuris 4, 5, 6, secundus figuris 7, 8, tertius figura 9 delineatur.[(16)]

(12) Denoted as *g'm'n'* in (our restored) Figures 4 and 6.

Now in the propounded triangle let the sides be denoted by the letters c and z, the base by the letter p, the segments of the base by the letters b and x, the angles at the base by the letters e and o. The propositions in the first case will then be resolved by the fifth figure by employing the triangle gmn for that proposed; those in the second case by employing the triangle $g^{[']}m^{[']}n^{[']}$ for the triangle[12] whose angles are, as delineated in the fourth figure, at the poles of the sides of the triangle proposed—that is, by employing angles in place of sides and sides in place of angles. For the sides of these triangles are either equal to their partner angles or, equivalently,[13] are their complements to two right angles.[14] In either case there are two operations: first is found either, in the first case, the base segment b or, in the second, the angle e at the *cathetus*; and then is ascertained, in either case, the required side or angle. The first operation is dealt with in the first column, the second in the four following ones.[15] However, the perpendicular falls sometimes within the triangle, sometimes outside it in the direction of u and z or of a and c, and in consequence the first mode is shown in figures 4, 5 and 6, the second in figures 7 and 8, the third in figure 9.[16] In

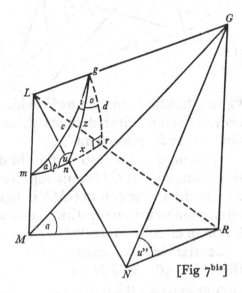

[Fig 7bis]

---

(13) Since $\mathrm{Cos}''x = -\mathrm{Cos}\,x$ and $\mathrm{Tan}''x = -\mathrm{Tan}\,x$ (or in modern terms

$$\cos(\pi - x) = -\cos x, \quad \tan(\pi - x) = -\tan x).$$

(14) We may confidently guess, lacking the manuscript of St John Hare's 'Trigonometria', that Cases 1 and 2 cover the complementary situations in which respectively the sides c and $p\,(= b + x)$ together with the included angle d, and the angles $i\,(= e + o)$ and u together with the adjacent side z are given. In the first instance the equalities

$$\frac{MG}{MR} = \frac{MG}{LM} : \frac{MR}{LM}, \quad \frac{LM}{LR} : \frac{LN}{LR} = \frac{LM}{LG} : \frac{LN}{LG} \quad \text{and} \quad \frac{NR}{LR} : \frac{MR}{LR} = \frac{GR}{MR} : \frac{GR}{NR}$$

yield $R : \mathrm{Cos}\,a = \mathrm{Tan}\,c\,\mathrm{Tan}\,x$, $\mathrm{Cos}\,b : \mathrm{Cos}\,x = \mathrm{Cos}\,c : \mathrm{Cos}\,z$ and $\mathrm{Sin}\,x : \mathrm{Sin}\,b = \mathrm{Tan}\,a : \mathrm{Tan}\,u$, which together completely determine the triangle. In the latter case, similarly, the same equalities applied to the polar triangle yield the equivalent determination by $R : \mathrm{Cos}\,z = \mathrm{Tan}\,u : \mathrm{Tan}\,o$, $\mathrm{Sin}\,o : \mathrm{Sin}\,e = -\mathrm{Cos}\,u : -\mathrm{Cos}\,a$ and $\mathrm{Cos}\,e : \mathrm{Cos}\,o = -\mathrm{Tan}\,z : -\mathrm{Tan}\,c$.

(15) Newton refers to some tabular layout in Hare's manuscript 'Trigonometria' not accessible to us.

(16) 'Fig 8' is manifestly identical in structure to 'Fig 7' but with the triangle gmn replaced by its polar triangle. 'Fig 9' is perhaps a variant on 'Fig 7' in which the angle at m is obtuse and so the cathetus gr falls outside the triangle gmn nearer to the side gm (or, in other words, 'Fig 9' is perhaps the mirror image of 'Fig 7').

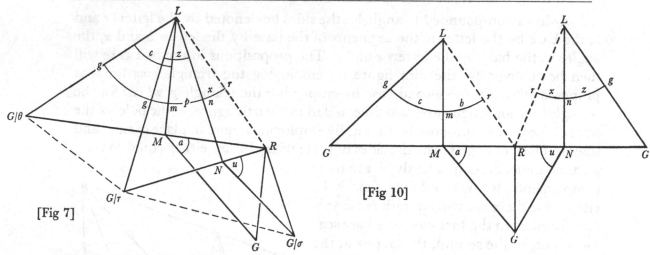

[Fig 7]

[Fig 10]

Figura decima & undecima[17] exhibentur plana laterum trianguli & sectionum transversarum aliter deducta & expansa. In his figuris cernuntur analogiæ omnes quæ in præcedentibus.

In triangulorum casu tertio ubi dicitur quod sit *GM* sinus anguli *u* ad *GN* sinum anguli *a* ut *GM* sinus anguli *c* ad *GN* sinum anguli *z*, sunt *GM* & *GN* in circulo circa triangulum *GMN* in figura [10] descripto chordæ quibus anguli illi ad circumferentiam constituti subtenduntur, et propterea sinus eorundem angulorum in circulo duplo.[18]

Sic etiam in casu quarto (fig. 5[19]) $\frac{1}{2}\theta\tau$ & $\frac{1}{2}R\tau$ in circulo circa triangulū $\theta R\tau$ descripto sunt sinus angulorum oppositorum *MRN*, & *Rθτ*. Angulus *MRN* ob angulos rectos *RML*, *RNL* complet angulum *MLN*, id est arcum *mn* seu *p* ad duos Rectos, adeoꝗ eundem habet sinum. Et si de angulo *LθM* seu complemento ipsius *c*

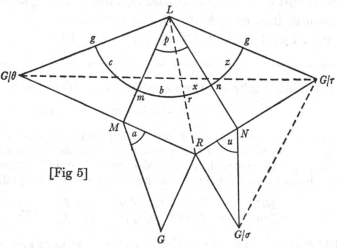

[Fig 5]

subducatur angulus *Lθτ* id est complementum semissis *θLτ* seu semisummæ laterum *c*, *p* & *z*, manebit angulus *Rθτ* æqualis $\dfrac{[-]c+p[+]z}{2}$.[20] Porrò in circulo qui radio *NG* describitur $NG\times R$ est *RR* et $\frac{1}{2}R\tau\times R$ est *ccss* $\frac{1}{2}u$[21] seu quadratum cosinus $\frac{1}{2}u$, propterea quod $R\tau\times 2NG$ est *στ* quadratum & $\frac{1}{2}\sigma\tau$ est cosinus *Rτσ* seu $\frac{1}{2}u$. Casus quintus idem est cum quarto nisi quod mutantur anguli in latera & latera in angulos per figuram quartam.[22]

the tenth and eleventh figures are shown the planes of the triangle's sides and of the cross-sections drawn and folded out another way.[17] In these figures may be discerned all the proportions exhibited in the preceding ones.

In the third case of the triangles—where it is said that GM, the sine of the angle u, is to GN, the sine of the angle z, as GM, the sine of the angle c, to GN, the sine of the angle $z - GM$ and GN are, in the circle described round the triangle GMN in figure 10, the chords subtended by those angles set at the circumference and consequently the sines of those angles in a circle twice the size.[18]

So also in the fourth case (figure 5[19]) $\frac{1}{2}\theta\tau$ and $\frac{1}{2}R\tau$ are, in the circle described round the triangle $\theta R\tau$, the sines of the opposite angles $\widehat{MRN}$ and $\widehat{R\theta\tau}$. The angle $\widehat{MRN}$ on account of the right angles RML, RNL is the complement to two right angles of the angle $\widehat{MLN}$, that is, of the arc $\widehat{mn}$ or p, and so has the same sine. And if from the angle $\widehat{L\theta M}$, or its complement c, be taken the angle $\widehat{L\theta\tau}$, that is, the complement of half $\theta\widehat{L}\tau$ or the half sum of the sides c, p and z, there will remain $\widehat{R\theta\tau} = \frac{1}{2}(-c + p + z)$.[20] Further, in the circle described with NG as radius, $NG \times R$ is R^2 and $\frac{1}{2}R\tau \times R$ is $\mathrm{Cos}^2 \frac{1}{2}u$, that is, the square of the cosine of $\frac{1}{2}u$, for the reason that $R\tau \times 2NG$ is $\sigma\tau$ squared and $\frac{1}{2}\sigma\tau$ is the cosine of $\widehat{R\tau\sigma} = \frac{1}{2}u$. The fifth case is the same as the fourth except that angles are changed into sides and sides into angles by means of the fourth figure.[22]

---

(17) That is, 'Fig 10' and 'Fig 11' are just layouts of the solid figures depicted in 'Fig 5bis' and 'Fig 7bis' whose nets differ from those shown in (our restored) 'Fig 5' and 'Fig 7'.

(18) Evidently in Case 3 the equality $\dfrac{GR}{GN} : \dfrac{GR}{GM} = \dfrac{GM}{GL} : \dfrac{GN}{GL}$ yields $\mathrm{Sin}\,u : \mathrm{Sin}\,a = \mathrm{Sin}\,c : \mathrm{Sin}\,z$.

(19) The argument holds *mutatis mutandis* for the complementary 'Fig 7'.

(20) The text has $\dfrac{`c + p - z'}{2}$, a faithful transcript by Humphrey Newton of a momentary lapse on Newton's part in his autograph draft (ULC. Add. 3959.4: 30^{v}).

(21) The notation 'cs' for *cosinus* is evidently borrowed from John Newton's *Institutio Mathematica* (London, 1654) if it is not Isaac's own present *ad hoc* invention. John Newton similarly employed 's', 't' and 'ct' for *sinus*, *tangens* and *cotangens*; see F. Cajori, *A History of Mathematical Notations*, **2** (Chicago, 1929): 162.

(22) On replacing the triangle *gmn* by its polar triangle, that is. Newton's Case 4 (surely not present in Hare's 'Trigonometria'?) evidently derives the standard cosine formula in the equivalent form

$$\mathrm{Cos}^2 \tfrac{1}{2}u / R^2 = \mathrm{Sin}\tfrac{1}{2}(p + z - c) \times \mathrm{Sin}\tfrac{1}{2}(p + z + c) / \mathrm{Sin}\,p \times \mathrm{Sin}\,z$$

by invoking the equality

$$\left(\frac{\frac{1}{2}\sigma T}{NT}\right)^2 = \left(\frac{\frac{1}{2}RT}{NT} \text{ or}\right) \frac{RT}{\theta T} \times \frac{\frac{1}{2}\theta T / LT}{NT / LT},$$

in which $RT/\theta T = \sin R\widehat{\theta}T / \sin\theta\widehat{R}T$, $\frac{1}{2}\theta T/LT = \sin L\widehat{T}\theta$ and $NT/LT = \sin L\widehat{T}N$. The equivalent modern result,

$$\cos u = \frac{\cos c - \cos(p + z)}{\sin p \sin z} - 1 = \frac{\cos c - \cos p \cos z}{\sin p \sin z},$$

In casu sexto (Fig. [12]) lineæ *LG, LM, LN* æquales sumuntur, & planis totidem ad puncta *G, M, N* normaliter erectis secantur producta circulorum plana *LGM, LMN, LNG*.
Plana illa transversa con-
currentia in *F, K* & *Q* hic
secantur secundum lineas
diagonales *HM, HN, HG* in
bina triangula *KHN* &
HNQ, QHG & *GHF, FHM*
& *MHK* rectangula ad *F, K*
& *Q*. Concipe plana illa
super planis circulorum
erigi ad perpendiculum ut
lineæ æquales *HF* et *HF*,
HK et *HK, HQ* et *HQ* coincidant, dein circulorum
plana duo *LGM, LG*[*N*]

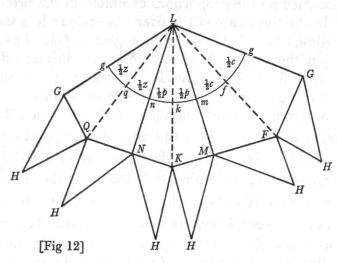

[Fig 12]

super tertio *LM*[*N*] erigi et ad invicem ita applicari, ut lineæ binæ æquales *LG* et *LG, HG* & *HG, HM* & *HM, HN* & *HN* coincidant, & figurā in hoc situ naturali intuendo patebit quod summa angulorum *FMK, KMH* angulo[23] trianguli ad *m* et summa *KNH, QNH* angulo[23] ad *n*, atqͻ summa *QGH, FGH* angulo[23] ad *g* æqualis sit. Ob æquales lineas *LG, LM, LN* anguli *GLM, MLN, NLG* a diagonalibus *LF, LK, LQ* bisecantur ideoqͻ respondent anguli *FLM, MLK, NLQ* semissibus laterum *gm, mn,* et *ng* trianguli sphærici,[24] suntqͻ lineæ binæ *GF* & *FM, MK* & *KN, NQ* & *QG* æquales, & inde etiam triangula bina *GFH* & *MFH, MKH* & *NKH, NK*[*Q*] & *QGH* æqualia,[25] et lineæ omnes *GH, HM, HN* æquantur inter se. Trianguli sphærici angulus ad *n* subducatur de ipsius angulo ad *g*, id est summa angulorum *KNH, QNH* seu *KMH, QNH* de summa

---

follows more straightforwardly in Eulerian manner (see note (9) above) by proving the equality

$$LO \times LM = LO^2 \text{ (or } LN^2 + NO^2) - LO \times MO \text{ (or } NO \times RO) = LN^2 + NO \times NR,$$

since then
$$\frac{NR}{NG} = \frac{LM/LG - (LN/LO) \times (LN/LG)}{(NO/LO) \times (NG/LG)}.$$

As an evident corollary Newton might well also have produced

$$\operatorname{Sin}^2 \tfrac{1}{2}u/R^2 = \operatorname{Sin}\tfrac{1}{2}(c+p-z) \times \operatorname{Sin}\tfrac{1}{2}(c+z-p)/\operatorname{Sin}p \times \operatorname{Sin}z.$$

The complementary Case 5 furnishes $\cos u = -\cos a \cos e + \sin a \sin e \cos c$ in the equivalent forms

$$\operatorname{Sin}^2 \tfrac{1}{2}c/R^2 = -\operatorname{Cos}\tfrac{1}{2}(a+e+u) \times \operatorname{Cos}\tfrac{1}{2}(a+e-u)/\operatorname{Sin}a \times \operatorname{Sin}e$$

and

$$\operatorname{Cos}^2 \tfrac{1}{2}c/R^2 = \operatorname{Cos}\tfrac{1}{2}(u+e-a) \times \operatorname{Cos}\tfrac{1}{2}(u+e-a)/\operatorname{Sin}a \times \operatorname{Sin}e.$$

The basic cosine formula, known to the tenth-century Arab astronomer and mathematician al-Battani, had already appeared in published form in Europe a century and a half before in

In the sixth case (Figure 12) the lines LG, LM and LN are taken equal, and by an equal number of planes erected normally at the points G, M, N are cut the extended circle planes LGM, LMN and LNG. Those cross planes, concurrent in F, K and Q, are there cut along the diagonal lines into pairs of triangles KHN and NHQ, QHG and GHF, FHM and MHK right-angled at F, K and Q. Imagine those planes to be raised perpendicularly upon the circle planes so that the equal lines HF and HF, HK and HK, HQ and HQ coincide, and then the two circle planes LGM and LGN to be raised upon the third LMN and so applied to one another

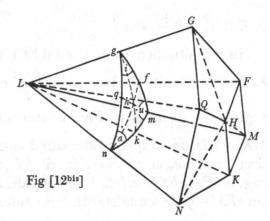

Fig [12bis]

that the pairs of equal lines LG and LG, HG and HG, HM and HM, HN and HN coincide. By inspecting the figure in this natural situation it will then be evident that the sum of the angles $F\widehat{M}H$ and $K\widehat{M}H$ is equal to the angle[23] of the triangle at m, that of $K\widehat{N}H$ and $Q\widehat{N}H$ to the angle[23] at n, and the sum of $Q\widehat{G}H$ and $F\widehat{G}H$ to the angle[23] at g. Because of the equal lines LG, LM, LN the angles $G\widehat{L}M$, $M\widehat{L}N$, $N\widehat{L}G$ are bisected by the diagonals LF, LK, LQ, and consequently the angles $F\widehat{L}M$, $M\widehat{L}K$, $N\widehat{L}Q$ correspond to the halves of the sides $\widehat{gm}$, $\widehat{mn}$ and $\widehat{ng}$ of the spherical triangle;[24] moreover, the pairs of lines GF and FM, MK and KN, NQ and QG are equal, and therefore also the pairs of triangles GFH and MFH, MKH and NKH, NKQ and QGH,[25] while the lines GH, HM and HN are all equal to one another. Let the angle at n in the spherical triangle be taken from its angle at g, that is, the sum of the angles $K\widehat{N}H$ and $Q\widehat{N}H$, or of

Regiomontanus' *De Triangulis Omnimodis* (Nuremberg, 1533: Liber v, Propositio III: 129) but its 'half-angle' forms are usually attributed to Euler. (Compare Johannes Tropfke, *Geschichte der Elementar-Mathematik*, **5** (Berlin/Leipzig, 1923): 137–43.) The polar standard form was known to Tycho Brahe about 1590 (see A. von Braunmühl, *Geschichte der Trigonometrie*, **1** (Leipzig, 1900): 201, note 1) and was derived as the polar analogue of the standard formula by Viète in his *Variorum de Rebus Mathematicis Responsorum Liber* viii (Tours, 1593 [= *Opera Mathematica* (Leyden, 1646): §§xv/xvi: 407–8]).

(23) $\hat{u}$, $\hat{a}$ and $\hat{i}$ respectively.

(24) This phrase replaces an equivalent following cancelled sentence 'Respondent igitur anguli *FLM*, *MLK* semissibus laterum trianguli; & summa angulorum *FMH*, *FMK* angulo incluso *m*, ac eorundem differentia differentiæ angulorum oppositorum æquatur' (...; and the sum of the angles $F\widehat{M}H$, $F\widehat{M}K$ is equal to the included angle *m*, their difference to the difference of the angles opposite).

(25) 'respectivè' (respectively) is cancelled.

$QGH+FGH$ seu QNH, FMH, et manebit differentia $FMH-KMH$.[26] Quare cum $FMH+KMH$ sit m et $FMH-KMH$ sit $g-n$ erit $FMH\dfrac{m+g-n}{2}$ seu $\dfrac{a+i-u}{2}$ et $KMH\dfrac{m-g+n}{2}$ seu $\dfrac{a-i+u}{2}$.

His expositis dico primò quod si FK bisecetur in O, in circulo circa triangulum KLF descripto sit OK sinus anguli $\dfrac{p+c}{2}$ et OM sinus anguli $\dfrac{p\|c}{2}$. Nam si in FK sumatur MD æqualis MK, et LD jungatur et producatur donec secet hunc circulum in d, et agatur Fd: erit Fd (ob æquales angulos FdL, FKL, KDL, FDd) æqualis FD et $\frac{1}{2}FD$ seu $\frac{1}{2}FM-\frac{1}{2}MK$ æqualis OM. Sunt autem FK et Fd chordæ angulorum FLK et FLd seu $\dfrac{c+p}{2}$ et $\dfrac{c-p}{2}$ ad circumferentiam, et chordarum semisses OK et OM sunt sinus eorundem angulorum ad centrum.[27]

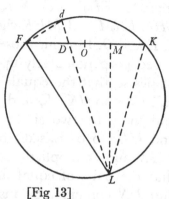

[Fig 13]

Dico jam secundò quod si circuli circa triangula MFH, MKH descripti concurrant in puncto X[28] et in FK demittatur perpendiculum XO, ad circulum centro X et radio XO descriptum erit OK cotangens anguli $\frac{1}{2}a$ et OM tangens anguli $\dfrac{i\|u}{2}$. Nam si jungantur FX, MX, KX erunt anguli FXM, MXK angulis FHM, MHK æquales respectivè; & quoniā circuli illi diametris æqualibus MH et MH descripti æquantur inter se et FX et KX in his circulis chordæ sunt quibus anguli FMX et XMK ad circumferentiam constitu[t]i sub-

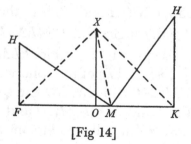

[Fig 14]

tenduntur, suntcҙ hi anguli simul sumpti rectis duobus æquales, erunt FX et XK æquales et propterea FK bisecabitur in O. Jam cum angulus FXM seu FHM sit $90^{\text{gr}}-FMH$ seu $90^{\text{gr}}-\dfrac{a+i-u}{2}$ et angulus MXK seu MHK sit $90^{\text{gr}}-KMH$ seu $90^{\text{gr}}-\dfrac{a+u-i}{2}$; patet quod angulorum FXM et MXK semisumma OXK sit $90^{\text{gr}}-\frac{1}{2}a$, et quod eorundem semidifferentia OXM sit $\dfrac{i-u}{2}$ ideocҙ OK et OM sunt tangentes angulorū $90^{\text{gr}}-\frac{1}{2}a$ et $\dfrac{i-u}{2}$ ad radium OX.[29]

(26) Newton first concluded: 'Est autem in circulo qui circa triangulum FLK describitur, FK chorda subtendens angulum FLK ad circumferentiam, & $FM-MK$ æqualis est chordæ

$K\widehat{M}H$ and $Q\widehat{N}H$, from the sum of $Q\widehat{G}H$ and $F\widehat{G}H$, or of $Q\widehat{N}H$ and $F\widehat{M}H$, and there will remain the difference $F\widehat{M}H - K\widehat{M}H$.[26] Hence, since

$$F\widehat{M}H + K\widehat{M}H = m \quad \text{and} \quad F\widehat{M}H - K\widehat{M}H = g-n,$$

then $F\widehat{M}H = \frac{1}{2}(m+g-n)$ or $\frac{1}{2}(a+i-u)$ and

$$K\widehat{M}H = \frac{1}{2}(m-g+n) \quad \text{or} \quad \frac{1}{2}(a-i+u).$$

With these things explained I assert, first, that if FK be bisected at O, then, in the circle described round the triangle KLF, OK shall be the sine of the angle $\frac{1}{2}(p+c)$ and OM the sine of the angle $\frac{1}{2}|p-c|$. For if in MK there be taken MD equal to MK and LD be joined and extended till it cut this circle in d, and Fd be drawn, then (because of the equal angles $F\widehat{d}L$, $F\widehat{K}L$, $K\widehat{D}L$ and $F\widehat{D}d$) Fd will be equal to FD with $\frac{1}{2}FD$, that is, $\frac{1}{2}FM - \frac{1}{2}MK$, equal to OM. However, FK and Fd are chords of the angles $F\widehat{L}K$ and $F\widehat{L}d$, namely, $\frac{1}{2}(c+p)$ and $\frac{1}{2}(c-p)$, at the circumference, and so the halves, OK and OM, of these chords are the sines of the same angles at the centre.[27]

I now say, secondly, that if the circles described round the triangles MFH, MKH be concurrent at the point X[28] and onto FK there be dropped the perpendicular XO, then, with regard to the circle described with centre X and radius XO, OK will be the cotangent of the angle $\frac{1}{2}a$ and OM the tangent of the angle $\frac{1}{2}|i-u|$. For if FX, MX and KX be joined, the angles $F\widehat{X}M$, $M\widehat{X}K$ will be equal to the angles $F\widehat{H}M$, $M\widehat{H}K$ respectively; and since the circles described with equal diameters MH and MH are equal to each other, while FX and KX in these circles are chords by which the angles $F\widehat{M}X$ and $X\widehat{M}K$ set at the circumference are subtended, and these angles taken together are equal to two right angles, FX and XK will be equal and accordingly FK will be bisected at O. Now because the angle $F\widehat{X}M$ or $F\widehat{H}M$ is $90° - F\widehat{M}H$, that is,

$$90° - \frac{1}{2}(a+i-u),$$

and the angle $M\widehat{X}K$ or $M\widehat{H}K$ is $90° - K\widehat{M}H$, that is, $90° - \frac{1}{2}(a+u-i)$, it is evident that the half sum $O\widehat{X}K$ of the angles $F\widehat{X}M$ and $M\widehat{X}K$ is $90° - \frac{1}{2}a$, and that their half difference $O\widehat{X}M$ is $\frac{1}{2}(i-u)$; accordingly OK and OM are the tangents of the angles $90° - \frac{1}{2}a$ and $\frac{1}{2}(i-u)$ with respect to the radius OX.[29]

---

qua angulus $FLM - MLK$ subtenditur. Quare...' (In the circle described round the triangle FLK, however, FK is a chord subtending the angle $F\widehat{L}K$ at the circumference, while $FM - MK$ is equal to the chord by which the angle $F\widehat{L}M - M\widehat{L}K$ is subtended. Hence...).

(27) In modern terms, $OK:OM = FK:(FD$ or$)$ $Fd = \operatorname{Sin}\frac{1}{2}(c+p):\operatorname{Sin}\frac{1}{2}(c-p)$.

(28) Evidently the mid-point of the line joining HH, since the two right triangles HMX are congruent.

(29) In modern terms, equivalently,

$$OK:OM = (OK/OX:OM/OX \text{ or}) \operatorname{Cot}\tfrac{1}{2}a:\operatorname{Tan}\tfrac{1}{2}(i-u).$$

Quare cùm OK et OM sint et sinus angulorum $\frac{c+p}{2}$ et $\frac{c-p}{2}$ et tangentes angulorum $90^{gr}-\frac{1}{2}a$ et $\frac{i-u}{2}$, patet quod hi sinus et tangentes sint in eadem ratione:[30] Quod erat primò demonstrandum.

Trianguli *aiu* producantur latera p et z donec compleantur ad semicirculos et in triangulo complementali $a"i"u$ per jam demon-
strata erit sinus semisummæ laterum c et $p"$ seu $90^{[gr]}-\frac{1}{2}p+\frac{1}{2}c$ ad sinum semidifferentiæ eorundem laterum seu $90^{[gr]}-\frac{1}{2}p-\frac{1}{2}c$ ut cotangens semissis anguli $a"$ ad tangentem semidifferentiæ angulorum reliquorum $i"$ et u, hoc est cosinus anguli $\frac{c-p}{2}$ ad

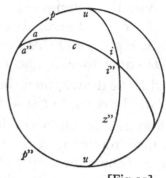

[Fig 15]

cosinum anguli $\frac{c+p}{2}$ ut tangens anguli $\frac{1}{2}a$ ad co-
tangentem anguli $\frac{i+u}{2}$, ideoq (cum tangens et cotangens sint reciproce proportionales) ut tangens anguli $\frac{i+u}{2}$ ad cotangentem anguli $\frac{1}{2}a$, et inverse cosin $\frac{c+p}{2}$ ad cosin $\frac{c-p}{2}$ ut cotang. $\frac{1}{2}a$ ad tang $\frac{i+u}{2}$.[31] Quod erat secundò demonstrandum.

Casus septimus idem fit cum sexto si modò[32] vertantur latera in angulos et anguli in latera.

---

(30) This is the so-called Napierean 'analogy' (see note (32))

$$\mathrm{Sin}\tfrac{1}{2}(c+p):\mathrm{Sin}\tfrac{1}{2}(c-p) = \mathrm{Cot}\tfrac{1}{2}a:\mathrm{Tan}\tfrac{1}{2}(i-u).$$

(31) The corresponding Napierean 'analogy'

$$\mathrm{Cos}\tfrac{1}{2}(c+p):\mathrm{Cos}\tfrac{1}{2}(c-p) = \mathrm{Cot}\tfrac{1}{2}a:\mathrm{Tan}\tfrac{1}{2}(i+u).$$

These two proportions were not, in fact, stated by Napier himself but by Henry Briggs in 'Annotationes aliquot' on the former's 'Propositiones quædam eminentissimæ ad triangula sphærica, mirâ facilitate resolvenda' (both published for the first time by Napier's son Robert in appendix to his edition of his father's *Mirifici Logarithmorum Canonis Constructio* (Edinburgh, 1619): 63–7 and 54–62 respectively). In Briggs' (unproved) statement (*ibid.*: 66–7) these appear as:
'1. prop. Sinus compl. $\frac{1}{2}$ summæ crurum [est ad] Sinu[m] compl. $\frac{1}{2}$ differentiæ crurū [ut] Tangens compl. $\frac{1}{2}$ anguli vertic. [ad] Tang. $\frac{1}{2}$ sum. angul. ad basim.
'2. prop. Sinus semisummæ laterum [est ad] Sinu[m] semidifferentiæ laterum [ut] Tangens com. semiang. vert. [ad] Tang. semidiff. ang. ad basim.' Newton's present demonstration

Hence since OK and OM are at once the sines of the angles $\frac{1}{2}(c+p)$ and $\frac{1}{2}(c-p)$ and the tangents of the angles $90° - \frac{1}{2}a$ and $\frac{1}{2}(i-u)$, it is evident that these sines and tangents are in the same ratio: the first thing to be demonstrated.[30]

Let the sides p and z of the triangle aiu be extended till they are complete semicircles, and in the complementary triangle $a''i''u$ by what has already been demonstrated there will be the sine of the half sum of the sides c and p'' (or $90° - \frac{1}{2}p + \frac{1}{2}c$) to the sine of the half difference of those same sides (or $90° - \frac{1}{2}p - \frac{1}{2}c$) as the cotangent of half the angle a'' to the tangent of the half difference of the remaining angles i'' and u; that is, the cosine of the angle $\frac{1}{2}(c-p)$ to the cosine of the angle $\frac{1}{2}(c+p)$ as the tangent of the angle $\frac{1}{2}a$ to the cotangent of the angle $\frac{1}{2}(i+u)$, and consequently (since tangent and cotangent are reciprocally proportional) as the tangent of the angle $\frac{1}{2}(i+u)$ to the cotangent of the angle $\frac{1}{2}a$. Inversely, the cosine of $\frac{1}{2}(c+p)$ to the cosine of $\frac{1}{2}(c-p)$ is as the cotangent of $\frac{1}{2}a$ to the tangent of $\frac{1}{2}(i+u)$.[31] This was the second thing to be demonstrated.

The seventh case becomes identical with the sixth if only[32] sides be converted into angles and angles into sides.

---

would appear to be their first direct proof, though Napier in his *Mirifici Logarithmorum Canonis Descriptio* (Edinburgh, 1614): Liber II, Caput VI: 49–52 had given an ingenious deduction, making use of the circle-preserving property of stereographic projection, of the corollary $\mathrm{Tan}\frac{1}{2}(c+p):\mathrm{Tan}\frac{1}{2}(c-p) = \mathrm{Tan}\frac{1}{2}(i+u):\mathrm{Tan}\frac{1}{2}(i-u)$. (Compare A. von Braunmühl, *Geschichte der Trigonometrie*, **2** (note (9) above): 15–16.)

(32) Namely, on replacing the spherical triangle by its polar dual. There result in this way the two complementary Napierean 'analogies'

$$\mathrm{Sin}\tfrac{1}{2}(i+u):\mathrm{Sin}\tfrac{1}{2}(i-u) = \mathrm{Tan}\tfrac{1}{2}z:\mathrm{Tan}\tfrac{1}{2}(c-p)$$

and

$$\mathrm{Cos}\tfrac{1}{2}(i+u):\mathrm{Cos}\tfrac{1}{2}(i-u) = \mathrm{Tan}\tfrac{1}{2}z:\mathrm{Tan}\tfrac{1}{2}(c+p),$$

first stated in cumbrous logarithmic form by Napier in his 'Propositiones quædam' (note (31) above): 61 (compare J. Tropfke, *Geschichte der Elementar-Mathematik*, **5**: 155, note 330) and restated by Briggs in his 'Annotationes' thereon in the more intelligible direct form (*Constructio* (note (31) above): 66) as:

'3. prop. Sinus $\frac{1}{2}$ aggregati angulorum [ad basin est ad] Sinu[m] semidifferentiæ angulorū [ut] Tangens semibasis [ad] Tangen[tem] $\frac{1}{2}$ differentiæ crurum';

'[4.] Proport. Sinus compl. $\frac{1}{2}$ summæ angulorū [est ad] Sin. comp. $\frac{1}{2}$ differētiæ angulor. [ut] Tangens semibasis [ad] Tangen[tem] semisummæ crurum.'

Apparently no explicit proof of these two proportions appeared in print till Oughtred demonstrated them in his *Trigonometria* (London, 1657: 26 ff.). The term 'analogies de Neper' was bestowed much later (by A. R. Mauduit in his *Principes d'Astronomie Sphérique; ou Traité complet de trigonométrie sphérique* (Paris, 1765): 76).

APPENDIX 1. NEWTON'S 'FUNDAMENTALS OF TRIGONOMETRY' (1683).[1]

From a transcript(?) in Lambeth Palace Library, London[2]

TRIGO[NO]METRIÆ FUNDAMENTA.[3]

R Radium significat. s sinum. t tangentem. s', t' sinum & Tangen[tem] complementi. ∠ Angulum. L Latus. = æqualia. :: proportionalia. ∞ s[ign]um proportionale. ∥ parallelam. ⊥ perpendicularem. + addendam. − subduc[endam.]

[Lemma.]

$t . R :: R . t'$. Nam $AT . AC :: BC . Bt$.

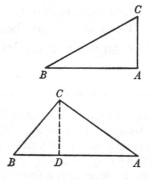

Propositiones.

1. In triangulo rectilineo ABC rectangulo ad A, ut

$$BC . CA :: R . s B. \quad \text{prop 1.}$$

$$BA . CA :: R . t B. \quad \text{prop 2.}$$

2. In triangulo quovis rectilineo ABC, sit $CD \perp AB$ et erit (per Prop 1)

$$BC . CD :: R . s B \ \& \ CD . CA :: s A . R,$$

et ex æquo $BC . CA :: s A . s B.$ pr 3.

---

(1) This self-styled 'Trigometriæ Fundamenta / a Viro Cl. Isaaco Newton, Matheseos Professore, anno 1683. data', carefully copied out in the hand of Henry Wharton (admitted as pensioner to Caius College, Cambridge, in February 1680, graduated B.A. *primus inter omnes* in March 1684, later to become a prolific writer of theological tracts and private chaplain to the Archbishop of Canterbury at Lambeth Palace), is now to be found among his manuscript 'Theoremata et Problemata quædā Anno 1683 a me conscripta Mensibus Julio, Augusto, Septembri Horis subsecivis. Liber Primus'. S. P. Rigaud, who first identified it (Bodleian. MS Rigaud 9: 43ʳ), later wrote that it 'contains the notes of common propositions...written out by [Wharton]' (*Historical Essay on the First Publication of Sir Isaac Newton's Principia* (Oxford, 1838): 97, note *e*). Twelve years later Joseph Edleston pointed out a passage in the 'Life of Mr. Hen. Wharton' prefixed to the second, posthumous edition of his sermons which suggested how this augmented version of Newton's 'Epitome' (§1 above) came to be in the possession of a young man destined to make his career in the Church: as an undergraduate Wharton 'pursued his Studies with an indefatigable Industry', reading widely in classics, French and philosophy, likewise attaining 'no mean Skill in *Mathematicks*. Which last was much encreased by the kindness of Mr. *Isaac Newton*,... who was pleased to give him further instructions in that noble Science, amongst a select Company in his own private Chamber' (*Fourteen Sermons Preach'd in Lambeth Chapel. Before...Dr. William Sancroft...In the Years MDCLXXXVIII. MDCLXXXIX.... With an Account of the Author's Life. The Second Edition Corrected* (London, 1700): A3ᵛ/A4ʳ; compare J. Edleston's *Correspondence of Sir Isaac Newton and Professor Cotes,*

Sit $BE = BC = BG$. et $EF \parallel AC$. erit (per 31. III element). $\angle ECG$ rectang. et $AG \cdot AE :: CG \cdot CF$.

id est $AB + BC \cdot AB - BC :: \mathrm{t}\,\dfrac{CBG}{2} \cdot \mathrm{t}\,\dfrac{C-A}{2}.$

[prop 4.]

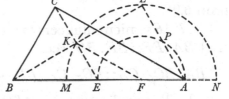

Sit $BE = BC$. [&] $KE \cdot CE :: 1 \cdot 2 :: KF \cdot CA :: EF \cdot EA$. MLN, EPA semicirculi centro F, radijs FK, FE descripti. [erit]

$$CKE \parallel APL \perp BKL.$$

tum

$$R \cdot \mathrm{s}' \frac{B}{2} :: BE \cdot BK :: BA \cdot BL :: BE \backsim BA \cdot BK \backsim BL = BM \backsim BN.$$

id est $R \cdot \mathrm{s}' \dfrac{B}{2} :: BC \backsim BA \cdot \dfrac{BA + BC - CA}{2} \backsim \dfrac{BA + BC + CA}{2}.$ [prop] 5.

(London, 1850): xlv, note (16)). We have seen (III: xviii) that it was Newton's statutory obligation, as Lucasian professor, to make himself freely available to students twice a week during term-time and the gathering in Newton's rooms could have been 'select' only in their common enthusiasm for the higher reaches of mathematics, but we may readily accept Edleston's conjecture (*Correspondence*: xcv, note †) that the present 'rules for the solution of plane and spherical triangles' were given to Wharton 'probably at one of those private lessons' rather than—as Rigaud would seem to have it—taken down by him during a lecture course of Newton's which has left no other recorded trace. Neither Rigaud nor Edleston had access to Newton's original paper, the 'Epitome Trigonometriæ', on which Wharton's 'Fundamenta' is based and were thus unable to determine the extent of the latter's innovations (though they might well have noticed the paper's narrow indebtedness to Seth Ward's popular undergraduate textbook, *Idea Trigonometriæ Demonstratæ* (*In usum Juventutis Oxoniensis*), published thirty years before). In hindsight, we may now see that Wharton's text slightly augments the first fourteen propositions of Newton's autograph, adding seven new theorems (Propositions 7 and 15–20) and the verbal enunciations of all twenty now listed, but is otherwise essentially unoriginal. Particular points of divergence between the two versions are discussed in following footnotes.

(2) Two sheets (unpaginated) in Wharton's 'Scripta Academica, viz Tentamina quædam Philosophica, Mathematica, Oratoria, Philologica, &c Annos inter 1682 et 1686 a me facta, Cum apud Academiam Cantabrigiensem literis operam Juven[is] nava[rem]...Necnon Scripta aliorum Mathematica, in usum proprium a me descripta, quædam' (Lambeth Palace, London. MS Codex H. Wharton 592), accurately described by Edleston (*Correspondence* (note (1) above): xcv, note †) as consisting of 'two folio leaves (i.e. of two pages and seven lines on the last page, the second being blank)'. The manuscript, rather larger than the surrounding sheets with which it is bound, is now battered and badly fragmented: there is now lacking, in particular, a gnomonic strip (about half an inch wide on its right-hand side and some half dozen lines deep at its bottom) which has completely eroded away. In the present reproduction of its text we have not hesitated to make wholesale restoration of the missing portions in line with Newton's original autograph 'Epitome' (§1).

(3) For Wharton's full title see note (1) above.

Item $R \cdot s\dfrac{B}{2} :: BE \cdot EK :: BA \cdot AL :: BE \infty BA \cdot EK \infty AL = PL \infty AL = AN \infty E[N.$ id est]

$$R \cdot s\frac{B}{2} :: BC \infty BA \cdot \frac{BC + AC - AB}{2} \infty \frac{AC + AB - BC}{2}.$$

prop 6.[4]

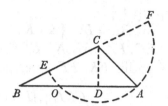

[Sit *EAF* semicirculus centro *C* radio *CA* descriptus. erit] $BA \cdot BF :: BE \cdot BO$, seu

$$BA \cdot BC + CA :: BC - CA \cdot BD - DA.^{[5]} \quad \text{prop 7.}$$

3. In Triangulis sphæricis *ABC*, *DBE*[6] rectangulis ad *A* et *E*, designent *DF*, *DA*, *DC*, *CF*, *CE* quadrantes circulorum in octavâ parte sphæræ, cujus centrum est *C*, cæterisꝗ ut in analemmate[7] constructis erit $CE \cdot EH :: CB \cdot [B]L.$ et $CD \cdot DM :: AK \cdot CA.$ id est

In triang ABC $\begin{cases} R \cdot s\, C :: s\, CB \cdot s\, AB. & \text{prop 8.} \\ R \cdot t'\, C :: t\, AB \cdot s\, AC. & \text{prop 9.} \end{cases}$

et

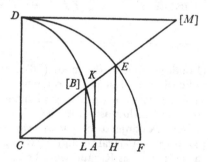

In triang DEB $\begin{cases} R \cdot s'\, DE :: s'\, BE \cdot s'\, BD. \\ \qquad\qquad\qquad\qquad \text{prop 10.} \\ R \cdot t\, DE :: t'\, DB \cdot s'\, D. \\ \qquad\qquad\qquad\qquad \text{prop 11.} \end{cases}$

---

(4) These first six propositions are a straightforward revise of their equivalents in Newton's 'Epitome'.

(5) A 'new' proposition, not present in the 'Epitome' but taken over with little alteration from Ward's *Idea* (note (1) above): 7: 'Fundamentum solutionis Trianguli, ubi dantur tria latera, & quæruntur anguli, (demissâ perpendiculari)'. As a corollary Ward's solution yields the 'cosine rule' (Euclid, *Elements*, II, 12/13; compare III: 406) in the form

$$BD \ (\text{or } BC\cos\hat{B}) = \tfrac{1}{2}[BA + (BC^2 - CA^2)/BA].$$

The insertion is probably Newton's but may just possibly be due to Wharton (whose surrounding 'Scripta Academica' includes a copy, in 19 sheets, of Ward's tract under the head 'Idea Trigonometriæ Demonstratæ. Authore Setho Wardo...1654. Additis e regione cujusꝗ Paginæ a me Commentariis'). To take up the extra 'prop 7', the numbers of the six following propositions (but not later references to them) have been increased by unity, while the general Napierean rule of circular parts no longer has the status of a separate 'Prop 13'.

(6) The point '*B*' is not marked in Wharton's accompanying figure, doubtless through an oversight on his part.

(7) Understand 'by orthogonal projection', a mapping (here of the sphere of centre *C* onto its diametral plane *CDEF*) first used systematically by Ptolemy in his *Analemma* to project the celestial sphere onto the three mutually perpendicular planes of the horizon, meridian and 'prime vertical'. For the exact significance of the word ἀνάλημμα consult T. L. Heath, *A History of Greek Mathematics*, **2** (Oxford, 1921): 287.

Quare (per prop 7 et 8)$^{(8)}$ in triang $DE[B]$ est $R . s B :: s DB . s DE$. et $R . t' B ::$
$t DE$. [$s BE$. Hoc est]

in triang $ABC \begin{cases} R . s B :: s' AB . s' C. & \text{prop 12.} \\ R . t' B :: t' C . s' BC. & \text{prop 13.} \end{cases}$

Ex his quinque terminis nimirum cruribus anguli recti, et complementis tum
hy[po]tenusæ, tum angulorum ad hypotenusam designet M quemlibet, P et p
duos huic ut[rinque] proximos, Q et q cæteros duos, et erit

$$R \times s M = s' Q \times s' q = t P \times t p.^{(9)}$$

quâ una prop[ositione] sex priores continentur.

4. In triangulo quovis sphærico ABC sit $CD [\perp] AB$ et (per prop $7^{(10)}$) erit
$s BC . R :: s CD . s B$ et $R . s AC :: s A . s CD$. et ex æquo

$s BC . s AC :: s A . s [B]$.   prop 14.

$^{(11)}$[Et mixtim

$s BC + s AC . s BC - s AC :: s A + s B . s A - s B$.   prop 15.

id est (per prop 4)

$$t \frac{BC+AC}{2} . t \frac{BC-AC}{2} :: t \frac{A+B}{2} . t \frac{A-B}{2} . \quad \text{prop 16.]}$$

$^{(12)}$s' $BC . s' BD (:: \text{per prop } 9,^{(13)} s CD . R) :: s' AC . s' AD$.   prop 17.

s' $B . s BCD (:: \text{per prop } 11,^{(13)} s' CD . R) :: s' A . s ACD$.   prop 18.

t' $B . s BD (:: \text{per prop } 8,^{(13)} R . t AB) :: t' A . s AD$.   prop 19.

t' $BC . s' BCD (:: \text{per prop } 10,^{(13)} R . t DC) :: t' AC . s' ACD$.   prop 20.

Quæ omnia verbis sic enunciantur.$^{(14)}$

Lemma, Ut Tangens ad Radium, ita Radius ad tangentem [complementi].

---

(8) Read 'per prop 8 et 9'; compare note (5) above.

(9) This is labelled 'Prop 13' in Newton's 'Epitome'. For the reason underlying its
suppression see note (5) above.

(10) Read 'per prop 8'; compare note (5) above.

(11) The remainder of this (first) page of Wharton's manuscript has flaked away. The
context does not allow a unique restoration of Propositions 15 and 16 following, but some
equivalent to our suggested versions would seem to be needed.

(12) We omit an unnecessary reminder (set at the top of the second page of Wharton's
manuscript) that 'In triangulo ABC, si $CD \perp AB$ erit' and an incorrectly lettered (but uncancelled) preliminary figure.

(13) In line with note (5) here read '10', '12', '9' and '11' respectively.

(14) The following twenty-two paragraphs enunciate in words the opening lemma, Propositions 1–20 and Napier's general rule of circular parts in the preceding mathematical text.

[1] In Triangulo rectilineo rectangulo ut Radius ad Sinum anguli alterius acuti, ita hypotenusa ad latus angulo illi oppositum.

Ut Radius ad Tangentem anguli, ita latus conterminum ad latus alt[erum].

[2] In Triangulis obliquangulis latera sunt sinubus angulorum oppositorum proportion[alia].

Ut summa laterum duorum ad differentiam eorum, ita Tangens complementi semissis anguli inclusi ad Tangentem semidifferentiæ reliquorum ang[ulorum].

Ut medium proportionale inter latera duo ad medium proportionale in[ter] semisummam omnium laterum et excessum illius semisummæ supra lat[us] tertium, ita Radius ad Sinum complementi semissis anguli inter latera du[o] prima.

Ut medium proportionale inter latera duo ad medium proportionale int[er] semisummam et semidifferentiam lateris tertij ad differentiam duorum later[um] priorum, ita Radius ad sinum semissis anguli inter latera duo priora.

Ut basis ad summam laterum, ita differentia laterum ad duplam dist[an]tiam perpendiculi a medio basis.

[3] In Triangulis Sphæricis Rectangulis ut Radius ad Sinum Hypotenusæ, ita sinus anguli cujusvis ad sinum lateris oppositi.

Ut Radius ad Tangentem complementi anguli alterutrius obliqui, vel Tangens eju[s ad] Radium, ita Tangens cruris oppositi ad sinum cruris alterius.

Ut Radius ad Sinum complementi cruris alterutrius, ita sinus compl[emen]ti cruris alterius ad sinum complementi Hypotenusæ.

Ut Radius ad Tangentem complementi Hypotenusæ, (vel Tangens ejus ad [Radi]um) ita Tangens cruris alterutrius ad sinum complementi anguli contermin[um].

Ut Radius ad sinum complementi lateris alterutrius, ita sinus anguli c[onter]-mini ad sinum complementi anguli alterius.

Ut Radius ad Tangentem complementi anguli alterutrius (vel Tangens [ad] Radium) ita tangens complementi anguli alterius ad sinum complemen[ti Hy]potenusæ.

Hâc unâ propositione complectimur sex priores. Spectentur hi quinq[ue ter]mini, crura duo, et complementa tum hypotenusæ, tum angulorum duor[um ad] Hypotenusam, et rectangula sub Radio et sinu termini cujusvis, sub T[angentibus terminorum] duorum is[ti proximorum et sub sinubus complementi reliquorum duorum erunt æqualia].[15]

---

(15) This badly eroded paragraph (entered at the bottom of the manuscript's second page) has required extensive restoration, but its essential structure is clear. The four following paragraphs (corresponding to Propositions 14–17 in the preceding mathematical text) have completely disappeared and our insertions serve only as an indication of their rough outline.

[4. In Triangulis sphæricis obliquangulis sinus laterum sunt sinubus angulorum oppositorum proportionales.

Ut summa sinuum hypotenusarum ad eorum differentiam, ita summa sinuum angulorum ad basim ad illorum differentiam.

Ut Tangens semisummæ hypotenusarum ad Tangentem earum semidifferentiæ, ita Tangens semisummæ angulorum ad basim ad Tangentem semidifferentiæ eorundem angulorum.

Sinus complementi hypotenusarum sunt sinubus complementi segmentorum basis proportionales.]

Sinus complementi angulorum ad basim sunt sinubus angulorum ad perpendiculum proportionales.

Sinus segmentorum basis sunt Tangentibus complementi angulorum ad basim directè et tangentibus angulorum ad basin reciprocè proportionales.

Sinus complementi angulorum ad perpendiculum sunt tangentibus complementi hypotenusarum directè, et tangentibus Hypotenusarum reciprocè proportionales.

APPENDIX 2. AN OUTLINE OF A CANON OF SINES.[1]

From the original autograph in the University Library, Cambridge[2]

TABULA SINUUM AD SEMIGRADUS.

0.00	00000	4.30	07846	9.00	15643	13.30	23345
0.30	00573	5.00	08716	9.30	16505	14.00	24192
1.00	01745	5.30	09585	10.00	17365	14.30	25038
1.30	02618	6.00	10453	10.30	18224	15.00	25882
2:00	03490	6.30	11320	11.00	19081	15.30	26724
2.30	04362	7.00	12187	11.30	19937	16.00	27564
3.00	05234	7.30	13053	12.00	20791	16.30	28402
3.30	06105	8.00	13917	12.30	21644	17.00	29237
4.00	06976	8.30	14781	13.00	22495	17.30	30071
4.30	07846	9.00	15643	13.30	23345	18.00	30902

(1) A table of $10^5\sin\theta$ listed at half-degree intervals, $0° \leqslant \theta \leqslant 90°$, and doubtless copied by Newton from one or other of the contemporary trigonometrical canons available to him. (At his death his library contained Adriaen Vlacq's table of logarithmic sines, *Trigonometria Artificialis, sive Magnus Canon Triangulorum Logarithmicus* (Gouda, 1633) but not the allied Briggsian table of natural trigonometrical functions, Gellibrand's edition of the *Trigonometria Britannica*, with which it was published; see R. de Villamil, *Newton: the Man* (London, 1931): 69. This, however, is most probably the 'Vlac's Trigonometry' for which Newton in October 1695 paid '8ˢ...wᵗʰ many thanks' to Halley (*Correspondence of Isaac Newton*, 4, 1967: 181).)

(2) ULC. Add. 3959.5:45ʳ. In §2: note (10) above we suggested that Newton intended this autograph list to be appended to his 'Trigonometria'.

18.00	30902	36.00	58779	54.00	80902	72.00	95106
18.30	31730	36.30	59482	54.30	81412	72.30	95372
19.00	32557	37.00	60181	55.00	81915	73.00	95630
19.30	33381	37.30	60876	55.30	82413	73.30	95882
20.00	34202	38.00	61566	56.00	82904	74.00	96126
20.30	35021	38.30	62251	56.30	83389	74.30	96363
21.00	35837	39.00	62932	57.00	83867	75.00	96593
21.30	36650	39.30	63608	57.30	84339	75.30	96815
22.00	37461	40.00	64279	58.00	84805	76.00	97030
22.30	38268	40.30	64945	58.30	85264	76.30	97237
23.00	39073	41.00	65606	59.00	85717	77.00	97437
23.30	39875	41.30	66262	59.30	86163	77.30	97630
24.00	40674	42.00	66913	60.00	86603	78.00	97815
24.30	41469	42.30	67559	60.30	87036	78.30	97992
25.00	42262	43.00	68200	61.00	87462	79.00	98163
25.30	43051	43.30	68835	61.30	87882	79.30	98325
26.00	43837	44.00	69466	62.00	88295	80.00	98481
26.30	44620	44.30	70091	62.30	88701	80.30	98629
27.00	45399	45.00	70711	63.00	89100	81.00	98769
27 30	46175	45.30	71325	63.30	89493	81.30	98902
28.00	46947	46.00	71934	64.00	89879	82.00	99027
28.30	47716	46.30	72537	64.30	90259	82.30	99144
29.00	48481	47.00	73135	65.00	90631	83.00	99255
29.30	49242	47.30	73727	65.30	90996	83.30	99357
30.00	50000	48.00	74319	66.00	91355	84.00	99452
30.30	50754	48.30	74896	66.30	91706	84.30	99540
31.00	51504	49.00	75471	67.00	92050	85.00	99619
31.30	52250	49.30	76041	67.30	92388	85.30	99692
32.00	52992	50.00	76604	68.00	92718	86.00	99756
32.30	53730	50.30	77162	68.30	93042	86.30	99813
33.00	54464	51.00	77715	69.00	93358	87.00	99863
33.30	55194	51.30	78261	69.30	93667	87.30	99905
34.00	55920	52.00	78801	70.00	93969	88.00	99939
34.30	56641	52.30	79335	70.30	94264	88.30	99966
35.00	57358	53.00	79864	71.00	94552	89.00	99985
35.30	58070	53.30	80386	71.30	94832	89.30	99996
36.00	58779	54.00	80902	72.00	95106	90.00	1.00000

APPENDIX 3. A FIRST VERSION OF SECTION 3 OF THE 'TRIGONOMETRIA', TOGETHER WITH A PRELIMINARY REVISE.

From the originals in the University Library, Cambridge[1]

[1]   SECT 3.

DE TRIANGULIS QUIBUSVIS PLANIS.

THEOREMATA.

1. Triangulorum latera sunt sinubus oppositorum angulorum proportionalia.

2. Ut medium proportionale inter crura[2] duo ad medium proportionale inter semisummam trium laterum & excessum illius semisummæ supra basem,[3] ita Radius ad co-sinum dimidij anguli verticalis.[4]

3. Ut summa duorum laterum ad differentiam eorundem ita tangens semisummæ angulorum reliquorum ad tangentem semidifferentiæ ipsorum.

DEMONSTRATIO.

1. In triangulo ABC demisso perpendiculo, est (per Th: 1, Sect: 2) $AC.DC::R.sA$, et $DC.BC::sB.R$. Quare ex æquo, $AC.BC::sB.sA$. Q.E.D.

2.[5] Sit ABC triangulum. In crure AB cape AD æqualem cruri AC. Actam CD bisecet recta AE, in quam productam incidat perpendiculum BF. Basi CB parallelam age EK secantem AB in K. Et quoniam ED dimidium est CD erit EK dimidium cruris CB et DK dimidium DB. Centro K radio KE vel KF describe circulum, qui secet crus AB in G et H, et erit AH semisumma trium laterum, et AG excessus ejus supra basem: et ob æqualia rectangula GAH, EAF idem erit medium proportionale inter AE et AF quod inter AG et AH. Sit

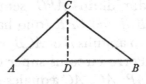

---

(1) ULC. Add. 3959.5: 39ʳ/49ʳ. This transitional text is reproduced both for its intrinsic interest and to reveal the closeness of the relationship between the 'Epitome Trigonometriæ' (§1 above) and its augmented revise as the following 'Compendium', a connection which is here (compare §2: note (42)) much more tenuous in the latter's final text.

(2) Newton first wrote 'latera'. Below, correspondingly, 'latus' is universally replaced by 'crus' (compare §2: note (37)).

(3) 'tertium latus' is cancelled; compare note (2).

(4) The equivalent adjectival phrase 'qui tertio lateri opponitur' is cancelled.

illud *AN*. Parallela *ED* acta *NL* secet *AB* in *L* et erit *AL* similiter medium proportionale inter *AD* et *AB*, id est inter *AC* et *AB*. Deniqʒ ob angulum *ANL* rectum est *AL* ad *AN* ut radius ad cosinum anguli *LAN* id est dimidij anguli *BAC*. Q.E.D.

     Coroll. Hinc etiam medium proportionale inter crura est ad medium proportionale inter summam et differentiam basis et differentiæ crurum (id est inter *DH* et *BD*) ut radius ad sinum dimidij anguli verticalis *A*. vizt

$$\sqrt{DAB}\,.\,\sqrt{DHB}::R\,.\,\text{s}\tfrac{1}{2}A.$$

Nam $\sqrt{DAB}=AL$. Cape *FP*=*ED* et ob angulum *DPB* rectum, circulus centro *K* radio *KB* vel *KD* descriptus transiret per punctum *P*. Quare cum *F* et *H* æqualiter distent a centro *K*, rectangula [*P*]*FB*, *DHB* æqualia erunt. Unde $\sqrt{DHB}(=\sqrt{BF[P]}=\sqrt{BF\times ED})=LN$. Sed $AL\,.\,LN::R\,.\,\text{s}\tfrac{1}{2}A$. Q.E.D.

     3.[6] In triangulo *ABC* in crure *AB* producto cape hinc inde *AP*[,] *AQ* æquales cruri *AC*. Junge *CP*, *CQ* et age *PR* paral- lelam *CB*, et erit *BQ* summa & *BP* differentia crurum, et *CAQ* summa angu- lorum *B* et *ACB* ad basem, et hujus dimidium *APC* semisumma, et angulus *BCP* vel *CPR* (quo hæc semisumma differt ab angulis *B* et *ACB*) eorum semidifferentia.

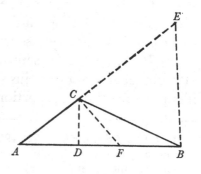

Porro circulus super diametro *PQ* descriptus transit per punctum *C*, eo quod *AP*, *AQ*, *AC* æquales sunt. Quare angulus *PCQ* rectus est et *CQ*, *CR* tangentes sunt angulorum *APC*, *CPR* respectu Radij *CP*. Est autem (ob parallelas *BC*, *PR*) *BQ* ad *BP* ut *CQ* ad *CR*. Q.E.D.

---

[2]   Propositiones

     $AC\,.\,CD::R\,.\,\text{s}A$.   & $CD\,.\,CB::\text{s}B\,.\,R$,   Ergo $AC\,.\,CB::\text{s}B\,.\,\text{s}A$. Prop 1. et mixtim $AC+CB$. $-AC+CB(::\text{s}B+\text{s}A\,.\,\text{s}A-\text{s}B)::$ (per Lem 2) $\text{t}\dfrac{A+B}{2}\,.\,\text{t}\dfrac{A-B}{2}$. Prop 2. Id est (si *BE*=*CE*) $AB+AE-BE$. $AB-AE+BE::\text{t'}E\,.\,\text{t'}ABE::$ $\text{t}ABE\,.\,\text{t}E$. Prop 3. [7]Et (si *BF*=*CF*)

$$AF+CF+AC\,.\,AF+CF-AC::\left[\text{t'}\frac{A}{2}\,.\,\text{t}\frac{ACF}{2}\right.$$

Prop 4. $\Big]$

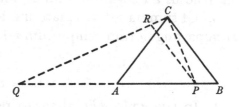

---

     (5) Compare Proposition 5 of the 'Epitome' (§1 : note (14) above).
     (6) Compare Proposition 4 of the 'Epitome' (§1: note (12) above).

$AB . AD :: R . s B$. $AD . AC :: s C . R$. Ergo ex æquo $AB . AC :: s C . s B$. Prop 1.

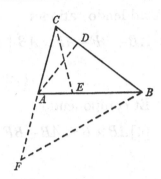

Et mixtim $AB + AC . AB - AC (:: s C + s B . s C - s B)$ :: (per Lem 2) $t \dfrac{C+B}{2} \left(t' \dfrac{A}{2} \right) . t \dfrac{C-B}{2}$. Prop. 2.

Id est (si $BE = EC$ et $CF = FB$) in tri ACE,

$$AE + EC + AC . AE + EC - AC :: t' \frac{A}{2} . t \frac{ACE}{2} ::$$

$$t' \frac{ACE}{2} . t \frac{A}{2}. \text{Prop 3.}$$

Et in tri ABF, $AB + BF - AF . AB - BF + AF :: t \dfrac{A}{2} . t \dfrac{ABF}{2} :: t' \dfrac{ABF}{2} . t' \dfrac{A}{2}$. Prop 4.

Et mixtim

$$AE + EC . AC :: \left(t' \frac{A}{2} + t \frac{ACE}{2} . t' \frac{A}{2} - t \frac{ACE}{2} :: \text{per Lem 1} \right) s' \frac{A - ACE}{2} . s \frac{AEC}{2}.$$

Prop 5.

Et

$$AB . BF - AF :: \left(t \frac{A}{2} + t \frac{ABF}{2} . t \frac{A}{2} - t \frac{ABF}{2} :: \text{per Lem 1} \right) s' \frac{F}{2} . s \frac{A - ABF}{2}. \text{Prop 6.}$$

Coroll.[8] Ut summa sinuum arcuum ad sinum dupli complementi semisummæ arcuum, ita sinus complementi semidifferentiæ ad sinū compl. semisummæ. &c.[9]

Sit A complementum angulorum B et C ad duos rectos et erit

1. $s A . s B + s C :: s' \dfrac{B+C}{2} . s' \dfrac{B-C}{2}$.

2. $s A . s B - s C :: s \dfrac{B+C}{2} . s \dfrac{B-C}{2}$.

Et rursus in tri ABF (per prop 3)

$$AF + BF + AB . AF + BF - AB :: t' \frac{A}{2} . t \frac{ABF}{2}^{[10]}$$

---

(7) Newton first continued 'Et rursus $AE + AB . AE - AB :: t \dfrac{B+E}{2} . t \dfrac{B-E}{2}$. id est

$$AC + BC + AB . AC + BC - AB :: t' \frac{A}{2} . t \frac{CBA}{2}. \text{Prop 4'}.$$

(8) A cancelled first version of the following reads 'Ut co-sinus differentiæ dimidiorum ad cosinum summæ ita summa sinuum arcuum duplorum ad sinum dupli complementi summæ arcuum'.

(9) Understand the complementary enunciation 'Ut differentia sinuum arcuum ad sinum dupli complementi semisummæ arcuum, ita sinus semidifferentiæ ad sinum semisummæ'.

(10) In his further revise Newton made this a separate 'Prop 5', increasing the numbers of the two following propositions by unity.

& addendo rationes

$$AB + BF - AF \times AB + BF + AF \, . \, AB - BF + AF \times AF + BF - AB$$

$$\left(::\text{t}' \frac{ABF}{2} \, . \, \text{t} \frac{ABF}{2} \right) :: \text{t}' \, \text{t}' \frac{ABF}{2} \, . \, R^q :: R^q \, . \, \text{tt} \frac{ABF}{2}. \quad \text{Prop 5.}$$

Et componendo

$$[4] AB \times BF \, . \, AB - BF + AF \times AF + BF - AB$$

$$:: R^q + \text{tt} \frac{ABF}{2} \, . \, \text{tt} \frac{ABF}{2} :: R^q \, . \, \text{ss} \frac{ABF}{2}. \quad \text{[Prop 6.]}^{(11)}$$

APPENDIX 4. TRIGONOMETRICAL EXTRACTS
FROM ST. JOHN HARE'S
'COMPENDIUM MATHEMATICOALGEBRAICUM' (1675).[1]

From the original in the University Library, Cambridge[2]

[1]   *Sphæricorum* △^{orū} *Rectangulorū.*[3]

1. $R \, . \, \dfrac{\text{s}\,B}{\text{s}\,C} :: \text{s}\,BC \, . \, \dfrac{\text{s}\,CA}{\text{s}\,BA} :: \dfrac{\text{sco}\,BA \, . \, \text{sco}\,C,}{\text{sco}\,CA \, . \, \text{sco}\,B,}$

2. $R \, . \, \text{sco}\,BA :: \text{sco}\,CA \, . \, \text{sco}\,BC,$

3. $R \, . \, \text{s}\,BA :: \dfrac{\text{t}\,B \, . \, \text{t}\,CA,}{\text{tco}\,CA \, . \, \text{tco}\,B,}$

4. $R \, . \, \text{tco}\,BC :: \dfrac{\text{t}\,CA \, . \, \text{sco}\,C,}{\text{t}\,BA \, . \, \text{sco}\,B,}$

5. $R \, . \, \text{sco}\,B :: \dfrac{\text{t}\,BC \, . \, \text{t}\,BA,}{\text{tco}\,BA \, . \, \text{tco}\,BC,}$

6. $R \, . \, \text{sco}\,BC :: \text{t}\,B \, . \, \text{tco}\,C.$[4]

In Quadrantali △° BCD ubi latus $DC \infty 90^{\text{gr}}$. Polo D descripto circulo CA, solvatur △^{m} ABC.[5]

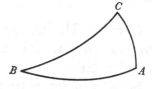

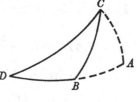

$DB + BA \infty 90$

---

(11) In the revised 'Sect 3' reproduced in §2 above Newton has done little more than to permute the vertices of his triangle ($ABC \to CAB$) and add a seventh proposition:

$$`4AB \times BF \, . \, AF + BF - AF \times AB + BF + AF :: R^q \, . \, \text{s's}' \frac{ABF}{2} \, `$$

(1) See §3: note (1) above for the general background to this extract. Both this and a preceding passage in Hare's pocket book which concerns itself with the 'Trigonometria △^{orū} planorum Rectangulorum/Obliquangulorum' (in effect, stating the sine and cosine rules for the plane scalene triangle) are narrowly based on Seth Ward's *Idea Trigonometriæ Demonstratæ* (*In usum Juventutis Oxoniensis*) (Oxford, 1654). The notation, however, of 'sco' and 'tco' for

[2]   [*Sphæricorum*] *Obliquangulorum* △^{orũ}.[3]

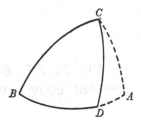

1. s B . s DC :: s D . s BC :: s C . s BD.

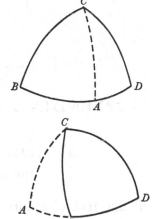

2. Datis B . BC . BD.
R . sco B :: t BC . t BA, tum
1°, sco BA . sco DA :: sco BC . sco DC.
2°, s DA . s BA :: t B . t D.

3. Datis B, C et BC.
R . sco BC :: t B . tco BCA, tum
1°, sco DCA . sco BCA :: t BC . t DC.
2°, s BCA . s DCA :: sco B . sco D.[6]

4. $BD \infty BA \pm DA$. $C \infty BCA \pm DCA$.

5. s BC . s $\dfrac{DC+BD-BC}{2}$ :: s $\dfrac{DC-BD+BC}{2}$. H [ubi] s BD . H :: R^q . Q : s $\frac{1}{2}B$.[7]

'sine complement' (Cos) and 'tangent complement' (Cot) is evidently borrowed from Oughtred's *Trigonometria* (London, 1657); compare F. Cajori, *A History of Mathematical Notations*, **2** (Chicago, 1929): 158–9.

(2) Add. 3998: 121^r. A first version may be found on a loose leaf in the terminal pocket of its companion, Add. 3997.

(3) Understand 'Trigonometria' or perhaps 'Solutio'.

(4) Compare Ward's *Idea* (note (1) above): **13**: 'Solutio Triangulorum Sphæricorum Rectangulorum'.

(5) The figure is taken from Ward's *Idea*: 9, where the text asserts more generally 'In obliquang. *BDC* dimittatur perpēdicularis [*CA*]'.

(6) Compare Ward's *Idea*: **14**: 'Solutio Triangulorum Sphæricorum obliquangulorum'.

(7) Ward's 'Axioma' for the 'Solutio Trianguli non Quadrantalis, ubi dantur tria latera & quæruntur anguli' (*Idea*: 20), there stated in the equivalent form

'$$\text{s, } CD \times \text{s, } BD. \ R^q :: \text{s, } \frac{BC}{2} + \frac{BD-CD}{2} \times \text{s, } \frac{BC}{2} \text{mi} \frac{BD-CD}{2}. \ Q : \text{s, } \frac{D}{2}$$'.

Hare does not mention Ward's lengthy Briggsian proof of this result (*ibid.*: 20–1; compare §1: note (21) preceding).

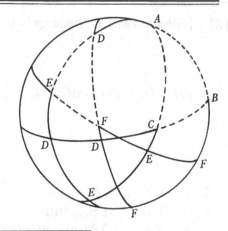

$A \infty EF$, $C \infty DE$. B acut ∞DF.
$\triangle^{\text{um}} ABC$ convertitur in $\triangle DEF$.[8]

APPENDIX 5. AN EARLY VERSION OF NEWTON'S AMPLIFICATION OF ST. JOHN HARE'S 'TRIGONOMETRY'.[1]

From the original in the University Library, Cambridge[2]

Plana tria circulorum trianguli sphærici plano quarto transversim secta continent Prisma cujus vertex est ad centrum sphæræ, & basis sectio illa transversa. Anguli autem trianguli sphærici ijdem sunt cum angulis horum planorum et latera respondent angulis ad verticem quos latera seu intersectiones planorum continent. Plana illa hic sunt *GML, GRL, MRL* quæ si super basibus *GM, GR, MR* erigi et ad invicem applicari concipiantur, constituent Prisma illud basem habens triangularem *GMR* et verticem *L*. Respondent igitur anguli *GLM, GLR, MLR* lateribus trianguli sphærici. Concipe angulum quem plana *GLR, MLR* ad lineam *RL* continent rectum esse et angulus *GLM* respondebit Hypotenusæ trianguli rectanguli, cæteriꝗ duo anguli *GLR, MLR* respondebunt lateribus continentibus angulum rectum. Concipe præterea basem *GMR* per-

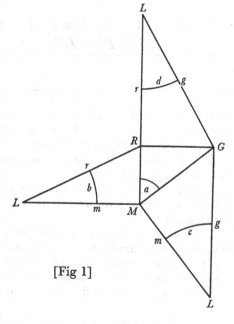

[Fig 1]

---

(8) Hare hints at the fundamental duality between the general spherical triangle and its corresponding polar triangle. Except for some renaming of points, the accompanying figure is essentially Ward's (*Idea*: 22).

pendicularem esse planis *GML*, *MRL* et ipsius angulus *GMR* æqualis erit angulo quem plana illa continent et angulus alter *GRM* evadet rectus atcҙ adeo hi duo anguli ijdem erunt cum angulis trianguli sphærici. Hoc modo trianguli cujuscuncҙ rectanguli tria latera cum duobus angulis recto et non recto in casu 1, 2, 3 et 4 designari concipe. Anguli vocalibus latera consonantibus passim designantur[3] nimirum latera circa angulum rectum literis *b*, *d*, latus tertium quod angulo recto opponitur et hypotenusa dici solet litera *c*, angulus lateri *b* conterminus litera *a* et angulus alter litera *e*. Et horum complementa ad angulum rectum sic notantur *b'*, *d'*, *c'*, *a'*, *e'*.

Si polo *e* in Fig [2] describatur circulus maximus *e'd'* trianguli hujus lateribus *b*, *c* (si opus est productis) occurrens, constituetur triangulum novum cujus hypotenusa est *b'*, latera *c'*, *e'*, angulus lateri *c'* conterminus *a* et angulus alter *d'*. Prisma in Fig. [3][3] designat hoc triangulum et adhibetur in casu 5 et 6 ubi angulus utercҙ [*a*] et *e* in quæstione est.

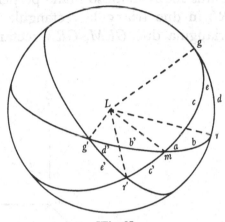

[Fig 2]

Quoniam recti sunt anguli *LMG*, *LMR*, *LRG*, *MRG*, si *MR* fiat radius circuli cujusvis erit *RG* tangens anguli *a* et *ML* cotangens anguli *b* in eodem circulo, sin *GM* fiat radius erit *GR* sinus et *MR* cosinus anguli *a* atcҙ *ML* cotangens anguli *c*. Si *ML* fiat radius erit *MR* tangens anguli *b*, et *MG* tangens anguli *c*. Si *GR* fiat radius erunt *MR*, *RL* cotangentes angulorum *a*, *d*. Si *RL* fiat radius erunt *MR* sinus et *ML* cosinus anguli *b* et *GR* tangens anguli *d*. Et vicissim si *MR* fiat cosinus anguli *a* erit *GM* radius, *GR* sinus anguli ejusdem *a* et *ML* cotangens anguli *c*. Unde Propositionem casus primi sic legas. Ut est *MR* radius in circulo quovis qui radio illo describitur ad *ML* cotangentem anguli *b*

---

(1) See §3: note (2) above. In this variant derivation of the basic properties of the spherical right-angled triangle *gmr*, the plane faces of the solid angle *LGMR* which it determines are collapsed into a corresponding plane net by 'opening up' the prism *LGMR* at its vertex *L*, not (as Newton found it more convenient subsequently to do) at the point *G* on the triangular transverse section *GMR*. Otherwise the differences between this and the equivalent revised text reproduced in §3 preceding are minimal.

(2) Add. 3959.4: 28ʳ/29ʳ, an amanuensis copy (in Humphrey Newton's hand) with a few autograph alterations. The manuscript is bereft of all figures and those here reproduced are our restorations. At the head of the first sheet a large gap has been left for the insertion of a title, doubtless 'Trigonometria / succinctè proposita ac nova methodo / demonstrata a Sᵗᵒ Johanne / Hareo Armʳᵒ'. A few minor slips in transcription on Humphrey's part have been silently corrected.

(3) Figure 3 is manifestly identical with Figure 1 in essential structure, the triangle *gmr* merely being replaced by the complementary triangle *g'mr'*.

in eodem circulo, ita est *MR* cosinus anguli *a* in circulo quovis qui radio *GM* describitur ad *ML* cotangentem anguli *c* in eodem circulo. Et sic in cæteris. Cum autem sinus et tangentes angulorum in circulo quovis sint inter se ut sinus et tangentes eorundem angulorum in circulo ad quem Canon conditur,[,] licebit vice sinuum et tangentium in his circulis usurpare sinus et tangentes in Canone. Imo verò si dividi intelliguntur radij circulorum omnium in partes 1000000, sinus et tangentes in singulis ijdem semper erunt qui in Canone.

Prismatis latera in Fig. [4][(4)] sunt *GLM*, *GLN*, *MLN* rectangula ad *M* et *N*. Latus *MLN* demisso plano perpendiculari *GRLG* scinditur secundum lineam *RL* in duo triangula rectangula *MLR*, *RLN*. Et loco basis unius habentur triangula duo *GRM*, *GRN* rectangula ad *R*,[,] quæ si latera Prismatis super

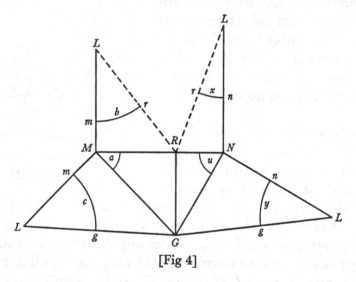

[Fig 4]

basibus *GM*, *GN*, *MN* erigantur et ita ad invicem applicentur ut puncta concurrentia *LLLL* fiant vertex figuræ, continebunt angulum ad lineam *GR* summæ angulorum *MLR*, *RLN* æqualem, plano *GRM* cum recta *ML* et plano *GRN* cum recta *NL* angulum rectum continente. Igitur si centro *L* & intervallo quovis describatur sphærica superficies triangularis ad latera prismatis terminata, trianguli hujus latera subtendent angulos *GLM*, *GLN*, *MLN* ad centrum. Et Partes in quas latus aliquod ceu basis, demisso perpendiculo secatur, subtendent angulos *RLM*, *RLN*; anguliꝗ ad basem ijdem erunt cum angulis *GMR*, *GNR*. Ponuntur *c*, *y* pro lateribus, *p* pro basi, *b*, *x* pro ejus segmentis, *a*, *u* pro angulis ad basem, *i* pro angulo ad verticem et *e*, *o* pro ejus partibus adjacentibus perpendiculo, suntꝗ *a*, *b*, *e* lateri *c* semper contermina.[(5)]

---

(4) The reader may, at this point, find it helpful to look at our Figure 5[bis] on page 177 above.

(5) This variant discussion is here abandoned without the trigonometrical applications of the present geometrical 'dissections' being indicated.

4

MISCELLANEOUS NOTES ON ANNUITIES AND ALGEBRAIC FACTORISATION[1]

[c. 1675–6]

From autograph worksheets in the University Library, Cambridge

§1. THE VALUE OF AN ANNUITY AFTER n YEARS.[2]

[1] $a. \quad \dfrac{3a}{100}. \quad b-\dfrac{3a}{100}. \quad a-b+\dfrac{3a}{100}=c.$

$$c-b+\frac{3c}{100}=d[=]a-2b+\frac{6a}{100}-\frac{3b}{100}+\frac{9a}{100^q}.$$

$$a-3b+\frac{9a}{100}-\frac{9b}{100}+\frac{27a}{100,100}-\frac{9b}{100,100}+\frac{27a}{100,100,100}. \quad [\&c]$$

[In general] a in $\overline{1+\dfrac{3}{100}}\Big|^{\,n}-e.$ [supposing that]

$$e=b \text{ in } \overline{1+\frac{3}{100}}\Big|^{\,n} \text{ in } \frac{100}{3}:-\frac{100b}{3}=\frac{100b}{3} \text{ in } \overline{1+\frac{3}{100}}\Big|^{\,n}-\frac{100b}{3}.$$

[that is] a in $\overline{1+\dfrac{3}{100}}\Big|^{\,n}+\dfrac{100b}{3}:-\dfrac{100b}{3}$ in $\overline{1+\dfrac{3}{100}}\Big|^{\,n}.$

---

(1) Edited transcripts of two preliminary worksheets on which Newton has jotted down in rough outline mathematical notes on computing the value of (terminating) annuities and on splitting simple algebraic polynomials into factors. Newton's handwriting style in his autograph texts is once more our main authority for the composition date conjecturally assigned but this dating is not at odds with the little we know of their background (see note (2) following and §2: note (1)). There seems little point in 'translating' these rough calculations into modern English equivalent, but in compensation we have made a few insertions in Newton's text for comprehension's sake and have summarized their content in an appended *Explanation* in each case.

(2) ULC. Add. 3964.8: 13ʳ. These unattached notes on determining the purchase price of an annuity (returning an agreed sum once each year over an allotted number of years when compound interest is paid, at a known rate, on the value of the annuity) corresponds to the 'æstimatio' of the 'usura usuræ' needed to maintain a 'pensio annua librarum a per quinque annos proxime sequentes solvenda [quæ] ematur parata pecunia' which Newton about this time inserted into 'Lect 6' of the ten mathematical (self-styled) lectures for 'Octob 1675' he was later to deposit in Cambridge University Library (Dd. 9.68: 52: Problema 17 [= *Arith-*

[When] $\overline{a-\dfrac{100}{3}\,b \text{ in } \overline{1+\dfrac{3}{100}}\,\Big|^{n}}:+\dfrac{100b}{3}=0.$ [then is]

$$\overline{100b-3a \text{ in } \overline{1+\dfrac{3}{100}}\,\Big|^{n}}=100b. \quad \text{[or]} \quad \dfrac{100b}{100b-3a}=\overline{1+\dfrac{3}{100}}\,\Big|^{n}. \quad {}^{(3)}$$

[2] $a.\ ca.\ b-ca.\ a-b+ca=d.$

$[cd=]ca+cca-cb.\quad [b-cd=]b+cb-ca-cca.$

$[d-b+cd=e=]a-b+ca-b-cb+ca+cca=a+2ca+cca-2b-cb.$

$[ce=]ca+2cca+c^{3}a-2cb-ccb.\quad [b-ce=]b+2cb+ccb-ca-2cca-c^{3}a.$

$[e+ce-b=f=]a+3ca+3cca+c^{3}a-3b-3cb-ccb.$

$[f+cf-b=]a+4ca+6cca+4c^{3}a+c^{4}a-4b-6cb-4ccb-c^{3}b.$

$a\times\overline{1+c}\,|^{n}-\dfrac{b}{c}\times\overline{1+c}\,|^{n}=\dfrac{-b}{c}=\dfrac{ac-b}{c}\times\overline{1+c}\,|^{n}.\quad \text{that is}\quad \dfrac{[-]b}{ac-b}=\overline{1+c}\,|^{n}.$

Explanation. In the first instance Newton determines the progressive depreciation in value year by year of an annuity, bought initially at the price a, which each year returns a fixed sum b while the outstanding capital accumulates interest at an annual (compound)[4] rate of 3%. He then proceeds analogously in the general case where the interest rate is taken to be c (not to be confused with his preceding use of this letter to denote the value of the annuity, appreciating annually by 3%, after the first year's repayment is made). Here the successively decreasing values of the annuity, after deducting the annual repayment, are in turn

$$d=(1+c)\,a-b \text{ after 1 year,}$$

$$e=(1+c)\,d-b=(1+c)^{2}\,a-(2+c)\,b \text{ after 2 years,}$$

$$f=(1+c)\,e-b=(1+c)^{3}\,a-(3+3c+c^{2})\,b \text{ after 3 years,}$$

metica Universalis; sive De Compositione et Resolutione Arithmetica Liber (Cambridge, ₁1707): 96: Problema 16]). We will return to this point in the next volume, but here may observe that Newton's solution of the latter case is essentially the particular instance $n = 5$ of his present one when it is transposed into the equivalent form $(1/c)\,(1-(1+c)^{-5}) = a/b$.

(3) In a following calculation, here omitted, Newton takes an initial price $a = 1,928,570$ for an annuity returning $b = 135,000$ at the same interest rate of 3%, from these computing $c = 1\cdot03a-b$ and $d = 1\cdot0609a-2\cdot06b$, the value of the annuity after the first and second years respectively.

(4) The comparable formula relating the purchase price a of an annuity 'discompting' simple interest at the rate c (that is, $100c\%$) to the number n of years for which it returns the fixed sum b annually is, as we have seen (III: xvi, note (20)), $a = \sum\limits_{1\leqslant i\leqslant n} b/(1+ic).$

and so on; hence after n years its value will have diminished to

$$(1+c)^n a - (1/c)((1+c)^n - 1) b,$$

that is, $(1+c)^n (a-b/c) + b/c$, as Newton computes. Evidently the annuity will terminate (with a suitable, slightly increased final payment) after $[n]$ years, where by the above $n = -\log(1-ac/b)/\log(1+c)$.[5] In practical terms the interest rate c will be determined by external economic factors but the skilful actuary will be able to offer a psychologically 'best buy' for given down-payment a by suitably varying the annual return b against the total period n of years over which the annuity is paid.

---

§2. QUADRATIC FACTORIZATION OF ALGEBRAIC BINOMIALS AND ITS APPLICATION TO INTEGRATION BY MEANS OF PARTIAL FACTORS.[1]

$$[1]^{(2)} \quad \frac{1}{aa+bb+2bx+xx} = y = \overset{(3)}{\underset{}{\frac{1}{cc+2bx+xx}}} = \frac{[e-fx]}{ecc+2ebx+exx-fx^3-fccx-2fbxx}.$$

---

(5) As a corollary, when the annuity is a perpetuity ($n = \infty$) and so repays the sum b annually for ever, it follows that $a = b/c$. This will explain the terms of a letter Newton wrote in late 1693 to a 'Mr Martin' on behalf of his recently widowed half-sister Hannah Barton: 'I paid lately (April 28) into the Exchequer by yor hands the sum of 833ᶫᶦ. 7ˢ. 0ᵈ for an Annuity of 100ᶫᶦ for two lives at ye rate of 12 per cent.... The first of the lives is my own for the whole summ. The second life I would have divided between three persons the children of Hannah Barton of Brigstock in ye county of Northampton widdow by her late husband Robert Barton, that is to say Robert Barton [Hannah's son] for 60ᶫᶦ, Katherine Barton [later to marry John Conduitt] for 20ᶫᶦ & Margaret Barton for 20ᶫᶦ. And after my own life I would divide the ann[uity] into three parcels to be paid during ye lives of [these] 3 persons...' (ULC. Add. 3965.18: 671ᵛ). Evidently £833. 7. 0 is £100/0·12 rounded off to the nearest shilling upwards.

(1) ULC. Add. 3970.3: 570ᵛ. On the same double sheet (f. 569ʳ, reproduced in *The Correspondence of Isaac Newton*, 2, 1960: 266–8) is Newton's draft of a letter (to Aubrey?) composed, by its own admission, some time 'after Mr Oldenburgs death [in September 1677]', probably in the early summer of 1678. The present calculations may well be of a slightly earlier date and certainly repeat—if they do not inspire—the series $1 + \frac{1}{3} - \frac{1}{5} - \frac{1}{7} + \dots$ for $\pi/2\sqrt{2}$ which Newton communicated to Oldenburg in October 1676 for onward transmission to Leibniz (see note (29) below).

(2) In an opening computation, here omitted, Newton set himself the problem of evaluating $\int y \,.\, dx$ where '$\frac{1}{b+x^n} = y$. $1 - by - x^n y = 0$. $y^{-1} - x^n - b = 0$'. An integration by parts yields $\int y \,.\, dx = yx - \int v \,.\, dz$ where '$\overline{z^{-1} - b}|^{\frac{1}{n}} [=v]$'. A final equation '$\frac{n-1}{n} = \tau$' suggests (compare III: 380) that Newton intended to evaluate $\tau \int x \,.\, dy = \int v \,.\, dz$ where $y^\tau = z$ and so $x = z^{1/(1-n)}v$: on completing the computation he would have found $\int y \,.\, dx = yx - (1/\tau) \int v \,.\, dz$ where $v^n = 1 - bz^{n/(n-1)}$. Compare III: 374–85.

(3) On setting $a^2 + b^2 = c^2$, that is. It will be evident from the series expansions for $\pi/4$ and $\pi/2\sqrt{2}$ invoked in the sequel (see note (18) below) that Newton's interest in this expression for y is directed to its integral $(1/a)\tan^{-1}[(x+b)/a]$.

[Si] $2eb = fcc$. $e = 2fb$. [fit] $4fbb = fcc$. $2b = c$. [adeoĝ scribendo a pro b et reducendo]

$$y = \frac{1}{4aa + 2ax + xx} = \frac{2a - x}{8a^3 - x^3} \cdot \left[y = \frac{1}{a + bx + xx} = \right] \frac{[c - dx + exx]}{\begin{array}{l} ca + cbx + cxx - dx^3 + ex^4 \\ -da - db + eb \\ + ea \end{array}} \cdot$$

[Si] $cb = da$. [&] $eb = d$. [erit] $c = ea$. [adeoĝ] $2ea = 2c = db = ebb$. [sive] $2a = bb$.

[Quare] $y = \dfrac{1}{2aa + 2ax + xx} = \dfrac{2eaa - 2eax + exx}{4ea^4 + ex^4}$. [(4)]

$$\left[y = \frac{1}{a + bx + xx} = \right] \frac{[c - dx + exx - fx^3 + gx^4]}{\begin{array}{l} ca + cbx + cxx - dx^3 + ex^4 - fx^5 + gx^6 \\ -da - db + eb - fb + gb \\ + ea - fa + ga \end{array}} \cdot$$

[Si] $f = gb$. [$e = fb - ga$. $d = eb - fa$. $c = db - ea$.] $cb = da$. [erit] $\dfrac{cb}{a} = d$. $fc = dga$.

$e - gbb + ga = 0$. $eb - \dfrac{cb}{a} = abg$. [seu] $ea - c = aag$. [adeoĝ] $aag + 2c = db = \dfrac{cbb}{a}$.

[Quare $c + ag = ea = gabb - gaa$ & $c = abbg - 2aag$. Deniĝ]

$$2abbg - 3aag = \frac{cbb}{a} = b^4 g - 2abbg.$$

[hoc est] $b^4 - 4abb + 3aa = 0$. [sive] $bb = a$. $bb = 3a$. [adeoĝ]

$$y = \frac{1}{aa + ax + x^2} = \frac{-a^4 + a^3 x * - ax^3 + x^4}{-a^6 + x^6} \cdot \text{ (5)}$$

[vel] $y = \dfrac{1}{3aa + 3ax + xx} = \dfrac{9a^4 - 9a^3 x + 6aax^2 - 3ax^2 + x^4}{27a^6 + x^6}$. [(6)]

---

[2] $y = \dfrac{1}{1 + nx + xx} \times \dfrac{1 - nx + pxx - qx^3 + rx^4 - sx^5 + tx^6[-vx^7 + wx^8]}{[1 - nx + pxx - qx^3 + rx^4 - sx^5 + tx^6 - vx^7 + wx^8]} \cdot$

[Denominator est]

$$\begin{array}{l} 1 + nx + xx - nx^3 + px^4 - qx^5 + rx^6 - s[x^7] + t[x^8] - v[x^9] + w[x^{10}]. \\ -nx - nnxx + pn - qn + rn - sn + tn - vn + wn \\ + p - q + r - s + t - v + w \end{array}$$

---

(4) Newton has neglected to cancel the common factor e top and bottom in this fraction.

(5) The numerator further factorizes into the quadratic pair $(-a^2 + x^2)(a^2 - ax + x^2)$.

(6) Similarly the numerator here further factorizes into $(3a^2 + x^2)(3a^2 - 3ax + x^2)$.

(7) The third root $n = 0$ corresponds to the factor $1 + x^2$ of $1 - x^4$.

(8) The numerator further factorizes, on evaluating $n = \frac{1}{2} \pm \sqrt{\frac{5}{4}}$, as $(1 - x)(1 - nx + x^2)$.

(9) And also $n = 0$. The roots $n = 0$ and $n = -\sqrt{3}$ yield the quadratic factorization $(1 + x^2)(1 - (\sqrt{3})x + x^2)$ of the following numerator.

[Quare si]

$$nn-1=p[=0]. \quad [\text{erit}] \ nn=1. \quad [\text{adeoq} \ y=] \ \frac{1-x}{1-x^3}.$$

$$n^3-2n[=pn-n]=q[=0]. \quad nn=2.^{(7)} \quad \frac{1-x\sqrt{2}+xx}{1+x^4}.$$

$$n^4-3nn+1[=qn-p]=r[=0]. \quad nn=\frac{3}{2}\pm\sqrt{\frac{5}{4}}.$$

$$\frac{1-\sqrt{\left(\frac{3}{2}\pm\sqrt{\frac{5}{4}}\right)}x+\frac{1\pm\sqrt{5}}{2}xx-[x^3]}{1-x^5}.^{(8)}$$

$$n^5-4n^3+3n[=rn-q]=s[=0]. \quad nn=1 \ \& \ 3.^{(9)} \quad \frac{1-\sqrt{3})\,x+2x^2-\sqrt{3})\,x^3+x^4}{1+x^6}.$$

$$n^6-5n^4+6nn-1[=sn-r]=t[=0]. \quad nn=^{(10)}$$

$$n^7-6n^5+10n^3-4n[=tn-s]=v[=0]. \quad nn=2 \ \& \ 2\pm\sqrt{2}.^{(11)}$$

$$\frac{1-\sqrt{2})\,x+xx*-x^4+\sqrt{2})\,x^5-x^6}{1-x^8}$$

[&] $$\frac{1-\sqrt{2\pm\sqrt{2}})\,x+\overline{1\pm\sqrt{2}})\,xx\mp\sqrt{4\pm\sqrt{8}})\,x^3+\overline{1\pm\sqrt{2}})\,x^4-\sqrt{2\pm\sqrt{2}})\,x^5+x^6}{1+x^8}.$$

$$n^8-7n^6+15n^4-10nn+1=w[=0].$$

$$n^9-8n^7+21n^5-20n^3+5n=0.^{(12)}$$

$$n^{10}-9n^8+28n^6-35n^4+15nn-1=0.$$

$$n^{11}-10n^9+36n^7-56n^5+35n^3-6n[=0]. \quad nn=1 \ \& \ 2 \ \& \ 3.^{(13)}$$

$$\frac{1-x*+x^3-x^4*+x^6-x^7*+x^9-x^{10}}{1-x^{12}}.$$

---

(10) A hopeful entry! In fact $1\pm x^7$ has no factors except $1\pm x$ in the field of rationals and quadratic surds.

(11) Together with $n=0$, of course. The roots $n=0$ and $\pm\sqrt{2}$ give

$$(1-x^8)/(1-x^2) \equiv (1+x^2)(1+(\sqrt{2})x+x^2)(1-(\sqrt{2})x+x^2),$$

while the roots $n=\pm\sqrt{[2\pm\sqrt{2}]}$ determine correspondingly $1+x^8 \equiv \prod_{1\leqslant i\leqslant 4}(1+n_i x+x^2)$ where $n_1=-n_4=\sqrt{[2+\sqrt{2}]}$ and $n_2=-n_3=\sqrt{[2-\sqrt{2}]}$.

(12) Newton evidently fails to see that this is the pair of equations

$$(n^5-5n^3+5n)(n^4-3n^2+1) = 0,$$

whose two sets of roots ($n=0, \pm\sqrt{[\frac{1}{2}(5\pm\sqrt{5})]}$ and $n=\pm\frac{1}{2}(\sqrt{5}\pm 1)$ respectively) yield corresponding quadratic factorizations of $1+x^{10}$ and $(1-x^{10})/(1-x^2)$.

(13) Along with $n=0$ and $n^2=2\pm\sqrt{3}$. The six roots $n=\pm\sqrt{2}, \pm\sqrt{[2\pm\sqrt{3}]}$ and the five remaining ones $n=0, \pm 1, \pm\sqrt{3}$ yield corresponding quadratic factors $1+nx+x^2$ of $1+x^{12}$ and $(1-x^{12})/(1-x^2)$ respectively.

$$[\&] \quad \frac{1-\sqrt{2})\,x+xx*-x^4+\sqrt{2})\,x^5-x^6*+x^8-\sqrt{2})\,x^9+x^{10}}{1+x^{12}}.$$

$$[\&] \quad \frac{1-\sqrt{3})\,x+2xx-\sqrt{3})\,x^3+x^4*-x^6+\sqrt{3})\,x^7-2x^8+\sqrt{3})\,x^9-x^{10}}{1-x^{12}}:$$

---

[3] $n^4-4nn+2[=0.\text{ fit}]\ nn=2\pm\sqrt{2}.$[14]

$n^8-8n^6+20n^4-16n^2+3[=0.\text{ fit vel }nn=1\text{ vel}]\ n^6-7n^4+13n^2-3[=0.\text{ hoc}$
est vel $nn=3$ vel] $n^4-4nn+1[=0.\text{ id est}]\ nn=2\pm\sqrt{3}.$[15]

---

[4] $AC=1.\ AB=\frac12\sqrt{2+\sqrt{2}}.\ BC=\frac12\sqrt{2-\sqrt{2}}.$

$n=1.5707963\ \&c=\dfrac{\text{Quadr}}{\text{Rad}}.$[16]

ang $BAC=22\frac12^{[\text{gr}]}=\frac14 n.$[17] $ACB=\frac34 n.$[17]

$DAC=BAE=\frac38 n.$[17]

[erit]

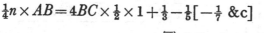

$$\tfrac14 n\times AB=4BC\times\tfrac12\times 1+\tfrac13-\tfrac15[-\tfrac17\ \&c]$$

$$+4\sqrt{2})\,BC\times\tfrac14\times\tfrac11-\tfrac13+\tfrac15-\tfrac17\ \&c.$$[18]

[Si] $AB=1.$ [erit $AB\times$] $n=2\times\tfrac11-\tfrac13+\tfrac15-\tfrac17\ [\&c]=2d.$

[ac si] $AB=\sqrt{\tfrac12}.\ AB\times n=\tfrac11+\tfrac13-\tfrac15-\tfrac17\ [\&c]=c.$[19]

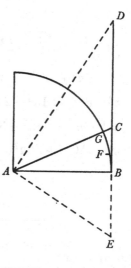

---

(14) The four roots $n=\pm\sqrt{[2\pm\sqrt{2}]}$ are the 'complementary subtenses' $2\cos\frac18(2i-1)\pi$, $i=1,2,3,4$, whose values yield the quadratic factors of $1+x^8$. See note (11) above.

(15) The eight roots $n=\pm1,\ \pm\sqrt{3}$ and $\pm\sqrt{[2\pm\sqrt{3}]}$ are $2\cos\frac{1}{12}(3i-2)\pi$, $i=1,2,3,\ldots,8$. By analogy with the preceding Newton may well have intended to factorize

$$n^8-8n^6+20n^4-16n^2+2=0,$$

whose eight roots $2\cos\frac{1}{16}(2i-1)\pi$ yield the quadratic factors of $1+x^{16}$.

(16) Understand $\dfrac{\text{'[arcus] Quadr[antalis]'}}{\text{Rad[ius]}}$, that is, $\frac12\pi$. In the preceding line, where the 'radius' AC is unity, evidently $AB=\cos\frac18\pi$ and $BC=\sin\frac18\pi$.

(17) That is, since $n=\frac12\pi$, $B\hat{A}C=\frac18\pi$, $A\hat{C}B=\frac38\pi$ and $D\hat{A}C=B\hat{A}E=\frac{3}{16}\pi$ (in radians) respectively.

Explanation. In general terms Newton here seeks to split off from the general binomial $\alpha^m \pm x^m$, m a positive integer, the quadratic factor $a+bx+x^2$. His technique is straightforwardly to set

$$(a+bx+x^2)\,(c-dx+ex^2\ldots[\pm x^{m-2}])=\alpha^m\pm x^m$$

and then, by equating $ac=\alpha^m$ and the remaining coefficients of powers x^i, $i=1, 2, 3, \ldots, m-1$, on the left-hand side to zero, to attempt to determine constants $a, b, c, d, e, \ldots$ which make the equation an identity. After successfully locating in this way single quadratic factors of $8a^3-x^3$, $4a^4+x^4$, $-a^6+x^6$ and $27a^6+x^6$, he sees that there is no loss of generality occasioned by directing his attention to the identity

$$(1+nx+x^2)\,(1-nx+px^2-qx^3+rx^4-\ldots[\pm x^{m-2}])\equiv 1\pm x^m.$$

By equating coefficients of x^i, $2\leqslant i\leqslant m-1$, to zero he is able to determine a pattern of equations which n must satisfy corresponding to $m=3, 4, 5, \ldots$, and 12 in the binomial instances $1-x^3$, $1+x^4$, $1-x^5$, $1+x^6$, $1\pm x^8$ and $1\pm x^{12}$ is able to identify single quadratic factors (two in the case of $1-x^{12}$), though he is unable to resolve the equally simple binomial $1\pm x^{10}$ in this manner.[20] At first he fails adequately to distinguish the binomial $1+x^m$ from its partner $1-x^m$, both which are covered indiscriminately by the equations derived, and indeed to see that the equation derived for the compound case $1\pm x^m$ splits universally into the pair $P(n)\times Q(n)=0$, where

$$\left.\begin{cases}P(n)=0,\ m\text{ even}\\Q(n)=0,\ m\text{ odd}\end{cases}\right\}\ \text{is satisfied by }n_i=2\cos\frac{2i-1}{m}\pi$$

and $\left.\begin{cases}P(n)=0,\ m\text{ odd}\\Q(n)=0,\ m\text{ even}\end{cases}\right\}\ \text{is satisfied by }n_i=2\cos\dfrac{2i}{m}\pi,\ i=1, 2, 3, \ldots, [\tfrac{1}{2}m].$

Thus, for instance, when $m=5$ Newton's equation $n^4-3n^2+1=0$ breaks into

$$(n^2-n-1)\,(n^2+n-1)=0,$$

where $P(n)\equiv n^2-n-1=0$ $(n_i=-2\cos\tfrac{2}{5}i\pi,\ i=1, 2)$ yields the factorization $1+x^5\equiv(1+x)\,(1+n_1x+x^2)\,(1+n_2x+x^2)$, while correspondingly

$$Q(n)\equiv n^2+n-1=0\quad(n_i=-2\cos\tfrac{1}{5}(2i-1)\,\pi,\ i=1, 2)$$

---

(18) This is wrong as it stands. Since (see note (19)) $1+\tfrac{1}{3}-\tfrac{1}{5}-\tfrac{1}{7}\ldots = n/\sqrt{2}$ and $1-\tfrac{1}{3}+\tfrac{1}{5}-\tfrac{1}{7}\ldots = n/2$, while $BC/AB = \tan\tfrac{1}{8}\pi = (\tfrac{1}{2}\sqrt{[2-\sqrt{2}]}/\tfrac{1}{2}\sqrt{[2+\sqrt{2}]}$ or$)\sqrt{2}-1$, Newton might have intended some equivalent to

'$\tfrac{1}{2}n\times AB = BC\times 1+\tfrac{1}{3}-\tfrac{1}{5}-\tfrac{1}{7}$ &c$+BC\times\tfrac{1}{1}-\tfrac{1}{3}+\tfrac{1}{5}-\tfrac{1}{7}$ &c'.

(19) For the justification of these expansions see the following *Explanation* and compare notes (29) and (30) below.

(20) See note (21) below and compare note (12) above.

gives $1-x^5 \equiv (1-x)(1+n_1 x+x^2)(1+n_2 x+x^2)$. Similarly the equation

$$n^9 - 8n^7 + 21n^5 - 20n^3 + 5n = 0,$$

left intact by Newton in his text,[21] breaks into the pair of

$$P(n) \equiv n^5 - 5n^3 + 5n = 0 \quad (n_i = -2\cos\tfrac{1}{10}(2i-1)\pi, \; i=1,2,3,4,5)$$

and $Q(n) \equiv n^4 - 3n^2 + 1 = 0$ $(n_i = -2\cos\tfrac{1}{5}i\pi, i=1,2,3,4)$, which together produce $1 \pm x^{10} \equiv (1\pm x^2)(1+Ax+x^2)(1-Ax+x^2)(1+Bx+x^2)(1-Bx+x^2)$ on taking $A = \sqrt{5}/B = \sqrt{[\tfrac{1}{2}(5+\sqrt{5})]}$ and $A = 1/B = \tfrac{1}{2}(\sqrt{5}+1)$ respectively.

The following section suggests that Newton rapidly surmounted this hurdle, for there we find the equation $n^4 - 4n^2 + 2 = 0$ (satisfied by $n_i = -2\cos\tfrac{1}{4}i\pi$, $i=1,2,3,4$) whose solution yields in explicit form the factorization

$$1+x^8 \equiv (1+n_1 x+x^2)(1+n_2 x+x^2)(1+n_3 x+x^2)(1+n_4 x+x^2).$$

In line with this we may suppose[22] that his second equation was intended to be $n^8 - 8n^6 + 20n^4 - 16n^2 + 2 = 0$ (satisfied by $n_i = -2\cos\tfrac{1}{8}i\pi$, $i=1,2,3,\ldots,8$), one which yields in a comparable way the quadratic factorization

$$1+x^{16} = \prod_{1 \leqslant i \leqslant 8} (1+n_i x+x^2).$$

From this it is but one more step to Cotes' formulas

$$\left.\begin{cases} (m \text{ even}) \; 1+x^m \\ (m \text{ odd}) \; (1+x^m)/(1+x) \end{cases}\right\} = \prod_{1 \leqslant i \leqslant [\frac{1}{2}m]} \left(1 - 2\left(\cos\frac{2i-1}{m}\pi\right)x + x^2\right)$$

and $\left.\begin{cases} (m \text{ even}) \; (1-x^m)(1-x)/(1+x) \\ (m \text{ odd}) \; 1-x^m \end{cases}\right\} = \prod_{1 \leqslant i \leqslant [\frac{1}{2}m]} \left(1 - 2\left(\cos\frac{2i}{m}\pi\right)x + x^2\right),$

but we know nothing to suggest that Newton was ever consciously aware of these before Robert Smith presented them (in geometrical form) to the world in his edition of Cotes' *Harmonia Mensurarum*.[23] Again, Newton must have given some thought to the reason why, for some m, $1 \pm x^m$ is readily expressible by quadratic factors whose coefficients are rational numbers or quadratic surds, but not apparently so for others; yet we must not credit him with any real inkling of Gauss' lemma[24] to his celebrated criterion of the Euclidean constructability of the angle $2\pi/m$: namely, the binomial $1-x^m$ splits into quadratic

(21) See note (12) above.

(22) Compare note (15) above.

(23) In the *Præfatio* to Smith's compilation from Cotes' papers, 'Theoremata tum Logometrica tum Trigonometrica Datarum Fluxionum Fluentes exhibentia, per Methodum Mensurarum Ulterius extensam' (*Harmonia Mensurarum, sive Analysis & Synthesis Per Rationum & Angulorum Mensuras promotæ* (Cambridge, 1722): 111–247, especially 113–14).

(24) First announced by Gauss in the *Intelligenzblatt der Allgemeinen Literaturzeitung* (Jena), Nr. 66 [for 1 July 1796], though the discovery had been made the previous 29 March. His full proof appeared only five years later in his *Disquisitiones Arithmeticæ* (Leipzig, 1801): §§335–66, especially 365 [= *Werke*, 1 (Göttingen, 1870): 461].

factors of this type if (and only if) m is an integral multiple of $2^\alpha . p_\beta . p_\gamma . p_\delta . \ldots$ where $\alpha, \beta, \gamma, \delta, \ldots$ are integers and p_i is the i-th Fermatian prime in sequence.[25]

The reason for Newton's present interest in splitting a binomial into quadratic factors is clarified by the last section. As Cotes demonstrated at length in his *Harmonia*[26] such a factoring of the denominator of an algebraic integrand allows it to be expressed much more conveniently as the analogous sum of partial fractions having one each of the component factors for their denominators. While Newton never went into this subject with the same intensity as Cotes he had, from the time of his first researches in calculus and above all in the 'Catalogus posterior' of integrals he listed in his 1671 tract,[27] made wide use of this reduction of an integrand to a sum of partial fractions. It will here be evident that the main goal of his calculations is to evaluate the integral

$$\int_0^x \frac{1}{1+nx+x^2} . dx = \frac{2}{\sqrt{[4-n^2]}} \tan^{-1} \frac{2x+n}{\sqrt{[4-n^2]}}$$

by setting it in the equivalent form

$$\int_0^x \frac{1-nx+px^2-qx^3+rx^4-\ldots+x^{2m-2}}{1+x^{2m}} . dx$$

in which $p=n^2-1 \, (m \neq 1)$, $q=np-n$, $r=nq-p \, (m \neq 2)$, $s=nr-q$, ... At once

$$\frac{1}{\sqrt{[4-n^2]}} \tan^{-1} \frac{x\sqrt{[4-n^2]}}{1-x^2} = \frac{1}{2} \int_{-x}^x \frac{1}{1+nx+x^2} . dx = \int_0^x \frac{1+px^2+rx^4 \ldots}{1+x^{2m}} . dx$$

and hence, on taking $x=1$,

$$\tfrac{1}{2}\pi / \sqrt{[4-n^2]} = \int_0^1 (1+px^2+rx^4 \ldots [+x^{2m-2}])/(1+x^{2m}) . dx.$$

Newton rounds off his worksheet by attempting to connect the two interesting particular cases of this, namely ($m=1$ and so $n=p=r=\ldots=0$)

$$\pi/4 = \int_0^1 1/(1+x^2) . dx = 1 - \tfrac{1}{3} + \tfrac{1}{5} - \tfrac{1}{7} + \ldots,\text{[28]}$$

(25) That is, the i-th prime in the sequence $2^{2^j}+1$, j integral. It is known that this expression is composite for $5 \leqslant j \leqslant 16$ (see G. H. Hardy and E. M. Wright, *An Introduction to the Theory of Numbers* (Oxford, $_4$1960): 15) and so the only low Fermat primes are $p_1 = 3$, $p_2 = 5$, $p_3 = 17$, $p_4 = 257$ and $p_5 = 65537$.

(26) In his 'Theoremata tum Logometrica tum Trigonometrica, quæ Datarum Fluxionum Fluentes exhibent per Mensuras. Adjiciuntur Theoremata de Continuatione Formarum et Fluentium ad Infinitum' (*Harmonia Mensurarum* (note (23) above): 46–76).

(27) See III: 244–54.

(28) The 'Gregory–Leibniz' series, first discovered about 1500 by the Hindu Nilakaṇṭha (see III: 34, note (5)), was communicated to Newton by Leibniz in his letter (to Oldenburg) of 17/27 August 1676 with the somewhat high-flung recommendation of it as the 'expressio [pro area circuli posito Quadrato circumscripto 1] jam triennio abhinc et ultra à me communicata amicis, [quæ] haud dubiè omnium possibilium simplicissima est, maximèque afficiens mentem' (*Correspondence of Isaac Newton*, **2**, 1960: 60).

and $(m=2$ and hence $p=1$, $r=\ldots=0$, $n=\sqrt{2})$

$$\pi/2\sqrt{2} = \int_0^1 (1+x^2)/(1+x^4)\,.\,dx = 1+\tfrac{1}{3}-\tfrac{1}{5}-\tfrac{1}{7}\ldots, ^{(29)}$$

(29) This elegant variant on the preceding series was returned by Newton to Leibniz in his *epistola posterior* of 24 October 1676 (*Correspondence*, **2**: 120) together with the hint that its proof depended on reducing an integrand—evidently $(1+x^2)/(1+x^4)$—into partial fractions of the form $d/(e+fx+gx^2)$, where $2eg = f^2$. The subtleties of this series completely passed their intermediary, Oldenburg, by and on 14 November Newton had to reassure him that in his *epistola* 'yᵉ signes of yᵉ series $1+\tfrac{1}{3}-\tfrac{1}{5}-\tfrac{1}{7}\ldots$ &c are rightly put two + and two − after one another, it being a different series from yᵗ of M. Leibnitz' (*ibid.*: 181). We may suspect that Leibniz fared no better in understanding Newton's hint, for twenty-five years afterwards in his 'Specimen Novum Analyseos Quadraturarum pro Scientia Infiniti, circa Summas et Quadraturas (*Acta Eruditorum* (May 1702): 210–19 [= (ed. C. I. Gerhardt) *Mathematische Schriften*, 5 (Halle, 1858): 350–61]), while correctly splitting x^4+a^4 into its complex linear factors '$x+a\sqrt{\sqrt{-1}}$, $x-a\sqrt{\sqrt{-1}}$, $x+a\sqrt{-\sqrt{-1}}$, $x-a\sqrt{-\sqrt{-1}}$', he failed to observe that

$$\sqrt{i} = \pm(1+i)/\sqrt{2} \quad \text{and} \quad \sqrt{-i} = \pm(1-i)/\sqrt{2}$$

and that the linear factors may be combined two at a time to produce the real quadratic pairs $x^2\pm(\sqrt{2})ax+a^2$ found by Newton in the present manuscript: on the contrary, Leibniz was encouraged by this failure to draw the false conclusion that '*Itaque* $\int dx:(x^4+a^4)$ *neque ex Circuli neque ex Hyperbolæ Quadratura per Analysin hanc nostram reduci potest, sed novam sui generis fundat. Et optarem*..., *ut* $\int dx:(x+a)$ *seu Quadraturam Hyperbolæ constat dare Logarithmos seu Sectionem Rationis, et* $\int dx:(xx+aa)$ *Sectionem Anguli, ita porro continuari posse progressionem, constareque cuinam problemati respondeant*

$$\int dx:(x^4+a^4), \ldots \int dx:(x^8+a^8), \quad \&c'$$

(*Acta*: 218–19). Fourteen years later, on 5 May 1716, Roger Cotes (who at that time knew better than anyone the truth of the matter) wrote forcefully to William Jones to redress the error: 'M. Leibnitz, in the Leipsic Acts of 1702 p. 218 and 219, has very rashly undertaken to demonstrate that the fluent of $\dfrac{\dot{x}}{x^4+a^4}$ cannot be expressed by measures of ratios and angles; and he swaggers upon the occasion (according to his usual vanity), as having by this demonstration determined a question of the greatest moment. Then he goes on thus: As the fluent of $\dfrac{\dot{x}}{x+a}$ depends upon the measure of a ratio, and the fluent of $\dfrac{\dot{x}}{xx+aa}$ upon the measure of an angle, so he has more than once expressed his wishes that the progression may be continued, and it be determined to what problem the fluents of $\dfrac{\dot{x}}{x^4+a^4}$, $\dfrac{\dot{x}}{x^8+a^8}$, &c may be referred. His desire is answered in my general solution, which contains an infinite number of such progressions....In truth I am inclined to believe that Leibnitz's grand question ought to be determined the contrary way, and that it will be found...that the fluent of any rational fluxion whatever does depend upon measures of ratios and angles, excepting those which may be had in finite terms, even without introducing measures' (S. P. Rigaud, *Correspondence of Scientific Men of the Seventeenth Century*, **1** (Oxford, 1841): 271–2).

(30) Compare note (18) above. In speaking of the application of the 'Series Leibnitij' $(1-\tfrac{1}{3}+\tfrac{1}{5}-\tfrac{1}{7}\ldots)$ and his own variant series $(1+\tfrac{1}{3}-\tfrac{1}{5}-\tfrac{1}{7}\ldots)/\sqrt{2}$ to the numerical evaluation

by some 'interesting' equation, probably intended to be an equivalent to $\frac{1}{4}\pi = (\pi/4 + \pi/2\sqrt{2})\tan\frac{1}{8}\pi$ $[= 2(1 - \frac{1}{7} + \frac{1}{9}\ldots)\tan\frac{1}{8}\pi]$, but does not succeed.[30]

---

of $\frac{1}{4}\pi$ ('area totius Circuli posito diametro 1') Newton in his *epistola posterior* of 24 October 1676 had remarked that 'si...alia quædam similia artificia adhibeantur, potest computum produci ad multas figuras' (*Correspondence*, **2**: 122), and two days later in a further letter to Oldenburg he suggested that a following phrase 'ut et ponendo summam terminorum

$$1 - \frac{1}{7} + \frac{1}{9} - \frac{1}{15} + \frac{1}{17} - \frac{1}{23} + \frac{1}{25} - \frac{1}{31} + \frac{1}{33} \quad \&c$$

esse ad totam seriem $1 - \frac{1}{3} + \frac{1}{5} - \frac{1}{7} + \frac{1}{9} - \frac{1}{11}$ &c ut $1 + \sqrt{2}$ ad 2' (*ibid.*: 162) be inserted. In framing this addition Newton may well have had the present (intended) variant in mind.

PART 2

RESEARCHES IN PURE AND ANALYTICAL GEOMETRY

(*c.* 1678-1680)

INTRODUCTION

On the background to the geometrical papers here reproduced contemporary report is wholly silent, and it would appear that at the time he wrote them Newton allowed no hint of their existence to escape him either by letter or by word of mouth. From identifying quirks in the handwriting style of the original autograph manuscripts we may, it is true, locate their date of composition in a fairly narrow interval of time, probably during 1678–9 and certainly not much after 1680,[1] but for all other illumination we must necessarily have recourse to essentially unprovable conjecture, relying extensively on personal prejudices which can be controlled only by the internal evidence of the texts themselves. Within these limitations let us dare to make a few general observations.

The first two sections present Newton's first extended researches into the geometry of the classical 'plane' and 'solid' locus, that is, of the circle (including the 'locus rectilineus' or straight line) and general conic respectively. This topic has not usually been well understood by previous students of his mathematical development who, unduly influenced by what they take to be the Grecian façade of his *Principia* and recalling Pemberton's remark shortly after Newton's death that 'Of [the antients'] taste, and form of demonstration Sir Isaac always professed himself a great admirer',[2] have been tempted to suppose (on little real evidence) that he had from the first a deep-felt sympathy with the rigour and elegance of Greek geometry, coupled with a wide knowledge and understanding of its technical niceties.[3] In regard to Newton's prevailing mathematical interests up to the time (October 1669) of his election to the Lucasian chair and for long afterwards this viewpoint is certainly invalid; for while it is true that in accordance with the demands of a scholastic curriculum he made in his undergraduate days a close study of portions of Euclid's *Elements*[4] and bore the impress of its logical structure through the rest of his days, there is no

---

(1) A unique confirmation of this otherwise circumstantially imposed dating is afforded by the presence of certain computations for determining the asymptotes of a general cubic, related to the paper reproduced as 3, §2.1 below, on the verso of the draft of a letter of Newton's (to Aubrey?) penned about June 1678. See 3, §2.2: note (32) below.

(2) Henry Pemberton, *A View of Sir Isaac Newton's Philosophy* (London, 1728): Preface: [a1ʳ]. The immediately following phrase 'I have heard him even censure himself for not following them yet more closely than he did' (compare I: 18, note (14)) is not so often quoted.

(3) G. L. Huxley, for instance, in a not unperceptive recent essay on 'Newton and Greek Geometry' (*Harvard Library Bulletin*, 13, 1959: 354–61) speaks of his 'deep understanding' and 'love of ancient mathematics' and states his opinion that 'many of his greatest achievements in creative mathematics were developments of Greek ideas'.

(4) Particularly its more arithmetical books II, v, vII and x (see I: 12, note (28)) though he also mastered the easier geometrical books as well and may well have lectured on their content; compare III: 402–6.

evidence to show that he ever acquired more than the bare working knowledge of Apollonius and Archimedes he needed to get by in his own researches and professorial lectures. Indeed, though his knowledge of Greek was adequate for the task, at no time in his life (so far as we know) was Newton tempted to study the mathematical classics of the 'ancients' in their original language, but rather made extensive use of such Latin editions and commentaries as Barrow's epitome of the *Elements* and (after 1675) perhaps also of his modernized condensations of Apollonius' *Conics* i–iv and the various *opera* of Archimedes then known.[5] What the geometrical manuscripts now published do reveal is that in his middle-thirties Newton developed an acute interest in the Συναγωγή ([*Mathematical*] *Collection*) of the late Alexandrian mathematician Pappus, minutely studying its seventh and eighth books in one or other of the available editions of Commandino's Latin translation.[6] Two stimuli may well have prompted and continued this new-found enthusiasm in the late 1670's for a relatively obscure Greek geometer.

In the first place, perhaps in a search for new material for his contemporary university course of lectures in algebra, Newton renewed his knowledge of the detail of the three books of Descartes' *Geometrie*—again in Schooten's second Latin edition of 1659—more than a decade after his first encounter with the work in the summer of 1664.[7] Where previously, however, his attention had been absorbed by the latter sections of the 'Essai' in which Descartes had developed his method for constructing the subnormal at a point on an algebraic

---

(5) *Euclidis Elementorum Libri XV breviter demonstrati, Operâ Is. Barrow* (Cambridge, ₁1655); *Archimedis Opera: Apollonii Pergæi Conicorum Libri IIII. Theodosii Sphærica: Methodo Nova Illustrata, & Succinctè Demonstrata Per Is. Barrow* (London, 1675). For information on Newton's library copies of these see 1: 11, note (26).

(6) *Pappi Alexandrini Mathematicæ Collectiones À Federico Commandino Vrbinate in Latinum conversæ, et Commentariis illustratæ* (Pesaro, ₁1588 → ₂1602), revised—with the addition to the preface of the seventh book of the passage, of disputed authenticity, containing the anticipation of 'Guldin's theorem'—by C. Manolessi in a new edition (Bologna, 1660) which hopefully but inaccurately to the title of the *editio princeps* adds the subhead '.... *In hac nostra editione ab innumeris, quibus scatebant mendis, & præcipuè in Græco contextu diligenter vindicatæ*'. Though Halley in his *Apollonii Pergæi De Sectione Rationis Libri Duo Ex Arabico MS^to. Latine Versi....Præmittitur Pappi Alexandrini Præfatio ad VII^mum Collectionis Mathematicæ nunc primum Græce edita* (Oxford, 1706): i–liii published an improved (Greek and Latin) version of the preface to Pappus' seventh book from two hitherto unused manuscripts in the Savilian library, his text in places remained greatly defective and at one point (p. xxxvii) he frankly confessed to his reader: '[*Euclidis*] *Porismatum* descriptio, nec mihi intellecta nec lectori profutura. Neque aliter fieri potuit: tam ob defectum Schematis...quàm ob omissa quædam ac transposita vel aliter vitiata in...expositione; unde quid sibi velit *Pappus* haud mihi datum est conjicere'. No accurate versions of the *Synagoge* were available in print till H. J. Eisenmann published the full Greek text (Paris, 1824) and F. Hultsch the standard modern (Greek/Latin) commented edition (2 volumes, Berlin, 1875/1878), complemented by P. Ver Eecke's French translation (Brussels, 1933). There is still no English version.

curve and given a concise exposition of the elementary theory of algebraic equations of up to sixth degree, Newton's eyes were now focused mainly on the second half of the first book and opening pages of the second wherein is sketched the derivation of the (not quite) general Cartesian equation of the conic, referred to oblique coordinates x and y, in the form

$$`y \backsim m - \frac{n}{z}x + \sqrt{mm + ox - \frac{p}{m}xx},$$

qui doit estre la longeur de la ligne *BC*, en laissant *AB*, ou x indeterminée'[8] along with its reduction to the canonical defining equations of its three component species.[9] As a rider to this analytical discussion Descartes introduced an 'Exemple tiré de Pappus', quoting Commandino's Latin version[10] of the passage in the prolegomenon to the *Mathematical Collection*'s seventh book in which Pappus enunciates the classical problem of the τόπος ἐπὶ τρεῖς καὶ τέσσαρας γραμμάς, the locus regarding three and four lines: namely, given three straight lines in position, to find the locus of a point such that, if lines are let fall from it to them at given angles, the ratio of the product of two of these lines to the square of the third is given; or more generally, given four straight lines in position, to determine the locus of a point such that, if lines are let fall from it to them at given angles, the ratio of the product of two of these lines to

(7) See I: 11, 146. The short *critique* Newton wrote at this time on 'Errores Cartesij Geometriæ' is reproduced in 3, §1 below. As we shall see in the next volume, much of the algebraic theory expounded in his Lucasian lectures during 1673–83 derives from Book 3 of the *Geometrie* by way of Mercator's Latin version of Kinckhuysen's *Algebra ofte Stelkonst* (Haerlem, 1661) and his earlier 'Observations' upon it (see II, 3, 1, §§1/2), but more direct influences from Descartes' work are also traceable: notably, the lengthy analysis in the first half of the *Geometrie* of the Greek 4-line locus is presented by Newton in a somewhat variant, drastically summarizcd form as the 'Prob. 12' which he subsequently claimed in the margin of his deposited copy to have read to his Cambridge audience in 'Lect. 6' of his 'Octob. 1676' series (ULC. Dd. 9.68: 73–4 [= *Arithmetica Universalis, sive De Compositione et Resolutione Arithmetica Liber* (Cambridge, ₁1707): 128–30]).

(8) *Geometrie* [= *Discours de La Methode* (Leyden, 1637): 297–413]: 326–7; compare Schooten's Latin rendering in his *Geometria à Renato Des Cartes Anno 1637 Gallicè edita*; *postea autem...in Latinam linguam versa...* (Amsterdam, ₂1659): 27.

(9) See *Geometrie*: 326–34 (= *Geometria*: 27–35). Descartes' reduction is effected by the axis transformation $\left\{ \begin{array}{l} x = (z/a)X \\ y = Y - (n/a)X + m \end{array} \right\}$, which yields the simplified defining equation

$$Y^2 = m^2 + (oz/a)X - (pz^2/a^2m)X^2$$

(that of an ellipse, parabola or hyperbola according as pz^2/a^2m is greater than, equal to or less than zero). Compare D. T. Whiteside, 'Patterns of Mathematical Thought in the later Seventeenth Century' (*Archive for History of Exact Sciences*, **1**, 1961: 179–388): 292–3.

(10) *Mathematicæ Collectiones* (note (6) above): 165^r; see *Geometrie*: 304–5 (= *Geometria*: 8) and 2, §1: note (4) below.

that of the other two is given.[11] In his analytical exploration of this example[12] Descartes was able, by introducing an appropriate pair of oblique coordinates x and y, to show that the defining condition on the locus may always be represented by an equation of the form $y^2 + 2\alpha xy + \beta x^2 + \gamma y + \delta x = 0$ (on taking the origin at the point A on the curve): since this is at once reducible to his general equation for the conic, he accurately concluded that the general 4-line locus (and the 3-line locus which is its particular case when two of the base lines coincide) is identical with the general conic. The recognition of this congruence of the *locus ad tres et quatuor lineas* with the 'solid' locus was not new: its proof had been attempted by Euclid in one of the books of his lost treatise of *Conics*, but was not adequate according to Apollonius who in the general preface to his own *Conics* asserted that 'Euclid had not worked out the synthesis of the locus with respect to three and four lines, but only a chance portion of it, and that not successfully; for the synthesis could not be completed without the theorems discovered by me'.[13] Commenting upon this passage Pappus later added, in

---

(11) We slightly paraphrase Pappus' enunciation so as to bring it into line with Descartes' restatement of the general 4-line case (*Geometrie*: 325, with insertions from 309–10): '[Soient] les 4 lignes *AB, AD, EF*, & *GH* données cy dessus, & qu'il faille trouuer vne autre ligne, en laquelle il se rencontre vne infinité de poins tels que *C*, duquel ayant tiré les 4 lignes *CB, CD, CF*, & *CH*, a angles [*CBA, CDA, CFE*, & *CHG*] donnés, sur les données, *CB* multipliée par *CF*, produist vne somme esgale a *CD*, multipliée par *CH*, [oubien qu'elles ayent quelque autre proportion donnée, car cela ne rend point la question plus difficile]'. Newton below closely follows the latter version; see 2, §1: note (20).

(12) *Geometrie*: 310–12/325–7 (= *Geometria*: 13–15/25–7).

(13) Ivor Thomas' translation in *Selections Illustrating the History of Greek Mathematics. II: From Aristarchus to Pappus* (London, ₂1951): 283. This portion of Apollonius' text is also quoted by Pappus in the preface to Book VII of his *Mathematical Collection*, where it serves to introduce the following comment.

(14) *Mathematical Collection*, Book VII: prefatory remarks on Apollonius' *Conics*. The translation is taken from Ivor Thomas, *Selections Illustrating the History of Greek Mathematics. I: From Thales to Euclid* (London, ₂1951): 487. Commandino's somewhat misleading Latin rendering of the not wholly authenticated original Greek text is given in the next note.

(15) *Geometrie* (note (8) above): 304; compare 2, §1: note (2) below. Commandino's unhappily punctuated Latin version of the original Greek text, flourished immediately afterwards in 'proof' by Descartes with an '& voycy ses mots' (tempered by the marginal remark that 'Ie cite plutost la version latine que le texte grec affin que chascun l'entende plus aysement'), reads: 'Quem autem dicit [Apollonius] in tertio libro locum ad tres, & quatuor lineas ab Euclide perfectum non esse, neque ipse perficere poterat, neque aliquis alius: sed neque paululum quid addere iis, quæ Euclides scripsit, per ea tantum conica, quæ usque ad Euclidis tempora præmonstrata sunt' (*Mathematicæ Collectiones* (note (6) above): 164ᵛ). (The phrase ': sed neque' (rendering ἀλλ' οὐδὲ) is essentially superfluous and may well mark a later interpolation in Pappus' text; compare F. Hultsch, *Pappi Alexandrini Collectionis quæ supersunt...*, **2** (Berlin, 1878): 676.) The original text of Pappus' preface was published only in 1706 (by Halley in his *Apollonii Pergæi De Sectione Rationis Libri Duo* (note (6) above): i–xvii, especially xv) but it seems unlikely that Descartes' mastery of Greek was strong enough to allow him to penetrate its subtleties even if a manuscript copy had been made available to him. Significantly,

mitigation of Euclid's insufficiency and also to show his low opinion of Apollonius' vaunt, the double-edged observation that 'neither Apollonius himself nor anyone else could have added anything to what Euclid wrote, using only those properties of conics which had been proved up to Euclid's time'.[14] When Descartes came to read Commandino's Latin rendering of this sentence, he took Pappus' words to mean that 'ny Euclide, ny Apollonius, ny aucun autre n'auoient sceu entierement resoudre':[15] which, if true,[16] presented him with an ideal opportunity, eagerly seized, for asserting the superiority of his own analytical *methode* over that of the ancients by applying it efficiently and successfully to a problem which no classical geometer had been able to solve. In the outcome later mathematicians readily allowed Descartes the essential novelty of his solution, but many—and Newton foremost among them—were less willing to sanction his boastful claim that 'ie pense auoir entierement satisfait a ce que Pappus nous dit auoir esté cherché en cecy par les anciens',[17] particularly since he had so clearly given merely a preliminary algebraic analysis

the 3/4-line locus problem had been brought to his notice not through his own untutored reading of the *Collection* but by the intervention in late 1631 of the Dutch classicist and orientalist Jakob Gool, who had earlier proposed it to Mydorge without success; see G. Milhaud, *Descartes Savant* (Paris, 1921): 124–5. What interpretation Gool put upon the passage does not seem to be recorded.

(16) We will not join the lists by attempting to formulate a definitive answer to the question as to whether or not Apollonius did construct a fully general proof that the 3/4-line locus is, in all cases, a conic. Indeed, within the limits of our present incomplete knowledge, that no one can do. There can be no doubt, however, that the solution of the 3-line problem is implicitly contained in his *Conics*, III, 54–6, while the general 4-line locus is derivable from the 3-line case in a simple, immediate way (see 2, §1: note (7)), and we need only a minimum of charity to allow that Apollonius himself consciously saw these connections. The topic has been well-nigh exhaustively examined during the past century—with regard to classical sources at least—in Charles Taylor, *An Introduction to the Ancient and Modern Geometry of Conics* (Cambridge, 1881): xlv–xlvi, 266–7; H. G. Zeuthen, *Die Lehre von den Kegelschnitten im Altertum* (Copenhagen, 1886): 126–62; T. L. Heath, *Apollonius of Perga: Treatise on Conic Sections.... With an Essay on the Earlier History of the Subject* (Cambridge, 1896): cxxxviii–cl; J. J. Milne, *An Elementary Treatise on Cross-ratio Geometry with Historical Notes* (Cambridge, 1911): 146–9; and the same author's 'Newton's Contributions to the Geometry of Conics' [= (ed. W. J. Greenstreet) *Isaac Newton, 1642–1727* (London, 1927): 96–114]: 105–8.

(17) *Geometrie* (note (8) above): 309. To be fair, Descartes here has in view not merely the Euclidean 3/4-line locus but Pappus' generalization of it to the analogous curve of n-th degree defined with respect to $(2n-1)/2n$ lines; he is, however, no less incisive in a later passage where, having reduced the defining equation of the simple conic locus to his standard form (see note (9)), he asserts: 'Au reste a cause que les equations...sont toutes comprises en ce que ie viens d'expliquer; non seulement le problesme des anciens en 3 & 4 lignes est icy entierement acheué; mais aussy tout ce qui appartient à ce qu'ils nommoient la composition des lieux solides....le plus haut but qu'ayent eu les anciens en cete matiere a esté de paruenir a la composition des lieux solides: Et il semble que tout ce qu'Apollonius a escrit des sections coniques n'a esté qu'à dessein de la chercher' (*ibid.*: 334–5).

of the locus and not at all the rigorous geometrical composition which the
ancients required.[18]

Whether or not such thoughts initially impelled Newton to turn from the
quoted Latin passages in the *Geometrie* to Commandino's edition itself, they
must rapidly have come to the forefront of his mind when he began to study the
seventh and eighth books of Pappus' *Collection*. Never one to stomach an over-
reacher and increasingly in a mood to be sharply critical of all things Cartesian,
Newton soon concluded that in supposing the Greeks not to have solved the
problem of the 3/4-line locus in all its generality Descartes had gone badly
wrong. In his preliminary paper on the 'Veterum Loca solida restituta' he
pointed, in particular, to the indubitable fact that the Greeks knew how to
construct a conic through five given points[19]—by itself enough, he argued, to
'compose' any given locus of this kind—while to refute Descartes' assertion of

---

(18) To paraphrase the comment with which Newton concluded the version of Propositions
1–4 of the 'Solutio Problematis Veterum de Loco solido' (2, §2) which he later added to his
self-styled lectures 'De Motu Corporum': namely, '...Problematis Veterum de quatuor
lineis ab Euclide incœpti et ab Apollonio continuati non calculu[m] sed compositio[nem]
Geometrica[m]...Veteres quærebant' (ULC. Dd. 9.46: 41ʳ [= *Philosophiæ Naturalis Principia
Mathematica* (London, ₁1687): Liber Primus, Lemma XIX, Corol. 2: 75]). But the same senti-
ment already transfuses through his 'Veterum Loca solida restituta' (2, §1 below).

(19) The case of the ellipse is dealt with in Pappus, *Mathematical Collection*, VIII, 13 and, as
Newton remarked (see 2, §1: note (6)), the two other conic species are constructible 'eadem
ratione'.

(20) Isaac Barrow, for example—in ignorance of his *Method*, still to lie unnoticed for
another two centuries in its Byzantine palimpsest—held the same view of much of Archimedes'
geometrical treatises, inserting after his modern analysis of Proposition 5 of Book 2 of the
Sphere and Cylinder the remark '...qui ipsissimus est analogismus ad quem rem deduxit
Archimedes, (quod, ut hic obiter moneam, satis prodit qualem is analysin usurpârit; nam huc
eum devenisse varias istas proportionum compositiones, divisiones, alternationes, & inversiones,
quas ostentat, adhibendo, penè supra fidem sit: quod si fecisset, casui potiùs imputandum
esset, quàm arti, quod in genuinas inciderit solutiones, & hoc ei constanter obtigisse, vix
concipi potest.)' (*Archimedis Opera*... (note (5) above): 33) before proceeding to the Archimedean composition of the problem 'Authoris insistentes vestigiis suppositâ hujus analogismi
effectione'. Similarly, in his lectures 'incerti temporis' on the same work Barrow began his
'Lect. I' with the words: 'Propositum est nobis methodum exponere, quâ Archimedes
præclara sua theoremata, libris qui extant comprehensa, adinvenit; subtilissimæ istius utcunq
vestigia persequendo. Conabimur autem id efficere singulas materias ad problemata revocando,
qualia nimirum ille sibi solvenda proponebat, & è quorum solutione cum theoremata sua,
tum ipsorum demonstrationes, deducebat. (Unde patebit qualem analysin, & quàm nostræ
modernæ similem exercuerit.)' (*Lectiones [IV. In quibus Theoremata & Problemata Archimedis De
Sphærâ & Cylindro Methodo Analyticâ eruuntur] Habitæ in Scholis Publicis Academiæ Cantabrigiensis*
(London, 1684) [= *Lectiones Mathematicæ XXIII; In quibus Principia Matheseôs generalia expo-
nuntur. Habitæ Cantabrigiæ A.D. 1664, 1665, 1666* (London, 1683): Appendix: 339–88]: 341).
See also 2, §1: note (9) below. As will appear in the seventh volume, Newton likewise came
later to believe that the 'hidden' analysis used by the Greeks was the geometrical equivalent
of contemporary algebraic (Cartesian) analysis in which indeterminate line-lengths departing
from given points take on the rôle of free variables.

the relative efficiency and power of his own algebraical analysis Newton invoked a widely popular thesis of the period that the ancients had clothed their mathematical deductions in the synthetic dress of proportions following a logical ideal which dictated that the prior analysis by which they were originally derived should be concealed.[20] (The polemical note here evident throughout was caught again almost exactly by Edmond Halley nearly thirty years afterwards and we might well suspect that the close similarity was not accidental but due to a preliminary briefing of Halley by Newton.[21]) Subsequently Newton widened the scope of his reading of the *Mathematical Collection* and he became deeply interested in the account given by Pappus in the preface to his seventh book of the lost works of Euclid, Aristæus and Apollonius (notably, Euclid's three books of *Porisms* and Apollonius' two of *Plane Loci*, three on *Vergings*, two on *Determinate Section* and two of *Contacts*), which, together with Euclid's extant *Data* and the full eight books of Apollonius' *Conics*, dealt with the 'analysed locus'.[22] Relying extensively at first on the somewhat shadowy summaries of

---

(21) Halley wrote in the preface (signature [a 4^r]) to his *Apollonii Pergæi De Sectione Rationis Libri Duo Ex Arabico MS^{to}. Latine Versi. Accedunt Ejusdem de Sectione Spatii Libri Duo Restituti*... (Oxford, 1706): 'Quin & alias ob causas expoliri & publicari meruerit hæc *Pappi* Præfatio [ad Librum VII *Collect. Math.*] Primo, ut ex eâ ostendatur *Cartesium* falso Veteres ignorantiæ insimulasse, quasi is primus mortalium Locum ad quatuor rectas ab *Euclide* incœptum componere noverit [see note (17) above]; cum tamen *Apollonius* hoc ipsum se effecisse non obscure indicaverit. Nam impossible esse dicit [note (13) above], perfectam ejus Compositionem exhibere, absque propositionibus quas ipse à se inventas prodidit in tertio *Conicorum*: quod idem est ac si dixisset, illis concessis facile & proclive fuisse *Euclidi* Locum composuisse. Et sane si quis contulerit solutionem illam operosam & immani calculo Algebraico perplexam, quam in principio *Geometriæ* suæ [see notes (9) and (12) above] dedit *Cartesius*, cum admiranda illa concinnitate qua res tota Geometrice & absque omni calculo absolvitur, per Lemmata XVII, XVIII, XIX. Lib. primi *Princip. Math. Naturalis Philosophiæ* [that is, Propositions 1–4 of the 'Solutio Problematis Veterum de Loco solido'; see 2, §2: note (1) below], adhibitis duabus propositionibus [17 and 18] Lib. III. *Conic.* minime dubitabit quin *Apollonius* ipse hac in re majus quiddam præstiterit, quam ab eo præstitum existimat *Cartesius*. Insuper adjicere licet, [compare 2, §1: note (19)] quod ad problema de *Sectione Determinatâ*, ab *Apollonio* plenissime resolutum, tota redeat difficultas inveniendi punctum quintum in Loco describendo. Datis autem quinque punctis docet *Pappus* Locum Ellipticum perficere, Lib. VIII. Prop. 13, 14 [see note (19) above]. Eodemque modo, nec difficilius, mutatis mutandis, Locus Hyperbolicus per data quinque puncta describitur'. It is difficult not to believe that this is a ghost-written Newtonian manifesto, but if the criticism was Halley's own he was in no position to accuse anyone of appropriating another's mathematical discovery—the hyperbolic 'Locus Geometricus' which underlay his restoration of the two books of Apollonius *De Sectione Spatii* (*ibid.*: 139–68, especially 163–8) was borrowed, without acknowledgement and with the implication that he himself had discovered it, as a particular case of the more general conic envelope described by Newton in Book 1, Lemma 25 of his *Principia*!

(22) A loose translation of Pappus' ὁ καλούμενος ἀναλυόμενος [*sc.* τόπος], literally 'the so-called resolved [locus]'. The phrase is rendered in Commandino's Latin version (see note (6) above) as 'locus qui vocatur ἀναλυόμενος, hoc est resolutus', a circumlocution followed by Hultsch in his *Pappi Alexandrini Collectionis quæ supersunt*, 2 (Berlin, 1878): 635 and by Ver

content and numerous lemmatical propositions for each treatise presented by Pappus, he began to list for his own use the enunciations of many simple 'rectilinear', 'plane', 'solid' and 'linear' loci (constructible by straight lines and as circles, conics and more general higher curves respectively) but soon extended these to cover a wider range of problems set down in sequence in his Waste Book under the general head of 'Quæstionum solutio Geometrica',[23] exhibiting many of their geometrical analyses at length together with a more rapid sketch of their synthetic composition. But we should not be misled by these classical sources of the content of Newton's arguments into mistaking the terse, essentially modern style in which they are presented and their several direct appeals to the limit value of a quantity (as a component geometrical variable tends to zero or infinity) for anything but what they are—typically seventeenth-century, indeed Newtonian, forms of proof. At no point does Newton consciously attempt a Euclidean or Apollonian restyling in the mould of such contemporary restorers of the ancients' analysis as Vincenzio Viviani[24] or Pierre de Fermat.

None the less we may conjecture with good reason that the posthumous publication in 1679 (nearly twenty years after his death) of Fermat's reconstructions of Apollonius' treatise on *Plane Loci* and of five of the propositions in Euclid's *Porisms*[25] was a significant factor in continuing Newton's new-found enthusiasm for the Greek 'Treasury of Analysis'. He never, so far as can be

Eecke in his French translation 'domaine ou champ de l'analyse (*Pappus d'Alexandrie: La Collection Mathématique*, **2** (Brussels, 1933): 477, note 1). The traditional English title of 'Treasury of Analysis' goes back to Newton, who (perhaps recalling the name which Oughtred gave to Caput XVIII of his *Clavis Mathematicæ*; compare 1: 25, note (3)) observed to David Gregory in May 1694 that 'Locus resolutus veterum est penus Analytica' (see *The Correspondence of Isaac Newton*, **3**, 1961: 331: 'Adnotata Math: ex Neutono. 1694. Maio', but note that 'penus' is there mistranscribed as 'prorsus').

(23) These are reproduced in 1, §1 and 1, §2 respectively. Not all the 'new' loci and locus problems are of his own *ad hoc* contrivance: they are taken in part from contemporary sources or borrowed from his own earlier unpublished researches. The most interesting example of the last is the straight-line locus presented in Proposition 20 of 1, §1.1, inspired by an earlier text in the same Waste Book which is reproduced in 1, Appendix.

(24) Notably in his *De Maximis et Minimis Geometrica Divinatio in Quintum Conicorum Apollonii Pergæi adhuc desideratum* (Florence, 1659). It is well known that Viviani succeeded in delaying publication of Borelli's edition of Abraham of Ecchelles' Latin version of the 'lost' Books 5–7 of the *Conics* (Florence, 1661: from the tenth-century paraphrase of Abu al-Fath of Ispahan) till after his own restoration of the fifth book appeared, and that he consistently refused to consult the Latin manuscript lest the finer purities of his restitution be contaminated by contact with the original.

(25) *Varia Opera Mathematica D. Petri de Fermat* (Toulouse, 1679): 28–43/116–19: 'Apollonii Pergæi Libri Duo de Locis Planis Restituti' / 'Porismatum Euclidæorum Renovata Doctrina, & sub formâ Isagoges recentioribus Geometris exhibita'. This compilation, edited by Fermat's son Samuel, also contains (pp. 74–88) an extension of Apollonius' problem of *Contacts* to the spherical case.

determined, owned a personal copy of Fermat's 'remains' nor does he ever refer to the *Varia Opera Mathematica* by name, but there is strong circumstantial evidence in the following geometrical papers[26] that he did, in fact, study these Apollonian and Euclidean restorations with some care. Above all, the two concluding circular loci upon which he exemplifies the general technique for determining the existence and species of a 'solid' locus expounded in his 'Solutio Problematis Veterum de Loco solido'[27] seem narrowly indebted to two of the propositions in Fermat's reconstructed second book of Apollonius' *Plane Loci*. In the thirteen main theorems and problems of this treatise Newton develops his synthetic composition of the 4-line locus problem of the ancients, rigorously establishing its identity with the general conic (Propositions 1 and 2) on the basis of Apollonius, *Conics*, III, 17/18,[28] and then (in Proposition 5) makes use of this equivalence to redefine the latter curve in a novel way[29] as the locus of meets of two line-pencils, each centred on a fixed pole, which pass through pairs of points cut off on the arms of a given angle by a line which moves parallel to itself (and will therefore, in modern terms, be through a fixed point at infinity): from this essentially projective definition he was at once able to deduce an elegant proof (Proposition 7) that the general conic may be organically generated from a straight line[30] and hence to construct conics required to pass through given points and touch straight lines given in position, while in culmination he produces the fundamental generalisation (Proposition 12) that a locus is 'solid' if and only if it is constructible as the intersection of two line-pencils, each through fixed poles, which are (as Newton puts it) 'mutually determinable by simple geometry', that is, are in 1, 1 correspondence.[31] This concept of 'simple' geometry, doubtless adapted from Descartes' slightly

---

(26) See, for instance, 1, §1: notes (13) and (36); 2, §2: notes (60), (76) and (82).

(27) Reproduced in 2, §2.

(28) These propositions, already familiar to him when he composed his fragment on 'Conick propertys to bee examined in other curves' a decade before (see II: 93, note (20)), were doubtless brought freshly to Newton's attention when he came to read Pappus' *Synagoge* (compare 2, §1: note (1)). This is the only occasion on which he makes an explicit citation from the *Conics* and he appears never to have acquired a detailed knowledge of Apollonius' work (whose first four books were readily available to him after 1675 in Barrow's Latin epitome, a second-hand copy of which he later bought; see I: 11, note (26) and note (5) above).

(29) Cavalieri had discovered a particular case of the construction in the 1630's (see 2, §2: note (19)) but Newton probably did not know this since he almost certainly never read any of Cavalieri's published works. He was, however, familiar with de Witt's later equivalent construction of the parabola (compare I: 35, 40).

(30) Compare II: 106–48.

(31) This is what we know as 'Chasles' theorem'. The simple anharmonic (cross-ratio) definition of the conic was probably known to Apollonius (if not to Euclid) and reappears in the seventeenth century in the geometrical researches of Fermat and Pascal; see 2, §2: note (60).

different usage of the term,[32] is developed at some length—not wholly clearly[33] —in Proposition 8, but in the sequel Newton narrows his interest to the 1, 1 correspondence between pairs of points in two straight lines which may, on taking x and y to be the respective distances of points from fixed origins in the lines, be analytically determined by an equation of the form

$$a+bx+cy+(d/e)\,xy = 0.$$

In particular, all pairs of points which are in perspective through a fixed point satisfy a relationship of this sort (Proposition 11) but, after being initially tempted to conclude that the converse holds true, Newton soon saw that the general relationship between pairs of points in two given straight lines may not in general be represented by a geometrical perspectivity.[34] All this is radically new and has no equal till more than half a century later William Braikenridge published his *Exercitatio Geometrica De Descriptione Linearum Curvarum*.[35] For all the surface classicism of Newton's 'Solutio Problematis Veterum'—and this is indeed more apparent than real—it is essential that we grasp its pioneering modernity and its radical viewpoint of the nature of a geometrical locus. Here is no antiquarian tract but a piece of geometry worthy to be included under Poncelet's analysis of general *propriétés projectives*.

The contemporary researches in analytical geometry presented in the last section may be dealt with here in a more summary fashion, for with the continuing, indeed overriding exception of the *Geometrie* traceable external influences are negligible and the papers for the most part form a logical sequence to earlier studies by Newton.

Following on his preceding discussion of the 4-line locus in the *Geometrie*, a short opening paper on 'Errores Cartesij *Geometriæ*'[36] locates two mistakes in Descartes' treatment of the generalised Pappus problem of constructing the $(2n-1)/2n$-line locus $(n>2)$, remarking that the Cartesian 'grade' of the construction required to determine an arbitrary point upon it is one less than stated and further—a much more important point though one somewhat shakily made[37]—that the general n-degree algebraic curve is not representable as such a locus: to this he adds an essentially qualitative criticism, sustained with pertinent counter-examples, of Descartes' choice of the 'simplest' 5/6-line locus.

(32) *Geometrie* (note (8) above): 307–8; compare 3, §1: note (5).
(33) See 2, §2: notes (40) and (43).
(34) See 2, §2: note (51).
(35) For details see 2, §2: note (55).
(36) Reproduced in 3, §1 below. As we might expect, the edition to which Newton refers his criticism is again the second Latin (Amsterdam, 1659); compare I: 20, note (6).
(37) See 3, §1: notes (27) and (29).
(38) Reproduced in II, **1**, 1: passim.

In sequel, a series of general observations (in English) regarding the asymptotes and diameters of cubics is largely a rephrasing of his equivalent conclusions of a decade before[38] without evidence of any fresh insight: not least, the nine 'cases' and sixteen 'species' into which Newton earlier subdivided the general cubic continue to be assumed as the present basis of classification.[39]

The most important paper in this last section, the untitled tract on the geometrical properties of cubics and their subdivision into component sub-species,[40] likewise borrows heavily from his comparable researches of the previous decade. The introductory paragraphs, defining terminology and pointing to the close analogies which exist between general properties of the conic and cubic, are essentially distinguishable from his earlier researches into the topic[41] only by a first appearance of 'Newton's theorem' affirming (in extension of Apollonius, *Conics*, III, 17/18)[42] that the product of the three ordinates at any point to a cubic bears a constant ratio to that of the three corresponding abscissas and the derivation of analogous but somewhat un-wieldy general defining 'symptoms' of the conic and cubic.[43] The following reduction of the Cartesian equation of the general cubic by coordinate trans-formation to its four canonical types is likewise an improved version of his preceding approach in 1667, now much simplified by being split into several more easily manageable stages, and his identification of the fifty-three species into which he subdivides these four standard equations but a refinement of his earlier, more idiosyncratically named classifications.[44] At the same time we should insist on the continuing novelty and subtlety of Newton's analysis of the cubic curve: his use of analytical transformation to simplify its Cartesian defining equation has no parallel at all in the researches of contemporary mathematicians and its generality was not matched till 1740 when, largely as an attempted complement to the variant geometrical approach later sketched by Newton in his published *Enumeratio*,[45] de Gua de Malves brought out his

---

(39) Newton's renewed study of the infinite branches of the cubic may possibly have been inspired by (or at least reflect) his contemporary interest in distinguishing the three non-degenerate species of the 'locus solidus' on a comparable basis; compare 3, §2: note (1).

(40) 3, §3 below.

(41) See II: 90–104.

(42) Compare note (28) above.

(43) See 3, §3: notes (18), (23) and (24).

(44) Compare II: 10–16 and 18–70 respectively.

(45) *Enumeratio Linearum Tertii Ordinis*, first published as the first of the 'Two Treatises of the Species and Magnitude of Curvilinear Figures' appended to his *Opticks: Or, A Treatise of the Reflexions, Refractions, Inflexions and Colours of Light* (London, $_1$1704): $_2$138–62. The autograph manuscript (ULC. Add. 3961.1: 38^r–50^r → [marked printer's fair copy] Add. 3961.2: 1^r–14^r), composed in midsummer 1695 in a straightforward revision of the present tract, will be reproduced in the seventh volume.

Usages de l'Analyse de Descartes.[46] What is, even for Newton, brand new in the present treatise is its concluding discrimination of a cubic's genus without prior reduction of its Cartesian equation to standard canonical form. His lengthy preliminary computations[47] reveal that this distinction into the sixteen Newtonian main species—a problem which was, it would appear, still beyond the attainments of James Stirling, Newton's most gifted and enlightened immediate successor in the analysis of the cubic curve, when he published his commentary on the *Enumeratio* in 1717[48]—was, to be sure, accomplished by a preliminary simplifying change of axes whose effect was subsequently allowed for by the equivalent definition of a new set of coefficients, in terms of which his following criteria for the various species and his test for their being diametral were later specified.[49] Despite the few imperfections and omissions which escaped Newton's notice when he gathered his several prior drafts into the condensed final version—first published by Rouse Ball more than two centuries afterward in his valuable essay 'On Newton's Classification of Cubic Curves'[50] —which we now reproduce, these concluding paragraphs on the discrimination of cubic genera are one of the highlights of his researches in the field of analytical geometry.

---

(46) Jean Paul de Gua de Malves, *Usages de l'Analyse de Descartes Pour découvrir, sans le secours du Calcul Différentiel, les Propriétés, ou Affections principales des Lignes Géométriques de tous les ordres* (Paris, 1740): especially Section III, Probleme cinquiéme. 'Faire les Divisions générales des Lignes du second, & du troisiéme Ordre, & indiquer des moyens pour en faire d'analogues dans les Ordres plus élevés': 421–52, particularly 436–45. In 1693 John Craige was still proud to publish his equivalent 'Nova Methodus Determinandi Loca Geometrica' (*Tractatus Mathematicus De Figurarum Curvilinearum et Locis Geometricis* (London, 1693): 63–76, Newton's library copy of which is now Trinity College, Cambridge. NQ.8.51), applying it uniquely to the case of the conic effectively dealt with nearly sixty years before by Descartes (see note (9) above), while even in 1717 James Stirling—manifestly unaware of Newton's unpublished researches—bemoaned the contemporary lack of knowledge of properties of curves higher than the conic with the words 'Proprietates Sectionum Conicarum à Geometris hactenus traditæ sufficiunt ad determinanda Loca quæ incidunt in Sectiones Coni: Consimiles vero Linearum superiorum ordinum proprietates traditas nondum habemus' (*Lineæ Tertii Ordinis Neutonianæ, sive Illustratio Tractatus D. Newtoni De Enumeratione Linearum Tertii Ordinis* (Oxford, 1717): 126).

(47) Reproduced, for the first time, in 3, §3, Appendixes 4–8.

(48) In the concluding section (on the 'Determinatio Locorum Geometricorum') of his *Lineæ Tertii Ordinis Neutonianæ* ((note (46) above): 120–8) Stirling accurately observed (pp. 120–1) that 'Postquam Species omnes Linearum alicujus ordinis enumerantur, convenit ut dignoscatur Species quam constituit Linea particularis æquatione quâlibet propositâ denotata. Innumeræ enim æquationes, quoad formam multum inter se discrepantes, eandem Lineam designare possunt. Lineæ quæ excurrunt in infinitum ex Asymptotis suis optime determinantur; Ovales vero ex Diametris'; but in the two simple examples he gave in sequel (the equations $y^3 = x^3 + a^3$ and $y(y-x)^2 = a^3$ of two cubic hyperbolas, elliptical and parabolic respectively) he resorted each time to a change of axes to determine their species.

(49) See 3, §3: note (79).

(50) *Proceedings of the London Mathematical Society*, **50**, 1891: 104–43, especially 125–8. Compare 3, §3: note (1).

The final fragment 'De solutione Problematum per Regulam trium',[51] dealing with the construction of a conic through five given points when the 'errors' of these in a given direction from a fixed straight line are known and the determination of its intersections with that base line, is a minor, somewhat pointless exercise of established technique which need not detain us.

(51) Perhaps Newton intended '...per Regulam falsæ positionis'; see 3, §4: note (2).

1

MISCELLANEOUS PROBLEMS IN ELEMENTARY PURE GEOMETRY[(1)]

[Late 1670's]

Extracted from Newton's Waste Book in the University Library, Cambridge

§1.[(2)] SIMPLE LOCUS PROBLEMS.

[1]   LOCA PLANA.[(3)]

1. Si datur linea AB, et angulus ACB, punctū [C] est in circulo per A et B transeunte.[(4)]

2. Si dantur puncta A, B et ratio AC ad BC punctum C est in circulo. Et si C est in circulo[(5)] recta BC vergit ad locum B.[(6)]

---

(1) Of the background to these problems, grouped together on ff. 91^v–96^v of Newton's Waste Book, nothing is accurately known. Their handwriting style suggests that they were entered up in this mathematical compendium at roughly the same period (but on different occasions) around or a little before 1680, a dating tentatively confirmed by the position of these autograph leaves in the Waste Book. It will be evident that here, after the lapse of more than a decade, Newton returns to a theme which was well to the forefront in his undergraduate annotations of Descartes, Schooten and de Witt: namely, the construction in various interesting ways of the different species of conic section. To this youthful acquaintance with the properties of the general 'solid' locus he now adds certain general insights gained over the preceding years, notably into higher curves and their possibilities of construction by his organic method. (See I: 28–45 and II: 106 ff.) Why Newton at this time came to renew his interest in elementary pure geometry will probably never be known with certainty—in the preceding introduction we suggest that his reading of Pappus and perhaps also of Fermat's *Opera Varia* was a prime impulse—but we may once more hazard the shrewd guess that he intended the present stockpile of problems as the basis of a course of instruction in synthetic geometry for his Cambridge students; indeed, several of these were (as we shall see in the fifth and sixth volumes) introduced into his contemporary lectures on algebra, there given analytical proof, or incorporated subsequently into the first book of his *Principia* in passages which he claimed to have read from the Lucasian chair in the autumns of 1684 and 1685. Newton's discussion of the problems and loci he lists remains highly individualistic and essentially modern, but his use of the somewhat restrictive terminology which distinguishes curves merely into 'plane', 'solid' and 'linear' classes and his penchant for questions already treated by classical Greek mathematicians underline his increasing knowledge of and growing respect for the geometrical attainments of

Translation

[1]   PLANE LOCI[3]

1. If there is given the line *AB* and the angle $\widehat{ACB}$, the point *C* is in a circle passing through *A* and *B*.[4]

2. If the points *A*, *B* and the ratio of *AC* to *BC* are given, the point *C* is in a circle. And if *C* is in a circle[5] then the straight line *BC* is centred on the position *B*.[6]

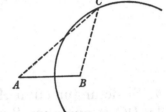

---

the 'ancients', particularly as epitomized in those books of Pappus' *Mathematical Collection* which were available to him.

(2) ULC. Add. 4004: 91ᵛ–93ᵛ, reproduced in strict sequence except for the final problem of section [1], which has been there inserted from f. 91ᵛ.

(3) In the classical Greek sense, outlined by Pappus in his *Mathematical Collection* (IV, 30, Scholium), of straight lines and circles as distinct from 'solid loci' (non-degenerate conics other than the circle) and 'linear loci' (all higher algebraical and all non-algebraical curves): for the exact significance of the terminology see T. L. Heath, *A History of Greek Mathematics*, **1** (Oxford, 1921): 117–18; and P. Ver Eecke, *Pappus d'Alexandrie: La Collection Mathématique*, **1** (Brussels, 1933): 206–8. Newton gives no antecedent for any of the twenty-seven circle loci he proceeds to elaborate, but it is overwhelmingly clear that their common source in large part is Pappus' description in Book VII of his *Mathematical Collection* of the still too little studied 'treasury' of analytical loci (ὁ καλούμενος ἀναλυόμενος [τόπος]) stored up by his predecessors Euclid, Aristæus and Apollonius in works which have now mostly disappeared. Problems 1–11 construct circles on the basis of Pappus' outline of the two lost books of Apollonius on *Plane Loci*, Problems 12–16 elaborate in the case of the circle the 3/4-line locus propounded by Euclid in his lost tract on *Solid Loci* and solved in the general conic case by Apollonius in a lost section of his *Conics*, while Problems 17 and 20–4 closely follow Pappus' commentary on and introductory lemmas to the three books of Euclid's treatise of *Porisms*, perhaps the most significant of those of his mathematical pieces which are known not to have survived.

(4) The second of Charmandrus' introductory propositions to the first book of Apollonius on *Plane Loci* which in Pappus' phrasing announces: 'If straight lines drawn from two given points intersect at a given angle, their common point is restricted to be on a circumference [of a circle] given in position.' The theorem is, in fact, Euclid, *Elements*, III, 21: 'In circulo... qui in eodem segmento sunt anguli...sunt inter se æquales' (Barrow, 1655). Evidently all angles subtended by the chord *AB* at the circle's circumference are half the (constant) angle subtended by it at the centre.

(5) Understand 'dato puncto *A* ut et ratione *AC* ad *BC*' (given the point *A* together with the ratio of *AC* to *BC*).

(6) This 'Apollonius' circle' is probably cited from Pappus, who locates it as an opening proposition of the second book of Apollonius' *Plane Loci* (see Ver Eecke's *Pappus* (note (3) above), **2**: 499), though it was known to earlier Greek mathematicians before the time of Euclid. Aristotle in his *Meteorologica*, III. v, 376a accurately fixed the centre, *D* say, of the locus in the line *AB* by dividing *AB* in the point *C'* such that $AC':C'B = AC:CB$ and then making $C'D^2 = AD \times BD$: at once the triangles *ACD*, *BCD* are similar and so *CD* is fixed in length. See 2, §2: note (76) for further historical details.

3. Si a dato puncto *A* ad rectam positione datam *BD* ducatur recta quævis *AB* et in ea sumatur punctum *C* ea lege ut detur rectangulum *BAC*[,] punctum *C* est in circulo, transeunte per *A*.[7]

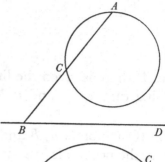

4. Si detur punctum *A* et rectangulum *BAC* vel *BA* × *DC* et punctum *B* ac *D* est ad circulū, etiam punctum *C* erit ad circulum.[8]

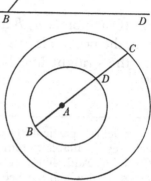

5. Si detur punctum *A* et proportio *AB* ad *AC* vel *AD* ad *DC* et punctum *B* est ad circulum[,] etiam *C* erit ad circulum.[9] Sin punctū *B* est ad rectam erit *C* ad rectam.[10]

6. Si dentur puncta duo *A*, *B* et differentia quadratorum $AC^q - BC^q$ punctum *C* est ad rectam.[11]

7. Si dentur puncta duo *A*, *B* et summa quadratorum $AC^q + BC^q$ [punctum] *C* erit ad Circulum.[12]

8. Si dentur puncta plura *A*, *B*, *D* et quadratorum ex lineis *AC*, *DC*, *BC* vel quadratorum quæ ad ipsa sunt in datis rationibus summa vel quod subducendo aliqua ab alijs restat vel proportio unius ad aggregatū aliud, punctū *C* erit in circulo.[13]

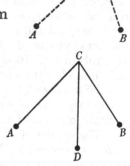

(7) This and the two following propositions are effectively included in Pappus' final, most general type of theorem contained in the second book of Apollonius' *Plane Loci* (see Ver Eecke's *Pappus* (note (3) above), **2**: 500). Here, if *c* is the point on the locus corresponding to the foot *b* of the normal let fall from *A* onto *BD*, it is immediate that the triangles *ABb*, *AcC* are similar and hence $A\widehat{C}c$ is right. In modern terms the circle (*C*) is the inverse (of 'power' $AB \times AC$) of the line *BD* with respect to centre *A*. The converse theorem is, of course, valid.

(8) Since *B* and *D* are on the same circle, the rectangle $BA \times AD$ is constant and so (in either case) the ratio of *AC* to *AD* is given: hence if *BD* meets the locus again in *E*, it follows that $EA:BA \, (= AC:AD)$ is given and therefore the rectangle $EA \times AC$ is constant. The point *A* is manifestly the centre of similitude of the two circles.

(9) When the ratio of *AD* to *DC* is given, so also is that of the constant rectangle $BA \times AD$ to the rectangle $BA \times DC$, which is therefore given in magnitude, and the proposition

3. If from the given point A to the straight line BD given in position there be drawn any straight line AB and in it be taken a point C with the stipulation that the rectangle $BA \times AC$ be given, then the point C is in a circle passing through A.[7]

4. If there be given the point A and the rectangle $BA \times AC$ or $BA \times DC$ and if the point B or D is on a circle, the point C too will be on a circle.[8]

5. If there be given the point A and the ratio of AB to AC or of AD to DC and if the point B is on a circle, C too will be on a circle.[9] But if the point B is on a straight line, C will be on a straight line.[10]

6. If the two points A, B and the difference of squares $AC^2 - BC^2$ be given, the point C is on a straight line.[11]

7. If the two points A, B and the sum of squares $AC^2 + BC^2$ be given, the point C will be on a circle.[12]

8. If several points A, B, D and also of the squares of the lines AC, BC, DC or of squares in given ratio to them the sum or the outcome of taking some away from others or the proportion of one to some aggregate be given, then the point C will be on a circle.[13]

---

effectively reduces to the previous one; similarly, when the ratio of AB to AC is given, we deduce that the rectangle $AD \times AC$ is constant for all pairs of points D and C, so that the circles (D) and (C) will be inverse with respect to the centre A. By Proposition 3, however, when the point A is on the circle BD the locus of C is a straight line.

(10) In either case, namely, the parallel through C to the previous straight line.

(11) The first proposition cited by Pappus in his epitome of the second book of Apollonius' *Plane Loci*: 'If the squares of two intersecting straight lines drawn from two points [understood to be fixed] differ by a given area, their common point is restricted to be in a straight line [perpendicular to the line joining the two points].' If the point D is determined in AB such that $AD^2 - BD^2 = AC^2 - BC^2$ given, it is an immediate deduction from Euclid, *Elements*, I, 47, that CD is perpendicular to AB.

(12) The particular case of the third proposition where the ratio is unity, quoted by Pappus from the second book of Apollonius' *Plane Loci*: 'If in intersecting straight lines drawn from two given points the square of one is proportionally greater than the square of the other by a given area, their common point is restricted to be in a circumference given in position.' In the present instance, since on bisecting AB in D there results $AC^2 + BC^2 = 2CD^2 + \frac{1}{2}AB^2$, it is clear that the length of CD is constant and so D is the circle centre.

(13) These generalizations of the preceding proposition are partly Pappus' quotation from the second book of Apollonius' *Plane Loci*, partly Newton's variants on Pappus' account. In modern form the proposition may be restated in this manner: given a number of fixed points A_i, $i = 1, 2, 3, \ldots$, the locus of a point C restricted by $\sum_i (\pm a_i . CA_i^2) = k$, where the a_i and k are constants, is a circle. The general geometrical proof is cumbrous but is evident analytically since, on setting the points A_i as (α_i, β_i) in a rectangular Cartesian coordinate system in which the point C is (x, y), the defining equation of the locus has the form

$$\sum_i \pm a_i [(x - \alpha_i)^2 + (y - \beta_i)^2] = k.$$

It is not clear whether Newton at the time he wrote up these 'loca plana' had yet seen Fermat's equivalent generalization in his *Apollonii Pergæi Propositiones de Locis Planis restitutæ* [= *Varia*

10.[14] Si dantur puncta B, D, E et productam BD secet quævis EC in A, et sit rectangulum EAC æquale rectangulo $BAD_{[,]}$ erit punctū C ad circulum.[15]

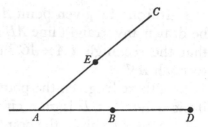

11. Si dantur puncta E, D et concur[r]ant rectæ $CE_{[,]}$ BD ad rectā positione datam AF et rectanguloru EAC, DAB differentia vel nulla est vel æqualis rectangulo sub AF et recta data: sit autem punctum B ad circulū, erit punctum C ad circulum.[16]

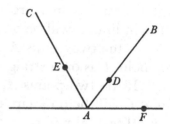

12. Si detur quadrilaterum $A[B]DE$ quod in circulo inscribi potest,[17] et a puncto C ad latera ejus ducantur in datis angulis rectæ quatuor CF, CG, CH, CI ita ut rectangulū sub duabus $CG \times CI$ datam habeat rationem ad rectangulū sub alijs duabus $CF \times CH$, et in uno aliquo casu punctū C est in circumferentia circuli transeuntis per $ABDE$, semper erit in circulo illo.[18]

Affinia sunt 1, 2, 6, 7, 8. Item 3, 4, 5. Item 10, 11.[19]

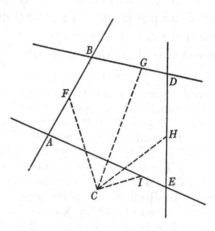

13. Si dentur puncta A, B. et circulus centro A radio dato AD descriptus utcunqჳ secetur in D et E a ducta AC ac demittatur normalis CF, sit autem rectangulum $DCE = 2ABF_{[,]}$ punctum C erit in circulo.[20]

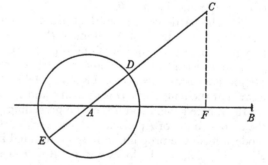

Opera Mathematica D. Petri de Fermat (Toulouse, 1679): 12–43]: Liber II, Propositio V: 33–41, where it is enunciated (p. 33) in the restricted form 'Si à quotcunque datis punctis ad punctum unum inflectantur rectæ lineæ, & sint species, quæ ab omnibus fiunt, dato spatio æquales, punctum continget positione datam circumferentiam' which narrowly repeats Pappus' phrasing.

(14) Newton lists no ninth locus.

10.[14] If the points B, D, E are given and any line EC shall cut BD produced in A, let the rectangle $EA \times AC$ be equal to the rectangle $BA \times AD$, and the point C will be on a circle.[15]

11. If the points E, D are given and the straight lines CE, BD be concurrent on the straight line AF given in position, and if the difference of the rectangles $EA \times AC$ and $DA \times AB$ is either zero or equal to the rectangle of AF and a given line-length, let the point B be on a circle and the point C will be on a circle.[16]

12. If the quadrilateral $ABDE$ inscribable in a circle[17] be given and from the point C to its sides four straight lines CF, CG, CH, CI be drawn at given angles so that the rectangle of the two $CG \times CI$ shall have a given ratio to the rectangle $CF \times CH$ of the other two, then if in some single instance the point C is in the circumference of the circle passing through A, B, D, E, it will always be in that circle.[18]

1, 2, 6, 7 and 8 are akin. Likewise 3, 4 and 5. And likewise 10 and 11.[19]

13. If the points A, B be given and the circle described with centre A and given radius AD be cut anywhere in D and E by AC when drawn, and if, on letting fall the normal CF, the rectangle $DC \times CE = 2AB \times BF$, the point C will be in a circle.[20]

---

(15) Understand 'per E, B, D transeuntem' (passing through E, B and D). The proposition is virtually the converse of Euclid, *Elements*, III, 36, on the products of the segments of intersecting circle chords.

(16) This is untrue in general unless we make the restriction that the circle (B) passes through D. In this case suppose that F is the origin of a rectangular Cartesian coordinate system in which AF is the line $y = 0$, A is the general point $(x, 0)$ and $O(a, b)$ is the centre of the circle (B) passing through $D(c, d)$: at once $AF = x$ and the rectangle

$$AB \times AD = (x-a)^2 + b^2 - (a-c)^2 - (b-d)^2 = x^2 - 2ax + 2ac + 2bd - c^2 - d^2.$$

Hence if the locus (C) is determined by the condition $AC \times AE = AB \times AD + 2k \times AF$, then on setting the given point E as (e, f) the locus will be a circle through E of centre

$$(a+k, \tfrac{1}{2}[2ac + 2bd - 2(a+k)e - c^2 - d^2 + e^2 + f^2]/f).$$

In particular, when $k = 0$, the line AF will be the radical axis of the two circles.

(17) Newton first wrote equivalently 'cujus anguli opp[ositi duos rectos conficiunt]' (whose opposite angles together make up two right angles).

(18) For the moment Newton evidently relies on Apollonius' assertion that the '4 line locus' defined by $CF \times CH = k \times CG \times CI$ is in general 'solid': a theorem first given analytical proof in Book 2 of Descartes' *Geometrie*, as he well knew (see the preceding introduction). The restriction to the circle is dealt with in more detail in Propositions 14 and 15 following.

(19) Indeed, Propositions 1, 2, 6, 7 and 8 give variant bipolar definitions of plane loci; Propositions 4, 5 and 6 construct plane loci as similar to or the inverse of given plane loci; while Proposition 11 evidently generalizes Proposition 10.

(20) Analytically, if A is the origin of a rectangular Cartesian coordinate system in which C is the general point (x, y) and AB is the line $y = 0$, then on taking $AB = a$ and $AD = AE = r$ the condition $CD \times CE$ (or $CA^2 - AD^2$) $= 2BA \times BF$ yields $x^2 + y^2 - r^2 = 2a(a-x)$ as the defining equation of the circle (C) of centre $(-a, 0)$ and radius $\sqrt{[3a^2 + r^2]}$. More generally, the locus defined by $CD \times CE = k \times BA \times BF$ will be a circle.

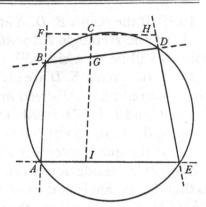

[14.] Si trapezij *ABDE* anguli *A* ac *D* recti sint et ad *AB* et *AE* demittantur perpendicula *CF*, *CI* a puncto quovis *C* secantia *BD* ac *DE* in *H* et *G* et fuerit rectangulum *FCH=GCI*, erit punctum *C* in circulo transeunte per puncta *ABDE*. Et vice versa. Idem eveniet si ad latera singula Trapezij a puncto *C* demittantur perpendicula.[21]

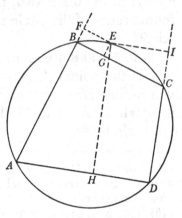

[15.] Si in circulo quovis *ABCD* inscribatur trapezium *ABCD*, et a circumferentiæ puncto quovis *E* ad latera trapezij ducantur lineæ *EF*, *EG*, *EH*, *EI* constituentes cum lateribus conterminis *AB*, *BC* parallelogrammum *EFBG* et cum alijs duobus lateribus conterminis *AD*, *DC* parallelogrammum *EHD[I]*, quod sub ductis ad opposita duo latera continetur rectangulum *GE × EH* æquale est rectangulo *EF × E[I]* sub ductis ad reliqua duo latera contento.[22]

Idem eveniet si a puncto *E* ad latera trapezij demittantur perpendicula. Ut et si ductæ ad duo latera contermina *EH*, *E[I]* ad *A[D, C]D* æquales angulos *EHD*, *E[I]D* vel *EHA*, *E[I]D* cum ipsis conficiant et ductæ ad altera duo latera *EF*[,], *EG* æquales angulos cum ipsis.[23]

---

(21) A simple instance of the circular case of the 4-line locus (compare note (18) above). A straightforward proof of both assertions is obtained by constructing the rectangle *AEδB*, extending *CG* and *CH* to meet *βδ*, *AE*, the circumcircle of *AEδB* and *Eδ* in *γ*, *I*, *C'* and *β*, and drawing *Cg*, *Ch* normal to *BD*, *DE*. At once, since the points *B*, *C*, *δ*, *C'* are concyclic, we have

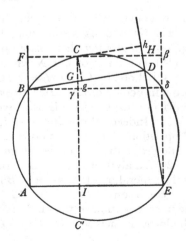

$$B\gamma : C\gamma \text{ (or } C'I) = E\beta \text{ (or } IC): (\delta\gamma \text{ or) } \beta C,$$

so that the quadrilaterals *BgCγ* and *EhCβ* are evidently similar and consequently *CG : Cg : Cγ = CH : Ch : Cβ*, while *CF × Cβ = Cγ ×* (*γC'* or) *CI* and hence *CF × CH = CG × CI*, *CF × Ch = Cg × CI*.

14. If the angles at A and D in the quadrilateral $ABDE$ are right and to AB and AE from any point C be dropped the perpendiculars CF, CI intersecting BD and DE in H and G, then if the rectangle $FC \times CH = GC \times CI$, the point C will be in the circle passing through the points A, B, D and E. And conversely so. The same will hold if from the point C perpendiculars are let slip to the individual sides of the quadrilateral.[21]

15. If in any circle $ABCD$ there be inscribed the quadrilateral $ABCD$ and from any point E in its circumference there be drawn to the quadrilateral's sides the lines EF, EG, EH, EI forming with two adjoining sides AB, BC the parallelogram $EFBG$ and with the other two adjoining sides AD, DC the parallelogram $EHDI$, the rectangle $GE \times EH$ contained by those drawn to two opposite sides is equal to the rectangle $EF \times EI$ contained by those drawn to the remaining two sides.[22]

The same will hold if from the point E to the quadrilateral's sides perpendiculars are let fall. So also if the lines, EH and EI, drawn to the two adjoining sides AD, CD shall make equal angles $\widehat{EHD}$, $\widehat{EID}$ or $\widehat{EHA}$, $\widehat{EID}$ with them and those, EF and EG, drawn to the other two sides equal angles with these.[23]

---

(22) More generally, draw $A\beta$ parallel to DC, meeting the circumcircle of $ABCD$ in β: at once, AD, βC are equal and equally inclined to DC and the same is true of their parallels $\phi E\iota$, EI, so that

$$E\phi \times EI \text{ (or } E\iota) = E\gamma \times (\gamma E' \text{ or)} EH,$$

and to complete the proof we need only show that the re-entrant quadrilaterals $AFE\phi$, $CGE\gamma$ are similar, since then $E\phi : EF = E\gamma : EG$.

(23) If in Newton's figure the perpendiculars Ef, Eg, Eh, Ei are let fall from E to AB, BC, AD, DC respectively, it will be obvious that, since $\widehat{fFE} = \widehat{gGE}$ and $\widehat{hHE} = \widehat{i\iota E}$, there is $EF : Ef = EG : Eg$ and $EH : Eh = EI : Ei$. (These proportions are evidently preserved when Ef, Eg make equal angles with AB, BC or when Eh, Ei make equal angles with AD, DC.) More directly, since we may easily show that

$$\frac{\triangle EAB \times \triangle ECD}{\triangle EBC \times \triangle EDA} = \frac{\sin \widehat{AEB} \times \sin \widehat{CED}}{\sin \widehat{BEC} \times \sin \widehat{DEA}} = \frac{AB \times CD}{BC \times DA},$$

it follows immediately that

$$\frac{Ef \times Ei}{Eg \times Eh} = \frac{2\triangle EAB \times 2\triangle ECD}{2\triangle EBC \times 2\triangle EDA} \bigg/ \frac{AB \times CD}{BC \times DA} = 1.$$

In the manuscript figure Newton carelessly entered the letters 'I' and 'C' in such a way that 'C' seems to mark both points C and I, and in consequence wrote 'C' in his text some half dozen times instead of 'I'.

Et hinc si ductæ quosvis angulos conficiant cum lateribus trapezij, et rectangula *GEH, FEI* sunt in data ratione[,] facile est cognoscere utrum punctum *E* sit in circūferentia circuli. Nam ad latera duo opposita *AB, CD* duc *EK, EL* in angulis *EKB, ELD* illis in quibus ad altera duo latera ductæ sunt lineæ[(24)] æqualibus. Verbi gratia *EK* in angulo *AKE*=ang *CGE* et *EL* in angulo *DLE*=ang *DHE*. Et si ratio rectanguli *FEI* ad rectangulum *GEH* componitur ex ratione *FE* ad *KE* et *EI* ad *EL* ita ut rectangula *KEL*[,] *GEH* æqualia sint, erit punctum *E* in circulo, secus non erit in circulo.

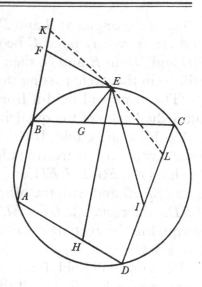

[16.] Si[(25)] *ABE* sit circulus et detur *A* & rectangulum *BAC*[,] erit punctum *C* in recta. Et si punctū *C* in recta sit converget recta *CA* ad datū punctū *A*. Et in fig 2 si datur rectang *BAC* erit *C* in circulo.[(26)]

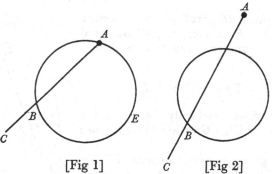

[Fig 1]   [Fig 2]

[17.] Si linearum *AD, BD, CE* poli *A B C* in linea recta sunt, et puncta intersectionum duo *D, E* lineas rectas describunt[,] tertia intersectio *F* lineam rectam[(27)] describet. Idem eveniet si linea *DE* parallela est lineæ *BC*.[(28)] Ut et si puncta *A B C* non sint in directum si modo loca punctorum *D, E* se secant in recta *BC*.[(29)]

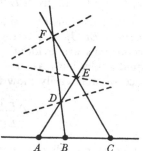

---

(24) Newton first wrote equivalently '...illis quos ductæ ad altera latera opposita faciunt'.

(25) Understand 'in fig 1 ' (in Figure 1).

(26) In modern terms the locus (C) is the inverse of the given circle, and so is a straight line (see note (7) above) when *A* is on the given circle, or an anti-homothetic circle of 'power' $AB \times AC$ when it is not (compare note (9) above).

And hence, if the lines drawn make any angles you like with the sides of the quadrilateral and the rectangles $GE \times EH$, $FE \times EI$ are in given ratio, it is easy to determine whether the point E is in the circumference of the circle. For to two opposite sides AB, CD draw EK, EL at angles equal to those at which lines are drawn to the other two sides—for instance, EK at the angle $A\widehat{K}E = C\widehat{G}E$ and EL at the angle $D\widehat{L}E = D\widehat{H}E$. Then if the ratio of the rectangle $FE \times EI$ to the rectangle $GE \times EH$ is compounded from the ratio of FE to KE and EI to EL such that the rectangles $KE \times EL$ and $GE \times EH$ are equal, the point E will be in the circle; otherwise not.

16. If[25] ABE be a circle and A, along with the rectangle $BA \times AC$, be given, the point C will be in a straight line. And if the point C be in a straight line, the line CA will be inclined through the given point A. And in Figure 2 if the rectangle $BA \times AC$ is given, then C will be in a circle.[26]

17. If the poles A, B, C of the lines AD, BD, CE are in a straight line and two points D, E of intersection describe straight lines, the third intersection F will describe a straight line.[27] This will hold true if the line DE is parallel to the line BC.[28] So also if the points A, B, C be not in line providing that the loci of the points D, E intersect on the straight line BC.[29]

---

(27) Namely, passing through the meet of the straight-line loci (D) and (E), though from the manuscript figure (here accurately reproduced) it is clear that Newton has not realized this.

(28) When, that is, the point A lies at infinity in the line BC. This variant on Desargues' theorem on perspective triangles is Pappus' codification of the first ten propositions of the first book of Euclid's lost treatise of *Porisms*: 'If three points situated on a unique straight line (or two points in the case of parallelism) are given and the [three] remaining points, one excepted, are restricted to be on straight lines given in position, then this single point, too, is restricted to be on a straight line given in position' (compare Ver Eecke's *Pappus* (note (3) above), **2**: 488).

(29) This has nothing to do with the preceding, but is rather an equivalent form of 'Pappus' theorem' on a hexagon inscribed in a line-pair, which first appears as Pappus' thirteenth and seventeenth lemmas on Euclid's *Porisms* (that is, *Mathematical Collection* VII, 139 and 143): for,

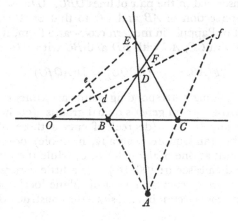

[18.] Si dati anguli *DBA*, *DCA* circa polos *B*, *C* volvan-
tur et angulis *ABD*, *ACD* æquales capiantur *CBF*[,] *BCF*
sitʒ punctum *D* in recta transeunte per punctum *F* vel
etiam in conica sectione per puncta tria *BCF*[,] erit punctum *A* in recta. Et si *D* in recta sit non transeunte per
punctū *F* aut in con. sectione transeunte per duo e tribus
punctis *B*, *C*, *F*, non autem per omnia tria, punctū *A* erit
in conica sectione. Si per unicum tantum e tribus punctis
BFC transit conica sectio, punctum *A* erit in curva primi
gradus tertij generis. Si per nullum erit primi [gradus]
quarti [generis].[30]

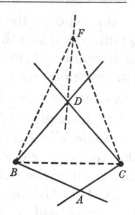

[19.] Si circuli duo *AC*[,] *AD* se secuerint
in *A* et per *A* agatur recta *ACDB*, et detur
ratio *CD* ad *DB* erit *B* in circulo transeunte
per intersectiones priorum.[31]

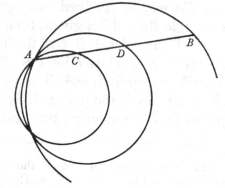

[20.] Si recta *CD* secet rectas positione datas[32] *AC*,
BD in data ratione et secetur in *E* in data ratione,
tanget *E* rectam cujus data est ratio ad *AC* vel *BD*.[33]

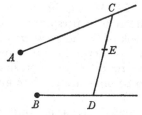

if the straight-line loci (*D*) and (*E*) meet in the point *O* on *BC*, then *eFf* is the 'Pappus' line
of the hexagon *ECABDO* inscribed in the pair of lines *OBC*, *ADE*, so that the locus of *F* is the
straight line joining the intersection of *AB* and *EO* to that of *AC* and *DO*. Newton's proof
would no doubt repeat that of Pappus: in modern cross-ratio form, if *P*, *Q*, *R* (not shown) are
the respective meets of *AB* with *CE*, *AC* with *BD* and *BC* with *AD*, then

$$E(ABeP) = (RBOC) = D(AQfC)$$

and so the lines *BQ*, *ef*, *PC* joining corresponding points are concurrent (namely, in *F*).

(30) Newton carelessly wrote 'primi generis quarti gradus' (of the first order of the fourth
grade) here. Evidently in any 'order' (degree) of curves those of first 'grade' have the
maximum number of double (and, more generally, multiple) points: here the constructed
cubic will have a double point at one of the poles *B*, *C*, while the quartic will have a double
point at each. (For further details see ɪɪ: 106–16.) It is a little contrary of Newton to include
his organic method of constructing curves in a list of 'plane loci' without bothering to show
how it can be applied to construct circles: in fact, since the constructed locus must pass through

18. If the given angles $D\widehat{B}A$, $D\widehat{C}A$ rotate round the poles B, C and to the angles $A\widehat{B}D$, $A\widehat{C}D$ there be taken equal $C\widehat{B}F$, $B\widehat{C}F$, let the point D be in a straight line passing through the point F, or even in a conic through the three points B, C and F, and the point A will be in a straight line. And if D be in a straight line not passing through the point F, or in a conic passing through two of the three points B, C, F but not all three, the point A will be in a conic. If the conic passes through only a single one of the three points B, F and C, the point A will be in a curve of the first grade of the third order. If through none of them it will be of the first [grade] of the fourth [order].[30]

19. If, when two circles AC, AD intersect in A and through A is drawn the straight line $ACDB$, the ratio of CD to DB be given, B will be in a circle passing through the intersections of the previous ones.[31]

20. Should the straight line CD cut the straight lines AC, BD (given in position)[32] in a given ratio and be itself cut at E in a given ratio, E will lie on a straight line whose ratio to AC or BD is given.[33]

---

the poles B, C the angle $B\widehat{A}C$ subtended at the circle (A) must be constant and hence the angle $B\widehat{D}C$ subtended at the describing curve (D) must also be fixed in magnitude, whence the describing curve must in all generations of circles be itself a circle through the poles B, C.

(31) For if a is the second intersection of the circles (C) and (D), the angles $A\widehat{C}a$, $A\widehat{D}a$ in the circle segments $\widehat{AC}$, $\widehat{AD}$ are constant and hence the figure $aCDB$ is given in species; in consequence, the angle $A\widehat{B}a$ is constant and so the locus of B is a circle through the points A and a. This locus, apparently Newton's own discovery in its present form, was a few years later introduced by him into his *Philosophiæ Naturalis Principia Mathematica* (London, ₁1687: 98–100) as a rider to Book 1, Lemma 27.

(32) As Newton already knew in 1665, these need not be in the same plane, though in context it is clear that this restriction is made in the present instance.

(33) Newton's earlier proof of the general case where AC and BD are not in the same plane is reproduced in appendix to the present section (pages 270–3 below). He subsequently gave a simple proof of the plane case in Lemma 23 of Book 1 of his *Principia* (note (31) above): 89–90, essentially by extending AC to its meet F with BD, constructing the points G and H in BD such that

$$AC\!:\!BD = FC\!:\!GD \quad \text{and} \quad FH\!:\!FG = CE\!:\!CD$$

(so that H is on the locus), setting off $KD = FG$ and drawing the parallel through E to BD to its meet I with CK: immediately $FH\!:\!(FG$ or$)$ $KD = CE\!:\!CD = IE\!:\!KD$, so that IE is equal and parallel to FH; but the ratios of $FG\!:\!(GD$ or$)$ FK and of $CI\!:\!IK$ ($= CE\!:\!ED$) are given, and consequently the locus of I is a straight line through F, making the locus of E the parallel to it through H.

[21.] Si detur punctum B & circulus AEC sintɋ AD, BD, CD continuè proportionales vel $AB.BC::AD.DC$ tanget D locum rectilineum.[34]

Sin detur punctum D et AC bisecetur in B tanget B circulum.[35]

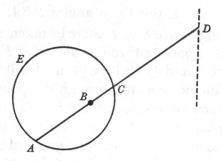

[22.] Datis positione lineis $AE\,BE$ et punctis $A\,B$: si recta quævis CD secet alteras in C, D ea lege ut rectangulum AC, BD æquetur dato rectangulo $AE \times BE$, comple parallelogrammum $AEBP$ et locus ad quem recta CD vergit erit punctum P.[36]

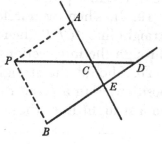

[23.] Si AB datur positione et longitudine & AD BC longitudine sintɋ CE DE æquales cape $AF.BF::AD.BC$ et CD verget ad datum locum puncti F.[37]

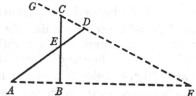

[24.] Si a datis punctis A, B ductæ AC, BC datam habeant summam vel differentiam N: Duc CD parallelam AB et in ratione ad AC quam habet N ad

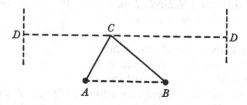

AB et punctum D erit in recta quæ perpendicularis est ad AB. Debet vero CD ad plagam versus A duci ubi datur summa $AC+BC$, ad plagam versus B ubi datur $AC-BC$ vel $BC-AC$.[38]

---

(34) Pappus' thirty-fifth lemma on Euclid's *Porisms* (*Mathematical Collection*, VII, 161). In the preceding lemma (VII, 160) Pappus had shown that, if b is the mid-point of AC, then

$$AB \times BC = bB \times BD:$$

for, since $AB \times CD = BC \times AD$, therefore

$$2AB \times CD = AB \times CD + BC(AB+BC+CD) = (AB+BC)(BC+CD)$$

21. If the point *B* be given together with the circle *AEC* and *AD*, *BD*, *CD* are continued proportionals, that is, $AB:BC = AD:DC$, *D* will trace a straight-line locus.[34]

But if the point *D* be given and *AC* be bisected at *B*, *B* will trace a circle.[35]

22. Given the lines *AE*, *BE* in position and the points *A* and *B*, if any straight line *CD* intersect the others in *C* and *D* subject to the restriction that the rectangle $AC \times BD$ be equal to the given rectangle $AE \times BA$, complete the parallelogram *AEBP* and the locus on which the line *CD* is centred will be the point *P*.[36]

23. If *AB* is given in position and length and *AD*, *BC* in length, and if *CE*, *DE* be equal, take $AF:BF = AD:BC$ and *CD* will be inclined through the given point-locus *F*.[37]

24. If the lines *AC*, *BC* drawn from the given points *A*, *B* have a given sum or difference *N*, draw *CD* parallel to *AB* and in ratio to *AC* as *N* to *AB*, and the point *D* will be in a straight line which is perpendicular to *AB*. When the sum, indeed, $AC+BC$ is given, *DC* ought to be drawn in direction towards *A*, but towards *B* when $AC-BC$ (or $BC-AC$) is given.[38]

---

and hence $AB \times CD = bC \times BD = bB \times BD + (BC \times BD)$ or $AB(CD-BC)$. As an immediate corollary, if *d* is the foot of the normal from any point *D* of the locus onto the diameter *ac* of the circle which is through *B*, then, when *O* is the circle's centre, $AB \times BC = aB \times Bc$ and $bB \times BD = OB \times Bd$, so that the position of *d* in *aBc* is fixed by $aB \times Bc = OB \times Bd$ for all points *D*. At once (*D*) is the straight line through *d* normal to *aBc*, that is, the polar of *B* with respect to the circle.

(35) For, if *O* is the circle's centre, *OB* will be perpendicular to *ACD* in all positions, and hence the locus of *B* is the circle on diameter *OD*.

(36) This is the first theorem, recast into more classical form, of the five restored *soi-disant* 'porisms' set out by Fermat in his *Porismatum Euclidæorum Renovata Doctrina, & sub formâ Isagoges recentioribus Geometris exhibita* [= *Varia Opera Mathematica* (note (13) above): 116–19, especially 117], but is only vaguely contained in Pappus' own listing of the contents of Euclid's treatise under the heading of those in which an 'area' is given (compare ver Eecke's *Pappus* (note (3) above), **2**: 491–2). The elementary proof is immediate, for by similar triangles $AC:AE = BE:BD$ determines *P* to lie in all transversals *CD*. In projective terms the proportion may be set as $(A \infty CE) = (\infty BDE)$, in which *E* is a self-corresponding point, so that *CD* is through the meet of the parallel to *BE* through *A* and of the parallel to *AE* through *B*.

(37) Namely, in the straight line *AB*. The proof is immediate on drawing the parallel to *CD* through *B*, for this meets *AD* in *H*, say, such that $HD = BC$ and $AH:HD = AB:BF$. This somewhat uninteresting theorem is evidently Newton's own discovery.

(38) The locus of *C* is evidently either an ellipse or hyperbola which has the points *A* and *B* for its foci, so that the locus of *D* is the pair of perpendiculars to *AB* which are its directrices. In context it would seem likely that Newton has taken this proposition over from the lemmas on Euclid's two lost books of *Surface Loci* (*Mathematical Collection*, VII, 235–7) in which Pappus first elaborated the focus-directrix definition of the general conic. (Compare T. L. Heath, *Greek Mathematics* (note (3) above), **2**: 119–21.)

[25.] Si datur circulus *ABD* et rectangulum *ACB*, punctū *C* erit in circulo idem habens centrum cum circulo *ABD*.[39]

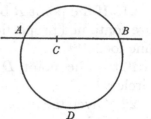

[40][26.] Si per data puncta *AB* transit circulus secans in *E* rectam [*CE*] ipsi *AB* perpendicularem et arcui *BE* æqualis sit arcus *EF*[,] erit *F* in circulo cujus centrum est *A*. Et si *F* in tali circulo sit et bisecetur [arcus] *BF* in *E*, erit *E* in recta.[41]

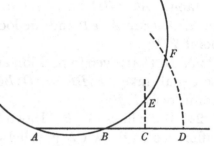

[27.] Punctum *B* circumferentiam contingit a quo si duæ agantur rectæ, una ad datum punctum *A* altera in dato angulo *C* ad rectam positione datam & ad datum punctū terminatam *DC*, quadratum prioris *AB^q* æquale sit rectangulo sub posteriore abscissa ac data *BC × E*.

Vel sic. Si *AB^q = FC × E* punctum *B* contingit circ.[42]

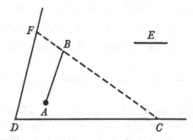

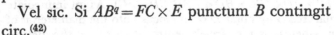

[2]   DE LOCO RECTILINEO.[43]

[1.] Si locus crura duo infinita opposita habet, et non plura, aut rectus est, aut tertij, quinti, septimi vel imparis alicujus generis curva linea.[44]

[2.] Si rectæ locum tangentis plaga determinata est rectus est locus.

[3.] Si recta nulla ad plagam infiniti cruris tendens potest locum secare[45] rectus est locus.

[4.] Si per datum loci punctum recta transiens non potest locū alibi secare rectus est locus.[46]

---

(39) It is not difficult to see that the distance of *C* from the centre, *O* say, of the circle is the constant quantity $\sqrt{[AO^2 - AC \times CB]}$.

(40) The two final propositions are later additions, the latter being found (see note (2) above) by itself on f. 91ᵛ.

(41) The proof of this proposition, evidently original with Newton, follows readily on noticing that the angles $B\hat{A}E$, $E\hat{A}F$ (subtended in the same circle by equal chords *BE*, *EF*) are equal, for then the obtuse triangles *AED*, *AEF* have a common side *AE* and equal corresponding sides *AD*, *AF* and angles $E\hat{A}D$, $E\hat{A}F$ and are therefore congruent, having their third sides *ED*, *EF* accordingly equal.

25. If the circle ABD together with the rectangle $AC \times CB$ is given, the point C will be in a circle having the same centre as the circle ABD.[39]

[40]26. If through the given points A, B passes a circle intersecting in E the straight line CE perpendicular to AB, and the arc $\widehat{EF}$ be equal to the arc $\widehat{BE}$, then F will be in a circle whose centre is A. And if F be in such a circle and the arc $\widehat{BF}$ be bisected at E, then E will be in a straight line.[41]

27. The point B traces a [circle] circumference: if from it two straight lines are drawn, one to the given point A, the other at the given angle C to the straight line DC given in position and terminating at a given point, the square AB^2 of the former shall be equal to the rectangle $BC \times E$ of the latter segment and a given length.

Or thus. If $AB^2 = FC \times E$, the point B traces a circle.[42]

---

[2]   ON THE STRAIGHT-LINE LOCUS[43]

1. If a locus has two opposite infinite branches, and no more, it is either rectilinear or a curve of third, fifth, seventh or some odd order.[44]

2. If the direction of a straight line tangent to the locus is fixed, the locus is rectilinear.

3. If no straight line tending in the direction of an infinite branch can intersect the locus,[45] the locus is rectilinear.

4. If a straight line passing through a given point of a locus can nowhere else intersect it, the locus is rectilinear.[46]

---

(42) Analytically, on drawing fc through A parallel to FC and letting fall the normals AP and Dp onto FC and fc respectively, we may define the rectangular Cartesian coordinate system in which $AP = x$, $PB = y$ and set $Ac = a$, $fc = b$. Where the ratio $pc/pD = \lambda$, the first locus $AB^2 = E \times BC$ has the defining equation $x^2 + y^2 = E(y + a + \lambda x)$, while on making $fc/pD = \mu$ the second locus is defined by $x^2 + y^2 = E(b + \mu x)$: both are circles whose synthetic construction follows readily from setting their equations in the form

$$(x - \tfrac{1}{2}E\lambda)^2 + (y - \tfrac{1}{2}E)^2 = aE + \tfrac{1}{4}E^2(\lambda^2 + 1)$$

and $(x - \tfrac{1}{2}E\mu)^2 + y^2 = Eb + \tfrac{1}{4}E^2\mu^2$ respectively.

(43) Newton lists eight necessary (but not always sufficient) conditions for a locus to be rectilinear.

(44) Pairs of roots of the n-degree equation defining the infinite points of an n-degree curve in direction may be conjugate complex. Thus, for example, the Cartesian folium $x^3 + y^3 = nxy$ has only 'duo crura infinita' (see I: 185).

(45) Without wholly coinciding with the locus, that is.

(46) Newton ignores the point-curves $ax^{2p} + by^{2p} = 0$ 'through' the origin $(0, 0)$.

[5.] Si a loci puncto quovis ad rectas duas positione datas in datis angulis demittantur aliæ duæ rectæ, et progrediendo per additionem subductionem et rationes datas, alterutra demissarum ex altera assumpta vel utracʒ ex assumpta tertia determinari potest rectus est locus.[47]

[6.] Si in recta quavis ad datam non infiniti cruris plagam tendente determinabile est loci punctū per simplicem Geometriā[48] rectus est locus.

[7.] Si rectæ per punctum datū extra locum transeuntis et loci intersectio determinabilis est per simplicem Geometriam, rectus est locus.[49]

[8.] Si rectæ cujusvis assignatæ et loci intersectio determinabilis est per simplicem Geometriam rectus est locus.[49]

---

[3][50]

1. A datis punctis *A*, *B* ductæ conveniant *ACBC* in *C* et si dentur ipsorum *A*[*C*], *B*[*C*] summa vel differentia (loc. *C* [erit] solid.).[51] proportio (loc. circ.)[52] differentia quadratorum (rect.)[53] summa quadratorum vel aliud quodvis compositum ex quadratis (circ).[54] rectangulum (lineare).[55] Area *ABC* (rect).[56] angulus *ACD* (circ).[57] differentia angulorum *B* − *A* vel summa 2*B* + *C* (Hyperb vel rect).[58] diff

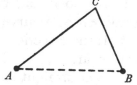

---

(47) If, in other words, on drawing *CA*, *CB* from the point *C* at given angles to two fixed straight lines *AA'*, *BB'* there is *a*.*AC* + *b*.*BC* = *c*, where *a*, *b*, *c* are constants, the locus (*C*) is a straight line (through the meet of the two lines given in position). For, on drawing *A'C*, *B'C* parallel to *BB'*, *AA'* respectively, the oblique Cartesian equation of the locus reduces to *a'*.*A'C* + *b'*.*B'C* = *c* since the ratios *A'C*/*AC* = *a*/*a'* and *B'C*/*BC* = *b*/*b'* are determined.

(48) Understand 'by a construction involving straight lines, directly or implicitly'. In a draft (ULC. Add. 3963.8: 64^r) of Problem 8 of 2, §2 below, Newton stressed that 'Per Geometriam simplicem determinabiles esse intelligo quæ per intersectiones solarum rectarum sine adminiculo circuli vel anguli dati hoc est per additionem subductionem & inventionem quartæ proportionalis'.

(49) This is not entirely true. For instance, the polar of an external point with respect to a given conic (already drawn) may be constructed merely by drawing straight lines through the point and connecting their intersections with the conic, so that the tangents from the point to the conic are readily determined 'per simplicem geometriam' to be through the intersection of the conic with its polar line.

(50) Newton classifies the types of locus which arise in three types of coordinate system (bipolars, 'focus-directrix' coordinates and a generalized Cartesian system) when the coordinate line-lengths are related in a variety of simple ways.

5. If from any point of a locus to two straight lines given in position two other straight lines be let fall at given angles, and through a succession of additions, subtractions and [multiplications by] given ratios either of the lines let fall can be determined from assuming the other, or if both are determinable in terms of some third line, then the locus is rectilinear.[47]

6. If in any straight line tending in a given direction, not that of an infinite branch, the point of a locus is determinable by simple geometry,[48] the locus is rectilinear.

7. If the intersection of a locus with a straight line passing through a given point outside the locus is determinable by simple geometry, the locus is rectilinear.[49]

8. If the intersection of a locus with any straight line assigned is determinable by simple geometry, the locus is rectilinear.[49]

---

[3][50]

1. Let lines AC, BC drawn from the given points A, B meet in C. Then if the sum or difference of AC and BC be given, the locus C is solid;[51] if their ratio be given, the locus is a circle;[52] if the difference of their squares, it is a straight line;[53] if the sum or any other compound of their squares, it is a circle;[54] if their product, the locus is linear;[55] if the area of [$\triangle$]ABC be given, it is a straight line;[56] if the angle $A\widehat{C}B$, it is a circle;[57] if the difference of the angles $B-A$ (or the sum $2B+C$), it is a hyperbola or a straight line;[58] if the difference $B-C$ or

---

(51) Namely, an ellipse or hyperbola. See, for example, 1: 29, note (15); 36, note (38); 295, note (89).

(52) See note (6) above.

(53) See note (11) above.

(54) See notes (12) and (13) above.

(55) The first recorded occurrence of Cassini's oval. More than a decade later Jacques Ozanam noted that 'Monsieur Cassini a inventé une nouvelle espece d'Ellipse, pour representer le mouvement des Planetes & de la Terre autour du Soleil. Cette Ellipse est une ligne du second genre, comme vous connoîtrez par sa description' (*Dictionaire Mathematique, ou Idée Generale des Mathematiques* (Paris, 1691): 436). Newton is not known to have explored the properties of this 'ligne du second genre' (in the Cartesian sense of a quartic oval) and he was doubtless as surprised as its 'second' inventor Cassini eventually to learn that this 'Ellipse' may on occasion split into two complementary ovals.

(56) Parallel to AB, namely, by Euclid, *Elements*, 1, 38.

(57) See note (4) above.

(58) This is 'Prob 35' of his contemporary Lucasian lectures on algebra (ULC. Dd. 9.68: 95–8 [= *Arithmetica Universalis* (Cambridge, ₁1707): 165–9]), read out, according to the none too trustworthy marginal assertion, in 'Lect 8' and 'Lect 9' of his 'Octob 1677' series. Newton's synthetic proof of this and the following 'Prob 36' will be reproduced in the next volume.

$B-C$ sive $C-B$ vel summa $2B+A$ vel $2C+A$ (lineare).[59] $2A=B$ vel dat. $3A+C$ (Hyperb.).[60] $2A=C$ (Lin). $2C=A$ (Lin). $A=B$ (rect).[61] $B=C$ (circ).[62]

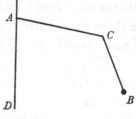

2. Detur AD positione et ang. DAC et punctū B. Et si dentur etiam ipsorum $AC\,BC$ summa, differentia (erit C in Parab).[63] Proportio (loc. sol.).[64] differentia quadratorum (Parab).[63] summa quadratorum vel aliud quodvis compositum ex quadratis (Loc solid). rectangulum (Lineare).[65]

3. Detur positione DA et punctum P. Et si detur AC (Locus [erit] Conchoid).[66]

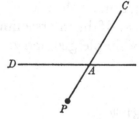

$$\left\{\begin{matrix}\text{[summa]}\\ \text{diff}\end{matrix}\right\} AC\pm AP\left(\text{loc}\left\{\begin{matrix}\text{circ.}\\ \text{linear.}\end{matrix}\right\}\right).\,^{(67)}$$

proportio PA ad AC (loc. rect.).[68] rectangulum APC (circ).[69] rectang PAC (lin.). rectang PCA (lin.). $AP^q\pm PC^q$. [vel] $AP^q\pm AC^q$. [vel] $PC^q\pm AC^q$ (lin.).[70]

---

(59) Newton first wrote 'Ellipsis vel lineare' (an ellipse or linear) here and then wisely changed his mind.

(60) 'Prob 36' of his contemporary professorial lectures on algebra (ULC. Dd. 9.68: 98–9 [=*Arithmetica Universalis* ($_1$1707): 169–71]). Compare note (59) above.

(61) Namely (since then $AC = BC$) the perpendicular bisector of AB.

(62) That is (since here $AC = AB$ is constant) of centre A. In general, if C is the general point (x, y) in the rectangular Cartesian coordinate system in which A, B are the fixed points $(-a, 0)$, $(a, 0)$, then $AB = 2a$, $BC = \sqrt{[(x-a)^2+y^2]}$, $AC = \sqrt{[(x+a)^2+y^2]}$, $\tan A = y/(a+x)$, $\tan B = y/(a-x)$ and $\tan C = 2ay/(x^2+y^2-a^2)$ and the defining equation of any locus containing these six elements is readily obtained: namely, if $AC = 2k \pm BC$, then the locus is the central conic $(k^2-a^2)x^2+k^2y^2+4ak^2x+k^2(a^2-k^2) = 0$, if $AC = k.BC$, then the locus is the circle $(k^2-1)(x^2+y^2)-2a(k^2+1)x-a^2 = 0$; if $AC \times BC = k^2$, then we have the Cassini oval $(x^2+y^2)^2+2a^2(-x^2+y^2)+a^4-k^4 = 0$; if $B-A = k = \tan^{-1}l$ or

$$\tan B - \tan A = l(1+\tan B\tan A),$$

there results the hyperbola $l(x^2-y^2)+2xy-la^2 = 0$; if $2A = B$ or $2\tan A = (1-\tan^2 A)\tan B$, then we have the pair of the line AB ($y = 0$) and the hyperbola $3x^2-y^2+2ax-2a^2 = 0$; and, lastly, if $B-C = \tan^{-1}m$ or $\tan B - \tan C = m(1+\tan B\tan C)$, there results the cubic

$$(lx+y)(x^2+y^2)-alx^2+2axy+aly^2-a^2lx-3a^2y+a^3l = 0,$$

which reduces to the pair of the line AB ($y = 0$) and the circle $(x+a)^2+y^2 = (2a)^2$ when ($m = 0$ or) $B = C$.

(63) Only when $D\hat{A}C$ is right; in general the locus is a hyperbola.

(64) Since CA is proportional to the perpendicular from C to DA, this is effectively the standard focus-directrix definition of the general conic. Compare note (38) above.

(65) As we have seen (III: 132–6) Newton had earlier introduced this general 'focus-directrix' system (in which the locus (C) is defined with regard to the coordinates AC, BC as

$C - B$ (or the sum $2B + A$ or $2C + A$), it is linear;[59] if it be given that $2A = B$ (or that $3A + C$ is given), the locus is a hyperbola;[60] if that $2A = C$, it is linear; if that $2C = A$, it is linear; if that $A = B$, it is a straight line;[61] if that $B = C$, it is a circle.[62]

2. Let there be given AD in position, the angle $D\widehat{A}C$ and the point B. Then if there also be given the sum or difference of AC and BC, C will be on a parabola;[63] if their ratio, the locus is solid;[64] if the difference of their squares, it is a parabola;[63] if the sum—or any other compound—of their squares, the locus is solid; if their product, it is linear.[65]

3. Let there be given DA in position and the point P. Then if AC be given, the locus will be a conchoid;[66] if the sum or difference $AC \pm AP$, the locus will be a circle or linear;[67] if the ratio of PA to AC be given, the locus is a straight line;[68] if the product $AP \times PC$, it is a circle;[69] if the product $PA \times AC$ it is linear; if the product $PC \times CA$, it is linear, if $AP^2 \pm PC^2$ or $AP^2 \pm AC^2$ or $PC^2 \pm AC^2$, it is linear.[70]

---

the particular case of the preceding bipolar system when one of the poles is at infinity) as 'Mod: 2' of Problem 4 of his 1671 tract. We may reduce to a standard rectangular Cartesian system in which C is the general point (x, y) by setting B as the origin $(0, 0)$ and the x-axis as parallel to the line AC given in direction, in which case $BC = \sqrt{[x^2 + y^2]}$ and $AC = a - x + by$, where a is the oblique distance, parallel to CA, of B from DA and $b = \cot D\widehat{A}C$. Hence, for example, when $AC \pm BC = a + k$, the locus is ($b \neq 0$) the hyperbola

$$(b^2 - 1)y^2 - 2bxy + 2k(x - by) + k^2 = 0$$

or ($b = 0$) the parabola $y^2 = 2k(x - \tfrac{1}{2}k)$; and when $\lambda \cdot AC^2 + \mu \cdot BC^2 = k^2$, it is the general conic $\lambda(x - by - a)^2 + \mu(x^2 + y^2) = k^2$—in particular, for $\lambda = -\mu = 1$ ($b \neq 0$) the hyperbola

$$(b^2 - 1)y^2 - 2bxy - 2a(x - by) + a^2 - k^2 = 0$$

or ($b = 0$) the parabola $y^2 = 2a(\tfrac{1}{2}a - \tfrac{1}{2}k^2/a - x)$.

(66) Strictly, as Newton draws his figure, the 'prima Conchoïdes Veterum' or upper 'shell' of the full conchoid. Compare 1: 502–4.

(67) When $(AC + PA$ or$)$ $PC = k$ it is obvious that the locus is a circle of centre P but it is not evident that when $AC - PA = k$ the locus is of fourth degree.

(68) Parallel to DA, namely.

(69) The inverse (through P) of the line DA which passes through points B, B' in DA where $PB^2 = PB'^2 = PA \times PC$. Compare note (7) above.

(70) If C is the general point (x, y) of a rectangular Cartesian coordinate system in which the foot of the normal from P to DA is the origin, distant b from P, then

$$PA = b\sqrt{[x^2 + (y + b)^2]}/(y + b) \quad \text{and} \quad AC = y\sqrt{[x^2 + (y + b)^2]}/(y + b).$$

Thus, when $AC = k$, the locus is the conchoid $x^2 y^2 = (k^2 - y^2)(y + b)^2$; when $AC - PA = k$, the locus is the quartic $x^2 = (y + b)^2(k^2/(y - b)^2 - 1)$; and when $PA(PA + AC) = k^2$, the locus is the circle $x^2 + y^2 + (2b - k^2/b)y + b^2 - k^2 = 0$ through $P(0, -b)$. All Newton's present 'linear' loci are quartics. The coordinate system is akin to that introduced in 'Mod: 4' of Problem 4 of the 1671 tract (III: 138–40), where PA and PC are chosen as the coordinates.

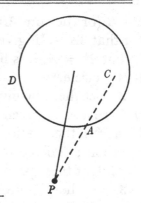

4. Detur circ AD, punctum P et si detur etiam APC vel $\dfrac{AP}{AC}$ (erit C in loc. Plan.).[71]

§2.[1] THE 'GEOMETRICAL SOLUTION OF PROBLEMS'.

[1]   QUÆSTIONUM SOLUTIO GEOMETRICA.

1. Datis trianguli cujusvis angulo latere et summa vel differentia reliquorum laterum[,] datur triangulum. Detur latus AB, reliquorum laterum AC & BC summa vel differentia AD. Si angulus datus dato lateri conterminus est, sit iste A. Et dabitur triangulum DAB. Angulorum vero CDB, ABD differentia in priori casu, summa in posteriori est ang ABC.

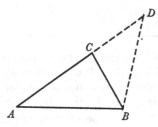

Sin angulus datus dato lateri op[p]onitur, sit iste C et dabitur triangulum CDB specie.[2] In triangulo autem ADB datis lateribus AB AD et ang D datur ang ABD. Unde datur Ang ABC ut ante.

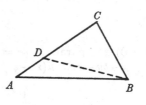

2. Data differentia segmentorum basis[,] summa vel differentia laterum et uno angulorum datur triangulum. Nam si datur summa laterum dabitur ratio differentiæ laterum ad basin[,] si differentia dabitur ratio summæ laterum ad basin.[3] Ex ratione utravis & uno angulorum[4] per problema superius datur triangulū specie et ex data ratione differentiæ segmentorum basis ad latera dantur latera.

3. Data summa vel differentia laterum[,] uno angulorum et ratione basis ad perpendiculum: ex duobus posterioribus dabitur triangulū specie[,] ex priori dabitur etiam magnitudine.

4. Data summa vel differentia laterum[,] uno angulorum et area: ex area datur rectangulum laterum datum angulum comprehendentiū.[5] Si istorum

---

(71) The generalization of the preceding coordinate system where the line DA (a 'circle of infinite radius') is replaced by a given circle: since PA meets the circle in two points A, the correspondence between A and C is (2, 1) and the greatest care is needed when cases other than that of inversion ($PA \times PC = $ constant; see note (9) above) and homothety ($PA:PC = $ constant; see note (8) above) are considered.

4. Let the circle AD and the point P be given. Then if there be given also $AP \times PC$ or AP/AC, C will be on a plane locus.[71]

Translation

[1]   THE GEOMETRICAL SOLUTION OF QUESTIONS

1. In any triangle given an angle, a side and the sum or difference of the remaining sides, the triangle is given. Let there be given the side AB and the sum or difference AD of the remaining sides AC and BC. If the given angle adjoins the given side, let it be A and the triangle DAB will be given. The angle $A\widehat{B}C$ is, indeed, the difference of the angles $C\widehat{D}B$, $A\widehat{B}D$ in the former case, but their sum in the latter.

If, however, the given angle is opposite the given side, let it be C and the triangle CDB will be given in species.[2] Now in the triangle ADB given the sides AB, AD and the angle D, the angle $A\widehat{B}D$ is given. Hence, as before, the angle $A\widehat{B}C$ is given.

2. Given the difference of the segments of the base, the sum or difference of the sides and one of the angles, a triangle is given. For if the sum of the sides is given, there will be given the ratio of the difference of the sides to the base; if the difference, there will be given the ratio of the sum of the sides to the base.[3] From either ratio and one of the angles[4] the triangle is, by the previous problem, given in species, and from the given ratio of the difference of the base segments to the sides there are given the sides.

3. Given the sum or difference of the sides, one of the angles and the ratio of the base to the perpendicular: from the two latter the triangle will be given in species, while by the first it will also be given in magnitude.

4. Given the sum or difference of the sides, one of the angles and the area: from the area is given the product of the sides enclosing the given angle.[5] If

---

(1) ULC. Add. 4004: 94^v – 96^v, three sets of geometrical 'questions', each pursuing a distinct theme, together with a final miscellaneous gathering of locus problems.

(2) Newton first wrote equivalently 'et dabuntur ang[uli] *CDB CBD*' (and the angles $C\widehat{D}B$, $C\widehat{B}D$ will be given). Evidently, since the triangle BCD is isosceles, there is

$$C\widehat{B}D = C\widehat{D}B = \tfrac{1}{2}(\pi - \widehat{C})$$

and hence BC (or CD) to BD is a given ratio.

(3) The difference of the squares of the base segments equals that of the squares of the sides. Compare §1: note (11).

(4) A first cancelled opening to the sentence reads 'In utroq̃ casu' (In either case).

(5) Namely, $AC \times BC = 2\triangle ABC / \sin \widehat{C}$.

summa vel diff. datur a quadrato summæ aufer duplum[6] rectangulū vel ad quadratum differentiæ laterū adde duplū[6] rectangulum et habebitur priori casu quadratū differentiæ, posteriori quadratū summæ laterum. Ex datis autem summa ac differentia laterum dantur latera. Si angulus datus basi conterminus est problema erit solidum.[7]

5. Datis angulo A latere AC vel BC et differentia segmentorum basis AD,[8] dabitur triangulum ADC ut et angulus B quod est complementū[9] anguli ADC.

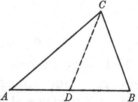

6. Si detur angulus verticalis C laterum alterutra AC vel BC[10] et segment[ū] basis $A[D]$: quiescant BC, AC et punctum D in circulo erit radio CB centro C descripto. ut et in Chonchoide Polo B, asymptoto AC intervallo AD descripta.[11]

Vel sic. Dato angulo ACB datur summa ang: $A+B$. seu $A+CDB$. Aufer hoc de duobus rectis ac dabitur differentia ang $ADC-A$. Unde datur triang. per sequ. Prob.[12]

7. Datis basi & differentia angulorum ad basin una cum latere alterutro vel summa differentia ratione laterum aut summa vel differentia □[i] laterum aut area, perpendiculo vel segmento basis aut summa vel differentia lateris alterutrius et perpendiculi vel segmenti basis. &c Datur triang. Nam data basi et angulorum ad Basem differentia, punctū C erit ad Hyperbolam;[13] et ex dato tertio, punctum C erit ad rectā aut circulū aut conicam aliquā sectionem.

8. Ubi datur angulus verticalis et differentia segmentorum basis[14] et

---

(6) Read 'quadruplum' (four times) correctly.

(7) Constructible, that is, by the meets of intersecting 'solid loci' (conics) and so of third or fourth degree. In fact, on taking E to be the foot of the perpendicular from C onto AB we may set $AE = x$, $EB = y$ and hence $BC = \lambda x$, $CE = \mu x$ where $\lambda = \sec \hat{B}$, $\mu = \tan \hat{B}$: the defining conditions $AC \pm BC = l$ and $AB \times CE = m^2$ then become

$$\sqrt{[x^2+\mu^2 y^2]} \pm \lambda y = l \quad \text{and} \quad (x+y)\mu y = m^2$$

and the point C may therefore be determined by the intersections of the hyperbolas

$$x^2 - y^2 \pm 2l\lambda y - l^2 = 0 \quad \text{and} \quad xy + y^2 - m^2/\mu = 0.$$

(8) Understand that $CB = CD$ in the accompanying figure, so that $AD = AE - EB$ where CE is the normal from C onto AB. In consequence $C\hat{D}B = C\hat{B}D$.

(9) 'ad duos rectos', that is.

(10) Understand 'puta BC' (BC say) to smooth the way into Newton's further argument.

(11) There will be four solutions (two of which may not be real) since, as Newton at once proceeds to show, construction of the meets of a conchoid with a circle through its pole is a 'solid' problem. A first conclusion to the present paragraph reads 'ut et $\dfrac{AC^q - BC^q}{AB} = AD$, ergo D determinatur per Geometriam simplicem [dato A]. Sed et ubi punctum D incidit in B positio rectæ AB determinatur. Ergo locus puncti D conica est sectio' (and also

$$(AC^2 - BC^2)/AB = AD,$$

the sum or difference of these is given, from the square of their sum take away twice[6] their product or to the square of the difference of the sides add twice[6] their product, and there will be obtained in the former case the square of the difference of the sides, in the latter the square of their sum. From the given sum and difference of the sides, however, the sides are given. If the given angle is adjacent to the base, the problem will be solid.[7]

5. Given the angle A, the side AC or BC and the difference AD of the segments of the base,[8] there will be given the triangle ADC along with the angle B which is the supplement of the angle $A\widehat{D}C$.

6. If there be given the vertex angle C, either of the sides AC or BC[10] and the base segment AD, let BC, AC be stationary and the point D will then be on a circle described with centre C and radius CB, and also on a conchoid described with pole B, asymptote AC and interval AD.[11]

Or thus. Given the angle $A\widehat{C}B$, there is given the sum of the angles $A+B$, that is, of $A+C\widehat{D}B$. Take this from two right angles and the angle difference $A\widehat{D}C-A$ will be given. The triangle is consequently given by the following problem.[12]

7. Given the base and difference of the angles at the base, together with one or other of the sides, or the sum, difference or ratio of the sides, or the sum or difference of the squares of the sides, or the area, perpendicular or a base segment, or the sum or difference of either side and the perpendicular or a base segment, and so on, then the triangle is given. For given the base and the difference of the angles at the base, the point C will be on a hyperbola;[13] while, depending on the third given element, the point C will be on a straight line, circle or some conic.

8. When there is given the vertex angle, the difference of the segments of the

---

so that [given A] D is determined by simple geometry. But also when the point D coincides with B the position of the straight line AB is determined. Consequently the locus of the point D is a conic section). This is true as far as it goes, but in fact the locus of D such that

$$AB \times AD = AC^2 - BC^2$$

determines that $CD = CB$ and so (D) is the same circle (on centre C and through B) as before! (More accurately, as analysis will show, the locus (D) is this circle together with the tangent at B to the circle—the line determined by Newton to be given in position.)

(12) Namely, by the (four) intersections of the hyperbola (D) determined (see next note) by the condition that D is in AB such that $C\widehat{D}A - \widehat{A}$ is constant, and of the circle (D) on centre C which passes through B.

(13) See §1: notes (59) and (63). The following statement evidently refers back for its proof likewise to the opening paragraph of §1.3 above.

(14) That is, given $\widehat{C}$ and AD in the preceding figure. In this generalization of the *idem aliter* of Problem 6 the locus of (D) is again a hyperbola (compare note (12) above) and the triangle ADC is determined by the 'tertium aliquod' which Newton does not specify.

tertium aliquod, habebitur aliud triangulum *ADC* ubi datur basis et differentia angulorum ad basem, et tertium aliquod.

9. Data basi ratione laterum et tertio quovis ut ⊥°,[15] segmento basis, angulo aliquo, ratione ⊥¹[15] ad latus vel ad segmentum Basis &c. datur triangulum. Nam ex data basi et rat. lat.[16] Datur circulus in quo vertex est.

[**2**][17]   QUÆSTIONUM SOLUTIO GEOMETRICA.

Prob 1.

Circulum *ABE* per data duo puncta *A*, *B* describere qu[i] rectam *FG* positione datam continget.

[18][Puta factum.] Sit *E* punctum contactus. Produc *AB* donec occurrat *FG* in *G* et erit *EG* medium proportionale inter datos *AG*, *BG*.[19]

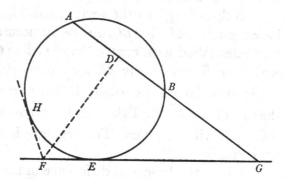

Prob 2.

Circulum *ABE* per datum punctum *A* describere qui rectas duas *FE*, *FH* continget.

[20]Recta *FD* biseca angulum *HFE*. Ad *FD* demitte normalem *AD* et produc ad *B* ut sit *DB*=*AD* et per puncta *A*, *B* describe circulum ut prius qui contingat rectam *FE*.[21]

---

(15) Read 'perpendiculo' and 'perpendiculi' respectively.

(16) Read 'ratione laterum'. For this Apollonian 'plane' locus see §1: note (6).

(17) Four problems whose solutions yield particular cases of Apollonius' general problem of 'contacts', given a Euclidean solution by him in the two books of his lost treatise on the topic. In Book VII of his *Mathematical Collection* Pappus enunciated it in most general form as: 'Given any three elements, either points, straight lines or circles, to describe a circle which, passing through each of the given points (when there are points given) shall touch each of the lines or circles' (compare P. Ver Eecke's *Pappus d'Alexandrie: La Collection Mathématique*, **2** (Bruges, 1933): 483). On the subsequent history of the problem see Ver Eecke's *Pappus*, **1**: lxvi-lxix. In 1600, in particular, François Viète published at Paris a celebrated restoration of Apollonius' lost work, his *Apollonius Gallus, Seu, Exsuscitata Apollonii Pergæi ΠΕΡῚ 'ΕΠΑΦΩΝ Geometria* [= *Opera Mathematica* (Leyden. 1646): 325–46]. Newton in his undergraduate annotations on Viète's mathematical works (see 1: 63–88) had made no notes on the latter's Apollonius restoration but it is not unlikely that he read it and that memory of Viète's tract lingered in his mind, to be recalled when a fresh reading of Pappus' *Mathematical Collection* brought the topic again to his attention. Whether Newton also at this time knew Fermat's generalization of Viète's solution to the four spheres' tangency problem in his *De Contactibus Sphæricis* [= *Varia Opera Mathematica D. Petri de Fermat* (Toulouse, 1679): 74–88] is not known.

base[14] and some third element, another triangle *ADC* will be obtained in which there is given the base, the difference of the angles at the base and some third element.

9. Given the base, the ratio of the sides and any third element (such as the perpendicular, a segment of the base, some angle, the ratio of the perpendicular to a side or base segment, and so on), the triangle is given. For from the base and ratio of the sides given[16] there is given the circle in which the vertex is situated.

---

[2][17]   THE GEOMETRICAL SOLUTION OF QUESTIONS

Problem 1.

To describe a circle *ABE* through two given points *A, B* which shall touch the straight line *FG* given in position.

[18][Suppose it done.] Let *E* be the point of contact. Produce *AB* till it meets *FG* in *G*, and then *EG* will be the mean proportional between *AG, BG* given.[19]

Problem 2.

To describe a circle *ABE* through the given point *A* which shall touch two straight lines *FE, FH*.

[20]By the straight line *FD* bisect the angle *HFE*. To *FD* let fall the normal *AD* and extend it to *B* so that *DB* = *AD* and through the points *A* and *B* describe, as before, a circle which shall touch the straight line *FE*.[21]

---

(18) Newton first began: 'Junge *A, B*. Biseca eam in *D*. Erige normalem *DF* occurrentem *FG* in *F*. Produc *AB* donec occurrat *FG* in [*G*]' (Join *A, B*. Bisect it in *D*. Raise the normal *DF* meeting *FG* in *F*. Extend *AB* till it meets *FG* in *G*). To fill the hiatus left by its omission we have, in line with 'Prob 3' below, inserted a bridging phrase.

(19) Essentially Viète's 'Problema II. Datis duobus punctis, & linea recta, per data puncta circulum describere, quem data linea recta contingat' (*Opera* (note (17) above): 326–7). Since the point *E* may lie in *FG* on either side of *G* there will be two solutions. A first cancelled state of the accompanying figure has lines drawn from *B* and *E* to the circle's centre *C* (in *FD*) but omits the line *FH*, exactly as in that illustrating an equivalent analytical proof of the construction (ULC. Dd. 9.68: 99–100 [= *Arithmetica Universalis* (Cambridge, ₁1707): 171–2], to be reproduced in the next volume) which Newton a little earlier inserted as 'Prob 37' of his contemporary professorial lectures on algebra and subsequently claimed to have given out in the last of his 'Octob. 1677' series.

(20) A cancelled opening began equivalently 'Centro *F*, radio *FA* describe [arcum qui secet rectam *FD* angulum *HFE* bisecantem in *B*?]'.

(21) The construction, all but self-evident, is that given earlier by Viète in his 'Problema IV. Datis duabus lineis, & puncto, per datum punctum circulum describere, quem datæ duæ lineæ rectæ contingant' (*Opera* (note (17) above): 328). Here again a cancelled accompanying diagram makes it clear that Newton is indebted to an earlier analytical proof of the construction (ULC. Dd. 9.68: 100–1 [= *Arithmetica Universalis* (note (19) above): 173–4]) which he subsequently claimed to have given out in the opening lecture of his 'Octob. 1678' series on algebra as 'Prob 38'.

Prob 3.

Circulum *ABE* per data duo puncta *A*, *B* describere qui alium circulum positione datum *EKL* continget.

Puta factum.[22] Sit punctum contactus *E*. Linea contingens *EM*. et erit $AM \times BM = EM^q = MK \times ML$. Divide ergo *BK* in *M* ut sit

$$AM . MK :: ML . MB.^{[23]}$$

Cape *ME* medium proportionale inter *AM* et *BM* et centro *M* radio *ME* describe circulum. Hic secabit

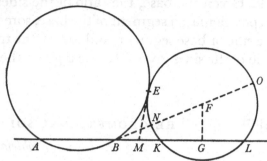

circulum *EKL* in puncto contactus *E*.[24] Recta autem *BK* sic secatur in *M*. Est $AM = AB + MB$. $MK = BK - MB$. $ML = BL - MB$.[25] Ergo

$$AB + MB . BK - MB :: BL - MB . MB.$$

et componendo $AB + BK(AK) . BK - MB :: BL . MB$. et inverse $AK . BL :: BK - MB . MB$. et rursus componendo $AK + BL . BL :: BK . MB$. [seu] $2AG^{[26]} . BL :: BK . MB$. Unde cum sit $BL . BO :: BN . BK$, erit $2AG^{[26]} . BO :: BN . BM$. Quæ solutio universalis est.[27]

Prob 4.

Circulum *BDE* per datum punctum *B* describere qui datum circulū *E* & rectam lineam *AD* pos[i]tione datam continget.

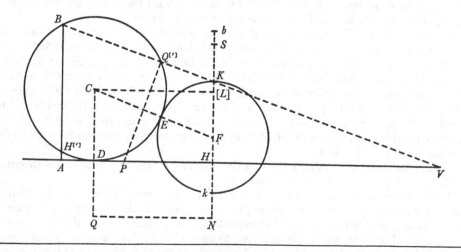

---

(22) We have trivially corrected the manuscript figure, where by mistake Newton has made *NFO* pass through *M* and not *B* (as it should do).

Problem 3.

To describe a circle *ABE* through two given points *A*, *B* which shall touch another circle *EKL* given in position.

Suppose it done.[22] Let the point of contact be *E*, the tangent line *EM*. Then will there be $AM \times BM = EM^2 = MK \times ML$. So divide *BK* at *M* so that

$$AM : MK = ML : MB,^{[23]}$$

take the mean proportional *ME* between *AM* and *BM*, and with centre *M* and radius *ME* describe a circle: this will intersect the circle *EKL* in the point *E* of contact.[24] The straight line *BK*, however, is cut at *M* in this manner. There is $AM = AB + MB$, $MK = BK - MB$, $ML = BL - MB$[25] and therefore

$$AB + MB : BK - MB = BL - MB : MB,$$

so that by compounding $AB + BK$ (or $AK) : BK - MB = BL : MB$ and inversely $AK : BL = BK - MB : MB$ and again *componendo* $AK + BL : BL = BK : MB$, that is, $2AG^{[26]} : BL = BK : MB$. Hence, since $BL : BO = BN : BK$, there will be $2AG^{[26]} : BO = BN : BM$. This solution is universal.[27]

Problem 4.

To describe a circle *BDE* through the given point *B* which shall touch both a given circle *E* and the straight line *AD* given in position.

---

(23) See below for this construction, which, as Newton must have known from reading the opening to Book VII of Pappus' *Mathematical Collection*, was discussed at length by Apollonius in his lost treatise on *Determinate Section*.

(24) It remains to construct the centre of the required circle as the meet of *FE* with the perpendicular bisector of *AB*.

(25) Newton takes the line-segment *BM* as independent variable.

(26) Read '2*A'G*' where *A'* is the mid-point of *AB*, since $AK + BL = AG + BG$.

(27) Including, that is, the case where the given circle does not intersect *AB* in real points *K*, *L*. In a cancelled preceding sentence Newton singled this out for special attention: 'Si *AB* non secat *EKL* pro $MK \times ML$ scribe $MG^q + FG^q - KF^q$. erit

$$AG - MG \times BG - MG (AM \times BM) = MG^q + FG^q - KF^q.$$

Hoc est $AG \times BG - FG^q + KF^q - \overline{AG + BG} \times MG[= 0]$', whence it follows that

$$MG = (AG \times BG + KG^2)/(AG + BG).$$

Much as before, an earlier analytical discussion of this problem was inserted as 'Prob 39' of his professorial lectures on algebra and asserted to have been given out in 'Lect 2' in October 1678 (ULC. Dd. 9.68: 101–2 [= *Arithmetica* (note (19) above): 174–6]).

[Puta factum.] AB est $2CD - AH^{[r]}$. NF est $2CQ - NS$ posito $CQ = CF = CS$.[28] [ideoꝗ] $AB - NF$ est $2CD - AH^{[r]} - 2CQ + NS$. Adde $2DQ$, erit

$$AB + DQ - HF = NS - AH^{[r]} = \frac{NQ^q}{NF} - \frac{AD^q}{AB}.^{[29]}$$

Dividenda est itaꝗ data AH in D ita ut $\dfrac{DH^q}{NF} - \dfrac{AD^q}{AB}$ dato æquale sit, nempe dato $AB(Hb) + DQ - HF$, seu dato bk. [Est autem]

$$DH = AH - AD. \quad DH^q = AH^q - 2DAH + AD^q.$$

[adeoꝗ] $AH^q - 2DAH + AD^q - \dfrac{HK \times AD^q}{AB} = HK, bk.$ [sive]

$$AH - 2DA + \frac{AB - HK}{BAH} AD^q = \frac{HK, bk}{AH}.$$

Fac $AB - HK \cdot AH :: BA \cdot AV$. & $AH \cdot HK :: bk \cdot HP$. [Erit]

$$AH - 2DA + \frac{AD^q}{AV} = HP.$$

[vel] $AD^q - 2DAV + AV^q = AV \times PV$. [hoc est] $-AD + AV = AV * PV^{[30]} = DV$. Age ergo BK occurrentem AH in V. In HA versus A si HK jacet versus b, aliter adversus A, cape HP ad bk [ut HK ad] AH, et $[V]D$ medium proport[io]nale inte[r VA et VP. Tum per punctum B describe circulum qui contingat rec]tam AH in D.

Nota etiam quod Problematis quatuor sunt casus[31] quorum duo sunt impossibiles ubi circulus datus et recta data se mutuò secant. Casus impossibiles sunt ubi punctum V cadit inter A et P.[32]

---

(28) Newton's figure has two points H and two points Q. To avoid confusion these are distinguished as H, $H^{[r]}$ and Q, $Q^{[r]}$ in the above text.

(29) For, where CL is the perpendicular from C to FK, there is

$$NQ^2 = CS^2 \text{ (or } CQ^2 = LN^2) - SL^2 = NS \times NF.$$

while $AD^2 = AH^{[r]} \times AB$. In the sequel Newton chooses the line-segment AD as his independent variable.

(30) Newton's '$*$' is evidently a further variant on Oughtred's 'm' (for 'medium proportionale'); compare 1, 3, §1: note (6) above. As Newton indicates in his figure the second meet $Q^{[r]}$ of BV with the circle to be constructed is the foot of the perpendicular from P to BV, for, since $AV \times PV = DV^2 = BV \times Q^{[r]}V$, the triangles BAV, $PQ^{[r]}V$ are similar.

(31) This will be intuitively obvious. In more precise terms each of the intersections K, k of HF with the given circle yield corresponding points P, p in AH defined by

$$HP = HK \times bk/AH \quad \text{and} \quad Hp = Hk \times bK/AH$$

Suppose it done. AB is $2CD - AH'$, NF is $2CQ - NS$ on putting

$$CQ = CF = CS,^{(28)}$$

and consequently $AB - NF$ is $2CD - AH' - 2CQ + NS$. Add $2DQ$ and then $AB + DQ - HF = NS - AH' = NQ^2/NF - AD^2/AB$.[29] Accordingly, the given line AH must be divided at D so that $DH^2/NF - AD^2/AB$ is equal to a given quantity, namely AB (or Hb) $+ DQ - HF$ or the given magnitude bk. However,

$$DH^2 = (AH - AD)^2 = AH^2 - 2AD \times AH + AD^2$$

and so $AH^2 - 2AD \times AH + AD^2 - (HK/AB)\,AD^2 = HK \times bk$, that is,

$$AH - 2AD + (AB - HK) \times AD^2/AB \times AH = HK \times bk/AH.$$

Make $AB - HK : AH = AB : AV$ and $AH : HK = bk : HP$, and there follows

$$AH - 2AD + AD^2/AV = HP \quad \text{or} \quad AD^2 - 2AD \times AV + AV^2 = AV \times PV,$$

that is, $(-AD + AV$ or$)$ $DV = \sqrt{[AV \times PV]}$.[30] Therefore draw BK meeting AH in V, and in AH (in the direction of A if HK lies towards b, otherwise the opposite way) take HP to bk as HK to AH, and DV the mean proportional between AV and PV. Then through the point B describe a circle which shall touch the straight line AH in D.

Note also that there are four cases of the problem,[31] two of which are impossible when the given circle and given straight line intersect one another. The impossible cases are when the point V falls between A and P.[32]

---

and so points $Q^{[\prime]}$, q in BK, Bk respectively through which the tangent circle must pass, while (by Problem 1) each point $Q^{[\prime]}$, q gives two solutions to the reduced problem of drawing a circle through B and $Q^{[\prime]}$ (or q) to touch the given line AH. On taking v to be the meet of Bk with AH, it will be evident that each pair of points P, V and p, v defines the four possible points of contact D, D' and d, d' by $AD^2 = AD'^2 = AV \times PV$, $Ad^2 = Ad'^2 = Ap \times Av$.

(32) Newton continues to assume that the point B is outside the given circle. As a result in the 'impossible' cases (when the given circle intersects AH and so the points B and q through which (see note (31)) the required circle must pass lie one within and one outside it) we have AH^2 (or Bb^2) $> kb \times bK$ (or $HK - AB$), whence $HK \times bk/AH (= Hp) < HK \times AH/(HK - AB)$ and so $Ap > AH \times AB/(HK - AB) = Av$, that is, v must lie between A and p (since it is clear that both v and p must lie in AH). As with the preceding problems Newton had earlier presented an equivalent analytical construction of the present 'Prob 4' in his contemporary Lucasian lectures on algebra (ULC. Dd. 9.68: 102–4: Problem 40 [= *Arithmetica* (note (19)): 176–9]), purportedly as 'Lect 3' of his 'Octob. 1678' series, but without touching on the number and possibility of particular cases constructible.

[3][33]   QUÆSTIONUM SOLUTIO GEOMETRICA.

1. Angulum datum DAB recta datæ longitudinis CB subtendere quæ ad datum punctum P converget.[34]

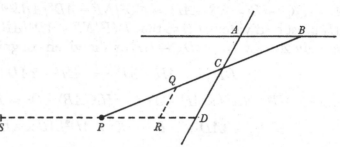

Cape $PQ = C[B]$ et Q erit in circulo cujus centrum P radius PQ. Age $QR \| AD$ et $PR \| AB$ et erit $PD . DC = AD - QR :: PR . QR$. Ergo Q in conica sectione est.[35] Pone QR infinitū et erit $AD - QR . QR :: PD . PR$[36] seu $PR = -PD$. Pone PR infinitū et erit[37] $PD + PR . PR :: AD . QR$. ergo $AD = QR$. et AB Asymptotos. Cape ergo $PS = PD$ et per S parallelam AD age alteram Asymptoton & his Asymptotis per punctū P describe Hyperbolam secantem circulum prædictū in Q.[38]

2. Inter circulum PDF et rectam DF ponere rectam[39] *datæ longitudinis BC quæ ad punctum P in circumferentia circuli datum converget.*

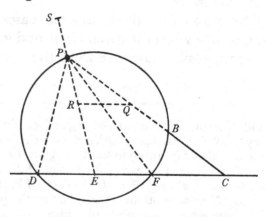

Biseca DF in E. Age PD, PE, PF. Cape $PQ = BC$. Age $QR \| DC$ & occurrentem PE in R. Et erit

$$PR . PQ :: PE . PC.$$

$$PR . PE :: PQ(BC) . PC :: RQ . EC.$$[40]

Et

$$BC . \frac{PE \times RQ}{PR} - EF(FC) ::$$

$$\frac{PE \times RQ}{PR} + EF(DC) . \frac{PE \times BC}{PR} (PC).$$

---

(33) Geometrical analyses of the two simplest cases (where one curve is a straight line and the other a straight line or circle) of the classical 'verging' problem: 'two curves being given in position, to set between them a straight line of given magnitude which shall be inclined through a given point' as Pappus enunciated it in his summary of the content of the two books of Apollonius' lost treatise *On Inclinations* in Book VII of his own *Mathematical Collection* (see Ver Eecke's *Pappus* (note (17) above), **2**: 501). We have earlier seen (1: 509, note (9)) that some time before May 1665 Newton had come upon the simplest instance (where $D\hat{A}B$ is right) of the first problem below in the third book of the 1659 Latin edition of Descartes' *Geometrie*, an example drawn ultimately from two of Pappus' lemmas (*Mathematical Collection*, VII, 71/72) on Apollonius' lost work; and that a little later (see II: 456–78, 508–12) he had reduced the solution of cubic and quartic algebraic equations to the construction, in geometrical equivalent, of several related types of straight line and circle vergings. In context it will be clear

[3]^(33) THE GEOMETRICAL SOLUTION OF QUESTIONS

1. In a given angle $D\widehat{A}B$ to place a straight line CB of given length which shall be inclined through the given point P.^(34)

Take $PQ = CB$ and Q will be in a circle of centre P and radius PQ. Draw $QR \parallel AD$ and $PR \parallel AB$, and there will be $PD:DC$ (or $AD-QR$) $= PR:QR$. Therefore Q is in a conic.^(35) Suppose QR to be infinite and then

$$AD - QR : QR = PD : PR,^{(36)}$$

that is, $PR = -PD$. Suppose PR to be infinite and then^(37)

$$PD + PR : PR = AD : QR,$$

so that $QR = AD$ and AB is an asymptote. Therefore take $PS = PD$ and through S parallel to AD draw the second asymptote, and with these asymptotes through the point P describe a hyperbola cutting the above-mentioned circle in Q.^(38)

2. Between the circle PDF and the straight line DF to set a straight line^(39) *BC of given length which shall incline towards the given point P in the circumference of the circle.*

Bisect DF in E, draw PD, PE, PF, take $PQ = BC$ and draw $QR \parallel DC$ meeting PE in R. Then $PR:PQ = PE:PC$, $PR:PE = PQ$ (or BC):$PC = RQ:EC$,^(40) and

$$BC:PE \times RQ/PR - EF \text{ (or } FC)$$

$$= PE \times RQ/PR + EF \text{ (or } DC):PE \times BC/PR \text{ (or } PC),$$

---

that this previous interest was now refreshed by Newton's study for the first time of Pappus' original text in one or other of the Latin editions available to him.

(34) Pappus' first general verging problem: 'given a parallelogram, one of whose sides is extended, to adjust a straight line given in magnitude set in the exterior angle so that it is inclined through the opposite corner' (compare Ver Eecke's *Pappus* (note (17) above), **2**:502).

(35) For PR, RQ are oblique 'Cartesian' coordinates, referred to P as origin, of the locus (Q) defined by $PD \times RQ = PR(DA - RQ)$ or $(PR + PD)(RQ - DA) = -PD \times DA$. As Newton goes on to show this is a hyperbola of asymptotes DA and the line $(PR = -PD)$ through S parallel to DA.

(36) The equivalent equation '$PR = \dfrac{PD \times QR}{AD - QR}$' is cancelled.

(37) Understand 'componendo' (by composition).

(38) As drawn, Newton's figure yields two possible solutions, the second (not shown) corresponding to a position of CB in the second 'exterior' angle vertically opposite $D\widehat{A}B$. When BC is large enough, there will be two further solutions corresponding to positions of C in AD below D. Compare Newton's figure on 1:509, which illustrates the particular problem for which $D\widehat{A}B$ is a right angle. (On taking $PD = a$, $DA = b$, $BA = x$ and $\cos D\widehat{A}B = k$, the length of $BC = x\sqrt{[(x-a)^2 + 2bck(x-a) + b^2]}/(x-a)$ attains a (local) minimum for $x = a + X$, where $X^3 - bkX(a - X) - ab^2 = 0$: when the given magnitude of BC is less than this minimum only two solutions are possible.)

(39) Two or four at most, depending on the given length of BC. In Newton's figure as drawn there will be a second solution in which C lies on the further side of D.

(40) A first, cancelled continuation reads 'et $PQ(BC).FC::DC.PC.$ ergo

$$PQ \times PC = FC \times DC = EC^q - EF^q\text{'}.$$

seu BC, $PR.PE$, $RQ-PR$, $EF::PE \times RQ+PR$, $EF.PE$, BC. [et] BC^q, PE, PR $=PE^q$, RQ^q-PR^q, EF^q.[41] Ergo Q locatur in Conica sectione cujus diameter PR, ordinata QR. Sit $RQ=0$, erit $PR=0$ et $=\dfrac{-BC^q, PE}{EF^q}$. In EP producta cape ergo $PS.\frac{1}{2}PE::BC^q.EF^q$ et erit S centrum et P vertex figuræ. Pone PR infinitū et erit PE^q, $RQ^q=PR^q$, EF^q, seu PE, $RQ=\pm EF$, PR. Quare per S ipsis PD, $P[F]$ age parallelas et hæ erunt Asymptoti figuræ. His igitur Asymptotis per punctum P describe Hyperbolam, ut et centro P radio PQ circulum & per eorum intersectionem Q age rectam PC.

Corol. Si ang. PEC rectus est Problema planū erit. Nam circuli centrum incidit in axem figuræ.[42]

3. A dato puncto P rectam PC ducere cujus pars BC inter circulum et productam diametrum DF æquabitur semidiametro EF.

Age EB ac demitte $\perp$ [a] PG, BH. Est $EH.HB::GC(GE+2EH).GP$. Ergo punctum B in Hyperbola est.[43] Pone $EH=0$ et erit $HB \times GE=0$ adeoφ

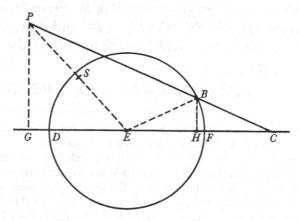

---

(41) The defining equation of the locus (Q) with regard to origin P and oblique coordinates PR and RQ. Since $BC = PQ = \sqrt{[PR^2+RQ^2-2k.\,PR \times RQ]}$ where $k = \cos P\hat{R}Q$, the locus is a nodal cubic having its double point at P and a unique real asymptote coincident with DF. But Newton is interested only in its intersection with the circle $PQ =$ constant on centre P and at once derives the condition that the four meets of this cubic and circle (other than the circular points at infinity) lie on the locus which is defined by setting $BC = PQ$, constant, in the present equation: namely, a hyperbola whose asymptotes are the line-pair

$$PE \times RQ = \pm EF(PR+\tfrac{1}{2}(BC^2/EF^2).PE)$$

through the centre S in PE (or $RQ = 0$) where $SP = \frac{1}{2}(BC^2/EF^2).PE$.

(42) Namely, PE. In this particular case the circle of centre P may be defined by

$$PR^2+RQ^2 = BC^2$$

and hence its intersections with the hyperbola are identical with its four meets with the pair of parallels Q_1Q_1', Q_2Q_2' to DF defined by $PR = -\alpha \pm \beta$, where

$$\alpha = (EF^2/PF^2).PS \quad \text{and} \quad \beta = \sqrt{[(PE^2/PF^2).BC^2+(EF^4/PF^4).PS^2]}$$

that is, $BC \times PR : PE \times RQ - PR \times EF = PE \times RQ + PR \times EF : PE \times BC$. And so $BC^2 \times PE \times PR = PE^2 \times RQ^2 - PR^2 \times EF^2$.[41] Consequently Q is located in a conic whose diameter is PR and ordinate QR. Let RQ be zero, and then $PR = 0$ or $-BC^2 \times PE/EF^2$. Therefore in EP produced take $PS : \frac{1}{2}PE = BC^2 : EF^2$ and S will be the figure's centre, P a vertex. Suppose PR to be infinite and then $PE^2 \times RQ^2 = PR^2 \times EF^2$, that is, $PE \times RQ = \pm PR \times EF$. In consequence, through S draw parallels to PD, PF and these will be the figure's asymptotes. With these asymptotes, therefore, through the point P describe a hyperbola, and also with centre P and radius PQ a circle, then draw the line PC through their intersection Q.

Corollary. If the angle $P\widehat{E}C$ is right, the problem will be plane. For the centre of the circle then falls in the figure's axis.[42]

3. *From the given point P to draw the straight line PC whose section BC between the circle and the extended diameter DF shall equal the radius EF.*

Draw EB and drop the perpendiculars PG, BH. Then is

$$EH : HB = GC \text{ (or } GE + 2EH) : GP$$

and therefore the point B is in a hyperbola.[43] Suppose EH equal to zero and

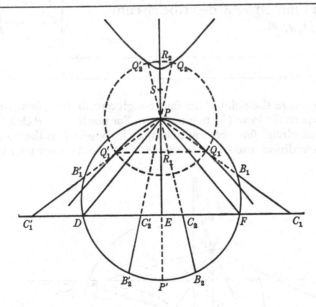

—or alternatively, of course, with either of the similarly definable line-pairs $Q_1 Q_2 \times Q_1' Q_2'$ or $Q_1 Q_2' \times Q_1' Q_2$.

(43) For with respect to origin E the segments EH, HB are rectangular coordinates of the locus B defined, for fixed P, E and varying radius EB ($= BC$) of the given circle by

$$GP \times EH = HB(GE + 2EH) \quad \text{or} \quad (EH + \tfrac{1}{2}GE)(HB - \tfrac{1}{2}GP) = -\tfrac{1}{4}GE \times GP :$$

manifestly a rectangular hyperbola referred to asymptotes through S, the mid-point of PE, parallel and perpendicular to GE.

$HB = 0$. Quare Hyperbola transit per punctum E. Pone EH infinitū et erit $EH \cdot HB :: 2EH \cdot GP$. Ergo $\frac{1}{2}GP = HB$. Pone HB infinitū et erit

$$EH \cdot GE :: HB \cdot GP - 2HB :: HB \cdot -2HB.$$

Ergo $\frac{1}{2}GE = -EH$. Quare biseca PE in S et per S age Asymptotos parallelas EH et HB et per punctum E vel P describe Hyperbolam secantem circulum in B. Et per B age PC.[44]

Corol. Hinc si ang PEG semirectus erit PE axis Hyperbolæ adeoꝗ Problema in eo casu planum.[45]

---

[4][46]

Ut in ang GED agatur GD data per A transiens posito AE quadrato, quære summam radicum Fd, FD. Ad AD erige $\perp DK$ [&] erit AK summa illa,[47] et $CD^q + CK^q (DK^q) + GD^q = GK^q$. Aufer BG^q seu CK^q et restabit $CD^q + GD^q = BK^q$. Datur ergo summa AK. Quare cum ang ADK rectus [sit]; super diametro AK describe circulū secantem FE in D, d.[48]

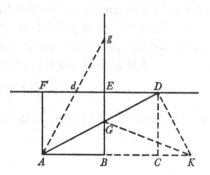

---

(44) In Newton's figure the point P lies far enough outside the given circle to allow only one branch of the hyperbolic locus (B) to intersect it. For positions of P closer to E (and always when P lies within the circle) four real solutions may be possible, as the accompanying figure shows. There are, in addition, two trivial solutions of the problem not mentioned by Newton:

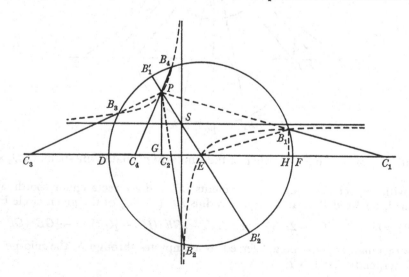

there will be $HB \times GE = 0$ and so $HB = 0$. Hence the hyperbola passes through the point E. Suppose EH infinite and then $EH:HB = 2EH:GP$, consequently $HB = \frac{1}{2}GP$. Suppose HB infinite and there will be $EH:GE = HB:GP - 2HB$, that is, $HB:-2HB$, consequently $EH = -\frac{1}{2}GE$. Hence bisect PE in S and through S draw asymptotes parallel to EH and HB, then through the point E or P describe a hyperbola cutting the circle in B and draw PC through B.[44]

Corollary. Hence if $P\widehat{E}G$ is half a right angle, PE will be the axis of the hyperbola and so the problem in this case is plane.[45]

---

[4][46]

To draw a given line GD passing through A in the angle $G\widehat{E}D$, supposing that AE is a square, seek the sum of the roots Fd, FD. At AD erect the perpendicular DK and AK will be that sum,[47] with $CD^2 + CK^2$ (or DK^2) $+ GD^2 = GK^2$. Take away BG^2, that is, CK^2, and there will remain $CD^2 + GD^2 = BK^2$. The sum AK is therefore given. Consequently, since the angle $A\widehat{D}K$ is right, upon the diameter AK describe a circle cutting FE in D and d.[48]

---

namely, when B lies at either end B_1', B_2' of the diameter through P, E and C coincides with the circle's centre E.

(45) This is intuitively evident since then both the circle and the hyperbolic locus (B) will be mirror-symmetric round PE. In this case, since now $GE = -GP$, the hyperbolic locus (B) is defined by $-GE \times EH = HB(GE + 2EH)$ and so its intersections with the given circle $EH^2 + HB^2 = BC^2$ lie on the locus $EH^2 + 2EH \times HB + HB^2 + GE(EH + HB) - BC^2 = 0$, which is the pair of perpendiculars to PE defined by $EH + HB + \frac{1}{2}GE \pm \sqrt{[\frac{1}{4}GE^2 + BC^2]} = 0$.

(46) Four miscellaneous problems: a simple Apollonian verging, two identifications of loci (both wrong) and an erroneous converse to the organic construction of a conic from a straight line.

(47) Since the triangles AFd, DCK are readily shown to be congruent.

(48) A straight summary of Pappus, *Mathematical Collection*, VII, 71/72, which expounds Heraclitus' construction of Apollonius' verging problem in the particular case where the given parallelogram is a square; see note (33) above and I: 509, note (9).

Super datis rectis tribus *AB*, *CD*, *EF* tria constituere triangula quorum vertices erunt ad idem punctum *G* et anguli ad vertices *AGB*, *CGD*, *EGF* æquales.[49] Super lineis *AB*, *CD*, *EF* describe similia segmenta quorumvis circulorum satis magnorum ita ut se mutuo secent. Comple segm. ad circulos & per intersectionem circulorum *AB*, *CD* age rectam, ut et aliam rectam per intersectiones circulorum *CD*, *FE*: nam hæ rectæ se secabunt in puncto *G*.[50]

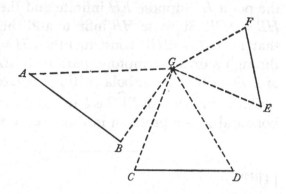

In triangulo dato *ABC* aliud triangulum *DEF* dato *def* simile inscribere cujus latus *EF* transibit per datum punctum *G*. Nempe verticis *D* trianguli *DEF* locus est linea recta.[51]

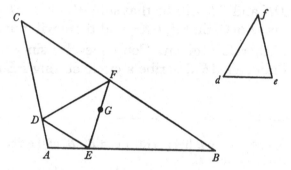

(49) Newton began a first, cancelled construction at this point: 'Junge *AD*, *BC*. Biseca eas in *r*, *s*. Secent *AB*, *CD* se mutuo in *t*. Age *tG*∥*sr*. Idem fac in lineis *CD*, *FE*' (Join *AD* and *BC*. Bisect them in *r* and *s*. Let *AB*, *CD* intersect one another in *t*. Draw *tG*∥*sr*. Do the same in the

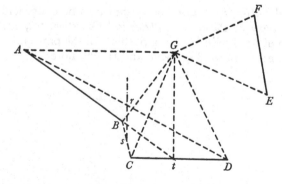

case of the lines *CD* and *FE*). It is not obvious what inspired Newton to guess that *tG* might be the locus of points *G* such that $A\hat{G}B = C\hat{G}D$: as he doubtless at once realized, not even the point *t* is upon it.

Upon the three given straight lines AB, CD, EF to construct triangles whose vertices will be at the same point G and vertex angles $A\widehat{G}B$, $C\widehat{G}D$, $E\widehat{G}F$ will be equal.[49] Upon the lines AB, CD, EF describe similar segments of any circles large enough to intersect one another. Complete the circle segments and through the intersections of the circles AB, CD draw a straight line, and likewise a second straight line through the intersections of the circles CD, FE: for these lines will cut one another in the point G.[50]

In the given triangle ABC to inscribe another triangle DEF, similar to the given one def, whose side EF shall pass through the given point G. Obviously the locus of the vertex D of the triangle DEF is a straight line.[51]

---

(50) Newton makes the plausible but erroneous conjecture that the locus (G) defined by $A\widehat{G}B = C\widehat{G}D$ is a straight line. If in modern terms we take G as the general point (x, y) in a rectangular Cartesian system in which the points A, B, C, D are respectively (a, b), (c, d), (A, B) and (C, D), the given condition becomes

$$\frac{(x-a)(y-d)-(x-c)(y-b)}{(x-a)(x-c)+(y-b)(y-d)} = \frac{(x-A)(y-D)-(x-C)(y-B)}{(x-A)(x-C)+(y-B)(y-D)},$$

defining the locus to be a circular cubic (evidently through each of the points A, B, C and D) of form

$$(x^2+y^2)((b-d-B+D)x+(c-a-C+A)y)+Ex^2+Fxy+Gy^2+Hx+Iy+K=0.$$

(Compare G. Salmon, *A Treatise on the Higher Plane Curves* (London, $_3$1879): §164:141–2.) However, when AB and CD are equal chords in the same circumscribing circle, the locus (G) breaks into the pair of that circle and the line round which AB and CD are symmetrically placed.

(51) Read 'hyperbola'. It should have been obvious to Newton that when FE is parallel to AB or BC, the corresponding points D are at infinity: the asymptotes to the locus are the corresponding positions of DF and DE respectively. If G is the origin of a rectangular Cartesian system in which D is the general point (x, y) and AB, BC the lines $y = \alpha$, $y = ax+b$ respectively, while $D\widehat{F}E = \tan^{-1}r$, $D\widehat{E}F = \tan^{-1}s$, we may set the general points E and F as (λ, α) and $(b\lambda/(\alpha-a\lambda), b\alpha/(\alpha-a\lambda))$ respectively and so obtain the equation of the locus parametrically in terms of λ as

$$\begin{cases} r = \left(\dfrac{\alpha}{\lambda}-\dfrac{b\alpha-y(\alpha-a\lambda)}{b\lambda-x(\alpha-a\lambda)}\right)\Big/\left(1+\dfrac{\alpha(b\alpha-y(\alpha-a\lambda))}{\lambda(b\lambda-x(\alpha-a\lambda))}\right), \\[2mm] s = \left(\dfrac{\alpha-y}{\lambda-x}-\dfrac{\alpha}{\lambda}\right)\Big/\left(1+\dfrac{\alpha(\alpha-y)}{\lambda(\lambda-x)}\right). \end{cases}$$

On eliminating λ there results a quartic which splits into a pair of hyperbolas, (D) and (D'), where D' is instantaneously the mirror-image in EF of D.

In data conica sectione *ABCDE*, trapezium *ACDB* inscribere cujus anguli duo oppositi *CAD*[3] *CBD* dantur et data puncta *A* et *B* consistunt. Viz^t si locus puncti *D* est Conica sectio locus puncti *C* erit linea recta.[52]

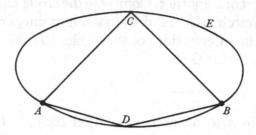

(52) A *non-sequitur*. When the locus (*D*) is a conic, the locus (*C*) is in general a second conic through the poles *A*, *B* since the homography defined by *A*(*D*) = *B*(*D*) determines a corresponding homography *A*(*C*) = *B*(*C*), the cross-ratios of the two sets of equiangular line-pencils *A*(*D*), *A*(*C*) and *B*(*D*), *B*(*C*) evidently being equal in each case. It follows that Newton's

In the given conic $ABCDE$ to inscribe a quadrilateral $ACBD$ whose two opposite angles $C\hat{A}D$, $C\hat{B}D$ are given and stand on the given points A and B. Evidently, if the locus of the point D is a conic, that of the point C will be a straight line.[52]

---

problem has, on discarding the two trivial meets A and B of the conics (C) and (D), two solutions which may be imaginary. If, however, some point (D' say) in the describing conic corresponds to the whole line AB (that is, if $C\hat{A}D = B\hat{A}D'$ and also $C\hat{B}D = A\hat{B}D'$ for some point D'), then the conic (C) breaks into the pair of AB and a second 'directrix' straight line. When Newton came to draft Problem 7 of his 'Solutio Problematis Veterum de Loco solido' (2, §2 following) he correctly asserted that 'si rectæ $[AD, BD]$ attingunt locum solidum per puncta $[A, B]$ transeuntem & ubi hæ concurrunt ad ejus punctum aliquod $[D']$ alteræ duæ $[AC, BC]$ coincidunt cum linea recta $[AB]$, punctum $[C]$ continget rectam positione datam' (see 2, §2: note (30) below).

APPENDIX. THE 'CENTER OF MOTION' OF TWO BODIES EACH MOVING UNIFORMLY IN A GIVEN STRAIGHT LINE.[1]

[Early 1665]

Extracts from Newton's Waste Book in the University Library, Cambridge[2]

Two bodys being uniformely moved in yᵉ same plaine their center of motion will describe a streight line.

28. If two bodys *b* & *c* move in the lines *br* & *cr*. The body *c* moveing through yᵉ space *cg* in yᵉ time *vs*, & th[r]ough *gk* in yᵉ time *nv*, & through *kr* in yᵉ time *nr*. & yᵉ velocity of yᵉ body *b* is to yᵉ velocity of *c* as *d*, to *e*, & as yᵉ line *cg* to yᵉ line *be*, or as *ck* to *br*, then when yᵉ body *c* is in yᵉ place *g*, *b* will bee in *e*, & when *c* is in *k*, *b* will be in *r*. to find yᵉ line wᶜʰ the center of their motion describes, viz *dfo*[3].

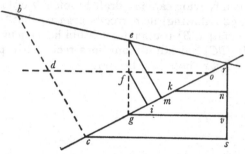

Then nameing yᵉ quantitys $br = a$, $cr = f$. $bc = g$. $e : d :: a : \dfrac{da}{e} = ck$. $kr = \dfrac{ef - da}{e}$. If

---

(1) See §1: note (33) above. This extract dealing with the rectilinear locus of the 'center of motion' (instantaneous centre of gravity) of two 'bodys' moving uniformly in a pair of straight lines which are, in general, skew is taken from a lengthy tract on the motion and impact of spherical bodies written up by Newton, from no longer existing preliminary worksheets, at about the time he received his Cambridge B.A. degree. The complete treatise—with the trivial omission of two concluding corollaries on 'compound force' (ULC. Add. 4004: 38ʳ)— is reproduced by J. W. Herivel in his *The Background to Newton's Principia: A Study of Newton's Dynamical Researches in the Years 1664–84* (Oxford, 1966): 133–5, 136–9, 141–50, 153–9, 162–79, 182.

(2) ULC. Add. 4004: 13ᵛ/14ʳ [= Herivel's *Background* (note (1) above): 163–8]. The date of the present extract might be set with a high degree of confidence as 'late January 1665' since on the opening folio (10ʳ) of the tract from which it is taken Newton has set in the margin 'Jan 20ᵗʰ 1664[/5]'. Since the purpose of the present appendix is not to establish an accurate text but merely to illuminate the background to a single rectilinear locus stated by Newton without proof above (Proposition 20 of §1.1), we have not reproduced cancelled first versions, present in the manuscript, of Proposition 28 and the *idem aliter* of Proposition '28 & 30ᵗʰ'. For the text of these we refer the reader to Herivel's *Background*.

(3) By the 'line' *dfo* understand the locus, not yet shown to be rectilinear, of the instantaneous 'center of motion' of the bodies *b* and *c*: namely, of the point *d* in the straight line *bc* which divides it inversely in the ratio of the weight of those bodies. (The 'gravity' acting on the bodies is assumed to be uniform.) In mathematical terms we are required to find the locus *dfo* such that, where *be* : *er* = *cg* : *gk*, then

$$bd : dc = ef : fg = ro : ok = \text{constant.}$$

o be y^e center of motion of y^e bodys at k & r; y^n, $b+c:b::\dfrac{ef-da}{e}:\dfrac{bef-abd}{eb+ec}=or.$

And y^e line df must passe through o.[4] againe making $gk=v$. y^n $d:e::v:\dfrac{ev}{d}=er.$

& if $bc\,\|\,fi\,\|\,em$, y^n $a:g::\dfrac{ev}{d}=er:\dfrac{gev}{ad}=em.$ &, $a:f::\dfrac{ev}{d}=er:\dfrac{fev}{ad}=mr.$

$$gr=gk+kr=v+\frac{ef-ad}{e}.\quad gm=gr-mr=\frac{ev+ef-ad}{e}\frac{-fev}{ad}$$

$$=\frac{adev+adef-aadd-feev}{ade}.$$

Since f is y^e center of motion in y^e bodys at g & e tis,

$$b+c:c::eg:fg::em=\frac{gev}{ad}:\frac{cgev}{abd+acd}=fi::gm=\&c^{(5)}:$$

$$gi=\frac{cadev+cadef-caadd-cfeev}{bade+cade}.$$

$$gr=gk+kr=\frac{ev+ef-ad}{e}.\quad go=gr-or=\frac{ev+ef-ad}{e}\frac{+abd-bef}{eb+ec}$$

$$=\frac{bev+cev+cef-cad}{eb+ec}.$$

$$go-gi=io=\frac{adbev+cfeev}{bade+cade}=\frac{abdv+cefv}{abd+acd}.\quad co=cr-or=\frac{fec+abd}{eb+ec}.$$

$b+c:c::g:\dfrac{cg}{b+c}=cd.$ Now if the lines $oi:if::oc:cd.$ Then y^e line od[6] must be a streight line. but $oi:if::adb+cfe:cge::oc:cd.$ therefore y^e line do[6] is a streight line, wch may bee found by y^e two points d & o.[7]

The demonstracon is y^e same if y^e body b moved from $[e]$ to b.[8]

29. If two bodys q & c be moved in divers pl[a]ines, then find y^e shortest line (pr) wch can bee drawne frome one line (cr) to y^e other line (qp) in wch those

They doe y^e same in divers plaines.

---

(4) Observe that Newton implicitly takes the constant ratio $b:c$ to be inversely as the weights of the bodies b and c. The convention is retained throughout.

(5) Understand the preceding value of gm.

(6) Read 'dfo'.

(7) Newton here omits the evident corollary that, since

$$do:fo = co:io = (d/e)\,a:v = ck:gk = br:er,$$

the motion of the 'center of motion' d of the bodies b and c is uniform.

(8) The manuscript has 'from a to b'. Newton has perhaps misread his carelessly formed Greek 'ϵ' in a preliminary worksheet.

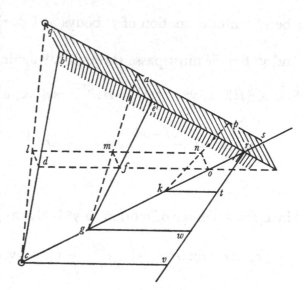

bodys are moved, & y^t line pr shall bee perpendicular to both y^e lines cr & qp,[9] viz $\angle qpr = \angle rps = [\angle]prc =$ recto. then draw qb equall & $\parallel pr$ & draw $br = qp$. Then shall y^e plaine $qbrp$ be perpendicular to y^e plaine bcr. Suppose also y^e body c moves over y^e spac[e] $\left.\begin{matrix} cg \\ gk \\ kr \end{matrix}\right\}$ in the time $\left\{\begin{matrix} vw \\ wt. \\ tr \end{matrix}\right.$

& y^t y^e body q moves over y^e space qa in y^e time vw, & ap in y^e time wt. Also suppose another body $b(=q)$ & equivelox to (q) y^t is to move over the space $\left\{\begin{matrix} be = qa \\ er = ap \end{matrix}\right\}$ in y^e time $\left\{\begin{matrix} vw \\ wt \end{matrix}\right.$.

Soe y^t when (c) is in (g) or (k), (b) will bee in $(e,)$ or (r) & (q) in (a) or (p). Then drawing the streight lines qc, ag, $[p]k$, vr if $b + c : c :: bc : cd :: eg : fg :: rk : ok$, the points d, f, & o, shall be y^e centers of motion[10] of y^e bodys b & c, when they are in y^e places b & c, e & g, r & k. & (prop 28) therefore y^e line (dfo) is a streight line. Likewise if it bee $q + c : c :: qc : lc :: ag : mg :: pk : nk$, then y^e points l, m, n are centers of motion to y^e bodys $(q$ & $c)$ being in y^e places (q) & (c), a & g, p & k. Then drawing y^e lines ld, mf, no, (twixt y^e neighbouring centers of motion) since $b + c : c :: q + c : c :: bc : cd :: q[c] : lc$. therefore $\angle qbc = \angle ldc$ & by y^e same reason $\angle gfm = \angle gea$. & $\angle krp = \angle kon$. Wherefore all the lines qb, ae, pr, ld, mf, no, are parallell to one another. And

$$b + c : c :: bc : dc :: qb : ld :: eg : fg :: ea(=qb) : mf :: kr : ko :: pr(=bq) : no,$$

soe y^t $ld = mf = no$. & since these line[s] ld, mf, no, are parallell, equall, in y^e same plaine $ldon$, & stand upon y^e same streight line do,[11] y^e line (lmn) in wch their other ends l, m, n, are terminated (i.e. in wch are all y^e centers of motion of y^e bodys $(c$ & $q)$) must bee a streight line.

The demonstracon is the same if (q) moved from (p) to (q). 28 & 30th.[12] Or

Or thus.   thus. The bodys $(b$ & $c)$ being in b & c, e & g, r & k, in y^e same times, & dn being

---

(9) Self-evidently so, since the shortest distance from any point p in qs to the line co is the perpendicular pr and conversely.

(10) Newton first wrote 'centers of gravity'.

(11) A cancelled first continuation reads 'their other ends (the centers of motion of c & q) must bee in y^e same streight line lmn, wch line [? is $=$ & $\parallel dfo$]'.

(12) Understand that this present '30th prop' is the '28th...done otherwise', as a cancelled marginal heading has it. A first preceding version of this proposition, not here reproduced (see note (2) above) is titled in the margin alongside 'Of y^e velocity of y^e center of motiõ'.

described by their centers of motion. Also making $dc \parallel fs \parallel ey$. & $mf \parallel cn$. Then

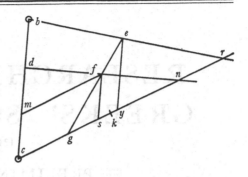

$$be:br::cy:cr::cg:ck$$

(for y^e motions of b & c are unifor̄)
$$::gy(=cy-cg):kr(=cr-ck)::gs:kn \text{ (for}$$

$$b+c:c::ge:gf::gy:gs::kr:kn)::$$

$$mf(=cs=cg+gs):cn(=ck+kn).$$

Againe $br:er::bc:ey::dc:fs=mc$ (for $b+c:c::ge:gf::ey:fs::bc:dc$ (prop 25)).[13] Therefore $be:br::dm:dc::mf:cn$. & consequently y^e points dfn are in one streight line. also since $be:br::df:dn$ y^e center of motions motion must bee uniforme.

---

(13) Proposition 25 'To find y^e centers of m[otion] in [moving] bodys' reads: 'Having [y^e] ce[nters] of motion of y^e 2 bodys ob & de to find y^e common center of both th[eire motions] draw a line oe from the centers of theire motions o, & e, & divid[e it in] a, soe y^t the body ob is to y^e body de as the line ae to y^e line oa: y^t is s[oe] y^t $ob \times oa = ae \times de$. For then if they move about y^e center a being always opposite to one another they have equall motion...& consequently have an equall endeavour from y^e center a...soe y^t...the one hindereth y^e other from forcing y^e center a any way soe y^t it shall stand in equilibrio betwixt them & ...is therefore theire center of motion' (ULC. Add. 4004: 11^v/12^r; compare Herivel's *Background* (note (1) above): 148).

Newton seemingly never attempted to identify the nature and properties of the locus of the general 'center of motion' l as the ratio $b:c$ varies, but it will be evident to the modern reader that the two families of straight lines (amg), (lmn) define this to be a ruled quadric surface (in fact, a hyperboloid of one sheet) of which they are the two systems of generators. The present theorem—and, with trivial variants, its proof—was independently rediscovered eighty years later by the Slav physicist Roger Boscovich while exploring possible indeterminacies of Newton's construction (ULC. Dd. 9. 68: 122–4 [= *Arithmetica Universalis* (Cambridge, $_1$1707): 205–7]: Problema 52) of a comet's path under the simplified Keplerian supposition that it is a straight line traversed uniformly; see Boscovich's *Dissertatio de Cometis, Habita...In Collegio Romano Anno 1746...* (Rome, 1746): §§ 21–5 (reprinted in *Rogeri Josephi Boscovich Opera pertinentia ad Opticam et Astronomiam...*, **3** (Venice, 1785): 316–68, especially 329–32). We will return to this point in the next volume when we reproduce the text of Newton's Cambridge lectures on algebra and analytical geometry.

2

RESEARCHES INTO THE GREEKS' 'SOLID LOCUS'[1]

[?Late 1670's]

§1. PRELIMINARY REMARKS.[2]

From Newton's Waste Book in the University Library, Cambridge[3]

VETERUM LOCA SOLIDA RESTITUTA.

Cartesius de hujus Problematis[4] confectione se jactitat quasi aliquid præstitisset a Veteribus tantopere quæsitum, cujus gratia putat Apollonium libros

---

(1) Researches into the pure geometry of the στερεοὶ τόποι of the 'ancients', that is, of the general non-degenerate conic, whose plane 'symptom' was derived by the classical Greek mathematicians (notably Menæchmus, Euclid and Apollonius) from its 'solid' definition as the plane section of a circular cone. Insofar as the internal evidence of the texts here reproduced offers a sure guide to Newton's inspiration we may firmly seize on Pappus' *Mathematical Collection* as his primary source of information, initially but obliquely through Descartes' quotation of Pappus' account of the locus *ad tres et quatuor lineas* in the first book of his *Geometrie*, later and more directly through his reading of Commandino's *editio princeps* of the Latin version of the extant portion of Pappus' work. In the opening proposition of the 'Solutio Problematis Veterum' (§2 following) Newton makes a unique quotation from Apollonius' *Conics* (namely of III, 17/18) but he may well have derived this in the first instance by way of Pappus, *Mathematical Collection*, VIII, 13 (see note (12) below). Contemporary documentary evidence is wholly silent regarding the background to these two autograph pieces, which were evidently written at the same period. The conjectured date of composition is assigned on the strength of our assessment of Newton's handwriting style, but it will be clear that a firm post-date for the latter tract is early 1685 when he incorporated a lightly revised version of its first seven propositions in his *Principia*, while the two circle loci discussed at length in its conclusion may well derive from his recent reading of Fermat's 'restitution' of Apollonius' *Plane Loci*, first published by Pierre's son Samuel in 1679 (see §2: notes (1), (76) and (82)).

(2) This polemical piece, virtually an admonitory preface to the lengthy 'Solutio Problematis Veterum de Loco solido' which follows (§2 below), mounts a reasoned attack on Descartes for boasting that, in establishing the identity of the Greek 3/4-line locus with the general 'locus solidus' (conic), he had done something which 'ny Euclide [in his lost treatise on conics], ny Apollonius, ny aucun autre n'auoient sceu entierement resoudre' (*Geometrie*: Livre Premier [= *Discours de la Methode* (Leyden, 1637): 297–314]: 304). For a more detailed account of the prehistory and justice of Descartes' assertion see the preceding introduction (pages 270–3 above).

(3) ULC. Add. 4004: 89ᵛ–90ᵛ.

(4) The celebrated 'locus ad tres et quatuor lineas', partially resolved by Euclid in the third book of his lost treatise on conics and whose general solution was effectively known to Apollonius: 'Si positione datis tribus rectis lineis ab vno, & eodem puncto ad tres lineas in datis angulis rectæ lineæ ducantur, & data sit proportio rectanguli contenti duabus ductis ad

Translation

RESTORATION OF THE ANCIENTS' SOLID LOCI.

Descartes, in regard to his accomplishment of this problem,[4] makes a great show as if he had achieved something so earnestly sought after by the Ancients

quadratum reliquæ: punctum contingit positione datum solidum locum, hoc est vnam ex tribus conicis sectionibus. Et si ad quattuor rectas lineas positione datas in datis angulis lineæ ducantur; & rectanguli duabus contenti ad contentum duabus reliquis proportio data sit: similiter punctum datum coni sectionem positione continget' (Pappus' enunciation in his prefatory remarks 'De conicis Apollonij' [*Pappi Alexandrini Mathematicæ Collectiones À Federico Commandino Vrbinate In Latinum Conversæ et Commentariis Illustratæ* (Pesaro, 1588): Liber Septimus: 165ʳ], quoted by Descartes in his *Geometrie* (note (2) above): 304–5). In the terms of Newton's

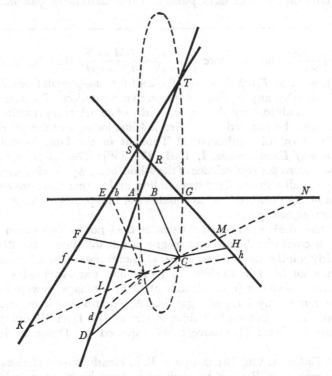

following diagram the general 4-line locus (*C*) is defined by $CB \times CF = k \times CD \times CH$, where *CB*, *CD*, *CF*, *CH* are lines let fall from the general point *C* at given angles to the fixed lines *AB*, *AD*, *EF* and *GH*. The locus will evidently pass through the meets *A*, *G* and *S*, *T* of *AB* and *EF* with *AD* and *GH*, since at these points one each of the pairs *CB*, *CF* and *CD*, *CH* vanishes. To verify that the locus is solid we may show that, if *c* is a given point on the locus (that is, such that $\dfrac{CB \times CF}{CD \times CH} = \dfrac{cb \times cf}{cd \times ch}$ where *cb*, *cf*, *cd*, *ch* are drawn parallel to their upper-case mates), then (*C*) is a conic through the five points *c*, *A*, *G*, *S*, *T*; for on extending *Cc* to meet *EF*, *AD*, *GH* and *AB* in *K*, *L*, *M* and *N* we readily deduce, since

$$CB:CN = cb:cN, \quad CF:CK = cf:cK, \quad CD:CL = cd:cL \quad \text{and} \quad CH:CM = ch:cM,$$

suos[5] de Conicis sectionibus scripsisse. Sed cum tanti viri pace rem Veteres neutiquam latuisse crediderim. Docet enim Pappus[6] modum ducendi Ellipsin per quincß data puncta et eadem est ratio in cæteris Con. sect. Et si Veteres norint ducere Conicam sectionem per quincß data puncta, quis non videt eos cognovisse compositionem loci solidi.[7] Imo vero eorum methodus longe elegantior est Cartesiana. Ille rem peregit[8] per calculum Algebraicum qui in verba (pro more Veterum scriptorum) resolutus adeo prolixu[s] et perplexu[s] evaderet ut nauseam crearet nec posset intelligi. At illi rem peregerunt per simplices quasdam Analogias, nihil judicantes lectu dignum quod aliter scriberetur, & proinde celantes Analysin per quam constructiones invenerunt.[9] Ut pateat hanc rem eos non latuisse, conabor inventum restituere insistendo vestigijs Problematis Pappiani. Incß eum finem propono hæc Problemata.[10]

1. *Conicam sectionem per tria data puncta A B C describere quæ datum centrum O*

---

that $\dfrac{CN \times CK}{CL \times CM} = \dfrac{cN \times cK}{cL \times cM}$ and therefore $\dfrac{CK \times cL}{CL \times cK} = \dfrac{CM \times cN}{CN \times cM}$, that is, in cross-ratio terms $(CcKL) = (CcMN)$, whence $T(CcSA) = G(CcSA)$ and the line-pencils from T, G to C, c, S, A are equi-cross. Manifestly, any equivalent projective definition of a conic—say, Pascal's *hexagrammum mysticum* condition that the opposite sides of an arbitrary inscribed hexagon meet in collinear points—may be reduced to the Greek 4-line locus, and conversely so. (Compare D. T. Whiteside, 'Patterns of Mathematical Thought in the later Seventeenth Century' [= *Archive for History of Exact Sciences*, **1**, 1961: 179–358]: 274–5.) In a *tour de force* in §2 following Newton will show the equivalence of the 4-line locus, Apollonius' *Conics*, III, 17/18, the locus described 'organically' from a fixed straight line, and the general homographic definition of a conic 'per simplicem Geometriam' as the meet of line-pencils through two fixed points which are in 1, 1 correspondence.

(5) 'tractatum' (treatise) is cancelled. Descartes does not in fact claim that Apollonius wrote his treatise on conics in the attempt to give a full solution of the 3/4-line locus, but merely observes fairly mildly that the *locus ad tres et quatuor lineas* was a 'question...qui auoit esté commencée a resoudre par Euclide, & poursuiuie par Apollonius, sans auoir esté acheuée par personne' (*Geometrie* (note (2) above): 306). Perhaps Newton here has in mind a far more hostile remark by Pappus regarding Apollonius' work on the locus, 'in quo magnifice se iactat, & ostentat nulla habita gratia ei [*sc.* Euclid], qui prius scripserat' (*Mathematicæ Collectiones* (note (4) above): 165$^\mathrm{r}$, quoted by Descartes in his *Geometrie*: 304).

(6) *Mathematical Collection*, VIII, 13; compare T. L. Heath, *Greek Mathematics*, **2** (Oxford, 1921): 434–7. The essence of Pappus' proposition is given by Newton in his two following lemmatical problems.

(7) Presuming, that is, that the 'ancients' had fully established the identity of the 3/4-line locus with the general 'locus solidus'. In point of fact the 3-line locus is an immediate corollary of Apollonius, *Conics*, III, 54–6 (see J. J. Milne, 'Newton's Contribution to the Geometry of Conics' [= (ed. W. J. Greenstreet) *Isaac Newton, 1642–1727* (London, 1927): 96–114]: 105–8; and compare T. L. Heath, *Apollonius of Perga: Treatise on Conic Sections edited in Modern Notation* (Cambridge, 1896): cxxxix), while the 4-line locus is readily derivable by compounding instances of the particular 3-line case: namely, if in Newton's following figure (see also note (4) above) $C\alpha$, $C\beta$, $C\gamma$, $C\delta$ are drawn from C at fixed angles to the tangents at A, G, S, T to the conic locus, then by the converse of the defining property of the 3-line locus each of the

and for whose sake he considers that Apollonius wrote his books[5] on conics. With all respect to so great a man I should have believed that this topic remained not at all a mystery to the Ancients. For Pappus informs us[6] of a method for drawing an ellipse through five given points and the reasoning is the same in the case of the other conics. And if the Ancients knew how to draw a conic through five given points, does any one not see that they found out the composition of the solid locus?[7] To be sure, their method is more elegant by far than the Cartesian one. For he achieved[8] the result by an algebraic calculus which, when transposed into words (following the practice of the Ancients in their writings), would prove to be so tedious and entangled as to provoke nausea, nor might it be understood. But they accomplished it by certain simple proportions, judging that nothing written in a different style was worthy to be read, and in consequence concealing the analysis by which they found their constructions.[9] To reveal that this topic was no mystery to them, I shall attempt to restore their discovery by following in the steps of Pappus' problem. And to this end I propose these problems:[10]

1. *To describe a conic through three given points A, B, C which shall have the given*

---

ratios $CB^2 : C\alpha \times C\beta$, $CF^2 : C\gamma \times C\delta$, $CD^2 : C\beta \times C\delta$ and $CH^2 : C\alpha \times C\gamma$ is given and so accordingly is $CB \times CF : CD \times CH$.

(8) Newton first wrote 'docuit' (explained).

(9) This plausible, if ultimately unacceptable, view that the ancients in formulating their synthetic proofs deliberately concealed a prior, arcane method of mathematical analysis which has not been communicated to the modern world—one widespread in the seventeenth century and but a facet of a wider tradition, supported not least by Newton, which clung tenaciously to belief in the previous existence of a *prisca sapientia*—was succinctly stated by Descartes in Rule 4 of his early *Regulæ ad Directionem Ingenii* (posthumously published in 1701): 'Et quidem hujus veræ Matheseos vestigia quædam adhuc apparere mihi videntur in Pappo & Diophanto, qui, licet non primâ ætate, multis tamen sæculis ante hæc tempora vixerunt. Hanc verò postea ab ipsis Scriptoribus perniciosâ quâdam astutiâ suppressam fuisse crediderim; nam sicut multos artifices de suis inventis fecisse compertum est, timuerunt fortè, quia facillima erat & simplex, ne vulgata vilesceret, malueruntque nobis in ejus locum steriles quasdam veritates ex consequentibus acutulè demonstratas, tanquam artis suæ effectus, vt illos miraremur exhibere, quàm artem ipsam docere, quæ planè admirationem sustulisset' (*Œuvres de Descartes* (ed. C. Adam and P. Tannery), **10** (Paris, 1908): 376–7). Newton would probably not have disagreed with Descartes' further opinion that 'nihil aliud esse videtur ars illa, quam barbaro nomine Algebram vocant, si tantùm multiplicibus numeris & inexplicabilibus figuris, quibus obruitur, ita possit exsolvi, vt nonampliùs ei desit perspicuitas & facilitas summa, qualem in verâ Mathesi debere esse supponimus' (*ibid.*: 377): a viewpoint not entirely consistent with his later observation regarding his solution of 'tous les Problesmes de la Geometrie ordinaire' by a uniform circle construction that 'ie ne croy pas que les anciens [l']ayent remarqué. car autrement ils n'eussent pas pris la peine d'en escrire tant de gros liures, ou le seul ordre de leurs propositions nous fait connoistre qu'ils n'ont point eu la vraye methode pour les trouuer toutes, mais qu'ils ont seulement ramassé celles qu'ils ont rencontrées' (*Geometrie* (note (2) above): 304). We will return to this point in the seventh volume.

(10) See note (6).

habebit. A duobus punctis $A\,B$ ad centrum O age rectas $AO\;BO$ et[11] produc AO ad P ut sit $OP=AO$. A tertio puncto C age CS parallelam AO et occurrentem OB in S et cape ST ad $\dfrac{SB\times SQ}{SC}::AO^q.BO^q$,[12] et erit punctum T ad curvam. Biseca TC in V, et ipsi OV parallelam age CR occurrentem AO in R, erit CR ordinatim applicata ad diametrum AP. Et latus rectum[13] erit ad AP ut RC^q ad $AR\times RP$, Figura existente Ellipsi si punctum R cadit intra A et P, aliter Hyperbola.[14]

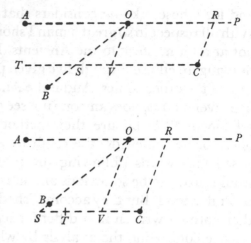

2. *Per data quinque puncta A, B, C, D, E Conicam Sectionem describere.* Junge duo puncta $A\,C$ et alia duo $B\,E$ sitꝗ jungentium intersectio K. Ipsis $AC\;BE$ age parallelas $DI\;DG$ occurrentes BE, AC in punctis H et F, et facto

$$BK\times KE.AK\times KC::$$

$$\frac{BH\times HE}{DH}.HI::FG.\frac{AF\times FC}{FD},$$

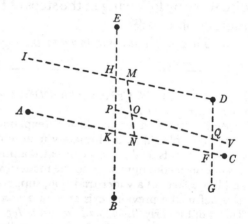

puncta G et I erunt ad curvam. Biseca ergo ID et AC in M et N ut et BE ac GD in P et Q et actarum MN, PQ intersectio O erit centrum Conicæ sectionis.[15] Habito autem centro, describe figuram per puncta A, B, C ut supra.[16] Quod si MN et PQ parallelæ sint Figura erit Parabola.[17] Quo casu produc PQ ad V ut sit $BP^q.GQ^q::PV.QV$. et erit PV diameter$_{[,]}$ V vertex et $\dfrac{BP^q}{PV}$ latus rectum figuræ.

---

(11) Newton first continued 'eas produc ad P, Q ut sit $OP = AO$, et $OQ = BO$, et puncta P et Q erunt ad curvam' (extend these to P and Q such that $OP = AO$ and $OQ = BO$, and the points P and Q will be on the curve). In his manuscript diagram, correspondingly, he first drew the extension of BO beyond the centre O to Q and then subsequently deleted it.

(12) Understand (compare note (11)) that the point Q is on the conic, lying diametrically opposite to B. Alternatively, in place of 'SQ' read '$BO+SO$'. The present theorem (equivalent to the projective definition of the general conic) is Apollonius, *Conics*, III, 17/18 but Pappus' text seems to assume that it was widely known among the classical writers on conics—enough so, at least, not to require the citation of Apollonius' text as authority.

(13) Corresponding to the ordination angle $A\hat{O}V = A\hat{R}C$ between conjugate diameters through O, that is. To determine the principal *latus rectum* we need still to construct the axes of

centre O. From two points A, B to the centre O draw the straight lines AO, BO and[11] extend AO to P so that OP = AO, from the third point C draw CS parallel to AO and meeting OB in S, and take ST to SB × SQ/SC as AO² to BO²:[12] the point T will then be on the curve. Bisect TC in V and parallel to OV draw CR meeting AO in R, and CR will be ordinate to the diameter AP. The *latus rectum*[13]

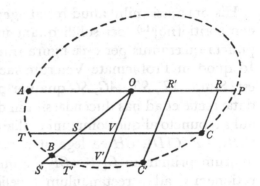

will then be to AP as RC² to AR × RP, the figure proving to be an ellipse if the point R falls between A and P, otherwise a hyperbola.[14]

2. *To describe a conic through the five given points A, B, C, D, E.* Join two points A, C and two others B, E and let the point of junction be K. Parallel to AC, BE draw DI, DG meeting BE, AC in the points H and F, then on making

$$BK \times KE : AK \times KC = BH \times HE/DH : HI$$
$$= FG : AF \times FC/FD$$

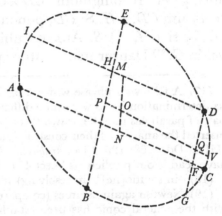

the points G and I will be on the curve. So bisect ID and AC in M and N, and likewise BE and GD in P and Q, and the intersection O of MN and PQ when drawn will be the conic's centre.[15] Having obtained the centre, however, describe the figure through the points A, B and C as above.[16] But should MN and PQ be parallel, the figure will be a parabola.[17] In this case extend PQ to V so that BP² : GQ² = PV : QV and PV will be the diameter, V the vertex and BP²/PV the *latus rectum* of the figure.

---

the conic (say, by one or other of Apollonius' *Conics*, VII, 1–23, according to the various cases which arise).

(14) The parabola has, of course, no finite centre but is readily constructed through four given points by the following problem. See note (17) below.

(15) The cancelled opening 'nisi ubi [? MN et PQ parallelæ sunt] (except when [MN and PQ are parallel]) to an abortive final clause probably anticipates the next sentence but one following.

(16) Newton here inserted and then deleted 'Quæ quidem Ellipsis erit si punctum [K] cadit inter A et [C], aliter Hyperbola' (This, indeed, will be an ellipse if the point K falls between A and C, otherwise an hyperbola). The assertion is true for Newton's configuration of points (where the conic through A, B, C, D, E is manifestly an ellipse) but is evidently false when, or instance, B lies between A and C on one branch of a hyperbola and E is on the opposite branch.

His præmissis nihil aliud restat agendum in compositio[ne][18] loci solidi quam ut quinqʒ puncta quæramus per quæ figura transibit.[19] Id quod in Problemate Veterum facillimum est. Sunto *AT, ST, AG, SG* quatuor positione datæ rectæ et ad has ducendæ sint in datis angulis a puncto aliquo communi *C* aliæ quatuor *C*[*B*], *CF*, *C*[*D*], *CH* ea lege ut rectangulum duarum primarum *C*[*B*] × *CF* datam habeat rationem ad rectangulum reliquarum *C*[*D*] × *CH*.[20] Curva in qua punctum *C* perpetim reperitur transibit per 4 intersectiones datarum *A, G, S, T,* nam ubi *FC* nulla est, rectangulum *FC* × *CB* nullum erit, adeoqʒ et rectangulum *CD* × *CH*, et una rectarum *CD, CH*. Si *CD*, punctum *C* incidet ad *T*, si *CH*, ad *S*. Atqʒ ita ubi *CB* nulla est punctum *C* incidet vel in *A* vel in *G*.[21] Dantur itaqʒ quatuor puncta *A, G, S, T* per quæ figura transibit et

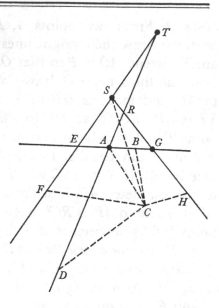

---

(17) A following phrase which adds little, 'cujus determinatio neutiquam difficilior' (and its determination in no way more difficult), has been deleted. In general, of course, a unique pair of parabolas may be drawn through four given points, as Newton well knew: indeed, he inserted the analysis of their construction as 'Prob. 54' of his contemporary lectures on algebra (ULC. Dd. 9.68: 125–7 [= *Arithmetica Universalis* (Cambridge, 1707): 209–11]), claiming to have read it out publicly as 'Lect 4' of his 'Octob. 1680' series.

(18) 'in solutio[ne]' (in resolving) is cancelled.

(19) Newton again assumes (compare note (7) above) that the identity of the 4-line locus with the general conic has been established. A stray remark noted down in May 1694 by David Gregory from conversation with him, 'Apollonij Liber de sectione lineæ determinata est in ordine ad Cartesianum probl: veterum, scilicet ex isto libro invenitur unum punctum, et per illud et quatuor data ducitur Conisectio' (*Correspondence of Isaac Newton*, **3**, 1961: 327), confirms that the identity followed self-evidently for Newton from the fact that a general line *KN* (see the figure accompanying note (4) above) will intersect the locus in a pair of points in involution with the pairs *K, N* and *L, M*; whence, given one of the points, *c* say, the other (*C*) may be constructed from it by the defining condition

$$CN \times CK/CL \times CM = cN \times cK/cL \times cM, \text{ constant,}$$

the general problem considered in the two books of Apollonius' lost treatise on *Determinate Section* (διωρισμένη τομή) according to Pappus. (The text is corrupt at this point but, with Hultsch, we may follow Snell's interpretation of the passage in his *Apollonius Batavus, seu Exsuscitata Apollonii Pergæi Geometrica* (Leyden, 1608), and take Pappus' summary in the prolegomenon to Book VII of his *Mathematical Collection* to read 'The unique proposition in the two books of *Determinate Section*...may be thus enunciated: to cut an indefinite straight line in a point such that...the rectangle comprised by two segments cut off from it up to given points may be in given ratio to the rectangle comprised by two other segments cut off (either way) up to given points'; compare P. Ver Eecke, *Pappus d'Alexandrie: La Collection Mathématique*, **2** (Brussels, 1933): 482.)

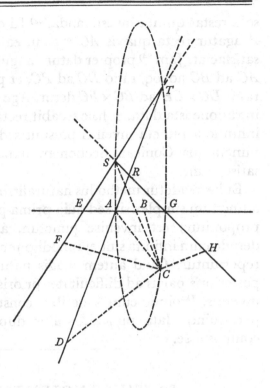

With these premissed nothing further remains to be done in composing[18] the *locus solidus* but to look for five points through which the figure shall pass.[19] In the Ancients' problem this is very easy. Let *AT*, *ST*, *AG*, *SG* be four straight lines given in position and to these from some common point *C* four more *CB*, *CF*, *CD*, *CH* are to be drawn at given angles subject to the condition that the product *CB* × *CF* of the first two shall bear a given ratio to the product *CD* × *CH* of the remainder.[20] The curve in which the point *C* is perpetually found will pass through the four intersections *A*, *G*, *S*, *T* of the given lines, for when *FC* is nil, the product *FC* × *CB* will be nil, and so also the product *CD* × *CH* along with one of the lines *CD*, *CH*. If it is *CD*, the point *C* will fall at *T*; if *CH*, at *S*. And thus when *CB* is nil the point *C* will fall either at *A* or *G*.[21] As a result the four points *A*, *G*, *S*, *T* through which the figure shall pass are given and it

---

(20) By a slip of his pen Newton twice inverted '*B*' and '*D*' in the manuscript. The accompanying figure is taken over from Descartes' *Geometrie* (note (2) above): 309, doubtless by way of its second Latin edition (*Geometria, à Renato des Cartes Anno 1637 Gallicè edita* (Amsterdam, ₂1659): 12), with the modification that *R*, *S*, *T* now mark the mutual intersections of *AD*, *EF*, *GH* and not (as in Descartes' figure) their respective meets with *CB*.

(21) Descartes' figures (*Geometrie* (note (2) above): 325 [where the locus (*C*) is a circle] and 331 [where it is drawn as an hyperbola]) are badly wrong in this respect, for in neither case is the conic drawn through *G* or would it pass accurately through the (there unmarked) intersections of *EF* with *AD* and *GH*. Soon after publication of the *Geometrie* in 1637 the error was brought to Descartes' attention by a certain 'Monsieur N.' [Roberval? or perhaps Godefroi de Haestrecht?] and accepted by him in a typically bad-tempered way in a letter (of about September 1639, most probably) to his editor and draughtsman Frans van Schooten: 'il ne faut pas auoir grande science, pour connoistre que la ligne courbe doit passer en cet exemple par les quatre intersections qu'il remarque: Car dans la figure de la page 325. on voit à l'œil, que puisque *CB* multipliée par *CF* doit produire vne somme égale [understand, in Newton's case, 'proportionnelle'] à *CD*, multipliée par *CH*, le point *C* se rencontre necessairement aux quatre intersections susdites; à sçauoir, en l'intersection *A*, pour ce qu'alors les lignes *BC* & *CD* sont nulles, & par consequent estant multipliées par les deux autres, elles composent deux riens, qui sont égaux entr'eux. Tout de mesme en l'intersection *G*, les lignes *CH* & *CB* sont nulles; & ainsi en l'vne des deux autres intersections, qui ne sont pas marquées dans la figure, *CD* & *CF*, & dans l'autre *CH* & *CF*, sont nulles. Mais on peut changer la question, en sorte que le mesme n'arriue point' (Claude Clerselier, *Lettres de Mr Descartes. Où il répond à plusieurs difficultez, qui luy ont esté proposées sur la Dioptrique, la Geometrie, & sur plusieurs autres sujets*, **3** (Paris,

sola restat quinta investiganda.[22] Id quod facillimum est. Nam per punctum *A* agatur recta quævis *AC*[23] et in ea quæratur punctum *C* quod Problemati satisfaciat. Jam[24] propter datum angulum *DAC* datur ratio *DC* ad *AC*, et ratio *AC* ad *BC* adeoʠ ratio *DC* ad *BC*, et proinde etiam ratio *CH* ad *FC* siquidem ratio *DC* × *CH* ad *BC* × *FC* detur. Age ergo rectam *SC* ea lege ut sit *CH* ad *CF* in ratione ista data, et hæc secabit rectam *AC* in puncto quæsito *C*. Eadem lege innumera puncta inveniri possunt sed uno aliquo invento habebimus quinʠ puncta quæ Conicam Sectionem juxta præcedentia determinando, Problemati satisfaciunt.

Et hæc videtur methodus naturalissima solvendi problema non tantum quod admodum simplex sit sed quia prima pars Problematis (prout ab ipso Cartesio proponitur) est invenire punctum aliquod datam habens conditionem,[25] deinde cum infinita sint ejusmodi puncta, determinare locum in quo ista omnia reperiuntur. Quid autem magis naturale quam reducere difficultates hujus posterioris partis ad difficultates prioris determinando locum ex paucis punctis inventis. Proinde cum Veteribus constiterit ratio ducendi conicam sectionem per quinʠ data puncta, nullus dubitaveri[t] eos hoc medio loca Solida composuisse.

---

§2 THE ANCIENTS' PROBLEM OF THE SOLID LOCUS SOLVED.[1]

From original autographs in the University Library, Cambridge[2]

Solutio Problematis Veterum de Loco solido.

Prop. 1.[3]

Si a Conicæ Sectionis puncto quovis P ad Trapezij alicujus ABDC in Conica ista sectione inscripti, latera quatuor infinitè producta AB, CD, AC, DB, totidem rectæ PQ, PR, PS,

---

1667): 469–70 [= (ed. Adam and Tannery) *Œuvres de Descartes*, **2** (Paris, 1898): 576]). Nevertheless, all subsequent editions of the *Geometrie*, in its original French and in Latin and English translation, repeat Schooten's inadequate figures without essential alteration.

(22) Read 'et solum restat quintum investigandum'! The Newtonian biographer with psychosexual leanings will doubtless find significance in the illogical attraction of the preceding fœminine 'figura' for the neutered 'puncta'.

(23) Newton first continued 'datum eff[iciens angulum cum *AB*]' (making a given [angle with *AB*]) but cancelled the phrase before completing it. Compare the sequel with note (19) above.

(24) The equivalent opening 'Est ergo' (Therefore) is deleted.

(25) Newton first completed his sentence with 'et secunda pars determinare locum in quo ejusmodi puncta omnia reperiuntur' (and the second part to determine the locus in which all points of the same nature are found). Descartes himself wrote: 'La question…estoit telle. Ayant trois ou quatre…lignes droites données par position; premierement on demande vn point, duquel on puisse tirer autant d'autres lignes droites, vne sur chascune des données, qui

remains only to discover a fifth.[22] Which is very easy. For through the point A let any straight line AC[23] be drawn and look for the point C in it which is to satisfy the problem. Now[24] because of the given angle $D\widehat{A}C$ there is given the ratio of DC to AC and that of AC to BC, and hence that of DC to BC, and consequently also the ratio of CH to FC seeing that the ratio of $DC \times CH$ to $BC \times FC$ is given. Draw therefore the straight line SC defined by CH to FC being in that given ratio, and this will cut the line AC in the required point C. By the same precept countless points may be found, but when we have found some single one we shall have the five points which, by determining a conic in line with the preceding, satisfy the problem.

And this seems the most natural method of solving the problem, not merely because it is relatively simple but since the first part of the problem (in the form propounded by Descartes himself) is to find some point having the given condition,[25] and thereafter, since there are an infinity of points of this sort, to determine the locus in which they are all found. What then is more natural than to reduce the difficulties of this latter part to those of the former by determining the locus from a few points after they are found? In consequence, since the Ancients did develop a procedure for constructing a conic through five given points, no one should have doubted that they composed solid loci by this means.

Translation

SOLUTION OF THE ANCIENTS' PROBLEM OF THE SOLID LOCUS
Proposition 1[3]

If from any point P of a conic to the four infinitely extended sides AB, CD, AC, BD of some quadrilateral ABDC inscribed in that conic an equal number of straight lines PQ,

---

façent auec elles des angles donnés, & que le rectangle contenu en deux de celles, qui seront ainsi tirées d'vn mesme point, ait la proportion donnée auec le quarré de la troisiesme, s'il n'y en a que trois; ou bien auec le rectangle des deux autres, s'il y en a quatre;.... Puis a cause qu'il y a tousiours vne infinité de diuers poins qui peuuent satisfaire a ce qui est icy demandé, il est aussy requis de connoistre, & de tracer la ligne, dans laquelle ils doiuent tous se trouuer. & Pappus dit que lorsqu'il n'y a que trois ou quatre lignes droites données, c'est en vne des trois sections coniques. mais il n'entreprend point de la determiner, ny de la descrire' (*Geometrie* (note (2) above): 306–7).

(1) Newton's general solution of the classical problem of the 'solid locus' (seen by him as that of constructing and identifying a conic which is required to pass through given points, touch lines given in position or satisfy other equivalent conditions) is built up gradually from the identification (in Propositions 1 and 2) of the 4-line locus with the general conic, culminating in the redefinition of the latter in homographic terms as the locus of intersections of two line-pencils which are in 1, 1 correspondence—'mutually determinable by simple geometry' as he phrases it. For his basic conic 'symptom' Newton refers to the semi-projective 'intercept-product' theorem (*Conics*, III, 17/18) which Apollonius himself drew upon for his own solution

PT in datis angulis ducantur, singulæ ad singula: rectangulum ductarum ad opposita duo latera PQ × PR erit ad rectangulum ductarum ad alia duo latera opposita PS × PT in data ratione.

Cas. 1. Ponamus imprimis lineas ad opposita latera ductas parallelas esse alterutri reliquorum laterum, puta *PQ* et *PR* lateri *AC* et *PS* ac *PT* lateri *AB*. Sintq̃ insuper latera duo ex oppositis, puta *AC* et *BD*, parallela. Et recta quæ bisecat parallela illa latera erit una ex diametris sectionis conicæ et bisecabit etiam *RQ*. Sit *O* punctum ubi *RQ* bisecatur, et erit *PO* ordinatim applicata ad diametrum illam. Produc *PO* ad *K* ut sit *OK* æqualis *PO* et erit *OK* ordinatim applicata ad

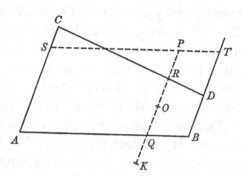

contrariam partem diametri. Cum igitur puncta *A, B, P* et *K* sint ad Conicam sectionem et *PK* secet *AB* in dato angulo, erit (per Prop. 17 & 18 lib. 3.[4] Apollonij) rectangulum *PQK* ad rectangulum *AQB* in data ratione. Sed *QK* et *PR* æquales sunt, utpote æqualium *OK, OP* et *OQ, OR* differentiæ; et inde etiam *PQK* et *PQ × PR* æqualia sunt: atq̃ adeo *PQ × PR* est ad *AQB* hoc est ad *PS × PT* in data ratione. Q.E.D.

---

of the *locus ad tres et quatuor lineas* (compare §1: notes (4) and (7) above), but this was probably familiar to him initially from his reading of Pappus' *Mathematical Collection* (see §1: note (1) above) rather than through direct contact with an edition of Apollonius' text itself, though at some later date he acquired a second-hand copy (see 1: 11, note (26) above) of Barrow's modernized epitome, *Apollonii Pergæi Conicorum Libri IIII* (London, 1675). The common source of the two loci, both circles somewhat disappointingly, on which Newton exemplifies his general solution of the 'locus solidus' we may tentatively identify as the second book of Pierre de Fermat's *Apollonii Pergæi Libri Duo de Locis Planis Restituti*, first published in his son Samuel's edition of his *Varia Opera Mathematica* (Toulouse, 1679): 28–43. The handwriting style of the manuscript suggests that the present tract was, in fact, composed about 1679–80 but there exists no contemporary documentary evidence to confirm this. A firm post-date is, as we shall see in the sixth volume, the spring of 1685 when Newton inserted a lightly revised, somewhat amplified version of the first seven propositions below in the first book of his *Principia* as Lemmas 17–21 and Propositions 22/23.

(2) In strict sequence, Add. 3963.8: 145^r–146^v/129–130^v/63^r–64^r/131^r–133^v/66[= 65a]^r–66^v/149^r–150^v/68^r/136^v–136^r. These sheets, together with the numerous first drafts interleaved with them, existed till 1960 scattered randomly in Add. 3963.8: 63^r–68^v, 3963.12: 127^r–149^v and 3963.13: 152^r and as a loose paper (now Add. 3963.8 : 149 bis^r–150^v) inserted in Newton's Waste Book (Add. 4004) between pages 72 and 73. Significant portions of the widely variant preliminary drafts are reproduced in appendix.

(3) Newton attempted at least four preliminary main drafts of this proposition. A first state of the first version (ULC. Add. 3968.8: 141^r–144^r, headed 'Tractatus de Compositione Locorum solidorum') is reproduced as Appendix 1 below: in this initial text 'Cas. 1' and 'Cas. 2' of the present proposition are essentially 'Cas. 1' of a first 'grade of demonstration' (when *BD* is parallel to *AC*) and 'Cas. 2' of a second grade (when *AC, BD* are no longer

PR, PS, PT are drawn at given angles, each to a separate one, the product PQ × PR of those drawn to two opposite sides will be to the product PS × PT of those drawn to the other two sides in a given ratio.

    Case 1. Let us suppose in the first instance that the lines drawn to opposite sides are parallel to one or other of the remaining sides, say *PQ* and *PR* to the side *AC*, *PS* and *PT* to the side *AB*. In addition, let two of the sides opposite, say *AC* and *BD*, be parallel. Then the straight line which bisects those parallel sides will be one of the diameters of the conic and will also bisect *RQ*. Let *O* be the point in which *RQ* is bisected, and *PO* will be ordinate to that diameter. Produce *PO* to

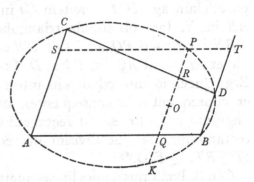

K so that *OK* equals *PO* and *OK* will be the ordinate on the opposite side of the diameter. Since, then, the points *A*, *B*, *P* and *K* are on the conic and *PK* intersects *AB* at a given angle, therefore (by Apollonius' [*Conics*][4] III, 17 and 18) the product *PQ* × *QK* will be to the product *AQ* × *QB* in a given ratio. But *QK* and *PR* are equal, being namely the difference of the equals *OK*, *OP* and *OQ*, *OR*, and consequently *PQ* × *QK* and *PQ* × *PR* are also equal. As a result, *PQ* × *PR* is to *AQ* × *QB*, that is *PS* × *PT*, in a given ratio. As was to be demonstrated.

---

parallel); conversely, 'Cas. 2' of the former grade and 'Cas. 1' of the latter are now subsumed into Proposition 2. This restructuring, already outlined in a first revise on ff. 141ᵛ/142ᵛ, was confirmed in a new draft 'De Compositione Locorum solidorum' (*ibid.*: 137ʳ–137ᵛ) where the separation of Proposition 1 into three component cases was established; but this was abandoned, uncompleted, for a further revise (*ibid.*: 127ʳ–127ᵛ) under the same title, to which Newton added a first version (*ibid.*: 127ᵛ–128ᵛ) of his final Propositions 2–4. Having achieved a roughly finished form of the opening of his tract, Newton entered a fair copy of Proposition 1, under the present final title 'Solutio Problematis Veterum de Loco solido', on a separate folio sheet (*ibid.*: 152ʳ) but this too he ultimately abandoned in favour of an insert (*ibid.*: 145ʳ–146ᵛ) in his second revise (127ʳ ff.), on which he wrote out the final version of Propositions 1–4 here reproduced. Subsequently, in the spring or early summer of 1685, he made a few alterations in a readily identifiable smaller hand in preparation for having it copied up into Book 1, Section 5, of his *Principia* (ULC. Dd. 9.46: 54–5/40–7/88–96 [= *Philosophiæ Naturalis Principia Mathematica* (London, ₁1687): 70–103]: 'Artic. v. Inventio Orbium ubi umbilicus neuter datur'), but these are not reproduced in the present text. Humphrey Newton's transcript of the revised 'Lemma xvɪɪ' (ULC. Dd. 9.46: 54–5 [= *Principia*: 70–1]) will be printed in the sixth volume.

    (4) In 1685 Newton inserted the clarification 'Conic.' at this place. In his original version (see note (6) of the following appendix) he had left the reference out completely. The theorem was, indeed, widely known among writers on conics before Apollonius, and Newton himself may well have derived it in the first instance from its (anonymous) quotation in Pappus' *Mathematical Collection*, vɪɪɪ, 13 (compare §1: notes (1) and (12) above). When Humphrey Newton transcribed the present text (see note (3)) he missed out this autograph insertion and the reading 'per Prop. 17 & 18 Lib. ɪɪɪ *Apollonii*' passed into print in the *Principia* (note (3) above): 70 without further ado.

Cas. 2. Ponamus jam Trapezij latera opposita *AC* et *BD* non esse parallela.[5]
Age *Bd* parallelam *AC* et occurrentem tum
rectæ *ST* in *t*, tum Conicæ Sectioni in *d*.
Junge *Cd* secantem *PQ* in *r*, & ipsi *PQ*
parallelam age *DM* secantem *Cd* in *M* et
AB in *N*. Jam ob similia triangula *BTt*,
DBN est *Bt* seu *PQ* ad *Tt* ut *DN* ad *NB*.
Sic et *Rr* est ad *AQ* seu *PS* ut *DM* ad *AN*.
Ergo ducendo antecedentes in antecedentes
et consequentes in consequentes, ut rect-
angulum *PQ* × *Rr* est ad rectangulum *Tt* × *PS* ita rectangulum *QPr* est ad
rectangulum *SPt*, ac divisim ita rectangulum *PQ* × *PR* est ad rectangulum
SP × *PT*. Q.E.D.[6]

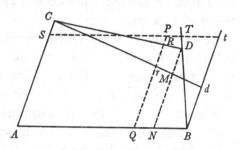

Cas. 3. Ponamus deniꝗ lineas quatuor *PQ*,
PR, *PS*, *PT* non esse parallelas lateribus *AC*,
AB, sed utcunꝗ inclinatas.[7] Earum vice age
Pq, *Pr* parallelas *AC* et *Ps*, *Pt* parallelas *AB* et
propter datos angulos triangulorum *PQq*,
PRr, *PSs*, *PTt* dabuntur rationes *PQ* ad *Pq*,
PR ad *Pr*, *PS* ad *Ps* et *PT* ad *Pt* atꝗ adeo
rationes compositæ *PQ* × *PR* ad *Pq* × *Pr* et
PS × *PT* ad *Ps* × *Pt*. Sed per superius demon-
strata ratio *Pq* × *Pr* ad *Ps* × *Pt* data est. Ergo
et ratio *PQ* × *PR* ad *PS* × *PT*. Q.E.D.

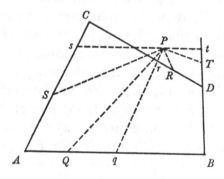

Prop. 2.[8]

Et [*e*] *contra,*[9] *si rectangulum ductarum ad opposita duo latera Trapezij PQ × PR sit
ad rectangulum ductarum ad reliqua duo latera PS × PT in data ratione: punctum P a quo
lineæ ducuntur attinget Conicam sectionem circa Trapezium descriptam.*

Per puncta *A*, *B*, *C*, *D* et aliquod infinitorum punctorum *P* puta *p* concipe
Conicam Sectionem describi: dico punctum *P* hanc semper attingere. Si negas,

---

(5) Preceding versions of the proposition (see note (3) above) have the variant opening
'Sin parallelogrammi[!] latus *BD* non sit parallelum lateri opposito *AC*' (But if the side *BD*
of the quadrilateral be not parallel to the opposite side *AC*), first drafted on ULC. Add.
3963.8: 141ᵛ.

(6) Earlier drafts of this second case (ULC. Add. 3963.6: 137ᵛ → 127ᵛ) concluded more
succinctly with 'Ergo *PQ* × *Rr*. *Tt* × *PS*::*DN* × *DM*. *AN* × *NB*::(per Cas. 1)

$$PQ \times Pr.\ PS \times Pt::(\text{divisim})\ PQ \times PR.\ PS \times PT.\ \text{Q.E.D.'}$$

The generality and power of this theorem is somewhat concealed by Newton's classically
modelled proof: to anticipate Proposition 5 following we deduce immediately from the equality
PQ × *PR*/*PS* × *PT* = *PQ* × *Pr*/*PS* × *Pt* that *PR*/*Pr* = *PT*/*Pt*, which when set in modern cross-
ratio form yields the homographic definition of the conic (*D*) through *A*, *B*, *C*, *d* and *P* by

Case 2. Let us now suppose that the opposite sides AC and BD are not parallel.[5] Draw Bd parallel to AC and meeting both the straight line ST in t and the conic in d. Join Cd cutting PQ in r and parallel to PQ draw DM cutting Cd in M and AB in N. Now because of the similar triangles BTt, DBN there is Bt (or PQ) to Tt as DN to NB. So also Rr is to AQ (or PS) as DM to AN. Therefore, on separately multiplying prior and posterior members of the ratios into one another, as the product $PQ \times Rr$ is to the

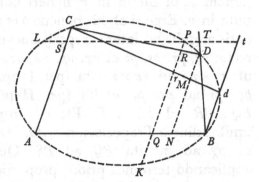

product $Tt \times PS$ so is the product $PQ \times Pr$ to the product $PS \times Pt$, and *divisim* so is the product $PQ \times PR$ to the product $PS \times PT$. As was to be demonstrated.[6]

Case 3. Let us suppose finally that the four lines PQ, PR, PS, PT are not parallel but inclined to one another in any manner.[7] In their place draw Pq, Pr parallel to AC and Ps, Pt parallel to AB, and then, because of the given angles of the triangles PQq, PRr, PSs and PTt, the ratios of PQ to Pq, PR to Pr, PS to Ps and PT to Pt will be given, and consequently the compound ratios $PQ \times PR$ to $Pq \times Pr$ and $PS \times PT$ to $Ps \times Pt$ also. But, by what is demonstrated above, the ratio of $Pq \times Pr$ to $Ps \times Pt$ is given. So also therefore that of $PQ \times PR$ to $PS \times PT$. As was to be demonstrated.

Proposition 2[8]

Conversely,[9] *if the product $PQ \times PR$ of lines drawn to two opposite sides of a quadrilateral be to the product $PS \times PT$ of those drawn to the remaining two in a given ratio, the point P from which the lines are drawn will belong to a conic described round the quadrilateral.*

Through the points A, B, C, D and some one, say p, of the infinity of points P imagine a conic to be described: I assert that the point P belongs always to the

---

$C(PDdA) = (PRr\,\infty) = (PTt\,\infty) = B(PDdA)$. The parallelism of RT, rt to BK is readily shown from first principles since these are the Pascal-lines of the respective inscribed hexagons $PKBDCL$, $PKBdCL$, where L is the second meet of PS with the conic and accordingly (since the Pascal-line of the 'Leibnizian' inscribed hexagon $PKBACL$ is at infinity) CL is paralle to BK.

(7) This replaces an earlier phrase (on f. 127$^\mathrm{v}$) 'in quibusvis datis angulis duc[tas]' (drawn at any given angles).

(8) Little variant prior versions of this proposition, a generalization (see note (3) above) of 'Cas. 2' of the first grade of the first state of Proposition 1 (reproduced in Appendix 1 below), exist on ULC. Add. 3963. 6: 141$^\mathrm{v}$/142$^\mathrm{v}$ and 127$^\mathrm{v}$–128$^\mathrm{r}$. Two slight alterations made by Newton in 1685, when his amanuensis Humphrey Newton transcribed it as Lemma XVIII of the lectures 'De Motu Corporum Liber Primus' (ULC. Dd. 9.46: 55 [= *Principia* (note (3) above): 72]), are noticed below.

(9) This replaces an earlier opening (on f. 141$^\mathrm{v}$) '*Ijsdem positis*' (*With the same suppositions*)— one, curiously, to which Newton returned in his 1685 revision of the proposition,

junge *AP* secantem hanc Conicam Sectionem alibi quam in *P* si fieri potest,
puta in *π*. Ergo si ab his punctis *p* et *π* in
datis angulis ad latera trapezij ducantur
rectæ *pq*, *pr*, *ps*, *pt* et *πχ*, *πρ*, *πσ*, *ππ*, erit
ut *πχ* × *πρ* ad *πσ* × *ππ* ita (per Prop. 1)
pq × *pr* ad *ps* × *pt*, et ita (per Hypoth)
PQ × *PR* ad *PS* × *PT*. Est et propter
similitudinem Trapeziorum *πχAσ*, *PQAS*,
ut *πχ* ad *πσ* ita *PQ* ad *PS*. Quare

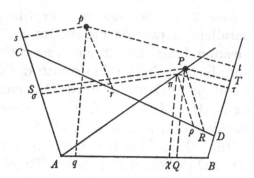

applicando terminos prioris proportionis ad terminos correspondentes hujus,
erit *πρ* ad *ππ* ut *PR* ad *PT*. Ergo Trapezia æquiangula *Dρππ*, *DRPT* similia
sunt, eorumcp diagonales *Dπ*, *DP* coincidunt. Cadit[10] itacp *π* in intersectionem
rectarum *AP*, *DP* adeocp coincidit cum puncto *P*. Quare punctum *P*, ubicuncp
sumatur, incidit in assignatam Conicam Sectionem. Q.E.D.

Prop 3.[11]

*Invenire punctum P a quo si quatuor rectæ PQ, PR, PS, PT ad totidem alias positione
datas rectas AB, CD, AC, BD singulæ ad singulas in datis angulis ducantur, rectangulum
sub duabus ductis PQ × PR sit ad rectangulum sub
alijs duabus PS × PT in data ratione.*

Ab aliquo angulorum trapezij *A*, age rectam *AP* secantem latera opposita *BD* in *H*
et *CD* in *I*,[12] et ob datos omnes angulos
figuræ dabuntur rationes *PQ* ad *PA* et *PA*
ad *PS*. adeocp ratio *PQ* ad *PS*. Auferendo
hanc a data ratione *PQ* × *PR* ad *PS* × *PT*
dabitur ratio *PR* ad *PT* et huic addendo
datas rationes *PI* ad *PR* et *PT* ad *PH* dabitur
ratio *PI* ad *PH* atcp adeò punctum *P*. Q.E.I.[13]

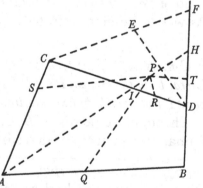

---

(10) This was converted in 1685 to read 'Incidit'—a trivial alteration which passed into
print soon afterwards.

(11) Preliminary versions of this and the following proposition, a first draft (ULC. Add.
3963.8: 144ʳ–144ᵛ) and a minor revise (*ibid.*: 128ʳ–128ᵛ, reproduced in Appendix 2), reveal
that Newton considerably altered its structure at a late stage, omitting a repetitive second
paragraph from the construction and suitably recasting the first. In 1685 this was transcribed
by Humphrey Newton as Lemma XIX and its Corollary 1 of the 'De Motu Corporum Liber
Primus' (ULC. Dd. 9.46: 40 [= *Principia* (note (3) above): 74–5]).

(12) In an afterthought, probably in 1685, Newton cancelled this opening in favour of the
version added to the draft on f. 128ʳ: 'Lineæ *AB*, *CD* ad quas rectæ duæ *PQ*, *PR* unum rectangulorum continentes ducuntur conveniant cum alijs duabus positione datis lineis in punctis
A, *B*, *C*, *D*: Ab horum punctorum aliquo *A* age rectā quamvis in qua velis punctum *P* reperiri.

latter. Should you deny it, join *AP* cutting
this conic elsewhere, if possible, than at
P, say at π. In consequence, if from these
points p and π the straight lines pq, pr, ps,
pt and $\pi\chi$, $\pi\rho$, $\pi\sigma$, $\pi\tau$ be drawn at given
angles to the sides of the quadrilateral,
there will be as $\pi\chi \times \pi\rho$ to $\pi\sigma \times \pi\tau$ so (by
Proposition 1) $pq \times pr$ to $ps \times pt$, and so
(by hypothesis) is $PQ \times PR$ to $PS \times PT$.
Furthermore, through the similarity of

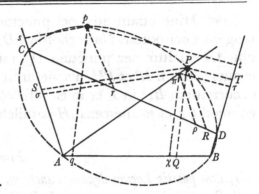

the quadrilaterals $\pi\chi A\sigma$ and *PQAS*, as $\pi\chi$ to $\pi\sigma$ so *PQ* to *PS*. Hence by dividing
the terms of the previous proportion by the corresponding terms of the
present one, there will be $\pi\rho$ to $\pi\tau$ as *PR* to *PT*. Therefore the equiangular
quadrilaterals $D\rho\pi\tau$, *DRPT* are similar and their diagonals $D\pi$, *DP* coincide.
Accordingly, π falls at the intersection of the lines *AP*, *DP* and so coincides with
the point *P*. Hence the point *P*, wherever it be taken, falls in the allotted conic.
As was to be demonstrated.

Proposition 3[11]

*To find a point P such that, if four straight lines PQ, PR, PS, PT be drawn from it at
given angles to an equal number of other straight lines AB, CD, AC, BD given in position,
one to each separately, the product PQ × PR of two of those drawn shall be to the product
PS × PT of the other two in a given ratio.*

From some corner *A* of the quadrilateral draw the straight line *AP* cutting
the opposite sides *BD* in *H* and *CD* in
I,[12] and then, because all angles in the
figure are given, the ratios of *PQ* to *PA*
and of *PA* to *PS* will be given, and
consequently that of *PQ* to *PS*. On eliminating this from the given ratio *PQ × PR*
to *PS × PT* there will be given the ratio
of *PR* to *PT*, and by adjoining to this
the given ratios *PI* to *PR* and *PT* to *PH*

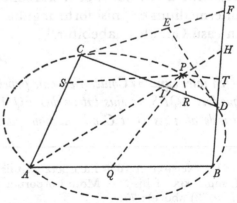

the ratio of *PI* to *PH* and hence the point *P* will be given. As was to be found.[13]

Secet ea lineas oppositas *BD*, *CD*, nimirum *BD* in *H* et *CD* in *I*' (Let the lines *AB*, *CD* to which
the two straight lines *PQ*, *PR* containing one of the products are drawn meet the other two
lines given in position in the points *A*, *B*, *C*, *D*: from some one of these points, *A*, draw any
straight line in which you would like the point *P* to be located, let it cut the opposing lines *BD*,
CD, namely *BD* in *H* and *CD* in *I*).

(13) In his preliminary versions of this proposition (f. 144ʳ → f. 128ʳ) Newton inserted an
'example' of the construction in a following paragraph. See note (23) of Appendix 2 below.

Corol. Hinc etiam ad Loci punctorum infinitorum *P* punctum quodvis *D* tangens duci potest. Nam chorda *PD* ubi puncta *P* ac *D* conveniunt[,] hoc est ubi *AH* ducitur per punctum *D*[,] tangens evadit. Quo in casu ultima ratio evanescentium *IP* et *PH* invenietur ut supra. Ipsi igitur *AD* duc parallelam *CF* occurrentem *BD* in *F* et in ea ultima ratione sectam in *E* et *DE* tangens erit, eo quod *CF* et evanescens *IH* parallelæ sint et in *E* et *P* similiter sectæ.

Prop 4.[14]

Ijsdem positis Locum definire punctorum infinitorū P.

[15]Per quodvis punctorum *A, B, C, D* puta *A* duc Loci tangentem *AE* & per aliud quodvis *B* duc tangenti parallelam *BF* occurrentem Loco in *F*. Biseca *BF* in *G* et acta *AG* diameter erit ad quam *BG* et *FG* ordinatim applicantur. Hæc *AG* occurrat Loco in *H* et erit *AH* latus transversum ad quod latus rectum est ut BG^q ad *AGH*. Si *AG* nullibi[16] occurrit Loco, linea *AH* existente infinita, Locus erit Parabola et latus rectum ejus $\dfrac{BG^q}{AG}$. Sin ea alicubi occurrit, Locus

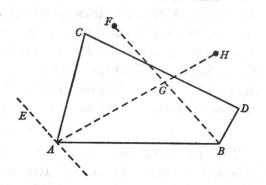

Hyperbola erit ubi *A* et *H* sita sunt ad easdem partes ipsius *G* et Ellipsis ubi *G* intermedium est, nisi forte angulus *AGB* rectus sit & insuper $BG^q = AGH$, quo in casu Circulus habebitur.[17]

Prop 5.[18]

Si inter cujusvis Conicæ Sectionis puncta duo data A et P inscribatur parallelogrammum quodvis AQPS, et ejus latera duo AQ AS producantur donec Curvæ occurrant in B et C, deinde ab istis B et C ad quintum quodvis Curvæ punctum D agantur rectæ BD, CD,

---

(14) Newton in 1685 incorporated this into the preceding proposition as Corollary 2 of Lemma XIX of his 'De Motu Corporum Liber Primus' (ULC. Dd. 9.46: 40–1 [= *Principia* (note (3) above): 75]).

(15) Newton first began to copy 'Sint *A, B, C, D* concursus...' from his revised draft on f. 128ᵛ, and then changed his mind, substituting the present, more pithy opening. (The full former version is reproduced in Appendix 2 below.) Several other cancellations in the sequel which tie the present text narrowly to the predraft on f. 128ᵛ are not specially noticed.

(16) At a finite distance by implication.

(17) In 1685 at this point in the revised transcript copied by Humphrey Newton as 'ARTIC. v' of the 'De Motu Corporum Liber Primus' (ULC. Dd. 9.46: 41 [= *Principia* (note (3) above): 75]) Newton added a late autograph insert to direct his readers' attention to what he had achieved: 'Atqȝ ita Problematis Veterum de quatuor lineis ab Euclide incœpti et ab Apollonio continuati non calculus sed compositio Geometrica qualem Veteres quærebant,

Corollary. Hence also at any point D of the locus of the infinity of points P a tangent may be drawn. For when the points P and D are coincident, that is, when AH is drawn through the point D, the chord PD comes to be tangent. In this case the ultimate ratio of the vanishing lines IP and PH will be found as above. Parallel to AD, therefore, draw CF meeting BD in F and cut at E in that ultimate ratio, and DE will be tangent for the reason that CF and the vanishing line IH are parallel and similarly cut at E and P.

Proposition 4[14]

With the same suppositions to define the locus of the infinity of points P.

[15]Through any of the points A, B, C or D, say A, draw the tangent AE to the locus and through any other point B draw BF parallel to the tangent, meeting the locus in F. Bisect BF in G and when AG is drawn it will be the diameter to which BG and FG are ordinate. Let this line AG meet the locus in H and AH will be the transverse diameter whose *latus rectum* is to it as BG^2 to $AG \times GH$. If AG nowhere[16] meets the locus, with the line AH proving to be infinite, the locus will be a parabola and BG^2/AG its *latus rectum*.

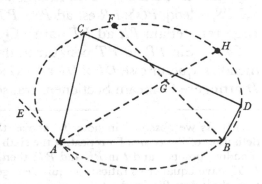

But if it does meet it somewhere, the locus will be a hyperbola when A and H are situated on the same side of G and an ellipse when G intervenes between them, unless perchance the angle $\widehat{AGB}$ is right and in addition $BG^2 = AG \times GH$, in which case a circle will be obtained.[17]

Proposition 5[18]

If between two given points A and P of any conic there be inscribed any parallelogram $AQPS$ and its two sides AQ, AS extended till they meet the curve in B and C, and subsequently from those points B and C to any fifth one D of the curve straight lines BD, CD

---

in hoc Corollario [*sc.* Prop. 4; see note (14) above] exhibetur.' The 'calculus [Algebraicus]' which he rejects is, of course, that of Descartes' in the first two books of his *Geometrie*; compare the opening paragraph of Newton's 'Veterum Loca solida restituta' (§1 preceding).

(18) Apart from a slightly changed enunciation this became, in 1685, Lemma xx of the 'De Motu Corporum Liber Primus' (ULC. Dd. 9.46: 41–2 [= *Principia* (note (3) above): 75–7]). Newton's draft of the three following corollaries exists on a stray sheet in private possession below the 'De solutione planorum Problematum per regulam trium' (3, §4 below) and is immediately followed by the first draft of Proposition 7 (titled 'Prop. 8') which is reproduced in Appendix 3. Also on other sides of this folded sheet are to be found the observations on cubic properties reproduced as 3, §2, and portions of the preliminary computation, for 3, §3 (especially those listed in Appendices 4 and 6 to it).

occurrentes lateribus oppositis PS PQ in T et R, erit semper PR ad PT in ratione data. Et vice versa si PR ad PT sit in ratione data punctum D attinget Conicam Sectionem per puncta quatuor ABPC transeuntem.[19]

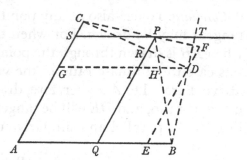

Cas. 1. Jugantur enim *BP, CP* et a puncto *D* age rectam *DG* parallelam *AB* et occurrentem *PB, PQ* et *CA* in *H, I* ac *G,* ut et rectam *DE* parallelam *AC* et occurrentem *PC, PS* et *AB* in *F, K* et *E.* Est ergo per Prop 1, $DE \times DF$ ad $DG \times DH$ in ratione data. Sed est $DE(IQ).PQ(::HB.PB)::DH.PT$, adeoqʒ $PQ \times DF$ est ad $DG \times PT$ in ratione data. Est et $DF.PR(::DC.RC)::DG.IG$ vel *PS,* adeoqʒ $PQ \times PR$ est ad $PS \times PT$ in ratione data. Sed *PQ* et *PS* dantur. Ergo ratio etiam *PR* ad *PT* datur. Q.E.D.

Cas. 2. Sin *PR* ad *PT* ponatur in data ratione, tunc simili ratiocinio regrediendo sequetur esse $DE \times DF$ ad $DG \times DH$ in ratione data, adeoqʒ[20] punctum *D* attingere Conicam Sectionem transeuntem per puncta *A, B, P, C.* Q.E.D.

---

(19) As we observed in note (6) above, this property is equivalent to the homographic definition of the conic. To repeat, if the sixth point *d* on the conic is constructed from corresponding points *r* and *t* in *PR* and *PT*, then *PR*:*Pr* = *PT*:*Pt* and so the line-pencils *C(R),* *B(T)* have equal cross-ratios: in equivalent geometrical terms, since *RT* is evidently parallel to *rt,* the points *R, T* and *r, t* are in perspective when seen from a unique point at infinity in the direction of *RT.* More generally, as Newton will go on to deduce in Proposition 12, if the line-pencils *C(R), B(T)* mark out corresponding ranges of points, (*R'*) and (*T'*), on any arbitrary fixed lines *PQ'* and *PS'* through *P,* then these ranges of points will be in 1, 1 correspondence (and hence related quadruples of points will have equal cross-ratios). This general insight is Newton's undivided discovery, but he had—whether he ever knew it is improbable—been to some extent anticipated by Bonaventura Cavalieri, who in the sixth book of his *Geometria Indivisibilibus Continuorum Nova quadam ratione promota* (Bologna, 1635: Liber Sextus, Theorema ix (Propositio ix), Corollarium/Scholium: ₆20–1) showed that, in the case where *B* (and hence *A* and *Q*) is at infinity and *CPD* is a parabola of vertex *C,* the ratio of *PR* to *PT* is constant, 'Ex quo novus, ni fallor, ac pulcherrimus describendi parabolam elicitur modus'. Subsequently on 15 June 1641 (N.S.) he communicated to Torricelli the improved result that the ratio of *PR* to *PT* remains constant if *B* and *C* are chosen to be two opposing vertices of a central conic with *PR* and *PT* drawn parallel to the diameter *BC* and its conjugate direction (*Opere di Evangelista Torricelli,* **3** (Faenza, 1919): 53–4) and his following proposition 'De modo facili describendi Sectiones Conicas, & in omnibus uniformi' was later published in the *Exercitatio Sexta* of his *Exercitationes Geometricæ Sex* (Bologna, 1647): 445–7, where he observed (p. 445) that 'Dum meæ *Geometriæ Indiv.* Librum contexerem incidi in modum describendæ parabolæ satis facilem, quem in eo Libro Scholio 2. Prop. 9 studiosis postmodum communicavi. Diù dolui non posse pariter hyperbolam, & ellipsem tam facili ratione describi. At deniqʒ animadverti idem in ijs quoqʒ perfici posse, quod mihi non parum attulit voluptatis'. Ten years afterwards Jan de Witt established a moving-angle construction of the three species of conic which is equivalent to Cavalieri's (see his *Elementa Curvarum Linearum* [appended by Schooten to *Renati Descartes Geometriæ Pars Secunda* (Amsterdam, 1661): 153–340]: Liber I, Cap. IV, 'Alia Parabolam, Hyperbolam, & Ellipsin in plano delineandi Methodus': 229–38)

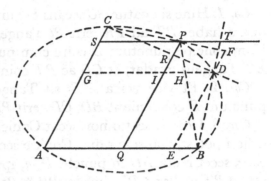

*be drawn meeting the opposite sides PS, PQ
in T and R, then will PR to PT be always
in a given ratio. Conversely, if PR to PT
be in a given ratio, the point D will belong to
a conic passing through the four points A, B,
P and C.*[19]

Case 1. For let BP, CP be joined and
from the point D draw the line DG
parallel to AB and meeting PB, PQ and
CA in H, I and G, as also the line DE
parallel to AC and meeting PC, PS and AB in F, K and E. In consequence, by
Proposition 1, $DE \times DF$ is to $DG \times DH$ in a given ratio. But

$$DE \text{ (or } IQ\text{)}:PQ(=HB:PB)=DH:PT$$

and so $PQ \times DF$ is to $DG \times PT$ in a given ratio. Also

$$DF:PR(=DC:RC)=DG:(IG \text{ or) } PS$$

and hence $PQ \times PR$ is to $PS \times PT$ in a given ratio. But PQ and PS are given.
Therefore the ratio of PR to PT is also given. As was to be demonstrated.

Case 2. But if PR be supposed to be to PT in a given ratio, then by a similar
argument in reverse it will follow that $DE \times DF$ is to $DG \times DH$ in a given ratio,
and hence[20] that the point D belongs to a conic passing through the points
A, B, P and C. As was to be demonstrated.

---

and Newton had noted the parabolic construction in his undergraduate annotations of the
Geometria (see I: 35, 40), but the underlying homographic property of the conic was there well
hidden. (Compare Michel Chasles, *Aperçu Historique sur l'Origine et le Développement des Méthodes
en Géométrie* (Paris, ₃1889): Chapitre III, §8: 100.) Independently of Cavalieri but about the
same time Pierre de Fermat had, in seeking to restore certain of the porisms in Euclid's lost
treatise on the topic on the basis of Pappus' account of the work, stumbled on a second
particular case of Newton's present theorem: namely, that in which the lines PQ and PS are
coincident and the resulting conic (in general, a hyperbola one of whose asymptotes is parallel
to PRT) is taken in the limit form of a parabola of which PRT is a diameter. This new porism,
communicated in June 1658 (by way of Kenelm Digby) to John Wallis and William Brouncker
as a challenge, was first published by Wallis in appendix to his *Commercium Epistolicum De
Quæstionibus quibusdam Mathematicis, nuper habitum* (Oxford, 1658): Epistola XVII: 188, and reappeared twenty years later in Fermat's posthumous *Porismatum Euclidæorum Renovata Doctrina
& sub formâ Isagoges recentioribus Geometris exhibita* [= *Varia Opera Mathematica* (Toulouse, 1679),
116–19]: 117: Porisma secundum; the former work Newton had annotated as a young man
(see I: 116–19), though perhaps in an issue which lacked the appendix, while it is very probable
(see note (1) above) that he had recently read portions at least of Fermat's *Opera*, but the
distance between Fermat's particular case and Newton's general proposition is great.

(20) Understand 'per Prop. 2' (by Proposition 2). In 1685 Newton made an equivalent
insertion, 'per Lemma XVIII', at this point in Humphrey Newton's transcript (ULC. Dd.
9.46: 41).

Cor 1. Hinc si agatur *BC* secans *PQ* in *r*, et in *PT* capiatur *Pt* in ratione ad *Pr* quam habet *PT* ad *PR*, erit *Bt* Tangens Conicæ sectionis ad punctum *B*.[21] Nam concipe punctum *D* coire cum puncto *B* ita ut Chorda *BD* evanescente *BT* Tangens evadat, et *CD* ac *BT* coincident cum *CB* et *Bt*.

Cor. 2. Et vice versa si *Bt* sit Tangens et ad quodvis Conicæ sectionis[22] punctum *D* conveniant *BD*, *CD*: erit *PR* ad *PT* ut *Pr* ad *Pt*.

Cor. 3. Conica sectio non secat Conicam sectionem in quinǫ punctis. Nam, si fieri potest, transeant duæ Conicæ sectiones per quinǫ puncta *A*, *B*, *C*, *D*, *P* easǫ secet recta *BD* in punctis *D*, *d*, ipsamǫ *PQ* secet recta *Cd* in *ρ*. Ergo *PR* est ad *PT* ut *Pρ* ad *PT*, hoc est *PR* & *Pρ* sibi invicem æquantur contra Hypoth.

Prop. 6.[23]

Conicam Sectionem definire quæ per data quinǫ puncta transibit.

Sunto quinǫ puncta *A*, *B*, *C*, *D*, *P*. Ab aliquo eorum *A*, ad alia duo quævis *B*, *C*[24] duc rectas *AB*, *AC*, & a quarto *P* duc alias duas rectas[,][25] unam *PT*

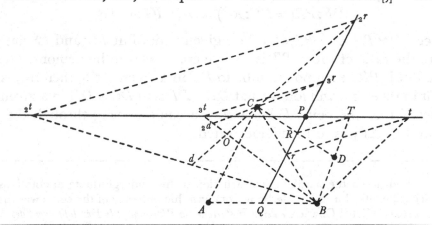

(21) Newton's first draft (see note (18) above) concludes: 'Concipe enim chordam *BD* minui donec punctum *D* incidat in ipsum *B* et *BT* ut notum est tangens evadet, sed in isto casu *CD* coincidet cum *CB* et punctū *R* cum puncto *r* adeoǫ etiam punctum *T* cum puncto *t* siquidem sit *PR.PT*::*Pr.Pt*. Quare *Bt* coincidit cum Tangente' (For imagine that the chord *BD* diminishes till the point *D* coincides with *B*, and then *BT*, as is known, will come to be tangent; but in that case *CD* will be coincident with *CB* and the point *R* with the point *r*, and consequently also the point *T* with the point *t* seeing that *PR*:*PT* = *Pr*:*Pt*. Hence *Bt* coincides with the tangent).

(22) Understand 'per puncta *A*, *B*, *C*, *P* transeuntis' (passing through the points *A*, *B*, *C* and *P*).

(23) A shortened version of this proposition, omitting the detail of the construction and the distinction of the conic's species (which, indeed, merely repeats Proposition 4), was transcribed by Humphrey Newton as 'Prop. xxii Prob. xiv. Trajectoriam per data quinǫ puncta describere' of Newton's 'De Motu Corporum Liber Primus' (ULC. Dd. 9.46: 43 [= *Principia* (note (3) above): 79–80]), while its Corollary 1 became 'Cas. 1' of the following 'Prop. xxiii Prob. xv. Trajectoriam describere quæ per data quatuor puncta transibit et rectam continget positione datam' (*ibid.*: 44 [= *Principia*: 81–2]).

Corollary 1. Consequently, if *BC* be drawn meeting *PQ* in *r*, and along *PT* be taken *Pt* in the ratio to *Pr* which *PT* has to *PR*, then *Bt* will be tangent to the conic at the point *B*.[21] For imagine the point *D* to coalesce with the point *B* with the result that, as the chord *BD* vanishes, *BT* comes to be tangent, and then *CD* and *BT* will coincide with *CB* and *Bt*.

Corollary 2. Conversely, if *Bt* be tangent and *BD*, *CD* meet at any point *D* on the conic,[22] then will *PR* be to *PT* as *Pr* to *Pt*.

Corollary 3. A conic does not intersect a conic in five points. For, if it is possible, let two conics pass through the five points *A*, *B*, *C*, *D* and *P*, and let the straight line *BD* cut these in the points *D*, *d* with *ρ* the intersection of the straight line *Cd* and *PQ*. Then *PR* is to *PT* as *Pρ* to *PT*, that is, *PR* and *Pρ* are equal to each other, contrary to hypothesis.

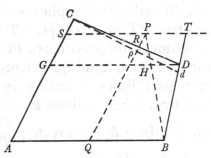

Proposition 6[23]

To define the conic which shall pass through five given points.

Let the five points be *A*, *B*, *C*, *D* and *P*. From some one of them, *A*, to any two others *B* and *C*[24] draw the straight lines *AB*, *AC*, and from a fourth *P* draw any two other straight lines,[25] one *PT* parallel to *AB* and the other *PR* parallel to

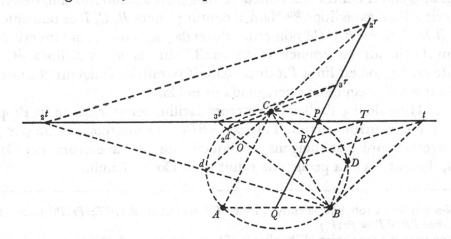

(24) In 1685 Newton introduced at this point the defining clause 'quæ poli nominentur' (let these be named 'poles'), doubtless having in mind his similar terminology for the organic construction of conics (ii: 118: '[puncta] circa quæ tanquam polos regulæ rotentur').

(25) A first, cancelled continuation reads equivalently 'complentes cum his parallelogrammum *AQPS*' (completing together with these the parallelogram *AQPS*). Newton first began this sentence 'Junge tria quævis *A*, *B*, *C*, et a quarto *P* ad trianguli *AB*[*C* latera duo quævis *AB*, *AC* parallelas duc duas rectas]' (Join any three *A*, *B*, *C* and from a fourth *P* [parallel to any two sides] of the triangle *AB*[*C* draw two straight lines]).

parallelam AB et alteram PR parallelam AC. Junge CB secantem PR in r, et ab ejus terminis C, B ad quintum punctum D age CD secantem PR in R et BD secantem PT in T. Age rt parallelam RT et occurrentem PT in t,[26] et acta Bt tanget Conicam sectionem in B, per Cor 1 Prop 5. Huic tangenti parallelam age Cd secantem PR in $_2r$. In PT cape $P_2t . P_2r :: PT . PR$. et ipsi RT parallelam age $_2r_2t$ occurrentem PT in $_2t$. Et punctum d ubi B_2t occurrit $_2rC$ erit ad Conicam sectionem per Prop 5. Biseca ergo Cd in O et acta BO erit diameter Figuræ, & CO, dO ordinatim applicatæ ad hanc diametrum. Insuper produc BO donec occurrat PT in $_3t$, et ipsi RT age parallelam $_3t_3r$ occurrentem PR in $_3r$, et punctum $_2d$ ubi $_3rC$ occurrit BO attinget curvam per Prop: 5, adeoq̃ B_2d latus erit transversum, ad quod latus rectum est ut $CO^{quadr.}$ ad $BO \times O[_2d]$, Figura existente Ellipsi si O cadit inter B ac $_2d$, aliter Hyperbola:[27] puta si $_3rC$ et BO concurrunt. Sed si lineæ istæ nullibi concurrunt, parallelæ existentes, Figura erit Parabola & latus rectum $\dfrac{CO^q}{BO}$.

Coroll. 1. Hinc si dentur quatuor puncta et tangens Conicæ sectionis ad unum istorum punctorum descriptio Curvæ expeditior evadet. Nam si e.g. dentur puncta A, B, C, P & tangens Bt positione lineæ CD, BD jam non ducendæ erunt, sed concipiendum erit quod punctum D coincidit cum puncto B, et lineæ CD, BD cum lineis CB, Bt, adeoq̃ puncta R ac T eadem esse cum punctis r et t, dein cætera peragenda ut supra.

Coroll. 2. Quod si dentur tria puncta & tangentes ad duo istorum, descriptio Curvæ erit adhuc expeditior.[28] Nam si dentur puncta B, C, P et tangentes ad puncta B & C, lineæ BA, CA non erunt ducendæ, sed concipiendum erit quod punctum A coincidit cum puncto C et linea AB cum linea CB ac linea AC cum Tangente ad C; hoc est linea PR ducenda erit parallela Tangenti et linea PT parallela ipsi CD, dein cætera peragenda ut in Cor 1.

Corol. 3. Hinc deniq̃ Problema Veterum facilius solvitur quàm in Prop. 4. Scilicet cum concursus laterum Trapezij $ABDC$ sint quatuor puncta per quæ Conica Sectio transit: ubi tangens ad aliquod istorum punctorum, per Corol: Prop: 3, ducitur, cætera peragenda erunt ut in Corol. 2 hujus.

---

(26) Newton first wrote equivalently 'Produc PT ad t ut sit $PR.PT::Pr.Pt$' (Extend PT to t such that $PR:PT = Pr:Pt$).

(27) Compare the conclusion of Problem 1 of §1 above, of which this argument is, *mutatis mutandis*, almost an exact copy.

AC. Join CB intersecting PR in r, and from its end-points C, B to the fifth point D draw CD intersecting PR in R and BD intersecting PT in T. Draw rt parallel to RT and meeting PT in t,[26] and Bt when drawn will (by Proposition 5, Corollary 1) touch the conic at B. Parallel to this tangent draw Cd intersecting PR in $_2r$. In PT take [P_2t where] $P_2t:P_2r = PT:PR$, and parallel to RT draw $_2r_2t$ meeting PT in $_2t$. The point d in which B_2t meets $_2rC$ will (by Proposition 5) be on the conic. Therefore bisect Cd in O and BO when drawn will be a diameter of the figure with CO, dO ordinate to it. Moreover, extend BO till it meets PT in $_3t$ and parallel to RT draw $_3t_3r$ meeting PR in $_3r$, and the point $_2d$ in which $_3rC$ meets BO will belong to the curve (by Proposition 5); and hence B_2d will be a transverse diameter with its *latus rectum* in proportion to it as CO^2 to $BO \times O_2d$, the figure proving to be an ellipse if O falls between B and $_2d$, otherwise a hyperbola[27]—presuming that $_3rC$ and BO meet. But should those lines meet at no point, proving to be parallel, the figure will be a parabola and its *latus rectum* CO^2/BO.

Corollary 1. Hence if four points and the tangent to the conic at one of those points be given, the description of the curve will prove to be more speedy. For if, in instance, there be given the points A, B, C, P and the tangent Bt in position, the lines CD, BD will not now have to be drawn: rather, you should conceive the point D to be coincident with the point B, the lines CD, BD with the lines CB, Bt and so the points R and T to be identical with the points r and t, and then the rest to be accomplished as above.

Corollary 2. But should three points and the tangents at two of these be given, the description of the curve will be yet more speedily achieved.[28] For if there be given the points B, C, P and the tangents at the points B and C, the lines BA, CA will not have to be drawn; rather, you should imagine that the point A coincides with the point C and the line AB with the line CB, while the line AC is coincident with the tangent at C—that is, the line PR will have to be drawn parallel to this tangent and the line PT parallel to CD, and then the rest must be performed as in Corollary 1.

Corollary 3. Hence, lastly, the Ancients' problem is more easily solved than in Proposition 4. For evidently, since the meets of the sides of the quadrilateral $ABDC$ are four points through which the conic passes, after the tangent at some one of those points is drawn (by the Corollary to Proposition 3) the rest will have to be accomplished as in the present Corollary 2.

---

(28) 'simplicior' (simpler) is cancelled.

Prop 7.[29]

Si rectæ duæ BM CM a datis punctis B, C ductæ concursu suo M contingunt tertiam positione datam rectam MN & aliæ duæ rectæ BD, CD cum prioribus duabus ad puncta illa data B, C datos angulos MBD MCD efficientes ducantur; dico quod hæ duæ concursu suo D attingunt locum solidum. Et vice versa si rectæ BD, CD attingunt locum solidum per puncta B, C, G transeuntem & ubi hæ concurrunt ad ejus punctum aliquod G, alteræ duæ BM CM coincidant cum linea BC,[30] punctum M continget rectam positione datam.[31]

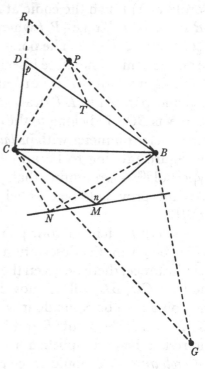

Nam in recta *MN* detur punctum *N*, et ubi punctum mobile *M* incidit in immotum *N*, incidat punctum mobile *D* in immotum *P*. Junge *CN, BN, CP, BP* et a puncto *P* age rectas *PT, PR* occurrentes ipsis *BD, CD* in *T* et *R* & facientes angulos *BPT* æqualem angulo *BNM* et *CPR* æqualem angulo *CNM*.[32] Cum ergo ex Hypothesi æquales sint anguli *MBD, NBP* ut et anguli *NCD, NCP*: aufer communes *NBD* et *MCP* et restabunt æquales *NBM*,

---

(29) Originally 'Prop. 6', a numbering which suggests that the preceding Proposition 6 was a late insertion. In 1685 Humphrey Newton transcribed this as Lemma xxi of Newton's 'De Motu Corporum Liber Primus' (ULC. Dd. 9.46: 42–3 [= *Principia* (note (3) above): 77–9]), while an expanded version of its Corollary 1 there appeared as the *Idem aliter* to Proposition xxii following (*ibid.*: 44–5 [= *Principia*: 82–3]). Two drafts of an abortive 'Prop 8' which essentially repeats its structure are reproduced in Appendices 3 and 4: the latter of these, later retitled 'Prob 7', was evidently intended at one time to replace the present proposition, but we take the version transcribed in 1685 into Newton's 'De Motu Corporum' to be his final preference. For his earlier, more general researches into the organic construction of curves see ii, **1**, 3 *passim*. The present proof that a straight line organically generates a general conic (and, conversely, that any conic can be so generated by a straight line) is, of course, wholly new.

(30) This to ensure that the whole line *BC* shall correspond to a single point on the described curve (*D*). In general, since the pairs of line-pencils *B(M)*, *B(D)* and *C(M)*, *C(D)* are manifestly equiangular and so have equal cross-ratios *B(M) = B(D)* and *C(M) = C(D)*, when the locus (*D*) is a conic through the poles *B* and *C* (and hence the pencils *B(D)*, *C(D)* are equicross) the describing curve (*M*) satisfies the cross-ratio equality *B(M) = C(M)* and is therefore likewise a conic through the poles *B*, *C*; in the present instance, however, where the line *BC* corresponds to a point *G* on the conic (*D*), the conic (*M*) breaks into the pair of *BC* and a second straight line *MN*, which will accordingly generate the complete conic (*D*). This subtlety was not well understood by J. L. Coolidge when he asserted with regard to the

Proposition 7[29]

If the two straight lines BM, CM drawn from the given points B, C should join at their meet M with a third straight line MN given in position, and if two further straight lines BD, CD be drawn, making given angles $M\widehat{B}D$, $M\widehat{C}D$ *with the two previous ones at those given points B, C; then I assert that the two latter trace by their meet D a solid locus. Conversely, if the lines BD, CD trace out a solid locus passing through the points B, C, G, and when these meet at some point G of it the two others BM, CM coincide with the line BC,*[30] *then the point M will belong to a straight line given in position.*[31]

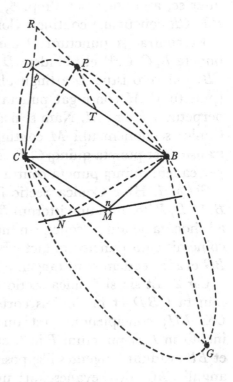

For in the line MN let the point N be given, and when the mobile point M falls at the stationary one N, let the mobile point D fall at the stationary one P. Join CN, BN, CP, BP and from the point P draw the straight lines PT, PR meeting BD, CD in T and R and making the angles $B\widehat{P}T$ equal to $B\widehat{N}M$ and $C\widehat{P}R$ equal to the angle $C\widehat{N}M$.[32] Since therefore by hypothesis the angles $M\widehat{B}D$, $N\widehat{B}P$ and also the angles $N\widehat{C}D$, $N\widehat{C}P$ are equal, take away the common ones $N\widehat{B}D$ and $M\widehat{C}P$ and there will remain the equal ones $N\widehat{B}M$, $P\widehat{B}T$ and $N\widehat{C}M$, $P\widehat{C}R$; in consequence the triangles NBM, PBT and also the

---

equivalent proposition later published in Newton's *Principia* (see note (29)) that 'I do not quite know what he means by his vice versa: usually if *D* describes a conic through *B* and *C*, *M* will do the same thing' (*A History of the Conic Sections and Quadric Surfaces* (Oxford, 1945): 46).

(31) Newton first concluded equivalently '...*per puncta B, C, G transeuntem et MBD, MCG æquentur angulis GBC, GCB si G ac D cadunt ad easdem partes rectæ BC, aliter complementis eorum ad duos rectos, punctum M attinget rectam positione datam*' (...*passing through the points B, C, G and M$\widehat{B}$D, M$\widehat{C}$G be equal to the angles G$\widehat{B}$C, G$\widehat{C}$B if G and D fall on the same side of the straight line BC, otherwise to their supplements, then the point M will be in a straight line given in position*).

(32) It is clearly enough to choose *PT* and *PR* through *P* parallel to lines drawn from *B* and *C* to any arbitrary point *A* on the conic locus, for then the homography *B*(*D*) = *C*(*D*) will determine ranges of points, (*R*) and (*T*), on *PR* and *PT* such that (*PR* ... ∞) = (*PT* ... ∞) and so the ratio *PR/PT* is constant. Newton's present choice of direction of *PR* and *PT* is equivalent to choosing *A* to be the point on the conic corresponding to the point at infinity on *MN*, for in this case $B\widehat{P}T$ = (exterior angle) $A\widehat{B}P$ = $B\widehat{N}M$ and $C\widehat{P}R$ = $A\widehat{C}P$ = $C\widehat{N}M$. It follows that the inclinations of *PR* to *CG*, of *PT* to *BG* and of *MN* to *BC* are equal.

PBT et NCM, PCR: adeoq triangula NBM, PBT similia sunt, ut et triangula NCM, PCR. Quare PT est ad NM ut PB ad NB, & PR ad NM ut PC ad NC. Ergo PT et PR datam habent rationem ad NM proindeq datam rationem inter se, atq adeo per Prop. 5, punctum D (perpetuus rectarum mobilium BT, CR concursus) contingit Conicam sectionem. Q.E.D.

Et contra, si punctum D contingit conicam sectionem transeuntem per puncta B, C, G,[33] et ubi idem D attingit punctum ejus G, M incidit in lineam CB: ubi vero istud D attingit alia duo quævis Conicæ sectionis puncta p, P, [punctum] M contingat puncta n, N. Agatur recta nN[34] et hæc erit locus perpetuus puncti M. Nam si negas liquet punctum D contingere[35] aliquam Conic: sectionem ubi M contingit rectam MN ut et hanc Coni sectionem transire per puncta quinq C, P, R, B, G;[36] hoc est duas Coni sectiones transire per eadem quinq puncta contra Corol: 3 Prop 5.

[37]*Cor 1.* Hinc Conica Sectio faciliùs describetur per quatuor data puncta B, C, D, P ita ut in uno istorum B contingat rectam positione datam BH. Nam ad hoc faciendum concipiendum est tantum quod punctum quintum G coincidit cum puncto contactus[38] B ita ut restent quatuor tantum puncta, et BG evadat eadem cum tangente BH, atq angulus CBG evanescat.

Cor 2. Et sic si Conica sectio describenda est, quæ transeat per tria data puncta CBD et in duobus eorum C, B tangat duas rectas positione datas CK, BH; concipiendum est quod punctum G incidit in B et punctum P in C et inde quod CG et BP evadunt tangentes illæ positione datæ, atq anguli CBG, BCP evanescunt: unde constructio evadet facillima. Nempe fac angulos CBM, DBN æquales angulo CBH et concurrat BM cum tangente CK in M, & BN cum CD in N. Actaq recta MN volvatur angulus datus [C]BM circa polum B et recta CN[39] circa polum C ita ut ea semper secet crus anguli istius BN ad rectam MN

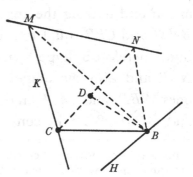

et punctum D ubi secat alterum crus BD describet conicam sectionem desideratam.

---

(33) Newton first continued 'sit N punctum quod M attingit ubi D attingit Conicæ Sectionis punctum quodvis P et punctum M semper attinget rectam positione datam MN' (let N be the point attained by M when D attains an arbitrary point P on the conic and the point M will lie always in the straight line MN given in position).

(34) This is logically displayed in Newton's figure as coincident with MN.

(35) The cumbrous equivalent 'manifestum est quod punctum D attinget' is cancelled.

(36) Newton first concluded: 'Attingat jam M concursum rectarum CB NM et ob angulos MBD MCD æquales angulis GBC GCB, lineæ BD CD coincident cum lineis BG CG, adeoq punctum D attinget punctū G. Transit itaq hæc Coni sectio per punctum G, et proinde duæ Coni sectiones transirent per eadem quinq puncta D, C, P, R, G contra Corol: 3 Prop 5' (Now

triangles *NCM*, *PCR* are similar. Hence *PT* is to *NM* as *PB* to *NB*, and *PR* to *NM* as *PC* to *NC*. Therefore *PT* and *PR* have a given ratio to *NM* and accordingly a given ratio to one another, so that by Proposition 5 the point *D* (the perpetual meet of the mobile straight lines *BT*, *CR*) belongs to a conic. As was to be proved.

Conversely, if the point *D* lies in a conic passing through the points *B*, *C*, *G*[33] and when this same point *D* arrives at the point *G* of it, *M* falls in the line *CB*, then to be sure when that point *D* reaches any two other points *p*, *P* of the conic, the point *M* shall attain the points *n*, *N*. For let the straight line *nN*[34] be drawn and this will then be the locus for ever of the point *M*. For if you deny this it is apparent that the point *D* belongs to some conic when *M* belongs to the straight line *MN* and also that this conic passes through the five points *C*, *P*, *R*, *B*, *G*;[36] that is, that two conics pass through the same five points contrary to Proposition 5, Corollary 3.

[37]*Corollary 1*. Hence will a conic more easily be described through four given points *B*, *C*, *D*, *P* so as to touch at one of these, *B*, a straight line *BH* given in position. For to do this you have merely to imagine that the fifth point *G* coincides with the point *B* of contact with the result that but four points remain, with *BG* proving to be identical with the tangent *BH* and the angle $\widehat{CBG}$ vanishing.

Corollary 2. And so if a conic is to be described which shall pass through three given points *C*, *B*, *D* and be tangent at two of them, *C* and *B*, to two straight lines *CK*, *BH* given in position, you should conceive that the point *G* passes into *B* and *P* into *C*, and consequently that *CG* and *BP* come to be those tangents given in position, the angles $\widehat{CBG}$, $\widehat{BCP}$ vanishing: whence the construction proves to be very easy. Namely, make the angles $\widehat{CBM}$, $\widehat{DBN}$ equal to the angle $\widehat{CBH}$ and let *BM* meet the tangent *CK* in *M* and *BN* the line *CD* in *N*. Then, when the straight line *MN* is drawn, let the given angle $\widehat{CBM}$ rotate round the pole *B* and the straight line *CN*[39] round the pole *C* so that it always intersects the leg *BN* of that angle on the line *MN*, and the point *D* in which it intersects the other leg *BD* will describe the desired conic.

let *M* attain the meet of the lines *CB*, *NM* and because the angles $\widehat{MBD}$, $\widehat{MCD}$ are equal to the angles $\widehat{GBC}$, $\widehat{GCB}$ the lines *BD*, *CD* will coincide with the lines *BG*, *CG* and consequently the point *D* will attain the point *G*. This conic accordingly passes through the point *G*, and as a result two conics would pass through the same five points *D*, *C*, *P*, *R* and *G* contrary to Corollary 3 of Proposition 5).

(37) The four corollaries following belong strictly with the rejected 'Prob 7' (reproduced in Appendix 4 below) but they form a natural sequel to the preceding proposition.

(38) This replaces the long-winded equivalent phrase 'cum puncto *B* quod tanget rectam positione datam'.

(39) Namely, an 'angulus datus' of two right angles.

Cor. 3. Dato centro Figuræ et tribus punctis in p[e]rimetro vel duobus punctis cum tangente ad unū eorum, potest etiam figura describi siquidem ex altera parte centri ad eandem distantiam simul dantur totidem alia puncta cum tangente.

Cor 4. Datis præterea quatuor punctis et plaga puncti infinite distantis potest figura describi. Sit enim punctum infinite distans *D* et punctum *M* invenietur ducendo parallelas *CD BD* versus plagam istam & faciendo angulos *DCM DBM* æquales *BCG CBG*, cæteraꝗ peragendo ut supra. Eodem modo res se habet ubi dantur tria puncta cum tangente vel duo puncta cum totidem tangentibus, & plaga puncti infinite distantis; vel etiam si dentur plagæ duorum punctorum infinite distantium, una cum alijs tribus punctis vel 2ᵇᵘˢ ac Tangente.

Hactenus ostendimus quomodo loca solida ex cognitis paucis punctis per quæ transibunt determinari ac describi possunt: restat ut quo pacto innotescat utrum locus aliquis propositus sit solidus necne, ostendamus.[40]

Prop 8.[41]

Si linea aliqua vel duæ aut plures parallelæ a datis punctis in data positione ductæ terminantur ad aliā quāvis rectam[42] quæ a dato puncto ducitur, et partes rectæ terminantis colligantur ex assumpta quavis parallelarum, quadrata partium istarum colligentur per simplicem Geom. et in operatione ista linea assumpta non ascendet ultra quadraticam dimensionem nec in denominatore fractionis alicujus reperietur.

Per Geometriam simplicem colligi[43] intelligo quæ per ductum solarum rectarum sine adminiculo circuli vel anguli dati, hoc est per additionem

---

(40) This replaces the more personal equivalent 'restat ut methodum ostendam quo pacto...' (it remains for me to reveal the method...).

(41) A much rewritten and extended version of 'Prop 8' on ff. 64ʳ–65ʳ preceding (reproduced in Appendix 6 below). A preliminary enunciation of the present revise on f. 133ʳ began 'Si linea mobilis secat rectas quotcunꝗ positione datas & ad data puncta terminatas et duæ earum quæ positione dantur sunt cognitæ, cæteræ omnes determinabiles erunt per simpl. Geom.' (If a mobile line intersects any number of straight lines given in position and terminated at given points, and two of these given in position are known, all the others will be determinable by simple geometry), while a following 'Prop 9' stated more generally that 'Si lineæ quotcunꝗ datas inter se positiones habentes, secant alias quotcunꝗ lineas datas intʳ se positiones habentes, hæ omnes posteriores una cum rationibus partium priorum a tribus quibusvis sive lineis posterioribus sive rationibus partium priorum cognitis determinabiles erunt per simp. Geom. dummodo tres illæ determinent positionem figurarum inter se' (If any number of lines having given positions one to another intersect any number of other lines having given positions one to another, all these latter ones together with the ratios of the parts of the former will be determinable by simple geometry from knowing any three either of the latter lines or ratios of the former ones). In a following paragraph Newton first went on with 'Ijsdem positis quadrata partium linearum priorū determinabiles erunt per simpl. Geom. Et ad ista determinanda lineæ posteriores nec ascendunt ultra quadraticam dimensionem neꝗ in denominatoribus fractionum reperientur' (With the same suppositions the squares of the parts of the former lines will be determinable by simple geometry. And to determine them the

Corollary 3. Given the centre of a figure and three points in its perimeter, or two points together with the tangent at one of them, the figure can also be described seeing that at equal distances on the other side of the centre there are at once given an equal number of other points along with a tangent.

Corollary 4. Furthermore, a figure can be described given four points of it and the direction of an infinitely distant point. For let the infinitely distant point be D and the point M will be found by drawing the parallels CD, BD in that direction and making the angles $D\widehat{C}M$, $D\widehat{B}M$ equal to $B\widehat{C}G$, $C\widehat{B}G$ and then by accomplishing the rest as above. The same method holds good when three points together with a tangent, or two points and the same number of tangents, are given along with the direction of an infinitely distant point; or also if there be given the directions of two infinitely distant points together with three further points, or two and a tangent.

Thus far we have shown how solid loci can be determined and described when a few points through which they shall pass are known: it remains for us to reveal the procedure[40] for ascertaining whether or not some proposed locus be solid.

Proposition 8[41]

If some line or two or more parallels drawn from given points in given position terminate at any other straight line[42] drawn from a given point, and the segments of the terminating line be gathered from assuming any one of the parallels, then the squares of those segments will be gathered by simple geometry and in that operation the assumed line will not rise above the quadratic dimension nor will it be found in the denominator of some fraction.

By 'gathered[43] by simple geometry' I understand what can be determined by drawing straight lines alone without the aid of a circle or given angle; that is,

---

latter lines will neither rise above the second power nor be found in the denominators of fractions), but then pulled himself up with 'Quære?' (Is it so?) and substituted 'Item quadrata linearū priorum sed non ipsæ lineæ determinabiles erunt per simp. Geom.' (Likewise the squares of the former lines but not the lines themselves will be determinable by simple geometry). Compare note (38) of the following Appendix 6.

(42) Newton first wrote 'secantur ab alia quavis recta' (be intersected by any other straight line).

(43) 'determinabiles' (determinable) is cancelled. To a modern eye this definition seems to confuse the geometrical (projective) 1,1 point-correspondence $x \leftrightarrow y$ (between two line-segments of magnitudes x and y) which is most generally determinable by an equation of form $A + Bx + Cy + Dxy = 0$, and the general 1,1 correspondence between magnitudes $x \leftrightarrow y$ (in which, for example,—to cite Newton's present allowable instance—y may be equal to x^2). It will be clear from the following propositions that Newton is essentially interested only in the former. Perhaps some vague recollection of Galileo's use of 1,1 correspondence to show the identity of the infinite set of positive integers with their squares (*Discorsi e Dimostrazioni Matematiche, intorno à due nuove scienze Attenenti alla Mecanica & i Movimenti Locali* (Leyden, 1638): Giornata prima: 33–4 [= *Opere*, 8 (Florence, 1898): 78–9]) tempted Newton to allow the correspondence $x \to x^2$ (but not $x^2 \to +x$) into his definition?

subductionem & inventionem quartæ proportionalis, vel ut jam loquuntur Geometræ[44] per multiplicationem et Divisionem sine extractione radicis determinari possunt.

Sint jam lineæ parallelæ *AD BE* terminatæ ad lineam *CD* per datum punctum *C* transeuntem, junge *AC* et [ad] eam demitte normalem *DG*. Cum ergo angulus *DAC* detur dabitur

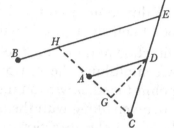

ratio *AD* ad *AG*. Sit ista *d* ad *e* et erit $AG = \frac{e}{d} AD$. Et

per [13,] 2 Elem est $AD^q + AC^q - 2AG \times AC = DC^q$.

adeoq $AD^q + AC^q - \frac{2e}{d} AD \times AC = DC^q$. Produc *CA*

donec occurrat *BE* in *H* et erit $AC^q . AH^q :: CD^q . DE^q$.

Ergo $DE^q = \frac{AH^q}{AC^q}$ in $AD^q + AC^q - \frac{2e}{d} A[D] \times AC$. Colliguntur ergo CD^q et

DE^q ex *AD* sine altiori ejus potestate quàm quadratica vel fractione in cujus denominatore *AD* reperitur. Q.E.D.

Prop 9.

Si recta aliqua secet alias quotcunq rectas datam inter se positionem habentes, hæ omnes una cum rationibus partium lineæ secantis et quadratis earundem, a duabus quibusvis[45] linearum positione datarum cognitis determinabiles erunt per simplicem Geometriam. At partes ipsæ lineæ secantis det[erminandæ sunt per Geometriam planam.][46]

Sint jam lineæ aliquot positione datæ *AB*, *AC*, *BC* et secet eas infinite productas *DF* in *D E* et *F*: dico primo quod ex assumptis duabus positione datis, cæteræ dabuntur per simpl. Geom. Assumantur enim *AE . AF*, et ad *AF* ducta *EK* parallelâ *DC*, propter datos angulos trianguli *AEK* dabitur ratio *AE* ad *AK* et *EK*, et proinde *AK* et *EK* dabuntur per simpl. Geom. Ab *AF* aufer *AK* et

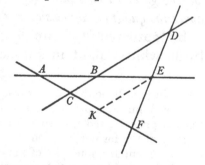

---

(44) Following Descartes, who wrote on the opening page of his *Geometrie* [= *Discours de la Methode* (Leyden, 1637): 297] that 'comme toute l'Arithmetique n'est composée, que de quatre ou cinq operations, qui sont l'Addition, la Soustraction, la Multiplication, la Division, & l'Extraction des Racines... : Ainsi n'at'on autre chose a faire en Geometrie touchant les lignes qu'on cherche, pour les preparer a estre connuës, que leur en adiouster d'autres [*sc.* equivalent geometrical operations]', adding that 'ie ne craindray pas d'introduire ces termes d'Arithmetique en la Geometrie, affin de me rendre plus intelligible' (p. 298). Thirty years later in his *Vera Circuli et Hyperbolæ Quadratura, In propria sua proportionis Specie, Inventa, & demonstrata* (Padua, 1667)—a work well known to Newton at this time (see III: 69, note (72)) —James Gregory neatly formalized this Cartesian insight in his definition of an analytical function (p. 9): 'Definitio 5. Quantitatem dicimus à quantitatibus esse compositam, cum à

by addition, subtraction and the finding of a fourth proportional—or, as geometers[44] now say, by multiplication and division without root extraction.

Let, now, the parallel lines AD, BE be terminated at the line CD passing through the given point C, join AC and to it let fall the normal DG. Since therefore the angle $D\widehat{A}C$ is given, there will be given the ratio of AD to AG. Let this be d to e, and then $AG = (e/d)\,AD$. Also by *Elements*, II, 13,

$$AD^2 + AC^2 - 2AG \times AC = DC^2,$$

and consequently $AD^2 + AC^2 - 2(e/d)\,AD \times AC = DC^2$. Extend CA till it meets BE in H and there will be $AC^2 : AH^2 = CD^2 : DE^2$. Therefore

$$DE^2 = \frac{AH^2}{AC^2}\,(AD^2 + AC^2 - 2(e/d)\,AD \times AC).$$

As a result CD^2 and DE^2 are gathered in terms of AD without entry of a higher power of it than its square or of a fraction in whose denominator AD is found. As was to be demonstrated.

Proposition 9

If some straight line intersect any number of others having their position one with regard to another given, all these latter, together with the ratios of the segments of the intersecting line and the squares of these, will be determinable by simple geometry from any two[45] of the lines given in position known. But the segments themselves of the intersecting line [will have to be determined by plane geometry].[46]

Let, now, a number of lines AB, AC, BC be given in position and let DF intersect their infinite extensions in D, E and F. I say, first, that from the assumption of two of the lines given in position the remainder will be given by simple geometry. For let AE, AF be assumed and EK drawn to AF parallel to DC, and because the angles of the triangle AEK are given, the ratios of AE to AK and EK will be given and consequently AK and EK will be given by simple geometry. From AF take away AK and there will be had KF, which is to the

---

quantitatum additione, subductione, multiplicatione, diuisione, radicum extractione, vel quacunque alia imaginabili operatione, fit alia quantitas' / 'Definitio 6. Quandò quantitas componitur ex quantitatum additione, subductione, multiplicatione, diuisione, radicum extractione; dicimus illam componi analyticè.' In Newton's present phrase there is perhaps the trace of a sneer.

(45) A first, cancelled continuation reads 'sive lineis datam inter se positionem habentibus sive rationibus partium lineæ secantis cognitis' (whether lines having a given position one to the other or ratios of parts of the intersecting line, when known). Compare note (41)

(46) The manuscript (f. 131ʳ) breaks off with this sentence, a late insertion, incomplete. 'Plane geometry' (construction by the intersection of a straight line and circle) will evidently construct the two possible square roots of the 'quadrata' determinable by 'simple geometry'.

habebitur *KF* quæ est ad datam *CF* ut *KE* ad *CD* et *FE* ad *FD*. Datur ergo *CD* et ratio *FE* ad *FD* per simpl. Geom.[47] Deniqȝ e datis *AE*, *AF*, *CD* datur *FE^q*, *FD^q*, *ED^q* per simp. Geom (prop 8). Q.E.D.

Prop 10.

Si rectæ quotcunqȝ datam inter se positionem habentes secent alias quotcunqȝ rectas datam itidem inter se positionem habentes: hæ omnes posteriores una cum rationibus partium priorum ad invicem, a trium punctorum per quæ aliquæ ex prioribus transeunt positione ad posteriores cognita, determinabiles erunt per simplicem Geometriam, dummodo puncta illa non in directum jaceant.[48]

S[int][49]

Prop 11.

Positis quibusvis duabus[50] *ab invicem per simplicem Geometriam determinabilibus quantitatibus, illæ exponi possunt per duas rectas a datis punctis in data positione ductas ac terminatas ad lineam circa datum punctum gyrantem.*[51]

Assumatur plenissima quævis relatio quantitatum quæ ab invicem per simplicem Geometriam determinabiles sunt,

qualis est hæc $a = + bx + cy + \dfrac{d}{e} xy$ ubi a, b, c

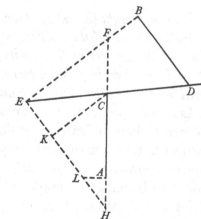

denotant quantitates datas cum signis suis + & − affectas et x et y quantitates incertas ex quarum alterutra cognita supponitur alteram posse determinari per simplicem Geometriam.

In linea quavis *AF* cape $AF = \dfrac{a}{b}$ & $AH = \dfrac{e}{d} c$.[52]

Ad *A* erige perpendiculum $AL = c$. Junge *LH* et ad eam productam demitte perpendiculum *FE* in quo cape $EB = b$, et a *B* age *BD* parallelam

---

(47) Newton first concluded: 'Eodem modo cæteræ lineæ et rationes colliguntur. Q.E.D.' (In the same manner the remaining lines and ratios are gathered. As was to be demonstrated).

(48) In this case, by the Euclid–Desargues theorem on perspective triangles (see 1, §1: note (28)), two of the points will uniquely determine the third and hence the three points are not independent bases of the configuration.

(49) Newton breaks off at this point, leaving half a page (on f. 131ᵛ) for a future insertion of the proof he never in fact entered.

(50) In a first draft of the enunciation of this problem on f. 135ᵛ Newton here added 'fluentibus' (fluent).

(51) In sequel Newton copied from his draft enunciation on f. 135ᵛ the converse assertion 'Et vice versa si lineæ a datis punctis in data positione ad lineam mobilem ductæ sunt ab invicem per simplicem Geometriam determinabiles, linea illa mobilis convolvetur circa datum

Prop 11

Positis quibusvis duabus ab invicem ~~datoris~~ per simplicem Geometriam determinabilibus quantitatibus, ~~illæ~~ exponi possunt per duas rectas a datis punctis in data positione ductas ac ~~ter~~ minatas ad lineam circa datum punctum gyrantem. ~~Et~~ ~~vice versa~~ si linea a datis punctis in ~~data~~ positione ad ~~li~~ ~~ram mobilem ductæ sunt ab invicem per simplicem Geometriam determinabiles, linea illa mobilis convolvetur circa datum punctum.~~

Assumatur plenissima quævis relatio quantitatum ^{quæ} ab invicem per simplicem Geometriam determinabiles ~~sunt~~, qualis est hæc $a = +bx + cy + \frac{d}{e}xy$ ~~= 0~~ ubi a, b, c denotant quantitates datas cum signis suis $+$ & $-$ affectas et x et y quantitates incertas ex quarum alterutra cognita ^{supponitur} altera posse determinari per simplicem Geometriam. In linea quavis ~~EB cape EB = b et~~ cape $AF = \frac{a}{b}$ & $AH = \frac{e}{d}c$. Ad A erige perpendiculum $AL = c$ Junge LH et ad eam productam demitte per pendiculum FE in quo cape $EB = b$, et a B age BD parallelam LH. Dico, quod ~~a dato puncto E~~ ~~age~~ rectam quavis secantem AF in C et BD in D, x ac y exponentur per AC et BD, ita ut si capi $^{semper fugit}$ atur $AC = x$ AF $BD = y$. ~~& vice versa~~. Nam a C demisso ad EH normali CK est $HL.LA :: CH.CK$ et $LHL.$ $AH :: FC . EK.$ Et proinde ~~CKxEK~~ $CK.EK (:: HL \times CK$. $ALxCH$ $HL \times EK$ ~~=~~ $AH \times FC) :: \frac{AL}{AH}$ ~~CH . FC~~ $CH . FC$ ~~=~~ sed $CK.EK$ :: $EB . BD.$ Ergo $\frac{AL}{AH} CH . FC :: EB . BD.$ hoc est ~~=~~ $c + \frac{d}{e}x . \frac{a}{b} - x ::$ $b . y.$ ~~=~~ vel $cy + \frac{d}{e}xy = a - bx.$ Q.E.D.

given length CF as KE to CD and FE to FD. Therefore by simple geometry there is given CD and the ratio of FE to FD.[47] Finally, from the given lengths AE, AF, CD there are given FE^2, FD^2 and ED^2 by simple geometry (by Proposition 8). As was to be demonstrated.

Proposition 10

If any number of straight lines having a given position one to another intersect any number of other straight lines likewise having a given position one to another, all those latter lines together with the ratios to each other of the former ones will be determinable by simple geometry from the known position with regard to the latter lines of three of the points through which some of the former pass, provided that those points do not lie in line.[48]

[Let...][49]

Proposition 11

Supposing any two[50] *quantities to be determinable each from the other by simple geometry, they can be exhibited by means of two straight lines drawn from given points in given position and terminated at a line revolving round a given point.*[51]

Let any fullest general relationship be assumed between quantities which are determinable from each other by simple geometry: it will be of this form, $a = bx + cy + (d/e)\, xy$, where a, b, c denote given quantities affected with their signs $+$ and $-$, while x and y are unknown quantities from either of which, when ascertained, the other can (it is supposed) be determined by simple geometry. In any line AF take $AF = a/b$ and $AH = (e/d)\,c$.[52] At A erect the perpendicular $AL = c$. Join LH, to its extension let fall the perpendicular FE

---

punctum' (And *vice versa* if lines drawn from given points in given position to a mobile line are mutually determinable by simple geometry, that mobile line will revolve round a given point). When he came to attempt the proof of this converse he must have seen its falsity, for (see note (55) below) he abandoned it after penning the words 'Et vice versa s...' and cancelled its enunciation. As we shall see in the seventh volume, Newton soon came to realize that the 'mobile' line CD which joins points C, D in 1, 1 correspondence in $AC = x$, $BD = y$ will, in general, envelop a conic: an insight probably already present in his mind when in early 1685 he composed Lemma xxv of his 'De Motu Corporum Liber Primus' (ULC. Dd. 9.46: 89 [= *Principia* (note (3) above): 91–2]). Only when the intersection of AC and BD is a self-corresponding point in the correspondence will the line CD pass through a single pole E, though it is all but self-evident that the correspondence may be generated in every case as the product of two such perspectivities. (In the next proposition, in effect, the 1, 1 correspondence $I \leftrightarrow K$ is generated by coupling the three perspectivities $B(I) = B(T)$, $C(K) = C(R)$ and $\omega(T) = \omega(R)$, where ω is the point at infinity in the direction TR.)

(52) Strictly, $AH = -(e/d)\,c$ since H is taken to lie on the opposite side of A to the points C and F.

LH.[53] Dico quod si a dato puncto *E* agatur recta quævis secans *AF* in *C* et *BD* in *D*, *x* ac *y* exponentur per *AC* et *BD*, ita ut si capiatur *AC*=*x* semper fuerit *BD*=*y*.[54] Nam a *C* demisso ad *EH* normali *CK* est *HL*.*LA*::*CH*.*CK* et *HL*.*AH*::*FC*.*EK*. Et proinde

$$CK.EK(::HL \times CK = AL \times CH . HL \times EK = AH \times FC) :: \frac{AL}{AH} CH.FC.$$

Sed *CK*.*EK*::*EB*.*BD*. Ergo $\frac{AL}{AH} CH.FC$::*EB*.*BD*. hoc est $c + \frac{d}{e} x. \frac{a}{b} - x :: b . y.$

vel $cy + \frac{d}{e} xy = a - bx.$ Q.E.D.[55]

Prop. 12.[56]

Si rectæ duæ CD BD circa data duo puncta C B gyrantes secent alias duas positione datas rectas HI HK, & longitudines earum quæ positione dantur sunt[57] *ex alterutra assumpta mutuo determinabiles per simplicem geometriam: locus intersectionis linearum mobilium D erit Conica sectio transiens per puncta duo C B circa quæ mobiles istæ volvuntur.*

Sint enim *A P* puncta duo quævis figuræ istius quam *D* attingit, et a *P* ipsis *CA BA* parallelas age *PR PT* secantes lineas mobiles *CD* in *R* et *BD* in *T*.[58] Et cum lineæ *PR PT* positione dentur, ex assumpta *PR* determinabilis erit *HK* per simplicem Geometriam per Prop 9 et inde ex Hypothesi dabitur *HI* per eandem Geometriam et inde etiam *PT* per eandem Geom. (per prop 9) et sic vice versa ex assumpta *PT* determinabilis erit *PR* per simpl. Geom. Convolvantur jam

---

(53) Without loss Newton might have taken $H\hat{A}L$ to be arbitrary and then set $B\hat{E}H = H\hat{A}L$, again drawing *BD* parallel to *EH*: for the points *F* and *H* (corresponding to $y = 0$ and $y = \infty$) here too yield $AF = a/b$ and $AH = -(e/d)c$ on taking $AC = x$, $BD = y$ to be related by $a = bx + cy + (d/e)xy$, while the triangles *HAL*, *HKC*, *HEF* and also *ECK*, *DEB* continue to be similar. Newton's choice of $BE = b$ and $AL = c$ evidently determines the meet, say *G*, of *AC* and *BD* to be a self-corresponding point: for, where $f = \sqrt{[d^2 + e^2 - 2de.\cos H\hat{A}L]}$, it is easily shown that $AG = (f/d)b - (e/d)c$ and

$$BG = (e/f)a/b - (e/d)b + (e^2/df)c.$$

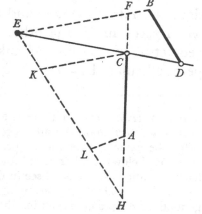

(54) '& vice versa' (and conversely) is cancelled. Compare note (51) above.

(55) This is the bottom line on f. 133[r]. At the top of the next page (f. 133[v]) Newton started to write 'Et vice versa s[i lineæ *AC*, *BD* a datis punctis *A*, *B* ad lineam mobilem *CD* ductæ sint ab invicem per simplicem Geometriam determinabiles, linea illa convolvetur circa datum punctum *E*?]' but—doubtless (compare note (51) above) realizing his error—abandoned this faulty converse in the middle of the fourth word, leaving the next three pages (ff. 133[v]–134[v]) blank for possible future improvements.

The present analytical expression for two magnitudes in 1,1 correspondence was never published by Newton during his lifetime and its rediscovery came only half a century later

and in this take $EB = b$, then from B draw BD parallel to LH.[53] I assert that, if from the given point E there be drawn any straight line cutting AF in C and BD in D, x and y will be exhibited by AC and BD and the result of taking $AC = x$ will always have been to make $BD = y$.[54] For after the normal CK is let fall from C to EH there follows $HL:LA = CH:CK$ and $HL:AH = FC:EK$ and accordingly

$$CK:EK(= HL \times CK \text{ or } AL \times CH:HL \times EK \text{ or } AH \times FC) = CH \times AL/AH:FC.$$

But $CK:EK = EB:BD$ and therefore $CH \times AL/AH:FC = EB:BD$; that is, $c + (d/e)\,x:a/b - x = b:y$ or $cy + (d/e)\,xy = a - bx$. As was to be demonstrated.[55]

Proposition 12[56]

If two straight lines CD, BD rotating round the two given points C, B intersect two other straight lines HI, HK given in position and the lengths of those given in position are [each][57] *mutually determinable by simple geometry from assuming the other, then the locus of the intersection D of the mobile lines will be a conic passing through the two points C, B round which those mobile lines revolve.*

For let A and P be any two points of the figure traced out by D and from P parallel to CA, BA draw PR, PT intersecting the mobile lines CD in R and BD in T.[58] Then, since the lines PR and PT are given in position, when PR is assumed HK will be determinable from it by simple geometry (by Proposition 9) and thence, by hypothesis, HI will be given by the same geometry and thence also PT by this same geometry (by Proposition 9); and so, conversely, when PT is assumed PR will be determinable from it by simple geometry. Let, now,

---

when William Braikenridge came to develop a form of Newton's Proposition 12 in his *Exercitatio Geometrica De Descriptione Linearum Curvarum* (London, 1733): Sectio III. 'Ubi describuntur Sectiones Conicæ ope plurium rectarum circa polos moventium': 60–8. Braikenridge, indeed, went slightly beyond Newton in making the transitivity of the 1, 1 correspondence a deducible theorem from the analytical defining equation: namely, if the correspondences $s \leftrightarrow t$ and $t \leftrightarrow r$ are expressed by 'æquationes hujusmodi... $ats + bt + cs + d = 0$, $frt + gr + ht + k = 0$, ... ubi quævis ex indeterminatis ... $s, ..., r, t$, est unius dimensionis ... & inde valor ejus exprimi potest per fractionem in cujus numeratore & denominatore occurrit alia variabilis etiam unius dimensionis, & hujus iterum valor exprimi potest per aliam fractionem in qua est alia variabilis unius tantum dimensionis' (pp. 67–8), on eliminating t there results an equation of form $Ars + Br + Cs + D = 0$, whence $s \leftrightarrow r$.

(56) Originally '11'. In Newton's first version this and the next proposition followed on Propositions 8–10 (reproduced in Appendix 6) on ff. 64^r–66 [= 65 a]^v. In his preliminary revise on f. 135^v he wrote '*Prop 12. Eadem 11* sup *Si rectæ duæ* &c'.

(57) 'omnes' (all) is cancelled.

(58) Since the line-pencils $B(D)$ and $C(D)$ are equicross, this determines that the ranges of points (R) and (T) in PR and PT have $C(PR \infty) = B(PT \infty)$, whence the ratio of PR to PT will be given. Compare note (32) above.

lineæ mobiles donec punctum D incidat in P et lineæ PR $P.T$ simul evanescent. Convolvantur eædem lineæ donec punctum D incidat in A et lineæ istæ congruentes cum CA BA evadent parallelæ lineis PR PT adeoʒ concursus earum R ac T recedunt in infinitum. Quare cum PR ac PT sunt ab invicem determinabiles per simplicem Geometriam & simul evanescunt, atʒ simul fiunt infinitæ, data erit earum ratio per Cor prop 10[59] et proinde Locus puncti D erit Conica Sectio transiens per B et C per Prop 5. Q.E.D.[60]

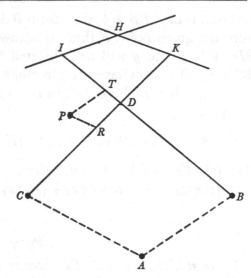

Coroll 1. Hinc si rectarum mobilium alterutra BD non gyret sed motu parallelo feratur, punctum D attinget etiam conicam sectionem: nam concipe punctum B circa [quod] BD volvitur amoveri ad infinitam distantiam et DB perpetuo tendens versus istud punctum non amplius gyrabit sed parallelè feretur.

Cor. 2. Hinc omnis figura est Conica sectio, cujus duo sunt puncta B, C per quæ si agantur utcunʒ rectæ, earum intersectiones cum reliquis partibus

---

(59) Of the first version (reproduced in Appendix 6), that is. In the present version Newton's text requires an equivalent corollary to Proposition 11, viz: 'duæ indeterminatæ quantitates ex alterutra cognita determinabiles per simplicem Geometriam, si possunt simul evanescere et simul evadere infinitæ, erunt in data ratione'. See also note (53) of the following Appendix.

(60) To derive this result from the preceding analytical treatment of 1, 1 point-correspondence it would have been enough to compute that, if the four pairs of points x_i, y_i, $i = 1, 2, 3, 4$, are in a 1, 1 correspondence defined by $a = bx_i + cy_i + (d/e) x_i y_i$, then on eliminating the constants a, b, c and d/e there results $\dfrac{(x_1 - x_3)(x_2 - x_4)}{(x_1 - x_4)(x_2 - x_3)} = \dfrac{(y_1 - y_3)(y_2 - y_4)}{(y_1 - y_4)(y_2 - y_3)}$, that is, in equivalent cross-ratio terms $(x_1 x_2 x_3 x_4) = (y_1 y_2 y_3 y_4)$. Newton, however, never took this final step, though as we shall see in the seventh volume he came later to appreciate the power of defining the general conic as the locus of meets of equicross line-pencils. In this, to be sure, he was not the first: indeed, in the case where the conic locus (D) is a circle, the line-pencils $B(D)$, $C(D)$ are equiangular and so superimposable, whence they are equicross—a result very likely known to Euclid and presented by him in his lost treatise on *Porisms*. (Compare Michel Chasles, *Les Trois Livres de Porismes d'Euclide, rétablis pour la première fois d'après la notice et les lemmes de Pappus* (Paris, 1860): 229–324: 'IIIᵉ Livre des Porismes', especially Porisme CXXVI: 230–1. The theorem that the cross-ratio of a line-pencil is invariant is Pappus' tenth/nineteenth lemma on Euclid's *Porisms* [= *Mathematical Collection*, **7**, 136/145].) By projecting this circular instance into the general conic case both Blaise Pascal and Fermat (the one relying on the insights of his mentor, Girard Desargues, the other seeking to interpret Pappus' rather vague clues regarding the content of Euclid's lost treatise) had, seemingly independently, come upon the general homographic definition of a conic. In his rare broadsheet, *Essay Povr Les Coniqves. Par B.P.* (Paris, 1640), Pascal stated explicitly as the first of his 'proprietez vniuerselles' that

the mobile lines revolve till the point D
coincides with P and the lines PR, PT
will vanish simultaneously. Let the same
lines revolve till the point D coincides
with A and those lines, here flush with
CA and BA, will prove to be parallel to
the lines PR and PT, and hence the
meets R and T of these will recede to
infinity. Consequently, since PR and PT
are determinable from each other by
simple geometry and both simultaneously
vanish and simultaneously become
infinite, their ratio will be given (by the
Corollary to Proposition 10),[59] and
accordingly (by Proposition 5) the locus

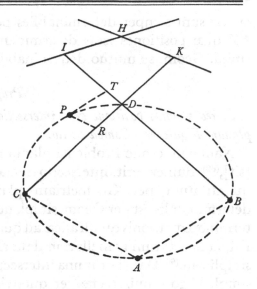

of the point D will be a conic passing through B and C. As was to be
demonstrated.[60]

Corollary 1. Hence if either one, BD, of the mobile lines should not rotate but
be carried along in a parallel movement, the point D will still trace a conic: for
imagine the point B round which BD revolves to move off to infinity and BD,
ever tending towards that point, will no longer rotate but be carried along
parallel to itself.

Corollary 2. Hence every figure is a conic where, if straight lines be drawn in
any way through two of its points B and C, their intersections with the remainder

---

'si dans le plan MSQ, dans la section de Cone, PKV, sont menées les droictes AK, AV,
atteignantes la section aux poincts PK, QV, & que de deux de ces quatre poincts qui ne sont
point en mesme droicte auec le poinct A, comme par les poincts K, V, & par deux poincts N,
O, pris dans le bord de la section sont menées quatre droictes KN, KO, VN, VO, coupantes les
droictes AV, AP, aux poincts $[S, T, L, M]$, ie dis que la raison composée des raisons de la
droicte PM, à la droicte MA, & de la droicte AS, à la droicte SQ, est la mesme que la composée des raisons de la droicte PL, à la droicte LA, & de la droicte AT, à la droicte TQ'
(compare R. Taton, 'L' *Essay pour les Coniques* de Pascal', *Revue d'Histoire des Sciences*, **8** (1955):
1–18, especially 13–14): in modern terms, if the points N, O, P, Q, K, V are on a conic, then
the line-pencils V ($PQON$), K ($PQON$) mark out quadruples of points on the lines APK, AQV
respectively whose cross-ratios ($PAML$), ($AQTS$) are equal. Correspondingly, Fermat in the
Porisma tertium of his *Porismata Euclidæorum Renovata Doctrina* (note (19) above) announced
(p. 117): 'Esto circulus cujus diameter recta AD, cui parallela [vel in dato angulo] utcunque
ducatur NM; circulo in punctis, N & M occurrens; & sint data puncta N & M, inflectatur
utcumque recta NBM, quæ secet diametrum in punctis O & V. Aio datam esse rationem
rectanguli sub AO in DV, ad rectangulum sub AV in DO [*sc.* the cross-ratio ($ADOV$) of the
line-pencil B ($ADNM$)]; ideoque si inflectatur NCM secans diametrum in punctis R [&] S,
erit semper ut...rectangulum sub AR in DS, ad rectangulum sub AS in DR [*viz.* the cross-
ratio ($ADRS$) of the line-pencil C ($ADNM$)], nec difficile est propositionem ad ellipses,
hyperbolas & oppositas sectiones extendere.'

curvæ sunt semper determinabiles per simplicem Geometriam, adeoϱ *PR* et *PT* quæ positiones istas determinant et ab illis vicissim determinantur per simpl. Geom. se mutuo determinabunt per eandem Geom.

Prop 13.[61]

Si rectæ omnis positione datæ intersectio cum curva determinabilis est per Geometriam planam Figura erit Conica Sectio.

Nam cum omne Problema planum ad æquationem quadraticam reducibilis [sit],[62] duplex erit intersectio rectæ et Curvæ propositæ. [63]Duarum vero quantitatum per Geometriam planam determinabilium summa semper determinabilis est per Geom. simpl, utpote dum sit cognita quantitas in secundo termino æquationis quadraticæ ad quam omne Problema planum reduci potest. Ergo una istarum gemellarum data altera determinabilis erit per Geometriam simplicem.[64] Data igitur una intersectione, altera determinabilis erit per Geom. simpl. Ergo omnis rectæ per quodvis hujus curvæ punctum datum ductæ[65] intersectio cum altera parte curvæ determinabilis erit per Geom. simp. et proinde per Cor. 2. Prop 1[2][66] Figura est Con. Sect.

[67]Jactis hisce fundamentis, ubi quæstio aliqua de loco proponitur, imprimis inquirendum est per Prop. 9, 10, 11, 12 & 13 num locus sit solidus. Et si solidus esse deprehenditur, tunc inquirenda sunt aliqua ejus puncta sive ad finitam sive ad infinitam distantiam. Plagæ punctorum infinitè distantium facilius inveniri solent, et ex istis species optimè definietur.[68] Nam si duo obvenerint ejusmodi puncta Curva erit Hyperbola, si unicum tantum Parabola, si nullum Ellipsis.

Ubi Hyperbola habetur, quærendo tangentes ad puncta infinite distantia invenies Asymptotos, ex quibus et puncto quovis figuræ ad finitam distantiam insuper invento Figuram datam esse satis notum est.[69]

---

(61) Originally '12'; compare note (56) above. A first version of this proposition (on f. 149 bis^r) is reproduced as Appendix 7.

(62) Newton first copied from his draft on f. 149 bis^r the opening 'Nam cum quantitates per Geometriam planam determinabiles semper prodeant duplices pro numero radicum æquationis construendæ intersectionúmve rectæ & circuli per quas determinantur' (For since quantities determinable by plane geometry always prove to be double in line with the number of roots of the equation to be constructed or of intersections of the straight line and circle by which they are determined). Compare 1: 492–4.

(63) A cancelled sequel at this point reads 'Duarum vero quantitatum per Geometriam planam determinabilium una data, altera semper determinabilis est per Geometriam simplicem siquidem omne Problema planum reduci potest ad æquationem quadraticam' (However, when one of two quantities determinable by plane geometry is given, the other is always determinable by simple geometry seeing that every plane problem can be reduced to a quadratic equation).

(64) This 'rule of twins'—earlier introduced by Newton in his 'observations' on Kinckhuysen's *Algebra* (II: 436–8) as one 'qua terminos ad ineundum calculum maximè accom-

of the curve are forever determinable by simple geometry, and consequently *PR* and *PT* (which determine the positions of these, and are in turn determined by them, by simple geometry) will mutually determine each other by the same geometry.

Proposition 13[61]

If the intersection of every straight line given in position with a curve is determinable by plane geometry, the figure will be a conic.

For, since every plane problem is reducible to a quadratic equation, the intersection of a straight line with the curve proposed is twofold. [63]However, the sum of two quantities determinable by plane geometry is always determinable by simple geometry inasmuch as it is the known quantity in the second term of the quadratic equation to which every plane problem can be reduced. Therefore, given one of those twin quantities, the other will be determinable by simple geometry.[64] Accordingly, given one intersection, the other will be determinable by simple geometry. Therefore in the case of every straight line drawn[65] through any given point of this curve its intersection will be determinable by simple geometry and consequently (by Proposition 12,[66] Corollary 2) the figure is a conic.

[67]The foundations have now been laid. When some question regarding a locus is proposed, you should in the first instance inquire (by Propositions 9, 10, 11, 12 and 13) whether the locus is solid. If the locus is discovered to be solid, you should then investigate some points on it either at a finite or infinite distance. The directions of infinitely distant points are usually found more easily, and from these the species will best be defined.[68] For if two points of this kind occur the curve will be a hyperbola, if but a single one it will be a parabola, if none an ellipse.

When a hyperbola is had, you will find its asymptotes by looking for the tangents at the infinitely distant points, and from these and any additional point of the figure found at a finite distance it is well enough known that the figure is determined.[69]

---

modatos prima fronte possis utplurimùm eligere'—was developed by him at about this time in a short paper on the 'Regula fratrum' (to be reproduced in the next volume).

(65) 'transientis' (passing) was here first copied from the preceding draft on f. 149 bis[r] and then deleted.

(66) The manuscript has '11'; compare note (56) above.

(67) A first version of the next four paragraphs is reproduced (from f. 150[v]) in Appendix 8 following. A preliminary revise on f. 149 bis[v] is little different from the final text (ff. 149[r]–149[v]) here given.

(68) 'hoc modo' (in this manner) is cancelled.

(69) Namely, if the given point has coordinates (a, b) in the Cartesian system in which the axes coincide with the asymptotes, then the hyperbola's defining equation is $xy = ab$. In his draft version (f. 149 bis[v]) Newton first wrote a little more forcefully that 'Hyperbolam determinari notissimum est' (...it is very well known that the hyperbola is determined).

Ubi Parabola occurrit, præter plagam puncti·infinite distantis quod situm diametrorum determinat[70] quærenda sunt duo puncta *A*, *B*, finitè distantia cum tangente ad unum eorum. Nam ea determinant figuram per Cor [1] Prop [6 vel 7].[71]

Ubi vero Genus Ellipticum obvenit inquirendum est ultra num Figura sit circulus. Id quod varijs argumentis discerni potest. Ut si ordinatim applicata ad axem sit media proportionalis inter segmenta axis, vel duo deprehendantur axes non ad rectos angulos, vel duo axes æquales ad rectos angulos, aut tres aut plures diametri æquales, concludes figuram esse circulum: ut et si angulus in aliquo segmento figuræ determinatus sit, tresve sint anguli æquales in eodem segmento, vel duo[72] anguli recti in semissi figuræ.[73] Et postquam per hæc aut similia criteria species figuræ innotescit si Ellipsis sit, determinabitur per Prop 6 per tria puncta inventa cum tangentibus ad duo eorum vel per quatuor puncta et tangentem vel deniꝗ per quinꝗ puncta si tot invenire facilius sit quam tangentes ducere. Quo pacto determinabitur etiam Hyperbola et Parabola ubi species Figuræ ignoratur[74] aut plagæ punctorum infinite distantium non facile inveniuntur.

Quemadmodum[75] si a duobus datis punctis *A B* ad tertium punctum *C* duæ rectæ *AC BC* datam haben-tes rationem inflectantur, et quæratur locus pu[n]cti *C*.[76] Imprimis duco lineam aliquam *GC* quam concipio positione datā esse & quæro utrum intersectio ejus *C* cum curva determinari possit

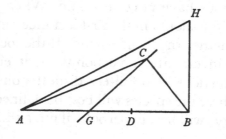

---

(70) In direction, that is.

(71) The references are left blank in the manuscript. In effect the present parabola is·determined by four points, one at *B*, two more (coincident) at *A* and a fourth (at infinity upon it) which determines the diametral direction and so that of the tangent at *A*. In first draft on f. 149 bis[v] Newton wrote more explicitly 'Nam *BD* acta parallela [tangenti] *AK* erit ordinatim applicata ad diametrum *AD* existente latere recto $\frac{BD^q}{AD}$' (For when *BD* is drawn parallel to the tangent *AK*, it will be ordinate to the diameter *AD*, the *latus rectum* being BD^2/AD).

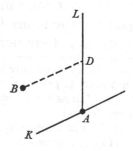

(72) Newton added 'aut plures' (or more) in first draft.

(73) In draft on f. 149 bis[v] Newton first continued 'vel plures quam duo æquales diametri vel deniꝗ alia quævis nota et convertibilis proprietas circuli. Et si per has aut similes pro-prietates circuli non depre[hendatur num figura sit circulus necne]' (or more than two equal diameters or finally any other known property of the circle directable to this end. And if by these or similar circle properties it should not be discovered [whether or not the figure is a circle]...).

When a parabola is met with, apart from the direction of the infinitely distant point which fixes the location of the diameters,[70] two finitely distant points *A*, *B* must be looked for together with the tangent at one of them. For (by Corollary 1 of Proposition 6 or 7)[71] they determine the figure.

When, indeed, the elliptical genus occurs, you should further inquire whether the figure is a circle. This can be distinguished on various grounds. Thus, if the ordinate to an axis be a mean proportional between the segments of the axis, or two axes not at right angles are discovered, or two equal axes at right angles, or three or more equal diameters, you will conclude that the figure is a circle: also if the angle in some segment of the figure be determinate, or if there be three equal angles in the same segment, or two[72] right angles in half of the figure.[73] After—by these or similar criteria—the species of the figure is known it will, should it be an ellipse, be determined (by Proposition 6) by means of three of its points, when found, together with the tangents at two of them, or by four points and a tangent, or finally by five points if it be easier to find so many than to draw tangents. In this manner, also, the hyperbola and parabola will be determined when the figure's species is not known[74] or the directions of infinitely distant points are not easily found.

For instance,[75] if from two given points *A*, *B* to a third point *C* two straight lines *AC*, *BC* having a given ratio are inclined and the locus of the point *C* be sought:[76] in the first place *I* draw some line *GC* which I conceive to be given in position and inquire whether its intersection *C* with the curve may be determined by plane geometry. Evi-

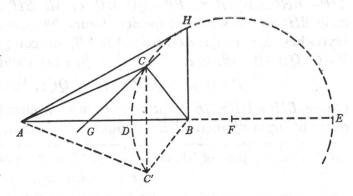

(74) This replaces the draft phrase 'non facile designaretur' (were not easily designated) on f. 149 bis[v].

(75) In the two following examples Newton in fact restricts his attention to two circular loci which illustrate the 'genus Ellipticum' alone.

(76) Compare 1, §1: note (6) above. That this locus is a circle was known at least as early as Aristotle, who in using it to account for the observed circular shape of the rainbow in his *Meteorologica*, III, v, 376*a* (see F. Poske, 'Die Erklärung des Regenbogens bei Aristoteles', *Zeitschrift für Mathematik und Physik*, 27 (1883): 134–8; T. L. Heath, *A History of Greek Mathematics*, 1 (Oxford, 1921): 340, and more especially *Mathematics in Aristotle* (Oxford, 1949): 181–90; C. B. Boyer, *The Rainbow: From Myth to Mathematics* (New York, 1959): 45–6) states the simple proof that, on taking points *D*, *E*, *F* in *AB* such that

$$AD:DB = AE:BE = AC:CB \quad \text{(constant)}$$

per Geom. planam. Nempe ex assumpto C datur AC^q et BC^q et ratio eorum ad invicem per simpl. Geom. Et ad hanc rationem investigandam GC ascendit tantum ad quadraticam dimens[ionem].[77] Ergo ex data illa ratione revertetur ad GC per solam extractionem radicis quadraticæ, hoc est punctum $[C]$ determinari potest per Geom. planā, et figura est Con. Sec.

Hoc animadverso quæro puncta figuræ quæ facilius occurrent. Et primum considero quid eveniet ponendo punctum ad infinitam distantiam. Nempe lineæ $A[C]$ $B[C]$ ‖ evadent & æquales[,] nam omnes infinitæ lineæ parallelæ et ad puncta finite distantia terminatæ rationem habent æqualitatis. Ergo ubi $A[C]$ et $B[C]$ non sunt æquales punctum C non potest recedere in infinitum sed Figura erit de genere Elliptico.

Hoc cognito quæro[78] quævis ejus puncta quæ occurrent, qualia sunt puncta D et E in linea AB quæ inveniuntur capiendo AD ad DB et AE ad EB in ista data ratione AC ad CB, ut patebit fingendo punctum C accedere ad D vel E. Interea vero animadverto quod si punctum C transferatur ad alteras partes lineæ AB eundem habebit situm ad puncta A B ubi AC est ejusdem longitudinis, et proinde curvæ portio similis et æqualis ad utramcɞ partem describitur adeocɞ DE esse axem figuræ. Erigo ergo ordinatam BH et in ea quæro aliud punctum H.[79] Nempe sit $AC.CB::P.Q::AH.BH$ et erit $AH^q.BH^q::PP.QQ.$ et $AH^q - BH^q (AB^q).BH^q::PP-QQ.QQ.$ et $AB.BH[::]\sqrt{PP-QQ}.Q.$ Jam cum ratio BH^q ad $DB \times BE$ det speciem figuræ,[80] quæro etiā $DB \times BE$. Nempe ex Hypothesi est $P.Q::AD.DB::AE.EB$, et com[ponendo] ac div[idendo] $P+Q.Q::AB.DB.$ & $P-Q.Q::AB.BE$. Ergo addendo rationes

$$P+Q \times P-Q \ (PP-QQ).QQ::AB^q.DB \times BE.$$

Quare $BH^q = DB \times BE$ et proinde figura est circulus.[81] Hoc cognito solum restat ut super diametro DE circulum describamus.

---

and F is the mid-point of DE, the triangles AFC, CFB are similar and consequently

$$CF \ (= \sqrt{[AF \times BF]})$$

is fixed in length as it rotates round its fixed end-point F. Both Pappus (in the seventh book of his *Mathematical Collection*) and Eutocius (in his commentary on Apollonius' *Conics*, I–IV) cite the proposition from Book 2 of Apollonius' lost treatise on *Plane Loci*: the latter's proof of the circularity of the locus is essentially that of Aristotle, 'which shows that the proposition was fully known and a standard proof of it was in existence before Euclid's time' (Heath, *Greek Mathematics*, **1**: 340). Newton may have derived his knowledge of the locus either from Pappus' account or perhaps from reading Fermat's restoration of *Apollonii Pergæi Libri Duo de Locis Planis* (see note (1) above), where it is set as Liber II, Propositio IV: 32–3: 'Si à duobus punctis datis rectæ lineæ inflectantur, & sit quod ab una efficitur eo quod ab altera dato majus quàm in proportione, punctum positione circumferentiam continget.' In his manuscript figure (where $AC < AH$) Newton has located the point C outside the right triangle ABH: this impossible site has been adjusted in our otherwise exact reproduction of it.

(77) Namely, $AC^2 = AG^2 - 2\lambda.AG \times GC + GC^2$ and $BC^2 = BG^2 + 2\lambda.BG \times GC + GC^2$, where $\lambda = \cos A\hat{G}C$.

dently, when C is assumed, AC^2 and BC^2 and their ratio to each other are given from it by simple geometry. And in discovering this ratio GC rises merely to its square power.[77] Therefore, when that ratio is given, the return to GC is made by extraction of a square root alone; that is, the point C may be determined by plane geometry and the figure is a conic.

This observed, I seek the points on the figure which are more easily met with. And first I consider what will happen by supposing a point to be at infinity. Evidently the lines AC, BC will prove to be parallel and equal, for all infinite lines which are parallel and terminate at finitely distant points have [to one another] a ratio of equality. Therefore when AC and BC are not equal, the point C cannot recede to infinity—rather, the figure will be of elliptical genus.

With this knowledge I seek[78] any points on it which come to suggest themselves, such as the points D and E in the line AB which are found by taking AD to DB and AE to EB in that given ratio AC to CB (as will be evident by imagining the point C to arrive at D or E). Furthermore, indeed, I notice that, if the point C be transposed to the further side of the line AB, it will have the same situation with regard to the points A, B when AC is of the same length: accordingly, an equal, congruent section of the curve is described on either side, and so DE is an axis of the figure. I therefore erect the ordinate BH and seek in it a further point H.[79] Namely, let

$$AC:CB = P:Q = AH:BH \quad \text{and then} \quad AH^2:BH^2 = P^2:Q^2$$

with $AH^2 - BH^2$ (or AB^2)$:BH^2 = P^2 - Q^2:Q^2$ and $AB:BH = \sqrt{[P^2 - Q^2]}:Q$. Now since the ratio of BH^2 to $DB \times BE$ yields the figure's species,[80] I seek also $DB \times BE$. Namely, by hypothesis $P:Q = AD:DB = AE:EB$, and so *componendo* and *dividendo* $P + Q:Q = AB:DB$ and $P - Q:Q = AB:BE$. Therefore, by adjoining the ratios, $(P + Q) \times (P - Q)$ (or $P^2 - Q^2$)$:Q^2 = AB^2:DB \times BE$. Hence $BH^2 = DB \times BE$ and the figure is accordingly a circle.[81] Once this is known it remains only for us to describe the circle on the diameter DE.

---

(78) Newton first continued 'insuper puncta ejus quæ determinatio figuræ occurrent, qualia sunt puncta in linea AB quæ inveniuntur secando AB in D et producendo ad E ut sit AD ad DB...' (in addition the points on it which more easily occur in the determination of its figure, such as the points in the line AB which are found by cutting AB in D and extending it to E such that there is AD to DB...).

(79) Of the locus, that is.

(80) Newton has just shown that DE is the ellipse's axis, and so its eccentricity is

$$\sqrt{[1 - BH^2/DB \times BE]}.$$

(81) A rather lengthy, tedious argument in comparison with the neat standard proofs (for example, that which shows CD, CE to be internal/external bisectors of $A\hat{C}B$ and so the angle $D\hat{C}E$ to be right) but Newton is seeking to illustrate his general method, not to furnish a concise and elegant demonstration.

Rursum, si a tribus datis punctis A, B, C ad commune punctum D inflectantur
tres lineæ *AD BD CD* quarum
quadratorum summa datur, et
quæratur locus puncti D.[82]
Imprimis duco rectam ali-
quam *ED* quam concipio
positione datam esse et curvam
secare alicubi in D et inqu[i]ro
num intersectio ejus cum curva
determinabilis sit per Geom.
planā. Nempe juncta *AC* et
producta ea donec secet *ED* in
E et demisso normali [D]H[83]

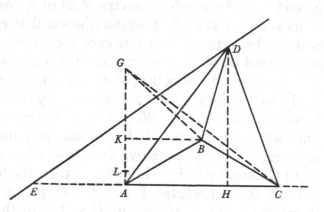

ad *AC* ex assumpta *ED* habebitur AD^q per simpl. Geom. (per 1[2,] 2 El) nec
ED ultra quadraticam dimensionem ascendet,[84] ut et DC^q et DB^q nec *ED*
ultra quadraticam dimensionem ascendet aut in denominatore fractionis alicujus reperietur. Ergo ad obtinendum $AD^q + BD^q + CD^q$, *ED* non ascendet
ultra quadraticam dimens. Et proinde ab ista summa data regredi potest ad
ED per solam extractionem radicis quadraticæ.[85] Quare punctum D determinabile est per Geom. plan.

Hoc animadverso sentio præterea punctum D non posse recedere ad infinitam
dist. quia summa quadratorum $AD^q + BD^q + CD^q$ definitur, et proinde figura
erit de genere Elliptico. Quæro igitur ejus puncta finite distantia quæ facilius
occurrent, puta in linea *AC* infinite producta vel in $\perp^{lo}$ ejus *AG*. Sit istud punctum
G et ex assumpta *AG* habebitur

$$GC^q = AG^q + AC^q \quad \text{et} \quad GB^q = AG^q[+]AB^q - 2AG \times AK^{(86)}$$

(82) This generalized Apollonian locus may (see 1, §1: notes (11) and (13) above) have
been derived either from Pappus' summary of the contents of the second book of Apollonius'
Plane Loci or from the corresponding Liber II of Fermat's *Apollonii Pergæi Libri Duo de Locis
Planis Restituti* (see note (1) above), where it is treated in general form (pp. 38–41) as the *idem
aliter* ('Propositio Altera') to Proposition V. In modern analytical equivalent, given the fixed
points $A_i \equiv (a_i, b_i)$, $i = 1, 2, ..., n$, Fermat constructs (pp. 40–1) the locus (D) defined by
$\sum\limits_{1 \leqslant i \leqslant n} (DA_i)^2 = k$ as the circle whose centre is the centroid $\left(\frac{1}{n}\Sigma a_i, \frac{1}{n}\Sigma b_i\right)$ of the points A_i, but
incorrectly takes its radius to be $\sqrt{[k/n]}$. (In fact, the locus-point $D \equiv (x, y)$ satisfies

$$\sum\limits_{1 \leqslant i \leqslant n} [(x-a_i)^2 + (y-b_i)^2] = k,$$

a circle of radius $\sqrt{\left[\left(\frac{1}{n}\Sigma a_i\right)^2 + \left(\frac{1}{n}\Sigma b_i\right)^2 - \frac{1}{n}\Sigma(a_i^2 + b_i^2) + \frac{k}{n}\right]}$.)

(83) The manuscript (f. 68^r) reads '*BH*' in error. In his accompanying figure (here
corrected) Newton likewise located H (in AC) as the foot of the perpendicular from B.

(84) Namely, $AD^2 = ED^2 - EA^2 - 2EA \times AH$.

Again, if from three given points A, B, C to the common point D three lines AD, BD, CD be inclined, the sum of whose squares is given, and the locus of the point D be sought:[82] in the first place I draw some straight line ED which I conceive to be given in position and to intersect the curve at some point in D, and then I inquire whether its intersection with the curve is determinable by plane geometry. Evidently, on joining AC and extending it till it cuts ED in E and on letting fall the normal DH[83] to AC, when ED is assumed AD^2 will be had

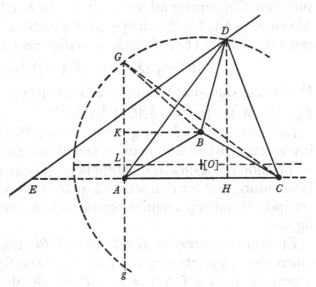

from it by simple geometry (by *Elements*, II, 12) nor will ED rise beyond its square power;[84] and likewise with both DC^2 and DB^2, ED will neither ascend beyond its quadratic dimension [n]or be found in the denominator of some fraction. Therefore, to obtain $AD^2 + BD^2 + CD^2$, ED will not rise beyond its square power, and accordingly regression may be made from that given sum to ED by means of square-root extraction alone.[85] Hence the point D is determinable by plane geometry.

Having noticed this, I perceive, moreover, that the point D cannot recede to an infinite distance, for the sum of the squares $AD^2 + BD^2 + CD^2$ is limited, and consequently the figure will be of elliptical genus. I therefore seek the more easily met with of its finitely distant points, say in the line AC infinitely extended or in its perpendicular AG. Let the latter point be G and then, when AG is assumed, from it will be had $GC^2 = AG^2 + AC^2$ and

$$GB^2 = AG^2 + AB^2 - 2AG \times AK,$$

so that the sum of the three squares will here be

$$3AG^2 + AC^2 + AB^2 - 2AG \times AK^{(86)}$$

---

(85) Most simply, if Ea, Eb, Ec are the projections on ED of EA, EB and EC respectively, then $AD^2 + BD^2 + CD^2 = 3ED^2 - 2(Ea + Eb + Ec)ED + EA^2 + EB^2 + EC^2$; hence the line ED meets the locus $AD^2 + BD^2 + CD = MN$ (constant) such that, on taking

$$E\lambda = \tfrac{1}{3}(Ea + Eb + Ec) \quad \text{and} \quad E\mu = \sqrt{[\tfrac{1}{3}(EA^2 + EB^2 + EC^2)]},$$

there follows $ED = E\lambda \pm \sqrt{[E\lambda^2 - E\mu^2 + \tfrac{1}{3}MN]}$, which is equivalent to Newton's following result.

(86) The text reads incorrectly '$AG^q - AB^q - 2AG \times AK$' and carries the wrong sign of AB^2 through the remaining computation.

adeoꝗ summa trium quadratorum erit $3AG^q + AC^q[+]AB^q - 2AG \times AK$, (si punctum G quæratur ad partes AC versus B sed si quæratur ad contrarias partes ista summa erit $3AG^q + AC^q[+]AB^q + 2AG \times AK$) quam cum ex Hypoth. detur ponamus $[M]N$ et resolvendo æquationem erit

$$AG = \tfrac{1}{3}AK \pm \sqrt{}: \overline{\tfrac{1}{9}AK^q[-]\tfrac{1}{3}AB^q - \tfrac{1}{3}AC^q + \tfrac{1}{3}MN}.$$

Hoc pacto duo simul investigantur puncta G et g capiendo $AL = \tfrac{1}{3}AK$ et LG, $Lg = \sqrt{\tfrac{1}{9}AK^q[-]\tfrac{1}{3}AB^q - \tfrac{1}{3}AC^q + \tfrac{1}{3}MN}.$[87]

Eodem modo si ad C erigeretur $\perp^{\text{lum}}$ ad AC et in eo quærerentur duo puncta[,] inveniretur distantiam medij puncti ab AC esse trientem ejusdem perpendiculi. Et proinde recta qua hæc parallela perpendicula utrinꝗ ad figuram terminata bisecantur[,] hoc est diameter ad quam LG ord[inatim] applic[atur,] parallela est ipsi AC adeoꝗ angulos rectos faciens cum ordinatim applicatis erit axis figuræ.

Et simili argumento si ad AB vel BC erigerentur perpendicula et in ijs quærerentur puncta figuræ, inveniretur axis figuræ ipsi AB vel BC parallela et a perpendiculo a C vel A ad AB vel BC demisso resecans tertiam partem. Quare Figura, tum quod axes habet plures duobus tum quod axes isti sunt ad angulos obliquos circulus erit, et centrum ejus[88] reperietur in intersectione linearum quæ ducuntur parallelæ laterib^s $\triangle^{\text{li}}$ ad distantiam ab ipsis æqualem trienti perpendiculorum quæ ab oppositis angulis ad illa demittuntur, hoc est in centro gravitatis trianguli.[89] Hoc igitur centro per punct[um] G in peripheria supra inventum, describatur circulus et habebitur fig. desiderata. Et eodem modo res se habet ubi linearum a pluribus datis punctis ad commune punctum inflexarum quadrata æquantur dato spatio.

Supponamus jam[90]

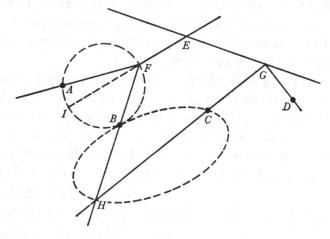

---

(87) Newton first concluded equivalently with the sentence 'Hoc pacto duo simul investigantur puncta, G ad partes AC versus B capiendo

$$AG = \tfrac{1}{3}AK + \sqrt{\tfrac{1}{9}AK^q[-]\tfrac{1}{3}AB^q - \tfrac{1}{3}AC^q + \tfrac{1}{3}MN}$$

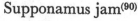

& g ad contrarias partes capiendo $Ag\,[= \tfrac{1}{3}AK - \sqrt{\quad}\,]$ '.

(88) We have inserted this in our diagram as the point O.

(89) See note (82) above in the case where $n = 3$.

(should the point G be sought on the side of AC on which B is; but if it be sought on the opposite side, the sum will be $3AG^2 + AC^2 + AB^2 + 2AG \times AK$). Since by hypothesis this is given, let us suppose it to be MN and by resolving the equation there will be $AG = \frac{1}{3}AK \pm \sqrt{[\frac{1}{9}AK^2 - \frac{1}{3}AB^2 - \frac{1}{3}AC^2 + \frac{1}{3}MN]}$. In this way two points G and g are simultaneously discovered by taking $AL = \frac{1}{3}AK$ and LG, $Lg = \sqrt{[\frac{1}{9}AK^2 - \frac{1}{3}AB^2 - \frac{1}{3}AC^2 + \frac{1}{3}MN]}$.[87]

In the same manner if the perpendicular at C to AC were erected and two points sought in it, it would be found that the distance of their midpoint from AC is one-third of the same perpendicular. Consequently, the straight line bisecting these parallel perpendiculars terminated in either direction at the figure, that is, the diameter to which LG is ordinate, is parallel to AC and so, as it makes right angles with the ordinates, is an axis of the figure.

And by a similar reasoning, if to AB or BC perpendiculars be erected and points of the figure be sought in them, an axis of the figure parallel to AB or BC would be found, cutting off one-third of the perpendicular let fall from C or A to AB or BC. Hence the figure, both because it has more than two axes and because those axes are at oblique angles, will be a circle and its centre[88] will be found at the intersection of lines drawn parallel to the triangle's sides at a distance from them equal to one-third of the perpendiculars let fall from the opposite corners to them, in other words, at the centroid of the triangle.[89] With this centre, therefore, through the point G in its circumference found above let a circle be described and the desired figure will be obtained. The procedure is the same when the squares of lines inclined from several given points to a common point are equal to a given area.

Now let us suppose[90]

---

(90) The manuscript here breaks off but from Newton's accompanying figure it is evident that the lines AF through the fixed point A (meeting either the conic AFB or the straight line IFE in F) and DG through the fixed point D (meeting EG in G) are in 1, 1 correspondence, and so consequently are the lines FB, GC through the fixed points B and C respectively, whence by Proposition 12 the locus (H) of their meets is a conic through B, C. The particular case where the points A, D are coincident, and the points F and G in the straight lines FE and GE are in line through the fixed point A, is in fact the converse of Pascal's theorem on the *hexagrammum mysticum* (compare R. Taton's 'L'œuvre de Pascal en géométrie projective', *Revue d'Histoire des Sciences*, **15**, 1962: 197–252, especially 240–3), first published as Lemma 1 of his *Essay Povr Les Coniqves* (Paris, 1640); for the locus (H), manifestly through the intersections, say b and c, of AB with GE and of AC with FE, is such that AFG is the Pascal-line of the inscribed hexagon $BbEcCH$. Newton himself never published his more general theorem but it is a comment on the slow publication and dissemination of mathematical knowledge in the seventeenth century that the particular Pascalian case could still, fifty years afterwards, excite (in total ignorance of its French antecedents) a bitter, full-fledged dispute regarding who could claim priority for its discovery. The theorem was first publicized at London in 1732 by William Braikenridge in papers presented to the Royal Society, and published a year later by him as the opening

APPENDIX. VARIANT DRAFTS FOR THE 'SOLUTIO PROBLEMATIS VETERUM DE LOCO SOLIDO'[1]

From the originals in the University Library, Cambridge and in private possession

[1][2]   TRACTATUS DE COMPOSITIONE LOCORUM SOLIDORUM.

Prop. 1.

Si a Conicæ Sectionis puncto aliquo indefinite spectato P ad Trapezij alicujus ABDC in Conica ista sectione inscripti latera quatuor infinite producta si opus est AB, CD, AC, BD, totidem rectæ PQ, PR, PS, PT in datis angulis ducantur, singulæ ad singula; rectangulum ductarum ad opposita duo latera, PQ × PR erit ad rectangulum ductarum ad alia duo opposita latera PS × PT in data ratione: Et vicissim si rectangula ista sint in data ratione punctum P continget Conicam Sectionem circa Trapezium[3] descriptum.

---

proposition of his *Exercitatio Geometrica de Descriptione Linearum Curvarum* (London, 1733: 1–15; compare note (55) above) with the claim that he had discovered it 'Anno 1726 cum *Edinburgi* degebam.... Et insequenti anno 1727, per tres menses commoratus Londini, [propositionem] impertivi... D. *J. Craig* Matheseos peritissimo.... Post paucos dies accidit ut Cl. D. *Maclaurin Edin. Prof* inviserem qui tunc Londini fuit, & ille...retulit se colloquium habuisse cum D. *Craigio*, qui Theoremata mea sibi narravit; dixitque insuper se quædam invenisse similia, & MSS ostendit quo innuebat inventa sua contineri; sed qua ratione ductus nescio'. Forewarned by friends of the content of Braikenridge's book in advance of its publication, Colin Maclaurin at once wrote on 21 December 1732 to John Machin (then Secretary of the Royal Society) 'A Letter...concerning the Description of Curve Lines' (*Philosophical Transactions*, **39**, No. 439 [for October to December 1735], §v: 143–8), asserting (p. 145) that he himself had found the theorem 'in *June* or *July*, 1722, upon the Hint I got from Mr. *Sympson* of Mr. *Pappus*'s Porisms' (compare Robert Simson's 'Pappi Alexandrini Propositiones duæ generales, quibus plura ex Euclidis Porismatis complexus est, Restitutæ', *Philosophical Transactions*, **32**, No. 377 [for May/June 1723], §vi: 330–40) and noting (p. 146) that 'In 1727 I added to a Chapter in my Algebra [published posthumously as *A Treatise of Algebra, in Three Parts* (London, 1748): Part 3, Chapter 2: 325–52, especially 346–8], which is very publick in this place [Edinburgh], an Algebraick Demonstration of the Locus, when three Poles are employed', while in appendix (*ibid.*: 163–5) he reproduced 'some Leaves' of a paper on the topic 'dated at Nancy, Novem. 27, 1722'. When two years later Simson supported his claim in the *Præfatio* to his own *Sectionum Conicarum Libri V* (Edinburgh, 1735: vi: 'Multis abhinc annis mecum communicavit Mathematicus eximius, vir doctissimus, mihique amicissimus, *Colinus Maclaurin*, ...hunc locum solidum') Maclaurin had all but won his case in his contemporaries' eyes. (See, for instance, Francis Hutcheson's aside on this point in his review of Simson's book in the *Bibliothèque Raisonnée des Ouvrages des Savans de l'Europe*, **14** (Amsterdam, 1735): Seconde Partie [for April to June 1735]: 476–83, especially 482.) We have earlier indicated that Newton himself produced, as an example of a rectilinear locus, the degenerate case of the Pascalian theorem (Pappus') where the conic locus (*H*) splits into a line-pair (see 1, §1: note (29)).

(1) See §2: note (2) above. These preliminary versions of Newton's general solution and construction of the classical problem of the 'locus solidus' are reproduced, in a severely edited version, above all for the light they shed on his still growing insight into the properties of the general conic and the 1, 1 point-correspondence even as the structure of his treatise hardened into its definitive form.

Demonstrationis gradus primus.[4]

Cas. 1. Ponamus imprimis lineas ad opposita latera ductas parallelas esse alterutri reliquorum laterum, puta *PQ* et *PR* lateri *AC* et *PS* ac *PT* lateri *AB*; sintcɜ insuper latera duo ex oppositis, puta *AC* ac *BD*, parallela; et bisecto *RQ* in *O*, in *OQ* cape *OK* æqualem *OP* et punctum *K* continget conicam sectionem, nam *OP* & *OK* erunt ordinatim applicatæ ad diametrum[5] qua parallelæ *AC BD* bisecantur. Est itacɜ per   [6] Apollonij *AQ* × *QB* ad *PQ* × *QK* in data ratione. Sed *QK* & *PR* æquales sunt, utpote æqualium *OK OP*, & *OQ OR* differentiæ, & proinde etiam *PQ* × *QK* & *PQ* × *PR* æquales sunt, atcɜ adeo *AQ* × *QB* hoc est *PS* × *PT* est ad *PQ* × *PR* in data ratione. Q.E.D.

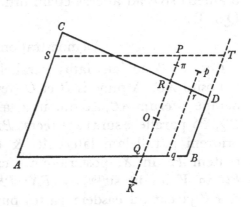

Cas. 2. E contra verò si hæc rectangula sint in data ratione,[7] per aliquod infinitorum punctorum *P* puta *p* ipsi *PQ* agatur parallela quævis *pq* secans rectas *AB* et *CD* in *q* et *r* &[8] punc[t]um *P* continget Conicam sectionem quæ per data quincɜ puncta *A B D p C* transit. Nam per jam demonstrata manifestum est quod ubicuncɜ ducatur recta *PQ* contingens hanc Conicam sectionem, puta in *π*, erit *AQ* × *QB* . *πQ* × *πR* :: *Aq* × *qB* . *pq* × *qr*, adeocɜ

$$AQ \times QB \,.\, \pi Q \times \pi R :: AQ \times QB \,.\, PQ \times PR \quad \text{et} \quad \pi Q \times \pi R = PQ \times PR.^{[9]}$$

---

(2) ULC. Add. 3963.8: 141ʳ–144ʳ. In later versions of this first opening to Newton's tract (see §2: note (3) above) 'Cas. 1' of the first 'gradus demonstrationis' and 'Cas. 2' of the second are collected together as Cases 1 and 2 of Proposition 1, while their converses ('Cas. 2' of the first grade and 'Cas. 1' of the second) are subsumed into the single general proof of Proposition 2 of the final version.

(3) The equivalent 'per angulos Tr[apezij]' is cancelled.

(4) When, that is, either of the pairs of opposite sides of the 'trapezium' *ABCD* are parallel.

(5) 'axem' was first written and is rightly cancelled.

(6) In revise on f. 137ʳ (compare §2: note (3) above) Newton completed the reference by inserting 'per Prop 17 & 18 lib 3 [*sc. Conicarum*]'. See also §2: note (4) above.

(7) Newton first wrote somewhat abruptly in sequel 'erit regrediendo *AQ* × *QB* ad *PQ* × *QK* in eadem data ratione [adeocɜ] punctum *P* continget Conicam Sectionem'.

(8) The preceding phrase 'per aliquod...punctorum *P* puta *p*' replaces a lengthier cancelled equivalent at this point which reads 'in ea capiatur punctum *p* ita ut sit *Aq* × *qB* ad *pq* × *qr* ut *PS* × *PT* vel *AQ* × *QB* ad *PQ* × *QR* et dico [quod]'.

(9) In continuation Newton first set out a preliminary version of his following argument: '*sive PQ . πQ :: πR . PR* & *PQ* − *πQ*(*Pπ*) :: *πR* − *PR*(−*Pπ*) . *PR* vel :: *PR* − *πR*(*Pπ*) . −*PR*. Adeocɜ *πQ* = −*PR*.' At once we deduce that *π* is coincident with the second meet, *K*, of *PQ* with the conic; however, Newton's conclusion presupposes that *Pπ* is not zero, that is, that *π* does not coincide with *P*.

Aufer utrincæ quadratum de OQ vel OR et restabunt $O\pi^{quad.} = OP^{quadr.}$.[10] Est itacæ $OP = O\pi$. At $O\pi$ ordinatim applicatur ad diametrum quæ bisecat rectas parallelas AC, BD et proinde OP ei æqualis, sive jaceat ad easdem partes hujus diametri sive ad alteras etiam ordinatim applicatur, et[11] figuram continget. Q.E.D.

Demonstrationis gradus secundus.[12]

Cas. 1. Pone jam latus Parallelogrammi BD non esse parallelum lateri opposito AC. A punctis B et C (terminis nempe laterum AB, AC quibus actæ rectæ PQ, PS parallelæ sunt) age rectas BE, CE, priorem parallelam lateri AC & occurrentem PS in X, posteriorem secantem PQ in V ita ut sit[13] $PV.PX::PR.PT$, & PV jaceat ad easdem partes puncti P cum PR si modo PX jaceat ad easdem partes cum PT, aliter ad contrarias. His constructis dico primo quod si $PQ \times PR$

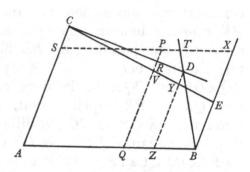

ponatur ad $PS \times PT$ in data ratione, punctum P continget Conicam Sectionem

---

(10) Newton first proceeded: 'Id quod in duobus tantum casibus contingere potest: nempe ubi punctū P incidit in punctum π vel jacet ad alteras partes puncti O & ad eandem ab eo distantiam, puta in K ita ut sit $OK = O\pi$. Est enim divisim $PQ - \pi Q$ (vel $\pi Q - PQ$) hoc est $P\pi$ ad PQ ut $\pi R - PR$ (vel $PR - \pi R$) hoc est etiam $P\pi$ ad PR, adeocæ πQ et PR [æquales sunt].' Compare note (9) above.

(11) Understand 'in utrocæ casu'.

(12) That is, when neither pair of opposite sides are parallel. Somewhat illogically, Case 1 is reduced to Case 2 of the preceding grade and doubtless, if Newton had completed it, Case 2 would have been reduced to Case 1 of the 'primus gradus demonstrationis'. In an immediately following redraft on ff. 141^v/142^v the present 'Cas. 1' is abandoned in favour of a separate 'Prop. 2', essentially that reproduced in §2 above, where the converse of both 'grades' of argument is developed in a single, relatively simple *reductio* form.

(13) Newton first continued more shortly but no less correctly '$RV.TX::PR.PT$, & V jaceat a parte R versus P si X jaceat a parte T versus P, aliter ad contrarias partes. His constructis dico primo quod si $PQ \times PR$ sit ad $PS \times PT$ in data ratione punctum P continget Conicam Sectionem. Cum enim sit $PR.PT::RV.TX$ erit

$$PQ \times PR.PS \times PT::PQ \times RV.PS \times TX'.$$

Exactly as below it then follows, since the ratios of PQ (or XB) to TX and hence of RV to PS (or QA) are given, that V lies in a unique straight line through C and consequently that E is a fixed point in the parallel through B to AC such that

'$PQ \times PR. PS \times PT::PQ \times RV. PS \times TX::$ (addendo antecedentes & consequentes)

$$PQ \times PV. PS \times PX'$$

is a given ratio. Whence by Case 2 of the first grade the locus of P is a conic through A, B, E, C, evidently (as Newton goes on to remark) also through the point D at which PR, PT become simultaneously zero.

per puncta *A B D C* descriptam. Cum enim sit $PR.PT::PV.PX::$ (divisim) $RV.TX$, erit

$$PQ \times PR.PS \times PT::PQ \times PV.PS \times PX::PQ \times RV.PS \times TX.$$

Quare cum $PQ \times PR$ sit ad $PS \times PT$ in data ratione, erit etiam $PQ \times PV$ ad $PS \times PX$ & $PQ \times [R]V$ ad $PS \times [T]X$[14] in data ratione, et praeterea cum (propter datos angulos trianguli *BTX*) PQ vel *XB* sit ad *TX* in data ratione, erit etiam *RV* ad *PS* in data ratione, & insuper cum *PS* vel *AQ* (propter parallelas *AC*, *QR*) sit ad *CR* in data ratione erit etiam *RV* ad *CR* in data ratione. Angulus itacȝ *RCV* datus est et ideo punctum *E* et Trapezium *ABEC* determinatum. Quamobrem cum hujus Trapezij latera *AC BE* parallela sint & insuper ostensum fuerit esse $PQ \times PV$ ad $PS \times PX$ in data ratione, sequitur (per prius demonstrata in secundo casu) punctum *P* contingere Conicam sectionem transeuntem per puncta *A, B, E, C*, ut et per punctum *D*, nempe in quod punctum *P* incidit ubi *PR* et *PT* simul evanescentes efficiunt rectangula $PQ \times PR$ & $PS \times PT$ nulla. Q.E.D.

Cas. 2. Et e converso si punctum *P* contingat Conicā sectionem[15]

[2][16]   *Prop 3.*

Invenire punctum P, a quo quatuor rectæ PQ, PR, PS, PT ad totidem alias positione datas AB, CD, AC, BD ducantur in datis angulis, ita ut duarum ductarum rectangulum $PQ \times PR$ sit ad aliarum duarum rectangulum $PS \times PT$ in ratione data.

Lineæ *AB, CD* ad quas rectæ duæ *PQ, PR* unum rectangulorum continentes ducuntur, conveniant cum alijs duabus positione datis *AC, BD* in punctis *A, B, C, D*: a quorum aliquo *A* age[17] rectam quamvis *AH* in qua vis punctum

---

(14) Newton first wrote '$PQ \times RV$ ad $PS \times TX$', correctly, and then for some obscure reason altered it to '$PQ \times PV$ ad $PS \times PX$'.

(15) The manuscript here breaks off, but we should understand 'per puncta *A, B, C, D* transeuntem, rectangulum $PQ \times PR$ erit ad rectangulum $PS \times PT$ in data ratione'. Newton's proof would evidently proceed much as in the revised 'Cas. 2' of 'Prop. 1' which replaces it in an immediately following draft on f. 141^v (little changed in all succeeding versions), which is indeed referred to 'the figure in pag 3 [= f. 143^r]': namely, 'age *BE* parallelam *AC* et occurrentem conicæ sectioni in *E* et rectæ *PQ* in *R*. Junge *CE* secantem *PQ* in *V*, et parallelam *PQ* age *DZ* secantem *AB* et *CE* in *Y* et *Z*. Jam ob sim. tri. *BXT*, *DBZ* erit *BX(PQ).TX*:: *DZ.ZB*. Sic et *RV.AQ(PS)::DY.AZ*. Ergo $PQ \times RV.TX \times PS::DZ \times DY.AZ \times ZB$:: (per Cas 1) $PQ \times PV.PS \times PX$:: (divisim) $PQ \times PR.PS \times PT$. Q.E.D.'.

(16) ULC. Add. 3963.8: 144^r – 144^v; compare §2: note (11) above. A somewhat shorter but otherwise trivially variant first draft exists on ff. 128^r–128^v preceding. It will be clear that Newton's final version of Proposition 3, reproduced (from the revised text on ff. 146^r–146^v) in §2 above, is a considerably altered account which omits the present 'Exemplum' entirely.

(17) In first draft (f. 144^r) Newton began this paragraph more tersely with 'A quovis punctorum *A, B, C, D*, puta *A* age...'.

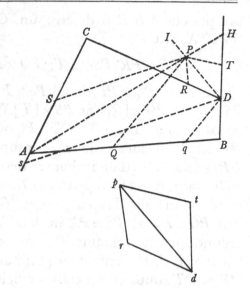

P reperiri. Et propter datos angulos triangulorum *APQ*, *APS*, dabitur ratio *PQ* ad *PA* et *PA* ad *PS* adeoǫ ratio *PQ* ad *PS*, et inde ratio etiam *PR* ad *PT*, quippe quæ una cum ratione *PQ* ad *PS* ex Hypothesi datur: Atǫ adeo Trapezium *DRPT* datur specie & diagonalis *DP* positione.[18] Age ergo hanc diagonalem et ea secabit *AH* in puncto quæsito *P*. Q.E.F.

Exempli gratia, sit $PQ \times PR . PS \times PT ::$ $d . e$ & $PQ . PS :: d . f$, et erit

$$d \times PR . f \times PT :: d . e,$$

sive $PR . PT \left(:: \dfrac{d}{d} . \dfrac{e}{f} \right) :: f . e.$ Quare extra figuram[19] duc rectas $pr = f$ & $pt = e$ continentes angulum *rpt* æqualem angulo dato *RPT*, ut et rectas *rd td* facientes cum priùs ductis angulos *prd*, *ptd* æquales angulis datis *PRD*, *PTD* & Trapezium *PRDT* erit simile huic dato Trapezio *prdt*. Age ergo diagonalem *dp* & recta *DI* faciens angulos *IDR*, *IDT* æquales angulis *pdr*, *pdt* secabit *AH* in puncto quæsito *P*.

Corol. Hinc etiam patet modus ducendi Tangentem ad quodvis punctum Curvæ lineæ quam puncta infinita *P* attingunt, puta ad punctum *D*. Nam chorda *PD* ubi puncta *P* ac *D* coincidunt, hoc est ubi *AH* ducitur per punctum *D*, Tangens evadet. Concipe ergo *AH* duci per *D* & perinde Actis *Dq*, *Ds* parallelis *PQ*, *PS*, pone $f . d :: Ds . Dq$, dein[20] constituto Trapezio *prdt* ut ante, hoc est cujus latera *pr*, *pt* sunt ut *f* ad *e*, fac angulum *CDI* æqualem angulo *rdp*, et *DI* tanget Curvam.

Prop. 4.

Iisdem positis, Locum definire quem infinita ista puncta P attingunt.

[21]Sint *A*, *B*, *C*, *D* concursus linearum positione datarum per quos Curva definienda (juxta Prop 2) transit. Per quemvis eorum *A* duc Tangentem

---

(18) Newton first concluded (f. 144ʳ) with 'cujus intersectio cum recta *AH* est punctum quæsitum *P*'.

(19) This replaces 'seorsim' (on its own) on f. 144ʳ.

(20) Newton first copied the opening 'Duc ergo *AH* per *D* et' from f. 144ʳ and then cancelled it for the present lengthier revise.

(21) On f. 144ᵛ Newton first began: 'Ab aliquo angulorum Trapezij *A* ad latera opposita *BD*, *CD* age rectas *At*, *Ar* rectis *PT*, *PR* ad eadem latera ductis parallelas.' In a cancelled accompanying figure, correspondingly, *Cq*, *Ar*, *Cs*, *At* were drawn parallel to lines *PQ*, *PR*, *PS*, *PT* let fall obliquely onto the sides *AB*, *CD*, *AC* and *BD* respectively of the general 'trapezium' *ABCD*.

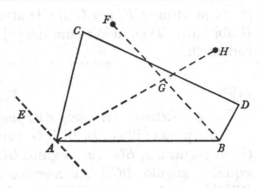

Curvæ *AE* & huic per alium quemvis *B*
parallel[a]m *BF* occurrentem Curvæ in
F. Biseca *BF* in *G*, et acta *AG* diameter
erit ad quam *BG* vel *FG* ordinatim
applicatur. Hæc *AG*, si opus est, producta, occurrat Curvæ in *H*, et erit *AH*
latus transversum:[22] ad quod latus rectum[22] est ut *BG^q* ad *AG* × *GH*. Si *AG*
non occurrat Curvæ in quovis puncto *H*
(lineis *AH*, *DI* in superiore Propositione
existentibus parallelis) figura erit Parabola: in quo casu latus rectum est $\dfrac{BG^q}{AG}$.

Sed si *AG* occurrat Curvæ, Figura erit Hyperbola ubi *A* et *H* sita sunt ad easdem
partes ipsius *G*, et Ellipsis ubi *G* intermedium est, nisi forte angulus *AGB* rectus
sit & insuper *BG^q* = *AG* × *GH* in quo casu circulus esse noscitur.

Tangens autem *AE* ducitur per Corollarium præcedentis Propositionis, et
puncta *F* et *H* per ipsam Propositionem determinantur.[23]

[3][24]   *Prop. 8.*

Conicam Sectionem per quinq; data puncta describere.

[Sunto] *B C D G P*. A duobus eorum *B C* ad reliqua
D G P duc rectas. Fac angulos *DBE PBH* æquales
angulo *GBC*[25] & angulos *DCF PCI* æqu. ∠^{lo} *GCB*.
Produc *BE . CF* donec concurrant in *M* et [*BH . CI*]
donec conc. in *N*, et per puncta *M* et *N* age rectam
MN. Deniq; convertantur anguli dati *MBD MCD*
circa polos *B*, *C* ita ut eorum crura *BM CM* perpetim conveniant ad rectam *MN*, et concursus
reliquorum crurum *BD CD* (per Prop 7) describet
con. sect.: quæ (ex constructione) transibit per puncta
quinq; *B C D G P*, nempe per *D* ipso motus initio, per

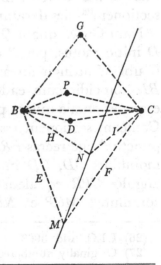

(22) That is, a general diameter of the conic locus and the *latus rectum* conjugate to it
respectively. Compare §1: note (13) above.

(23) This final paragraph, evidently a late addition, is not found on f. 144^v.

(24) A first draft (from the original in private possession) of an intended sequel to Proposition 7, using the organic construction of a conic there expounded in its solution. It is revised
by 'Prob 7' in Appendix 4 following.

(25) So making the point *G* of the described curve correspond to the polar line *BC*, whence
the line *MN* will correspond to the remainder of the conic locus (*D*) through *B*, *C*, *G*, *P*.
Compare §2: note (30) above.

P ubi *M* attingit *N*, per *G* ubi *M* attingit intersectionem rectarū *BC*, *MN*, per *B* ubi latus *CD* coincidit cum line[a] *CB*, et per *C* ubi latus *BD* coincidit cum eadem *BC*.

[4][26]   *Prob* 7.[27]

Conicam Sectionem per quinꝗ data puncta describere.

Sunto puncta illa *C*, *B*, *D*, *P*, *G*. Junge duo eorum *B*, *C* et inde[28] ad terti[u]m *G* duc rectas *CG*, *BG*. Fa cangulos *DCM*, *PCN* æquales angulo *BCG*, et angulos *DB*[*M*], *PB*[*N*] æquales angulo *CBG* si modo *G* cadit ad easdem partes lineæ *CB* cum punctis *D* et *P*, aliter fac angulos istos æquales complementis angulorum *BCG*, *CBG*. Et per puncta *M* et *N* ubi rectæ *CM*, *BM* & *CN*, *BN* concurrunt age rectam *MN*. Deniꝗ circa polos *C* et *B* convertantur anguli dati *DCM*, *DBM* ita ut eorum crura *CM*, *BM* perpetuo conveniant ad rectam *MN* et punctum *D* ubi reliqua duo crura concurrunt describet Conicam sectionem[29] desideratam.

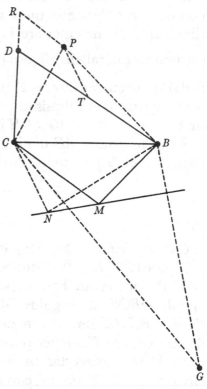

Nam Curva quam *D* describit transibit per *D* initio motus, per *P* ubi *M* attingit *N*, per *G* ubi *M* attingit lineam *CB*, per *C* ubi linea *BD* coincidit cum eadem linea *CB*. Transit ergo per omnia quinꝗ puncta data. Eam vero Conicam sectionem esse sic ostendo. A dato puncto *P* age rectas *PR*, *PT* occurrentes lineis mobilibus *BD*, *CD* in *R* ac *T*, & facientes angulos *CPR* æqualem angulo *CNM* & *BPT* æqualem angulo *BNM*. Aufer communes *MCP* et *NBD* et restabunt æquales *PBT NBM* & *PCR NCM*:

(26) ULC. Add. 3963.8: 63ʳ, the revise of 'Prop. 8' in Appendix 3 preceding.

(27) Originally numbered as 'Prob 8' [= Proposition 8]. This problem evidently began its existence as a revise of the abortive *addendum* to Proposition 7 in Newton's 'Solutio Problematis Veterum de Loco solido' and was, it is clear, at one time intended to replace it. We have earlier (§2: note (29) above) taken his preference in 1685 for the original 'Prop 7' (ff. 130ʳ–130ᵛ) as our authority for discarding this revise from the main text of Newton's treatise.

(28) Much as in Newton's preliminary draft, a first cancelled sequel reads 'ad reliqua duc rectas *CD*, *BD*, *CP*, *BP*, *CG*, *BG*'. The manuscript lacks an accompanying diagram: the figure reproduced is a simplified version of that of Proposition 7 of the main text (on f. 130ᵛ).

(29) Newton first concluded more verbosely with the words 'per quinꝗ illa data puncta *C*, *B*, *D*, *P*, *G* transeuntem'.

adeoɋ triangula *PBT*, *NBM* similia sunt ut et *PCR*, *NCM*. Quare *PT*.*NM*::
PB.*NB* & *PR*.*NM*::*PC*.*NC*. *PT* ergo et *PR* datam habent rationem ad *NM*
et proinde datam rationem inter se, atɋ adeo per Prop 5 punctum *D* (quod
in hac demonstratione suppono non determinatum esse ad punctum initio
datum, sed esse perpetuum rectarum mobilium *BT*, *CR* concursum) attingit
Conicam sectionem. Q.E.D.

---

[5][30]

1. Si recta mobilis *DE* secat tres rectas positione datas *AD*, *AE*[,] *BF*, et
earum duæ *AD*, *AE* cognitæ
sunt per simplicem Geom. tertia
BF determinabilis erit per simpl
Geom.

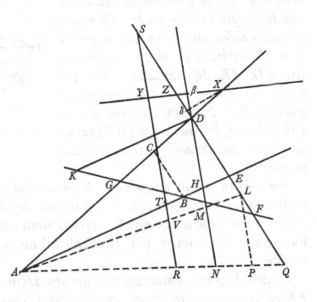

Nam age *BC*∥*D*[*E*] et erit
CG.*GB*::*GD*.*GF*.

2. Item ratio *DE* ad *EF* habebitur per simpl. Geom.

Nam age *DK*∥*BE* et erit
AG.*GB*::*DG*.*GK*. et *DE*.*EF*::
KB.*BF*.

3. Si ang *EDH* datur, dabitur
AH, *EH* per simpl. G. Nam fac
DAL=*EDH*=*EAQ*=*QLP*[31] et
erit *AD*.*AE*::*AL*.*AP* et

$$LQ.DQ::PQ(AQ-AP).NQ.$$

Unde habetur *AN*.[32] hæc et *AD* dat *AH* vel *EH*. per 1.

4. Si *SD* ad *DE* in relat. simp. et ang *DSR* dat: erit *AT*, *AV* simpl. Age
DH∥*SR* et erit *DE*.*ES*::*EH*.*ET* et *DL*.*LS*::*ML*.*VL*.

---

(30) ULC. Add. 3963.8:68ᵛ. (The recto side has two contemporary letter drafts, one to his
Woolsthorpe tenant [Storer?], one to his step-brother by marriage Robert Barton, both of
which are reproduced by H. W. Turnbull in *The Correspondence of Isaac Newton*, **3** (Cambridge,
1961): 393; however, his judgement (*ibid.*: note (1)) that the present verso is a page 'belonging
to the early 1690's' is evidently unsupportable, and the letter fragments themselves should
probably be dated as *c.* 1680.) In this preliminary draft Newton constructs a number of
examples of geometrical magnitudes mutually 'determinable by simple geometry', that is, in
modern terms (see §2: note (43) above) magnitudes in 1, 1 correspondence.

(31) In consequence *D*, *E*, *A*, *N* are concyclic and *DHN* is parallel to *LP*, whence

$$D\hat{E}A = D\hat{N}A = L\hat{P}A \quad \text{and} \quad D\hat{A}E = L\hat{A}P,$$

so that the triangles *DAE*, *LAP* are similar.

(32) 'et inde' is cancelled.

5. Si $AD.AE.XY$ simpl erunt AL, AT, AV simpl. Fac $ZX\delta = \text{ang } DSY$ et erit[33] $ZX.ZY::Z\delta.ZS$ simpl rat. Ergo[34]

---

[6][35]   *Prop* [8].[36]

Si Crura anguli cujusvis dati [DP]C[37] *circa punctū angulare P gyrantis secent rectas quotcunqȝ positione datas AB AC in punctis B, C, D, E et ab intersectionibus ad alias quasvis positione datas rectas ducantur rectæ ET BV in datis angulis et omnes quantitates pro datis habeantur. Dico quod ex quâvis indeterminatarum et non gyrantium AB cognita cæteræ omnes non gyrantes AC AD AE determinabiles erunt per simplicem Geometriam*[,] *sicut et gyrantium quadrata PB^q, PD^q, PC^q, PE^q et rationes PB, PC, PD, PE ad invicem*[;] *Gyrantes vero ipsæ solummodo per Geometriam planam.*[38]

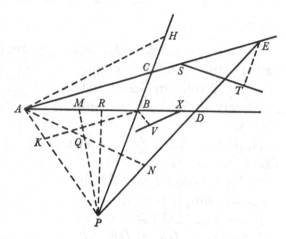

Per Geometriam simplicem determinabiles esse intelli[g]o quæ per intersectiones solarum rectarum sine adminiculo Circuli vel anguli dati, hoc est per additionem subductionem & inventionem quartæ proportionalis[,] vel ut jam loquuntur Geometræ per multiplicationem et divisionem sine extraxione radicis.

Imprimis enim assumatur non gyrans AB ut cognita et ipsi AC agatur parallela BK occurrens AP in K: et propter datos angulos trianguli ABK dabitur ratio AB ad AK et BK adeoqȝ AK et BK determinabiles sunt per simplicem Geometriam. A dato AP aufer AK et restabit PK quæ est ad BK ut AP ad AC & ad AK ut PB [ad] BC. Ergo et AC et ratio PB ad BC determinabilis est per simplicem Geom. Fac ang[ulum] BAH [æqualem] ang[ulo] BPD et triang[ula] ABH

---

(33) Since X, S, Y and δ are concyclic.

(34) The manuscript breaks off but understand 'Ergo $Z\delta$ datur per simplicem Geometriam' probably.

(35) ULC. Add. 3963.8: 64ʳ–66[= 65a]ᵛ, a considerably variant preliminary version of Propositions 8–10 in §2 above. Two first drafts of the present Proposition 10 follow in sequence on ff. 67ʳ–67ᵛ/67ᵛ–68ʳ.

(36) Evidently an improved version of the preceding Appendix 5.

(37) The manuscript reads '*BDC*'.

(38) Newton first concludes 'Gyrantium vero et ad polum P terminatarum sola quadrata per simplicem Geometriam determinabilia erunt'. A cancelled following sentence continues: 'Et e contra si una gyrantium PB assumatur cæteræ gyrantes PC, PD, PE determinabiles erunt per simplicem Geometriam, at non gyrantes solummodo per geometriam planam.'

PBD erunt sim[ilia]. Ergo $PB.PD::AB.AH_{[,]}$ hoc est [in] rat[ione] sim[plici].[39] Præterea ab *A* et *P* age *AN PM* facientes angulos *DAN APM* æquales dato angulo *BPD* et se secantes in *Q* et propter datos angulos trianguli *APQ* datamcg basem *AP* dabuntur latera *AQ PQ*. Ad æquales angulos *APM BPD* adde *MPB* et habebuntur æquales *APB QPN*. Item ad angulum *PAN* adde æquales *NAB APQ* et habebuntur æquales *PAB, PQN*. Et proinde triangula *PAB, PQN* similia sunt et *AB* erit ad *QN* sicut *AP* a[d] $PQ_{[,]}$ hoc est in ratione data. Quare *QN* et proinde etiam *AN* determinabilis est per simplicem Geometriam. Et eadem ratione qua ex cognita *AB* fuit *AC* determinabilis per simplicem Geometriam: ex cognita *AN* erunt *AD* et *AE* determinabiles per simplicem Geometriam. Aufer *AB, AC* et simul innotescent *BD CE* per simpl: Geom. Q.E.D.

Præterea cum sit[40] $PB.BC::PK.AK$ et *PK, AK* sint determinabiles per simplicem Geometriam, per eandem Geometriam determinabitur ratio *PB* ad *BC*. Et eodem argumento rationes *PN, ND, NE* ad invicem determinabiles erunt per simplicem Geometriam, adeocg ratio *PB* ad omnes cum propter similia *PAB PQN* sit *PB* ad *PN* in data ratione *PA* ad *PQ*. Quare[41] ratio omnium *PB, PC, PD, PE, CB, DE* ad invicem dantur per simplicem Geometriam. Q.E.D.

Insuper a *P* ad *AD* demitte perpendiculum *PR* et ab *AB* aufer datam *AR* et habebitur *RB* cujus quadratum una cum quadrato dati *PR* efficit quadratum *PB*. Ergo datur PB^q per simplicem Geometriam[42] sed [ad] inveniendum *PB* requiritur extractio radicis quadraticæ quæ non fit generaliter sine Geometria plana.[43] Cum ergo detur PB^q sed non *PB* per simplicem Geometriam, et *PB* ad *PC PD PE BC DE* rationem habeat per simplicem Geometriam determinabilem dabuntur $PC^q, PD^q, PE^{[q]}, BC^q, DE^q$ per simplicem Geometriam sed non ita dabuntur earum longitudines. Q.E.D.

Et cum pun[c]ta *S* et *X* dentur,[44] a cognitis *AE, AX* aufer cognitas *AS AB* et restabunt cognitæ *SE BX* quæ propter datos angulos triangulorum *SET*

---

(39) These two sentences were inserted hurriedly on f. 67ʳ.

(40) Since *AC* and *BK* are parallel.

(41) 'componendo' is cancelled.

(42) In a preceding, otherwise trivially variant draft of this paragraph Newton first went on: 'at ad PB universaliter determinandum requiritur extractio radicis quadraticæ, quod non fit sine Geometria plana. Et eodem argumento dabuntur quadrata linearum *PC PD PE* per simplicem Geometriam sed ad determinationem longitudinum eorum requiritur Geometria plana et multo magis ad determinationem longitudinum *BC DE* quæ sunt differentiæ radicum. Q.E.D.'

(43) Since Newton's definition of constructibility 'per simplicem Geometriam' allows the 'inventio quartæ proportionalis' but not an 'extraxio radicis' with its two possible (positive and negative) results, he is permitted to construct $PR^2 = PR \times PR/1$ but not *PB* (the positive root of $\pm \sqrt{[PB^2]}$).

(44) It is understood that *ST, XV* are in given position through the given points *S* and *X*, and that the triangles *EST, BXV* are accordingly given in species.

BXV, datas habebunt rationes ad *ET*, *ST*, *BV*, *VX*. Ergo et hæ dabuntur per simplicem Geometriam. Et vice versa ex harum aliqua cognita, dabuntur per retrogradum ratiocinium cæteræ omnes. Q.E.D.

Prop 9.

Supra e Prop 5.[45]

Prop 10.[46]

Si a dato puncto A in data positione ducatur recta quævis AB et ad quodvis ejus punctum B erigatur in dato angulo recta BC[47] *et ea sit relatio linearum AB, BC ut ex alterutra cognita altera [sit] determinabilis per simplicem Geometriam et ambæ etiam simul evadant infinitæ: punctum C attinget rectam positione datam.*

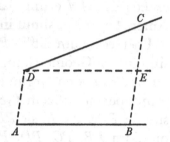

Cum enim *AB* et *BC* se mutuo determinent per simplicem Geometriam non requiritur quadratum aut altior dimensio alterutrius ad determinationem alterius. Nam si verbi gratia ad determinandam *BC* requiretur $AB^q_{[3]}$ tum ex cognita *BC* non possis regredi ad *AB* sine extractione radicis ejus. Assumatur itaɋ relatio quævis inter *AB* et *BC* qua mutuo determinabiles existant per [æquationem $P \mp Q \times AB \mp R \times BC \mp AB \times BC = 0$, datis quantitatibus *P*, *Q* et *R*. Nam hi sunt omnes termini integri qui componi possunt ex *AB* et *BC* sine ipsorum quadratis vel fractionibus.][48]

---

(45) Evidently an anticipation of Proposition 12 of the final tract (§2 preceding). We may perhaps restore its intended enunciation in the following way: '*Si rectarum CR, BT circa polos C, B gyrantium concursus D contingat sectionem Conicam per A & P transeuntem, et PR, PT rectis AC, AB parallelas secent CD, BD in R et T, erunt PR, PT mutuo determinabiles per simplicem Geometriam.*' (Understand the figure accompanying Proposition 5.)

(46) Appropriate extracts from two preliminary drafts of this proposition (on ff. 67ʳ–68ʳ) are reproduced in following footnotes.

(47) Newton first continued: 'et *AB*, *BC* sint mutuo determinabiles per simplicem Geometriam ex alterutra cognita et possunt simul evadere infinitæ:...'.

(48) The latter portion of this cancelled paragraph is restored from a first version on f. 67ʳ. Ignoring cancellations and afterthoughts Newton's preliminary draft (on ff. 67ʳ–67ᵛ) of the following text reads: 'Sit itaɋ simplicissima proportio qua ex cognita *AB* pervenitur ad *BC* istiusmodi $P \cdot Q :: R \cdot BC$. sive $\frac{QR}{P} = BC$ et manifestum est quod ubi $\frac{QR}{P}$ reducitur ad formam simplicissimam quam potest habere quærendo simplicissimos terminos rationum *Q* ad *P* et *R* ad *P* una quantitatum *Q* et *R* erit data quantitas, nam si utraɋ componeretur ex *AB* rectangulum *QR* componeretur ex AB^q. Sit ergo *R* data quantitas & ubi *BC* evadit inf. hoc est infinite major quam *R* erit *Q* infinite major quam *P*. Compona[n]tur igitur *P* et *Q* ex *AB* et assumptis quibusvis datis *k. l. M. N* sit $M \pm \frac{l}{k} AB = P$ et $N \pm AB = Q$. Ergo $N \pm AB$ infinite major quam $M \pm \frac{l}{k} AB$. Aufer finitas *M* et *N* et restabit *AB* infinite major [quam] $\frac{l}{k} AB$ sive 1

[49]Componatur enim BC ex quibuscunꝗ potest terminis et termini isti postquam ad communem denominatorem reducuntur constituant fractionem $\frac{M}{N}$, et manifestum est quod in M et N non reperietur quadratum de AB aut altior ejus dimensio[50] quia sic ex dato BC non posset vicissim determinari AB sine extractione rad[ic]is, contra Hypothesin. Quinimo in N non omninò reperietur AB quia sic ubi AB infinita est, N foret etiam infinita, et $\frac{M}{N}$, hoc [est] BC, non posset esse simul infinita, contra Hypothesin. Quod ut clarius ostendā, assumantur quævis datæ quantitates P, Q, R, S, T, V, et componatur M ex $P \times Q + R \times AB$ et N ex $S + \frac{T}{V} AB_{[,]}$ nam ex pluribus terminis sine altiori dimensione AB componi nequeunt. Ergo $\dfrac{P \times Q + R \times AB}{S + \dfrac{T}{V} AB} = BC$ sive

$P \times Q + R \times AB = S \times BC + \frac{T}{V} AB \times BC.$[51] Divide omnia per $AB \times BC$ et fiet $\frac{P \times Q}{AB \times BC} + \frac{R}{BC} = \frac{S}{AB} + \frac{T}{V}$. Fiant jam AB et BC infinitæ et omnes termini præter $\frac{T}{V}$ in nihilum redigentur, hoc est restabit $\frac{T}{V}$ æquale nihilo. Rejiciendus est itaꝗ terminus $\frac{T}{V} AB$ et restabit solum $\dfrac{P \times Q + R \times AB}{S} = BC$. Quare ab A duc AD parallelam BC & æqualem $\frac{PQ}{S}$ et age DE parallelam AB ut per eam de BC auferatur $BE = \frac{PQ}{S}$. et restabit $CE = \frac{R \times AB}{S}$. Quare est CE ad AB vel DE

---

infinite $\sqsubset \frac{l}{k}$. Quare $\frac{l}{k}$ nihil est et terminus $\frac{l}{k} AB$ deleri debet. ergo $M . N \pm AB :: R . BC$ et inverse $M . R :: N \pm AB . BC$. Componitur ergo BC ex dato $\frac{RN}{M}$ & indeterminato $\frac{R}{M} AB$. Quare ad AB erige AD parallelam BC et per D age DE parallelam AB et secantem BC in E et erit $EC = \frac{R}{M} AB$. Hinc ut $M . R :: AB$ vel $DE . EC$ hoc est EC ad AB vel DE in data ratione, et proinde per conversū 2.6 Elem DC quam C attingit recta est.' In sequel (on ff. 67v–68r) this was altered into a revised form little different from that here reproduced.

(49) Newton first began 'Quantitates per simplicem [mutuo determinabiles...]' but returned at once to the reading of his immediately prior draft on f. 67v.

(50) The equivalent 'potestas' is cancelled.

(51) In the prior draft on f. 67v Newton first continued: 'At ubi AB fit infinita etiam BC sive $\frac{M}{N}$ fit inf. hoc est M infinite $\sqsubset N$ et maximus terminus in M infinite major maximo termino in N id est $R \times AB$ [infinite $\sqsubset \frac{T}{V} AB$]'. Compare the preceding version quoted in note (48).

in data ratione R ad S et proinde per conversum 2^dæ Prop. lib. 6. Elem. locus puncti C est linea recta. Q.E.D.[52]

Corol. Hinc duæ indeterminatæ quantitates si possunt simul evanescere, erunt in data ratione[;] tunc enim puncta D et A coincident.[53]

---

[7][54]   *Prop 12.*

Si intersectio curvæ et omnis rectæ positione datæ determinabilis est per Geometriam planam figura erit Conica sectio.

Sit FC recta quævis positione data curvam secans in C, et cum intersectio ejus cum curva determinabilis sit per pl. Geom. et[55] in omni constructione per planam Geom intersectio rectæ et circuli vel duorum circulorum per quos Problema construitur duplex sit adeoꝗ quantitas per intersectionem istam determinata sit ambigua[,] intersectio rectæ FC & curvæ erit ambigua, hoc est FC curvam secabit in duobus punctis. Sint ista B et C.[56] Ergo si punctum B daretur punctum C determinabile esset per simplicem Geometriam. Nam omne[57] problema

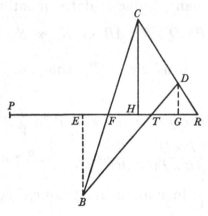

planum reducibile est ad æquationē quadraticam duas habentem radices per duas intersectiones circuli determinabiles et una istarum radicum cognita æquatio per divisionem deprimi[58] potest ita ut ad ejus constructionem postmodum sufficiat Geometria simplex. Omnis itaꝗ rectæ per punctum quodvis datum B in perimetro figuræ transientis intersectio C cum curva determinabilis est per simpl. Geom. et proinde[59] figura est con. sect.

---

(52) An ingenious reduction of the 1, 1 correspondence $AB \leftrightarrow BC$ defined by $0 \leftrightarrow (PQ/S)$ and $\infty \leftrightarrow \infty$ to the equation $BC = (PQ/S) + (R/S)\,AB$, which is then shown to define a straight line in the Cartesian system in which AB, BC are coordinates of the general point C with respect to A as origin. Newton's more general equation

$$P \times Q + R \times AB = S \times BC + (T/V)\,AB \times BC$$

represents an hyperbola referred to asymptotes parallel to $AB = 0$ and $BC = 0$: it will here be geometrically evident that AB and BC are 'mutually determinable by simple geometry but less obvious that this is the most general Cartesian curve to have its coordinates in 1, 1 correspondence. But Newton is not interested here in pursuing his geometrical model, requiring for his immediate purposes only the following corollary.

(53) A late addition on f. 66 [=65a]ᵛ. In modern terms, since two quadruples of points (R) and (T) in 1, 1 correspondence have the same cross-ratio, when the point P is self-corresponding and the points at infinity on PR and PT are related, then $(PRR'\infty) = (PTT'\infty)$ and accordingly (compare §2: note (19) above) the ratio PR/PT is given.

[8]$^{(60)}$

Jactis hisce fundamentis$^{(61)}$ ubi quæstio aliqua de loco proponitur, primum inquirendum est per Prop   num locus sit solidus necne. Et si solidus esse deprehenditur tunc inquirenda erunt aliqua ejus puncta sive ad finitam sive ad infinitam distantiam, & si non primo obvenerint quinꝗ vel 4 puncta sine constructione geometrica, satis est determinare tria & ad duo eorum tangentes ducere$_{[,]}$ dein per Prop [5 vel 6] curvam inde determinare et describere, nisi forte ex statu problematis aliqua obvenerit nota proprietas figuræ per quam determinatio facilior redditur. Quando vero figura est Hyperbola vel Parabola solet punctum infinite distans facile deprehendi: unde distinctio harum figurarum ab Ellipsi facilis redditur, sed Ellipsin a circulo distinguere paulo difficilior esse solet.$^{(62)}$ Et quamvis per solam inventionem punctorum satis commode fit, saltem si practica determinatio tantum requiritur, tamen ubi speculativa$^{(63)}$ requiritur determinatio plerumꝗ redditur$^{(64)}$ simplicior per alias notas proprietates circuli. Ut si duo deprehendantur axes non ad rectos angulos, vel duo axes æquales ad rectos angulos, vel angulus in periphcria determinatus, vel ordinatim applicata media proportionalis inter segmenta axis, vel data distantia peripheriæ a puncto aliquo,$^{(65)}$ vel alia proprietas convertibilis quam status problematis facilius manifestabit. Sed prima tamen inquisitio semper fieri debet de puncto aliquo per quam curva transit.

---

(54) ULC. Add. 3963.8: 66^v/149 bis^r, a first version of Proposition 13 in §2 preceding.

(55) Newton first wrote in sequel 'omnis constructio per planam Geom. exhibeat duas simul æquationis constructæ radices pro numero punctorum in quibus circulus sectus' and then 'in omni constructione per planam Geom. duæ semper prodeant quantitates problemati satisfacientes'.

(56) The broken lines *BE, CH, DG* in the figure are explained by a cancellation at this point, 'Agatur alia recta *PF* eaꝗ positione detur et propter datas utriusꝗ *PF FC* positiones Dabitur *PF* et angulus [*PFC*]. A puncto *B* in alia quavis data positione age rectam *BD* secantem curvam in *D* et junge *CD*': this became in first revise 'Deinde a dato quovis puncto *P* in data quavis positione age rectam *PF* secantem rectas *BC, BT, CD* in *F, T* et *R*, et ad eam demitte ⊥^{la} *BE, CH, DG*'.

(57) This replaces the long-winded equivalent 'Nam in omni æquatione qua constructio per'.

(58) 'reduci' is cancelled.

(59) By 'Prop 11' on f. 66^r, which became Proposition 12 in the revised tract (see §2: note (56) above).

(60) ULC. Add. 3963.8: 150^v. A preliminary revise on f. 149 bis^v is little different from the version reproduced (from f. 149^r) in §2 above (see pages 312–14).

(61) Newton first began with the phrase 'Hactenus fundamenta jecimus: Jam...'.

(62) The equivalent '...non est ita facile' is cancelled. Newton first continued: 'Illud tamen per solam inventionem punctorum alijsꝗ modis fieri potest. Nempe ubi rectæ a duabus punctis ad reliqua tria ductæ continent æquales angulos si ad easdem partes ducuntur vel angulos quorum suma æquatur duobus rectis figura erit circulus. Vel brevius, si circulus per

3

MISCELLANEOUS TOPICS IN ANALYTICAL GEOMETRY[1]

[Late 1670's]

§1. THREE MISTAKES IN DESCARTES' *GEOMETRIE*.[2]

From the original in the University Library, Cambridge[3]

Errores Cartesij[4] Geometriæ

Lib. 1 pag. 11[5] tradit puncta in Problemata Pappi[6] invenienda esse per solidorū Geometriam[,] hoc est per[7] Conicas sectiones ubi plures quam quinqɜ lineæ proponuntur, et per curvas uno gradu magis compositas ubi lineæ plures sunt quàm novem et quod curva adhuc magis composita adhiberi debet ubi sunt plures quàm tridecim. Cùm tamen[8] ejusmodi puncta per simplicem Geometriam uno gradu inferiorem[9] in his casibus semper reperiri possint nisi

---

tria p[uncta] descriptus transit per reliqua duo vel tangit rectas positione datas. Et similis est determinatio ubi quatuor puncta cum tangente, vel tria cum duabˢ tangentibus dantur.'

(63) 'demonstrativa' is cancelled.

(64) This replaces the less definite 'sæpe solet esse'.

(65) Newton has deleted in sequel 'vel proportionalitas linearum quæ intus decussantes vel a dato puncto ab extra ductæ ad peripheriam terminantur'.

(1) Of the background to these four contemporary pieces nothing is known beyond what may tentatively be surmised from the internal evidence of their texts. The date of composition is conjectured on the basis of Newton's handwriting style in the original autographs but is, for once, amply confirmed by the fact that the manuscript of §2.2 is penned on the verso of a letter written (probably to Aubrey) in the summer of 1678. In 1777 (compare ɪ: xxvii) Samuel Horsley found the third and first of the following sections, titled by him 'Fragments concerning lines of the 3ᵈ Order & some mistakes of Des Cartes', packaged together at Hurstbourne Castle and on a loose insert (now ULC. Add. 3961.4: 22ʳ) contributed his myopic judgement that they were 'not worth publishing. S. Horsley Octʳ 23ᵈ 1777'. No doubt the second and fourth papers—then in possession of the Earl of Macclesfield—would have received a similar consignment to oblivion if Horsley had seen them.

(2) When certain adjustment to the detail of their objections is made (see especially note (27) below) these three 'errors', all noticed here for the first time, are all valid and the second, in particular, is fundamental. As with his first studies of Descartes' work fifteen years before, the edition of the *Geometrie* (first published as the final appendix to the *Discours de la Methode* (Leyden, 1637): ₂295–413; see ɪ: 19–20) here used by Newton is Schooten's second Latin edition, *Geometria, à Renato des Cartes Anno 1637 Gallicè edita; postea autem...in Latinam linguam versa, & Commentariis illustrata.... Nunc demum...diligenter recognita...*, 1 (Amsterdam, 1659): 1–106. (See ɪ: 20, note (6) for a fuller bibliographical listing.)

Translation

ERRORS IN DESCARTES' *GEOMETRY*

On page 11 of Book 1[5] he recounts that the points in Pappus' problem[6] have to be found by the geometry of solids (that is, by means of[7] conics) when more than five lines are proposed, and by curves one degree more compound when the lines are more than nine in number, and that a curve still more complex must be employed when there are more than thirteen. Since, however,[8] points of this sort can always in these cases be found by a simple geometry one degree

---

(3) ULC. Add. 3961.4: 23^r–24^r. The manuscript, carefully written out (evidently from a preliminary worksheet) in the first instance, was subsequently somewhat recast in a more rough and ready manner.

(4) The equivalent 'CARTESIANÆ' is cancelled.

(5) *Geometria*, $_2$1659: 11 [= $_1$1649: 12]: *Responsum ad Quæstionem Pappi*: 'Primò autem inveni, quòd, dum hæc quæstio in tribus, quatuorve, aut quinque duntaxat lineis proponitur, puncta quæsita per simplicem semper Geometriam inveniri queant; hoc est, ut non nisi regulâ atque circino utamur; nec aliud quicquam, quàm quod jam traditum est, faciamus. Præterquam si quinque lineæ dantur, quæ omnes inter se parallelæ fuerint. [see note (10) below] Quo casu, ut & quum quæstio in 6, 7, 8, aut 9 lineis proponitur, quæsita puncta per Solidorum Geometriam inveniri possunt, hoc est, adhibendo, ad constructionem, aliquam ex tribus Conicis sectionibus. Excepto tantùm, si novem lineæ datæ fuerint, quæ omnes inter se parallelæ existant. Quo casu, ut & quum quæstio in 10, 11, 12, aut 12 lineis proposita est, quæsita puncta per curvam lineam, quæ uno tantùm gradu magis composita est, quàm sectiones Conicæ, inveniri possunt. Excepto in 13, quæ omnes inter se sint parallelæ, quo casu, ut & in 14, 15, 16, & 17 lineis, linea curva adhiberi debet, quæ uno gradu supra præcedentem composita est. Atque ita in infinitum.' (For the original French text of this passage see Descartes' *Geometrie* (note (2) above): 307–8: 'Response à la question de Pappus.') Observe that Descartes' usage of the phrase 'per simplicem Geometriam' corresponds to Newton's 'per planam Geometriam' (by a geometry of 'plane' loci, viz: straight lines and circles): in more modern style Newton himself restricts 'simple' geometry to be that of the straight line alone (compare 2, §2: note (43), and see note (9) below).

(6) Namely, Pappus' extension of the Euclidean 3/4-line locus (a conic) to the general $(2n-1)/2n$-line case (an algebraic curve of n-th degree). For historical details see the preceding introduction to the present Part 2.

(7) 'circulū et' (the circle and) is deleted.

(8) 'innumera' (innumerable) is cancelled but is none the less to be understood.

(9) Newton first wrote, in its Cartesian sense, 'per simplicem Geometriam' (see note (5) above), then replaced it with his own equivalent phrase 'per planam Geometriam' (by plane geometry) and finally cancelling this in favour of the present more general phrase: in modern terms, since the $(2n-1)/2n$-line locus is a curve of n-th degree, Newton asserts that any number of its points can be constructed by a 'simple' geometry making use of curves of, at most, $(n-1)$-th degree. Thus in the 6-line case (where the locus is a general cubic) the intersections of the locus with any arbitrary line through one of the nine meets of the two line-triples from which it is defined are, as Newton will show in detail below, determinable by a 'simple' (plane) geometry of second degree.

ubi[10] rectæ omnes, aut omnes præter unam sint inter se parallelæ[, erravit].[11] Nempe [ubi] punctum [locatur] in recta quavis[12] iungente intersectiones linearum diversæ conditionis,[13] hoc est linearum in quas rectæ diversa duo contenta per multiplicationem facientes demittuntur in datis angulis[,] linea illa jungens in punctis duobus quæ jungit secat curvam in qua puncta illa omnia reperiuntur [adeoꝗ curva] transit per duo puncta quæ recta illa jungit. Et proinde si curva non sit plusquam quadratici[14] generis reliqua duo puncta in quibus hæc recta jungens secat curvam, quæꝗ propterea quæstioni satisfaciunt inveniri possunt per planam Geom. et si curva non plusquam cubocubici generis reliqua quatuor puncta invenientur per solidam Geom.[15] Quinetiam semper ubi sex lineæ proponuntur[,] possunt innumera puncta per planam Geometriam[16] reperiri. Putemus enim quæstionem more Cartesiano ad finem perductam esse et ubi sex sunt lineæ, æquationem hanc emergere

$$
\begin{aligned}
x^3 + axx \quad + bx \quad - ey \\
+ cyxx \quad + dyx - gyy \\
+ fyyx - hy^3
\end{aligned} = 0.^{(17)}
$$

Juxta Cartesiu[m] Problema solvendum esset per hanc æquationem et proinde solidum fore.

Sed posito $AB = x$ et $BC = y$ & ang $ABC =$ dato, Age AD, DE ipsis CB, BA parallelas, sitꝗ AD unitas et si DE occurrens CA in E dicatur $z^{(18)}$ erit AB sive $x = zy$. Pro x igitur scribe zy in æquatione superiori et orietur[19]

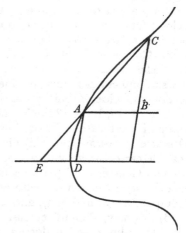

$$
\begin{aligned}
z^3 yy + azzy \quad + bz \quad - e \\
+ czzyy + dzy \quad - gy \\
+ fzyy - hyy
\end{aligned} = 0.
$$

Æquatio in qua y duas tantum habet dimensiones et proinde ex assumpto utcunꝗ z investigari possit per geometriam planam. Hoc est ex assumpto

---

(10) In these exceptional cases, where the base lines are (all but one) through the same point at infinity and the only finite meets of the two sets of line n-ples are collinear, the line joining any two of these meets can meet the locus in no new points. However, a general transversal through one of these meets will continue to meet the $(2n-1)/2n$-line locus in $n-1$ new points determinable by a 'simple' geometry of $(n-1)$-th degree.

(11) Newton's manuscript sentence lacks a main verb. We have not hesitated to fill several further lacunas in his following text in a similarly appropriate manner.

(12) A cancelled first continuation reads equivalently 'quæ jungit intersectiones linearum in quas rectæ se multiplicantes demittūtur fac[tas] cum alijs lineis in quas aliæ rectæ se multiplicantes demittuntur' (which joins intersections made by the lines onto which the first

lower[9] except when[10] all the straight lines, or all but one, are parallel to one another, [he is mistaken].[11] Evidently, [when] a point [is located] in any straight line[12] joining the intersections of lines of different condition,[13] that is, lines onto which are let fall at given angles straight lines forming by their multiplication two different contents, that joining line intersects the curve in which all those points are found in the two points it joins, [and so the curve] passes through the two points which that straight line joins. Accordingly, if the curve be of not more than quadratic[14] kind, the two remaining points in which this joining line intersects the curve (and which consequently satisfy the question) can be found by plane geometry; and if the curve be of not more than sextic kind, the remaining four points will be found by solid geometry.[15] Always, to be sure, when six lines are proposed, innumerable points can be found by plane geometry.[16] For let us consider that the question has been brought to a finish in Cartesian style: when there are six lines, this equation emerges,[17] $x^3 + (a+cy)\, x^2 + (b+dy+fy^2)\, x - (ey+gy^2+hy^3) = 0$. According to Descartes the problem would have to be solved by this equation and would consequently be solid.

However, putting $AB = x$ and $BC = y$ and setting the angle $\widehat{ABC}$ constant, draw AD, DE parallel to CB, BA and let AD be unity. Then if DE, meeting CA in E, be called z[18] there will be AB (or x) $= zy$. In place of x, therefore, in the previous equation write zy and there will arise[19]

$$y^2(z^3+cz^2+fz-h)+y(az^2+dz-g)+bz-e = 0,$$

an equation in which y has but two dimensions and consequently, when z is assumed arbitrarily, might be ascertained therefrom by plane geometry. In

group of straight lines multiplying one another are let fall with the other lines onto which the second set of straight lines multiplying each other are let fall). In the general case of the $2n$-line locus the two sets of line n-ples will intersect in n^2 such points.

(13) Newton first wrote 'diversi ordinis' and then 'diversi generis' (of different order/kind).

(14) Read 'quadratoquadratici' (quartic).

(15) Namely, by a 'simple' geometry of fourth degree whose constructions may be duplicated by the intersections of 'solid' loci (conics).

(16) Newton has cancelled in sequel 'uno gradu inferiorem quam Cartesius assignavit' (one degree lower than Descartes assigned it). A concluding phrase 'si modò calculus rectè ineatur' (if only the computation is correctly instituted) is likewise deleted.

(17) Observe that *more Cartesiano* Newton has taken one of the nine intersections of the two sets of base line-triples as origin of his coordinates, whence the cubic's equation necessarily lacks a constant term and the general transversal $y = zx$, of slope z, through the origin (on the cubic) will meet this 6-line locus in only two further points (which will therefore be constructible by a 'simple' geometry of degree 2).

(18) This will (compare previous note) evidently measure the slope of the transversal AC through the origin A (on the cubic).

(19) On ignoring the root $y = 0$ which yields the origin A itself.

‖ utcunᴄȝ *DE* potest *BC* quo Curvæ punctum *C* determinatur per Geometriam planam inveniri.[20]

Sed pro *y* scribe *xz* et emerget $\dfrac{\begin{array}{llll} xx & +ax & +b & -ez \\ & +czxx & +dzx & -gzzx \\ & & & +fzzxx-hz^3xx \end{array}}{}=0.$ æquatio ubi *x* est

duarum tantum dimens. et proinde per planam geometriã (ex assumpto *z*) investigatur.[21] Habito autem *x* habetur simul *xz* sive *y*. Et eodem modo si decem sint lineæ puncta invenientur per solidam Geometr[iam].

Sed cum in hac prima parte quæ[s]tionis non agatur de inventione infinitorum punctorum[,] hoc est de loco in quo puncta omnia reperiantur sed tantum de puncto aliquo quod quæstioni satisfaciat[,] dico præterea quod ejusmodi puncta quædam semper reperiri possunt per planam Geometriam ubi non sunt plures quam octo lineæ et per solidam Geometriã ubi non sunt plures quam duodecim & sic deinceps. &c[22]

Erravit præterea Cartesius in eo quod asseruit[23] omnes curvas quas Geometricas vocat utiles esse in Problemate Pappi. Nam si proponatur curva ordinis quadrato-cubici et agantur duæ rectæ quævis secantes eam in decem punctis et per hæc puncta agantur aliæ quinᴄȝ lineæ secantes eam in alijs quin-

---

(20) In sequel Newton began a following paragraph 'Ponatur jam *FG*=*z* et *GH*=*y*, *FI*=1, et *IK*=*v*, est si triangula [?*FGH*, *KIF* similia sunt, *z*.*y*::*v*.1 adeoᴄȝ *z*=*vy* ...]' but abandoned it without making its purpose clear.

(21) Newton first interchanged *x* and *y*, writing 'Sed pro *x* scribe *zy* et emerget $yyz^3\dfrac{\begin{array}{lll}+ay & +b & -e\\+cyy & +dy & -gy\\+fyy & -hyy\end{array}}{}zz\ z=0.$ æquatio ubi

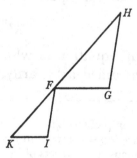

y duarum est dimensionum tantum et proinde per planam geometriam inveniri potest', and then sought to remove the constant term by a 'simplifying' substitution: 'præterea pro *z* scribatur *yv*+$\frac{e}{b}$. et orietur yyv^3 $\begin{array}{llll} +3\frac{e}{b}yy & +3\frac{ee}{bb}yy & +\frac{e^3}{b^3}yy \\ & vv & v \\ +ay & +2\frac{ae}{b}y & +\frac{ee}{bb}ay \\ +cyy & +2\frac{e}{b}cyy & +\frac{ee}{bb}cyy \\ & +b & [-gy] \\ & +dy & [-hyy] \\ +f[yy] \end{array}$ [=0]. æquatio in qua *y* deprimetur ad unam dimensionem [?ponendo *b*=0]' (further, in place of *z* let there be written *yv*+*e*/*b* and there will arise..., an equation in which *y* will be lowered to one dimension [? on putting *b* = 0]). The approach is evidently futile since *b* = 0 requires *e* = 0 also.

(22) Newton has left a long gap at this point in his manuscript, evidently intending to develop the criticism further.

other words, on assuming DE arbitrarily, from it BC—by which the point C on the curve is determined—is found by plane geometry.[20]

However, in place of y write xz and there will emerge

$$x^2(1+cz+fz^2-hz^3)+x(a+dz-gz^2)+b-ez=0,$$

an equation in which x is but of two dimensions and may consequently (from assuming z) be discovered by plane geometry.[21] When however x is had, there is obtained at once xz, that is, y. And in the same manner, should there be ten lines, the points will be found by solid geometry.

But since in this first part of the question it is not a matter of finding an infinity of points—in other words, the locus in which all those points are to be found—but merely of some point which is to satisfy the question, I assert moreover that certain points of the [required] sort can always be found by plane geometry when there are no more than eight lines, and by solid geometry when there are no more than twelve, and so on.[22]

Descartes was further mistaken in his assertion[23] that all curves he calls geometrical are of use in Pappus' problem. For if a curve of quintic order be propounded and there be drawn any two straight lines intersecting it in ten points, and through these points be drawn another five lines intersecting it in a

---

(23) Descartes sought to establish this by analogy; see [*Geometrie* (note (2) above): 308–9 =] *Geometria*, ₂1659: 11–12: 'Deinde inveni quoque, si tantùm tres aut quatuor lineæ datæ fuerint, quæsita puncta, non modò in aliqua trium Conicarum sectionum, sed interdum etiam in circuli circumferentia, aut in recta linea reperiri. Et si 5, 6, 7, aut 8 lineæ datæ fuerint, tum puncta illa incidere in aliquam ex lineis, uno gradu magis compositis, quàm sectiones Conicæ. Quarum quidem nullam, quæ ad hanc quæstionem non sit utilis, imaginari licet. Sed possunt rursus illa etiam in sectione Conica, aut in Circulo, aut linea recta reperiri. Similiter si 9, 10, 11, aut 12 lineæ datæ fuerint, reperientur hæc puncta in aliqua linea, quæ non nisi uno gradu supra præcedentes poterit esse composita: quemadmodum etiam nullam earum imaginari licet, quæ ibidem utilis non possit. Atque ita porrò in infinitum.' (Notice that in strict Cartesian terms a curve of n-th 'grade' is one of either $(2n-1)$-th or $2n$-th algebraic degree. See *Geometrie*: 323 [= *Geometria*, ₂1659: 24]: 'ie mets les lignes courbes qui font monter cete equation iusques au quarré de quarré, au mesme genre que celles qui ne la font monter que iusques au cube, & celles dont l'equation monte au quarré de cube, au mesme genre que celles dont elle ne monte qu'au sursolide, & ainsi des autres. Dont la raison est, qu'il y a reigle generale[!] pour reduire au cube toutes les difficultés qui vont au quarré de quarré, & au sursolide toutes celles qui vont au quarré de cube, de façon qu'on ne les doit point estimer plus composées.' Newton himself (in I, **2**, 1: *passim*) was the first person systematically to classify algebraic curves according to the dimension of their Cartesian defining equations: to avoid confusion with Schooten's rendering 'gradus' of Descartes' 'degré' he is careful in his present critique to talk of a curve of n-th 'order'.) Later, in the second book of his *Geometrie* Descartes concluded unhesitantly (p. 324) that 'il n'y a pas vne ligne courbe qui tombe sous le calcul & puisse estre receüe en Geometrie, qui n'y soit vtile pour quelque nombre de lignes' (= *Geometria*, ₂1659: 25: '...nulla curva linea, quæ sub calculum cadit, atque in Geometriam recipi potest, reperiatur, quæ ibidem ad aliquem linearum numerum non sit utilis').

decim punctis[24] deinde per sex horum punctorum agantur aliæ tres rectæ, hæ
transire debent per reliqua novem puncta si modo quinꝗ linearum quæ in has
tres et duas primo ductas demittuntur factus habeat rationem datam ad
factum demissarum in reliquas quinꝗ uti suppono. Novem igitur restant
conditiones pro determinanda positione decem rectarum: sed omnium positio[nes] ex positione duarum primo ductarum adimplentur, et ad harum
positionem [determinandam] sufficiunt quatu[o]r conditiones. Nam duæ
conditiones determinant positionem unius rectæ. Ergo quinꝗ restant conditiones adimplendæ post determinationem rectarum,[25] hoc est quinꝗ restant
intersectiones per quæ curva non transibit nisi forte ad hoc aptetur. Quis igitur
non videt partem longe maximam curvarū quadrato [cubici] ordinis inutiles
esse[26] in problemate Pappi. In generibus autem magis compositis pauciores
isti problemati inserviunt.[27]

  Potest etiam res alio modo ostendi. In plena æquatione defin[i]ente curvas
ordinis cubo cubici sunt 28 termini quorum 27 continent quantitates[28] datas
quæ omnimodo variari possunt. At in Problemate Pappi ad duodecim lineas
viginti duæ ejusmodi quantitates determinant positiones undecim linearum ad
duodecim[am] et alia ad determinandam rationem facti sex linearum ad factū
aliarum sex [requiritur].[29] Latior est itaꝗ natura curvarū hujus ordinis quàm

---

  (24) Newton first wrote 'et per quatuor horum punctorum agantur aliæ duæ lineæ secantes
eam in alijs sex punctis' (and through four of these points there be drawn two further lines
intersecting it in another six points).

  (25) A cancelled sequel reads 'quæ quidem nulla lege impleri possunt nisi curva aptetur ita
ut transeat per omnes' (which indeed cannot be filled unless the curve be adapted so as to
pass through all). There is a crucial *non-sequitur* in Newton's deduction here, for by 'Cramer's
paradox', if three of the four 'conditions' implicit in the free choice of the two original straight
lines are used to make the quintic pass through a total of 19 intersections of the two sets of line
5-ples, then it must pass in all cases through the remaining 6 intersections. See note (27) below.

  (26) Newton first wrote equivalently '…non videt perpaucas esse curvas…utiles' (…does
not see that there are very few curves…useful).

  (27) To paraphrase Newton's argument in the general
case where a curve of n-th degree is propounded, since $2n$
conditions fix the meets of the curve with two given straight
lines and a further $2(n-2)$ conditions its intersections
with two straight lines drawn through four of those meets,
to determine that the remaining $(n-2)^2$ meets of the two
sets of line n-ples shall lie on the curve there are apparently
only the 4 conditions afforded by the free choice of the two
original straight lines: from which he concludes that for
$n > 4$ the n-th degree curve is not in general representable
as a $2n$-line locus. However, by 'Cramer's paradox', if the
curve passes through $\frac{1}{2}n(n+3)-1$ meets of the line n-ples
it will pass through all the rest, so that this test shows that
the representation is impossible only when

$$\tfrac{1}{2}n(n+3)-1 > 2n+2(n-2)+4,$$

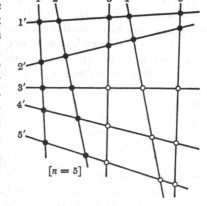

[n = 5]

further fifteen points,[24] and then through six of these points be drawn a further three points, these ought to pass through the remaining nine points provided that, as I suppose, the product of five of the lines which are let fall onto these last three and the two first drawn has a given ratio to the product of those let fall onto the remaining five. Nine conditions, therefore, remain for determining in position ten of the straight lines. But the positions of all will be fully fixed from the position of the two first drawn, while to determine the latter's position four conditions suffice: for two conditions determine the position of a single line. Therefore five conditions remain to be filled after the determination of the straight lines,[25] in other words there remain five intersections, through which the curve will not pass unless by chance it is suited to do so. Who, therefore, does not see that by far the greatest portion of curves of the quintic order are useless[26] in Pappus' problem? In more complex classes, indeed, fewer still will be of service to that problem.[27]

The point can also be shown another way. In the complete equation defining curves of sextic order there are 28 terms, 27 of which contain given quantities[28] capable of being varied arbitrarily. But in Pappus' twelve-line problem twenty-two quantities of this sort determine the positions of eleven lines with regard to the twelfth and a further one [is needed] to determine the ratio of the product of six of the lines to the product of the other six.[29] The character of curves of this order is consequently broader than those which may wholly be

---

that is, for (integral) $n > 5$. We may be a little disappointed that Newton failed to appreciate this subtlety, but his instinct that the general representation of a curve of n-th degree as a $2n$-line locus is impossible held true. (The paradox was not, in fact, noticed till forty years afterwards when Colin Maclaurin observed in his *Geometria Organica: sive Descriptio Linearum Curvarum Universalis* (London, 1720: Pars Secunda, Sectio v. 'De Descriptione Linearum Geometricarum per data Puncta', Lemma III, Corollary II: 137) that 'Linea Ordinis (n) occurrere potest aliæ ejusdem Ordinis in punctis n^2. Proinde duæ Lineæ Ordinis (n) per eadem puncta n^2 transire nonnunquam possunt; adeoque puncta data quorum numerus est $\frac{1}{2}n^2 + 3n$ non sufficiunt ad Lineam Ordinis (n) ita determinandum ut unica sit curva quæ per ea data puncta duci possit: Cum vero coefficientes in æquatione generali ad Lineam Ordinis (n) sint $\frac{1}{2}n^2 + 3n$, patet si plura dentur puncta, Lineam Ordinis (n) per ea forsan duci non posse & Problema reddi posse impossibile'. See also C. B. Boyer, *History of Analytic Geometry* (New York, 1956): 196, 246.)

(28) 'literas' (algebraic constants) is cancelled.

(29) Newton forgets that two further conditions are needed to fix the first line in position. However, his argument is unaffected: since the general $2n$-line locus,

$$\prod_{1 \leqslant i \leqslant n} P_i = \lambda \times \prod_{n+1 \leqslant i \leqslant 2n} P_i \quad \text{where} \quad P_i \equiv x + a_i y + b_i,$$

requires $4n + 1$ conditions to determine the constants λ and $a_i, b_i, i = 1, 2, \ldots, 2n$, while the general curve of n-th degree needs $\frac{1}{2}n(n+3)$ conditions for its determination, the latter is not generally representable as one of the former when $\frac{1}{2}n(n+3) > 4n + 1$, that is, for $n > 5$. Compare note (27).

quæ per Problema Pappi omnimodo designentur. Nam anguli in quibus rectæ se multiplicantes demittuntur ad alias positione datas rectas nihil mutant præter proportionem factorum et proinde pro alijs conditionibus non sunt habendi.

Præterea quod asserat[30] primam et simplicissim. post Con. sect. esse Parabolam illam per quam construxit cubo cubica problemata, non assentior ei. Nam eodem ratiocinio Hyperbola et Parabola essent simpliciores quam circulus siquidem locus puncti est ad istas figuras ubi duæ sint parallelæ rectæ et tertia ijs ⊥^{lis}.[31] Imò simplicior est casus Cartesiano, quando tres ∥ normaliter secantur a quarta et parallelip.[32] sub 3 demissis æquale parallelip.[32] sub 4ta demissa et recta data.[33] Et adhuc simplicior ubi 2 ∥ secantur a 3ª normaliter et parallelip. sub tribus demissis æquatur cubo dato.[34]

---

(30) *Geometrie* (note (2) above): 335–6: 'si la question des anciens est proposée en cinq lignes,...dont il y en ait quatre qui soient paralleles, & que la cinquiesme les couppe a angles droits, & mesme que toutes les lignes tirées du point cherché les rencontrent aussy a angles droits, & enfin que le parallelepipede composé de trois des lignes ainsi tirées sur trois de celles qui sont paralleles, soit esgal au parallelepipede composé des deux lignes tirées l'vne sur la quatriesme de celles qui sont paralleles & l'autre sur celle qui les couppe a angles droits, & d'vne troisiesme ligne donnée. ce qui est ce semble le plus simple cas qu'on puisse imaginer aprés le precedent [le problesme en 3 & 4 lignes]; le point cherché sera en la ligne courbe, qui est descrite par le mouuement d'vne parabole en la façon cy-dessus [*Geometrie*: 322; compare II: 363, note (125) above] expliquée.' (An equivalent Latin version of the passage will be found on *Geometria*, ₂1659: 35.) Earlier Descartes had spoken of this cubic curve—of equation $(x-a)(x^2-b^2) = cxy$ when the four parallels are taken to be $x = 0$, $x = a$, $x = \pm b$ and the fifth line perpendicular to these is $y = 0$—as 'la premiere, & la plus simple de toutes aprés les sections coniques' since 'on peut [la] descrire par l'intersection d'vne Parabole, & d'vne ligne droite' (*Geometrie*: 309 [= *Geometria*: 12]). On Descartes' use of his 'trident' to construct the real roots of the general sextic consult *Geometrie*: 402–12 [= *Geometria*: 97–105]; see also I: 495. The descriptive name is Newton's (II: 70).

(31) Newton's meaning should be clear from the accompanying diagram, in which the two verticals are $x = 0$, $x = a$ and the horizontal is $y = 0$. The defining conditions

$$CP \times CR = k.CQ \quad \text{and} \quad CQ \times CR = k.CP$$

determine the locus (C) to be the hyperbolas $x(y+k) = ak$ and $(x-a)(y+k) = -ak$ respectively, while the condition $CP \times CQ = k.CR$ yields the parabola

$$(x - \tfrac{1}{2}a)^2 = -k(y - \tfrac{1}{4}a^2/k).$$

(32) Read 'parallelipipedon' and 'parallelipipedo' respectively.

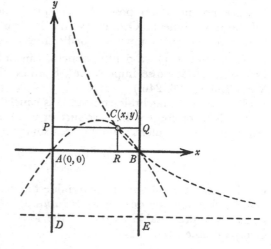

defined by means of Pappus' problem. For the angles at which the straight lines multiplying each other are let fall onto the other straight lines given in position alter nothing except the ratio of the products and accordingly must not be considered as further conditions.

His further assertion[30] that first and simplest after the conic is the parabolic curve by which he constructed sextic problems, I cannot admit. For by the same reasoning the hyperbola and parabola would be simpler than the circle inasmuch as the locus of the point is one of those figures where two of the straight lines are to be parallel and the third perpendicular to them.[31] It is, to be sure, a simpler case than Descartes' when three parallels are cut at right angles by a fourth and the cubic product of three of the lines let fall is equal to that of the fourth and a given line.[33] Still simpler is the case where two parallels are cut at right angles by a third and the product of the three lines let fall equals a given cube.[34]

---

(33) When the three parallels are taken to be $x = 0$, $x = a$, $x = b$ and the base is $y = 0$, the defining condition $CP \times CQ \times CR = k.CS$ determines the locus (C) to be the general Wallis cubic of equation $x(x-a)(x-b) = ky$. Compare II: 449, note (20).

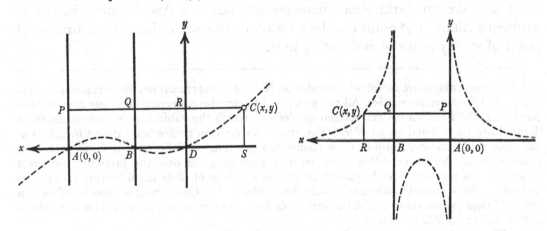

(34) When the parallels are $x = 0$, $x = a$ and the base line is $y = 0$, the defining condition $CP \times CQ \times CR = k^3$ determines the locus (C) to be the cubic $x(x-a)y = k^3$, a hyperbolic hyperbola or (as Newton will name it in §3) hyperbolism of a hyperbola.

§2. OBSERVATIONS ON ASYMPTOTES AND DIAMETERS IN THE GENERAL CUBIC.[1]

From originals in private possession and in the University Library, Cambridge

[1][2]

1 To find y^e Asymptote, 2 the Diameter.

1. To find ye plaga or position of an Asymptote or infinite tangent by a Cubic Æquation &c wch is easily done wthout Algebra by supposing all finite distances to be a point, all infinite ones to be finite.[3]

2. To find ye intersection of the Asymptotes wth any given line, by supposing ye motion of ye intersection of ye tangent to be 0.[4]

3. To find ye intersections of ye Curve & Asymptotes[,] wch is a simple problem. Note that if there be 3 intersections they all ly in a straight line.

4. If there be no more points of the Curve given by the nature of the Problem, To find any other point[5] whether in one of ye other Asymptotes if there be others, or in a line drawn through ye former point parallel to one of ye other Asymptotes, or if there be no Asymptotes in a line passing through ye former point parallel to ye diameter of ye Parabola.[6] All this is a simple Probleme but if the line have no Parabolical diameter nor but one Assymptote, then is ye Probleme plane & ye point is to be sought in a line wch either passes through ye point of ye Asymptot or is draw[7] ∥ to it.

---

(1) These remarks are manifestly founded on Newton's first researches into the general cubic (see II, **1**, 1/2) a decade earlier. All the analytical theory there developed is here assumed: in particular, the nine 'cases' and sixteen species into which the cubic is now understood to be divided are those listed on II: 86/88. The reason for Newton's refreshed interest in cubics at this time is not known. From its close connection (see note (2) following) with his preceding exploration of the Greek 4-line locus we may perhaps guess that the present manuscript represents his renewed zeal in classifying curves in terms of their infinite branches; for, as he had previously asserted when explaining his method for distinguishing the species of a given conic, 'Plagæ punctorum infinitè distantium facilius inveniri solent, et ex istis species optimè definietur' (page 312 above).

(2) This section is entered on three sides of a folded quarto sheet (in private possession), below the preliminary calculations reproduced in appendix and preceding calculations (given in edited form in §3, Appendixes 4/6) for determining the case of a given cubic from its unreduced Cartesian equation. On its fourth side occur the 'De Solutione planorum Problematum per regulam trium' (§4 below) together with drafts of Corollaries 1–3 of Proposition 4 (see 2, §2: note (18)) and a first version of Proposition 7 (reproduced in 2, §2, Appendix 3) of Newton's 'Solutio Problematis Veterum de Loco solido'.

(3) In other words, by considering only the infinite branches of the cubic.

(4) For 'tangent' read 'subtangent'. The remark is applicable to any smoothly continuous curve which does not have an inflexion point at its meet with the asymptote; compare III: 126, note (200).

(5) Newton first wrote 'To find the intersection or other points'.

5. to find 3 other points. If yᵉ two first points lye in two Asymtotes draw a line through them & that shall cut yᵉ 3ᵈ in a 3ᵈ point.[8] If there be 3 Asymptotes, draw a line through yᵉ two points & a 3ᵈ point shall be in it as far from yᵉ 3ᵈ Asymptote on one side as yᵉ 2ᵈ is from its Asymptote on yᵉ other.[9]

[10]Now if there be 3 Asymptotes, no two parallel, all intersected, the Curve is of the first Case yᵉ first species[,] if only two be intersected it is of yᵉ second, if none of the third. In yᵉ 1ˢᵗ & third case any assymptote may be chosen for yᵉ ordinate, in yᵉ second that wᶜʰ is not intersected. In all thre[11] cases the line wᶜʰ bisects all yᵉ lines drawn from one inordinate[12] Asymptote to another parallel to yᵉ ordinate is yᵉ diameter or base & also yᵉ bisecant diameter in yᵉ 2ᵈ & 3ᵈ case but in yᵉ first the bisecant is a conic Hyperbola thus found. Draw through yᵉ intersection of either inordinate Assymp. a line ∥ to yᵉ ordin. Asymp. [&] find its other intersection point. Bisect it & through yᵉ middle point wᵗʰ yᵉ base & ordinate Asymptote for Asymptotes de[s]cribe a conic Hyperbola.

If 2 Asymptotes be ∥ the 3ᵈ is yᵉ ordinate Asymptote & the base is æquidistant between yᵉ other two. And yᵉ figure is of yᵉ 8ᵗʰ species if yᵉ ordin. Asympt. be inters[ected], otherwise of the 9ᵗʰ. In yᵉ former case draw any line ∥[13] yᵉ base. Find its intersection. Thence draw a line ∥ to yᵉ Ordinate, bisect it, And yᵉ Conic Hyperb wᶜʰ passes through yᵉ middl point as in yᵉ 1ˢᵗ species is yᵉ bisecant but in yᵉ second case yᵉ base is bisecant.

If there be two Asymptotes, the double or intersected one is base[,] yᵉ other ordinate & yᵉ figure is of yᵉ 12ᵗ species if yᵉ Ordinate be intersected[,] otherwise of yᵉ 13ᵗʰ. In yᵉ first case yᵉ bisecant is a conic Hyperbola found as in yᵉ 8ᵗʰ species, in yᵉ latter the Base is bisecant.

If there be but one Asymptote, examin if it have a Parabolic diameter. If it have & yᵉ applicate As:[14] be intersected,[15] it is of yᵉ sixt species, if not intersected tis of the 7ᵗʰ. in yᵉ latter case yᵉ diameter is bisecant, in yᵉ former draw a

---

(6) A cancelled first version reads '. . . one point in a line passing through yᵉ former point parallel to yᵉ diameter of yᵉ Parabola, & another point in a line parallel to yᵗ'.

(7) Read 'drawn'.

(8) On the cubic, that is. In proof see II: 44, note (22).

(9) This is Theorem 2 on II: 102.

(10) In the sequel Newton understands (as on II: 86/88) that the general cubic is divided into nine Cases, comprising Species 1–3, 4/5, 6/7, 8/9, 10/11, 12/13, 14, 15 and 16 respectively. The latter subdivision is, of course, made according to whether the cubic has 0, 1 or 3 diameters.

(11) 'three'.

(12) That is, 'non-ordinate'. In the figure on II: 48 (which illustrates Case 1, Species 1), for example, the segment *Lλ* is 'ordinate' to the asymptote *Dδ* to which it is parallel but joins points in the 'inordinate' asymptotes *δd*, *Dd*. In this instance the 'bisecant' locus (*x*) is the Apollonian hyperbola *Φfφ*.

(13) 'to' is cancelled, somewhat unnecessarily.

(14) 'Asymptote.'

(15) Newton first wrote 'cut'. He has also cancelled a following adverb 'then'.

tangent to ye intersection point or[16] a line from it ‖ to the base & from its intersection another line ‖ to ye ordinate, bisect this & through its middl an Ellipsis described as in ye 1st species[17] shall be ye bisecant.

If ye Asymptote & ordinate diameter be coincident it is of ye 15th species[18] & has no bisecant.

If the figure have an Asymptote & no parabolic diam. it is of y^e 4th or 10 species if y^e Asymptote be intersected, otherwise of y^e fift or 11th. In y^e former case draw a tangent to y^e intersection point & from its intersection another line ‖ to y^e ordinate &c.[19] Or in both find y^e locus of $\frac{1}{2}$ y^e summ of y^e two applicates to any given line, & that shall be y^e bisecant in both cases, w^{ch} [in] y^e first case will be an Hyperbola whose Asymptote is y^e base, in y^e 2d it will be y^e base it self. If the intersection of a line drawn ‖ to y^e base be determinable by a simple[20] æquation it is of 10th or 11 species, if by a cubic one, of ye 4th or 5th: whence in ye latter case is opened a way of finding ye position of y^e base.[21]

If it have no Asymptote it is of y^e 14th or 16 species, of y^e 14th if y^e Parabolic[22] leggs ly on one side of y^e diameter, otherwise of y^e 16th. in y^e first case y^e lines ‖ to y^e Parabolic diameter are ordinates bisected by the base, so that the base is y^e Locus of y^e $\frac{1}{2}$ sum of y^e applicates to any given line, or may be determined by y^e ordinates of any one applicate wth y^e tangents to its intersection points but in y^e second case there is no bisecant.

---

(16) The preceding phrase 'a tangent to y^e intersection point or' needs to be deleted. The tangent $(bl^2/h^2 - k)x = hy + l$ to the parabolic cubic $bxy^2 = gx^2 + hy + kx + l$ at the point $(0, -l/h)$ in which it intersects its asymptote $x = 0$ meets it again in the point

$$(2bh^2l\lambda/(b\lambda^2 - gh^6),\ ^r(b\lambda^2 + gh^6)\,l/(b\lambda^2 - gh^6)\,h),$$

where $\lambda = bl^2 - h^2k$; whence it follows that the mid-point

$$(bh^2l\lambda/(b\lambda^2 - gh^6),\ bl\lambda^2/(b\lambda^2 - gh^6)\,h)$$

does not in general lie on the bisecant hyperbola $2bxy = h$.

(17) That is, through the mid-points of all chords parallel to the parabolic diameter; compare Figures 1 and 2 on II: 80.

(18) 'of wch hereafter' is cancelled. Species 15 (the Cartesian trident $dx^3 + fxy + kx + l = 0$) manifestly has the vertical $x = 0$ both for its asymptote and as (Newtonian) diameter ordinate to parallels $y = p$ to the base, but this also holds true for the general cubic of second 'grade' (see II: 47, note (26) and the figure on II: 49) having for Cartesian equation

$$bxy^2 = dx^3 + hy + kx + l.$$

(19) Much as above (see note (16)) it seems enough here to have written 'In y^e former case from y^e intersection point draw a line ‖ to y^e ordinate &c'. Compare the figures on II: 78 and 82.

(20) That is, linear.

How the Base is determined in these species.

In y^e 1st, 2^d, 3^d, 8th, 9th the base lies in ye midd way between ye inordinate Asymptotes bisecting ye ordinates to them. In ye 12th & 13th species the double Asymptote is base, In ye sixt & 7th ye Parabolic diameter is base. In ye 5^t, 11 & 14 the base bisects ye ordinates, in ye 4th & 10th it is ye Asymptote of a conic Hyperbola wch bisects ye ordinates. In ye 16 it is a tangent to ye vertex[23] of ye figure. In ye 15th ye base is tangent to ye vertex of a conic Parabola wch ad infinitum falls in wth its thighs.[24] Neglect ye terms y^t never become infinite[25] & you have ye Parab.

How ye bisecant is determined.

In ye 1st through the intersection of either inordinate Asymptote draw a line ‖ to the ordinate Asymptote, bisect it, through ye bisection point de[s]cribe a Conic Hyperbola wch shall be bisecant. In y^e 2^d, 3, 9th & 13 y^e base is bisecant. in y^e 8th, 10, 11, 12 draw any line parallel to y^e base, & from its intersection another parallel to y^e ordinate, bisect it & the bisecant (wch in y^e 10th is Base) shall pass through y^e p$^{t[26]}$ of bisection. In y^e 10th & 11 you must find y^e tangent to y^e intersection points & thence determin y^e base & bisecant. In the 4th, 5, 10, 11 & 14 the mean inclination of y^e two tangents is y^e tangent to y^e bisecant.[27] & this & y^e half sum of y^e ordinates (or wch is all one the locus of the half sum of y^e ordinates[)] gives y^e bisecant.

---

(21) In Species 4/5 ($bxy^2 = -dx^3 + gx^2 + hy + kx + l$) the meets of the general parallel to the base, say $y = p$, with the cubic are given by the equation $dx^3 - gx^2 + (bp^2 - k)x - (hp + l) = 0$. In the particular instance of Species 10/11 $d = g = 0$ and so at once $x = (hp + l)/(bp^2 - k)$. In the former case it is not obvious—except in the trivial sense that $p = 0$ yields a simpler equation—how the position of the base is to be found in terms of the roots of the cubic.

(22) Newton first continued with the words 'Diameter go between its crura', a phrase evidently cancelled because Species 16 (the Wallis cubic) has no diameter of any kind, being symmetrical round its sole finite point of inflexion. But the following revised phrase is equally nonsensical and we should probably read '. . . ly[e] on one side of y^e tangent to the vertex'.

(23) Namely, the point A in the figure illustrating Species 16 on II: 84. Compare note (22) preceding.

(24) That is, 'crura' (infinite branches) or 'leggs' as he has previously called them.

(25) Namely, all but fxy and dx^3 in the equation ($fxy = dx^3 + l$) for the trident on II: 70. These yield, the pair of the asymptote $x = 0$ and the parabola $fy = dx^2$ as approximations to the infinite branches.

(26) 'point'.

(27) The slope of the cubic $bxy^2 = -dx^3 + gx^2 + hy + kx + l$ at the general point (x, y) is

$$-\frac{3dx^2 - 2gx - k + by^2}{2bxy - h} = -\frac{h}{2bx^2} \pm \frac{4bdx^4 - 2bgx^3 + 2blx + h^2}{2bx^2\sqrt{[4bx(-dx^3 + gx^2 + kx + l) + h^2]}},$$

the mean of which, $-h/2bx^2$, is the slope of the diametral hyperbola $2bxy = h$ at the point having the same abscissa.

How yᵉ line is defined.

If there be 3 Asymptotes[28] & yᵉ curve cut yᵉ ordinate Asymptote [in *S*], draw a right line *ST* tangent to yᵉ point & also a Parabolic one whose diameter is ∥ to yᵉ [base *AT*] & latus rectum *L*.[29] Put $b = 1$. [Erit] $-\dfrac{2Ad^q}{AD} = g$.

$-\dfrac{g}{2AD} = d$. $AS = -\dfrac{l}{h}$. $AS \cdot AT :: AS^q - k \cdot h$. (Si $AS = 0$ erit $l = 0$, & $AS \cdot AT :: -k \cdot h$. Sin $AS = \dfrac{1}{0}$ erit $h = 0$ et ultimum $AS^q \times AT = -l$.) rectangulum xy (ubi $x = 0$, $y = \text{infin}$) $= h$ in primis 6 casibus[30] & $AS \times h = -l$. Sed si $h = 0$, erit ultimum rectangulum $xyy = l$. Item semper est $k = \dfrac{xyy - hy - l}{x}$ ubi y infin. et $x = 0$.[31]

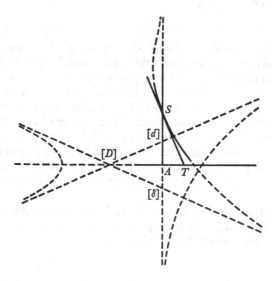

---

(28) This opening phrase is a late insertion which is inconsistent with Newton's following demand that the cubic should have a parabolic asymptotal curve.

(29) This final phrase should probably be deleted, for Newton's following argument requires that the cubic be taken in the more general form $bxy^2 = dx^3 + gx^2 + hy + kx + l$. Only when $d = 0$ will the infinite branches be given by (the pair of $x = 0$ and) the parabola $y^2 = Lx$,

$L = g/b$. Newton's rough sketch lacks the asymptote *Dd* referred to in the sequel, but the accompanying figure—one of several amplifications possible—, in which (compare Figure 18 on II: 74) the full cubic and its three asymptotes have been entered in broken line, should serve to remedy the deficiency.

(30) That is, Species 1–13.

Or better, at once find y^e æquation $bxyy = dx^3 + gxx + hy + kx + l$ by y^e nature of y^e Problem.

---

[2]$^{(32)}$ [$xyy = dx^3 + gxx + hy + kx + l$. Capiendo fluxiones]

$$myy + 2nxy = 3dmxx + 2gmx + hn + km.$$

[Sed] $n \cdot m :: y \cdot AT$. [adeoq;] $AT = \dfrac{hy - 2xyy}{yy - 3dxx - 2gx - k}$. $^{(33)}$ [hoc est]

$$AT = -2x[+]\frac{h}{y} + \frac{2kx}{yy} - \frac{hk}{y^3} \ [\&c].^{(34)}$$

[Pone] $x = 0$. [fit] $myy = km + hn$. [adeoq;] $yy - k \cdot h :: [n \cdot m :: y \cdot AT]$. maximum rectangulum$^{(35)}$ $xy = h$.

---

(31) On this last stretch Newton breaks unconsciously into Latin. In English translation: 'Then $-2Ad^2/AD = g$, $-g/2AD = d$, $AS = -l/h$ and $AS : AT = (AS^2 - k) : h$. (If $AS = 0$, then $l = 0$ and $AS : AT = -k : h$. But if $AS = 1/0$, then $h = 0$ and ultimately $AS^2 \times AT = -l$.) The rectangle xy (in which $x = 0$, $y = \infty$) $= h$ in the first six cases [13 species] and $AS \times h = -l$. But if $h = 0$, then the ultimate rectangle is $xy^2 = l$. Likewise it is always $k = (xy^2 - hy - l)/x$ when $y = \infty$ and $x = 0$'. In these computations Newton, taking the equation of the cubic in its simplest form $xy^2 = dx^3 + gx^2 + hy + kx + l$, assumes the results established on II: 18 ff. and in [2] below. The asymptote Dd, of equation $y = \sqrt{(d/b)} \, (x + g/2d)$ (see II: 20, note (9)), determines $AD = -g/2d$, $Ad = g/2\sqrt{(bd)}$, while the cubic meets the asymptote $Ad(x = 0)$ in the point $S(0, -l/h)$, whence $AS = -l/h$ is zero when $l = 0$ $(h \neq 0)$ and infinite when $h = 0$ $(l \neq 0)$. (In the last case S will be an inflexion point at infinity on the cubic.) The ratio $AS/AT = (y^2 - k)/h$ (see note (33) in the case where $x = 0$) determines the slope of the cubic at A; in the limit, however, when S (and so T) coincides with A, this ratio is $-k/h$, while if AS becomes infinite (and so $h = 0$) in the limit

$$AS^2 \times AT = [AS^3 \times h/(AS^2 - k) \ \text{or}] - AS^2 \times l/(AS^2 - k) = -l.$$

Infinite branches in the direction of the x-axis (for infinite y) are, in Species 1–13, determined by $(h \neq 0) \, y(xy - h) = 0$ or $(h = 0) \, xy^2 - l = 0$; more accurately, by $xy^2 = hy + kx + l$.

(32) These calculations (ULC. Add. 3970.3: 570^v), evidently related to the last paragraphs of [1] preceding, are entered on the back of a draft of a letter from Newton, possibly to Aubrey, which mentions 'Mr Oldenburgs death' (in early September 1677) and is reproduced, under the date of '?June 1678', in *The Correspondence of Isaac Newton*, **2** (Cambridge, 1960): 266–7.

(33) Whence the slope of the cubic at its general point (x, y) is

$$-(y^2 - 3dx^2 - 2gx - k)/(2xy - h).$$

Compare note (27) above. The letters m and n denote, of course, the fluxions ($\dot{x}$ and $\dot{y}$) of x and y.

(34) On expanding $AT = -(2xy - h)y^{-2}(1 - y^{-2}(k + \ldots))^{-1}$ as an infinite series in ascending powers of $1/y$.

(35) This 'greatest rectangle' is the 'ultimum rectangulum' (ultimate rectangle) of [1] above.

$$k = yy - dxx - gx \frac{-hy - l}{x}. \text{ [hoc est ubi } y \text{ infin. et } x = 0] \; k = \frac{xyy - hy - l}{x}. \text{ [vel si}$$

$$AS = 0] \; \frac{xy^2 - hy}{x}. \; ^{(36)}$$

APPENDIX. MISCELLANEOUS COMPUTATIONS WITH REGARD TO THE GENERAL CUBIC.[1]

From the original in private possession

$$[\text{Sit}] \begin{array}{l} ax^3 + bx^2 + cx + d \\ + exxy + fxy + gy \\ + hxyy + kyy \\ + ly^3 \end{array} = 0. \quad 2ex + f + 2hy = 0.^{(2)} \quad \begin{array}{l} 6axx + 4bx + 2c \\ + 2exy + fy \end{array} = 0.^{(3)}$$

$$[\text{adeoq}]^{(4)} \begin{array}{l} 6axx + 4bx + 2c \\ -4\dfrac{ee}{2h}xx - 4\dfrac{ef}{2h}x - \dfrac{ff}{2h} \end{array} = 0. \quad \begin{array}{l} 6lyy + 4ky + 2g \\ + 2hxy + fx \end{array} [=0.]^{(3)}$$

$$[\text{Si}] \quad e = 0. \quad \begin{array}{l} ax^3 + bxx + cx + d \\ + fxy + gy \\ + hxyy + kyy \\ + ly^3 \end{array} [=0.] \quad f + 2hy = 0 \quad [\text{seu}] \quad -\frac{f}{2h} = y. \quad [\text{At}]$$

$$3lyy + 4ky + 2g = 0.^{(5)} \quad [\text{adeoq}] \quad 3lff + 8kfh + 8ghh = 0. \quad [\text{vel}] \quad -\frac{3lff}{8hh} - \frac{kf}{h} = g. \quad [\text{ut et}]$$

$$12ahxx + 8bhx \begin{array}{l} +4ch \\ -ff \end{array} = 0. \quad [\text{seu}] \quad x = \frac{b}{3a} \pm \sqrt{\frac{bb}{9aa} + \frac{ff}{12ah} - \frac{c}{3a}}.$$

(36) Since $AS = -l/h$ and so when AS is zero $(h \neq 0)$ $l = 0$.

(1) These calculations are found at the head of the first page of the manuscript sheet on which Newton entered his preceding observations on cubics (see §2: note (2)). Their exact purpose is not explained and, while Newton was always prepared to give his equations a geometrical interpretation, he may have here been unable to do so. On denoting the general cubic $ax^3 + ex^2y + hxy^2 + ly^3 + bx^2 + fxy + ky^2 + cx + gy + d = 0$ by ϕ, the condition

$$\phi_{xx} = \phi_{xy} = \phi_{yy} = 0$$

determines the point (x, y) to be the centre of the cubic, while the further stipulation that $\phi_x = \phi_y = 0$ determines it to be also a triple point (so that the cubic degenerates into a line-triple) but Newton's restriction $\phi_x = \phi_y = \phi_{xy} = 0$ by itself seems to have no geometrical meaning (though it would have been obvious to him that $\phi_x = \phi_y = 0$ determines (x, y) to be a double point on the cubic).

(2) Where the preceding cubic equation is denoted by $\phi = 0$, this is the derivative condition that $\phi_{xy} = 0$.

(3) The conditions $2\phi_x - y\phi_{xy} = 0$ and $2\phi_y - x\phi_{xy} = 0$ respectively. In conjunction with the preceding equation (see note (2)) these determine $\phi_x = \phi_y = \phi_{xy} = 0$.

Si $l = 0$ erit $2ex + f + 2hy = 0$ et $ky + 2g - 2exx = 0$ [adeoɋ

$$4ehxx + 2ekx \begin{matrix} +fk \\ -4gh \end{matrix} = 0.]^{(6)}$$

(4) On substituting $-(2ex+f)/2h$ for y in the preceding equation, whence the following condition is equivalent to setting $\phi_x = \phi_{xy} = 0$.

(5) That is, $\phi_y - \frac{1}{2}x\phi_{xy} = 0$.

(6) The text breaks off at this point.

§3. REVISED ANALYTICAL INVESTIGATION OF THE GENERAL CUBIC CURVE.[1]

From original drafts in the University Library, Cambridge[2]

Conicarum sectionum communes proprietates magna ex parte conveniunt curvis magis compositis.[3] Eæ sunt hæ.[4]

Si ductis duabus rectis parallelis terminatis ad Conicam sectionem[5] recta eas bisecans bisecat alias omnes parallelas terminatas ad curvam, ea proinde dicitur *diameter* et rectæ bisectæ *ordinatim applicatæ* ad diametrum, et concursus omnium diametrorum *centrum* figuræ, et diameter illa *axis*, cui ordinatim applicatæ insistunt ad angulos rectos. Eodem modo in curvis trium dimensionum, si rectæ duæ quævis parallelæ ducantur occurrentes curvæ in tribus punctis: recta quæ ita secat has parallelas ut summa duarum partium ex una plaga ad curvam terminatarum æquetur parti tertiæ ex altera plaga ad curvam terminatæ, eodem modo secabit omnes alias his parallelas & curvæ in tribus punctis occurrentes[,] hoc est ita ut summa partium ex uno latere æquetur parti ex altero latere. Has itaqɜ partes quæ hinc inde æquantur *ordinatim applicatas*, et rectam secantem cui ordinatim applicantur *diametrum* nominare licebit. Et diameter rectangula si modo aliqua est etiam *axis* dici potest. Et si forte omnes diametri coeunt ad idem punctum, istud erit *centrum*; et *vertex* erit ubi diameter secat curvā. Cæterum distinctionis gratia Diametros cui duæ rectæ æquales hinc inde ordinatim applicantur, dicemus *Diametros bisectionis*[6] vel *diametros simplices*, et eas cui tres ordinatim applicantur *diametros trisectionis*[6] et sic [in] infinitum. Et eodem modo centra distingui possunt. Nempe[7] *centrū bisectionis*

---

(1) This revised paper on the general properties of cubics and their classification into 'genera' on the basis of their Cartesian defining equation (and a sketch of their further subdivision into species) is evidently based on his similar researches of a decade earlier (II, **1**, 1, §§1–3/2, §2). What, if anything, stimulated Newton to return to the topic is not known and we know too little of his life at this period (*c.* 1678–9) to be firm in conjecture. Quickly passed over by Horsley during his hurried survey of the Newton papers at Hurstbourne in October 1777 (see §1: note (1) above), the manuscript lapsed into oblivion for another century till its importance was at length recognized by Rouse Ball, who in an excellent article 'On Newton's Classification of Cubic Curves' (*Proceedings of the London Mathematical Society*, **22** (1891): 104–43, especially 109–12/120–2/125–8) gave a detailed account of the piece, along with extensive excerpts from it. All subsequent accounts (for example, the German summary of Rouse Ball's paper in *Bibliotheca Mathematica*, ₂5 (1891): 35–40 and C. B. Boyer's *History of Analytical Geometry* (New York, 1956): 138–40) restrict themselves to his findings.

(2) Add. 3961.4: 19ʳ–21bisᵛ/3961.1: 18ʳ–19ʳ. (The division in the manuscript already existed in 1727 when Pellet packaged the first portion as '12/K' except for the sheet now marked as f. 21 bis (originally f. 23), which he used to wrap up later drafts (Add. 3961.2) in the *Enumeratio* sequence as 'Math. MSS 40/K'.) The more interesting of Newton's preliminary drafts for the piece are reproduced in the Appendix, while other variants are noticed in following footnotes to the present text. Substantial extracts from the manuscript were first published

Translation

Properties held in common by conics obtain for the most part in more complex curves.[3] These are they.[4]

If, when two parallel straight lines terminated at a conic are drawn, the straight line bisecting them bisects all other parallels terminated at the curve, it is for that reason called a 'diameter' and the bisected straight lines 'ordinates' to the diameter, while the meet of all the diameters is the figure's 'centre' and that diameter is an 'axis' whose ordinates are set at right angles to it. In the same way in curves of third power if any two parallel straight lines be drawn, meeting the curve [each] in three points, the straight line which cuts these parallels such that the sum of its two parts terminated at the curve in one direction are equal to its third part terminated at the curve on the other will cut all other lines parallel to these, meeting the curve in three points, in the same manner—namely, such that the sum of the parts on one side are equal to the part on the other. It will accordingly be permissible to name these parts which are equal on either hand 'ordinates' and the intersecting straight line to which they are ordinate their 'diameter'. And a rectangular diameter, should there be one present, may also be called an 'axis', while if all diameters happen to meet in the same point, that will be the 'centre', and a 'vertex' will be where the diameter intersects the curve. In addition, for the sake of making a distinction, diameters to which two equal straight lines are ordinate on either side we shall call 'bisection diameters'[6] or 'simple diameters', those to which three are ordinate 'trisection diameters',[6] and so on indefinitely. And in the same way centres can be distinguished. Evidently [7] a 'bisection centre' will be the

by Rouse Ball in 1891 in his essay 'On Newton's Classification of Cubic Curves' (note (1) above): Appendices 1/2: 132–43. In 1695 Newton extensively revised the opening paragraphs in preparation to including them in introduction to his 'Enumeratio Curvarum tertij Ordinis' (Add. 3961.2: 38ʳ–50ᵛ, especially 38ʳ–39ᵛ), but his later hand is clearly discernible and we here give our edited version of the pristine text.

(3) In 1695 (see note (2)) Newton corrected this phrase to read '...mutatis mutandis ad Curvas magis compositas applicari possunt' (may be applied *mutatis mutandis* to more complex curves). The four following paragraphs were also numbered in sequence 1–4.

(4) Compare II: 90–6.

(5) In 1695 this became (compare note (3) above) '1. Si rectæ duæ parallelæ ad Conicam sectionem utrinqɜ terminatæ ducantur...' with corresponding slight adjustment in the sequel. These and other later changes essentially transform the text into the 1695 revise, which will be reproduced in the seventh volume, and are not systematically noticed here.

(6) Newton has cancelled '*Diametros ad duas ordinatas*' and '*diametros ad tres ordinatas*' (two/three-ordinate diameters) respectively.

(7) A first, cancelled sequel reads '*centrū bisecans* erit quod bisecat rectas omnes hinc inde ad curvam terminatas' (a 'bisecting centre' will be one which bisects all straight lines terminating at the curve on either side).

erit ubi duæ diametri bisectionis conveniunt, *centrum trisectionis* ubi duæ diametri trisectionis conveniunt.

Præterea hoc etiam observandum est,[8] quod sicut omnes Diametri Parabolæ conicæ sunt parallelæ sic fit in Parabolis secundi generis. Et sicut partes rectæ cujusvis inter Hyperbolam Conicam et Asymptotos ejus sunt hinc inde æquales, sic in Hyperbolis secundi generis ducta recta quavis secante curvam et tres ejus Asymptotos in tribus punctis,[9] summa duarum partium istius rectæ quæ a duobus quibusvis Asymptotis versus eandem plagam ad duo puncta curvæ extenduntur æquales erunt tertiæ parti quæ a tertia Asymptoto versus contrarium plagam ad tertium curvæ punctum extendetur.[10]

Adhæc sicut in Conicis Sectionibus Quadrata Ordinatim applicaratū, hoc est rectangulum earum quæ ad contrarias partes diametri ducuntur, sunt ad rectangula partium diametri ad vertices Ellipseos vel Hyperbolæ extensarum ut data quædam linea quæ dicitur *latus rectum* ad partem diametri quæ inter vertices jacet et dicitur *latus transversum*, sic in[11] figuris non Parabolicis secundi ordinis Parallelepipedon ex tribus ordinatim applicatis est ad parallelepipedum ex partibus diametri ad tres vertices figuræ extensis in ratione quadam data; in qua ratio[ne] si sumantur tres rectæ[12] ad tres partes diametri inter vertices figuræ sitas, singulæ ad singulas, tunc illæ tres rectæ dici possunt *latera recta* figuræ et partes diametri inter vertices *latera transversa*. Et sicut in Parabola Conica quæ non nisi unicam habet verticem ad eandem diametrum rectangulum ordinatarum æquatur rectangulo diametri ad verticem extensæ et rectæ cujusdam datæ quæ *latus rectum* dicitur, sic in Parabolis secundi ordinis,[13] Parallelipipedum ex ordinatis æquatur parallelipipido ex duabus partibus diametri ad vertices curvæ extensæ et recta quadam data, quæ proinde *latus rectum* dici potest modo Parabola secat diametrum in duobus punctis.[14] Vel si curva tangat diametrum[15] tunc in hoc et superiori theoremate, punctum contactus pro duobus habendum est[,] hoc est quadratum diametri ad punctum contactus terminatæ adhiberi debet.

---

(8) In 1695 Newton added a '†' at this point and a 'Quære' (Is it so?) in the margin opposite, then cancelled the whole sentence. He need not have worried, for the set of parallels $y = mx + z$ (z free) of constant slope m meet the divergent parabola $y^2 = ax^3 + bx^2 + cx + d$ in triples of points given by $ax^3 - (m^2 - b) x^2 + (c - 2mz) x - (z^2 - d) = 0$, whence the corresponding 'trisection' diameter is $x = \frac{1}{3}(m^2 - b)/a$, parallel to the ordinate direction. Newton may perhaps have been disappointed because the analogy is not exact, since the same parallels determine as the locus of mid-points of their intersections with the Apollonian parabola $y^2 = kx$ the 'bisection' diameter $y = \frac{1}{2}k/m$, parallel to the base.

(9) Newton first wrote in sequel 'partes istius rectæ inter duo diversa puncta curvæ et totidem puncta Asymptotorum, æquales erunt' (the parts of that line between two diverse points on the curve and an equal number of points on the asymptotes will be equal...).

(10) This is Theorem 2 on II: 102.

(11) 'Hyperbolis et' (hyperbolas and) is cancelled.

point where two bisection diameters meet, a 'trisection centre' that where two trisection diameters meet.

Furthermore, this also should be noticed:[8] that, just as all diameters of a conic parabola are parallel, so also in second-order parabolas; and, just as the parts of any straight line between the conic hyperbola and its asymptotes are equal on either hand, so in second order hyperbolas, when any straight line is drawn cutting the curve and its asymptotes in three points,[9] the sum of the two parts of that line which extend in the same direction from any two asymptote to two points of the curve will be equal to the third part which shall extend from the third asymptote in the opposite direction to a third point on the curve.[10]

In addition, just as in conics the squares of ordinates—that is, the product of those drawn on opposite sides of a diameter—are to the products of the portions of the diameter extending (in an ellipse or hyperbola) to the vertices as a certain given line, called the '*latus rectum*', to the portion of the diameter between the vertices, called the 'transverse diameter', so in[11] non-parabolic figures of second order the cubic product of the three ordinates is to that of the parts of the diameter extending to the three vertices of the figure in a certain given ratio; and if three straight lines[12] be taken in this ratio to the three portions of the diameter situated between the figure's vertices, one to each separately, the three former lines may be called '*latera recta*' of the figure and the portions of the diameter between the vertices 'transverse diameters'. Also, just as in the conic parabola (which has but a single vertex corresponding to the same diameter) the product of the ordinates is equal to that of the diameter extended to its vertex and a certain given straight line which is called its '*latus rectum*', so in parabolas of second order[13] the cubic product of the ordinates is equal to that of the two portions of the diameter extending to its vertices and a certain given straight line which may be called the '*latus rectum*' (providing the parabola intersects the diameter in two points).[14] Or should the curve touch the diameter,[15] then in this and the previous theorem the point of contact must be considered as double; that is, the square of the diameter terminated at the point of contact should be employed.

---

(12) Newton first wrote '*tres rectæ quædam datæ*' (any three given straight lines).

(13) This replaces '*tertij ordinis*' (of third order).

(14) Newton cancelled this unnecessary final phrase in 1695.

(15) The condition for $y = 0$ to be both a diameter of a cubic through $(0, 0)$ and also a tangent to it there is that the cubic's defining equation be of the form

$$y^3 = (ax^2 + bx + c)y + dx^2(x + e).$$

There is no strict parallel in the general conic except in the degenerate case of the line-pair.

Deniꝗ sicut in conicis sectionibus, ubi duæ parallelæ ad curvam utrinꝗ terminatæ secantur a duabus parallelis utrinꝗ ad curvam terminatis, prima a tertia et secunda a quarta, rectangulum partium primæ est ad rectangulū partium tertiæ ut rectangulum partium secundæ ad rectangulum partium quartæ;[16] sic ubi quatuor tales lineæ occurrunt curvæ secundi ordinis singulæ in tribus punctis, parallelipipidum partium[17] primæ lineæ erit ad parallelipipidum partium tertiæ ut parallelipipedum partium secundæ ad parallelipipedū partium quartæ.[18]

Hæ sunt insigniores proprietates Conicarum sectionum, et videtis quod curvis etiam superioris ordinis conveniunt.

[19]Cæterum cum hujusmodi proprietates eo præsertim spectent ut sciamus commode exprimere Curvas, eo ut de ijsdem commode ratiocinemur dicam breviter qua ratione id fiat in Curvis cubicis[20] æque ac in Conicis sectionibus.

Sit primo ABC con. sectio. Eam secent rectæ duæ parallelæ $AG\,BH$ in punctis A, G, B, H et tertia quævis recta non parallela AL in puncto L. Fiat $AG \cdot EH - BE :: AM \cdot EM$.[21] Et sit $d \cdot e :: AG \cdot AM$ ut et $b \cdot e :: EB \times EH \cdot EA \times EL$. Et ad AE erectâ quavis $TV \parallel BE$ et occurrente curvæ in V, dicatur $[TV] = y$ et erit

$$yy - \frac{d}{e}\,TM \times y - \frac{b}{e}\,TA \times TL = 0,$$

expressio conicæ sectionis.[22]

Eodem modo se res habet in curvis cubicis. Dentur enim positione tres rectæ parallelæ AG, BH, CI secantes curvam propositam singulæ in totidem datis punctis A, D, G; B, E, H; C, F, I. Et per extremum unius rectæ punctum A et alterius medium punctum E agatur quarta recta AE, eaꝗ producta secet

---

(16) Apollonius, *Conics*, III, 17/18. Compare 2, §1: note (12) and 2, §2: note (4).

(17) The equivalent phrase 'sub trib[us partibus]' is cancelled.

(18) The first appearance of this celebrated Newtonian theorem (which has an evident generalization in the case of higher curves). The proof is immediate, for the general ordinate through $(x, 0)$ to the cubic $y^3 - (ax + b)y^2 + (cx^2 + dx + e)y - (fx^3 + gx^2 + hx + k) = 0$ meets it in points (x, y_i), $i = 1, 2, 3$, such that $y_1 \times y_2 \times y_3 = fx^3 + gx^2 + hx + k$ while its intersections $(X_i, 0)$, $i = 1, 2, 3$, with the base $y = 0$ yield $fx^3 + gx^3 + hx + k = f(x - X_1)(x - X_2)(x - X_3)$, whence $y_1 \times y_2 \times y_3 / (x - X_1)(x - X_2)(x - X_3) = f$, constant.

(19) A cancelled first version of the following geometrical argument is reproduced in Appendix 1 below. A draft revise (on f. 22 bis[r]) is little different from the following final text (ff. 19[v]/20[r]).

(20) 'curvis secundi generis' (curves of second kind) was first written.

(21) This opening replaces 'Eam secent rectæ parallelæ AG, BH, CI in punctis A, G, B, H, C, I et quævis recta non parallela AL in puncto L. Fiat $EH - EB \cdot CK - KI :: EM \cdot KM$' (Let

Lastly, just as in conics when two parallels terminated at each end at the curve are intersected by two parallels terminated at each end at the curve, the first by the third and the second by the fourth, the product of the segments of the first is to that of the segments of the third as the product of the segments of the second to that of the segments of the fourth;[16] so when four such lines meet a curve of second order, each one in three points, the cubic product of the segments of the first line will be to that of the segments of the third as the cubic product of the segments of the second to that of the segments of the fourth.[18]

These are the more outstanding properties of conics and you see that they hold also for curves of higher order.

[19]However, since properties of this kind particularly regard our knowing how fittingly to represent curves in order to reason appropriately concerning them, I shall tell briefly the way that is to be done in cubic curves[20] no less than in conics.

First, let ABC be a conic. Let two parallel straight lines AG, BH intersect it in the points A, G, B, H and any third line AL, not parallel, in the point L. Make $AG:EH-BE=AM:EM$[21] and let $AG:AM=d:e$, also

$$EB\times EH:EA\times EL=b:e.$$

Then, having raised to AE any TV parallel to BE and meeting the curve in V, call $TV=y$ and $y^2-(d/e)\,TM\times y-(b/e)\,TA\times TL=0$ will be the conic's expression.[22]

The procedure is the same in cubic curves. For let there be given in position three parallel straight AG, BH, CI intersecting the proposed curve, each in as many given points A, D, G; B, E, H; C, F, I respectively. Then through the end-point A of one line and the middle point E of another draw a fourth straight line

---

the parallel lines AG, BH, CI intersect it in the points A, G, B, H, C, I and any other non-parallel straight line AL meet it in the point L. Make $EH-EB:CK-KI=EM:KM$). Newton has correspondingly deleted a parallel CKI between BEH and VT in his manuscript figure.

(22) On taking $AT=x$ to be the corresponding abscissa the Cartesian equation of the conic $V(x,y)$ through the five given points $A(0,0)$, $B(AE,EB)$, $G(0,-GA)$, $H(AE,-HE)$ and $L(AL,0)$ can be put as $y^2-(ax+b)y+cx^2+dx=0$. When $x=0$, then

$$y=(0\text{ or})-GA=AG,$$

so that $b=AG$. When $x=AE$, then $y=EB$ or EH, so that $EH+EB=a$. $AE+AG$ or $a=-AG/AM$ on making $EM/AM=(EH+EB)/AG$; also $EH\times EB=c$. AE^2+d. AE. Lastly, when $y=0$, $x=(0\text{ or})\,AL$, so that $c.AL^2+d.AL=0$ and $-d/c=AL$, whence

$$EH\times EB/AE\times EL=c,$$

constant. (Compare note (16) above.) When these values are substituted, the conic's equation becomes $y^2+(AG/AM)(x-AM)y+(EH\times EB/EA\times EL)x(x-AL)=0$, which agrees with Newton's expression since $-TM=MT=x-AM$, $-TA=AT=x$ and $TL=x-AL$.

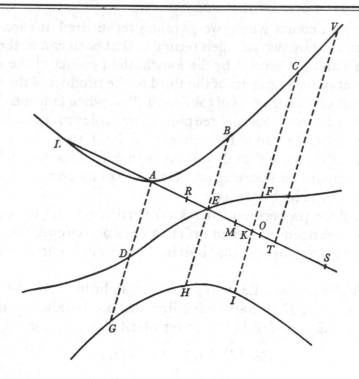

curvam in *L* et rectam *CI* in *K*. Præterea in *AE* capiatur punctum *M* ita ut sit
$AD+AG.EH-EB::AM.EM$, et sit $d.e::AD+AG.AM$ ut et

$$b.e::KC\times KF\times KI.KL\times KA\times KE.$$

Deinde facto $KI.KF::KC-KI.KM$, capiantur etiam in *AE* puncta *R* et *S* ita
ut sint

$$AD\times AG.AR\times AS::EB\times EH.ER\times ES::$$

$$KC\times KF-KC\times KI-KF\times KI.KR\times KS::c.e.$$

Et acta quavis recta *TV* ipsi *AG* parallela, et occurrens rectæ *AE* in *T* et curvæ
ABC in *V*, si *TV* dicatur *y* erit

$$y^3-\frac{d}{e}TM\times yy+\frac{c}{e}TR\times -TS\times y-\frac{b}{e}TA\times TL\times TE=0.^{(23)}$$

Estꝗ generalis regula exprimendi curvas omnes ordinis cubici, si modo signorū
+ et − mutatio pro diversa positione linearum probe fiat.

---

(23) Here, similarly, on taking $AT = x$ the Cartesian equation of the cubic $V(x, y)$ through
the nine given points $A(0, 0)$, $B(AE, EB)$, $C(AK, KC)$, $D(0, -DA)$, $E(AE, 0)$, $F(AK, KF)$,
$G(0, -GA)$, $H(AE, -HE)$ and $I(AK, -IK)$ can be put as

$$y^3-(ax+b)y^2+(cx^2+dx+e)y-(fx^3+gx^2+hx) = 0.$$

AE and let it intersect the curve in *L* and the line *CI* in *K*. Moreover, let a point *M* be taken in *AE* such that $AD + AG : EH[+]EB = AM : EM$ and put

$$AD + AG : AM = d : e,$$

also $KC \times KF \times KI : KL \times KA \times KE = b : e$. Then, having made

$$KI : KF = KC - KI : KM,$$

take also in *AE* points *R* and *S* such that

$$AD \times AG : AR \times AS = EB \times EH : ER \times ES$$

$$= KC \times KF[+]KC \times KI[+]KF \times KI : KR \times KS = c : e.$$

Then, when any straight line *TV* is drawn parallel to *AG*, meeting the line *AE* in *T* and the curve *ABC* in *V*, if *TV* be called *y* there will be

$$y^3 - (d/e)\,TM \times y^2 + (c/e)\,TR \times - TS \times y - (b/e)\,TA \times TL \times TE = 0.^{(23)}$$

The rule is a general one for expressing all curves of cubic order provided that appropriate interchange of the signs + and − be made for differing positions of the lines.

When $x = 0$, then $y = 0$, *AD* or *AG*, so that $b = AD + AG$ and $e = AD \times AG$. When $x = AE$, then $y = 0$, *EB* or *EH*, so that $EB + EH = a.AE + AD + AG$, or $a = -(AD + AG)/AM$ on making $EM/AM = (EB + EH)/(AD + AG)$; also

$$EB \times EH = c.AE^2 + d.AE + AD \times AG \equiv c.ER \times ES,$$

say, whence $-d/c = AR + AS$ and $e/c = AR \times AS$ or $c = AD \times AG/AR \times AS$. Next, when $x = AK$, then $y = KC$, *KF* or *KI* and consequently

$$KC + KF + KI = a.AK + AD + AG = -(AD + AG)MK/AM;$$

also

$$KC \times KF + KF \times KI + KI \times KC = c.(AK - AR)(AK - AS)$$

and so

$$(KC \times KF + KF \times KI + KI \times KC)/KR \times KS = AD \times AG/AR \times AS.$$

Finally, $fx^3 + gx^2 + hx = 0$, $x = AE$, *AL*, whence $-g/f = AE + AL$ and $h/f = AE \times AL$, and so $fx^3 + gx^2 + hx \equiv fx(x - AE)(x - AL)$; accordingly $f.AK \times EK \times LK = KC \times KF \times KI$ (compare note (18) above). When these values are substituted the cubic's equation assumes the form

$$y^3 + \frac{AD + AG}{AM}(x - AM)y^2 + \frac{AD \times AG}{AR \times AS}(x - AR)(x - AS)y$$
$$+ \frac{KC \times KF \times KI}{KA \times KE \times KL}x(x - AE)(x - AL) = 0,$$

which yields Newton's expression on substituting $x = AT = -TA$ (and so $x - AM = -TM$, $x - AR = -TR$, $x - AS = -TS$, $x - AE = -TE$, $x - AL = -TL$). We cannot account for the superfluous phrase 'facto *KI.KF*::*KC − KI.KM*', which is surely a *non-sequitur* in context.

Puncta autem R et S per planam Geometriam investigantur. Scilicet faciendum est

$$\frac{AD \times AG \times EK}{AE} + \frac{EB \times EH \times KA}{AE} + KC \times KI - KC \times KF + KI \times KF$$

$$. AK \times EK :: [c] . e,$$

et erit $[c] . e :: AD \times AG . AR \times AS :: BE \times EH . ER \times ES.$[24] dantur igitur rectangula $AR \times AS$ et $ER \times ES$. Sint ista $AE \times p$, et $AE \times q$ et erit

$$p + q + AE^{[25]} = AR + AS.$$

Ejus summæ dimidium esto AO et erunt $[O]R$ et $[O]S = \sqrt{A[O]^q - AE \times p}$. Ubi $A[O]^q$ semper major erit quam $AE \times p$,[26] si modo recta AE [ducatur] non per duo extrema linearum puncta A et B vel G et H nec per media D et E sed per extremum et medium uti A et E.[27] Unde constat hanc regulam nulli exceptioni obnoxiam esse.[28]

Cæterùm cum hæc sit æquatio cubica quæ non nisi per solidorū[29] geometriam construi possit docebimus quomodo hæ curvæ redigendæ sunt ad æquationes quæ per planam Geometriam construi possunt.[30] Et in hunc finem sit basis $AB = z$, ordinatim incedens $BC = v$ [applicata in dato angulo ABC &][31] proponatur æquatio plena
$$\begin{array}{l} av^3 + bzvv + czzv + dz^3 \\ \;\;\;\;+ evv \;\;+ fzv + gzz \\ \;\;\;\;\;\;\;\;\;\;\;+ hv \;\;+ kz \\ \;\;\;\;\;\;\;\;\;\;\;\;\;\;\;+ l \end{array} = 0.$$ Construatur quomodocunqȝ hæc æquatio $a[-]bp + cpp[-]dp^3 = 0$[32] et invento p cape BD ad BC ut

---

(24) For, on taking $KE/AE = r$ and $KA/AE = s \, (= r - 1)$, there results

$$KR \times KS = r . AR \times AS - s . ER \times ES + KA \times KE,$$

so that
$$\frac{c}{e} = \frac{AD \times AG}{AR \times AS} = \frac{EB \times EH}{ER \times ES} = \frac{KC \times KF + KF \times KI + KI \times KC}{r . AR \times AS - s . ER \times ES + KA \times KE}$$

$$= \frac{KC \times KF + KF \times KI + KI \times KC - r . AD \times AG + s . EB \times EH}{KA \times KE}.$$

(25) Read '$p - q + AE$' accurately.

(26) Newton first began, in line with his draft on f. 22 bis[r], 'Quæ quidem quantitas radicalis semper possibilis [erit]' (The quantity within the radical will indeed always be possible).

(27) This rule guarantees that $AD \times AG$ and $EB \times EH$ have opposite signs, but in fact, since $AO = \frac{1}{2}(AR + AS)$ and $AE \times p = AR \times AS$, always $AO^2 - AE \times p = \frac{1}{4}(AR - AS)^2 \geqslant 0$.

(28) In first draft on f. 22 bis[r] Newton concluded much to the same purpose 'adeo ut hæc regula generalis exprimendi omnes curvas hujus ordinis nullam exceptionem pati[atur]' (so that this general rule for expressing all curves of this order suffers no exception).

(29) The equivalent adjective 'solidam' is cancelled. Understand, as elsewhere, by geometry involving 'solid' loci (conics) and hence by the quartic equations whose roots may be constructed by their meets.

(30) Newton first wrote '...quomodo ea semper reduci poss[it]' (how it might always be reduced). Immediately after this in the manuscript there follows a cancelled first version of

The points R and S are, however, discovered by plane geometry. Plainly, you must make

$$(AD \times AG \times EK + EB \times EH \times KA)/AE$$
$$+ (KC \times KI[+]KC \times KF + KI \times KF) : AK \times EK = c : e$$

and then $c : e = AD \times AG : AR \times AS = BE \times EH : ER \times ES.$[24] The products

$$AR \times AS \quad \text{and} \quad ER \times ES$$

are therefore given. Let these be $AE \times p$ and $AE \times q$, and there follows

$$p[-]q + AE = AR + AS.$$

Let AO be half this sum, and then $OR, OS = \sqrt{[AO^2 - AE \times p]}$. Here AO^2 will always be greater than $AE \times p$[26] provided the straight line AE be drawn, not through two end-points, A and B or G and H, of the lines nor through the middle ones, D and E, but through an end-point and a middle one such as A and E.[27] It is hence established that this rule is guilty of no exception.[28]

For the rest, since this is a cubic equation which it is impossible to construct except by the geometry of solids,[29] we shall explain how these curves are to be reduced to equations constructible by plane geometry.[30] To this end let the base be $AB = z$ with the ordinate $BC = v$ applied at the given angle $A\widehat{B}C$, and let the full equation $av^3 + (bz + e)v^2 + (cz^2 + fz + h)v + dz^3 + gz^2 + kz + l = 0$ be proposed. Let this equation $a - bp + cp^2 - dp^3 = 0$ be constructed some way or other,[32] and when p has been found take BD to BC as p to 1 and draw CD. Let

his improved reduction of the general cubic's equation to standard canonical form by co-ordinate transformation (ff. 20^r–21^r, reproduced in Appendix 3 below). A stray fragment (ULC. Add. 3965.18: 705^v) on which Newton at about this time began a first revise of his first reduction a decade earlier (see II: 10–12) is given in Appendix 2.

(31) This necessary phrase has been inserted by us from Newton's first version (see Appendix 3).

(32) By a slip of Newton's pen the manuscript equation reads '$a + bp + cpp + dp^3 = 0$'. In sequel he first continued 'et invento p fiat BD ad BC ut p ad 1, et dicto $AD = s$ erit$AB = s - py$, quo scripto in æquatione pro z, termini per y^3 multiplicati evanescent. Sit æquatio emergens $bs\,yy + css\,y + ds^3 = 0$, retentis scilicet literis (a, b, c &c) cum significatione mutata. Produc BC
+e   +fs   +gss
+h   +ks
+l

ad F ut sit $2e.c::AB.BF$ et dicto $CF = t$ erit $BC = t - [\frac{c}{2e}z]$' (and when p has been found make BD to BC as p to 1, and on calling $AD = s$ there will be $AB = s - py$: once this is written in the equation in place of z the terms multiplied by y^3 will vanish. Let the emergent equation be $(bs + e)y^2 + (cs^2 + fs + h)y + ds^3 + gs^2 + ks + l = 0$, where, of course, the letters $a, b, c, \ldots$ are retained in an altered meaning. Extend BC to F so that $2e : c = AB : BF$ and, on calling $CF = t$, there will be $BC = t - (c/2e)z$). Since the reduced equation relates, by implication, the co-ordinates $AD = s$ and $DC = (1/q)y$, the next step in the operation—as Newton at once realized—must be to extend (not CB but) CD.

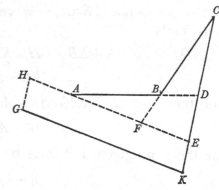

p ad 1, et agatur CD quæ sit ad BC ut [1] ad [q] et dicto $AD=r$ et $CD=s$ erit $BC=qs$ et $AB=r-pqs$.[33] Quibus scriptis in æquatione pro v et z orietur nova æquatio in qua terminus multiplicatus per s^3 deerit. Sit ista

$$\begin{aligned}brss+crrs+dr^3\\+ess\ +frs+grr\\+hs\ +kr\\+l\end{aligned}=0,^{(34)}$$ et si terminus $brss$ non

desit produc CD ad E ut sit $2b\,.\,c::AD\,.\,DE$ et propter datos angulos trianguli ADE simul dabitur ratio AD ad AE. Sit ista n ad 1, et dictis $AE=t$ et $CE=w$ erit

$AD=nt$ et $CD=w-\dfrac{cn}{2b}\,t$.[35] Quibus scriptis in æquatione pro r et s emerget[36]

æquatio sine termino multiplicato per ttw. Sit ista $$\begin{aligned}btww+ftw+dt^3\\+eww\ +hw+gtt\\+kt\\+l\end{aligned}=0,$$

retentis denuò literis cum mutato significatu[,] et produc CE ad K ut sit $EK=\dfrac{f}{[2]b}$,$^{(37)}$ et EA ad H ut sit $AH=\dfrac{e}{b}$ et completo parallelogrammo $HEKG$

dic $GK=x$ et $CK=y$ et erit $AE=x-\dfrac{e}{b}$ et $CE=y-\dfrac{f}{[2]b}$.$^{(38)}$ quibus scriptis pro t

et w orietur æquatio sine terminis yy et xy, exprimens relationem inter basem GK et ordinatam KC, & non ultra reducibilis. Sit ista

$$bxyy+hy+dx^3+gxx+kx+l=0.^{(39)}$$

Quod si terminus $br[s]s$ supra defuerit, non autem $crrs$.[40] fac $c\,.\,d::AD\,.\,DE$,

et pro r et s scriptis nt, et $w-\dfrac{dn}{c}\,t$, emerget æquatio sine termino t^3. Sit ista

$$\begin{aligned}eww+cttw+gtt\\+ftw\ +kt\\+hw\ +l\end{aligned}=0$$ et positis $EK=\dfrac{g}{c}$, et $AH=\dfrac{f}{2c}$, & pro t et w scripto $x-\dfrac{f}{2c}$ &

---

(33) This represents Newton's first Cartesian transformation from $C(z, v)$ to $C(r, s)$ by $\left\{\begin{matrix}z = r-pqs\\v = qs\end{matrix}\right\}$: evidently the origin A is unchanged. The same transforming equations are invoked in a preliminary draft (reproduced in Appendix 4) of his later computations for discriminating the genus of a cubic from its unreduced defining equation.

(34) Understand, as in Newton's first draft (note (32)) that the coefficients $b, c, d, ..., l$ are retained 'cum significatione mutata' (with their designation changed).

(35) This second transform $\left\{\begin{matrix}r = nt\\s = w-(cn/2b)\,t\end{matrix}\right\}$ of $C(r, s)$ to $C(t, w)$ likewise leaves the origin A unaltered.

this be to BC as 1 to q and on calling $AD = r$, $CD = s$ there will be $BC = qs$ and $AB = r - pqs$.[33] When these are entered in the equation in place of v and z there will arise a new equation in which the term multiplied by s^3 is lacking. Let that be $(br + e) s^2 + (cr^2 + fr + h) s + dr^3 + gr^2 + kr + l = 0$,[34] and should the term brs^2 not be lacking extend CD to E so that $2b:c = AD:DE$ and then, because the angles of the triangle ADE are given, there will at once be given the ratio of AD to AE. Let that be as n to 1, and on calling $AE = t$ and $CE = w$ there will be $AD = nt$ and $CD = w - (\frac{1}{2}cn/b) t$.[35] And when these are written in the equation in place of r and s there will emerge an[36] equation without a term multiplied by t^2w. Let that be $(bt + e) w^2 + (ft + h) w + dt^3 + gt^2 + kt + l = 0$, once more retaining letters with an altered meaning, then extend CE to K so that $EK = \frac{1}{2}f/b$[37] and EA to H so that $AH = e/b$, and after completing the parallelogram $HEKG$ call $GK = x$ and $CK = y$, and there will be

$$AE = x - e/b, \quad CE = y - \tfrac{1}{2}f/b.\text{[38]}$$

And when these are entered in place of t and w there will result an equation without terms in y^2 and xy which expresses the relationship between the base GK and ordinate KC and is not further reducible. Let that be

$$bxy^2 + hy + dx^3 + gx^2 + kx + l = 0.\text{[39]}$$

But should the term brs^2—not, however, cr^2s—be lacking besides,[40] make $c:d = AD:DE$ and, when nt and $w - (dn/c) t$ are written in place of r and s, there will emerge an equation without a term in t^3. Let that be

$$ew^2 + (ct^2 + ft + h) w + gt^2 + kt + l = 0$$

and, on setting $EK = g/c$ and $AH = \frac{1}{2}f/c$ and with $x - \frac{1}{2}f/c$ and $y - g/c$ written in

---

(36) 'nova' (a new) is cancelled. Compare note (41).

(37) By a trivial oversight Newton here simply wrote '$\frac{f}{b}$' (and '$CE = y - \frac{f}{b}$' in the next line correspondingly).

(38) The effect of this axis translation $\begin{Bmatrix} t = x - e/b \\ w = y - \frac{1}{2}f/b \end{Bmatrix}$ is to remove the origin from A to $G(-e/b, -\frac{1}{2}f/b)$.

(39) Again understand the phrase 'retentis literis b, g, h, k, l cum mutato significatu'. Contrast the elegance with which Newton achieves his present reduction to standard canonical form—by splitting the total transformation $C(z, v) \to C(x, y)$ into the simple components (note (33)) $C(z, v) \to C(r, s)$, (note (35)) $C(r, s) \to C(t, w)$ and (note (38)) $C(t, w) \to C(x, y)$—with the bulldozing directness of his pioneering technique (II: 12–16) for simplifying the equation of 84 terms which results on substituting

$$z = x' + qy' + r, \quad v = px' + y' + s,$$

where x', y' are the oblique projections, parallel to the original coordinates z, v, of the transformed pair $GK = x$, $KC = y$.

(40) 'tum' (then) is cancelled. The preceding phrase is a late insertion.

$y - \dfrac{g}{c}$ orietur[41] æquatio sine terminis x^3, xyy, y^3, xx, xy hujus formæ

$$eyy + cxxy + kx \atop + hy \; + l = 0.^{(42)}$$

quæ quidem easdem habet conditiones cum æquatione supra inventa

$$bxyy + hy + gxx + kx + l = 0$$

quando terminus dx^3 nullus est.

Sin terminus etiam *crrs* defuerit, non autem *ess*, fac $2e . f :: AD . DE$, et perinde scriptis nt et $w - \dfrac{f}{2e} \, nt$ pro r et s orietur æquatio sine terminis w^3, tww, ttw, tw. Sit ista $eww + hw + dt^3 + gtt + kt + l = 0.^{(43)}$ Deinde positis $EK = \dfrac{h}{2e}$ et $AH = \dfrac{g}{3d}$ et pro t et w perinde scriptis $x - \dfrac{g}{3d}$ et $y - \dfrac{h}{2e}$, orietur æquatio hujus formæ

$$eyy + dx^3 + kx + l = 0.^{(44)}$$

Quod si terminus *ess* etiam defuerit, non autem terminus *frs*.[45] produc DA ad H ut sit $AH = \dfrac{h}{f}$, et dicto $DH[=]t$ erit $AD = t - \dfrac{h}{f}$, quo scripto pro r orietur æquatio hujus formæ

$$fts + dt^3 + gtt + kt + l = 0.$$

Dein produc CD ad E et K ut sit $f . g :: HD . DE$

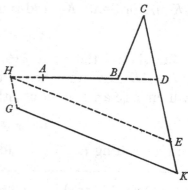

---

(41) 'nova' (a new) is again cancelled. Compare note (36).

(42) When, that is, the change of axes $\begin{Bmatrix} z = r - pqs \\ v = qs \end{Bmatrix}$ produces a cubic (in r and s) which lacks the term brs^2, reduction to the standard reduced form (lacking correspondingly a term in x^3) is made by the axis transformation $\begin{Bmatrix} r = nt \\ s = w - (dn/c)\,t \end{Bmatrix}$ and then the translation to new origin $(-\tfrac{1}{2}f/c, -g/c)$ determined by $\begin{Bmatrix} t = x - \tfrac{1}{2}f/c \\ w = y - g/c \end{Bmatrix}$. The following clause is a late insertion (on f. 21 bis^v) which replaces the more abrupt equivalent 'Vel quod perinde est

$$bxyy + hy + gxx + kx + l = 0'$$

(or, what is all but the same, $bxy^2 + hy + gx^2 + kx + l = 0$).

(43) This sentence, too, is a late insertion on f. 21 bis^v.

(44) In other words, when the transformation of axes $\begin{Bmatrix} z = r - pqs \\ v = qs \end{Bmatrix}$ yields a cubic lacking terms brs^2 and cr^2s, reduction to this canonical form (the Cartesian equation of a divergent

place of t and w, there will arise an[41] equation without terms in x^3, xy^2, y^3, x^2 and xy of this form $ey^2 + (cx^2 + h)y + kx + l = 0$.[42] This has, indeed, the same nature as $bxy^2 + hy + gx^2 + kx + l = 0$, the above found equation when the term dx^3 is not present.

Should, however, the term cr^2s but not es^2 be lacking, make $2e : f = AD : DE$ and likewise, after writing nt and $w - (\frac{1}{2}fn/e)t$ in place of r and s, there will arise an equation without terms in w^3, tw^2, t^2w and tw. Let that be

$$ew^2 + hw + dt^3 + gt^2 + kt + l = 0.\text{[43]}$$

Then, on setting $EK = \frac{1}{2}h/e$ and $AH = \frac{1}{3}g/d$ and likewise writing $x - \frac{1}{3}g/d$ and $y - \frac{1}{2}h/e$ in place of t and w, there will result an equation of this form

$$ey^2 + dx^3 + kx + l = 0.\text{[44]}$$

But should the term es^2—not, however, the term frs—also be lacking,[45] extend DA to H so that $AH = h/f$ and on calling $DH = t$ there will be

$$AD = t - h/f,$$

and with this written in place of r there will arise an equation of this form

$$fts + dt^3 + gt^2 + kt + l = 0.$$

Next, produce CD to E and K so that $f : g = HD : DE$ and $EK = k/f$, then, having

---

parabola) is effected by the axis change $\begin{Bmatrix} r = nt \\ s = w - \frac{1}{2}(f/e)nt \end{Bmatrix}$ followed by the translation $\begin{Bmatrix} t = x - \frac{1}{3}g/d \\ w = y - \frac{1}{2}h/e \end{Bmatrix}$.

(45) A cancelled first continuation at this point, 'fac $f.g :: AD.DE$ et scripto nt & $w - \frac{g}{f}nt$ pro r et s orietur æquatio simplicior sine [terminis w^3, tww, ttw, ww, tt] hujus formæ

$$[ftw + hw + dt^3 + kt + l = 0]'$$

(make $f : g = AD : DE$ and on writing nt and $w - (g/f)nt$ in place of r and s there will arise a simpler equation, without terms in w^3, tw^2, t^2w, w^2 and t^2, of this form

$$ftw + hw + dt^3 + kt + l = 0),$$

suggests that Newton intended a reduction to canonical form on the preceding lines, namely by compounding a change of axes which leaves the origin unchanged with a simple translation. However, when $\begin{Bmatrix} r = nt \\ s = w - (g/f)nt \end{Bmatrix}$ has reduced the equation to the form

$$dt^3 + ftw + hw + kt + l = 0,$$

a further compound transformation $\begin{Bmatrix} t = nx - h/f \\ w = y + 3(dh/f^2)nx - k/f \end{Bmatrix}$ is needed to produce the canonical equation of form $dx^3 + fxy + l = 0$: this is doubtless Newton's reason for contriving his elegant alternative approach outlined in the sequel. He might, alternatively, have kept his previous reduction scheme, invoking the axis change $\begin{Bmatrix} r = nt \\ s = w - (g/f - 3dh/f^2)nt \end{Bmatrix}$ and then applied the translation $\begin{Bmatrix} t = x - h/fn \\ w = y - (g/f - gh/f^2) \end{Bmatrix}$ to the resulting equation.

& $EK=\dfrac{k}{f}$ et Acta *HE* comple parallelogrammum *HEKG* sitɋ $HD.HE::n.1$ et dicto *GK* vel $HE=x$ & $KC=y$ erit $HD=nx$, et $CD=y-\dfrac{k}{f}-\dfrac{gnx}{f}$. quibus scriptis pro *t* et *s* emerget æquatio hujus formæ $fxy+dx^3+l=0.^{(46)}$

Sin deniɋ terminus etiam *frs* desit; fac $AH=\dfrac{g}{3d}$ & scripto $t-\dfrac{g}{3d}$ pro *r* orietur æquatio hujus formæ $hs+dt^3+kt+l=0$. Deinde posito $h.k::HD.DE$, et $EK=\dfrac{l}{k}$, scriptisɋ nx pro *t*, et $y-\dfrac{l}{h}-\dfrac{k}{h}nx$ pro *s*, orietur æquatio $hy+dn^3x^3=0.^{(47)}$

Atɋ ita videre est quod omnes æquationes ad hasce quatuor formas reduci possint[₃] $bxyy=dx^3+gxx+hy+kx+l.$ $eyy=dx^3+kx+l.$ $fxy=dx^3+l.$ et $hy=dx^3$, ubi tamen signa + et − omnimodò variari possint, & aliqui termini deesse. Atɋ hæ formæ non sunt ultra reducibiles. Jam vero in prima harum formarum posito *bxyy* affirmativo si terminus dx^3 sit etiam affirmativus, curva erit Hyperbola triformis$^{(48)}$ cum tribus Asymptotis quarum nullæ sunt parallelæ, sin termin[us] ist[e] dx^3 sit negativus$^{(49)}$ duæ ex Hyperbolis mutabuntur in Ellipsin quæ tamen non semper habet ovalem formā sed aliquando conjungitur reliquæ figuræ Hyperbolicæ, aliquando evadit punctum, et aliquando imaginaria est. Inter Hyperbolam et Ellipsin media existit Parabola cum Hyperbola conjugata ubi dx^3 nullus est dummodo terminus etiam *gxx* non desit. Nam si iste etiam desit figura rursum evadet omni ex parte Hyperbolica. Et quidem triplex Hyperbola$^{(50)}$ cum tribus Asymptotis quarum duæ sunt parallelæ si terminus *k* affirmativus est, Ubi vero *k* evanescit una Hyperbola evanescit, & parallelæ Asymptoti coincidunt, et ex his hyperbolis alia evanescit ubi *k* fit negativus.$^{(51)}$

In secunda forma curva semper Parabola$^{(52)}$ est sine Hyperbola conjugata et aliquando habet Ellipsin conjugatam juxta verticem. Tertia forma exprimit Parabolam Cartesianam et quarta Parabolam simplicem cubicam.

---

(46) Namely, the canonical defining equation of a Cartesian trident. Newton's procedure is equivalent to the single transformation $\left\{\begin{array}{l}r=nx-h/f\\s=y-(g/f-3dh/f^2)nx-(k/f+3dh^2/f^3)\end{array}\right\}$ of the equation $dr^3+frs+gr^2+hs+kr+l=0$.

(47) Of canonical form $dx^3+hy=0$, that is. Proceeding by his first method Newton would have derived an equivalent result by using the transformation $\left\{\begin{array}{l}r=nt\\s=w-(k/h-g^2/3dh)nt\end{array}\right\}$ to reach an equation of form $dt^3+gt^2+hw+(g^2/3d)t+l=0$, whence the translation

$$\left\{\begin{array}{l}t=x-g/3d\\w=y-(l/h-g^3/27d^2h)\end{array}\right\}$$

produces a second equation of the desired form $dx^3+hy=0$ (which represents, of course, a general Wallis cubic).

(48) In the following tabulation this designation, evidently taken over from II: 88, was changed to 'Hyperbola Hyperbolica' (hyperbolic hyperbola); see note (59) below.

drawn *HE*, complete the parallelogram *HEKG* and let $HD:HE = n:1$; on calling *GK* (or *HE*) $= x$ and $KC = y$ there will be $HD = nx$ and

$$CD = y - (gn/f)\,x - k/f,$$

and with these written in place of *t* and *s* there will emerge an equation of this form $fxy + dx^3 + l = 0.$[46]

Should, finally, the term *frs* also be lacking, make $AH = \tfrac{1}{3}g/d$ and when $t - \tfrac{1}{3}g/d$ is written in place of *r* there will arise an equation of this form

$$hs + dt^3 + kt + l = 0.$$

Then, on setting $h:k = HD:DE$ and $EK = l/k$ and with *nx* written in place of *t* and $y - (kn/h)\,x - l/h$ for *s*, there will arise the equation $hy + dn^3x^3 = 0.$[47]

And in this way it is evident that all equations might be reduced to these four forms: $bxy^2 = dx^3 + gx^2 + hy + kx + l$, $ey^2 = dx^3 + kx + l$, $fxy = dx^3 + l$ and $hy = dx^3$. (Here, however, the signs $+$ and $-$ may be interchanged in any manner and some terms may be lacking.) Nor are these forms further reducible. If now, in fact, in the first of these forms on supposing bxy^2 to be positive dx^3 should also be positive, the curve will be a three-limbed hyperbola[48] with three asymptotes, none of which are parallel, but if the term dx^3 be negative[49] two of the hyperbolas will change into an ellipse, which will not, however, always have an oval shape but is sometimes joined onto the remaining hyperbolic figure, sometimes proves to be a point and is sometimes imaginary. Midway between the hyperbola and ellipse, when dx^3 is nil and so long as the term gx^2 is not also lacking, there exists a parabola with a conjugate hyperbola. For should the latter term also be missing, the figure will again turn out to be hyperbolic on every hand. It is, indeed, a threefold hyperbola[50] with three asymptotes, two of which are parallel, if the term in *k* is positive; but when *k* vanishes one of the hyperbolas vanishes and the parallel asymptotes coincide, while a further one of these hyperbolas vanishes when *k* comes to be negative.[51]

In the second form the curve is always a parabola[52] without any conjugate hyperbola, though on occasion it has a conjugate ellipse close to its vertex. The third form represents Descartes' parabola, and the fourth the simple cubical parabola.

(49) The text has the illogical plural 'termini isti'. Newton first wrote 'sin vero terminorum istorum signa sint eadem' (but if indeed the signs of those terms be the same), and then altered this to read 'sin termini isti sint ejusdem signi'.

(50) Renamed the 'Hyperbolismus Hyperbolæ' (hyperbolism of a hyperbola) in the following table of species.

(51) These first six 'kinds' of cubic are Cases 1, 2, 3, 4, 6, 5 respectively on II: 88. Compare note (53) below.

(52) The 'Parabola divergens' (divergent parabola) of the following table.

Atcp ita patet novem esse genera[53] curvarum ordinis cubici. Sub hisce tamen multæ[54] comprehenduntur species. Nam aliæ curvæ non habent aliquam diametrum bisectionis, aliæ habent quas bifidas[55] nominare licebit. Aliquæ nullam habent partem similem alij parti,[,] aliæ habent partes omnino similes et æquales a centro intermedio hinc inde [in] infinitum procurrentes quas ideo æquicruras[56] nominare licet. In aliquibus tres Asymptoti et omnes diametri trisectionis concurrunt ad idem punctum, in alijs non. Aliquæ habent punctum conjugatum quas punctatas voco, aliæ acutè terminantur ad vertices ad modum Cissoidis quas ideo Cissoidales vel cuspidatas voco,[57] et aliæ se decussant ad modum crucis. Hæ sunt præcipuæ differentiæ harum curvarum. Et secundum has diff[er]entias possumus enumerare species harum Curvarum juxta sequentem Tabulā.[58]

Hyperbola Hyperbolica[59]

non bifida[60]
- cum Ovali
- deficiens, ad angulos oppositos, non æqui[crura][61]
- deficiens, ad angulos oppositos, æqui[crura][61]
- deficiens, ad angulos non oppositos
- conjugatam decussans
- se decussans
- Cuspidata[62]
- punctata

bifida[60]
- cum Ovali
- deficiens, ad angulos oppositos[63]
- deficiens, ad angulos non oppositos
- Conjugatam decussans
- se decussans
- Cuspidata
- punctata

trifariam bifida[60]

---

(53) These are identical with the nine 'casus' (cases) on II: 88.

(54) The exaggeration 'innumeræ' (innumerable) is replaced.

(55) Newton had earlier used 'fissus' in this sense in describing the Cartesian trident (see II: 86/88) and has deleted the equivalent adjective 'bisectus' in his following table (see note (60)).

(56) That is, with its (infinite) branches symmetrically situated. The adjective 'æqui-later[as]' (equilateral) is cancelled; compare note (61).

(57) In the following table, in fact, Newton came to prefer the designation 'cuspidata' (cusped); see note (62) below.

(58) Altogether Newton here lists 53 separate species. Observe that the 1667 'grading' by possession or lack of coincident asymptotes and a 'centrum [generale]' (see II: 46) is here abandoned, but that in compensation clear distinction is now made (namely in species 4 and 8; 11 and 15; 18 and 20; 24 and 26; 30 and 33, though not in species 49, the divergent parabola

And so it is evident that there are nine kinds[53] of curve of cubic order. Under these, however, many[54] species are comprehended. For some curves do not possess a bisection diameter, others—which it will be permissible to name 'cleft'[55]—do. Some have no portion similar to any other, others have exactly congruent portions running out from an intermediate centre either way to infinity—these it is consequently permissible to name 'equal-branched'.[56] In some the three asymptotes and all trisection diameters are concurrent at the same point, in others not so. Some have a conjugate point—these I call 'punctate'—, others terminate sharply at their vertices in the manner of a cissoid—these I accordingly name 'cissoidal' or 'cusped'[57]—, and yet others intersect themselves in the manner of a cross. These are the principal differentiations between these curves. Following them we can enumerate the species of these curves as laid out in the following table.[58]

The hyperbolic hyperbola[59]

uncleft[60]
- with an oval
- deficient, at opposite angles, not equal-branched[61]
- deficient, at opposite angles, equal-branched[61]
- deficient, at non-opposite angles
- crossing its conjugate
- crossing itself
- cusped[62]
- punctate

cleft[60]
- with an oval
- deficient, at opposite angles[63]
- deficient, at non-opposite angles
- crossing its conjugate
- crossing itself
- cusped
- punctate

cleft[60] three ways.

'sine Ovali') between the pure cubic ('deficiens') and that ('punctata') having a conjugate point. In a first draft of the list (on f. 21 bis[v]) the earlier subdivision between cubics 'sine centro generali' and 'cum centro generali' was still retained, so doubling species 2, 4, 6, 10, 11, 12, 16 and thereby increasing the total number of species to 60.

(59) 'Hyperbola triformis' (the three-limbed hyperbola) is cancelled. Compare note (48) above.

(60) These 'cleft' and 'uncleft' divisions were originally (on f. 21 bis[v]) distinguished as 'simplex' (simple) and 'circumflexa' (around-bent)—compare Newton's use of 'directa' and 'circumflexa' on II: 86—and then as 'bisectus' (bisected) and 'non bisectus'.

(61) 'æquilatera' (equilateral) is cancelled; see note (56). The 'angles' are understood to be those of the triangle formed by the three (real, non-coincident) asymptotes.

Hyperbola Elliptica [60]

- non bifida [60]
  - Cum Ovali [64]
  - Deficiens
  - se decussans
  - punctata
  - Cuspidata

- bifida [60]
  - Ovalem habens [65] ad concavitatem
  - Ovalem habens [65] ad convexitatem
  - deficiens
  - se decussans
  - punctata
  - Cuspidata, sive [66] Cissoides Veterum

Hyperbola Parabolica

- non bifida [60]
  - Cum Ovali [67]
  - Deficiens, aperta [68]
  - Deficiens[,] clausa
  - se decussans
  - Conjugatam decussans
  - Punctata
  - Cuspidata

- bifida [60]
  - Parabolam habens ad concavitatem
  - Parabolam habens ad convexitatem
  - Parabolam apperiens
  - Conjugatam decussans

Hyperbolismus [69] Hyperbolae

- non bifidus [60]
  - ad angulos non oppositos
  - ad angulos oppositos, non æquis cruribus [70]
  - ad angulos oppositos, æquis cruribus [70]
- bifidus [60]

---

(62) Originally (on f. 21 bis^v) 'Cuspidata sive Cissoidalis' (cusped or cissoidal). Compare note (57) above.

(63) 'non æquilatera' (non-equilateral) is cancelled. Compare notes (56) and (61) above.

(64) In draft on f. 21 bis^v Newton first tried the names 'redundans' (redundant), 'Ovalem habens' (having an oval) and 'plena' (full) in quick succession.

(65) 'redundans, Ellipsin habens' (redundant, having an ellipse) was Newton's first choice on f. 21 bis^v.

(66) This replaces 'ipsa nempe' (namely) on f. 21 bis^v.

(67) 'redundans' (redundant) is cancelled.

(68) This replaces 'fissa' (cleft) on f. 21 bis^v.

(69) In his 1695 revise of the present paper, titled by him 'Enumeratio Curvarum tertij Ordinis' (ULC. Add. 3961.1: 38^r–50^v [= *Opticks* (London, 1704): 138–62]) Newton defined this term as follows: 'Hyperbolismum figuræ notæ voco cujus ordinata prodit applicando

The elliptical hyperbola

- uncleft[60]
  - with an oval[64]
  - deficient
  - crossing itself
  - punctate
  - cusped
- cleft[60]
  - having an oval[65] on its concave side
  - having an oval[65] on its convex side
  - deficient
  - crossing itself
  - punctate
  - cusped, in other words[66] the Ancients' cissoid.

The parabolic hyperbola

- uncleft[60]
  - with an oval[67]
  - deficient, open[68]
  - deficient, shut
  - crossing itself
  - crossing its conjugate
  - punctate
  - cusped
- cleft[60]
  - having a parabola on its concave side
  - having a parabola on its convex side
  - opening its parabola
  - crossing its conjugate.

The hyperbolism[69] of a hyperbola

- uncleft[60]
  - at non-opposite angles
  - at opposite angles, not equal-branched[70]
  - at opposite angles, equal-branched[70]
- cleft.[60]

---

ordinatam figuræ notæ ad abscissam communem. Hac ratione linea recta vertitur in Hyperbolam conicam et sectio omnis conica vertitur in aliquam figurarum secundi generis quas hic Hyperbolismos sectionū conicarū voco. Nam æquatio ad figuras de quibus agimus, nempe

$$xyy + ey = cx + d, \quad \text{seu} \quad y = \frac{e \pm \sqrt{ee + 4dx + 4cxx}}{2x},$$

producitur applicando ordinatam sectionis conicæ $\frac{e \pm \sqrt{ee + 4dx + 4cxx}}{2}$ ad ejus abscissam x'
(ff. 44^r–44^v [= pp. 154]). In other words, if the Cartesian equation of a curve is $y = f(x)$, then $xy = f(x)$ is its hyperbolism: thus in the present instance the conic $y^2 = hy + kx^2 + lx$ yields $(xy)^2 = h(xy) + kx^2 + lx$, that is, $xy^2 = hy + kx + l$, as its hyperbolism (hyperbolic, parabolic or elliptical according as $k > 0$, $k = 0$ or $k < 0$).

(70) Newton has cancelled 'non æquilateros/æquilateros' (non-equilateral/equilateral) respectively. In first draft on f. 21 bis^v he first wrote 'dissimilaris/similaris' (dissimilar/similar) and then, in the first instance, 'ex adverso dissimilis/similis'.

Hyperbolismus[69] Ellipseos $\begin{cases} \text{non bifidus,}^{[60]} \text{ nec æquis cruribus}^{[70]} \\ \text{non bifidus,}^{[60]} \text{ sed æquis cruribus}^{[70]} \\ \text{bifidus}^{[60]} \end{cases}$

Hyperbolismus[69] Parabolæ $\begin{cases} \text{non bifidus}^{[60]} \\ \text{bifidus}^{[60]} \end{cases}$

Parabola divergens[71] $\begin{cases} \text{cum Ovali} \\ \text{sine Ovali} \\ \text{se decussans} \\ \text{Cuspidata,}^{[62]} \text{ viz: cujus longitudinem Neilus invenit.}^{[72]} \end{cases}$

Parabolismus Hyperbolæ[73] sive Parabola Cartesiana

Parabola cubica.[74]

[75]Enumeratis harum Curvarum speciebus et ostenso quo pacto æquationes quibus quomodocunœ exprimuntur reduci possint ad formulas quarum subsidio definivimus ipsas et distinximus in species, videor determinationem horum locorū ad finem perduxisse. Sed quoniam reductio æquationum ad formulas istas sæpenumero tædio esse possit, ex abundanti jam docebo quomodo speciem[76] curvæ sine istiusmodi transmutatione dignoscere liceat. Supponamus igitur quæstionem aliquam de hujus generis Loco ad æquationem pe[r]ductam esse, sitœ æquatio illa ut supra[77] $y^3 \begin{smallmatrix} +bx \\ +e \end{smallmatrix} yy \begin{smallmatrix} +cxx \\ +fx \\ +h \end{smallmatrix} y \begin{smallmatrix} +dx^3 \\ +gxx \\ +kx \\ +l \end{smallmatrix} = 0$ in qua x denotat

basem AB, y ordinatam BC, et $+b$, $+c$, $+d$ &c datas quasvis quantitates cum suis signis affectas. Construatur hæc æquatio $z^3 + bzz + cz + d = 0$, et ponatur $3zz^{[78]} + 2bz + c = m$. $3z + b = n$. $ezz + fz + g = p$. $2ez + f = q$. & $hz + k = r$.[79] Ubi si

___

(71) 'Parabola bifida' (cleft parabola) is cancelled. On f. 21 bis^v Newton first tried in quick succession 'Parabolismus Conicæ Sectionis' (conic parabolism), then 'Parabola reflexa' (back-bent parabola) and, lastly, 'Parabola semicubica, bisecta' (bisected, semicubic parabola).

(72) See II: 68, note (88) above. Newton evidently supported Wallis in his recent squabble with Huygens over whether William Neil or Henrik van Heuraet first constructed this semicubical parabola. (Compare Wallis' 'Epistola...Primam Inventionem & Demonstrationem Æqualitatis lineæ Curvæ *Paraboloïdis* cum Recta, anno 1657. factam, Dn. *Guilielmo Neile* p.m. asserens', *Philosophical Transactions*, **8**, 1673, No. 98 [for 17 November 1673]: 6146–9; and the two following letters by Brouncker and Wren, *ibid.*: 6149–50.)

(73) Newton nowhere explains how this 'hyperbolic parabolism' $fxy + dx^3 + l = 0$ (of the hyperbola $fxy + dx^2 + l = 0$?) is to be derived and the name was dropped from future cubic enumerations, as we shall see in the seventh volume.

(74) This replaces 'Parabolismus simplex' (simple parabolism) on f. 21 bis^v.

(75) Newton may well have intended to follow his preceding table with a detailed analytical justification (on the lines of his earlier enumeration on II: 38–70) of his present subdivision of

The hyperbolism[69] of $\left\{\begin{array}{l}\text{uncleft}^{(60)} \left\{\begin{array}{l}\text{not equal-branched}^{(70)} \\ \text{but equal-branched}^{(70)}\end{array}\right. \\ \text{cleft.}^{(60)}\end{array}\right.$
    an ellipse

The hyperbolism[69] of $\left\{\begin{array}{l}\text{uncleft}^{(60)} \\ \text{cleft.}^{(60)}\end{array}\right.$
    a parabola

The divergent parabola[71] $\left\{\begin{array}{l}\text{with an oval} \\ \text{without an oval} \\ \text{crossing itself} \\ \text{cusped,}^{(62)} \text{ namely, that whose length Neil found.}^{(72)}\end{array}\right.$

The parabolism of a hyperbola,[73] in other words, the Cartesian parabola.

The cubical[74] parabola.

[75]Having enumerated the species of these curves and shown how the equations by which they are arbitrarily represented can be reduced to the formulas with whose assistance we defined them and distinguished them into species, I appear to have brought my determining of these loci to an end. But, seeing that reduction of equations to those formulas may frequently be irksome, as an extra I shall now explain a procedure which will allow us to discover the species[76] of a curve without a transformation of that sort. Let us suppose, therefore, that some question regarding a locus of this kind has been brought to an equation, and let that equation be, as above,[77]

$$y^3 + (bx+e)\,y^2 + (cx^2+fx+h)\,y + dx^3 + gx^2 + kx + l = 0,$$

in which x denotes the base AB, y the ordinate BC and b, c, d, ... are any given quantities affected with their signs. Let this equation $z^3 + bz^2 + cz + d = 0$ be constructed and put $3z^{2(78)} + 2bz + c = m$, $3z + b = n$, $ez^2 + fz + g = p$, $2ez + f = q$ and $hz + k = r$.[79] Here, if in constructing the aforesaid equation there proved

---

the general cubic into nine *genera* and fifty-three *species*. What seems to be a complete set of his preliminary drafts and computations for the following final section on discriminating a cubic's genus from its unreduced Cartesian equation still survives (partly in private possession): lavish extracts from these are reproduced in Appendixes 4–7.

(76) Understand by this one of the sixteen major species (compare II: 86) into which the cubic subdivides under Newton's system of classification.

(77) The 'æquatio plena' of the cubic on page 362 above with variables x, y substituted for z, v and with a set equal to unity.

(78) In line with the preceding draft on f. 21ʳ (reproduced in Appendix 7 below) the manuscript reads '$3azz$'. Newton has momentarily forgotten that in the present revise he has set $a = 1$.

(79) Whence the substitution $y = zx+v$ converts the 'æquatio plena' into the reduced form $mx^2v + nxv^2 + v^3 + px^2 + qxv + ev^2 + rx + hv + l = 0$ since $\phi \equiv z^3 + bz^2 + cz + d$ is zero. To the

in construenda præfata æquatione tres obvenerint reales radices z illæ dabunt totidem quantitates m, n, p, q, r. Et si forte d non habeatur tunc duæ radicum ex æquatione $zz+bz+c=0$ elici debent, et 0 pro tertia usurpari. hoc est pro m, n, p, q, r scribendum erit $c, b, g, f,$ & k, neglectis terminis in quibus z sive 0 reperitur. Eodem modo[80] si c et d simul desunt duæ reputandæ sunt radices et sibi mutuo et nihilo æquales, et tertia radix erit $-b$:[81] Et sic si omnes tres termini b, c, d desunt, tres radices reputandæ sunt[82] æquales et pro illis scribendum erit 0.

Jam igitur si tres obvenerint[83] inæquales radices, et nulla earum in colligendis quantitatibus[84] m, n, p, q, r usurpata efficiat terminos $npp-qpm+rmm$ sese destruere[,] Figura erit Hyperbola triformis non bifida, sed si aliqua earum efficiat terminos istos $npp-qpm+rmm$ evanescere Hyperbola illa triformis erit uno modo bifida, et trifariam bifida si termini illi in unoquoq́ trium casuum evanescunt.[85]

Quod si æquatio illa non nisi unicam [h]abeat[83] radicem z, cæteris duabus existentibus imaginarijs, Figura erit Hyperbola Elliptica modo non sit $nn-2m=0$,[86] nam in hoc casu habebitur Hyperbolismus Ellipseos.

Adhæc si trium radicum z duæ sunt æquales[87] et tertia usurpata pro z non efficiat $p=0$, Figura erit Hyperbola Parabolica. Sed si efficiatur $p=0$, habebitur Hyperbolismus Hyperbolæ, & Hyperbolismus Parabolæ si insuper

---

modern reader this is a Taylor expansion in which $m = d\phi/dz$, $n = \frac{1}{2}d^2\phi/dz^2$ and $q = dp/dz$, but, though this was no doubt evident to Newton also, he makes no explicit reference to the fact and we should beware in present context of interpreting these coefficients as derivatives, as Rouse Ball freely does in his 'Newton's Classification of Cubic Curves' (note (1) above): 121. The detailed exploration of this approach had to await another sixty years till Jean Paul de Gua de Malves developed his 'Méthode abrégée, & analogue à celles du Calcul Différentiel, pour transporter l'Origine d'une Courbe dans un point quelconque du Plan sur lequel cette Courbe est décrite' (*Usages de l'Analyse de Descartes Pour découvrir, sans le secours du Calcul Différentiel, les Proprietés, ou Affections principales des Lignes Géométriques de tous les ordres* (Paris, 1740): Section II, Lemme Quatriéme: 236–347).

(80) 'Pariter' (equivalently) is cancelled.

(81) In this case $z^3+bz^2 = 0$.

(82) 'et sibi mutuo et nihilo' (both to one another and to zero) is cancelled.

(83) Newton here copied 'reales et' and 'realem' respectively from his preceding draft on f. 21^r (see Appendix 7) and then deleted them, clearly considering the reality of the roots to be a self-evident assumption.

(84) This replaces 'valoribus' (values).

(85) On replacing the reduced cubic equation (note (79) above) by the approximation mx^2v+px^2 when x and v are indefinitely large there results $v = -p/m+t$, and when this is substituted in the original equation there is $x(r-qp/m+np^2/m^2)+mx^2t+O(x^0) = 0$, whence the infinite branches of the cubic are highly approximated, to $O(x^{-2})$, by

$$y = zx-p/m-m^{-3}(np^2-qpm+rm^2)\,x^{-1}.$$

to be three real roots z, they will yield an equal number of the quantities m, n, p, q, r. And should d chance not to be present, then two of the roots ought to be derived from the equation $z^2 + bz + c = 0$, with zero used in place of the third: in other words, in place of m, n, p, q and r you must write c, b, g, f and k (neglecting the terms in which z, that is, zero, is found). In the same manner[80] if c and d are simultaneously lacking, two roots are to be reckoned as equal both to one another and to nothing, and the third root will be $-b$.[81] And so if all three terms b, c and d are missing, the three roots are to be counted as equal[82] and in their place there will need to be written zero.

Now, therefore, if the three roots prove to be[83] unequal and none of them, when employed in evaluating the quantities[84] m, n, p, q and r, makes the terms $np^2 - qpm + rm^2$ destroy one another, the figure will be a three-limbed uncleft hyperbola, but should any of them make those terms $np^2 - qpm + rm^2$ vanish, then that three-limbed hyperbola will be cleft singly, and it will be cleft threefold if those terms vanish in each of the three cases.[85]

But should that equation possess only a unique[83] root z, the two others turning out to be imaginary, the figure will be an elliptical hyperbola provided $n^2 - 2m = 0$[86] does not hold, for in this case a hyperbolism of an ellipse will be obtained.

Further, if two of the three roots z are equal[87] and the third, when employed in place of z, does not produce $p = 0$, the figure will be a parabolic hyperbola. But if $p = 0$ is produced, there will be had a hyperbolism of a hyperbola, and a hyperbolism of a parabola if in addition there be $q^2 - 4nr = 0$.[88] And

---

(For Newton's computation see Appendix 6.) The necessary condition for this to have the (bisection) diameter $y = zx - p/m$ is clearly that the coefficient of x^{-1} in this defining equation be zero ($m \neq 0$).

(86) This should read '$nn - 4m = 0$ [compare notes (37), (38), (75) and (89) of the following appendix] ut et $p = 0$'. (Though Newton does not say so, this is also the condition for the 'Hyperbola triformis' to reduce to a 'Hyperbolismus Hyperbolæ' when all three roots z are real.) When $n^2 = 4m$ the reduced cubic (note (79) above) becomes

$$v(\tfrac{1}{2}nx + v)^2 + px^2 + qxv + ev^2 + rx + hv + l = 0,$$

whence the 'æquatio plena' when changed to new axes by the transformation $\begin{cases} x = \lambda u \\ y = \lambda zu + v \end{cases}$ is $v(\tfrac{1}{2}n\lambda u + v)^2 + p\lambda^2 u^2 + q\lambda uv + ev^2 + r\lambda u + hv + l = 0$; which is in turn reducible ($p \neq 0$) to $XY^2 + RX^2 + SY + TX + V = 0$ or ($p = 0$) to $XY^2 + S'Y + T'X + V' = 0$, the canonical forms of a parabolic hyperbola and conic hyperbolism respectively, by an appropriate second transformation to new axes $\begin{cases} u = Y - (2\mu/\lambda n)X + \alpha \\ v = \mu X - \beta \end{cases}$.

(87) And so $m = 0$, whence the reduced cubic (note (79) above) takes on the simplified form $nxv^2 + v^3 + px^2 + qxv + ev^2 + rx + hv + l = 0$.

sit $qq-4nr=0$.[88] Et in omnibus hisce casibus figura erit bifida si tertia illa radix z efficiat $npp-qpm+rmm=0$: aliter erit non bifida.[89]

Deniꝗ si omnes tres radices z æquales esse contigerit & non sit $p=0$ Figura erit Parabola divergens[,] sed si sit $p=0$ habebitur Parabolismus Hyperbolæ nisi insuper sit $q=0$, in quo casu Figura erit Parabola Cubica.[89]

Cognita specie Curvæ, si præterea desideretur positio Asymptotorum et plaga ad quam crura Parabolica tendunt, In ordinata BC quam y designat cape $BS=zx-\dfrac{p}{m}$[90] et punctum S attinget Asymptoton si

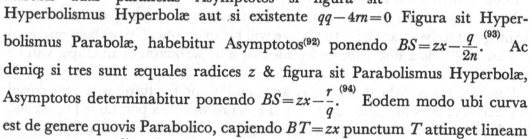

modo z non sit una duarum vel trium æqualium radicum.[91] Hoc igitur pacto tot invenies Asymptotos quot sunt radices z non æquales. Sed si z sit una duarum æqualium radicum cape $BS=zx-\dfrac{q}{2n}\pm\sqrt{\dfrac{qq-4rn}{4nn}}$ & habebis duas parallelas Asymptotos si figura sit Hyperbolismus Hyperbolæ aut si existente $qq-4rn=0$ Figura sit Hyperbolismus Parabolæ, habebitur Asymptotos[92] ponendo $BS=zx-\dfrac{q}{2n}$.[93] Ac deniꝗ si tres sunt æquales radices z & figura sit Parabolismus Hyperbolæ, Asymptotos determinabitur ponendo $BS=zx-\dfrac{r}{q}$.[94] Eodem modo ubi curva est de genere quovis Parabolico, capiendo $BT=zx$ punctum T attinget lineam rectam quæ tendit ad eandem plagam cum infinitis cruribus Parabolicis figuræ, si modo pro z sumatur una æqualium radicum.[95]

In præfatis determinationibus supposui primum terminum æquationis y^3 non deesse. Quamobrem si terminus ille desit & dx^3 non desit,[96] debet y fieri basis figuræ & x ordinata, & cætera peragi ut supra. Sed si uterꝗ terminus simul desit sed b et c non desunt Figura erit Hyperbola triformis cujus una

---

(88) The fictitious equation for indefinitely large x and v is now $x(nv^2+qv+r+px)=0$. Accordingly, the infinite branches are highly approximated, when $p\neq 0$, by the parabola $y=zx-q/2n\pm\sqrt{[(q^2-4nr)/n^2-(p/n)x]}$; but, when $p=0$, by the parallel asymptotes

$$y=zx-q/2n\pm\sqrt{[q^2-4nr]}/n$$

(coincident for $q^2=4nr$) which define the degenerate case of the conic hyperbolism.

(89) In this case $m=n=0$, whence the reduced cubic is

$$v^3+px^2+qxv+ev^2+rx+hv+l=0$$

and the 'æquatio plena' may be reduced to the corresponding transformed equation

$$v^3+p\lambda^2u^2+q\lambda uv+ev^2+r\lambda x+hv+l=0 \quad (\lambda\neq 0)$$

by the change of axes $\begin{cases} x=\lambda u \\ y=\lambda zu+v \end{cases}$. When $p\neq 0$, this in turn is reducible to the form $ey^2+dx^3+kx+l=0$ by a second coordinate change (compare note (44) above); but, when

in all these cases the figure will be cleft if the third root z should make $np^2 - qpm + rm^2 = 0$: otherwise it will not be.[89]

Finally, if all three roots z should chance to be equal and there be not $p = 0$, the figure will be a divergent parabola; but if there should be $p = 0$ a parabolism of a hyperbola will be obtained, unless in addition $q = 0$, in which case the figure will be a cubical parabola.

Once the species of a curve is found out, if besides there be desired the location of the asymptotes and the direction to which the parabolic branches tend, in the ordinate BC (which y denotes) take $BS = zx - p/m$[90] and the point S will lie in the asymptote provided z is not one of two or three equal roots.[91] In this way, accordingly, you will find as many asymptotes as there are unequal roots z. But if z be one of two equal roots take $BS = zx - \frac{1}{2}q/n \pm \sqrt{[\frac{1}{4}(q^2 - 4rn)/n^2]}$ and you will have two parallel asymptotes if the figure be a hyperbolism of a hyperbola; or if (there being $q^2 - 4rn = 0$) the figure should be a hyperbolism of a parabola, the asymptote[92] will be obtained by putting $BS = zx - \frac{1}{2}q/n$.[93] And, lastly, if there are three equal roots z and the figure be a parabolism of a hyperbola, the asymptote will be obtained by putting $BS = zx - r/q$.[94] In the same manner, when the curve is of any parabolic kind, on taking $BT = zx$ the point T will trace a straight line tending in the same direction as the infinite parabolic branches of the figure provided one of the equal roots be assumed in place of z.[95]

In the above determinations I have presumed that the first term y^3 of the equation is not lacking. Consequently, if that term is missing and dx^3 is not,[96] y ought to be made the figure's base and x the ordinate, with the rest accomplished as before. But if each term be simultaneously lacking but not those in b and c, the figure will be a three-limbed hyperbola, one of whose asymptotes is

---

$p = 0$, to $fxy + dx^3 + l = 0$ (compare note (46) above) and, when $p = q = 0$, to $hy + dx^3 = 0$ (compare note (47)).

(90) See note (85) above. Understand that $AB = x$ in the accompanying figure.

(91) That is, if $m \neq 0$. Compare note (87) above.

(92) 'inter duo crura' (between two infinite branches) is cancelled.

(93) See note (88).

(94) In this case $y = zx + v$ is determined, for large x and y (and so v), by the fictitious equation $qxv + rx = 0$.

(95) Newton first began a following paragraph: 'Deniɕ si Curva sit Hyperbolarum aliqua vel Parabola reflexa et scire velis utrum habeat punctum aliquod duplex...' (If, finally, the curve be some one of the hyperbolas or a divergent parabola and you should wish to know whether it possesses a double point...). In the case of the 'æquatio plena'

$$F(x, y) \equiv y^3 + bxy^2 + cx^2y + dx^3 + ey^2 + fxy + gx^2 + hy + kx + l = 0$$

the condition for the point (x, y) on it to be double (namely, a node or conjugate point) is that $F_x = F_y = 0$, since the slope of the cubic there ($-F_x/F_y$) is necessarily indeterminate.

(96) A cancelled first sequel reads 'mutanda sunt nomina quantitatum x et y' (the designations of the quantities x and y are to be interchanged).

Asymptotos determinatur capiendo $AB = -\frac{e}{b}$ & ad B erigendo lineam parallelam ordinatis, altera determinatur capiendo $BS = -\frac{g}{c}$ et ducendo per S lineam parallelam Basi. Nam istæ parallelæ erunt Asymptoti. Tertia attingetur a puncto S sumendo $BS = -\frac{cx}{b} + \frac{ec}{bb} - \frac{f}{b} + \frac{g}{c}$. [97] Deniq̃ si terminorum etiam b et c alteruter puta c desit Figura vel Hyperbola Parabolica erit vel Hyperbolismus aliquis cujus determinationem supra satis explicuimus. [98]

---

(97) In this case, where the 'æquatio plena' of the cubic is

$$bxy^2 + cx^2y + ey^2 + fxy + gx^2 + hy + kx + l = 0,$$

the direction of the asymptotes is given by $bxy^2 + cx^2y = 0 = xy(by + cx)$. Those,

$$(B) \equiv bx + e = 0 \quad \text{and} \quad (S) \equiv cy + g = 0,$$

parallel to $x = 0$ and $y = 0$ are readily determined from the corresponding fictitious equations

determined by taking $AB = -e/b$ and erecting at B a line parallel to the ordinates, another by taking $BS = -g/c$ and drawing through S a line parallel to the base (for they will be parallel to an asymptote), while the third will be traced by the point S on assuming $BS = -(c/b)x + ec/b^2 - f/b + g/c$.[97] Finally, if one or other of the terms b and c—say c—should also be lacking, the figure will be either a parabolic hyperbola or some hyperbolism whose determination we have adequately explained above.[98]

$bxy^2 + ey^2 = 0$, $cx^2y + gx^2 = 0$ respectively. The third, $by + cx = ce/b - f + bg/c$, is determined by substituting $y = -(c/b)x + v$ in the cubic and equating the resulting coefficient of x^2 to zero. (Newton's computations for this are reproduced in Appendix 8 below.)

(98) A note of tiredness creeps into the text. Newton has had enough!

APPENDIX. VARIANT DRAFTS OF PORTIONS OF THE REVISED TREATISE ON CUBICS.[1]

From the originals in the University Library, Cambridge and in private possession

[1][2]

Deniꝗ figuræ compositæ imitantur Con. sect.[3] non tantum respectu diametrorū[4] sed relatione ad alias quasvis rectas. Sit enim in fig 1 *AKB* quævis Conica sectio, *BE KG* rectæ parallelæ secantes conicam sect. in punctis *B E G K*, sitꝗ *AB* tertia quævis recta positione data quæ secat figuram in punctis *A*, *B*. [5]Bisectis *BE KG* in *Q* et *S* age *QS* occurrentem ipsi *AB* in *V*, sitꝗ

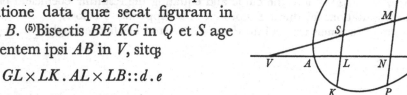

Fig 1

$$GL \times LK . AL \times LB :: d . e$$

& acta quavis ordinata $NO = x$ erit

$$xx - 2MN \times x + \frac{d}{e} AN \times NB = 0. \quad \text{vel} \quad x = MN \pm \sqrt{MN^q - \frac{d}{e} AN \times NB}$$

æquatio ad Conicam Sectionem ex quinꝗ datis punctis *A, B, E, G, K* eruta.[6] Et eodem modo in fig 2 si dentur positione tres rectæ *EBH FCI GDK* secantes curvam in novem punctis *E B H F C I G D K* et quarta *BC* ducatur secans curvam in alio puncto *A*. Age diametrum *RVQ* ad quam rectæ parallelæ *EBH*,

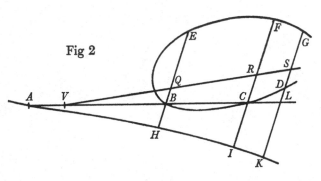

Fig 2

---

(1) These preliminary versions of and computations for §3 preceding (notably for the concluding section on discriminating the sixteen species of general cubic from its unreduced defining equation) are reproduced for their intrinsic interest and for the light they cast on the development of Newton's exposition into its final form in the finished tract.

(2) ULC. Add. 3961.4: 19ᵛ, a first version of the equivalent passage in the preceding treatise on cubics (see §3: note (19) above).

(3) 'sicut Conicæ sectiones exprimi possunt' is cancelled.

(4) The equivalent 'quoad diametro[s]' was first written.

(5) An opening phrase 'Hinc *AB* ordinētur rectæ parallelæ *BE* et quævis *LGK* in dato angulo, et' is deleted.

(6) Evidently *M* will be the mid-point of the chord *OP*, while by Apollonius, *Conics* III, 17, there is $ON \times NP : AN \times NB = GL \times LK : AL \times LB = d:e$, so that the segments $NO = x_1$ and $NP = -x_2$ are determined by $x_1 + x_2 = NO - NP = 2NM$ and $x_1 \times x_2 = (d/e) AN \times NB$.

FCI ordinatim applicantur facto sc. $HQ = \dfrac{HB + HE}{3}$, et $IR = \dfrac{IC + IF}{3}$. Præterea rectangula $BE \times BH$, et $CF \times CI$, et $LG \times LK + LD \times LK - LD \times LG$ applicentur ad datam aliquam lineam β et latera tria capiantur[7]

---

[2][8]

Proposita quacunq curva *EC* norunt mathematici relationem inter ejus Abscissam[9] et Ordinatim Applicatam invenire. Sit ejus Abscissa $AB = z$ et Ordinata $BC = v$ et sit

$av^3 + bzvv + czzv + dz^3 + evv$

$\quad + fzv + gzz + hv + kz + l = 0$

æquatio qua relatio inter Abscissam et Ordinatam definitur existentibus a, b, c, d, e &c numeris quibuscunq datis cum signis suis $+$ et $-$.

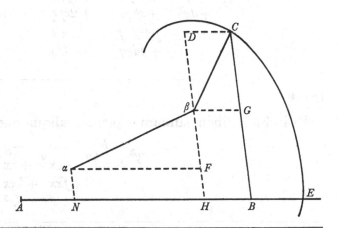

---

(7) The manuscript breaks off at this point, but we should understand some equivalent to '…et latera tria prodeuntia capiantur esse *k*, *l* et *m*'. In these terms if the point *A* is taken to be the origin of a system of Cartesian coordinates in which *AB* is the base and the ordinate direction is parallel to *BE*, then the defining equation of the cubic may be set as

$$y^3 - (ax + b)y^2 + (cx^2 + dx + e)y - (fx^3 + gx^2 + hx) = 0.$$

When $x = AB$, $y = 0$, *BE* or $-BH$, so that $3BQ = BE - BH = a \times AB + b$; similarly, when $x = AC$, $y = 0$, *CF* or $-CI$, and so $3CR = CF - CI = a \times AC + b$: whence

$$a = -3(BQ - CR)/BC, \quad b = 3(BQ \times AC - CR \times AB)/BC.$$

Again, when $x = AL$, $y = LD$, *LG* or $-LK$ and so the coefficients c, d, e may be determined from the equalities

$$BE \times -BH = \beta k = c \times AB^2 + d \times AB + e, \quad CF \times -CI = \beta l = c \times AC^2 + d \times AC + e$$

and $LD \times LG + LD \times -LK + LG \times -LK = \beta m = c \times AL^2 + d \times AL + e$. Finally the equalities $fx^3 + gx^2 + hx = 0$, $x = AB$, *AC*, and $f \times AL^3 + g \times AL^2 + h \times AL = LD \times LG \times -LK$ yield $-g/f = AB + AC$, $h/f = AC^2$ and lastly $f = LD \times LG \times -LK/AL \times BL \times CL$ (compare §3: note (18) above). The equation of the cubic through the nine points *B*, *C*, *D*, *E*, *F*, *G*, *H*, *I*, *K* is then found by substituting for a, b, c, d, e, f, g, h their values found above.

(8) ULC. Add. 3965.18: 705^v. This unfinished fragment, evidently a revise of II: 10–12, may (compare §3: note (30) above) represent Newton's first attempt at an improved analytical reduction of the general cubic to standard form as a preliminary to inserting it in the preceding treatise. The handwriting style is consistent with a date of c. 1678–9. There is no figure accompanying the fragment: evidently that on II: 10/12 is to be understood and accordingly a simplified version of the latter diagram is here repeated.

(9) 'quamcunq' is cancelled.

Nam hæc comprehendit curvas omnes secundi generis. Sit insuper $\alpha\beta = $ Abscissa nova[10] et βC ordinata nova[10] ejusdem Curvæ. Compleantur parallelogramma βGBH, βGCD, αNHF & sint $AN = r$, $N\alpha = s$, $AF = x$,[11] $CG = y$, $F\beta = px$. $CB = y + px + s = v$. $\beta G = qy$ & $BA = qy + x + r = z$, et in æquatione superiore substituantur ipsorum v et z valores inventi, et orietur æquatio nova[12]

$$
\begin{array}{lllll}
a & +3ap & +3app & +ap^3 & +3as \\
{}_{+bq}y^3 & {}_{+b}xyy & {}_{+2bp}xxy & {}_{+bpp}x^3 & {}_{+2bqs}y^2 \\
+cqq & +2bpq & +bppq & +cp & +cqqs \\
+dq^3 & +cpqq & +2cpq & +d & +br \\
+2cq & +c & & & +2cqr \\
+3dqq & +3dq & & & +3dqqr \quad {}^{(13)}
\end{array}
$$

[3][14]

Et in hunc finem sumamus plenam aliquam æquationem

$$
\begin{array}{llll}
ay^3 & +b & +d & +g \\
& {}_{+cx}yy & {}_{+ex}y & {}_{+hx} \\
& +fxx & +kxx & \\
& & & +lx^3
\end{array} = 0.
$$

(10) We would normally expect the datives 'Abscissæ novæ' and 'ordinatæ novæ' after ' = ' (that is, 'æqualis'). The manuscript shows many other signs of being hurriedly composed. Compare the next note.

(11) This should be '$\alpha F = x$' (as on II: 12).

(12) Understand 'exprimens relationem inter αF et GC'.

(13) Newton breaks off his transcript of the 84 term equation on II: 12 in the middle of its fifth column.

(14) ULC. Add. 3961.4: 20$^\mathrm{r}$–21$^\mathrm{r}$, a first version of the analytical reduction of the general cubic equation on pages 362–8 above (see §3: note (30)). It will be evident that, while his geometrical interpretation of the following argument is seriously in error, Newton's analytical reduction itself is unimpeachable. By the transformation $AD = z = x + py$, $p = BD/BC$ constant, of the coordinates $AB = x$, $BC = y$ of $C(x, y)$ he is able without effort to simplify the general equation of the cubic into the form

$$(b + cz)y^2 + (d + ez + fz^2)y + (g + hz + kz^2 + lz^3) = 0$$

but fails to produce the second equation $BD = w = p'y$ ($p'^2 = p^2 + 2p\cos A\hat{B}C + 1$), which completes the transform $C(x, y) \to C(z, w)$. Thereafter he compounds this basic mistake, introducing the second 'untied' parameter $CF = v = y + qz$, $q = BF/AD$ constant, and then from the variables z and v defining the new pair $s = (\beta/\gamma)(z + m)$, $t = (\beta/\delta)(v + n)$ where $DG = m$, $FH = n$: these he would have to be the coordinates AK, KC of $C(s, t)$. But this clearly cannot be, for the origin $A(0, 0)$ in $C(x, y)$ determines the corresponding parameters z, v likewise to be zero, so that A must be the point $((\beta/\gamma)m, (\beta/\delta)n)$ in the new coordinate system in which C is the point (s, t). When Newton came to see this inconsistency, he realized that his analytical argument might be retained with little change, and in his revise (ff. 21$^\mathrm{r}$–21$^\mathrm{v}$, reproduced in §3 preceding) he added a 'Note. These reductions are to be adapted to this figure after this method' before proceeding to elaborate the correct argument.

ubi x esto basis AB et y incedens BC applicata in dato angulo ABC. age quamvis CD
occurrentem AB in D et sit $AD = z$ et $1 . p$::
$BC . BD$, et erit AB sive $x = z - py$. Pro x itaqȝ
in æquatione scribe $z[-]py$ et prodeuntes
terminos ubi y est trium dimensionum[,]
hoc est terminos $-lp^3y^3 + fppy^3 - cpy^3 + ay^3$,
per y^3 divisos pone æquales nihilo,
et constructio ejus æquationis quomodocunqȝ instituta[15] dabit p. Quo invento
scribe denuò $z[-]py$ pro x et terminus
cubicus y^3 evanescet, linea DC evadente

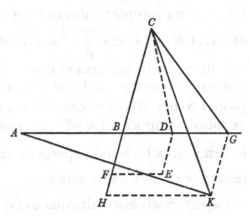

parallela Asymptoto curvæ, si modo curva Asymptoton habeat.[16] Sit itaqȝ

prodiens æquatio $\begin{matrix} b \\ +cz \end{matrix}yy \begin{matrix} +d \\ +ez \\ +fzz \end{matrix}y \begin{matrix} +g \\ +hz \\ +kzz \\ +lz^3 \end{matrix}=0$ retentis literis b, c, d &c licet mutato

significatu.[17] Præterea in BC producto sumpto quovis puncto F comple
pg͞rmum[18] $DBFE$[,] junge AE, et dicto $CF = v$ positoqȝ[19] $2c . f$:: $AD . BF$ erit BC

sive $y = v - \dfrac{f}{2c}z$. Scribe ergo $v - \dfrac{f}{2c}z$ pro y et evanescet terminus multiplicatus

per vzz. Sit itaqȝ æquatio prodiens $\begin{matrix} b \\ +cz \end{matrix}vv \begin{matrix} +d \\ +ez \\ +kzz \\ +lz^3 \end{matrix}v \begin{matrix} +g \\ +hz \end{matrix}=0$, retentis denuo literis

sub mutato significatu, produc AD ad G ut sit $DG = \dfrac{b}{c}$, & CB ad H ut sit

$FH = \dfrac{e}{2c}$,[20] comple paralle[lo]grammum $HBGK$, junge AK, CK, pone $AK = s$,

(15) That is, by any geometrical or algebraical method.

(16) A strange observation: all cubic equations have at least one real root and so all cubic
curves have at least one real asymptote (which may, it is true,—in the case of the divergent
parabolas and the Wallis cubic—be at infinity).

(17) Newton first continued: 'Agatur recta quævis AC occurrens BC in E et dicto $EC = v$,
positoqȝ $1 . q$::$x . EB$, erit BC sive $y = v + qx$. Quare scribe $v + qx$ [pro y...]'.

(18) 'parallelogrammum.'

(19) A cancelled first sequel reads '$1 . q$::$AD . DE$ erit BC sive $y = v + qz$. quare pro y scribe
$v + qz$ et emergentes terminos ubi reperitur vzz nempe $2cqvzz + fvzz$ divisos per vzz pone $[= 0]$'.

(20) Newton first wrote '...ut sit $FH = \dfrac{d}{2b}$' and then went on: 'et positis $AG = s$ et $CH = t$

scribe $s - \dfrac{c}{b}$ pro z et $t - \dfrac{d}{2b}$ pro y et emerget æquatio in qua termini multiplicati per $[st]$ et $[tt]$
(non per $[stt]$) evanescent. Sit ista...'.

$CK = t$, sitcp (propter datos angulos triangulorum AGK, CHK[)] $AK . AG :: \beta . \gamma$ et $CK . CH :: \beta . \delta$. erit $\frac{\gamma s}{\beta} - \frac{b}{c} = AD$, et $\frac{\delta}{\beta} t - \frac{e}{2c} = CF$, et scriptis hisce pro z et v orietur æquatio relationem exprime[ns] inter basem positione datam AK et ordinatim incedentem KC[21] in qua æquatione termini multiplicati per st et tt (non dico per stt) evanescent. Sit ista $cstt + dt + g + hs + kss + ls^3 = 0$. Et hæc erit[22] ex formis simplicissimis ad quas æquationes hujus generis reduci possunt, nisi forte aliqui ex his terminis per se evanescant, aut supra ubi posuimus $y = v - \frac{f}{2c} z$, terminus c nullus esse contingat.

In posteriori casu[23] sit æquatio $byy \begin{array}{l} +d \\ +ez \end{array} y \begin{array}{l} +g \\ +hz \end{array} = 0$. et ponatur $f . l :: AD . DE$
$$+fzz \quad +kzz$$
$$+lz^3$$

vel BF. et erit CB sive $y = v - \frac{lz}{f}$. Quare scribe $v - \frac{lz}{f}$ pro y et emerget æquatio in qua cubica dimensio ipsius z evanescet. Sit ista $bvv \begin{array}{l} +d \\ +ez \end{array} v \begin{array}{l} +g \\ +hz \end{array} = 0$. Fac $DG = \frac{e}{2f}$,
$$+fzz \quad +kzz$$

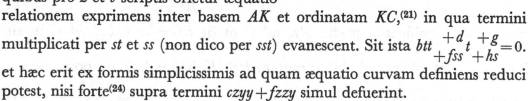

$FH = \frac{k}{f}$ et erit $\frac{\gamma}{\beta} s - \frac{e}{2f} = AD$, $\frac{\delta}{\beta} t - \frac{k}{f} = CF_{[,]}$ quibus pro z et v scriptis orietur æquatio relationem exprimens inter basem AK et ordinatam KC,[21] in qua termini multiplicati per st et ss (non dico per sst) evanescent. Sit ista $btt \begin{array}{l} +d \\ +fss \end{array} t \begin{array}{l} +g \\ +hs \end{array} = 0$. et hæc erit ex formis simplicissimis ad quam æquatio curvam definiens reduci potest, nisi forte[24] supra termini $czyy + fzzy$ simul defuerint.

---

(21) Not so, for the angle $A\hat{K}C$ will vary with z. Compare note (14) above.

(22) 'una' is deleted.

(23) On taking $c = 0$, that is.

(24) Newton first continued 'aliquos ex his terminis deesse contingat'. On writing this reduced form as $ftss + hs + g + dt + btt = 0$ it is readily seen to be merely a particular case of the preceding one.

(25) When $c = f = 0$, that is.

(26) Newton first went on erroneously 'sitcp $DE = GK$ punctis F et H, lineis $FE\ HK$ coincidentibus. Fac etiam $\frac{d}{2b} = DG$ et punctis D ac G coincidentibus erit $\frac{\gamma}{\beta} s = AD$ ac $v - \frac{e}{2b} z = CB$, quibus pro z et y scriptis orietur æquatio exprimens relationem inter basem AK et ordinatam $K[C]$'.

(27) '(non dico per [vv])' is cancelled.

Quo casu[25] sit æquatio $byy \begin{smallmatrix} +d \\ +ez \end{smallmatrix} y \begin{smallmatrix} +g \\ +hz \end{smallmatrix} =0$. Et si terminus *byy* non defuerit,
$$+kzz$$
$$+lz^3$$

fac $2b.e::AD.DE$ vel BF,[26] et erit $v-\dfrac{e}{2b}[z]=CB$, quo pro *y* scripto orietur

æquatio in qua terminus multiplicatus per zv[27] evanescet. Sit ista

$$bvv+dv+g+hz+kzz+lz^3=0,$$

[28]fac $DG=\dfrac{k}{3l}$ et $FH=\dfrac{d}{2b}$ et erit $\dfrac{\gamma}{\beta}s-\dfrac{k}{3l}=AD$, et $\dfrac{\delta}{\beta}t-\dfrac{d}{2b}=CF$. quibus pro *z* et *v*

scriptis orietur æquatio exprimens relationem inter basem *AK* et ordinatam
KC, in qua termini multiplicati per *ss* et *t* (non dico s^3 et *tt*) evanescent, quæcɞ
non erit ultra reducibilis. Sit ista $btt+g+hs+ls^3=0$.

Quod si terminus *byy* defuerit et terminus *ezy* non defuerit,[29] fac $DG=\dfrac{d}{e}$, et

erit $AD=\dfrac{\gamma}{\beta}s-\dfrac{d}{e}$. Quo scripto pro *z* emerget æquatio in qua terminus multi-

plicatus per *y*[30] deerit. Sit ista $esy+g+hs+kss+ls^3=0$. et fac $e.k::A[D].BF$,

et cape $FH=\dfrac{h}{e}$ et erit $\dfrac{\delta}{\beta}t-\dfrac{h}{e}-\dfrac{ks}{e}=BC_{[,]}$ quo scripto pro *y* emerget æquatio non

amplius reducibilis exprimens relationem inter basem *AK* et ordinatam *CK*, in
qua termini multiplicati per *s* et *ss* deerunt. Sit ista $est+g+ls^3=0$.

Sin deniɋ terminus uterɋ *byy* et *ezy* defuerit.[31] Fac $DG=\dfrac{k}{3l}$ et erit

$\dfrac{\gamma}{\beta}s-\dfrac{k}{3l}=AD$, quo scripto pro *z* emerget æquatio in qua terminus multiplicatus

per *ss* deerit. Sit ista $dy+g+hs+ls^3=0$ et fac $d.h::AK.BH$ et cape $FH=\dfrac{g}{d}$ et

erit $BC=\dfrac{\delta}{\beta}t-\dfrac{hs}{d}-\dfrac{g}{d}$. Quo scripto pro *y* orietur æquatio simplex $\dfrac{\delta d}{\beta}t+ls^3=0$,

exprimens relationem inter basem AK[32] et ordinatam CK.[33]

---

(28) Newton first continued in sequel 'et posito insuper...'.
(29) That is, when $b=c=f=0$.
(30) Understand '(non dico per *sy*)'.
(31) On taking $b=c=e=f=0$.
(32) Newton has corrected this in the manuscript to read '*GK*'. Was this the point at which
he saw for the first time the fallacy in his preceding geometrical argument?
(33) A final phrase 'simplicissima quæ potest' is cancelled. To recapitulate, in the terms of
note (14) above the preliminary transformation $x=z-py$ reduces the cubic equation to the
form $(b+cz)y^2+(d+ez+fz^2)y+(g+hz+kz^2+lz^3)=0$, which is then further reduced to

[4]$^{(34)}$

[$av^3 + bzvv + czzv + dz^3 + evv + fzv + gzz + hv + kz + l = 0$.   Cape $z = r - pqs$,

$v = qs^{(35)}$ et erit]
$$\begin{aligned}
&aq^3s^3 \\
&+ bqqrss - bpq^3s^3 \\
&+ cqsrr \quad - 2cpqqssr + cppq^3s^3 \\
&+ dr^3 \quad\;\; - 3pqdrrs \quad + 3ppqqdrss - [d]p^3q^3s^3[\ldots = 0.]
\end{aligned}$$

[Pone $a - bp + cpp - dp^3 = 0.^{(36)}$ fit $\begin{array}{l} b \\ -2cp \\ +3dpp \end{array} qqrss \begin{array}{l} +c \\ -3dp \end{array} qrrs + dr^3 = 0.$

hoc est $\dfrac{r}{d} \times \overline{\dfrac{cq - 3dpq}{2} s + dr}\Big|^2 = 0$ posito quod]

$$d \times bqq - 2cpqq + 3dppqq = \overline{\dfrac{cq - 3dpq}{2}}\Big|^2. \quad ^{(37)}$$

[id est] $2d.\; c - 3dp.\; b - 2cp + 3dpp \;\div\!\!\div$

[Simili argumento prodit] $2a.\,r.\,n\!\div\!\!\div^{(38)}$

---

standard canonical form by means of the transform $\begin{Bmatrix} z = (\gamma/\beta)s - m \\ v = (\delta/\beta)t - n \end{Bmatrix}$ where $y = v - qz$: here specifically (when $c \neq 0$) $m = b/c$, $n = e/2c$, $q = f/2c$; (when $c = 0$) $m = e/2f$, $n = k/f$, $q = l/f$; (when $c = f = 0$) $m = k/3l$, $n = d/2b$, $q = e/2b$; (when $b = c = f = 0$) $m = d/e$, $n = h/e$, $q = k\beta/e\gamma$; and (when $b = c = e = f = 0$) $m = k/3l$, $n = g/d$, $q = h\beta/d\gamma$.

(34) From the original in private possession (see §2: note (2) above). Newton seeks the condition for the general cubic to reduce to a conic hyperbolism (§3: note (69) above).

(35) Newton makes a preliminary change of axes, using his earlier transformation
$$\begin{Bmatrix} z = r - pqs \\ v = qs \end{Bmatrix};$$
see §3: note (33) above. Since he is interested only in the terms of cubic power in the variables r and s, the remaining ones (r^2, rs, s^2, r, s and constant term) are discarded in his further computation.

(36) This cubic equation determines, of course, the direction of the three asymptotes.

(37) Read '$\overline{\dfrac{cq - 3dpq}{2}}\Big|^2$' and in the sequel '$4d.c - 3dp.b - 2cp + 3dpp \div\!\!\div$'. (This simple slip, never noticed by Newton, is carried through all subsequent drafts on the discrimination of a cubic's genus.) When $4d(b - 2cp + 3dp^2) = (c - 3dp)^2$, that is, $3d^2p^2 - 2ccdp + 4bd - c^2 = 0$ (where p is a root of $a - bp + cp^2 - dp^3 = 0$), the cubic evidently reduces to
$$dr(As + r)^2 + Bs^2 + Crs + Dr^2 + Es + Fr + G = 0,$$
which is in turn reducible to the canonical form $xy^2 = \alpha x^2 + \beta y + \gamma x + \delta$ of a hyperbolic parabola (or, when $B \equiv (e - fp + gp^2)q^2 = 0$, to the conic hyperbolism $xy^2 = \beta'y + \gamma'x + \delta'$) by the second transformation to new axes $\begin{Bmatrix} r = mx + \mu \\ s = y - (m/A)x + \nu \end{Bmatrix}$. Compare §3: note (88) above.

[5]$^{(39)}$

$$\begin{array}{r}
ax^3 + bxx \;\; + cx \;\; + d \\
+ eyxx + fyx + gy \\
+ hyyx + kyy \\
+ ly^3
\end{array} = 0.$$

$ax^3 + eyxx + hyyx + ly^3 = 0$ [id est $an^3 + enn + hn + l = 0$] si $\dfrac{x}{y} = n$. $^{(40)}$

$$([x=]\, ny - \frac{bnn + fn + k \maltese \dfrac{cn+g}{y}}{3ann + 2en + h \maltese \dfrac{2bn}{y} + \dfrac{f}{y} + \dfrac{3anz}{y} + \dfrac{ez}{y}}. \,^{(41)}$$

[Pone] $ny + z = x.$

$$\left|\begin{array}{l}
an^3y^3 + 3annyyz + 3anyzz + az^3 \\
bnnyy \;\;\;\; + 2bnyz \\
cny \\
enny^3 + 2enyyz \;\; + eyzz \\
fnyy \;\;\;\;\;\; + fyz \\
+ gy \\
hny^3 \;\;\;\;\; + hyyz \\
kyy \\
ly^3
\end{array}\right.$$

$-\dfrac{r}{t} + v = z.$ $\qquad \left| tyv - a\dfrac{r^3}{t^3} + \dfrac{3arr}{tt}v[=0]\; \left(v = \dfrac{ar^3}{yt^4}\right.\right. \,^{(42)}$

---

(38) In line with the preceding note this should be '$4a.r.n \div$'. This late addition was made when Newton came to draft Appendix 6 following, where the general defining equation $ay^3 + bxy^2 + cx^2y + dx^3 + ey^2 \ldots + l = 0$ is first reduced by the substitution

$$y = zx + v \;\; (\text{with } az^3 + bz^2 + cz + d = 0),$$

so that the derived equation becomes $av^3 + rxv^2 + nx^2v + ev^2 \ldots + l = 0$ on taking $3az + b = r$ and $3az^2 + 2bz + c = n$. This assumes the form $av(v + (r/2a)x)^2 + ev^2 \ldots + l = 0$ desired by Newton when $r^2 = 4an$.

(39) ULC. Add. 3961.1: 22$^\text{v}$. Newton makes a first sustained attempt at discriminating the cubic's genus from its unreduced Cartesian equation.

(40) The roots n determine the slope of the three asymptotes (two of which may not be real). In sequel, for each root n Newton sets $x = ny + z$, finding

$$z = -(bn^2 + fn + k + O(y^{-1}))/(3an^2 + 2en + h + O(y^{-1})),$$

whence $x = ny - (bn^2 + fn + k)/(3an^2 + 2en + h)$ will be the equation of the corresponding asymptote.

(41) Newton constructs the root x of the preceding cubic as an infinite series in powers of y^{-1}, rounding off his result to $O(y^{-2})$. (For the method compare III: 54 ff.) The terms '$bzz + cz + d$' are omitted from the computation because they do not affect the value of x to $O(y^{-2})$. The elaborate '$\maltese$'s are merely variant signs of addition used (instead of brackets) to distinguish groups of terms.

[Sit rursus $ax^3 \begin{array}{l} +b \\ +ey \\ +hyy \end{array} xx \begin{array}{l} +c \\ +fy \\ +kyy \end{array} x \begin{array}{l} +d \\ +gy \\ +ly^3 \end{array} = 0.$] $n =$ roots of this æquat.

$an^3 + enn + hn + l = 0.$ $bnn + fn + k = p.$

$3ann + 2en + h = q.$ $cn + g = r.$

$2bn + f + 3an\dfrac{p}{q} + \dfrac{ep}{q} = s.^{(43)}$ [Cape] $AB = 1.\ BC = n.$

$CD = \dfrac{p}{q}.^{(44)}$ Age $DE \parallel AC$ et erit ED Asymptotos.

vel si $q = 0$ erit $AC \parallel$ diametro Parabolæ. Cape

$AM = y$ et erit $MN^{(45)} = ny - \dfrac{py + r}{qy + s}$ & punctum

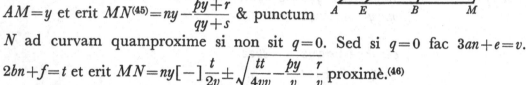

N ad curvam quamproxime si non sit $q = 0$. Sed si $q = 0$ fac $3an + e = v$.

$2bn + f = t$ et erit $MN = ny[-]\dfrac{t}{2v} \pm \sqrt{\dfrac{tt}{4vv} - \dfrac{py}{v} - \dfrac{r}{v}}$ proximè.$^{(46)}$

Si n habet duas æqu. rad.$^{(47)}$ fac

$$MN = ny[-]\frac{qy + t}{2v} \pm \sqrt{\frac{qqyy + 2tqy + tt}{4vv} \ \frac{-py - r}{v}}.^{(48)}$$

sed tunc erit $q = 0$, ergo sume n pro rad. dupl.$^{(49)}$ fac

$$MN = ny[-]\frac{t}{2v} \pm \sqrt{\frac{tt - 4rv - [4]pyv}{4vv}},$$

---

(42) At this point Newton breaks off and starts afresh. It is assumed, much as in the sequel, that $r = cn + g$, $t = 2bn + f$ to $O(y^{-1})$, whence, since $ry + tyz = 0$ is the 'fictitious' equation at the second stage, to that order $z = v - r/t$. The last term is computed by substituting this in $az^3 + bz^2 + \ldots + tvy \ldots = 0$ and ignoring all but the significant terms $-a(r^3/t^3)$, $+tvy$ in it.

(43) This should be '$2bn + f - 3an\dfrac{p}{q} - e\dfrac{p}{q} = s$'.

(44) Since $n = \lim\limits_{x,\,y \to \infty} \left(\dfrac{x}{y}\right)$, on taking $AM = y$ and $MN(\parallel BD) = x$ the asymptote NDE in the accompanying figure will have the equation $x = ny - p/q$.

(45) That is, x. If n satisfies $an^3 + en^2 + hn + l = 0$, then

$$qy^2 z + vyz^2 + az^3 + py^2 + tyz + bz^2 + ry + cz + d = 0$$

on setting $3an + e = v$, $2bn + f = t$, whence $y((qy + t)z + py + r) \approx 0$.

(46) When $q = 0$ (that is, when $an^3 + en^2 + hn + l = 0$ has a double root n), then

$$y(vz^2 + tz + py + r) \approx 0 \quad \text{and so} \quad z \approx -t/2v \pm \sqrt{[(t/2v)^2 - (py + r)/v]}.$$

(47) 'æquales radices.'

(48) $MN = ny + z$, where $vz^2 + (qy + t)z + py + r = 0$, $q = 0$!

(49) 'radice duplici.'

(50) If $p = t^2 - 4rv = 0$, then the term under the radical sign vanishes, leaving two coincident values $ny - t/2v = ny - 2r/t$ for MN.

et habebis duos parallelos Asymptotos si $p=0$, vel magnitudinem & situm parabolæ si p non sit $=0$, aut duplicem Asymptoton si $tt=4rv$. & $p=0$.[50] Dein sume n pro rad. simpl.[51] et inde cognosces utrum figura sit bifida necne.

Si tres n æquales sunt Figura erit vel Parabolismus Hyperbolæ[52] vel Parabola divergens vel deniqʒ Parabola cubica, et erit $q=v=0$. Si sit præterea $p=0$ erit Parabolismus Hyperb.[52] et $MN = ny - \dfrac{r}{t}$ dabit Asymptoton et $MN = ny \pm \sqrt{\dfrac{-ty}{a}}$ crura parabolica & $MN = ny - \dfrac{r}{t} + \dfrac{ar^3}{t^4 y}$. [53]

---

[6][54]

$$[y^3 + bxyy + cxxy + dx^3 + eyy + fxy + gxx + hy + kx + l = 0.]$$

$$y = zx + v.$$

$$
\begin{aligned}
& z^3x^3 + 3zzxxv + 3zxvv + v^3 \\
& + bzzx^3 + 2bzxxv \; + bxvv \\
& \quad + czx^3 \quad + cxxv \\
& \quad + dx^3 \\
& + ezzxx \; + 2ezxv \quad + evv \\
& + fzxx \quad\; + fxv \\
& + gxx \\
& + hzx \quad\;\; + hv \\
& + kx \\
& + l
\end{aligned}
\quad [=0.]
$$

---

(51) 'radice simplici.' The criterion for a cubic to have a diameter is developed in Appendix 6 following.

(52) The Cartesian trident in Newton's present terminology; see §3: note (73) above.

(53) When all three roots of $an^3 + en^2 + hn + l = 0$ are equal, then $q = v = 0$ and the transformed cubic becomes $az^3 + py^2 + tyz + bz^2 + ry + cz + d = 0$. In general this may (see §3: notes (43) and (44) above) be reduced by a transformation of axes to the divergent parabola; while if $p = 0$ it may (see §3: note (46) above) be reduced to the trident $fxy + dx^3 + l = 0$; and if $p = t = 0$ it is (see §3: note (47) above) reducible to the Wallisian cubic parabola. From a geometrical viewpoint in all three cases the cubic's asymptotes will coincide, though only in the case of the trident will they be at a finite distance from the origin. When $p = 0$ the root z of the equation $az^3 + tyz + bz^2 + ry + cz + d = 0$ is readily determined by Newton's method (using the fictitious equations $tyz + ry = 0$ and then $az^3 + tyz + ry = 0$) to be

$$z = -r/t + ar^3/t^4y \dots,$$

whence the corresponding asymptote is $(MN\text{ or})\, x = ny - r/t$ (at infinity when $t = 0$); while the parabolic branches are given by the 'fictitious' equation $az^3 + tyz = 0$ as

$$(MN\text{ or})\; x = ny \pm \sqrt{[-ty/a]}, \quad \text{that is,} \quad a(x-ny)^2 + ty = 0.$$

(54) Edited extracts from Newton's main computations for his method of discriminating a cubic's genus and the first verbal draft to which they led, reproduced from the original worksheets in Cambridge (ULC. Add. 3961.1: 22r/19 bisr–20v) and in private possession (see §2: note (2) above).

[hoc est positis $ezz+fz+g=m$. $3zz+2bz+c=n$. $hz+k=p$. $2ez+f=q$. $3z+b=r$.]

$$
\begin{array}{l}
mxx \quad +nxxv \quad +rxvv+v^3 \\
+px \quad\;\; +qxv \quad +evv \\
+l \quad\quad +hv
\end{array}
$$

$v=-\dfrac{m}{n}+t.$

$$
\begin{array}{l}
px \quad +nxxt \quad +rxtt+t^3 \\
+l \quad\;\; +qxt \quad +ett \\
-\dfrac{mq}{n}x \quad\;\; +ht-3\dfrac{m}{n}tt \\[4pt]
-\dfrac{mh}{n}-\dfrac{2rmx}{n}t \\[4pt]
+\dfrac{rmm}{nn}x-\dfrac{2em}{n}t \\[4pt]
+e\dfrac{mm}{nn}+3\dfrac{mm}{nn}t \\[4pt]
-\dfrac{m^3}{n^3}
\end{array}
$$

[Prodit] $t=\dfrac{-pnn+qmn-rmm}{n^3x}.$ [55] vel si hoc$=0$ erit

$$t=\dfrac{-ln^3+hmnn-emmn+m^3}{n^4xx+qn^3x-2rmnnx}.\quad{}^{[56]}$$

Si $n=0$,[57] erit [$mxx+rxvv=0$ proxime. Fac ergo]

$v=\sqrt{\dfrac{-mx}{r}}+t.$

$$
\begin{array}{lll}
px+qx\sqrt{}) & +qxt & +rxtt+t^3 \\
+l\;+h\sqrt{}) & +ht & +ett \\
-\dfrac{emx}{r} & +2rx\sqrt{})\,t+3\sqrt{})\,tt \\[6pt]
-\dfrac{mx}{r}\sqrt{}) & [+]2e\sqrt{})\,t \\[6pt]
& -3\dfrac{mx}{r}t
\end{array}
$$

(55) To $O(x^{-2})$, that is, on selecting the appropriate fictitious equation

$$x(p+nxt-qm/n+rm^2/n^2) = 0.$$

(56) Here the corresponding fictitious equation is chosen to be

$$l-hm/n+em^2/n^2-m^3/n^3+nx^2t+(q-2rm/n)\,xt = 0.$$

However, since x is supposed to be indefinitely great, the term $(q-2rm/n)\,xt$ is superfluous and we may therefore write $t = (m^3-em^2n+hmn^2-ln^3)/n^4x^2\,[+O(x^{-3})]$ simply.

(57) On taking $z^3+bz^2+cz+d = 0$ to have a double root.

[id est] $t = \dfrac{m}{2rr} - \dfrac{q}{2r}$. $^{(58)}$ Vel [posito $\begin{aligned} mxx + qxv + rxvv + v^3 \\ {}+ px \end{aligned} = 0$ proxime fit]

$$v = -\frac{q}{2r} \pm \sqrt{\frac{qq - 4mrx - 4pr - \dfrac{v^3}{x}}{4rr}}.$$

Si $m = 0 = n$ erit $[px + qxv + rxvv = 0$ proxime. Cape ergo]

$$v = -\frac{q}{2r} \pm \sqrt{\frac{qq - 4pr}{4rr}} \quad ([\text{puta}]\, s) + t.$$

$$[v = s + t.] \qquad \left| \begin{array}{l} l \;\; + qxt + rxtt + t^3 \\ {}+ 2rxst \\ {}+ hs \;\;\; + ht \;\; + ett \\ {}+ ess \;\; + 2est + 3stt \\ {}+ s^3 \;\;\; + 3sst \end{array} \right.$$

[prodit] $t = -\dfrac{l + hs + ess + s^3}{qx + 2rxs}$. $^{(59)}$ [At si $qq = 4pr$ adeoȝ $q + 2rs = 0$ erit]

$$t = \sqrt{\frac{-l - hs - ess - s^3}{rx}}.\,^{(60)} \quad [\text{Hoc est } \tfrac{1}{2}tt =] \; \frac{\dfrac{q^3}{8r^3} - \dfrac{eqq}{4rr} + \dfrac{hq}{2r} - l}{2rx} = \frac{\dfrac{pq}{2rr} - \dfrac{ep}{r} + \dfrac{hq}{2r} - l}{2rx}.\,^{(61)}$$

[Si $m = n = r = 0$ erit] $px + qxv + v^3 = 0$. [Cape $px + qxv = 0$ proxime seu

---

(58) The fictitious equation is here $(q - m/r + 2rt)\, x\sqrt{[-(m/r)\, x]} = 0$.

(59) Since $l + hs + es^2 + s^3 + (q + 2rs)\, xt[+ O(x^{-1})] = 0$ is the fictitious equation for x taken indefinitely large. Because by definition $s = (-q \pm \sqrt{[q^2 - 4pr]})/2r$, the denominator is $\pm \sqrt{[q^2 - 4pr]}\, x$ and the numerator may similarly be evaluated.

(60) Here $l + hs + es^2 + s^3 + rxt^2[+ O(x^{-1})] = 0$, where now $s = -q/2r$.

(61) On making appropriate substitution of p in place of $q^2/4r$. The extraneous factor 2 in the denominators was probably the result of a trivial slip on Newton's part in computing t^2. In first draft Newton first set $rxv^2 + v^3 \approx 0$ and then, on substituting '$v = -rx + t$', found '$\begin{aligned} p \;\; + qt - 2rtt \\ -qrx + rrxt \end{aligned}[= 0$ proxime. unde] $t = \dfrac{q + rrx}{4r} \pm \sqrt{\dfrac{qq + 8pr - 6qrrx + r^4xx}{16rr}}$'. This was manifestly a false trail. At a second attempt Newton rightly isolated the fictitious equation

$$px + qxv + rxv^2 = 0$$

and then made direct substitution of '$-\dfrac{q}{2r} + \sqrt{\dfrac{qq - 4rp}{4rr}} + t = v$' but failed to introduce the terms ev^3, hv and l into his resulting array and, in addition, made two arithmetical errors in his evaluation of the other terms, deriving the incorrect equations

$$`\frac{qq}{2rr}\sqrt{\frown}) - \frac{p}{r}\sqrt{\frown}) - \frac{q^3}{2r^3} + \frac{3pq}{2rr} + \sqrt{\frown})\, xt[= 0]'$$

and, when $\sqrt{\frown}) = 0$, '$-\dfrac{q^3}{2r^3} + \dfrac{3pq}{2rr} + rxtt[= 0]$'. (The symbol $\sqrt{\frown})$ is Newton's contraction for '$\sqrt{qq - 4rp}$'. Compare his similar usage on pages 207–8 above.)

$y =]zx - \dfrac{p}{q}$. [Pone] $-\dfrac{p}{q} + t = v$. [prodit] $\begin{aligned} -\dfrac{p^3}{q^3} + qxt \\ + \dfrac{3pp}{qq}t \end{aligned}$ [= 0].[62]   [Vel cape]

$qxv + v^3$ [= 0 proxime. ponendo] $\sqrt{-qx} + t = v$. [erit æquatio]

$$px \pm qx\sqrt{-qx}\,^{(63)} \;\; \begin{aligned} + qxt \\ - 3qxt \end{aligned} \; [= 0.$$

Unde $t = \dfrac{p}{2q}$ ut et $y =]zx \pm \sqrt{-qx} + \dfrac{p}{2q}$.

[Si deniqȝ $n = r = 0$ erit] $\begin{aligned} mxx + qxv + v^3 \\ + px \end{aligned}$ [= 0. Concipe $mxx + v^3 = 0$ proxime et

prodit $y =]zx - \sqrt[3]{mxx} + \dfrac{q\sqrt[3]{\dfrac{x}{m}}}{3} + \dfrac{9pm + q^3}{9qm}$.[64]

$-\sqrt[3]{mxx} + t = v.$	$\begin{aligned} px \quad\quad\quad + qxt \\ - qx\sqrt[3]{mxx} + 3\sqrt[3]{mmx^4})\, t - 3\sqrt[3]{mxx}\, tt + t^3 \end{aligned}$
$\dfrac{q\sqrt[3]{\dfrac{x}{m}}}{3} + s = t.$	$px + \dfrac{qqx\sqrt[3]{\dfrac{x}{m}}}{3} \quad\quad - qxs + 3\sqrt[3]{mmx^4})\,s$
	$+\dfrac{q^3 x}{9m} \quad\quad\quad +\dfrac{qq\sqrt[3]{\dfrac{xx}{mm}}}{3}s$

Enumeratis harum curvarum speciebus ut revertamur ad determinationem loci solidi: cum problema aliquod de his locis proponitur imprimis[65] de cruribus infinitis prospiciendum est et plaga versus quam serpunt in infinitum ut et Asymptotis crurum Hyperbolicorum. Nam species Curvæ hinc innotescit.

---

(62) This yields $y = zx - p/q + p^3/q^4x + O(x^{-2})$, the equation of the asymptotic hyperbola. Newton proceeds to find the equation of the asymptotic parabola correspondingly.

(63) Understand '$+ qx\sqrt{-qx} - qx\sqrt{-qx}$, that is, two terms which mutually cancel.

(64) This last term should read '$-\tfrac{1}{3}e$' (to the order of $x^{-\frac{1}{3}}$, implicitly). Newton has ignored the effect of the term ev^2 in substituting $v = -m^{\frac{1}{3}}x^{\frac{2}{3}} + \tfrac{1}{3}qm^{-\frac{1}{3}}x^{\frac{1}{3}} + t$ in

$$v^3 + mx^2 + qxv + ev^2 + px + hv + l = 0.$$

The correct fictitious equation at the last stage is $em^{\frac{2}{3}}x^{\frac{4}{3}} + 3m^{\frac{2}{3}}x^{\frac{2}{3}}s = 0$.

(65) Newton first continued 'prospiciendum est de tangentibus infinitis et situs eorum determinandus' and then altered this to 'tangentium infinitarum plaga et positio determinanda est, et ex quo latere tangunt crura infinita Curvæ'.

Nempe si tres sunt[66] Asymptoti Figura erit Hyperbola triformis ubi nullæ
Asymtotων sunt parallelæ, aut Hyperbolismus Hyperbolæ ubi duæ sunt
parallelæ, & Hyperbolismus Parabolæ si duæ coincidunt. Coincidere autem
cognoscas ubi duo habent Curvæ crura ex utroq latere[67] in infinitum serpentia.
Adhæc si unica tantum reperiatur Asymptotos & nullum crus Parabolicum
Figura erit vel Hyperbola Elliptica [vel] Hyperbolismus Ellipseos. Nempe
posterior si rectæ lineæ circa datum quodvis quod extra Curvam est punctum
convolventis intersectio cum Curva in aliqua positione determinari potest per
planam Geometriam, sed prior si ad hujus intersectionis determinationem solida
Geometria in omni lineæ positione requiritur. Præterea si unica sit Asymptotos
et duo crura parabolica non convergentia ad Asymptoton illam Figura erit
Hyperbola Parabolica. Sed si convergunt ad eam[,] hoc est si tendunt ad eandem
plagam cum alterutra extremitate Asymptoti[,] figura erit Parabolismus
Hyperbolæ.[68] Insuper si nulla reperiatur Asymptotos sed crura duo Parabolica
ad oppositas plagas tendentia[69] quarum convexitas eandem plagam respicit,
curva erit Parabola divergens, sed si convexitas respicit diversas plagas Parabola
erit Cubica.

Præterea ut innotescat utrum Figura bifida sit necne spectandus est situs
crurum Hyperbolicorum respectu Asymptoti, nam si tres reperiantur Asymptoti
crura habentes ex eodem latere figura erit trifariam bifida, sin unica sit ejusmodi
Asymptotos erit unico modo bifida, si nulla non erit bifida nisi forte sit Parabola
divergens quæ nullam[70] habet Asymptoton.

Postquam species principalis Figuræ cujusvis propositæ sic innotuit, si
ulterior determinatio speciei desideretur investiganda est Tangens Figuræ ad
punctum indeterminatum,[71] et Tangens illa ponenda parallela Asymptoto illi,
nempe habet crura Hyperbolica ex eodem latere si curva est Hyperbola
triformis bifida, aliter cuivis Asymptoto, aut lineæ cui crura Parabolica ad infinitum pergentia parallela evadunt si figura sit Parabola divergens. Hoc pacto
invenies puncta *T*, *t*, 7, τ, ex quorum situ supra[72] distinximus has curvas in
species.

---

(66) A cancelled first sequel at this point reads 'ejusmodi tangentes Hyperbolicæ positione
datæ quarum nullæ parallelæ sunt'.

(67) 'tangentis' is cancelled.

(68) Newton first wrote: 'Præterea si tres sunt infinitæ tangentes quarum unica tantum
positione datur [et] cæteræ duæ semper recedunt [in infinitum], Figura erit Parabolismus
Hyperbolæ.'

(69) This replaces 'ad easdem partes lineæ [jacentia]'.

(70) Understand 'finitam'.

(71) 'aliquod incertum' is deleted.

(72) See II: 46 ff. This adds weight to our earlier suggestion (§3: note (75) above) that
Newton intended to add a résumé of his previous detailed distinction of cubics into species to
his present treatise.

Habita Æquatione $y^3 \begin{matrix} +bx \\ +e \end{matrix} yy \begin{matrix} +cxx \\ +fx \end{matrix} y \begin{matrix} +dx^3 \\ +gxx \\ +h \end{matrix} \begin{matrix} \\ +kx \\ +l \end{matrix} = 0$ ad curvam aliquam ubi $AB = x_{[,]}$

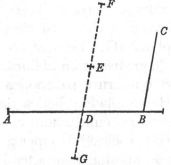

$BC = y$ quære radices hujus $z^3 + bzz + cz + d = 0$. Et si omnes tres reales sint, cape $AD = 1$, et sunto DE, DF, DG tres radices. Dein pone $ezz + fz + g = m$. $3zz + 2bz + c = n$. $hz + k = p$. $2ez + f = q$. $3z + b = r$, et erit

$$zx - \frac{m}{n} + \frac{-pn^2 + qmn - rmm}{nnnx}$$

æquatio ad crus infinitum si tres radices sunt reales et inæquales vel si z non sit una duarum æqualium vel si z sit unica realis radix$_{[,]}$ & figura erit bifida si

$pn^2 = qmn - rmm$, aliter non.[73] Et $zx - \frac{m}{n}$ erit [æquatio] ad Asymptoton. Et erit trifariam bifida si p q et $r = 0$.[74] Et tunc erit

$BC = zx - \frac{m}{n} + \frac{m^3}{n^4 xx}$ proxime[75] sive sit trifariam bifida sive simpliciter bifida. In his casibus figura erit Hyperb triformis si omnes radices sunt inæq: & real: vel Hyperb Ellipt[ica] aut Hyperbolismus Ellipseos si unica tantum realis est: nempe Hyperbolismus si sit $2a \cdot r :: r \cdot n$ id est $rr = 2n$,[76] Aliter Hyperbola.

Quod si duæ radices sint æquales et habetur m,[77] figura erit Hyperbola Parabolica existente $BC = zx \dfrac{[-]q \pm \sqrt{qq - 4pr - 4mrx}}{2r}$ et $zx [-] \dfrac{q}{2r}$ æquatione ad diametrum proxime.[78]

Sed si m desit erit $zx - \dfrac{q}{2r} \pm \sqrt{\dfrac{qq - 4rp}{4rr}}$ ad Asymptotos parallelas, figura existente Hyperbolismo Hyperbolæ nisi sit $qq = 4rp$, tunc enim Asymptotis coincidentibus figura erit Hyperbolismus Parabolæ. Fac $\dfrac{q}{2r} \pm \sqrt{\dfrac{qq - 4rp}{4rr}} = s$ et erit $zx - s \mp \dfrac{2rs^3}{x \sqrt{qq - 4rp}}$ [79] ad curvam proxime in priori casu$_{[,]}$ in posteriori vero erit $zx - \dfrac{q}{2r} \pm \sqrt{\dfrac{3pqr - q^3}{2r^4 x}}$ [80] ad curvam proxime

---

(73) See note (55) above.

(74) In this case $pn^2 - qmn + rm^2$ vanishes for all values of m and n.

(75) See note (56) above. Observe that Newton neglects the terms $-ln^3 + hmn^2 - em^2 n$ in this approximation to $O(x^{-3})$.

(76) Read '$4 \cdot r :: r \cdot n$ id est $rr = 4n$'. Newton carries through his earlier error (note (38) above) and also forgetfully retains the coefficient a (which he set as unity above).

(77) This replaces 'sint istæ z et erit'.

Si z habet tres æquales valores, et insuper sit $m=0$ sed habetur $q_{[,]}$ figura erit Parabolismus Hyperbolæ, et erit $BC=zx-\dfrac{p}{q}+\dfrac{ap^3}{q^4x}$ proxim[e][81] æquatio ad crus Hyperbolicum et $zx-\dfrac{p}{q}$ exacte ad Asymptoton & $BC=zx[-]\dfrac{p}{2q}\pm\sqrt{-qx}$[81] ad crura Parabolica.

Quod si non sit $m=0$, figura erit Parabola divergens et æquatio ad eam erit

$$zx-\sqrt[3]{mxx}+\frac{q}{3m}\sqrt[3]{mmx}+\frac{p}{q}+\frac{qq}{9m}.^{(82)}$$

Deniꝗ si m et q simul $=0$ figura erit Parabola Cubica et æquatio ad eam erit

$$zx-\sqrt{3:px+l}.^{(83)}$$

[7][84]

Enumeratis harum Curvarum speciebus & ostenso quomodo æquationes quibus quomodocunꝗ exprimuntur reduci possin[t] ad formulas quarum subsidio definivimus ipsas & distinximus in species, videor[85] determinationem horum locorum ad finem perduxisse: cum tamen reductio æquationum ad formulas illas sæpenumero perquam tædiosa esse possit, ex abundanti jam docebo quomodo species curvæ sine istiusmodi transmutatione discerni possit. Supponamus igitur quæstionem aliquam de hujus generis Loco ad æquationem perductam esse sitꝗ æquatio illa ut supra $ay^3+bx\ \pm cxx\ \pm dx^3=0.$ in qua[86]

$$\begin{array}{rrr} \pm e^{yy} & \pm dx^{y} & \pm gxx \\ \pm e & \pm kx & \\ \pm l & & \end{array}$$

siquis terminorum hic positorum desit terminus ille in sequenti calculo negligendus erit. Supono autem AB designare Basem x et BC Ordinatam [y]. Imprimis igitur construo hanc æquationem $az^3+bzz+cz+d=0.^{(87)}$ dein pono

(78) Newton omits the term $-v^3/x$ from his earlier computation as being negligible for indefinitely large x.

(79) See note (59) above. The terms $l+hs+es^2$ are omitted from the numerator while in 'compensation' the remaining term s^3 is multiplied by $2r$.

(80) Compare note (61) above. Correctly, $y = zx-q/2r+t$ where
$$t^2 = (pq/2r^2-ep/r+hq/2r-l)/rx.$$

(81) See note (63) above.

(82) The constant terms should be '$-\dfrac{e}{3}$' simply. See note (64) above.

(83) On taking the fictitious equation to be $v^3+px+l = 0$.

(84) ULC. Add. 3961. 1: 21ʳ–22ʳ. Newton revises the text of Appendix 6 preceding.

(85) 'satis' is cancelled, removing some hesitancy in Newton's affirmation.

(86) An unnecessary following phrase 'licet omnes terminos posuerim' is deleted.

(87) A cancelled first continuation reads: 'Et si obvenerint tres reales & inæquales radices concludo figuram esse Hyperbolam triformem. Si duæ earum sunt æquales, erit vel Hyperbola Parabolica vel Hyperbolismus Hyperbolæ Parabolæ've, si omnes tres æquales sunt erit vel parabolismus Hyperbolæ vel Parabola convexa [= divergens] vel Parabola Cubica. At si duæ

m $p^{(88)}$ $ezz + fz + g = m.$ $3azz + 2bz + c = n.$

n m $hz + k = p.$ $2ez + f = q.$ & $3az + b = r.$

p r Si tres obvenerint æquationis

q q $az^3 + bzz + cz + d = 0$ radices reales

r n & inæquales et nulla earum in colligendis valoribus quantitatum m, n, p, q & r usurpata efficiat terminos[89] $pnn - qmn + rmm$ sese destruere Figura erit Hyperbola triformis non bifida, sed si aliqua earum efficiat terminos istos $pnn - qmn + rmm$ evanescere Hyperbola triformis erit uno modo bifida, et trifariam bifida si termini illi in unoquoqƺ trium casuum evanescunt.

Quod si æquatio illa non nisi unicam habeat realem radicem z, Figura erit Hyperbola Elliptica nisi fuerit $rr - 2an = 0$,[90] Tunc enim erit Hyperbolismus Ellipseos.[91]

Adhæc si tres sunt reales radices z et earum duæ æquales Figura erit Hyperbola Parabolica nisi fuerit $m = 0_{[,]}$ in quo casu habebitur Hyperbolismus Hyperbolæ vel etiam Hyperbolismus Parabolæ si insuper sit $qq - 4rp = 0$. Et in omnibus hisce quinqƺ[92] casibus figura erit bifida si sit $pnn - qmn + rmm = 0$, aliter non erit bifida.

Quod si omnes tres radices z æquales esse contigerit et non sit $m = 0$ Figura erit Parabola divergens[,] sed si sit $m = 0$, habebitur Parabolismus Hyperbolæ nisi insuper sit $q = 0$. Id enim denotat figuram esse Parabolam Cubicam.

---

earum sunt imaginariæ habebitur aut Hyperbola Elliptica aut Hyperbolismus Ellipseos. Sed ut figura penitus defini[a]tur sumo aliquam radicem z et pono $ezz \pm fz \pm g = m$.

$$3azz \pm 2bz \pm c = n. \quad hz + k = p. \quad 2ez + f = q. \quad \& \quad 3[a]z + b = r.$$

et in Ordinata BC sumpta $BD = zx$ (vel $= z$ si AB sit 1) et $DG = \dfrac{m}{n}$ duco EG parallelam AD : nam erit As[ymptotos].'

(88) These two marginal columns define, passing from left to right, the cyclic permutation which transforms the present analytical argument into the final version in §3 above.

(89) 'exhibeat quantitatem' is cancelled.

(90) This should be '$rr - 4an = 0$'; compare notes (38) and (76) above. See also §3: note (88) preceding.

(91) Newton has deleted a following sentence 'Et in utroqƺ casu figura erit bifida si sit $pnn - qmn + rmm = 0$, aliter erit non bifida': in compensation a word 'quinqƺ' is added at the end of the next paragraph (see note (92)).

Ubi species figuræ sic innotuit[,] si Asymptotos, situm crurum infinitorum, cæteraqȝ determinare animus est In basi Figuræ capta *AB* vel *x* cujusvis longitudinis, in ordinata *BC* quam *y* designat cape $BG = zx - \frac{m}{n}$ et *BG* semper terminabitur ad Asymptoton figuræ nisi *z* fuerit una duarum vel trium æqualium radicum. Hoc pacto invenies[93] tres Asymptotos si sunt tres inæquales radices *z*, et si præterea positiones crurum infinitorum respectu Asymptotorum cognoscere desideras in *BC* cape $BE = zx - \frac{m}{n} - \frac{+pnn - qmn + rmm}{n^3 x}$ et punctum *E* attinget Hyperbolam Conicam quæ eandem habet Asymptoton[94] cum crure Figuræ et ad easdem partes[95] Asymptoti jacet quæqȝ cum crure figur[æ] quamproxime coincidit, adeo ut inde non tantum de situ cruris sed etiam de distantia ejus ab Asymptoto judicare possis[,] puta si non fuerit

$$pnn - qmn + rmm = 0,$$

tunc enim capienda est $BE = zx - \frac{m}{n} + \frac{m^3 - emmn + hmnn - ln^3}{n^4 xx}$ et *BE* attinget Hyperbolam simplicissimam[96] secundi generis quamproximè coincidentem cum crure figuræ.

Hoc pacto pro numero radicum *z*[97] determinantur tres Asymptoti Hyperbolæ triformis, et una Asymptotos Hyperbolæ Ellipticæ et Hyperbolismi Ellipseos, ut et Asymptotos Hyperbolæ Parabolicæ et illa Asymptotos Hyperbolismi Hyperbolæ & Parabolæ qui non est ex parallelis Asymptotis nec habet crura Hyperbolica ex utroqȝ latere[98] nam istæ Asymptoti ut et crura Parabolica figurarum determinantur per æquales radices *z*.[99] Igitur si habeatur Hyperbola Parabolica et situm ac magnitudinem crurum ejus cognoscere desideras, collige terminos *m*, *p*, *q*, *r* per unam æqualium radicum *z* et cape

---

(92) A late insertion; see note (91).

(93) Newton first went on with the equivalent phrase 'tot Asymptotos quot sunt inæquales radices *z*'.

(94) The line *EG*, namely.

(95) 'ad eandem plagam' is here replaced.

(96) Newton first wrote 'simplicem'. The distinction here is not very clear. Observe that he has now corrected his preceding error (see note (75) above).

(97) This replaces 'per tres radices *z*'. The three asymptotes in question are the lines *EG*, $_2E_2G$ and $_3E_3G$ in Newton's figure.

(98) 'ad utramqȝ partem' is cancelled.

(99) A cancelled first sequel reads: 'Igitur si habeatur Hyperbolismus Hyperbolæ, pone unam æqualium radicum pro *z* et cape $BG = zx - \frac{q}{2r} \pm \sqrt{\frac{qq - 4rp}{4rr}}$ & punctum *G* semper [a]ttinget unam vel alteram parallelarū Asymptotωn prout addas vel auferas $\sqrt{\frac{qq - 4rp}{4rr}}$'.

$BE = zx \pm \sqrt{-\dfrac{mx}{r} + \dfrac{m}{2rr} - \dfrac{q}{2r}}^{(100)}$ & punctum E attinget Parabolam Conicam quamproxime coincidentem cum Parabolicis cruribus figuræ. Quod si existente $m=0$ habeatur Hyperbolismus Hyperbolæ, ad determinandas ejus $\|$las[101] Asymp[totos] fac $\dfrac{q}{2r} + \sqrt{\dfrac{qq-4rp}{4rr}} = s$ & $\dfrac{q}{2r} - \sqrt{\dfrac{qq-4rp}{4rr}} = t$ et cape $BG = zx - s$

& $BE = zx - s - \dfrac{+s^3 - ess + hs - l}{x\sqrt{qq-4rp}}^{(102)}$ et punctum G attinget unam Asymptoton et E Hyperbolam quamproxime coincidentem cum crure figuræ cujus est Asymptotos. Dein cape $BG = zx - t$ & $BE = zx - t + \dfrac{t^3 - ett + ht - l}{x\sqrt{qq-4rp}}^{(102)}$ et G attinget alteram parallelarum Asymptotôn et E Hyperbolam quamproxime coincidentem cum crure Figuræ ad istam Asymptoton pertinente.

Quod si existente $qq - 4rp = 0$ figura deprehendatur esse Hyperbolismus Parabolæ capienda est $BG = zx - \dfrac{q}{2r}$ pro determinatione Asymptoti et $BE = zx - \dfrac{q}{2r} \pm \sqrt{\dfrac{pq - 2epr + hqr - 2lrr}{4r^3 x}}^{(103)}$ pro determinatione Figuræ ex utroɋ latere quæ quamproxime congruit cum utrovis crure Figuræ.

Præterea si figura sit Parabola divergens fac $BG = zx$ et G attinget lineam rectam quæ tendit ad plagas crurum infinitorum et figura ipsa quamproxime attingetur a puncto E quando $BE = zx - \sqrt[3]{mxx} + \dfrac{q}{3m}\sqrt[3]{mmx} + \dfrac{p}{q} + \dfrac{qq}{9m}.^{(104)}$ Sed si existente $m=0$ figura sit Parabolismus Hyperbolæ, fac

$$BG = zx - \dfrac{p}{q} \quad \& \quad BE = zx - \dfrac{p}{q} + \dfrac{ap^3}{q^4 x}^{(105)}$$

& G attinget ejus Asymptoton Hyperbolicum et E Hyperbolam conicam attinget eandem habentem Asymptoton et quamproxime coincidentem cum Hyperbolico crure figuræ. Fac insuper $BE = zx \pm \sqrt{-qx + \dfrac{p}{2q}}^{(105)}$ et E attinget

---

(100) See note (58) above.

(101) Read 'parallelas'.

(102) For the preceding computation see note (59) above; compare also note (79) above.

(103) The final term was first set erroneously as '$\pm \sqrt{\dfrac{3pqr - q^3}{2r^4 x}}$' as in the preceding version (see note (80) above). For the preliminary computation see note (61) above.

(104) In place of the two final terms read '$-\dfrac{e}{3}$', as at the corresponding place in the preceding version (compare notes (64) and (82) above).

(105) See notes (62) and (81) above.

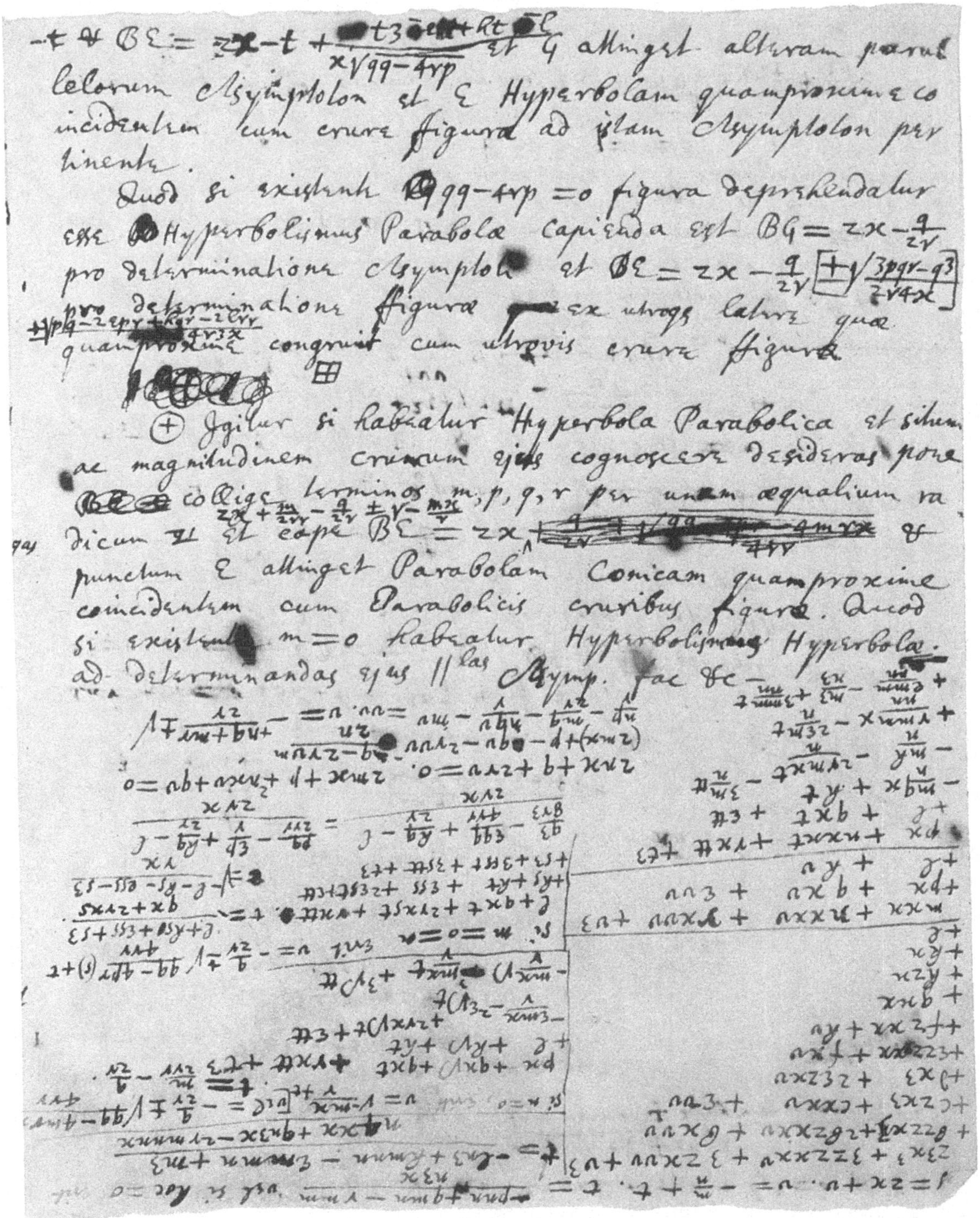

$-\not{t} \not{A} \; \mathcal{B}\mathcal{E} = zx - t + \dfrac{t3 - \overline{a} + kt - \overline{a}}{x\sqrt{qq-4rp}}$ Et $\not{q}$ attinget alteram paral

lelorum Asymptoton et $\mathcal{E}$ Hyperbolam quamproxime co

incidentem cum crure figuræ ad istam Asymptoton per

tinente.

Quod si existente $qq-4rp = o$ figura deprehendatur

esse Hyperbolismus Parabolæ capienda est $\mathcal{B}G = zx - \dfrac{q}{2r}$

pro determinatione Asymptoti et $\mathcal{B}\mathcal{E} = zx - \dfrac{q}{2r} \boxed{+\sqrt{\dfrac{3pqr-q3}{r^2 \cdot 4x}}}$

pro determinatione figuræ ex utroq latere quæ

quamproxime congruit cum utrovis crure figuræ

$\oplus$ Igitur si habeatur Hyperbola Parabolica et situm

ac magnitudinem crurum ejus cognoscere desideras pone

$\mathcal{B}\mathcal{E} =$ collige terminos m, p, q, r per unam æqualium ra

dicum x Et cape $\mathcal{B}\mathcal{E} = zx$ et

punctum $\mathcal{E}$ attinget Parabolam Conicam quamproxime

coincidentem cum Parabolicis cruribus figuræ. Quod

si existente $m = o$ habeatur Hyperbolismus Hyperbolæ.

ad determinandas ejus $\parallel^{las}$ Asymp. fac $\mathcal{B}\mathcal{E} -$

Plate III.   The discrimination of a cubic's genus: a
preliminary draft (**2**, 3, §3, Appendix).

Parabolam Conicam quamproxime coincidentem cum cruribus Parabolicis figuræ. Deniꝗ si terminis *m* et *q* simul deficientibus habeatur Parabola Cubica Fac $BE = zx$ et E continget rectam quæ tendit ad plagas crurum [infinitorum].

---

[8]$^{(106)}$

$$[bxyy + cxxy + eyy + fxy + gxx + hy + kx + l = 0 \quad \text{seu} \quad bxyy + cxxy = 0$$

adeoꝗ $y = -\dfrac{c}{b}x$ proxime. Pone] $-\dfrac{cx}{b} + v = y.$ [erit]

$$-cxxv + bxvv + \frac{ecc}{bb}xx - \frac{2ec}{b}xv - \frac{fc}{b}xx + fxv + gxx[= 0$$

sive $-cxxv + \dfrac{ecc}{bb}xx - \dfrac{fc}{b}xx + gxx = 0$ neglectis negligendis. unde]

$$\frac{ecc - bfc + bbg}{bbc} = v = \frac{ec}{bb} - \frac{f}{b} + \frac{g}{c}.^{(107)}$$

---

(106) ULC. Add. 3961.1: 19 bis^{v}, a late computation for the final paragraph of §3 above.

(107) Newton computes the asymptote of the cubic

$$bxy^2 + cx^2y + ey^2 + fxy + gx^2 + hy + kx + l = 0$$

which is parallel to $by + cx = 0$ by substituting $y = -(c/b)x + v$ and then equating the coefficient of x^2 in the resulting equation to zero (since the asymptote shares two points at infinity with the cubic and the term in x^3 is not present).

§4. THE SOLUTION OF PLANE PROBLEMS BY THE 'RULE OF THREE'.[1]

From the original in private possession

De solutione planorum Problematum per regulam trium.[2]

Sint assumptæ quinɋ GA, GB, GC, GD, GE; errores AH, $[B]I$, $[C]K$, $[D]L$, $[E]M$.[3] Fac $AH - ME \cdot AE :: AH \cdot AQ = HQ$. [ut et] $Ox = x$. $xy = y$. $OP = a$.

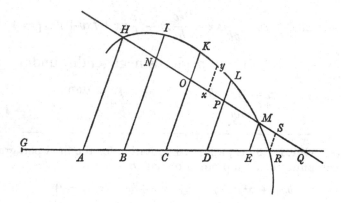

$OK = b$. $PL = c$. $IN = d$. [Posita æquatione] $xx + qxy \genfrac{}{}{0pt}{}{-4aa}{+ry} = 0$[4] [quæ designet

$$+syy$$

Conicam sectionem HMR, prodit] $4aa - sbb = rb$. $3aa$[5]$+ qac + rc + scc = 0$.

$3aa$[5]$- qad + rd + sdd = 0$. [adeoɋ si] $\dfrac{aa^{(6)}}{c} = f$. $\dfrac{aa^{(6)}}{d} = g$. $\dfrac{aa}{b} = e$. [fit] $\dfrac{3f + 3g + 8e^{(7)}}{b - c - d} = s$.

---

(1) Nothing is known of the reasons which stimulated Newton to compose these two variants on a 'regula quinɋ errorum' for constructing the meets of a straight line with a given conic. The date of this scrap is evidently little different from that of 2, §2 and 3, §§2/3 since it was written on the same sheet as (drafts of) these; see §2: note (2) above.

(2) Understand any variant on the 'rule' $\delta = \beta\gamma/\alpha$ which connects four quantities in the proportion $\alpha : \beta = \gamma : \delta$, though this is not an apt summary of the method used in sequel. (Indeed, Newton's present use of the terminology 'error' suggests that he might have intended to write '...PER REGULAM FALSÆ POSITIONIS' (BY THE RULE OF FALSE POSITION); compare Oughtred's *Clavis Mathematicæ Denuo Limata, sive potius Fabricata. Cum aliis quibusdam... Commentationibus* (Oxford, ₃1652): Appendix: 45–6: 'Regula Falsæ Positionis'.) The two intersections of a straight line with a given conic are, of course, determinable by a 'plane' construction (involving the meets of straight lines and circles).

(3) The (unstated) problem is to construct the meets R of the straight line GQ with the conic HIK. (The manuscript figure shows only one point R of intersection but we have extended the conic to meet GQ in the second point R' in that accompanying the English version.) Newton's method is to draw a second straight line HMQ through two given points on the conic,

Translation

ON THE SOLUTION OF PLANE PROBLEMS BY
THE RULE OF THREE[2]

Let there be assumed any five lines *GA*, *GB*, *GC*, *GD*, *GE* with errors *AH*, *BI*, *CK*, *DL*, *EM*.[3] Make $AH - ME : AE = AH : AQ$, that is, *HQ*; also $Ox = x$, $xy = y$, $OP = a$, $OK = b$, $PL = c$, $IN = d$. Then, on supposing the equation

$$x^2 + qxy + (-4a^2 + ry + sy^2) = 0^{(4)}$$

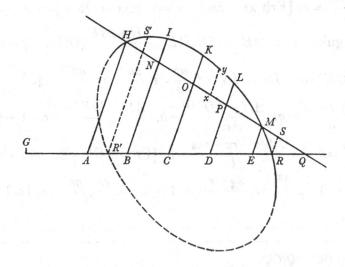

to denote the conic *HMR*, there will result $4a^2 - sb^2 = rb$, $3a^{2\,(5)} + qac + rc + sc^2 = 0$, $3a^{2\,(5)} - qad + rd + sd^2 = 0$; consequently, if $a^2/c^{(6)} = f$, $a^2/d^{(6)} = g$ and $a^2/b = e$, there comes to be $(3f + 3g + 8e)/(b - c - d)^{(7)} = s$, $4e - bs = r$ and $(3g + r + sd)/a = q$. Now,

---

draw the parallels *HA*, *ME* such that the triangles *QHA*, *QME* are isosceles, quadrisect the segment *AE* in *B*, *C*, *D* and raise further parallels *BI*, *CK*, *DL* to the conic. From the five 'errors' by which the conic diverges at the points *H*, *I*, *K*, *L*, *M* from the line *GQ* Newton will determine the points *R* on the conic at which the 'error' is nil, and so *R* is the intersection of the conic with *GQ*.

(4) Newton first added the term ' $+px$' to the equation but cancelled it when he realized that the conic passes through $H(-2a, 0)$ and $M(2a, 0)$. The following equations result by determining the conic to be through $K(0, b)$, $L(a, c)$ and $I(-a, b)$ respectively.

(5) Read ' $-3aa$'.

(6) Read ' $-\dfrac{aa}{c}$' and ' $-\dfrac{aa}{d}$' respectively; compare the previous note. In a first cancelled version these substitutions are not made.

(7) The denominator should be ' $2b - c - d$' on correctly eliminating between

$$4e - sb = r, \quad 3f + qa + r + sc = 0 \quad \text{and} \quad 3e - qa + r + sd = 0.$$

$4e-bs=r.\ \dfrac{3g+r+sd}{a}=q.$ [Jam ponendo] $OQ=w.\ OC=v.$ [erit]

$$w.v::w-x.v-\frac{vx^{(8)}}{w}=SR.$$

[adeoꝗ][9] $\quad xx+qxv-\dfrac{qvxx}{w}-4aa+rv-\dfrac{rv}{w}x+svv-\dfrac{2svvx}{w}+s\dfrac{vv}{ww}xx[=0]$ [seu]

$$\begin{array}{c}ww\\+svv\\-qvw\end{array}xx\ \begin{array}{c}+qvw^2\\-rvw\\-2svvw\end{array}x\ \begin{array}{c}-4aaww\\+rvww\\+svvww\end{array}=0.$$ [Fac] $\dfrac{v}{w}=t.\ \ 1+[s]tt-qt=l.\ \ \dfrac{rt+2svt-qv}{2l}=m.$

[&] $\dfrac{4aa-rv-svv}{l}=n.$ [Erit $xx-2mx-n=0$, adeoꝗ $OS=]x=m\pm\sqrt{mm+n}.$[10]

[Unde] Regula. Fac $AH-ME.AE::\dfrac{AH+ME}{2}(OC).CQ=OQ.$ [ut et]

$CD=OP=a.\ OK=b.\ PL=c.\ IN=d.\ \dfrac{aa}{b}=e.\ \dfrac{aa^{(6)}}{c}=f.\ \dfrac{aa^{(6)}}{d}=g.\ \dfrac{3f+3g+8e^{(7)}}{b-c-d}=s.$

$4e-bs=r.\ \dfrac{3f+r+sd}{a}=q.\ \dfrac{AH-ME^{(11)}}{AE}=h.\ \dfrac{AH+ME^{(12)}}{2}=n.\ \ 1+[s]hh-gh=l.$

$\dfrac{rh+2sn[h]-qn}{2l}=m.$ [&] $\dfrac{4aa-rn-snn}{l}=p.$ [Prodit $OS=]x=m\pm\sqrt{mm+p}.$

Sed si $s=0.$ erit $\dfrac{4aa^{(13)}}{b}=r.\ \dfrac{3g+r}{a}=q.\ 1-gh=l.\ \dfrac{rh-qn}{2l}=m.$ [&] $\dfrac{4aa-rn}{l}=p.$

---

(8) Since $SR = OC \times SQ/OQ$.

(9) Taking now the intersection point R to be $(x, v-(v/w)x)$.

(10) Whence the length of $OS = x$ is determined as the meet of the line $y = 0$ and the 'plane' (circle) locus $(x-m)^2+y^2 = \rho^2$, $\rho = \sqrt{[m^2+n]}$. See note (2) above.

on setting $OQ=w$ and $OC=v$, there will be $w:v=w-x:(SR$ or$)$ $v-(v/w)\,x$,[8] so that[9] $x^2+qvx-(qv/w)\,x^2-4a^2+rv-(rv/w)\,x+sv^2-2(sv^2/w)\,x+(sv^2/w^2)\,x^2=0$ or $(w^2+sv^2-qvw)\,x^2+(qvw^2-rvw-2sv^2w)\,x+(-4a^2+rv+sv^2)\,w^2=0$. Make

$$v/w=t, \quad 1+st^2-qt=l, \quad \tfrac{1}{2}(rt+2svt-qv)/l=m$$

and $(4a^2-rv-sv^2)/l=n$. Then $x^2-2mx-n=0$ and so

$$OS \text{ (or } x) = m\pm\sqrt{[m^2+n]}.^{(10)}$$

Hence the rule. Make $AH-ME:AE=\tfrac{1}{2}(AH+ME)$ (or OC)$:CQ$ (or OQ); also $CD=OP=a$, $OK=b$, $PL=c$, $IN=d$, $a^2/b=e$, $a^2/c^{(6)}=f$, $a^2/d^{(6)}=g$, $(3f+3g+8e)/(b-c-d)^{(7)}=s$, $4e-bs=r$, $(3g+r+sd)/a=q$,

$$(AH-ME)/AE^{(11)}=h, \quad \tfrac{1}{2}(AH+ME)^{(12)}=n, \quad 1+sh^2-gh=l,$$

$\tfrac{1}{2}(rh+2snh-qn)/l=m$ and $(4a^2-rn-sn^2)/l=p$: there results

$$OS \text{ (or } x) = m\pm\sqrt{[m^2+p]}.$$

But if $s=0$, then $4a^2/b^{(13)}=r$, $(3g+r)/a=q$, $1-qh=l$, $\tfrac{1}{2}(rh-qn)/l=m$ and $(4a^2-rn)/l=p$.

---

(11) That is, OC/CQ.
(12) Namely, OS.
(13) That is, $4e$ in the above designation.

PART 3

THE 'GEOMETRIA CURVILINEA'
(*c.* 1680)
AND THE
'MATHESEOS UNIVERSALIS
SPECIMINA'
(1684)

INTRODUCTION

In this final part of the present volume of Newton's mathematical papers we reproduce, from the original autographs, the extant portion of two intended treatises on fluxional analysis and computation by infinite series, neither of which he completed or communicated publicly, in their unfinished form, to his contemporaries. Except for a few opening pages of the two versions of the latter work, given by him about 1710 to William Jones—then planning the edition of Newton's mathematical tracts on calculus and series which subsequently appeared as *Analysis Per Quantitatum Series, Fluxiones ac Differentias*[1]—, they have remained till the past decade buried in the main corpus of Newton's scientific papers, scattered, neglected and their very existence forgotten. Both have their share of ill-polished crudities and tiresome elaborations of the obvious (Newton's two chief failings as a popular expositor) but also contain passages which significantly widen our insight into the subtlety of thought and immense technical skill of a man on the eve of encapsulating the mathematical principles underlying a wide range of natural phenomena in one of the most profound and enduring of scientific classics.

Of the background to the first treatise, the 'Geometria Curvilinea',[2] nothing is known: more exactly, an intensive search of contemporary sources has not only not brought to light any comment on its purpose or adequacy but has also failed to unearth any reference to its existence. In default we must rely heavily in assessment of its character and significance on the internal evidence of the text—notably the content of Newton's introductory remarks—and for the rest upon controlled guesswork.[3] As that preface makes clear, the general 'geometry of curved lines' announced in the title is one in which, analogously to Euclid's treatment of the elements of the theory of the straight line and circle, classical geometrical analysis is applied to the study of higher curves, notably to the derivation of their elementary differential properties. With a heavy-handed rigidity of outlook which was henceforward to become increasingly familiar in his scientific writings Newton here berates the equivalent Cartesian algebraic analysis of higher geometry for its lack of rigour, its ponderous

---

(1) London, 1711; see 1: xvi–xvii.

(2) Reproduced in 1, §1 (main text) and §2 (two outline schemes for an augmented revision of its first book).

(3) The date (in or about 1680) assigned to the 'Geometria' is conjectured uniquely on a visual estimate of Newton's handwriting style, though it is more broadly clear that the manuscript lies, in sequence of composition, midway between the 1671 tract (whose section on geometrical fluxions it expands) and Lemma II of the *Principia*'s second book (which is based on Propositions 3–10 of its 'Lib. 1'); see 1, §1: note (1).

inelegance, its 'scarcely tolerable' computational complexities and its wide-spread reliance on infinitesimals: in its place he will make basic appeal to quantities growing (or decreasing) continuously in time according to a given law and their 'fluxions' (instantaneous speeds of growth—in his eyes the 'natural fount' for measuring curvilinear magnitudes because of the clarity and brevity of its arguments and the simplicity of its conclusions). Of the four books in which it was his intention to elaborate this refined method of fluxions[4] only the first (on the general theory applied to simple algebraic, geometrical and trigonometrical magnitudes) appears in a substantially complete form in existing manuscript, while the opening problems of the second (where the direct method is applied to treat questions of maxima and minima and to construct normals and tangents) are but briefly outlined; Book 3 (on the inverse method of fluxions) and Book 4 ('on the nature of curves in general and the construction of problems through their intersections'[5]) remained still-born foetuses.

True to his promise Newton begins his 'Lib. 1' in Euclidean style with sets of 'Definitions', 'Axioms' and 'Postulates', though (as with his Greek model) the distinction between these is not always evident: above all the crucial sixth 'axiom' that 'fluxions are as the first (or last) ratios of their nascent (or evanescent) parts' seems to modern eyes to define the fluxion of a quantity in a not immediately coherent way, though Newton himself was clearly content to elevate it to the status of a self-evident truth. The fundamental appeal here—and equivalently in the following propositions—to the concept of a limit is glossed over by the use of such conventional, none too precisely defined verbal forms as 'first', 'beginning', 'last', 'vanishing'[6] and especially (instantaneous) 'speed', which Newton hopes will be intuitively understood by his reader. Nevertheless, these definitions and axioms of motion and the measure of its flux are a praiseworthy, however unsuccessful, attempt on his part to codify ideas of which he had made continual and increasingly sophisticated technical use since the autumn of 1665,[7] especially in the two extended treatises on calculus he had written in 1666 and 1671. To the section on geometrical fluxions

---

(4) See 1, §1: note (18).

(5) No doubt an intended revision of Newton's earlier 'Problems for construing æquations' (II: 450–516).

(6) In Axiom 6, for example, Newton's argument has recourse to the undefined notion of a 'principium generationis'; and in Proposition 3 following to an 'initium fluxionis ubi puncta... coincidunt & ab ipsis tantum incipiunt discedere'. The modern student of Newton's fluxional thought who hopes to find in the 'Geometria' a more precise linguistic—or indeed mathematical—analysis of the structural complexities of a 'first (or last) ratio' will be disappointed.

(7) See I: 343–4, and also 146–7.

(8) This dependence is most evident in the preliminary draft reproduced in 1, Appendix 1.1 below, which quotes *verbatim* from the 1671 addendum (III: 330–8).

appended to the latter tract the first ten propositions of the 'Geometria' are narrowly related,[8] though in preference to his earlier algebraical derivation of the fluxion of a product (now relegated to a scholium) Newton now gives pride of place in Propositions 1–3 to a geometrical argument which achieves an equivalent end by considering the limit-motion of the intersections of a rotating right angle with a stationary hypotenuse. In sequel Propositions 4–10 deal algebraically with particular cases and trivial extensions of the product theorem, applying it in particular (Proposition 8) to deriving the fluxion of the general power of a variable magnitude. Propositions 11—25 next following treat at tedious length[9] of the fluxional derivatives of the basic trigonometrical functions (defined geometrically in Propositions 17 and 18 with regard to the traditional model of a circle) and their application in the analysis of the limit-motions of the elements of a general triangle divided up by its altitudes—one afterwards scheduled to be discussed still more minutely in the numerous problems 'De Triangulis Rectangulis' and 'De Triangulis Obliquangulis' enunciated by Newton in his projected revise.[10] The remaining Propositions 26–30 concern themselves with the construction of the 'determination' (instantaneous direction) and centres of 'motion' (curvature) of curves defined in various generalized Cartesian and polar coordinate systems: these find their *raison d'être* in the concluding Problems 3–11 (conjecturally assigned by us to the proposed 'Liber 2'), which determine normals, tangents and (implicitly) curvature centres[11] in the case of the Apollonian ellipse, Cartesian oval and trident, and also more general classes of conchoids, cissoids, spirals and quadratrixes. Problems 1 and 2 deal straightforwardly with Fermatian problems of maxima and minima by equating the fluxion of the variable magnitude to zero.

Though the 'Geometria Curvilinea' had no direct impact upon Newton's contemporaries and successors,[12] the modified summary[13] of Propositions 3–10 of its first book which he incorporated in 1686 in his *Principia* as Lemma II of Book 2 was soon to become a document of primary importance in his dispute with the Leibnizians over calculus priority; indeed, though ostensibly introduced to justify two minor assertions in following propositions, this first public announcement (in July 1687) of his mature method of fluxions was manifestly

---

(9) Newton himself realized the ineffectiveness of much of his elaborate analysis—scarcely more than a monotonous display of empty technical virtuosity—for at a late stage he decided to eliminate two of the four particular cases of Proposition 20 which he originally discussed in the 'Geometria'; see 1, §1: note (87). The omitted cases are reproduced, for completeness' sake rather than because of any intrinsic significance, in 1, Appendix 1.5.

(10) 1, §2.2.

(11) See 1, §1: note (162).

(12) Not even in the revised version which, as we shall see in the seventh volume, Newton incorporated about 1693 in his still unpublished 'Geometriæ Liber Primus'.

(13) Reproduced in 1, Appendix 3 below.

intended as a counterblast to Leibniz' recently appeared 'singular calculus' for dealing with problems of maxima and minima and of tangents.[14] In ignorance of Newton's earlier dotted and literal notations for derivatives[15] and the elaborate researches in calculus which he had systematized in his two fluxional treatises of October 1666 and again in 1671, the continental mathematicians of the day—notably Leibniz himself—were persuaded into concluding that this single *Principia* lemma typified, both in its minuscule notation for fluxions[16] and in its conceptual inadequacies,[17] the form and qualities of his unpublished researches in calculus. In hindsight we may see how a little more forthrightness on Newton's part when he first presented his fluxional method to the world would have saved him a great deal of the bitterness and sense of frustration he experienced twenty-five years afterwards when he fought tenaciously to safeguard his priority of discovery.

Why Newton never completed his 'Geometria Curvilinea' is difficult to pinpoint. In its later propositions—and still more in the two hurried schemes for revising its first book which he subsequently outlined—there are signs of growing strain and of increasing dissatisfaction with the long-windedness of its diction and the patent sterility of much of its content; no doubt, when a suitable

(14) G. W. Leibniz, 'Nova Methodus pro maximis & minimis, itemque tangentibus, quæ nec fractas, nec irrationales quantitates moratur, & singulare pro illis calculi genus', *Acta Eruditorum* (October 1684): 467–73 [= C. I. Gerhardt, *Leibnizens Mathematische Schriften* (2), **1** (Halle, 1858): 220–6]. Compare 1, Appendix 3: note (1).

(15) See ı: 146, note (6) in particular.

(16) See 1, Appendix 3: note (5). Newton's employment of the notation originally introduced in Proposition 3 of the 'Geometria' (see 1, §1: notes (31) and (32)), by which lower-case letters denote the fluxions of corresponding upper-case fluents, was subsequently criticized by Leibniz in a letter to Huygens on 13 October 1690: '...ce que j'appelle *dx* ou *dy*, vous le pouvés designer par quelque autre lettre....Cependant je m'imagine qu'il y a certaines vues qui ne viennent pas aussi aisement que par mon expression [*dx* ou *ddx*].... Il me semble donc qu'il est plus naturel de les designer en sorte qu'elles fassent connoistre immediatement la grandeur dont elles sont les affections. Et cela paroist sur tout convenable, quand il y a plusieurs lettres, et plusieurs degres de differences à combiner,...car il y a alors à observer une certaine loy d'homogenes toute particuliere, et la seule vue decouvre ce qu'on ne demêleroit pas si aisement par des notes vagues, comme sont des simples lettres. Je voy que Mᵣ Newton se sert des minuscules pour les differences, mais quand on vient aux differences des differences, et au delà,...il faudra encor changer, de sorte qu'il me semble qu'on fait mieux de se servir d'une expression qui s'etend à tout. Cependant quand on est accoustumé à une methode on a raison de ne la pas changer aisement, quoyque on conseilleroit peut estre à d'autres, qui n'en ont encor aucune, de se servir de celle qui paroist la plus naturelle' (*Œuvres complètes de Christiaan Huygens*, **9** (The Hague, 1910): 516–17). If we recall the somewhat confusing expedient adopted by Newton in his 1671 tract (see ııı: 160–2 and 170–2, for example) of writing *r* (or *ż*) for the second fluxion (*ÿ*) of a quantity *y* whose first fluxion (*ẏ*) is taken to be *z*, we can but agree with Leibniz. Newton never extended the alphabetical parallel of his present literal notation to cover second fluxions (by writing, say, *α* for the second fluxion of a quantity *A* whose first fluxion is *a*) 'in such a way as to make known at once the magnitude whose affections [cognate quantities] they are'.

opportunity presented itself, he was happy enough to divert his energy to a more rewarding topic.[18] There is nothing to show that he returned to a study of the intertwined methods of infinite series and fluxions till mid-June 1684 when out of the blue he received a small package from James Gregory's nephew David, appointed months before to his uncle's old chair of mathematics at Edinburgh. The covering letter was brief but carefully worded:

Altho I have not yᵉ honour of your acquaintance, yet yᵉ character and place yee bear in the learned World I presume gives me a title to adress you especially in a matter of this sort, which is humbly to present you with a treatise latly published by me heir,[19] which I am sure contains things new to the greatest parte of the geometers. Sʳ I perceive by severall letters from Mʳ Collins to my Uncle, from whose remains this is for yᵉ most parte taken, that your selfe have of a long time cultivate this methode, and that yᵉ World have long expected your discoveries therein. and I hope that if yee doe me yᵉ honour to glance it over yee will see that I have according to justice acknowledged yᵉ same. Sʳ yee will exceedingly oblige me, if yee will spare so much time from your Philosophical and Geometrical studies, as to allow me your free thoughts and character of this expectation, which I assure you I will justly value more then that of all the rest of yᵉ World. and if yee please to add how farr Dʳ Wallis will pursue his designe, which wee wer acquainted of a good time agoe,[20] of explaining your methode of turning yᵉ

(17) However, the sophisticated fallacy in the Lemma's 'Cas. 1'—where (see 1, Appendix 3: note (6)) Newton sought to pass directly to his desired limit without appearing to discard non-zero infinitesimal quantities at a crucial stage in his argument—was not noticed by his contemporaries and has, indeed, equally evaded the notice of most modern students of his fluxional thought.

(18) If our dating of the piece as composed in 1680 is accurate, the sudden appearance of a comet in November of that year—the first seen in England for fifteen years—could well have provided such a diversion. During the four months from mid-December 1680 Newton entered into an active exchange of letters with Flamsteed (see *The Correspondence of Isaac Newton*, **2**, 1960: 315–20, 336–67) which suggests that for a while the topic of cometary orbits and the nature of their substance wholly monopolized his creative vigour.

(19) Gregory's presentation copy of his newly published *Exercitatio Geometrica de Dimensione Figurarum, sive Specimen Methodi Generalis Dimetiendi Quasvis Figuras. Auctore Davide Gregorio in Academiâ Edinburgensi Matheseos Professore*. (The *imprimatur* is dated 'Edinb: 29. Martii, 1684'.) As his letter went on to say, 'I sent a [second] coppy some weeks agoe to London adressed for you but learned it may not come so soon to your hands'. The present whereabouts of either of these Newtonian presentations is not known: doubtless one or other remained in Newton's library till his death, though the work is not found individually listed in either the Huggins or Musgrave catalogue. (The latter has an entry of 'Tracts Mathematical, 2 vols. by Borelli, Gregory, &c. 4to'; see R. de Villamil, *Newton: the Man* (London, 1931): 62.)

(20) Gregory refers to Wallis' announcement in his *A Proposal About Printing a Treatise of Algebra, Historical and Practical: ...containing not only a History, but an Institution of Algebra, according to several Methods hitherto in practice* that he will treat 'lastly, the method of infinite *Series*, or continual *Approximations* (grounded on the same Principles), arising principally from *Divisions* and Extraction of *Roots*, in *Species*, infinitely continued; invented by Mr. *Isaac Newton*, and pursued by Mr. *Nicholas Mercator*, and others, which is of great use for the rectifying of *Curve Lines*, squaring of *Curve-lined Figures*, and other abstruse Difficulties in *Geometry*' (ff. 2ʳ/2ᵛ [= *Algebra* (London, 1685): A2ʳ/A2ᵛ]; the corrected proof copy in Bodleian. B.1.14. Art is

roote of ane æquation to ane infinitie series, which is infinitly troublesome and tædious to me.[21]

Neither this diffident self-introduction nor the title, 'A Geometrical Exercise in the Measuring of Figures', of the little 50-page tract which accompanied it would quite have prepared Newton for the jolt he was to receive.

Gregory's *Exercitatio Geometrica de Dimensione Figurarum* opens mildly enough with a résumé of his uncle James' construction of the area of a general conic sector as the common 'termination' of a 'certain converging [double] sequence' and his ensuing attempt to demonstrate that the sector cannot be analytically compounded (in Descartes' sense) from its two first terms,[22] briefly touches in sequel on Mercator's derivation in his *Logarithmotechnia* of his variant quadrature of the hyperbola by a single series,[23] and then refers to Collins' transmission to Scotland in March 1670 of Newton's series: one which James in turn announced the following December to be derivable from his previously communicated

---

dated '1683'). In the last paragraph of his printed *Exercitatio* (note (19)) he had written similarly (p. 48) that 'jam primùm publico scripto moniti speramus [D: *Wallisium*], inter alia multa præclara sua aliorumque inventa, hanc etiam de æquationum radicibus in series resolvendis doctrinam, ad mentem *Clarissimi Summique Analystæ D: Isaaci Neutoni* aliquando enucleaturum'.

(21) ULC. Add. 3980.1 [= *Correspondence of Isaac Newton*, **2**: 396]. The letter is dated 'Edenburgh 9 of June 1684', so that Newton would have had it (in Cambridge) by the end of the month.

(22) *Vera Circuli et Hyperbolæ Quadratura, In propria sua proportionis Specie, Inventa, & demonstrata A Iacobo Gregorio Abredonensi Scoto* (Padua, 1667): Propositions I–IV/XI: 11–14/25–9. Compare III: 141, note (236).

(23) See II: 166.

(24) See 2, §1: note (6) below. James Gregory wrote to Collins on 19 December 1670 that 'In my last to you, I had not taken notice, that Mr Newtons series for the zone of a circle... togither with an infinite number of series of the like nature may be an consectarie to that which I sent you concerning logarithmes; viz Dato logarithmo invenire ejus numerum vel radicem potestatis cujuscumque puræ in infinitam seriem permutare. I admire much my own dulnes, that, in such a considerable time, I had not taken notice of this; nevertheless that I had taken much pains to find out that series: but the truth is, I thought always (if so be it wer a series) that I might fal upon it by some combination of my series for the circle, seing I had such infint numbers of them, not so much as once desiring any other methode' (*James Gregory Tercentenary Memorial Volume* (London, 1939): 148 [= *Correspondence of Isaac Newton*, **1**: 49–50]). Gregory's solution, effectively by a binomial expansion, of the problem 'To find the Number of a Logarithme' (reproduced, from Collins' contemporary copy in Shirburn Castle 101.H.2: 118r/118v, in S. P. Rigaud, *Correspondence of Scientific Men of the Seventeenth Century*, **2** (Oxford, 1841): 209–10) had been enclosed with his letter to Collins on the previous 23 November (*Gregory Volume*: 118–22). How much of the theory of infinite (sum) series he knew before March 1670 is a moot point: it is, for example, particularly tempting to suppose that the lavish inequalities published by Gregory in his *Exercitationes Geometricæ* (London, 1668) were derived by rounding off the 'notes' of a sine, cosine or tangent series of some kind (see H. W. Turnbull, 'The published works of James Gregory. II. The *Exercitationes Geometricæ*' [= *Gregory Volume*: 459–65]: 460–2; and J. E. Hofmann, 'Über Gregorys systematische Näherungen für den Sektor eines Mittelpunktkegelschnittes', *Centaurus*, **1**, 1950: 24–37). If Turnbull is right

logarithmic form of the general binomial series.[24] During the following year, David Gregory went on, his uncle extended his method of series, using it to extract the roots of any given 'literal' equation and applying it to various problems,[25] but death came prematurely to him before he publicly formulated his technique for so doing. Having since come into possession of certain of his uncle's *adversaria* (?worksheets) which listed several developments in series without general indication of the procedures by which they had been derived, he had worked hard to rediscover them: in his *Exercitatio* he would now present for the scrutiny of his fellow geometers his restoration of the Gregorian doctrine of series, completing it with applications to a number of the more elegant problems currently of major interest in the geometry of curves.[26] Thus forewarned, Newton was well prepared for Gregory's following sketch of the principles of exact algebraic integration and their exemplification in problems of the quadrature and rectification of curves and the mensuration of their solids and surfaces of revolution.[27] But, as he read on, a growing feeling of the *déjà vu*

(*Correspondence of Isaac Newton*, 1: 52, note (1)), Gregory had already found the general binomial expansion, as a deduction from the advancing-differences interpolation formula, in 1668 and applied it to determining the number corresponding to a given logarithm by 1669. Certainly, it is all but impossible to support Newton's belief—framed on inadequate evidence—that James Gregory derived the wealth of series he discovered during the late autumn and winter of 1670–1, many by repeated differentiation (see *Gregory Volume*: 347–67), solely on the basis of the single expansion, the simple integral of a square-root series, sent to him scarcely six months before.

(25) See 2, §1: note (6). David refers by implication to a passage in James' letter to Collins on 17 January 1672: 'I am confident that the tables of logarithmes & sines can not resolve all æquations, albeit they can verie manie, yet not without great preparation, so as an surdesolid equation wch can be reduced to a pure one, must first ascend to the 20th not without extraordina[r]ie work; the onlie universal method I know is the infinit serieses: ther can be given one wch wil serve for al cubick æquations, another for al biquadraticks, another for all surdesolids, &c and I suppose that tables of these series's wer the best of anie' (Shirburn Castle 101.H.2: 129^r, published in S. P. Rigaud's *Correspondence of Scientific Men* (note (24) above), 2: 230).

(26) See 2, §1: note (7) below.

(27) *Exercitatio* (note (19) above): 4–17. As the book's contemporary reviewer (Wallis?) put it: 'Mr. *Gregorie*...doth here assume (though in other words) the Doctrine of Indivisibles, and the Arithmetick of Infinites as already known.... And applies this to particular cases, in this manner; Supposing a streight line or Axis, which he calls *X*, cut into parts infinitely small, and the respective values of each *L* (which he calls *Elementum,*) or small part of the Curve, Plain, or Solid which is to be measured; answering to each of those particles of *X*; (or at least somewhat so near the values of *L*, as that the difference may be neglected; as when [*Exercitatio*: 5] a short Subtense or Tangent, is taken as coincident with a Curve;) he doth (according to the Doctrine of Infinites) collect the Aggregates of all such *L*; which Aggregate is the Magnitude sought. Of this he gives diverse examples in Parabola's, Hyperbola's, Ellipses, Spirals, Cycloids, Conchoids, Cissoids, and some other Curves, or Curve-lined Figures; as to their Area's, and Curve-lines, with the Solids, and Curve-surfaces, made by conversion of them, or otherwise derived from them' (*Philosophical Transactions*, 14, 1684: No. 163 [for 'Sept. 20 1684']: 730–2, especially 731).

must have come over him, for Gregory devoted the remainder of his tract to elaborating (with numerous well-chosen examples) two of the three methods of reducing quantities to series—by division, extraction of the square root of a binomial and the solution of affected equations—which Newton himself had set down in 'Reg. III' of his *De Analysi* fifteen years before;[28] in addition he gave, without proof, two instances (seemingly taken from his uncle's papers) of extracting the roots of a literal equation 'almost in the manner of Viète's *Exegetice Numerosa*'[29] and declared his intention of publishing a full explanation of the method 'on another occasion' unless John Wallis, as he had just announced in his *Proposal* for his forthcoming *Treatise of Algebra*, should forestall him by printing his 'enucleation' of the equivalent Newtonian doctrine.[30] The challenge thrown out to Newton himself was unspoken but no less real: publish or be published. After more than a decade of cautious reluctance to make known his discoveries in calculus and infinite series he was suddenly faced with the agonizing choice of doing so without delay or allowing a young Scots mathematician to take public credit for many of them. For who could tell what other rediscoveries Gregory would produce in his promised sequel?

(28) See II: 210–26. Gregory applies the approach *dividendo* ('Mercator' division) to derive the series for $\log(1+X)$ and $R\tan^{-1}(X/R)$ (*Exercitatio*: 17–18, 40–2); that *radicem extrahendo* to the direct extraction of $\sqrt{[c^2+X^2]}$ as an infinite series (*ibid.*: 19–20), which is in turn in following pages used as a model for determining the quadrature of various elliptical, hyperbolic and conchoidal areas.

(29) See I: 63, note (1); and compare II: 218, note (45). Gregory's exact words are 'extrahendo æquationum radices in Speciebus ad modum ferè *Exegetices Numerosæ Vietææ* (*Exercitatio* (note 19) above): 46). His first example, the 'Gregorian' *conchois Nicomedis* whose pole and vertex are equidistant from its asymptotic *norma* (see III: 271, note (608)), has the Cartesian equation $X^2L^2 = (L^2-a^2)(L+a)^2$, whence

$$\text{`Resolutâ hâc æquatione erit } L = a - \frac{a^{-1}X^2}{8} + \frac{a^{-3}X^4}{128} + \&c\text{';}$$

his second, constructing the '*Kepleri Problema*' (see note (38) of the concluding appendix to this volume) of drawing through the point $(BR/(B-R), 0)$ a transversal which cuts the semicircle $L = \sqrt{[X(2R-X)]}$ in the proportion P to Q, yields the implicit equation

$$L + (B-R)\sin^{-1}(L/R) = (B-R)P/(P+Q) = a,$$

whence on eliminating L there results $a = \sqrt{2}(B(X/R)^{\frac{1}{2}} + \frac{1}{12}(B-4R)(X/R)^{\frac{3}{2}} ...)$ and, on inverting this (and correcting two slight misprints in Gregory's text),

$$X/R = \frac{1}{2}B^{-2}a^2 - \frac{1}{24}(B-4R)B^{-5}a^4 + \frac{1}{720}(1-...)B^{-6}a^6 ....$$

(30) See note (20) above. Wallis' *Treatise of Algebra, Both Historical and Practical* (London, 1685) duly appeared the next year with five chapters (XCI–XCV: 330–46) devoted to 'The Doctrine of INFINITE SERIES, further prosecuted by Mr. Newton', but these merely repeat (in English) the relevant portions of Newton's two letters to Leibniz in 1676. (It seems unlikely that permission to publish was sought from Cambridge; very probably Newton was presented

Newton's reaction was swift. Whether or not he responded to Gregory's request to 'doe me yᵉ favour of a letter',[31] even as he continued to study the *Exercitatio* his thoughts turned towards the two bulky letters in which, eight years earlier, he had set down for Leibniz' benefit the essence of his mathematical achievements up to that time. Using this storehouse as basis and also borrowing appropriately from the chapters of his 1671 fluxional tract he would draft a new treatise on analysis which would effectively supplement Gregory's introduction and so render its intended sequel superfluous: at the same time he could set the historical record straight by stressing that David's uncle James had learnt the method of extracting infinite series (as he thought) from that of Newton's for the general circle zone. Such was the initial plan for the 'Matheseos Universalis Specimina'.[32] But as he began to gather his material Newton's purpose subtly changed: the impulse to outpace a mathematical junior became transmuted into the need to reply, years after they were made, to the deeper, more penetrating criticisms of his contemporary in Hanover. And then in turn this latter compulsion, too, was sublimated and the 'Specimina' changed into an abstract technical treatise 'De computo serierum'[33] in which the names of neither David

with a *fait accompli*.) In particular the 'enucleation' of Newton's 'new Method of Extracting Roots in Simple and Affected Equations...very different from that of *Vieta*, *Oughtred* and *Harriot*, which is commonly received' was presented in Chapter xcɪv (pp. 338–40). Wallis allowed himself a late concluding observation (p. 347): 'And (while these things are printing) comes out this present year 1684 (at *Edinburg*) a Treatise (of like nature,) of Mr. *David Gregorie*...Entituled, *Exercitatio nova de Dimensione Figurarum*. In which we have more Examples of the like Process'.

(31) Below Gregory's request that Newton should 'adresse [your letter] for me to be found in yᵉ Colledge of Edenburgh' the letter bears in Newton's hand the cryptic phrase 'By Sʳ Cartwright'. It may well be that Cartwright was intended to take Newton's reply to Scotland, but whether any letter did in fact leave Cambridge for Edinburgh is not known. The next letter which is recorded to have passed between the two, that of Gregory to Newton on 2 September 1687 (*Correspondence*, **2**: 484), makes no mention of the *Exercitatio*, while its hesitant concluding affirmation that 'you may beleive that I am very glade to have this occasion to write to you and wish for many such' is clear evidence that no regular correspondence had previously taken place between them.

(32) See the introductory paragraphs in 2, §1.1 below. Besides substantially complete versions of his letters of 13 June and 24 October 1676 to Oldenburg for Leibniz, Newton intended to publish important excerpts from Leibniz' replies to Oldenburg on 17/27 August 1676 and on 11/21 June and 12/22 July 1677 regarding 'pars illa quæ de seriebus infinitis agit'; see 2, §1: note (11). The alterations scheduled to be made to the public version of the *epistola posterior* (compare 2, §1: note (14)) are not considerable, serving more to paste over passages excised in the original letter and tighten up stray vagaries of phrasing than to 'improve' (that is, falsify) its content: the reader may check for himself by comparing the edited version reproduced in 2, Appendix 1 with the text of the letter sent to Oldenburg for onward transmission (*Correspondence*, **2**: 110–29).

(33) This—or its extant portion at least—is reproduced as 2, §2 below, while a considerably variant first state of its Chapter 2 is given in 2, Appendix 3.

Gregory nor Leibniz appear. With the passion quenched, however, so was the fire and the latter tract was soon abandoned in the middle of its third chapter.

In preface to the technical content of the 'Specimina' and its revise we need say little. For the most part Newton merely elaborates his earlier techniques for extracting a given quantity term by term as an infinite series—by use of the general binomial expansion or his tabular method for computing the root of an affected equation[34] where possible, but failing all else by

indefinitely assuming the first term of a series and making that term fulfil the conditions of the problem as far as can be done, by next assuming the second term indefinitely and making that fulfil the conditions of the problem as closely as possible, then assuming the third term and adapting that also to the conditions of the problem, and so on infinitely.[35]

The applications he gives of these procedures in solving geometrical and fluxional problems of various kinds are not notably profound or, indeed, always adequate, while the series expansions he derives in their solution are often carelessly stated, having the numerical coefficients of one or more of their later terms in error. What is truly fresh in either treatise are the second chapters in which Newton attacks the general problem of determining the limit-sums of infinite series, the coefficients of whose terms are numerical or simple algebraic quantities, and also of quickening the rate at which 'slowly convergent' or 'divergent' series[36] approach their 'terminations'. Both the methods he adduces[37] depend, in their different ways, on computing the various finite-differences of the coefficients: the one on identifying the 'progression' of those differences and extending them in consequence—exactly or approximately by an appropriate scheme of interpolation—one stage further back, the other on transforming the given series by an Eulerian 'transmutation' to a new series whose coefficients are those differences. Full commentary on the details of Newton's argument is given in sequel at appropriate points in footnote to the text.

Neither the Leibnizian criticisms of the 'Specimina' nor the technical procedures of the 'De computo serierum' were ever publicly communicated, and for more than two hundred and fifty years their manuscript has been disunited. Why the text of the 'De computo' was summarily abandoned in midsummer (July or early August?) 1684 we can only conjecture, but the near concurrence of that event with the date[38] of Halley's visit to Cambridge to

---

(34) This is most clearly enunciated in the first chapter of the 'De computo serierum' (2, §2).

(35) See 2, §1: note (151).

(36) The meaning Newton attaches to 'convergence' and 'divergence' is not quite the modern one; see 2, §2: note (48).

(37) See 2, §1: note (67) and 2, §2: note (40) respectively.

(38) Conventionally given as August; see page 657, note (4) below.

persuade Newton to turn his attention to the dynamics of elliptical planetary motion is scarcely mere coincidence. For the next three years he was to be completely absorbed in writing first his 'De Motu Corporum' and then in preparing his mighty *Principia* for the press, with his little spare time fully devoted to tending the furnaces in his chemical 'elaboratory' at Trinity; indeed, after Wallis had published the essence of his two *epistolæ* to Leibniz (in his 1685 *Algebra*) there must have seemed little immediate point in completing either of his two treatises for publication. Not till late 1691 was external circumstance—a second jolt from David Gregory, as it happened—to turn his attention back to the topic of fluxions and infinite series. But this must wait till our sixth volume.

1

THE 'GEOMETRY OF CURVED LINES'[1]

[c. 1680]

From autograph drafts in the University Library, Cambridge

§1. THE PRINCIPAL VERSION.[2]

GEOMETRIA CURVILINEA

Nuperi veterum inventis addere studentes, Arithmeticam speciosam[3] conjunxerunt cum Geometria. Ejus beneficio longe lateꝗ progressum est, si copiam rerum spectes, sed minus commodè si perplexitatem conclusion[u]m.[4] Nam hæc computa per operationes Arithmeticas solummodo progressa, sæpissime per ambages haud ferendas exprimunt quantitates quæ in Geometriâ

---

(1) Nothing at all is known of the reaons which led Newton to draft this intended treatise—more technically ingenious than effective—on geometrical fluxions and to abandon it unfinished. No doubt he was as conscious as any modern reader of the increasing sterility of its unnecessarily elaborate revise (§2 below) and, in time, only too happy to be diverted to another interest. The highly tentative date of composition is conjectured from our intuitive analysis of the character of Newton's handwriting, but it cannot be too far out in time since the present 'Geometria' is evidently a straightforward (if considerably augmented) revise of the closely similar piece (III: 328–52) which he added as an afterthought to his lavish fluxions tract of a decade earlier, while a précis of its introductory propositions was afterwards (about mid-1684) incorporated into the second book of his *Principia Mathematica* as Lemma II (here reproduced in Appendix 3). Once more following Barrow's lead (compare II: 71–2, notes (80), (81), (82) and (84)) Newton again throws much of the strain of the ill-defined notion of limit-value implicit in his concept of a fluxion onto such equally vague verbal equivalents as 'first', 'last', 'beginning' and 'end'. The applications of geometrical fluxions to problems of maxima and minima and the construction of tangents and normals sketchily outlined in 'Liber 2' mark no significant advance upon the corresponding sections (elaborated in terms of limit-increments, to be sure) of his 1671 tract.

(2) In sequence (gathering the two now separated portions of the autograph in logical order) ULC. Add. 3963.7: 46^r/Add. 3960.5: 49–52/Add. 3963.7: 48^v–61^v. Variant preliminary drafts are given in Appendix 1 below. An inserted slip bears the judgement, for once favourable, of Newton's eighteenth-century editor: 'A fragment concerning fluxions, which contains a geometrical demonstration of the theorem for the fluxion of a power ($\dot{x^n} = nx^{n-1}\dot{x}$) very different from any yet made public. I think this fragment very proper to be publishd in so much as it relates to proportionals. S. Horsley. Octr 22^d 1777.' In the last sentence 'as an appendix to the introduction to the [Quadrature of Curves]' is cancelled; as we shall see in the seventh volume the published version of the revised *De Quadratura Curvarum* [= *Opticks* (London, 1704): $_2$163–211, especially 165–9], whose main text Newton drafted in 1693, was

Translation

THE GEOMETRY OF CURVED LINES

Men of recent times, eager to add to the discoveries of the ancients, have united the arithmetic of variables[3] with geometry. Benefiting from that, progress has been broad and far-reaching if your eye is on the profuseness of output but the advance is less of a blessing if you look at the complexity of its conclusions.[4] For these computations, progressing by means of arithmetical operations alone, very often express in an intolerably roundabout way quantities which in geometry are designated by the drawing of a single line. Thus the

---

completed ten years after by a short preface incorporating brief excerpts from the 'Geometria Curvilinea'. Also to be found with Newton's autograph manuscript is a stray sheet (Add. 3960.5: 67–8), again in Horsley's hand, on which are written out six—of a projected eight— 'demonstrationes' of Newton's Propositions 3, 11 and 12: these are referred partly to the manuscript text, partly to Theorems I and IV of Roger Cotes' *Æstimatio Errorum in Mixta Mathesi per Variationes Partium Trianguli Plani et Sphærici* [= *Opera Miscellanea* (Cambridge, 1722): 1–22]. No such supplement was, in fact, published by Horsley in his *Isaaci Newtoni Opera quæ exstant Omnia* (London, 1779–85) though he had no qualms in appending to its 'Volumen primum' his wholly new 'Samuelis Horsleii de Geometria Fluxionum Liber Singularis. Sive Additamentum Tractatus Newtoniani de Methodo Rationum Primarum et Ultimarum' (pp. 571–92), a pompous commentary which owed more to Robert Simson's recently published posthumous *De Limitibus Quantitatum et Rationum. Fragmentum* [= *Opera Quædam Reliqua* (Glasgow, 1776): ₃1–33] than to any Newtonian text.

(3) That is, the algebraic analysis of the free 'specious' variable, which Descartes 'conjoined' with pure coordinate geometry in his Cartesian *analyse*. Indeed, Newton first wrote 'Analysin' (analysis); see note (4) following.

(4) Newton first began his introduction: 'Ut veterum inventis adderent, Analysin excogitarunt & coluerunt hodierni, & ejus beneficio longe lateç progressum est, sed non semper absç circuitu' (So as to add to the findings of the ancients, men of today have pondered and cultivated [algebraic] analysis and with its help progress has been deep and broad, but not always free from circuitousness). In sequel he continued: 'Nam quæ synthetice resolvenda erant his computis sæpenumero sunt aggressi, calculum ['fastidiosum' is cancelled] satis compositum ineuntes & conclusiones perplexas præferentes ubi res per ductum linearum more veterum nullo fere negotio peragi poterat. Per ambages verò incedere non minùs vitium est in ratiocinijs quā in constructione problematum. Sed cum ejusmodi computa satis composita esse soleant & non pertineant ad Geometriam nisi quatenus ejus conclusiones omni ex parte postmodum geometricè construantur ac demonstrentur' (For what should have been resolved synthetically they have very often attacked by the latter computational approach, engaging in a [loathsome] elaborate enough calculation and preferring an intricate finish when the object might have been attained by means of a linear construction in the manner of the ancients with scarcely any trouble. To proceed by digression is, indeed, no less a fault in the case of deductive reasoning than in the construction of problems. But since computations of this nature are usually complex enough and relate to geometry only to the extent its conclusions do, these should afterwards be constructed and demonstrated in an entirely geometrical way). On Newton's hardening view of the superiority of a geometrical synthesis 'more veterum' over mere repetition of the preliminary (algebraical) analysis in expounding theorems in geometry see the preceding introduction and compare **2**, 2, §1: note (9).

ductu unius lineæ designantur. Sic segmenta perp[endicu]lorū trianguli obliquanguli ex datis lateribus; vel segmenta diagonalium Trapezij ex datis lateribus et una diagonali per computum ægerrime derivantur et exprimuntur. Imo et perpendiculi ac segmentorū basis trianguli obliquanguli atcȝ partiū rectarum fere omnium oblique concurrentium vel circulo occurrentium derivatio et designatio ex datis sat composita est. Unde non mirum est si conclusiones[5] compositæ oriantur quas fastidio esset ratione Geometrica per totum[6] construere. Nec Analystæ[7] hoc semper faciunt sed satis esse putant si quamvis complexionem datarum quantitatum pro uno dato habeant.[8] Bona quidem est hæc Arithmetica sed inelegans Geometria ubi methodus Synthetica[9] locum obtinet, imò vitiosa si modò ratiocinia nimis complexa æque ac constructiones problematum nimis compositæ vitiū dici[10] meruerint. Animadvertens igitur nonnulla Problematum genera quæ per Analysin resolvi solent, posse simplicius resolvi (maxima saltem ex parte) per Synthesin, de ea scripsi sequentem tractatum. Sed interim cum elementa[11] Euclidea huic negotio vix sufficiant speculationi curvarum, adactus sum alia compingere. Ille fundamenta rectilineæ Geometriæ tradidit. Qui[12] curvilineas figuras dimensi sunt, eas tanquam ex partibus infinite parvis & multis constantes contemplari solent.[13] Ego vero eas considerabo tanquam crescendo generatas, argumentatus eas majores, æquales vel minores esse prout ab initio celerius, æque celeriter vel tardius crescunt. Et hanc crescendi[14] celeritatem vocabo fluxionem quantitatis. Sic ubi linea describitur per motum puncti, velocitas puncti[,] hoc est celeritas generationis lineæ erit fluxio ejus. Genuinum hunc fontem esse men[sur]andi[15] quantitates continuo fluxu juxta certam legem generatas crediderim,[16] tum ob perspicuitatem & brevitatem ratiocinij,[17] tum ob simplicitatem conclusionum & schematum quæ requiruntur. Hujus autem Geometriæ elementa

---

(5) In sequel 'adeo compositæ oriantur ut Analystæ in constructionibus earum nonnunqu[am desperent?]' (such complicated [results] should arise that algebraists would despair of their construction) was first written. An adverb 'valdè' (extremely) was subsequently inserted and then likewise cancelled.

(6) 'ple[nè]' (fully) is deleted.

(7) The (Cartesian) algebraists whose approach is wholly analytical.

(8) Newton has cancelled the concluding clause 'etsi talium quantitatū plena constructio qualem geometria requirit intollerabili esset fastidio' (even though the full construction of such quantities in the manner required by geometry would be an insufferably intricate matter).

(9) Evidently a slip of the pen for the converse 'Analytica' (analytic) required by the context.

(10) 'tribui' (conceded) was first written.

(11) The equivalent 'principia' (principles) is cancelled.

(12) Those geometers—notably Cavalieri, Torricelli, Pascal, Fermat, Huygens (post-1657; see III: 389, note (3)), Wallis and James Gregory—who, unconsciously following the lead of Archimedes in his *Method*, approached the topic of geometrical integration from the viewpoint of the theory of indivisibles, either from philosophical belief or (more usually) out of practical utility.

segments of the perpendiculars of a scalene triangle, given its sides, or the segments of the diagonals of a quadrilateral, given the sides and one diagonal, are derived and expressed by computation with the greatest difficulty. Indeed, both with the perpendicular and base segments of a scalene triangle and with the portions of almost all straight lines concurrent at an oblique angle or meeting a circle their derivation and designation in terms of given magnitudes are complex enough. Hence it is not to be wondered if[5] complicated results should arise which it would be loathsome to construct in their entirety[6] by a geometrical procedure. Nor do analysts[7] always do this but reckon it enough if they treat any combination of given quantities, however compound, as a single given magnitude.[8] This arithmetical approach is indeed fine—and a geometrical one inelegant—where the synthetic[9] method has the whiphand, but is at fault, conversely, should ever an over-elaborate analysis deserve to be called[10] no less of a defect than the over-complicated construction of a problem. Observing therefore that numerous kinds of problem which are usually resolved by [an algebraic] analysis may (at least for the most part) be more simply effected by synthesis, I have written the following treatise on the topic. At the same time, since Euclid's elements[11] are scarcely adequate for a work dealing, as this, with curves, I have been forced to frame others. He has delivered the foundations of the geometry of straight lines. Those[12] who have taken the measure of curvilinear figures have usually viewed[13] them as made up of infinitely many infinitely-small parts. I, in fact, shall consider them as generated by growing, arguing that they are greater, equal or less according as they grow more swiftly, equally swiftly or more slowly from their beginning. And this swiftness of growth[14] I shall call the fluxion of a quantity. So when a line is described by the movement of a point, the speed of the point—that is, the swiftness of the line's generation—will be its fluxion. I should have believed that this is the natural source for measuring quantities generated by continuous flow according to a precise law,[16] both on account of the clarity and brevity of the reasoning[17] involved and because of the simplicity of the conclusions and the illustrations required. The elements of this geometry, I may add, will be pre-

---

(13) 'finxerunt' (imagined) is here deleted.

(14) 'generandi' (generation) was first written.

(15) The manuscript reads 'menusandi'.

(16) A cancelled first continuation at this point reads 'tum quod nullis cedat certitudine & cæteras perspicuitate demonstrationum, alias brevitate computi, simplicitate conclusionum & constructionum & schematum ferè superet' (both because it yields to none in its precision, and since it all but surpasses others in the clarity of its proofs and still others in the conciseness of its analysis, the simplicity of its conclusions and of its constructions and diagrams).

(17) This replaces 'computi' (analysis); compare note (16).

libro primo tradentur, secundus continebit problemata quibus usus Theorematum libri primi patebit in derivatione fluxionum a quantitatibus. Tertius e contra continebit problemata in quibus quantitates derivantur a fluxionibus, quartus erit de natura curvarum in genere decg constructione problematum per earum intersectiones.[18]

Lɪʙ. 1.[19]

Dᴇғɪɴɪᴛɪoɴᴇs.[20]

1. Fluens est quod continua mutatione augetur vel diminuitur.

2. Fluxio est celeritas mutationis illius.

3. Fluxio affirmativa, sive Profluxio est celeritas tum incrementi rei affirmativæ tum decrementi negativæ sive ablativæ.

4. Fluxio negativa sive Defluxio est celeritas tum decrementi rei affirmativæ tum incrementi negativæ.

Schol.[21] Per negativam rem intellige quicquid subducitur vel restat subducendo majus e minori. Sic in negotijs humanis, Debita sunt negativa quia rem diminuunt, restantcg subducendo majorem substantiam de minori. Et incrementum Debiti est Defluxio substantiæ realis, decrementum vero profluxio ejus. Sic et in composito quovis $A - B$ incrementum negativi B defluxio est quia diminuit compositum.

5. Eodem modo fluere dicuntur quæ omnia fluunt affirmativè vel omnia negativè: diverso modo quorum unum fluit affirmativè alterum negative.

6. Permanens voco et stabile vel etiam datum ac determinatum quod non fluit.

---

(18) His still extant manuscripts, reproduced in sequel, reveal that Newton drafted and extensively revised a virtually complete version of the first book (on the direct method of fluxions, treated in a geometrical framework) and briefly outlined the opening eleven problems of the second (applications to tangents, normals and minimax properties). There is no evidence to suggest—and we may conclude it to be highly unlikely—that he ever mapped out the content of the projected third and fourth books (on the inverse geometrical method of fluxions and on the nature of curves and their use in constructing problems respectively). As on III: 350–2 the former would no doubt make heavy use of Axioms 2 and 4 following, while the latter would probably incorporate a revised version of II, 3, 2, §2.

(19) This subhead is crowded in after the main title 'Geometria Curvilinea' in the manuscript (Add. 3963.7: 46ʳ) but is here more logically resited. The distinction of the 'Geometria' into separate books is manifestly a late decision whose effects have not been allowed for in the following text. In agreement with Newton's preceding scheme we have split off the final problems as a separate 'Liber 2', but it should be understood that references there to Propositions are to those of the present 'Lib. 1'. All of the first three books in Newton's scheme clearly depend narrowly on the following definitions, axioms and postulates which are, as he has earlier asserted, designed to complement their Euclidean equivalents. A preliminary version of these 'Definitiones', 'Axiomata' and 'Petitiones' together with Propositions 1–15 is reproduced in Appendix 1.1 below.

sented in the first book. The second will contain problems in which the use of the theorems of the first book in the derivation of fluxions from quantities will be plain. The third, conversely, will contain problems in which quantities are derived from their fluxions, while the fourth will be on the nature of curves in general and the construction of problems by their intersections.[18]

Book 1[19]

Definitions[20]

1. A fluent is what is increased or diminished by continuous change.

2. Its fluxion is the swiftness of that change.

3. A positive fluxion, or profluxion, is the swiftness both of increment of a positive thing and of decrement of a negative or detractive one.

4. A negative fluxion, or defluxion, is the swiftness both of decrement of a positive thing and of increment of a negative one.

Scholium.[21] By a negative thing understand any that is subtracted or remains after subtracting a greater from a less. So, in human affairs, debts are negative because they diminish a thing and are the remnant of subtracting a greater substance from a less. And an increment of a debt is a defluxion of real substance, while indeed a decrement is its profluxion. So also, in any compound $A - B$, increment of the negative thing B is a defluxion because it diminishes the compound.

5. Things are said to flow in the same manner, all of which flow positively or all negatively; and in a different manner when one of them flows positively, the other negatively.

6. I call permanent and stable, or also given and determined, what does not flow.

---

(20) In general these codify (and often make explicit) the equivalent definitions invoked in the 1671 fluxional tract (see especially III: 72), though Definitions 7, 8 and 9 owe something to those which open the contemporary 'Problems for construing æquations' (see II: 450–2).

(21) In line with his preliminary draft (see Appendix 1.1: note (3)) Newton first entered the content of this scholium as separate definitions '5. Affirmativum autem est quod rem positivam auget' (A positive, however, is that which increases a positive thing) and '6. Et negativum seu privatum quod diminuit' (And a negative, that is, privation, that which diminishes it), to which he then began to add 'Schol. Sic si linea AB pro positiva habeatur' (Scholium. Thus if the line AB be taken as positive...) but broke off. His accompanying sketch (here accurately reproduced) suggests that Newton intended to illustrate the effect of adding AD to AB to form the sum $AB + (AD$ or$) BC = AC$ and of taking away AD (to yield DB) or its negative equal $AE = -AD$ (to form

$$AD - (-AE) = ED).$$

7. Polus lineæ rectæ mobilis est punctum in quo recta illa circumacta incipit locum linearem quem modo occupavit secare, vel in quo desinit secare si in locum istum revertitur.

8. Locus puncti moventis est linea sive recta sive curva quam punctum istud motu suo describit.

9. Determinatio vel Plaga motûs puncti istius[22] est positio rectæ lineæ tangentis Curvam illam ad punctum illud movens.

[23]10. Quantitas per quantitatem exaltari dicitur quando multiplicatur per rationem ejus ad unitatem.

11. Quantitas per quantitatem deprimi dicitur quando dividitur per rationem ejus ad unitatem.

Axiomata.[24]

1. Quæ sunt perpetim æqualia, fluxionibus æqualibus generantur.
2. Quæ fluxionibus æqualibus simul generantur sunt æqualia.
3. Quæ sunt perpetim in ratione data, fluxiones habent in eadem ratione.
4. Quæ fluxionibus in data ratione simul generantur sunt in ratione fluxionum.

Nota simul generari intelligo quæ tota eodem tempore generantur.[25]

5. Fluxio totius æquatur fluxionibus partium simul sumptis.
6. Fluxiones quantitatum sunt in prima ratione partium nascentium, vel quod perinde est, in ultima ratione partium istarum per defluxionem vicissim evanescentium.[26] Sint *AB, CD* fluentes quanti-tates. Fluant hæ donec evadant *AE, CF*. Dico fl: *AB* esse ad fl: *CD* in prima ratione quam partes generatæ *BE, DF* habebant, id est in ratione quam partes istæ habebant in principio

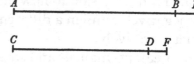

generationis, vel quod perinde est, in ultima ratione quam possunt habere ubi *AE* et *CF* vicissim defluentes evadant *AB* et *CD*.

Petitiones.

1. Lineas quasvis quacunꝗ[27] ratione Geometrica movere. Per rationem Geometricam intelligo talem rationem movendi in qua quævis positio lineæ motæ potest geometricè designari.

---

(22) 'moventis' (moving) is cancelled.
(23) These two final definitions following, not present in Newton's preliminary version (Appendix 1.1), are a late insertion in the manuscript.
(24) The first five axioms are taken over with little change from III: 330, while Axiom 6 borrows heavily from an earlier portion of the same 1671 tract (see III: 72, 114).
(25) This is to avoid any difficulties over a possible added constant in the 'integration' of a fluxional quantity.

7. The pole of a mobile straight line is the point in which that line, as it goes round, begins to intersect the linear position it occupied a moment before or in which it ceases to intersect it if it returns to that position.

8. The locus of a moving point is the line, straight or curved, which that point describes in its movement.

9. The determination or direction of motion of that[22] point is the position of the straight line touching that curve at that moving point.

[23]10. A quantity is said to be heightened by a quantity when it is multiplied by its ratio to unity.

11. A quantity is said to be lowered by a quantity when it is divided by its ratio to unity.

AXIOMS

1. What are perpetually equal are generated by equal fluxions.

2. What are simultaneously generated by equal fluxions are equal.

3. What are perpetually in a given ratio have their fluxions in the same ratio.

4. What are simultaneously generated by fluxions in a given ratio are in the ratio of the fluxions.

Note: I understand to be simultaneously generated things which are wholly generated in the same time.[25]

5. The fluxion of the whole is equal to the fluxions of its parts taken simultaneously.

6. Fluxions of quantities are in the first ratio of their nascent parts or, what is exactly the same, in the last ratio of those parts as they vanish by defluxion.[26] Let AB, CD be fluent quantities, and let them flow till they come to be AE, CF. I say that fl (AB) is to fl (CD) in the first ratio which the parts BE, DF have [to each other]—that is, in the ratio which those parts have at the beginning of their generation—or, what is exactly the same, in the last ratio they can have when AE and CF, mutually defluent, come to be AB and CD.

POSTULATES

1. That any lines may move in any[27] geometrical fashion whatever. By 'geometrical fashion' I understand such a fashion of moving that any position of a line moved in it can be geometrically designated.

---

(26) On this Barrovian concept of a 'first' or 'last' ratio of the 'nascent' or 'evanescent' fluent increments of two related quantities, here for the first time explicitly made the basis of his concept of a fluxion, see III: 70–2, especially 71, note (81).

(27) 'data' (given) is cancelled.

2. Datas esse Lineas quas puncta vel intersectiones linearum ratione geo-metrica motarum describunt.[28]

Prop 1.

Si trium semper continuè proportionalium quantitatum media detur et extremæ fluant, erit profluxio unius extremæ ad defluxionem alterius ut prior extrema ad illam alteram. Sint $B.C::C.D$, et dato C, erit $B.D::$ defl: $B.$ profl: D.

Exponantur enim hæ quantitates per tres lineas AB, AC, AD, quarum extremæ in directum jaceant, et media ijs ad communem terminum[29] perpendicu-lariter insistat. Fluant jam extremæ AB, AD donec evadant AG, AE, et erunt etiam ex Hypothesi AG AC AE continue propor-tionales. Quare actis BC, DC, GC, EC, triangula BCD, GCE erunt rectan-

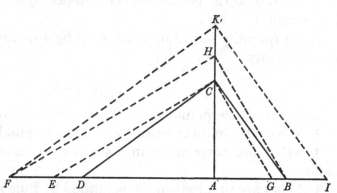

gula ad C, adeoq̃ dempto communi angulo DCG, qui restabunt anguli BCG, DCE æquales erunt. Age ipsi $[G]C$ parallelam BH et occurrentem AC in H et triangula BCH, CDE erunt similia, ut et triangula CAG, HAB, adeoq̃ est $DE.CH::CD.BC::AD.AC$ & $CH.BG::AC.AG$, & ex æquo $DE.BG::AD.AG$. Refluant jam AG et AE dum coeant cum AB et AD, et ultima ratio evanes-centium partium DE & BG erit AD ad AB. Quare per ax 6, fluxiones ipsarum AD et AB sunt in eadem ratione. Q.E.D.

Cor. Si quatuor proportionalium duo media dantur vel etiam si detur rectangulum mediorum, erit defluxio unius extremi ad profluxionem alterius, ut extremum prius ad hoc alterum. Sit $A.B::C.D$ et dato $B\times C$ erit

$$A.D::\text{defl}\,A.\text{profl}\,D.$$

Nam medium proportionale in[ter] B et C hoc est inter A ac D dabitur.

Prop 2.

Si trium semper proportionalium quantitatum una extrema datur, erit fluxio mediæ ad fluxionem alterius extremæ ut extrema data ad duplum mediæ. Sint $A.B::B.C$ et dato A, erit $A.2B::$ fl: $B.$ fl: C.

---

(28) These two postulates, an addition to the 'static' set given by Euclid in his *Elements* which Newton in his preceding introduction has observed to be inadequate for his present purposes, are evidently founded in part on the three 'postulata' of his 'Problems for construing æquations' (see II: 452). The narrow indebtedness of Propositions 1–10 following to Theorems 1

2. That there are given the lines described by points or the intersections of lines moved in a geometrical fashion.[28]

Proposition 1

If of three quantities ever in continued proportion the middle one be given and the extremes flow, then the profluxion of one extreme will be to the defluxion of the other as the former extreme to the other. Let $B:C = C:D$ and, when C is given, there will be

$$B:D = -\mathrm{fl}\,(B):+\mathrm{fl}\,(D).$$

For let these quantities be exhibited by the three lines AB, AC, AD, the extreme ones of which are to lie in a straight line, while the middle one shall stand perpendicularly to them at their common end-point.[29] Now let the extremes AB, AD flow till they come to be AG, AE, and then also by hypothesis will AG, AC, AE be continued proportionals. Consequently, on drawing BC, DC, GC and EC, the triangles BCD, GCE will be right-angled at C and hence, when the common angle $D\widehat{C}G$ is taken away, the angles $B\widehat{C}G$, $D\widehat{C}E$ which remain will be equal. Draw BH parallel to GC, meeting AC in H, and the triangles BCH, CDE will be similar, also the triangles CAG, HAB; accordingly

$$DE:CH = CD:BC = AD:AC \quad \text{and} \quad CH:BG = AC:AG,$$

and *ex æquo* $DE:BG = AD:AG$. Now let AG and AE flow back till they coincide with AB and AD, and the last ratio of the vanishing parts DE and BG will be AD to AB. Hence by axiom 6 the fluxions of AD and AB are in the same ratio. As was to be demonstrated.

Corollary. If of four proportionals the two middle ones are given, or also should the product of the middle ones be given, then the defluxion of one extreme will be to the profluxion of the other as the former extreme to the latter one. Let $A:B = C:D$ and, when $B \times C$ is given, there will be

$$A:D = -\mathrm{fl}\,(A):+\mathrm{fl}\,(D).$$

For the mean proportional between B and C, that is, between A and D, will be given.

Proposition 2

If of three ever proportional quantities one extreme is given, then the fluxion of the middle one will be to the fluxion of the other extreme as the given extreme to twice the middle one. Let $A:B = B:C$ and, when A is given, there will be $A:2B = \mathrm{fl}\,(B):\mathrm{fl}\,(C)$.

---

and 2 of the geometrical addendum to his 1671 tract (III: 330–8) is more immediately obvious in the preliminary version of the 'Geometria' reproduced in Appendix 1.1 below.

(29) The point A in Newton's figure.

Exponantur enim hæ quantitates per lineas $AB.AC.AD$, nempe perpendiculum et segmenta basis trianguli rectanguli BCD ut ante, et fluant AC et AD donec evadant AH et AF. Age CE parallelam HF & propter similia triangula $BCH\ CDE$ & $CAE,\ HAF$, erit

$$CH.DE(::BH.CD)::AB.AC. \quad \& \quad CH.EF(::AH.AF)::AB.AH.$$

Quare componendo $CH.DF::AB.AC+AH$. Defluant jam AH, AF donec evadant[30] AC et AD, et $AC+AH$ fiet $2AC$, adeoq ultima ratio partium evanescentium CH et DF erit AB ad $2AC$. Quare per ax 6 fluxiones ipsarum AC et AD sunt in ista ratione. Q.E.D.

Prop 3.

Si sint tres fluentes et semper proportionales quantitates: fluxio mediæ exaltata per duplum mediæ, æqualis erit fluxionibus extremarum reciprocè exaltatis per extremas. Sint $A.B::B.C$ et erit $2Bb=Ac+Ca.$[31]

Exponantur enim quantitates per lineas AB, AC, AD, et constituatur triangulum BCD rectangulum ad C ut ante. Fluant hæ donec evadant AI, AK, AF. Sitq AH semper media proportionalis inter AB et AF, et actis rectis BH, HF, IK, KF, per Prop 2 erit $2AH.AB::af.ah$,[32] sive $\dfrac{AB\times af}{2AH}=ah$. Item per ea quæ ibi demonstra[ta] sunt, $AH+AK.AF::BI.HK$, atq adeo sub initio fluxionis ubi AH & AK æquales erant $2AH.AF::BI.HK::$ (per ax. 6) bi vel $ai.hk$. Quare $\dfrac{AF}{2AH}ai=hk$. Sed per ax 5 est $ah+hk=ak$. Ergo

$$\frac{AB}{2AH}af+\frac{AF}{2AH}ai=ak. \quad \text{hoc est} \quad \frac{AB}{2AC}ad+\frac{AD}{2AC}ab=ac:$$

nam de initio fluxionis agimus ubi puncta F, I et K cum punctis $D\ B$ et C coincidunt & ab ipsis tantum incipiunt discedere. Exalta igitur fluxiones omnes per $2AC$ et fiet $AB\times ad+AD\times ab=2AC\times ac.$[33] Q.E.D.

Schol.[34] Hæc omnia sic brevius demonstrari possent. Sint tres fluentes et perpetim proportionales quantitates A, B, C. Fluant hæ donec evadant

---

(30) This replaces 'conveniant cum' (coincide with).

(31) Without any prior warning Newton introduces a new notation, taking capitals to represent fluent magnitudes and the corresponding lower-case letters their fluxions. Here understand that a, b, c are the fluxions of the corresponding quantities A, B, C. In his more usual notation (compare III: 334: Theorem 1, 'Cor: 1') Newton would write this enunciation as '$2B\times$ fl $B=A\times$ fl $C+C\times$ fl A'.

(32) By an extension of the preceding notation (see note (31) above) Newton now takes af and ah to be the geometrical fluxions of the fluent line-segments AF and AH. Below, correspondingly, understand that ab, ac, ad, ai, ak, bi and hk are the respective fluxions of AB, AC, AD, AI, AK, BI and HK.

For let these quantities be exhibited by the lines *AB*, *AC*, *AD*—namely, the perpendicular and the base segments of the right-angled triangle *BCD* as before—and let *AC* and *AD* flow till they come to be *AH* and *AF*. Draw *CE* parallel to *HF* and because of the similar triangles *BCH*, *CDE* and *CAE*, *HAF* there will be

$$CH:DE\,(\text{or } BH:CD) = AB:AC \quad \text{and} \quad CH:EF\,(\text{or } AH:AF) = AB:AH.$$

Hence, by compounding, $CH:DF = AB:AC+AH$. Now let *AH*, *AF* flow back till they come to be[30] *AC* and *AD*, and $AC+AH$ will become $2AC$; accordingly, the last ratio of the vanishing parts *CH* and *DF* will be *AB* to $2AC$. Hence by axiom 6 the fluxions of *AC* and *AD* are in that ratio. As was to be demonstrated.

Proposition 3

Should there be three fluent, ever proportional quantities, the fluxion of the middle one heightened by twice the middle will be equal to the fluxions of the extremes heightened reciprocally by the extremes. Let $A:B = B:C$ and there will be $2Bb = Ac+Ca$.[31]

For let the quantities be exhibited by the lines *AB*, *AC*, *AD* and the triangle *BCD* right-angled at *C* be set up as before. Let these flow till they come to be *AI*, *AK*, *AF* and let *AH* be ever the mean proportional between *AB* and *AF*. Then, after the straight lines *BH*, *HF*, *IK* and *KF* are drawn, by Proposition 2 there will be $2AH:AB = af:ah$,[32] that is, $ah = (AB/2AH)\,af$. Likewise, by what are there demonstrated, $AH+AK:AF = BI:HK$ and so at the start of fluxion, where *AH* and *AK* were equal, $2AH:AF = BI:HK$, that is, by axiom 6, (*bi* or) $ai:hk$. Consequently $hk = (AF/2AH)\,ai$. But by axiom 5 $ah+hk = ak$. Therefore $(AB/2AH)\,af+(AF/2AH)\,ai = ak$, that is,

$$(AB/2AC)\,ad+(AD/2AC)\,ab = ac:$$

for we are speaking of the beginning of fluxion when the points *F*, *I* and *K* are coincident with the points *D*, *B* and *C* and are merely starting to depart from them. Heighten, therefore, all fluxions by $2AC$ and there will prove to be $AB \times ad+AD \times ab = 2AC \times ac$.[33] As was to be demonstrated.

Scholium.[34] All these things might be given a shorter demonstration in the following way. Let *A*, *B*, *C* be three fluent and ever proportional quantities, and

---

(33) In his earlier semi-verbal notation (see III, **1**, **2**, §4 and compare note (31) above) Newton would write this, as indeed he did in his immediately preceding draft (Appendix 1.1), in the form '$AB \times \text{fl } AD + AD \times \text{fl } AB = 2AC \times \text{fl } AC$'.

(34) Observe that, to avoid an intolerable confusion with his newly invented notation for fluxions (see note (31) above), Newton has had to convert the lower-case fluents in his draft version (Appendix 1.1 below) into more cumbersome upper-case equivalents.

$A+D$. $B+E$. $C+F$. Ergo $AC=BB$, & $A+D\times C+F=B+E\times B+E$, sive $AC+AF+DC+DF=BB+2BE+EE$. aufer æquales AC & BB et restabunt $AF+DC+DF=2BE+EE$, sive $\frac{F}{E}A+\frac{D}{E}\times C+F=2B+E$. Defluant jam quantitates $A+D$. $B+E$ et $C+F$ donec evadant $A.B.C$, et $\frac{D}{E}\times C+F$ fiet $\frac{D}{E}\times C$ atɋ $2B+E$ fiet $2B$. Quare $\frac{F}{E}A+\frac{D}{E}C=2B$, sive per ax 6, $\frac{c}{b}A+\frac{a}{b}C=2B$. hoc est $cA+aC=2bB$, ut in hac tertia propositione. Unde si B datur[,] hoc est b sit nulla erit $cA+aC=0$, sive A. $C::-a$. $+c$ ut in Prop. 1, & si A datur[35] erit $cA=2bB$ sive A. $2B::b$. c [ut in Prop 2].

Prop 4.[36]

Positis tribus continue proportionalibus $A.B.C$ si summa extremorum stabilis[37] est, fluxio minoris extremæ erit ad fluxionem mediæ, ut duplum mediæ ad differentiam extremarum. $2B$. $C-A::a$. b.

Nam per Prop 3 est $Ac+Ca=2Bb$. Et præterea cum $A+C$ stabile sit, erit $a+c=0$ & $Aa+Ac=0$. Aufer hoc de $Ac+Ca$ $[=2Bb]$ et restabit $Ca-Aa=2Bb$. Unde $2B$. $C-A::a$. b.

Prop 5.

Positis tribus continue proportionalibus A, B, C si differentia extremarum datur, fluxio alterutrius extremæ[38] erit ad fluxionem mediæ ut duplum mediæ ad summam extremarum. $2B$. $A+C::a$. b.

Demonstratur sicut Propositio superior.[39]

Prop 6.

Positis tribus continue proportionalibus $A.B.C$, si summa primæ et secundæ detur, erit fluxio secundæ ad fluxionem tertiæ, ut prima ad duplum secundæ auctum tertia. A. $2B+C::b$. c.

Nam per Prop 3 est $Ac+Ca=2Bb$, et ex hypothesi $A+B$ stabile est, adeoɋ $a+b=0$ et $Ca+Cb=0$. Aufer hoc de $Ac+Ca$ $[=2Bb]$ et restabit $Ac-Cb=2Bb$. Adde Cb utrinɋ et erit $Ac=2Bb+Cb$. Unde A. $2B+C::b$. c.

---

(35) And so $a = 0$.

(36) Propositions 4–8 following are, as Newton's preliminary draft emphasizes (see Appendix 1.1: note (16)) straightforward adaptations of Corollaries 2–6 of Theorem 1 of his earlier addendum to the 1671 tract (see III: 334–6).

(37) That is, non-fluent (constant); see Definition 6.

(38) If the difference of the fluents A and C is constant, their fluxions a and c will be equal.

(39) Namely, on suitably substituting $-A$ for A and $-a$ for a.

let them flow till they come to be $A+D$, $B+E$, $C+F$. In consequence $AC = B^2$ and $(A+D)(C+F) = (B+E)^2$, that is,

$$AC+AF+DC+DF = B^2+2BE+E^2.$$

Take away the equals AC and B^2 and there will remain

$$AF+DC+DF = 2BE+E^2 \quad \text{or} \quad (F/E)A+(D/E)(C+F) = 2B+E.$$

Now let the quantities $A+D$, $B+E$ and $C+F$ flow backward till they end as A, B and C: $(D/E)(C+F)$ will then become $(D/E)C$ while $2B+E$ will become $2B$. Hence $(F/E)A+(D/E)C = 2B$; in other words (by axiom 6)

$$(c/b)A+(a/b)C = 2B, \quad \text{that is,} \quad cA+aC = 2bB$$

as in the present third proposition. From this if B is given—that is, should b be zero—, then $cA+aC = 0$ or $A:C = -a:+c$ (as in Proposition 1); and if A is given,[35] then $cA = 2bB$ or $A:2B = b:c$ (as in Proposition 2).

Proposition 4[36]

Positing three continued proportionals A, B, C, if the sum of the extremes is stable[37] the fluxion of the lesser extreme will be to the fluxion of the middle one as twice the middle to the difference of the extremes: $2B:C-A = a:b$.

For by Proposition 3 there is $Ac+Ca = 2Bb$. Further, since $A+C$ is stable, there will be $a+c = 0$ and $Aa+Ac = 0$. Take this from $Ac+Ca = 2Bb$ and there will remain $Ca-Aa = 2Bb$. Whence $2B:C-A = a:b$.

Proposition 5

Positing three continued proportionals A, B, C, if the difference of the extremes is given, the fluxion of either extreme[38] will be to the fluxion of the middle one as twice the middle to the sum of the extremes: $2B:A+C = a:b$.

The demonstration is similar to that of the previous proposition.[39]

Proposition 6

Positing three continued proportionals A, B, C, if the sum of the first and the second be given, the fluxion of the second will be to the fluxion of the third as the first to twice the second increased by the third: $A:2B+C = b:c$.

For by Proposition 3 $Ac+Ca = 2Bb$, while by hypothesis $A+B$ is stable, so that $a+b = 0$ and $Ca+Cb = 0$. Take this from $Ac+Ca = 2Bb$ and there will remain $Ac-Cb = 2Bb$. Add Cb to each side and there will be $Ac = 2Bb+Cb$. Whence $A:2B+C = b:c$.

Prop 7.

Positus tribus continue proportionalibus $A . B . C$ si differentia primæ et secundæ stabilis est, erit fluxio alterutrius ad fluxionem tertiæ ut prima ad duplum secundæ diminutum tertia. $A . 2B - C :: b . c.$

Demonstratur ut Prop 6.[40]

Prop 8.

Positis quotcunq̃ continuè proportionalibus quarum una sit stabilis et cæteræ fluant: fluxiones fluentium erunt inter se ut fluentes illæ ductæ in numerum terminorum quibus distant a dato illo termino.

Sint $A . B . C . D . E . F . G$ continuè proportionales et si datur C, erit

$$-2A . -B . D . 2E . 3F . 4G :: a . b . d . e . f . g.$$

Nam in tribus continue proportionalibus $C . D . E$ propter stabilem C est $D . 2E :: d . e$ per prop 2. Insuper per prop 3 est $Df + Fd = 2Ee.$ et inde cum fuerit $d . e (:: D . 2E) :: E . 2F$ sive $2Fd = Ee$ et $4Fd = 2Ee$, erit $Df + Fd = 4Fd.$ hoc est $Df = 3Fd$ sive $D . 3F :: d . f.$ Præterea $C . E . G$ sunt continue proportionales et proinde per Prop 2 est $e . g :: C . 2E :: 2E . 4G.$ Et sic in infinitum. Ex altera autem parte est $D . C :: C . B$ adeoq̃ per Prop 1 defluxio $-b$ ad profluxionem d ut B ad D vel $b . d :: -B . D.$ Præterea est $C . B :: B . A$ et inde per prop 2,

$$[b] . [a] :: C . 2B :: B . 2A :: -B . -2A.$$

Et sic in infinitum.[41]

Coroll. Fluxio Potestatis est ad fluxionem radicis suæ, ut Potestas multiplicata per indicem suam ad radicem. Verbi gratia fluxio cubi A^3 est ad fluxionem radicis A ut $3A^3$ ad A: nam $1 . A . A^2 . A^3$ sunt continuè proportionales. Sic et fl$: A^{\frac{2}{3}} .$ fl$: A :: \frac{2}{3} A^{\frac{2}{3}} . A.$ Nam $1 . A^{\frac{1}{3}(42)} . A^{\frac{2}{3}} . A$ sunt continuè proportionales, adeoq̃ fluxiones earum ut $0 . A^{\frac{1}{3}(42)} . 2A^{\frac{2}{3}} . 3A.$

Prop 9.

Si quatuor proportionalium prima datur: fluxio quartæ exaltata per hanc primam æquabitur fluxionibus mediarum mutuò exaltatis per medias.

Sint $A . B :: C . D.$ et detur A; sitq̃ E media proportionalis inter B et C, et eadem erit etiam media proportionalis inter A ac D. Quare per Prop 2 & 3 est $A . 2E :: e . d$ sive $Ad = 2Ee$ et $Bc + Cb = 2Ee.$ adeoq̃ $Ad = Bc + Cb.$ Q.E.D.

---

(40) On replacing B and b by their negatives $-B$ and $-b$ at appropriate places.

(41) See III: 336, note (23).

(42) Read '$A^{\frac{1}{3}}$'.

(43) Newton introduces a new modification in his previous notation (see note (31) above) so as to comprehend the fluxions of compound quantities: to clarify the sense we 'translate'

Proposition 7

Positing three continued proportionals A, B, C, if the difference of the first and second is stable, then the fluxion of either one will be to the fluxion of the third as the first to twice the second diminished by the third: $A:2B-C=b:c$.

The demonstration is as in Proposition 6.[40]

Proposition 8

Positing any number of continued proportionals, one of which shall be stable and the others fluent, the fluxions of the fluent ones will be to one another as those fluents multiplied by the number of terms they are distant from the given term.

Let A, B, C, D, E, F, G be the continued proportionals and, if C is given, then $-2A:-B:D:2E:3F:4G = a:b:d:e:f:g$. For in the case of the three continued proportionals C, D, E, because C is stable there is $D:2E = d:e$ by Proposition 2. Moreover, by Proposition 3 $Df+Fd = 2Ee$ and accordingly, since

$$d:e \text{ (or } D:2E) = E:2F, \quad \text{that is,} \quad 2Fd = Ee \quad \text{and} \quad 4Fd = 2Ee,$$

there will be $Df+Fd = 4Fd$—that is, $Df = 3Fd$ or $D:3F = d:f$. Further, C, E, G are continued proportionals and consequently by Proposition 2 there is $e:g = C:2E = 2E:4G$. And so on indefinitely. On the other hand, however, it is $D:C = C:B$, so that by Proposition 1 the defluxion $-b$ is to the profluxion $+d$ as B to D, or $b:d = -B:D$. Further, $C:B = B:A$ and thence, by Proposition 2, $b:a = C:2B = B:2A = -B:-2A$. And so on indefinitely.[41]

Corollary. The fluxion of a power is to the fluxion of its root as the power multiplied by its index to its root. For instance, the fluxion of the cube A^3 is to the fluxion of its root A as $3A^3$ to A; for 1, A, A^2, A^3 are continued proportionals. Likewise also, fl $(A^{\frac{2}{3}}):\text{fl }(A) = \frac{2}{3}A^{\frac{2}{3}}:A$; for 1, $A^{[\frac{1}{3}]}$, $A^{\frac{2}{3}}$, A are continued proportionals and so their fluxions are as 0, $A^{[\frac{1}{3}]}$, $2A^{\frac{2}{3}}$, $3A$.

Proposition 9

If of four proportionals the first is given, then the fluxion of the fourth heightened by this first one will equal the fluxions of the middles heightened reciprocally by the middles.

Let $A:B = C:D$ and A be given; again, let E be the mean proportional between B and C, and the same will also be the mean proportional between A and D. In consequence, by Propositions 2 and 3, it is $A:2E = e:d$ or $Ad = 2Ee$, and $Bc+Cb = 2Ee$, so that $Ad = Bc+Cb$. As was to be demonstrated.

---

his fluxional rectangles by his more usual operator 'fl'. This usage of an enveloping square or rectangle must sharply be distinguished from its early employment in the 'De Analysi' (compare II: 226, note (78)) to represent the converse operation of 'integrating' a fluxion to obtain its fluent.

Corol. Hinc si duæ vel plures fluentes quantitates sese multiplicent, fluxio facti æquabitur fluxionibus multiplicantium mutuo exaltatis per multiplicantes si duæ sunt, vel per factos reliquorum multiplicantium si sunt plures. [Sic][43] $\boxed{AB}=Ba+Ab.$ nam $1.A::B.AB.$ Item $\boxed{ABC}=BCa+ACb+ABc.$ nam

$$\boxed{ABC}=BCa+A\times\boxed{BC} \quad \text{et} \quad \boxed{BC}=Cb+Bc.$$

ergo $\boxed{ABC}=BCa+ACb+ABc.$ Et sic de pluribus multiplicantibus.

[Prop 10.

Positis quatuor proportionalibus fluxiones extremarum mutuò ductæ in extremas æquantur fluxionibus mediarum mutuò ductis in medias.

Sint $A.B::C.D$ sitcʒ E media proportionalis inter B & C ut et inter A & D et per Prop 3 erit $Ad+Da=2Ee=Bc+Cb.$][44]

Prop 11.

[45]*Si Trianguli perpetim rectanguli hypotenusa datur, profluxio unius lateris erit ad defluxionem alterius, ut illud alterum ad latus primum, et demisso perpendiculo ab angulo recto in Hypotenusam, fluxio perpendiculi erit ad fluxionem alterutrius*[46] *lateris ut differentia segmentorum basis ad latus alterum, & ad fluxionem vero alterutrius segmenti basis ut differentia segmentorum ad duplum perpendiculi.*

Sit istud triangulum ABC rectangulum ad B cujus latus datum AC[47] bisecetur in E et demisso ad hoc latus perpendiculo $BD,$ erit ED semidifferentia segmentorum $AD,$ $DC.$ Dico jam esse

flBD.flDC.flBC.defl$AB::2ED.2BD.AB.BC.$

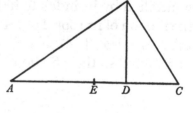

Nam quia $DC.BD::BD.AD$ et summa extremarum DC & AD datur erit per prop 4 flBD.fl$DC::2ED.2BD.$ Deinde quia $AC.BC::$ $BC.DC$ et AC datur erit per prop 2 flDC.fl$BC::2BC.AC::2BD.AB.$

---

(44) We borrow this Proposition 10, in a suitably recast form, from the preliminary draft on Add. 3963.7: 48 (reproduced in Appendix 1.1). The manuscript revise which now omits it is clearly incomplete, for the next theorem is 'Prop 11'. It may well be that the present proposition, which in sequence would have been entered at the top of a new sheet of paper, was inserted on a loose separate slip which has now disappeared.

(45) In line with the summary sketch in his preliminary draft (see Appendix 1.1: note (22) below) Newton first began with a straightforward transcript of 'Theor. 2' on III: 338, namely: '*In Triangulo quovis perpetim rectangulo*[...]' (*In any perpetually right-angled triangle*...). He then abruptly changed his mind, bringing forward instead the next two propositions (see Appendix 1.2: notes (32) and (33)) which are its particular cases. Compare note (53) below.

Corollary. Hence if two or more fluent quantities be multiplied into one another, the fluxion of the product will equal the fluxions of the multiplying quantities mutually heightened (if there are two) by the multiplying quantities, or (if there are more) by the products of the remaining multiplying quantities. Thus[43] fl $(AB) = Ba + Ab$; for $1:A = B:AB$. Likewise

$$\text{fl}\,(ABC) = BCa + ACb + ABc; \quad \text{for} \quad \text{fl}\,(ABC) = BCa + A \times \text{fl}\,(BC)$$

while fl $(BC) = Cb + Bc$, and therefore fl $(ABC) = BCa + ACb + ABc$. And so in the case of more multiplying quantities.

[*Proposition 10*

Positing four proportionals, the fluxions of the extremes reciprocally multiplied into the extremes equal the fluxions of the middles reciprocally multiplied into the middles.

Let $A:B = C:D$ and E be the mean proportional between B and C, as also between A and D, and by Proposition 3 there will be

$$Ad + Da = 2Ee = Bc + Cb.][44]$$

Proposition 11

[45]*If in a perpetually right-angled triangle the hypotenuse is given, the profluxion of one side will be to the defluxion of the other as that second one to the first; and, when a perpendicular is let fall from the right angle onto the hypotenuse, the fluxion of the perpendicular will be to the fluxion of either*[46] *side as the difference of the base segments to the other side, and to the fluxion, indeed, of either segment of the base as the difference of the segments to twice the perpendicular.*

Let that triangle be ABC, right-angled at B, and let its given side AC[47] be bisected in E; then, when the perpendicular BD is let fall to this side, ED will be the half-difference of the segments AD, DC. I now assert that

$$\text{fl}\,(BD):\text{fl}\,(DC):\text{fl}\,(BC):-\text{fl}\,(AB) = 2ED:2BD:AB:BC.$$

For, because $DC:BD = BD:AD$ and the sum of the extremes DC and AD is given, there will (by Proposition 4) be fl $(BD):\text{fl}\,(DC) = 2ED:2BD$. Next, because $AC:BC = BC:DC$ and AC is given, there will (by Proposition 2) be fl $(DC):\text{fl}\,(BC) = 2BC:AC = 2BD:AB$. Consequently, by joining the ratios it

(46) Newton hesitantly first altered this to 'majoris' (the greater) and then returned to his first choice of word.

(47) Doubtless seeing that AC is self-evidently the triangle's hypotenuse Newton has cancelled a following phrase 'subtendens angulum rectum ABC' (subtending the right angle $A\hat{B}C$).

Quare connectendo rationes est fl BD.fl BC::2ED.AB. Et simili ratiocinio erit fl BD.defl AB[::]2ED.BC. Adeoꝗ ex æquo fl BC.defl AB::AB.BC.(48) Q.E.D.

Prop 12.

Si Trianguli perpetim rectanguli latus angulo recto conterminum detur erit fluxio alterius lateris ad fluxionem Hypotenusæ ut Hypotenusa ad latus illud fluens. Et demisso perpendiculo in Hypotenusam,(49) *fluxio basis erit ad defluxionem segmenti dato lateri contermini ut basis ad segmentum illud, e[t] ad fluxionem alterius segmenti ut basis ad summam basis et segmenti prioris, e[t] ad fluxionem perpendiculi ut latus fluens ductum in Hypotenusam ad latus alterum ductum in conterminum segmentum basis.*

In triangulo ABC rectangulo ad B, detur latus AB. Et demisso perpendiculo $B[D]$ dico esse fl BC.fl AC.defl AD.fl DC.fl BD:: AC^q.$BC\times AC$.$BC\times AD$.$BC\times\overline{AC+AD}$.$AB\times AD$. (50)Nam ad AC erige normalem CE occurrentem AB in E et erunt AB.BC.BE∹ adeoꝗ per Prop 2 $\frac{1}{2}AB$.BC::fl BC.fl BE. Item AB.AC.AE∹ & inde per Prop 2, $\frac{1}{2}AB$.AC::fl AC.fl AE. Jam cum AB non fluat adeoꝗ fl AE et fl BE æquales sint, erit ex æquo perturbatè BC.AC::fl AC.fl BC. Q.E.D. Præterea est AC.AB::AB.AD. adeoꝗ per Prop 1, AC.AD::fl AC.defl AD vel $-$fl:AD et componendo AC.$AC+AD$[::]fl AC.fl DC. [Q.E.D.] Deniꝗ per prop 11 est BD.AD::fl AD(51).fl BD::BC.AB. sed erat AC.AD::fl AC.fl AD.(51) Quare connectendo rationes est $BC\times AC$.$AB\times AD$:: fl AC.fl BD. Q.E.D.

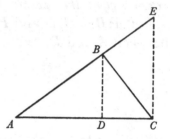

Cor. Est

fl BC.[fl] AC.[defl] AD.[fl] DC.[fl] BD::AE.EC.BD.$EC+BD$.AD.(52)

---

(48) Compare the predraft ('13') in Appendix 1.2 below.

(49) A cancelled first sequel reads '*fluxio ejusdem lateris erit ad fluxionem perpendiculi ut quadratum Hypotenusæ ad idem latus ductum in oppositum segmentum basis, et ad fluxionem istius segmenti ut quadratum Hypotenusæ ad latus datum ductum in segmentum*' (*the fluxion of the same side will be to the fluxion of the perpendicular as the square of the hypotenuse to the same side multiplied into the opposite segment of the base, and to the fluxion of that segment as the square of the hypotenuse to the given side multiplied into the segment*).

(50) A verbally more jejune but otherwise little variant preliminary draft of the next three sentences exists on f. 52ʳ.

(51) Read 'defl AD'.

(52) For AE:AC = AC:AB and EC:AC = BD:AD = BC:AB.

is fl (BD):fl $(BC) = 2ED:AB$. And by a similar reasoning there will be

$$\text{fl }(BD):-\text{fl }(AB) = 2ED:BC.$$

As a result, *ex æquo* fl $(BC):-$fl $(AB) = AB:BC$.[48] As was to be demonstrated.

Proposition 12

If in a perpetually right-angled triangle a side bordering on the right angle be given, then the fluxion of the other side will be to the fluxion of the hypotenuse as the hypotenuse to that fluent side. And, when a perpendicular is let fall onto the hypotenuse, [49]*the fluxion of the base will be to the defluxion of the segment adjoining the given side as the base to that segment; to the fluxion of the other segment as the base to the sum of the base and the former segment; and to the fluxion of the perpendicular as the fluent side multiplied into the hypotenuse to the other side multiplied into the adjoining segment of the base.*

In the triangle ABC right-angled at B let the side AB be given. Then, having let fall the perpendicular BD, I assert that

$$\text{fl }(BC):\text{fl }(AC):-\text{fl }(AD):\text{fl }(DC):\text{fl }(BD)$$
$$= AC^2:BC\times AC:BC\times AD:BC(AC+AD):AB\times AD.$$

[50]For to AC raise the normal CE meeting AB in E and there will be AB, BC, BE in continued proportion, so that by Proposition 2

$$\tfrac{1}{2}AB:BC = \text{fl }(BC):\text{fl }(BE).$$

Likewise AB, AC, AE are in continued proportion and therefrom, by Proposition 2, $\tfrac{1}{2}AB:AC = \text{fl }(AC):\text{fl }(AE)$. Now since AB does not flow and accordingly fl (AE) and fl (BE) are equal, there will be *ex æquo perturbatè*

$$BC:AC = \text{fl }(AC):\text{fl }(BC).$$

As was to be proved. Furthermore, $AC:AB = AB:AD$, so that by Proposition 1 $AC:AD = \text{fl }(AC):(\text{defluxion } AD \text{ or}) -\text{fl }(AD)$, and on compounding

$$AC:AC+AD = \text{fl }(AC):\text{fl }(DC).$$

As was to be proved. Finally by Proposition 11 there is

$$BD:AD = [-]\text{fl }(AD):\text{fl }(BD) = BC:AB.$$

But it was $AC:AD = \text{fl }(AC):[-]\text{fl }(AD)$. Therefore by adjoining ratios

$$BC\times AC:AB\times AD = \text{fl }(AC):\text{fl }(BD).\quad \text{As was to be proved.}$$

Corollary. At once

$$\text{fl }(BC):\text{fl }(AC):-\text{fl }(AD):\text{fl }(DC):\text{fl }(BD) = AE:EC:BD:EC+BD:AD.\text{[52]}$$

Prop 13.

Si trianguli perpetim rectanguli latera omnia fluunt, summa laterum ductorum in suas fluxiones æquatur hypotenusæ ductæ in fluxionem suam.[53]

[In triangulo *ABC* rectangulo ad *B* dico esse

$$AB \times \mathrm{fl}\, AB + BC \times \mathrm{fl}\, BC = AC \times \mathrm{fl}\, AC.$$

Nam per Prop 8 est $\mathrm{fl}\, AB . \mathrm{fl}\, AB^q (:: AB . 2AB^q) :: 1 . 2AB$ adeoȝ

$$\mathrm{fl}\, AB^q = 2AB \times \mathrm{fl}\, AB.$$

Eadem ratione $\mathrm{fl}\, BC^q = 2BC \times \mathrm{fl}\, BC$ et $\mathrm{fl}\, AC^q = 2AC \times \mathrm{fl}\, AC$. Quare cum per 47, 1 Elem sit $AB^q + BC^q = AC^q$ adeoȝ per Ax 5 $\mathrm{fl}\, AB^q + \mathrm{fl}\, BC^q = \mathrm{fl}\, AC^q$. fit dimidiando $AB \times \mathrm{fl}\, AB + BC \times \mathrm{fl}\, BC = AC \times \mathrm{fl}\, AC$. Q.E.D.]

Prop 14.[54]

[55]*In circulo dato fluxio arcus est ad fluxionem sinus ut radius ad cosinum et ad fluxionem tangentis ut cosinus ad secantem,*[56] *et ad fluxionem secantis ut cosinus ad Tangentem.*

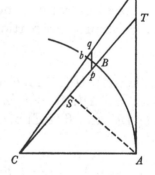

Sit *AB* circulus centro *C* radio *AC* descriptus sitȝ *AB* arcus, *AT* recta tangens eum in dato puncto *A*. Age secantem [*C*]*T* arcui occurrentem in *B* et ad hanc demitte sinum *AS*. fluan[t] jam arcus et Tangens donec evadant *Ab* et *At*. Et cum sector *CBb* sit $\frac{1}{2}CA \times Bb_{[,]}$ ut demonstravit Archimedes[,][57] et triangulum *CTt* sit $\frac{1}{2}CA \times Tt$ erit pars arcus *Bb* ad partem tangentis *Tt* ut sector *CBb* ad triangulū *CTt*.[58] Concipe jam aliud triangulum *Cpq* inter easdem lineas *CT* et *Ct*

---

(53) As before (see note (45) above) Newton first began a straightforward transcription of Theorem 2 on III: 338, beginning '*In Triangulo quovis per*[*petim rectangulo*...], and then slightly reworded it as it is here enunciated. A proof is not given (compare 'Prop 11' in Appendix 1.1 below): that which follows is founded on the demonstration given by Newton in his earlier addendum to the 1671 tract.

(54) A first version of this proposition is reproduced in Appendix 1.3 following.

(55) Newton later altered the following enunciation tentatively to read, in line with Proposition 15, '*In circulo dato fluxiones arcus sinus cosinus tangentis cotangentis secantis et cosecantis sunt inter se &c*' (*In a given circle the fluxions of an arc and its sine, cosine, tangent, cotangent, secant and cosecant are to one another as*...): understand '*...sunt inter se ut radius, cosinus, sinus, quadratum secantis, quadratum cosecantis, factum ex secante et tangente applicatum ad radium & factum ex cosecante et cotangente applicatum ad radium*'.

(56) The equivalent phrase '*ut radius ad quadratum secantis*' (*as the radius to the square of the secant*) is deleted.

Proposition 13

If in a perpetually right-angled triangle all the sides are fluent, the sum of the sides multiplied into their fluxions equals the hypotenuse multiplied into its fluxion.[53]

[In the triangle ABC right-angled at B I say that

$$AB \times \mathrm{fl}\,(AB) + BC \times \mathrm{fl}\,(BC) = AC \times \mathrm{fl}\,(AC).$$

For by Proposition 8 $\mathrm{fl}\,(AB) : \mathrm{fl}\,(AB^2) = (AB : 2AB^2$ or$)\ 1 : 2AB$ so that

$$\mathrm{fl}\,(AB^2) = 2AB \times \mathrm{fl}\,(AB).$$

For the same reason $\mathrm{fl}\,(BC^2) = 2BC \times \mathrm{fl}\,(BC)$ and $\mathrm{fl}\,(AC^2) = 2AC \times \mathrm{fl}\,(AC)$. Consequently, since by *Elements* I, 47, there is $AB^2 + BC^2 = AC^2$ and hence (by Axiom 5) $\mathrm{fl}\,(AB^2) + \mathrm{fl}\,(BC^2) = \mathrm{fl}\,(AC^2)$, there comes to be after halving

$$AB \times \mathrm{fl}\,(AB) + BC \times \mathrm{fl}\,(BC) = AC \times \mathrm{fl}\,(AC).$$

As was to be demonstrated.]

Proposition 14[54]

[55]*In a given circle the fluxion of an arc is to the fluxion of its sine as the radius to its cosine; to the fluxion of its tangent as its cosine is to its secant;*[56] *and to the fluxion of its secant as its cosine to its tangent.*

Let AB be a circle described on centre C and with radius AC, and let AB be an arc, AT a straight line tangent to it at the given point A. Draw the secant CT meeting the arc in B and to this let fall the sine AS. Now let the arc and tangent flow till they come to be Ab and At. Then, since the sector CBb is $\frac{1}{2}CA \times \widehat{Bb}$ (as Archimedes has demonstrated)[57] and the triangle CTt is $\frac{1}{2}CA \times Tt$, the portion Bb of the arc will be to the portion Tt of the tangent as the sector CBb to the triangle CTt.[58] Now conceive a second triangle Cpq to be set up between the same lines CT and Ct, similar indeed to the triangle CTt but equal [in area] to

---

(57) This is an immediate corollary of Proposition 1 of Archimedes' *Measure of a Circle*. Newton is probably thinking of his variant *reductio* proof in 'Prop. 1' on III: 408–10.

(58) This replaces the syncopated equivalent phrase 'erit $Bb \,.\, Tt :: CBb \,.\, CT[t]$'. In line with the draft version (Appendix 1.3) Newton first continued, inserting corresponding concentric arcs $\widehat{TR}$, $\widehat{tr}$ (later deleted) in his figure: 'Centro C radijs CT Ct describe duos arcus TR tr occurrentes rectis Ct CT in R et r et similes sectores [CTR, Crt ...]' (With centre C and radii CT, Ct describe two arcs $\widehat{TR}$, $\widehat{tr}$ meeting the lines Ct, CT in R and r, and the similar sectors [CTR, Crt] ...). In first revision he drew parallels BR, br to ATt through B and b, and then wrote in (a likewise cancelled) sequel: 'A punctis B et b age rectas BR br parallelas AT et occurrentes Cb et CB in R et r. Et per 19 VI Euc triangula similia Cbr, CBR CTt erunt ut

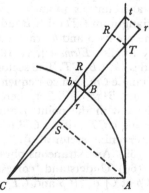

co[n]stitui, simile quidem triangulo CTt sed æquale sectori [CBb], sitq ang. Cpq æqualis angulo CTt[59] & per 19 VI Eucl erunt hæc triangula ut quadrata laterū[60] hoc est erit arc $Bb . Tt :: Cp^{quad} . CT^{quad}$. Insuper linearum $Cp Cq$ una Cp minor est altera Cq major quam radius CB, aliter triangulum vel contineret totum sectorem vel totum in eo contineretur & sic non posset ei æquale esse. Quare si triangula defluunt donec puncta T et t coincident, puncta etiam p et q similiter coincidentia convenient ad arcum intermedium et Cp evadet æqualis CB. Itaq ultima ratio Cp^{quad} ad CT^{quad} hoc est ultima ratio evanescentis arcus Bb ad evanescentem rectam Tt est CB^q ad CT^q id est CS ad CT. Quare per ax 6 fluxio arcus AB est ad fluxionem tangentis AT in eadem ratione. Q.E.D. Præterea per prop 12 est $CT . AT :: \text{fl} AT . \text{fl} CT$. ergo connectendo rationes $CS . AT :: \text{fl} AB . \text{fl} CT$. [Q.E.D.] Est et per Prop 12 $CT . CS :: \text{fl} CT . \text{defl} CS$. adeoq $CT . AT :: \text{fl} AB . \text{defl} CS :: CA . AS$. Deniq per pro[p] 11 $AS . CS :: \text{defl} CS . \text{fl} AS$. Quare $CA . CS :: \text{fl} AB . \text{fl} AS$. Q.E.D.

Prop 15.[61]

Si anguli duo æqualiter fluant datam habentes differentiam, fluxiones sinuum istorū angulorum erunt ut cosinus, et generaliter fluxiones tangentis, secantis, sinus versi, & sinus recti unius anguli, necnon sinus recti, sinus versi, secantis et tangentis alterius anguli, erunt inter se ut sunt ratio quadrati secantis ad radium, ratio rectanguli tangentis et secantis ad radium, sinus rectus, & cosinus prioris anguli, atq[62] cosinus, sinus rectus, ratio rectanguli tangentis et secantis ad radium & ratio quadrati secantis ad radium. Et eædem sunt etiam proportiones ubi summa duorum angulorū datur.

Sint anguli isti ACB, ACD; eorum data differentia BCD; arcus AB, AD; sinus AS, AR; tangentes AT, AV, & secantes CT, CV.[63] Ad CT et CV erige perpendicula TP[5] VQ occurrentia radio CA producto in P et Q, et erit

---

quadrata laterum $Cr. CB. CT$. Sector vero CBb minor est triangulo CBR & major triangulo Crb adeoq ad triangulum CTt minorem habet rationem quam Cr quadratum habet ad CT^q & majorem quam CB quadratum habet ad CT^{quad}. Defluant ergo sector et triangula donec evanescunt et punctis r et B coincidentibus, ultima ratio utriusq trianguli Crb CBR ad triangulum CTt erit CB^q ad CT^q. Quare et sectoris intermedij eadem erit ultima ratio' (From the points B and b draw the lines BR, br parallel to AT and meeting Cb, CB in R and r. Then by Euclid [*Elements*], VI, 19, the similar triangles Cbr, CBR, CTt will be as the squares of the sides Cr, CB, CT. The sector CBb, however, is less than the triangle CBR and greater than the triangle Crb, and consequently bears to the triangle CTt a ratio less than Cr^2 to CT^2 but greater than CB^2 to CT^2. Let, therefore, the sector and the triangles flow back till they disappear and then, with the points r and B coinciding, the final ratio of either triangle Crb, CBR to the triangle CTt will be CB^2 to CT^2. This accordingly will be the selfsame final ratio [to it] of the intervening sector also).

(59) An extraneous phrase since pq is drawn parallel to Tt.

(60) Understand 'correspondentium' (corresponding). Newton has in fact here cancelled 'Cp et $C[T]$' (Cp and CT).

the sector CBb, and let the angle $\widehat{Cpq}$ be equal to the angle $\widehat{CTt}$:[59] then by Euclid VI, 19, these triangles will be as the squares of their[60] sides, in other words there will be $\widehat{Bb}:Tt = Cp^2:CT^2$. Moreover, of the lines Cp, Cq one (Cp) is less than the radius CB, the other (Cq) greater than it, otherwise the triangle would wholly contain the sector or wholly be contained in it and thus could not be equal to it. Consequently, if the triangles flow back till the points T and t coincide, the points p and q, likewise coinciding, will coalesce along the intervening arc and Cp will prove to be equal to CB. As a result, the last ratio of Cp^2 to CT^2—in other words, the last ratio of the vanishing arc $\widehat{Bb}$ to the vanishing straight line Tt—is CB^2 to CT^2, that is, CS to CT. Consequently, by axiom 6 the fluxion of the arc $\widehat{AB}$ is to the fluxion of the tangent AT in the same ratio. As was to be proved. Further, by Proposition 12, $CT:AT = \text{fl}\,(AT):\text{fl}\,(CT)$, therefore by linking the ratios $CS:AT = \text{fl}\,(AB):\text{fl}\,(CT)$. As was to be demonstrated. Also by Proposition 12 $CT:CS = \text{fl}\,(CT):-\text{fl}\,(CS)$, and accordingly $CT:AT = \text{fl}\,(AB):-\text{fl}\,(CS) = CA:AS$. Finally, by Proposition 11

$$AS:CS = -\text{fl}\,(CS):\text{fl}\,(AS)$$

and consequently $CA:CS = \text{fl}\,(AB):\text{fl}\,(AS)$. As was to be demonstrated.

Proposition 15[61]

Should two angles flow equally, preserving a given difference, the fluxions of the sines of those angles will be as their cosines; and, generally, the fluxions of the tangent, secant, versine and right sine of one angle and those, again, of the right sine, versine, secant and tangent of the other angle will be to one another as are the ratio of the square of the secant to the radius, the ratio of the product of the tangent and secant to the radius, the right sine and the cosine of the former angle and as the[62] cosine, right sine, the ratio of the product of its tangent and secant to the radius and the ratio of the square of its secant to the radius. The same propositions hold also when the sum of two angles is given.

Let those angles be $\widehat{ACB}$, $\widehat{ACD}$; their given difference $\widehat{BCD}$; their arcs $\widehat{AB}$, $\widehat{AD}$; their sines AS, AR, tangents AT, AV and secants CT, CV.[63] To CT and CV erect the perpendiculars TP, VQ meeting the radius CA produced in P and

---

(61) In the margin alongside Newton has written 'Vide pag vers'. (See the previous page [viz: f. 50ʳ]), an instruction that 'Cor 2' below is to be inserted from the other side of the sheet where (for want of room at the end of the present proposition) he had set it lengthways in the margin.

(62) Understand 'alterius anguli' (other angle's).

(63) Newton first continued: 'Dico primo quod sit fl AS.fl $AR::CS.CR$. Nam per Prop 14 est fl $[AS.\text{fl}\,AB::CS.CA$ & fl $AR.\text{fl}\,AD$ seu fl $AB::CR.CA]$' (I assert first that

$$\text{fl}\,(AS):\text{fl}\,(AR) = CS:CR.$$

For by Proposition 14 there is

$$\text{fl}\,[(AS):\text{fl}\,(AB) = CS:CA \quad \text{and} \quad \text{fl}\,(AR):\text{fl}\,(AD) \quad \text{or} \quad \text{fl}\,(AB) = CR:CA]).$$

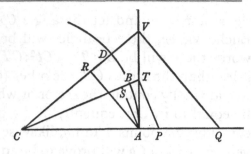

$$CP = \frac{CT^q}{CA}, \quad CQ = \frac{CV^q}{CA}. \quad TP = \frac{AT \times CT}{CA} \quad \text{et}$$

$$VQ = \frac{AV \times CV}{CA}. \text{ hoc est } CP, CQ, TP \text{ et } VQ$$

sunt rationes tangentium quadratorum et factorum tangentium ductarum in secantes ad radium. Quo pacto dico fluxiones quantitatum $AT.CT.BS.AS.$ arc AB vel $AD.AR.DR.CV.AV$ esse inter se ut sunt[64] $CP.$ (per Prop 12) $TP.$ (Prop 12) $AS.$ (Prop 11) $CS.$ (Prop 14) $CA.$ (Prop 14) $CR.$ (Prop 11) $AR.$ (Prop 12) $VQ.$ (Prop 12) $CQ.$ Q.E.D.

[65]Quod si summa angulorum ACB ACD detur tunc inspice secundam figuram et eodem ratiocinio (signis profluxionum ac defluxionum nempe + et − probe observatis) patebit fluxiones quantitatum

$$AT.CT.BS.AS. \text{ arc } AB \text{ vel } AD.AR.DR.CV.AV$$

esse inter se ut

$$CP.TP.AS.CS. \pm CA.^{[66]} - CR. - AR. - VQ. - CQ.$$

Cor [1]. Valent et proportiones hæ in circulis inæqualibus quoniam aucto circulo sinus tangentes & secantes & eorum fluxiones in eadem ratione augentur.

Cor 2.[67] Hinc in omni triangulo unum stabilem angulum habent[e], ex cognita fluxione sinus tangentis vel secantis unius e flue[n]tibus angulis cognoscitur etiam fluxio sinus tangentis vel secantis alterius.

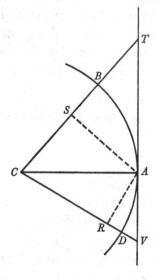

Prop 16.

Si trium angulorum data est vel summa omnium, vel excessus quo duo superant tertiū aut ab eo superantur: affirmativæ & negativæ fluxiones sinuum divisæ per[68] cosinus sibi mutuò æquales erunt.

Sunto tres arcus $PQR_{[,]}$ eorum sinus $A, B, C,$ cosinus D E et F respective. [69]Et posito C sinu anguli vel defluentis vel subductitij quo reliqui duo superant datum

---

(64) Much as in his preliminary draft (see Appendix 1.1: note (29) below) Newton first concluded: '$CT^q.$ (per Prop 12) $AT \times CT.$ (Prop 12) $-AT \times CS,$ vel $-CA \times AS.$ (Prop 11) $CA \times CS.$ (Prop 14) $CA^q.$ (Prop 14) $CA \times CR.$ (Prop 11) $CA \times AR,$ vel $-AV \times CR.$ (Prop 12) $AV \times CV.$ (Prop 12) $CV^q.$ Q.E.D.'

(65) In a preceding cancellation the following paragraph was (compare Appendix 1.1: note (30) below) made a separate 'Prop 16', viz: '*Eædem sunt proportiones præfatarum linearum*

Q, and then will there be $CP = CT^2/CA$, $CQ = CV^2/CA$, $TP = AT \times CT/CA$ and $VQ = AV \times CV/CA$: that is, CP, CQ, TP and VQ are the ratios of the squares of the tangents and of the products of the tangents multiplied into the secants to the radius. Stipulating this, I say that the fluxions of the quantities AT, CT, BS, AS, $\widehat{AB}$ or $\widehat{AD}$, AR, DR, CV and AV are to one another as are[64] CP, TP (by Proposition 12), AS (by Proposition 12), CS (by Proposition 11), CA (by Proposition 14), CR (by Proposition 14), AR (by Proposition 11), VQ (by Proposition 12) and CQ (by Proposition 12). As was to be demonstrated.

[65]But should the sum of the angles $\widehat{ACB}$, $\widehat{ACD}$ be given, then inspect the second figure and by the same reasoning (with the signs, namely $+$ and $-$, of profluxions and defluxions taken appropriately into account) it will be manifest that the fluxions of the quantities AT, CT, BS, AS, $\widehat{AB}$ or $\widehat{AD}$, AR, DR, CV and AV are to one another as CP, TP, AS, CS, $\pm CA$,[66] $-CR$, $-AR$, $-VQ$ and $-CQ$.

Corollary 1. These proportions are also valid in unequal circles seeing that, when a circle is increased, its sines, tangents and secants along with their fluxions increase in the same ratio.

Corollary 2.[67] Hence in every triangle having one angle stable, once there is known the fluxion of the sine, tangent or secant of one of the fluent angles, the fluxion of the sine, tangent or secant of the other is also known.

Proposition 16

If of three angles there is given either their total sum or the excess by which two surmount the third or are surmounted by it, the positive and negative fluxions of their sines divided by their cosines will be mutually equal to one another.

Let there be three arcs P, Q, R and let their sines be A, B, C and cosines D, E, F respectively. [69]Then, on putting C to be the sine either of the defluent angle or of the subtractive difference by which the two others surmount a given angle,

---

ubi summa fluentium angulorum datur' (*The proportions of the above-mentioned lines are the same when the sum of the fluent angles is given*), adding in proof that 'Eadem est utriusq̃ Propositionis demonstratio signis profluxionum et defluxionum nempe $+$ et $-$ probe observatis' (*The demonstration of either proposition is the same when due attention is paid to the signs of the positive and negative fluxions, namely '$+$' and '$-$'*).

(66) According as $\widehat{AB}$ is greater or less than $\widehat{AD}$.

(67) This is a late insertion 'in pag. vers.' (on the preceding page), as Newton notes. See note (61) above.

(68) The equivalent 'applicatæ ad' is cancelled.

(69) A cancelled sequel reads: 'Et in priori casu cum e tribus duo simul augeantur, sint illi A et B et ubi augentur erit $\dfrac{\mathrm{fl}\,A}{D} + \dfrac{\mathrm{fl}\,B}{E} = \dfrac{\mathrm{fl}\,C}{F}$,' (*And in the former case, since two of the three increase together, let these be A and B and when the increase takes place there will be* $\mathrm{fl}\,(A)/D + \mathrm{fl}\,(B)/E = \mathrm{fl}\,(C)/F$).

angulum, erit [in priori casu] $\dfrac{\text{fl}\,A}{D}+\dfrac{\text{fl}\,B}{E}+\dfrac{\text{fl}\,C}{F}=0$. Est enim per prop [14]

$D\,.\,\text{radius}::\text{fl}\,A\,.\,\text{fl}\,P$. adeoꝗ $\dfrac{\text{rad.}\times\text{fl}\,A}{D}=\text{fl}\,P$. Et eodem modo $\dfrac{\text{rad}\times\text{fl}\,B}{E}=\text{fl}\,Q$ &

$\dfrac{\text{rad}\times\text{fl}\,C}{F}=\text{fl}\,R$. Jam vero cum summa arcuum P, Q, R detur adeoꝗ summa

fluxionum nulla sit erit $\dfrac{\text{rad}\times\text{fl}\,A}{D}+\dfrac{\text{rad}\times\text{fl}\,B}{E}+\dfrac{\text{rad}\times\text{fl}\,C}{F}=0$. Divide omnia

per radium et erit $\dfrac{\text{fl}\,A}{D}+\dfrac{\text{fl}\,B}{E}+\dfrac{\text{fl}\,C}{F}=0$. hoc est si C negative fluat erit

$\dfrac{\text{fl}\,A}{D}+\dfrac{\text{fl}\,B}{E}=\dfrac{\text{neg}:\text{fl}\,C}{F}$. Aut si B et C negative fluant erit $\dfrac{\text{fl}\,A}{D}=\dfrac{\text{neg}\,\text{fl}\,B}{E}+\dfrac{\text{neg}\,\text{fl}\,C}{F}$.
Q.E.D.

Cor 1. Hinc patet etiam[70] fluxiones affirmativas et negativas cosinuum divisas per sinus esse sibi mutuò æquales. Nam cosinus sunt sinus arcuum complementalium ad quadrantem et sinus sunt cosinus.[71]

Cor 2. Patet etiam fluxiones affirmativas et negativas sinuum multiplicatas per secantes eorundem angulorum esse sibi mutuo æquales. Nam cum cosinus

sit ad radium sicut radius ad secantem erit $\dfrac{\text{rad}}{\text{cosin}}=\dfrac{\text{sec}}{\text{rad}}$. Sint itaꝗ trium angu-

lorum secantes G, H, I et erit $\dfrac{G\times\text{fl}\,A}{\text{rad}}+\dfrac{H\times\text{fl}\,B}{\text{rad}}+\dfrac{I\times\text{fl}\,C}{\text{rad}}=0$. hoc est si C negativè

fluat et omnia multiplicentur per radium, $G\times\text{fl}\,A+H\times\text{fl}\,B=I\times[\text{neg}]\,\text{fl}\,C$.

Cor 3. Hinc in omni triangulo si sinus cosinus vel secantes duorum angulorum innotescant, simul habetur sinus cosinus vel secans tertij.[72]

Prop 17.

Fluxio sinus anguli cujusvis est ad fluxionem sinus anguli perpetim dupli ut cosinus anguli minoris ad duplum cosinus anguli majoris. Et universaliter si anguli fluant in data ratione, fluxio sinus anguli minoris est ad fluxionem sinus anguli majoris ut cosinus minoris ad cosinum majoris ductum in rationem majoris anguli ad minorem.

Nam per prop 14 fluxio sinus minoris est ad fluxionem arcus ejus ut cosinus ad radium et per ax: 3 fluxio minoris est ad fluxionem majoris in ratione angulorum[,] hoc est ut unitas ad rationem quam major angulus habet ad minorem et per prop 14 fluxio majoris est ad fluxionem sinus ejus ut radius ad

---

(70) This replaces a first opening 'Eodem modo probatur' (In the same manner it is proved).

(71) Again understand 'arcuum complementalium ad quadrantem'.

(72) On taking the trigonometrical radius to be unity, here $D=\sqrt{[1-A^2]}$, $E=\sqrt{[1-B^2]}$, $F=\sqrt{[1-C^2]}$ with $A=BF+CE$, $D=EF-BC$. There seems little point in introducing such fluxional equivalents as fl $(A)/D+$ fl $(B)/E+$ fl $(C)/F=0$ to evaluate A or D given B and C or E and F.

there will (in the former case) be $\text{fl}(A)/D + \text{fl}(B)/E + \text{fl}(C)/F = 0$. For, by Proposition 14, $D : \text{radius} = \text{fl}(A) : \text{fl}(P)$, so that $\text{fl}(P) = \text{radius} \times \text{fl}(A)/D$. And in the same manner $\text{fl}(Q) = \text{radius} \times \text{fl}(B)/E$ and $\text{fl}(R) = \text{radius} \times \text{fl}(C)/F$. Now, however, since the sum of the arcs P, Q and R is given and so the sum of their fluxions is nil, there will be

$$\text{radius} \times \text{fl}(A)/D + \text{radius} \times \text{fl}(B)/E + \text{radius} \times \text{fl}(C)/F = 0.$$

Divide throughout by the radius and then

$$\text{fl}(A)/D + \text{fl}(B)/E + \text{fl}(C)/F = 0.$$

Thus, should C flow negatively, then $\text{fl}(A)/D + \text{fl}(B)/E = -\text{fl}(C)/F$; or if B and C should flow negatively, then $\text{fl}(A)/D = [-\text{fl}(B)/E] + [-\text{fl}(C)/F]$. As was to be demonstrated.

Corollary 1. It is hence evident also[10] that positive and negative fluxions of cosines divided by sines are mutually equal to one another. For cosines are sines of complementary arcs and sines are cosines [of them].

Corollary 2. It is evident, too, that positive and negative fluxions of sines multiplied by the secants of the same angles are mutually equal to one another. For since cosine to radius is as radius to secant, there will be

$$\text{radius}/\text{cosine} = \text{secant}/\text{radius}.$$

Accordingly, let the secants of the three angles be G, H, I and then

$$G \times \text{fl}(A)/\text{radius} + H \times \text{fl}(B)/\text{radius} + I \times \text{fl}(C)/\text{radius} = 0.$$

That is, should C flow negatively and if we multiply throughout by the radius, $G \times \text{fl}(A) + H \times \text{fl}(B) = I \times -\text{fl}(C)$.

Corollary 3. Hence in every triangle if the sines, cosines or secants of two angles be known, there is obtained at once the sine, cosine or secant of the third.[72]

Proposition 17

The fluxion of the sine of any angle is to the fluxion of the sine of an angle perpetually twice it as the cosine of the lesser angle to twice the cosine of the greater angle. And universally, should the angles flow in given ratio, the fluxion of the sine of the lesser angle is to the fluxion of the sine of the greater angle as the cosine of the lesser one to the cosine of the greater multiplied into the ratio of the greater angle to the lesser one.

For by Proposition 14 the fluxion of the sine of the lesser is to the fluxion of its arc as its cosine to the radius, while by Axiom 3 the fluxion of the lesser is to that of the greater one in the ratio of the angles—that is, as unity to the ratio which the greater angle bears to the less—and by Proposition 14 the fluxion of the greater is to the fluxion of its sine as the radius to its cosine. Combine all these

cosinum ejus. Connecte omnes rationes et erit fluxio sinus minoris ad fluxionem sinus majoris ut cosinus minoris ad cosinum majoris ductum in rationem ejus ad minorem.

Schol. Eodem fundamento fluxiones tangentium et secantium prodeunt. Quinetiam si anguli[73] aliam quamvis relationem habeant, (puta si sint continue proportionales a tertio dato angulo, vel habe[a]nt datum angulum pro medio proportionali,) potest inde ratio fluxionum angulorum colligi, dein ex ratione istarum fluxionum ad fluxiones sinuum tangentium et secantium innotescet ratio fluxionum sinuum tangentium et secantium inter se.

Prop 18.[74]

Si trianguli cujusvis basis et angulus ad verticem dentur cæteraꝗ motu quocunꝗ fluant & perpendiculum demittatur ad basem,[75] *dico fluxionem lateris alterutrius esse ad fluxionem sinus anguli oppositi ut latus est ad sinum. Et præterea si a vertice ad basem agatur recta continens angulum cum alterutro latere æqualem angulo quem perp[endiculum] continet cum altero latere et a medio puncto hujus actæ rectæ demittantur perpendicula ad latera basem et perpendiculum trianguli, fluxiones laterum, majoris segmenti basis, & perpendiculi trianguli erunt inter se ut perpendicula quæ in ea demittuntur.*

Sit *ABC* triangulum, *C* datus angulus ad verticem[,] *AB* data basis & *CD* perpendiculum et erit

fl *AC* . defl *BC* ::

$$AC \times DB . BC \times AD.^{(76)}$$

Præterea agatur *CE* occurrens *AB* in *E* & constituens ang *ACE* æqualem angulo *BCD* et bisecta ea in *X*[77] demitte ad *AC*, *CB*, *AB*, *CD* perpendicula *XF*, *XG*, *XH*, *XI*, & erit fl *AC* . defl *CB* . fl *AD* . defl *CD* :: *XF* . *XG* . *XH* . *XI*.

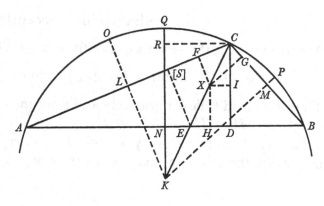

---

(73) 'arcus' (arcs) is cancelled.

(74) A late addition to the text founded on the draft version reproduced in Appendix 1.4, for the manuscript has, immediately before, a first 'Pro[p] 18' which is essentially Proposition 19 following. The text of the latter reads: 'Si trianguli cujusvis datum angulum ad basem habentis, latus dato angulo oppositum convertatur circa terminum basis, erit profluxio lateris *BC* dato angulo [*A*] oppositi ad fluxionem *C* :: sin *C*. *AC*.' (The second ratio should be ' :: tan C . 1/*BC*' since, where *Cc* is the limit-increment of *AC*, $c\gamma = C\gamma$. tan *C* and $C\gamma/BC$ are the contemporaneous increments of *BC* and $\hat{C} = (\pi - \hat{A}) - \hat{B}$ (in radians) on letting fall *Cγ* normal from *C* to *Bc*.)

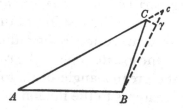

ratios and then the fluxion of the sine of the lesser will be to the fluxion of the sine of the greater as the cosine of the lesser one to the cosine of the greater multiplied into its ratio to the lesser.

Scholium. On the same basis the fluxions of tangents and secants are forthcoming. Indeed, should the angles[73] have any other relationship (should they, for instance, be in continued proportion from a third given angle or have a given angle for their mean proportional), the ratio of the fluxions of the angles can be gathered from this, and then from the ratio of those fluxions to the fluxions of the sines, tangents and secants will be ascertained the ratios of the fluxions of the sines, tangents and secants to each other.

Proposition 18[74]

If in any triangle the base and the angle at the vertex be given, its other elements flowing in any manner whatever, and the perpendicular be let fall to the base,[75] *I assert that the fluxion of either side is to the fluxion of the sine of the angle opposite as the side is to that sine. Furthermore, if from the vertex to the base there be drawn a straight line containing with either side an angle equal to that contained by the perpendicular with the other side, and from the mid-point of this drawn side there be dropped perpendiculars to the sides, the base and the perpendicular of the triangle, then the fluxions of the sides, the greater base segment and the triangle's perpendicular will be to one another as the perpendiculars let fall onto them.*

Let ABC be the triangle, C the given angle at its vertex, AB the given base and CD the perpendicular, and there will be

$$\text{fl}\,(AC) : -\text{fl}\,(BC) = AC \times DB : BC \times AD.^{(76)}$$

Further, let there be drawn CE meeting AB in E and constituting the angle $\widehat{ACE}$ equal to the angle $\widehat{BCD}$, bisect it in X[77] and to AC, CB, AB, CD let fall the perpendiculars XF, XG, XH, XI: there will then be

$$\text{fl}\,(AC) : -\text{fl}\,(CB) : \text{fl}\,(AD) : -\text{fl}\,(CD) = XF : XG : XH : XI.$$

---

(75) Newton first continued, in anticipation of Corollary 1 following, with the equivalent statement 'dico fluxiones laterum esse ut latera ducta in segmenta opposita' (I assert that the fluxions of the sides are as the sides multiplied into the opposing [base] segments). In equivalent limit-increment form, since $AC/\sin B = BC/\sin A$, constant and also $\cos A = AD/AC$, $\cos B = DB/BC$, there follows

$$d(AC)/d(BC) = d(\sin B)/d(\sin A) = (AC \times DB/BC \times AD) \times (dB/dA).$$

Compare Newton's preliminary version (Appendix 1.4: note (40) below).

(76) This last phrase should probably be cancelled, for it evidently takes up a portion of the original proposition which (see note (75)) is delayed till Corollary 1 of the present revise.

(77) The manuscript reads 'Q' here and also at a number of corresponding points, but we have silently corrected the text to agree with Newton's figure. (This originally had two points 'Q', one of which—evidently to avoid confusion—he altered to 'X' at a late stage without everywhere changing his text to suit.)

Quod ut pateat, super basi *AB* describe (per 32.3 Eucl) circuli segmentum *ACB* quod capiet datum angulum *ACB* sitᴂ *K* centrum hujus circuli, et a *K* ad latera trianguli *AC BC* & basem *AB* demitte perpendicula *KL KM KN* bisecantia arcus *AC*, *CB*, *AB* in *O*, *P* et *Q*, et ad *KQ* demitte etiam a vertice trianguli perpendiculum *CR*. Cum itaᴂ vertex trianguli semper reperiatur in peripheria circuli hujus, erunt *CL*, *CM*, *CR* sinus arcuum *CO*, *CP*, *CQ*. et *KL*. *KM*. *KR* cosinus. Quare cum summa arcuum *CO*, *CP* detur utpo[te] dimidia totius arcus *ACB*, erit (per prop [14]) fl *CL* ad fl *CM* hoc est fl *CA* ad fl *CB* ut *KL* ad *KM* id est ut *XF* ad *XG*. Q.E.D. Præterea cum arcus *QB* hoc est summa arcuum *CQ*, *CB* detur, erit profluxio arcus *QC* æqualis defluxioni arcus *CB* hoc est æqualis duplo defluxionis arcus dimidij *CP*, ergo per prop 17 fluxio sinus *CR* vel *ND* est ad defluxionem sinus *CM* ut cosinus *KR* ad dimidium cosinus *KM*, et ad defluxione[m] chordæ *CB* ut cosinus *KR* ad cosinum *KM*. Sed *RC ND* et *AD* æqualiter fluunt et est *KR*. *XH* :: *KC*. *XE* vel *XC* :: *KM*. *XG*. hoc est *KR*. *KM* :: *XH*. *XG*. Quare fl *AD*. fl *CB* :: *XH*. *XG*. Q.E.D. Deniᴂ per prop [11] est fl *CR*. [de]fl *KR* :: *KR*. *CR* :: *CI* vel *XH*. *XI*. Sed *CR*, *DN*, *DA* æqualiter fluunt, ut et *CD*, *NR*, *KR*. Ergo fl *DA*. [de]fl *CD* :: *X*[*H*]. *X*[*I*]. Q.E.D

Cor 1. Profluxio unius lateris & defluxio alterius sunt ut cosinus oppositorum anguloū ad basem. Vel fl *AC*. [de]fl *BC* :: $\dfrac{BD}{BC} \cdot \dfrac{AD}{AC}$ [superscript](78)

Cor 2. fl *AC*. fl *AD* :: *AE*. *AC*.[superscript](79) Nam ad *AC* demitte perp[endiculum] *ES* et erit *XF*. *XH* :: *E*[*S*]. *CD* :: *AE*. *AC*.

Cor. 3. fl *AD*. fl *DC* :: *DC*. *ED*.[superscript](80)

Prop 19.[superscript](81)

Si trianguli cujusvis angulus ad verticem et latus alterutrum angulo isti conterminum detur cæteraᴂ motu quocunᴂ fluunt, demittatur perpendiculum a vertice ejus ad basem: demitte etiam in latus non datum duo perpendicula[,] unum a termino basis alterum a communi termino segmentorum ejus[,] & fluxiones lateris non dati, basis, perpendiculi & segmenti basis dato lateri contermini erunt inter se ut sunt basis, latus fluens productum ad perpendiculum secundum, differentia duorum ultimorum perpendiculorum[superscript](82) & segmentum lateris inter verticem trianguli et perpendiculum ultimum.[superscript](83) Et præterea fluxio basis est ad defluxionem sinus anguli dato lateri oppositi ut basis ad sinum istum.

---

(78) See notes (75) and (76) above.

(79) Compare Appendix 1.4: note (43) below. Observe that $XF = \frac{1}{2}ES$, $XH = \frac{1}{2}CD$.

(80) Since $XH = \frac{1}{2}CD$, $XI = \frac{1}{2}ED$.

(81) The manuscript has a cancelled 'Prop 18' which is a first version of this; see note (74) above.

(82) 'quæ a basi ad latus illud demitt[unt]ur' (which are let fall from the base to that side) is cancelled.

(83) This replaces the more precise equivalent 'perpendiculum ultimo demissum' (the perpendicular last let fall).

To make this obvious, on the base AB describe (by Euclid, III, 32) the segment ACB of a circle which shall take the given angle $A\widehat{C}B$ and let K be the centre of this circle; then from K to the triangle's sides AC, BC and base AB let fall the perpendiculars KL, KM, KN bisecting the arcs $\widehat{AC}$, $\widehat{CB}$, $\widehat{AB}$ in O, P and Q, and to KQ let fall also from the triangle's vertex the perpendicular CR. Since, then, the vertex of the triangle is always to be found in the circumference of this circle, CL, CM, CR will be the sines of the arcs $\widehat{CO}$, $\widehat{CP}$, $\widehat{CQ}$ and KL, KM, KR their cosines. In consequence, since the sum of the arcs $\widehat{CO}$ and $\widehat{CP}$ is given (being, namely, half the whole arc $A\widehat{C}B$) there will, by Proposition 14, be $\mathrm{fl}(CL)$ to $\mathrm{fl}(CM)$—that is, $\mathrm{fl}(CA)$ to $\mathrm{fl}(CB)$—as KL to KM, and so as XF to XG. As was to be proved. Moreover, since the arc $\widehat{QB}$ (or the sum of the arcs $\widehat{CQ}$, $\widehat{CB}$) is given, the profluxion of the arc $\widehat{QC}$ will equal the defluxion of the arc $\widehat{CB}$, that is, twice the defluxion of the half arc $\widehat{CP}$: therefore, by Proposition 17, the fluxion of the sine CR (or ND) is to the defluxion of the sine CM as the cosine KR to half the cosine KM, and to the defluxion of the chord CB as the cosine KR to the cosine KM. But RC (or ND) and AD flow uniformly and $KR:XH = KC:XE$ (or XC) $= KM:XG$, that is, $KR:KM = XH:XG$. Consequently, $\mathrm{fl}(AD):\mathrm{fl}(CB) = XH:XG$. As was to be demonstrated. Finally, by Proposition 11 $\mathrm{fl}(CR):-\mathrm{fl}(KR) = KR:CR = CI$ (or XH)$:XI$. But CR (or DN) and DA flow uniformly, likewise also DC (or NR) and KR. Therefore $\mathrm{fl}(DA):-\mathrm{fl}(CD) = XH:XI$. As was to be demonstrated.

Corollary 1. The profluxion of one side and the defluxion of the other are as the cosines of the angles opposite the base. Equivalently,

$$\mathrm{fl}(AC):-\mathrm{fl}(BC) = BD/BC:AD/AC. \quad [78]$$

Corollary 2. $\mathrm{fl}(AC):\mathrm{fl}(AD) = AE:AC.$ [79] For to AC let fall the perpendicular ES and then $XF:XH = ES:CD = AE:AC$.

Corollary 3. $\mathrm{fl}(AD):\mathrm{fl}(DC) = DC:ED.$ [80]

Proposition 19 [81]

If in any triangle the angle at the vertex and either side bordering that angle be given, the other elements flowing with any motion whatever, and there be let fall the perpendicular from its vertex to the base, let fall also two perpendiculars onto the side not given, one from the end-point of the base, the other from the common end-point of its segments, and the fluxions of the side not given, the base, the perpendicular and the base segment adjoining the given side will be to one another as are the base, the fluent side extended to the second perpendicular, the difference of the two latter perpendiculars [82] and the segment of the side between the vertex of the triangle and the last perpendicular. [83] Furthermore, the fluxion of the base is to the defluxion of the sine of the angle opposite the given side as the base to that sine.

Sit ABC triangulum istud[,] AC latus datum[,]
AB basis[,] CD perpendiculum demissum in
basem: A punctis A ac D ad latus fluens CB
demitte etiam perpendicula AE, DF, et erunt

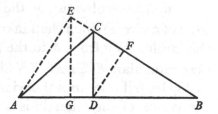

$$\text{fl}\,AB\,.\,\text{fl}\,CB\,.\,\text{defl}\,CD\,.\,\text{fl}\,AD$$

$$::EB\,.\,AB\,.\,AE-DF\,.\,CF.$$

Demitte enim ad AB normalem EG, et in triangulo rectangulo AEG cum
detur latus AE[(84)] erit (per prop 12[)] fl EB vel CB. fl $EG::AB^q\,.\,AE\times AG$.
Præterea cum angulus $[E]AC$ hoc est differentia istorum angulorum BAE,
BAC detur, erunt per Prop 15 fluxiones sinuum angulorum BAE, BAC ut sunt
cosinus. hoc est[(85)] per Cor Prop 15 fl EG. fl $CD::AG$. AD. Adeoœ connectendo
rationes fl CB. fl $CD::AB^q\,.\,AE\times AD::AB\,.\,\dfrac{AE\times AD}{AB}$. Sed $\dfrac{AE\times AD}{AB}=AE-DF$.
Ergo fl CB. fl $CD::AB$. $AE-DF$. Q.E.D. Præterea per Prop 12 est fl EB vel
CB. fl $AB::AB$. EB. Et per prop 11 fl CD. defl $AD::AD$. $CD::AE-DF$. CF.
Q.E.D.

Cor. Est fl CB. fl $AB::CB$. BD.[(86)]

Possem omnes casus triangulorum percurrere sed mallem[(87)] rem totam
paucis generalibus complecti ut sequitur.

Prop 20.

In quovis triangulo acutangulo,[(88)] *demisso ad basem perpendicul[o], differentia
fluxionum laterum ductarum in latera æquatur differentiæ fluxionis basis ductæ in alterutrum
segmentum ejus & fluxionis segmenti alterius ductæ in basem.*

---

(84) That is, $AC.\sin C$ where AC and $A\hat{C}B$ are given.

(85) Newton first went on:

$$\text{`fl}\,\frac{EG}{AE}\,.\,\text{fl}\,\frac{CD}{AC}::\frac{AG}{AE}\,.\,\frac{A[D]}{AC}.$$

Duc antecedentia in datum AE et consequentia in datum AC et erit fl EG. fl $CD[::AG.AD]$'
(fl (EG/AE):fl $(CD/AC) = AG/AE$:AD/AC. Multiply the antecedents [of the ratios] by the
constant AE and the consequents by the constant AC and there will be fl (EG):fl $(CD) =$
AG:AD).

(86) For, since the right triangles ABE, CBD are similar, AB:$EB = CB$:BD.

(87) Newton first wrote 'præstat' (it is better to). The manuscript contains in fact, immediately preceding the present paragraph, two variant Propositions 20 and 21—manifestly to
be omitted from the final version of the 'Geometria'—which consider two further particular
cases, one where two sides are given, the second where an angle is given. The text of these is
reproduced in Appendix 1.5 below.

(88) Newton explains this restriction in the scholium following Proposition 25 below: in his
view, to consider the general scalene case would introduce a proliferation of changes in sign

Let *ABC* be that triangle, *AC* the given side, *AB* the base, *CD* the perpendicular dropped onto the base: also, from the points *A* and *D* to the fluent side *CB* let fall the perpendiculars *AE*, *DF*, and there will be

$$\mathrm{fl}\,(AB):\mathrm{fl}\,(CB):-\mathrm{fl}\,(CD):\mathrm{fl}\,(AD) = EB:AB:AE-DF:CF.$$

For let fall the normal *EG* to *AB*, and then in the right-angled triangle *AEG*, since the side *AE*[84] is given, there will (by Proposition 12) be

$$\mathrm{fl}\,(EB)\ [\text{or }\mathrm{fl}\,(CB)]:\mathrm{fl}\,(EG) = AB^2:AE\times AG.$$

Moreover, since the angle $\widehat{EAC}$—namely, the difference of those angles $\widehat{BAE}$, $\widehat{BAC}$—is given, by Proposition 15 the fluxions of the sines of the angles $\widehat{BAE}$, $\widehat{BAC}$ will be as their cosines: whence[85] by the Corollary to Proposition 15 $\mathrm{fl}\,(EG):\mathrm{fl}\,(CD) = AG:AD$. In consequence, on combining the ratios,

$$\mathrm{fl}\,(CB):\mathrm{fl}\,(CD) = AB^2:AE\times AD, \quad\text{that is,}\quad AB:AE\times AD/AB.$$

But $AE\times AD/AB = AE-DF$. Therefore $\mathrm{fl}\,(CB):\mathrm{fl}(CD) = AB:AE-DF$. As was to be demonstrated. Furthermore, by Proposition 12 there is

$$\mathrm{fl}\,(EB)\ [\text{or }\mathrm{fl}\,(CB)]:\mathrm{fl}\,AB = AB:EB,$$

and by Proposition 11 $\mathrm{fl}\,(CD):-\mathrm{fl}\,(AD) = AD:CD = AE-DF:CF$. As was to be demonstrated.

Corollary. There is $\mathrm{fl}\,(CB):\mathrm{fl}\,(AB) = CB:BD$.[86]

I could run through all the cases of triangles but I would rather[87] embrace the whole topic in a few generalities, as follows.

Proposition 20

In any acute-angled[88] *triangle, on letting fall the perpendicular to the base, the difference of the fluxions of the sides multiplied into the sides equals the difference of the fluxion of the base multiplied into either of its segments and the fluxion of the other segment multiplied into the base.*

---

in the basic proposition to no other purpose than making their structure unnecessarily obscure. In sequel he first went on with the phrase 'demissis ad basem et latera perpendiculis se mutuo secantibus' (on letting fall mutually intersecting perpendiculars to the base and sides), implicitly invoking the theorem—one known at least as early as Archimedes but still not to be found in any contemporary English geometrical textbook (see note (90) below)—that the altitudes of a triangle are concurrent. Realizing that this theorem was neither self-evident nor commonly known, he at once decided to eliminate mention of it in the present proposition, deleting the perpendiculars *AE* and *BF* from his manuscript figure (along with the point *X* in *CG* of their concurrence) but omitting in his alteration of the text to change 'perpendiculis' into 'perpendiculo', and then reintroduced it with full proof as the Lemma to Proposition 21 following.

In triangulo ABC ad basem AB demittatur perpendicularis CG et (per $13_{[,]}2$ Elem Eucl) erit

$$AC^q - BC^q = 2AB \times AG - AB^q:$$

Adeoꝗ per Prop [8]

$$AC \times \text{fl} \, AC - BC \times \text{fl} \, BC$$
$$= AB \times \text{fl} \, AG + AG \times \text{fl} \, AB - AB \times \text{fl} \, AB$$

hoc est $= AB \times \text{fl} \, AG - BG \times \text{fl} \, AB.$[89] Q.E.D.

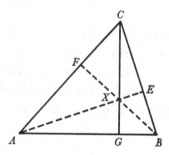

Cor 1. Si dentur latera duo, tunc fluxionibus $AC \; BC$ evanescentibus erit $AB \times \text{fl} \, AG - BG \times \text{fl} \, AB = 0$ sive $AB.BG :: \text{fl} \, AB . \text{fl} \, AG.$

Cor 2. Si detur basis et latus alterutrum puta BC erit $AB.AC :: \text{fl} \, AC . \text{fl} \, AG.$

Cor 3. Si detur segmentum et latus oppositum [puta AG et BC] erit $AC.BG :: \text{fl} \, AB . \text{defl} \, AC.$

Eodem modo si detur unum latus vel basis vel segmentum prodeunt Theoremata nonnihil simpliciora quàm fundamentali[a].

Lemma.[90]

Perpendiculorum AE, BF, CG ad omnia latera trianguli alicujus ABC demissorum communis est intersectio.

Sit enim X intersectio duorum AE, CG et propter sim tri BGC, AGX erit $CG.GB :: AG.GX$ et vicissim $CG.AG :: GB.GX.$ Sit autem Y[91] intersectio duorum CG et BF et propter sim tri CGA, BGY erit $CG.AG[::]BG.GY.$ Quare GX et GY æquantur$_{[,]}$ hoc est $\perp^{1a} AE$ et BF secant CG in eodem puncto. Q.E.D.[92]

---

(89) That is, '$= AG \times \text{fl} \, AG - BG \times \text{fl} \, BG$'. The proposition follows more simply from the equality $AC^2 - BC^2 = AG^2 - BG^2$ (see **2**, 1, §1: note (11) above).

(90) That Newton felt compelled to introduce a lemma at this point to demonstrate the existence of the orthocentre may surprise the modern reader. Nearly two thousand years before, in fact, Archimedes had cited the theorem from a lost 'tractatus de triangulis rectangulis' in Proposition 5 of his *Book of Lemmas* (first published by John Twysden as *Tractatus XI. Lemmata Archimedis...e vetusto Codice M.S. Arabico* of his edition of Samuel Foster's *Miscellanea sive Lucubrationes Mathematicæ* (London, 1659), and again two years later by G. A. Borelli in the Latin version by Abraham of Ecchelles, *Archimedis Liber Assumptorum Interprete Thabit Ben-Kora*, which he appended to his edition of al-Fath's paraphrase of *Apollonii Pergæi Conicorum Lib. V. VI. VII* (Florence, 1661): especially 391-2). The proposition was also afterwards quoted by Pappus in his *Mathematical Collection*, VII, 62, but again without proof: Commandino's attempted reconstitution of one in his Latin *editio princeps* (Pesaro, 1588) was both prolix and erroneous. (See P. Ver Eecke, *Pappus d'Alexandrie: La Collection Mathématique*, **2** (Brussels, 1933): 589, note 1.) Its first public appearance in Europe would seem to be Regiomontanus' incidental (and unproven) remark in his *De Triangulis Omnimodis Libri Quinque* (Nuremberg, 1533): Liber I, Theorema XXXII: 28 that 'Tres autem perpendiculares [ductæ a verticibus angulorũ versus latera ipsis angulis opposita] in eodem puncto se intersecabunt, quod alio in loco

In the triangle ABC to the base AB let fall the perpendicular CG and (by Euclid's *Elements*, II, 13) there will be $AC^2 - BC^2 = 2AB \times AG - AB^2$. In consequence, by Proposition 8,

$$AC \times \mathrm{fl}(AC) - BC \times \mathrm{fl}(BC) = AB \times \mathrm{fl}(AG) + AG \times \mathrm{fl}(AB) - AB \times \mathrm{fl}(AB),$$

that is, $AB \times \mathrm{fl}(AG) - BG \times \mathrm{fl}(AB)$.[89] As was to be demonstrated.

Corollary 1. If there be given the two sides, then, with the fluxions of AC and BC vanishing, there will be

$$AB \times \mathrm{fl}(AG) - BG \times \mathrm{fl}(AB) = 0 \quad \text{or} \quad AB:BG = \mathrm{fl}(AB):\mathrm{fl}(AG).$$

Corollary 2. If there be given the base and one or other of the sides, say BC, then $AB:AC = \mathrm{fl}(AC):\mathrm{fl}(AG)$.

Corollary 3. If there be given a [base] segment and the side opposite, say AG and BC, then $AC:BG = \mathrm{fl}(AB): -\mathrm{fl}(AC)$.

In the same manner, should there be given one side, the base or a [base] segment several theorems arise which are simple rather than basic.

Lemma.[90]

The perpendiculars AE, BF, CG dropped to each of the sides of any triangle ABC have a common intersection.

For let X be the intersection of the two AE, CG and because of the similar triangles BGC, AGX there will be $CG:GB = AG:GX$ and alternately

$$CG:AG = GB:GX.$$

Let Y,[91] however, be the intersection of the two CG and BF, and then because of the similar triangles CGA, BGY there will be $CG:AG = BG:GY$. Consequently GX and GY are equal, in other words the perpendiculars AE and BF cut CG in the same point. As was to be demonstrated.[92]

---

demonstratum tradidimus'—where, he nowhere specified. The observation was echoed almost a century later by Ludolph van Ceulen in *Zetema* 31 of his 'De Zetematum Geometricorum Epilogismo' [= *Fundamenta Arithmetica et Geometrica, cum eorundem usu in varijs problematis Geometricis....E vernaculo in Latinam translata a Wil*[*lebrod*] *Sn*[*ell*] (Leyden, 1615): 137–84]: 165–7, where he stated that 'Perpendiculares [trianguli] in eodem puncto se intersecare alias demonstratum est'. The first published proof known to us was that given by Samuel Marolois in his *Geometrie* [the opening tract of his *Opera Mathematica. Ou Œuvres Mathematicques traictans de Geometrie, Perspective, Architecture et Fortification* (The Hague, 1619)]: signature K^v: 'Theoreme 5. En tous triangles Oxigones les trois perpendiculaires qui tombent des angles sur les costez opposez s'entrecoupent mutuellement en un point'.

(91) Not shown in Newton's figure.

(92) In its present *reductio* form Newton's proof is evidently original but has considerable structural similarity with the contemporary adaptation of Marolois' proof (see note (90) above) given by Antoine Arnauld in his *Nouveaux Elemens de Geometrie* (Paris, 1667): Livre XIII,

Prop 21.

In quovis Triangulo acutangulo demissis ad basem et latera perpendiculis se mutuo secantibus: summa fluxionum laterum ductarū in opposita segmenta perpendiculorum quæ in ea demittuntur, æqualis est aggregato fluxionis basis ductæ in conterminum segmentum perpendiculi in eam demissi, & fluxionis perpendiculi istius ductæ in basem.

In Triangulo ABC, ad latera AB, BC, AC demitte perpendicula AE BF CG se mutuò secantibus in X. Et erit

$$BX \times \text{fl} \, AC + AX \times \text{fl} \, BC = GX \times \text{fl} \, AB + AB \times \text{fl} \, GC.$$

Erat enim supra[93] $AC \times \text{fl} \, AC - BC \times \text{fl} \, BC = AB \times \text{fl} \, AG - BG \times \text{fl} \, AB$. Est et (per prop [13]) $AC \times \text{fl} \, AC = AG \times \text{fl} \, AG + GC \times \text{fl} \, GC$. [Itacɜ] æqualia priora ducta in AG aufer a posterioribus ductis in AB et restabit

$$AC \times BG \times \text{fl} \, AC + BC \times AG \times \text{fl} \, BC = BG \times AG \times \text{fl} \, AB + GC \times AB \times \text{fl} \, GC.$$

Atqui propter sim tri ACG, XBG est $GC \cdot AC :: BG \cdot BX$,[94] et sic propter sim tri BCG, XAG est $GC \cdot BC :: AG \cdot AX$,[94] et $GC \cdot BG :: AG \cdot GX$. Quare applica æqualia novissima ad[95] GC et ea evadent

$$BX \times \text{fl} \, AC + AX \times \text{fl} \, BC = GX \times \text{fl} \, AB + AB \times \text{fl} \, GC \quad \text{Q.E.D.}$$

Cor 1. Si datur basis et latus alterutrum AC erit $AB \cdot AX :: \text{fl} \, BC \cdot \text{fl} \, GC$.

Cor 2. Si dantur latera erit $AB \cdot GX :: \text{fl} \, AB \cdot \text{defl} \, GC$.

Cor 3. Si datur basis et perpendiculum erit $AX \cdot BX :: \text{fl} \, AC \cdot \text{defl} \, BC$.[96]

[97]Hactenus comparavimus partes rectilineas triangulorum, jam vero de fluxi[onibus] angulorum eorum etiam agemus, exponendo angulum per arcum subtendentem dato aliquo Radio descriptum. Arcum vero hunc designabimus scribendo arc A pro arcu subtendente angulum A, arc ABC pro arcu subtendente angulum ABC & sic in cæteris.

---

§xxix: 273: 'Theoreme. Si de tous les angles d'un triangle oxygone on tire des perpendiculaires aux costez, elles se couperont en un même point.' (Arnauld effectively shows that, since the points B, E, X, G and also A, C, E, G are concyclic, therefore $E\hat{B}X = C\hat{G}E = F\hat{A}X$ and the triangles EBX, FAX are accordingly similar.) The correspondence is, however, probably fortuitous: Newton's French was weak (see 1: 549, note (1)) and there is no indication that he ever knew of the existence of Marolois' or Arnauld's treatises. In the manuscript Newton originally entered his lemma after Proposition 21 following, but it is here located in agreement with his dictate 'Lemma præponenda præced[enti] Prop' (The lemma [is] to be set before the preceding proposition).

(93) See Proposition 20.

(94) Here respectively are cancelled in sequel 'adeocɜ $\dfrac{BG \times AC}{GC} = BX$' and 'et inde $\dfrac{AG \times BC}{GC} = AX$'.

(95) This replaces the equivalent phrase 'divide...per'.

(96) Newton began to enter a 'Cor 4' on the next line but broke off after writing 'Si' (If...).

Proposition 21

In any acute-angled triangle, on letting fall mutually intersecting perpendiculars to the base and sides, the sum of the fluxions of the sides multiplied into the opposite segments of the perpendiculars let fall onto them is equal to the aggregate of the fluxion of the base multiplied into the adjoining segment of the perpendicular let fall onto it and the fluxion of that perpendicular multiplied into the base.

In the triangle ABC to the sides AB, BC, AC let fall the perpendiculars AE, BF, CG mutually concurrent at X: then will there be

$$BX \times \text{fl}\,(AC) + AX \times \text{fl}\,(BC) = GX \times \text{fl}\,(AB) + AB \times \text{fl}\,(GC).$$

For above[93] there was

$$AC \times \text{fl}\,(AC) - BC \times \text{fl}\,(BC) = AB \times \text{fl}\,(AG) - BG \times \text{fl}\,(AB).$$

Also (by Proposition 13) $AC \times \text{fl}\,(AC) = AG \times \text{fl}\,(AG) + GC \times \text{fl}\,(GC)$. Accordingly, take the former equals multiplied by AG from the latter ones multiplied by AB and there will be left

$$AC \times BG \times \text{fl}\,(AC) + BC \times AG \times \text{fl}\,(BC)$$
$$= BG \times AG \times \text{fl}\,(AB) + GC \times AB \times \text{fl}\,(GC).$$

Here, however, on account of the similar triangles ACG, XBG there is

$$GC:AC = BG:BX^{[94]}$$

and, because of the similar triangles BCG, XAG, likewise $GC:BC = AG:AX^{[94]}$ and $GC:BG = AG:GX$. Consequently, divide the most recent equality through by GC and it will become

$$BX \times \text{fl}\,(AC) + AX \times \text{fl}\,(BC) = GX \times \text{fl}\,(AB) + AB \times \text{fl}\,(GC).$$

As was to be demonstrated.

Corollary 1. If there are given the base and one or other side AC, then

$$AB:AX = \text{fl}\,(BC):\text{fl}\,(GC).$$

Corollary 2. If the sides are given, then $AB:GX = \text{fl}\,(AB):-\text{fl}\,(GC)$.

Corollary 3. If there is given the base and its perpendicular, then

$$AX:BX = \text{fl}\,(AC):-\text{fl}\,(BC).^{[96]}$$

[97]So far we have compared the rectilinear parts of triangles; now indeed let us go on to discuss also the fluxions of their angles, specifying an angle by its subtending arc described with some given radius. This arc we shall, in fact, denote by writing 'arc A' for the arc subtending the angle A, 'arc $A\widehat{B}C$' for the arc subtending the angle $A\widehat{B}C$, and likewise in other cases.

Prop. 22.[98]

In quovis Triangulo acutangulo excessus quo fluxio basis ducta in Radium superat fluxiones laterum ductas in cosinus conterminorum angulorum ad basem, æqualis est fluxioni arcûs angulum verticalem subtendentis ductæ in perpendiculum ab angulo isto ad basem demissum.

In triangulo ABC demissis ad latera omnia $\perp^{\text{lis}}$ AE, BF, CG sit C verticalis angulus et erit

$$\text{Rad} \times \text{fl}\,AB - \cos A \times \text{fl}: AC$$
$$-\cos B \times \text{fl}: BC = CG \times \text{fl}: \text{arc}\,C.^{[99]}$$

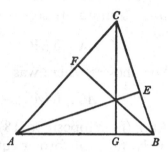

Est enim rad. cosin $A :: AC . AG$. Ergo per prop [8] $\cos A \times \text{fl}\,AC + AC \times \text{fl}\cos A = \text{Rad} \times \text{fl}\,AG$. Est et per prop [14] $\text{fl}\cos A . \text{defl arc}\,A :: \sin A . \text{rad} :: CG . AC$. Quare $AC \times \text{fl}\cos A = C[G] \times \text{defl arc}\,A$. et

$$\cos A \times \text{fl}\,AC + [C]G \times \text{defl arc}\,A = \text{Rad} \times \text{fl}\,AG.$$

Eodem modo est $\cos B \times \text{fl}\,BC + CG \times \text{defl arc}\,B = \text{rad} \times \text{fl}\,BG$. Horum æqualium summa est $\cos A \times \text{fl}\,AC + \cos B \times \text{fl}\,BC + CG \times \text{defl arc}\,A + B = \text{Rad} \times \text{fl}\,AB$. sed defl arc $A+B$ est fl arc C. Ergo

$$\cos A \times \text{fl}\,AC + \cos B \times \text{fl}\,BC + CG \times \text{fl arc}\,C = \text{rad} \times \text{fl}\,AB.$$

Aufer utrincæ $\cos A \times \text{fl}\,AC + \cos B \times \text{fl}\,BC$ et restabit

$$CG \times \text{fl arc}\,C = \text{rad} \times \text{fl}\,AB - \cos A \times \text{fl}\,AC - \cos B \times \text{fl}\,BC. \quad \text{Q.E.D.}$$

[100]*Cor 1.* Si dentur latera AB, AC angulum[101] comprehendentia erit $AE . \text{Rad} :: \text{fl}\,BC . \text{fl arc}\,A$.

Cor. 2. Si detur basis[102] et latus alterutrum AC, erit

$$AE . \cos B :: \text{fl}\,AB . \text{defl arc}\,A.$$

Cor 3. Si detur angulus aliquis sit iste $A^{[103]}$ et erit

$$\text{Rad} \times \text{fl}\,BC = \cos C \times \text{fl}\,AC + \cos B \times \text{fl}\,AB,$$

vel quod perinde est $BC \times \text{fl}\,BC = CF \times \text{fl}\,AC + BG \times \text{fl}\,AB$. Unde ex datis fluxionibus duorum laterum si modo ambo fluant datur fluxio tertij.

---

(97) The following paragraph is clearly an afterthought since Newton first continued with an exact copy of the opening of Proposition 22, viz: '*Prop* . *In quovis acutangulo triangulo excessus quo fluxio basis ducta in . . .*'.

(98) A variant first proof of this proposition, cancelled in the manuscript, is reproduced in Appendix 1.6 below.

(99) Understand that Newton's *cosinus A* is $R.\cos A$ in modern terms. Compare 1, 3, §1: note (4).

Proposition 22[98]

In any acute-angled triangle the excess of the fluxion of the base multiplied by the radius over the fluxions of the sides multiplied by the cosines of the adjoining base angles is equal to the fluxion of the arc subtending the vertical angle multiplied by the perpendicular let fall from that angle to the base.

In the triangle ABC, where the perpendiculars AE, BF, CG have been dropped to each of the sides, let C be the vertical angle and then

$$\text{radius} \times \text{fl}(AB) - \text{Cos}\, A \times \text{fl}(AC) - \text{Cos}\, B \times \text{fl}(BC) = CG \times \text{fl}(\text{arc}\, C). \quad [99]$$

For radius: $\text{Cos}\, A = AC : AG$, and therefore by Proposition 8

$$\text{Cos}\, A \times \text{fl}(AC) + AC \times \text{fl}(\text{Cos}\, A) = \text{radius} \times \text{fl}(AG).$$

Also, by Proposition 14, $\text{fl}(\text{Cos}\, A) : -\text{fl}(\text{arc}\, A) = \text{Sin}\, A : \text{radius} = CG : AC$. Consequently $AC \times \text{fl}(\text{Cos}\, A) = CG \times -\text{fl}(\text{arc}\, A)$, and so

$$\text{Cos}\, A \times \text{fl}(AC) + CG \times -\text{fl}(\text{arc}\, A) = \text{radius} \times \text{fl}(AG).$$

In the same way there is $\text{Cos}\, B \times \text{fl}(BC) + CG \times -\text{fl}(\text{arc}\, B) = \text{radius} \times \text{fl}(BG)$. The sum of these equalities is

$$\text{Cos}\, A \times \text{fl}(AC) + \text{Cos}\, B \times \text{fl}(BC) + CG \times -\text{fl}(\text{arc}\,[A+B]) = \text{radius} \times \text{fl}(AB).$$

But $-\text{fl}(\text{arc}\,[A+B])$ is $\text{fl}(\text{arc}\, C)$, and therefore

$$\text{Cos}\, A \times \text{fl}(AC) + \text{Cos}\, B \times \text{fl}(BC) + CG \times \text{fl}(\text{arc}\, C) = \text{radius} \times \text{fl}(AB).$$

Take away $\text{Cos}\, A \times \text{fl}(AC) + \text{Cos}\, B \times \text{fl}(BC)$ from either side and there will be left $CG \times \text{fl}(\text{arc}\, C) = \text{radius} \times \text{fl}(AB) - \text{Cos}\, A \times \text{fl}(AC) - \text{Cos}\, B \times \text{fl}(BC)$. As was to be demonstrated.

[100]*Corollary 1.* If the sides AB, AC bounding the angle[101] be given, then $AE : \text{radius} = \text{fl}(BC) : \text{fl}(\text{arc}\, A)$.

Corollary 2. If there be given the base[102] and one or other side AC, then $AE : \text{Cos}\, B = \text{fl}(AB) : -\text{fl}(\text{arc}\, A)$.

Corollary 3. If some angle be given let it be A,[103] and then

$$\text{radius} \times \text{fl}(BC) = \text{Cos}\, C \times \text{fl}(AC) + \text{Cos}\, B \times \text{fl}(AB);$$

or, equivalently, $BC \times \text{fl}(BC) = CF \times \text{fl}(AC) + BG \times \text{fl}(AB)$. Hence, given the fluxions of two sides, from these (provided both are fluent) is given the fluxion of the third.

---

(100) These corollaries strictly relate to the analogous theorem

'$\text{Rad} \times \text{fl}\, BC = \cos B \times \text{fl}\, AB + \cos C \times \text{fl}\, AC + AE \times \text{fl}\, \text{arc}\, A$'

demonstrated in the first version of the proof (see note (98) and compare Appendix 1.6).

(101) That is, 'angulum verticalem' (vertical angle), here $\hat{A}$.

(102) Understand this to be BC; compare note (100).

(103) Newton first concluded: 'et si basis BC detur, erit $\cos C . \cos B :: \text{fl}\, AB . [\text{de}]\, \text{fl}\, AC$' (and should the base BC be given, then $\text{Cos}\, C : \text{Cos}\, B = \text{fl}(AB) : [-]\text{fl}(AC)$).

Cor 4. Si angulus [A] et latus aliquod AC detur erit $BC.BG::\mathrm{fl}\,AB.\mathrm{fl}\,BC$.

Cor 5. Si angulus [A] et basis [BC] detur erit[104]

Prop 23.

In Triangulo quovis acutangulo perpendiculo ad basem demisso differentia fluxionum laterum ductarum in sinus conterminorum angulorum ad basem, æqualis est differentiæ fluxionum arcuum subtendentium angulos ad basem ductarum in contermina segmenta basis.

In triangulo ABC demisso ad basem $AB \perp^{\mathrm{lo}} CG$, est $\sin B \times \mathrm{fl}\,BC - \sin A \times \mathrm{fl}\,AC = AG \times \mathrm{fl}\,\mathrm{arc}\,A - BG \times \mathrm{fl}\,\mathrm{arc}\,B$. Nam demiss[is] ad reliqua latera perpendiculis AE BF est $\sin A . \sin B :: BF . AE :: BC . AC$. Ergo per prop [8] $\sin A \times \mathrm{fl}\,AC + AC \times \mathrm{fl}\,\sin A = \sin B \times \mathrm{fl}\,BC + BC \times \mathrm{fl}\,\sin B$. Sed per prop [14] est $\mathrm{fl}\,\sin A . \mathrm{fl}\,\mathrm{arc}\,A[::]\cos\mathrm{in}\,A . \mathrm{rad}:: AG . AC$. Ergo $AC \times \mathrm{fl}\,\sin A = AG \times \mathrm{fl}\,\mathrm{arc}\,A$. Et eodem modo est $BC \times \mathrm{fl}\,\sin B = BG \times \mathrm{fl}\,\mathrm{arc}\,B$. Quare

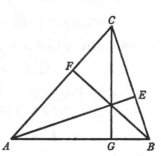

$$\sin A \times \mathrm{fl}\,AC + AG \times \mathrm{fl}\,\mathrm{arc}\,A = \sin B \times \mathrm{fl}\,BC + BG \times \mathrm{fl}\,\mathrm{arc}\,B$$

et ablatis utrobiꝗ $BG \times \mathrm{fl}\,\mathrm{arc}\,B$ et $\sin A \times \mathrm{fl}\,AC$ erit

$$AG \times \mathrm{fl}\,\mathrm{arc}\,A - BG \times \mathrm{fl}\,\mathrm{arc}\,B = \sin B \times \mathrm{fl}\,BC - \sin A \times \mathrm{fl}\,AC. \quad \text{Q.E.D.}$$

Cor 1. Si dentur latera erit $BG . AG :: \mathrm{fl}\,\mathrm{arc}\,A . \mathrm{fl}\,\mathrm{arc}\,B :: \mathrm{fl}\,\mathrm{ang}\,A . \mathrm{fl}\,\mathrm{ang}\,B$.

[*Cor*] *2.* Si detur angulus B et latus oppositum erit $AG . \sin B :: \mathrm{fl}\,BC . \mathrm{fl}\,\mathrm{arc}\,A$.

[*Cor*] *3.* Si detur angulus B et latus conterminum BC erit

$$AG . \sin A :: \mathrm{fl}\,AC . \mathrm{de\,fl}\,\mathrm{arc}\,A \text{ vel } \mathrm{fl}\,\mathrm{arc}\,C.$$

[*Cor*] *4.* Si detur angulus[105] sit iste B et erit

$$\sin B \times \mathrm{fl}\,BC - \sin A \times \mathrm{fl}\,AC = AG \times \mathrm{fl}\,\mathrm{arc}\,A.$$

[*Cor*] *5.* In omni vero triangulo rectang. cum angulus rectus detur sit iste B et erit

$$\mathrm{rad} \times \mathrm{fl}\,BC - \sin A \times \mathrm{fl}\,AC = AB \times \mathrm{fl}\,\mathrm{arc}\,A.\text{[106]}$$

Unde si latus BC detur erit

$$AB . \sin A :: \mathrm{fl}\,AC . \mathrm{de\,fl}\,\mathrm{arc}\,A \text{ vel } \mathrm{fl}\,\mathrm{arc}\,[C].$$

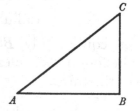

Sin Hypotenusa AC detur erit $AB . \mathrm{rad} :: \mathrm{fl}\,BC . \mathrm{fl}\,\mathrm{arc}\,A$.

(In left margin:) Vide Coroll. Prop. sequ.

---

(104) Understand '$BC.CF::\mathrm{fl}\,AC.\mathrm{fl}\,BC$'. The given angle is once more assumed to be the 'angulus verticalis' $\hat{A}$.

(105) The parenthesis 'ut fit in omni triangulo rectangulo' (as happens in every right triangle) is deleted.

Corollary 4. If the angle A and some side AC be given, then

$$BC:BG = \text{fl}\,(AB):\text{fl}\,(BC).$$

Corollary 5. If the angle A and base BC be given, then[104]

Proposition 23

In any acute-angled triangle, when the perpendicular is let fall to the base, the difference of the fluxions of the sides multiplied by the sines of the angles adjoining the base is equal to that of the fluxions of the arcs subtending the base angles multiplied by the adjoining base segments.

In the triangle ABC, on letting fall the perpendicular CG to the base AB, there is $\text{Sin}\,B \times \text{fl}\,(BC) - \text{Sin}\,A \times \text{fl}\,(AC) = AG \times \text{fl}\,(\text{arc}\,A) - BG \times \text{fl}\,(\text{arc}\,B)$. For, when the perpendiculars AE, BF are dropped onto the remaining sides, there is $\text{Sin}\,A:\text{Sin}\,B = BF:AE = BC:AC$. Therefore by Proposition 8

$$\text{Sin}\,A \times \text{fl}\,(AC) + AC \times \text{fl}\,(\text{Sin A}) = \text{Sin}\,B \times \text{fl}\,(BC) + BC \times \text{fl}\,(\text{Sin}\,B).$$

But by Proposition 14 $\text{fl}\,(\text{Sin}\,A):\text{fl}\,(\text{arc}\,A) = \text{Cos}\,A:\text{radius} = AG:AC$, and therefore $AC \times \text{fl}\,(\text{Sin}\,A) = AG \times \text{fl}\,(\text{arc}\,A)$. And in the same way

$$BC \times \text{fl}\,(\text{Sin}\,B) = BG \times \text{fl}\,(\text{arc}\,B).$$

Consequently

$$\text{Sin}\,A \times \text{fl}\,(AC) + AG \times \text{fl}\,(\text{arc}\,A) = \text{Sin}\,B \times \text{fl}\,(BC) + BG \times \text{fl}\,(\text{arc}\,B),$$

and on taking away $BG \times \text{fl}\,(\text{arc}\,B)$ and $\text{Sin}\,A \times \text{fl}\,(AC)$ from either side there is $AG \times \text{fl}\,(\text{arc}\,A) - BG \times \text{fl}\,(\text{arc}\,B) = \text{Sin}\,B \times \text{fl}\,(BC) - \text{Sin}\,A \times \text{fl}\,(AC)$. As was to be proved.

Corollary 1. If the sides be given, then

$$BG:AG = \text{fl}\,(\text{arc}\,A):\text{fl}\,(\text{arc}\,B) = \text{fl}\,(\hat{A}):\text{fl}\,(\hat{B}).$$

Corollary 2. If there be given the angle B and the side opposite, then

$$AG:\text{Sin}\,B = \text{fl}\,(BC):\text{fl}\,(\text{arc}\,A).$$

Corollary 3. If there be given the angle B and the adjoining side BC, then $AG:\text{Sin}\,A = \text{fl}\,(AC):-\text{fl}\,(\text{arc}\,A)$, that is, $\text{fl}\,(\text{arc}\,C)$.

Corollary 4. If an angle be given,[105] let it be B and then

$$\text{Sin}\,B \times \text{fl}\,(BC) - \text{Sin}\,A \times \text{fl}\,(AC) = AG \times \text{fl}\,(\text{arc}\,A).$$

Corollary 5. In every right-angled triangle, indeed, since the right angle is given let it be B and there will be

$$\text{radius} \times \text{fl}\,(BC) - \text{Sin}\,A \times \text{fl}\,(AC) = AB \times \text{fl}\,(\text{arc}\,A).^{(106)}$$

Hence, should the side BC be given, there will be

$$AB:\text{Sin}\,A = \text{fl}\,(AC):-\text{fl}\,(\text{arc}\,A), \text{ that is } \text{fl}\,(\text{arc}\,C).$$

But if the hypotenuse AC be given, then

$$AB:\text{radius} = \text{fl}\,(BC):\text{fl}\,(\text{arc}\,A).$$

See the Corollaries to the following proposition.

Prop 24.

In triangulo quovis acutangulo demisso ad Basem perpendiculo differentia fluxionum laterum ductarum in sinus conterminorum angulorum ad basem æquatur aggregato fluxionis arcûs subtendentis alterutrum angulum ad basem ductæ in basem et fluxionis arcus subtendentis angulum ad verticem ductæ in segmentum basis priori angulo conterminum.

In triangulo ABC demisso ad basem $AB \perp^{\text{lo}} CG$ erit
$\sin B \times \text{fl } BC - \sin A \times \text{fl } AC = AB \times \text{fl arc } A + BG \times \text{fl arc } C$.
Erat enim supra[107]

$$\sin B \times \text{fl } BC - \sin A \times \text{fl } AC = AG \times \text{fl arc } A - BG \times \text{fl arc } B.$$

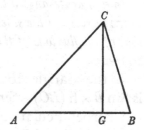

Quare cum summa trium angulorum detur, adeoൻ aggregatum fluxionum omnium nullum sit, vel duorum fluxio æqualis defluxioni tertij pro $-BG \times \text{fl arc } B$ scribe $+BG \times \text{fl arc } \overline{A+C}$ et ipsi $AG \times \text{fl arc } A$ adde $BG \times \text{fl arc } B$ et fiet

$$\sin B \times \text{fl } BC - \sin A \times \text{fl } AC = AB \times \text{fl arc } A + BG \times \text{fl arc } C. \quad \text{Q.E.D.}$$

Cor 1. Si dētur latera erit $AB \cdot BG :: \text{fl arc } C \cdot \text{defl arc } A$.

Cor 2. Si detur angulus A et latus oppositum BC erit
$$BG \cdot \sin A :: \text{fl } AC \cdot \text{defl arc } C.^{[109]}$$

Cor 3. Si detur idem angulus A et latus conterminum AC erit
$$BG \cdot \sin B :: \text{fl } BC \cdot \text{fl arc } C.$$

Cor 4. Si alter angulus C inter duo latera detur una cum lateri AC utriൻ angulo contermino erit $AB \cdot \sin B :: \text{fl } BC \cdot \text{fl arc } A$.

Cor 5. Si detur ille angulus C et latus BC alteri angulo oppositum erit
$$AB \cdot \sin A :: \text{fl } AC \cdot \text{defl arc } A \text{ vel fl arc } [\overline{B+C}].$$

Hæc Corol. potius referantur ad prop. præced.[108]

Prop 25.

In triangulo quovis acutangulo demissis ad tria[110] latera perpendiculis, fluxiones angulorum sunt inter se ut excessus quo fluxiones laterum oppositorum ductæ in latera illa superant fluxiones reliquorum laterum ductas in eorum segmenta alteri lateri contermina.

---

(106) Since $AB = AC \cdot \cos A$, this follows directly from the equality
$$\text{'rad} \times BC = AC \times \sin A\text{'}$$
on taking fluxions.

(107) In Proposition 23 preceding.

(108) With A and B interchanged Corollaries 2 and 3 are identical with their namesakes in Proposition 23, while with B and C interchanged Corollaries 4 and 5 are particular instances of the preceding Corollary 4. The whole present Proposition 24 is, in fact, no more than an *idem aliter* to Proposition 23 and in redraft (§2.2 following) is rightly omitted.

Proposition 24

In any acute-angled triangle, when the perpendicular is let fall to the base, the difference of the fluxions of the sides multiplied by the sines of the adjoining base angles equals the aggregate of the fluxion of the arc subtending either base angle multiplied by the base and the fluxion of the arc subtending the vertex angle multiplied by the base segment adjoining the former angle.

In the triangle ABC, when the perpendicular CG is let fall to the base AB, there will be

$$\mathrm{Sin}\,B \times \mathrm{fl}\,(BC) - \mathrm{Sin}\,A \times \mathrm{fl}\,(AC) = AB \times \mathrm{fl}\,(\mathrm{arc}\,A) + BG \times \mathrm{fl}\,(\mathrm{arc}\,C).$$

For above[107] there was

$$\mathrm{Sin}\,B \times \mathrm{fl}\,(BC) - \mathrm{Sin}\,A \times \mathrm{fl}\,(AC) = AG \times \mathrm{fl}\,(\mathrm{arc}\,A) - BG \times \mathrm{fl}\,(\mathrm{arc}\,B).$$

Consequently, since the sum of the three angles is given, so that the total of the fluxions of all of them is nil—in other words, the fluxion of two of them is equal to the defluxion of the third—, in place of $-BG \times \mathrm{fl}\,(\mathrm{arc}\,B)$ write

$$+BG \times \mathrm{fl}\,(\mathrm{arc}\,[A+C])$$

and to $AG \times \mathrm{fl}\,(\mathrm{arc}\,A)$ add $BG \times \mathrm{fl}\,(\mathrm{arc}\,B)$: there will then result

$$\mathrm{Sin}\,B \times \mathrm{fl}\,(BC) - \mathrm{Sin}\,A \times \mathrm{fl}\,(AC) = AB \times \mathrm{fl}\,(\mathrm{arc}\,A) + BG \times \mathrm{fl}\,(\mathrm{arc}\,C).$$

As was to be demonstrated.

Corollary 1. If the sides be given, then $AB:BG = \mathrm{fl}\,(\mathrm{arc}\,C): -\mathrm{fl}\,(\mathrm{arc}\,A)$.

Corollary 2. If there be given the angle A and the side opposite BC, then $BG:\mathrm{Sin}\,A = \mathrm{fl}\,(AC): -\mathrm{fl}\,(\mathrm{arc}\,C)$.[109]

Corollary 3. If there be given the same angle A and the adjoining side AC, then $BG:\mathrm{Sin}\,B = \mathrm{fl}\,(BC):\mathrm{fl}\,(\mathrm{arc}\,C)$.

Corollary 4. If the other angle C between the two sides be given along with the side AC adjoining either angle, then $AB:\mathrm{Sin}\,B = \mathrm{fl}\,(BC):\mathrm{fl}\,(\mathrm{arc}\,A)$.

Corollary 5. If that angle C be given together with the side BC opposite the other angle, then $AB:\mathrm{Sin}\,A = \mathrm{fl}\,(AC): -\mathrm{fl}\,(\mathrm{arc}\,A)$, that is, $\mathrm{fl}\,(\mathrm{arc}\,[B+C])$.

These Corollaries are better referred to the preceding proposition.[108]

Proposition 25

In any acute-angled triangle, when perpendiculars are let fall to[110] the three sides, the fluxions of the angles are to one another as the excess of the fluxions of the sides opposite multiplied by those sides over the fluxions of the remaining sides multiplied by their segments adjoining the other side.

(109) That is, 'fl arc B' since $\hat{B}+\hat{C} = \pi-\hat{A}$, constant.
(110) 'omnia' (all) is cancelled.

In triangulo ABC demissis perpendiculis AE, BF, CG erit

$$\text{fl:ang:}C.\text{fl:ang:}B::AB\times\text{fl}\,AB-AF\times\text{fl}\,AC-BE\times\text{fl}\,BC$$

$$.AC\times\text{fl}\,AC-CE\times\text{fl}\,BC-AG\times\text{fl}\,AB.$$

Erat enim in Prop [22]

$$\text{rad}\times\text{fl}\,AB-\cos A\times\text{fl}\,AC-\cos B\times\text{fl}\,BC=CG\times\text{fl arc }C.$$

Sed sunt rad. $\cos A.\cos B::AB.AF.BE.$ Quare æqualia

illa ducta in $\dfrac{AB}{\text{rad}}$ evadent

$$AB\times\text{fl}\,AB-AF\times\text{fl}\,AC-BE\times\text{fl}\,BC=\frac{AB\times CG}{\text{rad}}\times\text{fl arc }C$$

hoc est $=\dfrac{CB\times AE}{\text{rad}}\times\text{fl arc }C$, nam $AB\times CG$ et $CB\times AE$ æqualia sunt. Eodem

modo erit $AC\times\text{fl}\,AC-CE\times\text{fl}\,BC-AG\times\text{fl}\,AB=\dfrac{CB\times AE}{\text{rad}}\times\text{fl arc }B$. Quare cum

sit $\dfrac{[CB\times AE]}{\text{rad}}\times\text{fl arc }C.\dfrac{[CB\times AE]}{\text{rad}}\times\text{fl arc }B::\text{fl arc }C.\text{fl arc }B$, liquet proposi-

tum. Q.E.D.

Corol 1. Si dentur latera duo quævis AC et BC erunt

$$\text{fl ang }A.\text{fl ang }B.\text{fl ang }C::BG.AG.-AB.$$

Corol 2. Si detur latus AB[111] fac ut $\text{fl}\,BC.\text{fl}\,AC::CF.R::AF.S$,[112] et erunt
$R-BC.CE-R-S.S+BE::\text{fl ang }A.\text{fl ang }B.\text{fl ang }C$.

Schol. [113]Hæc de triangulis acutangulis solummodo declaravimus ne per determinationes juxta varietatem angulorum obscuræ redderentur Proposi-tiones. Ut eadem vero[114] applicentur ad Triangula obtusangula, [mutandum erit] signum cosinus anguli obtusi ut et conterminorum segmentorum laterum continentium ipsum, quippe quæ sunt cosinus anguli istius ad circulos alternis lateribus descriptos et crescente angulo cosinus diminuitur[115] evanescens ubi angulus evadit rectus & postea cadens ad contrarias partes centri, id est nega-tivus evadens. Præterea ubi angulus vel linea aliqua[116] defluit tunc ejus fluxio auferenda est a quantitatibus quibus juxta Theorema addenda esset, vel addenda quibus esset auferenda. His probe observatis præcedentia Theoremata

---

(111) Newton first continued in the more direct way 'erunt

$$BC-\frac{\text{fl}\,AC}{\text{fl}\,BC}\times CF.CE-\frac{\text{fl}\,AC}{\text{fl}\,BC}\times AC.BE+\frac{\text{fl}\,AC}{\text{fl}\,BC}\times AF::\text{fl ang }A.\text{defl ang }B.\text{defl ang }C'.$$

(112) That is, '$::AC.R+S$'.

(113) See note (88) above. Newton first began: 'In hisce posuimus triangulum acutan[gulum esse]' (In these [propositions] we have taken the triangle to be acute-angled).

(114) 'facile' (easily) is cancelled.

In the triangle ABC, after the perpendiculars AE, BF, CG are let fall, there will be

$$\mathrm{fl}\,(\hat{C}):\mathrm{fl}\,(\hat{B}) = AB \times \mathrm{fl}\,(AB) - AF \times \mathrm{fl}\,(AC)$$
$$- BE \times \mathrm{fl}\,(BC):AC \times \mathrm{fl}\,(AC) - CE \times \mathrm{fl}\,(BC) - AG \times \mathrm{fl}\,(AB).$$

For in Proposition 22 there was

$$\text{radius} \times \mathrm{fl}\,(AB) - \mathrm{Cos}\,A \times \mathrm{fl}\,(AC) - \mathrm{Cos}\,B \times \mathrm{fl}\,(BC) = CG \times \mathrm{fl}\,(\text{arc } C).$$

But radius: $\mathrm{Cos}\,A:\mathrm{Cos}\,B = AB:AF:BE$. Consequently, when the equality is multiplied by AB/radius there will prove to be

$$AB \times \mathrm{fl}\,(AB) - AF \times \mathrm{fl}\,(AC) - BE \times \mathrm{fl}\,(BC) = AB \times CG \times \mathrm{fl}\,(\text{arc } C)/\text{radius},$$

that is, $CB \times AE \times \mathrm{fl}\,(\text{arc } C)/\text{radius}$, for $AB \times CG$ and $CB \times AE$ are equal. In the same manner there will be

$$AC \times \mathrm{fl}\,(AC) - CE \times \mathrm{fl}\,(BC) - AG \times \mathrm{fl}\,(AB) = CB \times AE \times \mathrm{fl}\,(\text{arc } B)/\text{radius}.$$

Consequently, since

$$CB \times AE \times \mathrm{fl}\,(\text{arc } C)/\text{radius}:CB \times AE \times \mathrm{fl}\,(\text{arc } B)/\text{radius} = \mathrm{fl}\,(\text{arc } C):\mathrm{fl}\,(\text{arc } B),$$

the proposition is manifest. As was to be demonstrated.

Corollary 1. If any two sides AC and BC be given, then

$$\mathrm{fl}\,(\hat{A}):\mathrm{fl}\,(\hat{B}):\mathrm{fl}\,(\hat{C}) = BG:AG:-AB.$$

Corollary 2. If there be given the side AB,[111] make

$$\mathrm{fl}\,(BC):\mathrm{fl}\,(AC) = CF:R = AF:S^{[112]}$$

and then $\mathrm{fl}\,(\hat{A}):\mathrm{fl}\,(\hat{B}):\mathrm{fl}\,(\hat{C}) = R - BC:CE - (R+S):S + BE$.

Scholium. [113]We have pronounced these theorems for acute-angled triangles alone so as not to render the propositions obscure by directions according with the variety of angles possible. To apply these[114] in fact to obtuse-angled triangles the sign of the cosine of the obtuse angle will have to be altered, and likewise those of the adjoining segments of the sides containing it, seeing that they are cosines of that angle in circles described on the alternate sides and as the angle increases the cosine diminishes,[115] vanishing when the angle comes to be right and thereafter falling on the opposite side of the centre—that is, coming to be negative. Moreover, when an angle or some line[116] is defluent, then its fluxion must be taken away from the quantities to which, according to the theorem, it should have been added or added to those from which it should have been taken away. When these instructions are duly observed the preceding

---

(115) Newton has quickly cancelled a first following clause 'usqɜ dum angulus evanescit' (till the angle vanishes)!

(116) This replaces 'latus aliqu[od]' (some side).

ad omnes Triangulorum casus se extendent, adeo ut non opus fuerit casus istos sigillatim prosequi.

Patet igitur quomodo in triangulo quovis ex fluxionibus laterum fluxiones arcuum subtendentium angulos derivari possint & vice versa, & quomodo per fluxiones perpendiculorum et segmentorum basium vel ex fluxionibus laterum per prop [20] & Cor [1, 2, 3] Prop [21] vel ex fluxionibus arcuum subtendentium angulos per prop [22 et 23] derivari liceat, & proinde cum omnes figuræ rectilineæ in triangula resolvi possint patet quomodo ex fluxionibus partium unius trianguli progredi liceat ad fluxiones aliarum linearum usqɜ dum perveniatur ad fluxionem lineæ quam quærimus. Id quod primo ostendendum suscepi. Restat ut determinemus centra motuum angularium et plagas ad quas puncta qu[æ]vis mota directè tendunt.

Prop 26.[117]

Si recta mobilis insistat rectæ positione datæ, habeatur autem fluxio rectæ positione datæ, & fluxio arcus angulum earum ad datam distantiam subtendentis: distantia centri motûs angularis a concursu linearum erit ad sinum anguli comprehensi, ut fluxio rectæ positione datæ ad fluxionem arcus angulum subtendentis.

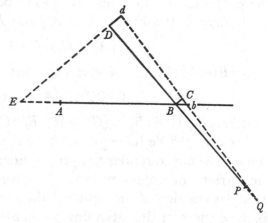

Sit *AB* recta positione data, *BD* recta incedens. Cape *BP* . sin *ABD* :: fl *AB* . fl arc *ABD*[118] et erit *P* polus motus angularis rectæ *BD*. Nam moveatur *BD* aliquantulum donec veniat ad locum *bd* et sit *Q* concursus linearum *BD*, *bd*, sitqɜ *QD* radius dati circuli et *Dd* arcus ejus subtendens angulum *DQd*. Age chordam *dD* et produc eam donec occurrat rectæ *AB* in *E*, et ei parallelam duc *BC* occurrentem *Qd* in *C*. Et propter sim tri *BbC, Ebd*, ut et *QDd, QBC* erit *Bb* . *BC* :: *Eb* . *Ed*. et *BC* . *Dd* :: *BQ* . *DQ*. Et connectendo rationes *Bb* . *Dd* :: *BQ* . *DQ* × $\frac{Ed}{Eb}$. Redeat jam recta *dQ* in locum priorem *DQ* et[119] centro dati circuli *Dd* semper existente ad *Q* ultima ratio rectæ evanescentis *Bb* ad chordam evanescentem *Dd* erit [*BQ* ad] *DQ* × $\frac{ED}{EB}$, siquidem *Ed* & *Eb* tunc evadant *ED* et *EB*. Sed est *EB* . *ED* :: rad . sin *ABD*. Atqɜ adeo *BQ* ad sin *ABD* est ultima ratio rectæ *Bb* ad chordam *Dd* hoc est

---

(117) Manifestly a generalization of 'Theor. 4' of the 1671 tract's addendum on geometrical fluxions; see III: 342–4.

theorems extend to all cases of triangles, so that there is no need to pursue those cases separately.

It is therefore evident how in any triangle, once the fluxions of the sides are given, from them may be derived the fluxions of the arcs subtending the angles, and the converse; and how by means of the fluxions of the perpendiculars and base segments we are at liberty, either (by Proposition 20 and Corollaries 1, 2, 3 to Proposition 21) from the fluxions of the sides or (by Propositions 22, 23) from the fluxions of their subtending arcs, to derive the angles: accordingly, since all rectilinear configurations can be resolved into triangles, it is evident how from the fluxions of the parts of one triangle we are allowed to proceed to the fluxions of other lines till we attain the fluxion of the line we seek. This is what I undertook first to show. It remains for us to determine the centres of angular motion and the directions in which any moving points tend instantaneously.

Proposition 26[117]

If a mobile straight line be ordinate to a straight line given in position, while there is had both the fluxion of the straight line given in position and that of the arc subtending their angle at a given distance, then the distance of the centre of angular motion from the meet of the lines will be to the sine of the angle they comprehend as the fluxion of the straight line given in position to the fluxion of the arc subtending their angle.

Let AB be the straight line given in position, BD the ordinate line. Take $BP : \operatorname{Sin} \widehat{ABD} = \mathrm{fl}\,(AB) : \mathrm{fl}\,(\text{arc } \widehat{ABD})$ and P will be the pole of angular motion of the straight line BD. For let BD move slightly till it comes to the position bd and let Q be the intersection of the lines BD, bd with QD the radius of the given circle and Dd its arc subtending the angle $\widehat{DQd}$. Draw the chord dD and extend it till it meets the straight line AB in E, then parallel to it draw BC meeting Qd in C. And because of the similar triangles BbC, Ebd and also QDd, QBC there will be $Bb : BC = Eb : Ed$ and $BC : Dd = BQ : DQ$, and on combining the ratios $Bb : Dd = BQ : DQ \times (Ed/Eb)$. Now let the straight line dQ retreat to its former position DQ and,[119] with the centre of the given circle Dd lastingly at Q, the final ratio of the vanishing line Bb to the vanishing chord Dd will be BQ to $DQ \times (ED/EB)$, insofar as Ed and Eb then come to be ED and EB. But

$$EB : ED = \text{radius} : \operatorname{Sin} \widehat{ABD}$$

and consequently BQ to $\operatorname{Sin} \widehat{ABD}$ is the final ratio of the straight line Bb to the

---

(118) Newton first wrote this proportion out in full as 'Cape BP ad sinum anguli ABD ut fluxio rectæ AB ad fluxionem arcus subtendentis angulum ABD'.

(119) A first sequel 'angulo Q evanescente' (with the angle $\widehat{Q}$ vanishing) is cancelled.

(per Lem)[120] arcu[m] *Dd*. Adeoქ per ax [6] fluxio rectæ *AB* est ad fluxionem arcus subtendentis angulum ipsi *ABD* æqualem in eadem ratione. Itaქ *BQ* et *BP* eandem habentes rationem ad sinum ang *ABD* æquales sunt, hoc est ultima intersectio linearum coeuntium *bd BD* est ad punctum *P*, quod proinde per [Def 7] erit polus[121] motûs angularis rectæ *BD*. Q.E.D.

Corol. Hinc si recta *PB* quomodocunქ mota secet alias quotcunქ positione datas rectas *AB*, *AC*, [*E*]*D* fluxiones[122]

Agendum est hic de rectis insistentibus alijs rectis in angulis incertis.

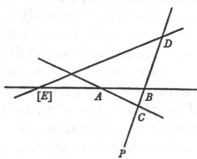

Prop 27.[123]

Si recta mobilis *BD* rectæ positione datæ *AB* et ad datum punctum *A* terminatæ tanquam basi insistat et habeantur fluxiones Basis illius *AB*[,] rectæ insistentis *BD* et arcûs subtendentis angulum *ABD* quem hæ rectæ comprehendunt;[124] sint hæ fluxiones ut sunt rectæ *R*, *S*, *T* respective. Produc *BD* ad *E* ita ut sit *DE*=*S*. Erige normalem *EF* quæ sit ad *T* ut *BD* ad radium.[125] Age *FG* parallelam *AB* et æqualem *R*. Et acta *DG* exhibebit determi-

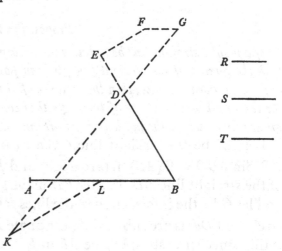

---

(120) This first version of the 'Geometria' contains no corresponding lemma but Newton evidently here appeals to the principle 'Chordæ et arcus evanescentium ultima ratio est ratio æqualitatis' (The final ratio of an arc and its chord as they vanish is one of equality) which he set down in his preliminary revise (§2.1 below) and was a few years later to insert in his *Principia* as Book 1, Lemma VII: '...dico quod ultima ratio arcus chordæ et tangentis ad

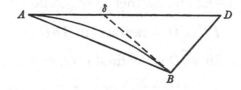

invicem est ratio æqualitatis' (ULC. Dd. 9.46: 24ʳ [= *Philosophiæ Naturalis Principia Mathematica* (London, ₁1687): 30]). To rephrase the somewhat woolly argument he gave in proof in 1685, if the tangent and normal at *B* to the arc $\widehat{AB}$ meet the tangent at *A* in δ and *D*, then (by

chord *Dd*—that is, (by Lemma[120]) to the arc $\overset{\frown}{Dd}$. In consequence, by Axiom 6 the fluxion of the straight line *AB* is to that of the arc subtending an angle equal to $\overset{\frown}{ABD}$ in the same ratio. As a result *BQ* and *BP*, having the same ratio to the sine of $\overset{\frown}{ABD}$, are equal: in other words, the final intersection of the lines *bd*, *BD* as they come to coincide is at the point *P*, and this will accordingly (by Definition 7) be the pole[121] of angular motion of the straight line *BD*. As was to be demonstrated.

Corollary. Hence if the straight line *PB*, moving in any manner whatever, should intersect any number of other straight lines *AB*, *AC*, *ED* given in position, the fluxions[122]

We must here deal with straight lines ordinate to other straight lines at unspecified angles.

Proposition 27[123]

If the mobile straight line *BD* stands on the straight line *AB*, given in position and terminated at the given point *A*, as base and there be had the fluxions of that base *AB*, the standing line *BD* and the arc subtending the angle $\overset{\frown}{ABD}$ comprised by these lines,[124] let these fluxions be respectively as the lines *R*, *S*, *T*. Extend *BD* to *E* so that *DE* = *S*, erect the normal *EF* so as to be to *T* as *BD* to the radius,[125] draw *FG* parallel to *AB* and equal to *R*, and *DG*, when drawn, will

---

Archimedes' convexity lemmas) $AB < \overset{\frown}{AB} < A\delta + \delta B$ and it is evident that $\delta B < \delta D$, whence $AB < \overset{\frown}{AB} < AD$. In the limit therefore as *B* (and so *D*) coincides with *A*, since the 'final' ratio of *AB*/*AD* is unity, so is that of $AB/\overset{\frown}{AB}$. Compare James Gregory, *Geometriæ Pars Universalis, Inserviens Quantitatum Curvarum transmutationi & mensuræ* (Padua, 1668): 'Prop. I. Theorema': 1–3, where the accurate restriction to a 'curva...simplex & non sinuosa' is made.

(121) 'centrū' (centre) is cancelled.

(122) Understand '...fluxiones rectarum *AB*, *AC*, *ED* erunt inter se ut

$$PB, \quad PC \times \frac{AC}{AB} \quad \text{et} \quad PD \times \frac{ED}{EB},$$

(the fluxions of the lines *AB*, *AC*, *ED* will be to one another as

$$PB, \quad PC \times AC/AB \quad \text{and} \quad PD \times ED/DB).$$

Since he cites this cancelled corollary as a 'lemma' in Proposition 30 (see note (143) below) it would appear to have been Newton's intention to amplify rather than discard it. A large space immediately following in the manuscript is blank of writing except for an isolated 'Prop'.

(123) Construction of the 'determination of motion' (instantaneous tangential direction) of a curve (*D*) defined with respect to generalized Cartesian coordinates *AB* and *BD* whose ordination angle $\overset{\frown}{ABD}$ is no longer fixed but determined in some relationship to *AB*, *BD*.

(124) 'vel relatio harum fluxionum ad invicem' (or the relationship of these fluxions to one another) is cancelled.

(125) Understand this to be *DO*, where *O* is the limit-meet of ordinates indefinitely near to *BD*. At once $EF = BD \times \text{fl}(\overset{\frown}{ABD})$.

nationem motus pu[n]cti D hoc est tanget curvam quam punctum D describit.[126]

Vel sic. In AB cape BL quæ sit ad BD ut fluxio rectæ AB ad fluxionem rectæ BD.[127] Dein cape lineam T ad BL ut fluxio arcus subtendentis ang: ABD, ad fluxionem rectæ AB. Et ad rectos angulos cum BD age LK quæ sit ad T ut BD ad radium circuli dati cujus arcus subtendit angulum B[128] et acta KD tanget curvam. Sed in hac constructione fluxio BD non debet esse nulla. Quare constructio prior melior.[129]

Prop 28.

Si rectæ DP DQ datum angulum QDP servantes quocunꝗ motu ferantur, fuerint[130] autem P et Q poli earum, et ad hos polos erigantur normales PE, QE concurrentes in E: erit E polus communis motarum linearum $QD[, D]P$; et rectæ omnes ED EF EG EH ab hoc polo ad puncta quævis D, F, G, H datam positionem ad motas rectas DP DQ servant[ia],[131] perpendiculares erunt ad curvas quas mota illa puncta describunt in plano immobili.[132]

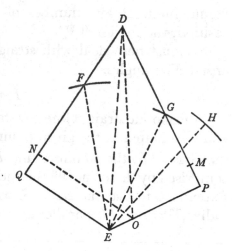

---

(126) In equivalent limit-increment terms, if the arc of the curve is instantaneously increased by the infinitesimal segment Dg, this is compounded of the imcrement $Bb = fg$ of AB, De of BD and

$$ef = BD \times D\hat{B}f \quad \text{of} \quad BD \times A\hat{B}D,$$

so that

$$FG:DE:EF = \text{fl}(AB):\text{fl}(BD):BD \times \text{fl}(A\hat{B}D)$$

$$= fg:De:ef,$$

and accordingly DG coincides with the 'determination' Dg of motion of the locus-point D.

(127) The equivalent phrase 'ut S ad R' (as S to R) is deleted. In more direct terms

$$BL:BD = \text{fl}(AB):\text{fl}(BD) = FG:DE.$$

(128) That is,

$$LK:BD = BD \times \text{fl}(A\hat{B}D):\text{fl}(BD) = EF:DE.$$

It follows $DB:BL:LK = DE:EH:HG$ on completing the paralleogram $EFGH$; hence the quadrilaterals $DBLK$, $DEHG$ whose sides BL, EH and LK, HG (both perpendicular to BDE) are parallel are in homothety round D, and so KDG is a straight line.

exhibit the direction of motion of the point D, that is, it will touch the curve described by the point D.[126]

Or thus. In AB take BL, which is to be to BD as the fluxion of the straight line AB to the fluxion of the straight line BD.[127] Next take the line T to BL as the fluxion of the arc subtending the angle $A\widehat{B}D$ to the fluxion of the line AB, and at right angles to BD draw LK—this to be to T as BD to the radius of the given circle whose arc subtends the angle B[128]—, and then KD when drawn will touch the curve. But in this construction the fluxion of BD ought not to be nil. Consequently the former construction is better.[129]

Proposition 28

If the straight lines DP, DQ travel, keeping the angle $Q\widehat{D}P$ constant, with any motion whatever, while P and Q were[130] their poles, and if at these poles there be erected the normals PE, QE concurrent in E, then E will be the common pole of the moving lines QD, DP and all straight lines ED, EF, EG, EH from this pole to any arbitrary points D, F, G, H which maintain a given position with respect to the moving straight lines DP, DQ[131] will be perpendicular to the curves which those moving points describe in the stationary plane.[132]

---

(129) This, doubtless Newton's reason for cancelling the present paragraph, seems a little harsh. It is true that when $\mathrm{fl}(BD) = 0$ (and so $DE = 0$) BL and LK cannot be set in a finite proportion to BD, but the derived proportion $BL:LK = \mathrm{fl}(AB):BD \times \mathrm{fl}(A\widehat{B}D)$ (that is, now, $FG:DF$ since the points D, E are coincident) holds true and we need only construct the 'determination of motion' DG as the parallel there to BK. In sequel Newton began, but almost at once cancelled, the opening to an abortive following paragraph: 'Hæc propositio ad omnes casus Tangentium sufficit sed res [?alijs adhuc patebit exemplis]' (This proposition suffices in all cases of tangency, but the matter [?will be clarified by yet further examples]).

(130) Understand 'initio motus a D' (at the start of motion from D): P and Q are instantaneously the poles round which the arms PD, QD of the fixed angle $P\widehat{D}Q$ rotate.

(131) And so fixed in the plane of the moving angle $P\widehat{D}Q$.

(132) If the angle $P\widehat{D}Q$ moves instantaneously into the position $P\widehat{d}Q$ and E is the limit-meet of the perpendiculars at D and d to the infinitesimal arc Dd of the curve (D), it will be evident, since

$$D\widehat{P}d = D\widehat{Q}d = D\widehat{E}d,$$

that DE is the diameter of the circle through D, P, Q and consequently that $EP \perp PD$, $EQ \perp QD$. Further, since E is the centre of curvature at the point D on the curve (D), all points F, G, H fixed in the plane of the angle $\widehat{D}$ move instantaneously at right angles to the corresponding *radii vectores* EF, EG and EH; compare 1: 264, note (52).

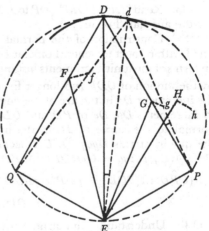

Prop. 29.

Sin angulus *QDP* instabilis fuerit: tunc

Cas 1. [ubi fluxiones angulares ipsarum *DP* et *DQ* habentur] in recta *DP* cape *DM* ad *DP*[133] ut motus angularis ipsius *DP* ad motum angularem ipsius *DQ*, deinde in *DQ* cape *DN . DM :: DQ . DP* et erecta *NO* perpendiculari ad *DQ* erit *OD* perpendicularis ad curvam quam punctum describit.[134]

Cas 2. ubi fluxiones *PD QD*[135] a punctis *P, Q* positione datis habentur.[136]

Cas 3. ubi fluxio *PD* et fluxio angularis [ipsius] *QD* datur.[137]

Prop [30].[138]

Si ad mobile punctum *D* rectæ alicujus [*K*]*D* positione datæ conveniant rectæ quotvis *AD, BD, CD* quarum aliquæ ut *AD, BD* ad puncta *A, B* terminatæ sunt, aliæ, ut *CD*, normaliter insistant rectis positione datis *CE*;[139] centro quovis in recta *KD* posito describe circulum per punctum *D*, secantem rectas *AD, BD, CD, KD* in punctis *E, F, G, H* & fluxiones harum linearum *AD, BD, CD, KD* erunt ut earum partes *ED, FD, GD, HD*, a circulo comprehensæ.[140] Et præterea[141] fluxiones arcuum subtendentium angulos quos rectæ ad data puncta terminatæ faciunt cum recta quavis *CE* vel *HD* positione data erunt ut[142]

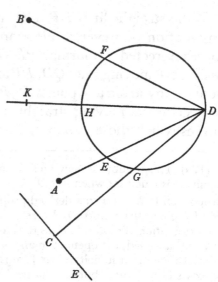

---

(133) Read '*DP* ad *DM*' (*DP* to *DM*). Newton first continued with 'ut fluxio angularis...' (as the angular fluxion...).

(134) Construction of the normal to a curve (*D*) defined with respect to general bipolar coordinates *PD, QD*. In terms of the limit-increments instantaneously generated by the motion of $Q\hat{D}P$ (no longer fixed in magnitude), if the increment *Dd* in the arc of the curve (*D*) corresponds to increments *Dp, Dq* of *PD* and *QD*, then, if *OD* is the normal to the curve at *D*, in the limit as *Dd* vanishes the quadrilaterals *Dpdq, OPDN* become similar, so that *Dd:Dp:Dq:pd:qd = OD:OP:ON:PD:ND*. Accordingly

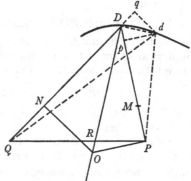

$$\mathrm{fl}(Q\hat{P}D):\mathrm{fl}(P\hat{Q}D) = pd/PD:qd/QD$$

$$= QD:ND = PD:MD.$$

(135) Understand 'ductarum' (drawn). In sequel Newton first wrote 'a perpendiculis *QE PE*' (from the perpendiculars *QE, PE*).

Proposition 29

But were the angle $Q\widehat{D}P$ not stable, then

Case 1, when the angular fluxions of *DP* and *DQ* are had: in the straight line *DP* take *DM* to *DP*[133] as the angular motion of *DP* to the angular motion of *DQ*, then in *DQ* take $DN:DM = DQ:DP$ and, when *NO* is erected perpendicular to *DQ*, *OD* will be perpendicular to the curve which the point describes.[134]

Case 2, when the fluxions of *PD* and *QD*[135] from the points *P*, *Q* given in position are had: ...[136]

Case 3, when the fluxion of *PD* and the angular fluxion of *QD* are given: ...[137]

Proposition 30[138]

If at the mobile point *D* of any straight line *KD* given in position there meet any number of straight lines *AD*, *BD*, *CD*, some of which (as *AD*, *BD*) are terminated at points (*A*, *B*), while others (such as *CD*) stand perpendicularly to straight lines (*CE*) given in position,[139] with any centre situated in the line *KD* describe a circle through the point *D* cutting the lines *AD*, *BD*, *CD*, *KD* in the points *E*, *F*, *G*, *H* and the fluxions of these lines *AD*, *BD*, *CD*, *KD* will then be as their segments *ED*, *FD*, *GD*, *HD* included by the circle.[140] Furthermore,[141] the fluxions of the arcs subtending the angles which the lines terminated at given points make with any straight line *CE* or *HD* given in position will be as[142]

---

(136) In this case $\mathrm{fl}(PD):\mathrm{fl}(QD) = Dp:Dq = OP:ON$. If *PQ* meets *OD* in *R* it is an immediate corollary that $PR:QR$ (or $PD \times OP:QD \times ON) = PD \times \mathrm{fl}(PD):QD \times \mathrm{fl}(QD)$; compare I: 297, note (96).

(137) Here

$$\mathrm{fl}(PD):QD \times \mathrm{fl}(P\widehat{Q}D) = Dp:qd = OP:ND.$$

(138) In revise this cancelled proposition was added as a corollary to Proposition 13. See §2: note (4) below.

(139) In these cases the given points are at infinity in effect.

(140) In the corresponding argument by limit-increments, if the moving point *D* instantaneously slides along *KH* to *d*, then the corresponding increments of *AD*, *BD* and *CD* will evidently be αd, βd and γd where α, β, γ are the meets of the infinitesimal circle on diameter *Dd* with *Ad*, *Bd* and *cd*: at once, since the concyclic configurations $D\alpha\gamma d\beta$, *HEGDF* are similar, there results

$\mathrm{fl}(AD):\mathrm{fl}(BD):\mathrm{fl}(CD):\mathrm{fl}(KD)$

$= \alpha d:\beta d:\gamma d:Dd = ED:FD:GD:HD.$

(141) Newton first continued: 'si circulus iste *HGD* permaneat magnitudine datus licet non

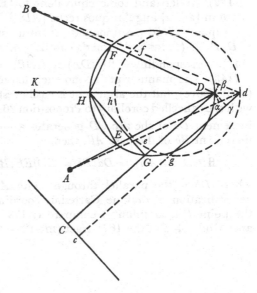

Prop 30.

Si recta [B]D quomodocunq; mota occurrat rectæ *AB* et Curvæ *CD* positione datis; sitq; *P* polus rectæ *BD*, et *DE* recta tangens curvam in *D* et occurrens rectæ *AB* in *E*: erit fluxio curvæ *CD* ad fluxionem rectæ *AB* ut $ED \times DP$ ad $EB \times BP$.

Probatur per Lemma ad Prop [26].[143]

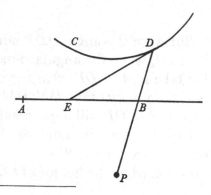

[LIBER 2. PROBLEMATA.][144]

Prob 1.

Rectam datam ita secare ut prima quatuor medie proportionaliū[145] sit quam maxima.

Sunto proportionales *A. B. C. D. E. F* quarum *A & F* constituunt rectam, & *B* maxima est. Jam si *B* crescit evadet major[,] si decrescit fuit major contra Hypoth. Permanet[146] igitur et proinde per Prop [8] est fluxio *A* ad fluxionem *F* ut $-A$ ad $4F$. quare cum $A+F$ data sit[147] adeoq; *A* defluit quantum *F* profluit, erit $-A+4F=0$ sive $A=4F$ et $A+F$ (tota nempe linea) $=5F$. Quare *F* erit quinta pars totius lineæ.[148]

---

positione, fluxiones arcuum *G*[*H*], *E*[*H*], [*FH*] subtendentium...' (if the circle *HGD* remains given in magnitude though not in position, the fluxions of the arcs *GH, EH, FH* subtending...).

(142) Understand some equivalent to 'Et præterea fluxiones arcuum [*EH, FH*] subtendentium [ad *D*] angulos quos rectæ [*AD, BD*] ad data puncta [*A, B*] terminatæ faciunt cum recta quavis [*HD*] positione data erunt ut istorum chordæ [*EH, FH*] applicatæ ad has rectas [*AD, BD*]' (...as the former's chords...divided by the latter lines). In the diagram of note (140), correspondingly, $\text{fl}(H\hat{D}A):\text{fl}(H\hat{D}B) = D\alpha/AD:D\beta/BD = HE/AD:HF/BD$.

(143) The manuscript has no such lemma (see note (122) above), but the argument is easily restored in line with the cancelled corollary to Proposition 26. If the limit-increment *Dd* of the arc $\overset{\frown}{CD}$ generates a corresponding increment *Bb* of the base *AB*, then

$$\text{fl}(\overset{\frown}{CD}):\text{fl}(AB) = Dd:Bb = ED/EB:BP/DP$$

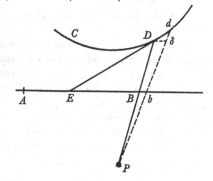

where *Dδ* is the parallel through *D* to *AB*. For this generalization of oblique Cartesian coordinates (where the point *P* is at infinity) compare I: 173: '*Examp* 3ᵈ' and Modes 4–6 of the 1671 tract (III: 138–40).

Proposition 30

If the straight line BD moving in any manner whatever should meet the straight line AB and the curve CD, both given in position, let P be the pole of the line BD and DE a straight line tangent to the curve at D and meeting the line AB in E: the fluxion of the curve CD will then be to the fluxion of the straight line AB as $ED \times DP$ to $EB \times BP$.

This is proved by the Lemma to Proposition 26.[143]

[BOOK 2. PROBLEMS][144]

Problem 1

To cut a given straight line in such a way that the first of the four mean proportionals[145] *shall be as great as possible.*

Let the proportionals be A, B, C, D, E, F, of which A and F make up the straight line and B is a maximum. Now if B is increasing it will come to be greater, if decreasing it was greater: both contrary to hypothesis. It is therefore stable[146] and accordingly, by Proposition 8, the fluxion of A is to the fluxion of F as $-A$ to $4F$. Consequently, since $A+F$ is given[147] and hence A is defluent by as much as F is profluent, there will be $-A+4F = 0$, that is, $A = 4F$ and $A+F$ (the whole line, namely) equals $5F$. Consequently F will be the fifth part of the whole line.[148]

---

(144) This title is inserted in agreement with the breakdown of the contents of the 'Geometria' outlined by Newton in his preface; compare note (18) above.

(145) Understand 'inter duo segmenta' (between the two segments). This is the first occurrence in Newton's papers of this standard Fermatian problem, though he had broached an equivalent case in the first of the additional problems appended to 'Prob 3' of his 1671 tract (see III: 118, note (178)). Ten years after Newton is now probably familiar (compare **2**, Introduction: note (26)) with the recently published collection of Fermat's posthumous 'remains', which for the first time printed the Latin 'escrit' on maxima and minima, originally sent by Fermat to Mersenne in early January 1638 in a letter no longer extant. (See Samuel de Fermat's *Varia Opera Mathematica D. Petri de Fermat Senatoris Tolosani* (Toulouse, 1679): 63–4 [= *Œuvres de Fermat*, **1** (Paris, 1891): 133–6]: 'Methodus Ad disquirendam maximam et minimam.')

(146) That is, when $B = \sqrt[5]{[A^4F]}$ is a maximum, its instantaneous increment—and so its fluxion—is zero. It follows (by Proposition 8) that $4A^3F \times \mathrm{fl}(A) + A^4 \times \mathrm{fl}(F) = 0$.

(147) Newton first wrote 'data quantitas' (a given quantity).

(148) In modern equivalent, if $A = x$ and $F = a-x$, then $B = \sqrt[5]{[x^4(a-x)]}$ is a maximum when $\mathrm{fl}(x) : [\mathrm{fl}(a-x)$ or$] - \mathrm{fl}(x) = -x : 4(a-x)$, so that $x = 4(a-x)$ and $a-x = \frac{1}{5}a$.

Prob 2.

Si recta BC datū angulū BAC subtendit invenire maximam vel minimam distantiā quā punctum aliquod ejus habere potest ab A.

Est igitur AD eo tempore linea non fluens, Adeoꝗ per [Cor 1] Prop [22] est in triangulo BAD fl: arc: ABD . fl AB :: rad. DH & fl arc ABD vel defl arc ADC . defl AC :: rad. DI. Dein per [Cor 1] prop [18] est in triangulo BAC fl AB . [de]fl AC :: cosin C . cosin B. Quare

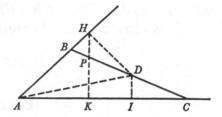

$$\frac{\text{fl arc } ABD \times DH}{\text{rad}} \cdot \frac{\text{defl arc } ADB \times DI}{\text{rad}}$$

$$:: \cosin C . \cosin B$$

[hoc est] $DH . - DI :: \dfrac{IC}{DC} . - \dfrac{BH}{DB}$. Vel $DH \times BH \times \dfrac{DC}{DB} = DI \times IC$. Vel $DH \times BH . DI \times CI :: DC . DB :: \text{tri } DIC . \text{tri } DHB$. Sive H et I æquidistantia [sunt] a recta BC. Demitte ergo $HK \perp AC$ occurrentem BC in P et erit $HP = DI$.[149]

Prob 3.

[*Ad punctum quodvis Ellipsis educere normalem.*]

Si sit $AB \times BD . BC^q :: DD . EE$[150] hoc est $AB . \dfrac{BC \times D}{E} . BD \dot{\div}$ erit per prop 4

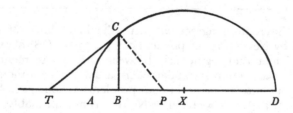

$\dfrac{2BC \times D}{E} . BD - AB :: \text{fl } AB . \text{fl } \dfrac{BC \times D}{E}$. Vel $\dfrac{2BC \times DD}{EE} . BD - AB :: \text{fl } AB . \text{fl } BC ::$ $BT . BC :: BC . BP$. sive $BP = \dfrac{EE}{DD} \times BX$.[151]

---

(149) Since the locus of D is an ellipse (see 1: 32, note (25)), this problem effectively determines the condition $HP = DI$ that AD shall be the instantaneously greatest or least distance from A to the ellipse (D), that is, that D shall be the foot of one of the four normals from A to the ellipse. Compare Problem 3 following.

(150) Newton supposes the ordinate BC to be perpendicular to the main axis ABD and that BP, BT are the corresponding subnormal and subtangent at the general point C.

Problem 2

If the straight line BC subtends the given angle B$\widehat{A}$C, to find the greatest or least distance which some point in it can have from A.

At that instant, therefore, AD is a non-fluent line and hence in the triangle BAD, by Corollary 1 to Proposition 22, fl (arc $A\widehat{B}D$) : fl (AB) = radius : DH and [fl (arc $A\widehat{B}D$) or] $-$ fl (arc $A\widehat{D}C$) : $-$ fl (AC) = radius : DI. Next, by Corollary 1 to Proposition 18, there is in the triangle BAC

$$\text{fl } (AB) : -\text{fl } (AC) = \text{Cos } C : \text{Cos } B.$$

Consequently

$$\text{fl (arc } A\widehat{B}D) \times DH/\text{radius} : -\text{fl (arc } A\widehat{D}B) \times DI/\text{radius} = \text{Cos } C : \text{Cos } B,$$

that is, $DH : -DI = IC/DC : -BH/DB$ or $DH \times BH \times DC/DB = DI \times IC$ and so $DH \times BH : DI \times CI = DC : DB = $ triangle DIC : triangle DHB; in other words H and I are equidistant from the line BC. So let fall HK perpendicular to AC meeting BC in P and then $HP = DI$.[149]

Problem 3

[*To draw the normal at any point on an ellipse*]

If there be $AB \times BD : BC^2 = D^2 : E^2$,[150] that is, if AB, $BC \times [D/E]$ and BD be in continued proportion, then by Proposition 4

$$2BC \times [D/E] : BD - AB = \text{fl } (AB) : \text{fl } (BC \times [D/E])$$

or $2BC \times D^2/E^2 : BD - AB = \text{fl } (AB) : \text{fl } (BC) = BT : BC = BC : BP$, that is,

$$BP = [E^2/D^2] \times BX.\text{[151]}$$

---

(151) A neat fluxional derivation of Apollonius, *Conics*, v, 10. Since D^2/E^2 is the ratio of the ellipse's *latus rectum*, say r, to its transverse axis AD, on taking $AD = q$, $BP = v$, $AB = x$ and $BC = y$ Newton's result is equivalent to Descartes' derivation of the subnormal of the ellipse $y^2 = rx - (r/q)x^2$ as $v = \frac{1}{2}r - (r/q)x$; see his *Geometrie* [= *Discours de la Methode* (Leyden, 1637): 297–413]: 347; and compare I: 237: '*Example*, [1st]'.

Prob 4.

[*Ad punctum quodvis Ellipsis educere tangentem.*]

Si Ellipsis describatur per filum *ACB* per focos[(152)] ductum, tunc *BC* tantum defluit quantum *AC* fluit. quare per Prop [29] cape *CE=CB* et erectis ⊥[lis] *EF BG* erit [*CF*] tangens. Hoc est ang *BCE* bisecandus est.[(153)]

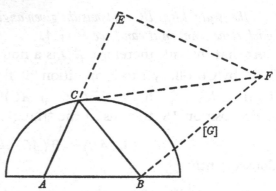

Prob 5.

Idem fit in Ellipsibus 2[di] generis. [Vide] Geom. Chart. lib. 2.[(154)] Et inde refractio eorum facile demonstratur.

Prob 6.

[Item] In Parabola Cartesiana 2[di] generis[(155)] & infinitis similibus.

---

(152) Namely, *A* and *B*.

(153) See I: 380, note (18) and 387, note (21); III: 138, note (229).

(154) See III: 138, note (228).

(155) The Cartesian 'trident', as Newton was later to name it (see II: 70, note (92)). Descartes in his *Geometrie* (note (151) above): 343 (compare II: 263, note (125)) constructed this cubic parabola as the locus of the intersections *C* (and *F*) of an Apollonian parabola *CDF*, moving vertically along its main axis *DH*, with a rotating line *AE* joining the fixed pole *A* to a point *E* in that axis fixed in the translating plane of the parabola (so that *DE* is a constant length). As we shall see in the seventh volume, Newton amplified his intended construction of the trident's tangent more than a decade afterwards in his unpublished 'Geometriæ Liber Primus' (ULC. Add. 4004: 129ʳ–150ʳ, especially 149ʳ–150ʳ). Most simply, in the equivalent terms of limit-increments, if the parabola passes instantaneously into the position *cdf* and *Cγ* is drawn parallel to *ED*, meeting *cdf* in γ and *Ace* in γ′, then is *Cγ = Ee* and so, on drawing the tangents *CL*, *CM* at *C* to the two parabolas, at once *LE:LM = Cγ′:(Cγ* or) *Ee = AC:AE*. Similarly, where *FN*, *FO* are the two corresponding tangents at *F*, *OE:ON = AF:AE*. Unknown to Newton already in the late 1630's Roberval had given an equivalent construction by 'compounding motions' in the concluding example of his 'Observations sur la Composition des Mouvemens, et sur le Moyen de trouver les Touchantes des Lignes Courbes', first published in 1699 in his *Divers Ouvrages* [= *Memoires de l'Académie Royale des Sciences Depuis 1666 jusqu'à 1699* (Paris, [₁1699 →]₂1730): 1–478]: 1–89, especially 86–9: 'Treiziéme exemple, de la Parabole de M. des Cartes'. Taking *AH = b*, *DE = c* and the parameter of the conic parabola to be *d*, Descartes himself had rapidly derived (*Geometrie*: 343–4) the analytical defining equation of the trident to be $dxy = (b-y)(y^2-cd)$ where *HG = x*, *GC = y* and then computed the length of the subnormal *HP* to be (see *Geometrie*: 348–9)

$$v = y - (b-y)(y^2-cd)(-2y^3+by^2+bcd)/d^2y^3.$$

Problem 4

[To draw the tangent at any point on an ellipse]

If the ellipse be described by means of the thread *ACB* drawn through the foci,[152] then *BC* is defluent exactly as much as *AC* is fluent. In consequence, by Proposition 29 take *CE* = *CB* and, after the perpendiculars *EF*, *BG* have been erected, *CF* will be tangent. That is, the angle *BCE* is to be bisected.[153]

Problem 5

The same holds in ovals of second order. See Book 2 of Descartes' *Geometry*.[154] And from this their refractive property is easily demonstrated.

Problem 6

Likewise in the Cartesian parabola of second order[155] and an infinity of similar curves.

This agrees with Newton's construction, in which $LE:EM$ (or $c+y^2/d$) $= (b-y):y$ and so $LG(= (v-y)y/x) = LE-(y^2/d-c)$.

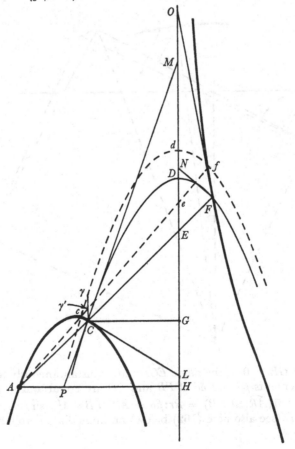

Prob 7.

[*Ad punctum quodvis Conchoidis vel Cissoidis educere normalem.*]

In Conchoide [puncto E descripta] fl AR . fl PB vel PD :: AP . AB :: DC . DB.[156]

In [Cissoide][157] puncto E descripta est fl AR . fl AB :: AP . PB :: PF . AP.[158] et fl AB . fl BD vel BG :: AP . DC. et fl BD . fl BC :: BP . AB. Ergo fl AR . fl BC :: AP^q . $AB \times DC$ :: PF . DC. et

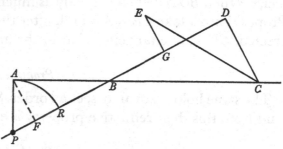

fl AR . fl AC :: PF . $AP + DC$.[159]

Habita autem fluxione ipsius AC et ang APD vel ACE[160] habetur motus puncti [E].[161]

---

(156) In the accompanying figure the angle $E\widehat{G}D$ (here drawn right) is rigid, while its arm DG is constrained always to pass through the fixed pole P. In the present instance the conchoid (E) is generated by keeping BD of fixed length: the 'prima conchois Veterum' (upper shell of the simple conchoid; see I: 502, note (1) and III: 124, note (195)) is the particular case of this

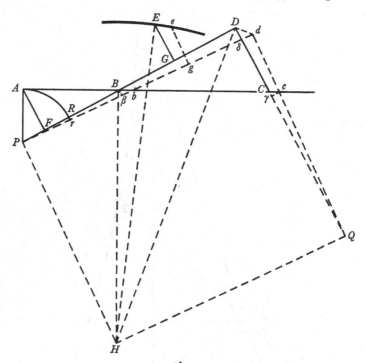

locus in which $DG = GE = 0$. If the angle $E\widehat{G}D$ moves instantaneously to the position $e\widehat{g}d$, it is clear that the increments βb and δd of PB and PD are equal and so (as Newton asserts) fl(PB) = fl(PD); but fl$(\widehat{AR})$: fl(PB) = Rr : βb = AP^2 : $PB \times AB$, since $Rr/B\beta = AP/PB$ and $B\beta/\beta b = AP/AB$. It is (see also note (162) below) an immediate corollary of Proposition 28

Problem 7

[*To erect the normal at any point on a conchoid or cissoid*]

In the conchoid described by the point E

$$\text{fl} (AR) : (\text{fl} (PB) \text{ or}) \text{fl} (PD) = AP : AB = DC : DB. \text{[156]}$$

In the cissoid[157] described by point E

$$\text{fl} (AR) : \text{fl} (AB) = AP : PB = PF : AP, \text{[158]}$$

$\text{fl} (AB) : (\text{fl} (BD) \text{ or}) \text{fl} (BG) = AP : DC$ and $\text{fl} (BD) : \text{fl} (BC) = BP : AB$. Therefore

$$\text{fl} (AR) : \text{fl} (BC) = AP^2 : AB \times DC = PF : DC$$

and $\text{fl} (AR) : \text{fl} (AC) = PF : AP + DC.$[159] When, however, the fluxion of AC is had together with the angle $\widehat{APD}$ or $\widehat{ACE}$,[160] there is obtained the motion of the point E.[161]

---

that, if Q is the limit-meet of the perpendiculars DC, dc at D and d to PD, Pd and the rectangle $PDQH$ is completed, the point H (that is, the meet of the perpendiculars at B and P to AB and PB) is the instantaneous centre of rotation—and so centre of curvature—of the simple conchoid (D), while EH is the normal to the general conchoidal locus (E).

(157) By a slip of his pen Newton again wrote 'In Conch.' (In the conchoid). In this case the whole 'Engine' $CDGE$ (as Newton would have called it in May 1665; compare 1: 264, note (52)) is rigid, supported by its fixed arm DC on the base ABC, while its axis DG is again constrained to pass through the fixed pole P. As Newton proved in his contemporary Lucasian lectures on algebra (ULC. Dd. 9.68: Problem 20: 81–2 [= *Arithmetica Universalis*; *sive De Compositione et Resolutione Arithmetica Liber* (Cambridge, ₁1707): 142–4], claimed in the margin of this deposited copy to have been given out in 'Lect. 1' of his 'Octob. 1677' series), the Dioclean 'cissois Veterum' results when E is taken to be the mid-point of $DC = AP$. As before (see note (156)) H is the centre of curvature of the locus (D) and EH the normal to the general cissoid (E).

(158) This is mistaken. In equivalent limit-increment terms,

$$\text{fl} (\widehat{AR}) : \text{fl} (AB) = Rr : Bb = AP^2 : AB^2 = PF : PB \quad \text{since} \quad Rr/B\beta = B\beta/Bb = AP : AB.$$

The error is carried through the remainder of the paragraph.

(159) It is correct that $\text{fl} (BD) = \text{fl} (BG)$ (since GD is fixed in length), also that

$$\text{fl} (AB) : \text{fl} (BD) = AP : DC$$

(since the ratio $AB : BD = AP : DC$ is constant) and that $\text{fl} (BD) : \text{fl} (BC) = BP : AB$ (since $BC^2 - BD^2 = DC^2$, constant, while $BC : BD = BP : AB$), but these final proportions incorporate the erroneous result $\text{fl} (AR) : \text{fl} (AB) = AP : PB$ (see note (158)). Read

'$\text{fl} AR . \text{fl} BC :: AP \times PF . AB \times DC$'

accurately, with corresponding correction to the ratio

$$\text{fl} (\widehat{AR}) : \text{fl} (AC) [= \text{fl} (AB) + \text{fl} (BC)].$$

(160) $\widehat{APD} - \widehat{ACE}$, that is.

(161) In effect, this reduces the problem to the particular instance of the generalized coordinate system of Proposition 27 in which the 'determination of motion' at E is constructed from the fluxions of AC and $\widehat{ACE}$ (since the second coordinate CE is non-fluent).

Cæterum uno opere ducuntur tangentes ad infinitas curvas si modo polus rectæ *DC* inveniatur, deinde et communis polus regulæ *PDC*.[162]

Prob 8.

Si a punctis quotvis datis rectæ ad idem punctum mobile datam habentes summam [*ducantur*][163]

Prob 9.

Si rectæ quotvis in datis angulis ad totidem rectas positione datas ducantur[164] Ut in problemate veterum.[165]

---

(162) Namely, by Proposition 28. In terms of the diagram of note (156) above, Q (the limit-meet of *DC* and *dc*) is instantaneously the 'polus rectæ *DC*' and P (through which its arm *DG* is always constrained to pass) the 'communis polus regulæ *PDC*', so that H (the meet of the perpendiculars at P and Q to PD and QD) is the centre of curvature of the locus (D) and hence (compare note (132)) EH is normal to the locus (E). In the case of the simple conchoid of Nicomedes, the locus (D) such that BD is constant (and so the increments βb and δd are equal), the curvature centre H is readily determined by

$$PH:PB = \beta b:\beta B = AB:AP \quad \text{or} \quad PH = AB \times PB/AP;$$

while in the general cissoidal case (where *DC* is fixed in length), since $AC^2 - PD^2 = CD^2 - AP^2$, constant, and so $\delta d:Cc = AC:PD$, there results

$$PH:(PH-DC) = \delta d:\gamma c = AC \times PB:PD \times AB,$$

whence $PH = AB \times AC \times DC/PF \times BD$.

(163) This locus reappears in Chapter 3 of the 'Matheseos Universalis Specimina' below (see 2, §1: note (94)). When n points are given it will in general be of degree $2n-2$ but, given

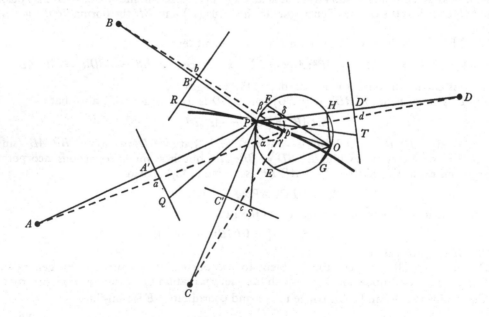

But tangents are drawn at one go to an infinity of curves if only the pole of the straight line *DC* may be found, along with the common pole of the rule *PDC*.[162]

Problem 8

If from any number of given points straight lines having a given sum [be drawn] to the same mobile point...[163]

Problem 9

If any number of straight lines be drawn to an equal number of straight lines given in position...[164] *As in the ancients' problem...*[165]

---

the curve, its tangent is readily determined by the cancelled Proposition 30 above (see note (140)). In equivalent limit-increment terms, if *Pp* is an infinitesimal arc of the curve (*P*) defined, say, by $PA+PB = PC+PD+k$, where *A, B, C, D* are given fixed points, then the circle on diameter *Pp* cuts off from *pA, pB, pC, PD* corresponding increments *pα, pβ, pγ, Pδ* of *PA, PB, PC* and *PD*: namely, such that $p\alpha+p\beta = p\gamma+P\delta$. The 'direction of motion' *PO* at *P* is then constructed as the diameter of a circle through *P* such that, if *OE, OF, OG, OH* are chords parallel to *PA, PB, PC, PD*, then

$$OE:OF:OG:OH = \mathrm{fl}(PA):\mathrm{fl}(PB):\mathrm{fl}(PC):\mathrm{fl}(PD).$$

(164) This is essentially the same type of problem as that preceding except that here, in effect, the given points lie at infinity. If, when given the lines *A'Q, B'R, C'S* and *D'T* and the angles $P\hat{Q}A', P\hat{R}B', P\hat{S}C', P\hat{T}D'$, the locus (*P*) is defined by some relationship

$$\phi(PQ, PR, PS, PT) \equiv \phi'(PA', PB', PC', PD') = 0$$

on introducing the perpendicular distances $PA' = PQ.\sin\hat{Q}, PB' = PR.\sin\hat{R}, PC' = PS.\sin\hat{S}$ and $PD' = PT.\sin\hat{T}$, then much as before the 'direction of motion' *PO* at *P* is constructible as the diameter of the circle *PEGOHF* in which

$$OE:OF:OG:OH = p\alpha:p\beta:p\gamma:p\delta = \mathrm{fl}(PA'):\mathrm{fl}(PB'):\mathrm{fl}(PC'):\mathrm{fl}(PD').$$

(165) The Greek 'locus ad tres et quattuor lineas'; see **2**, Introduction: passim. Here the relationship between *PQ, PR, PS* and *PT* is defined by an equation of form

$$PQ\times PR = k.PS\times PT$$

and the locus of *P* is a general conic through four of the six intersections of *A'Q, B'R, C'S* and *D'T*: whence, on putting $k' = k.\sin\hat{Q}\times\sin\hat{R}/\sin\hat{S}\times\sin\hat{T}$, there results

$$PA'\times PB' = k'.PC'\times PD'$$

and consequently $PA'\times\mathrm{fl}(PB')+PB'\times\mathrm{fl}(PA') = k'.(PC'\times\mathrm{fl}(PD')+PD'\times\mathrm{fl}(PC'))$. Where *P* is one of the four intersections lying on the conic, the tangent *PO* is very simply constructible; see Proposition 5, Corollary 1 of Newton's 'Solutio Problematis Veterum de Loco solido' (**2**, 2, §2 above).

Prob 10.

In Spiralibus[166]

Prob 11.

In Quadratricibus[167]

§2. TWO SCHEMES FOR REVISING BOOK 1 OF THE 'GEOMETRIA'.[1]

[1][2]

Propositiones decem priores non mutentur.

Prop 11. De collectione fluxionum ex æquationibus.[3]

Prop 12 13 14. Sunto tres priores de Triangulis rectangulis. Sed ad Prop 13 adjice hoc *Corollariū.* Si rectæ plures a datis punctis conveniant ad aliam rectam positione datam erunt fluxiones ut...[4]

Lemma

Chordæ et arcus evanescentium ultima ratio est ratio æqualitatis.[5]

Prop 15. In triangulo quovis rectangulo cujus Hypotenusa datur fluxio lateris alterutrius est ad fl: arcûs subtendentis oppositum angulum ut[6]

Prop 16. Dato autem latere fluxio alterius lateris est ad fluxionem arcus subtendentis oppositum angulum ut[7]

---

(166) Compare Mode 7 of Problem 4 of the 1671 tract (III: 140–2).

(167) Likewise, compare the following Mode 8 (III: 144–6).

(1) The handwriting of these two autograph manuscripts, indistinguishable from that used to pen the preceding 'Geometria' whose text they revise and augment, suggests very strongly that they were composed at the same time as Newton abandoned his first scheme. Nothing is known of the reasons which impelled him to modify the structure of the first version of this treatise on geometrical fluxions, but the revise sketched in [2] below is a logically clearer and stylistically more finished text. Whether its elaborations of minor points of detail represents a significant improvement on the previous text (§1) is doubtful and partially, no doubt, a matter of individual mathematical taste.

(2) Add. 3963.7: 61ᵛ, a hurriedly written memorandum indicating preliminary alterations to be made to the preceding text of the 'Geometria' (§1 above).

(3) Newton evidently intends to introduce at this point a summary of Problem 1 of his 1671 tract (III: 74–82), now couched in the language of geometrical fluxions. As a result the numbers of the three propositions following (Propositions 11, 12, 13) have to be increased by unity in the revised scheme.

Problem 10

In spirals...[166]

Problem 11

In quadratrixes...[167]

Translation

[1][2]

The first ten propositions are not to be altered.
Proposition 11. On the gathering of fluxions from equations.[3]
Propositions 12, 13, 14. Let these be the first three on right-angled triangles. But to Proposition 13 add this *Corollary*: If several straight lines from given points meet on a further straight line given in position, their fluxions will be as...[4]

Lemma

The final ratio of a vanishing chord and its arc is one of equality.[5]

Proposition 15. In any right-angled triangle whose hypotenuse is given, the fluxion of either side is to the fluxion of the arc subtending the opposite angle as...[6]

Proposition 16. Given a side, however, the fluxion of the other side is to the fluxion of the arc subtending the opposite angle as...[7]

---

(4) This is, of course, a particular case of the cancelled Proposition 30 in the first version of the 'Geometria'; see §1: note (138) above.

(5) This lemma, later to be subsumed into the *Principia* as Lemma VII of its first book, is already implicitly invoked in Proposition 26 of the first version; see §1: note (120) above.

(6) Understand '...ut latus alterum ad unitatem' (as the second side to unity), an all but trivial new proposition. In the case of the right triangle ABC whose hypotenuse AC is given, if it instantaneously passes into the position AbC, then $\mathrm{fl}(BC):\mathrm{fl}(\hat{A}) = B\beta:B\beta/AB = AB:1$.

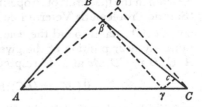

(7) Here understand '...ut quadratum hypotenusæ ad datum latus' (...as the square of the hypotenuse to the given side). For if the side AB of the right triangle ABC is given and the triangle instantaneously passes to the position $A\beta\gamma$ when the vertex A and direction of the hypotenuse AC is fixed, at once, since in the limit $Cc:\gamma c = \gamma c:B\beta = BC:AB$, there follows $\mathrm{fl}(BC):\mathrm{fl}(\hat{A}) = (B\beta+Cc):B\beta/AB = AC^2:AB$.

Cor 1. Si recta circa datum punctum convoluta secet alias quotcunᗡ positione datas rectas erunt fluxiones[8]

Cor 2. Si rectæ duæ datum angulum comprehendentes circa punctum angulare convolutæ secent alias duas positione datas rectas erunt fluxiones[9]

Prop 17. In omni Triangulo rectangulo[10]

Prop 18. Si Trianguli cujusvis basis et angulus ad verticem dentur

Prop 19. Si Trianguli cujusvis angulus ad verticem et latus alterutrum angulo isti conterminū dentur

Schol. Possem omnes casus triang &c.

Prop 20, 21, 22, 23, 24, 25, 26 non mutentur.[11] Sed ad Prop 22 vel 23 corollaria quædam adde de inventione ⊥[11] et segmenti basis ubi ang[ulus] ad verticem datur.

Prop 27[12]

---

[2][13]   DE PROPORTIONALIBUS[14]

Prop. 1. Si trium semper continuè proportionalium quantitatum media detur et extremæ fluant, erit profluxio unius extremæ ad defluxionem alterius ut prior extrema ad illam alteram. Sint $A. B. C \div\!\!\cdot$ et dato B erit $A . C :: a. - c$.

Prop 2. Si trium semper[15] proportionalium una extrema datur, erit fluxio mediæ ad fluxionem alterius extremæ ut extrema data ad duplum mediæ. Sint $A. B. C \div\!\!\cdot$, et dato A, erit $A . 2B :: b . c$.

---

(8) The cancelled corollary to Proposition 26 of the first version; see §1: note (122) above. If the vector *AFC* rotating round the given pole *A* meets the lines *BC*, *EF* in *C* and *F*, then on taking *Afc* to be an infinitely near vector and drawing *Fφ* parallel to *BC* there results

$$\mathrm{fl}(BC):\mathrm{fl}(EF) = Cc:Ff$$

$$= (Cc/F\phi \text{ or}) \ AC/AF : (F\phi/Ff \text{ or}) \ EC/EF.$$

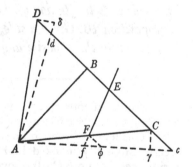

(9) This generalization of the preceding corollary was perhaps suggested by the moving angle mechanism invoked by Newton in the first draft of Proposition 8 of his contemporary 'Solutio Problematis Veterum de Loco solido' (see **2**, 2, §2, Appendix 6 above). If the fixed angle $C\hat{A}D$ rotates round the pole *A* to intersect the given line *CD* in *C* and *D* and if $c\hat{A}d$ is an infinitely near position of the given angle, then on letting fall the normals *AB*, *Cγ*, *Dδ* from *A*, *C*, *D* to *CD*, *Ac* and *Ad* respectively, it is readily shown that

$$\mathrm{fl}(AC):\mathrm{fl}(BC):\mathrm{fl}(AC \times B\hat{A}C) = \gamma c:Cc:C\gamma = BC:AC:AB$$

and similarly that $\mathrm{fl}(AD):\mathrm{fl}(BD):\mathrm{fl}(AD \times B\hat{A}D) = \delta d:Dd:D\delta = BD:AD:AB$.

(10) Another new proposition, evidently a variant on Proposition 13 (renumbered here as '14') of the preceding 'Geometria' (see §1: note (53) above) which introduces fluxional equivalents of the relations $AB = AC.\cos\hat{A} = AC.\sin\hat{C} = BC.\tan\hat{C} = BC.\cot\hat{A}$ between the

Corollary 1. If a straight line revolving round a given point should intersect any number of other straight lines given in position, then the fluxions...[8]

Corollary 2. If two straight lines embracing a given angle should, as they revolve round their corner point, intersect two other straight lines given in position, then the fluxions...[9]

Proposition 17. In every right-angled triangle...[10]

Proposition 18. If in any triangle the base and the angle at the vertex be given,...

Proposition 19. If in any triangle the angle at the vertex and either side bordering that angle be given,...

Scholium. I could [run through] all the cases of triangles...

Propositions 20, 21, 22, 23, 24, 25 and *26* are not to be altered.[11] But to Proposition 22 or 23 add a corollary or two on finding the perpendicular and base segment when the angle at the vertex is given.

Proposition 27....[12]

---

[2][13]   ON PROPORTIONALS[14]

Proposition 1. If of three quantities ever in continued proportion the middle one be given while the extremes are fluent, the profluxion of one extreme will be to the defluxion of the other as the former extreme to the other. Let A, B, C be in continued proportion and, when B is given, there will be $A:C = a:-c$.

Proposition 2. If of three ever [continued] proportionals one of the extremes is given, then the fluxion of the middle one to the fluxion of the other extreme will be as the given extreme to twice the middle. Let A, B, C be in continued proportion and, when A is given, there will be $A:2B = b:c$.

---

sides, hypotenuse and angles of a general right triangle ABC (right-angled at $\hat{B}$). Compare Propositions 2–10 in the section 'De Triangulis rectangulis' in [2] following.

(11) From the partial enunciations given by Newton it is evident that Propositions 18 and 19 are likewise to continue without essential change in this first revision.

(12) Presumably some change is to be made in Proposition 27 (and Propositions 28–30?) of the 'Geometria' or it is to be repositioned. At this point, however, the manuscript breaks off.

(13) Add. 3960.5: [in sequence] 53–4/66/57/60/59/64, taking the jumbled sheets of the manuscript in their most logical order. Abandoning the relatively mild revision of the 'Geometria' sketched in [1] preceding, Newton now launches himself into a more drastic programme of reconstruction.

(14) A considerably augmented revise of Propositions 1–10 of the first state of the 'Geometria'. Variant preliminary drafts of the new proofs now adduced are printed in Appendix 2 below. Again without warning Newton introduces 'lower-case' fluxions; compare § 1: note (31).

(15) Understand 'continuè'.

Prop 3. Positis tribus semper continue proportionalibus, si summa extremarum datur, fluxio minoris extremæ erit ad fluxionem mediæ ut duplum mediæ ad differentiam extremarum. Sint $A \,.\, B \,.\, C \dotplus$, et dato $A + C$, erit $2B \,.\, C - A :: a \,.\, b$.

Prop 4. Sin differentia extremarum datur, fluxio alterutrius extremæ erit ad fluxionem mediæ ut duplum mediæ ad summam extremarum. Detur $A - C$ et erit $2B \,.\, A + C :: a \,.\, b$.

Prop 5. Quod si summa primæ et secundæ detur, erit fluxio secundæ ad fluxionem tertiæ ut prima ad duplum secundæ auctum terti[â]. Detur $A + B$ et erit $A \,.\, 2B + C :: b \,.\, c$.

Prop 6. Si deniæ differentia primæ et secundæ detur, erit fluxio alterutrius ad fluxionem tertiæ ut prima ad duplum secundæ diminutum tertia. Detur $A - B$ et erit

$$A \,.\, 2B - C :: b \,.\, c.$$

Prop 7. Positis tribus quibuscunæ semper continue proportionalibus, fluxio mediæ componitur ex duabus fluxionibus quarum una est ad fluxionem primæ ut tertia ad duplum secundæ, et altera ad fluxionem tertiæ ut prima ad duplum secundæ. Sint $A \,.\, B \,.\, C \dotplus$. Fac $2B \,.\, C :: a \,.\, m$. et $2B \,.\, A :: c \,.\, n$ et erit $m + n = b$.

Vel sic brevius: *Fluxio mediæ ducta in duplum mediæ æquatur fluxionibus extremarum reciprocè ductis in extremas.* Sint $A \,.\, B \,.\, C \dotplus$ et erit $Ca + Ac = 2Bb$.

Quantitas autem hic duci dicitur in aliam quantitatem quando augetur vel diminuitur in ratione quam ista alia habet ad unitatem vel ad quamvis quantitatem sui generis[16] quæ ut unitas spectatur. Sic Ca denotat fluxionem quæ sit ad a ut C ad unitatem linearem. In ratiocinijs mathematicis extra Arithmeticam speciosam[17] et genesin Superficierum a lineis minùs usitata est hæc locutio: sed quia modus brevior est et commodior exprimendi quartam proportionalem ab unitate quam per nuncupationem proportionis,[18] non verebimur in sequentibus ubi opus est, adhibere.

Prop 8. Si quatuor semper proportionalium quantitatum prima una cum summa mediarum datur, erit fluxio secundæ ad fluxionem quartæ ut prima ad excessum tertiæ super secundam. Sint $A \,.\, B :: C \,.\, D$, et datis A et $B + C$, erit $A \,.\, C - B :: b \,.\, d$.

Prop 9. Ijsdem positis si prima una cum differentia mediarum datur, erit fluxio secundæ ad fluxionem quartæ ut prima ad summam secundæ ac tertiæ. Datis A, et $B - C$, erit $A \,.\, B + C :: b \,.\, d$.

---

(16) Dimensionally, that is.

(17) 'Universal' arithmetic, in which numerical quantities are replaced by undetermined algebraic constants and variables (compare §1: note (3) above). The following 'genesis of surfaces [that is, areas] from lines' refers to corresponding contemporary approaches to integration by conceiving the elements of a continuum to be indivisible *quanta*. In the third 'Exercise' of his *Exercitationes Geometricæ Sex* (Bologna, 1647): 3ᵃ *Exercitatio*, Caput xv. 'In quo soluitur quædam difficultas, quæ contra Indivisibilia fieri poterat': 238–41 Cavalieri had, in the course of his reply to Paul Guldin's *Centrobaryca, seu De Centro Gravitatis Trium Specierum Quantitatis Continuæ* (Vienna, 1635/1641), introduced (pp. 239–40) the metaphorical comparison of indivisible lines in a plane surface to threads in woven cloth (*tela filis contexta*).

Proposition 3. Supposing three ever continued proportionals, if the sum of the extremes is given, the fluxion of the lesser extreme will be to the fluxion of the middle one as twice the middle to the difference of the extremes. Let A, B, C be in continued proportion and, when $A+C$ is given, there will be $2B:C-A = a:b$.

Proposition 4. But if the difference of the extremes is given, the fluxion of either extreme will be to the fluxion of the middle as twice the middle to the sum of the extremes. Let $A-C$ be given and there will be $2B:A+C = a:b$.

Proposition 5. If, however, the sum of the first and second be given, the fluxion of the second will be to the fluxion of the third as the first to twice the second increased by the third. Let $A+B$ be given and there will be $A:2B+C = b:c$.

Proposition 6. If, finally, the difference of the first and second be given, the fluxion of either one will be to the fluxion of the third as the first to twice the second diminished by the third. Let $A-B$ be given and there will be $A:2B-C = b:c$.

Proposition 7. Supposing any three quantities ever in continued proportion, the fluxion of the middle one is compounded of two fluxions, one of which is to the fluxion of the first as the third to twice the second, while the other is to the fluxion of the third as the first to twice the second. Let A, B, C be in continued proportion. Make $2B:C = a:m$ and $2B:A = c:n$ and there will be $b = m+n$.

Or more shortly thus: *the fluxion of the middle multiplied into twice the middle equals the fluxions of the extremes reciprocally multiplied into the extremes.* Let A, B, C be in continued proportion and there will be $2Bb = Ca+Ac$.

Here, however, a quantity is said to be multiplied into another quantity when it is increased or diminished in the ratio which that second one has to unity, or to any arbitrary quantity of its kind[16] which is regarded as unity. Thus Ca denotes the fluxion which shall be to a as C to the linear unity. In mathematical reasoning, outside of the arithmetic of free variables[17] and the genesis of surfaces from lines, this phrasing is comparatively little used: yet because this is a shorter and also more serviceable way of expressing their fourth proportional to unity than by formal statement of the proportion,[18] we shall not shrink from employing it, when need be, in the following.

Proposition 8. If of four quantities ever in proportion the first together with the sum of the middle ones is given, the fluxion of the second will be to the fluxion of the fourth as the first to the excess of the third over the second. Let $A:B = C:D$ and, when A and $B+C$ are given, there will be $A:C-B = b:d$.

Proposition 9. With the same supposition, if the first together with the difference of the middle ones is given, the fluxion of the second will be to the fluxion of the fourth as the first to the sum of the second and third. Given A and $B-C$, there will be $A:B+C = b:d$.

---

(18) This replaces the more pithy phrase 'quia nihil aliud designat quam quartam proportionalem ab unitate' (because it denotes nothing else than their fourth proportional to unity). In place of 'nuncupationem' Newton first started to write 'deno[minationem]'.

Prop. 10. Quod si prima una cum summa tertiæ et quartæ detur erit summa primæ et secundæ ad tertiam ut fluxio secundæ ad fluxionem quartæ. Dentur A et $C+D$ et erit $A+B.C::b.d$.

11. Sin prima cum differentia tertiæ et quartæ datur, fluxio secundæ erit ad fluxionem quartæ ut excessus primæ super secundam ad tertiam.[19]

12. Si quarumvis quatuor proportionalium prima datur, fluxio quartæ ducta in hanc primam æquabitur fluxionibus mediarum mutuò ductis in medias. Sint $A.B::C.D$ et dato A erit $Ad=Bc+Cb$.

13. Positis quatuor fluentibus proportionalibus, fluxiones extremarum mutuò ductæ in extremas æquantur fluxionibus mediarum mutuò ductis in medias. Sint $A.B::C.D$ et erit $Ad+Da=Bc+Cb$.

14. Positis quotcunɔ continuè proportionalibus quarum una datur et cæteræ fluant: fluxiones fluentium erunt inter se ut fluentes illæ ductæ in numerum terminorum quibus distant a dato illo termino. Sint $A.\ B.\ C.\ D.\ E.\ F.\ G\dotplus$ et si datur C erunt $-2A.-B.D.2E.3F.4G::a.b.d.e.f.g$.

[20]*Dem[onstratio] pr[op] 1.* [Posito] $A.B::B.C$. [erit] $A\pm e.B::B.C+f$. Ergo ex æqu[o] $A.A\pm e::C+f.C$. et div[isim] $\pm e.A\pm e::f.C$. sive $\pm e.f::A\pm e.C::$ (init[io] fl[uxionis]) $A.C$. Quare per ax 6 [prodit propositio].

[21]*Dem[onstratio prop]* 2 et 8, 9, 10, 11, 12. Si $NA.NB::NC.ND$ fluant [NB, NC, ND] donec evadant Nb, Nc, Nd. et age $Ce\parallel Ab$. Ergo $NA.NC(::AB.CD)::Bb.De$ et $NA.Nb(::NC.Ne)::Cc.ed$. Quare initio fluxionis $NA.NC::\mathrm{fl}\,Bb$ (sive fl$[N]B$)$.\mathrm{fl}\,De$ & $NA.NB::\mathrm{fl}\,Cc$ (sive fl$[N]C$)$.\mathrm{fl}\,ed$. Quare si $NB=NC$ ut in Prop 2 erit

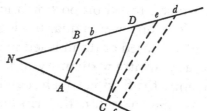

$$NA.2NB::\mathrm{fl}\,[N]B.\mathrm{fl}\,De+ed=\mathrm{fl}\,N[D].\text{[22]}$$

---

(19) In line with the preceding Newton doubtless intended to add 'Dentur A et $C-D$ et erit $B-A.C::b.d$' (Let A and $C-D$ be given and there will be $B-A:C=b:d$).

(20) The following proofs are entered in a rough and exceedingly abrupt manner on an inserted worksheet which Newton clearly intended to revise and polish before exposing it to public view. For the reader's better understanding we have not hesitated to make *verba* interpolations in the manuscript text.

(21) Newton first began this paragraph: 'Dem. 2. Si $NA.NB=NC.ND\dotplus$fluant NB, NC, ND donec evadant Nb, Nc, Nd. et age $Ce\parallel Ab$. Ergo $NA.NC(NB)(::AB.CD)::Bb.$ De et $NA.Nb(::NC.Ne)::Cc$, vel $Bb[.ed]$.'

(22) Out of the blue, without prior warning or explanation, Newton sneaks in the notion of the fluxion of a quantity which is itself vanishingly small and hence identical with its instantaneous increment. Since fl(Bb) is the speed of motion of $B\to b$ *initio fluxionis* as NB passes into Nb it is, of course, legitimate to set fl$(NB)=$ fl$(Nb)=$ fl(Bb) and correspondingly fl$(NC)=$ fl$(Nc)=$ fl(Cc) in the limit as B, b and C, c coincide. It would nevertheless have been more consistent—and certainly less worrying, logically—to have argued straightforwardly that, since $Dd=(NC/NA)\times Bb+(NB/NA)\times Cc$, therefore

$$\mathrm{fl}\,(ND)=(NC/NA)\times\mathrm{fl}\,(NB)+(NB/NA)\times\mathrm{fl}\,(NC),$$

Proposition 10. But if the first together with the sum of the third and fourth be given, the sum of the first and second will be to the third as the fluxion of the second to the fluxion of the fourth. Let A and $C+D$ be given and there will be $A+B:C = b:d$.

[*Proposition*] *11. If, however, the first along with the difference of the third and fourth is given, the fluxion of the second will be to the fluxion of the fourth as the excess of the first over the second to the third.*[19]

[*Proposition*] *12. If of any four proportionals the first is given, the fluxion of the fourth multiplied into this first one will equal the fluxions of the middles mutually multiplied into the middles.* Let $A:B = C:D$ and, when A is given, there will be $Ad = Bc+Cb$.

[*Proposition*] *13. Supposing there to be four fluent proportionals, the fluxions of the extremes mutually multiplied into the extremes equal the fluxions of the middles mutually multiplied into the middles.* Let $A:B = C:D$ and there will be $Ad+Da = Bc+Cb$.

[*Proposition*] *14. Supposing any number of continued proportionals, one of which is given while the rest are fluent, the fluxions of the fluents will be to one another as those fluents multiplied by the number of terms they are distant from the given term.* Let A, B, C, D, E, F, G be in continued proportion and, if C is given, then

$$-2A:-B:D:2E:3F:4G = a:b:d:e:f:g.$$

[20]*Demonstration of Proposition 1.* Supposing $A:B = B:C$, there will be

$$A\pm e:B = B:C+f.$$

Therefore *ex æquo* $A:A\pm e = C+f:C$ and *dividendo* $\pm e:A\pm e = f:C$ or

$$\pm e:f = A\pm e:C,$$

that is, (at the start of fluxion) $A:C$. Consequently by Axiom 6 the proposition results.

[21]*Demonstration of Proposition 2, also of 8, 9, 10, 11 and 12.* If

$$NA:NB = NC:ND,$$

let NB, NC, ND flow till they come to be Nb, Nc, Nd and draw Ce parallel to Ab. Accordingly,

$$NA:NC \text{ (or } AB:CD) = Bb:De \quad \text{and} \quad NA:Nb \text{ (or } NC:Ne) = Cc:ed.$$

Hence at the start of fluxion $NA:NC = (\text{fl}\,(Bb) \text{ or}) \text{ fl}\,(NB):\text{fl}\,(De)$ and

$$NA:NB = (\text{fl}\,(Cc) \text{ or}) \text{ fl}\,(NC):\text{fl}\,(ed).$$

Consequently, if $NB = NC$ (as in Proposition 2), there will be

$$NA:2NB = \text{fl}\,(NB):(\text{fl}\,(De+ed) \text{ or}) \text{ fl}\,(ND).[22]$$

---

so avoiding this tangled concept. Further to confuse, as in his preliminary draft (see Appendix 2: note (11) below) Newton sets the proportion between the fluxions and their respective limit-increments to be unity, whence $\text{fl}\,(Cc) = Cc$ and $\text{fl}\,(Dd) = Dd$.

Sed si NB et NC non sunt $=$[23] erit $NA \times$ fl$De = NC \times$ fl$[N]B$ &

$$NA \times \text{fl}\, ed = NB \times \text{fl}\,[N]C \quad \text{adeoq} \quad NA \times \text{fl}\, De + ed$$

sive $NA \times$ fl $ND = NC \times$ fl $NB + NB \times$ fl NC ut in Prop 12. Et si præterea $NB + NC$ detur adeoq sit fl $NB +$ fl $NC = 0$, de $NC \times$ fl $NB + NB \times$ fl NC aufer

$$NC \times \text{fl}\, NB + \text{fl}\, NC = 0$$

et restabit $NB \times$ fl $NC - NC \times$ fl $NC = NA \times$ fl $ND_{[,]}$ unde

$$NA \,.\, NB - NC :: \text{fl}\, NC \,.\, \text{fl}\, ND$$

ut in Prop 8. Et eodem modo prop 9, 10 & 11 demonstrantur.

Dem. prop. 3, 4, 5, 6, 7, 13. Sint $A \,.\, B :: C \,.\, D$. Cape $N \,.\, A :: D \,.\, P$ et erit etiam $N \,.\, B :: C \,.\, P$. Ergo si N datur erit[24] $Ad + Da = Np = Bc + Cb$. Quod est prop. 13. Unde si $B = C$ ut in Prop 7 erit $Ad + Da = 2Bb$. Et si præterea $A + D$ datur adeoq $a + d$ sit $= 0$, de $Ad + Da[= 2Bb]$ aufer $Da + Dd = 0$ et restabit

$$Ad - Dd = 2Bb_{[,]} \quad \text{unde} \quad A - D \,.\, 2B :: b \,.\, d$$

ut in Prop 3. Eodem fere[25] modo prop 4 5 & 6 demonstrantur.

Dem[onstratio] prop 14. Sint $A \,.\, B \,.\, C \,.\, D \,.\, E \,.\, F \,.\, G \text{⁝⁝}$ et dato C erit per Prop 2[26]

Schol. Potuerunt hæ omnes Propositiones ordine[27] demonstrari et illustrari per s[c]hemata. Sic ad demonstrandam tertiam super diametro AD describe circulum. Ad puncta quævis diametri erige $\perp$[1a] BC, GE. Junge CE et biseca eam in $F_{[,]}$ age FK ad centrū circuli et demitte $\perp$[1a] FL ad[28] AD, et CH ad EG. et erunt $\triangle CEH\ FLK$ similia adeoq $CH(BG) \,.\, EH :: FL \,.\, LK$. Coeant jam BC et EG et ultima ratio BG ad EH erit CB ad BK. proinde per ax 6 fluxiones ipsarum AB et BC sunt in ea ratione. &c.

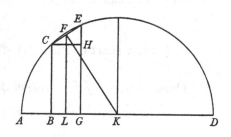

[Aliter in hac figura]

$$WY(RS) \,.\, VY :: PR \,.\, RW :: WR \,.\, RT.\text{[29]}$$

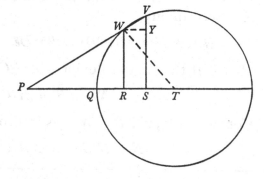

But if NB and NC are not equal, then $NA \times \text{fl}\,(De) = NC \times \text{fl}\,(NB)$ and

$$NA \times \text{fl}\,(ed) = NB \times \text{fl}\,(NC), \quad \text{so that} \quad NA \times \text{fl}\,(De+ed)$$

or $NA \times \text{fl}\,(ND) = NC \times \text{fl}\,(NB) + NB \times \text{fl}\,(NC)$, as in Proposition 12. If, more-over, $NB+NC$ be given (so that $\text{fl}\,(NB) + \text{fl}\,(NC) = 0$) from

$$NC \times \text{fl}\,(NB) + NB \times \text{fl}\,(NC) \quad \text{take away} \quad NC \times (\text{fl}\,(NB) + \text{fl}\,(NC)) = 0$$

and there will remain $NB \times \text{fl}\,(NC) - NC \times \text{fl}\,(NC) = NA \times \text{fl}\,(ND)$, whence $NA:NB-NC = \text{fl}\,(NC):\text{fl}\,(ND)$, as in Proposition 8. And in the same manner Propositions 9, 10 and 11 are demonstrated.

Demonstration of Propositions 3, 4, 5, 6, 7 and 13. Let $A:B = C:D$. Take $N:A = D:P$ and there will also be $N:B = C:P$. If therefore N is given, then[24] $Ad+Da = Np = Bc+Cb$. This is Proposition 13. Hence if (as in Proposition 7) $B = C$, there will be $Ad+Da = 2Bb$. And if in addition $A+D$ is given (so that $a+d = 0$), from $Ad+Da = 2Bb$ take away $Da+Dd = 0$ and there will remain $Ad-Dd = 2Bb$, whence $A-D:2B = b:d$, as in Proposition 3. In the same manner, almost,[25] Propositions 4, 5 and 6 are proved.

Demonstration of Proposition 14. Let A, B, C, D, E, F, G be in continued proportion and then, given C, by Proposition 2...[26]

Scholium. All these propositions could have been demonstrated in order[27] and illustrated by means of figures. So to demonstrate the third, on the diameter AD describe a circle and at any points on this diameter raise the perpendiculars BC, GE. Join CE, bisecting it in F, draw FK to the circle's centre and let fall the perpendiculars FL to[28] AD and CH to EG. Then will the triangles CEH, FLK be similar, so that CH (or BG): $EH = FL:LK$. Now let BC and EG coalesce and the final ratio of BG to EH will be CB to BK. Accordingly, by Axiom 6 the fluxions of AB and BC are in that ratio. And so on.

---

(25) A wise late insertion!

(26) In line with Proposition 8 of the first version of the 'Geometria', understand '...erit per Prop 2 $D.2E::d.e$. Insuper per Prop. 3 est $Df+Fd = 2Ee$ et inde cum fuerit

$$d.e(::D.2E)::E.2F \quad \text{sive} \quad 2Fd = Ee, \quad \text{erit} \quad Df = 3Fd \quad \text{sive} \quad D.3F::d.f.$$

Præterea cùm sit $C.E.G \div$ per Prop. 2 erit $2E.4G::C.2E::e.g$. Et sic in infinitum. Ex altera autem parte est $D.C::C.B$ adeoꝗ per Prop. 1 $B.D::-b.d$ vel $-B.D::b.d$. Præterea est $C.B::B.A$ et inde per Prop. 2 $-2A.-B::2A.B::b.c$. Et sic in infinitum'.

(27) As in the first version of the 'Geometria' (§1 above).

(28) 'diametrum' (the diameter) is cancelled. The structure of this proof is very close to that of Archimedes, *Sphere and Cylinder*, I, 28 ff.

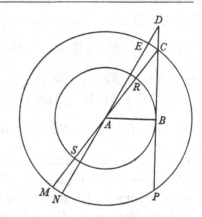

Sic in Prop 4 [vide figuram]

$$DC.DE::DN.DP::DA.DB.$$

hoc est fl: CB . fl CR :: $CR + CS$. [2]CB.[30]

De Triangulis rectangulis.[31]

1. Si Trianguli ad verticem rectanguli basis[32] *datur erunt fluxiones laterum alterne ut latera.* [In triangulo ABC rectangulo ad A] detur BC et erit $ab.ac::AC.AB$.[33]

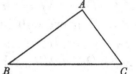

2. Eadem data fluxio lateris alterutrius erit ad fluxionem arcus subtendentis angulum oppositum ut latus illud ad tangentem anguli istius[34] *vel ut latus alterum ad radium.* Detur BC et erit fl:AC . fl:arc:B:: sin C . BC.[35]

3. Eâdem datâ erit fluxio lateris alterutrius ad fluxionem perpendiculi ut latus alterum ad differentiam segmentorū basis.

4. Si trianguli ad verticem rectanguli latus alterutrum recto angulo conterminum detur, erit fluxio alterius lateris ad fluxionem basis, ut basis ad illud latus fluens. In triangulo ABC rectangulo ad A si detur latus AB, erit $ac.bc::BC.AC$.

5. Eodem latere dato erit fluxio alterius lateris ad fluxionem arcus subtendentis angulum oppositum ut basis ad cosinum anguli istius. Detur AB et erit fl AC . fl:arc:B:: BC . sin:C.[36]

6. Eodem dato erit fluxio Basis ad fluxionem arcus subtendentis angulum dato lateri conterminum, ut latus fluens ad cosinum anguli istius. Detur AB et erit

$$\text{fl } BC . \text{fl:arc:}B::AC.\sin C.^{[36]}$$

7. Eodem dato erit fluxio lateris alterius ad fluxionem perpendiculi ut quadratum Basis ad rectangulum sub latere dato et contermino segmento basis.

8. In omni triangulo rectangulo summa laterum ductorum in fluxiones suas æquatur basi ductæ in fluxionem suam.

---

(29) Understand 'initio fluxionis' (at the start of fluxion). On combining the analysis of the two figures, if $\left\{\begin{matrix} BA \\ RQ \end{matrix}\right\} = A$, $\left\{\begin{matrix} BC \\ RW \end{matrix}\right\} = B$ and $\left\{\begin{matrix} BD \\ RX \end{matrix}\right\} = C$ (where in the second figure X denotes the point on the circle diametrically opposite Q), then $A + C = \left\{\begin{matrix} AD \\ QX \end{matrix}\right\}$ is given and $AC = B^2$, while fl(A):fl$(B) = a{:}b = 2B{:}C - A$ in either case.

Alternatively, in this figure WY (or RS): $VY = PR:RW = WR:RT$.[29]

So in Proposition 4 (see the figure). $DC:DE = DN:DP = DA:DB$; in other words, fl (CB):fl $(CR) = CR+CS:2CB$.[30]

ON RIGHT-ANGLED TRIANGLES[31]

1. If in a triangle right-angled at the vertex the base[32] is given, the fluxions of the sides will be inversely as the sides. In the triangle ABC right-angled at A let BC be given and there will be $ab:ac = AC . AB$.[33]

2. Given the same, the fluxion of either side will be to the fluxion of the arc subtending the opposite angle as that side to the tangent of that angle,[34] or as the other side to the radius. Let BC be given and there will be fl (AC):fl $(\text{arc } B) = \text{Sin } C:BC$.[35]

3. Given the same, the fluxion of either side will be to the fluxion of the perpendicular as the other side to the difference of the base segments.

4. If in any triangle right-angled at the vertex either side bordering the right angle be given, the fluxion of the other side, will be to the fluxion of the base as the base to that fluent side. In the triangle ABC right-angled at A, if the side AB be given, then $ac:bc = BC:AC$.

5. The same side given, the fluxion of either side will be to the fluxion of the arc subtending the opposite angle as the base to the cosine of that angle. Let AB be given and there will be fl (AC):fl $(\text{arc } B) = BC:\text{Sin } C$.[36]

6. The same given, the fluxion of the base will be to the fluxion of the arc subtending the angle bordered by the given side as the fluent side to the cosine of that angle. Let AB be given and there will be fl (BC):fl $(\text{arc } B) = AC:\text{Sin } C$.[36]

7. The same given, the fluxion of either side will be to the fluxion of the perpendicular as the square of the base to the rectangle formed by the given side and the adjoining base segment.

8. In every right-angled triangle the sum of the sides multiplied into their fluxions equals the base multiplied into its fluxion.

---

(30) Here again, on setting $CS = A$, $CB = B$ and $CR = C$, there is $AC = B^2$ but now $A-C = RS$, given, while $a:b = 2B:C+A$.

(31) A much augmented revise of Propositions 11, 12 and 13 of the preceding version of the 'Geometria'. The manuscript places the following propositions in the sequence 4, 5, 6, 1, 2, 8, 9, 10, 3, 7 but we have rearranged them to accord with Newton's revised numbering of them.

(32) In a cancelled following sentence Newton originally added: 'Per basem vero in his propositionibus semper intellige latus angulo recto vel cuivis angulo dato oppositum' (By 'base' in these propositions always understand in fact the side opposite the right angle or any given angle).

(33) Read '$-AB$' more precisely.

(34) This replaces '...*ut cosinus anguli istius ad basem*' (*as the cosine of that angle to the base*).

(35) Understand '$::AC.\tan B::AB.\text{rad}$' ($= AC:\text{Tan } B = AB:R$).

(36) That is '$\cos B$' ($\text{Cos } B$).

9. In omni triangulo rectangulo fluxio arcus subtendentis angulum alterutrum ad basem ducta in latus angulo conterminum æquatur excessui quo fluxio lateris angulo isti oppositi ducta in radium superat fluxionem basis ductam in sinum ejusdem anguli.

10. In omni triangulo rectangulo fluxio arcus subtendentis angulum alterutrum ad basem ducta in basem æquatur excessui quo fluxio lateris angulo isti oppositi ducta in radium superat fluxionem alterius lateris ductam in tangentem ejusdem anguli.[37]

DE TRIANGULIS OBLIQUANGULIS DATUM ANGULUM AD VERTICEM HABENTIBUS.[38]

1. Si Basis datur fluxiones laterum erunt ut cosinus oppositorum angulorum ad basem, id est ut rectangula sub lateribus conterminis et oppositis segmentis basis.

2. Eadem data fluxio lateris alterutrius erit ad fluxionem arcus subtendentis angulum oppositum,[39] *ut latus illud ad tangentem anguli istius.*

3. Eâdem datâ erit fluxio lateris alterutrius ad fluxionem ⊥^li ut cosinus anguli oppositi ad sinum differentiæ angulorum.

4. Eâdem datâ erit fluxio lateris alterutrius ad fluxionem contermini segmenti basis ut cosinus anguli oppositi ad cosinum differentiæ angulorum.

5. Dato alterutro latere erit fluxio alterius lateris ad fluxionem basis ut radius ad cosinum anguli contermini.

6. Eodem dato erit fluxio alterius lateris ad fluxionem arcûs subtendentis angulum oppositum ut basis ad sinum alterius anguli.

---

(37) Sic! Newton for some reason has cancelled the correct enunciation '...æquatur excessui quo fluxio lateris angulo isti oppositi ducta in latus conterminum superat fluxionem lateris eidem angulo contermini ductam in sinum anguli istius' (equals the excess of the fluxion of the side opposite that angle multiplied into the adjoining side over the fluxion of the side adjoining the same angle multiplied into the sine of that angle).

This group of ten propositions is most easily justified by letting fall the perpendicular AD from the right angle $\hat{A}$ onto the base (that is, hypotenuse) BC and then, for brevity's sake, employing a modified Leibnizian 'd' to represent instantaneous increments (proportional to the corresponding fluxions). In the case of Propositions 1–3 (which expand Newton's earlier Proposition 11) the base BC is given, whence the equalities $AB^2 + AC^2 = BC^2$, $AB = BC.\sin C$ and $BC \times AD = AB \times AC$ yield in turn the theorems $dAB/dAC = -AC/AB$ [1],

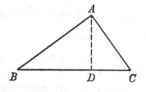

$$dAB/dC = BC.\cos C = AC = AB/\tan C \quad [2]$$

and $BC.dAD = (AB.dAC \text{ or}) - (AB^2/AC).dAB + AC.dAB$, so that

$$dAB/dAD = AC \times BC/(DC^2 - BD^2) = -AC/(BD - DC) \quad [3].$$

In Propositions 4–7 (which augment the earlier Proposition 12), however, the side AB is given and accordingly $dAC/dBC = (BC/AC \text{ or}) \, 1/\cos C$ [4],

$$dAC = (-BC.\sin C.dC + \cos C.dBC \text{ or}) \, BC.\sin C.dB + \cos^2 C.dAC$$

9. *In every right-angled triangle the fluxion of the arc subtending either base angle multiplied into the side bordering the angle equals the excess of the fluxion of the side opposite that angle multiplied into the radius over the fluxion of the base multiplied into the sine of this same angle.*

10. *In every right-angled triangle the fluxion of the arc subtending either base angle multiplied into the base equals the excess of the fluxion of the side opposite that angle multiplied into the radius over the fluxion of the other side multiplied into the tangent of the same angle.*[37]

ON OBLIQUE-ANGLED TRIANGLES HAVING THEIR VERTEX ANGLE GIVEN[38]

1. *If the base is given, the fluxions of the sides will be as the cosines of the opposite base angles; that is, as the products of the adjoining sides and the opposite base segments.*

2. *Given the same, the fluxion of either side will be to the fluxion of the arc subtending the angle opposite*[39] *as that side to the tangent of that angle.*

3. *Given the same, the fluxion of either side will be to the fluxion of the perpendicular as the cosine of the opposite angle to the sine of the difference of the angles.*

4. *Given the same, the fluxion of either side will be to the fluxion of the adjoining base segment as the cosine of the angle opposite to the cosine of the difference of the angles.*

5. *Given either side, the fluxion of the other side will be to the fluxion of the base as the radius to the cosine of the adjoining angle.*

6. *Given the same, the fluxion of the other side will be to the fluxion of the arc subtending the angle opposite as the base to the sine of the other angle.*

---

and so $dAC/dB = BC/\sin C$ [5] or $cdBC/dB = AC/\sin C$ [6], while lastly

$$BC.dAD + AD.dBC \text{ (or } (AD \times AC^2/BC^2).dAC) = AB.dAC,$$

whence $dAC/dAD = BC^2/AB \times BD$ [7]. In general, in Propositions 8–10 (compare Proposition 13 of §1) $AB.dAB + AC.dAC = BC.dBC$ [8],

$$AB.\cos B.dB + (-AC.\cos C.dC \text{ or}) AC.\cos C.dB = \sin C.dAC - \sin B.dAB$$

or $BC.dB = (AB.dAC - AC.dAB)/BC$ [10] and finally

$$AB.dB = (AB^2.dAC - AC \times AB.dAB)/BC^2 = dAC - (AC/BC).dBC \quad [9].$$

(38) A much amplified revise of Propositions 18 and 19 of the first version of the 'Geometria' (§1 preceding).

(39) Newton first concluded: 'ut cosinus istius anguli ad radium, posito radio $\dfrac{AB \times AC}{2BC}$. hoc est ut $\dfrac{\text{segm. con.}}{\text{lat. conterm.}} \times \dfrac{AB \times AC}{2BC}$ ad rad :: $\dfrac{AB \times \text{segm.}}{2BC} . \dfrac{\text{rad} \times \sin A}{\sin C}$' (as the cosine of that angle to the radius on putting the radius $AB \times AC/2BC$; that is, as

$$(\text{adjacent segment/adjacent side}) \times (AB \times AC/2BC)$$

to the radius, so is the segment $\times AB/2BC$ to the radius $\times \mathrm{Sin}\, A/\mathrm{Sin}\, C$).

7. *Eodem dato erit fluxio basis ad fluxionem arcus subtendentis angulum dato lateri conterminum, ut latus fluens productum usꝗ ad perpendiculum ab angulo opposito in illud demissum, ad sinum anguli dato lateri oppositi.*

8. *Eodem dato erit fluxio lateris alterius ad fluxionem perpendiculi ut rectangulum basis et lateris fluentis ad rectangulum perpendiculi et segmenti basis quod dato lateri conterminum est.*

9. *Eodem dato erit flux[io] lateris alterius ad fluxionem segmenti basis quod dato lateri conterminum est, ut rectangulum basis et lateris fluentis ad quadratum perpendiculi.*

10. *In omni triangulo acutangulo datum angulum ad verticem habente fluxio basis ducta in radium æquatur fluxionibus laterum ductis in cosinus conterminorum angulorum ad basem,* vel quod perinde est, dato ∠*A*, est

$$BC \times \mathrm{fl} : BC = BD \times \mathrm{fl} : AB + CE \times \mathrm{fl}\, AE.$$

11. *In omni triangulo acutangulo datum angulum ad verticem habente differentia fluxionum laterum ductarum in sinus conterminorum angulorum ad basem, æquatur fluxioni arcus subtendentis alterutrum angulum ad basem ductæ in basem.*
12.[40]

DE TRIANGULIS OBLIQUANGULIS NULLUM DATUM ANGULUM HABENTIBUS.[41]

1. *Si dentur duo latera erit fluxio basis ad fluxionem arcûs angulum verticalem*[42] *subtendentis ut perpendiculum*[43] *ad radium.*

2. *Si dentur basis et latus alterutrum erit fluxio lateris alterius ad fluxionem arcûs*[44] *subtendentis ut perpendiculum ad cosinum anguli dato lateri oppositi.*

---

(40) The manuscript breaks off at this point (on p. 60) though a considerable gap is left for insertion of this intended Proposition 12. In all the theorems of this section it is supposed that the vertex angle $\hat{A}$ is given, so that (in the Leibnizian terms of note (37)) $dB = -dC$ throughout. In Propositions 1–4 the base BC is also given, whence, since $AB.\sin B = AC.\sin C$ and $AB.\cos B + AC.\cos C = BC$, at once $\cos B.dAB + \cos C.dAC = 0$ or

$$dAB/dAC = -\cos C/\cos B \quad [1];$$

further, because $AB.\sin A = BC.\sin C$, at once $dAB/dC = AB/\tan C$ [2]; and, because $BC \times AF = AB \times AC.\sin A$, also $BC.dAF = \sin A(AB.dAC + AC.dAB)$, so that

$$dAB/dAF = \cos C/(-(AF/AC)\cos B + (AF/AB)\cos C) = \cos C/\sin(B-C) \quad [3];$$

while lastly, since $AB^2 = AF^2 + BF^2$ and so $AB.dAB = AF.dAF + BF.dBF$, there results

$$dAC/dBF = \cos B \cos C/(\cos C - \sin B \sin(B-C)) = \cos C/\cos(B-C) \quad [4].$$

In Propositions 5–9 we may suppose the adjacent side AC to be given, whence first, since $BC^2 = AB^2 + AC^2 - 2AB \times AD$ and so $BC.dBC = (AB-AD).dAB$, there is

$$dAB/dBC = (BC/BD \text{ or}) \ 1/\cos B \quad [5];$$

7. Given the same, the fluxion of the base will be to the fluxion of the arc subtending the angle bordered by the given side as the fluent side, extended as far as the perpendicular let fall from the opposite angle onto it, to the sine of the angle opposite the given side.

8. Given the same, the fluxion of either side will be to the fluxion of the perpendicular as the product of the base and the fluent side to the product of the perpendicular and the base segment adjoining the given side.

9. Given the same, the fluxion of either side will be to the fluxion of the base segment adjoining the given side as the product of the base and the fluent side to the square of the perpendicular.

10. In every acute-angled triangle having its vertex angle given, the fluxion of the base multiplied into the radius equals the fluxions of the sides multiplied into the cosines of the adjoining base angles. Equivalently, given $\hat{A}$, there is

$$BC \times \mathrm{fl}\,(BC) = BD \times \mathrm{fl}\,(AB) + CE \times \mathrm{fl}\,(AE).$$

11. In every acute-angled triangle having its vertex angle given, the difference of the fluxions of the sides multiplied into the sines of the adjoining base angles equals the fluxion of the arc subtending either base angle multiplied into the base.

12. ...[40]

ON OBLIQUE-ANGLED TRIANGLES HAVING NO ANGLE GIVEN[41]

1. If the two sides be given, the fluxion of the base will be to the fluxion of the arc subtending the angle at the vertex[42] *as the perpendicular*[43] *to the radius.*

2. If the base and one or other side be given, the fluxion of the second side will be to the fluxion of the arc subtending[44] *as the perpendicular to the cosine of the angle opposite the given side.*

---

and again, because $BC.\sin B = AC.\sin A$ or $BC.\cos B.dB + \sin B.dBC = 0$, also

$$dAB/dC \text{ (or } -dBC/\cos B.dB) = BC/\sin B \quad [6];$$

further, as an immediate consequence, $dBC/dC = (BC.\cot B \text{ or}) BD/\sin B$ [7]; while, since $BC \times AF = AB \times CD$ (where $CD = AC.\sin A$ is given) and so

$$BC.dAF + (AF.dBC \text{ or}) AF.\cos B.dAB = CD.dAB,$$

there follows $dAB/dAF = (BC/(DC - AF \times BF/AB) \text{ or}) AB \times BC/AF \times FC$ [8]; and finally, because $AF^2 + FC^2 = AC^2$ or $AF.dAF + FC.dFC = 0$, there is $dAB/dFC = -AB \times BC/AF^2$. In general (Propositions 10 and 11) the equality $BC^2 = AB^2 + AC^2 - 2AB \times AC.\cos A$ gives

$$BC.dBC = ((AB - AC.\cos A).dAB + (AC - AB.\cos A).dAC \text{ or}) BD.dAB + CE.dAC \quad [10];$$

while, because $AB.\sin B = AC.\sin C$, there follows

$$\sin B.dAB - \sin C.dAC = (AC.\cos C.dC - AB.\cos B.dB \text{ or}) BC.dC = -BC.dB \quad [11].$$

(41) An amplified revision of Propositions 20–25 in the first version of the 'Geometria' (§1).
(42) Newton first wrote equivalently but less consistently with the preceding 'a datis lateribus comprehensum' (comprehended by the given sides).
(43) 'ad basem demi[ssum]' (let fall to the base) is cancelled.
(44) Understand 'angulum verticalem' (the vertex angle).

3. *Si dentur duo latera, erit fluxio basis ad fluxionem alterutrius segmenti ejus ut basis ad alterum segmentu[m].*

4. *Si dentur basis et latus alterutrum erit fluxio alterius lateris ad fluxionem segmenti contermini, ut basis ad latus fluens.*

5. *Si dentur latera erit fluxio basis ad fluxionem perpendiculi ut* $\begin{cases} \textit{rectangulum sub basi} \\ \textit{basis ad partem per-} \end{cases}$ *et perpendiculo ad rectangulum sub segmentis basis.*
pendiculi quæ inter basem et intersectionem perpendiculorum ad reliqua latera demis-sorum[45] *jacet.*

6. *Si dentur basis et latus alterutrum erit fluxio lateris alterius ad fluxionem perpendiculi* $\begin{cases} \textit{ut rectangulum sub basi et perpendiculo ad rectangulum sub latere fluente \& opposito} \\ \textit{ut basis ad partem alterius perpendiculi ad latus fluens demissi quæ inter angulum} \end{cases}$ *segmento basis.*
oppositum a quo demittitur & perpendiculum ad basem demissum intercipitur.

7. *Si detur basis et perpendiculum erunt fluxiones laterum ut segmenta laterum quæ inter basem et perpendicula ad latera ista jacent.*[46]

8. *Si detur segmentum basis et latus oppositum erit fluxio alterius lateris ad defluxionem basis ut alterum segmentum ad latus fluens.*

[9.] *Si dentur latera duo, fluxiones angulorum ad basem erunt ad defluxionem anguli verticalis ut opposita segmenta basis ad basem.* [id est] fl: B . fl: C . fl: A :: DC . DB . $-BC$.

[10.] *Si detur basis BC, fluxiones angulorum ad Basem erunt inter se ut excessus quo fluxiones laterum oppositorum ductæ in latera superant fluxiones laterum conterminorum ductas in eorum segmenta basi contermina.* [hoc est]

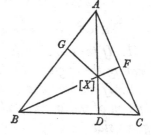

$$\text{fl } B \text{ . fl } C :: AC \times \text{fl } AC - AG \times \boxed{ab} \text{ . } AB \boxed{ab} - AF \boxed{ac}.^{(47)}$$

---

(45) The orthocentre (marked as 'X' on Newton's following diagram), in other words.

(46) This replaces 'terminantur' (are terminated).

(47) Newton collapses the variant fluxional notations used in Propositions 3 and 9 of the first version of the 'Geometria' (see §1: notes (32) and (43) above), now representing the fluxion of an upper-case fluent line-segment by its lower-case equivalent placed in a rectangular box. This denotation is found nowhere else in Newton's mathematical papers.

These first ten propositions, representing a much augmented revise of Propositions 20 and 21 of the earlier 'Geometria', are most easily proved in the Leibnizian terms of notes (37) and (40) above. In Propositions 1, 3, 5 and 9 the sides AB and AC are given: hence

$$BC^2 = AB^2 + AC^2 - 2AB \times AC \text{ . cos } A$$

yields the limit-equality $BC \text{ . } dBC = (AB \times AC \text{ . sin } A \text{ . } dA$ or) $AB \times CG \text{ . } dA$ and so

$$dBC/dA = AD \quad [1];$$

further, since $AB^2 + BC^2 - 2BC \times BD = AC^2$, $(BC - BD) \text{ . } dBC - BC \text{ . } d(BD) = 0$ or

$$dBC/dBD = BC/DC \quad [3]; \quad \text{while} \quad BC \times AD = AB \times AC \text{ . sin } A$$

3. *If the two sides be given, the fluxion of the base will be to the fluxion of either of its segments as the base to the other segment.*

4. *If the base and one or other side be given, the fluxion of the second side will be to the fluxion of the adjoining segment as the base to the fluent side.*

5. *If the sides be given, the fluxion of the base will be to the fluxion of the perpendicular as the product of the base and the perpendicular to the product of the base segments; or, alternatively, as the base to the portion of the perpendicular which lies between the base and the intersection of the perpendiculars let fall to the remaining sides.*[45]

6. *If the base and one or other side be given, the fluxion of the second side will be to the fluxion of the perpendicular as the product of the base and the perpendicular to the product of the fluent side and the opposite base segment; or, alternatively, as the base to the portion of a second perpendicular let fall to the fluent side which is intercepted between the opposite angle it was dropped from and the perpendicular let fall to the base.*

7. *If there be given the base and the perpendicular, the fluxions of the sides will be as the segments of the sides which lie*[46] *between the base and the perpendiculars to those sides.*

8. *If there be given a base segment and the opposite side, the fluxion of the other side will be to the defluxion of the base as the other segment to the fluent side.*

9. *If the two sides be given, the fluxions of the angles at the base will be to the defluxion of the vertex angle as the opposite base segments to the base.* That is,

$$\mathrm{fl}\,(\hat{B}):\mathrm{fl}\,(\hat{C}):\mathrm{fl}\,(\hat{A}) = DC:DB:-BC.$$

10. *If the base BC be given, the fluxions of the angles at the base will be to one another as the excess of the fluxions of the opposite sides multiplied into the sides over the fluxions of the adjoining sides multiplied into their adjoining base segments.* That is,

$$\mathrm{fl}\,(\hat{B}):\mathrm{fl}\,(\hat{C}) = AC\times\mathrm{fl}\,(AC)-AG\times\mathrm{fl}\,(AB):AB\times\mathrm{fl}\,(AB)-AF\times\mathrm{fl}\,(AC).^{(47)}$$

---

gives $BC.dAD+AD.dBC = (AB\times AC.\cos A.dA$ or) $AB\times AG.dBC/AD$, whence, since

$$AB\times AG = AD\times AX, \quad \text{there is} \quad dBC/dAD = BC/XD = BC\times AD/BD\times DC \quad [5];$$

and lastly, since $AB.\sin B = AC.\sin C$ and so $AB.\cos B.dB = AC.\cos C.dC$, there results $dB/dC = DC/BD$, whence dA/dC (or $-d(B+C)/dC) = -BC/BD$ [9]. In Propositions 2, 4 and 6 a side, say AB, and the base BC are given: accordingly, because

$$AB^2+AC^2-2AB\times AC.\cos A = BC^2,$$

there is $(AC-AB.\cos A).dAC+AB\times AC.\sin A.dA = 0$ and therefore

$$dAC/dA = (-AB\times CG/FC \text{ or}) -BC\times AD/FC = -AD/\cos C \quad [2];$$

also, since $AC^2+BC^2-2BC\times DC = AB^2$, there is $AC.dAC-BC.dDC = 0$ or

$$dAC/dDC = BC/AC \quad [4];$$

and finally, because $BC\times AD = AB\times AC.\sin A$,

$$BC.dAD = (AC\times AB.\cos A.dA+AB.\sin A.dAC \text{ or}) AC\times AF.dA+BF.dAC,$$

[*11.*] *In Triangulo quovis acutangulo excessus quo fluxio basis ducta in radium &c.*
Vide prop 22.

[*12.*] *In tri. quovis acut.* ⊥^{lo} *ad basem demisso, differentia fluxionum laterum &c.*
Vide prop. 23.

[*13.*] *In triang. &c.* Vide Prop: 25.

[*14*] *In triang &c.* Vide Prop 26.[48]

[*15.*] *In triang &c.* Vide Prop 20.

[*16.*] *In triang &c.* Vide Prop 21.[49]

D[E FLUXIONIBUS LINEARUM][50]

1. Si rectæ duæ alijs duabus positione datis rectis ad datos angulos insistentes conveniant ad tertiam positione datam lineam erunt fluxiones[51]

2. Si rectæ duæ quarum una ducitur a dato puncto, altera rectæ positione datæ in dato angulo insistit conveniant ad aliquam positione datam lineam[52]

---

so that, since $AF \times FC = BF \times XF$ and $BX \times AD = AC \times BD$, there comes

$$dAC/dAD = (BC \times BF/(BF^2 - AF \times FC) \text{ or}) BC/BX = BC \times AD/AC \times BD \quad [6].$$

In Proposition 7 BC and AD are given: here $AB \times AC . \sin A = BC \times AD$ yields

$$\sin A . d(AB \times AC) + AB \times AC . \cos A . dA = 0,$$

while $AB^2 + AC^2 - 2AB \times AC . \cos A = BC^2$ gives

$$AB . dAB + AC . dAC + AB \times AC . \sin A . dA - \cos A . d(AB \times AC) = 0,$$

whence $d(AB \times AC)/\cos A = AB . dAB + AC . dAC$, so that

$$(AB - AC . \cos A) . dAC + (AC - AB . \cos A) . dAB = 0 \quad \text{or} \quad dAB/dAC = -BG/CF \quad [7].$$

In Proposition 8, where AC and BD are given, the equality $AB^2 + BC^2 - 2BC \times BD = AC^2$ gives at once $AB . dAB + (BC - BD) . dBC = 0$, whence $dAB/dBC = -DC/AB$ [8]. Most generally in Proposition 10, where BC alone is given, $AC^2 = AB^2 + BC^2 - 2AB \times BC . \cos B$ yields $AC . dAC = AB . dAB(-BC . d(AB . \cos B) \text{ or}) + BC . d(AC . \cos C)$ since

$$AB . \cos B + AC . \cos C = BC,$$

so that, on introducing $AB . \sin B = AC . \sin C$,

$$AB . dAB - (AC - BC . \cos C) . dAC = BC \times AB . \sin B . dC$$

and similarly $AC . dAC - (AB - BC . \cos B) . dAB = BC \times AC . \sin C . dB$; accordingly, because again $AB . \sin B = AC . \sin C$, $dB/dC = (AC . dAC - AG . dAB)/(AB . dAB - AF . dAC)$.

(48) This is evidently a ghost entry and should be deleted: Proposition 26 of the first version of the 'Geometria' is concerned with the motion of a rotating radius vector! Observe that Newton omits the unnecessary 'Prop. 24' from his revised scheme; compare §1: note (108) above.

11. *In any acute-angled triangle the excess of the fluxion of the base multiplied into the radius.... See Proposition 22.*

12. *In any acute-angled triangle, when the perpendicular is let fall to the base, the difference of the fluxions of the sides.... See Proposition 23.*

13. *In...triangle.... See Proposition 25.*

14. *In...triangle.... See Proposition 26.*[48]

15. *In...triangle.... See Proposition 20.*

16. *In...triangle.... See Proposition 21.*[49]

[On the fluxions of lines][50]

1. *If two straight lines standing at given angles on two other straight lines given in position should meet on a third line given in position, then the fluxions...*[51]

2. *If two straight lines, one of which is drawn from a given point, while the other stands at a given angle on a straight line given in position, should meet on some line given in position...*[52]

---

(49) On transposing Newton's enunciations to fit his present figure, we may set Propositions 11, 12, 13, 15 and 16 as:

'11. $\mathrm{rad} \times \mathrm{fl}\,BC - \cos B \times \mathrm{fl}\,AB - \cos C \times \mathrm{fl}\,AC = AD \times \mathrm{fl}\,\mathrm{arc}\,A.$

12. $\sin C \times \mathrm{fl}\,AC - \sin B \times \mathrm{fl}\,AB = BD \times \mathrm{fl}\,\mathrm{arc}\,B - DC \times \mathrm{fl}\,\mathrm{arc}\,C.$

13. $\mathrm{fl}\,\mathrm{ang}\,A . \mathrm{fl}\,\mathrm{ang}\,C :: BC \times \mathrm{fl}\,BC - BG \times \mathrm{fl}\,AB - CF \times \mathrm{fl}\,AC$
$$. AB \times \mathrm{fl}\,AB - AF \times \mathrm{fl}\,AC - BD \times \mathrm{fl}\,BC.$$

15. $AB \times \mathrm{fl}\,AB - AC \times \mathrm{fl}\,AC = BC \times \mathrm{fl}\,BD - DC \times \mathrm{fl}\,BC.$

16. $CX \times \mathrm{fl}\,AB + BX \times \mathrm{fl}\,AC = DX \times \mathrm{fl}\,BC + BC \times \mathrm{fl}\,AD.$'

(50) Our conjectural title, amplifying Newton's single capital '*D*', for this schematic revise of Propositions 26 to 30 of the first state of the 'Geometria' (§1 above). The following propositions evidently deal with the construction of tangents (instantaneous 'directions of motion') and normals at given points on curves defined in a variety of systems of coordinates.

(51) The particular case of bipolars in which the curve (*C*) is defined by some relationship $f(A'C, B'C) = 0$ between lines $A'C$, $B'C$ drawn at given angles to the fixed lines DA', EB'. This and the two following theorems are essentially amplifications of the cancelled 'Prop 30' of the first version of the 'Geometria' (see §1: note (138) above) in which the curve (*C*) is defined by some relationship between its angled distances CA', CB', ... from given lines DA', EB' ... fixed in position and also between its distances CA, CB, ... from fixed poles A, B, ...

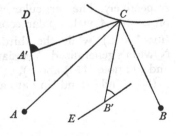

(52) The 'focus-directrix' case in which (*C*) is determined by some equation

$$f(AC, A'C) = 0.$$

Compare Mode 2 of Problem 4 of Newton's 1671 tract (see III: 132–6).

3. Si rectæ duæ circa data puncta volventes[53] *conveniant ad lineam positione datam*[54]

4. Si recta circa datum punctum volvens secet alias duas positione datas lineas[55]

5. Si recta circa datum punctum volvens secet alias duas positione datas lineas[55]

6. Si recta circa datum punctum volvens secet alias duas positione datas lineas —[55]

7. Si duæ rectæ quarum una circa datum punctum volvitur[,] *altera rectæ positione datæ*[56] *insistit conveniant ad lineam positione datam et secent aliam positione datam lineam* —[57]

8. Si duæ rectæ circa data puncta volventes conveniant ad lineam positione datam et secent aliam lineam positione datam[58]

9. Lineæ cujusvis mobilis invenire centrum gyrationis.[59]

10. Figuræ cujusvis mobilis invenire centrum gyrationis.[59]

11. Pūcti cujusvis mobilis invenire plagam motus.[60]

De fluxionibus superficierum.[61]

1.

2.

3.

4.

---

(53) Newton first wrote equivalently 'a duobus positione datis punctis ductæ' (drawn from two points given in position)—a phrase he forgot to cancel in the manuscript.

(54) The general bipolar case in which the curve (C) is determined by some relationship between the *radii vectores AC, BC*. This is essentially Proposition 29 of the preceding 'Geometria' (§1).

(55) Observe that the enunciations of Propositions 4, 5 and 6 are, as far as they go, identical. By analogy with Modes 4, 5 and 6 of Problem 2 of the 1671 tract (see III: 138–40) we may suppose that here the curve (C) is determined with respect to a line *ABC* rotating round the pole A and intersecting the given curve αB in B: namely, by some equation

$$f_1(AB, AC) = 0, \quad f_2(\widehat{\alpha B}, AC) = 0 \quad \text{or} \quad f_3(\widehat{\alpha B}, BC) = 0$$

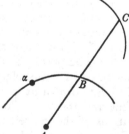

respectively. The particular case of the second when $\widehat{\alpha B}$ is a circle on centre A yields polar coordinates—Mode 7 of the 1671 tract (III: 140–2), while the third, when $\widehat{\alpha B}$ is a straight line, gives Newton's generalized Cartesian coordinates (see I: 173: '*Examp* 3ᵈ'; and §1: note (143) above).

(56) Understand 'in dato angulo' (at a given angle).

3. If two straight lines revolving round given points[53] *should meet on a line given in position*...[54]

4. If a straight line revolving round a given point should intersect two other lines given in position...[55]

5. If a straight line revolving round a given point should intersect two other lines given in position...[55]

6. If a straight line revolving round a given point should intersect two other lines given in position...[55]

7. If two straight lines, one of which revolves round a given point, while the other stands[56] *on a straight line given in position, should meet on a line given in position and intersect another line given in position*...[57]

8. If two straight lines revolving round given points should meet on a line given in position and intersect another line given in position...[58]

9. Of any mobile line to find the centre of gyration.[59]

10. Of any mobile figure to find the centre of gyration.[59]

11. Of any mobile point to find the direction of its motion.[60]

On the fluxions of surfaces[61]

1.

2.

3.

4.

---

(57) Here the curve (C) is understood to be defined by some equation $f(Aa, B'b') = 0$, probably, though the arc-lengths $\overset{\frown}{aa}$, $\overset{\frown}{ab'}$ could conceivably be introduced.

(58) The curve (C) is most probably here understood to be defined by some equation $f'(Aa, Bb) = 0$; compare note (57) above.

(59) The centre of curvature (defined as the point which is instantaneously stationary in the moving plane of the curve), that is.

(60) Namely, the 'determinatio motus' in Proposition 27 of the first version of the 'Geometria'; compare §1: note (123) above.

(61) Understand (in context) plane surfaces. The following propositions were doubtless intended to elaborate the theorems on the fluxions of geometrical areas earlier expounded in Newton's addendum to his 1671 tract (see III: 338–40, 342–6).

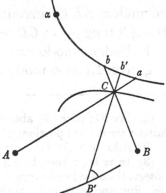

APPENDIX 1. VARIANT DRAFTS OF THE 'GEOMETRIA CURVILINEA'.[1]

From the originals in the University Library, Cambridge

[1][2]   DEFINITIONES

Affirmativum est quod additur ad nihil.

Negativum est quod subducitur a nihilo.[3]

1. Fluens est quod continuo augetur vel diminuitur.

2. Fluxio est celeritas mutationis illius.

3. Fluxio affirmativa sive profluxio est celeritas incrementi rei affirmativæ vel decrementi negativæ.

4. Fluxio negativa sive defluxio est celeritas decrementi rei affirmativæ vel incrementi negativæ.

5. Determinatum est quod permanet sine incremento vel decremento.

6. Voces Radij, chordæ, sinus, tangentis et secantis[4] usurpo ut sunt in Trigonometria. Nempe si centro quovis *C* & radio *CA* describatur circulus *EAF*, et agatur[5] quælibet recta *EF* circulo occurrens in *E* et *F*, erit *EF* chorda arcus *EAF*. Biseca eas in *D* et *A* et ejus[6] dimidium *ED* vel *DF* est sinus dimidij arcus *EA* vel *FA*. Junge *CA*, *CE*, & idem dimidium *ED* dicetur etiam sinus anguli cui semichorda [*D*]*E* respondet. Erit etiam sagitta *DA* sinus versus ejusdem anguli et alterum radij segmentū *CD* erit cosinus ejus. Deniꝗ ad *CA* erige normalem *AB* occurrentem radio producto *CE* in *B* et erit *AB* tangens & *CB* secans ejusdem anguli.

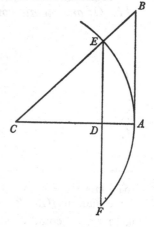

7. Eodem modo fluere dicuntur quæ omnia fluunt affirmativè vel omnia negative. Diverso modo quorum unum fluit affirmative alterum negative.

---

(1) See §1: note (2) above. As the handwriting of these autographs confirms, these preliminary versions of portions of the 'Geometria' do not significantly precede their revise in time.

(2) Add. 3963.7: 47^r–49^r; compare §1: note (19) above.

(3) In redraft (see §1: note (21) above) Newton initially made these two complementary observations—a late insertion in the manuscript—new Definitions 5 and 6, but at once changed his mind and introduced them in a following scholium in the version reproduced.

(4) This replaces an original opening 'Voces sinuum rectorum et versorum, cosinuum, tangentium et secantium'. The following definitions of sine (chord), tangent and secant are virtually those expounded in the 'Prænoscenda' to Newton's *Trigonometria* (**1**, 3, §2) and in his preface to St John Hare's *Trigonometria* (**1**, 3, §3), both which likewise avoid introducing the complementary functions of cosine, cotangent and cosecant.

(5) An omitted sequel reads 'quilibet radius *CE* et ab ejus termino *E* ad alterum radium *AC* demittatur perpendicularis *ED*, hæc *ED* erit sinus anguli *ACE* quem subtendit'.

8. Polus lineæ rectæ mobilis est punctum in quo recta illa circumacta incipit locum linearem quem modo occupavit secare vel in quo desinit secare si in locum istum revertitur.

9. Locus puncti moti est linea sive recta sive curva quam punctū istud motu suo describit.

10. Determinatio motus puncti ejus est positio rectæ tangens curvam illam ad punctū motum.[7]

Axiomata.

1. Quæ fluxionibus æqualibus simul generantur sunt[8] æqualia.

2. Quæ sunt perpetim æqualia, fluxionibus æqualibus generantur.

3. Quæ fluxionibus in data ratione simul generantur sunt in ratione fluxionum.

4. Quæ sunt perpetim in ratione data, fluxiones habent in eadem ratione.
Nota simul generari intelligo quæ tota eodem tempore generantur.

5. Fluxio totius æquatur fluxionibus partium simul sumptis.

6. Fluxiones quantitatum sunt in prima ratione partium nascentium, vel quod perinde est, in ultima ratione partium istarum per defluxionem evanescentium. Sint *AB, CD* fluentes quantitates. Fluant hæ donec evadant *AE, CF*: dico fl. *AB* esse ad fl: *CD* in prima ratione quam partes generatæ *BE, DF* habebant in principio generationis, vel quod perinde est in ultima ratione quam partes habere possunt ubi *AE* et *CF* vicissim defluentes evadunt *AB* et *CD*.

Petitiones.

Pet. 1. Lineas quasvis quacunq̃ ratione geometrica movere. Per rationem geometricam intelligo rationem movendi in qua positio lineæ motæ potest geometricè designari.

Pet: 2. Alias lineas per puncta vel intersectiones priorum describere.[9]

---

(6) Of *EF*, that is.

(7) In redraft Definition 6 was omitted (doubtless because it was part of common mathematical knowledge of the period and perhaps also because it is so obviously out of place in present context), while slightly revised versions of Definitions 5 and 7 were inverted in order (becoming Definitions 6 and 5) respectively; the ordinal numbering of Definitions 8, 9 and 10 was then correspondingly reduced by unity.

(8) Newton began to insert 'perp[etim]' as his next word and then reverted to the form he had earlier used in his 1671 tract (see Axiom 1 on III: 330).

(9) These 'Axiomata' and 'Petitiones' reappear without essential change in the revised version (reproduced in §1 preceding). Here (and widely below) we have not reproduced certain late changes entered by Newton into the manuscript which convert its text into an exact forerunner of the revise.

Prop 1

*Si trium semper proportionalium quantitatum media datur[,] erit profluxio unius extremæ
ad defluxionem alterius ut prior extrema ad illam alteram.*

Exponantur enim hæ quantitates per tres lineas *AB AC AD* quarum media
perpendiculariter insistat extremis ad
communem terminum.[10] Fluant jam
AB AD donec evadant *Ab* et *Ad* et erunt
etiam ex Hyp *Ab*, *AC*, *Ad* continue
proportionales. Quare actis *BC*, *DC*; *bC*,
dC triangula *BCD*, *bCd* erunt rectangula
ad *C*. Erige normalem *bE* occurrentem
[*B*]*C* in *E* et (per [29], 1 *Elem*) erit
ang *BEb* = ang *BCA* = ang *ADC* adeoꝗ

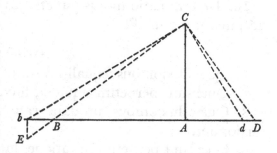

triangula *CEb*, *CDd* similia sunt ut et *BEb*, *ACD*. Quare *Dd*.*Eb*::*Cd*.*Cb*::
Ab.*AC*. et *Eb*.*Bb*::*AC*.*AD*. Adeoꝗ per 9[,] 5 Eucl *Dd*.*Bb*::*Ab*.*AD*. Refluant
jam *Ab* et *Ad* dum coeant cum *AB* et *AD* et in ultima ratione evanescentiū
partium *Dd* et *Bb* erit *AB* ad *AD*. Quare per ax 6 fluxiones ipsarum *AD* et *AB*
sunt in eadem ratione. Q.E.D.

Prop 2.

Si trium semper proportionalium quantitatum una extrema[11] *datur, erit fluxio mediæ ad
fluxionem alterius extremæ ut extrema data ad duplum mediæ.*

Sint istæ *AB AC AD*
nempe perpendiculum et
segmenta basis trianguli
rectanguli *BCD* ut ante, et
flu[a]nt *AC* et *AD* donec
evadant *Ac* et *Ad*. Age *CE*
parallelam *cd* et propter
similia triangula *BCc*,
CDE, & *CAE*, *cAd* erit
Cc.*DE*::*BC*.*CD*::*AB*.*AC*
& *Cc*.*Ed*::*Ac*.*Ad*::*AB*.*Ac*.

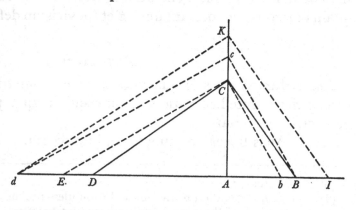

Quare componendo est *Cc*.*Dd*::*AB*.*AC*+*Ac*. Conveniant jam *Ac* et *Ad* cum
AC et *AD* et ultimo *AC*+*Ac* evadet 2*AC* adeoꝗ ultima ratio partium *Cc* et *Dd*
evanescentium erit *AB* ad 2*AC*. Quare per ax 6 fluxiones ipsarum *AC* et *AD*
sunt in ista ratione. [Q.E.D.]

---

(10) Namely, *A*. A terminal following phrase 'et agantur *BC DC* constituentes triangulum
[*BCD*]' is deleted.

(11) This replaces 'prima', an all but trivial correction.

Prop 3.[12]

Sin trium semper proportionalium fluentium quantitat[um nulla datur], fluxiones extremarum reciproce ductæ in extremas æquales erunt fluxioni mediæ ductæ in duplum mediæ.

Sint AB, AC, $A[D]$ tres quantitates ut ante, et fluant hæ donec evadant AI, AK, Ad. Sitcȝ Ac semper media proportionalis inter AB et Ad et agantur rectæ Bc, cd, IK, Kd. Et per prop 2 erit $2Ac . AB :: \text{fl}\,Ad . \text{fl}\,Ac$ sive $\dfrac{AB \times \text{fl}\,Ad}{2Ac} = \text{fl}\,Ac$.

Item per ea quæ Ibi demonstrata sunt erit $Ac + AK . Ad :: BI . cK$ adeocȝ sub initio fluxionis[13] ubi Ac & AK æquales erant

$$2Ac . Ad :: BI . cK :: (\text{per ax 6}) \text{ fl}\,BI\,(\text{vel }[\text{fl}]\,AI) . \text{fl}\,cK.$$

Quare $\dfrac{Ad \times \text{fl}\,AI}{2Ac} = \text{fl}\,cK$. Sed per ax 5 est $\text{fl}\,Ac + \text{fl}\,cK = \text{fl}\,AK$. Quare

$$\frac{AB \times \text{fl}\,Ad}{2Ac} + \frac{Ad \times \text{fl}\,AI}{2Ac} = \text{fl}\,AK.$$

Coincidant jam AI cum AB, Ac et AK cum AC et Ad cum AD ut fit sub initio fluxionis et omnia ducendo in $2Ac$ vel $2AC$ erit

$$AB \times \text{fl}\,AD + AD \times \text{fl}\,AB = 2AC \times \text{fl}\,AC.$$

Schol. Hæc omnia sic brevius demonstrari possent. Sint tres quantitates a, b, c [&] fluant [hæ] donec evadant $a+e$, $b+f$ & $c+g$. Ergo

$$ac = bb\ [\&]\ \overline{a+e} \times \overline{c+g}\,[= \overline{b+f} \times \overline{b+f}].$$

hoc est $ac + ag + ec + eg = bb + 2bf + ff$. Subduc ac et bb et erunt

$$ag + ec + eg = 2bf + ff.$$

defluant jam quantitates $a+e$, $b+f$, $c+g$ donec evadunt, a, b, c[14] et erit $a\dfrac{g}{f} + \dfrac{e}{f}c = 2b$. sive per ax 6 $a \times \dfrac{\text{fl}\,c}{\text{fl}\,[b]} + c \times \dfrac{\text{fl}\,a}{\text{fl}\,[b]} = 2b$ hoc est

$$a \times \text{fl}\,c + c \times \text{fl}\,a = 2b \times \text{fl}\,b.$$

Prop 4. Si trium proportionalium summa[15] *datur*—[*Prop*] *5. Si differentia*—[*Prop*] *6. Si extremæ et mediæ summa detur,*—[*Prop*] *7. Si differentia*[—].[16]

---

(12) Halfway through this proposition Newton decided to rename the points c and d in the figure 'H' and 'F' respectively, an alteration confirmed in the revise (§1). For logical consistency we here retain the earlier denotation throughout, making silent 'correction' of occurrences of H and F in the manuscript text where necessary.

(13) 'motus' is cancelled.

(14) The equivalent phrase 'donec e, f et g evanescunt' is deleted.

(15) Understand 'extremarum'.

Prop 8. Si quotcunꝗ continue proportionalium una datur—.

Cor. Fluxio dignitatis est ad fluxionem radicis ut dignitas ducta in indicem suam ad radicem.[17]

Prop 9. Si quatuor proportionalium prima datur[,] ea ducta in fluxionem quartæ æquabitur fluxionibus mediarum mutuo ductis in medias.[18]

Sint $A.B::C.D$, et detur A.[19] Sitꝗ E media proportionalis inter B et C et eadem erit media proportionalis inter A et D. Quare per prop 2 et 3

$$A \times \text{fl}\, D = 2E \times \text{fl}\, E = B \times \text{fl}\, C + C \times \text{fl}\, B.$$

Cor 1, 2, 3 de mult. divis. & æquat.[20]

Prop 10. Positis quatuor proportionalibus flux[i]ones extremarum mutuo ductæ in extremas æquantur fluxionibus medioru̅ mutuo ductis in medias.[21]

Sint $A.B::C.D$ sitꝗ E media proportionalis inter B & C et per prop 3 erit
$$A \times \text{fl}\, D + D \times \text{fl}\, A = 2E \times \text{fl}\, E = B \times \text{fl}\, C + C \times \text{fl}\, B.$$

Prop 11. In Triangulo quovis perpetim rectangulo Cor 1 et 2.[22]

Prop 12. In circulo dato fluxio sinus est ad fluxionem tangentis ut quadratum cosinus ad secantem ductum in radium, ad fluxionem secantis vero ut quadratum cosinus ad tangentem ductum in radium.[23] *Defluxio cosinus autem est ad fluxionem tangentis ut sinus in cosinū ad secantem in radium et ad fluxionem secantis ut cosinus ad secantem. Deniꝗ defluxio cosinus* [*est*] *ad fluxionem chordæ ut chorda ad radium.*

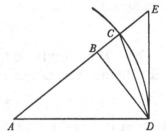

Est enim $\text{fl}\, DB.[\text{fl}] - AB.[\text{fl}]\ AE.[\text{fl}]\ ED::AB^q$. (Cor [2] pr 11) $AB \times BD$. (Prop 1) $AE \times BD$, vel $AC \times ED$. (cor [1] pr 11) $AC \times AE$. Est et $2AC.DC::DC.BC$ adeoꝗ per Prop 2 est

$$AC.DC::\text{fl}\, DC.\text{fl}\, BC.$$

---

(16) The full enunciations of these four propositions and their proofs are manifestly to be entered from Corollaries 2–5 of Theorem 1 of his earlier addendum to the 1671 tract (see III: 334–6). Likewise 'Prop 8' following is to be completed from Corollary 6 (III: 336) by understanding 'fluxiones fluentium erunt inter se ut fluentes illæ ductæ in numerum terminorum quibus distant a dato illo'.

(17) This is effectively 'Cor 9' on III: 338.

(18) A more general version of 'Cor. 7' on III: 336, which is now set as its first corollary (see note (20) below).

(19) In sequel a cancelled first conclusion reads: 'Cum ergo A, $\sqrt{BC}$ & D ut et B, $\sqrt{BC}$, C sint continue proport. erit per prop 2 et 3 $A \times \text{fl}\, D = 2\sqrt{BC} \times \text{fl}\sqrt{BC} = B \times \text{fl}\, C + C \times \text{fl}\, B$'. The revise merely substitutes E^2 for BC.

(20) Read 'de multiplicatione, divisione & æquatione'. We must evidently understand that the first two corollaries (much as 'Cor. 7' and 'Cor 8' on III: 336–8) deal with the particular instance $A = 1$ of the preceding proposition when the proportion is set out in the equivalent forms $1:B = C:(D$ or$)\ BC$ and $B:1 = C:C/B$. The third corollary, by analogy, was intended no doubt to cover the corresponding equality $BC = D$, yielding

$$\text{fl}(BC) = B \times \text{fl}(C) + C \times \text{fl}(B).$$

Prop 13.[24] *Si angulorū duorū fluentium differentia datur, fluxiones sinuum erunt ut cosinus, fluxiones tangentium ut quadrata secantium,*[3] *fluxiones secantium ut rectangula sinu[u]m et secantium, fluxio sinus unius ad fluxionem tangentis alterius ut radius ductus in cosinum prioris ad tangentem ductam in secantem alterius & fluxio tangentis unius ad fluxionem secantis alterius ut quadratum secantis prioris ad tangentem ductam in secantem posterioris.*

Sint anguli isti *DAE DAG* quorum Differentia *EAG* detur. Sintꝗ sinus e[o]rum *DB, DF,*[25] tangentes *DE DG* et secantes *AE AG*: dico primo quod sit fl *DB* . fl *DF* :: *AB* . *AF*. Nam ad *AF* et *AK* demitte *BH* & *BK* normales.[26]

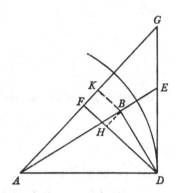

Et propter similia triangula *DBH, ABK*[27] erit *DB* . *DH* :: *AB* . *AK* :: (per ax 3) fl *DB* . fl *DH*. Et *AB* . *BK* vel *FH* :: (per ax 3) defl *AB* . defl *FH* :: *DB* . *BH*. Sed per prop [11 Cor 2] est *AB* . *BD* :: fl *BD* . defl *AB*. Ergo connectendo rationes[28] *AB* . *BH* :: fl *BD* . defl *FH*. vel *AB* . — *BH* :: fl *DB* . fl *FH*. Et componendo

$$AB . AK - BH(AF) :: \text{fl } DB . \text{fl } DH + \text{fl } FH.$$

Sed fl *DH* + fl *FH* componunt fl *DF* per ax 5 adeoꝗ est *AB* . *AF* :: fl *DB* . fl *DF*. Q.E.D.

Hoc ostenso dico præterea quod sint

fl *DE* . fl *AE* . defl *AB* . fl *DB* . fl *DF* . defl *AF* . fl *AG* . fl *DG* ::

$$AE^q . (\quad) DE \times AE . (\quad) DE \times AB, \text{ vel } AD \times DB . (\quad) AD \times AB.$$

(per modo ostensa)

$$AD \times AF . (\quad) AD \times DF \quad \text{vel} \quad DG \times AF . (\quad) DG \times AG . (\quad) AG^q.[29]$$

Q.E.D.

(21) Theorem 1 on III: 330. Its proof, however, is now made to depend on a twin appeal to Proposition 3 instead of being developed from first principles (see III: 330–4).

(22) This proposition and its two corollaries are manifestly a word for word repetition of Theorem 2 on III: 338. Here Newton completes his borrowing from the appendage to the 1671 tract (III, **1**, 2, §3) and the two remaining propositions in this preliminary draft are new.

(23) This replaces a first opening: '*Fluxio sinus est ad fluxionem tangentis ut cosinus ductus in radium ad quadratum secantis, ad fluxionem secantis ut cosinus ductus in radium ad tangentem × secantem.*'

(24) Newton first began equivalently 'Si anguli duo æqualiter fluant, fluxiones....'.

(25) Where, that is, *DB* and *DF* are perpendiculars from *D* to *AE* and *AG* respectively.

(26) A cancelled first continuation at this point reads: 'et erit ut radius ad sinum anguli dati *EAG* ita *AB* ad *FH* :: (per ax 3) fl *AB* . fl *FH*. et ut radius ad cosinum ejusdem anguli dati *EAG* vel *BDH* ita *BD* ad *DH* :: (per ax 3) fl *BD* . fl *DH*.'

(27) Newton first wrote 'propter datos et æquales angulos *EAG* [et *BDH*]'.

(28) 'componendo' is rightly cancelled.

(29) The parentheses are left blank for a future insertion of references to previous propositions which Newton never, in fact, made.

Eædem proportiones obtinent ubi summa flu-entium angulorum datur. [30]Ut patebit relegenti demonstratione[m] præcedentis propositionis et simul inspicienti hanc figuram, si modo angu-lorū alteruter *DAB* una cum ejus sinu et tan-gente considerentur jam ut defluens, & cosinus ut profluens.

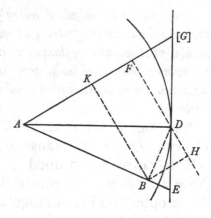

Prop [14]

Si anguli fluunt in ratione data[31]

[2][32]

13. Dato *AC* est

$$\text{fl } AB . [\text{fl}] BC . [\text{fl}] DC :: BC \times AC . AB \times AC . \text{ (prop 2) } 2BC \times AB.$$

Præterea est $AC . AB :: BC . BD$.[34] Biseca *AC* in

Dele prop 12[33] et scribe hæc duo.

E et Age *EB*. hæc æqualis erit *AE* vel *EC* adeoɋ data, quare $BD . ED :: \text{fl } ED (AD) . \text{fl } BD$. [Hoc est]

$$\text{defl } BD . \text{fl } ED . \text{fl } AB . \text{fl } BC ::$$
$$2ED \times BA . 2BD \times AB . \text{ (prop 2) } BD \times AC$$

vel $AB \times BC . AB^q$. vel defl $BD . \text{fl } ED$ vel

$$\text{def}[l] DC . \text{fl } BC . \text{fl } AB :: 2ED, BC . 2BD, BC . BD, AC \text{ vel } AB \times BC . BC^q.[35]$$

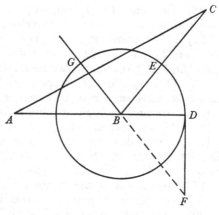

(30) Newton began a direct proof of this comple-mentary case, in fact, but abandoned it unfinished: 'Nam in Triangulo quovis *AB*[*C*] detur summa angulorum *B* et *C* et dabitur angulus *A* qui est differentia angulorum *C* et *CBD*, latere *AB* producto ad *D*. Prædictæ igitur proportiones obtinent inter sinus tangentes & secantes angulorum *C* et *CBD* sed angulus *CBA* eosdē habet sinus tangentes et secantes cum angulo *CBD*, nam centro *B* radio quovis *BD* describe circulum occurrentem *BC* in *E* et cape angulum *DBG* æqualem angulo [*ABE*].' In Newton's figure the radius *GB* is manifestly extended to meet the tangent at *D* to the circle in *F*, whence (on taking the trigonometrical 'radius' to be that of the circle) it follows that the (equal) distances of *E* and *G* from *ABD* are sinus $A\hat{B}C$ = sinus $C\hat{B}D$, $DF = -\text{tangens } A\hat{B}C$ and $BF = -\text{secans } A\hat{B}C$.

Dato AB est $AD.AB::\text{fl}\,AB.$ [fl] AC. Ergo $AC.AB::\text{defl}\,AC.\text{defl}\,AD$. [Item]
defl $AD.\text{fl}\,BD::BD.AD$.   Ergo fl$AC.\text{fl}\,BD::AC\times BD.AD^q$. [Insuper est]
fl$BC.\text{fl}\,AC::AC.BC$. ergo fl$BC.\text{fl}\,BD::AC^q\times BD.BC\times AD^q::AC^q.AB\times AD$.
et $AD+AC.AD::\text{fl}\,DC.\text{def}[\text{l}]\,AD$. [Itacɜ hoc Theorema prodibit.]

   14. Dato AB [est]

$$\text{fl}\,BC.[\text{fl}]\,AC.[\text{fl}]\,AD.[\text{fl}]\,BD::AC^q.BC\times AC.BC\times AD.AB\times AD.$$

Præterea cum sit $AC.AD::\text{fl}\,AC.\text{defl}\,AD$ erit

$$AC.AC+AD.AD::\text{fl}\,AC.\text{fl}\,DC.\text{fl}\,AD.^{(36)}$$

[**3**]$^{(37)}$

   $^{(38)}$[Est] $EF.[K]C::AO\times EF.AO\times DC::AFE.ACD$. sed

$$AFE.ACD\sqsubset AFE.ACG$$

hoc est $\sqsubset AF^q.AC^q$ & $\sqsupset AFE.AH[K]$ hoc est $\sqsupset AF^q.AD^q$. Coeant ergo $AC\ AD$

---

(31) Newton breaks off this trivial extension of the preceding proposition in mid-sentence (at the top of f. 49^v in the manuscript) and thereupon abandons this preliminary version of the 'Geometria' for the revise reproduced in §1 above.

(32) Add. 3963.7: 52^r, preliminary drafts of the proofs of Propositions 11 and 12 (here numbered 13 and 14) in §1.

(33) Understand in the preliminary version of the 'Geometria' reproduced above. Newton subsequently decided to delay his Proposition 11 till after these two, which accordingly became Propositions 11 and 12 in the revise; see §1: note (45), and compare the preceding note.

(34) As before, it is taken that $A\hat{B}C$ is right and that BD is the altitude, so that the right triangles ABC, ADB, BDC are similar. Newton first continued: 'Ergo

$$\frac{BC\times\text{fl}\,AB+AB\times\text{fl}\,BC}{AC}=\text{fl}\,BD.\quad\text{sed}\quad AB\times\frac{\text{fl}\,AB}{BC}=\text{fl}\,BC.$$

Ergo $\dfrac{BC^q+AB^q}{BC\times AC}\times\text{fl}\,AB=\text{fl}\,BD=\dfrac{AC^q}{BC\times AC}\,[\times\text{fl}\,AB]=\dfrac{AC}{BC}\,[\times\text{fl}\,AB].$'

(35) That is, '$::2ED.2BD.BC.AB$'!

(36) A final unfinished phrase 'Et $AD.AC+AD$' is left hanging.

(37) Add. 3963.7: 52^r, a first draft of the demonstration of Proposition 14 in the main text of the 'Geometria' (§1 above). The sheet was first used by Newton to draft an unidentified letter to an unknown correspondent, abandoned after its opening phrase 'S^r / The last wednesday I had packed up your[...]'.

(38) A cancelled preliminary opening reads: '[Ratio triangulorum] $AFE.ACD$

   $\sqsubset AFE.AGC$ hoc est $AF^q.AC^q$. & $\sqsupset AFE.AHD$ hoc est $AF^q.AD^q$.

Coeant igitur $AD\ AC$ et ultima ratio [triangulorum AFE,ACD] erit $AF^{[q]}.AC^{[q]}$. Quare fl $A[D]\,F.\text{fl}\,A[D]\,C::AF^q.AC^q::AB.AC$.' The following revised proof is substantially that of Proposition 14 of the 'Geometria' after considerable verbal rephrasing and a new naming of points in the accompanying figure.

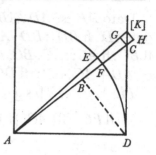

et ultima ratio $EF . [K]C$ erit $AF^q . AC^q$ sive $AB . AC$. Quare fluxiones [ipsorum] $DF \, DC$ sunt in ista ratione. Q.E.D. Præterea est fl DC . fl AC :: $AC . DC$. ergo fl DF . fl AC :: $AB . DC$. Q.E.D. Item fl AC . defl AB :: $AC . AB$. Ergo fl DF . defl AB :: $AC . DC$. Deniqg

defl AB . fl BD :: $BD . AB$.

ergo fl DF . fl BD :: $AC \times BD . AB \times DC$ :: $AD . AB$. [Q.E.D.]

---

[4][39]

Pr[*op*] *18. Si trianguli cujusvis basis et angulus ad verticem dentur [cæteraq motu quocunq fluant].*

Super basi AB describe (per 32.3[40] Eucl[i]d) circuli segmentum ACB quod capiet datum angulum ACB, sitqg GH circuli hujus diameter parallela AB et perpendiculo [CD] trianguli si opus est producto occurrens in E. Quare cum per 20[,] 3 Eucl vertex trianguli semper sit ad peripheriam hujus circuli, Differentia inter[41]

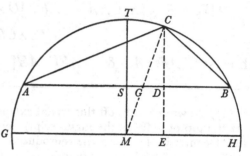

Est $\left.\begin{array}{l} AB . AC :: \sin C . \sin B. \\ AB . BC :: \sin C . \sin A. \end{array}\right\}$ Ergo[42] $AB . \sin C :: $ fl AC . fl $\sin B :: $ fl BC . fl $\sin A$.

Sed cum summa ang A et B datur est per pro[p 15]

fl $\sin A$. fl $\sin B :: \cos A . -\cos B$.

Ergo fl AC . fl $BC :: -\cos B . \cos A :: -\dfrac{BD}{BC} . \dfrac{AD}{AC}$.[43]

---

(39) Add. 3963.7: 52^v, a preliminary version of Proposition 18 of the 'Geometria'; see §1: note (74).

(40) 'Elem' is deleted as unnecessary.

(41) Understand probably 'Differentia inter angulos A et B est angulus GCD', though this is evident enough for Newton not to finish his phrase.

(42) Since $AB/\sin C = AC/\sin B$, constant.

(43) A cancelled following sentence reads 'Præterea cum detur summa angulorum ACD BCD est fl $\sin ACD$. fl $\sin BCD :: \cos ACD . \cos BCD$'.

(44) Newton spoilt a first version of this paragraph by erroneously deducing that

'$AC \times$ fl $AC + BC \times$ fl $BC = AB \times$ fl AD.   sed $\dfrac{BD, AC}{AD}$, fl $AC = BC \times$ defl BC[!].

Ergo $\dfrac{BD + AD}{AD} \times$ defl $BC = AB \times$ fl AD. hoc est $AC \times$ fl $AC = AD \times$ fl AD'.

[44]Insuper [cum per 13, 2 Eucl sit $AC^q - BC^q$ æqualis $2AB \times AD - AB^q$, est]

$$AC \times \text{fl}\, AC - BC \times \text{fl}\, BC = AB \times \text{fl}\, AD. \quad \text{sed} \quad BC \times \text{fl}\, BC \quad \text{est} = \frac{AD \times BC^q}{-BD \times AC} \times \text{fl}\, AC.$$

ergo $\dfrac{AC^q \times BD + BC^q \times AD}{AC \times BD} \times \text{fl}\, AC = AB \times \text{fl}\, AD.$[45]

---

fl BC . fl arc BC [::] cosin A . rad. et fl arc BC vel [defl] AC . fl AC :: cosin B . rad. Ergo [de]fl AC . fl BC :: cosin B . cosin A.[46] Item

$$\text{fl}\, BC . \text{fl}\, CE \text{ vel } CD :: \text{rad. cosin } CH = A + \frac{ACB - \text{rect}}{2} :: CM . ME^{[47]} :: CG . GD.$$

Fl AC . Fl BC :: [−]cosin $\frac{1}{2}$ Arc AC sive ang B . cosin $\frac{1}{2}$ arc BC sive ang A. Item fl CD (vel CE) . [de]fl AD(SD vel ME) :: (pr[op 9])[48] $ME . CE :: GD . DC.$

---

[5][49]

Prop 20.

Si trianguli cujusvis duo latera dentur cæteraꝗ fluant ac demittatur perpendiculum ad tertium latus duo ejus segmenta constituens, fluxio basis erit ad fluxionem alterutrius segmenti ejus ut ipsū[50] *est ad alterum segmentum; ad fluxionem vero perpendiculi ut rectangulum basi[s] et perpendicul[i] ad rectangulū segmentorum ipsius basis.* [51]*Sin*

---

(45) Whence $\dfrac{\text{fl}\, AD}{\text{fl}\, AC} = \dfrac{CD^2 + AD \times BD}{AC \times BD} = \dfrac{\cos(\hat{B} - \hat{A})}{\cos \hat{B}} = \dfrac{AC}{AG}$,

since $AC^2 = BC^2$ (or $BD^2 + CD^2$) $+ AB^2 - 2AB \times BD$ and $\hat{B} = \frac{1}{2}\pi - A\hat{C}G$. Compare §1 : note (79) above.

(46) That is, ' $:: \dfrac{BD}{BC} . \dfrac{AD}{AC}$' as above. Two illogical transpositions are here corrected.

(47) This is doubly incomprehensible: for, first,

$$C\hat{M}H = \tfrac{1}{2}\pi - (\hat{B} - \hat{A}) = 2\hat{A} + \hat{C} - \tfrac{1}{2}\pi \ne \hat{A} + \tfrac{1}{2}(A\hat{C}B - \tfrac{1}{2}\pi), \quad \text{that is,} \quad \tfrac{1}{4}\pi + \tfrac{1}{2}(\hat{A} - \hat{B});$$

while $BC = 2CM . \sin \hat{A}$, $CE = CM . \sin C\hat{M}H$ and so

$$\text{fl}(BC) : \text{fl}(CE) = 2\cos \hat{A} \times \text{fl}(\hat{A}) : \cos C\hat{M}H \times \text{fl}(C\hat{M}H) = \cos B\hat{C}G : \cos C\hat{M}H \ne 1 : \cos C\hat{M}H.$$

(48) Since $CE^2 + ME^2 = CM^2$, constant.

(49) Add. 3963.7: 53ᵛ, two further particular cases of Proposition 20 of the 'Geometria', omitted (see §1 : note (87)) because of their lack of generality and relative triviality.

(50) The preceding 'tertium latus'.

(51) A cancelled first continuation reads: 'Sin perpendiculum demittatur ab alterutro termino basis ad latus oppositum, erit fluxio tertij lateris ad fluxionem segmenti contermini ut summa perpendiculi et segmenti oppositi est ad basem; ad fluxionem vero perpendiculi ut tertiū latus est ad summam perpendiculi et quadrati ejus applicati ad segmentum oppositum : puta si perpendiculum cadit ad partes lateris dati versus latus fluens, nam si cadit ad alteras partes tunc vice summæ differentia inter perpendiculum ad quadratū ejus applicatum ad hoc segmentum in priori analogia et differentia inter perpendiculum et hoc segmentū in posteriori

alterutrū latus datum co[n]stituatur basis perpendiculo ad ıd demisso, erit fluxio lateris fluentis ad fluxionem segmenti contermini ut basis ad latus fluens; ad fluxionem vero perpendiculi ut rectangulum basis et perpendiculi est ad rectangulum lateris fluentis et oppositi segmenti.

Sit ABC triangulum, CD perpendiculum, AC, CB latera data et erit fl AB. fl AD::AB.DB, et

$$\text{fl } AB \,. \text{fl } CD :: AB \times DC \,. AD \times DB.$$

Nam per prop [11] est fl DB. defl CD::CD.DB & defl CD. fl AD::AD.CD. Adde rationes et fit,

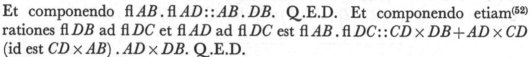

$$\text{Fl } DB \,. \text{fl } AD :: AD \,. DB.$$

Et componendo fl AB. fl AD::AB.DB. Q.E.D. Et componendo etiam[52] rationes fl DB ad fl DC et fl AD ad fl DC est fl AB. fl DC::$CD \times DB + AD \times CD$ (id est $CD \times AB$). $AD \times DB$. Q.E.D.

[53]Quod si perpendiculum AE demittatur ad alterutrum latus datum BC: dico fore fl AB. fl EB::[CB. AB] et

$$[\text{fl}] AB \,. [\text{fl}] AE :: CB \times AE \,. AD \times CE.$$

Nam per prop [11] est fl AE. defl CE (vel fl EB)::CE.AE. adeoꝗ

$$CE \times \text{fl } EB = AE \times \text{fl } AE = (\text{per prop 13}) \, AB \times \text{fl } AB - EB \times \text{fl } EB.$$

Quare $AB \times \text{fl } AB = CE \times \text{fl } EB + EB \times \text{fl } EB = CB \times \text{fl } EB$[54] et resolvendo

$$\text{fl } AB \,. \text{fl } EB :: CB \,. AB. \quad \text{Q.E.D.}$$

Erat autem fl EB. fl AE::AE.CE. Ergo connectendo rationes

$$\text{fl } AB \,. \text{fl } AE :: CB \times AE \,. AB \times CE. \quad \text{Q.E.D.}$$

Coroll. Dantur simul fluxiones sinu[u]m et cosinuum omnium angulorum. Nam perpendicula CD AE respectu radij AC sunt sinus angulorū A et C et segmenta AD, CE cosinus.

---

sumendum est.' On taking one of the given sides AC, BC to be the base, say BC, this enunciation yields the erroneous proportions

$$\text{fl}(AB):\text{fl}(BE) = (AE + CE):BC \quad \text{and} \quad \text{fl}(AB):\text{fl}(AE) = AB:(AE + AE^2/CE).$$

(52) 'rursus addendo' was first written. Newton's result follows more directly from the Pythagorean equalities $CD^2 + DB^2 = BC^2$ and $AB^2 - 2AB \times DB = AC^2 - BC^2$ (Euclid, *Elements*, II, 12).

(53) Newton first went on: 'Quod si latera AC et AB dentur et perpendiculum CD demittatur a C ad AB erit fl AB. fl EB::AE.CE.'

(54) More directly, this follows at once from the equality $AB^2 = 2BC \times BE + AC^2 - BC^2$; compare note (52) above.

Prop 21.

Si latera cujuscunɢ Trianguli acutanguli datum angulum habentis utcunɢ fluant[,] Ab angulis omnibus[55] *demitte perpendicula ad opposita latera, et si segmenta laterum angulis non datis contermina ducantur in fluxiones laterum quorum sunt segmenta, summa factorum ubi ambo anguli isti acuti sunt vel differentia ubi alter eorū obtusus est æqualis erit tertio lateri ducto in fluxionem suam. Præterea fluxio sinus anguli alterutrius fluentis multiplicata per latus dato angulo oppositum æquatur excessui quo fluxio lateris angulo isti fluenti oppositi ducta in sinum anguli dati superat fluxionem lateris angulo isti fluenti oppositi*[56] *duct[am] in sinum anguli fluentis. Insuper fluxio sinus alterutrius anguli fluentis ducta in quadratum lateris dato angulo oppositi æquatur fluxioni lateris angulo isti fluenti oppositi ductæ in factum sinus anguli dati & istius segmenti lateris oppositi quod angulo fluenti conterminum est, & fluxioni lateris [inter] angulū datum et fluentem jacentis ductæ in factum sinus anguli fluentis et istius segmenti lateris hujus quod angulo fluenti conterminum est. Deniɢ fluxiones cosinuum angulorum fluentium sunt ad invicem ut latera opposita. Eadem valent*[57]

---

(55) This replaces 'non datis'. Observe that Newton here assumes it to be known that the altitudes of a triangle are concurrent, evidently not yet having had second thoughts on the matter; compare §1: notes (87) and (90).

(56) An obvious slip of Newton's pen for 'angulo dato opposti'; see the next note.

(57) The text breaks off in mid-sentence but we should understand something like 'Eadem valent in triangulis obtusangulis si modò mutentur signa cosinus anguli obtusi ut et conterminorum segmentorum laterum eum continentium, quippe quæ sunt cosinus anguli istius respectu alternorum laterum ut radiorum'. (Compare the scholium to Proposition 25 in the main text of the 'Geometria' above.) On taking the angle $B\hat{A}C$ in the diagram of Appendix 1.6 following to be the 'angulus datus' and correcting an evident slip (see note (56)) Newton's four results are, in sequence,

$$BC \times \text{fl}(BC) = BG \times \text{fl}(AB) + FC \times \text{fl}(AC),$$

$$BC \times \text{fl}(\sin B) = \sin A \times \text{fl}(AC) - \sin B \times \text{fl}(BC),$$

$$BC^2 \times \text{fl}(\sin B) = BE.\sin A \times \text{fl}(BC) - BG.\sin B \times \text{fl}(AB),$$

and
$$AB \times \text{fl}(\cos B) = [-]AC \times \text{fl}(\cos C).$$

The first follows at once as the fluxional form of $BC^2 = AB^2 + AC^2 - 2AB.AC.\cos A$ on putting $BG = AB - AC.\cos A$ and $FC = AC - AB.\cos A$; on putting $\sin B = GC/BC$ the second is immediate from evaluating $BC \times \text{fl}(GC/BC) = \text{fl}(GC) - (GC/BC) \times \text{fl}(BC)$ since

$$\text{fl}(GC) = \sin A \times \text{fl}(AC);$$

but the third needs emending to

$$BC^2 \times \text{fl}(\sin B) = \sin A(BC \times \text{fl}(AC) - AC \times \text{fl}(BC)$$
$$= (\sin A/\cos C)(BE \times \text{fl}(BC) - BG \times \text{fl}(AB)).$$

The fourth proportion results from the equality $\hat{B} + \hat{C} = \pi - \hat{A}$, constant, whence

$$\text{fl}(\hat{B}) = -\text{fl}(\hat{C})$$

and so $AB \times \text{fl}(\cos B) + AC \times \text{fl}(\cos C) = -(AB.\sin B - AC.\sin C) \times \text{fl}(\hat{B}) = 0.$

[6][58]

[*Prop. 22*]

In triangulo *ABC* demissis ad latera omnia ⊥^lis *AE*, *BF*, *CG* Dico esse Rad. Cosin: ang: *A*::*AC*.*AG*. Ergo per prop [14] $\cos: A \times \mathrm{fl}\, AC + AC \times \mathrm{fl} \cos A = \mathrm{Rad} \times \mathrm{fl}\, AG$. Est et per prop [14] fl cos *A*. defl arc *A*::sin *A*.rad:: *CG*.*AC*. Quare $AC \times \mathrm{fl} \cos A = CG \times \mathrm{defl}\, \mathrm{arc}\, A$. et $\cos A \times \mathrm{fl}\, AC = \mathrm{Rad} \times \mathrm{fl}\, AG + CG \times \mathrm{fl}\, \mathrm{arc}\, A$. Erat autem in Prop [20]

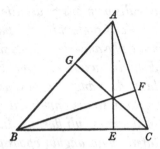

$$AC \times \mathrm{fl}\, AC - BC \times \mathrm{fl}\, BC = AB \times \mathrm{fl}\, AG - BG \times \mathrm{fl}\, AB.$$

Quare si posteriora æqualia ducta in radium auferantur de prioribus ductis in *AB* restabunt

$$\overline{AB \times \cos A - AC \times \mathrm{Rad}}\ \mathrm{in}\ \mathrm{fl}\, AC : + BC \times \mathrm{rad} \times \mathrm{fl}\, BC$$
$$= BG \times \mathrm{rad} \times \mathrm{fl}\, AB + AB \times CG \times \mathrm{fl}\, \mathrm{arc}\, A.$$

Sed est Rad. cos *A*::*AB*.*AF*. Quare $AB \times \cos A = AF \times \mathrm{Rad}$. Aufer $AC \times \mathrm{Rad}$, et derit $FC \times \mathrm{Rad}$.[59] hoc est restabit

$$BC \times \mathrm{rad} \times \mathrm{fl}\, BC - FC \times \mathrm{Rad} \times \mathrm{fl}\, AC = BG \times \mathrm{rad} \times \mathrm{fl}\, AB + AB \times CG \times \mathrm{fl}\, \mathrm{arc}\, A.$$

Deniꝗ est *BC*.*FC*::rad.cos *C* et *BC*.*BG*::rad.cos *B*. et *BC*.*CG*::*AB*.*AE*. Quare applica æqualia novissima ad *BC* et subducto utrobiꝗ cos $B \times \mathrm{fl}\, AB$, ea evadent $\mathrm{Rad} \times \mathrm{fl}\, BC - \cos C \times \mathrm{fl}\, AC - \cos B \times \mathrm{fl}\, AB = AE \times \mathrm{fl}\, \mathrm{arc}\, A$. Q.E.D.

APPENDIX 2. PRELIMINARY CALCULATIONS FOR THE REVISED SECTION 'DE PROPORTIONALIBUS'.[1]

Extracts from the original worksheet[2] in the University Library, Cambridge

[1][3] *A*.*B*::*C*.*D*. [Dato *A* erit] $Bc + Cb = Ad$.[4] [Et si præterea] $\begin{matrix} b \pm c = 0. \\ b \pm d = 0. \end{matrix}$ [erit]

$\begin{matrix} Cb \mp Bb = Ad. \\ Bc \mp Cd = Ad. \end{matrix}$ Ergo $\begin{matrix} C \mp B\,.\,A :: d\,.\,b. \\ B\,.\,A \pm C :: d\,.\,c. \end{matrix}$

---

(58) Add. 3963.7: 55ʳ, a variant first proof of Proposition 22 of the 'Geometria'; see §1: note (98).

(59) In other words 'prodibit $-FC \times \mathrm{Rad}$'.

(1) See §2.2: note (14). These variant proofs are reproduced mainly for their intrinsic interest and to illustrate Newton's ingrained habit of developing fluxional arguments on the (often unstated) basis of equivalent limit-increment proofs.

[Erit generaliter] $Ad+Da=Bc+Cb$. [adeoᛩ si]

$$d\pm a=0=b\pm c. \qquad Da\mp Aa=Bc\mp Cc. \begin{cases} D\mp A\,.\,B\mp C::c\,.\,a::b\,.\,d. \\ \text{vel } D\mp A\,.\,C\mp B::b\,.\,a::c\,.\,d. \end{cases}$$

[erit]

$$d\pm c=0=b\pm a. \qquad Da\mp Ac=Bc\mp Ca. \begin{cases} D\pm C\,.\,B\pm A::c\,.\,a::d\,.\,b. \\ \text{vel } D\pm C\,.\,A\pm B::d\,.\,a::c\,.\,b. \end{cases}$$

Prop. Si quatuor proportionalium *A. B. C. D* prima datur et

$$\left.\begin{array}{l} \left.\begin{array}{l}\text{sum̄a}\\ \text{differentia}\end{array}\right\}\text{mediarum}\\ \left.\begin{array}{l}\text{summa}\\ \text{differentia}\end{array}\right\}\text{tertiæ et }4^{\text{tæ}} \end{array}\right\} \text{erit fluxio quartæ ad fluxionem secundæ}$$

$$\text{ut}\begin{cases}\begin{cases}\text{excessus tertiæ super secundam}\\ \text{sum̄a secundæ ac tertiæ}\end{cases}\text{ad primam.}\\ \text{tertia ad}\begin{cases}\text{summam primæ et secundæ.}\\ \text{excessum primæ super secundam.}\end{cases}\end{cases}$$

---

[2]

Demonstr: 1.[5] $A\,.\,B::C\,.\,D::B\,.\,N::B+C\,.\,N+D$. Ergo,
$$A\,.\,B+C::B\,.\,N+D::b\,.\,n+d.$$

Sed $A\,.\,B::B\,.\,N$ ergo $A\,.\,2B::b\,.\,n$. et $A\,.\,C-B::b\,.\,d$.

Demonstr 2.[6] $A\,.\,B::C\,.\,D$ et $A\,.\,A+B::C\,.\,C+D$. Sit N medium proportionale inter A et $C+D$ et erit $A+B\,.\,N::N\,.\,C$. Quare cum A et $C+D$ adeoᛩ medium proportionale N dentur erit $A+B\,.\,C::c\,.\,a+b::c\,.\,b$,[7] nam a nullum est.

---

[3][8]

[Posito] $A\,.\,B::C\,.\,D$ [ubi datur A, cape]

$$A\,.\,B+b::C+c\,.\,D+d::F\,.\,D::C+c-F\,.\,d$$

[et erit] Ad[9] $=Bc+Cb+bc$. [adeoᛩ] $d\,.\,b::C+c+\dfrac{Bc}{b}\,.\,A$.

---

(2) Add. 3960.5: 63, the recto side of an otherwise blank folio sheet.

(3) A first version of Propositions 8–12.

(4) Proposition 12 of the 'De proportionalibus'.

(5) Understand 'ubi A et $B+C$ dantur'. This is revised as Proposition 8 of the 'De proportionalibus'.

(6) Here, correspondingly, understand 'ubi A et $C+D$ dantur' in this preliminary version of Proposition 10.

(7) Since $(A+B)C=N^2$ and hence $(A+B)c+C(a+b)=0$, read '$\ldots::-c\,.\,a+b::-c\,.\,b$'.

(8) A new attempt at a proof of Proposition 12 of the 'De proportionalibus'.

[Vel generalius fac] $A.B+b::C+c.D+d::G.H::C+c+G.D+d+H.$ [et] $A.C+c+G::B+b.D+d+H::B.K.$ [Erit] $A.C+c+G::b.D+d+H-K.$ [Hic est] $G=\dfrac{Bc}{b}$ [ubi] $K=D+H.$

---

[4]$^{(10)}$

[Si] $AB.AC::AH.AE::AI.AF.$ [et] $AB.AD::AI.AG.$ [erit] $AB.AC::HI.EF.$ [et] $AB.AI::CD.FG.$ Ergo [dato AB] initio fluxionis [punctis $H,\ I$ ut et $C,\ D$; $E,\ G$ convenientibus prodit] $AB.AH::$ $CD.FG::\mathrm{fl}\,AC.\mathrm{fl}\,FG=\mathrm{fl}\,AG-AF^{(11)}$ et

$AB.AC::HI.EF[::]\mathrm{fl}\,AH.\mathrm{fl}\,AF.^{(12)}$

[Insuper] $AC.AB::CD.LE::LE.KH.$ ergo $AC.AH^{(13)}$

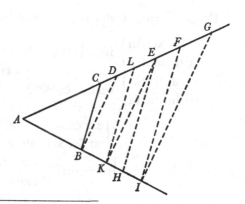

---

[5]

[Dato AB si] $AB.AC::AD.AE.$ [erit] $AB.AF::AG.AH.$ [adeoϛ]

$AB.AD(::CF.EK)::\mathrm{fl}\,AF.\mathrm{fl}\,AK.^{(14)}$

[et] $AB.AF::DG.KH::\mathrm{fl}\,AG.\mathrm{fl}:AH-AK.^{(14)}$

[ut et] $\dfrac{AD\times CF+AF\times DG}{AB}=EH.^{(15)}$

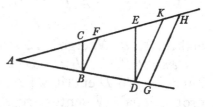

---

(9) That is, $(B+b)(C+c)-(AD$ or$)\,BC.$

(10) A first essay at a geometrical proof of the preceding. It is understood that AB is given in magnitude.

(11) Not so, since $FG = EG-(EF$ or$)\dfrac{AC}{AB}\times HI$ where, on taking (with Newton) the proportion between increment and fluxion to be unity, $EG = \mathrm{fl}(AE)$ and $HI = \mathrm{fl}(AH).$

(12) Again untrue, since (compare note (11)) $EF = (AC/AB)\times\mathrm{fl}(AH) \neq \mathrm{fl}(AF)$ in general.

(13) Newton breaks off at this point and starts afresh.

(14) These fluxional equivalents are again mistaken since, in Newton's present terms, $\mathrm{fl}(AK) = \mathrm{fl}(AE) = EH.$

(15) Abandoning the erroneous fluxional equivalents in his preceding proportions Newton combines the various limit-increments, finding

$$AD\times CF+AF\times DG = (AB\times EK+AB\times KH \text{ or})\ AB\times EH.$$

In the limit therefore as $F,\ G,\ H$ come respectively to coincide with $C,\ D,\ E$ and accordingly $CF = \mathrm{fl}(AC),\ DG = \mathrm{fl}(AD),\ EH = \mathrm{fl}(AE)$ it follows that

$$AD\times\mathrm{fl}(AC)+AC\times\mathrm{fl}(AD) = AB\times\mathrm{fl}(AE).$$

[6]

$a \cdot b \cdot c \vdots$ [dato b erit] $a + e \cdot b \cdot c + f \vdots$ [Ergo initio fluxionis]

$$a \cdot b :: e \cdot d :: b \cdot c :: h \cdot f.^{(16)}$$

[7]$^{(17)}$

[Si] $a \cdot b :: b \cdot n - a$. [erit] $f + a \cdot b + e :: b + e \cdot n - a - f$. [hoc est multiplicando in crucem et deletis æqualibus $an - aa$ & bb]

$$-2af + nf - ff = 2be + ee.$$

[sive] $f \cdot e :: 2b + e \cdot n - 2a - f$.

[In figura]$^{(18)}$

$BD \cdot FE :: BC + DF \cdot BG - AD$

$\quad\quad = GD - AB = DH + BH.$

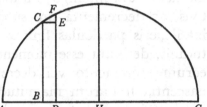

APPENDIX 3. NEWTON'S FIRST PUBLIC ANNOUNCEMENT OF FLUXIONS

[Late 1686?]$^{(1)}$

Extracted from his *Philosophiæ Naturalis Principia Mathematica*$^{(2)}$

Lemma II

Momentum Genitæ æquatur momentis Terminorum singulorum generantium in eorundem laterum indices dignitatum & coefficientia continue ductis.

---

(16) Read '$\ldots :: h \cdot -f$' strictly.

(17) A first draft of the scholium to 'De proportionalibus'.

(18) A semicircle, namely, of centre H and diameter AG, in which $AB = a \perp BC = b$, $AG = 2AH = n$, $EF = e$ and $BD = f$.

(1) Since no autograph draft of the second book of the *Principia* is now preserved, a more precise date of composition is impossible: that suggested is the period when Newton, rejecting his prior 'De Motu Corporum Liber Secundus' (posthumously published by Conduitt in 1728 as *De Mundi Systemate Liber*; see 1: xx, note (12)), composed his revised Liber II. (As he told Halley on 13 February 1686/7 (*Correspondence of Isaac Newton*, 2, 1960: 464), 'The second book I made ready for you in Autumn [last], having wrote to you...that it should come out wth ye first'.) This general fluxional lemma, manifestly based on Propositions 3–10 of the main version of his 'Geometria Curvilinea' (§1 above), occupies a fairly incongruous position in the second book of Newton's published *Principia*. It is there ostensibly introduced to justify two statements, in Propositions 8 and 9 following, that the '*momentum*' of AP^2 is $2AP \times PQ$ where PQ is the instantaneous increment of AP, but there are several places in the preceding first book (drafted in its essential structure by spring 1685) where it might have been cited to equal

Genitam voco quantitatem omnem quæ ex[3] Terminis quibuscunꝗ in Arithmetica per multiplicationem, divisionem & extractionem radicum; in Geometria per inventionem vel contentorum & laterum, vel extremarum & mediarum proportionalium absꝗ additione & subductione generatur. Ejusmodi quantitates sunt Facti, Quoti, Radices, rectangula, quadrata, cubi, latera quadrata, latera cubica & similes. Has quantitates ut indeterminatas & instabiles, & quasi motu fluxuve perpetuo crescentes vel decrescentes hic considero, & e[a]rum incrementa vel decrementa momentanea sub nomine momentorum intelligo; ita ut incrementa pro momentis additiis seu affirmativis, ac decrementa pro subductitiis seu negativis habeantur. Cave tamen intellexeris particulas finitas. Momenta, quam primum finitæ sunt magnitudinis, desinunt esse momenta. Finiri enim repugnat aliquatenus perpetuo eorum incremento vel decremento.[4] Intelligenda sunt principia jamjam nascentia finitarum magnitudinum. Neꝗ enim spectatur in hoc Lemmate magnitudo momentorum, sed prima nascentium proportio. Eodem recidit si loco momentorum usurpentur vel velocitates incrementorum ac decrementorum, (quas etiam motus, mutationes & fluxiones quantitatum nominare licet) vel finitæ quævis quantitates velocitatibus hisce proportionales. Termini autem cujusꝗ Generantis coefficiens est quantitas, quæ oritur applicando Genitam ad hunc Terminum.

—indeed better—advantage: notably in Proposition 39, where (p. 123) the Leibnizian derivative $d(V^2)/dAD$ is determined, on taking I and DE to be corresponding limit-increments of V and AD, as the limit-ratio $[(V+I)^2 - V^2]/DE = 2V \times (I/DE)$ 'si primæ quantitatum nascentium rationes sumantur'; and in Proposition 45, where (pp. 139–41) the value of the derivative $d(A^n)/dA$ is found, by setting $A = T - X$, expanding $(T-X)^n$ 'per Methodum nostram Serierum convergentium' and taking X to be infinitesimal, to be

$$[T^n - (T-X)^n]/X = nA^{n-1}$$

'ob factas R, T æquales, atꝗ X in infinitum diminutam'. When the restrained *scholium* which Newton added to the published lemma (see note (11) below) is taken into account, it is hard to resist the impression that Newton's overriding interest in publicizing this revise of the opening theorems of his 'Geometria' was less to establish a rigorous foundation for the limit-increment arguments which he widely employed in his treatise than to assert—in riposte to Leibniz' recently printed account of algorithmic calculus, 'Nova Methodus pro maximis & minimis, itemque tangentibus, quæ nec fractas, nec irrationales quantitates moratur, & singulare pro illis calculi genus' (*Acta Eruditorum* (October 1684): 467–73)—his own unchallengeable priority in its discovery.

(2) *Philosophiæ Naturalis Principia Mathematica Autore Is. Newton* (London, ₁1687): Liber II ['De Motu Corporum Liber Secundus']: 250–3. The only preliminary manuscript version now extant, Humphrey Newton's press copy lightly corrected in Newton's own hand (now Royal Society MS. LXIX), is essentially identical with the printed text here reproduced.

(3) In the second edition of his *Principia* (Cambridge, 1713) Newton here added 'lateribus vel'. In the last line of the present paragraph, correspondingly, 'Termini' was replaced by 'Lateris'.

Igitur sensus Lemmatis est, ut si quantitatum quarumcunq perpetuo motu crescentium vel decrescentium A, B, C, &c momenta, vel mutationum velocitates dicantur a, b, c, &c,[5] momentum vel mutatio rectanguli AB fuerit Ab+aB, & contenti ABC momentum fuerit ABc+AbC+aBC: & dignitatum A^2, A^3, A^4, $A^{\frac{1}{2}}$, $A^{\frac{3}{2}}$, $A^{\frac{1}{3}}$, $\frac{1}{A}$, $\frac{1}{A^2}$, & $\frac{1}{A^{\frac{1}{2}}}$ momenta $2Aa$, $3aA^2$, $4aA^3$, $\frac{1}{2}aA^{-\frac{1}{2}}$, $\frac{3}{2}aA^{\frac{1}{2}}$, $\frac{1}{3}aA^{-\frac{2}{3}}$, $\frac{2}{3}aA^{-\frac{1}{3}}$, $-aA^{-2}$, $-2aA^{-3}$, & $-\frac{1}{2}aA^{-\frac{3}{2}}$ respective. Et generaliter ut dignitatis cujuscunq $A^{\frac{n}{m}}$ momentum fuerit $\frac{n}{m}A^{\frac{n-m}{m}}$. Item ut Genitæ $A^{quad} \times B$ momentum fuerit $2aAB+A^2b$; & Genitæ $A^3B^4C^2$ momentum

$$3aA^2B^4C^2+4A^3bB^3C^2+2A^3B^4Cc;$$

& Genitæ $\frac{A^3}{B^2}$ sive A^3B^{-2} momentum $3aA^2B^{-2}-2A^3bB^{-3}$: & sic in cæteris. Demonstratur vero Lemma in hunc modum.

Cas. 1. Rectangulum quodvis motu perpetuo auctum AB, ubi de lateribus A & B deerant momentorum dimidia $\frac{1}{2}a$ & $\frac{1}{2}b$, fuit $A-\frac{1}{2}a$ in $B-\frac{1}{2}b$,[6] seu $AB-\frac{1}{2}aB-\frac{1}{2}Ab+\frac{1}{4}ab$; & quam primum latera A & B alteris momentorum dimidiis aucta sunt, evadit $A+\frac{1}{2}a$ in $B+\frac{1}{2}b$ seu $AB+\frac{1}{2}aB+\frac{1}{2}Ab+\frac{1}{4}ab$. De hoc rectangulo subducatur rectangulum prius, & manebit excessus $aB+Ab$. Igitur laterum incrementis totis a & b generatur rectanguli incrementum $aB+Ab$.[7] Q.E.D.

---

(4) 'For to be bounded is to some extent in contradiction to their perpetual increment or decrement'—a curious phrase (due to Halley) replaced, along with the preceding sentence, by 'Particulæ finitæ non sunt momenta, sed quantitates ipsæ ex momentis genitæ' in 1713.

(5) Newton takes up the notation he had earlier introduced in Proposition 3 of the first version of his 'Geometria' (see §1: note (31) above).

(6) A celebrated *non-sequitur*; compare D. T. Whiteside, 'Patterns of Mathematical Thought in the later Seventeenth Century' [*Archive for History of Exact Sciences*, 1, 1961: 179–388]: 373: note **. Newton here has clearly deluded himself into believing he has contrived an approach which avoids the comparatively messy appeal to the limit-value of $(A+a)(B+b)-AB$ as the increments a, b vanish that he had earlier made in his 1671 fluxions addendum (see 'Th. 1' on III: 330–4, especially 330, note (9)). This eminently plausible circumvention is, in fact, irreparably a false trail: for, if o is the increment of the base variable x and $A \equiv A(x)$, then $\frac{1}{2}a = A(x+o)-A(x) = o. A' + \frac{1}{2}o^2.A'' + \frac{1}{6}o^3.A''' \ldots$ and so, when $A'' \neq 0$,

$$A-\tfrac{1}{2}a(= A-o.A'-\tfrac{1}{2}o^2.A'' \ldots) \neq A(x-o),$$

as Newton requires. Similarly, if $B \equiv B(x)$ and $\frac{1}{2}b = B(x+o)-B(x)$, then

$$B-\tfrac{1}{2}b(= 2B(x)-B(x+o)) \neq B(x-o) \quad \text{when} \quad B'' \neq 0.$$

(7) It is worth observing that even when $A'' = B'' = 0$, and so $A \equiv ax+b$, $B \equiv cx+d$, Newton's result is more subtle than it appears. For if $\alpha\beta = x$ and $\beta C = y$ are the abscissa

Cas. 2. Ponatur AB æquale G, & contenti ABC seu GC momentum (per Cas. 1) erit $gG + Gc$, id est (si pro G & g scribantur AB & $aB + Ab$)

$$aBC + AbC + ABc.$$

Et par est ratio contenti sub lateribus quotcunqʒ. Q.E.D.

Cas. 3. Ponantur A, B, C æqualia; & ipsius A^2, id est rectanguli AB, momentum $aB + Ab$ erit $2aA$, ipsius autem A^3, id est contenti ABC, momentum

$$aBC + AbC + ABc$$

erit $3aA^2$. Et eodem argumento momentum dignitatis cujuscunqʒ A^n est naA^{n-1}. Q.E.D.

Cas. 4. Unde cum $\frac{1}{A}$ in A sit 1, momentum ipsius $\frac{1}{A}$ ductum in A, una cum $\frac{1}{A}$ ducto in a erit momentum ipsius 1, id est nihil. Proinde momentum ipsius $\frac{1}{A}$ seu A^{-1} est $\frac{-a}{A^2}$. Et generaliter cum $\frac{1}{A^n}$ in A^n sit 1, momentum ipsius $\frac{1}{A^n}$ ductum in A^n una cum $\frac{1}{A^n}$ in $naA^{n-1\,(8)}$ erit nihil. Et propterea momentum ipsius $\frac{1}{A^n}$ seu A^{-n} erit $-\frac{na}{A^{n+1}}$. Q.E.D.

---

and ordinate of the parabola $y = (ax+b)(cx+d) \equiv A.B$ and $b\beta = \beta b' = o$, then the ratio of the 'moment'

$$A.b + B.a = 2o.[(ax+b)c + a(cx+d)]$$

of y to the corresponding 'moment' $bb' = 2o$ of x, or the slope of the parabola at C, is determined by Newton as that,

$$[(a(x+o)+b)(c(x+o)+d) - (a(x-o)+b)(c(x-o)+d)]/2o,$$

of the parallel chord cc'.

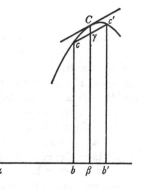

(8) The 'momentum ipsius A^n', that is.

(9) The 1687 text (duly corrected in 1713) here misprints '$2aA^{-\frac{1}{2}}$', 'æqualem' and '$maA^{m-1} + nbB^{n-1}$' respectively.

(10) See Proposition 8 of the 'Geometria Curvilinea' (§1).

(11) In an immediately following scholium (*Principia* (note (2) above): 253–4 Newton appended the autobiographical remark: 'In literis [*sc.* his *epistola posterior* to Oldenburg for Leibniz of 24 October 1676 (= *Correspondence*, **2**: 110–29, especially 129); see III: 25, note (24) above] quæ mihi cum Geometra peritissimo *G. G. Leibnitio* annis abhinc decem intercedebant, cum significarem me compotem esse methodi determinandi Maximas & Minimas, ducendi Tangentes, & similia peragendi, quæ in terminis surdis æque ac in rationalibus procederet & literis transpositis hanc sententiam involventibus *Data æquatione quotcunqʒ fluentes quantitates involvente, fluxiones invenire, & vice versa* eandem celarem: rescripsit Vir Clarissimus [in Leibniz to Oldenburg for Newton, 11 June 1677 (= *Correspondence*, **2**: 212–19, especially 213–15)] se quoque in ejusmodi methodum incidisse, & methodum suam communicavit a mea vix abludentem præterquam in verborum & notarum formulis. Utriusqʒ fundamentum continetur in hoc Lemmate.' Compare Newton's more extensive observations, hitherto unpublished, on

Cas. 5. Et cum $A^{\frac{1}{2}}$ in $A^{\frac{1}{2}}$ sit A, momentum ipsius $A^{\frac{1}{2}}$ in $2A^{\frac{1}{2}}$ erit a, per Cas. 3: ideoꝗ momentum ipsius $A^{\frac{1}{2}}$ erit $\dfrac{a}{2A^{\frac{1}{2}}}$ sive $[\frac{1}{2}]aA^{-\frac{1}{2}}$.[9] Et generaliter si ponatur $A^{\frac{m}{n}}$ æquale[9] B, erit A^m æquale B^n, ideoꝗ maA^{m-1} æquale nbB^{n-1}, & maA^{-1} æquale nbB^{-1} seu $\dfrac{nb}{A^{\frac{m}{n}}}$, adeoꝗ $\dfrac{m}{n}aA^{\frac{m-n}{n}}$ æquale b, id est æquale momento ipsius $A^{\frac{m}{n}}$. Q.E.D.

Cas. 6. Igitur Genitæ cujuscunꝗ A^mB^n momentum est momentum ipsius A^m ductum in B^n, una cum momento ipsius B^n ducto in A^m, id est

$$maA^{m-1}[B^n]+nbB^{n-1}[A^m]\,;^{[9]}$$

idꝗ sive dignitatum indices m & n sint integri numeri vel fracti, sive affirmativi vel negativi. Et par est ratio contenti sub pluribus dignitatibus. Q.E.D.

Corol. 1. Hinc in continue proportionalibus, si terminus unus datur, momenta terminorum reliquorum erunt ut iidem termini multiplicati per numerum intervallorum inter ipsos & terminum datum. Sunto A, B, C, D, E, F continue proportionales: & si detur terminus C, momenta reliquorum terminorum erunt inter se ut $-2A$, $-B$, D, $2E$, $3F$.[10]

Corol. 2. Et si in quatuor proportionalibus duæ mediæ dentur, momenta extremarum erunt ut eædem extremæ. Idem intelligendum est de lateribus rectanguli cujuscunꝗ dati.

Corol. 3. Et si summa vel differentia duorum quadratorum detur, momenta laterum erunt reciproce ut latera.[11]

---

his exchange of letters with Leibniz during 1676–7, which are reproduced in the immediately following 'Matheseos Universalis Specimina' (2, §1: passim).

Some time after the appearance of the second edition of his *Principia*, in (or soon after) 1714 probably, Newton introduced into his personal interleaved copy of the 1713 printing (now ULC. Adv. b. 39.2) an intended postscript, never published in any subsequent edition, with the instruction that it be added 'To the Scholium in *Princip. Philos.* [Cambridge, $_2$1713] pag. 227 after the words *continetur in hoc Lemmate*', namely: 'Sunto quantitates datæ a, b, c; fluentes x, y, z; fluxiones p, q, r; & momenta op, oq, or: et proponatur æquatio quævis flu[entes] involvens $x^3-2xxy+bxx-bbx+byy-y^3+czz=0$. Et per hoc Lemma si sola fluat x, momentum totius erit $3xxop-4xopy+2bxop-bbop$; si sola fluat y momentum totius erit

$$-2xxoq+2byoq-3yyoq[;]$$

si sola fluat z, momentum totius erit $+2czor$; si fluant omnes momentum totius erit

$$3xxop-4xopy+2bxop-bbop-2xxoq+2byoq-3yyoq+2czor.$$

Et quoniam totum semper æquale est nihilo erit momentum ipsius æquale nihilo. Dividatur totum per momentum o, et prodibit æquatio quæ ex fluentibus dat fluxiones, vizt

$$3xxp-4xpy+2bxp-bbp-2xxq+2byq-3yyq+2crz=0.$$

Exhibet igitur Lemma solutionem Propositionis hujus, *Data æquatione fluentes quotcunꝗ quantitates involvente fluxiones invenire.* Dixi autem anno 1676, in Literis præfatis [see [G_2] on III: 25 and

2

'SPECIMENS OF A UNIVERSAL SYSTEM OF MATHEMATICS'

[*c.* late June/July 1684][1]

From autograph drafts in the University Library, Cambridge and in private possession

§1. THE FIRST VERSION OF THE 'SPECIMINA'.[2]

[1]   MATHESEOS[3] UNIVERSALIS
                          SPECIMINA.

Methodum quandam resolvendi problemata per series convergentes quam ante annos octodecim[4] excogitaveram, per ea ferè tempora significarat D. Joannes Collinsius amicus meus integerrimus[5] D. Jacobo Gregorio Matheseos in Academia Edinburgensi apud Scotos Professori celeberrimo, penes me esse. Eandem D. Gregorius pro ingenio quo pollebat, ex unica ejusmodi serie ad

---

[G_5] on III: 26], hanc Propositionem esse fundamentum methodi generalis de qua scripseram tunc ante annos quinq̃, seu anno 1671. In ijs quæ tunc scripsi explicui solutionem Analytice, in hoc Lemmate demonstravi eandem synthetice'. (Except for the added term '$+czz$' Newton has drawn his example from his letter to Collins on 10 December 1672; see *Correspondence*, 1, 1959: 247.)

(1) The two unfinished papers on series reproduced in this final section, the 'Matheseos Universalis Specimina' and its revise 'De Computo Serierum', were evidently the main by-product of Newton's study in late June 1684 of David Gregory's *Exercitatio Geometrica de Dimensione Figurarum, sive Specimen Methodi Generalis Dimetiendi Quasvis Figuras*, fresh from its Edinburgh press. To summarize the details given in the preceding introduction, in his letter (ULC. Add. 3980.1, printed in *The Correspondence of Isaac Newton*, 2, 1960: 396) accompanying the no longer traceable copy of his book which Gregory sent Newton on 9 June David had both stressed his overriding debt to the towering genius of his deceased uncle, James, and yet not been afraid to make a modest claim to have discussed in his work 'things new to the greatest parte of the geometers': in his *Exercitatio*, indeed, Gregory solidly based his summary of contemporary techniques of quadrature and rectification on passages quoted from James Gregory's *Geometriæ Pars Universalis* and *Exercitationes Geometricæ* of 1668, but later went on to discuss approximate methods of integration by expansion into infinite series in a manner which—almost certainly in ignorance of Newton's prior researches—virtually duplicated the developments *dividendo, radicem* [*quadraticam*] *extrahendo* and *per resolutionem æquationum affectarum* which the latter had fifteen years before made fundamental in his 'De Analysi' (see II: 206–46, especially 212–32). Whether or not Newton replied in person to Gregory's polite request to be allowed his 'free thoughts and character of this exercitation' we do not know, but it will be

Translation

[1]   SPECIMENS OF A UNIVERSAL
               [SYSTEM OF] MATHEMATICS.[3]

A certain method of resolving problems by convergent series devised by me about eighteen years ago[4] had, by my very honest friend[5] Mr John Collins, around that time been announced to Mr James Gregory, the renowned Professor of Mathematics at Edinburgh University in Scotland, as being in my possession. This method Mr Gregory learnt—a measure of the power of his intellect—from but a single series of this type which had been passed on to him, but not much

---

clear that he was speedily encouraged to review contemporary advances in the method of infinite series on a broad front—not only those propounded by the Gregorys, uncle and nephew, but those promoted in his still unpublished 1671 fluxional tract and above all those announced and discussed in the extensive correspondence centred on this topic which he had, by way of Oldenburg, conducted with Leibniz during 1676–7. In a way which was to become familiar not quite thirty years later Newton's initial instinct to restrict himself to technical criticism of Gregory's treatise rapidly became subordinated to a dominant urge to establish in the public arena his own absolute priority of discovery over all 'second' inventors, James and David Gregory and Leibniz alike. As it happened, the present annotated 'Commercium Epistolicum' was never completed, nor did either of its versions reach public knowledge till the present day. We may readily guess that, when Edmond Halley paid Newton his famous visit at Cambridge in August 1684 (or so well-based tradition has it), Newton's attention was effectively distracted from this somewhat sterile obligation to claim his rights of first discovery to the demanding, infinitely more rewarding task of writing his *Principia*. The composition date of late June or July 1684 is amply confirmed by the internal evidence of the following text; compare notes (4) and (132) below.

(2) The autograph originals of the introduction to this piece (portions of which were published in 1960 by H. W. Turnbull in his edition of *The Correspondence of Isaac Newton*, **2**: 400–1) and of its first chapter exist in private possession; preliminary drafts—partially published in Appendix 2—and the main text of the remaining chapters are now to be found in ULC. Add. 3964.3: 7ʳ–20ᵛ. It would seem that the division was made during Newton's lifetime, probably during 1709–10 when William Jones (into whose possession the introduction and opening chapter later passed) actively contemplated a full edition of Newton's unpublished mathematical papers (compare ɪ: xvii).

(3) 'ANALYSEOS' (ANALYSIS) is cancelled. A first title for the piece is heavily cancelled, but appears to read 'MATHESEOS UNIVERSALIS SPECIMINA ANALYTICA' (ANALYTICAL SPECIMENS OF A UNIVERSAL MATHEMATICS). It has not been possible to confirm the reading by examining the manuscript under ultra-violet light.

(4) That is, in 1665–6. Newton here seems to refer to his discovery of the general binomial expansion early in 1665 (see ɪ: 104–11, 122–34) and perhaps the 'mechanichal' extraction of the root of an algebraic equation as an infinite series according to 'Vieta's Analyticall resolution of powers' adumbrated in his October 1666 tract (ɪ: 413–14) but not, of course, developed at length till he came to write his 1671 treatise.

(5) This replaces the more fulsome phrase 'rerum Mathematicarum cultor eximius' (that outstanding cultivator of things mathematical). Newton knew very well the moderate limits of Collins' mathematical talents and not even the latter's recent death (in November 1683) would provoke him more than momentarily to an undeserved epithet.

eum transmissa didicit nec multò post morte præmatura sublatus est.[6] Ex ejus chartis ubi computa quædam absᴄᶾ methodi delineatione extabant, didicit etiam clarissimus ejus in Professione Math. successor David Gregorius ipsam computandi methodum eamᴄᶾ t[r]actatu concinno et ingenioso complexus est, quo non solum solertiam suam in rebus Mathematicis sed etiam probitatem summam ostendit, candide agnoscens quæ ipse ab Antecessore suo, Antecessor a Collinsio acceperat.[7] Ista dum legerem[8] cogitare statim cœpi quod cum sæpe rogatus essem ut aliquid in lucem ederem, importunitati amicorum et alioru̅[9] expectationi jam minus resistere liceret et consultiùs foret citò morem gerere quam serò cum molestia tempore minus opportuno vinci. Intercesserant quondam inter me et D. Leibnitium jam celebrem Ducis Hannoveræ in negotijs publicis ministrum epistolæ quædam de his seriebus procurante[10] Domino H.O. Et his edendis Lectori certè magis gratificabor quam si rem omnem post D. Gregorium de novo componerem: præsertim cùm in his contineatur perelegans Leibnitij methodus a mea longè diversa perveniendi ad easdem series, quam nec honestum foret reticere dum meam in lucem edo; mea verò sic describatur ut Liber Cl. Gregorij fusiore explicatione eorum quæ hic carptim attinguntur &

---

(6) James Gregory died in late October 1675, when he was not quite 37 years old and still in the prime of life, within a few hours of having had a severe stroke accompanied by blindness and paralysis. Newton's source of information is David Gregory who wrote in his *Exercitatio Geometrica* (note (1) above): that 'Circa initium anni 1670 [in fact, on 24 March 1669/70], vir Doctissimus & ad Geometriam promovendam natus D. *Joannes Collinseus*, literis ei [*sc. Jac. Gregorio*] significavit Clarissimum virum D. *Isaacum Newtonum* methodi quadraturarum generalis compotem esse, quæ *Mercatoris* Hyperbolæ quadraturæ [as presented in his *Logarithmotechnia* of 1668; see II: 166] analoga esset: videlicet si detur radius *R*, & latitudo Zonæ *B*, erit

$$\text{Zona} \left[\int_{-B}^{B} \sqrt{[R^2 - x^2]} \, . \, dx \right] = 2RB - \frac{B^3}{3R} - \frac{B^5}{20R^3} - \frac{B^7}{56R^5} - \frac{5B^9}{576R^7} - \&\text{c.}$$

[see III: 567–8] quam suis congruere, ac ex priùs communicatis de inveniendo numero ex dato Logarithmo; vel radicem potestatis cujusvis puræ in infinitam seriem permutando, apertè consequi, eidem D. *Collinseo* postea circa mensem *Decembris* insequentem [on the 19th] rescripsit. Interea harum serierum infinitarum doctrinam variis problematis ex occasione applicuit, Geometriam illius ope multò auctiorem factam esse asserens, cum multis de causis, tum hâc etiam, quòd in quâvis æquatione literali, facile esset seriem infinitam ejusdem radici accommodatam exhibere, et quod unica cuivis gradui sufficeret; quodque harum tabulæ radicibus expeditè inveniendis admodum conducerent, ut ineunte anno 1672 [on 17 January] *Collinseum* literis monuit. Hisce cogitationibus detentum mors præmatura eripuit'. The question of James Gregory's debt to Newton in the realm of infinite series during 1670–1 is briefly discussed—with reference to passages in the letters of Collins and Gregory here referred to—in the preceding introduction, but we may here reiterate the point that Newton continued to believe till the end of his life—inaccurately we now know—that Gregory discovered the general method of convergent (sum) series by equating the present series for the circle zone to its exact value $\int_{-B}^{B} \sqrt{[R^2 - x^2]} \, . \, dx$ and hence, by differentiation, deriving the series expansion of the binomial square root. See Introduction: note (24).

aliorum afterwards he was snatched away by an untimely death.[6] From his papers, extant in which were certain calculations though without a description of the method, his celebrated successor in the Mathematical Chair, David Gregory, also learnt this method of calculation and developed it in a neat and stimulating tract: in this he revealed not only his expertise in mathematical topics but also the utmost probity, candidly acknowledging what he himself had taken from his predecessor and what his predecessor had received from Collins.[7] While still reading it I began at once to reflect that, since I had often been asked to publish something, I had now less excuse to resist the entreaties of my friends and thwart the expectation of others,[9] and that I were better advised to be swiftly acquiescent rather than have to submit with annoyance at a later, less opportune time. Through the agency[10] of Mr H.O. certain letters regarding series of this sort once passed between me and Mr Leibniz, now the Duke of Hanover's eminent minister in public affairs. By publishing these I shall certainly oblige the reader more than if (after what Mr Gregory has done) I should write up the whole subject afresh, and especially so since in them is contained Leibniz' extremely elegant method, far different from mine, of attaining the same series—one about which it would be dishonest to remain silent while publishing my own. Mine shall, indeed, be described in such a way that Mr Gregory's book, with its more lavish explanation of points which are here touched upon piecemeal and omission of others which are here more

---

(7) Compare David Gregory's *Exercitatio Geometrica* (note (1) above): 4: '...ejus adversaria (saltem quæ in manus nostras delapsa) præter methodi hujus exempla quædam, ipsâ tamen methodo et operandi formâ destituta, nihil quod ad hanc rem spectaret continebant. Invenienda itaque erat methodus istis seriebus producendis apta, quam etiam illustravimus elegantioribus aliquot exemplis figurarum, quarum apud Geometras frequentior et celebrior est contemplatio, quamque ingenuis candidorum Geometrarum et benevolis studiis hic exponimus.' An overwhelming number of the examples to which David applies his rediscovered Gregorian method in the body of his book come in fact (though this is not always made clear in the case of still unpublished *adversaria*) from the papers and printed works of his uncle James, though he at one point refers to Nicolaus Mercator's *Logarithmotechnia* for the series expansion of $\log(1+x)$ and twice cites René-François de Sluse's *Miscellanea* (Liège, 1668), once as source for the general Slusian 'pearl', once as inventor of the Slusian conchoid.

(8) The subjunctive mood is manifestly a slip on Newton's part: the act of reading is coterminous with the reaction it inspired.

(9) In the present mathematical context Newton would probably rank Barrow and Collins (now both dead) as friends, Oldenburg (also deceased seven years before) and John Wallis as 'other' acquaintances!

(10) 'mediante' (through the mediation) is cancelled. 'H.O.' is, of course, 'H[enrico] O[ldenburgo]' (Henry Oldenburg). The designation of Leibniz—after the spring of 1676 Librarian (and unofficial legal adviser) to the Duke of Hanover—as his 'minister in negotijs publicis' (minister of state?) probably points to Newton's lack of awareness of contemporary political realities at this period.

omissione quæ hic fusius describuntur vicem introductionis commode suppleat. Sunt autem epistolæ hujusmodi.[11]

Cl. viro H.O. Newtonus Junij 13. 1676.

Quanquam D. Leibnitij modestia &c.[12]

Cl. viro H.O. Gothofredus Gulielmus Leibnitius 27 Aug 1676.

Literæ tuæ (die 26 Julij datæ) plura...meditationum plenus.—Quod dicere videmini plerascȝ...nunc satis.[13]

---

(11) In sequel Newton indicates extracts relating to infinite series from the five letters which passed (by way of Oldenburg) between himself and Leibniz during the period June 1676 to July 1677. We relate these brief citations of terminal phrases (quoted by Newton, in the case of the three Leibniz letters, from copies furnished him by Oldenburg and Collins) to the reproductions of the originals as sent, given by H. W. Turnbull in his edition of *The Correspondence of Isaac Newton*, **2**, 1960. While rightly choosing not to tamper with the text of his *epistola prior* of 13 June 1676—nor, of course, with Leibniz' words—Newton has slightly revised and clarified passages in his *epistola posterior* of the following 24 October. To show the full force of these changes we have incorporated them in the revised version of the letter reproduced in Appendix 1 below. Some more general observations on this Newton–Leibniz correspondence are made in the concluding appendix to this volume (see pages 666–74).

(12) *Correspondence*, **2**: 20 [opening line of the letter]. Newton indicates no close to the quotation but it seems that he intended to cite the whole letter as passed on to Leibniz (as far, that is, as *ibid.* **2**: 31, l. 20), neglecting the concluding note meant for Oldenburg's eyes alone. The content of this *epistola prior* of Newton's to Leibniz (whose original (now ULC. Add. 3977.2) was returned by Oldenburg to him after he passed on its 'apographum' to Leibniz on 26 July) is discussed at some length in the concluding appendix below (pages 666–70). To be very brief, Newton here, in response to Leibniz' request of 2 May to Oldenburg for information on this point, outlined his general method of series—introducing an improved exposition of his 1665 binomial expansion—and then, in illustration, gave some dozen or more particular logarithmic and trigonometrical series.

(13) *Correspondence*, **2**: 57[sixth line of letter]—62[end of first paragraph]/64, ll. 4–16. (Leibniz' original letter—in Oldenburg's possession at his death in 1677—was some time after acquired by Hans Sloane and is now British Museum. Add. 4294: 67–71. Newton's present quotation from the letter is evidently made from a transcript of Collins', now ULC. Add. 3971.1: 34^r–36^v: 'Leibnitz to Mr Oldenburg 27 Augst 1676', sent to Cambridge on 9 September 1676.) Leibniz opened his letter with the remark that 'Neutoni...methodus inveniendi radices æquationum et areas figurarum per series infinitas, prorsus differt à mea; ut mirari libeat diversitatem itinerum per quæ eodem pertingere licet.' Though he had nothing equivalent to Newton's method of binomial root extraction 'per seriem infinitam' to offer, Leibniz described two simple types of integral transform by which a surd integrand might be reduced to one involving a simple algebraic fraction and hence, by a simple division 'more Mercatoris', to a geometrical progression whose terms can then be integrated separately one at a time. According to his first approach the area $\int_0^x y \, . \, dx$ under a given locus $D(x, y)$ is reduced to its

copiously described, may profitably fill the rôle of an introduction. The letters, however, are to this effect.[11]

Newton to the worthy H.O., 13 June 1676.
'Though the modesty of Mr Leibniz...'.[12]

Gottfried Wilhelm Leibniz to the worthy H.O., 27 August 1676.
'Your letter (dated 26 July) [contains] more...abounding in [splendid] ideas.'
'What you seem to say, that most...enough [of this] for the present.'[13]

---

equal $-\int_{x=0}^{z} v \cdot dz$ by a Barrovian transform $z = ry/x$ (whence $v = -r^{-1}xz(dx/dz)$): in Leibniz' example of the circle quadrant $y^2 = 2rx - x^2$, $0 \leqslant x, y \leqslant r$, the area

$$\int_0^x \sqrt{[2rx - x^2]} \cdot dx = -\int_\infty^z 8r^5 z^2/(r^2 + z^2)^3 \cdot dz$$

on computing $dx/dz = -4r^3 z/(r^2 + z^2)^2$, $y = 2r^2 z/(r^2 + z^2)$, but the resulting series is not specified nor is its convergence discussed—unfortunately so since the whole quadrant $\frac{1}{4}\pi r^2$ cannot be evaluated directly by this means. (The derivative dx/dz is calculated from first principles as $\lim_{\beta \to 0} \left[\frac{1}{\beta} \left(\frac{2r^3}{r^2 + z^2} - \frac{2r^3}{r^2 + (z-\beta)^2} \right) \right]$.) Leibniz' second general method is to evaluate the general sector $\frac{1}{2}\int_0^x [y \cdot dx - x \cdot dy]$ under $D(x, y)$ as $\frac{1}{2}\left[t \cdot f(t) - \int f(t) \cdot dt \right]_{x=0}^t$ by introducing the Robervallian transform $x = f(t)$, $t = y - x(dy/dx)$: in his loosely sketched example of the general conic $y^2 = rx \mp (r/q)x^2$, at once $t = \sqrt{[\frac{1}{4}qrx/(q \mp x)]}$ and so

$$x = qt^2/(\tfrac{1}{4}qr \pm t^2), \quad y = \tfrac{1}{2}rx/t = \tfrac{1}{2}qrt/(\tfrac{1}{4}qr \pm t^2),$$

whence on adding the triangle $\frac{1}{4}qy$ to the general sector the central sector of the conic (cut off between the main axis and the central radius through D) is

$$\tfrac{1}{2}\left(qt \mp \int_0^t qt^2/(\tfrac{1}{4}qr \pm t^2) \cdot dt \right) = \tfrac{1}{2}q \int_0^t 1/(1 \pm (4/qr)t^2) \cdot dt.$$

In the particular case of the circle $y^2 = 2x - x^2$ ($q = r = 2$) this yields, as Leibniz immediately goes on to assert, the Gregorian series $\tan^{-1} t = \int_0^t 1/(1 + t^2) \cdot dt = t - \tfrac{1}{3}t^3 + \tfrac{1}{5}t^5 - \dots$. (On the analogy of the 'area circuli' $\frac{1}{4}\pi = 1 - \frac{1}{3} + \frac{1}{5} - \frac{1}{7} \dots$ he somewhat irrelevantly digressed to discuss a 'Harmony' in which infinite series of simple patterns of unit-reciprocals compound to make up numerical fractions and the complementary hyperbolic area $\frac{1}{2}\log 2$. Newton was singularly unimpressed, adding to a neat exposition of Leibniz's 'Harmonia' in his review of his own *Commercium Epistolicum D. Johannis Collins et Aliorum de Analysi promota* (London, 1712)— where (pp. 58–65) Leibniz' letter was republished from Wallis' *Opera Mathematica*, **3** (Oxford, 1699): 629–33 with Newtonian footnotes—the ironic plaudit 'See the Mystery!' (*Philosophical Transactions*, **29**, No. 342 [for January/February 1714/15]: 183).) In the concluding portion of the present extract Leibniz propounded the series expansions for $e^l - 1$, $\cos a$ and $\sin a$ 'ex seriebus regressuum'. At this point in his *Commercium Epistolicum* in 1712 Newton added (p. 62) a bitter footnote that 'Methodum perveniendi ad has Series *Leibnitius a Newtono* jam modo

Cl. viro H.O Newtonus Octob 24 1676.[14]

Quanta cum voluptate... (p. 2) in quadraticis radicibus hæc erat.[15] Quibus perspectis et inventa subinde resolutione affectarum æquationum qua regredi licebat ab areis aut arcubus ad ordinatim applicatas, neglexi penitus interpolationem serierum et has operationes tanquam fundamenta magis genuina in posterum colui.[16]

(f. 3.1.) Incidi postea in aliorum generum series infinitas quæ in finitas degenerant quoties cu[r]vilinea figura Geometricè quadrari potest. Harum unam in rei specimen[17] subjungere suffecerit.

Ad curvam aliquam....*Parabolam*[18] *describere quæ per data quotcunq puncta transibit.* (f. 4.1.)[19]

Seriei $\frac{t}{1} \pm \frac{t^3}{3} + \frac{t^5}{5} \pm \frac{t^7}{7}$ &c quam D. Leibnitius ad quadraturas con. sect. usurpat,[20] affinis est hæc, $1 + \frac{1}{3} - \frac{1}{5} - \frac{1}{7} + \frac{1}{9} + \frac{1}{11} - \frac{1}{13} - \frac{1}{15}$ &c longitudinē arcus quadrantalis exprimens cujus chorda est unitas. Differentia solum in signis est ubi t est unitas at inventio diversa. Illa longitudinem arcus quadrantalis ex

---

acceperat, idque ex ipsius rogatu', understanding that Leibniz had derived these series from the expansions of $l = \log(1+x)$ and $a = \sin^{-1}y$ by his own method of series inversion. From still unpublished notes at Hanover in Leibniz' autograph draft of the letter (Niedersächsische Landesbibliothek 35, II, 1: 47r–50v) we now know that he, in fact, derived these series recursively from the schemes $x = e^l - 1 = \int_0^l x.dl + l$ and $z = \cos a = 1 - \int_0^a \int_0^x z.dx\,dx$. (See J. E. Hofmann, *Die Entwicklungsgeschichte der Leibnizschen Mathematik während des Aufenthaltes in Paris (1672–1676)* (Munich, 1949): 155–6; compare also II: 237, note (112).) Leibniz, indeed, was still ignorant of Newton's method of series inversion and went on to request Newton to explain in more detail 'quomodo in Methodo regressuum se gerat, ut cum ex logarithmo quærit numerum' (*Correspondence*, 2: 62).

(14) The celebrated *epistola posterior* to Oldenburg for Leibniz in 1676. We have incorporated the minor alterations now indicated in the full version of Newton's extract restored, on the basis of the annotated amanuensis copy of the letter (now ULC. Add. 3977. 4) he retained for reference, in Appendix 1 below. Newton himself seems to have had a further text (perhaps the original draft letter, no longer traceable, which he gave to Wickins to transcribe early in October 1676?) before him as he wrote his present revisions: the unexplained citations 'f. 3.1', 'f. 4.1' and 'f. 6.2' (by which understand 'f[olij] 3 [paginâ] 1'—that is, as we would putit, 'f. 3r'—and so on) suggest that this unknown text was less densely written than Wickins' transcript (for which the corresponding references would be 'f. 2.2 [inf.]', 'f. 4.1 [sup.]' and 'f. 5.1 [inf.]'). Because of the imbalance in Newton's quotation from his letter we here restrict our commentary in the main to textual points, deferring our remarks on its technical content to Appendix 1. Some more general historical observations on the letter as sent by Oldenburg to Leibniz are presented in the general appendix which concludes this volume (pages 671–3 below).

(15) *Correspondence*, 2: 110 [opening line of letter]–112 [last line of text]. Newton evidently meant also to include in his quotation the immediately following algebraic computation of the square root of $1 - x^2$ as '$1 - \frac{1}{2}xx - \frac{1}{8}x^4 - \frac{1}{16}x^6$ &c' (compare III: 40).

(16) This sentence is a considerably reordered version of *Correspondence*, 2: 113, ll. 1–4. The

Newton to the worthy H.O., 24 October 1676.[14]

'[I can hardly tell] with what pleasure.... This was [the form of the working] in square roots.[15] When these matters were thoroughly clear to me and having found straightaway after a resolution of affected equations by means of which I was free to go back from areas or arcs to the ordinates, I completely neglected series interpolation and subsequently cultivated these procedures as being more genuinely fundamental.[16]

'I afterwards fell upon infinite series of yet other kinds, which degenerate into finite ones each time a curvilinear figure can be geometrically squared. It will be enough if I subjoin an example of this.[17]

'For any curve..., *To describe a parabola*[18] *which shall pass through any number of given points.*[19]

'To the series $t \pm \frac{1}{3}t^3 + \frac{1}{5}t^5 \pm \frac{1}{7}t^7 \dots$ which Mr Leibniz employs[20] for the quadratures of conics there is related this, $1 + \frac{1}{3} - \frac{1}{5} - \frac{1}{7} + \frac{1}{9} + \frac{1}{11} - \frac{1}{13} - \frac{1}{15} \dots$, which expresses the arc-length of a quadrant whose chord is unity. The difference lies only in the signs when t is unity, but the way they are found is not the same: the former determines the arc-length of a quadrant from the tangent of

---

short paragraph which follows fills the gap left by the omission of *ibid.*: 113, l. 4–115, l. 16 in which Newton in 1676 described for Leibniz his use of the 'Mercator' series expansion of $\pm \log(1 \pm x)$ to compute 'Logarithmi [of low primes] in Tabulam inserendi' and then outlined how he wrote his 'De Analysi' about the time (1668) that Mercator's *Logarithmotechnia* came out (see II: 165–7), following it with the more lavish revised 1671 tract—though in it 'series infinitæ non magnam partem obtinebant'—but suppressing its publication, along with that of his optical lectures, 'subortæ statim per diversorum epistolas Objectionibus alijsꝗ refertas crebræ interpellationes' (see III: 8–9). Two final paragraphs reminded Oldenburg (and informed Leibniz!) that 'Sub id tempus [1671] Gregorius ex unica tantùm serie quadam e meis quam D. Collinsius ad eum transmiserat, post multam considerationem, ut ad Collinsium rescripsit, pervenit ad eandem methodum' (compare note (6) above) and that he had developed his variant on the general algebraic method of tangents in the 1660's 'two or three years before Sluse communicated it to you' (in 1672). A final anagram in which Newton locked up the 'Fundamentum' of his methods of tangents and of maxima and minima yields, when the component letters are set in correct sequence, 'Data æquatione quotcunque fluentes quantitates involvente, fluxiones invenire, et vice versâ' (see III: 25, note (24)).

(17) 'in sequenti serie' (in the following series) is cancelled.

(18) This replaces the cancelled phrase '*Curvam geometricam* (*geometric curve*) which Newton first copied from his 1676 letter (see *Correspondence*, **2**: 119, l. 3). As we earlier saw (**1**, **3**: passim) his finite-difference method depends on approximating a given arbitrary curve $y = f(x)$, tabulated for a number of particular values of the argument x, by the general parabola $y = a + bx + cx^2 + dx^3 \dots$ (and not, for instance, by the hyperbola $y = a + b/x + c/x^2 + d/x^3 \dots$).

(19) *Correspondence*, **2**: 115, l. 17–119, l. 4. The cancellation '....ex pulcherrimis quæ solvere desiderem' (...among the most beautiful I could wish to solve) shows that Newton initially meant to continue the quotation for another seven lines to the end of the paragraph: as he doubtless saw, this would include remarks on his organic construction of conics and cubics which were irrelevant to a treatise on infinite series.

(20) Newton first wrote 'proponit ut omnium simplicissimam' (proposes as simplest of all).

tangente semissis, hæc ex chorda totius[21] definit. Utinam hæ series convergerent celerius, necg opus esset longitudinem[22] areamve circuli aliunde computare. Eam una cum area Hyperbolæ rectangulæ aliquando sic assecuti sumus. Posito axe transverso$=1$....peripheriam 3.1415926535897928.[23] Sic paucis horis area circuli et Hyperbolæ ad loca figurarum decimalia sexdecim colligitur quam vix anni pauci per series superiores darent. Nec tamen Series Leibnitiana minoris facienda est,[24] quæ uticg varijs speculationibus ansam ministrare potest. Quinimo per seriem illam (f. 6.2)....$+\dfrac{a^{28}}{9}$ [&c].[25]

Credo Cl. Leibnitium vix animum advertisse ad seriem meam pro determinatione sinus versi ex arcu dato, ubi ponit seriem pro determinatione sinus complementi ex eodem arcu eamcg aliunde derivat. Nec tamen dubitem quin seriem illam aliascg a me positas, a se verò prætermissas ipse sua dudum methodo invenerat. Quis enim methodi generalis comp^{dae}[26] hærebit in exemplis? Ex occasione verò quod[27] ponit seriem $\dfrac{l}{1}-\dfrac{l^2}{1\times2}+\dfrac{l^3}{1\times2\times3}-\dfrac{l^4}{1\times2\times3\times4}$ &c tanquam a mea $\left[\dfrac{l}{1}+\dfrac{l^2}{1\times2}+\dfrac{l^3}{1\times2\times3}+\dfrac{l^4}{1\times2\times3\times4}\text{ \&c}\right]$ diversam, notaverim quod in theorematîs[28] a me positis usurpo passim species singulas pro quantitatibus[29] sub signis suis sive $+$ sive $-$, id adeo, ne signorum casus singulos enumerando Theoremata cum tædio multiplicem. Sic enim v.g. Theorema illud unicum superius quo curvas quadravi Geometricè, resolvendum esset in 32 Theoremata, præter casus 130 ubi species una vel plures deficiunt. Præterea quæ habentur de...segmenta in tabulam referenda.[30]

Quæ Cl. Leibnitius a me desiderat explicanda, ex parte descripsi. Quod verò attinet ad...decretum est.[31] Verbi gratia, pro primo illo termino seligatur

---

(21) The clarification 'unitate assumpta' (taken to be unity) is deleted.

(22) 'quadrantis' (of a quadrant) is cancelled.

(23) *Correspondence*, 2: 121, l. 6–122, l. 3. (Newton copies the erroneous 15th/16th places in the decimal value of π—which should be '32'—without remark.) The effect of the present revision is drastically to summarize the somewhat rambling observations in the two preceding paragraphs (*ibid.*: 119–21), omitting the précis of the 1671 tract's 'Catalogus posterior' of integrals evaluable as conic areas (see III: 244–54; and compare *Correspondence*, 2: 119).

(24) Newton first wrote equivalently 'Nec tamen Seriei Leibnitianæ laus debita denegan-[da est]' (Nor, however, should due praise be denied to Leibniz' series).

(25) *Correspondence*, 2: 122, ll. 9–20. In a first revision of the next paragraph but one in his 1676 letter Newton here first continued with 'Mitto quod series Leibnitij qua ex arcu dato determinatur sinus complementi coincidit cum mea qua ex eodem arcu sinū versum exhibui' (I put aside that Leibniz' series by which from a given arc is determined its cosine coincides with mine by which from the same arc I exhibited its versine), and then wrote (sticking more closely to his earlier text) 'Credo Cl. Leibnitium ijs quæ habuit [pro determinatione cosinus ex arcu dato, vix animum advertisse ad seriem meam pro determinatione sinus versi ex eodem arcu,] siquidem hæ eædem sunt'.

its half arc, the present one from the chord of the whole.[21] If only these series would converge more swiftly! There would then be no need to compute the length[22] or area of a circle in any other manner. This, along with the area under a rectangular hyperbola we once obtained as follows. The transverse axis being taken = 1 ... the [whole] circumference, 3·1415926535897928.[23] In this way in a few hours we acquire the area of the circle and hyperbola to sixteen decimal places when by means of the previous series a few years would scarcely give it. Leibniz' series should not, however, be the less valued[24] for it can certainly be a useful prising tool in various enquiries. Indeed, by that series... $+\frac{1}{9}a^{28}[-]$....[25]

'I believe that Mr Leibniz has scarcely taken note of my series for determining the versed sine from its arc given when he sets down the series for determining the cosine from the same arc and then derives it from another source. I should not, however, doubt that he had some time before found by his own method both that series and others set down by me, though indeed passed over by him. For who, when constructing[26] a general method, shall be bogged down in particular instances of it? On the occasion, indeed, when he[27] sets down the series $l - \dfrac{l^2}{1\times2} + \dfrac{l^3}{1\times2\times3} - \dfrac{l^4}{1\times2\times3\times4}$ ... as though it differed from mine, I should have noted that in[28] the theorems set down by me I everywhere employ single variables for[29] quantities under their signs, whether $+$ or $-$, so as not, by enumerating the separate cases of signs, tediously to multiply theorems. Thus, for example, that single theorem above by which I squared curves geometrically would have to be split into 32 theorems, apart from 130 cases in which one or more of the variables were lacking. Moreover, the statements about...the segments [will be left], to be entered in a table.[30]

'What Mr Leibniz wishes me to explain, I have in part described. But as to...has been decided [to extract one or other of the roots].[31] For instance, let

(26) Reading 'componendæ'. Newton could well have meant 'complendæ' (filling out).

(27) 'Cl. Leibnitius' (Mr Leibniz) is cancelled.

(28) 'omnibus' (all) is deleted.

(29) 'significandis' (signifying) has been struck out.

(30) *Correspondence*, 2: 123, l. 12–124, l. 11. Newton first began the next paragraph 'Terminorum *y.p.q.r* &c in extractione radicum affectarum, inventionem Cl. Leibnitius a me postulat. Primum *y* sic eruo....' (Mr Leibniz requests me for the way of finding the terms *y, p, q, r*, ... in the extraction of affected roots. The first, *y*, I hunt out in this way) but almost at once returned to the original phrasing of his 1676 letter (*ibid.*: 126, l. 20 ff). The two omitted pages deal with the tabulation of logarithmical and trigonometrical functions (see 1, 1, §2 above) and are, at best, only marginally relevant in the present context of infinite series expansions.

(31) *Correspondence*, 2: 126, l. 20–127, l. 13. The effect of Newton's revision in sequel is to add a sentence clarifying the solution of $y^6 - 5xy^5 + a^{-1}x^3y^4 - 7a^2x^2y^2 + 6a^3x^3 + b^2x^4 = 0$ in the case of the root $y \approx a^{\frac{1}{3}}x^{\frac{2}{3}}$.

$+\sqrt{ax}$ et pro reliquis in infinitum terminis nondū cognitis scribatur p, et erit $\sqrt{ax}+p=y$. Sic æquationem $y^3+axy+aay-x^3-2a^3=0$ resolvendo ut in priori epistola, parallelogramma da[n]t $-2a^3+aay+y^3=0$ et inde $y=a$. Cum itacꝫ a sit primus terminus valoris y...prodire solent[32] ex terminis duobus uno infimo in columna prima altero infimo in columna secunda, (ut in hoc exemplo ipse p ex terminis $aax+4aap$, ipse q ex terminis $-16axx+4aaq$, ipse r ex terminis $-\frac{131}{128}x^3[+4aar]$,) dividendo priorem (ut aax, $-16axx$, vel$-\frac{131}{128}x^3$) per coefficientem illius p, q, r in posteriore (ut hic per $4aa$) et mutando signum Quoti.

Intellexti credo...altera generaliori.[33] Sed et abscꝫ his methodis solvitur inversum illud de tangentibus problema quando tangens inter punctum contactus et axem figuræ est datæ longitudinis, quo casu prodit curva quædam Mechanica cujus determinatio pendet ab area Hyperbolæ Apollonianæ. Et ejusdem generis est etiam...responsum darem.[34]

D^{no} O: *Leibnitius* [*21 Junij 1677*].

Amplissime D^{ne}

Accepi literas tuas &c...p^r tangentes sit Demonstrata.[35]—Pulcherrimæ sunt illæ series Newtonianæ...necꝫ curvam Ellipseos quantum memini.[36]

Hannovero. Jun 21 1677.

---

(32) *Correspondence*, **2**: 127, ll. 16–21. Newton first continued, much as in his 1676 letter, with 'dividendo terminum infimum columnæ primæ per coefficientem p, q, r vel s in termino infimo columnæ secundæ, ut in hoc exemplo terminum aax per ipsius p coefficientem $4aa$ ...' (by dividing the lowest term of the first column by the coefficient of p, q, r or s in the lowest term of the second column, as in this example the term a^2x by the coefficient $4a^2$ of p) before making the following slight amplification.

(33) *Correspondence*, **2**: 127, l. 23–129, l. 10. In sequel Newton omits his celebrated fluxions anagram 'literis transpositis', which when unscrambled reads (see II: 191, note (25); and compare III: 26, note (30)) 'Una methodus consistit in extractione fluentis quantitatis ex æquatione simul involvente fluxionem ejus; altera tantum in assumptione seriei pro quantitate qualibet incognita ex qua cætera commode derivari possunt, & in collatione terminorum homologorum æquationis resultantis ad eruendos terminos assumptæ seriei'.

(34) *Correspondence*, **2**: 129, ll. 19–33. Newton concluded his 1676 letter with the more personal remark 'Et volui hac vice copiosior esse quia credidi amœniora tua negotia severiori hocce scribendi genere non deberi a me crebrò interpellari' (*ibid.*: ll. 33–5). Due in the main to Oldenburg's ignorance of his new German address—the *epistola posterior* reached London a few days after he left England in late October 1676, unfortunately—Leibniz received Newton's letter only at the beginning of the following June but made haste to reply 'immediately' on the 11th of that month (old style).

(35) *Correspondence*, **2**: 212, the opening paragraph of Leibniz' response. Making apologies for noting only what he observed 'festinante oculo' in Newton's preceding letter, he remarked that what pleased him most was the latter's description 'qua via in nonnulla sua elegantia sanè theoremata inciderit et quæ de Wallisianis interpolationibus habet, vel ideò placent, quia hac ratione obtinetur harum interpolationum demonstratio, cum res antea quod sciam sola

$+\sqrt{[ax]}$ be chosen for that first term and in place of the infinity of remaining ones not yet known let there be written p, and then $y = \sqrt{[ax]} + p$. So, by resolving the equation $y^3 + axy + a^2y - x^3 - 2a^3 = 0$ as in my previous letter, the parallelograms give $-2a^3 + a^2y + y^3 = 0$ and thence $y = a$. And so since a is the first term of the value of y ... commonly result[32] from two terms, one the lowest in the first column, the other the lowest in the second column, (as in this example p from the terms $a^2x + 4a^2p$, q from the terms $-16ax^2 + 4a^2q$, r from the terms $-\frac{131}{128}x^3 + 4a^2r$) on dividing the former (here a^2x, $-16ax^2$, or $-\frac{131}{128}x^3$) by the coefficient of p, q, r in the latter (here by $4a^2$) and then changing the sign of the quotient.

'You will, I think, have understood...the other more general.[33] But that inverse problem of tangents is solved without these methods when the tangent between its point of contact and the figure's axis is of given length: in this case there results a certain mechanical curve whose determination depends on the area of an Apollonian hyperbola. And [the problem] is also of the same kind...I should give [it this more extended] reply.'[34]

Leibniz to Mr O., 21 June 1677.

'Most honourable Sir,

'I have received your [long awaited] letter...proved by means of tangents.[35]

'Those series of Newton's [which degenerate from infinite to finite] are most beautiful...nor the curve of an ellipse, as far as I remember.[36]

Hanover, 21 June 1677.'

---

inductione niteretur...'. In his quotation of Leibniz' letter Newton omits Leibniz' account of his variant on the general Slusian method of tangents which follows immediately afterwards (*ibid.*: 213–15).

(36) *Correspondence*, **2**: 215, l. 22–218, l. 27. In this passage Leibniz, while generally laudatory of Newton's 'very fine series', was a little cautious in accepting their universal effectiveness in extracting the roots of an algebraic equation: '...si in ipsius generali illa æquationis affectæ indefinitæ extractione...præstari posset ut...inter extrahendum radices ex æquationibus... invenire liceret radices rationales finitas, quando eæ insunt, vel etiam irrationales, tunc dicerem methodum serierum infinitarum ad summam perfectionem esse productam. Opus esset tamen...discerni posse varias æquationis ejusdem radices, item necesse esset ope serierum discerni æquationes possibiles ab impossibilibus. Quod si hæc nobis obtinuerit [Neutonus] atque effecerit...ut possimus seriem infinitam convertere in finitam quando id fieri potest aut saltem agnoscere ex quanam finita sit deducta, tunc in methodo serierum infinitarum quæ divisione atǫ extractione inveniuntur vix quicquam amplius optandum restabit....velim...nosse an per extractiones in seriebus discernere possit æquationes possibiles ab impossibilibus; nam si generalis ejusmodi extractio procederet, sequeretur nullam æquationem fore impossibilem: item quomodo inveniat diversas ejusmodi æquationis radices, ita ut ex pluribus radicibus eam possit invenire quam quærimus: item an tales habeat series quarum ope extrahendo æquation[is radic]es inveniuntur valores finiti quando tales insunt; deniǫ quid sentiat de resolutione æquationum quales paulò ante posui [he cites in exempli-

Dⁿᵒ O. Leibnitius [12 Julij 1677].

Amplissime Domine

Nuperas meas credo...nondum mihi liquet.[37]

Hannovero 12ᵐᵒ Julij 1677.

Hactenus Epistolarum pars illa quæ de seriebus infinitis agit. Namqꝫ his ultimis nihil respondi morte D.O.[38] præventus. Quare jam supplenda sunt quæ deesse videantur. Id sub certis capitibus qua possim brevitate perficiam.[39]

Cap. 1. De extractione et natura radicum affectarum.[40]

Cap 2. Ex seriebus minus convergentibus vel etiam ex divergentibus quantitates quæsitas elicere.

Cap 3. Series appropinquantes ex lineis pluribus Geometricè inveniendis componere.[41]

Cap 4. Methodus generalior reducendi problemata[42] *ad series infinitas.*

Cap 5. De migratione serierum infinitarum in finitas.

---

fication $x^y + y^x = xy$ and $x^x + y^y = x + y$], ubi scilicet incognita ingreditur in exponentem' (*ibid.*: 215–16/218). (To the first two of these very real difficulties in the way of the universal application of infinite series expansions in mathematical analysis Newton will attempt some answer in the first chapter of his present 'Specimina'.) In the possibility of representing a given continuous function by a 'parabolic' curve (the basis of Newton's finite-differences method) he found further difficulty: while agreeing that such a 'curva analytica, seu æquationis capax' can always be made to pass through a given finite set of points—he cites a remark made to him by Johann Hudde at Amsterdam in November 1676 that 'posse se curvam describere, Analyticam seu certa æquatione uniformi constantem quæ faciei hominis cujusdam noti lineamenta designet' (*ibid.*: 216)—, he queried whether this were wholly true when an infinite set of points were given. (In his example an indefinite number of points are supposed to be given on two 'analytic' curves: can a third 'curva analytica' be drawn to pass alternately in some sequence through both sets of points?) With regard to Newton's claim (*ibid.*: 129: 'inversa de tangentibus Problemata sunt in potestate, aliaqꝫ illis difficiliora') to have mastered the general inverse problem of tangents—by which he meant at least the general method of algebraic integration—Leibniz held still more reservations: 'Quod ait problemata methodi tangentium inversæ esse in potestate, hoc arbitror ab eo intelligi per series...infinitas. Sed à me ita desiderantur ut curvæ exhibeantur geometricè quatenus id fieri potest suppositis (minimum) quadraturis.... Cum ait Neutonus inventionem curvæ quando tangens vel intervallum tangentis et ordinatæ in axe sumtum est recta constans non indigere his methodis [in his *epistola posterior* in 1676 Newton wrote (*ibid.*: 129) that 'Est...Curva illa Mechanica, cujus determinatio pendet ab area Hyperbolae'; compare III: 26, note (31)] innuit, credo, se intelligere methodum tangentium inversam generalem in potestate esse per methodos serierum appropinquativas,...ego verò methodum quærebam quæ accuratè curvam quæsitam exhibeat saltem ex suppositis quadraturis et cujus ope ejus æquationem si quam habet aut aliam primariam proprietatem possumus invenire.... Generalem...methodum tangentium inversam nondum quod sciam habemus' (*ibid.*: 216–17). In sequel Leibniz went on to make certain remarks *à propos* of the particular case where 'datur relatio inter duo ex lateribus...trianguli quod ego characteristicum (ob crebros usus) vocare soleo' (*ibid.*: 217): these are taken up by Newton at the close of the fourth chapter of his present 'Specimina' (see note (147) below).

Leibniz to Mr O., 12 July 1677.

'Most honoured Sir,

'[You will,] I think, [have received] my recent letter...not yet clear to me.[37]

Hanover, July 12th, 1677.'

This concludes the portion of the correspondence which has to do with infinite series. For to these last letters I made no reply, being forestalled by Mr O.'s death.[38] Accordingly, what may seem to be missing must now be supplied. This I shall accomplish with all possible brevity under certain heads.[39]

Chapter 1. On the extraction and nature of affected roots.[40]

Chapter 2. To elicit required quantities from less convergent series or even divergent ones.

Chapter 3. To construct approximating series from several lines to be found geometrically.[41]

Chapter 4. A more general method of reducing[42] *problems to infinite series.*

Chapter 5. On the passage of infinite series into finite ones.

---

(37) *Correspondence*, **2**: 231 [opening line of letter]–232, l. 18. Leibniz here added a few complementary observations to his previous letter, stressing the difficulty he had raised previously about the employment of infinite series 'circa æquationes impossibiles' (*ibid.*: 231: 'Necdum verò illa [difficultas] sublata est, et meretur res excuti diligentius') and again emphasizing the current lack of an adequate 'geometrical' inverse method of tangents 'saltem suppositis curvarum analyticarum quadraturis', adding that 'si demonstrari potest, (ut arbitror) quasdam figuras non esse quadrabiles nec per circulum nec Hyperbolam, restat ut alias quasdam figuras primarias altiores constituamus, ad quarum quadraturam reducantur cæteræ omnes, quando id fieri potest: hoc quamdiu non fit hæremus, et sæpè per seriem infinitam particularem quærimus' (*ibid.*: 232).

(38) This is the exact truth since Oldenburg died a week after Collins forwarded to Newton (on 30 August 1677; see *Correspondence*, **2**: 237) his transcript of the 'scarce legible' original of Leibniz' letter of 12 July (now Royal Society MS LXXXI, No. 70: his preceding communication of 11/21 June is No. 56 in the same 'Commercium Epistolicum' volume). After Oldenburg's death Newton's correspondence with Collins soon ceased also, and there remained no contact between Cambridge and Hanover short of the direct epistolary exchange on which Newton was unwilling to embark.

(39) 'caput' here has the double meaning (impossible to convey in a single English word) of both 'heading' and 'chapter'. The first four 'capita' in the following outline correspond to Chapters 1, 2, 3 and 5 in the succeeding main text of the *Specimina*, while 'Cap 6' is dealt with in Chapter 3 of the revised tract 'De Computo Serierum' (§2 below). There is nothing in the extant corpus of Newton's mathematical writings to show that he ever implemented his projected 'Cap 5. De migratione serierum infinitarum in finitas' (on determining the 'finite' magnitudes to which a given infinite series converges?). The *Specimina*'s fourth chapter, dealing with the solution of problems by fluxional methods, is evidently a late insertion made in afterthought.

(40) Originally 'De extractione radicum affectarum ex æquationibus' (On the extraction of affected roots from equations).

(41) Newton first wrote simply 'Quantitates imperatas appropinquare per series' (To approximate required quantities by series).

(42) 'difficiliora' (more difficult) is cancelled.

Cap 6. De Solutione Problematum per lineam Parabolicam.[43]
Conclusio.[44]

---

[2]   CAP 1.

DE RADICIBUS AFFECTARUM ÆQUATIONUM.[45]

Difficultatem movit Cl. Leibnitius circa æquationum radices impossibiles quæ per series nostras[46] inveniuntur et exhibentur more possibilium.[47] Verum sciendum est quod series ipsæ perinde possibiles et impossibiles sunt atcʒ radices quas exhibent. Radix possibilis per seriem possibilem, impossibilis per impossibilem semper exprimitur. Series autem possibilis est cujus omnium in infinitum terminorū quantitas aliqua certa et determinata est,[48] impossibilis cujus quantitas certa et determinata non est. Quantitatis autem incertitudo et indeterminatio in seriebus exprimitur per infinitatem. Quoties radix æquationis est impossibilis series illam designans fit infinita.[49]

Sic in semicirculo *AEB*[50] series ordinatam *ED* designans finita est ubi *CD* semidiametro *CB* minor sumitur, et ex aucta *CD* minuitur uscʒ in nihilum[,] deinde ex nihilo per saltum fit infinita quamprimum *CD* fit major quam *CD*. Sic et series sinum versum[51]

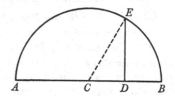

BD designans ex finita fit infinita per saltum quamprimum ordinata *DE* crescendo fit semidiametro *CE* major.

Quærit etiam Vir Celeberrimus[52] quomodo inveniam diversas ejusdem æquationis radices, ita ut ubi plures sint imperatam quamvis eruam.[53] Nimirum termini per parallelogrammum inventi ex quibus primum seriei terminū

---

(43) A following entry, '*Cap 7. Miscellanea*', is struck out.

(44) A cancelled following phrase 'de tangentibus et alijs Leibnitianis' (on tangents and other Leibnizian matters) suggests the topics intended to be covered in Newton's conclusion. We have seen earlier (note (35) above) that he omitted from his quotation of Leibniz' letter to Oldenburg on 11/21 June 1677 the passage (*Correspondence*, **2**: 213–15) in which Leibniz elaborated his variant of the general tangent rule.

(45) This title replaces the cancelled heading 'DE EXTRACTIONE ET NATURA RADICUM AFFECTARUM' (ON THE EXTRACTION AND NATURE OF AFFECTED ROOTS) under which it is listed in Newton's preliminary scheme in [1] (see note (40) above).

(46) 'infinitas' (infinite) was first written.

(47) See Leibniz' letter to Oldenburg of 11/21 June 1677 (note (36) above). Newton first continued: 'Quæritur unde radix quæ in æquatione impossibilis est per seriem exhibetur tanquam si possibilis esset. Numquid quantitas impossibilis rectè exhibetur per seriem tanquam si possibilis esset?' (The query concerns the circumstances in which a root, impossible in its equation, may be displayed by a series as though it were possible. Is it at all correct that an impossible quantity may be displayed by a series as though it were possible?)

(48) A following amplification, 'ea nempe quæ est radicis quam expr[imit]' (that, namely, of the root which it expresses), is cancelled.

Chapter 6. On the solution of problems by a parabolic line.[43]
Conclusion.[44]

[2]   CHAPTER 1.
               ON THE ROOTS OF AFFECTED EQUATIONS.[45]

Mr Leibniz has raised a difficulty about impossible roots of equations which
are found by our[46] series and displayed in the manner of possible ones.[47] But
it should be known that these series are alike possible and impossible according
as the roots they display. A possible root is always expressed by a possible series,
an impossible one always by an impossible. A series, however, is possible when
some quantity of all its terms to infinity is precisely determined,[48] impossible
when its quantity is not precisely determined. Imprecision and indeterminacy,
indeed, of a quantity is expressed in the series by infinity. Each time an equation's root is impossible the series denoting it becomes infinite.[49] Thus in the
semicircle *AEB*[50] the series denoting the ordinate *ED* is finite when *CD* is taken
less than the radius *CB*, and as *CD* increases it diminishes accordingly till it is
nothing, then from nothing at a bound it comes to be infinite as soon as *CD*
becomes greater than *CD*. So also, the series denoting the versed sine[51] *BD*
from being finite comes at a bound to be infinite as soon as the ordinate *DE* by
its growth comes to be greater than the radius *CE*.

My renowned[52] correspondent also requests how I shall find the different
roots of the same equation so as, where there are several, to obtain any appointed
one.[53] Here, of course, the terms found by the parallelogram—from which I
elicit the first term of the series—will yield as many first terms of series as there

(49) This argument that a divergent series 'represents' a non-real quantity (with its
imaginary part lurking, at any finite stage, in the infinitely great remaining terms?) is repeated
from Newton's 1671 tract (see III: 69, note (73)). As we there remarked, the hyperbolic
function $\log(1+x)$ and the 'Mercator' series $x - \frac{1}{2}x^2 + \frac{1}{3}x^3 - ...$, $x > 1$, are together a simple
confutation of this half-baked, if superficially plausible, idea.

(50) In sequel Newton first added the explanatory phrase 'cujus centrum *C*, radius *CB* = *r*,
ordinata *DE* = *y* et pars axis ubi *ED* ad diametrum *AB* ordinatim applicatur [*CD*=*x*]'
(whose centre is *C*, radius *CB* = *r*, ordinate *DE* = *y* and the portion of the axis at which *ED*
is applied at right angles to the diameter *AB* is *CD* = *x*). In these terms the defining equation
of the circle is $x^2+y^2 = r^2$ and Newton's further argument (compare III: 68) is that when the
ordinate *DE* is imaginary ('impossible')—that is, when *CD* > *CB* or $|x| > r$—the corre-
sponding binomial expansion $y = \sqrt{[r^2-x^2]} = r - \frac{1}{2}x^2/r - \frac{1}{8}x^4/r^3 ...$ is divergent to infinity.

(51) 'segmentum' (segment) was first written.

(52) Evidently this epithet is the *mot juste* in Newton's eyes, for he has cancelled the equally
empty equivalent 'Clarissimus' (most brilliant).

(53) Newton first concluded with '...ita ut ex pluribus dignitatum indicibus eam possim
eruere quæ quæritur' (so as to be able, out of several indices of powers, to lay hold of the one
required).

elicio, tot dabit primos serierum terminos quot sunt radices æquationis per seriem designabiles, et ex singulis primis terminis subinde colligentur singulæ series. Sic ad resolvendā æquationē $y^3 + axy + aay - x^3 - 2a^3 = 0$ parallelogrammum dat æquationem fictitiā $y^3 + aay - 2a^3 = 0$ cujus tres radices

$$a, \quad [-]\tfrac{1}{2}a + \sqrt{-\tfrac{7}{4}aa} \quad \& \quad [-\tfrac{1}{2}]a - \sqrt{-\tfrac{7}{4}aa}$$

sunt primi termini totidem serierum. Et ad resolvendam

$$y^6 - 5xy^5 + \frac{x^3}{a}y^4 - 7aaxxyy + 6a^3x^3 + bbx^4 = 0.$$

parallelogram̄um dat fictitiā $y^6 - 7aaxxyy + 6a^3x^3 = 0$, cujus sex radices $+\sqrt{ax}$, $-\sqrt{ax}$, $+2\sqrt{ax}$, $-2\sqrt{ax}$,[54] $+\sqrt{-3ax}$, $-\sqrt{-3ax}$ sunt termini initiales serierum totidem. Elige jam terminum initialem radici quæsitæ congruentem et consequetur series optata. Quod si terminus aliquis initialis sit impossibilis[55] id arguit seriem impossibilem esse.[56] Quo in casu augenda vel minuenda est longitudo quantitatis indefinitæ[,] non illius cujus radix extrahi debet sed alterius ex cujus dignitatibus componitur series. Oritur enim seriei impossibilitas ex impossibilitate radicis ubi quantitas illa indefinita valde parva est. Sic in æquatione $y^3 + axy + aay - x^3 - 2a^3 = 0$ ubi x valde parva est duo valores ipsius y[57] sunt impossibiles et inde series istis competentes sunt etiam impossibiles. Quære igitur longitudinē quamvis ipsius x ubi tres sunt ipsius y radices possibiles. Sitçç longitudo illa n et dato n augeatur vel minuatur indefinita x.[58] Sic et ubi serierum termini aliqui initiales deficiunt augenda vel minuenda est indefinita illa quantitas x vel etiam mutanda in ejus reciprocum. Ut si resolvenda esset æquatio $y^5 - byy + 9bxx - x^3 = 0$, inde per Paralle[lo]grammum

---

(54) A trivial slip here: read '$+\sqrt{2ax}$, $-\sqrt{2ax}$', since

$$y^6 - 7a^2x^2y^2 + 6a^3x^3 \equiv (y^2 - ax)(y^2 - 2ax)(y^2 + 3ax).$$

See III: 52.

(55) 'vel deficit' (or lacking) is cancelled. In sequel Newton first wrote '...seriem respondentem' (the corresponding series).

(56) A cancelled first continuation reads: 'Sic ad resolvendam æquationem

$$y^5 - byy + 9bxx - x^3 = 0$$

parallelogrammum dabit æquationem fictitiam $-byy + 9bxx = 0$ cujus radices sunt $+3b$ et $-3b$[!]. Tertia quarta et quinta radix deest' (So, to resolve the equation

$$y^5 - by^2 + 9bx^2 - x^3 = 0,$$

the parallelogram will yield the fictitious equation $-by^2 + 9bx^2 = 0$ whose roots are $+3[x]$ and $-3[x]$. The third, fourth and fifth roots are lacking). A corrected version of this example reappears below; see note (59) below.

(57) Namely $-\tfrac{1}{2}(1 \pm \sqrt{-7})a$.

are roots of the equation denotable by series, and from each of the first terms separate series will be developed forthwith. Thus to resolve the equation

$$y^3 + axy + a^2y - x^3 - 2a^3 = 0$$

the parallelogram yields the fictitious equation $y^3 + a^2y - 2a^3 = 0$, the three roots a, $\frac{1}{2}(-1+\sqrt{-7})\,a$ and $\frac{1}{2}(-1-\sqrt{-7})\,a$ of which are the first terms of an equal number of series. While to resolve

$$y^6 - 5xy^5 + a^{-1}x^3y^4 - 7a^2x^2y^2 + 6a^3x^3 + b^2x^4 = 0$$

the parallelogram yields the fictitious $y^6 - 7a^2x^2y^2 + 6a^3x^3 = 0$, the six roots $+\sqrt{[ax]}$, $-\sqrt{[ax]}$, $+2\sqrt{[ax]}$, $-2\sqrt{[ax]}$,[54] $+\sqrt{[-3ax]}$, $-\sqrt{[-3ax]}$ of which are the initial terms of an equal number of series. Now choose the initial term fitting the required root and the desired series will ensue. But if some initial term be impossible,[55] this declares that the series is impossible.[56] In that case you must increase or diminish the length of the indefinite quantity, not that whose root is due to be extracted but the other one from whose powers the series is composed. For the impossibility of the series arises in consequence of the impossibility of the root when that indefinite quantity is exceedingly small. So, in the equation $y^3 + axy + a^2y - x^3 - 2a^3 = 0$, when x is exceedingly small two values of y[57] are impossible and accordingly the series corresponding to those are also impossible. Search, therefore, for any length of x for which the three roots of y are possible. Let that length be n and then increase or diminish the indefinite quantity x by the given amount n.[58] So also, when one or more initial terms of the series are lacking, the indefinite quantity x should be increased or diminished, or alternatively changed into its reciprocal. For instance, if the equation

$$y^5 - by^2 + 9bx^2 - x^3 = 0$$

were to be resolved, from this by means of the parallelogram there would result

---

(58) In the present instance, for example, we might transfer the origin to $(-\sqrt[3]{2}a, 0)$, whence, on substituting $x \rightarrow x - \sqrt[3]{2}a$, the defining equation of Newton's unicursal hyperbolic hyperbolism (of asymptote $y = x - \frac{1}{3}a$) becomes

$$y^3 - x^3 + axy + 3\sqrt[3]{2}ax^2 - (\sqrt[3]{2}-1)\,a^2y - 3\sqrt[3]{4}a^2x = 0,$$

whose corresponding fictitious equation $y^3 - (\sqrt[3]{2}-1)\,a^2y = 0$ yields the three real roots $y = 0$, $\pm\sqrt{[\sqrt[3]{2}-1]}\,a$. Following Newton's procedure the general ordinate y may then be expressed dually as an infinite series in x whose first term is one or other of the two latter roots. This is no real help, however, since when $x = \sqrt[3]{2}a$ these expansions—'representing' the 'impossible' ordinates $-\frac{1}{2}(1\pm\sqrt{-7})\,a$—will diverge to infinity, as Newton well realizes. Nor is it possible in general to construct an ordinate to a given curve of n-th degree which shall meet it in n real points: thus, for no k will the general ordinate $x = k$ of the Wallis cubic $y = x^3$ meet it in more than one real point, whence Newton's present argument here fails.

prodiret æquatio fictitia $-byy+9bxx=0$ ubi duo sunt tantum valores y nimirum $+3x$ et $-3x$ pro terminis initialibus serierum totidem.[59]

Cap 2.

De serierum proprietatibus.[60]

Serierū proprietates de quibus hic ago sunt progressio terminatio et intercalatio. Progressionem sic eruo. Esto series $A. B. C. D. E. F. G. H$ &c. Finge terminum unumquemꝗ subsequentem produci multiplicando antecedentem per $\dfrac{p+nq+n^2r \,\&c}{s+nt+n^2v \,\&c}$ ubi p, q, r, s, t, v &c designent datas quantitates et n numerum terminorum seriei numeratorum a tertio C, id est numeros $-2. -1. 0. 1. 2. 3.$ &c successive perinde ut multiplicator iste successivè ducitur in $A. B. C. D. E. F$ &c. Sic erit $\dfrac{p-2q+4r}{s-2t+4v}A=B.$ $\dfrac{p-q+r}{s-t+v}B=C.$ $\dfrac{p}{s}C=D.$ $\dfrac{p+q+r}{s+t+v}D=E.$ $\dfrac{p+2q+4r}{s+2t+4v}E=F.$[61] Ex his æquationibus collatis et reductis[62] invenio quantitates q, r, t, v, positis p numeratore et s denominatore fractionis reductæ $\dfrac{D}{C}$. Dein tento si seriei terminus quilibet subsequens prodeat multiplicando terminum præcedentem per $\dfrac{p+nq+n^2r}{s+nt+n^2v}$, id est si fuerit $\dfrac{p+3q+9r}{s+3t+9v}F=G.$ $\dfrac{p+4q+16r}{s+4t+16v}G=H$ & sic deinceps. Hoc pacto progressiones inveni quas posui in epistola mea priori[63] ut et sequentes.[64]

Serierum progressiones et terminationes invenire est unum et idem problema. Per terminationem[65] intelligo valorem seriei cujusvis seu terminorum omnium $A. b. c. d. e. f.$ [&c] aggregatum. Sunto differentiæ primæ $B. \,^2b. \,^3b$ &c. secundæ

---

(59) In sequel Newton has cancelled 'Desunt ergo tres. Ut reliquos obtineam scribo reciprocum $\dfrac{1}{z}$ pro y in æquatione prima [? et prodit $1-bz^3+9bxxz^5-x^3z^5=0$ adeoꝗ æquatio fictitia $1-bz^3=0$, unde $z=\sqrt[3]{\dfrac{1}{b}}$]' (So three are lacking. To obtain the rest I write its reciprocal $1/z$ in place of y in the first equation and there results $1-bz^3+9bx^2z^5-x^3z^5=0$, so that the fictitious equation is $1-bz^3=0$, whence $z=\sqrt[3]{[1/b]}$). The three resulting 'valores' $y=b^{\frac13}\omega$, where $\omega^3=1$, are directly obtainable from the fictitious equation $y^5-by^2=0$ and there is clearly no point in introducing the reciprocal of y to determine them.

(60) Newton first wrote 'CULTURA' (CULTIVATION). Earlier drafts, reproduced in Appendix 2. 2/3 below, are titled 'DE PROGRESSIONE [ET TERMINATIONE]/POLITIONE SERIERUM' (ON THE PROGRESSION [AND TERMINATION]/REFINING OF SERIES). Following on his original title Newton first began in sequel: 'Serierum cultura consistit in invenienda progressione aggregato et intercalatione terminorum' (The cultivation of series consists in finding the progression and aggregate of their terms and in their interpolation).

the fictitious equation $-by^2 + 9bx^2 = 0$, where there are but two values of y (namely $+3x$ and $-3x$) for the initial terms of the same number of series.[59]

CHAPTER 2
ON THE PROPERTIES[60] OF SERIES

The properties of series about which I am here concerned are their progression, termination and intercalation. Their progression I derive this way. Let the series be $A, B, C, D, E, F, G, H, \ldots$. Imagine that each subsequent term is produced by multiplying the preceding one by $\dfrac{p + nq + n^2r \ldots}{s + nt + n^2v \ldots}$, in which $p, q, r, s, t, v, \ldots$ are to denote given quantities and n the number of terms counted from the third one C; that is, in succession the numbers $-2, -1, 0, 1, 2, 3, \ldots$ correspondingly as that multiplier is drawn successively into $A, B, C, D, E, F, \ldots$. There will thus be $\dfrac{p - 2q + 4r}{s - 2t + 4v}A = B$, $\dfrac{p - q + r}{s - t + v}B = C$, $\dfrac{p}{s}C = D$, $\dfrac{p + q + r}{s + t + v}D = E$, $\dfrac{p + 2q + 4r}{s + 2t + 4v}E = F$.[61] After these equations have been set in unison and reduced,[62] from them I find the quantities q, r, t, v, setting p to be the numerator and s the denominator of the fraction D/C when reduced. I then test if any random subsequent term of the series results on multiplying the preceding term by $\dfrac{p + nq + n^2r}{s + nt + n^2v}$; that is, if there proves to be $\dfrac{p + 3q + 9r}{s + 3t + 9v}F = G$, $\dfrac{p + 4q + 16r}{s + 4t + 16v}G = H$ and so on in turn. By this approach I found the progressions which I set down in my first letter,[63] and also the following.[64]

To find the progressions and terminations of series is one and the same problem. By 'termination'[65] I understand the value of any series; in other words, the aggregate of all its terms $A, b, c, d, e, f, \ldots$. Let the first differences

---

(61) Since only the ratios of p, q, r, s, t, v in the particular 'progressio'

$$(p + nq + n^2r)/(s + nt + n^2v)$$

here assumed are significant, five equations are sufficient to determine it.

(62) Compare Newton's preliminary computations reproduced in Appendix 2.1 following.

(63) In his letter to Oldenburg for Leibniz on 13 June 1676 Newton had elaborated— probably by visual inspection rather than by the intricate method here expounded—the progression of the series expansions of $r\sin^{-1}(x/r)$, $r\sin(z/r)$, $d\sin[n\sin^{-1}(x/d)]$,

$$\int_0^x \sqrt{[(c^2r - (c-r)x^2)/c(cr - x^2)]} \, . \, dx \quad \text{and} \quad 2\int_0^x \int_0^y z \, . \, dy \, dx$$

where $x^2/a^2 + (y^2 + z^2)/c^2 = 1$; see *Correspondence*, 2: 25–9 and compare pages 668–9 below.

(64) Half a page following is left blank in the manuscript for the intended future insertion of these additional expansions into infinite series. Presumably these would involve simple algebraic, circular and hyperbolic functions singly or in combination.

$C.$ $^2c.$ ^3c &c. tertiæ $D.$ $^2d.$ ^3d &c et sic deinceps: id est ita ut sit

$$A-b=B. \quad b-c={}^2b. \quad c-d={}^3b \quad \text{&c.}$$

$Z.$	$^2Z.$	$^3Z.$	$^4Z.$	$^5Z.$	$^6Z.$	^7Z
$A.$	$b.$	$c.$	$d.$	$e.$	$f.$	
B	^2b	^3b	^4b	^5b		
C	^2c	^3c	^4c			
D	^2d	^3d				
E	^2e					
F						

$B-{}^2b=C.$ ${}^2b-{}^3b={}^2c$ &c. $C-{}^2c=D$ &c subductis
semper sequentibus de præcedentibus[66] & signis
residuorum notatis. Manifestum est quod omnium
$B.$ $^2b.$ $^3b.$ ^4b &c aggregatum sit A, quodcß B sit
aggregatū omnium $C.$ $^2c.$ ^3c &c et sic in infinitum atcß adeo si aggregatorum
series $A.$ $B.$ $C.$ $D.$ $E.$ F &c producatur versus A præfixo uno termino Z iste Z
erit aggregatū omnium A, b, c, d, e &c. Et vice versa si series A, B, C, D &c
augenda est termino novo Z, colligendæ erunt terminorum A B C D &c
differentiæ primæ $b, {}^2b, {}^2c, {}^2d$ &c, secundæ $c, {}^3b, {}^3c, {}^3[d]$ &c, tertiæ $d, {}^4b, {}^4c$ &c
cæteræcß et aggregatum differentiarū A, b, c, d, e &c habendum pro Z.[67]
Igitur si progressio seriei $A.$ $B.$ $C.$ D &c versus A exactè inveniri potest terminabitur series $A.$ $b.$ $c.$ d &c exactè et vice versa:[68] sin minus approximatio adhi-
benda est. Ea fit ducendo lineam regularem per puncta quotcuncß data. Ad
rectam quamvis positione datam ZE æqualibus intervallis $ZA, AB, BC, CD,$
DE erigantur normales ZQ, AP, BP, CP, DP &c æquales terminis seriei $Z, A,$

---

(65) One more Newtonian borrowing from James Gregory's *Vera Circuli et Hyperbolæ Quadratura* (Padua, 1667): 19: 'imaginando...seriem in infinitum continuari, possumus imaginari vltimos terminos cōuergentes esse æquales, quos terminos æquales appellamus seriei terminationem'.

(66) Contrary to modern convention but in line with Newton's standard practice (compare **1, 1**: passim).

(67) This would seem to be the first historical occurrence of a summation by inverting a difference scheme. Retaining the sequence in which Newton takes his differences (see previous note) we may, in modern equivalent, set $A = f_0 (= \Delta_0^0)$, $b = f_1$, $c = f_2$, $d = f_3$, ... and from these define in succession $(i = 1, 2, 3, ...)$ $B = f_0 - f_1 = \Delta_0^1$, $b_i = f_i - f_{i+1} = \Delta_i^1$;

$$C = \Delta_0^1 - \Delta_1^1 = \Delta_0^2, \quad c_i = \Delta_i^1 - \Delta_{i+1}^1 = \Delta_i^2; \quad D = \Delta_0^2 - \Delta_1^2 = \Delta_0^3, \quad d_i = \Delta_i^2 - \Delta_{i+1}^2 = \Delta_i^3;$$

and so on: whence the series $A, B, C, D, ...$ is $\Delta_0^j = \sum_{0 \leqslant i \leqslant \infty} \Delta_i^{j+1}$, $j = 0, 1, 2, 3, ...$, and accordingly the prefixed term $Z = \Delta_0^{-1} = \sum_{0 \leqslant i \leqslant \infty} f_i$. At the close of the present chapter Newton cites

an example in which $f_i = \dfrac{i!}{1.3.5....(2i+1)} \left[= \int_0^1 (\tfrac{1}{2}[1-x^2])^i . dx \right]$, from which

$$\Delta_i^j = \int_0^1 (\tfrac{1}{2}[1-x^2])^i (\tfrac{1}{2}[1+x^2])^j . dx$$

and consequently $Z = \int_0^1 (\tfrac{1}{2}[1+x^2])^{-1} . dx = 2\tan^{-1}1 = \tfrac{1}{2}\pi$. This exact summation of the series expansion of $2\sqrt{[z/(1+z)]} \tan^{-1}\sqrt{[z(1+z)]}$, $z = \tfrac{1}{2}$ (see §2: note (44) below) seems here to elude Newton (who later broaches a second finite-difference method for approximating Z to desired numerical accuracy): in this case, to be sure, the general 'progressio' $\Delta_0^{j+1}/\Delta_0^j$ of the series $A, B, C, D, ...$ of differences is, when j is not a positive integer, non-algebraic and con-

be $B, b_2, b_3, \ldots$; the second ones $C, c_2, c_3, \ldots$; $Z \quad Z_2 \quad Z_3 \quad Z_4 \quad Z_5 \quad Z_6 \quad Z_7[\ldots]$
the third ones $D, d_2, d_3, \ldots$, and so forth: $A \quad b \quad c \quad d \quad e \quad f \; [\ldots]$
that is, so as to make $A - b = B, \; b - c = b_2,$ $B \quad b_2 \quad b_3 \quad b_4 \quad b_5$

$c - d = b_3, \quad \ldots; \quad B - b_2 = C,$ $C \quad c_2 \quad c_3 \quad c_4$
$\quad b_2 - b_3 = c_2, \quad \ldots; \quad C - c_2 = D, \quad \ldots,$ $D \quad d_2 \quad d_3$
     $E \quad e_2$
taking always the following from the preceding ones[66] F
and noting the signs of the remainders. It is manifest
that the total of all the $B, b_2, b_3, b_4, \ldots$ must be A, and that B must be the aggregate of all the $C, c_2, c_3, \ldots$, and so on indefinitely; consequently, if the series $A, B, C, D, E, F, \ldots$ of totals be extended in A's direction by prefixing a term Z, that quantity Z will be the aggregate of all the $A, b, c, d, e, \ldots$. And conversely, if the series $A, B, C, D, \ldots$ is to be increased by the new term Z, of the terms $A, B, C, D, \ldots$ there must be gathered the first differences $b, b_2, c_2, d_2, \ldots$; the second ones $c, b_3, c_3, d_3, \ldots$; the third ones $d, b_4, c_4, \ldots$, and the rest, and the total of the differences $A, b, c, d, e, \ldots$ is to be accepted for Z.[67] ‖ Therefore, if the progression of the series $A, B, C, D, \ldots$ in the direction of A can be exactly ascertained, the series $A, b, c, d, \ldots$ will have an exact termination, and conversely so:[68] but if not, an approximation must be employed. This is accomplished by drawing a regular line through any number of given points. To any straight line ZE given in position at equal intervals ZA, AB, BC, CD, DE let there be erected the normals $ZQ, AP, BP, CP, DP, \ldots$ equal respectively to the

sequently not to be discovered by the simple algorithm proposed in the previous paragraph. (In a comparable instance twenty years earlier Newton had likewise, it would appear, failed to identify the integral pattern $\int_0^1 (x+x^2)^i \, . \, dx$, $i = 0, 1, 2, 3, \ldots$, which underlies the series $1, \frac{5}{6}, \frac{31}{30}, \frac{209}{140}, \ldots$ communicated by Wallis to Kenelm Digby in November 1657; see 1: 116, note (94).)

The remainder of this paragraph and the two following ones, cancelled by Newton in favour of the numerical finite-difference argument which concludes the chapter, are here retained because of their general interest. (Study of the manuscript shows that the cancellation was made at a late stage, perhaps in preparation for the revise treatise 'De Computo Serierum' (§2 below), in which the following argument is subsumed into a separate 'Cap 3. De serierum interpolatione'.)

(68) Newton first wrote 'Sive igitur seriei $A.b.c.d.e$ &c terminatio definienda sit sive series A, B, C, D, E &c producenda, res eo redit' (Whether therefore the termination of the series $A, b, c, d, e, \ldots$ is to be delimited or the series $A, B, C, D, E, \ldots$ extended, the circumstance is one and the same), and then to much the same effect 'Eodem igitur recidit seriei A, b, c, d, e &c terminationem Z definire et seriei $A, B, C,$ &c terminū novum producere'. In a first version of the preceding diagram the row '$A.b.c.d.e.f$' and the oblique column '$A.B.C.D.E.F$' were originally interchanged, but Newton has not everywhere in his manuscript made the necessary revisions to allow for the inversion: these, where pertinent, we have silently incorporated in the text reproduced. The change-over was evidently made so that in the sequel Newton's standard upper case denotation of the interpolation arguments $Z, A, B, C, D, \ldots$ would be preserved.

B, C, D &c respectivè et linea regularis transiens per puncta omnia data *P, P, P, P* &c abscindet longitudinem *ZQ* æqualem[69] quæsito seriei termino *Z*. Eadem ratione intercalatur series inserto ubivis termino *RS*.[70] Ea sic fiunt.

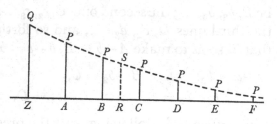

Ordinatarum *A, B, C, D, E, F* &c collige differentias primas $b, {}^2b, {}^2c, {}^2d$, &c. secundas $c, {}^3b, {}^3c, {}^3d$ &c. tertias $d, {}^4b, {}^4c$ [&c] cæterasq̃. Deinde ut ex assumpta utcunq̃ longitudine *AR* innotescat longitudo ordinatæ *RS*, pone intervalla *AB, BC, CD, DE* unitates esse et dic $AP = a$.[71]

$$-AR = k. \quad \tfrac{1}{2}kx - BR = l. \quad \tfrac{1}{3}lx + CR = m. \quad \tfrac{1}{4}mx + DR = n. \quad \tfrac{1}{5}nx + ER = o. \quad \&c$$

pergendo usq̃ ad penultimum terminum *PE* et signis probe observatis erit $RS = a^{(71)} + bk + cl + dm + en + fo$ &c.[72]

Linea ad quam ordinata *RS* terminatur Parabolici est generis[73] atq̃ adeo non satis quadrat cum progressionibus Hyperbolicis serierum convergentium.[74] Aptando hanc Parabolam ad puncta aliquot curvarum linearum approximantur areæ curvarum omnium. Potest enim semper quadrari.[75] Tabulæ cujuscunq̃ generis ex datis paucis earum terminis eadem ratione complentur, motus siderum[76] ex paucis observationibus innotescunt in locis intermedijs, I[n]tercalantur series illæ nobilissimæ quibus Wallisius noster aream circuli et Hyperbolæ dudum exhibuit[77] et approximationum corriguntur errores.[78] Sed ad series convergentes si calculus exactior esse debet, aptanda est linea hyperbolica cujus Asymptotos sit linea [*ZE*]. Esto series convergens *A. B. C. D. E* &c. Ducatur Parabola per terminos ordinatarum $\dfrac{1}{A} \cdot \dfrac{1}{B} \cdot \dfrac{1}{C} \cdot \dfrac{1}{D} \cdot \dfrac{1}{E}$ &c et erit $\dfrac{1}{RS}$

---

(69) Newton tentatively altered this to 'respondentem' (corresponding) and then returned to his original choice of phrase.

(70) An unnecessary following affirmation that 'Ducitur verò hæc linea regularis per puncta omnia' (For, to be sure, this regular line is drawn through every one of the points) is deleted.

(71) That is, of course, *A*.

(72) The simple 'Harriot–Briggs' formula for intercalating the general 'ordinate' *RS* in the sequence of terms *A, B, C, D, E*, ... (given at unit-intervals of argument) by means of the advancing finite differences $b, c, d, e, f, ...$ If, on setting $AR = x$ (whence $AB = 1$, $AC = 2$, $AD = 3$, ...), the 'linea regularis' (*P*) is defined by $RS = f_x$ (so that $AP = A(= a) = f_0$, $BP = B = f_1$, $CP = C = f_2$, ...), at once $-AR = -x$, $-BR = -(x-1)$, $CR = -(x-2)$, $DR = -(x-3)$, ... as drawn, while Newton's differences are

$$b = f_0 - f_1 = -\Delta_0^1, \quad {}^2b = f_2 - f_1 = -\Delta_1^1, \quad ...; \quad c = b - {}^2b = +\Delta_0^2,$$

$$^2c = {}^2b - {}^3b = +\Delta_1^2, \quad ...; \quad d = c - {}^2c = -\Delta_0^3, \quad ...;$$

terms $Z, A, B, C, D, \ldots$ of the series, and the regular line passing through all the given points $P, P, P, P, \ldots$ will cut off the length ZQ equal[69] to the required term Z of the series. By the same method the series is intercalated by a term RS inserted at any point.[70] That is accomplished in the following way.

Of the ordinates $A, B, C, D, E, F, \ldots$ gather the first differences $b, b_2, c_2, d_2, \ldots$; the second ones $c, b_3, c_3, d_3, \ldots$; the third ones $d, b_4, c_4, \ldots$, and the rest. Then, so that from any length AR taken arbitrarily the length of the ordinate RS may be found out, set the intervals AB, BC, CD, DE to be units and call $AP = a$,[71] $-AR = k$, $\frac{1}{2}kx - BR = l$, $\frac{1}{3}lx + CR = m$, $\frac{1}{4}mx + DR = n$, $\frac{1}{5}nx + ER = o$, and so on, proceeding as far as the last term but one PE: when the signs are appropriately observed, there will then be $RS = a$[71]$+ bk + cl + dm + en + fo \ldots$.[72]

The line at which the ordinate RS terminates is of parabolic kind[73] and consequently does not square adequately with hyperbolical progressions of convergent series.[74] On fitting this parabola to a set of points on curved lines the areas of all curves are approximated: for it can always be squared.[75] By the same method tables of every shape and form are, when a few of their entries are given, completed therefrom, from a few observations the movements of comets[76] are ascertained in the intermediate positions, intercalation is made of those most noble series by which our own Wallis once exhibited the area of the circle and hyperbola,[77] and errors in approximation are corrected.[78] But as for converging series, if the calculation ought to be more precise, to them must be fitted a hyperbolic line whose asymptote shall be ZE. Let the convergent series be $A, B, C, D, E, \ldots$, draw a parabola through the end-points of the ordinates $1/A, 1/B, 1/C, 1/D, 1/E, \ldots$ and $1/RS$ will then be the ordinate of this hyperbola.

---

and so on. Accordingly, Newton's expansion becomes, in modern equivalent,

$$f_x \equiv f_0 + x\Delta_0^1 + \tfrac{1}{2}x(x-1)\Delta_0^2 + \tfrac{1}{6}x(x-1)(x-2)\Delta_0^3 \ldots.$$

A few months afterwards Newton inserted a little variant revise of this interpolation formula in the third book of his *Principia* as Case 1 of Lemma V (reproduced in **1**, 1, Appendix above).

(73) On putting $b = a - \alpha$, $c = a - 2\alpha + \beta$, $d = a - 3\alpha + 3\beta - \gamma$, $e = a - 4\alpha + 6\beta - 4\gamma + \delta$, $\ldots$, Newton's preceding formula reduces to $RS = a + \alpha x + \beta x^2 + \gamma x^3 + \delta x^4 \ldots$, where $AR = x$. Compare **1**, 1, Appendix: note (4).

(74) Which may, of course, be equally useful on occasion, as James Stirling was later to see. See **1**, 1, §3: note (3).

(75) See **1**, 1, Appendix: note (6) above.

(76) 'stars' in Descartes' sense (*Principia Philosophiæ* (Amsterdam, 1644): Pars III: 144–64). As we shall see in the next volume Newton had spent much time during the period 1681–4 in seeking a viable method of accurately computing the orbits of comets, spurred on by the recent appearance of the two great comets of 1680–1 and of 1682 ('Halley's'). A few months later he was to set his general divided differences interpolation formula in his *Principia* as an introduction to Book 3, Lemma VI. 'Ex observatis aliquot locis Cometæ invenire locum ejus ad tempus quodvis intermedium datum'; see **1**, 1, Appendix: note (1).

(77) See I: 99, note (23) and 116, note (94). A following phrase 'aliæꝗ omnes' (and all others) is very rightly cancelled.

(78) Newton illustrates this point with an example at the close of the present chapter.

ordinata hujus hyperbolæ. Loquor de seriebus debitè præparatis. Nam variæ sunt præparationes quibus differentiæ terminorum ejus minuuntur.[79] Augere minuere multiplicare vel dividere licet terminos ejus per terminos progressionis cujusvis regularis ut fecit Wallisius in seriebus suis.[80]

Sed quid tandem agendū est ubi nec seriei A, B, C, D, E &c progressio innotescit nec series A, b, c, d, e &c convergit celeriter? Transmutanda est series[81] Z, A, B, C, D, E, F in hanc $\frac{1}{Z}\cdot\frac{1}{A}\cdot\frac{1}{B}\cdot\frac{1}{C}\cdot\frac{1}{D}\cdot\frac{1}{E}$ &c. Dein colligendæ ejus differentiæ primæ, puta $\frac{1}{A}-\frac{1}{Z}$. k. [^2k. ^3k. ^4k] &c. secundæ, puta $k-\frac{1}{A}+\frac{1}{Z}$. [$l$]. ^2l. [^3l. ^4l. &c.] cæteræq̃ ut in annexo diagrammate, et ponendum

$$\frac{1}{Z}=\frac{1}{A}-k+l-m \quad \text{&c.}^{(82)}$$

$\frac{1}{Z}$.	$\frac{1}{A}-\frac{1}{Z}$.	$k-\frac{1}{A}+\frac{1}{Z}$.	$l-k+\frac{1}{A}-\frac{1}{Z}$.	
$\frac{1}{A}$.	k.	l.	m.	n.
$\frac{1}{B}$.	^2k.	^2l.	^2m.	^2n.
$\frac{1}{C}$.	^3k.	^3l.	^3m.	^3n.
$\frac{1}{D}$.	^4k.	^4l.	^4m.	^4n.

Si hæc series nondum convergit satis celeriter, capienda est secunda series differentiarum convergentium, id est si series differentiarum primarum k, ^2k, ^3k, ^4k &c convergit capienda est series differentiarum secundarum $k-\frac{1}{A}+\frac{1}{Z}$. l. ^2l. ^3l. ^4l &c. Si series differentiarum primarum divergit et ea secundarum l. ^2l. ^3l, ^4l [&c] convergit, capienda est series tertiarum[83] aut ea quartarum si fortè secundæ divergunt et tertiæ convergunt et sic in cæteris.

Demus seriem tertiarum $l-k+\frac{1}{A}-\frac{1}{Z}$. m. ^2m. ^3m. ^4m &c capiendam esse, et pro

---

(79) The equivalent 'red[d]untur exiguæ' is cancelled, along with the sequel 'ut multiplicando vel dividendo terminos per progressionē quamvis geometricam vel arithmeticam quæ exactè intercalari et in infinitum produci potest' (for instance, by multiplying or dividing the terms by any geometrical or arithmetical series which can be exactly interpolated and extended to infinity). A further sentence 'Dabo exemplum' (I shall give an example) was deleted without being implemented.

(80) Newton refers to John Wallis' various attempts to find the general 'progression' of certain types of integral sequence in his *Arithmetica Infinitorum, sive Nova Methodus Inquirendi in Curvilineorum Quadraturam, aliaq̃ difficiliora Matheseos Problemata* (Oxford, 1656). Thus, having computed the particular values $g_0 = 1$, $g_1 = \frac{1}{6}$, $g_2 = \frac{1}{30}$, $g_3 = \frac{1}{140}$, ... of the sequence $g_s = \int_0^1 (x-x^2)^s\,.\,dx$ and hence inferring that $g_s = \frac{s}{2(2s+1)}\times g_{s-1}$ (Propositio CXXXIII: 106), Wallis subsequently examined the derived series $g'_s = (2s+1)\times g_s\left[=\frac{s}{2(2s-1)}\times g'_{s-1}\right]$ and

I speak of series which have been duly prepared. For there are various preparatory procedures by which the differences of its terms may be made smaller.[79] To enlarge, diminish, multiply or divide its terms by the terms of any regular progression (as Wallis did in the case of his series) is permissible.[80]

But what action, at length, is to be taken when neither the progression of the series $A, B, C, D, E, \ldots$ is known nor the series $A, b, c, d, e, \ldots$ convergent swiftly enough? [81]The series $Z, A, B, C, D, E, F \ldots$ must be transformed into this $1/Z, 1/A, 1/B, 1/C, 1/D, 1/E, \ldots$. Then you must gather its first differences, say $1/A-1/Z, k, k_2, k_3, k_4, \ldots$; its second ones, say $k-1/A+1/Z, l, l_2, l_3, l_4, \ldots$, and its remaining ones (as in the adjoining diagram) and then you must put $1/Z = 1/A-k+l-m \ldots$.[82]

If this series is still not swiftly convergent enough, a second series of convergent differences must be taken; that is, if the series of first differences converges [slowly], the series $k-1/A+1/Z, l, l_2, l_3, l_4, \ldots$ of second differences must be taken. If the series of first differences diverges while that of the second

$1/Z$				
	$1/A-1/Z$			
$1/A$		$k-1/A+1/Z$		
	k		$l-k+1/A-1/Z$	
$1/B$		l		n
	k_2		m	
$1/C$		l_2		n_2
	k_3		m_2	
$1/D$		l_3		n_3
	k_4		m_3	
		l_4		n_4
			m_4	

ones $l, l_2, l_3, l_4, \ldots$ converges, you must take the series of third differences,[83] or that of fourth ones should the second ones diverge and the third ones converge, and so on in other cases. Let us grant that the series $l-k+1/A-1/Z, m, m_2,$

---

showed that for $s = 0, 1, 2, 3, \ldots g'_s = 1 \Big/ \binom{2s}{s}$ (Propositio CLXVI: 133; compare Propositio

CXXXII: 105), readily concluding that $g_s = 1 \Big/ \prod_{1 \leqslant i \leqslant s} (2i+1) \times \binom{2s}{s}$, s positive integral.

In sequel Newton began the first word of a new sentence 'Multipl...' but never finished it, evidently deciding at that point to cancel the whole of the preceding passage. At the head of his cancellation he then drafted the opening of a new paragraph, '*Hoc pacto solvitur problema si series A, b, c, d, e celeriter convergit. Sed ubi secus transm[utanda est]*' (In this manner is the problem solved if the series $A, b, c, d, e, \ldots$ speedily converges. But when it is not so, it must be transformed), which then made way for the revised version which follows.

(81) A first continuation reads '*alterutra in aliam formam. Seriei A, B, C, D, E &c transmutatio talis est*' (One or other [series must be transformed] into another shape. This is such a transformation of the series $A, B, C, D, E, \ldots$).

(82) Assuming, that is, that this alternating series of differences is convergent.

(83) '$l-k+\dfrac{1}{A}-\dfrac{1}{Z}.m.^2m.^3m.^4m$ &c' is cancelled.

ejus primo termino ignoto $l-k+\dfrac{1}{A}-\dfrac{1}{Z}$ scribamus y, transmutemus eam in hanc

$\dfrac{1}{y} \cdot \dfrac{1}{m} \cdot \dfrac{1}{2m} \cdot \dfrac{1}{3m} \cdot \dfrac{1}{4m}$ &c, deinde ex differentijs terminorum $\dfrac{1}{m} \cdot \dfrac{1}{2m} \cdot \dfrac{1}{3m} \cdot \dfrac{1}{4m}$ &c

quæramus terminū primum $\dfrac{1}{y}$ ut supra. Sit is $=r$, seu $\dfrac{1}{r}=y$ et erit $1-k+\dfrac{1}{A}-\dfrac{1}{Z}=\dfrac{1}{r}$

adeoop $\dfrac{1}{l-k+\dfrac{1}{A}-\dfrac{1}{r}}=Z$. Eadem ratione si Z nondū satis exactè prodiret, posset

opus tertiò institui,[84] nam pergitur hac methodo in infinitum. Sed regulam illustremus exemplo.

Invenienda sit terminatio seriei $1+\dfrac{1}{1,3}+\dfrac{1,2}{1,3,5}+\dfrac{1,2,3}{1,3,5,7}$ &c. Differentiæ terminorū loco seriei Z. A. B. C. D. E &c scribendæ, sunt

$$Z. \quad 1. \quad \tfrac{2}{3}. \quad \tfrac{7}{15}. \quad \tfrac{12}{35}. \quad \tfrac{83}{315}. \quad \tfrac{146}{693} \quad \&c.^{[85]}$$

et reciproce $\dfrac{1}{Z}$. 1. $\tfrac{3}{2}$. $\tfrac{15}{7}$. $\tfrac{35}{12}$. &c. id est in numeris decimalibus

$\dfrac{1}{Z}$

1	−0.5	0.14285714	0.01290476[86]
1,5	−0.64285714	0.13095238	0.02624784
2,14285714	−0.77380952	[0.]10470454	
2,91666666	−0.87851406		
3,79518072		Inveniend[87]	

---

(84) As in his preliminary draft (see Appendix 2.3 below) Newton first concluded with the words 'quartoop et sæpius, sed prima vel secunda vice prodire solet [terminus ille satis exactè]' (and a fourth and even more often, but the term usually comes out exact enough the first or second time round).

(85) The sequence is, in fact, $\{f_j\}$ where (see note (67) above)

$$f_j = \int_0^1 (\tfrac{1}{2}[1+x^2])^j . dx, \quad j = -1, 0, 1, 2, 3, 4, 5, \dots.$$

For positive integer j the general term f_j is easily derived from the recursion

$$f_j = (1+j \cdot f_{j-1})/(2j+1).$$

The allied sequence $\int_0^1 (1+x^2)^j . dx$, $j = 0, 1, 2, 3, \dots$, was dealt with by Wallis in his letter to Leibniz on 21 November 1696; see i: 116–17, note (94).

m_3, m_4, ... is to be taken and, writing y in place of its unknown first term $l-k+1/A-1/Z$, let us transform it into this one $1/y$, $1/m$, $1/m_2$, $1/m_3$, $1/m_4$, ... and then from the differences of the terms $1/m$, $1/m_2$, $1/m_3$, $1/m_4$, ... hunt out the first term $1/y$, as above. Let this be equal to y, that is, $1/r = y$, and there will be $l-k+1/A-1/Z = 1/r$ and so $Z = 1/(l-k+1/A-1/r)$. Should a sufficiently precise value for Z still not result, the same method could be worked through a third time,[84] for you can go on using this procedure indefinitely. But let us illustrate the rule by an example.

Suppose the termination of the series $1 + \dfrac{1}{1.3} + \dfrac{1.2}{1.3.5} + \dfrac{1.2.3}{1.3.5.7} \ldots$ is to be found. The differences of the terms (to be written in place of the series Z, A, B, C, D, E, ...) are Z, 1, $\frac{2}{3}$, $\frac{7}{15}$, $\frac{12}{35}$, $\frac{83}{315}$, $\frac{146}{693}$, ...[85] and reciprocally $1/Z$, 1, $\frac{3}{2}$, $\frac{15}{7}$, $\frac{35}{12}$, ...; that is, in decimals,

$1/Z$			
1			
1·5	−0·5		
2·14285714	−0·64285714	0·14285714	
2·91666666	−0·77380952	0·13095238	0·01290476[86]
3·79518072	−0·87851406	0·10470454	0·02624784

... to be found[87]

---

(86) This should be 0·01190476. Rather dauntingly, the fourth difference −0·01434308 is larger than this in magnitude but the differences again begin to decrease numerically from the fifth (−0·00876694) onwards, though the rate of convergence to zero is slow. By a rough computation $1/Z = 1-0·5+0·1428 \ldots +0·0119 \ldots -0·0143 \ldots -0·0087 \ldots$ is very nearly 0·637 and so $Z \approx 1·57$. (Accurately $Z = \frac{1}{2}\pi$; see note (67) above and §2: note (44) following.)

(87) The manuscript breaks off abruptly at this point. Evidently Newton would have gone on to compute $1/Z$ as the sum of the top row of his difference scheme and then Z as its reciprocal. He may well have been dissuaded from going on with his calculation when he noticed that the next (fourth) difference was numerically larger than the third; see the previous note. Before he made the trial he could not have suspected that his initial series $1+\frac{1}{3}+\frac{2}{15}+\frac{2}{35}+\frac{8}{315}$ ... (the particular case $z=\frac{1}{2}$ of the 'transmuted' expansion of $2\sqrt{[z/(1-z)]}\tan^{-1}\sqrt{[z/(1-z)]}$; see page 608 below) in fact converges more swiftly than the difference-series to which it is here transformed.

CAP 3.

OSTENDITUR QUÀM LATÈ PATET ANALYSIS PER ÆQUATIONES INFINITAS.[88]

Scripsi in epistola mea prima[89] [*Ex his videre est quantum fines Analyseos per hujusmodi infinitas æquationes ampliantur: quippe quæ earum beneficio, ad omnia, pene dixerim, problemata (si numeralia Diophanti et similia excipias) sese extendit. Non tamen omninò universalis evadit, nisi per ulteriores quasdam methodos eliciendi series infinitas. Sunt enim quædam Problemata in quibus non liceat ad series infinitas per divisionem vel extractionem radicum simplicium affectarúmve pervenire: sed quomodo in istis casibus procedendum sit jam non vacat dicere, ut neğ alia quædam tradere quæ circa reductionem infinitarum serierum in finitas, ubi rei natura tulit, excogitavi.*] Quorum verborum sensus est methodum resolvendi problemata in seriebus infinitis valde generalem esse sed per solas divisiones & extractiones radicum satis generalem non esse, me aliquid amplius excogitasse et præterea nonnulla meditatum esse de reductione serierum infinitarū ad æquationes finitas sed hac in parte me methodum adeo generale invenisse minime asserui. Quoniam igitur Cl. Leibnitius in his hærere videatur,[90] primum exponam problematum genera quæ reducuntur ad series infinitas per divisiones et extractiones radicum; Dein reliqua problematum genera quorum solutio postulat methodū ulteriorem (sic enim constabit quam generalis sit hæc methodus) & ultimo nonnihil addam de reductione æquationum infinitarum ad finitas.[91]

Resolu[t]ionis[92] vulgarium Problematum quæ per Analysin Vulgi[93] difficulter solvuntur accipe hoc specimen.

---

(88) 'CONVERGENTES' (CONVERGENT) was earlier written (on f. 17ʳ). A cancelled first opening to this chapter is reproduced in Appendix 2.4 below. As we have seen (note (41) above) Newton's provisional title in his scheme had been the more explicit 'Series appropinquantes ex lineis pluribus Geometricè [now understand 'et Mechanicè' also] inveniendis componere'.

(89) A blank which follows in the manuscript is just large enough to take the following inserted extract from Newton's letter of 13 June 1676 (*Correspondence*, 2: 29; compare III: 24, note (21)), manifestly that to whose 'verba' he goes on to refer.

(90) In his reply to Newton's *epistola prior* Leibniz wrote back on 17/27 August 1676: 'Quod dicere videmini plerasҀ difficultates (exceptis problematis Diophanteis) ad series infinitas reduci id mihi non videtur. Sunt enim multa usҀ adeò mira et implexa, ut neҀ ab æquationibus pendeant, neҀ ex quadraturis, qualia sunt (ex multis aliis) problemata methodi tangentium inversæ quæ etiam Cartesius in potestate non esse fassus est' (*Correspondence*, 2: 64). To this Newton had returned in his *epistola posterior* on 24 October the firm answer that 'inversa de tangentibus Problemata sunt in potestate, aliaҀ illis difficiliora: ad quæ solvenda usus sum duplici methodo, una concinniori, altera generaliori' (*Correspondence*, 2: 129) but chose to transmit the outline of both these latter methods in anagram form (see note (33) above). Leibniz very rightly felt this rebuff to be too definite and in his further reply on 11/21 June 1677 accurately concluded that 'Quod ait problemata methodi tangentium inversæ esse in potestate hoc arbitror ab eo intelligi per series scilicet infinitas. Sed à me ita desiderantur ut curvæ exhibeantur geometricè quatenus id fieri potest suppositis (minimum) quadraturis'

Chapter 3

The broad expanse of analysis by infinite[88] equations is revealed

I wrote in my first letter:[89]

From these instances it may be seen how far the bounds of analysis are widened by infinite equations of this sort: indeed, I could almost have said it extends, with their support, to all problems (making exception for numerical Diophantine ones and the like). It does not, however, prove to be completely universal until certain further methods of deriving infinite equations are deployed. For there are certain problems in which we are not free to arrive at infinite series by means of division or extraction of simple or affected roots. But how one should proceed in those cases I have now no time to say, nor again to pass on some other thoughts I have had about the reduction of infinite series to finite ones where the nature of the topic allows.

The sense of these words is that the method of resolving problems in infinite series is exceedingly general but to do so by divisions and root extractions is not general enough, and that I had developed something more adequate and had, moreover, given some consideration to the reduction of infinite series to finite equations (though in this regard I in no way asserted that I had found any very general method). Seeing, therefore, that Mr Leibniz appears to be stuck in these matters,[90] I shall first expound the types of problem which are reducible to infinite scrics by divisions and root extractions, and then the types of problem whose solution demands a more sophisticated method (for in this way the generality of this method will be settled); and, lastly, I shall add something on the reduction of infinite equations to finite ones.[91]

Of the resolution[92] of common problems which are solved with difficulty by common analysis[93] take this as a specimen.

(*Correspondence*, **2**: 216; see also note (36) preceding, where more lavish quotation from Leibniz' letter is made). In his 'Specimina' Newton briefly summarized this thrust and parry in a cancelled amplification of his present remark: 'Minus generalem esse methodum nostram suspicabatur Cl. Leibnitius, nec ad problemata de tangentibus inversa extendere. Respondi problematū illorum solutionem in potestate esse. Postulat vir Clarissimus talem problematum illorum solutionem ut quando id fieri potest series infinitæ vertantur in æquationes finitas. Id me facturum esse minime asserui' (Mr Leibniz suspected that my method is not so general, and does not extend to inverse problems of tangents. I replied that the solution of those problems is in my power. He now demands a solution of them such that, when it may be done, infinite series shall be turned into finite equations. That I could do that I asserted not at all).

(91) This last topic is not developed in the extant version of the 'Specimina'.

(92) The manuscript reads 'Resolusionis' and then 'Resolusitionis'. Newton first wrote 'Computi' (computation).

(93) Newton understands the conventional algebraic analysis—that of Viète and Descartes— which reduces problems to the solution of sets of finite equations.

In curva quacunꝗ linea regulari PQ invenire punctū P a quo si rectæ quotcunꝗ HP, IP, KP, LP ad totidem data puncta H I K L ducantur summa ductarum omniū HP+IP+KP+LP æquetur dato N.[94]

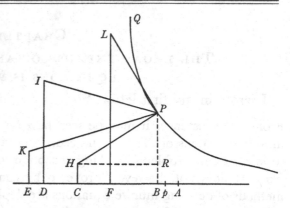

A punctis *P, H, I, K, L* ad rectam aliquam positione datam *EF* demitte perpendicula *PB, HC, ID, KE, LF* & in recta illa cape punctum aliquod *A* quod sit puncto *B* proximum quantum ex conjectura liceat assequi, id adeo ut indefinita quantitas *AB* fiat quam minima.[95] Eam dic x et *BP* y et ex natura curvæ *PQ* quære relationem inter x et y. Sit ea $\frac{1}{10}x^3-3x+2xy-y-6+y^3=0$.[96] Extrahe radicem y et fiet $y=2-\frac{1}{11}x+\frac{16}{1331}xx-\frac{16131}{1610510}x^3$ &c et inde $yy=4-\frac{4}{11}x+\frac{75}{1331}xx-\frac{68396}{1610510}x^3$ &c.[97] Ponantur jam $AC=3, CH=1, AD=5, DI=4. AE=6. EK=2. AF=-2. FL=6$ & summa $HP+IP+KP+LP=20$ et erit *BC* vel $HR=3-x. PR=y-1$ et inde $HP=\sqrt{10-6x+xx-2y+yy}$. Hic pro y et yy scribantur eorum valores et fiet $HP=\sqrt{10-\frac{68}{11}x+\frac{1374}{1331}xx-\frac{36134}{1610510}x^3}$ &c.[98] Et radice extracta

$$HP=\sqrt{10}-\frac{34}{11\sqrt{10}}x+\frac{512}{13310\sqrt{10}}xx+\frac{16318}{16105100\sqrt{10}}x^3 \text{ \&c.}^{[98]} \text{ Simili argumento}$$

invenientur $IP=\sqrt{29}-\frac{53}{11\sqrt{29}}x+\frac{18813}{66098\sqrt{29}}xx$ [&c].

$$KP=6-x+\frac{1}{1452}xx-\frac{16577}{115956720}x^3 \text{ \&c.}$$

$LP=\sqrt{20}-\frac{7}{11\sqrt{20}}x+\frac{1889}{53240\sqrt{20}}xx$ [&c].[99] Quorum omnium summa in numeris decimalibus est $19.01962-3.01444x+0.07361xx$ &c.[100] Est autem hæc summa

---

(94) In effect, the point *P* is constructed as the meet of the 'linea regularis' *PQ* and the curve (*P*) defined by $HP+IP+KP+LP = N$ (compare 1, §1: note (163) above).

(95) Newton tentatively changed the last phrase to read '...adeo ut series emergentes celeriter convergant' (so that the emergent series shall rapidly converge) but at once cancelled this revise.

(96) In geometrical terms a conchoidal cubic (hyperbolic hyperbolism) 'twisted' round its unique asymptote $y+10^{-\frac{1}{3}}x=\frac{2}{3}10^{\frac{1}{3}}$. Newton's first choice of curve *PQ* was the Cartesian folium '$x^3-3xy+y^3=0$. Extrahe radicem y et fiet $y=\frac{1}{3}xx \ldots$'. (Alternatively, on choosing $y^3-3xy = 0$ for fictitious equation, $y = \pm\sqrt{3}x^{\frac{1}{2}}+O(x^{\frac{3}{2}})$.)

(97) The coefficient of the term in x^3 should be $-\frac{68044}{1610510}$. The error throws the detail of the remaining computations correspondingly out.

(98) Correctly (see note (97)) the numerators of the terms in x^3 should be 35782 and 12578 respectively.

In any regular curved line PQ to find a point P such that, if any number of straight lines HP, IP, KP, LP be drawn to an equal number of given points H, I, K, L, the sum of all those drawn (HP+IP+KP+LP) shall be equal to a given magnitude N.[94]

From the points P, H, I, K, L to some straight line EF given in position let fall the perpendiculars PB, HC, ID, KE, and in that line take some point A which shall be as near to the point B as a crude guess allows you to attain: this so that the indefinite quantity AB shall become as small as possible.[95] Call it x and BP y and from the nature of the curve PQ seek the relationship between x and y. Let it be $\frac{1}{10}x^3+y^3+2xy-3x-y-6 = 0$.[96] Extract the root y and it will be $y = 2-\frac{1}{11}x+\frac{16}{1331}x^2-\frac{16131}{1610510}x^3 \dots$ and from this

$$y^2 = 4-\frac{4}{11}x+\frac{75}{1331}x^2-\frac{68396}{1610510}x^3 \dots \text{[97]}$$

Now put $AC = 3$, $CH = 1$, $AD = 5$, $DI = 4$, $AE = 6$, $EK = 2$, $AF = -2$, $FL = 6$ and the sum $HP+IP+KP+LP = 20$: there will then be

$$BC \text{ (or } HR) = 3-x, \quad PR = y-1$$

and thence $HP = \sqrt{[10-6x+x^2-2y+y^2]}$. Here in place of y and y^2 let their values be written in and there will come

$$HP = \sqrt{[10-\tfrac{68}{11}x+\tfrac{1374}{1331}x^2-\tfrac{36134}{1610510}x^3 \dots]},\text{[98]}$$

and after the root is extracted

$$HP = \sqrt{10}-\frac{34}{11\sqrt{10}}x+\frac{512}{13310\sqrt{10}}x^2+\frac{16318}{16105100\sqrt{10}}x^3 \dots \text{[98]}$$

By a similar deduction there will be found

$$IP = \sqrt{29}-\frac{53}{11\sqrt{29}}x+\frac{18813}{66098\sqrt{29}}x^2 \dots, \quad KP = 6-x+\frac{1}{1452}x^2-\frac{16577}{115956720}x^3 \dots$$

and $LP = \sqrt{20}-\frac{7}{11\sqrt{20}}x+\frac{1889}{53240\sqrt{20}}x^2 \dots$.[99] The sum of all these, in decimals, is $19{\cdot}01962-3{\cdot}01444x+0{\cdot}07361x^2 \dots$.[100] This sum, however, is 20, so that

---

(99) Each of these expansions has an erroneous last term. Accurately

$$IP = \sqrt{[(4-y)^2+(5-x)^2]} = \sqrt{29}-\frac{53}{11\sqrt{29}}x+\frac{6163}{77198\sqrt{29}}x^2 \dots,$$

$$KP = \sqrt{[(2-y)^2+(6-x)^2]} = 6-x+\frac{1}{1452}x^2-\frac{71}{175692}x^3 \dots,$$

and $\quad LP = \sqrt{[(6-y)^2+(2+x)^2]} = \sqrt{20}+\frac{26}{11\sqrt{20}}x+\frac{4211}{13310\sqrt{20}}x^2 \dots$

(100) The coefficients of the second and third terms are considerably out (see notes (98) and (99) above) but the general structure of the argument is unaffected. Observe that Newton has chosen $N = 20$ to approximate $\sqrt{10}+\sqrt{29}+6+\sqrt{20} = 19{\cdot}019 \dots$ to within a unit.

$=20$ adeoɋ reducta æquatione fit $0.98038 = -3.01444x + 0.07361xx$ &c et extracta radice $x = 0.323$. Et opere continuato valorem x ad plura figurarum[101] loca producere licet. Vel etiam posito $Ap = 0.323$, et errore $Bp = z$ licebit scribere $0.323 + z$ pro x & per operis repetitionem valorem z ad decem[102] vel plura decimalium figurarum producere.

Simili computo possunt intersectiones curvarum quarumvis regulariũ inveniri, et inde Problemata omnia solvi quæ ad inventionem talium intersectionum deduci possunt. Ut si Problema aliquod ad æquationes duas $x^4 - ax^3 + b^4 + cxyy - dy^3 = 0$ et $3ax^3 - d^3x + e^4 - f^3y + g^2xy + hxxy - k^2yy - 2y^4 = 0$, aliasve[103] magis compositas deduceretur[,] Fingerem curvas duas *EC, ED* his definiri ea lege ut y designet earũ ordinatas *BC* & *BD* & x basem communem *AB*, et quærerem intersectionem *E* seu basis longitudinem *AG* ubi ordinatæ *BC* & *BD* seu y in utraɋ æquatione evadunt æquales. Et primo quidem ubi quæstio determinatur ponerem numeros pro quantitatibus datis *a b c* &c & longitudinem *AG* ratione quacunɋ seu conjecturali seu mechanica quantum liceret sine molestia calculi assequerer. Sit

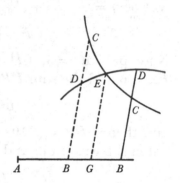

ea numerus aliquis *N* et error nominetur z. Tum in æquationibus scriberem $N + z$ pro x et ex æquationibus duabus prodeuntibus extraherem seorsim radices y, easɋ sibi mutuo æquales ponendo haberem æquationem novam ex qua extraherem radicem z.

Quinetiam ad curvas lineas vulgo dictas Mechanicas[104] inveniri possunt æquationes convergentes et earum beneficio problemata in his curvis non aliter solvi quam in curvis simplicioribus.[105] In angulis item valet hæc methodus æque ac in lineis ubi hi in quæstione sunt. Ut si determinandum esset Quadrilaterũ *ABCD* cujus latera $AB = 3, BC = 4, CD = 5$, $DA = 6$[106] una cum angulorum *BAC*, *CAD* vel differentia vel proportione dantur, inveniatur utcunɋ longitudo aliqua proxima diagonali *AC* puta 4 sitɋ error

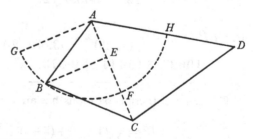

---

(101) 'decimalium' (decimals) is cancelled.

(102) That is, twice the $5D$ to which the preceding calculation is carried.

(103) A cancelled first continuation reads 'aliasve similes deduceretur et in harum ulteriore reductione hæreret Anal[ysis]' (or . . . to similar ones and in their further reduction [algebraic] analysis should become stuck).

(104) By Descartes, that is. See III: 141, note (236); and compare I: 369, note (2).

(105) A first, rejected following illustration in the case of the Archimedean spiral (f. 17ᵛ) is reproduced in Appendix 2.5.

when the equation is reduced there proves to be

$$0{\cdot}98038 = -3{\cdot}01444x + 0{\cdot}07361x^2 \ldots$$

and, when the root is extracted, $x = 0{\cdot}323$. By continuing the operation we are at liberty to extend the value of x to further places of figures.[101] Or again, on putting $Ap = 0{\cdot}323$ and the error $Bp = z$, it will be permissible to write $0{\cdot}323 + z$ in place of x and by a repetition of the operation extend the value of z to ten[102] or more decimal places.

By a similar computation the intersections of any regular curves may be found, and in consequence all problems solved which can be brought down to the finding of such intersections. So if some problem were reduced to the two equations $x^4 - ax^3 + b^4 + cxy^2 - dy^3 = 0$ and

$$3ax^3 - d^3x + e^4 + (-f^3 + g^2x + hx^2)\,y - k^2y^2 - 2y^4 = 0,\text{[103]}$$

or to others more complicated, I should imagine the two curves EC, ED to be defined by these when y is stipulated to denote their ordinates BC and BD, x the common base AB, and I should then seek out the intersection E (or its base length AG) where the ordinates BC and BD—that is, the values of y in either equation—come to be equal. And in the first place, indeed, where a question is determinate I should put numbers in place of the given quantities a, b, c, ... and obtain the length of AG by any method, guesswork or a rough construction, as far as permissible without troublesome calculation. Let this be some number N and the error be named z. In the equations I would then write $N + z$ in place of x and from the two resulting equations would extract the roots y separately: on setting these equal one to the other I would have a new equation from which I could extract the root z.

To be sure, convergent equations can be found for the curved lines commonly dubbed 'mechanical',[104] and with their assistance problems on these curves are solved no differently than in simpler curves.[105] In the case of angles, likewise, this method is valid, and equally so in lines where these are in question. For instance, if you had to determine the quadrilateral $ABCD$ whose sides $AB = 3$, $BC = 4$, $CD = 5$, $DA = 6$ are given[106] together with either the difference or ratio of the angles $B\widehat{A}C$, $C\widehat{A}D$, let some length—4, say—which is close to that of the diagonal AC be found in some manner and let its error be x, supposing

---

(106) Newton first added in sequel 'et angulus BAD graduum 125 et invenienda sit diagonalis AC: ponatur $AC = 5$ proximè sitcȝ error x' (and $B\widehat{A}D = 125°$, and let its diagonal AC be to be found: suppose $AC = 5$ approximately and let the error be x ...). In this case by direct computation $BD = 8{\cdot}10239 \ldots$ and so $B\widehat{C}D \approx 128° \, 2\tfrac{1}{2}'$, whence $AC = 3{\cdot}942 \ldots$ and $B\widehat{A}C \approx 69°$: accordingly Newton's following approximation $AC \approx 4$ is much nearer to the truth. Conversely, the requirement that $AC \approx 4$ be a good approximation may be met by the condition $B\widehat{A}C + C\widehat{A}D = 125°$ or alternatively, for instance, $B\widehat{A}C : C\widehat{A}D \approx 5:4$.

x posito $AC = 4 + x$. Tum a B ad AC demisso perpendiculo BE ex lateribus trianguli ABC inveniatur $AE = \dfrac{9 + 8x + xx}{8 + 2x}$ seu $= \frac{9}{8} + \frac{21}{32}x - \frac{5}{128}xx + \frac{5}{512}x^3$ &c.[107] Sin centro A radio AB describatur circuli Quadrans AFG et posito radio $AB = r$ et[108] arcu $BG = z$ habebitur iterum $AE = z - \dfrac{z^3}{6rr} + \dfrac{z^5}{120r^4}$ &c[109] ut in epistola mea prima docetur, et inde si pro r scribatur 3 & ex æquatione duorum valorum ipsius AE extrahatur radix z eaçç subducatur de dato arcu quadrantali GF habebitur arcus BF.[110] Et simili argumento invenietur arcus FH.[111] Tum ex data angulorum BAC CAD vel differentia vel proportione dabitur vel differentia vel proportio arcuum inventorum BF FH et inde habebitur æquatio cujus radice extracta x Problema solvitur.

In hujusmodi problematis prima difficultas est rudis aliqua approximatio quantitatis quæsitæ. Pro varijs problematum generibus ea variè institui potest. Suffecerit hic regula quadam generali rem complecti. Pro quantitate quæsita assumatur numerus quilibet OA et inde colligatur error quilibet AP. Dein pro eadem quantitate quæsita assumantur alij numeri OA^2, OA^3, OA^4, OA^5 &c et inde similiter colligantur errores A^2P^2, A^3P^3, A^4P^4, A^5P^5. Tum per puncta omnia

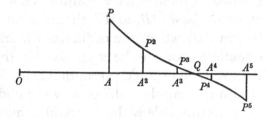

P, P^2, P^3, P^4, P^5 regulari[112] manus motu ducatur curva $PP^2P^3P^4P^5$ secans rectam OA in Q et erit OQ approximatio quæsita.[113] Sic in primo problematum hujus capitis pro longitudine AP assumpto numero quovis OA, inde colligo longitudinem BP, dein circino capiendo longitudines HP, IP, KP, LP colligo

---

(107) This should read '$\frac{9}{8} + \frac{23}{32}x - \frac{7}{128}xx + \frac{7}{512}x^3$ &c'.

(108) Newton first proceeded to set the 'sinus rectus' AE of arc $\widehat{BG}$ equal to z, thence determining '$GB[= r\sin^{-1}(z/r)] = z + \dfrac{z^3}{6rr} + \dfrac{3z^5}{40r^4}$ &c' where $r = 3$ and

$$'z[=(9 + 8x + x^2)/(8 + 2x)] = \frac{9}{8} + \frac{21}{32}x - \frac{5}{128}xx \text{ &c [!]}. \quad \frac{z^3}{6rr} = \frac{27}{512} + \frac{81 \times 21}{64 \times 64}x \text{ [!!]} \dots'$$

(compare note (107) above). On introducing the correct series expansions of AE and its powers in terms of x, this approach yields (analogously to the following argument)

$$\widehat{BG}[= 3 . B\widehat{A}G] = 3\left(\frac{3}{8} + \frac{1}{3}\left(\frac{3}{8}\right)^3 + \frac{3}{40}\left(\frac{3}{8}\right)^5 \dots + \frac{23\sqrt{55}}{288}x \dots\right).$$

(109) The series expansion of $r\sin(z/r)$, derived by Newton from inversion of that of $r\sin^{-1}(x/r)$ and communicated to Leibniz in his *epistola prior* of 13 June 1676; see page 668 below.

$AC = 4+x$. Then, letting fall the perpendicular BE from B to AC, from the sides of the triangle ABC there is to be found

$$AE = (9+8x+x^2)/(8+2x), \quad \text{that is,} \quad \tfrac{9}{8}+\tfrac{21}{32}x-\tfrac{5}{128}x^2+\tfrac{5}{512}x^3 \dots^{(107)}$$

If, however, with centre A and radius AB the circle quadrant AFG be described, then on setting radius $AB = r$ and[108] the arc $\widehat{BG} = z$ there will again be obtained $AE = z-\tfrac{1}{6}z^3/r^2+\tfrac{1}{120}z^5/r^4 \dots$,[109] as I explained in my first letter. Accordingly, if in place of r there be written 3 and from the equation between the two values of AE the root z be extracted, take this from the given quadrantal arc $\widehat{GF}$ and the arc $\widehat{BF}$ will be obtained.[110] And by a similar argument the arc $\widehat{FH}$ will be found.[111] Then from the given difference or ratio of the angles $\widehat{BAC}$, $\widehat{CAD}$ will be given the difference or ratio of the arcs $\widehat{BF}$, $\widehat{FH}$ found and from this an equation will be found by means of whose root x, when it is extracted, the problem is solved.

In problems of this sort the primary difficulty is to determine some rough approximation to the quantity required. According to the various kinds of problem that can be effected in a variety of ways. It will here be sufficient to cover the topic with an inclusive general rule. For the quantity required let any number OA be assumed at will and from it gather any error AP. Next, for the same required quantity let other numbers OA_2, OA_3, OA_4, OA_5, ... be assumed and from these similarly collect the errors A_2P_2, A_3P_3, A_4P_4, A_5P_5, ... Then through all the points P, P_2, P_3, P_4, P_5, ... with a regular[112] movement of the hand draw a curve $PP_2P_3P_4P_5$ cutting the straight line OA in Q and OQ will then be the required approximation.[113] Thus in the first of the problems in this chapter, on assuming any number OA for the length AP I gather the length of BP from it and then, taking the lengths of HP, IP, KP and LP with dividers,

---

(110) From identifying the two values, $3\sin^{-1}(\tfrac{1}{3}z)$ and $(9+8x+x^2)/(8+2x)$, of AE there comes (compare note (108) above) $\widehat{BF} = \tfrac{3}{2}\pi - \widehat{GB} = 3\cos^{-1}\dfrac{3}{8} - \dfrac{23\sqrt{55}}{96}x \dots$. Newton has cancelled a mysterious final phrase 'cujus tertia pars est' (whose third part it is): the restriction $\widehat{BAC} = \tfrac{1}{3}\pi$ itself serves to determine the problem and will, in general, be inconsistent with any additional posited relationship of that angle to $\widehat{CAD}$ which makes the latter differ from

$$\cos^{-1}[(21+\sqrt{165})/4(3\sqrt{3}+\sqrt{55})].$$

(111) On identifying $\cos\widehat{CAD}$ with $(27+8x+x^2)/12(4+x)$ it follows that

$$\widehat{CAD} = \cos^{-1}\dfrac{9}{16} - \dfrac{1}{12\sqrt{7}}x \dots.$$

(112) 'æqua' (even) is cancelled.

(113) The general rule of false position, based on the assumption that the 'error' curve (P) relating, for any given problem, the error AP to the base variable OA is a smoothly continuous 'linea regularis'. In the next chapter Newton will assume that this curve (P) may be approximated to within any required limits as the general parabola defined by

$$AP = a+b.OA+c.OA^2+d.OA^3 \dots.$$

earum summam et inter hanc summam et summam quam debent constituere differentiam pono errorem esse AP. Tum rursus pro [OA] adhibendo alios numeros OA^2, OA^3, OA^4 &c colligo iterum alia[s] summarum differentias A^2P^2, $A^3P^{[3]}$, A^4P^4 &c. Sic in Problemate secundo pro basi Curvarum AB assumpto numero quovis OA, colligo Ordinatas BC, BD et earum differentiam CD quæ nulla esse debet pono errorem esse AP, et simili opere repetito colligo errores A^2P^2, A^3P^3 &c. Si[c] in Problemate tertio pro longitudine AC assumpto numero quovis OA describo quadrilaterum $ABCD$ et anguloru̅ BAC, DAC differentiam proportionemve invenio. Sit ista differentia K, proportio 1 ad M, differentia requisita[114] k, proportio requisita 1 ad m, et error AP erit $K-k$ vel $M-m$. Atcꝫ ita in reliquis. Pro errore AP adhiberi potest quælibet quantitas quæ in legitima longitudine quæsitæ quantitatis evanescit, et suffecerit hunc errorem ratione quacunꝗ seu Geometrica seu Mechanica seu conjecturali præterpropter colligere. Verum quo exactius eam assecutus fueris eo celerius convergent series in opere sequente, et quantitas quæsita pauciori terminorum numero exactius colligetur.

Et hinc prodit alia methodus generalis[115] solvendi hujusmodi problemata. Computo satis exacto collige errores aliquot AP, A^2P^2, A^3P^3, A^4P^4 &c et per puncta P, P^2, P^3, P^4 &c methodo in præcedente capite[116] exposita age curvam regularem quæ secet rectam OA in puncto Q et erit OQ quantitas quæsita. Convenit autem intervalla AA^2, A^2A^3, A^3A^4 &c æqualia assumi.[117] Primum quære methodo aliqua rudiori limites satis strictos AA^5 inter quos necesse est punctum Q inveniri, Dein spatium AA^5 in æqualia aliquot intervalla AA^2, A^2A^3 &c divide. Si noveris rectas ducere quæ curvam ducendā $PP^2P^3P^4$ in datis punctis P, P^2, P^3, P^4, &c tangent, suffecerit intervallu̅ unicum tantum vel ad plurimum duo.

Proponi possunt problemata ubi opus fuerit duas vel plures ignotas quantitates assumere ad errorem colligendum. Ut si dentur tres æquationes

$$x^4 - axyz + byyz - z^4 + 2y^4 - ab^3 = 0. \quad zx^4 - cxy^3 + dxyzz + yyz^3 \ \&c = 0.$$

& $zx^4 - 3xyyzz - 2ayz^3 + y^5$ &c $= 0$, quæ tres involvunt ignotas quantitates, pro earum aliqua x quam inveniri vellem, assumo numerum aliquem OA et sic habeo tres æquationes quæ duas involvunt ignotas quantitates z et y. Per primam et secundam æquationem quæro quantitatem ignotam z perinde ut in superioribus expositum est,[118] percꝫ secundam ac tertiam rursus quæro

---

(114) 'quam habere debent' (due) and then 'legitima' (legitimate) was first written.

(115) The general method of finite differences, namely.

(116) In the long central passage which Newton subsequently cancelled; see note (67) above.

(117) When, of course, the conditions of the given problem easily allow this equisection of the base interval OA.

(118) Understand by eliminating y between the two. In the text of the 'Specimina' as we have it the general method of eliminating a variable between two given (algebraic) equations

I collect their sum, and the difference between this sum and the sum they ought to total I set as the error AP. Again, by employing in place of OA other numbers OA_2, OA_3, OA_4, ... I gather yet other differences of sums A_2P_2, A_3P_3, A_4P_4, ... So in the second problem, on assuming any number OA for the base AB of the curves I gather the ordinates BC, CD and set their difference CD, which ought to be nil, as the error AP, then by similarly repeating the procedure I collect the errors A_2P_2, A_3P_3, ... Likewise in the third problem, on assuming any number OA for the length of AC I describe the quadrilateral $ABCD$ and find the difference or ratio of the angles $\widehat{BAC}$, $\widehat{DAC}$. Let that difference be K or ratio 1 to M, the difference required[114] k or ratio required 1 to m, and the error AP will be $K-k$ or $M-m$. And so in the remaining instances. For the error AP there can be used any quantity vanishing in the right length of the quantity sought and it will be enough to collect this error somewhere thereabouts by any method, mathematical or constructional, or by guess. But the more accurately you achieve it, the swifter will the series converge in the following work and the fewer the number of terms needed to ascertain the required quantity more accurately.

And from this results another general method[115] of solving problems of this sort. With an adequately exact computation gather a number of errors AP, A_2P_2, A_3P_3, A_4P_4, ... and through the points P, P_2, P_3, P_4, ... by the method revealed in the preceding chapter[116] draw a regular curve which shall intersect the straight line OA in the point Q and OQ will be the quantity sought. It is, however, convenient to take the intervals AA_2, A_2A_3, A_3A_4, ... equal.[117] First by some cruder method seek out adequately tight limits A, A_5 within which it is necessary that the point Q be found, then divide the region AA_5 into a number of equal intervals AA_2, A_2A_3, ... If you should know how to draw straight lines which shall touch the curve $PP_2P_3P_4$ to be described at the given points P, P_2, P_3, P_4, ..., it will be enough to have but a single interval, or at most two.

Problems can also be proposed in which it will be necessary to assume two or more unknown quantities in order to gather the error. For instance, if there be given the three equations

$$x^4 - axyz + by^2z - z^4 + 2y^4 - ab^3 = 0, \quad zx^4 - cxy^3 + dxyz^2 + y^2z^3 \ldots = 0$$

and $zx^4 - 3xy^2z^2 - 2ayz^3 + y^5 \ldots = 0$ involving three unknown quantities, in place of some one of them, x, which I would like to find I assume some number OA and thus have three equations involving the two unknown quantities z and y. By means of the first and second equations I seek out the unknown quantity z exactly as was revealed above,[118] and again seek it out by means of the second

---

is not broached. As we shall see in the next volume, however, Newton discussed the topic at some length in his contemporary Lucasian lectures on algebra (ULC. Dd. 9.67: 35–8: 'Exterminatio quantitatis incognitæ quæ plurium in utraq æquatione dimensionum existit'

quantitatem *z* et differentiam inter utramcɓ *z* pono errorem esse *AP*. Denicɓ ex ejusmodi pluribus erroribus A^2P^2 A^3P^3 &c determino quantitatem *OQ* seu *x* ut supra. Et simili operis progressu conficitur problema in æquationibus quatuor vel pluribus. Onerosa sunt hæc computa sed aliquid est Problemata vel sic solvere quæ per Algebram vulgarem tractari nequeunt.[119]

In æquationibus etiam de maximis et minimis utilis est hæc methodus. Quantitas ignota quam maximam vel minimam esse velis reducatur ad seriem infinitam. Sit ea $y = a + bx - cxx + dx^3 + ex^4$ &c. Juxta methodum Huddenij[120] fac $0 = 1b - 2cx + 3dxx + 4ex^{[3]}$ &c. Hinc extrahe radicem *x* et extractam substituendo in æquatione priore, habebis maximam vel minimam *y*. Hoc modo procedi potest in difficilioribus ubi æquationes finitæ in promptu non sunt.

His addantur Problemata de inveniendis curvarum areis, longitudinibus, tangentibus, curvatura, centris gravitatum, solidorū contentis et superficiebus et similia, et complebitur ferè systema problematum[121] quæ per divisiones et extractiones radicum reduci possunt ad series infinitas. Pergendū jam est ad difficiliora in quibus divisiones et extractiones illæ sine ulteriori artificio parum prosunt. Ea sunt regressus ad curvas ab ipsarū tangentibus, longitudinibus, areis, curvatura, centris gravitatum et horum complicationibus varijs et siqua alia sunt quæ cum his affinitatem habent. Ad horum solutionem dixi me duplici uti methodo una concinniori altera generaliori.[122] Posteriorem exponere suffecerit.

Cap 4.

Exponitur methodus resolvendi Problemata per fluxionem quantitatum.[123]

In Epistola mea secunda celavi methodum tangentiū[124] his literis transpositis 6*a cc d æ* 13*e ff* 7*i* 3*l* 9*n* 4*o* 4*q rr* 4*s* 8*t* 12*v x*: quæ recto ordine hanc continent sententiam. *Data æquatione quotcuncɓ fluentes quantitates involvente fluxiones invenire; et vice versa.*[125] Prior pars sententiæ methodum tangentium directam altera

---

[= *Arithmetica Universalis*; *sive De Compositione et Resolutione Arithmetica Liber* (Cambridge, ₁1707): 72–6], whose marginal dating claims it to have been the topic of the last two lectures of his 'Octob. 1674' series). Compare also II: 404–8.

(119) This last sentence was originally inserted at the end of the previous paragraph. Evidently the present one was a last-minute insertion added on the spur of the moment.

(120) See I: 213–15.

(121) Newton is again, no doubt, thinking of the full version of the fluxional tract he had intended to complete thirteen years before; compare Appendix 2.4: note (31) below.

(122) At the close of his *epistola posterior*; see notes (33) and (90) preceding. Immediately following Newton has cancelled the sentence 'Ad prioris explicationem conducit methodus tangentium, quam ideo jam prius exponenda' (In developing the former the method of tangents contributes, and it must now accordingly be expounded in preliminary). Despite the deletion the method of deriving the fluxion of a given fluent quantity is discussed at length in the next chapter, 'Cap 4' (a late insertion not allowed for in Newton's preliminary scheme; see note

and third, and then I set the difference between the two z's as the error AP. Lastly, from several errors A_2P_2, A_3P_3, ... of this sort I determine the quantity OQ, that is x, as above. And by a similar sequence of operations the problem in four or more equations is completed. These computations are indeed oppressive, but it is something to solve even in this fashion problems which it is impossible to treat by common algebra.[119]

In equations regarding maxima and minima, too, this is a useful method. Should you wish an unknown quantity to attain a maximum or minimum, reduce it to an infinite series. Let this be $y = a + bx - cx^2 + dx^3 + ex^4$ .... Following Hudde's method[120] make $0 = b - 2cx + 3dx^2 + 4ex^3$ ..., extract the root x from this and on substituting this when extracted in the former equation you will have maximum or minimum y. This procedure can be followed in more difficult cases where finite equations are not readily available.

To these may be added problems on finding the areas, lengths, tangents, curvature and centres of gravity of curves, the volumes and surfaces of solids and other like matters, and the system of problems[121] reducible by means of divisions and root extractions to infinite series will almost be complete. We must now pass on to more difficult ones in which those divisions and extractions are, without further artifice, inadequate. These deal with the regression to curves from their tangents, lengths, areas, curvature, centres of gravity and various combinations of these, and anything else which is related to that reversion. For their solution I said that I use a 'twofold method, one part of which is neater, the other more general'.[122] It will be sufficient to expound the latter.

CHAPTER 4
A METHOD OF RESOLVING PROBLEMS BY MEANS OF THE FLUXION OF QUANTITIES IS EXPOUNDED.[123]

In my second letter I concealed my method of tangents[124] in this transposition of letters [6a cc d æ 13 e ff 7i 3l 9n 4o 4q rr 4s 8t 12v x]: in their right order they contain this sentence [*Data æquatione quotcunq fluentes quantitates involvente fluxiones invenire; et vice versa*].[125] The first part of the sentence

---

(39) above), while the elaboration of the latter general method—consisting (to unscramble his October 1676 anagram) 'in assumptione seriei pro quantitate qualibet incognita ex qua cætera commodè derivari possunt, et in collatione terminorum homologorum æquationis resultantis ad eruendos terminos assumptæ seriei' (II: 191, note (25))—is delayed till Chapter 5.

(123) Newton elaborates the 'methodus concinnior' of fluxions which he had announced to Leibniz in 1676 at the close of his *epistola posterior* (see previous note).

(124) The equivalent phrase 'ducendi tangentes' (of drawing tangents) is cancelled.

(125) The anagram, accurately copied from ULC. Add. 3977.4: 2^v, should have '...9*t*...'. The error was repeated in the letter sent to Oldenburg on 24 October 1676 (British Museum. Add. 4294.119 [= *Correspondence*, **2**: 115]) but corrected by the latter in the version subse-

inversam continet. Per fluentem quantitatem intelligo quantitatem quamcunꝗ quæ sensim augetur vel minuitur et per fluxionem ejus celeritatem augmenti vel decrementi, augmenti quidem si fluxio affirmativa est decrementi si negativa.[126] Sic et si punctum aliquod motu continuo lineam describit, celeritatem descriptionis indifferenter appello tum fluxionem puncti generantis tum fluxionem lineæ genitæ. Exponantur jam fluentes quotcunꝗ quantitates x, y, z &c[127] et exprimatur harum relatio ad invicem æquatione quacunꝗ $x^3 - axy + byy - czz + ddz + e^3 = 0$. Fluant autem hæ quantitates minimo temporis spatio o, et[128] fluxione illa evadant $x + op$, $y + oq$, $z + or$ sumptis incrementis op, oq, or in ratione p, q, r, et illa p, q, r designabunt etiam rationem fluxionum ad invicem eo quod fluxiones sunt ut incrementa momentanea. Cum igitur x, y et z jam sint $x + op$, $y + oq$, $z + or$, scribantur hæc in æquatione pro x, y,

et z et emerget æquatio talis
$$\begin{aligned} &x^3 &&+ 3opxx &&+ 3ooppx &&+ o^3p^3 \\ &- axy &&- aopy &&- aoqx &&- aoopq \\ &+ byy &&+ 2boqy &&+ booqq \\ &- czz &&- 2corz &&- coorr \\ &+ ddz &&+ ddor \\ &+ e^3 \end{aligned} = 0.$$
Supra erat

$x^3 - axy + byy - czz + ddz + e^3 = 0$: his igitur deletis et reliquis per o divisis, restant $3pxx - apy + 2bqy - 2crz + ddr + 3ppox - aqx + boqq - corr + oop^3 - aopq = 0$. Defluant jam quantitates $x + op$, $y + oq$, $z + or$ et fiant x, y, z momento temporis o in nihilum diminuto,[129] et æquatio novissima evadet

$$3pxx - apy + 2bqy - 2crz + ddr - aqx = 0$$

cæteris terminis evanescentibus.

Hoc et similibus exemplis colligitur terminos omnes non multiplicatos per o semper evanescere ut et omnes multiplicatos per o plusquam unius dimensionis, cæteros verò ex æquationibus propositis sic colligi.[130]

---

quently passed on to Leibniz; see II: 191, note (25). Newton's sentence, 'Given an equation involving any number of fluent quantities, to find the fluxions; and conversely so', summarizes the two opening problems of his 1671 tract, 'Relatione quantitatum fluentium inter se datâ, fluxionum relationem determinare' / 'Exposita æquatione fluxiones quantitatum involvente, invenire relationem quantitatum inter se' (see III: 74, 82). In sequel he adapts the content of the former 'Prob: 1'.

(126) This definition, clearer than that given in the 1671 tract (III: 72), essentially repeats the four opening 'Definitiones' of the 'Geometria Curvilinea' (compare 1, §1: note (20) above).

(127) Newton first displayed his following argument in terms of the general fluent line-segments '$AB = x$, $CD = y$, $EF = z$ &c' departing from the origins A, C and E respectively. Compare I: 383–5, III: 70–2 and Axiom 6 of the preceding 'Geometria Curvilinea' (1, §1 above).

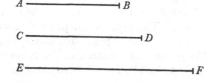

contains the direct method of tangents, the second the inverse one. By a fluent quantity I understand any quantity which grows or diminishes gradually, and by its fluxion the swiftness of its increase or decrease—of its increase, indeed, if its fluxion is positive, of its decrease if negative.[126] So also if some point describes a line by a continuous motion, the swiftness of its description I call indifferently both the fluxion of the generating point and the fluxion of the line generated. Now let any number of fluent quantities $x, y, z, \ldots$[127] be exhibited and let their relationship to one another be expressed by any equation

$$x^3 - axy + by^2 - cz^2 + d^2z + e^3 = 0.$$

Let these quantities flow for a very small space of time o and[128] by that fluxion let them come to be $x + o\dot{x}$, $y + o\dot{y}$, $z + o\dot{z}$, taking the increments $o\dot{x}$, $o\dot{y}$, $o\dot{z}$ in the ratio $\dot{x} : \dot{y} : \dot{z}$, and those quantities $\dot{x}$, $\dot{y}$, $\dot{z}$ will denote also the ratio of the fluxions to one another for the reason that fluxions are as the momentary increments. Since, therefore, x, y and z are now $x + o\dot{x}$, $y + o\dot{y}$, $z + o\dot{z}$, let these be entered in the equation in place of x, y and z and there will emerge an equation of this form

$$x^3 - axy + by^2 - cz^2 + d^2z + e^3 + o(3\dot{x}x^2 - a\dot{x}y - a\dot{y}x + 2b\dot{y}y - 2c\dot{z}z + d^2\dot{z})$$
$$+ o^2(3\dot{x}^2x - a\dot{x}\dot{y} + b\dot{y}^2 - c\dot{z}^2) + o^3\dot{x}^3 = 0.$$

Above there was $x^3 - axy + by^2 - cz^2 + d^2z + e^3 = 0$: with these, therefore, deleted and the remainder divided by o there is left

$$3\dot{x}x^2 - a\dot{x}y - a\dot{y}x + 2b\dot{y}y - 2c\dot{z}z + d^2\dot{z} + o(3\dot{x}^2x - a\dot{x}\dot{y} + b\dot{y}^2 - c\dot{z}^2) + o^2\dot{x}^3 = 0.$$

Now let the quantities $x + o\dot{x}$, $y + o\dot{y}$, $z + o\dot{z}$ flow back, becoming x, y, z as the moment of time o diminishes to nothing,[129] and the most recent equation will prove to be $3\dot{x}x^2 - a\dot{x}y - a\dot{y}x + 2b\dot{y}y - 2c\dot{z}z + d^2\dot{z} = 0$, the other terms vanishing.

It is gathered from this and similar examples that all terms not multiplied by o always vanish, as do all those multiplied by a higher dimension of o than the first, while indeed the others are gathered from the equations proposed in this way.[130]

---

(128) 'completo illo tempore' (on the completion of that time) is cancelled. This discussion closely follows the 'Demonstratio' for the example $x^3 - ax^2 + axy - y^3 = 0$ on III: 80.

(129) Newton first began to write 'evanescente tem[pore o]' (as the time o vanishes).

(130) The following rule is the 'Solutio' on III: 74 trivially restated and applied to slightly variant examples; namely, his previous equation $x^3 - axy + by^2 - cz^2 + d^2z + e^3 = 0$ and a revised 'Exempl 2' $2y^3 - x^2y + axy - a^2y + b^3 = 0$.

Æquationem dispone secundum dimensiones quantitatis cujusvis fluentis x et terminos multiplica per arithmeticam progressionem unitatum ac loco lateris unius x in quolibet termino scribe fluxionem ejus. Idem fac sigillatim in singulis quantitatibus fluentibus & summa productorum omnium componet æquationem novam qua fluxionum p, q, r relatio ad invicem exprimitur. Sic in æquatione allata terminos dispono primò secundum dimensiones x dein secundum dimensiones y tertio secundum dimensiones z et multiplico ut sequitur.

$$
\begin{array}{ccc}
\begin{array}{ccc} 3 & 1 & 0 \end{array} & \begin{array}{ccc} 2 & 1 & 0 \end{array} & \begin{array}{ccc} 2 & 1 & 0 \end{array} \\[2pt]
x^3 \;-axy+byy & byy-axy+x^3 & -czz\;+ddz+x^3 \\
-czz & -czz & -axy \\
+ddz & +ddz & +byy \\
+e^3 & +e^3 & +e^3 \\
\hline
3pxx-apy+0 & 2byq-axq+0 & -2czr+ddr+0.
\end{array}
$$

Et summa productorum $3pxx-apy+2byq-axq-2czr+ddr=0$ est æquatio quæsita relationem fluxionum $p\ q\ r$ ad invicem definiens. Sic habita æquatione $2y^3-xxy+axy-aay+b^3=0$, terminos multiplicare licebit in hunc modum

$$
\begin{array}{cc}
\begin{array}{ccc} 2 & 0 & -1 \end{array} & \begin{array}{ccc} 2 & 1 & 0 \end{array} \\[2pt]
2y^3-xx\,y+b^3 & -xxy+axy+2y^3 \\
+ax & -aay \\
-aa & +b^{[3]} \\
\hline
4yyq\;+0\;-\dfrac{b^3q}{y} & -2xpy+apy+0.
\end{array}
$$

Et summa productorum $4yyq-\dfrac{b^3q}{y}-2xpy+apy=0$ est æquatio quæsita relationem fluxionum p, q ad invicem definiens.[131] Quod si æquatio surdas aliquas quantitates vel fractas compositorum denominatorum involvat operor ut in sequente exemplo. Detur $x^3-ayy+\dfrac{by^3}{a+y}-xx\sqrt{ay+xx}=0$. pono z pro $\dfrac{by^3}{a+y}$ & v pro $xx\sqrt{ay+xx}$ et inde nactus sum tres æquationes $x^3-ayy+z-v=0$. $az+yz-by^3=0$ et $ax^4y+x^6-vv=0$. prima dat $3px^2-2aqy+r-s=0$ scripto nimirum s pro fluxione v, secunda dat $ar+ry+qz-3bqyy=0$, tertia $4apx^3y+ax^4q+6px^5-2sv=0$. Ipsorum r et s valores per secundam ac tertiam inventos $\left(\text{nempe } \dfrac{3bqyy-qz}{a+y} \text{ et } \dfrac{4apx^3y+ax^4q+6px^5}{2v}\right)$ substituo in prima et oritur

$3pxx-2aqy+\dfrac{3bqyy-qz}{a+y}-\dfrac{4apx^3y+ax^4q+6px^5}{2v}=0$. Et pro z et v scriptis $\dfrac{by^3}{a+y}$ &

---

(131) To be exact this is the fluxional derivative of $y^{-1}(2y^3-x^2y+axy-a^2y+b^3)=0$, so that y must now not be zero.

Arrange the equation according to the dimensions of any one, x, of the fluent quantities, multiply the terms by an arithmetical progression of units, and then in replacement for a single unit-power of x in every term write in its fluxion. Do the same separately for each of the fluent quantities and the sum of all the products will make up a new equation in which is expressed the relationship of the fluxions $\dot{x}$, $\dot{y}$, $\dot{z}$ to each other. Thus in the equation adduced I arrange the terms first according to the dimensions of x, then according to the dimensions of y, and thirdly according to the dimensions of z, and then I multiply as follows:

$$
\begin{array}{ccc}
\begin{array}{ccc}
3 & 1 & 0 \\
x^3 & -axy+by^2 & \\
 & -cz^2 & \\
 & +d^2z & \\
 & +e^3 & \\
\hline
3\dot{x}x^2 & -a\dot{x}y+0 &
\end{array}
&
\begin{array}{ccc}
2 & 1 & 0 \\
by^2 & -axy+x^3 & \\
 & -cz^2 & \\
 & +d^2z & \\
 & +e^3 & \\
\hline
2b\dot{y}y & -a x\dot{y}+0 &
\end{array}
&
\begin{array}{ccc}
2 & 1 & 0 \\
-cz^2 & +d^2z+x^3 & \\
 & -axy & \\
 & +by^2 & \\
 & +e^3 & \\
\hline
-2cz\dot{z} & +d^2\dot{z}+0. &
\end{array}
\end{array}
$$

The sum $3\dot{x}x^2 - a\dot{x}y + 2by\dot{y} - ax\dot{y} - 2cz\dot{z} + d^2\dot{z} = 0$ of the products is then the required equation defining the relationship of the fluxions $\dot{x}$, $\dot{y}$, $\dot{z}$ to one another. So when the equation $2y^3 - x^2y + axy - a^2y + b^3 = 0$ is obtained, it will be permissible to multiply its terms in this manner:

$$
\begin{array}{cc}
\begin{array}{ccc}
2 & 0 & -1 \\
2y^3 + (-x^2+ax-a^2)\,y & +b^3 \\
\hline
4y^2\dot{y} & +0 & -b^3\dot{y}/y
\end{array}
&
\begin{array}{ccc}
2 & 1 & 0 \\
-x^2y & +axy + (2y^3-a^2y+b^3) \\
\hline
-2x\dot{x}y & +a\dot{x}y & +0
\end{array}
\end{array}.
$$

The sum $4y^2\dot{y} - b^3\dot{y}/y - 2x\dot{x}y + a\dot{x}y = 0$ of the products is then the required equation defining the relationship of the fluxions $\dot{x}$, $\dot{y}$ to one another.[131] But should the equation involve one or more surd quantities or fractions with composite denominators, I operate as in the following example. Let there be given $x^3 - ay^2 + by^3/(a+y) - x^2\sqrt{[ay+x^2]} = 0$. I put z for $by^3/(a+y)$ and v for $x^2\sqrt{[ay+x^2]}$, and accordingly obtain the three equations $x^3 - ay^2 + z - v = 0$, $az + yz - by^3 = 0$ and $ax^4y + x^6 - v^2 = 0$. The first gives $3\dot{x}x^2 - 2a y\dot{y} + \dot{z} - \dot{v} = 0$ (on writing $\dot{v}$ for the fluxion of v, of course), the second yields

$$
a\dot{z} + \dot{z}y + \dot{y}z - 3by\dot{y}^2 = 0,
$$

the third $4a\dot{x}x^3y + ax^4\dot{y} + 6\dot{x}x^5 - 2\dot{v}v = 0$. The values of $\dot{z}$ and $\dot{v}$ found by means of the second and third—namely, $(3by\dot{y}^2 - \dot{y}z)/(a+y)$ and

$$
(4a\dot{x}x^3y + ax^4\dot{y} + 6\dot{x}x^5)/2v
$$

—I substitute in the first equation and there arises

$$
3\dot{x}x^2 - 2a y\dot{y} + \frac{3by\dot{y}^2 - \dot{y}z}{a+y} - \frac{4a\dot{x}x^3y + ax^4\dot{y} + 6\dot{x}x^5}{2v} = 0,
$$

$xx\sqrt{ay+xx}$, prodit $3pxx-2aqy+\dfrac{3abqyy+2bqy^3}{a^2+2ay+yy}-\dfrac{4apx^3y+ax^4q+6px^5}{2\sqrt{ay+xx}}=0$ æquatio

definiens relationem fluxionū p et q ad invicem. Atcɋ hæc ante annos novendecim[132] meditatus sum conferendo inventa   [133] & Huddenij inter se.

His intellectis[134] in promptu est et id genus alia perficere. Si ad curvæ cujusvis *DCH* punctum quodvis *C* ducenda sit tangens *CT*, considero quod ubi ordinata quævis *CB* super basi *AB* positione data incedit, fluxio ordinatæ *CB* est ad fluxionem basis *AB* ut ipsa ordinata *CB* ad lineam *BT* quæ ad tangentem terminatur: proinde habita relatione inter *AB* et *CB* quæro inde relationem fluxionum et capio *TB* ad *CB* ut fluxio *AB* ad fluxionem *CB*.[135]

[Fig 1]

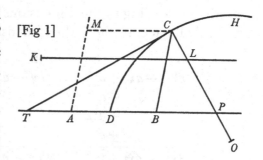

Ut si dictis *AB*=x & *BC*=z relatio inter *AB* et *BC* definiretur hisce duabus æquationibus $x^3-axy+byy-czz+ddz+e^3=0$ & $2y^3-xxy+axy-aay+b^3=0$. Per æquationem posteriorem quærerem relationem inter fluxiones ipsarum x et y seu inter p et q ut supra: Quo pacto assumpta utcuncɋ fluxione [p] haberem fluxionem q. Deinde per æquationem priorem quærerem relationem inter fluxiones ipsarum x, y et z seu inter p q et r: quo pacto ex notis fluxionibus p et q haberem fluxionem r. Denicɋ caperem *TB* ad *CB* ut p ad r, et agerem tangentem *TC*.

Conveniunt hæc cum Leibnitianis[136] fundamentaliter, verum tamen non nisi particula sunt methodi cujusdam generalioris quæ a speculatione locorum planorum derivatur.[137] Conantur Analystæ omnia deducere ad æquationes. Hac methodo vix tractantur æquationes nisi ubi hæ in statu quæstionis in-

---

(132) That is, about mid-1665. (For confirmation see I: 212–389 passim, especially 216–33, 236–8, 272–8.) Newton first wrote far more vaguely that 'hæc dudum inveneram' (I had found these things out some while ago).

(133) Understand 'Cartesij' (Descartes), of course. It is curious that less than twenty years after the event Newton cannot immediately recall the name of the man who contributed more than anyone to his own growth to mathematical—and indeed scientific—enlightenment. Perhaps already, in an unconscious rationalization of the past, he is trying to forget the magnitude of his debt to Descartes in mathematics as vigorously as he sought henceforth to conceal it in physics?

(134) The less expectant participle 'expositis' (expounded) is cancelled.

(135) A fluxional restatement of Problem 4, Mode 1, of the 1671 tract (III: 120–2) which eliminates the latter's plea to infinitesimals. Newton had communicated this general tangent construction, for an algebraic curve defined in oblique Cartesian coordinates, to Collins on 10 December 1672 (see *Correspondence*, 1: 247–8; compare also III: 122, note (191)).

(136) Leibniz had communicated his variant of the general Slusian tangent-method to Oldenburg on 11/21 June 1677 (*Correspondence*, 2: 213–15); compare note (35) above.

and on replacing z and v by $by^3/(a+y)$ and $x^2\sqrt{[ay+x^2]}$ there results

$$3\dot{x}x^2 - 2a\dot{y}y + \frac{3ab\dot{y}y^2 + 2b\dot{y}y^3}{(a+y)^2} - \frac{4a\dot{x}x^3y + ax^4\dot{y} + 6\dot{x}x^5}{2\sqrt{[ay+x^2]}} = 0,$$

the equation defining the relationship of the fluxions $\dot{x}$ and $\dot{y}$ to one another. On these matters I pondered nineteen years ago,[132] comparing the findings of [133] and Hudde with one another.

Once these [fundamentals] are understood[134] we are ready to accomplish other matters of the kind also. If at any point C on any curve DCH the tangent CT has to be drawn, I consider that, when any ordinate CB advances along the base AB given in position, the fluxion of the ordinate CB is to that of the base AB as that ordinate CB to the line BT terminating at the tangent: in consequence, when the relationship between AB and CB is had, I ascertain the relationship of the fluxions from it and take TB to CB as the fluxion of AB to the fluxion of CB.[135]

So if, on calling $AB = x$ and $BC = z$, the relationship between AB and BC were to be defined by these two equations

$$x^3 - axy + by^2 - cz^2 + d^2z + e^3 = 0 \quad \text{and} \quad 2y^3 - x^2y + axy - a^2y + b^3 = 0,$$

by the latter equation I should seek the relationship between the fluxions of x and y—that is, between $\dot{x}$ and $\dot{y}$—as above: in this way on assuming the fluxion $\dot{x}$ arbitrarily I would have the fluxion $\dot{y}$. Then by means of the first equation I should ascertain the relationship between the fluxions of x, y and z—that is, between $\dot{x}$, $\dot{y}$ and $\dot{z}$—and in this way from the known fluxions $\dot{x}$ and $\dot{y}$ I would gain the fluxion $\dot{z}$. Finally, I should take TB to CB as $\dot{x}$ to $\dot{z}$ and draw the tangent TC.

This approach agrees basically with that of Leibniz,[136] yet is, however, but a small part of a more general method which is derived from an examination of plane loci.[137] Analytical mathematicians attempt to bring everything down to equations. In the present method equations are hardly handled at all except when they are involved in the structure of a problem. Here the operations

(137) Newton first wrote 'a tangentibus derivatur' (...from tangents) and in sequel went on: 'tamₒ simplex est et latè patens ut descriptio Curvæ per motum localem vel relatio ejus ad lineas rectas proponi vix possit quin statim prodeat tangentis determinatio simplex. Rem totam exponere non est hujus loci. Dabo tantum instantiā virium hujus methodi' (and is so simple and broad in expanse that scarcely a description of a curve by local motion or its relationship to straight lines can be propounded without there immediately ensuing a simple determination of the tangent. This is not the place to develop the whole topic. I shall merely give an instance of the power of this method). He evidently here has in view the variant Modes 2–9 of constructing tangents which he had expounded in Problem 4 of his 1671 fluxional tract (see III: 132–48).

volvuntur. Operationibus facilioribus hic pergi solet et ad conclusiones simpliciores deveniri, idcg in quæstionibus difficillimis ut ex sequenti hujus methodi specimine constare potest.[138]

Sed redeamus ad methodum fluxionum.

Si determinanda est quantitas maxima vel minima considero quod in eo casu fluxio ejus nulla est. Maxima non augetur quia major non erit, non diminuitur quia major non fuit. Quæro igitur quantitatum indefinitarum fluxiones et fluxionem ejus quæ maxima vel minima esse debet pono nullam. Ut si fuerit $2y^3 - xxy + axy - aay + b^3 = 0$. et maxima y quæratur: quæro relationem fluxionum $4yyq - \dfrac{b^3 q}{y} - 2xpy + apy = 0$, et ponendo $q = 0$, restat $apy - 2xpy = 0$ seu $a = 2x$. Maxima est ergo vel minima y ubi x valet $\frac{1}{2}a$. Atcg hinc constat methodus Huddenij.[139]

Si curvæ alicujus ignotæ area $BCD = z$ ex longitudine baseos $AB = x$ æquatione quavis definiatur et inde invenienda sit longitudo ordinatæ CB, produco CB ad F ut sit perpendiculum $FG = 1$ et completo parallelogrammo $AEF[B]$ imaginor fluxionem hujus parallelogrammi seu $1 \times x$, exponi per lineam BF et fluxionem areæ BDC seu z per ordinatam CB. Tum per æquationem qua relatio z ad x definitur quæro relationem fluxionum p et r et capio BC ad BF ut r ad p.[140] Hoc modo pro arbitrio inveniuntur curvæ lineæ quarum areæ per finitas æquationes definiuntur.[141]

[Fig 2]

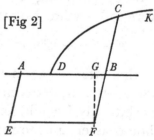

Si &c[142]

Si invenienda est longitudo Curvæ alicujus DC (Fig [1]) considero quod fluxio ejus est ad fluxionem rectæ AB ut CT ad BT. Quare (Fig [2]) constituo

(138) Some three-quarters of a page following in the manuscript (on f. 16[r]) has been left blank for the intended later insertion of this extra-difficult example.

(139) See 1: 214–15 and note (131) above; compare also Problem 3 of the 1671 tract (III: 116–20).

(140) The basic theorem of Problem 7 of the 1671 tract (see III: 194–6) generalized to cover the case where the angle between the coordinates, here AB and BC, is no longer right. In effect, on taking $BC = y$ and FG to be unity the area under the curve (C) is $z = \int y \sin A\hat{B}C \,.\, dx$, whence

$$r \text{ (or } \dot{z}) : p \text{ (or } \dot{x}) = y : \operatorname{cosec} A\hat{B}C = BC : BF.$$

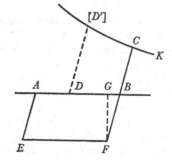

The general curve CK need not, of course, meet the base AB but it is a trivial extension of Newton's argument to introduce the initial ordinate DD' and then consider the area $(BDD'C) = z$.

usually proceed rather easily and the conclusions arrived at tend to be simple: this even in the case of the most difficult questions, as may be established from the following specimen of this method.[138]

But let us return to the method of fluxions.

If the maximum or minimum of a quantity is to be determined, I consider that its fluxion in this case is nil. At a maximum it neither increases (since it will not come to be greater) nor diminishes (since it was not greater before). I therefore seek the fluxions of the indefinite quantities and the fluxion of that which ought to be a maximum or a minimum I set to be nothing. If, for instance, there was $2y^3 - x^2y + axy - a^2y + b^3 = 0$ and the maximum y be sought, I ascertain the relationship, $4y^2\dot{y} - b^3\dot{y}/y - 2x\dot{x}y + a\dot{x}y = 0$, of the fluxions and on setting $\dot{y} = 0$ there remains $a\dot{x}y - 2x\dot{x}y = 0$ or $a = 2x$. The maximum or minimum of y therefore occurs when the value of x is $\frac{1}{2}a$. And on this Hudde's method is based.[139]

If the area $BCD = z$ of some unknown curve be defined in relation to the length of its base $AB = x$ by any equation and from this the length of the ordinate BC is to be found, I extend CB to F such that the perpendicular $FG = 1$ and, having completed the parallelogram $AEFB$, I imagine the fluxion of this parallelogram (or $1 \times x$) to be represented by the line BF and the fluxion of the area BDC (or z) by the ordinate BC. Then by means of the equation by which the relationship of z to x is defined I seek the relationship of the fluxions $\dot{x}$ and $\dot{z}$ and take BC to BF as $\dot{z}$ to $\dot{x}$.[140] In this way curved lines are found at will whose areas are defined by means of finite equations.[141]

If...[142]

If the length of some curve $\stackrel{\frown}{DC}$ (Figure 1) has to be found, I consider that its fluxion is to the fluxion of the straight line AB as CT to BT. Consequently, I set

---

(141) Namely, given $f(x, y, z) = 0$, by determining its fluxional derivative

$$\phi(x, y, z, p, q, r) = 0;$$

whence by eliminating z between the two and setting $r = py\sin A\widehat{B}C$ there results

$$F(x, y, p, q) = 0,$$

the required equation relating x, y and their fluxions p (or $\dot{x}$), q (or $\dot{y}$). Integration of the last will yield a family of curves (depending on choice of the constant of integration) sharing the property $f(x, y, z) = 0$.

(142) A late insertion squashed between the two surrounding paragraphs. Newton's added instruction, 'Here add y^e finding of lines whose areas are comparable to others', reveals unmistakably that he here intended to summarize his 1671 tract's Problem 8, 'Curvas pro arbitrio multas invenire quarum areæ ad aream datæ alicujus Curvæ relationem habent per finitas æquationes designabilem'.

curvam aliam *DCK* cujus basis *AB*
æqualis sit basi alterius *AB*, ordinata
verò *CB* sit ad datam *BF* ut *CT* ad *BT*
et aream *DCB* applicatam ad unitatem
GF pono æqualem longitudini quæsitæ
DC.[143]

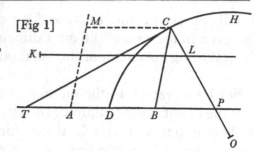

[Fig 1]

Si curvæ *DC* desideretur curvatura ad
punctum aliquod *C*, ipsi perpendiculare
agatur *CO* secans basem *AB* in *P* et aliam quamvis positione datam basicꝗ
parallelam rectam *KL* in *L*. Quære fluxiones ipsarū *AP* et *KL* et in ratione
earundem fluxionum cape *PO* et *LO*. Erit *CO* radius curvaturæ, seu radius
circuli qui æque curvus est ac linea [*D*]*CH* in puncto *C*. Idem *CO* longitudinem
exprimit alterius cujusdam curvæ quā punctū *O* describit[144] ut explicuit
Hugenius.[145]

Si æquatione aliqua definiatur relatio inter *AP* et *CP* et inde determinanda
est curva linea *DCH* cui *PC* perpetuò perpendicularis est, constitue triangulum
rectangulum *BPC* ea lege ut sit *BP* ad *CP* sicut fluxio *CP* ad fluxionem *AP* et
habebis Curvæ punctum *C*.[146]

Si ex data relatione laterum trianguli *BCT* (Fig [1]) definienda est curva illa
quam latus *CT* perpetuo tangit, considero quod fluxio Basis *AB* sit ad fluxionem
ordinatæ *BC* ut *BT* ad *BC*, id est si compleatur parallelogrammum *ABCM*
fluxio ordinatæ *CM* est ad fluxionē basis *AM* ut *BT* ad *AM*. Expono igitur basem
AM et ordinatam *MC* (Fig [1]) per areas *ABFE* et
DCB (Fig [2]) et illarum fluxiones per harum fluxiones
BF et *CB* faciendo ut sit *CB* ad *BF* (Fig [2]) sicuti *BT*
ad *BC* (Fig [1]) et tum demum inventa area *CD*[*B*]
ex relatione ejus ad aream *AEFB* seu ordinatæ
MC ad Basem *MA* determinabit Curvā *CD*.
Quoniam igitur hæc area semper inveniri potest per
divisiones et extractiones radicum et nonnunquam
per finitas æquationes dixi in epistola mea secunda
inversum de tangentibus problema hoc in casu nunquā indigere ulterioribus

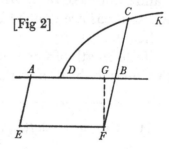

[Fig 2]

---

(143) Problems 11/12 of the 1671 tract (see especially III: 304–6, 314–16) generalized, much
as above, to the case where the 'ordination' angle $A\hat{B}C$ is no longer right. Here the fluxion of
orbit $d(\hat{DC})/dt = (CT/BT).d(AB)/dt$ is now $\sqrt{[p^2+q^2-2pq\cos A\hat{B}C]}$ and Newton sets
$$z = \int (CT/BT).dx.$$
(144) The evolute of the curve (*C*), that is.
(145) See Huygens' *Horologium Oscillatorium sive De Motu Pendulorum ad Horologia Aptato
Demonstrationes Geometricæ* (Paris, 1673): Pars Tertia, 'De linearum curvarum evolutione &
dimensione': 59–90. Implicitly obeying his own later maxim that 'second Inventors have no
Right' ('Observations upon [Leibniz' letter to Conti, 9 April 1716 (N.S.)]' [= J. Raphson,

up (Figure 2) a second curve *DCK* whose base *AB* shall be equal to the other's base *AB*, while its ordinate *BC* is to be to the given length of *BF* as *CT* to *BT*, and I then set the area *DCB*, divided by the unit *GF*, equal to the required length of $\widehat{DC}$.[143]

If the curvature of the curve *DC* at some point *C* should be desired, draw *CO* perpendicular to it, cutting the base *AB* in *P* and any other straight line *KL* given in position parallel to the base in *L*. Ascertain the fluxions of *AP* and *KL* and take *PO* and *LO* in the ratio of those fluxions. Then *CO* will be the radius of curvature: in other words, the radius of a circle which is equally as curved as the line *DCH* at the point *C*. This same *CO* expresses the length of a certain other curve described by the point *O*,[144] as Huygens has found.[145]

If the relationship between *AP* and *CP* be defined by some equation and from it is to be determined the curved line *DCH* to which *PC* is perpetually normal, set up the right-angled triangle so restricted that *BP* shall be to *CP* as the fluxion of *CP* to the fluxion of *AP* and you will obtain the point *C* on the curve.[146]

If from the given relationship of the sides of the triangle *BCT* (Figure 1) the curve perpetually touched by the side *CT* is to be defined, I consider that the fluxion of the base *AB* is to be to the fluxion of the ordinate *BC* as *BT* to *BC*; that is, if the parallelogram *ABCM* be completed, the fluxion of the ordinate *CM* is to that of the base *AM* as *BT* to *AM*. I therefore exhibit the base *AM* and ordinate *MC* (Figure 1) by the areas *ABFE* and *DCB* (Figure 2), and the fluxions of the former pair by the fluxions *BF* and *CB* of the latter, making *CB* to *BF* (Figure 2) as *BT* to *BC* (Figure 1); and then the area *CDB*, when at length found from its relationship to the area *AEFB* (that is, from the ratio of the ordinate *MC* to the base *MA*), will determine the curve *CD*. Seeing therefore that this area can always be found by divisions and root extractions and not infrequently by means of finite equations, I said in my second letter that the inverse problem of tangents in this case stood at no time in need of the ulterior

---

History of Fluxions (London, 1715): Appendix (1718?, not found in all English and Latin versions): 111–19]: 115) Newton makes no mention of his own independent, later discovery of Huygens' 'nova curvarum linearum consideratio..., earum scilicet quæ sui evolutione alias curvas generant. Vnde comparatio inter se longitudinis curvarum cum rectis nascitur' (*Horologium*: 2) in late 1664 (see 1: 245–9) or of the many improvements he introduced in the spring of 1665 (1: 259–71, 280–94), the next November (1: 387–9, essentially the modified fluxional approach now sketched in the case of a curve defined in standard Cartesian co-ordinates), in October 1666 (1: 419–40) and above all in Problems 5 and 6 of his 1671 tract (III: 150–92).

(146) As we have seen Newton invented this semi-intrinsic coordinate system in October 1666 (1: 434–6) and later introduced it in Problem 10 of his 1671 tract (III: 298–302). On Newton's implicit assumption that these coordinates uniquely define the locus (*P*) see III: 299, note (685).

methodis de quibus ibi locutus sum, at alijs in casibus rem aliter se habere solere[,] id est methodis ulterioribus non semper (ut verba mea interpretatus est Cl. Leibnitius) sed maxima ex parte indigere.[147]

Cap 5.[148]

Exponitur methodus generalior reducendi Problemata ad æquationes infinitas.

Consistit hæc methodus in assumptione indefinita primi termini seriei et faciendo[149] ut terminus iste impleat conditiones Problematis quoad fieri possit, deinde in assumptione indefinita secundi termin[i] et faciendo ut iste conditiones problematis quam proximè[150] impleat, tum in assumptione tertij termini et istum quoq; aptando ad conditiones Problematis, et sic in infinitum.[151] Res melius constabit exemplis.

---

(147) Compare note (90) above. In the closing page of his *epistola posterior* in October 1676 Newton had asserted (*Correspondence*, **2**: 129): 'Inversum hoc Problema de tangentibus quando tangens inter punctum contactus et axem figuræ est datæ longitudinis, non indiget his methodis. [Et] quando in triangulo rectangulo quod ab illa axis parte & tangente ac ordinatim applicata constituitur, relatio duorum quorumlibet laterum per æquationem quamlibet definitur, Problema solvi potest absq; mea methodo generali, sed ubi pars axis ad punctum aliquod positione datum terminata ingreditur vinculum tunc res aliter se habere solet.' In his present terms, given some 'relatio' (doubtless assumed to be algebraic and homogeneous) which relates the fluxions p (or $\dot{x}$), q (or $\dot{y}$) and r (or $\dot{z}$) of $AB = x$, $BC = y$ and $\widehat{DC} = z$ Newton asserts that the curve (C) may be determined without using the general series method which he will expound in the next paragraph. Since

$$TB:BC \text{ (or } y):CT = p:q:r \text{ (or } \sqrt{[p^2+q^2-2pq\cos A\widehat{B}C]}),$$

the defining relationship between any two sides of the 'characteristic' triangle TBC (say $f(y, py/q, ry/q) = 0$) may be expressed in the form $f'(y, p/q) = 0$, which in turn by algebraic extraction of its root p/q ($= \dot{x}/\dot{y}$) will, 'absque methodo generali', yield $p/q = \phi(y)$ and so immediately $x = \int \phi(y) . dy$. Thus in Newton's example of the tractrix (on making $A\widehat{B}C = \frac{1}{2}\pi$) $f \equiv ry/q - c, f' \equiv (p/q)y - \sqrt{[c^2-y^2]}$ and $\phi \equiv \sqrt{[c^2-y^2]}/y$, whence x is determinable by Ordo 5.1 of the 1671 tract's 'catalogus posterior' of integrals (III: 248) on setting $d = 1$, $e = c^2$, $f = -1$, $\eta = 2$: alternatively (compare III: 26, note (31))

$$x = \sqrt{[c^2-y^2]} - c\log[(c+\sqrt{[c^2-y^2]})/y]$$

in explicit form. Evidently when the abscissa x enters the 'vinculum', and consequently the defining 'relatio' has the more general form $f(x, y, py/q, ry/q) = 0$, this simple reduction is no longer possible. Leibniz seemingly did not comprehend Newton's argument when he first received Oldenburg's copy of the *epistola posterior* about late May 1677, for in its margin he scribbled three equations—on recasting in Newton's present terms $py/q = f(y)$, $ry/q = f(y)$ and $py/q = f(x)$ respectively—which he took to be an exact representation of Newton's assertion: since the result $\log y = \int 1/f(x) . dx$ follows from the third equation with no more difficulty than $x = \int f(y)/y . dy$ and $z = \int f(y)/y . dy$ are deducible from the first two he thought he had caught Newton out and accordingly wrote back on 11/21 June 1677 (*Correspondence*, **2**:

methods of which I there spoke, though in other cases the situation is apt to be otherwise; in other words, ulterior methods are for the most part—but not (as Mr Leibniz took my words to mean) always—needed by it.[147]

CHAPTER 5[148]

A MORE GENERAL METHOD OF REDUCING PROBLEMS TO
INFINITE EQUATIONS IS EXPOUNDED

This method consists in indefinitely assuming the first term of a series and making[149] that term fulfil the conditions of the problem as far as can be done, in next assuming the second term indefinitely and making that fulfil the conditions of the problem as closely as possible,[150] then assuming the third term and adapting that also to the conditions of the problem, and so on infinitely.[151] The point will better be taken in examples.

---

217): 'si habeatur valor ipsius $T[B]$ ex BC haberi poterit curva [DC suppositis quadraturis figurarum analyticarum;] quod verò addit...Neutonus non æquè rem procedere si detur relatio ipsius TB ad partem axis seu ad AB ..., ad hoc respondeo mihi æquè facile esse invenire curvæ naturam vel æquationem si detur relatio ipsius TB ad AB, quam si...detur relatio ad BC'. Newton's words, however, make explicit that his 'relatio' shall be between two (or all three?) of the sides TB, BC, CT—and also, in the examples which require his general method, the general abscissa AB—and Leibniz' counter-example $TB = f(AB)$ is therefore excluded. (See also C. J. Scriba's comments in his 'The Inverse Method of Tangents: A Dialogue between Leibniz and Newton (1675–1677)', *Archive for History of Exact Sciences*, **2**, 1964: 113–37, especially §§4/5: 125–33. Leibniz' *marginalia* were published by C. I. Gerhardt in his edition of *Der Briefwechsel von G. W. Leibniz mit Mathematikern*, **1** (Berlin, 1899): 224, note ∗ [= *Correspondence*, **2**: 211].)

In sequel Newton has cancelled in his manuscript a bridging passage to the (final) chapter which follows: 'Allatis instantijs constet credo quomodo problemata ad contemplationem fluxionū reducenda sunt et solvenda, derivando fluxiones a fluentibus quantitatibus. Restat ut regressus a fluxionibus ad fluentes quantitates, dein quod ulterius est exponatur' (Now that these instances have been adduced it should, I think, be established how problems are to be reduced to the survey of fluxions and solved by deriving fluxions from fluent quantities. It remains to expound the return from fluxions to fluent quantities and then what is beyond).

(148) '4' was first written: to be sure this chapter elaborates the theme of 'Cap 4. Methodus generalior reducendi problemata ad series infinitas' in Newton's preliminary outline of the 'Specimina'. Compare notes (39) and (122) above.

(149) 'operando' (contriving that) is cancelled.

(150) This replaces 'quoad licuerit' (as far as permissible).

(151) Exactly the second 'methodus generalior' communicated by Newton (in anagrammatic form) in his *epistola posterior* to Leibniz in October 1676, viz: 'altera [methodus constitit] tantum in assumptione seriei pro quantitate qualibet incognita ex qua cætera commodè derivari possunt, et in collatione terminorum homologorum æquationis resultantis ad eruendos terminos assumptæ seriei' (see II: 191, note (25)). No attempt is made in the two following examples to repeat the neat tabular arrays by which Newton promoted his method in the 1671 tract (III: 98–112) in the series solution of first-order fluxional equations. As a consequence the text is something of a jumble.

Ex. 1. Curvæ alicujus ignotæ *DC*, area *DCB* ad unitatem linearem applicata dicatur *z*, basis *AB x*, et ordinata *BC y*, tum æquatione quavis, puta

$$azz - 2xxz + 3x^3 - 2aay + y^3 - 4a^3 = 0.$$

definiatur relatio inter *x y* et *z*,[152] & invenienda sit curva quæ hanc conditionem impleat. Jam quærendo æquationē infinitam qua ex assumpta utcunɋ basi *x* determinetur ordinata *y*: pro primo seriei termino assumo *ex*^{n} fingens esse $y = ex^n$ &c unde consequens est fore[153] $z = \dfrac{e}{n+1} x^{n+1}$

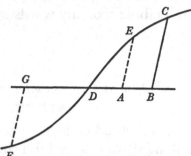

&c. Quamobrem in æquatione resolvenda scribo ex^n pro *y* et $\dfrac{e}{n+1} x^{n+1}$ pro *z* et emergit $\dfrac{eea}{nn+2n+1} x^{2n+2} - \dfrac{2e}{n+1} x^{n+3} + 3x^3 - 2aaex^n + e^3x^{3n} - [4]a^3 = 0.$ [154]Duo sunt hic genera terminorum, unū notorum[155] seu in quibus ignotum ex^n non reperitur, alterum ignotorum seu involventium ignotum ex^n. Utriusɋ generis terminos illos seligo qui si indefinita parva quantitas *x* in infinitum diminuitur evadunt cæteris infinitè majores, id est terminos ubi *x* est minimarum dimensionum, ut in hoc casu $-2aaex^n + e^3x^{3n}$ & $-4a^3$. Hos igitur[156] tanquam si cæteri in infinitum facti infinitè minores evanescerent solos retineo et conjunctim pono $= 0$.[157] Incertum est utrum dignitas x^n vel dignitas x^{3n} depressior sit et ideo terminum utrumɋ $-2aaex^n + e^3x^{3n}$ retineo. Si index *n* supponatur affirmativus altior erit dignitas ipsius x^{3n} quam ipsius x^n et ideo terminus e^3x^{3n} delendus. Quo facto manebit $-2aaex^n - [4]a^3 = 0$ seu $ex^n = -[2]a$. Unde sequitur indicem *n* nullum esse debere contra Hypothesin.[158] Si index ille *n* supponatur negativus altior erit dignitas ipsius x^n quàm ipsius x^{3n} et ideo

---

(152) Newton first continued 'et inde derivanda est relatio inter solas *x* et *y* per æquationē infinitam' (from this is to be derived the relationship between *x* and *y* alone by means of an infinite equation).

(153) More generally, as Newton will observe below, we might set $z = r + ex^{n+1}/(n+1)$ to $O(x^{n+2})$, where *r* is a constant of integration. The present series solution is the particular one for which $r = 0$.

(154) A cancelled first version of the following reads: 'Hic seligo terminum maximum cui ex^n non inest ut et maximum cui $[e]x^n$ inest. Maximos duco qui si indefinite parva *x* in infinitum minuatur minime omnium minuuntur, id est in quibus *x* est minimarum dimensionum' (Here I select the greatest term in which ex^n is not present and the greatest in which it is. I regard as greatest those which, if the indefinitely small quantity *x* should diminish infinitely, diminish least of all—those, that is, in which *x* is of least dimensions).

(155) Newton first wrote equivalently 'cognitorum' and then 'verorum' (true ones).

(156) 'ut in eo casu solos manentes' (as being in that case the sole remaining ones) is cancelled.

Example 1. Let the area *DCB* of some unknown curve (after division by the linear unit) be called *z*, the base *AB* *x* and the ordinate *BC* *y*, then let any equation, say $az^2 - 2x^2z + 3x^3 - 2a^2y + y^3 - 4a^3 = 0$, define the relationship between *x*, *y* and *z*:[152] we are required to find the curve which shall fulfil this condition. Now, in seeking the infinite equation by which the ordinate *y* is to be determined from any base *x* arbitrarily chosen, for the first term of the series I assume ex^n, conceiving that $y = ex^n \ldots$, and from this in consequence there will be[153]

$z = \dfrac{e}{n+1} x^{n+1} \ldots$ Therefore in the equation to be resolved I write ex^n in place of *y* and $\dfrac{e}{n+1} x^{n+1}$ for *z* and there emerges

$$\frac{e^2a}{(n+1)^2} x^{2n+2} - \frac{2e}{n+1} x^{n+3} + 3x^3 - 2a^2ex^n + e^3x^{3n} - 4a^3 = 0. \text{[154]}$$

There are here two kinds of term: one of knowns,[155] in which, namely, the unknown ex^n is not found, and the other of unknowns, or ones involving the unknown ex^n. From either kind I select those terms which, if the indefinite (small) quantity *x* diminishes indefinitely, prove to be infinitely greater than the rest—that is, terms in which *x* is of least dimension, as in the present case $-2a^2ex^n$, $+e^3x^{3n}$ and $-4a^3$. These consequently,[156] with the infinity of others, when they come to be infinitely less, vanishing as it were, I alone retain and set them collectively equal to 0.[157] It is not clear which of the powers x^n or x^{3n} is the lower and as a result I retain both terms $-2a^2ex^n$ and $+e^3x^{3n}$. If the index *n* be supposed positive the power x^{3n} will be higher than x^n and accordingly the term e^3x^{3n} is to be deleted. But when this is done there will remain

$$-2a^2ex^n - 4a^3 = 0 \quad \text{or} \quad ex^n = -2a,$$

from which it follows that the index *n* is zero, contrary to hypothesis.[158] If the index *n* be supposed negative, the power x^n will be higher than x^{3n} and accord-

---

(157) Newton first continued: 'et inde elicio $ex^n = -\frac{1}{2}a$ [!]. Est igitur $-\frac{1}{2}a$ primus terminus seriei quæsitæ. Ut secundum jam inveniam pro eo assumo ignotum quemvis ex^n, id est pono $y = -\frac{1}{2}a + ex^n$ et inde prodit $z = -\frac{1}{2}ax + \dfrac{e}{n+[1]} x^{n+1}$. Hos igitur ipsarum *y* et *z* valores substituo in æquatione prima et neglectis interea terminis qui se mutuo destruunt quive ratione jam exposita evasuri sunt cæteris infinite minores, emergit hæc æquatio' (and thence I derive $ex^n = -\frac{1}{2}a$, which is therefore the first term of the required series. Now to find the second in its place I assume any unknown quantity ex^n—that is, I put $y = -\frac{1}{2}a + ex^n$—and from this there results $z = -\frac{1}{2}ax + ex^{n+1}/(n+1)$. These values of *y* and *z* I therefore substitute in the first equation and, on neglecting at the same time terms which destroy one another or which on the principle just now expounded will turn out to be infinitely less than the rest, there emerges this equation...).

(158) In sequel Newton has cancelled the unnecessary corollary 'Non est igitur index ille [affirmativus]' (That index is therefore not positive).

terminus $-2aaex^n$ delendus erit. Quo facto manebit $e^3x^{3n}-[4]a^3=0$. Unde rursus colligitur indicem n nullum esse debere contra Hypothesin. Sit igitur index ille nullus et fiet $-2aaex^0+e^3x^{3\times0}-4a^3=0$ id est $-2aae+e^3-4a^3=0$. Cujus æquationis resolutione invenitur $e=2a$.[159] Primus igitur seriei quæsitæ terminus ex^n valet $2ax^0$ id est $2a$. Ut secundum jam inveniam, pro eo assumo ignotum quemvis puta ex^n, id est pono $y=2a+ex^n$ et inde prodit

$$z=2ax+\frac{e}{n+1}x^{n+1} \quad \&\text{c}.$$

Quibus pro y et z in æquatione resolvenda $azz-2xxz$ &c $[=0]$ substitutis et neglectis interea terminis qui se mutuò destruunt quive ratione jam ante exposita evasuri sunt cæteris infinitè minores, ex Hypothesi quod index n affirmativus est emergit hæc æquatio $+10aaex^n+4a^3x^2=0$.[160] seu $ex^n=-\dfrac{2ax^2}{5}$.

Invento seriei secundo termino pro tertio itidem assumo ignotum quemvis ex^n ponendo $y=2a-\dfrac{2ax^2}{5}+ex^n$ &c, et inde prodit $z=2ax-\dfrac{2ax^3}{15}+\dfrac{e}{n+1}x^{n+1}$ &c, et his pro y et z in æquatione resolvenda scriptis neglectiscꝫ negligendis, ex hypothesi quod index n affirmativus est prodit $10aaex^n+3x^3-4ax^3=0$, seu $ex^n=\dfrac{4a-3}{10aa}x^3$. Quo invento fiat $y=2a-\dfrac{2a}{5}x^2+\dfrac{4a-3}{10aa}x^3+ex^n$ &c[161] et con-sequetur $z=2ax-\dfrac{2a}{15}x^3+\dfrac{4a-3}{4[0]aa}x^4+\dfrac{e}{n+1}x^{n+1}$ &c. His in æquatione resolvenda pro y et z scriptis et rejectis rejiciendis obveniet $10aaex^n-\dfrac{[16]a^3}{75}x^4=0$ seu $ex^n=\dfrac{[16]a}{750}x^4$. Et eadem methodo pergi potest ad terminum quintum cæteroscꝫ in infinitum. Est igitur æquatio interminata quam invenire oportuit

$$y=2a-\frac{2a}{5}xx+\frac{4a-3}{10aa}x^3+\frac{[16]a}{750}x^4 \quad \&\text{c}.^{[162]}$$

Hæc est igitur una curvarum quæ conditionem problematis implent. Sunt

---

(159) Ignoring the non-real roots $e=(-1\pm\sqrt{-1})a$.

(160) On omitting terms in x^3 and higher powers of x, namely. In successive later stages the equations result 'neglectis negligendis' to $O(x^4)$ and 'rejectis rejiciendis' to $O(x^5)$.

(161) Newton originally set the denominator of the term in x^3 to be '12'; while here correcting it to '10' he has not bothered to make the three alterations of '112' into '16' which are thereby necessitated in the next few lines.

ingly the term $-2a^2ex^n$ will need to be deleted. When this is done there will remain $e^3x^{3n}-4a^3=0$, and from this it is again gathered that n ought to be zero, contrary to hypothesis. So let that index be zero and there will come to be $-2a^2ex^0+e^3(x^0)^3-4a^3=0$, that is, $-2a^2e+e^3-4a^3=0$. From the solution of this equation it is found that $e=2a$.[159] The value, therefore, of the first term ex^n of the series sought is $2ax^0$, that is, $2a$. To find the second one now, I assume any unknown term, say ex^n, for it; that is, I put $y=2a+ex^n$ and from this there

results $z=2ax+\dfrac{e}{n+1}x^{n+1}$ .... When these are substituted in place of y and z in

the equation $az^2-2x^2z\ldots=0$ to be resolved and on neglecting, moreover, those terms which mutually destroy one another or (for the reason already discussed above) prove to be infinitely less than the rest, from the hypothesis of a positive index n there emerges this equation $10a^2ex^n+4a^3x^2=0$[160] or $ex^n=-\frac{2}{5}ax^2$. Having found the second term of the series, for the third likewise I assume any unknown ex^n, putting $y=2a-\frac{2}{5}ax^2+ex^n\ldots$, and as a result

$$z=2ax-\tfrac{2}{15}ax^3+\frac{e}{n+1}x^{n+1}\ldots.$$

Once these are entered in the equation to be resolved in place of y and z and ignoring what is to be ignored, from the hypothesis that the index n is positive there results $10a^2ex^n+3x^3-4ax^3=0$, that is, $ex^n=\frac{1}{10}(4a^{-1}-3a^{-2})x^3$. When this is found, make $y=2a-\frac{2}{5}ax^2+\frac{1}{10}(4a^{-1}-3a^{-2})x^3+ex^n\ldots$[161] and there will be

obtained $z=2ax-\tfrac{2}{15}ax^3+\tfrac{1}{40}(4a^{-1}-3a^{-2})x^4+\dfrac{e}{n+1}x^{n+1}$ .... Once these are

entered in the equation to be resolved in place of y and z and on rejecting what is to be rejected there will fall out $10a^2ex^n-\frac{16}{75}a^3x^4=0$, that is, $ex^n=\frac{16}{750}ax^4$. And by the same approach you can advance to a fifth term and to an infinity of others. Therefore the non-terminating equation it was required to find is

$$y=2a-\tfrac{2}{5}ax^2+\tfrac{1}{10}(4a^{-1}-3a^{-2})x^3+\tfrac{16}{750}ax^4\ldots.\text{[162]}$$

This, therefore, is one of the curves which fulfil the condition of the problem.

(162) To summarize in the style of the 1671 tract (III: 98–112):

$$\begin{cases} y=2a * & -\tfrac{2}{5}ax^2+\tfrac{1}{10}(4a-3)a^{-2}x^3+\tfrac{8}{375}ax^4\ldots \\ z(=\int y\,.\,dx)=2ax & * \qquad -\tfrac{2}{15}ax^3+\tfrac{1}{40}(4a-3)a^{-2}x^4\ldots \end{cases}$$

$-2a^2y=$	$-4a^3 *$	$+\tfrac{4}{5}a^3x^2-\tfrac{1}{5}(4a-3)x^3-\tfrac{16}{375}a^3x^4\ldots$		
$+y^3=$	$8a^3 *$	$-\tfrac{24}{5}a^3x^2+\tfrac{6}{5}(4a-3)x^3+\tfrac{216}{375}a^3x^4\ldots$		
$-2x^2z=$		$-4ax^3 \qquad * \qquad \ldots$		
$+az^2=$	$4a^3x^2$	$* \qquad -\tfrac{8}{15}a^3x^4\ldots$		

$$+4a^3 * \qquad * \qquad -3x^3 \qquad * \qquad \ldots.$$

enim ejusmodi cu[r]væ numero infinitæ.[163] Initio superioris operationis assumpsi $y = ex^n$ &c et inde deduxi $z = \frac{e}{n+1} x^{n+1}$ &c. Deductio vera est sed non unica. Assumi enim potest nota[164] quævis quantitas puta r pro primo termino z et scribi $z = r + \frac{e}{n+1} x^{n+1}$ &c, id adeo, quia ex hoc valore z consequitur esse $y = ex^n$. Hisce autem pro y et z in æquatione resolvenda scriptis, oritur

$$arr + \frac{2are}{n+1} x^{n+1} + \frac{aeex^{2n+2}}{nn+2n+1} - 2rxx - \frac{2e}{n+1} x^{n+3} + 3x^3 - 2aaex^n + e^3x^{3n} - 4a^3 = 0. \quad \text{Et}$$

rejectis rejiciendis $e^3x^{3n} - 2aaex^n = 4a^3 - arr$.[165] Dignitas x eadem utrobiqʒ esse debet. Nulla est ad dextram ergo nulla erit ad sinistram. Est itaqʒ $e^3 - 2aae = 4a^3 - arr$. Resolutione hujus æquationis cubicæ invenienda est radix e. Factum puta, sitqʒ radix illa p, et erit px^n id est px^0 seu p primus terminus valoris y. Pro secundo termino jam assume ex^n et erit $y = p + ex^n$ &c et $z = r + px + \frac{e}{n+1} x^{n+1}$ [&c]. Quibus in æquatione resolvenda loco y et z scriptis ac deletis tum terminis $arr - 2aap + p^3 - 4a^3$ se mutuo destruentibus tum cæteris juxta superiora delendis orietur $2aaex^n - 3ppex^n = 2aprx$ seu $ex^n = \frac{2apr}{2aa - 3pp} x$. Est igitur $y = p + \frac{2apr}{2aa - 3pp} x$ &c et simili progressu producetur series[166] quoad usqʒ computi molestia sinit, quam ubi gravis esse incipit minuere licebit per numeralem determinationem quantitatum a, p, r & similium[167] quibus indefinita quantitas x afficitur.

[168]Ut si curvam DC transire velis per datum quodvis punctum E: Detur punctum illud E in recta AE quæ ordinatæ BC parallela est, et quoniam BC seu y coincidit cum AE ubi AB nulla est, in eo casu dabitur $y = AE$. Sed in eo casu extracta series evadit $y = p$. ergo datur $p = AE$. Pro datis p et a scribantur numeri quibus sunt æquales puta 1, et æquatio $e^3 - 2aae = 4a^3 - arr$ (ubi e idem erat cum p)

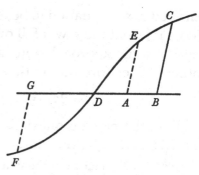

---

(163) Newton first continued: 'Nam deest una conditio quo minus problema determinatum sit' (For one condition is lacking to prevent the problem being determinate). A cancelled following sentence 'Proponatur jam curvam lineam per datum punctum ducere quæ conditionem præfatam complebit' (Let it now be proposed to draw a curved line through a given point, which shall fulfil the aforesaid condition) is taken up again in the next paragraph.

(164) That is, of constant magnitude. Compare note (153) above.

(165) To $O(x)$, that is.

(166) Newton first wrote equivalently 'Et simili computo colligitur tertius terminus cæteriqʒ' (And by a similar computation there is gathered the third term and the rest).

For there are an infinite number of curves of this sort.[163] At the start of the above operation I assumed $y = ex^n \ldots$ and deduced from this that

$$z = \frac{e}{n+1} x^{n+1} \ldots .$$

The deduction is a true one, but not unique. For any known[164] quantity, r say, can be assumed for z's first term and we may write $z = r + \frac{e}{n+1} x^{n+1} \ldots$, the reason being that $y = ex^n$ is a consequence of this value of z. With these, however, entered in place of y and z in the equation to be resolved, there arises

$$ar^2 + \frac{2aer}{n+1} x^{n+1} + \frac{ae^2}{(n+1)^2} x^{2n+2} - 2rx^2 - \frac{2e}{n+1} x^{n+3} + 3x^3 - 2a^2 ex^n + e^3 x^{3n} - 4a^3 = 0,$$

and on rejecting what is to be rejected $e^3 x^{3n} - 2a^2 ex^n = 4a^3 - ar^2$.[165] The power of x on either side should be the same. It is zero on the right and will therefore be zero on the left. Accordingly $e^3 - 2a^2 e = 4a^3 - ar^2$, and the root e is to be found by resolving this cubic equation. Consider it done and let that root be p, and then px^n, that is, px^0 or p, will be the first term in the value of y. For the second term now assume ex^n and there will be

$$y = p + ex^n \ldots \quad \text{with} \quad z = r + px + \frac{e}{n+1} x^{n+1} \ldots .$$

When these are entered in the equation to be resolved in place of y and z, and after deleting both the terms $ar^2 - 2a^2 p + p^3 - 4a^3$ which mutually destroy one another and also the others to be deleted in agreement with the above, there will arise $2a^2 ex^n - 3p^2 ex^n = 2aprx$ or $ex^n = \frac{2apr}{2a^2 - 3p^2} x$. Therefore $y = p + \frac{2apr}{2a^2 - 3p^2} x \ldots$ and by a similar procedure the series will be extended[166] as far as the troublesomeness of the computation allows. When this begins to be serious, you will be free to reduce it by determining numerically the quantities a, p, r and the like[167] which are coefficients of the indefinite quantity x.

[168]Thus should you want the curve DC to pass through any given point E: let that point E be given in the straight line AE which is parallel to the ordinate BC, and seeing that BC (or y) coincides with AE when AB is nil, in that case y will be given equal to AE. But in that case the series proves, when extracted, to be $y = p$ and it is therefore given that $p = AE$. In the place of the given quantities p and a let there be written the numbers they are equal to, say 1, and the equation $e^3 - 2a^3 e = 4a^3 - ar^2$ (where e was the same as p) by appropriate

---

(167) The equivalent phrase 'substituendo numeros pro a, p, r et similibus' (by substituting numbers in place of a, p, r and the like) is cancelled.

(168) A first opening to the paragraph reads 'Ut si a puncto A agatur AE ordinatæ BC parallela et in AE detur punctum quodvis E per quod curva transire debet' (If, for instance, from the point A there be drawn AE parallel to the ordinate BC and in AE there be given any point E through which the curve DC ought to pass...). See also note (163).

per debitam reductionem dabit $r = \sqrt{5}$. His numeris scriptis pro a p et r et resolutionis opere continuato prodibit $y = 1 - x\sqrt{20} \, \genfrac{}{}{0pt}{}{-4\sqrt{5}}{-72} xx$ [&c].[169] Quod si punctū per quod curvam DC transire velis locaretur extra lineam AE puta in F, agenda esset FG Ordinatæ BC parallela et dicto $GB = x$ scribenda esset $x - GB$ pro x in æquatione resolvenda et tum demum resolutio ineunda. Nam licet longitudinem AB dato quovis augere vel minuere manentibus ordinata y & area z.[170]

EXEMPL. 2.[171] Detur æquatio quæ Curvæ alicujus longitudinem z Basem x et ordinatam z simul involvit, [puta] $x^3 - 3xyz + y^3 + z^3 - a^2x - a^2y = 0$.[172] et requiratur Curva quæ hanc conditionem impleat serie infinita ex dignitatibus x conflata. Assumatur seriei initium $y = ex^n$ &c et posita 1 fluxione basis x fluxio curvæ lineæ invenietur $\sqrt{1 + nneex^{2n-2}}$ [&c] id est $\sqrt{1 + ee}$ [&c] si $n - 1$ nihil est[,] vel extracta radice $1 + \dfrac{nnee}{2} x^{2n-2}$ &c si $n - 1$ affirmativum est, vel alias extracta radice $nex^{n-1} + \dfrac{1}{2nex^{n-1}}$ &c si $n - 1$ negativum est. Juxta triplicem hunc casum longitudo curvæ z ex fluxione collecta erit vel $x\sqrt{1 + ee}$ [&c] vel $x + \dfrac{nnee}{2} x^{2n-1}$ &c[173] vel deniq̃ $nex^n + \dfrac{1}{2nex^{n-2}}$ &c.[173] Scribo igitur ex^n pro y in æquatione resolvenda et prodit $x^3 - 3zex^{n+1} + e^3x^{3n} + z^3 - aax - aaex^n$ [$= 0$]. dein tento tres valores z. Is retinendus est quo depress[iss]imæ dimensiones ipsius x in terminis cum ignota ex^n affectis et in terminis cum ignota ex^n non affectis æquentur inter se. Et hic est solus $x\sqrt{1 + ee}$. Nimirum scribendo $x\sqrt{1 + ee}$ pro z[174] et negligendo

---

(169) The coefficient of x^2 should be ' $+ 2\sqrt{5} - 51$ '. The equation

$$z^2 - 2x^2z + 3x^3 - 2y + y^3 - 4 = 0$$

is readily resolvable to any desired accuracy by the following Newtonian scheme (compare note (162) above):

$$\begin{cases} y = 1 - 2\sqrt{5}x + (2\sqrt{5} - 51)x^2 \ldots \\ z(= \int y.dx) = \sqrt{5} + x - \sqrt{5}x^2 \ldots \end{cases}$$

$-2y =$	$-2 + 4\sqrt{5}x - (4\sqrt{5} - 102)x^2 \ldots$
$+y^3 =$	$+1 - 6\sqrt{5}x + (6\sqrt{5} - 93)x^2 \ldots$
$-2x^2z =$	$-2\sqrt{5}x^2 \ldots$
$+z^2 =$	$+5 + 2\sqrt{5}x \qquad -9x^2 \ldots$

$$+4 \qquad * \qquad\qquad * -3x^3 \ldots.$$

(170) Depending, that is, on the lower bound taken to the 'area' $(z =) \int y.dx$.

reduction will yield $r = \sqrt{5}$. With these numbers replacing a, p and r there will, when the resolution procedure is continued, result

$$y = 1 - x\sqrt{20} - (72 + 4\sqrt{5})\, x^2 \ldots {}^{(169)}$$

But if the point through which you would like the curve DC to pass had been located outside the line AE, say at F, you would have had to draw FG parallel to the ordinate BC and, calling $GB = x$, write $GB - x$ for x in the equation to be resolved: only then could you have begun the resolution. For it is permissible to increase or diminish the length of AB by any given quantity while maintaining the ordinate y and area z.[170]

Example 2.[171] Let there be given an equation simultaneously involving the length z of some curve, its base x and ordinate y—say

$$x^3 - 3xyz + y^3 + z^3 - a^2x - a^2y = 0^{(172)}$$

—and let there be required a curve which shall fulfil this condition by an infinite series built up from powers of x. Let the beginning of the series be taken as $y = ex^n \ldots$ and, presupposing a unit fluxion of the base x, the fluxion of the curve's arc will be found to be $\sqrt{[1 + n^2e^2x^{2n-2} \ldots]}$; that is, $\sqrt{[1 + e^2]} \ldots$ if $n-1$ is zero, or on extracting the root $1 + \frac{1}{2}n^2e^2x^{2n-2} \ldots$ if $n-1$ is positive, or on extracting the root another way $nex^{n-1} + 1/(2nex^{n-1}) \ldots$ if $n-1$ is negative. In accordance with these three cases the length of the curve z gathered from its fluxion will be either $x\sqrt{[1 + e^2]} \ldots$ or $x + \frac{1}{2}n^2e^2x^{2n-1} \ldots {}^{(173)}$ or lastly $nex^n + 1/(2nex^{n-2}) \ldots {}^{(173)}$ I therefore write ex^n in place of y in the equation to be resolved, producing

$$x^3 - 3zex^{n+1} + e^3x^{3n} + z^3 - a^2x - a^2ex^n = 0,$$

and then test the three values of z. That one is to be kept which will make the lowest powers of x in the terms affected with the unknown quantity ex^n and in those not affected with it equal to one another. Here there is only $x\sqrt{[1 + e^2]} \ldots$ To be precise, on writing $x\sqrt{[1 + e^2]} \ldots$ in place of $z^{(174)}$ and neglecting terms in

---

(171) Newton first continued: 'Ex relatione inter Curvæ alicujus longitudinem z, basem x et ordinatam y per æquationem aliquā definita invenire Curvam ill[am]' (From the relationship between the length z of some curve, its base x and ordinate y defined by some equation, to determine that curve).

(172) The terms '$-a^2x - a^2y$' are a late insertion; see note (174). Newton in sequel will assume $a = 1$.

(173) These should read '$x + \dfrac{nnee}{4n-2} x^{2n-1}$' and '$ex^n + \dfrac{1}{4n - 2\,ex^{n-2}}$' respectively.

(174) In the first instance Newton went on with 'æquatio resolvenda fit

$$x^3 = 3x^3e\sqrt{1 + ee} - e^3x^3 - x^3\sqrt{1 + ee} - eex^3\sqrt{1 + ee}.$$

Unica et ideo depressissima dimensio ipsius x hic est x^3. Reductione autem æquationis hujus prodit $e = 1[!]$, adeoꞯ seriei terminus primus ex^n est x. Fingatur jam secundus terminus ex^n posit[o] $y = x + ex^n$ &c et fluxio curvæ lineæ erit $\sqrt{2 + 2nex^{n-1} + nneex^{2n-2}}$ seu extracta radice

terminos ubi x est plurimarum dimensionum prodibit $aax + aaex^n = 0$ seu $e = -1$,[175] adeoqʒ primus seriei terminus ex^n est $-x$. Ad inveniendum secundum terminum assumamus $y = -x + ex^n$ et fluxio curvæ lineæ erit

$$\sqrt{2 - 2nex^{n-1} + nneex^{2n-2}}$$

& extracta radice $= \sqrt{2} - \dfrac{ne}{\sqrt{2}} x^{n-1}$ &c. Unde curvæ longitudo z est $x\sqrt{2} - \dfrac{e}{\sqrt{2}} x^n$ &c et scriptis his pro y et z deletisqʒ delendis manet $5\sqrt{2} \times x^3 = ex^n$. Fingo jam $y = -x + 5\sqrt{2} \times x^3 + ex^n$ et fluxio curvæ erit $\sqrt{2 - 30\sqrt{2} \times xx + 450x^4 - 2nex^{n-1}}$ &c seu extracta radice $\sqrt{2} - 15xx + \dfrac{225}{2\sqrt{2}} x^4 - \dfrac{ne}{\sqrt{2}} x^{n-1}$ &c. Unde curvæ longitudo z est $x\sqrt{2} - 5x^3 + \dfrac{45}{2\sqrt{2}} x^5 - \dfrac{e}{\sqrt{2}} x^n$ &c. Et ipsarum y et z substitutis valoribus deletisqʒ delendis manet $15\sqrt{2} \times x^5 - 75x^5 - ex^n = 0$, seu $15x^5\sqrt{2} - 75x^5 = ex^n$. Et simili progressu prodibit quartò $\dfrac{4915}{2\sqrt{2}} x^7 - 555x^7 = ex^n$.[176] Est ergo series quæsita

$$y = -x + 5\sqrt{2} \times x^3 + \dfrac{15\sqrt{2}}{-75} x^5 + \dfrac{\frac{4915}{2\sqrt{2}}}{-555} x^7 \ \&c.$$

Atqʒ hoc exemplo constet quo modo methodus in hujusmodi casibus exercenda sit, verùm tamen falsa[177] est series

$\sqrt{2} + \dfrac{nex^{n-1}}{\sqrt{2}}$ &c. Unde Curvæ fluentis longitudo z erit $x\sqrt{2} + \dfrac{ex^n}{\sqrt{2}}$ &c. Jam pro y et z substitutis his eorum valoribus æquatio resolvenda evadet ' (the equation to be resolved becomes

$$x^3 = 3e\sqrt{[1 + e^2]}\, x^3 - e^3 x^3 - (\sqrt{[1 + e^2]})^3 x^3.$$

The unique, and consequently lowest, dimension of x here is x^3. By reduction of this equation, however, there results $e = 1$, so that the first term ex^n of the series is x. Now suppose the second term to be ex^n, putting $y = x + ex^n \dots$, and the fluxion of the curve's arc will be

$$\sqrt{[2 + 2enx^{n-1} + e^2 n^2 x^{2n-2} \dots]},$$

that is, on extracting the root, $\sqrt{2} + (en/\sqrt{2})\, x^{n-1} \dots$. Whence the length z of the fluent curve will be $\sqrt{2}x + (e/\sqrt{2})\, x^n \dots$. Now when these present values are substituted in place of y and z the equation to be resolved will prove to be...). It will be clear that Newton here works with his original equation $x^3 - 3xyz + y^3 + z^3 = 0$ (see note (172) above), but abruptly increases it with the further terms $-a^2 x - a^2 z$ when he sees his mistake in supposing $e = 1$ to be a root of $x^3 = (-e^3 + (3e - 1 - e^2)\sqrt{[1 + e^2]})\, x^3$.

(175) And, of course, $n = 1$.

(176) This should be '$\dfrac{3435}{2\sqrt{2}} x^7 - 600x^7 = ex^n$', yielding the corrected coefficient '$\dfrac{3435}{2\sqrt{2}} - 600$' for the term in x^7 in the series expansion of y. In Newtonian style (compare note (162) above) we may extract the root y of $x^3 - 3xyz + y^3 + z^3 - x - y = 0$ as follows (supposing, with Newton, the integration constant to be zero):

which x is of very many dimensions there will prove to be $a^2x + a^2ex^n = 0$, or $e = -1$[175] and consequently the first term ex^n of the series is $-x$. To find the second term let us assume $y = -x + ex^n$, and the fluxion of the curve's arc will be $\sqrt{[2 - 2nex^{n-1} + n^2e^2x^{2n-2}]}$, that is, when the root has been extracted,

$$\sqrt{2} - \frac{1}{\sqrt{2}}nex^{n-1}\ldots.$$

Hence the length z of the curve is $x\sqrt{2} - \frac{1}{\sqrt{2}}ex^n\ldots$, and after these have been written in place of y and z and on deleting what is to be deleted there remains $ex^n = (5\sqrt{2})x^3$. I now conceive that $y = -x + (5\sqrt{2})x^3 + ex^n$ and the fluxion of the curve will be $\sqrt{[2 - (30\sqrt{2})x^2 + 450x^4 - 2nex^{n-1}\ldots]}$ or, after the root is extracted, $\sqrt{2} - 15x^2 + \frac{225}{2\sqrt{2}}x^4 - \frac{1}{\sqrt{2}}nex^{n-1}\ldots.$ Hence the length z of the curve is $x\sqrt{2} - 5x^3 + \frac{45}{2\sqrt{2}}x^5 - \frac{1}{\sqrt{2}}ex^n\ldots$, and after the values of y and z have been substituted and on deleting what is to be deleted there remains

$$(15\sqrt{2})x^5 - 75x^5 - ex^n = 0 \quad \text{or} \quad ex^n = (15\sqrt{2} - 75)x^5.$$

And by a similar procedure there will result in the fourth place

$$ex^n = \left(\frac{4915}{2\sqrt{2}} - 555\right)x^7.[176]$$

The required series is therefore

$$y = -x + (5\sqrt{2})x^3 + (15\sqrt{2} - 75)x^5 + \left(\frac{4915}{2\sqrt{2}} - 555\right)x^7\ldots.$$

This example should establish how the method is to be practised in cases of this kind, though, for all that, the series which here results is false.[177] There is

$$y = -x + 5\sqrt{2}x^3 + \lambda x^5 + \mu x^7\ldots$$
$$z(= \int\sqrt{[1 + (dy/dx)^2]}.dx) = \sqrt{2}x - 5x^3 + [(45 - 2\lambda)/2\sqrt{2}]x^7\ldots$$

$-y =$	$x - 5\sqrt{2}x^3$	$-\lambda x^5$		$-\mu x^7\ldots$
$+y^3 =$		$-x^3 + 15\sqrt{2}x^5$	$(-130 + 3\lambda)x^7\ldots$	
$-3xyz =$		$+3\sqrt{2}x^3 \quad -45x^5 + [(435 - 18\lambda)/2\sqrt{2}]x^7\ldots$		
$+z^3 =$		$+2\sqrt{2}x^3 \quad -30x^5 \quad +[(285 - 6\lambda)/\sqrt{2}]x^7\ldots$		

$$+x \quad -x^3 \quad * \qquad * \qquad \ldots$$

where $\lambda = 15\sqrt{2} - 75$ and $\mu = 3435/2\sqrt{2} - 600$.

(177) As we have seen (note (49) above) Newton's following argument that, because the series expansion of y in terms of x is divergent ('false'), the quantity which it represents must be non-real is strictly a *non-sequitur*. In the present case, to be sure, we may reduce the cubic $z^3 - 3xyz + p = 0$, $p = x^3 + y^3 - x - y$, to the explicit form

$$z = \sqrt[3]{[-\tfrac{1}{2}p + \sqrt{(\tfrac{1}{4}p^2 - xy)}]} + \sqrt[3]{[-\tfrac{1}{2}p - \sqrt{(\tfrac{1}{4}p^2 - xy)}]};$$

quæ hic prodijt. Nulla est curva linea quæ conditiones Problematis implere potest et ex falsis positis falsa consequi necesse erat. Falsam verò hanc seriem esse ex ipsius forma fere colligitur: quippe in qua coefficientes dignitatum *x* quæ tandem decrescere deberent augentur magis et magis in immensum idcq in proportione plusquam Geometrica. Analogiam habent hæ operationes cum extractione radicum ex æquationibus affectis.[178] Si residua quæ radicum erroribus respondent augentur in perpetuum, radices prodeunt falsæ, & ubi radix impossibilis est quæ extrahitur coefficientes dignitatum *x* majores et majores evadunt in infinitum. Eodem modo coefficientes hic augentur et in æquatione resolvenda subductis terminis negativis de affirmativis residuum relinquitur majus ubi longius extracti sunt valores substituendi ipsarum *x* et *y*. Et quamvis residua illa ubi *x* valde parva assumitur aliquamdiu decrescere possint[,] tandem tamen ob incrementum coefficientium incip[i]ent increscere. Ut hæc exponerem dedi exemplum Problematis in casu impossibili. Sin exemplum in casu possibili desideres, pone $zz = xx + yy + ax - by$ et prodibit $y =$[179]

---

whence, if $y = -x + \epsilon$ (ϵ small), since $p^2 \approx \epsilon^2(3x^2 - 1)^2 < -4xy \approx 4x^2$ all three corresponding values of z will be real and Newton's conclusion that 'Nulla est curva linea quæ conditiones Problematis implere potest' is badly wrong.

(178) See III: 42–64: 'De affectarum æquationum reductione.' (This introductory passage from the 1671 tract is taken up again in Chapter 1 of the tract 'De Computo Serierum' which follows; see pages 592–604 below.)

(179) The text breaks off at this point. The example is still not a good illustration of a resolution into infinite series since, when the constant of integration is zero, the two particular solutions—namely $y = (a/b)x$ and $y = b - (a/b)x$—are both finite, while for non-zero values

no curve line which can fulfil the conditions of the problem and from false premisses it is a necessary truth that false deductions follow. This series is, indeed, gathered to be a false one almost from its very form, seeing that the coefficients of the powers of x in it, which ought eventually to decrease, increase more and more without bound and this in a proportion which is more than geometrical. These operations bear some analogy to the extraction of roots out of affected equations.[178] If the remainders—corresponding to the errors of the roots—increase perpetually, the roots turn out to be false, and when extraction of an impossible root is performed, the coefficients of the powers of x grow to be indefinitely larger and larger. In the same way here the coefficients increase, and when in the equation to be resolved negative terms are taken from positive ones the remnant left is greater the farther the extractions of the values of x and y to be substituted have been taken. And although those remnants might considerably diminish when x is assumed to be exceedingly small, they will nonetheless, through the increasing size of the coefficients, begin to grow larger. To illustrate this point I have given an example of the problem in an impossible case. But should you desire an example in a possible one, set

$$z^2 = x^2 + y^2 + ax - by$$

and there will result $y =$ [179]

---

of the constant the operation of extracting the series expansions of y and z in powers of x rapidly becomes tedious and unwieldy. In any case it seems better to eliminate the variable $z = \int \sqrt{[1 + (dy/dx)^2]} \, . \, dx$ by differentiating $z = \sqrt{[x^2 + y^2 + ax - by]}$ and seeking to resolve the first-order fluxional equation

$$\frac{q}{p}\left(\text{or}\,\frac{dy}{dx}\right) = \frac{(2x+a)(2y-b) \pm \sqrt{[(a^2+b^2)(x^2+y^2+ax-by)]}}{4x^2+4ax-b^2}.$$

§2. THE REVISED 'COMPUTATION OF SERIES'.[1]

DE COMPUTO SERIERUM

CAP 1.[2]

QUO MODO ÆQUATIONES RESOLVENDÆ SINT
IN SERIES INFINITAS.

I

Resolutio
quantitatum
simplicium
in series
infinitas.

Initio epistolæ meæ prioris regulam[3] dedi qua binomium quodvis ut $\overline{P+PQ}^{\frac{m}{n}}$ convertitur in seriem convergentem. Per binomium hic intelligo quantitatem quamvis ex partibus quotcunɔ constantem quarum maxima vel maximarum aggregatum sit P, & minima vel minimarum aggregatum PQ. Talium binomiorum multiplicationes divisiones et radicum extractiones hac unica regula complectuntur. [4]Si dignitatis index numerus integer est et affirmativus docet regula multiplicare et producere potestates binomij. Si index ille integer et negativus est, docet eadem regula dividere. Radices extrahuntur ubi index fractus obvenit et nonnunquā diversæ[5] multiplicandi dividendi et radices extra[hendi] operationes simul vice[6] peraguntur. Potest index ille surda esse quantitas et tunc producetur series quam nec multiplicatio nec divisio nec

---

(1) The autograph text of the first chapter is in private possession (compare H. W. Turnbull's note in *The Correspondence of Isaac Newton*, **2**: 402, note (1)), that of the two following ones is now ULC. Add. 3964.3: 13ʳ–14ʳ (with preliminary drafts and computations for Chapter 2 on ff. 11ʳ–12ᵛ; see note (38) below). In their original state the manuscript sheets, still in excellent physical condition, were fairly carefully composed, with outer rules (drawn freehand) to the pages and the numbered marginal titles here reproduced, but in revise Newton was less patient: his first title for the piece is now completely obliterated by heavy, superimposed horizontal, vertical and diagonal cancellations, while several of his scrawled interlineations are barely readable. There is little else which can usefully be said. Why Newton abandoned his previous 'Specimina' for the present 'De computo serierum', in which all references to Leibniz and David Gregory are suppressed, is not known. Equally, if he ever completed the unfinished third chapter on interpolation or the following ones he doubtless intended, in line with those of the 'Specimina', to compose, their text is no longer traceable and—with its obvious possible locations checked in vain—it seems now unlikely to exist.

(2) Like the analogous first chapter of the 'Specimina' (§1 preceding) this, apart from its discussion of the binomial expansion (paragraph I), borrows heavily from the 1671 tract's section 'De affectarum æquationum reductione' (III: 42–66, especially 50–64).

(3) 'gen[eralem]' (general) is cancelled. In his *epistola prior* to Oldenburg for Leibniz on 13 June 1676 (compare §1: note (12) and page 667 below) Newton had stated the series expansion of the general binomial $[P(1+Q)]^{m/n}$ as

$$`\overline{P+PQ}^{\frac{m}{n}} = P^{\frac{m}{n}} + \frac{m}{n}AQ + \frac{m-n}{2n}BQ + \frac{m-2n}{3n}CQ + \frac{m-3n}{4n}DQ + \&c.$$

Ubi $P+PQ$ significat quantitatem cujus radix vel etiam dimensio quævis vel radix dimensionis investiganda est, P primum terminum quantitatis ejus, Q reliquos terminos divisos per primum, & $\frac{m}{n}$ numeralem indicem dimensionis ipsius $P+PQ$ sive dimensio illa integra sit,

Translation

ON THE COMPUTATION OF SERIES
CHAPTER 1[2]
HOW EQUATIONS ARE TO BE RESOLVED
INTO INFINITE SERIES

At the beginning of my first letter I gave a[3] rule by which any binomial, such as $(P+PQ)^{\frac{m}{n}}$, is converted into an infinite series. By a binomial I here understand any quantity consisting of a number of parts, the greatest, individually or in total, of which is taken to be P and the least, individually or in total, of which is PQ. With such binomials, multiplications, divisions and root extractions are embraced in this single rule. [4]If the index of the power is a positive integer the rule teaches us how to multiply and extend the powers of a binomial. If that index is a negative integer, the same rule instructs us how to divide. Roots are extracted when the index proves to be a fraction and not infrequently different[5] operations of multiplication, division and root extraction are performed simultaneously at [one] go.[6] The index can be a surd quantity and a series will

I. Resolution of simple quantities into infinite series.

---

sive (ut ita loquar) fracta, sive affirmativa, sive negativa. . . . Deniɋ pro terminis inter operandum inventis in Quoto, usurpo A, B, C, D &c nempe A pro primo termino $P^{\frac{m}{n}}$, B pro secundo $\frac{m}{n}AQ$, & sic deinceps' (ULC. Add. 3977.2: 1ʳ [= *Correspondence*, **2**: 21]). Newton now explicitly makes the restriction (necessary for convergence of the binomial series) that $Q < 1$.

(4) A cancelled version of the following reads: 'Si dignitatis index $\frac{m}{n}$ numerus integer est et affirmativus docet regula multiplicare et invenire potestates binomij. Si index integer et negativus est docet eadem regula dividere unitatem per binomium aut per aliquam potestatē ejus. Si fractio est cujus numerator m sit unitas docet regula radicem aliquam extrahere, ut et unitatem per radicem illam dividere si modo fractio illa negativa sit' (If the index m/n of the power is integral and positive, the rule informs how to multiply and find powers or 'dignities' of the binomial; if the index is integral and negative, the same rule teaches how to divide unity by the binomial or by some power of it; if it is a fraction whose numerator m be unity, the rule tells how to extract some root, and also how to divide unity by that root should the fraction be negative). To Newton's contemporaries, of course, this was no mere elaboration of the obvious: indeed Newton was the first systematically to use the full range of the general (rational) coefficient in this way, and had felt compelled to explain the usage both in his 1671 tract (see III: 38) and in his *epistola prior* in June 1676 (see *Correspondence*, **2**: 21: 'Nam sicut Analystæ pro aa, aaa, &c scribere solent a^2, a^3, sic ego pro $\sqrt{a}$, $\sqrt{a^3}$, $\sqrt{c}:a^5$ scribo $a^{\frac{1}{2}}$, $a^{\frac{3}{2}}$, $a^{\frac{5}{3}}$, &

pro $\frac{1}{a}$, $\frac{1}{aa}$, $\frac{1}{a^3}$ scribo a^{-1}, a^{-2}, a^{-3}. Et sic pro $\dfrac{aa}{\sqrt{c}:a^3+bbx}$ scribo $aa \times \overline{a^3+bbx}\}^{-\frac{1}{3}}$ . . .').

(5) 'plures' (several) was first written.
(6) Probably to be deleted as a stray; Newton's 'simul' replaces 'una vice' but only 'una' is cancelled in the manuscript.

radicis extractio quævis daret.[7] Superest extractio radicum ex æquationibus affectis.

II
Resolutio radicum affectarum in series infinitas.

Hæc operatio superiores omnes in se complectitur[8] et in eo consistit ut mediante tabula rectangula[9] in cellulas rectangulas distincta sciamus ex æquatione resolvenda eruere terminum primum radicis: Deinde pro reliqua parte radicis ponentes terminum aliquem indefinitum ut p scribamus terminum inventum $+p$ pro radice in æquatione resolvenda. Quo pacto orietur æquatio secunda resolvenda cujus radix est p. Tum similiter ex hac æquatione per eandem tabulam tesselatam eruendus est primus terminus radicis p et pro reliqua radice ponendus terminus aliquis indefinitus ut q & scribendus inventus terminus $+q$ pro p in æquatione secunda resolvenda. Qua ratione prodibit æquatio tertia resolvenda cujus radix est q. Inde per tabulā tesselatam eruendus est primus terminus ipsius q & pro q scribendus terminus iste $+r$. Et sic in infinitum. Deniꝗ termini omnes inventi conjunctim component radicem quæsitam æquationis primæ. Sic in exemplo epistolæ prioris[10] ubi æquatio resolvenda erat $y^3 + axy + aay - x^3 - 2a^3 = 0$ et radix extrahenda erat y, tabula tessellata dedit primum radicis terminum $+a$. Quare pro reliqua parte radicis posui p et in æquatione resolvenda scripsi $a+p$ [pro] y. Sic prodijt æquatio secunda resolvenda $p^3 + 3app + 4aap + axp + aax - x^3 = 0$. Inde per tabulam tessellatam obtinui primum terminum radicis p nimirum $-\frac{1}{4}x$, tum pro reliqua parte radicis illius p ponendo q scripsi $-\frac{1}{4}x + q$ pro p in hac secunda æquatione resolvenda et prodijt æquatio tertia resolvenda

$$q^3 - \tfrac{3}{4}xqq + \tfrac{3}{16}xxq - \tfrac{65}{64}x^3 + 3aqq - \tfrac{1}{2}axq + 4aaq - \tfrac{1}{16}aax = 0.^{(11)}$$

Unde per tabulam tessellatam primus terminus radicis q inventus est $+\dfrac{xx}{64a}$

[adeoꝗ] pro reliqua parte radicis ponendo r et perinde scribendo $\dfrac{xx}{64a} + r$ pro q prodijt æquatio quarta resolvenda.[12] &c. Erat itaꝗ radix

$$y = a + p = a - \tfrac{1}{4}x + q = a - \tfrac{1}{4}x + \frac{xx}{64a} + r.$$

et sic in infinitum.

---

(7) In sequel Newton has cancelled 'neꝗ alias facile obtineri posset quam per hanc regulam. Sola hæc regula his in casibus nobis subvenit' (nor might it easily be obtained otherwise than by this rule. In these cases this rule alone is of assistance to us).

(8) For instance, the series expansion of the binomial $(1+x)^{m/n}$, m, n integral, may be derived as $1 + px$, where

$$(1+x)^m = 1 + mx + \tfrac{1}{2}m(m-1)x^2 \ldots \equiv (1+px)^n = 1 + npx + \tfrac{1}{2}n(n-1)p^2x^2 \ldots;$$

whence $p = m/n + qx$, where $\tfrac{1}{2}m(m-1)x^2 \ldots \equiv [nq + \tfrac{1}{2}m^2(n-1)/n]x^2 \ldots$ or

$$q = \tfrac{1}{2}(m/n)(m/n - 1) + rx;$$

and so on.

then be produced which no multiplication, division or root extraction could give.[7] There remains the extraction of roots from affected equations.

This operation embraces within itself all the above ones[8] and consists in knowing how, by means of a rectangular array[9] marked out into rectangular cells, to dig out the first term in the root; and then, after putting some indefinite quantity, as p, for the remaining part of the root, in entering the term $+p$ so found in replacement of the root in the equation to be resolved. In this manner a second equation to be resolved, of root p, will arise. Next, in a similar fashion the first term of the root p must be dug out from this equation with the aid of the same squared-off array and, with some indefinite term, as q, put for the remainder of the root, the term $+q$ so found must be entered in replacement of p in the second equation to be resolved. In this way a third equation to be resolved, of root q, will result. From this by means of the squared-off array the first term of q must be dug out and that first term, $+r$, entered in place of q. And so on infinitely. All the terms, lastly, when found will jointly compose the required root of the first equation. So in the example in my first letter,[10] where the equation to be resolved was $y^3 + axy + a^2y - x^3 - 2a^3 = 0$ and y the root to be extracted, the squared array gave $+a$ as the first term of the root. Consequently, for the remaining part of the root I put p and in place of y in the equation to be resolved I wrote $a+p$. In this way there resulted

II.
Resolution
of the roots
of affected
[equations]
into infinite
series.

$$p^3 + 3ap^2 + 4a^2p + axp + a^2x - x^3 = 0$$

for the second equation to be resolved. From this by means of the squared-out array I obtained the first term, namely $-\frac{1}{4}x$, of the root p, then on putting q for the remaining part of that root p I wrote $-\frac{1}{4}x+q$ in place of p in this second equation to be resolved and there resulted

$$q^3 - \tfrac{3}{4}xq^2 + \tfrac{3}{16}x^2q - \tfrac{65}{64}x^3 + 3aq^2 - \tfrac{1}{2}axq + 4a^2q - \tfrac{1}{16}a^2x = 0^{(11)}$$

for the third equation to be resolved. From this by means of the squared-out array the first term of the root q was found to be $\frac{1}{64}a^{-1}x^2$ and hence, on putting r for the remaining part of the root and likewise writing $\frac{1}{64}a^{-1}x^2+r$ in place of q, the fourth equation to be resolved[12] ensued. And so on. Accordingly the root was $y = a + p = a - \frac{1}{4}x + q = a - \frac{1}{4}x + \frac{1}{64}a^{-1}x^2 + r = \ldots$, and so on indefinitely.

---

(9) Newton first wrote 'parallelogrammo' (rectangle) and then 'figura rectangula [in minora rectangula distincta]' (rectangular figure [marked out into lesser rectangles]).

(10) See *Correspondence* **2**: 24. This passage in Newton's *epistola prior* to Leibniz in June 1676 is taken over virtually unchanged from his 1671 tract (III: 54–6). On 24 October following Newton slightly amplified his procedure at Leibniz' request (see *Correspondence*, **2**: 127).

(11) In his *epistola prior* in 1676 (see previous note). Newton slightly rounded this equation off, omitting the first two tems '$q^3 - \frac{3}{4}xqq$'. In his 1671 tract (III: 54) he wrote merely '$q^3 - \frac{3}{4}xqq + 3aqq$ &c'.

(12) '$x^3 + \dfrac{3xx}{64a}r + \dfrac{3x^4}{4096aa}r +$' is cancelled.

III
Quænam
series veræ
sunt et
quænam
falsæ.

Et hinc liquet operis demonstratio. Nam si p, q, r, s &c qui sunt errores seriei pe[r]petuo decrescunt sic, ut tandem dato quovis minores evadant, manifestū est seriem extractam legitime convergere et in infinitum appropinquare quantitatem y quam exprimit.[13] Id vero semper fit ubi quantitas indefinita (ut x) e cujus dignitatibus series conflatur satis parva est. Quod si ob magnitudinem indefinitæ illius quantitatis x vel alia de causa errores illi p, q, r, s &c non sic decrescunt, series nunquam appropinquabit quantitatem quæsitam y[14] sine errore nimio et ideo falsa dici meretur. Prodeunt autem semper falsæ series ubi radices impossibiles extrahimus.[15]

IV
De radicibus
pluribus
ejusdem
æquationis
et quomodo
radix
imperata
extrahi possit.

Cæterùm æquationis ejusdem radices sunt plures. Singularum extractionem tabula tessellata exhibet.[16] Termini per hanc tabulam selecti ex quibus æquationem fictam componentibus primi serierum termini eruendi sunt tot dab[unt] primos terminos quot sunt radices per seriem designabiles et ex singulis primis terminis subinde colligentur singulæ series. Sic ad resolvendam æquationem $y^3 + axy + aay - x^3 - 2a^3 = 0$ tabula dat æquationem fictitiam $y^3 + aay - 2a^3 = 0$ cujus tres radices $a, -\frac{1}{2}a + \sqrt{-\frac{7}{4}aa}$ & $-\frac{1}{2}a - \sqrt{-\frac{7}{4}aa}$ sunt primi termini totidem serierum, et ad resolvendam

$$y^6 - 5xy^5 + \frac{x^3}{a}y^4 - 7aaxxyy + 6a^3x^3 + bbx^4 = 0$$

tabula dat fictitiam $y^6 - 7aaxxyy + 6a^3x^3 = 0$, cujus sex radices $+\sqrt{ax}, -\sqrt{ax}, +2\sqrt{ax}, -2\sqrt{ax}$,[17] $+\sqrt{-3ax}, -\sqrt{-3ax}$ sunt termini initiales serierum totidem. Ex initialibus possibilibus consequuntur series possibles[,] ex impossibilibus impossibiles. Impossibilitatis causa est quod in æquatione resolvenda radices illæ aut semper impossibiles sunt, aut minuendo indefinitam quantitatem x possibilitatem amittunt. Nonnunquam tabula tessellata pro varia regulæ applicatione dabit varias æquationes fictas. Potest enim regula modo quocunꝗ applicari sic, ut cellulas omnes terminis æquationis respondentes ex eodem sui latere positas habeat, et extimas sibi vicinas attingat, parallela modò non sit tabulæ lateri sinistro AB.[18] Nam termini cellulis attactis respondentes component æquationem fictam cujus radices singulæ pro serierum terminis initialibus adhiberi

---

(13) Newton first wrote '...vere ['perpetuo' is deleted] convergere ad quantitatem y et errorem ejus tandem dato quovis minorem evadere' (truly [perpetually] converges to the quantity y and the error comes at length to be less than any given amount). The requirement that the series of 'errors' $p, q, r, s, \ldots$ should decrease to zero in the limit does indeed determine the convergence of the aggregate $a + (p-q) + (q-r) + (r-s) \ldots$, but this is scarcely more than a definition of (uniform) convergence. The following assertion is likewise true but adds little: the important question of what bounds need to be set on x to preserve the convergence of a given series expansion $a + bx + cx^2 + dx^3 \ldots$ is not broached.

(14) The amplification 'quam debet exprimere' (which it ought to express) is cancelled.

(15) A deleted conclusion reads 'et veræ ubi radices possibiles sunt et primus terminus seriei partem bene magnā radicis constituit' (and true ones when the roots are possible and the first term of the series constitutes a fairly large part of the root). On Newton's identification

And from this the demonstration of the procedure is clear. For if p, q, r, s, ...,
which are the errors of the series, perpetually decrease so as at last to come to be less than any given quantity, it is manifest that the series extracted justly converges and approaches infinitely near to the quantity y which it expresses.[13] That, in truth, always happens when the indefinite quantity (here x) from whose powers the series is made up is small enough. But if on account of the size of that indefinite quantity x or for some other reason those errors p, q, r, s, ... do not diminish in this way, the series will never approach the required quantity y[14] to within a sufficiently small error and is on this account deservedly called a false one. False series, however, always result when we extract impossible roots.[15]

For the rest, the same equation has several roots. The extraction of each is IV.
On the several
roots of the
same equation
and how an
appointed
root may be
extracted. displayed by the squared array. Terms selected by this array, making up a fictitious equation from which the first terms of the series are to be elicited, will yield as many first terms as there are roots which can be designated by a series, and from each separate first term a separate series will thereafter be gathered.[16] Thus to resolve the equation $y^3 + axy + a^2y - x^3 - 2a^3 = 0$ the array yields the fictitious equation $y^3 + a^2y - 2a^3 = 0$ whose three roots a, $\frac{1}{2}(-1 + \sqrt{-7})\,a$ and $\frac{1}{2}(-1 - \sqrt{-7})\,a$ are the first terms of an equal number of series, while to resolve $y^6 - 5xy^5 + a^{-1}x^3y^4 - 7a^2x^2y^2 + 6a^3x^3 + b^2x^4 = 0$ the array yields the fictitious

$$y^6 - 7a^2x^2y^2 + 6a^3x^3 = 0,$$

whose six roots $+\sqrt{[ax]}$, $-\sqrt{[ax]}$, $+\sqrt{[2ax]}$, $-\sqrt{[2ax]}$, $+\sqrt{[-3ax]}$, $-\sqrt{[-3ax]}$ are the initial terms of an equal number of series. From possible initial terms are obtained possible series, from impossibles impossible ones. The reason for the impossibility is that in the equation to be resolved those roots either are always impossible or lose their possibility on diminishing the indefinite quantity x. Not infrequently the squared array will yield different fictitious equations according as application of the ruler is differently made. For the ruler can be applied in any way at all so as to have all the cells corresponding to terms in the equation in position on the same side of it, the outermost neighbouring ones in contact with it, provided it is not parallel to the left side AB of the array.[18] The terms corresponding to the cells in contact will then compose the fictitious equation whose separate roots might be employed as initial terms of the series.

of 'false' (divergent) series with the 'impossible' (complex) roots of an equation see §1: note (49) above.

(16) The sequel is a light revision of the equivalent passage in Chapter 1 of the 'Specimina' (§1 preceding); see pages 542–4 above.

(17) As before (see §1: note (54) above) read ' $+\sqrt{2ax}$, $-\sqrt{2ax}$ '.

(18) Understand the 'parallelogram' on III: 50 (communicated to Leibniz in the *epistola posterior* in October 1676; see *Correspondence*, **2**: 126 and compare page 629 below). In this case the fictitious equation will be of the form $f(x) . y^p = 0$, whose roots y are, trivially, zero.

possint. Ut si resolvenda sit æquatio $y^5 - byy + 9bxx - x^3 = 0$ regula inferius applicata attinget uno situ cellulas terminorum $-byy + 9bxx$, alio situ cellulas terminorum $y^5 - byy$. Priores component æquationem fictam $-byy + 9bxx = 0$ cujus radices y sunt $+3x$ et $-3x$: Posteriores æquationem fictam $y^5 - byy = 0$ cujus radices y sunt $b^{\frac{3}{2}}$, $-\frac{1}{2}b^{\frac{3}{2}} + \sqrt{-\frac{3}{4}b^{\frac{3}{2}}}$ et $-\frac{1}{2}b^{\frac{3}{2}} - \sqrt{-\frac{3}{4}b^{\frac{3}{2}}}$. Quincp sunt igitur termini initiales serierum juxta numerum dimensionum æquationis resolvendæ. Contingit autem aliquando ut ex terminis initialibus duo vel plures prodeant æquales, quo in casu per æquationem secundam resolvendam tertiāve aut aliquā sequentem termini totidem secundi tertijve aut sequentes inæquales[19] invenientur.

V

Quomodo radices extra-hendæ sunt ubi indefinita quantitas valde magna est cujus dignitatibus conflatur series.

VI

Quomodo æquationes præparari possint ad varias extractiones radicum.

Hi sunt casus radicum ubi quantitas indefinita ex cujus dignitatibus series conflatur valde parva est. Quod si eandem valde magnam esse velis, applicanda est regula ad cellulas superius per totam operationem.[20] Ut si resolvenda esset æquatio $y^3 + axy + aay - x^3 - 2a^3 = 0$, regula superius applicata attingeret cellulas terminorum y^3 et $-x^3$. Hi component æquationem fictam $y^3 - x^3 = 0$ cujus tres sunt radices, una x, alteræ duæ[21] impossibiles. Ex priori per ulteriorem resolutionem obtinebitur tandem series $y = x - \dfrac{1}{3}a - \dfrac{aa}{3x} + \dfrac{37a^3}{81x^2}$ &c[22] cujus termini propter magnitudinem x convergunt.

Cæterum hoc genus extractionis eodem recidit ac si in æquatione resolvenda scriberem $\dfrac{1}{z}$ pro x et æquationem prodeuntem $y^3z^3 + azzy + aaz^3 - 1 - 2a^3[z]^3 = 0$ resolverem applicando regulam ad cellulas inferius. Et tum demum in serie prodeunte $y = \dfrac{1}{z} - \dfrac{1}{3}a - \dfrac{1}{3}aaz + \dfrac{37a^3z^2}{81}$ &c[23] vicissim scriberem $\dfrac{1}{x}$ pro z. Dato etiam augeri vel minui potest indefinita quantitas x vel z et sic radices modis infinitis extrahi, nec non radices impossibiles sæpe reddi possibiles. Ad curvæ alicujus lineæ basem $AB = x$[24] ordinatim applicetur BD vel $BC = y$. Huic ordinatæ agantur parallelæ duæ infinitæ AG non secans curvam et aF secans eandem

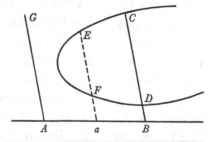

(19) 'diversi' (differing) was first written.

(20) Newton first went on: 'Sic enim prodibunt series quarum termini applicantur ad dignitates. Superiori casu termini serierum prodeuntium multiplicantur posteriori dividuntur per dignitates indefinitæ quantitatis' (For in this way result the series whose terms are applied to the powers. In the former case the terms of the resulting series are multiplied by the powers of the indefinite quantity, in the latter they are divided by them).

(21) Namely $\frac{1}{2}(-1 \pm \sqrt{-3})x$.

(22) 'pergendo in opere resolutione' (on proceeding with the solution procedure) is cancelled. The last term should read ' $+\dfrac{55a^3}{81x^2}$ ' correctly. When $y = x - \frac{1}{3}a - \frac{1}{3}a^2/x + \alpha/x^2 \dots$ is

For instance, should the equation $y^5 - by^2 + 9bx^2 - x^3 = 0$ need to be solved, when applied from below the ruler will be in contact in one position with the cells of the terms $-by^2 + 9bx^2$, and in another with the cells of the terms $y^5 - by^2$. The former will make up the fictitious equation $-by^2 + 9bx^2 = 0$ whose roots y are $+3x$ and $-3x$, the latter the fictitious equation $y^5 - by^2 = 0$ whose roots y are

$$b^{\frac{1}{3}}, \quad \tfrac{1}{2}(-1 + \sqrt{-3})\, b^{\frac{1}{3}} \quad \text{and} \quad \tfrac{1}{2}(-1 - \sqrt{-3})\, b^{\frac{1}{3}}.$$

There are therefore five initial terms of series, exactly the number of dimensions of the equation to be resolved. It sometimes chances, however, that two or more of the initial terms turn out to be equal: in this event the same number of unequal[19] second, third or following terms will be found through the second, third or some following equation to be resolved.

These are the cases of roots when the indefinite quantity from whose powers the series is made up is exceedingly small. But should you want it to be exceedingly large, the ruler must be applied to the cells from above throughout the operation.[20] If, for instance, the equation $y^3 + axy + a^2y - x^3 - 2a^3 = 0$ were to be resolved, the ruler when applied from above would contact the cells of the terms y^3 and $-x^3$. These will form the fictitious equation $y^3 - x^3 = 0$ which has three roots, one being x and the other two[21] impossible. From the first by further resolution there will at length be obtained the series $y = x - \tfrac{1}{3}a - \tfrac{1}{3}a^2/x + \tfrac{37}{81}a^3/x^2 \dots$,[22] whose terms converge on account of the size of x.

Yet this type of extraction falls out to be the same as if in the equation to be resolved I were to write $1/z$ for x and resolve the resulting equation

$$y^3z^3 + az^2y + a^2z^3 - 1 - 2a^3z^3 = 0$$

by applying the ruler to the cells from below. And then, lastly, in the resulting series $y = 1/z - \tfrac{1}{3}a - \tfrac{1}{3}a^2z + \tfrac{37}{81}a^3z^2 \dots$[23] I should in turn write $1/x$ in place of z. Increasing or diminishing the indefinite quantity x or z by a given amount is also possible, and in this manner roots may be extracted in an infinity of ways, while impossible roots, too, can often be made possible. To the base $AB = x$ of some curved line[24] let there be ordinately applied BD or $BC = y$. Parallel to this ordinate draw two unterminated lines AG, not intersecting the curve, and

(sidenote:) V. How roots are to be extracted when the indefinite quantity from whose powers the series is made up is exceedingly large.

VI. How equations might be prepared for various sorts of root extraction.

---

substituted in the preceding equation, on equating the constant term to zero there follows $\alpha = \tfrac{1}{3}(-\tfrac{1}{3} + \tfrac{64}{27})$: Newton has perhaps taken this as $-\tfrac{1}{3} + \tfrac{1}{3}(\tfrac{64}{27})$.

(23) The last term should be ' $+\dfrac{55a^3z^2}{81}$ ' (see previous note).

(24) 'dato angulo' (at a given angle) is cancelled.

in F et E. Manifestum est quod minuendo AB ordinata y possibilitatem tandem amittet [et] proinde per seriem convergentem in qua basis x ut indefinite parva spectatur exprimi nequit. De basi AB auferatur data Aa ponendo $aB = z$ et in æquatione resolvenda scribendo $Aa + z$ pro x et radicem y jam licebit extrahere. Et hoc modo series ad curvarum partes imperatas FD vel EC exactiùs definiendas aptantur.

VII
Quid facien-dum sit ubi primus seriei terminus commode inveniri nequit.

Siquando seriei terminus initialis sit surda quantitas vel inextricabilis radix æquationis multipliciter affectæ, extrahendus est is in numeris[25] fractis vel forte designandus per speciem. Ut in exemplo superiore

$$y^3 + axy + aay - x^3 - 2a^3 = 0,$$

si radix æquationis $y^3 + aay - 2a^3 = 0$ per tabulam tessellatam inventæ fuisset surda vel ignota, pro ea scripsissem speciem quamlibet b[26] et resolutionem ut sequitur perfecissem.

$$y = b - \frac{abx}{cc} + \frac{a^4bxx}{c^6} + \frac{x^3}{cc} + \frac{a^3b^3x^3}{c^8} - \frac{a^5bx^3}{c^8} + \frac{6a^5b^3x^3}{c^{10}} \quad \&\text{c.}$$

$b+p=y.$	$+y^3$	$b^3+3bbp+3bpp+p^3$
	$+axy$	$abx \quad +axp$
	$+aay$	$aab \quad +aap$
	$-x^3$	$-x^3$
	$-2a^3$	$-2a^3$

$-\dfrac{abx}{cc}+q=p.$	p^3	$-\dfrac{a^3b^3x^3}{c^6} \quad \&\text{c}$
	$+3bpp$	$+\dfrac{3aab^3xx}{c^4} - \dfrac{6abbx}{cc}q \quad \&\text{c}$
	$+axp$	$-\dfrac{aabxx}{cc} \qquad +axq$
	$+ccp$	$-abx \qquad +ccq$
	$-x^3$	$-x^3$
	$-abx$	$-abx$

$$cc+ax-\frac{6abbx}{cc}\bigg)\frac{a^4bxx}{c^4}+x^3+\frac{a^3b^3x^3}{c^6}\bigg(\frac{a^4bxx}{c^6}+\frac{x^3}{cc}+\frac{a^3b^3x^3}{c^8}-\frac{a^5bx^3}{c^8}+\frac{6a^5b^3x^3}{c^{10}} \quad \&\text{c.}$$

Scribens b in serie, pono $b + p = y$ et ex æquatione prima resolvenda substituendo $b+p$ pro y oritur secunda

$$p^3 + 3bpp + 3bbp + b^3 + axp + abx + aap + aab - x^3 - 2a^3 = 0.$$

Hic termini $-2a^3 + aab + b^3$ se mutuò destruunt propterea quod b supponitur

---

(25) Understand 'decimalibus' (decimals).

aF, cutting it in *F* and *E*. It is plain that by diminishing *AB* the ordinate *y* will eventually lose its possibility and then, consequently, cannot be expressed by a converging series in which the base *x* is regarded as being indefinitely small. From the base *AB* take away the given length *Aa* by setting $aB = z$ and in the equation to be resolved writing $Aa + z$ in place of *x*, and it will now be legitimate to extract the root *y*. And in this way series are adapted to define appointed portions *FD* or *EC* of curves more narrowly.

If at any time the initial term should be a surd quantity or the inextricable root of a multiply affected equation, it must be extracted in[25] fractions or perhaps be denoted by a general constant. So in the above example

$$y^3 + axy + a^2y - x^3 - 2a^3 = 0$$

VII.
What to do when the first term of a series cannot conveniently be found.

if the root of the equation $y^3 + a^2y - 2a^3 = 0$ found by the squared array had been surd or not known, in its place I should have written any general form b[26] and completed the resolution as follows:

$$y = b - abc^{-2}x + a^4bc^{-6}x^2 + (c^{-2} + a^3b^3c^{-8} - a^5bc^{-8} + 6a^5b^3c^{-10})\,x^3 \ldots$$

$b + p = y.$	y^3	$b^3 + 3b^2p + 3bp^2 + p^3$	
	$+axy$	$abx\ +axp$	
	$+a^2y$	$a^2b\ +a^2p$	
	$-x^3$	$-x^3$	
	$-2a^3$	$-2a^3$	
$-abc^{-2}x + q = p.$	p^3	$-a^3b^3c^{-6}x^3 \ldots$	
	$+3bp^2$	$+3a^2b^3c^{-4}x^2 - 6ab^2c^{-2}q$	$\ldots$
	$+axp$	$-a^2bc^{-2}x^2\ \ +axq$	
	$+c^2p$	$-abx\ \ +c^2q$	
	$-x^3$	$-x^3$	
	$-abx$	$-abx$	

$$c^2 + (a - 6ab^2c^{-2})\,x \big) a^4bc^{-4}x^2 + (1 + a^3b^3c^{-6})\,x^3 \Big(a^4bc^{-6}x^2 +$$

$$(c^{-2} + a^3b^3c^{-8} - a^5bc^{-8} + 6a^5b^3c^{-10})\,x^3 \ldots$$

Entering *b* in the series, I put $y = b + p$ and from the first equation to be resolved there arises, on substituting $b + p$ for *y*, the second one

$$p^3 + 3bp^3 + 3b^2p + b^3 + axp + abx + a^2p + a^2b - x^3 - 2a^3 = 0.$$

Here the terms $-2a^3 + a^2b + b^3$ mutually destroy one another for the reason that

---

(26) Where, that is, $b^3 + a^2b - 2a^3 = 0$. The present example is a slightly amplified revise of the equivalent passage in the 1671 tract (III: 62–4), largely repeating it word for word.

radix æquationis hujus $-2a^3 + aay + y^3 = 0$. Deleantur et æquatio secunda resolvenda evadit $p^3 + 3bpp + axp + ccp^{(27)} - x^3 - abx = 0$. Hujus termini

$$[-]abx + aap + 3bbp$$

per tabulam tessellatam selecti et positi nihilo æquales, dant $-\dfrac{abx}{aa+3bb}$ in serie

scribendum et $-\dfrac{abx}{aa+3bb} + q$ substituendum in æquatione resolvenda pro p.

Brevitatis autem gratia scribo cc pro $aa+3bb$, cavendo tamen ut $aa+3bb$ ubi commodum fuerit restituatur. Et sic prodit æquatio tertia resolvenda

$-\dfrac{a^4bxx}{c^4} - x^3 - \dfrac{a^3b^3x^3}{c^6} + ccq + axq - \dfrac{6abbx}{cc}q$ &c $= 0$. Unde tertium seriei terminū

$\dfrac{a^4bx^2}{c^6}$ elicio et sic opus quousqʒ libuerit produco. Quo tandem completo si radix

æquationis $y^3 + aay - 2a^3 = 0$ utcunqʒ inveniatur vel si ea ponendo numerum quemvis pro a extrahatur in numeris, scribenda erit eadem pro b in serie.[28]

VIII
Resolutiones abbreviare docetur.

Interea verò dum[29] æquatio quævis resolvitur omittendi sunt semper termini omnes æquationum resolvendarum qui nulli deinceps usui futuri sunt. Hoc pacto labor computandi plurimū minuitur estqʒ regula elidendorum terminorum talis. Ipso operis initio statuo seriem ad usqʒ certam dignitatem[30] indefinitæ quantitatis producere. Ad indicem illius dignitatis addo indicem dignitatis ejusdem indefinitæ quantitatis in cellula ad sinistrā penextima quam regula sic,[31] ut dictum est applicata inferius attingit et summam indicum pro indice dignitatis altissimæ æquationum resolvendarum retineo. Tum[32] negligo in æquationibus resolvendis terminos omnes ubi indefinita quantitas x transcendit illam dignitatem altissimam vel etiam transcenderet eandem scribendo pro y, p, q, r &c terminum primum valoris cujusqʒ. Et ultimò ubi deventum est ad æquationem resolvendam in qua indefinita quantitas p, q, r vel s &c non nisi simplicis est dignitatis, elicio reliquos seriei terminos per divisionem.[33] Ut si

---

(27) Read ' $+aap + 3bbp$ '. Newton anticipates his following replacement of $a^2 + 3b^2$ by c^2.

(28) A cancelled final sentence reads: 'Potuit etiam radix b in numeris extrahi priusquam æquatio resolveretur in seriem et numerus ille loco b pro primo seriei termino usurpari' (The root b could also have been extracted numerically before the equation was resolved into a series and that number b employed in place of the first term of the series).

(29) Newton earlier continued 'hæc aguntur et radix deinceps extrahitur consulendum est semper de contractione operis. Negligendi sunt enim termini omnes æquationū resolvendarum qui postea nulli deinceps usui esse possunt' (this is going on and the root successively extracted, continual consideration must be given to contracting the work. All terms in the resolvend equations which can afterwards be of no subsequent use must of course be omitted).

(30) 'certam periodum' (certain period) was first written.

(31) 'in secunda tessellatarum columna' (in the second column of tessellations) is cancelled.

(32) Newton first went on: 'fingendo y ejusdem esse dimensionis cum x in primo termino, p ejusdem cum x in secundo, q ejusdem cum x in tertio, r ejusdem cum x in quarto, et ita deinceps' (imagining y to be of the same dimension as x in the first term, p of the same as x in the second, q of the same as x in the third, r of the same as x in the fourth, and so on in turn).

b is supposed a root of this equation $-2a^3 + a^2y + y^3 = 0$. Delete these and the second equation to be resolved comes out as

$$p^3 + 3bp^2 + axp + [a^2 + 3b^2]p - x^3 - abx = 0.$$

Of this the terms $(a^2 + 3b^2)p - abx$ selected by the squared array yield, when set equal to nothing, $-ab(a^2 + 3b^2)^{-1}x$ to be entered in the series and

$$-ab(a^2 + 3b^2)^{-1}x + q$$

to be substituted for p in the equation to be resolved. However, for brevity's sake I write c^2 in place of $a^2 + 3b^2$, taking care nonetheless to restore $a^2 + 3b^2$ when convenient. And so there results the third equation to be resolved

$$-a^4bc^{-4}x^2 - (1 + a^3b^3c^{-6})x^3 + c^2q + (a - 6ab^2c^{-2})xq \ldots = 0.$$

From this I elicit the third term $a^4bc^{-6}x^2$ of the series and in this way extend the operation as far as I will. If, after this is at last completed, the root of the equation $y^3 + a^2y - 2a^3 = 0$ be found in any way at all or should it, on putting any number for a, be extracted in numbers, it will need to be written in place of b in the series.[28]

All the while, of course,[29] any equation is being resolved all terms in the [interim] equations to be resolved which can from then on be of no use are continually to be omitted. In this manner the labour of calculation is very much diminished. For striking out terms my rule is as follows. At the very start of the operation I settle that the series is to be extended as far as a certain power[30] of the indefinite quantity. To the index of that power I add the index of the power of the same indefinite quantity in the next outermost cell to the left which is in contact with the ruler when applied from below[31] in the manner stated, and the sum of the indices I keep as the index of the highest power in the equations to be resolved. I then[32] neglect in the equations to be resolved all terms in which the indefinite quantity x rises above that highest power, or again would transcend it on replacing $y, p, q, r, \ldots$ by the first term of each of their values. Finally, when I have arrived at an equation to be resolved in which the indefinite quantity $p, q, r, s, \ldots$ exists only as a simple power, I elicit the remaining terms of the series by division.[33] If, for instance, the equation

VIII.
Abbreviation of solutions is explained.

(33) Newton unnecessarily complicates an essentially simple idea: if the value of y is required to $O(x^{n+1})$, then all terms of order higher than x^n in its value, or of order higher than x^{m+n} in any equation whose primary 'fictitious' equation is $\alpha x^l + \beta x^m y = 0$, can safely be omitted. In the following inversion of the series expansion of $x = \log(1+y)$ into that of $y = e^x - 1$ (a slight expansion of the equivalent passage in his 1671 tract; see III: 58), Newton's desired evaluation of the latter to $O(x^6)$ allows omission of all powers of x and y higher than the fifth—and correspondingly of terms in p and q (of respective orders x and x^2)—since the primary fictitious equation is $-x + y = 0$ ($l = 1$, $m = 0$ with $n = 5$).

resolvenda esset æquatio $x - y + \frac{1}{2}yy - \frac{1}{3}y^3 + \frac{1}{4}y^4 - \frac{1}{5}y^5 + \frac{1}{6}y^6$ &c $= 0$ & series ad usqȝ quintam ipsius x dimensionem producenda: quoniam regula quærendo primum seriei terminum attingeret primam seu infimam cellulam earum quæ sunt in secunda seu penextima columna ad sinistram, et in hac cellula quantitas x nullius est dimensionis, summam indicum 5 & 0 id est 5 retineo ut indicem altissimæ dimensionis indefinitarum quantitatum. Tum in æquatione primâ resolvendâ terminos $\frac{1}{6}x^6$ cæterosqȝ altiores negligo ut dignitatem illam altissimam transcendentes. Et quoniam x in valore marginali ipsius p vel quod perinde est in secundo seriei termino $\frac{1}{2}xx$ duarum est dimensionum et in valore ipsius $[q]$ seu in tertio seriei termino $\frac{1}{6}x^3$ est trium, negligo terminos omnes æquationum subsequentium qui dignitatem x^5 scribendo x^2 pro p et x^3 pro q transcenderent, id est omnes post notā *&c* in sequente operis paradigmate.

$$y = x + \tfrac{1}{2}xx + \tfrac{1}{6}x^3 + \tfrac{1}{24}x^4 + \tfrac{1}{120}x^5 \quad \text{\&c.}$$

$x+p=y.$	$\frac{1}{5}y^5$	$\frac{1}{5}x^5$ &c	
	$-\frac{1}{4}y^4$	$-\frac{1}{4}x^4$ $-x^3 p$ &c	
	$+\frac{1}{3}y^3$	$+\frac{1}{3}x^3$ $+x^2 p + xpp$ &c	
	$-\frac{1}{2}yy$	$-\frac{1}{2}x^2$ $-xp - \frac{1}{2}pp.$	
	$+y$	$+x$ $+p.$	
	$-x$	$-x$	

$\frac{1}{2}xx + q = p.$	xpp	$\frac{1}{4}x^5$ &c	
	$-\frac{1}{2}pp$	$-\frac{1}{8}x^4 - \frac{1}{2}xxq$ &c	
	$-x^3 p$	$-\frac{1}{2}x^5$ &c	
	$+x^2 p$	$[+]\frac{1}{2}x^4 + xxq.$	
	$-xp$	$-\frac{1}{2}x^3 - xq.$	
	$+p$	$+\frac{1}{2}xx +q.$	
	$+\frac{1}{5}x^5$	$+\frac{1}{5}x^5.$	
	$-\frac{1}{4}x^4$	$-\frac{1}{4}x^4.$	$\frac{1}{6}x^3 + r = q.$
	$+\frac{1}{3}x^3$	$+\frac{1}{3}x^3.$	
	$-\frac{1}{2}x^2$	$-\frac{1}{2}x^2.$	

$$1 - x + \tfrac{1}{2}xx) \tfrac{1}{6}x^3 - \tfrac{1}{8}x^4 + \tfrac{1}{20}x^5 (\tfrac{1}{6}x^3 - \tfrac{1}{24}x^4 + \tfrac{1}{120}x^5 \quad \text{\&c.}$$

IX
Quomodo
æquationes
resolvendæ
sunt ubi
dignitatum
Siquando regula non tangit cellulam penextimam in angulo infimo sinistro sed eam in latere sinistro secat, sumendus est numerus fractus segmento proportionalis pro indice. Ut si resolvenda esset æquatio $x^3 - 2x^4 + xy^4 - y^7 = 0$, secaret regula tertiam cellulam secundæ columnæ in medio sinistri lateris,[34] et index respondens medio lateri tertiæ cellulæ est $2\frac{1}{2}$. Igitur ut habeatur index

---

(34) Namely, the point in the 'parallelogram' which would have represented the term in $x^{2\frac{1}{2}}y$ if it had been present: somewhat frustratingly, the series $y = \sqrt[4]{[-1]}x^{\frac{1}{2}} + O(x^{\frac{3}{2}})$ yielded by the corresponding fictitious equation $x^3 + xy^4 = 0$ is not real. However, the alternative fictitious equation $xy^4 - y^7 = 0$ gives the more manageable expansion $y = x^{\frac{1}{2}} + \frac{1}{8}x - \frac{5}{8}x^{\frac{3}{2}} \dots$

$$x - y + \tfrac{1}{2}y^2 - \tfrac{1}{3}y^3 + \tfrac{1}{4}y^4 - \tfrac{1}{5}y^5 + \tfrac{1}{6}y^6 \ldots = 0$$

were to be resolved and the series extended as far as the fifth dimension of x, seeing that the ruler for seeking out the first term of the series would contact the first—or lowest—of the cells which are in the second, next outermost column on its left, while in this cell the quantity x is of zero dimension, I keep the indices' sum $5 + 0 = 5$ as the index of the highest dimension of the indefinite quantities. Then in the first equation to be resolved I neglect the terms $\tfrac{1}{6}x^6$ and the other higher ones as transcending that highest power. Then since in the marginal value of p—or, what is the same, in the second term $\tfrac{1}{2}x^2$ of the series—x is of two dimensions while in the value of q—or in the third term $\tfrac{1}{6}x^3$ of the series—it is of three, I neglect all terms in the subsequent equations which would, on writing x^2 in place of p and x^3 for q, rise above the power x^5, that is, all after the symbol '...' in the following pattern of operation:

$$y = x + \tfrac{1}{2}x^2 + \tfrac{1}{6}x^3 + \tfrac{1}{24}x^4 + \tfrac{1}{120}x^5 \ldots$$

$x + p = y.$	$\tfrac{1}{5}y^5$	$\tfrac{1}{5}x^5$ $\ldots$
	$-\tfrac{1}{4}y^4$	$-\tfrac{1}{4}x^4$ $-x^3p$ $\ldots$
	$+\tfrac{1}{3}y^3$	$+\tfrac{1}{3}x^3$ $+x^2p + xp^2$ $\ldots$
	$-\tfrac{1}{2}y^2$	$-\tfrac{1}{2}x^2$ $-xp - \tfrac{1}{2}p^2$
	$+y$	$+x$ $+p$
	$-x$	$-x$

$\tfrac{1}{2}x^2 + q = p.$	xp^2	$\tfrac{1}{4}x^5$ $\ldots$
	$-\tfrac{1}{2}p^2$	$-\tfrac{1}{8}x^4 - \tfrac{1}{2}x^2q$ $\ldots$
	$-x^3p$	$-\tfrac{1}{2}x^5$ $\ldots$
	$+x^2p$	$+\tfrac{1}{2}x^4 + x^2q$
	$-xp$	$-\tfrac{1}{2}x^3 - xq$
	$+p$	$+\tfrac{1}{2}x^2 + q$
	$+\tfrac{1}{5}x^5$	$+\tfrac{1}{5}x^5$
	$-\tfrac{1}{4}x^4$	$-\tfrac{1}{4}x^4$
	$+\tfrac{1}{3}x^3$	$+\tfrac{1}{3}x^3$
	$-\tfrac{1}{2}x^2$	$-\tfrac{1}{2}x^2$

$$\tfrac{1}{6}x^3 + r = q.$$

$$1 - x + \tfrac{1}{2}x^2)\tfrac{1}{6}x^3 - \tfrac{1}{8}x^4 + \tfrac{1}{20}x^5 (\tfrac{1}{6}x^3 - \tfrac{1}{24}x^4 + \tfrac{1}{120}x^5 \ldots.$$

Whenever the ruler does not touch the next outermost cell at its bottom left corner but intersects it in its left side, the fraction proportional to the section cut off must be taken for the index. So if the equation $x^3 - 2x^4 + xy^4 - y^7$ were to be resolved, the ruler would intersect the third cell of the second column in the mid-point of its left side[34] and the index corresponding to the mid-side of the third cell is $2\tfrac{1}{2}$. Therefore to have the index of the highest power in the equations

IX.
How equations are to be resolved when the indices of the powers are

indices sunt
fracti vel
surdi numeri. dignitatis altissimæ æquationum resolvendarum addo $2\frac{1}{2}$ ad indicem altissimæ dignitatis quantitatis x ad quam seriem produci cupiam. Simili indicum et sectorum laterum analogia resolvuntur æquationes ubi indices dimensionum quantitatum indefinitarum (x et y) sunt fracti vel surdi numeri; nisi fortè mavis fractos reducere ad integros: ut in $x^{\frac{3}{2}} - 3xy + y^{\frac{1}{2}} = 0$, scribendo z pro $x^{\frac{1}{2}}$ et v pro $y^{\frac{1}{2}}$ et resolvendo

$$zz - 3z^3vv + v = 0.$$

Surdos reducere licet[35] ad fractos decimales. Et substituendo valores ipsorum y, p, q, r, s &c recurrendum est ad Regulam[36] quam initio epistolæ meæ prioris exposui.

Cæterum lex[37] de superfluis æquationum terminis elidendis hic allata non obtinet ubi duo vel plures initiales serierum termini prodeunt æquales. Eo casu expectandū est donec regula ad tabulam tessellatam applicata dederit ipsius p, q, r vel s valores omnes inæquales, et tum demum lex illa valebit.

CAP 2.[38]

DE TRANSMUTATIONE SERIERUM.

I
Describitur
transmuta-
tionū Regula. Ad copiam serierum spectat earum transmutatio. Sic ex una inventa derivantur aliæ innumeræ. Proponatur series quævis

$$v = At \pm Bt^2 + Ct^3 \pm Dt^4 + Et^5 \pm Ft^6 \quad \text{&c.}$$

$$b \quad {}^2b \quad {}^3b \quad {}^4b \quad {}^5b$$
$$c \quad {}^2c \quad {}^3c \quad {}^4c$$
$$d \quad {}^2d \quad {}^3d$$
$$e \quad {}^2e$$
$$f$$

Ubi A, B, C &c denotent terminorū coefficientes quarum ratio ultima si infinitè producatur series est æqualitatis et 1 ad t est ratio reliqua[39] terminorum; signum vero $\pm$ ambiguum est et signo $\mp$ contrarium. Terminorum A, B, C collige differentias primas $b, {}^2b, {}^3b$ &c. secundas $c, {}^2c, {}^3c$ &c, tertias $d, {}^2d, {}^3d$ &c et sequentes quotcunq. Collige autem eas subducendo semper terminum

---

(35) 'convenit' (it is convenient) was first written.

(36) The general binomial expansion; see note (3) above.

(37) This replaces 'regula' (rule).

(38) A first version of this chapter and preliminary notes for the present revise are reproduced in Appendix 3 below. We there conjecture that Newton may have been inspired to develop the following Eulerian 'transmutatio' by learning, through Collins, of 'An approach to the summe of [an] infinite series' (by means of an approximating geometrical series) which had been communicated by James Gregory in 1672; see Appendix 3.1: note (9) below. A few minor variants in a draft of the opening paragraph on ULC. Add. 3964.3: 12^r are given in following footnotes.

to be resolved I add $2\frac{1}{2}$ to the index of the highest power of the quantity x to fractions or which I should desire the series to be extended. By a similar proportion between surds. the indices and intersected sides equations are resolved in which the indices of the dimensions of the indefinite quantities (x and y) are fractions or surds. But you may perhaps prefer to reduce the fractions to integers: as, in the case of $x^{\frac{3}{2}}-3xy+y^{\frac{1}{2}}=0$, by writing z for $x^{\frac{1}{2}}$ and v for $y^{\frac{1}{2}}$ and resolving $z^2-3z^3v^2+v=0$. Surds you are free[35] to reduce to decimals. Then on substituting the values of $y, p, q, r, s, \dots$ you should have recourse to the rule[36] which I expounded at the beginning of my first letter.

For the rest, the precept[37] for striking out superfluous terms in equations which is here invoked does not obtain when two or more initial terms of series prove to be equal. In that case you must wait till the ruler, when applied to the squared array, has yielded values of $p, q, r, s, \dots$ which are all unequal: only then will the precept be valid.

CHAPTER 2[38]

ON THE TRANSFORMATION OF SERIES

The transformation of series has regard to their profusion. So, when one is I. found, countless others derive from it. Let any series A rule for transformation is described.

$$v = At \pm Bt^2 + Ct^3 \pm Dt^4 + Et^5 \pm Ft^6 \dots$$
$$b \quad b_2 \quad b_3 \quad b_4 \quad b_5$$
$$c \quad c_2 \quad c_3 \quad c_4$$
$$d \quad d_2 \quad d_3$$
$$e \quad e_2$$
$$f$$

be proposed. Here $A, B, C, \dots$ are to denote the coefficients of the terms whose ultimate ratio, if the series be extended infinitely, is one of equality, and 1 to t is the remaining[39] ratio of the terms; while the sign $\pm$ is ambiguous and the converse of the sign $\mp$. Collect the first differences $b, b_2, b_3, \dots$ of the terms $A, B, C, \dots$; then their second ones $c, c_2, c_3, \dots$, third ones $d, d_2, d_3, \dots$, and any number of following ones. Collect these, however, by always taking a latter

(39) On ignoring the coefficients of the terms, that is. Newton first wrote 'ubi [A, B, C] &c denotant seriei coefficientes numerales quarum ratio si longissimè producantur acced[it] ad rationem æqualitatis quamproximè, t vero est quantitas quæ secundū proportionem geometricam progreditur incipientem ab unitate seu consequens rationis ultimæ terminorum totorum posita unitate antecedente' (where $A, B, C, \dots$ denote the numerical coefficients of the series, whose ratio, should they be extremely far prolonged, approaches one of unity very closely; while t is a quantity which progresses following a geometrical proportion starting from unity, that is, the latter member of the ultimate ratio of all the terms on taking unity for its prior part).

posteriorem de priore, B de A, C de B &c. ^2b de b, ^3b de ^2b &c. ^2d de d, ^3d de ^2d &c.

Dein fac $\dfrac{t}{1 \mp t} = z$ et signis probe observatis erit $v = Az \mp bz^2 + cz^3 \mp dz^4 + ez^5 \mp fz^6$ &c.[40]

Ut si fuerit $y = x - \frac{1}{2}x^2 + \frac{1}{3}x^3 - \frac{1}{4}x^4 + \frac{1}{5}x^5$ &c, quoniam terminorum coefficientes 1, $\frac{1}{2}$, $\frac{1}{3}$, $\frac{1}{4}$, $\frac{1}{5}$ &c rationem æqualitatis perpetuo appropinquant & reliqua terminorum ratio est 1 ad x, colligo ipsorum 1, $\frac{1}{2}$, $\frac{1}{3}$ &c differentias primas $\dfrac{1}{1,2} \cdot \dfrac{1}{2,3} \cdot \dfrac{1}{3,4} \cdot \dfrac{1}{4,5}$ &c. secundas $\dfrac{1,2}{1,2,3} \cdot \dfrac{1,2}{2,3,4} \cdot \dfrac{1,2}{3,4,5}$ &c seu $\frac{1}{3}$, $\frac{1}{12}$, $\frac{1}{30}$ &c. tertias $\dfrac{1,2,3}{1,2,3,4} \cdot \dfrac{1,2,3}{2,3,4,5} \cdot \dfrac{1,2,3}{3,4,5,6}$ &c seu $\frac{1}{4}$, $\frac{1}{20}$, $\frac{1}{60}$ &c. quartas $\dfrac{1,2,3,4}{1,2,3,4,5} \cdot \dfrac{1,2,3,4}{2,3,4,5,6}$ &c seu $\frac{1}{5}$, $\frac{1}{30}$ &c. Et posito $t = x$ et inde $\dfrac{x}{1+x} = z$ scribo $y = z + \frac{1}{2}z^2 + \frac{1}{3}z^3 + \frac{1}{4}z^4$ &c.[41]

Si fuerit $y = x - \dfrac{x^3}{3aa} + \dfrac{x^5}{5a^4} - \dfrac{x^7}{7a^6}$ &c. Quoniam coefficientes 1, $-\frac{1}{3}$, $+\frac{1}{5}$ producendo seriem[42] appropinquant rationem æqualitatis et reliqua terminorum ratio est $\dfrac{xx}{aa} = t$,[43] facio ut termini seriei multiplicentur (ut in regula) per digni-

---

(40) In sequel Newton began to enter a final phrase 'series nova ejusdem valoris cum priore' (a new series identical in value to the former) in line with his predraft on f. 12^r (note (38) above), but at once cancelled it. The derivation of this 'transmutatio' is straightforward (compare Newton's preliminary computations reproduced in Appendix 3. 2/3 below): since $z = t/(1 \mp t)$, on taking the successive differences

$$b \text{ (or } ^1b) = A - B, \quad ^2b = B - C, \quad ^3b = C - D, \ldots; \quad c \text{ (or } ^1c) = b - ^2b,$$

$^ic = {}^ib - {}^{i+1}b$ $(i = 2, 3, 4, \ldots)$; d (or $^1d) = c - {}^2c$, $^id = {}^ic - {}^{i+1}c$ $(i = 2, 3, 4, \ldots)$; and so on, we deduce successively that

$$v = z(1 \mp t)(A \pm Bt + Ct^2 \pm Dt^3 + Et^4 \ldots)$$
$$= Az \mp z^2(b \pm {}^2bt + {}^3bt^2 \pm {}^4bt^3 \ldots)$$
$$= Az \mp bz^2 + z^3(c \pm {}^2ct + {}^3ct^2 \pm {}^4ct^3 \ldots)$$
$$= Az \mp bz^2 + cz^3 \mp z^4(d \pm {}^2dt + {}^3dt^2 \pm {}^4dt^3 \ldots),$$

and so on indefinitely. In a following cancelled paragraph on f. 12^r Newton did not succeed in expressing these 'partial' expansions correctly, declaring rather confusedly that 'Ubi hanc seriem ad periodum aliquem produxeris, pergere licet per terminos sequentes seriei prioris retinendo coefficientes termini periodici' (When you have extended this series to some period, it is permissible to proceed by the terms following in the former series, retaining the coefficients of the periodic term). In attempted clarification in sequel he wrote (when some silent adjustment of notation is made): 'Ut si seriem posteriorem produxeris ad terminum d seu $A - 3B + 3C - D$, pergere licet ponendo

$$B - 3C + 3D - E = {}^2d. \quad C - 3D + 3E - F = {}^3d. \quad D - 3E + 3F - G = {}^4d. \quad \text{&c},$$

deinde $v = Az \mp bz^2 + cz^3 \mp dz^4 : \mp dz^4 \pm {}^2dz^5 + {}^3dz^6 \pm {}^4dz^7 + {}^5dz^8$ &c:.'

Newton's transmutation scheme was used twenty years later by Jean Christophe Fatio de Duillier to tighten the convergence of the Gregory–Leibniz series $\frac{1}{4}\pi = 1 - \frac{1}{3} + \frac{1}{5} - \frac{1}{7} \ldots$ (see

term from the previous one: B from A, C from B, ...; b_2 from b, b_3 from b_2, ...; d_2 from d, d_3 from d_2, ..., and so on. Then make $t/(1 \mp t) = z$ and when the signs are appropriately observed there will be

$$v = Az \mp bz^2 + cz^3 \mp dz^4 + ez^5 \mp fz^6 \ldots.^{(40)}$$

If, for instance, there were $y = x - \frac{1}{2}x^2 + \frac{1}{3}x^3 - \frac{1}{4}x^4 + \frac{1}{5}x^5 \ldots$, seeing that the coefficients $1, \frac{1}{2}, \frac{1}{3}, \frac{1}{4}, \frac{1}{5}, \ldots$ of the terms perpetually approach the ratio of equality while the remaining ratio of the terms is 1 to x, from $1, \frac{1}{2}, \frac{1}{3}, \ldots$ I collect the first differences $\dfrac{1}{1.2}, \dfrac{1}{2.3}, \dfrac{1}{3.4}, \dfrac{1}{4.5}, \ldots$; second ones $\dfrac{1.2}{1.2.3}, \dfrac{1.2}{2.3.4}, \dfrac{1.2}{3.4.5}, \ldots$, that is, $\frac{1}{3}, \frac{1}{12}, \frac{1}{30}, \ldots$; third ones $\dfrac{1.2.3}{1.2.3.4}, \dfrac{1.2.3}{2.3.4.5}, \dfrac{1.2.3}{3.4.5.6}, \ldots$, that is, $\frac{1}{4}, \frac{1}{20}, \frac{1}{60}, \ldots$; fourth ones $\dfrac{1.2.3.4}{1.2.3.4.5}, \dfrac{1.2.3.4}{2.3.4.5.6}, \ldots$, that is, $\frac{1}{5}, \frac{1}{30}, \ldots$, and so on. Then, putting $t = x$ and accordingly $x/(1+x) = z$, I write

$$y = z + \tfrac{1}{2}z^2 + \tfrac{1}{3}z^3 + \tfrac{1}{4}z^4 \ldots.^{(41)}$$

If there were $y = x - \frac{1}{3}x^3/a^2 + \frac{1}{5}x^5/a^4 - \frac{1}{7}x^7/a^6 \ldots$, seeing that the coefficients $1, -\frac{1}{3}, \frac{1}{5}, \ldots$ on extending the series[42] approach the ratio of equality while the remaining ratio of the terms is $x^2/a^2 = t$,[43] I arrange that the terms of the series

---

note (44) below) but no general account of the 'transmutatio' was given in print till Euler discussed it in his *Institutiones Calculi Differentialis cum eius Vsu in Analysi Finitorum ac Doctrina Serierum* (St Petersburg, 1755): Pars Posterior, Caput I. 'De Transformatione Serierum': §§2–11: 281–96 [= *Leonhardi Euleri Opera Omnia* (1) **10**, Leipzig/Berlin, 1913: 217–28]. As Euler remarks (*ibid.*: §§6/11: 287/294) since $z = t/(1-t) > t$ for $0 < t < 1$, unless the 'transmuted' series $v = Az - bz^2 + cz^3 - dz^4 \ldots$ terminates (all k-th order finite differences of the coefficients $A, B, C, \ldots$ of the original series $v = At + Bt^2 + Ct^3 \ldots$ being zero) it will normally be more slowly convergent than the one from which it is derived; but, since

$$z = t/(1+t) < t \quad \text{for} \quad t > 0,$$

the procedure is effective in speeding the convergence of the alternating sequence

$$v = At - Bt^2 + Ct^3 - \ldots.$$

The particular case of the latter (essentially Fatio's in 1704) in which $t = 1$ yields the elegant transformation $v = A - B + C - D + E \ldots = \frac{1}{2}A + \frac{1}{4}b + \frac{1}{8}c + \frac{1}{16}d \ldots$, where

$$b = A - B, \quad c = A - 2B + C, \quad d = A - 3B + 3C - D, \quad \ldots:$$

one known widely in modern texts—not unjustly so—as 'Euler's'.

(41) The 'transmutation' $x/(1+x) = z$ converts the series expansion of $y = \log(1+x)$ into that of $y = \log(1/(1-z))$. For Jakob Bernoulli's and Hermann's derivation in 1705 of the case $x = 1, z = \frac{1}{2}$ see note (44) below.

(42) 'magis et magis' (more and more) is deleted.

(43) Newton has cancelled the following phrase 'ut primus seriei terminus formam debitam sortiatur' (so that the first term of the series may gain an appropriate form).

tates t scribendo $\dfrac{yx}{aa} = \dfrac{xx}{aa} - \dfrac{x^4}{3a^4} + \dfrac{x^6}{5a^6}$ &c. Dein colligo coefficientium differentias

primas $\dfrac{2}{1,3} \cdot \dfrac{2}{3,5} \cdot \dfrac{2}{5,7}$ &c. secundas $\dfrac{2,4}{1,3,5} \cdot \dfrac{2,4}{3,5,7} \cdot \dfrac{2,4}{5,7,9}$ &c. tertias $\dfrac{2,4,6}{1,3,5,7} \cdot \dfrac{2,4,6}{3,5,7,9}$

&c. Et posito $\dfrac{xx}{aa+xx} \left(= \dfrac{t}{1+t} \right) = z$ scribo $\dfrac{yx}{aa} = z + \dfrac{2}{1,3} z^2 + \dfrac{2,4}{1,3,5} z^3 + \dfrac{2,4,6}{1,3,5,7} z^4$

&c.[44]

Eodem modo si fuerit $y = x + \dfrac{x^3}{3aa} + \dfrac{x^5}{5a^4} + \dfrac{x^7}{7a^6}$ &c ponendum erit

$$\frac{x^2}{aa-xx} \left(= \frac{t}{1-t} \right) = z$$

et scribendum $\dfrac{yx}{aa} = z - \dfrac{2}{1,3} z^2 + \dfrac{2,4}{1,3,5} z^3 - \dfrac{2,4,6}{1,3,5,7} z^4$ &c.[45]

Siquando termini bini affirmativi et bini negativi per vices ponuntur ut in hac serie $1 + \frac{1}{3} - \frac{1}{5} - \frac{1}{7} + \frac{1}{9} + \frac{1}{11}$ &c, resolvenda est series in duas $1 - \frac{1}{5} + \frac{1}{9} - \frac{1}{13}$ &c et $\frac{1}{3} - \frac{1}{7} + \frac{1}{11}$ &c nisi mavis terminos binos conjungere.[46] Et sic in ternis, quaternis et pluribus.

Convenit aliquando transmutationem incipere in medio seriei. Ut in serie $y = x - \frac{1}{2}x^2 + \frac{1}{3}x^3 - \frac{1}{4}x^4$ &c si posueris $v = \frac{1}{3}x - \frac{1}{4}x^2 + \frac{1}{5}x^3$ &c et inde per

---

(44) Here the transformation $x^2/(a^2+x^2) = z$ converts the series expansion of

$$y = a\tan^{-1}(x/a)$$

into that of $xy/a^2 = \sqrt{[z/(1-z)]}\tan^{-1}\sqrt{[z/(1-z)]} = \sqrt{[z/(1-z)]}\sin^{-1}\sqrt{z}$. The particular series $\frac{1}{4}\pi = \dfrac{1}{2}\left(1 + \dfrac{1}{1.3} + \dfrac{1.2}{1.3.5} + \dfrac{1.2.3}{1.3.5.7} \cdots \right)$ which results on setting $a = x = 1$ and so $z = \frac{1}{2}$ was communicated by Jakob Bernoulli to Leibniz on 2 August 1704 (N.S.) as the recent discovery of Jean Christophe Fatio de Duillier (Nicolas' elder brother) 'ipse Geometra insignis, [qui] nuper cum Hermanno nostro communicavit ingeniosum ipsius inventum de Transmutatione seriei tuæ Cyclicæ $\dfrac{1}{1} - \dfrac{1}{3} + \dfrac{1}{5} - \dfrac{1}{7}$ &c, quæ terminos habet alternatim affirmativos et negativos, in hanc aliam pure affirmativam

$$\frac{1}{2} + \frac{1}{2.3} + \frac{1}{3.5} + \frac{1}{5.7} + \frac{4}{5.7.9} + \frac{4.5}{5.7.9.11} + \frac{4.5.6}{5.7.9.11.13} \quad \&c$$

quæ celerrime convergit, utpote in qua terminus quilibet minor est quam subduplus præcedentis, addiditque, quod sibi constitutum sit proximam hyemem impendere in provehendis seriei hujus ope numeris Ludolphinis [the digits in the decimal expansion of π] ad centum usque notas' (C. I. Gerhardt, *Leibnizens Mathematische Schriften* (1) **3**.1 (Halle, 1855): 91–2). The explanation of the 'Tetragonistica Fatiana' given by Jakob Hermann to Leibniz six months afterwards on 21 January 1705 (N.S.) (*Leibnizens Mathematische Schriften* (1) **4** (Halle, 1859): 267–8; see also J. E. Hofmann, 'Ueber Jakob Bernoullis Beiträge zur Infinitesimalmathematik' [= *L'Enseignement Mathématique. Monographie 3* (Geneva, 1957)]: 53) agrees exactly with the derivation in note (40) above on setting $+t = 1$, $A = 1$, $-B = -\frac{1}{3}$, $C = \frac{1}{5}$,

shall (as in the rule) be multiplied by the powers of t by writing

$$xy/a^2 = x^2/a^2 - \tfrac{1}{3}x^4/a^4 + \tfrac{1}{5}x^6/a^6 \dots$$

I next collect the first differences $\dfrac{2}{1.3}$, $\dfrac{2}{3.5}$, $\dfrac{2}{5.7}$, ... of the coefficients; their

second ones $\dfrac{2.4}{1.3.5}$, $\dfrac{2.4}{3.5.7}$, $\dfrac{2.4}{5.7.9}$, ...; their third ones $\dfrac{2.4.6}{1.3.5.7}$, $\dfrac{2.4.6}{3.5.7.9}$, ...,

and so on. Then, putting $x^2/(a^2+x^2)$ (or $t/(1+t)$) $= z$, I write

$$xy/a^2 = z + \frac{2}{1.3}z^2 + \frac{2.4}{1.3.5}z^3 + \frac{2.4.6}{1.3.5.7}z^4 \dots \tag{44}$$

In the same way if there were $y = x + \tfrac{1}{3}x^3/a^2 + \tfrac{1}{5}x^5/a^4 + \tfrac{1}{7}x^7/a^6 \dots$, I should have to put $x^2/(a^2-x^2)$ (or $t/(1-t)$) $= z$ and write

$$xy/a^2 = z - \frac{2}{1.3}z^2 + \frac{2.4}{1.3.5}z^3 - \frac{2.4.6}{1.3.5.7}z^4 \dots \tag{45}$$

Whenever the terms are set a pair positive and a pair negative alternately, as in this series $1 + \tfrac{1}{3} - \tfrac{1}{5} - \tfrac{1}{7} + \tfrac{1}{9} + \tfrac{1}{11} \dots$, the series must be resolved into the two $1 - \tfrac{1}{5} + \tfrac{1}{9} - \tfrac{1}{13} \dots$ and $\tfrac{1}{3} - \tfrac{1}{7} + \tfrac{1}{11} \dots$, unless you prefer to unite the pairs of terms.[46] And so in the case of triples, quadruples and higher groupings.

Sometimes it is convenient to begin the transformation in the middle of a series. So in the series $y = x - \tfrac{1}{2}x^2 + \tfrac{1}{3}x^3 - \tfrac{1}{4}x^4 \dots$, should you have set

$$v = \tfrac{1}{3}x - \tfrac{1}{4}x^2 + \tfrac{1}{5}x^3 \dots$$

$-D = -\tfrac{1}{7}$, ... and hence $z = \tfrac{1}{2}$, $b = \dfrac{2}{1.3}$, $^2b = \dfrac{2}{3.5}$, ...; $c = \dfrac{2.4}{1.3.5}$, $^2c = \dfrac{2.4}{3.5.7}$, ... and so on.

As an afterthought Hermann added that 'Ita mutavi quoque hanc seriem pro hyperbola $1 - \tfrac{1}{2} + \tfrac{1}{3} - \tfrac{1}{4} + \tfrac{1}{5} - \tfrac{1}{6}$ &c in hanc $\dfrac{1}{1.2} + \dfrac{1}{2.4} + \dfrac{1}{3.8} + \dfrac{1}{4.16} + \dfrac{1}{5.32} + \dfrac{1}{6.64}$ &c', which is likewise the particular case $x = 1$, $z = \tfrac{1}{2}$ of Newton's preceding 'transmutatio' (see note (41) above): the same deduction was made a month later by Jakob Bernoulli who, remarking to Leibniz on 28 February that 'Artificium commutandi seriem Tuam in Facianam leviculum est, et consistit in sola additione continua primi et secundi, secundi et tertii, tertii et quarti &c termini, prout uberius Tibi ab ipso Hermanno explicitum esse credo', added that 'ejusdem artificioli ope series hyperbolica $1 - \tfrac{1}{2} + \tfrac{1}{3} - \tfrac{1}{4} \dots$ &c convertatur in hanc $\dfrac{1}{1.2} + \dfrac{1}{2.4} + \dfrac{1}{3.8} \dots + $&c quæ eadem mihi e alio fundamento se obtulerat (vid. Prop. LIX *de Seriebus* [*Earumque Usu In quadraturis Spatiorum & rectificationibus Pars Quinta* (Basel, 1704) = *Opera*, **2** (Geneva, 1744): Nº CI: 969]) et in qua per 18 primos terminos tantundem [logarithmus 2] approximatur, quantum per mille terminos alterius' (*Leibnizens Mathematische Schriften* (1), **3**, 1: 96).

(45) The comparable 'transmutation' of $y = a\tanh^{-1}(x/a)$ into

$$xy/a^2 = \surd[z/(1+z)]\tanh^{-1}\surd[z/(1+z)] = \surd[z/(1+z)]\sinh^{-1}\surd z.$$

(These hyperbolic functions would, of course, have been expressed by Newton on the geometrical model of the rectangular hyperbola.)

(46) 'scribendo $\tfrac{4}{3} - \tfrac{12}{35} + \tfrac{20}{99}$ &c' (by writing [it as] $\tfrac{4}{3} - \tfrac{12}{35} + \tfrac{20}{99} \dots$) is cancelled. Newton first concluded his paragraph with 'Et eadem est ratio ternorum & plurium terminorum ejusdem signi' (And the method is the same for terms three or more at a time with the same sign).

WMP

transmutationem[47] obtinueris $v = \frac{1}{3}z + \frac{1}{3,4}z^2 + \frac{1,2}{3,4,5}z^3 + \frac{1,2,3}{3,4,5,6}z^4$ &c habebis

$$y = x - \tfrac{1}{2}x^2 + x^2 \text{ in } \overline{\frac{1}{3}z + \frac{1}{3,4}z^2 + \frac{1,2}{3,4,5}z^3} \text{ &c.}$$

II

Harum transmuta-tionum usus.

Harum transmutationum usus principalis est ut series divergentes in convergentes, et convergentes in magis convergentes vertantur. Series quarum termini omnes ejusdem sunt signi divergere nequeunt quin simul infinite magnæ et eo nomine falsæ evadant.[48] Hæ itacɞ opus non habent ut in convergentes vertantur. Illæ autem quarum termini per vices mutant signa et regulariter[49] progrediuntur, ita per alternas terminorum illorum additiones et subductiones temperantur ut etiam divergendo maneant veræ.[50] At in forma divergentium quantitas earum computari nequit. Vertendæ sunt in convergentes per allatam regulam et ubi tarde convergunt applicanda est regula ut convergant celerius. Sic series $y = x - \tfrac{1}{3}x^3 + \tfrac{1}{5}x^5$ &c ubi convergit vel tardè satis divergit et versa est in hanc[51]

$$y = x - \tfrac{1}{3}x^3 + \tfrac{1}{5}x^5 - \tfrac{1}{7}x^7 + \tfrac{1}{9}x^9 - \tfrac{1}{11}x^{11} + \tfrac{1}{13}x^{13} - \tfrac{1}{15}x^{15}$$

$$+ x^{15} \text{ in } \overline{\frac{1}{17}z + \frac{2}{17,19}z^2 + \frac{2,4}{17,19,21}z^3 + \frac{2,4,6}{17,19,21,23}z^4} \quad \text{&c}$$

(47) On setting $z = x/(1+x)$, that is.

(48) Newton cannot yet (compare note (15) above) free himself from the idea that a series divergent to infinity 'represents' a 'false' (complex) quantity, though he now restricts his assertion to the monotonically divergent series only. But, by the same law of continuity according to which he below assumes that the series expansion of $\tan^{-1}x$ continues to denote the inverse tangent even when $x > 1$, surely the series $1/(1-x)$ must continue to be 'represented' by

$$1 + x + x^2 + x^3 \dots \text{ when } \quad x > 1?$$

Further, since (as Newton knew) the Leibnizian 'harmonic' series

$$[-\log(1/0) =] 1 + \tfrac{1}{2} + \tfrac{1}{3} + \tfrac{1}{4} \dots$$

is divergent, under the present head simple 'real' infinity must be classed equally as a 'false' quantity. (In his 'Extracts from Mr Gregories Letters'—the so-called 'Historiola'—, sent to Oldenburg about late May 1676 and probably known to Newton soon afterwards, John Collins had noted Pietro Mengoli's proof of the divergency of the sum to infinity of the simple 'Musicall' progression in 'the Præface [+4^r/4^v] of his [*Novæ*] *Quadraturæ Arithmeticæ* [*seu De Additione Fractorum*] printed at Bononia [Bologna] A° 1650 … by this Argumentation. $\tfrac{1}{2} \tfrac{1}{3} \tfrac{1}{4}$ … the sum of all three tearmes are greater than triple of the meane which is an Unit. Also … $\tfrac{1}{5} \tfrac{1}{6} \tfrac{1}{7}$ … by the same Argumentation are greater than $\tfrac{1}{2}$ which is triple of the meane, and the next three $\tfrac{1}{8} \tfrac{1}{9} \tfrac{1}{10}$ greater than $\tfrac{1}{3}$ the triple of the meane, and the next three $\tfrac{1}{11} \tfrac{1}{12} \tfrac{1}{13}$ are greater than $\tfrac{1}{4}$ the triple of the meane. But $\tfrac{1}{2} \tfrac{1}{3} \tfrac{1}{4}$ … were greater than an Unit, and therefore the following Nine are greater than an Unit. [Likewise] the ensuing 27 … are greater than an Unit, and so ad infinitum' (Royal Society MS LXXXI, No. 46: 1–2). Oresme's similar proof three centuries before in his *Questiones super Geometriam Euclidis* (ed. H. L. L. Busard, Leyden, 1961), making repeated use of the inequality $1/(2n-1) + 1/2n > 1/n$, was not known in Newton's day.)

and from this by transformation[47] have obtained

$$v = \frac{1}{3}z + \frac{1}{3.4}z^2 + \frac{1.2}{3.4.5}z^3 + \frac{1.2.3}{3.4.5.6}z^4 \cdots,$$

you will get $y = x - \frac{1}{2}x^2 + x^2\left(\frac{1}{3}z + \frac{1}{3.4}z^2 + \frac{1.2}{3.4.5}z^3 \cdots\right)$.

The chief use for these transformations is to turn divergent series into convergent ones, and convergent series into ones more convergent. Series in which all terms are of the same sign cannot diverge without simultaneously coming to be infinitely great and on that account false.[48] These, consequently, have no need to be turned into convergent ones. Those, however, in which the terms alternate in sign and proceed regularly are so moderated by the successive addition and subtraction of those terms as to remain true even in divergence.[50] But in their divergent form their quantity cannot be computed and they must be turned into convergent ones by the rule introduced, while when they are sluggishly convergent the rule must be applied to make them converge more swiftly. Thus the series $y = \frac{1}{3}x^3 + \frac{1}{5}x^5 \cdots$, when it converges or diverges slowly enough and has been turned into this one[51]

II
Uses of these
transforma-
tions.

$$y = x - \frac{1}{3}x^3 + \frac{1}{5}x^5 - \frac{1}{7}x^7 + \frac{1}{9}x^9 - \frac{1}{11}x^{11} + \frac{1}{13}x^{13} - \frac{1}{15}x^{15}$$
$$+ x^{15}\left(\frac{1}{17}z + \frac{2}{17.19}z^2 + \frac{2.4}{17.19.21}z^3 + \frac{2.4.6}{17.19.21.23}z^4 \cdots\right),$$

---

(49) The equivalent phrase 'regulari lege omnium' was first written, followed by 'secundū legem aliquam communem' (following some common law).

(50) Newton is evidently impressed by the fact that the 'transmutations' $z = x/(1+x)$ and $z = x^2/(1+x^2)$, which (for all x) convert $\log(1+x)$ and $\tan^{-1}x$ into $-\log(1-z)$ and

$$\sqrt{[z/(1-z)]}\sin^{-1}\sqrt{z},$$

likewise transform the series expansions of the former (divergent when $x > 1$) into those of the latter (convergent for $z < 1$ and hence for all x positive): a similar comparison holds between any function $f(x)$ whose series expansion is divergent for $x > k$, say, and the 'transmuted' function $\phi(z)$, $z = x^p/(1+x^p)$. He has, however, ignored the unpleasant fact that no unique sum is assignable to a divergent alternating series, as Euler was later to discover in similar circumstances (see his *Institutiones Calculi Differentialis* (note (40) above): Pars Posterior, Caput 1: 288 ff., where the series $1-2+3-4+5-6 \ldots$ is shown to yield either $(1-2)+(3-4)+(5-6)+\ldots = -\infty$ or $1-(2-3)-(4-5)-\ldots = +\infty$ as its aggregate to infinity on suitable pairing of its terms). Fortunately for Newton's underlying instinct (or was it merely good fortune?) in identifying patterns of continuity in mathematical structures, in the case of the simple, 'well behaved' circular and hyperbolic functions which he considered in his researches into series the value to which the transformed function $\phi(z)$ converges is in fact the 'transmutation' of that 'represented' by a divergent original function $f(x)$.

(51) On setting $z = x^2/(1+x^2)$, it is understood. The series under the vinculum (whose successive coefficients are the k-th order Newtonian differences of the terms $\frac{1}{17}, \frac{1}{19}, \frac{1}{21}, \frac{1}{23}, \ldots$, $k = 0, 1, 2, \ldots$) is evidently convergent for $0 \leqslant z < 1$ and hence for all positive values of x.

ad multa decimalium figurarum loca sat citò computabitur. Eadem si celeriter divergit vertenda est in convergentem $yx = z + \frac{2}{1,3} z^2 + \frac{2,4}{1,3,5} z^3$ &c,[52] deinde per ea quæ in sequente capite[53] habentur computari potest. Convenit autem sæpenumero coefficientes A, B, C &c in numeros fractos decimales ipso operis initio reducere.

III
Transmuta-
tiones aliæ
prioribus
affines, et
earum usus.

Sunt et aliæ[54] his affines transmutationum formulæ quas longum esset prosequi. Formari potest series una ex differentijs primis, alia ex secundis, tertia ex tertijs, aliæ ex his differentijs vari[e] mixtis, sed harum vis et usus omnis in præcedentibus includitur. Formantur et aliae multæ aliunde. Nam data serie quavis $y = ax \pm bx^2 \pm cx^3 \pm dx^4 \pm ex^5$ &c si ex indefinita x et alia quavis indefinita z una cum earum dignitatibus in datas quasvis ductis formetur utcunqɜ æquatio et resolvatur hæc in seriem æquipollentem ipsi x, vel si ad significandum x assumatur series quælibet conflata ex dignitatibus z, sive finita ea sit, sive in infinitum progressa: deinde si series illa pro x, quadratū ejus pro x^2, cubus pro x^3 cæteræcɜ ejus dignitates pro cæteris dignitatibus ipsius x in serie data scriba[n]tur, prodibit nova series ex dignitatibus indefinitæ z conflata si modò seriei substitutæ termini omnes vel ducebantur in z et ejus dignitates vel omnes applicabantur ad easdem. Ut si assumatur $x = z + z^3$ et perinde pro x et ejus dignitatibus scribantur z et ejus dignitates, series $y = ax \pm bx^2 \pm cx^3$ &c

evadet $y = az \begin{smallmatrix} +a \\ \pm b \end{smallmatrix} z^2 \begin{smallmatrix} \pm 2b \\ \pm c \end{smallmatrix} z^3 \begin{smallmatrix} \pm b \\ \pm 3c \\ \pm d \end{smallmatrix} z^4$ &c. Confingi etiam potest series ex dignitatibus

ipsius x et ejus beneficio exterminari x juxta methodum in epistola mea secunda expositam.[55] Ut si fingatur $z = x + xx$, æquatio infinita $y = ax \pm bx^2 \pm cx^3$ &c

exterminato x evadet $y = az \begin{smallmatrix} -a \\ \pm b \end{smallmatrix} z^2 \begin{smallmatrix} +2a \\ \mp 2b \\ \pm c \end{smallmatrix} z^3$ &c.[56] Et per hujusmodi transmuta-

---

(52) See note (44) above.

(53) This method, not preserved in the opening paragraphs of 'Cap 3' in sequel which alone are now traceable, is manifestly the finite-difference technique for determining the 'terminatio' of a series which Newton expounded at the close of Chapter 2 of his preceding 'Specimina' (see §1: note (67)): indeed the example there instanced of a summation 'proxime in numeris fractis decimalibus' is essentially—a factor of 2 apart—the particular case $x = 1$ (and so $z = \frac{1}{2}$) of the present expansion (see §1: notes (85) and (86); and compare note (44) of the present 'De Computo Serierum').

(54) The more optimistic announcement 'multæ' (many) is cancelled. An earlier Gregorian 'approximatio' to the sum of an infinite series, elaborated by Newton in a first version (on ff. 11ʳ–12ʳ) of the present chapter, is reproduced in Appendix 3.1 below.

(55) Newton evidently intends that the given equation

$$z = f(x) \equiv \alpha x + \beta x^2 + \gamma x^3 + \delta x^4 \dots$$

will speedily enough be computed to many places of decimals. If the same series proves swiftly divergent it must be turned into the convergent

$$xy = z + \frac{2}{1 \cdot 3} z^2 + \frac{2 \cdot 4}{1 \cdot 3 \cdot 5} z^3 \ldots \text{(52)}$$

and then by what is presented in the following chapter[53] it can be computed. It is, however, frequently convenient to reduce the coefficients A, B, C, ... to decimal fractions at the very start of the work.

There are also[54] other related transformation formulas which it would be tedious to pursue. One series can be formed from the first differences, another from the second ones, a third from the third and others from various mixtures of these differences, but all the force and use of these is included in the preceding. Many others, also, are formed on other principles. For, given any series

$$y = ax \pm bx^2 \pm cx^3 \pm dx^4 \pm ex^5 \ldots,$$

let an equation be formed in any manner from the indefinite quantity x and any other indefinite one z together with their powers multiplied into any constants and let this be resolved into a series equipollent to x, or to signify x let any series made up from powers of z—whether finite or carrying on to infinity—be assumed: then if that series should be written in place of x, its square in place of x^2, its cube for x^3 and its other powers for the other powers of x in the given series, there will result a new series made up of powers of the indefinite quantity z provided all the terms in the substitute series were either multiples of z and its powers or the quotients of these. Should it be assumed, for instance, that $x = z + z^3$ and if correspondingly in place of x and its powers there be written z and its powers, the series $y = ax \pm bx^2 \pm cx^3$ ... will come to be $y = az + (a \pm b) z^2 + (\pm 2b \pm c) z^3 + (\pm b \pm 3c \pm d) z^4$ .... A series can also be fabricated from powers of x and by its aid x may be eliminated following the method disclosed in my second letter.[55] So if it be imagined that $z = x + x^2$, the infinite equation $y = ax \pm bx^2 \pm cx^3$ ... will, when x is eliminated, come to be $y = az + (-a \pm b) z^2 + (2a \mp 2b \pm c) z^3$ ....[56] And by transformations of this kind

III.
Other trans-
formations
related to the
previous ones
and their
uses.

be inverted by means of the two general theorems communicated in October 1676 in his *epistola posterior* (*Correspondence*, **2**: 128; see pages 631–2 below) and the inverse series

$$x = f^{-1}(z) \equiv (1/\alpha) z - (\beta/\alpha^3) z^2 \ldots$$

then substituted in the given 'æquatio infinita' $y = ax + bx^2 + cx^3$ ..., powers of z being subsequently collected.

(56) Since $x = z - x^2 = z/(1+x) = \dfrac{z}{1+} \dfrac{z}{1+} \dfrac{z}{1+} \ldots$, at once to $O(z^5)$

$$x = z - z^2/(1+z)^2 = z - z^2 + 2z^3 - 3z^4 \ldots$$

whence $y = a(z - z^2 + 2z^3) \pm b(z - z^2 \ldots)^2 \pm c(z \ldots)^3 \ldots.$

tiones sæpe prodeunt series magis convergentes quam per methodum differentiarum. Sed harum usus præcipuus est ut in problemate quovis habita serie ex dignitatibus unius indefinitarū quantitatum conflata, sciamus inde condere seriem ex dignitatibus alterius cujusvis.

IV
Transmuta-
tionū genus
aliud, et
earum usus. Sunt rursus aliæ transmutationes quærendo dignitates et radices seriei propositæ et has inter se variè componendo: qua methodo series infinitæ sæpe convertuntur in æquationes finitas, ut alibi[57] commodius ostendetur.

Cap 3.

De serierum interpolatione.[58]

I
Qua ratione
series inter-
polandæ sunt. Usus serierum augetur inventione terminorum novorum in locis imperatis. Id fit modis diversis. Meliorem nondū reperi quam ducendo lineam regularem per puncta quotcunꞯ data.[59] Dentur puncta quotcunꞯ P, P^2, P^3, P^4, P^5, P^6, P^7, P^8 &c et inde ad rectam quamvis positione datam OA (in qua detur punctum quodvis O) demittantur perpendicula PA, P^2A^2, P^3A^3, P^4A^4 &c serierum

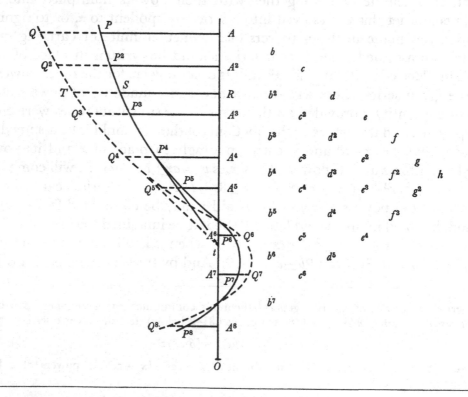

---

(57) This was doubtless to form the theme of a later chapter on the lines of 'Cap 5. De migratione serierum infinitarum in finitas' in Newton's preliminary scheme for the 'Matheseos Universalis Specimina' (see §1: note (39) above). In certain cases the preceding 'transmutation' will yield a finite series as the equal of an infinite one (where, namely, the latter is

there often result series more convergent than ones derived by the method of differences. But the pre-eminent use for these is that, once we have obtained in any problem a series made up of the powers of one indefinite quantity, we may thereby know how to build up a series from the powers of any other.

There are, again, yet other transformations by seeking out the powers and roots of the series proposed and compounding these one with another in various ways: by this method infinite series are often converted into finite ones, as will more suitably be revealed elsewhere.[57]

IV.
A further type of transformation and its use.

CHAPTER 3.

ON THE INTERPOLATION OF SERIES.[58]

The usefulness of series is increased by the discovery of new terms at appointed places. This comes about in different ways. I have not yet found a better one than by drawing a regular line through a number of given points.[59] Let any number of points P, P_2, P_3, P_4, P_5, P_6, P_7, P_8, ... be given, and from these to any straight line OA given in position (in which let there be given any point O) drop the perpendiculars PA, P_2A_2, P_3A_3, P_4A_4, ... corresponding to terms in the

I.
By what procedure series are to be interpolated.

$\sum_{1 \leqslant i \leqslant \infty} a_i x^i$, $a_i = r + si + ti^2 + vi^3 + ... wi^k$, k a finite positive integer, when $\Delta^{k+1}a_i = 0$): for example, if $y = x + 2x^2 + 3x^3 + 4x^4 ...$, then the substitution $z = x/(1-x)$ produces

$$y = z + z^2 = x/(1-x)^2.$$

More generally, if $y = x^p \pm \binom{m/n}{1} x^{p+q} + \binom{m/n}{2} x^{p+2q} \pm \binom{m/n}{3} x^{p+3q} ...$, then $y^{n/m} = x^{np/m}(1 \pm x^q)$.

(58) A stray remark in the previous chapter (see note (53) above) makes it certain that the general scheme of interpolation outlined in the opening paragraph following is intended to introduce a revise of the finite-differences technique for summation of series outlined in Chapter 2 of the preceding 'Specimina' (see §1: note (67) above).

(59) Newton proceeds to sketch his earlier technique for inserting the general ordinate $RS = f(OR)$ to a 'linea regularis' which is made to pass through a number of given points P^i, where $A^i P^i = f(OA^i)$. The accompanying figure, with the omission of the broken line $QQ^2 ... Q^8$ and the tangents P^4t, Q^4t at P^4 and Q^4 is the mirror image of that accompanying 'Prob 2. Curvam Geometricam describere quæ per data quotcunq puncta transibit' in 1, 1, §4.2 (page 62 above): that the tangents at P^4 and Q^4 intersect on the base OA confirms the suggestion afforded optically by the fairly carefully drawn manuscript figure that the proportion of the ordinates $A^i P^i$, $A^i Q^i$ is taken to be constant. The triangular array on the right follows Newton's earlier (1676) notation in representing the general divided differences

$$b^i = \frac{A^i P^i - A^{i+1} P^{i+1}}{A^i A^{i+1}}, \quad c^i = \frac{b^i - b^{i+1}}{A^i A^{i+2}}, \quad d^i = \frac{c^i - c^{i+1}}{A^i A^{i+3}}, \quad ...$$

by the primary differences b, c, d, ... [that is, b^1, c^1, d^1, ...] in each row increased by a corresponding posterior superscript integer i [$= 2, 3, 4, ..., 7$]. In imitation of his usage in Chapter 2 of the 'Specimina' (§1 preceding), however, Newton first began to enter the first row as 'b, ^2b, ^3b, ^4b, ^5b, ...'. but quickly emended it to that reproduced.

terminis respondentia. Tum ab ejusdem rectæ puncto quovis R erigatur normalis RS, eamcp concipe terminari ad lineam regularem $PP^2P^3P^4$ [&c] per puncta illa omnia data P, P^2, P^3, P^4 &c transeuntem. Ex assumpta utcuncp longitudine OR invenienda est longitudo RS.

Reg. 1. Ex longitudine indefinita OR, quam dic x, et quantitatibus quibuscuncp datis i, k, l [&c] formetur utcuncp quantitas v. Tum in valore v scribendo OA pro x fac[60]

---

(60) Understand 'fac $AP=v$....' (make $AP = v$). The last word 'fac' is the catchword at the bottom of f. 14^r but the continuation of this rule (on a further sheet no longer traceable) seems to be irreparably lost. Very possibly it was straightaway destroyed by Newton himself when he cancelled the opening lines of 'Reg. 1' here reproduced. It was evidently his initial intention to set up a divided differences scheme, much as in **1**, 1, §4 (see previous note), for interpolating the general ordinate $RS = v = i + kx + lx^2 \dots$, where $OR = x$: at once if the curve (S) is to pass through n given point P^j, $j = 1, 2, 3, \dots, n$, then

$$A^j P^j = i + k \cdot (OA^j) + l \cdot (OA^j)^2 + \dots + m \cdot (OA^j)^{n-1}.$$

We may guess that 'Reg. 1' and 'Reg. 2' dealt with the cases of central-difference interpolation arising according as n is odd or even; compare 'Cas. 1' and 'Cas. 2' of Problem 2 of **1**, 1, §4.2.

series. Then from any point R of the same straight line erect the normal RS and conceive it to terminate at the regular line $PP_2P_3P_4\ldots$ passing through all those given points P, P_2, P_3, P_4, ... On assuming the length of OR arbitrarily, the length of RS has to be found from it.

Rule 1. Call the indefinite length of OR x and from it and any given quantities i, k, l, ... let the quantity v be formed arbitrarily. Then on writing OA in place of x in the value of v, make[60]

No portion of the following chapters Newton doubtless intended to add has turned up after prolonged search on our part, nor indeed is it known that they were ever written. (Somewhat pessimistically perhaps, we think they were not, but would be happy to be wrong.) In content they would evidently revise the later Chapters 3–5 of the preceding 'Specimina' (see pages 554–88), while it was also Newton's declared intention (see note (57) above) to introduce some discussion of the theme 'De migratione serierum infinitarum in finitas'. In the present state of knowledge all else must be unsupported conjecture.

APPENDIX 1. THE EDITED VERSION OF NEWTON'S 'EPISTOLA POSTERIOR' TO LEIBNIZ INTENDED FOR PUBLICATION IN HIS 'SPECIMINA' (1684).[1]

Adapted from the corrected copy retained by Newton in 1676, now in the University Library, Cambridge[2]

Quanta cum voluptate legi Epistolas Clarissimorum virorum D. Leibnitij & D. Tschurnhausij vix dixerim. Perelegans sane est Leibnitij methodus perveniendi ad series convergentes, & satis ostendisset ingeniū Authoris etsi nihil aliud scripsisset. Sed quæ alibi per Epistolam sparguntur suo nomine dignissima, efficiunt etiam ut ab eo speremus maxima. Diversitas modorū quibus eodem tenditur, eo magis placuit,[3] quod mihi tres methodi perveniendi ad ejusmodi series innotuere,[4] adeò ut novam nobis communicandam vix expectarem. Unam e meis priùs[5] descripsi, jā addo aliam, illā sc: quâ primùm incidi in has series: nam incidi in eas antequam scirem divisiones et extractiones radicū quibus jam utor. Et hujus explicatione pandendum est fundamentū Theorematis sub initio Epistolæ prioris positi quod D. Leibnitius a me desiderat.

[6]Sub initio studiorū meorū Mathematicorū ubi incideram in opera Celeberrimi Wallisij nostri, considerando series quarum intercalatione ipse exhibit aream circuli, et Hyperbolæ, utpote quod in serie curvarū quarū basis sive axis communis sit x et ordinatim applicatæ $\overline{1-xx}|^{\frac{0}{2}}$. $\overline{1-xx}|^{\frac{1}{2}}$. $\overline{1-xx}|^{\frac{2}{2}}$. $\overline{1-xx}|^{\frac{3}{2}}$. $\overline{1-xx}|^{\frac{4}{2}}$. $\overline{1-xx}|^{\frac{5}{2}}$. &c, si areæ alternarū quæ sunt

$$x. \quad x-\tfrac{1}{3}x^3. \quad x-\tfrac{2}{3}x^3+\tfrac{1}{5}x^5. \quad x-\tfrac{3}{3}x^3+\tfrac{3}{5}x^5-\tfrac{1}{7}x^7 \quad \&c$$

interpolari possent haberemus areas intermediarū quarū prima $\overline{1-xx}|^{\frac{1}{2}}$ est circulus: ad has interpolandas notabam quod in omnibus primus terminus

---

(1) See §1: notes (11) and (14) above. In this slightly revised abridgment of his letter sent to Oldenburg on 24 October 1676 for onward transmission to Leibniz (*Correspondence of Isaac Newton*, **2**, 1960: 110–29) Newton has straightforwardly retained the portions—forming the bulk of the letter in fact—which relate to the construction, convergence and application of the infinite series expansions of elementary circular and hyperbolic functions, pasting over the omissions with suitable transitional passages and adding one or two verbal refinements (which in no way distort the content of the letter received by Leibniz seven years before). The present reconstruction is made primarily to allow the reader visually to appreciate the shortened letter which Newton intended to quote in the introduction to his 'Matheseos Universalis Specimina', but we have seized the opportunity to insert a few editorial observations on what is, beyond doubt, singly his most important mathematical communication to a contemporary.

(2) Add. 3977.4, a transcript made by Newton's chamber-mate John Wickins but with autograph corrections and additions by Newton himself. (The autograph letter sent to Oldenburg is now British Museum MS Add. 4294:119. See pages 666–74 below for a précis of the historical circumstances in which the letter was composed.)

(3) As in the case of the autograph sent to Oldenburg (compare the photocopy inserted in *Correspondence*, **2**: Plate I [facing p. 110]) this replaces the less restrained verb 'miror'.

esset x, quodꝗ secundi termini $\frac{0}{3}x^3$. $\frac{1}{3}x^3$. $\frac{2}{3}x^3$. $\frac{3}{3}x^3$ &c essent in Arithmeticâ progressione, & proinde quod duo primi termini serierū intercalandarū deberent esse $x - \frac{\frac{1}{2}x^3}{3}$. $x - \frac{\frac{3}{2}x^3}{3}$. $x - \frac{\frac{5}{2}x^3}{3}$ &c. Ad reliquas intercalandas considerabam quod denominatores 1.3.5.7 &c erant in arithmeticâ progressione, adeoꝗ solæ numeratorū coefficientes numerales restabant investigandæ. Hæ autem in alternis datis areis erant figuræ potestatū numeri 11 nempe harum $\overline{11}|^0$. $\overline{11}|^1$. $\overline{11}|^2$. $\overline{11}|^3$. $\overline{11}|^4$. hoc est primò 1. dein 1,1. tertiò 1,2,1. quartò 1,3,3,1. quintò 1,4,6,4,1. &c. Quærebam itaꝗ quomodo in his seriebus ex datis duabus primis figuris reliquæ derivari possent, et inveni quod positâ secundâ figura m, reliquæ producerentur per continuam multiplicationē terminorū hujus seriei

$$\frac{m-0}{1} \times \frac{m-1}{2} \times \frac{m-2}{3} \times \frac{m-3}{4} \times \frac{m-4}{5}$$

&c. E. gr. sit $m = 4$, et erit $4 \times \frac{m-1}{2}$ hoc est 6 tertius terminus, & $6 \times \frac{m-2}{3}$ hoc est 4 quartus, et $4 \times \frac{m-3}{4}$ hoc est 1 quintus, & $1 \times \frac{m-4}{5}$ hoc est 0 sextus, quo series in hoc casu terminatur. Hanc regulam itaꝗ applicui ad series interserendas et cùm pro circulo secundus terminus esset $\frac{\frac{1}{2}x^3}{3}$, posui $m = \frac{1}{2}$, et prodierunt termini $\frac{1}{2} \times \frac{\frac{1}{2}-1}{2}$ sive $-\frac{1}{8}$, $-\frac{1}{8} \times \frac{\frac{1}{2}-2}{3}$ sive $+\frac{1}{16}$, $\frac{1}{16} \times \frac{\frac{1}{2}-3}{4}$ sive $-\frac{5}{128}$, & sic in infinitū. Unde cognovi desideratā aream segmenti circularis esse $x - \frac{\frac{1}{2}x^3}{3} - \frac{\frac{1}{8}x^5}{5} - \frac{\frac{1}{16}x^7}{6} - \frac{\frac{5}{128}x^9}{9}$ &c. Et eadem ratione prodierunt etiam interserendæ areæ reliquarum curvarū, ut et area Hyperbolæ et cæterarū alternarū in hac serie $\overline{1+xx}|^{\frac{0}{2}}$. $\overline{1+xx}|^{\frac{1}{2}}$. $\overline{1+xx}|^{\frac{2}{2}}$. $\overline{1+xx}|^{\frac{3}{2}}$ &c. Et eadem est ratio intercalandi alias series idꝗ per intervalla duorū pluriumve terminorū simul deficientium.[7]

---

(4) As in the version sent off to Oldenburg, Wickins' transcript originally went on: 'et tamen illa Leibnitij ante lectas literas ejus penitus me latuit.' Newton's natural caution would evidently not allow him to be lavish in his praise of a still little known, younger German contemporary.

(5) Namely, in his *epistola prior* to Oldenburg for Leibniz on 13 June 1676, where he described his general binomial expansion. See §1: note (12) above.

(6) In the letter as sent to Oldenburg in 1676 (see note (2) above) Newton first began this paragraph with the words 'Cùm primùm appuli ad studia Mathematica et cœperam ea mediocriter callere, incidi in opera...Wallisij nostri, & considerando...'. The following version of his mathematical baptism—written more than a decade after the sequence of events it describes and so not necessarily to be trusted *prima facie*—is, in fact, broadly consistent with the evidence of Newton's extant annotations on Wallis' *Arithmetica Infinitorum* during his last undergraduate year (see 1: 104–11).

(7) Compare 1: 122–34.

Ubi verò hæc didiceram mox considerabam terminos $\overline{1-xx}|^{\frac{0}{2}}$. $\overline{1-xx}|^{\frac{2}{2}}$. $\overline{1-xx}|^{\frac{4}{2}}$. $\overline{1-xx}|^{\frac{6}{2}}$ &c hoc est 1. $1-xx$. $1-2xx+x^4$. $1-3xx+3x^4-x^6$ &c eodem modo interpolari posse ac areas ab ipsis generatas: et ad hoc nihil aliud requiri quàm omissionem denominatorū 1, 3, 5, 7 &c in terminis exprimentibus areas; hoc est coefficientes terminorū quantitatis intercalandæ $\overline{1-xx}|^{\frac{1}{2}}$, vel $\overline{1-xx}|^{\frac{3}{2}}$, vel generaliter $\overline{1-xx}|^{m}$, prodire per continuam multiplicationē hujus seriei $m\times\dfrac{m-1}{2}\times\dfrac{m-2}{3}\times\dfrac{m-3}{4}$ &c. Adeo ut e.g. $\overline{1-xx}|^{\frac{1}{2}}$ valeret $1-\frac{1}{2}x^2-\frac{1}{8}x^4-\frac{1}{16}x^6$ &c et $\overline{1-xx}|^{\frac{3}{2}}$ valeret $1-\frac{3}{2}xx+\frac{3}{8}x^4+\frac{1}{16}x^6$ &c et $\overline{1-xx}|^{\frac{1}{3}}$ valeret $1-\frac{1}{3}xx-\frac{1}{9}x^4-\frac{5}{81}x^6$ &c. Sic itacʒ innotuit mihi generalis reductio radicalium in infinitas series per regulam illam quam posui initio Epistolæ prioris[8] antequam scirem extractionem radicum. Sed hac cognita non potuit altera me diu latere: nam ut probarem has operationes multiplicavi $1-\frac{1}{2}x^2-\frac{1}{8}x^4-\frac{1}{16}x^6$ &c in se, et factum est $1-xx$, terminis reliquis in infinitū evanescentibus per continuationem seriei. Atcʒ ita $1-\frac{1}{3}xx-\frac{1}{9}x^4-\frac{5}{81}x^6$ &c bis in se ductum produxit etiam $1-xx$. Quod ut certa fuit harum conclusionū demonstratio, sic me manu duxit ad tentandū e converso num hæ series quas sic constitit esse radices quantitatis $1-xx$ non possent inde extrahi more Arithmetico. et res bene successit. Operationis forma in quadraticis radicibus hæc erat.[9]

$$1-xx \quad (1-\tfrac{1}{2}xx-\tfrac{1}{8}x^4-\tfrac{1}{16}x^6 \quad \&c$$
$$\underline{1}$$
$$0-xx$$
$$\underline{-xx+\tfrac{1}{4}x^4}$$
$$-\tfrac{1}{4}x^4$$
$$\underline{-\tfrac{1}{4}x^4+\tfrac{1}{8}x^6+\tfrac{1}{64}x^8}$$
$$0-\tfrac{1}{8}x^6-\tfrac{1}{64}x^8.$$

Quibus perspectis et inventa subinde resolutione affectarum æquationum qua regredi licebat ab areis aut arcubus ad ordinatim applicatas, neglexi penitus interpolationem serierum et has operationes tanquam fundamenta magis genuina in posterum colui.

---

(8) Of 13 June 1676, that is. Newton refers to his reformulation there (ULC. Add. 3977. 2: 1ʳ [= *Correspondence*, 2: 21]) of the binomial expansion of $[P(1+Q)]^{m/n}$ as, in effect,

$$P^{m/n}\left(1+\frac{m}{n}Q+\frac{m(m-n)}{n.2n}Q^2+\frac{m(m-n)(m-2n)}{n.2n.3n}Q^3\ldots\right)$$

'ubi...$\dfrac{m}{n}$ [significat] numeralem indicem dimensionis'. See also page 667, note (33) below.

(9) Though Newton does not indicate that the following root extraction is to be included in the present excerpt from the *epistola posterior*, we take this to be his intention (compare §1: note (15) above).

Incidi postea in aliorum generum series quæ in finitas degenerant quoties curvilinea figura Geometricè quadrari potest. Harum unam in rei specimen subjungere suffecerit.

Ad Curvam aliquam sit $dz^\theta \times \overline{e+fz^\eta}|^\lambda$ ordinatim applicata termino diametri seu basis z normaliter insistens: ubi literæ d, e, f denotant quaslibet quantitates datas, & θ, η, λ indices potestatū sive dignitatum quantitatū quibus affixæ sunt.

Fac $\dfrac{\theta+1}{\eta}=r$. $\lambda+r=s$. $\dfrac{d}{\eta f}\times\overline{e+fz^\eta}|^{\lambda+1}=Q$. & $r\eta-\eta=\pi$, & area Curvæ erit

Q in $\dfrac{z^\pi}{s}-\dfrac{r-1}{s-1}\times\dfrac{eA}{fz^\eta}+\dfrac{r-2}{s-2}\times\dfrac{eB}{fz^\eta}-\dfrac{r-3}{s-3}\times\dfrac{eC}{fz^\eta}+\dfrac{r-4}{s-4}\times\dfrac{eD}{fz^\eta}$ &c literis A, B, C, D &c

denotantibus terminos proximè antecedentes, nempe A terminum $\dfrac{z^\pi}{s}$, B ter-

minum $-\dfrac{r-1}{s-1}\times\dfrac{eA}{fz^\eta}$ &c.[10] Hæc series ubi r fractio est vel numerus negativus, continuatur in infinitum: ubi verò r integer est et affirmativus continuatur ad tot terminos tantùm quot sunt unitates in eodem r, et sic exhibet geometricam quadraturam Curvæ. Rem exemplis illustro.

Ex. 1. Proponatur Parabola cujus ordinatim applicata sit $\sqrt{az}$. Hæc in formam regulæ reducta fit $z^0\times\overline{0+az^1}|^{\frac{1}{2}}$. Quare est $d=1$. $\theta=0$. $e=0$. $f=a$. $\eta=1$.

---

(10) This 'Theorema primum generale' is an immediate generalization of Ordo 2 of the 1671 tract's 'Catalogus prior Curvarum' (see III: 237, note (540) and compare 260, note (574)). The following examples are evidently concocted *ad hoc* for the present letter. There is an easy proof of the theorem by repeated integration by parts (see J. E. Hofmann, *Studien zur Vorge-schichte des Prioritätstreites zwischen Leibniz und Newton um die Entdeckung der höheren Analysis. I. Abhandlung: Materialien zur ersten mathematischen Schaffensperiode Newtons (1665–1675)* [= *Abhand-lungen d. Preuss. Akademie der Wissenschaften*, Jahrgang 1943. Math.-naturw. Klasse Nr. 2 (Berlin, 1943)]: 64, note 253): in brief, where $I_\theta \equiv d\int z^\theta(e+fz^\eta)^\lambda.dz$, then on taking (with Newton) $(\theta+1)/\eta = r$ and $r+\lambda = s$ there follows

$$I_{\theta-k\eta} = \frac{d}{\eta f(s-k)}z^{\theta+1-(k+1)\eta}(e+fz^\eta)^\lambda - \frac{r-(k+1)}{s-k}\cdot\frac{e}{f}I_{\theta-(k+1)\eta}$$

and hence, when k is taken successively to be 0, 1, 2, ... and the intervening $I_{\theta-k\eta}$ are successively eliminated,

$$I_\theta = -\frac{d}{\eta fr}z^{\theta+1-\eta}(e+fz^\eta)^{\lambda+1}\cdot\sum_{0\leqslant j\leqslant k-1}\prod_{0\leqslant i\leqslant j}\left(\frac{-(r-i)e}{(s-i)fz^\eta}\right)+R_kI_{\theta-\lambda\eta}, \quad R_k=\prod_{0\leqslant i\leqslant k}\left(-\frac{(r-i-1)e}{(s-i)f}\right).$$

(The series will terminate if $R_k = 0$ for some k, otherwise a necessary condition for convergence is that $\lim\limits_{k\to\infty}(R_k) = 0$.) This way of deduction, however, seems alien to Newton's mathematical temperament: the theorem may well have been merely 'induced' by a Wallisian extrapolation from Ordo 2 of the 1671 'Catalogus prior' and further (if at all) justified by differentiation. Observe that Newton here obliterates in his verbal rephrasing the suggestion, strongly implied in his original letter to Leibniz in 1676, that this 'Theorema' was primary among the 'Theoremata quædam generalia' which (compare III: 236–8, 260–2) he stated 'candide' (but erroneously?) to have derived earlier in his 'speculationes de Quadratura curvarum' (see *Correspondence*, **2**: 115).

[$\lambda=\frac{1}{2}$.] Adeoq $r=1$. $s=1\frac{1}{2}$. $Q=\frac{1}{a}\times\overline{az}|^{\frac{3}{2}}$. $\pi=0$. et area quæsita $\frac{1}{a}\times\overline{az}|^{\frac{3}{2}}$ in $\frac{1}{1\frac{1}{2}}$, hoc est $\frac{2}{3}z\sqrt{az}$. Et sic in genere si cz^η ponatur ordinatim applicata, prodibit area $\frac{c}{\eta+1}z^{\eta+1}$.

Ex 2. Sit ordinatim applicata $\frac{a^4z}{c^4-2cczz+z^4}$. Hæc per reductionem fit $a^4z\times\overline{cc-zz}|^{-2}$, vel etiam $a^4z^{-3}\times\overline{-1+ccz^{-2}}|^{-2}$. In priori casu est $d=a^4$. $\theta=1$. $e=cc$. $f=-1$. $\eta=2$. $\lambda=-2$. Adeoq $r=1$. $s=-1$. $Q=\frac{a^4}{-2}\times\overline{cc-zz}|^{-1}$ hoc est $=\frac{-a^4}{2cc-2zz}$. $\pi=0$. Et area Curvæ $=Q$ in $\frac{z^0}{-1}$ id est $=\frac{a^4}{2cc-2zz}$. In secundo autem casu est $d=a^4$. $\theta=-3$. $e=-1$. $f=cc$. $\eta=-2$. $\lambda=-2$. $r=1$. $s=-1$. $Q=\frac{a^4}{-2cc}\times\overline{-1+ccz^{-2}}|^{-1}$ id est $=\frac{-a^4zz}{2c^4-2cczz}$. $\pi=0$. et area $=Q$ in $\frac{1}{z^0}$ hoc est $=\frac{a^4zz}{2c^4-2cczz}$. [11] Area his casibus diversimodè exhibetur quatenus computatur a diversis finibus quorum assignatio per hos inventos valores arearum facilis est.

Exempl. 3. Sit Ordinatim applicata $\frac{a^5}{z^5}\sqrt{bz+zz}$, hoc est per reductionem ad debitam formam, vel $a^5z^{-\frac{9}{2}}\times\overline{b+z}|^{\frac{1}{2}}$, vel $a^5z^{-4}\times\overline{1+bz^{-1}}|^{\frac{1}{2}}$. Et erit in priori casu $d=a^5$. $\theta=-\frac{9}{2}$. $e=b$. $f=1$. $\eta=1$. $\lambda=\frac{1}{2}$ adeoq $r=-\frac{7}{2}$ &c. Quare cum r non sit numerus affirmativus, procedo ad alterū casū. Hic est $d=a^5$. $\theta=-4$. $e=1$. $f=b$. $\eta=-1$. $\lambda=\frac{1}{2}$ adeoq $r=3$. $s=3\frac{1}{2}$. $Q=\frac{a^5}{-b}\times\overline{1+bz^{-1}}|^{\frac{3}{2}}$, seu

$$=-\frac{a^5z+a^5b}{bzz}\sqrt{zz+bz}. \quad \pi=-2.$$

$$\text{et area}=Q \text{ in } \frac{z^{-2}}{3\frac{1}{2}}-\frac{2}{2\frac{1}{2}}\times\frac{z^{-1}}{3\frac{1}{2}b}+\frac{1}{1\frac{1}{2}}\times\frac{2}{2\frac{1}{2}}\times\frac{z^0}{3\frac{1}{2}bb},$$

$$\text{hoc est}=\frac{-30bb+24bz-16zz}{105bbzz}\times\frac{a^5z+a^5b}{bzz}\sqrt{zz+bz}.$$

Exempl. 4. Sit deniq ordinatim applicata

$$\frac{bz^{\frac{1}{2}}}{\sqrt{⑤c^3-3accz^{\frac{2}{3}}+3aacz^{\frac{4}{3}}-a^3z^2}}.$$

(11) These two 'variant' integrals are fundamentally the same, their constant difference $\frac{1}{2}a^4/c^2=\int_{-\infty}^0 a^4z(c^2-z^2)^{-2}.dz$ being that of $\frac{1}{2}(a^4/c^2)c^2/(c^2-z^2)=\int_{-\infty}^z a^4z(c^2-z^2)^{-2}.dz$ and $\frac{1}{2}(a^4/c^2)z^2/(c^2-z^2)=\int_0^z a^4z(c^2-z^2)^{-2}.dz$.

Hæc ad formam Regulæ reducta fit $bz^{\frac{2}{3}} \times \overline{c-az^{\frac{2}{3}}}\rvert^{-\frac{2}{3}}$. Indeq3 est $d=b$. $\theta=\frac{1}{3}$. $e=c$. $f=-a$. $\eta=\frac{2}{3}$. $\lambda=-\frac{2}{3}$. $r=2$. $s=\frac{7}{5}$. $Q=\dfrac{3b}{-2a}\times\overline{c-az^{\frac{2}{3}}}\rvert^{\frac{2}{3}}$. $\pi=\frac{2}{3}$. Et

$$\text{area}=Q\times\frac{5z^{\frac{2}{3}}}{7}-\frac{5}{2}\times\frac{5c}{-7a},$$

id est $\dfrac{-30abz^{\frac{2}{3}}+75bc}{28aa}\times\overline{c-az^{\frac{2}{3}}}\rvert^{\frac{2}{3}}$. Quod si res non successisset in hoc casu, existente r vel fractione vel numero negativo, tunc tentassem alterum casum purgando terminū $-az^{\frac{2}{3}}$ in ordinatim applicata a coefficiente $z^{\frac{2}{3}}$, hoc est reducendo ordinatim applicatam ad hanc formam $bz^{-\frac{4}{15}}\times\overline{-a+cz^{-\frac{2}{3}}}\rvert^{-\frac{2}{3}}$. Et si r in neutro casu fuisset numerus integer et affirmativus conclusissem Curva ex earum numero esse quæ non possunt Geometricè quadrari. Nam quantū animadverto, hæc Regula exhibet in finitis æquationibus areas omniū Geometricam quadraturā admittentiū Curvarū, quarū ordinatim applicatæ constant ex potestatibus, radicibus, vel quibuslibet dignitatibus binomij cujuscunq3.

At quando hujusmodi Curvæ aliqua non potest Geometricè quadrari sunt ad manus alia Theoremata pro comparatione ejus cum Conicis Sectionibus, vel saltem cū alijs figuris simplicissimis quibuscū potest comparari, ad quod sufficit etiam hoc ipsum unicum jam descriptum Theorema. Pro trinomijs etiam et alijs quibusdam Regulas quasdem concinnavi. Sed in simplicioribus vulgoq3 celebratis figuris vix aliquid relatu dignū reperi quod evasit aliorū conatus nisi fortè longitudo Cissoidis ejusmodi censeatur. Ea sic construitur....[12] Demonstratio perbrevis est. Sed ad infinitas series redeo.

Quamvis multa restent investiganda circa modos approximandi, et diversa serierum genera quæ possunt ad id conducere: tamen vix cum D. Tschurnhausio[13] speraverim dari posse aut simpliciora aut magis generalia funda

---

(12) We omit Newton's following rectification of the general cissoid arc (ULC. Add. 3977. 4: 3^r/3^v [= *Correspondence*, 2: 117, ll. 22–30]: 'Sit *VD* Cissois...erit sextupla altitudinis *VN*'): this is copied virtually word for word from the second paragraph of Example 5 of the 1671 tract's 'Prob 12' (see III: 320, ll. 12–24). In his reply on 11/21 June 1677 Leibniz remarked: 'Oblitus eram dicere pulchram mihi videri Cyssoidis extensionem in rectam quam Neutonus invenit, ex supposita quadratura Hyperbolæ. Ego mihi videor eodem modo etiam metiri posse curvam Hyperbolæ æquilateræ[!]; sed nondum omnis, neq3 curvam E[l]lipseos quantum memini' (*Correspondence*, 2: 218; compare also 234, note (5)).

(13) In his letter to Oldenburg on 1 September 1676 (N.S.); compare J. E. Hofmann, *Die Entwicklungsgeschichte der Leibnizschen Mathematik während des Aufenthaltes in Paris (1672–1676)* (Munich, 1949): 162. The relevant passage was printed by Newton in his *Commercium Epistolicum D. Johannis Collins et Aliorum de Analysi promota* (London, 1712): 66, from the transcript (now Royal Society MS LXXXI, No. 54, partially reproduced in *The Correspondence of Isaac Newton*, 2: 84–5) sent to him by Collins on 9 September 1676 since 'Mr Oldenburgh is gone into the Country for 10 dayes' (*ibid.*: 99).

menta reducendi quantitates ad hoc genus serierū de quo agimus quàm sunt divisiones et extractiones radicū quibus Leibnitius et ego utimur: saltem non generaliora quia pro Quadraturâ et Euthunsi[14] Curvarū ac similibus, nullæ possunt dari series ex hisce simplicibus terminis Algebraicis, unicā tantū indefinitā quantitatem involventibus constantes quas non licet hac methodo colligere. Nā non possunt esse plures hujusmodi convergentes series ad idem determinandū quàm sunt indefinitæ quantitates ex quarū potestatibus series conflentur, et ego quidem ex adhibita quacunꝗ indefinita quantitate seriem novi colligere: et idem credo Leibnitio in potestate esse. Nam quamvis mea methodo liberū sit eligere pro conflanda serie quantitatem quamlibet indefinitā a qua quæsitū dependeat, et methodus quam ipse nobis communicavit determinata videatur ad electionem talium indefinitarum quantitatū quibus opus commodè deduci potest ad fractiones quæ per solam divisionem evadant series infinitæ: tamen aliæ quæcunꝗ indefinitæ quantitates pro seriebus conflandis adhiberi possunt per methodum istam qua affectæ æquationes resolvuntur, dummodo resolvantur in proprijs terminis hoc est conficiendo seriem ex solis terminis quos æquatio involvit. Præterea non video cur dicatur[15] his divisionibus et extractionibus Problemata resolvi per accidens siquidem hæ operationes eodem modo se habeant ad hoc genus Algebræ ac vulgares operationes Arithmeticæ ad Algebram vulgo notam. Quod autem ad simplicitatem methodi attinet, nolim fractiones et radicales absꝗ præviâ reductione semper resolvi in series infinitas. Sed ubi perplexæ quantitates occurrunt, tentandæ sunt omnimodæ reductiones, sive id fiat augendo, minuendo, multiplicando, vel dividendo quantitates indefinitas sive per methodū transmutatoriam Leibnitij aut alio quocunꝗ modo qui occurrat. Et tunc resolutio in series per divisionē et extractionem optimè adhibebitur. Hîc autem præcipuè nitendum est ut Denominatores fractionū et quantitates in vinculo radicū reducantur ad quam paucissimas et minimè compositas, et ad tales etiam quæ in seriem abeant citissimè convergentē, etsi radices neꝗ convertantur in fractiones neꝗ deprimantur. Nam per regulā initio alterius Epistolæ, extractio altissimarū radicum æque simplex et facilis est ac extractio radicis quadraticæ vel divisio; et series quæ per divisionē eliciuntur solent minimè omniū convergere. Hactenus de seriebus unicā indefinitam quantitatem involventibus locutus sū: sed possunt etiam perspecta methodo series ex duabus vel pluribus assignatis definitis quantitatibus pro arbitrio confici. Quinetiam beneficio ejusdem methodi

---

(14) That is, εὐθύνσει. The letter sent to Oldenburg transliterates 'Euthunsei'.

(15) Tschirnhaus had written: 'Non obstante quod existim[e]m ad quantitatem quamvis ad infinitam Seriem æquipollentem reducend[am], fundamenta adhuc dari et simpliciora et universaliora, quam sunt Fractionum et irrationalium reductio[nes] ad tales Series, ope divisionis aut Extractionis, quæ mihi tale quid non nisi per accidens præstare videntur' (*Correspondence*, **2**: 84).

possunt series ad omnes figuras efformari Gregorianis ad Circulum et Hyperbolā editis affines,[16] hoc est quarū ultimus terminus exhibebit quæsitā aream. Sed calculū hic onerosiorem nolim lubens subire. Possunt deniɋ series ex terminis compositis eadem methodo constitui. Quemadmodū si sit $\sqrt{aa-ax+\dfrac{x^3}{a}}$ ordinatim applicata curvæ alicujus, pone $aa-ax=zz$ et ex binomio $zz\left[+\dfrac{x^3}{a}\right]^{(17)}$ extracta radice, prodibit $z+\dfrac{x^3}{2az}-\dfrac{x^6}{8aaz^3}$ &c cujus seriei omnes termini quadrari possunt per Theorema jam ante descriptū. Sed hæc minoris facio quod ubi series simplices non sunt satis tractabiles, aliam nondum communicatam methodum habeo qua pro lubitu acceditur ad quæsitū. Ejus fundamentum est commoda expedita et generalis Solutio hujus Problematis. *Parabolam*[18] *describere quæ per data quotcunɋ puncta transibit.*

Seriei $\dfrac{t}{1}\pm\dfrac{t^3}{3}+\dfrac{t^5}{5}\pm\dfrac{t^7}{7}$ &c quam D. Leibnitius ad quadraturas con[icarum] sect[ionum] usurpat, affinis est hæc, $1+\frac13-\frac15-\frac17+\frac19+\frac{1}{11}-\frac{1}{13}-\frac{1}{15}$ &c longitudinē arcus quadrantalis exprimens cujus chorda est unitas.[19] Differentia solum in signis est ubi t est unitas at inventio diversa. Illa longitudinem arcus quadrantalis ex tangente semissis, hæc ex chorda totius definit. Utinam hæ series convergerent celerius, neɋ opus esset longitudinem areamve circuli aliunde computare. Eam una cum area Hyperbolæ rectangulæ aliquando[20] sic assecuti sumus.

---

(16) A late correction (in Newton's hand) of 'persimiles' which was communicated only as an afterthought to Oldenburg on 26 October 1676 (see *Correspondence*, **2**: 162, and compare 155, note (38)). The reference is to the double 'series' of converging upper and lower bounds to the area of a given elliptical or hyperbolic sector constructed by James Gregory in his *Vera Circuli et Hyperbolæ Quadratura* (Padua, 1667[→ ₂1668]); see III: 68, note (72) and compare M. Dehn and E. Hellinger, 'On James Gregory's *Vera Quadratura*' (*James Gregory Tercentenary Memorial Volume* (London, 1939): 468–78).

(17) Both Wickins' transcript (ULC. Add. 3977.4: 4ʳ) and the letter sent to Oldenburg (*Correspondence*, **2**: 118, note (39)) have the momentary slip of the pen '$zz-x^3$'.

(18) Newton's replacement in 1684 for the vaguer phrase '*Curvam geometricam*' transmitted to Leibniz eight years earlier; see §1: note (18) above.

(19) That is, $\displaystyle\int_0^1 \dfrac{1+x^2}{1+x^4}.dx = \dfrac{1}{\sqrt2}\left[\tan^{-1}(x\sqrt2+1)\right]_{-1}^1 = \dfrac{\pi}{2\sqrt2}$. More generally,

$$t+\tfrac13 t^3-\tfrac15 t^5-\tfrac17 t^7+\tfrac19 t^9+\tfrac{1}{11}t^{11}-\ldots = \dfrac{1}{\sqrt2}\tan^{-1}\dfrac{t\sqrt2}{1-t^2}.$$

Despite Newton's broad hint, in the *epistola posterior* as sent, that exploration of this neat variant on the simple Gregorian series for $\frac14\pi$ earlier submitted by Leibniz (see §1: note (13) above) depended on reducing an integrand into partial fractions of the form $d/(e+fx+gx^2)$ where $2eg = f^2$, the latter never, it would appear, succeeded in penetrating its subtleties. Compare **1**, 4, §2: note (29).

(20) In 'Prob: 9' of the 1671 tract (ULC. Add. 3960.14: [75g/75h]); see III: 222–6).

Posito axe transverso $= 1$, et sinu verso seu segmenti sagitta $= x$. erit semi-segmentū $\left.\begin{matrix}\text{Hyperbolæ}\\\text{Circuli}\end{matrix}\right\}^{(21)} = x^{\frac{1}{2}}$ in $\overline{\frac{2}{3}x \pm \frac{xx}{5} - \frac{x^3}{28} \pm \frac{x^4}{72}}$ &c. Hæc autem series sic in infinitū producitur. Sit $2x^{\frac{3}{2}} = a.$ $\frac{ax}{2} = b.$ $\frac{bx}{4} = c.$ $\frac{3cx}{6} = d.$ $\frac{5dx}{8} = e.$ $\frac{7ex}{10} = f$ &c et erit semisegm. $\left.\begin{matrix}\text{Hyperbolæ}\\\text{Circuli}\end{matrix}\right\} = \frac{a}{3} \pm \frac{b}{5} - \frac{c}{7} \pm \frac{d}{9} - \frac{e}{11} \pm \frac{f}{13}$ &c eorumq̃ semisumma $\frac{a}{3} - \frac{c}{7} - \frac{e}{11}$ &c et semidifferentia $\frac{b}{5} + \frac{d}{9} + \frac{f}{13}$ &c. His ita præparatis suppono x esse $\frac{1}{4}$ quadrantem nempe axis et prodit $a(\frac{1}{4}) = 0.25.$ $b\left(= \frac{ax}{2} = \frac{0.25}{8}\right) = 0.03125.$ $c\left(= \frac{bx}{4} = \frac{0.03125}{2 \times 8}\right) = 0.001953125.$ $d\left(= \frac{3cx}{6} = \frac{0.001953125}{8}\right) = 0.000244140625.$ Et sic procedo usq̃ dum venero ad terminū depress[iss]imum qui potest ingredi opus. Deinde hos terminos per 3, 5, 7, 9, 11 &c respectivè divisos dispono in duas tabulas, ambiguos cū primo in unam et negativos in aliam et addo ut hic vides - - - - - - - -   - - - - - - - .(22)

Tunc a priori summa [0.0896109885646618] aufero posteriorem

[0.0002825719389575]   et restat   0.0893284166257043

area semisegmenti Hyperbolici. Addo etiam easdem summas & aggregatum aufero a primo termino duplicato

0.1666666666666666,   et restat   0.0767731061630473

area semisegmenti circularis. Huic addo triangulū istud quo completur in sectorem, h.e. triangulū $\frac{1}{32}\sqrt{3}$ seu 0.0541265877365274, & habeo sectorem sexaginta graduū 0.1308996938995747, cujus sextuplum 0.7853981633974482 est area totius circuli, quæ divisa per $\frac{1}{4}$ quadrantem diametri dat totā peripheriam 3.1415926535897928. Sic paucis horis area circuli et Hyperbolæ ad loca figurarum decimalia sexdecim colligitur quam vix anni pauci per series superiores darent.(23) Nec tamen Series Leibnitiana minoris facienda est, quæ utiq̃ varijs speculationibus ansam ministrare potest. Quinimo per seriem illam

---

(21) That is, the area $\int_0^x y \cdot dx$ under the conic $y^2 = x \pm x^2$ (a rectangular hyperbola/circle of unit *latus rectum* and main axis).

(22) This calculation, which repeats the equivalent computation (III: 224) in the 1671 tract, is here omitted.

(23) Though the simple Gregorian series is slowly convergent, Newton's estimate of the time needed to compute $\frac{1}{4}\pi$ to $16D$ by its aid seems considerably exaggerated. As Newton at once proceeds to observe, with a little juggling the general inverse-tangent series is easily manipulable into a more rapidly converging form.

si ultimo loco dimidiū termini adjiciatur,[24] et alia quædam similia artificia adhibeantur, potest computum produci ad multas figuras[;] ut et ponendo summam terminorum $1-\frac{1}{7}+\frac{1}{9}-\frac{1}{15}+\frac{1}{17}-\frac{1}{23}+\frac{1}{25}$ &c esse ad totam seriem $1-\frac{1}{3}+\frac{1}{5}-\frac{1}{7}$ &c ut $1+\sqrt{2}$ ad 2. Sed optimus ejus usus videtur esse quando vel conjungitur cū duabus alijs persimilibus et citissimè convergentibus seriebus, vel sola adhibetur ad computandū arcum 30 grad. posita tangente $\sqrt{\frac{1}{3}}$. Tunc

enim series illa evadit $\dfrac{1-\dfrac{1}{3\times 3}+\dfrac{1}{5\times 9}-\dfrac{1}{7\times 27}+\dfrac{1}{9\times 81}\ \&c}{\sqrt{3}}$, quæ citò convergit.

vel si conjunges cum alijs seriebus, pone circuli diametrum $=1$ & $a=\frac{1}{2}$ & area totius circuli erit $\frac{a}{1}-\frac{a^3}{3}+\frac{a^5}{5}-\frac{a^7}{7}+\&,\ +\frac{aa}{1}+\frac{a^5}{3}-\frac{a^8}{5}-\frac{a^{11}}{7}+\frac{a^{14}}{9}+\frac{a^{17}}{11}-\&c,$

$+\frac{a^4}{1}-\frac{a^{10}}{3}+\frac{a^{16}}{5}-\frac{a^{22}}{7}+\frac{a^{28}}{9}\ \&c.$

Credo Cl. Leibnitium vix animum advertisse ad seriem meam pro determinatione sinus versi ex arcu dato, ubi ponit seriem pro determinatione sinus complementi ex eodem arcu eamq̃ aliunde derivat. Nec tamen dubitem quin seriem illam aliasq̃ a me positas, a se verò prætermissas ipse sua dudum methodo invenerat. Quis enim methodi generalis comp[onend]æ[25] hærebit in exemplis?

Ex occasione verò quod ponit seriem $\frac{l}{1}-\frac{l^2}{1\times 2}+\frac{l^3}{1\times 2\times 3}-\frac{l^4}{1\times 2\times 3\times 4}$ &c

---

(24) The error in the Leibniz series $1-\frac{1}{3}+\frac{1}{5}-\frac{1}{7}+\ldots$ after n terms is

$$\frac{1}{2n+1}-\frac{2}{(2n+3)(2n+5)}-\frac{2}{(2n+7)(2n+9)}-\ldots=\frac{2}{(2n+1)(2n+3)}+\frac{2}{(2n+5)(2n+7)}+\ldots,$$

that is, on taking the arithmetic mean of these,

$$\frac{1}{2}\left(\frac{1}{2n+1}\right)+\frac{4}{(2n+1)(2n+3)(2n+5)}+\frac{4}{(2n+5)(2n+7)(2n+9)}+\ldots$$

$$\approx\frac{1}{2}\left(\frac{1}{2n+1}\right)+\frac{8}{(2n+1)(2n+3)(2n+5)}-\frac{48}{(2n+1)(2n+3)(2n+7)(2n+9)}.$$

For large n, therefore, the error in setting

$$\tfrac{1}{4}\pi = 1-\tfrac{1}{3}+\tfrac{1}{5}-\ldots\pm\frac{1}{2n-1}\mp\frac{1}{2}\left(\frac{1}{2n+1}\right)$$

will be of the order of n^{-3}; for $n=9$, in fact, the error is $0\cdot0014\ldots$ (see *Correspondence*, **2**: 156, note (52)). The approximation is intuitively evident since successive terms

$$\pm\frac{1}{2i-1},\quad \mp\frac{1}{2i+1},\quad \pm\frac{1}{2i+3},\quad \ldots$$

are very nearly equal and so uniformly straddle the error; compare Newton's remark in 1665 on the convergence of Wallis' infinite product for $4/\pi$ (1: 104, note (43)).

(25) Our preferred expansion of Newton's contracted form 'comp^{dæ}'; see §1: note (26) above.

tanquam a mea $\left[\dfrac{l}{1}+\dfrac{l^2}{1\times 2}+\dfrac{l^3}{1\times 2\times 3}+\dfrac{l^4}{1\times 2\times 3\times 4}\ \&c\right]$ diversam, notaverim quod in theorematîs a me positis usurpo passim species singulas pro quantitatibus sub signis suis sive + sive −, id adeo, ne signorum casus singulos enumerando Theoremata cum tædio multiplicem. Sic enim v.g. Theorema illud unicum superius quo curvas quadravi Geometricè, resolvendum esset in 32 Theoremata, præter casus 130$^{(26)}$ ubi species una vel plures deficiunt.

Præterea quæ habentur de inventione numeri unitate majoris per datū Logarithmum Hyperbolicum ope seriei $\dfrac{l}{1}-\dfrac{ll}{1\times 2}+\dfrac{l^3}{1\times 2\times 3}-\dfrac{l^4}{1\times 2\times 3\times 4}+\&c$ potius quam ope seriei $\dfrac{l}{1}+\dfrac{ll}{1\times 2}+\dfrac{l^3}{1\times 2\times 3}+\dfrac{l^4}{1\times 2\times 3\times 4}+\&c$ mihi quidem haud ita clara sunt. Nam si unus terminus adjiciatur ampliùs ad seriem posteriorem, quam ad priorem posterior magis appropinquabit. Et certè minor est labor computare unam vel duas primas figuras adjecti hujus termini quā dividere unitatem per prodeuntem logarithmū Hyperbolicū ad multa figurarū loca extensum, ut inde obtineatur Logarithmus Hyperbolicus quæsitus. Utraꝗ series igitur (si duas dicere fas sit) officio suo fungatur. Potest tamen

$$\frac{l}{1}+\frac{l^3}{1\times 2\times 3}+\frac{l^5}{1\times 2\times 3\times 4\times 5}\quad \&c$$

series ex dimidia parte terminorū constans optimè adhiberi, siquidem hæc dabit $\tfrac{1}{2}$ differentiam duorū numerorū ex qua et rectangulo dato uterꝗ datur. Sic & ex serie $1+\dfrac{ll}{1\times 2}+\dfrac{l^4}{1\times 2\times 3\times 4}\ \&c$ datur semisumma numerorum indeꝗ etiam numeri. Unde prodit relatio serierum inter se quâ ex unâ datâ dabitur altera.

Theorema de inventione arcûs ex dato cosinu, ponendo radium 1, cosinum c, & arcum $\sqrt{6-\sqrt{24c+12}}$,$^{(27)}$ minus appropinquat quàm prima fronte videtur. Posito quidem sinu verso v, error erit $\dfrac{v^3}{90}+\dfrac{v^4}{194}$ $^{(28)}$ $+\&c$. Potest fieri ut $120-27v$

---

(26) According to the sign, plus or minus, of the coefficients and indices e, f, θ, η and λ the integrand will have the following particular instances, it would appear: ($\lambda = 0$: integrand $dz^{\pm\theta}$) 2 cases only; ($\lambda \neq 0$: integrand $dz^{\pm\theta}(\pm e \pm fz^{\mp\eta})^{\pm\lambda}$) $4.2^4 = 64$, $6.2^3 = 48$ and $4.2^2 = 16$ cases arising depending on whether one, two or three of θ, η, e, f are simultaneously taken to be zero.

(27) The root a, $0 \leqslant a \leqslant 1$, of $c = 1 - \tfrac{1}{2}a^2 + \tfrac{1}{24}a^4$.

(28) So corrected by Newton from Wickins' transcribed value '$\dfrac{v^3}{140}$'. A similar last-minute change was made in the letter sent to Oldenburg though, to be sure, he added the correction again in the list of *errata* sent on 26 October (see *Correspondence*, **2**: 162, and compare 157, note (57)).

ad $120-17v$ ita chorda $(\sqrt{2v})$ ad arcum, et error erit tantū $\dfrac{61v^3\sqrt{2v}}{44800}$ circiter[29] qui semper minor est quàm $5\tfrac{1}{4}$ minuta secunda, dum arcus non sit major quàm 45^{gr}, & singulis etiam bisectionibus diminuitur 128 vicibus.

Series $\dfrac{a^3}{1\times2\times3}-\dfrac{a^5}{1\times2\times3\times4\times5}+\dfrac{a^7}{1\times2\times3\times4\times5\times6\times7}$ &c applicari posset ad computationem Tabulæ segmentorum ut observat Vir Clarissimus,[30] sed res optimè absolvitur per Canonem sinuum. Utpote cognitâ Quadrantis areâ, per continuam additionē nonæ partis ejus habebis sectores ad singulos decem gradus in semicirculo, dein per continuam additionem decimæ partis hujus habebis sectores ad gradus; et sic ad decimas partes graduum et ultra procedi potest. Tunc radio existente 1, ab unoquocꝗ sectore et ejus complemento ad 180^{gr} aufer dimidium communis sinus recti & relinquentur segmenta in Tabulam referenda.

NB. Eodem modo Leibnitius collegit seriem

$$\dfrac{a^3}{1\times2\times3}-\dfrac{a^5}{1\times2\times3\times4\times5}$$

[&c].

Quæ Cl. Leibnitius a me desiderat explicanda, ex parte descripsi.[31] Quod verò attinet ad inventionem terminorum p, q, r in extractione radicis affectæ, primum p sic eruo. Descripto angulo recto BAC, latera ejus BA, AC divido in

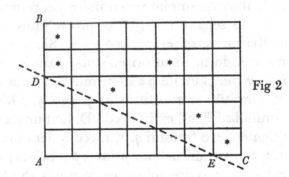

Fig 1   Fig 2

partes æquales, et inde normales erigo distribuentes angulare spatium in æqualia parallelogramma vel quadrata, quæ concipio denominata esse a

(29) To $O(v^{4\frac{1}{2}})$, namely. The corresponding error in Leibniz' computation of the arc $a\ [=\sqrt{2}(v^{\frac{1}{2}}+\tfrac{1}{12}v^{\frac{3}{2}}+\tfrac{7}{288}v^{\frac{5}{2}}\ldots)]$ from $v\ (=1-c)=\tfrac{1}{2}a^2-\tfrac{1}{24}a^4$ is $\tfrac{11}{1440}\sqrt{2}v^{\frac{9}{2}}+O(v^{\frac{11}{2}})$.

(30) Leibniz to Oldenburg, 17/27 August 1676; see *Correspondence*, **2**: 61, l. 10 ff.

(31) In his preceding letter Leibniz had written: 'Sed desideraverim ut Clarissimus Neutonus nonnulla amplius explicet, ut originem theorematis [the general binomial expansion] quod initio [epistolæ prioris] ponit, item modum, quo quantitates $p.q.r$ in suis operationibus invenit, ac deniꝗ quomodo in Methodo regressuum se gerat, ut cum ex logarithmo quærit numerum; necꝗ enim explicat, quomodo id ex methodo sua derivetur' (*Correspondence*, **2**: 62). The two following paragraphs, slightly retouched in 1684 for public view, are taken over essentially unchanged from the introduction to the 1671 tract (see III: 50–2).

dimensionibus duarum indefinitarū specierum, puta x et y, regulariter ascendentium a termino A prout vides in fig. 1 inscriptas: ubi y denotat radicem extrahendā et x alteram indefinitam quantitatem ex cujus potestatibus series constituenda est. Deinde cùm æquatio aliqua proponitur, parallelogramma singulis ejus terminis correspondentia insignio nota aliqua: et Regulâ ad duo vel fortè plura ex insignitis parallelogrammis applicatâ quorū unum sit humillimū in columna sinistrâ juxta AB, et alia ad regulam dextrorsū sita, cæteracꝫ omnia non contingentia Regulam supra eam jaceant: seligo terminos æquationis per parallelogramma contingentia Regulam designatos et inde quæro quantitatem Quotienti addendam.

Sic ad extrahendam radicem y ex $y^6 - 5xy^5 + \dfrac{x^3}{a}y^4 - 7a^2x^2y^2 + 6a^3x^3 + bbx^4 = 0$;

parallelogramma hujus terminis respondentia signo notā aliqua $*$ ut vides in fig. 2. Dein applico Regulam DE ad inferiorem e locis signatis in sinistrâ columna, eamcꝫ ab inferioribus ad superiora dextrorsū gyrare facio donec alium similiter vel fortè plura e reliquis signatis locis cœperit attingere, videocꝫ loca sic attacta esse x^3, $xxyy$ et y^6. E terminis itacꝫ $y^6 - 7aaxxyy + 6a^3x^3$ tanquam nihilo æqualibus (et insuper si placet reductis ad $v^6 - 7vv + 6 = 0$ ponendo $y = v\sqrt{ax}$) quæro valorem y et invenio quadruplicem $+\sqrt{ax}$, $-\sqrt{ax}$, $+\sqrt{2ax}$, & $-\sqrt{2ax}$, quorū quemlibet pro primo termino Quotientis accipere licet prout e radicibus quampiam extrahere decretum est. Verbi gratia, pro primo illo termino seligatur $+\sqrt{ax}$ et pro reliquis in infinitum terminis nondū cognitis scribatur p, et erit $\sqrt{ax} + p = y$. Sic æquationem $y^3 + axy + aay - x^3 - 2a^3 = 0$ resolvendo ut in priori epistola, parallelogramma dabat $-2a^3 + aay + y^3 = 0$ et inde $y = a$. Cum itacꝫ a sit primus terminus valoris y, pono p pro cæteris omnibus in infinitū, et substituo $a + p$ pro y. Obvenient hic aliquando difficultates nonnullæ,[32] sed ex ijs credo D. Leibnitius se proprio Marte extricabit. Subsequentes verò termini q, r, s, &c eodem modo ex æquationibus secundis, tertijs cæteriscꝫ eruuntur quo primus p e prima; sed cura leviori, quia cæteri termini valoris y prodire solent ex terminis duobus uno infimo in columna prima, altero infimo in columna secunda, (ut in hoc exemplo ipse p ex terminis $aax + 4aap$, ipse q ex terminis $-16axx + 4aaq$, ipse r ex terminis $-\frac{131}{128}x^3 + 4aar$,) dividendo priorem (ut aax, $-16axx$, vel $-\frac{131}{128}x^3$) per coefficientem illius p, q, r in posteriore (ut hic per $4aa$) et mutando signum Quoti.

Intellexti credo ex superioribus regressionem ab areis curvarū ad lineas rectas fieri per hanc extractionem radicis affectæ. Sed duo alij sunt modi quibus idem perficio. Eorum unus affinis est computationibus quibus colligebam approximationes sub finem alterius epistolæ,[33] et intelligi potest per hoc ex-

---

(32) Compare III: 62–6.

(33) Newton's *epistola prior* of 13 June, that is. See *Correspondence*, **2**: 25, where the series expansion of $z = r\sin^{-1}(x/r)$ is inverted to yield that of $x = r\sin(z/r)$.

emplum. Proponatur æquatio ad aream Hyperbolæ $z = x + \frac{1}{2}xx + \frac{1}{3}x^3 + \frac{1}{4}x^4 + \frac{1}{5}x^5$ &c. et partibus ejus multiplicatis in se emerget $zz = xx + x^3 + \frac{11}{12}x^4 + \frac{5}{6}x^5$ &c. $z^3 = x^3 + \frac{3}{2}x^4 + \frac{7}{4}x^5$ &c. $z^4 = x^4 + 2x^5$ &c. $z^5 = x^5$ &c. Jam de z aufer $\frac{1}{2}zz$ et restat $z - \frac{1}{2}zz = x - \frac{1}{6}x^3 - \frac{5}{24}x^4 - \frac{13}{60}x^5$ &c.[34] Huic addo $\frac{1}{6}z^3$ et fit

$$z - \frac{1}{2}zz + \frac{1}{6}z^3 = x + \frac{1}{24}x^4 + \frac{3}{40}x^5 \quad \&\text{c.}$$

Aufero $\frac{1}{24}z^4$ et restat $z - \frac{1}{2}zz + \frac{1}{6}z^3 - \frac{1}{24}z^4 = x - \frac{1}{120}x^5$ &c. Addo $\frac{1}{120}z^5$ et fit $z - \frac{1}{2}zz + \frac{1}{6}z^3 - \frac{1}{24}z^4 + \frac{1}{120}z^5 = x$ quamproximè, sive $x = z - \frac{1}{2}zz + \frac{1}{6}z^3 - \frac{1}{24}z^4 + \frac{1}{120}z^5$ &c.

Eodem modo series de una indefinita quantitate in aliam transferri possunt. Quemadmodum si posito r radio circuli, x sinu recto arcus z, et

$$x + \frac{x^3}{6rr} + \frac{3x^5}{40r^4} + \&\text{c}^{[35]}$$

longitudine arcus istius, atꝗ hanc seriem e sinu recto ad Tangentem vellē transferre: quæro longitudinem Tangentis[36] $\dfrac{rx}{\sqrt{rr-xx}}$ & reduco in infinitam seriem $x + \dfrac{x^3}{2rr} + \dfrac{3x^5}{8r^4} + \&$c: Qua dicta t, colligo potestates ejus $t^3 = x^3 + \dfrac{3x^5}{2rr} + \&$c. $t^5 = x^5 + \&$c. Aufero jam t de z et restat [posito $r=1$] $z - t = -\dfrac{x^3}{3} - \dfrac{3x^5}{10} - \&$c.

Addo $\frac{1}{3}t^3$ et fit $z - t + \frac{1}{3}t^3 = \frac{1}{5}x^5 + \&$c. Aufero $\frac{1}{5}t^5$ et restat $z - t + \frac{1}{3}t^3 - \frac{1}{5}t^5 = 0$ quamproxime. Quare est $z = t - \frac{1}{3}t^3 + \frac{1}{5}t^5 - \&$c. Sed siquis in usus Trigonometricos me jussisset exhibere expressionem arcus per Tangentem, eam non hoc circuitu sed directa methodo[37] quæsivissem.

Per hoc genus computi colliguntur etiam series ex duabus vel pluribus indefinitis quantitatibus constantes: et radices affectarum æquationū magna ex parte extrahuntur, sed ad hunc posteriorem usum adhibeo potiùs mcthodū in alterâ epistola descriptam tanquam generaliorem, &, (Regulis pro elisione superfluorū terminorum adhibitis,) paulo magis expeditam. Pro regressione

---

(34) The coefficient of the last term was originally transcribed by Wickins—evidently a slip of his copyist's pen—as '$-\frac{16}{60}$' and then the '6' in the numerator was altered by Newton into '3', but only after he himself had copied the erroneous coefficient into the autograph fair copy he sent to Oldenburg on 24 October (compare *Correspondence*, **2**: 159, note (69)). Due correction was made in the version subsequently transmitted by Oldenburg to Leibniz (reproduced by C. I. Gerhardt in *Der Briefwechsel von Gottfried Wilhelm Leibniz mit Mathematikern*, **1** (Berlin, 1899): 203–25, especially 222).

(35) That is, $r\sin^{-1}(x/r) = z$.

(36) Namely, $r\tan(z/r)$.

(37) Evidently by Leibniz' own technique of setting $z = \tan^{-1}t = \int_0^t 1/(1+t^2)\,.\,dt$. Compare §1: note (13) above.

verò ab areis ad lineas rectas et similibus, possunt hujusmodi Theoremata adhiberi.

THEOREM. 1. Sit $z = ay + byy + cy^3 + dy^4 + ey^5$ &c et vicissim erit

$$y = \frac{z}{a} - \frac{bzz}{a^3} + \frac{2bb - ac}{a^5}z^3 + \frac{5abc - 5b^3 - aad}{a^7}z^4$$

$$+ \frac{3a^2c^2 - 21abbc + 6aabd + 14b^4 - a^3e}{a^9}z^5 + \&c.$$

Ex. gr. Proponatur æquatio ad aream Hyperbolæ $z = y - \frac{yy}{2} + \frac{y^3}{3} - \frac{y^4}{4} + \frac{y^5}{5}$ &c et substitutis in Regula 1 pro a, $-\frac{1}{2}$ pro b, $\frac{1}{3}$ pro c, $-\frac{1}{4}$ pro d, et $\frac{1}{5}$ pro e, vicissim exurget $y = z + \frac{1}{2}zz + \frac{1}{6}z^3 + \frac{1}{24}z^4 + \&c.$

THEOREM. 2. Sit $z = ay + by^3 + cy^5 + dy^7 + ey^9 + \&c$, et vicissim erit

$$y = \frac{z}{a} - \frac{bz^3}{a^4} + \frac{3bb - ac}{a^7}z^5 + \frac{8abc - aad - 12b^3}{a^{10}}z^7$$

$$+ \frac{55b^4 - 55abbc + 10aabd + 5aacc - a^3e}{a^{13}}z^9 + \&c.$$

Ex. gr. Proponatur æquatio ad arcum circuli $z = y + \frac{y^3}{6rr} + \frac{3y^5}{40r^4} + \frac{5y^7}{112r^6} + \&c.$ Et substitutis in Regula 1 pro a, $\frac{1}{6rr}$ pro b, $\frac{3}{40r^4}$ pro c, $\frac{5}{112r^6}$ pro d, &c; orietur

$$y = z - \frac{z^3}{6rr} + \frac{z^5}{120r^4} - \frac{z^7}{5040r^6} + \&c.$$ Alterū modum regrediendi ab areis ad lineas rectas celare statui.

Ubi dixi omnia pene Problemata solubilia[38] existere, volui de ijs præsertim intelligi circa quæ Mathematici se hactenus occuparunt vel saltem in quibus ratiocinia mathematica locum aliq[u]em obtinere possunt. Nam alia sanè adeo perplexis conditionibus implicata excogitare liceat ut non satis comprehendere valeamus et multò minùs tantarum computationū onus sustinere quod ista requirerent. Attamen ne nimium dixisse videar, inversa de Tangentibus

---

(38) In his following letter to Oldenburg two days later Newton requested that this word be changed to read 'solutilia', apologizing for allowing this careless 'scape' (*Correspondence*, 2: 162). Though the latter form occurs classically (in Suetonius' *Vita Neronis*: §34), the former variant is found widely in such post-classical authors as Ammianus and Prudentius, so that the blemish—if any—is of minuscule importance.

(39) See §1: note (33) above. The two methods to which Newton here obliquely refers are Cases 1 and 2 of 'Prob 2' of his 1671 tract (III: 90–4/94–112), which deal respectively with the first-order fluxion which is reducible to integrable form by extracting the root of an algebraic equation, and with the more general type in which both fluent variables and their fluxions are connected in a complicated manner (and for whose solution Newton recommends an expansion as an infinite series). Observe that Newton is still unwilling to publish the unscrambled form of his 1676 anagram in which the two cases are briefly distinguished.

Problemata sunt in potestate aliaꝗ illis difficiliora ad quæ solvenda usus sum duplici methodo, unâ concinniori, alterâ generaliori.[39] Sed et absꝗ his methodis solvitur inversum illud de tangentibus problema quando tangens inter punctum contactus et axem figuræ est datæ longitudinis, quo casu prodit curva quædam Mechanica cujus determinatio pendet ab area Hyperbolæ Apollonianæ.[40] Et ejusdem generis est etiam Problema quando pars axis inter tangentem et ordinatim applicatam datur longitudine.[41] Sed hos casus vix numeraverim inter ludos naturæ:[42] nam quando in triangulo rectangulo quod ab illa axis parte et tangente ac ordinatim applicata constituitur, relatio duorū quorumlibet laterum per æquationem quamlibet definitur, Problema solvi potest absꝗ mea methodo generali; sed ubi pars axis ad punctū aliquod positione datū terminata ingreditur vinculum tunc res aliter se habere solet.

Communicatio Resolutionis affectarū æquationū per methodum Leibnitij pergrata erit, juxta et explicatio quomodo se gerat ubi indices potestatum sunt fractiones ut in hac æquatione $20 + x^{\frac{2}{3}} - x^{\frac{2}{3}}y^{\frac{2}{3}} - y^{\frac{7}{11}} = 0$, aut surdæ quantitates ut in hac $\overline{x^{\sqrt{2}} + x^{\sqrt{7}}}\,|^{\sqrt{}\textcircled{3}\frac{2}{3}} = y$, ubi $\sqrt{2}$ et $\sqrt{7}$ non designant coefficientes ipsius x sed indices potestatū seu dignitatum ejus et $\sqrt{\textcircled{3}\frac{2}{3}}$ indicem dignitatis binomij $x^{\sqrt{2}} + x^{\sqrt{7}}$. Res credo mea methodo patet, aliter descripsissem. Sed meta tandem prolixæ huic Epistolæ ponenda est. Literæ[43] sanè excellentissimi Leibnitij valde dignæ erant quibus fusiùs hocce responsum darem.

---

(40) This curve of constant tangent-length is, of course, the tractrix; see III: 26, note (31).

(41) The 'curva logarithmica'; see I: 376, note (47).

(42) John Collins, in the copy of Leibniz' letter of 17/27 August 1676 to Oldenburg which he made for Newton (ULC. Add. 3971.1: 34ʳ–36ᵛ; see §1: note (13) above), had inadvertently (*ibid.*: 36ᵛ [= *Correspondence*, **2**: 64, l. 10]) mistranscribed 'ludus' (a sport!) for the neutral article 'hujus' in Leibniz' remark regarding the 'logarithmica' that 'In Tomo IIIᵗⁱᵒ Epistolarum [ed. Clerselier, Paris, 1667] una habetur ad Bcaunium [No. LXXI = Descartes to de Beaune, 20 February 1639 (N.S.)], in qua propositas à Beaunio curvas quasdam invenire conatur, quarum una est hujus naturæ, ut intervallum inter tangentem, ad directricem (axem) usꝗ productam, et ordinatim applicatam ex curva ad directricem, sit semper idem, recta scilicet constans. Hanc curvam nec Cartesius nec Beaunius, nec quisquam alius quod sciam invenit; Ego verò qua primum die, imò hora cœpi quærere, statim certa analysi solvi'. (A manuscript of Leibniz' dated 'Jul. 1676', published by Gerhardt in his *Briefwechsel* (note (34)): 201–3 reveals that he had been less successful in his mastery of de Beaune's problem: in particular, he had not then succeeded in effecting Descartes' reduction (see III: 84, note (109)) of the defining condition $dx/dy = a/(y-x)$ to the particular solution $x/a = \log[(x-y+a)/a]$, the Cartesian equation of a logarithmic curve through the origin and having $y = x+a$ for its asymptote.) Newton ought to have seen that Leibniz' 'ludus' was merely the errant prank of a copyist's pen, but even if the remark had been made by Leibniz it cried out for a more playful rejoinder than his present over-solemn aside.

(43) Leibniz' preceding letter of 17/27 August, that is.

APPENDIX 2.
THE 'MATHESEOS UNIVERSALIS SPECIMINA': VARIANT DRAFTS AND CANCELLED PASSAGES.[1]

From the originals in the University Library, Cambridge

[1][2]

[Esto seriei] *a. b. c. d. e. f.* [terminus quilibet] $\dfrac{g+hn+in^2}{k+ln+mn^2}$.[3] [Fit]

$$g-2h+4i=ak-2al+4am. \quad \text{[adeoq$\mathfrak{z}$]}$$
$$g\ -h\ +i=bk\ -bl\ +bm.$$
$$g\qquad\quad=ck.$$
$$g\ +h\ +i=dk\ +dl\ +dm.$$
$$g+2h+4i=ek+2el+4em.$$

$$2h=\begin{matrix}d\\-b\end{matrix}k\begin{matrix}+d\\+b\end{matrix}l\begin{matrix}+d\\-b\end{matrix}m. \qquad 0=\begin{matrix}a\\-2b\\+2d\\-e\end{matrix}k\begin{matrix}-2a\\+2b\\+2d\\-2e\end{matrix}l\begin{matrix}+4a\\-2b\\+2d\\-4e\end{matrix}m.$$

$$4h=\begin{matrix}e\\-a\end{matrix}k\begin{matrix}+2e\\+2a\end{matrix}l\begin{matrix}+4e\\-4a\end{matrix}m.$$

$$2g+2i=\begin{matrix}b\\+d\end{matrix}k\begin{matrix}+d\\-b\end{matrix}l\begin{matrix}+d\\+b\end{matrix}m. \qquad 0=\begin{matrix}4b\\+4d\\-a\\-e\\-6c\end{matrix}k\begin{matrix}+4d\\-4b\\+2a\\-2e\end{matrix}l\begin{matrix}-4a\\+4b\\+4d\\-4e\end{matrix}m.$$

$$2g+8i=\begin{matrix}a\\+e\end{matrix}k\begin{matrix}+2e\\-2a\end{matrix}l\begin{matrix}+4a\\+4e\end{matrix}m.$$

[Hoc est ponendo] $a-2b+2d-e=p$. $a-b-d+e=q$. $a-4b+[6]c-4d+e=r$. $2a-b+d-2e=s$. [erit] $pk-2ql+2sm=0=-rk+2pl-2qm$.[4] [adeoq$\mathfrak{z}$]

$$pqk-2qql-rsk+2psl=0. \quad \text{[seu]} \quad \dfrac{pq-2rs}{qq-2ps}k=l.$$

---

(1) Edited versions of Newton's preliminary schemes and drafts for Chapter 2 of the 'Specimina' (§1 above) together with a first version (unfinished) of Chapter 3 and a cancelled passage from its revise.

(2) ULC. Add. 3964.3: 10ᵛ. In these rough computations (which we have, for comprehension's sake, slightly filled out with additions—largely synsemantic—in square brackets) Newton seeks to identify the general term of the sequence $\{f_n\}$ by taking f_n to be of various fractional algebraic forms. The simplest case in which $f_n = (p+qn+rn^2)/(s+tn+vn^2)$ is incorporated in the first draft of Chapter 2 following (Appendix 2.2), but then abandoned. In his second draft of Chapter 2 (Appendix 2.3), Newton introduces the modified assumption that the ratio of two successive terms in the sequence (that is, f_{n+1}/f_n) may be represented by an algebraic fraction $(p+qn+rn^2 \ldots)/(s+tn+vn^2 \ldots)$, and with little change this revised approach was incorporated into the opening of the final text of Chapter 2.

[Cape] $\dfrac{q}{s}=t$.   [fit] $\dfrac{p-2rt}{q-2pt}k=l$. $\dfrac{2ql-pk}{2s}=m$. $\dfrac{d-b}{2}k+\dfrac{d+b}{2}l+\dfrac{d-b}{2}m=h$.
$dk+dl+dm-g-h=i$.

[Terminus quilibet] $\dfrac{p+rn+tn^2+vn^3}{q+sn+tn^2+vn^3}$.   [Fit]

$$p-2r+4t-8v=aq-2as+4at-8av.$$
$$p\ -r\ +t\ -v=bq\ -bs\ +bt\ -bv.$$
$$p\qquad\quad=cq.$$
$$p\ +r\ +t\ +v=dq\ +ds\ +dt\ +dv.$$
$$p+2r+4t+8v=eq+2es+4et+8ev.$$

$$\underset{-2c}{\underset{+d^q-b^s+b^t-b}{b+d\ _s+d\ _t+d}}\,v=0.\qquad \underset{-2c}{\underset{+a^q-2a^s+4a^t-8a}{e+2e\ _s+4e\ _t+8e}}\,v=0.$$
$$\qquad\qquad -2 \qquad\qquad\qquad -8$$

$$\underset{-2}{\underset{-b^q+b^s-b^t+b}{d+d\ _s+d\ _t+d}}\,v=2r.\qquad \underset{-16}{\underset{-a^q+2a^s-4a^t+8a}{e+2e\ _s+4e\ _t+8e}}\,v=4r.$$

$$\underset{-3\ -1}{\underset{-b^q+b^s-b^t+b}{a-2a\ _s+4a\ _t-8a}}\,v=0.\ ^{(5)}\qquad \underset{+3\ -1}{\underset{-e^q-2e^s-4e^t-8e}{d+d\ _s+d\ _t+d}}\,v=0.\ ^{(5)}$$

[Terminus] $\dfrac{p+rn+sn^2}{q+rn+sn^2}$.   [Fit]$^{(6)}$ $p-r+s=aq-ar+as$.
$$p\qquad=bq.$$
$$p+r+s=cq+cr+cs.$$

$$\underset{-2}{\underset{-a^q+a^r-a}{c+c\ _r+c}}\,s=0.\qquad \underset{+a\quad-2}{\underset{-2b^q-a^r+a}{c+c\ _r+c}}\,s=0.$$

---

(3) In sequel, tacitly supposing that $a=f_{-2}$, $b=f_{-1}$, $c=f_0$, $d=f_1$ and $e=f_2$, Newton equates $(g+hn+in^2)=f_0(k+ln+mn^2)$, $n=-2,\ -1,\ 0,\ 1,\ 2$, and then systematically determines the ratios—alone significant—of g, h, i, k, l, m to each other.

(4) The last term should be '$-4qm$'. In compensation the sequel should read

'$2pqk-4qql-rsk+2psl=0$.   seu $\dfrac{2pq-rs}{4qq-2ps}k=1$. ...   fit $\dfrac{2p-rt}{4q-2pt}k=l$. ...'.

(5) Read '$-r-8v$' in each case.

(6) Where $a=f_{-1}$, $b=f_0$, $c=f_1$ Newton supposes the general term f_n to be

$$(p+rn+sn^2)/(q+rn+sn^2).$$

Since only the ratios of p, q, r and s are significant, three equations

$$p+r+sn^2=f_n(q+rn+sn^2),\quad n=-1,\,0,\,1,\quad\text{suffice.}$$

[Pone] $a-c=g$. $a+c-2=h$. $a-2b+c=k$. $\dfrac{h}{g}=l$. [Erit] $-gq+hr-gs=0$.

$kq-gr+hs=0$. [adeoꝗ] $\dfrac{-hg\ +hh}{+kg\ -gg}q-r=0$. $\dfrac{k-h}{1+l\times1-l}q=r$. [$l$]$r-q=s$. [Rursus]

$\dfrac{a\ -a\ +a}{-b\ +1\ -1}q\ r\ s=0$. $\dfrac{c\ +c\ +c}{-b\ -1\ -1}q\ r\ s=0$. [adeoꝗ] $\dfrac{a-1}{a-b}q=r-s$. $\dfrac{b-c}{c-1}q=r+s$.

[ubi] $bq=p$.

[Terminus] $\dfrac{p+qn+rn^2+sn^3+tn^4+n^6}{v\ \ \ \ \ \ \ \ \ \ \ \ +n^6}$. [7]

[Esto terminus quilibet] $\dfrac{p+qn+n^m}{r+n^m}$. [Fit] $\dfrac{p}{r}=b$. $\dfrac{p+q+1}{r+1}=c$. $\dfrac{p-q+1}{r+1}=a$.[8]

[adeoꝗ] $2p+2=\dfrac{a\ +a}{+c\ +c}r=2br+2$. [vel] $\dfrac{2-a-c}{a-2b+c}=r$, si m par. Sin m impar est

$\dfrac{p-q+1}{r-1}=a$. $2p=\dfrac{c}{+a}r+c-a=2br$. [seu] $\dfrac{a-c}{a-2b+c}=r$. Casu primo

$$\frac{2-2b}{a-2b+c}=r+1 \quad \text{et} \quad \frac{2b-2c}{a-2b+c}=r-1.$$

[Aliter][9] $\dfrac{p}{r}=a$. $\dfrac{p+q+1}{r+1}=b$. $\dfrac{p+2q+2^m}{r+2^m}=c$. [adeoꝗ]

$$rc+c\times2^m-2rb-2b=2^m-ra-2.$$

[hoc est] $r=\dfrac{2b-2+\overline{1-c}\times2^m}{a-2b+c}$.

---

(7) Two further terms '$+tn$' and '$+n^4$' were entered separately in the denominator of this fraction and then cancelled. No attempt is made in the manuscript to apply this general term to any sequence.

(8) Understanding 'si m par', a phrase set rather isolatedly by Newton at the end of his next line. Here three terms (taken to be $a=f_{-1}$, $b=f_0$, $c=f_1$) of the given sequence are needed to determine p, q and r in absolute magnitude, since the coefficient of n^m is set as unity.

(9) The slight variant which results on setting $a=f_0$, $b=f_1$, $c=f_2$. In the light of the simple series developments considered elsewhere in the 'Matheseos Universalis Specimina' examination of the sequence whose general term is $f_n=(p+qn+n^m)/(r+n^m)$ seems elaborately artificial.

(10) ULC. Add. 3964.3: 9^r, a first draft of Chapter 2 of the 'Specimina'. Newton develops a method, similar in structure to that used in Appendix 2.1 preceding, for determining the general term $f_n=(p+qn+rn^2\ldots)/(s+tn+vn^2\ldots)$ of any sequence $a=f_{-1}$, $b=f_0$, $c=f_1$, $d=f_2$, ... which is representable in this way.

(11) Newton has cancelled the equivalent phrase 'terminum unumquemꝗ per subsequentem divide[ndo]'.

[2]⁽¹⁰⁾   Cᴀᴘ [2]

De serierum progressione.

Inventa progressionis lege continuatur series utrinꝗ. Eam sic eruo. Esto series Z. A. B. C. D. E. F. G. V &c ubi terminus Z inveniendus sit, cæteri dentur. Ex terminorum rationibus⁽¹¹⁾ compone novam seriem $\dfrac{Z}{A} \cdot \dfrac{A}{B} \cdot \dfrac{B}{C} \cdot \dfrac{C}{D}$ &c seu $\dfrac{Z}{A}$. a. b. c. d. e. f [&c]. Finge horum terminorum valorem communē esse $\dfrac{p+qn+rn^2 \text{ \&c}}{s+tn+vn^2 \text{ \&c}}$ ea lege ut p, q, r, s, t, v sint datæ quantitates, n vero significet numerum terminorum numeratorū a tertio b, id est -2. -1. 0. 1. 2. 3. 4 &c successivè perinde ut valor iste successivè scribatur pro $\dfrac{Z}{A}$. a. b. c. d. e. f &c.⁽¹²⁾

Pendent quantitates r et v a quantitate ultimorum terminorū seriei $\dfrac{Z}{A}$. a. b. c. d. e. f &c in infinitum productæ. Si termini illi infinitè decrescunt debet vn^2 deleri et r poni $=[1]$; si infinitè decrescunt in duplicata circiter ratione numeri terminorum numeratorum ab c⁽¹³⁾ debet etiam tn deleri; si infinite augentur debet rn^2 deleri et v poni $=1$: si in infinitum decrescunt in duplicata circiter ratione numeri terminorum, debet etiam qn deleri; si convergunt ad unitatem⁽¹⁴⁾ debet r poni $=1=v$: si convergunt ad aliam quāvis quantitatem pone illam $=v$ et $1=r$. Verbi gratia si series esset $\dfrac{Z}{A} \cdot \dfrac{x}{e} \cdot \dfrac{x}{3e} \cdot \dfrac{x}{5e} \cdot \dfrac{x}{7e} \cdot \dfrac{x}{9e}$ &c delerem v et ponerem $r=1$. Si ea esset $\dfrac{Z}{A} \cdot \dfrac{x}{1,3e} \cdot \dfrac{x}{3,5e} \cdot \dfrac{x}{5,7e} \cdot \dfrac{x}{7,9e}$ &c delerem quoꝗ tn. Si $\dfrac{Z}{A} \cdot \dfrac{x}{e} \cdot \dfrac{3x}{e} \cdot \dfrac{5x}{e} \cdot \dfrac{7x}{e}$ &c delerem r et ponerem $v=1$. Si $\dfrac{Z}{A} \cdot \dfrac{1,3x}{e} \cdot \dfrac{3,5x}{e} \cdot \dfrac{5,7x}{e}$ &c delerem quoꝗ qn. Si deniꝗ $\dfrac{Z}{A} \cdot \dfrac{x}{4e} \cdot \dfrac{3x}{6e} \cdot \dfrac{5x}{8e} \cdot \dfrac{7x}{10e}$ &c ponerem $v=\dfrac{x}{e}$ et $r=1$.⁽¹⁵⁾ Inventis hoc modo r et v

---

(12) Henceforth, in fact, Newton will restrict his attention to the sequence $a = f_{-1}$, $b = f_0$, $c = f_1$, ... whose general 'progressio' (denoting the $(n+2)$th ratio in the series A/B, B/C, C/D, ... derived from the primary sequence A, B, C, D, ...) is
$$f_n = (p+qn+rn^2)/(s+tn+vn^2).$$

(13) Read 'b'. In a first version of this paragraph Newton counted n 'a quarto termino c' in the sequence Z, a, b, c, ...

(14) 'si perpetuo appropinquant ad æqualitatem' was first written.

(15) Newton has cancelled in sequel 'Nam coefficientes $\frac{1}{4}$. $\frac{3}{6}$. $\frac{5}{8}$. $\frac{7}{10}$ vergunt ad æqualitatem atꝗ adeo termini seriei vergunt ad quantitatem $\frac{x}{e}$'. The 'progressiones' in the preceding series are, in his preferred terms, $\dfrac{\frac{1}{2}x/e}{\frac{3}{2}+n}$, $\dfrac{\frac{1}{4}x/e}{\frac{3}{4}+2n+n^2}$, $\dfrac{\frac{3}{2}+n}{\frac{1}{2}e/x}$, $\dfrac{\frac{3}{4}+2n+n^2}{\frac{1}{4}e/x}$ and $\dfrac{(\frac{3}{2}+n)\,x/e}{3+n}$ respectively.

cæteræ quantitates p, q, s ac t (si forte q vel t non desit) invenientur conferendo terminorū a, b, c, d &c aliquos cum valore suo $\frac{p+qn+rn^2}{s+tn+vn^2}$, id est ponendo $\frac{p-q+r}{s-t+v}=a$. $\frac{p}{s}=b$. $\frac{p+q+r}{s+t+v}=c$. $\frac{p+2q+4r}{s+2t+4v}=d$. &c. Ex quibus æquationibus debitè reductis prodit regula talis. Pone $a-3b+3c-d=g$. $a-3c+2d=h$. $a+3c-4d=i$. $\frac{a-2b+c}{g}=k$. $a-c[+]hk=l$. $a+c-ik=m$. et erit $\left[\frac{mv-2r}{l}=t.\right.$ $\frac{2r+\overline{a-c}\times t-\overline{a+c}\times v}{gk}=s$. $bs=p$. $p+r-\overline{s-t+v}\times a=q.\Big]$ Hæc ita sunt ubi neutrum q vel t deest. Sin t desit fac $\left[-\frac{i}{g}v=s. \ bs=p.\right.$ et $p+r-\overline{s-t+v}\times a=q.\Big]$ Sin q deest pone $\left[\frac{mv-2r}{l}=t.\right.$ $\frac{2r+\overline{a-c}\times t-\overline{a+c}\times v}{gk}=s$. $bs=p.\Big]^{(16)}$

Cap 2.

De politione$^{(18)}$ serierum.

Inventarum serierum cultura consistit in inveniendis progressionibus, terminorum omnium aggregatis, et intercalationibus terminorum. Progressionis legem sic eruo. Esto series A. B. C. D. E. F. G. H &c cujus terminus quilibet consequens produ[c]atur multiplicando præcedentem per

$$\frac{p+qn+rn^2 \ \&c}{s+tn+vn^2 \ \&c}$$

ubi p, q, r, s, t, v &c significant datas quantitates et n numerum terminorum numeratorum a tertio $\left[\frac{D}{C}\text{ in serie composita } \frac{B}{A}\cdot\frac{C}{B}\cdot\frac{D}{C}\cdot\frac{E}{D}\cdot\frac{F}{E} \ \&c\right]$, id est numeros -2. -1. 0. 1. 2. 3 &c successive perinde ut valor iste scribatur pro $\frac{B}{A}\cdot\frac{C}{B}\cdot\frac{D}{C}$ &c.

---

(16) We have filled in some gaps in the manuscript text. Newton first wrote: 'Pone $b-2c+d=g$. $b-d=h$. $b+d=i$. $\frac{a-2b+c}{g}=[-]k$. $2a-2b+hk=l$. $ik+4a-2b=m$. $2+2k=o$, et erit $\frac{mv-or}{l}=t$. $\frac{ht-iv+2r}{g}=s$. $cs=p$. $p+r[-]\overline{s-t+v}\times b=q$. ... Sin t desit fac

$$\frac{2bv-4av+2r}{a-2b+c}=s. \quad cs=p. \quad \text{et} \quad p+r:[-]\overline{s-t+v}\times[b]=q.'$$

(17) ULC. Add. 3964.3: 9^{v}–10^{v}, an immediate revise of the preceding augmented by a discussion of summation considered as the operation inverse to that of taking (finite) differences. Extensive cancelled variants which show the continuity of this draft with that in Appendix 2.2 above are not noted.

(18) This replaces 'PROGRESSIONE ET TERMINATIONE'.

Sic erit $\dfrac{B}{A}=\dfrac{p-2q+4r}{s-2t+4v}$. $\dfrac{C}{B}=\dfrac{p-q+r}{s-t+v}$. $\dfrac{D}{C}=\dfrac{p}{s}$. $\dfrac{E}{D}=\dfrac{p+q+r}{s+t+v}$. $\dfrac{F}{E}=\dfrac{p+2q+4r}{s+2t+4v}$. Ex his æquationibus debitè collatis et reductis invenio quantitates q, r, t, v positis p numeratore & s denominatore fractionis $\dfrac{D}{C}$ ad minimos terminos reductæ. Dein tento si seriei primæ terminus quilibet posterior prodeat multiplicando terminum priorem per fractionem $\dfrac{p+qn+rn^2}{s+tn+vn^2}$, id est si fuerit $\dfrac{p+3q+9r}{s+3t+9v}F=G$. $\dfrac{p+4q+16r}{s+4t+16v}G=H$ & sic in infinitum. Hoc pacto progressiones inveni quas posui in epistola mea priori, ut et sequentes.

His quilibet plures suo marte facile adjunget.

Usus harum progressionum est ut series expeditè producamus ad computandum aggregatū terminorum. At in tardè convergentibus labor iste computandi crescit in immensum, et præterea in seriebus nimis compositis difficile est regulā progressionis invenire.[19] His igitur in casibus requiritur ex paucis seriei terminis initialibus invenire terminorum omnium in infinitum pergentium aggregatum. Nisi hoc fiat, inutiles erunt series quam plurimæ. Esto igitur series quælibet A. B. C. D. E. F &c et sunto terminorū differentiæ primæ b. ^2b. ^3b &c. secundæ c. ^2c. ^3c &c. tertiæ d. ^2d. ^3d &c. quartæ e. ^2e. ^3e &c et sic in infinitum: id est ita ut sit $A-B=b$. $B-C=^2b$. $C-D=^3b$. [&c]. $b-^2b=c$. $^2b-^3b=^2c$. &c. $c-^2c=d$ &c subductis semper sequentibus de præcedentibus et signis residuorum notatis. Manifestum est quod A sit aggregatū[20] omnium b, ^2b, ^3b, ^4b &c in infinitum, b aggregatum omnium c, ^2c, ^3c, ^4c &[c], c omnium d, ^2d, ^3d &c, d omnium e, ^2e, ^3e &c; atꝗ adeo quod si series A. b. c. d. e. f &c producatur versus A præfixo uno termino Z, erit ille Z aggregatum omnium A, B, C, D, E, F &c.[21] Præfigetur autem ille Z[22] inveniendo progressionem seriei ut supra si forte exactè inveniri potest, sin minus pergendum erit ut sequitur.

$$
\begin{array}{llllll}
 & & & & Z & \\
Y. & A. & B. & C. & D. & E. & F \ \&c \\
 & b. & ^2b. & ^3b. & ^4b. & ^5b \\
 & c. & ^2c. & ^3c. & ^4c. \\
 & d. & ^2d. & ^3d \\
 & e. & ^2e \\
 & f
\end{array}
$$

(19) Given any finite number of members of a sequence $\{f_n\}$, no unique algebraic expression for the general term f_n can be found, but Newton understands that the 'simplest' representation $f_n = (p+qn+rn^2 \ldots)/(s+tn+vn^2 \ldots)$ is to be chosen.

(20) 'summa differentiarum' is cancelled.

(21) Compare §1: note (67) above.

(22) Newton first concluded his paragraph with the words 'inveniendo ut supra progressionis legem $\dfrac{p+qn+rn^2}{s+tn+vn^2}$ et ponendo $\dfrac{s-3t+9v}{p-3q+9r}A=Z$. Si lex illa exacta est habebitur Z exactè,

Seriei terminum unumquemꝗ divide per terminum subsequentem, constituendo seriem novam $\frac{Z}{A} \cdot \frac{A}{b} \cdot \frac{b}{c} \cdot \frac{c}{d} \cdot \frac{d}{e}$ &c. Hoc opus repete semel vel sæpius.

Nimirum si termini ulteriores $\frac{c}{d} \cdot \frac{d}{e}$ &c appropinquant unitatem, repete semel constituendo seriem $\frac{Zb}{A^2} \cdot \frac{Ac}{b^2} \cdot \frac{bd}{c^2} \cdot \frac{ce}{d^2}$ &c. Sin illi non appropinquant unitatem sed seriei novissimæ termini ulteriores $\frac{bd}{c^2} \cdot \frac{ce}{d^2}$ &c appropinquant unitatem[,] repete adhuc semel constituendo seriem $\frac{Zb^3}{A^3c} \cdot \frac{Ac^3}{b^3d} \cdot \frac{bd^3}{c^3e} \cdot \frac{ce^3}{d^3f}$ &c. Et si termini ulteriores $\frac{bd}{c^2} \cdot \frac{ce}{d^2}$ &c non appropinquant unitatem repete adhuc tertiò constituendo seriem $\frac{Zb^6d}{A^4c^4} \cdot \frac{Ac^6e}{b^4d^4} \cdot \frac{bd^6f}{c^4e^4}$. &c. Tandem ubi veneris ad seriem cujus termini ulteriores appropinquant unitatem & opere repetito seriem novam collegeris esto series illa nova $\frac{Z}{F}$. *G. H. I. K. L* &c. Fac $\frac{G-H}{G-1}=p \cdot \frac{H-I}{I-1}=q. \frac{p+q}{2}=r. \frac{-p+q}{2}=s.$ et erit $\frac{H-2r+4s}{1-2r+4s}=\frac{Z}{F}$ quamproximè.[23] Sed quantus est error? In quantitate $\frac{H+rn+sn^2}{1+rn+sn^2}$ scribendo successive $-2. -1. 0. 1. 2. 3$ &c collige seriem novam $\frac{H-2r+4s}{1-2r+4s} \cdot$ *G. H. I.* $\frac{H+2r+4s}{1+2r+4s} \cdot \frac{H+3r+9s}{1+3r+9s}$ &c. Hujus terminos subduc a terminis seriei superioris $\frac{Z}{F}$. *G. H. I. K. L* &c. Residua, quæ sunto

---

sin minùs quamproximè. At malui sic pergere'. In the margin alongside at this point he made a baffling late addition, ordering that 'Ex terminorum differentijs auctis unitate componatur nova series $Z-A+1. A-b+1. b-c+1. c-d+1$' &c—that is,

'$Z-A+1. B+1. {}^2b+1. {}^2c+1$ &c'

since, as he went on to note, 'Differentiæ primæ in schemate annexo sunt $Z-A. B. {}^2b. {}^2c. {}^2d.$ &c'. Is this done to minimize the latter 'series errorum'?

(23) This replaces the more general phrase 'et valorem progressionis constitue $\frac{H+rn+sn^2}{1+rn+sn^2}$ quam proxime ponendo n numerum terminorum numeratorum a secundo H'. Newton's statement is easily justified since, on setting $G=f_{-1}$, $H=f_0$, $I=f_1$, the 'progressio' $f_n = (H+rn+sn^2)/(1+rn+sn^2)$ must satisfy

$$G = (H-r+s)/(1-r+s) \quad \text{and} \quad I = (H+r+s)/(1+r+s),$$

whence $r-s = (G-H)/(G-1) = p$ and $r+s = (H-I)/(I-1) = q$.

(24) '$T.V$' replaces the more exact expression '$K-\frac{H+2r+4s}{1+2r+4s} \cdot L-\frac{H+3r+9s}{1+3r+9s}$' which Newton originally wrote.

$\dfrac{Z}{F} - \dfrac{H-2r+4s}{1-2r+4s}$. 0. 0. 0. T. V.[24] W &c erunt series errorum. Et error primus

T ex una parte[25] erit ad errorem primum $\dfrac{Z}{F} - \dfrac{H-2r+4s}{1-2r+4s}$ ex altera parte[25] ut

$\dfrac{H+2r+4s}{H-2r-4s}$ quadratum ad $\dfrac{1-2r+4s}{1+2r+4s}$[26] quadratum circiter, sed afficietur signo

contrario.

Si error nimius est, opus totum sic repete. Addatur unitati series errorum ad

componendam novam seriem quæ sit $\dfrac{Z}{F} + \dfrac{1-H}{1-2r+4s}$.[27] 1. 1. 1. $T+1$. $V+1$.

$W+1$. Hujus seriei $\dfrac{Z}{F} + \alpha$. 1. 1. 1. β. γ. δ &c terminos antecedentes divide per

consequentes. Id fac iterum ac tertiò. Prodeat series $\dfrac{Z}{F} + \alpha \cdot \dfrac{1}{\beta} \cdot \dfrac{\beta^3}{\gamma} \cdot \dfrac{\gamma^3}{\beta^3\delta}$ &c.[28] Hujus

quære terminum primum $\dfrac{Z}{F} + \alpha$ eadem ratione qua prius. Et sic operis totius

perpetua repetitione pergitur in infinitum. Sed prima vel secunda vice solet

terminus ille primus exactè satis prodire.

---

[4][29]

CAP 3.

QUÀM LATÈ PATET SERIERUM CONVERGENTIUM METHODUS.

Beneficio harum serierum solvuntur Arithmeticè[30] problemata vulgaria

quarum reductio ad æquationes vulgares permolesta est; computa peraguntur

in angulis æque ac in lineis; curvarum areæ longitudines centra gravitatum et

---

(25) The more graphic 'ad dextram...ad sinistram' was first written.

(26) Newton intends the reciprocal $\dfrac{\text{'}1+2r+4s\text{'}}{1-2r+4s}$. The remark suggests that he here assumes

the sequence $\{F_n\}$, $Z/F = F_{-2}$, $G = F_{-1}$, $H = F_n$, $I = F_1$, $K = F_2$, ... to approximate the

sequence $\{f_n\}$, $f_n = (H+rn+sn^2)/(1+rn+sn^2)$ parabolically; that is, such that

$$F_n = f_n + \epsilon \cdot f_n^2 + O(\epsilon^2), \quad \epsilon \text{ small.}$$

It follows that $Z/F - f_{-2} \approx \epsilon \cdot f_{-2}^2$ and T (or $K - f_2$) $\approx \epsilon \cdot f_2^2$, whence

$$T : (Z/F - f_{-2}) \approx -f_2^2 : f_{-2}^2.$$

(27) That is, $-(H-2r+4s)/(1-2r+4s) + 1$.

(28) Newton first continued 'seu $\dfrac{Z}{F} + [\alpha] \cdot g \cdot h \cdot i \cdot k$ &c. Fac ut supra $\dfrac{g-h}{g-1} = p$. $\dfrac{h-i}{i-1} = q$.

$\dfrac{p+q}{2} = r$. $\dfrac{q-p}{2} = s$. et [erit] $\dfrac{h-2r+[4]s}{1-2r+[4]s} = \dfrac{Z}{F} + [\alpha \text{ quamproxime}]$'.

(29) ULC. Add. 3964.3: 17^r, a first opening to Chapter 3 of the 'Specimina'; see §1:
note (88) above.

(30) Invoking, that is, only the elementary arithmetical operations of addition, subtraction,
multiplication and division 'sine resolutione æquationum' (as a cancelled sequel has it).

similia pro arbitrio computantur; vicissim ex areis longitudinibus centris gravitatum et tangentibus regreditur ad curvas, et id genus alia multa peraguntur quæ omnia satis explicare volumen[31] requireret nos pauca festinanter attingimus.

Computi vulgariū problematū sine resolutione affectarum æquationum accipe hoc specimen. *Inter rectas AD, DB positione datas ponenda est recta AB quæ convergat ad datum punctum P.*[32] A puncto illo P ad *CD* age tum *PC* parallelam *DB* tum *PF* perpendicularem *DC* sitcz *PC*=2, *DC*=3, *AB*=4, *CF*=$\frac{1}{2}$ et quæratur *CA*. Esto *CE* quant[it]as quævis data quam putes ignotæ *AC* proximè æqualem esse, puta 1,[33] et nominetur *EA*=x, ita ut quæsita *CA*=1+x. Tum $PC^q + CA^q + 2ACF$ id [est] $6+3x+xx$ valet PA^q et extracta radice est

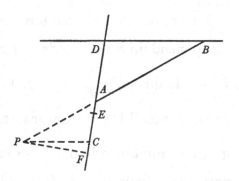

$$PA = \sqrt{6} + \frac{3}{2\sqrt{6}}x + \frac{5}{16\sqrt{6}}xx - \frac{5}{128\sqrt{6}}x^3 \ \&c.^{(34)}$$

[5][35]

Quinetiam Problemata circa curvas jam vulgo dictas Mechanicas eadem methodo solvuntur. Ad ejusmodi curvas inveniendæ sunt æquationes convergentes, dein cætera peragenda ut in curvis simplicioribus. Ut si Spiralis[36]

---

(31) Newton has his 1671 tract squarely in mind at this point; see in particular his preliminary scheme for that treatise (III: 28–30).

(32) A classical Apollonian neusis earlier solved geometrically by Newton through the intersection of a circle and hyperbola (see **2**, 1, §2.3: Problem 1 above).

(33) Since the triangles *ABD*, *APC* are similar, at once $DA \times AP = BA \times AC$ and so $(2-x)\sqrt{[6+3x+x^2]} = 4(1+x)$, whence $x^4 - x^3 - 18x^2 - 44x + 8 = 0$ whose least positive real root is $x \approx \frac{1}{6}$. In the first instance Newton doubtless set *CA* (= z, say) as his unknown, so determining $(3-z)\sqrt{[4+z+z^2]} = 4z$ or $z^4 - 5z^3 - 9z^2 - 13z + 36 = 0$ and isolating the root $z \approx 1\frac{1}{6}$.

(34) The text breaks off at this point. In sequel Newton still has to solve

$$\sqrt{6}(2-x)(1+\tfrac{1}{4}x+\tfrac{5}{96}x^2 \ldots) = 4(1+x)$$

'sine resolutione æquationis'. This may evidently be done by multiplying out the left side and then again extracting the square root, when $\sqrt[4]{\tfrac{3}{2}}(1-\tfrac{1}{8}x-\tfrac{17}{384}x^2 \ldots) = 1+\tfrac{1}{2}x-\tfrac{1}{8}x^2 \ldots$ and hence $x \approx (\sqrt[4]{\tfrac{3}{2}}-1)/\tfrac{1}{8}(\sqrt[4]{\tfrac{3}{2}}+4) \approx 0\cdot17$. This, however, is immeasurably more tedious than extracting this value as the least positive root of the quartic $x^4 - x^3 - 18x^2 - 44x + 8 = 0$ (see previous note) and no doubt explains why Newton here abandoned his intended exemplification of the superiority of expansions into infinite series over 'vulgar' algebraic analysis by solving equations.

(35) ULC. Add. 3964.3: 17^v, a second cancelled attempted illustration of the thesis of Chapter 3 that expansions into infinite series are widely superior to conventional techniques of analysis. See § 1: note (105) above.

in Quæstione esset, cujus partes circa DE contemplari vellem: a centro ejus C duco rectam quamvis CA secantem spiralem in A in quo detur punctum quodvis A et a spiralis puncto quovis D ad hanc rectam demitti conciperem ordinatam DB normaliter. Tum dicto $AC=a$, $AB=x$, $BD=y$, erit $BC=a-x$ et

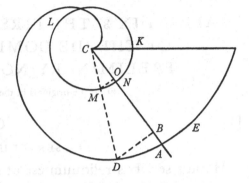

$$CD=\sqrt{aa-2ax+xx-yy}\,;^{(37)}$$

atcɣ hæc est longitudo arcus KLM circuli illius ad quem spiralis aptatur et ablata de longitudine CA relinquit arcum $MN^{(38)}=a-\sqrt{aa-2ax+xx-yy}$. Si arcus ille dicatur z & circuli radius r erit sinus $MO=z-\dfrac{z^3}{6rr}+\dfrac{z^5}{120r^4}$ &c. Est autem MO ad CM ut DB ad CD hoc est $z-\dfrac{z^3}{6rr}+\dfrac{z^5}{120r^4}$ &c ad r ut y ad $\sqrt{aa-2ax+xx-yy}$. Et inde ductis extremis et medijs in se, restituto valore z extracta radice ex $aa-2ax+xx-yy^{(39)}$ ordinatis omnibus et extracta ultimò radice y, habebitur æquatio convergens$^{(40)}$ qua definitur relatio inter [AB et BD].$^{(41)}$

---

(36) Understand 'Spiralis Archimedæa'. In his further argument Newton assumes this to be the particular 'helix' defined by $CA = \overset{\frown}{KLN}$.

(37) Read '$\sqrt{aa-2ax+xx+yy}$': the trivial error of mistaking '$+yy$' for '$-yy$' is carried through into the sequel but does not, of course, distort the general argument. Newton has deleted a following parenthesis '$seu\ a-x-\dfrac{yy}{2a}-\dfrac{xyy}{2aa}-\dfrac{y^4}{8a^3}$ &c' in which the erroneous root is extracted.

(38) For (see note (36) above) it is assumed that $CA = \overset{\frown}{KLN}$, $CD = \overset{\frown}{KLM}$.

(39) Read '$aa-2ax+xx+yy$' correctly; see note (37) above.

(40) 'ad spira[lem]' is deleted.

(41) In this intrinsic coordinate system (which has no precedent, Newtonian or otherwise), since $MO = r\sin(z/r)$ where $z = a-\sqrt{[(a-x)^2+y^2]}$, the exact 'relatio' which refers the general point $A(x, y)$ to origin $D(0, 0)$ will be $a-\sqrt{[(a-x)^2+y^2]} = r\tan[y/(a-x)]$, where $AB = x \perp BD = y$ and $C(a, 0)$ is the spiral's pole. In general an infinite number of values of AB or BD will correspond to any given length of the other coordinate (BD or AB respectively) but the series expansion dictated by Newton, convergent only for $|x| < a$ and $y^2 < (a-x)^2$, is effective only for the principal value of $M\overset{\frown}{C}O(= z/r+2k\pi)$ which obeys $|M\overset{\frown}{C}O| < \frac{1}{4}\pi$. In his projected application of these 'helical' coordinates Newton would no doubt keep x and y small enough to ensure fairly rapid convergence, but we may guess that the unwanted elimination of the multiple solutions possible which expansion into series of the spiral's defining equation imposed was the reason which led him to cancel this interesting intended 'mechanical' exemplification of the superiority of analysis by infinite series over conventional algebraic techniques throughout a wide area of mathematics.

APPENDIX 3. THE FIRST VERSION OF CHAPTER 2 OF THE 'DE COMPUTO SERIERUM' AND PRELIMINARY NOTES FOR ITS REVISE.[1]

From the originals in the University Library, Cambridge[2]

[1]

CAP 2.

DE SERIERUM QUANTITATE.[3]

I
De serierū
valoribus.

Habitis seriebus reliquum est ut sciamus inde quantitates computare quas exprimunt.[4] In celeriter convergentibus sufficit quantitatem primorum aliquot terminorum colligere, sed in tardè convergentibus labor ille crescit in immensum. Nunc centeni termini primi, nunc milleni, nunc milleni millies et amplius[5] non appropinquabunt satis, nunc verò series divergunt sic ut quo longius pergas eo magis erres.[6] His incommodis remediū aliquod afferre conatus [sum] per serierum transmutationes ut sequitur.

---

(1) See §2: note (38) above. In these preliminary drafts and worksheet computations Newton (in [1]) approximates the sum of a (uniformly convergent) 'series regularis'

$$ax^n \pm bx^{2n} + cx^{3n} \pm \ldots + d(\pm x^n)^k + \ldots$$

by means of a generalized geometrical progression

$$(ax^n + \beta x^{2n} + \gamma x^{3n} + \ldots + \delta x^{kn})(1 \pm kx^n + \tfrac{1}{2}k(k+1)x^{2n} \pm \ldots)$$
$$= (ax^n + \beta x^{2n} + \gamma x^{3n} + \ldots + \delta x^{kn})/(1 \mp x^n)^k,$$

subsequently (in [2]) achieving the insight that the latter 'approximatio' may be set as

$$ax^n/(1 \pm x^n) \mp (a-b)x^{2n}/(1 \mp x^n)^2 + (a-2b+c)x^{3n}/(1 \mp x^n)^3 \mp \ldots$$
$$+ (\mp 1)^k(a - kb + \tfrac{1}{2}k(k-1)c - \ldots + (\mp 1)^k d)x^{kn}/(1 \mp x^n)^k.$$

In sequel (in [3]) Newton concocts half a dozen simple examples of 'transmuting' a given series $ax^n \pm bx^{2n} + cx^{3n} \pm \ldots$ by means of the substitution $z = x^n/(1 \mp x^n)$ into the (hopefully more convergent) equal $az \mp (a-b)z^2 + (a-2b+c)z^3 \mp \ldots$, and finally (in [4]) composes a first draft of the revised Chapter 2 for the 'De computo'.

(2) ULC. Add. 3964.3: 11ʳ–12ᵛ.

(3) 'TERMINATIONE' was first written.

(4) This replaces the phrase 'ad quas convergunt', evidently unnecessarily restrictive in Newton's eyes. It is, of course, an *idée fixe* with him that a divergent series may also 'express' a unique quantity; compare §2: note (48) above.

(5) Understand 'serierum valores'; 'nedū longe plures' was first written.

(6) Newton first continued: 'Danda est regula qua serierum tardè convergentium valores etiam in his casibus expeditè ['celeriter' is cancelled] assequamur'.

(7) 'ad certum terminorum numerum producta' is cancelled. In sequel he first wrote '...ubi dignitates quantitatis indefinitæ *x* progressione arithmetica assurgant, sitꝗ *n* indicum illorū differentia'.

(8) Whence $z = 1 \mp x^n$.

Proponatur series quælibet regularis[7]

$$x^m \times a + bx^n + cx^{2n} + dx^{3n} + ex^{4n} + fx^{5n} + gx^{6n} + hx^{7n} \ \&c$$

ubi *a*, *b*, *c* &c designent coefficientes dignitatum indefinitæ quantitatis *x* cum signis suis $+$ et $-$ [affectis] et *n* differentiam indicum dignitatum. In se ducatur aliquoties $1 - x^n$ si termini omnes affirmativi sunt vel omnes negativi: sin alternatim affirmativi et negativi sint ducatur in se $1 + x^n$. Ducatur autem $1 - x^n$ vel $1 + x^n$ in se cubicè si regulam ad tres ultimos seriei terminos applicare velis, quadrato-quadraticè si ad quatuor ultimos, quadrato-cubicè si ad quinǫ et sic in infinitum. Aggregatum verò duorum primorum terminorum prodeuntium cum signo suo dicatur *p*, trium primorum *q*, quatuor *r*, quinǫ *s*, sex *t* &c et omniū *z*.[8] Deinde constituatur fractio cujus denominator sit *z*, numerator verò componatur ex ultimo seriei termino $+ p$ in penultimum $+ q$ in antepenultimum $+ r$ in tertium ab ultimo et ita deinceps, per omnes *p*, *q*, *r*, *s*, *t* &c præter *z* perǫ totidem seriei terminos pergendo: et hæc fractio una cum reliquis seriei terminis initialibus appropinquabit valorem seriei.[9]

II
Regula
computandi
valores
serierum tarde
convergentiū.

---

(9) The result is immediate since, to $O(x^{kn})$, *k* even,

$$(a \pm bx^n + cx^{2n} \pm \ldots \pm fx^{(k-3)n} + gx^{(k-2)n} \pm hx^{(k-1)n} + ix^{kn} \pm \ldots)z$$
$$= a\alpha \pm b\beta x^n + c\gamma x^{2n} \pm \ldots \pm fqx^{(k-3)n} + gpx^{(k-2)n} \pm hx^{(k-1)n}$$

with an error of $+\Delta^k a \cdot x^{kn} \pm \Delta^k b \cdot x^{(k+1)n} + \Delta^k c \cdot x^{(k+2)n} \pm \ldots$, where $\alpha, \beta, \gamma, \ldots, q, p$ are the first $k, k-1, k-2, \ldots, 3, 2$ terms of the expansion

$$z = (1 \mp x^n)^k = 1 \mp kx^n + \tfrac{1}{2}k(k-1)x^{2n} \mp \ldots + (\mp 1)^k x^{kn}$$

and the Δ^k are the successive *k*-th order finite-differences of the coefficients $a, b, c, \ldots, f, g, h, i, \ldots$ (namely

$$\Delta^k a = a - kb + \tfrac{1}{2}k(k-1)c - \ldots + (-1)^k i, \quad \Delta^k b = b - kc + \tfrac{1}{2}k(k-1)d \ldots,$$

$\Delta^k c = c - kd + \tfrac{1}{2}k(k-1)e \ldots$, and so on). Similarly, when *k* is odd

$$(a \pm bx^n + cx^{2n} \pm \ldots + fx^{(k-3)n} \pm gx^{(k-2)n} + hx^{(k-1)n} \pm ix^{kn} \ldots)z$$
$$= a\alpha \pm b\beta x^n + c\gamma x^{2n} \pm \ldots + fqx^{(k-3)n} \pm gpx^{(k-2)n} + hx^{(k-1)n}$$

with an error of $\mp \Delta^k a \cdot x^{kn} - \Delta^k b \cdot x^{(k+1)n} \mp \Delta^k c \cdot x^{(k+2)n} - \ldots$. The simplest case $k = 1$ yields the 'approximatio' a/z, that is, $a/(1 \mp x^n)$ to the series $a \pm bx^n + \ldots$, but here it would be better to set $a/(1 \mp (b/a)x^n)$ as the approximation. The latter had, in fact, been communicated by James Gregory to the London mathematical computer Michael Dary on 9 April 1672 as 'An approach to the summe of the infinite Series whereby the Logarithmes are made. Let the penultimate tearme be $= a$, the last tearme be $= b$. The sum of all the tearmes before the penultimate togeather with $\dfrac{a^2}{a-b}$ is $=$ to the aggregate of the whole infinite series' (from Collins' contemporary copy in Shirburn Castle 101. H. 2: 135^r, first published by S. P. Rigaud in his *Correspondence of Scientific Men of the Seventeenth Century*, **2** (Oxford, 1841): 240): in Gregory's example the series expansion of $\tfrac{1}{2}\log[(c+n)/(c-n)]$, namely $n/c + \tfrac{1}{3}n^3/c^3 + \tfrac{1}{5}n^5/c^5 + \ '\&c$ in infinitum' is readily shown to be a little 'more' than the first three terms increased by

$$\tfrac{1}{7}(n^7/c^1)(1 - \tfrac{7}{9}n^2/c^2)^{-1} = 9n^7/(63c^7 - 49c^5n^2).$$

Newton may well have been made aware of Gregory's observation about August 1676; if so

Ut si regulam ad terminos quinqȝ ultimos $dx^{3n}+ex^{4n}+fx^{5n}+gx^{6n}+hx^{7n}$ applicari velis sintqȝ termini omnes affirmativi, ducendo $1-x^n$ in se quadrato-cubicè, prodibit $1-5x^n+10x^{2n}-10x^{3n}+5x^{4n}-x^{5n}$. et inde fiet $1-5x^n=p$. $p+10x^{2n}=q$. $q-10x^{3n}=r$. $r+5x^{4n}=s$. $s-x^{5n\,(10)}=z$. Sin termini d, e, f, g, h alternatim affirmativi et negativi sint, ducendo $1+x^n$ in se quadrato-cubicè prodibit $1+5x^n+10x^{2n}+10x^{3n}+5x^{4n}+x^{5n}$. atqȝ adeo $1+5x^n=p$, $p+10x^{2n}=q$, $q+10x^{3n}=r$, $r+5x^{4n}=s$ & $s+x^{5n\,(10)}=z$. Deinde in utroqȝ casu (signis probe observatis) fiet x^m in $a+bx^n+cx^{2n}+\dfrac{sdx^{3n}+rex^{4n}+qfx^{5n}+pgx^{6n}+hx^{7n}}{z}$ approximatio seriei quam invenire oportuit.[11]

Exemplum demus in Hyperbola. Sit ea *EDG*, ejus Asymptoti *CB*, *CH* rectum angulū ad *C* capientes, axis *CEF*[,] vertex principalis *E*, ordinatæ ad axem *F*[*D*], applicatæ normaliter ad Asymptoton *EA*, *DB*, *db*. Dictis $AB=x$. $BD=y$. $EF=z$. $FD=v$. $CA=AE=a$, latere recto r et transverso q:[12]

erit area $ABDE=ax-\dfrac{x^2}{2}+\dfrac{x^3}{3a}-\dfrac{x^4}{4a^2}+\dfrac{x^5}{5a^3}$ &c

et area $E[D]F=r^{\frac12}[z]^{\frac32}$ in $\frac23$[13]

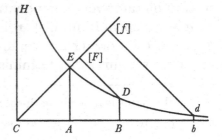

---

we may readily believe that it played some small part in leading Newton to his present more general technique of approximation. (John Collins included a *verbatim* repeat of this passage in the 'Extracts from Mr Gregories Letters to be lent Monsr Leibnitz to peruse' (Royal Society MS LXXXI, No. 46: 16; see note (30) of the concluding appendix to this volume) which he compiled at Oldenburg's request in May 1676. On 5 September 1676 Newton returned to Collins certain 'manuscript papers' which he had recently been lent, in particular 'your paper about Mr Gregory [in which] I have presumed to race out two things...: the last...in my generall method mentioned in your 4th section' (*Correspondence*, 2: 95). This 'paper about Mr Gregory' cannot have been the English original (Royal Society MS LXXXI, No. 45) of the 'Abridgement' of his '*Historiola*' which Collins sent to Oldenburg on 14 June, for the latter has no such '4th section', but it may have been a shortened version of his Gregory 'Extracts'.)

(10) That is, $(1-x^n)^5$ and $(1+x^n)^5$ respectively.

(11) In the terminology of note (9), since $k=5$ is odd,

$$d\pm ex^n+fx^{2n}\pm gx^{3n}+hx^{4n}=(d\alpha\pm e\beta x^n+f\gamma x^{2n}\pm g\delta x^{3n}+hx^{4n})/z$$

on taking α, β, γ, δ to be the first 5, 4, 3, 2 terms of the expansion of $z=(1\mp x^n)^5$, with a negative error of $\pm\Delta^5 d\,.\,x^{5n}+\Delta^5 e\,.\,x^{6n}\pm\Delta^7 f\,.\,x^{7n}+\dots$ where

$$\Delta^5 d=d-5e+10f-10g+5h\dots, \quad \Delta^5 e=e-5f+10g\dots,$$

and so on.

(12) Accordingly $y=a^2/(a+x)$ and, since $q=r=2\sqrt{2}a$, $v^2=rz+z^2$.

(13) Understand '$r^{\frac12}z^{\frac32}$ in $\dfrac{2}{3}+\dfrac{z}{5r}-\dfrac{z^2}{28r^2}-\dfrac{z^3}{72r^3}$ &c' (compare III: 292, l. 7). In general

$$(ABDE)=\int_0^x y\,.\,dx=a^2\log(1+x/a)$$

Dicatur jam inventa approximatio A et seriei termini d, e, f, g &c qui fractionem ingressi sunt[14] cæteriœ in infinitum ducantur gradatim in terminos præfatæ potestatis $\mp x^{5n}+5x^{4n}\mp 10x^{3n}+10x^{2n}\mp 5x^{n}+1$ et producta dividas per z ponendo[15]

$$\frac{\mp d+5e\mp 10f+10g\mp 5h+i}{z}\,x^{8n}=B.$$

$$\frac{\mp e+5f\mp 10g+10h\mp 5i+k}{z}\,x^{9n}=C.$$

$$\frac{\mp f+5g\mp 10h+10i\mp 5k+l}{z}\,x^{10n}=D. \quad \text{\&c}$$

Et ex his inventis terminis $A+B+C+D$ &c (signis probe observatis) componetur nova series quæ celerius converget[16] et ejusdem erit valoris cum priore.

Quod si eosdem seriei terminos d, e, f, g &c multiplices per terminos dignitatis $1\mp x^{n}$ uno gradu altioris, id est per hosce terminos

$$x^{6n}\mp 6x^{5n}+15x^{4n}\mp 20x^{3n}+15x^{2n}\mp 6x^{n}+1,$$

et per omnium aggregatum[17] (quod jam dicatur z) producta dividas, ponendo[18]

$$\frac{d\mp 6e+15f\mp 20g+15h\mp 6i+k}{z}\,[x^{9n}]=B^{2}.$$

$$\frac{e\mp 6f+15g\mp 20h+15i\mp 6k+l}{z}\,[x^{10n}]=C^{2}. \quad \text{\&c}$$

---

and $\qquad (EDF) = \displaystyle\int_{0}^{z} v\,.\,dz = \tfrac{1}{2}v(z+\tfrac{1}{2}r)-\tfrac{1}{8}r^{2}\log\left(1+2(v+z)/r\right).$

For x and z not small Newton's series are very slow to converge (and are of course divergent when $x>a$, $z>r$) and it was evidently his intention to apply his preceding approximation to obtain a rough approach to truth: indeed, the case of the former in which $x=a=1$ and so $(ABDE) = \log 2$ is taken up as his concluding 'Exemplum' in §V below. (The coefficients of the latter series $\dfrac{2}{2i+1}\dbinom{\frac{1}{2}}{i}$, $i=0,1,2,3,\ldots$ do not have conspicuously small k-th order finite differences Δ^{k} and hence the preceding 'approximatio' cannot usefully there be applied.)

(14) Newton first wrote 'ad quibus regula applicata fuit'. In an extension of his preceding 'approximatio' (see note (11) above) $A = (d\alpha \pm e\beta x^{n}+f\gamma x^{2n}\pm g\delta x^{3n}+hx^{4n})\,x^{3n}/(1\mp x^{n})^{5}$ to the terms $dx^{3n}\pm ex^{4n}+fx^{5n}\pm gx^{6n}+hx^{7n}\pm ix^{8n}+kx^{9n}\pm lx^{10n}+\ldots$ (where α, β, γ, δ are, as before, the first 5, 4, 3, 2 terms of the expansion of $(1\mp x^{n})^{5} = z$), he now refines its accuracy by successively incorporating the terms $B = \mp\Delta^{5}d\,.\,x^{8n}/z$, $C = -\Delta^{5}e\,.\,x^{9n}/z$, $D = \mp\Delta^{5}f\,.\,x^{10}/z$, $\ldots$, where the Δ^{5} are the fifth differences $\Delta^{5}d = d-5e+10f-10g+5h-i$, and so on.

(15) For all signs ' $+$ ' in the following read ' $\pm$ '.

(16) This will be true only if the coefficients $\mp\Delta^{5}d$, $-\Delta^{5}e$, $\mp\Delta^{5}f$, $\ldots$ decrease (on average) more rapidly to zero than those ($\pm i$, $+k$, $\pm l$, $\ldots$) of the original series.

(17) Namely $(1\mp x^{n})^{6}$.

(18) Here again (compare note (15) above) in place of all signs ' $+$ ' (including those understood to be attached to the initial coefficients d, e, $\ldots$) read ' $\pm$ '.

et pro $A+B^{(19)}$ scrib[endo] A^2 fiet $A^2+B^2+C^2$ &c series adhuc celerius convergens.[20]

Et simili ratione ponendo $A^2+B^{2(19)}=A^3$ et ex dignitate ipsius $1 \mp x^n$ uno gradu altiore colligendo B^3, C^3, D^3 [&c] obtinebitur series.[21]

Sic possunt innumeræ celeriùs convergentes series[22] ex una tardius convergente derivari. Sed sufficit eorum terminus primus A.

Cæterum anteq̄[23] calculus ineatur sciendus est numerus terminorum seriei primæ ex quibus approximatio A exactè satis prodeat. Si computare velis fractionē præfatā ex terminis d, e, f & sequentibus, collige $\dfrac{\mp 3dx^{3n}+ex^{4n}\,^{(24)}}{1 \mp 2x^n+x^{2n}}=\alpha,$

$$c.\ d.\ \frac{d^2}{c}\cdot\frac{d^3}{cc}\cdot\frac{d^4}{c^3}\cdot\frac{d^5}{c^4}.\ \&\mathrm{c}.$$

ut et progressiones geometricas $\quad \alpha.\ \alpha^2.\ \alpha^3.\ \alpha^4.\ \alpha^5.\ \alpha^6.\ \&\mathrm{c}.\quad$ In progressione inferiore quære numerum α^2, α^3, α^4, α^5 vel α^6 &c qui proximè minor est quam error tollerandus approximationis inveniendæ A. Et seriem produc usq̃ ad terminū illum cujus coefficiens proximè minor est quam terminus progressionis alterius capiti hujus numeri impositus.[25] Ut si numerus α^5 proxime minor sit quam error tollerandus et coefficiens g minor quam $\dfrac{d^4}{c^3}$, ad computū suffecerit series usq̃ ad g producta.[26]

---

(19) These should read '$A+\dfrac{B}{1 \mp x^n}$' and '$A^2+\dfrac{B^2}{1 \mp x^n}$' respectively.

(20) Here $A^2 = (d\alpha' \pm e\beta'x^n+f\gamma'x^{2n} \pm g\delta'x^{3n}+h\epsilon'x^{4n} \pm ix^{5n})\,x^{3n}/z$ on taking α', β', γ', $\delta'\ \epsilon'$ to be the first 6, 5, 4, 3, 2 terms of the expansion of $z = (1 \mp x^n)^6$, while

$$B^2 = +\Delta^6 d.x^{9n}/z, \quad C^2 = \pm\Delta^6 e.x^{10n}/z, \quad \ldots$$

where $\Delta^6 d\ [= d-6e+15f-20g+15h-6i+k]$, $\Delta^6 e$, ... are the successive 6-th order finite differences of the coefficients of $dx^{3n} \pm ex^{4n}+fx^{5n} \pm gx^{6n}+hx^{7n} \pm ix^{8n}+kx^{9n} \pm lx^{10n}+\ldots.$ At once, since $\alpha' = (1 \mp x^n)\alpha \mp x^{5n}$, $\beta' = (1 \mp x^n)\beta+5x^{4n}$, $\gamma' = (1 \mp x^n)\gamma \mp 10x^{3n}$, $\delta' = (1 \mp x^n)\delta+10x^{2n}$ and $\epsilon' = (1 \mp x^n) \mp 5x^n$, there follows

$$d\alpha' \pm e\beta'x^n+f\gamma'x^{2n} \ldots \pm ix^{5n} = (1 \mp x^n)\,(d\alpha \pm e\beta x^n+f\gamma x^{2n} \ldots +hx^{4n}) \mp \Delta^5 a.x^{5n}$$

and hence, on multiplying through by $x^{3n}/(1 \mp x^n)^6$, $A^2 = A+B/(1 \mp x^n)$. Likewise, since $\Delta^6 d = \Delta^5 d-\Delta^5 e$, there results $B^2 = (\mp Bx^n+C)/(1 \mp x^n)$ and similarly $C^2 = (\mp Cx^n+D)/(1 \mp x^n)$ and so on. —

(21) On taking α'', β'', γ'', δ'', ϵ'', ζ'' to be the first 7, 6, 5, 4, 3, 2 terms of the expansion of $z = (1 \mp x^n)^7$, since

$$\alpha'' = (1 \mp x^n)\alpha'+x^{6n}, \beta'' = (1 \mp x^n)\beta' \mp 6x^{5n}, \gamma'' = (1 \mp x^n)\gamma'+15x^{4n}, \ldots, \zeta'' = (1 \mp x^n) \mp 6x^n,$$

there follows $A^3 = A^2+B^2/(1 \mp x^n)$ where

$$A^3 = (d\alpha'' \pm e\beta''x^n+f\gamma''x^{2n} \pm g\delta''x^{3n}+h\epsilon''x^{4n}+i\zeta''x^{5n}+kx^{6n})\,x^{3n}/z.$$

Likewise $B^3 = \mp\Delta^7 d.x^{10n}/z = (\mp B^2 x^n+C^2)/(1 \mp x^n)$ and so on.

(22) This replaces 'Sic potest infinitus serierum celeriter convergentiū numerus'.

(23) Read 'antequam'.

Sic in serie ad aream Hyperbolæ $x - \frac{x^2}{2} + \frac{x^3}{3} - \frac{x^4}{4}$ &c, si duodecim terminos V
Exemplum.

primos ad computum adhiberi velis et regulam ad ultimos septem applicari, emerget appropinquatio

$$x - \frac{x^2}{2} + \frac{x^3}{3} - \frac{x^4}{4} + \frac{x^5}{5} + \frac{-\frac{vx^6}{6} + \frac{1}{7}tx^7 - \frac{1}{8}sx^8 + \frac{1}{9}rx^9 - \frac{1}{10}qx^{10} + \frac{1}{11}px^{11} - \frac{1}{12}x^{12}}{z} \quad \&c$$

ubi p significat duos primos terminos potestatis[27]

$$1 + 7x + 21x^2 + 35x^3 + 35x^4 + 21x^5 + 7x^6 + x^7,$$

q tres primos & sic deinceps. Pone $x = 1$ et approximatio illa fiet

$$1 - \tfrac{1}{2} + \tfrac{1}{3} - \tfrac{1}{4} + \tfrac{1}{5} - \frac{+\frac{127}{6} - \frac{120}{7} + \frac{99}{8} - \frac{64}{9} + \frac{29}{10} - \frac{8}{11} + \frac{1}{12}}{128},$$

id est in fractis decimalibus

$$0.783333333333 - 0.090185617334 = 0.69314771599.\text{[28]}$$

---

(24) '$\frac{\mp 3dx^n + \epsilon}{1 \mp 2x^n + x^{2n}}$' was first written, followed by '$\frac{\mp dx^{3n}}{1 \mp 2x^n + x^{2n}}$'. All three expressions are seemingly illogical. Perhaps Newton meant to write '$\frac{d \mp 2d \mp e \times x^n}{1 \mp 2x^n + x^{2n}} = \alpha c$'? On this supposition the series $a + bx^n + cx^{2n} + dx^{3n} + ex^{4n} + fx^{5n} \pm \dots$ could be replaced by its equal

$$a + bx^n + cx^{2n} + c\alpha x^{3n} + (1 \mp x^n)^{-2}(\Delta^2 d \cdot x^{4n} \pm \Delta^2 e \cdot x^{5n} + \dots).$$

It would then be reasonable in the case of Newton's simple 'series regulariter progredientes' to compare the rate of convergence of the two by means of the 'approximating' geometrical series
$$\left.\begin{cases} a + bx^n + cx^{2n} + dx^{3n} + (d^2/c)\,x^{4n} + (d^3/c^2)\,x^{5n} + \dots \\ a + bx^n + c(x^{2n} + \alpha x^{3n} + \alpha^2 x^{4n} + \alpha^3 x^{5n} + \dots) \end{cases}\right\}.$$
Some such intended comparison was clearly in Newton's mind, but the details recorded in this paragraph seem too inaccurate and elusive for us to be more definite in our judgement.

(25) '...than the term set over the top of this number in the other progression'. Newton first wrote 'huic numero imminens'.

(26) In sequel Newton has cancelled: 'De seriebus regulariter procedentibus loquor. Siquando obvenerint irregulares, hæ longius producendæ sunt. Termini ulteriores irregularitatem paulatim amittent'. A following paragraph he broke off after writing the opening phrase 'Ut si serie'.

(27) Namely, $z = (1+x)^7$.

(28) See note (13) above. Here the infinite series $-\frac{1}{6}x^6 + \frac{1}{7}x^7 - \frac{1}{8}x^8 + \frac{1}{9}x^9 - \dots$ is approximated by $-x^6(1+x)^{-7}(\frac{1}{6}v - \frac{1}{7}tx + \frac{1}{8}sx^2 - \frac{1}{9}rx^3 + \frac{1}{10}qx^4 - \frac{1}{11}px^5 + \frac{1}{12}x^6)$, where v, t, s, r, q, p are the first 7, 6, 5, 4, 3, 2 terms in the expansion of $z = (1+x)^7$; compare note (9) above in the case when $k = 7$ on taking $v = \alpha$, $t = \beta$, $s = \gamma$, ..., In particular, when $x = 1$, then $p = 1 + 7 = 8$, $q = p + 8 = 29$, $r = q + 35 = 64$, $s = r + 35 = 99$, $t = s + 21 = 120$ and $v = t + 7 = 127$. Newton's computations are accurate to the twelve places to which they are taken, but as he

[2]$^{(29)}$

$$c+d[x]+e[x^2]+f[x^3]+g[x^4]. \qquad c+\frac{d}{1-x}. \qquad c+\frac{\substack{1\\ -2x}\,d+e}{\overline{1-x}\,|^2}. \qquad c+\frac{\substack{1\\ -3x\\ +3x^2}\,d\ \substack{+1\\ -3x}\,e+f}{\overline{1-x}\,|^3}. \qquad [\&c]$$

$$c-d[x]+e[x^2]-f[x^3]+g[x^4]. \qquad c-\frac{d}{1+x}. \qquad c-\frac{\substack{1\\ +2x}\,d-e}{\overline{1+x}\,|^2}. \qquad c-\frac{\substack{1\\ +3x\\ +3x^2}\,d\ \substack{-1\\ -3x}\,e+f}{\overline{1+x}\,|^3}. \qquad [\&c]$$

$$-c+d[x]-e[x^2]+f[x^3]-g[x^4]. \qquad -c+\frac{d}{1+x}. \qquad -c+\frac{\substack{1\\ +2x}\,d-e}{\overline{1+x}\,|^2}. \qquad -c+\frac{\substack{1\\ +3x\\ +3x^2}\,d\ \substack{-1\\ -3x}\,e+f}{\overline{1+x}\,|^3}.$$

$$-c+\frac{\substack{1\\ +4x\\ +6x^2\\ +4x^3}\,d\ \substack{-1\\ -4x\\ -6x^2}\,e\ \substack{+1\\ +4x}\,f-g}{\overline{1+x}\,|^4}. \qquad [\&c]$$

1 Diff $\quad \dfrac{d}{1-x}. \quad \dfrac{-dx+e}{\overline{1-x}\,|^2}. \quad \dfrac{dx^2-2ex+f}{\overline{1-x}\,|^3}. \quad \dfrac{-dx^3+3ex^{[2]}-3fx+g}{\overline{1[-]x}\,|^4}. \quad [\&c]$

2 Diff $\quad \dfrac{-d}{1+x}. \quad \dfrac{-dx+e}{\overline{1+x}\,|^2}. \quad \left[\dfrac{-dx^2+2ex-f}{\overline{1+x}\,|^3}. \quad \dfrac{-dx^3+3ex^2-3fx+g}{\overline{1+x}\,|^4}. \quad \&c\right]$

---

observed in a cancelled sentence at this point 'Verius area illa [namely log 2] alias quæsita est $\dfrac{69314718}{100000000}$, adeo ut error prioris computi tantū $\dfrac{53}{100000000}$': more precisely, the error is about 53.54×10^{-8} too much.

(29) For three simple series of the general type $\pm(c\pm dx+ex^2\pm fx^3+gx^4\pm...)$, the first monotonic, the two latter alternating in sign, Newton sets down successively more refined approximations of the type developed in the preceding section: namely

$$A_k = \pm(c\pm(d\alpha\pm e\beta x+f\gamma x^2+g\delta x^3+...)x/(1\mp x)^k),$$

where $\alpha,\beta,\gamma,\delta,...$ are the first $k, k-1, k-2, k-3, ...$ terms of the expansion of $(1\mp x^n)^k$, $k=1,2,3,4,...$. He then computes successively the 'Diff'

$$A_k-A_{k-1} = \mp\Delta^{k-1}d.(\mp x/(1\mp x))^k,$$

$k=1,2,3,4,...$, where $\Delta^0 d=d, \Delta^1 d=d-e, \Delta^2 d=d-2e+f$, and so on; and so is led to develop the expansion of the general 'approximatio' A_k in powers of $x/(1\mp x)$, viz:

$$A_k = \pm(c\pm dz-\Delta^1 d.z^2\pm\Delta^2 d.z^3-\Delta^3 d.z^4\pm...), \quad \text{where} \quad z=x/(1\mp x).$$

Observe that in all but the last line Newton's 'd', 'e', 'f', 'g', ... are contractions for $dx, ex^2, fx^3, gx^4, ...$ respectively: this we have spelt out in the opening enunciation of the various series with appropriate insertions in square brackets.

3 Diff $\dfrac{d}{1+x}$. $\dfrac{dx-e}{|1+x|^2}$. $\dfrac{dx^2-2ex+f}{|1+x|^3}$. $\dfrac{dx^3-3ex^2+3fx-g}{|1+x|^4}$.   [&c]

$c+dx+ex^2+fx^3+gx^4$. Diff. $\dfrac{d}{1-x}x+\dfrac{-d+e}{|1-x|^2}x^2+\dfrac{d-2e+f}{|1-x|^3}x^3+\dfrac{d-3e+3f-g}{|1-x|^4}x^4$. [30]

---

[3][31]

$y=x-\frac{1}{2}x^2+\frac{1}{3}x^3-\frac{1}{4}x^4$ &c. Fac $\dfrac{x}{1+x}=z$. erit $[y=]z+\frac{1}{2}z^2+\frac{1}{3}z^3+\frac{1}{4}z^4$ &c.

$y=x-\frac{1}{3}x^3+\frac{1}{5}x^5-\frac{1}{7}x^7$ &c. Fac $\dfrac{x^2}{1+x^2}=z$. erit

$$yx=z+\frac{2}{1,3}z^2+\frac{2,4}{1,3,5}z^3+\frac{2,4,6}{1,3,5,7}z^4 \text{ &c.}$$

$y=x+\frac{1}{6}x^3+\frac{3}{40}x^5+\frac{5}{112}x^7$ &c$=x+\dfrac{1}{2,3}[x^3]+\dfrac{1,3}{2,4,5}[x^5]+\dfrac{1,3,5}{2,4,6,7}[x^7$ &c$]$.

   Fac $\dfrac{x^2}{1-x^2}=z$, erit $yx=z$[32]

$y=\sqrt{x^2-x^4}=x-\frac{1}{2}x^3[-]\frac{1}{8}x^5$ &c. Fac $\dfrac{x^2}{1+x^2}=z$, erit $yx=z+\frac{1}{2}z^2[-]\frac{1}{8}z^3$ [&c].

$\sqrt{1+x^2}=y=1+\frac{1}{2}x^2+\dfrac{1,-1}{2,4}x^4+\dfrac{1,-1,-3}{2,4,6}x^6$ &c. Fac $\dfrac{x^2}{1+x^2}=z$. [erit]

$$y=1+\frac{1}{2}z+\frac{1,3}{2,4}z^2+\frac{1,3,5}{2,4,6}z^3 \text{ &c.}$$

$y=x+\dfrac{1}{2,3}x^3+\dfrac{1,-1}{2,4,5}x^5+\dfrac{1,-1,-3}{2,4,6,7}x^7$ [&c]. Fac[33]

---

(30) The general 'Diff' $A_4-(A_0$ or$)$ $c=dz-\Delta^1 d.z^2+\Delta^2 d.z^3-\Delta^3 d.z^4$, $z=x/(1-x)$ which adds the terms in the preceding '1 Diff' (which lists A_k-A_{k-1}, $k=1,2,3,4$). It follows therefore (compare previous note) that $c+dz-\Delta^1 d.z^2+\Delta^2 d.z^3-\Delta^3 d.z^4$ is the 'approximatio' $c+([1-4x+6x^2-4x^3]dx+[1-4x+6x^2]ex^2+[1-4x]fx^3+gx^4)/(1-x)^4$ to the series

$$c+dx+ex^2+fx^3+gx^4 \ldots.$$

(31) Generalizing his new-found series 'transmutation' to the case

$$z=x^n/(1\mp x^n), \quad n=1,2,\ldots.$$

Newton now computes the transforms of six simple algebraic, circular and logarithmic (hyperbolic) series.

(32) Understand '$yx=z-\frac{5}{6}z^2+\frac{89}{120}z^3-\frac{381}{560}z^4$ &c': Newton has entered the first three coefficients '$1. \frac{5}{6}. \frac{89}{120}$' in the margin alongside.

(33) Read 'Fac $\dfrac{x^2}{1+x^2}=z$, erit $yx=z+\frac{7}{6}z^2+\frac{157}{120}z^3+\frac{803}{506}z^4$ &c', since the original series alternates in sign after the first term.

To summarize Newton's results, the 'transmutation' $z=x/(1+x)$ converts the expansion

[4]$^{(34)}$

Proponatur series quælibet dx^n. ex^{2n}. fx^{3n}. gx^{4n}. hx^{5n} &c$^{(35)}$ ubi d, e, f, g, &c significant datas terminorū coefficientes quæ si satis longe producuntur accedunt ad rationem æqualitatis. Si termini omnes ejusdem sunt signi pone $\dfrac{x^n}{1-x^n}=z$, sin signa $+$ et $-$ alternatim mutantur pone $\dfrac{x^n}{1+x^n}=z$. Fac etiam $d-e=p$. $d-2e+f=q$. $d-3e+3f-g=r$. $d-4e+6f-4g+h=s$.

$$d-5e+10f-10g+5h-i=t \quad \text{&c}^{(36)}$$

producendo seriem continuo per numerales coefficientes potestatum binomij. Dein loco prioris seriei scribe hanc$^{(37)}$ dz. pz^2. qz^3. rz^4. sz^5. tz^6 &c$^{(38)}$ ubi terminus d ejusdem sit signi ac in serie priori; cæterorū quoq; terminorum signa eadem sint si in serie priore alternatim mutabantur, at alternatim mutentur si illic erant eadem. Posteriori casu series convergentes in divergentes$_{[,]}$ priori divergentes in convergentes et minùs convergentes in magis convergentes abeunt.$^{(39)}$ Et hic est casus qui transmutatione$^{(40)}$ maximè indiget. Nam series$^{(41)}$ quarū termini omnes sunt ejusdem signi, divergere nequeunt quin simul amittant possibilitatem. Hæ igitur non opus habent ut in convergentes vertantur.$^{(42)}$

---

of $\log(1+x)$ into the series for $-\log(1-z)$, while $z = x^2/(1 \pm x^2)$ converts those of $\tan^{-1}x$, $\sin^{-1}x$, $x\sqrt{[1-x^2]}$, $\sqrt{[1+x^2]}$ and $\int_0^x \sqrt{[1+x^2]}\,.dx = \frac{1}{2}(x\sqrt{[1+x^2]}+\log(x+\sqrt{[1+x^2]}))$ into the series for $\sqrt{[z/(1-z)]}\sin^{-1}\sqrt{z}$, $\sqrt{[z/(1+z)]}\tan^{-1}\sqrt{z}$, $\sqrt{[z^2(1-2z)/(1-z)^3]}$, $\sqrt{[1-z]}$ and

$$\tfrac{1}{2}(z/(1-z)^{\frac{3}{2}}+z^{\frac{1}{2}}(1-z)^{-\frac{1}{2}}\log[(1+z^{\frac{1}{2}})/(1-z)^{\frac{1}{2}}]).$$

(34) A first rough draft of Chapter 2 of the 'De computo', now incorporating his new technique for series transformation.

(35) Descartes' notation in his *Geometrie* for a polynomial when the signs of the coefficients d, e, f, g, h, ... of its component terms are not known, one widely used by continental mathematicians in the later seventeenth century (for instance by Kinckhuysen in his *Stelkonst* (Haerlem, 1661); compare Mercator's reversion to the notation in his Latin version on II: 334, bottom line). The points, probably in the first place a mere printing convenience to separate the individual terms, are space-fillers (and not a primitive notation for the compound ' $\pm$ ' sign since the opening term ' dx^n ' is never written ' $.dx^n$ '). Newton's present employment of the convention is virtually unprecedented in his mathematical papers and here seems uselessly idiosyncratic.

(36) Newton first continued 'et sic in infinitum pergas multiplicando terminos d, e, f, g, &c per numerales coefficientes potestatum binomij'.

(37) An earlier qualifying phrase 'nova[m] celeriter convergen[tem]' is deleted.

(38) In sequel Newton first wrote 'ubi termini omnes signis affirmativis conjungendi sunt si habeatur $+dx^n-ex^{2n}+fx^{3n}$ &c. negativis si $-dx^n+ex^{2n}-fx^{3n}$ &c. variantibus hoc modo $+dz-pz^2+qz^3-rz^4$ &c si $+dx^n+ex^{2n}+fx^{3n}$ [&c] et hoc $-dz+pz^2-qz^3$ &c si $-dx^n-ex^{2n}-fx^{3n}$ &c'.

(39) The more accurate 'abire solent' is replaced.

(40) 'hac regula' was first written.

(41) 'legitim[æ]' is cancelled.

Ut si fuerit $y = x - \frac{1}{2}x^2 + \frac{1}{3}x^3 - \frac{1}{4}x^4$ &c.[43] Quoniam signa $+$ et $-$ mutantur per vices pono $\frac{x}{1+x} = z$, dein $1 - \frac{1}{2}$ seu $\frac{1}{2} = p$. $1 - \frac{2}{2} + \frac{1}{3}$ seu $\frac{1}{3} = q$. $1 - \frac{3}{2} + \frac{3}{3} - \frac{1}{4}$ seu $\frac{1}{4} = r$ &c.[44] ac tandem $y = z [+] \frac{1}{2}z^2 + \frac{1}{3}z^3 [+] \frac{1}{4}z^4 + \frac{1}{5}z^5$ &c.

Si fuerit $y = ax - \frac{1}{3a}x^3 + \frac{1}{5a^3}x^5 - \frac{1}{7}\frac{x^7}{a^5}$ &c id est $\frac{xy}{aa} = \frac{xx}{aa} - \frac{x^4}{3a^4} + \frac{x^6}{5a^6}$ &c pono

$\frac{[xx]}{aa} = v$. $\frac{[v]}{1+[v]} = z$. $1 - \frac{1}{3} = p = \frac{2}{3}$. $1 - \frac{2}{3} + \frac{1}{5} = q = \frac{8}{15} = \frac{2, 4}{3, 5}$.

$$1 - \frac{3}{3} + \frac{3}{5} - \frac{1}{7} = r = \frac{16}{35} = \frac{2, 4, 6}{3, 5, 7} \quad \text{&c}[45]$$

et $\frac{[x]y}{aa} = z + \frac{2}{3}z^2 + \frac{2, 4}{3, 5}z^3 + \frac{2, 4, 6}{3, 5, 7}z^4 + \frac{2, 4, 6, 8}{3, 5, 7, 9}z^5$ &c.[46]

---

(42) See our editorial comment in §2: note (48) above. In sequel Newton first began a new paragraph 'Porrò si seriem novam usq̃ ad certum terminum puta rz^4 producamus, deinde', there breaking off to cancel it. Perhaps he initially intended to introduce the remark that the remainder could then be set (compare note (9) above) as

$$\mp (\Delta^3 d.x^{3n} \pm \Delta^3 e.x^{4n} + \Delta^3 f.x^{5n} \pm \ldots)/(1 \mp x^n)^3,$$

that is (in Newton's present terms), $\mp z^3 (r \pm (r-s)x^n + (r-2s+t)x^{2n} \pm \ldots)$.

(43) 'quæ series est ad aream Hyperbolæ' [viz: $y = 1/(1+x)$] is cancelled. This example is taken straightforwardly from [3] preceding.

(44) $1 - \frac{1}{2}\binom{n}{1} + \frac{1}{3}\binom{n}{2} - \ldots + (-1)^n/(n+1) = \int_0^1 (1-x)^n.dx = 1/(n+1)$. In a stray draft higher on the same manuscript sheet Newton convinced himself by outright computation that

$$`\frac{1}{a} \frac{-3}{a-1} \frac{+3}{a-2} \frac{-1}{a-3} [=] \frac{4a-2}{aa-2a}[-]\frac{4a-10}{aa-4a+3}[=]\frac{[-]6}{a^4-6a^3+11aa-6a}`$$

(by multiplying $(4a-2)(a^2-4a+3)$ and $(4a-10)(a^2-2a)$, and subtracting the results $`4a^3 \begin{smallmatrix} -16 \\ -2 \end{smallmatrix} aa \begin{smallmatrix} +12 \\ +8 \end{smallmatrix} a - 6`$ and $`4a^3 \begin{smallmatrix} -8 \\ -10 \end{smallmatrix} aa + 20a`$) and further that

$$\sum_{0 \leqslant i \leqslant 5} (-1)^i/(a+i) = `\frac{1, 2, 3, 4, 5}{a, a+1, a+2, a+3, a+4, a+5}`.$$

(45) $1 - \frac{1}{3}\binom{n}{1} + \frac{1}{5}\binom{n}{2} - \ldots + (-1)^n/(2n+1) = \int_0^1 (1-x^2)^n.dx = \prod_{1 \leqslant i \leqslant n} [2i/(2i+1)]$.

(46) This example, in which the 'transmutation' $z = x^2/(a^2+x^2)$ converts the series expansion of $y = a\tan^{-1}(x/a)$ into that of $xy/a^2 = \sqrt{[z/(1-z)]}\sin^{-1}z$ is a slight generalization of the equivalent example in [3] preceding. For the particular case $x = a$, $z = \frac{1}{2}$ see §2: note (44) above.

APPENDIX

MATHEMATICAL TOPICS IN NEWTON'S CORRESPONDENCE
1674-1676

MATHEMATICAL TOPICS IN NEWTON'S CORRESPONDENCE, 1674-1676

Following on the precedent established in the previous volume we here briefly summarize—and in some particulars complement—points of mathematical significance which are discussed by Newton in extant letters to his contemporaries, notably Collins and Oldenburg, during the two-year period from June 1674 to November 1676.[1] These two terminal dates mark natural pauses in his correspondence. Between the time of his reply to Collins on 17 September 1673, polite in tone but with its firm refusal to be further involved in optical dispute,[2] and his next response on 20 June 1674 to a renewed attempt on Collins' part to establish contact, Newton would appear to have severed all ties with his scientific contemporaries in London, devoting himself uninterruptedly to his optical and chemical researches.[3] Correspondingly, the transmission of his second letter for Leibniz in October 1676 was the prelude to an interval of eight years during which Newton revealed a marked unwillingness to share his mathematical insights with his fellow men, and which terminated only at the very end of the period covered by the present volume with David Gregory's letter to him on 9 June 1684 and Halley's celebrated visit to Cambridge two months later.[4] On 8 November 1676 Newton wrote his last extant letter to Collins and thereafter, apart from Collins' communication in March and

---

(1) See the concluding appendix to the third volume (III: 559–71). As before, our basic text is H. W. Turnbull's edition (for the Royal Society) of *The Correspondence of Isaac Newton*, I: *1661–1675*; II: *1676–1687* (Cambridge, 1, 1959; 2, 1960), to which we refer by the convention typified in '[2: 179]': by this we understand a reference to page 179 of Volume 2 of the *Correspondence* (and hence to the opening page of Newton's letter to Collins of 8 November 1676). A more lavish discussion of certain points of detail and of the contemporary mathematical scene is given by J. E. Hofmann in the later sections (especially pages 95–100) of his *Studien zur Vorgeschichte des Prioritätstreites zwischen Leibniz und Newton um die Entdeckung der höheren Analysis. I: Materialien zur ersten mathematischen Schaffensperiode Newtons (1665–1675)* [= *Abhandlungen der Preuss. Akad. der Wissenschaften. Jahrgang 1943. Mathem.-Naturwiss. Klasse Nr. 2*], Berlin, 1943.

(2) 'I shall not trouble you any further wth discourses about y^e Perspective [*sc.* his 'Catadioptricall' telescope]' [1: 307].

(3) Compare the general introduction to the present volume, especially note (35).

(4) See **3**, Introduction: note (38) above. We accept the traditional view that this visit took place in August 1684: recent attempts to advance the date to May are not convincing, and deny Halley's assertion only two years later, on 29 June 1686, that the visit was paid in August [**2**: 442].

August 1677 of transcripts of two letters from Leibniz[5] and a last, seemingly fruitless attempt in October 1678 to interest Newton in the 'rumb Spirall',[6] their relationship ceased to be active.[7] To Oldenburg on 19 February 1676/7 Newton wrote a friendly letter, adding in postscript that he was 'going out of town for a few days.... I would not have you write to me till you hear y^t I am returned' [**2**: 194]: he was back in Cambridge three weeks later but Oldenburg died the following September without renewing the correspondence and contact with Leibniz was broken for more than a decade. Over the next seven years Newton rarely broke his silence, and only once on a topic of mathematical interest—namely, in the brief flurry of letters which passed between him and Hooke in the early winter of 1679/80 regarding the paths (notably those of the planets) traversed by a freely moving body under the action of constant and inverse-square centripetal forces. The details of this correspondence we postpone till we come, in the sixth volume, to present mathematical aspects of the mature Newtonian theory of motion in a central-force field to which it is an essential prelude. Here, therefore, we may justly confine our attention to mathematical points arising in Newton's letters during the two and a half years from June 1674.

---

(5) See notes (54) and (64) below.

(6) Collins to Newton, 12 October 1678 [**2**: 286]. The rhumb (logarithmic) spiral is the stereographic projection, from the South pole onto the equatorial plane, of a general spherical loxodrome: namely, the skew curve which makes some constant angle (α, say) with every meridian. At once, by the conformal property of the projection, the *radii vectores* make this same constant angle with the projected spiral, yielding the differential defining property $dr/r\,d\phi = \tan\alpha$. Some seventy years before Thomas Harriot had, in Collins' words, seen that the spiral's 'rayes' (r) are 'the tang^{ts} of halfe the Complements of the Latitudes' (θ, say) while the central angle (ϕ) is the 'difference of Longitude' of the loxodrome's end-points, and deduced a near-equivalent to the integral $\int_\theta^0 \sec x\,.\,dx = (\phi\text{ or}) - \log[\tan\tfrac{1}{2}(\tfrac{1}{2}\pi - \theta)]$ in the case where $\alpha = \tfrac{1}{4}\pi$: he subsequently went on to rectify the general arc ($s = r\sec\alpha$) of numerical instances of his proportional 'helix', thence deriving the correct generalization. (See J. V. Pepper, 'Harriot's Calculation of the Meridional Parts as Logarithmic Tangents', *Archive for History of Exact Sciences*, **4**. 5, 1968: 359–413, especially 366–75, 399–402; and compare **1**: 474, note (35).) The insight that the spiral may be 'streightened' by unwrapping it along its tangent as far as the subtangent perpendicular to the radius vector—known earlier to Descartes (1639) and Torricelli (*c.* 1645) and perhaps to Harriot also—was first published in 1659 by John Wallis (probably second-hand from Wren) in his *Tractatus Duo* (**1**: 24, note (22)): 106–8, together with the dampening observation 'fatendum erit, Tangentem spiralis hujus, non magis duci posse Geometrice, quam spiralis Archimedeæ, (ut quæ ex quadratura circuli dependeat...)' (*ibid.*: 107); in other words, ''tis presumed [to draw a touch line to the Spirall] cannot be done without streightening the Curve' into its tangent, which in turn depends on our being able to construct 'geometrically' the constant slope (α) of a given spiral to its radius vector ($r = e^{\phi\tan\alpha}$, where any number of points (r, ϕ) are given). How far Newton went in acceding to Collins' request that 'he consider this most usefull Probleme [of drawing a 'touch line to such Spirall']...of great use in Navigation [since it 'gives the angle of the rumbe from one place to the other'] and no Construction or good approach for it yet in lines' is not known: not very

In a letter to Newton in mid-June 1674, now lost, Collins had sought criticism of an enclosed copy of Robert Anderson's newly published *Genuine Use and Effects of the Gunne*—a tract propounding, in modification of the simple Galilean theory of projectile motion in a parabola, the Tartaglian supposition of an initially rectilinear flight path—and also (perhaps at the request of Michael Dary, then his assistant at the Farthing Office) had tried to draw Newton out on 'extracting yᵉ roots of literal equations'. To the first the reply was made [1: 309] that

> Mʳ Andersons book is very ingenious, & may prove as usefull if his principles be true. But I suspect one of them, namely yᵗ yᵉ bullet moves in a Parabola. This would be so indeed were yᵉ horizontal celerity of yᵉ bullet uniform, but I should think its motion decays considerably in yᵉ flight....if it were so, yᵉ celerity of yᵉ bullet would increas... whereas I should rather think yᵗ yᵉ celerity decreases very considerably.

However, Newton's proposed alternative 'rule for its decreasing [wᶜʰ] may pretty nearly approach yᵉ truth' (presupposing, in effect, that during flight the horizontal acceleration and speed are reduced uniformly to zero by air resistance) also, after a certain limit, increases the bullet's velocity indefinitely.[8] In amplification of the other topic, already outlined in the *De Analysi* in 1669 under the heading 'Literalis æquationum affectarum resolutio' and further developed in the revised tract he wrote two years later,[9] Newton laid out the series solution of $y^3 + a^2y - b^3 = 0$ ('wᶜʰ may be a form for all cubic equations') by means of the partial quotients $y = l + p$, $p = -m/n + q$, $q = -3lm^2/n^3 + r$ with r rounded off to $O(m^4)$ as $m^3(1/n^4[-18l^2/n^5])$, where

$$m = l^3 + a^2l - b^3 \qquad \text{and} \qquad n = a^2 + 3l^2:$$

thus in particular, according as a is 'considerably' less or greater than b, substitution of $l = b$ ($m = a^2b$, $1/n = \frac{1}{3}b^{-2} - \frac{1}{9}a^2b^{-4} + \frac{1}{27}a^4b^{-6} \ldots$) or of

$$l = b^3/a^2 \ (m = a^{-6}b^9, \ n = a^{-2} - 3a^{-4}b^2 + 9a^{-6}b^4 \ldots)$$

---

far, we may guess, for he knew enough about the logarithmic spiral (compare III: 186, 192) to realize that $\tan\alpha = (\log r)/\phi$ is not algebraically constructible in Wallis' (and Descartes') 'geometrical' sense.

(7) Their last known contact was in April 1682 when, out of the blue it would appear, Newton wrote to Collins for his support of Paget's candidacy as prospective 'Lecturer of Navigation' at Christ's Hospital (see Collins to (?) Flamsteed, 16 May 1682 [2: 376–7]).

(8) This remark will be justified when Newton's present ballistic curve is examined in detail in the sixth volume.

(9) See II: 222–30, 234 and III: 48–68. Writing to Gregory in October 1675 regarding 'the method of Series for finding the rootes of adfected æquations...by extracting [them] in Species', Collins added that 'Mʳ Newton...communicated some such Series to me, with a method of extracting them, much different from that of Vieta in Numbers, were it applyed to Species' [1: 355–6]. In context this probably relates to the section 'Numeralis æquationum affectarum resolutio' of Newton's *De Analysi* (see II: 218–22) but no doubt Collins also had the June 1674 letter in mind.

yields a suitably convergent series solution. Correspondingly, the series solution of the pure cubic $y^3 - a^3 - b^3 = 0$ where 'a is bigger then b' (by

$$y = a + p, \quad p = \tfrac{1}{3}b^3/a^2 + q, \quad q = -\tfrac{1}{9}b^6/a^5 + r,$$

and so on) yields the expansion of the binomial $a(1 + b^3/a^3)^{\frac{1}{3}}$, which 'els...may be done as in numerall Arithmetic'.[10] This affords a convenient way of rapidly approximating the cube root of any given number c, for 'if $y^3 = c$, suppose ... a to be pretty nearly equall to y^e cube root of c, & putting $c - a^3 = b^{[3]}$ ... extract y^e cube root out of $a^3 + b^{[3]}$ as in y^e former example'. Furthermore, this cube root being once extracted may be kept as a rule for extracting numeral cube roots. But yet (as you suggest) these infinite series are only usefull where y^e roots of equations cannot be attained accurately. For when they can be accurately attaind recours must be had to other methods, this only performing it by approximation [1: 311].

Collins evidently lost no time in passing the content of this letter on to Dary, and in the summer or early autumn of 1674 the latter began a brief but brisk exchange of letters with Newton, many of them now lost. From the earliest we have, written by Newton from Cambridge on 6 October [1: 319–20] it is evident that the solution of equations continued as the central topic of interest. 'The terme in my last letter of wch you are doubtful should be $\dfrac{8v - s}{28v}$.[11] I have run over y^e problem in universall terms as you desired & y^e result is this....' Newton then summarizes his series solution of 'literal' equations in the case of Dary's propounded equation $z^n + bz + R = 0$: namely, on setting $n - 1 = p$, $n - 2 = q$, $n - 3 = r$, ... and taking c 'as nearly as can be guessed, equall to y^e root z', the substitution $z = c + u$ yields $dg + gu + sc^q u^2 + tc^r u^3 \ldots = 0$, where $g = b + nc^p$ and $dg = c^n + bc + R$; then $u = -d + v$ gives

$$v + ed^2 - 2edv - (tc^r/g)\, d^3 \ldots = 0,$$

---

(10) Compare the equivalent extraction of the square root $a(1 + x^2/a^2)^{\frac{1}{2}}$ in the *De Analysi* (II: 214) and the 1671 tract (III: 40). In the latter, as we have seen (III: 214, note (466)), the corresponding cube-root series expansion of $a(1 + x^3/a^3)^{\frac{1}{3}}$ is introduced *en passant* without prior explanation.

(11) Turnbull's transcription on [1: 319] from Collins' copy of the letter (now in private possession). We should probably read '$^{28}\overline{8v - 5}$', understanding the iterative solution of $v^{28} - 8v + 5 = 0$. Later in the same letter Newton iterates the analogous example of the equation $z^{30} = 8z - 5$.

(12) As Newton put it, 'so shall y^e last found terme e, f, or g &c be y^e desired root z'. It is just possible that Collins had already told him of an equivalent procedure, communicated by James Gregory the previous April, for finding the root a of $a^{n+1} = b^{n-1}(b + c)a - b^n c$, in which '$b$ [is an] annuitie [discompting compound interest], c the present Worth, n [years] the time of continuance and $\dfrac{b^2}{a} - b$ an year's interest of the annuitie' (James Gregory to Collins, 2 April 1674, first published by S. P. Rigaud in his *Correspondence of Scientific Men of the Seventeenth Century*, **2** (Oxford, 1841): 255–6 [= *James Gregory Tercentenary Memorial Volume*, London,

where $e = sc^q/g$; while $v = -ed^2 + w$ yields $w + fd^3 \ldots = 0$ on putting $f = 2e^2 - tc^r/g$; and so on indefinitely. Similarly, the 'æquation

$$z^n + bz^p + R = 0 \quad [viz. \; z^n + bz^{n-1} + R = 0],$$

if you write $\dfrac{1}{x}$ for z becomes $\dfrac{1}{R} + \dfrac{b}{R}x + x^n[=0]$ whose root x being found by y^e

former approximation, make $\dfrac{1}{x} = z'$. This method, however, is none too efficient

in numerical cases in quickly yielding a good approximation to a root, and to achieve a more rapid result Newton now invoked the idea of attaining the root to desired accuracy by an iterative process—of necessity without any discussion of the convergence or uniqueness of such a procedure: 'When you have thus found y^e valor of z or x, if it be not exact enough you may write y^e valor for c & repeat y^e worke. But yet I conceive these roots may be easilier extracted by logarithmes.' He then requires Dary, given some equation $z^n = bz + R$, to find 'c as nearly as you can guesse equall to y^e root z' and then will $d = \sqrt[n]{[bc + R]}$, $e = \sqrt[n]{[bd + R]}$, $f = \sqrt[n]{[be + R]}$, $g = \sqrt[n]{[bf + R]}$, ... be ever closer approximations to z. Correspondingly a root of $z^n + bz^{n-1} = R$ which is near to c is approached by iterating $z = \sqrt[n-1]{[R/(z + b)]}$, so deriving the sequence c, d, e, f, g. ... converging to z.[12] In his reply a month later [1: 326] Dary announced the trivial extension of this iterative method to extracting the root of an equation

$$z^p = az^q + n \quad (p > q)$$

'consisting of two severall powers...and an absolute number; *per* an approximation easily performed by logarithmes'.[13]

1939: 278–9]). Such iterative methods had been employed at least as early as the fifteenth century when (according to Chelebi) the Samarkand mathematician Jamšīd al-Kāšī resolved the 'trisection' cubic $x = \frac{1}{3}(a + 4x^3)$ in this way, deriving $x = \sin 1°$ to eighteen places from $a = \sin 3°$. (See A. Aaboe, 'Al-Kāšī's Iterative Method for the Determination of $\sin 1°$', *Scripta Mathematica*, **20**, 1954: 24–9; and compare A. P. Yushkevich, *Geschichte der Mathematik im Mittelalter* (Leipzig, 1964): 321–3).

(13) 'The rule is thus: First guess at the root as nearly as you can, the nearer the better (not for nescessity but for accomodation,) and suppose that guess to be z. Then observing the following series $[z_{i+1} = \sqrt[p]{(az_i^q + n)}, \; z_i = z, b, c, d, e, \ldots$ in turn], you shall approach...toward the true z which is sought' (Dary to Newton, 15 October 1674 [1: 326]). Writing to Collins a month later, on 17 November, Newton was not much impressed: 'The rules wch I lately sent [Mr Dary] for resolving æquations by Logarithms to be communicated to you, he applys to Quadratic equations: whereas they are only to be applied to æquations wch have many intermediate terms wanting, four or five at least. And ye more intermediate terms are wanting, the sooner they approach to truth. He is desirous of ye best way of determining Logarithms by ye Hyperbola & ye solution of another Probleme of ye same kind, but I have nothing valuable to communicate therein wch you are not already acquainted wth' [1: 327–8]. As we have seen (1: 97, note (1)), Collins had earlier taken a copy of the Wallis annotations (1, 1, 3, §3) in which, among other applications of his binomial theorem, Newton had used the series expansion of $\log(1 + x)$ to 'Square the Hyperbola' $(1 + x)y = 1$.

Their subsequent correspondence is lost except for a letter written by Newton on 22 January 1674/5, extant in a copy made by Dary for Collins.[14] Here he restricted himself to answering Dary's request that he extend two infinite series he had previously sent, one rectifying the general ellipse arc, the other unspecified but no doubt taken likewise from his 1671 tract [1: 332–3]:

I should have sent you a Continuation of ye two series you desire but I have not any Computations of them by mee, & perceive them so tedious to compute y^t I am constrained to deferr them to another t[i]me: The series for the length of the Ellipsis I computed when I sent it you, & sent you so much as I computed of it. But at present in stead of these I have sent you an approximation for the length of the Ellipsis...instead of that you propounded....

Suppose[15] *AB*, *AD*, rectangular conjugate semidiameters of the Ellipsis, *BCD* a quadrant of it....if you would know the length of any...arch, as *BGC*, bisect its chord in *K*, Draw *AKG* [meeting the 'Ellipsis' in *G*], and y^e chords *BG* & *GC*: and make $\frac{4BG+4GC-BC}{3}$ y^e length of the arch *BGC*.

As Newton remarked, 'This [approximation] is derived from Hugenius' Quadrature of y^e Circle, and I believe approaches y^e Ellipsis as near as his doth the Circle'.[16] We should not, however, take his remarks about not having 'any Computations...by mee' for strict historical truth, since on page 113 of his 1671 fluxional tract[17] he had, as we have seen, calculated the general arc of the ellipse $y^2 = a^2 - bz^2$ to be $z + \frac{1}{6}a^{-2}b^2z^3 + \frac{1}{40}a^{-4}(4b^3 - b^4)\,z^5\,....$ From this the Huygenian approximation here propounded for Dary is an immediate deduction.[18] In the case of the whole quadrantal arc $\widehat{BD}$ Newton specified the

---

(14) Since Newton mentions, among other things, a gift copy of 'Dr Barrow's *Euclid*...safe enough now I know it was delivered', sent by him to Dary some time before, while thanking him 'for your problem', it is evident that a number of intervening—and subsequent?—letters between the two have disappeared.

(15) Understand the figure reproduced on [1: 333].

(16) See Christiaan Huygens' *De Circuli Magnitudine Inventa* (Leyden, 1654): Theorem VII, Proposition VII: 9–11 [= *Œuvres complètes de Christiaan Huygens*, 12 (The Hague, 1910): 93–181, especially 133–5]. Compare notes (43) and (45) below.

(17) That is, Example 8 of Problem 12 (see III: 326).

(18) Namely, on putting $A \equiv (0, 0)$, $B \equiv (a, 0)$, $G \equiv (y, z)$ and $C \equiv (v, x)$, at once $y/z = (v+a)/x$ with

$$v = \sqrt{[a^2 - bx^2]} = a - \tfrac{1}{2}a^{-1}bx^2 \,..., \quad z = ax/\sqrt{[(v+a)^2 + bx^2} = \tfrac{1}{2}x(1 + \tfrac{1}{8}a^{-2}bx^2 \,...)$$

and so $y = \sqrt{[a^2 - bz^2]} = a - \tfrac{1}{8}a^{-1}bx^2 \,....$ Hence

$$BC = \sqrt{[(v-a)^2 + x^2]} = x + \tfrac{1}{8}a^{-2}b^2x^3 \,..., \quad BG = \sqrt{[(y-a)^2 + z^2]} = \tfrac{1}{2}x + \tfrac{1}{64}a^{-2}(4b + b^2)\,x^3 \,...$$

and $GC = \sqrt{[(v-y)^2 + (x-z)^2]} = \tfrac{1}{2}x + \tfrac{1}{64}a^{-2}(-4b + 9b^2)\,x^3 \,...,$ so that

$$BG + GC = x + \tfrac{5}{32}a^{-2}b^2x^3 \,...$$

resulting construction, drawing '*BHD* its chord bisected in *H*; Draw *AH* abutting upon the Ellipsis at *C*, Joyne *BC* and *CD*, Take

$$BE = BC + CD, \quad \& \quad EF = \tfrac{1}{3}DE:$$

and *BF* shall bee the length of the quadrant *quam proximè*'. For greater accuracy, however, 'you may find *BC* and *CD* severally and the summe of them will give yᵉ Quadrant *BCD*, exacter then before' [**1**: 333].

Although Newton may have met Collins in London at this time (in late February or early March 1675), we possess no letter which passed between them during the next six months.[19] In May, however,—at Collins' request no doubt—he spent some time helping John Smith, 'late clerk [*sc.* accountant] at Brook House' under the Commissioner of Accompts, in calculating 'tables of Square rootes and Cube rootes...for all numbers from 0 to 10000, [he] having imparted a ready method for maintaining the last difference true'.[20] Newton's 'best way of performing it, yᵗ I can think of', communicated to Smith on 8 May [**1**: 342–4], required that 'yᵉ roots of every hundredth number be [first] extracted to ten decimal places', after which each intermediate set of 99 numbers could be interpolated (to eight place accuracy) by a scheme[21] in which third differences are assumed constant, 'the whole work being perform'd by Addition & Subduction excepting yᵗ in yᵉ computation of every 100ᵗʰ number, there is required yᵉ Extraction of one root'. This straight extraction of every hundredth root was, as Smith soon saw, the chief practical drawback in applying Newton's method. Could this difficulty be overcome? In reply to a lost request of Smith's for help Newton wrote back on 24 July following that, having

stay'd to think of something yᵗ might satisfy your Desire,...though I can not hitherto doe it to my owne liking, yet that I may not wrack your patience too much I have here

---

and finally $\overparen{BC}$ ($= x + \tfrac{1}{6}a^{-2}b^2x^3 \ldots$) $\approx \tfrac{1}{3}[4(BG + GC) - BC]$. The construction holds true for a general central conic to the same approximation; see J. E. Hofmann, *Studien* (note (1) above): 97, note 443.

(19) Collins spoke to Gregory on 1 May of 'Mʳ Newton being lately here' [**1**: 341] though on 19 October following he asserted inconsistently that 'I have not writt to or seene [him] this 11 or 12 Months, not troubling him as being intent upon Chimicall studies and practises' [**1**: 356]. Collins' draft of his letter to Newton of 'c. July 1675' [**1**: 346–7], perhaps never sent, is probably to be dated in the late autumn of that year, for it communicates 'an Account of some Errors observed by Dʳ Wallis in Dʳ Barrows *Archimedes*' (now ULC. Add. 3964.7: 9ᵛ/10ʳ: 'Errata Archimedis Typographica', preceded by (7ᵛ/8ʳ) Collins' list of 'Errors in Dʳ Barrowe's Conicks [and] In the Schemes') sent to him only about September. (See Collins to Gregory, 21 September 1675 [*Gregory Memorial Volume* (note (12) above): 333].) His remark in this draft about the 'Probleme, improperly proposed, about breaking [any Biquadratick æquation]... into two rationall quadratick æquations...which I troubled you with when here' [**1**: 346] may well refer to a London visit made by Newton between 14 and 23 October of that year (J. Edleston, *Correspondence of Sir Isaac Newton and Professor Cotes* (London, 1850): lxxxv).

(20) Collins to Gregory, 29 June 1675 [**1**: 345].

(21) This is reproduced as **1**, 1, §1 above.

w[r]itt you what occurs to mee, wch is only about facilitating ye Extrac[t]ion of [rootes]. The former Method might be applyed to determin all by every 1000th, as well as by every 100th [root], but not with advantage, for it will require the Extrac[t]ion of [rootes] to 14 or 15 places, besides a greater number of Addit[i]ons, Subduc[t]ions & Divisions in those greater numbers [1: 348].

In their place he proposed that the *n*-th root of any number A $(n = 2, 3, 4)$ should be extracted 'by common Arithmetick' or 'by Logarithms' to 5 decimal places, so obtaining the approximate 'Quotient' B, when $n^{-1}[(n-1)B + A/B^{n-1}]$ will approach the desired root to twice as many places:[22] 'But I think you will doe well to lett the Table of [squaresquare roots] alone, til you have done the other two, and then, if you finde your time too short, print the [square & cube roots] without troubling your selfe any further'.[23] Of this approximate rule Newton gave no justification, but evidently if $A = (B+b)^n \approx B^n + nbB^{n-1}$, b/B small, then $\sqrt[n]{A} \approx B + n^{-1}(A - B^n)/B^{n-1}$, which is his result.[24] The sequel did not repay Newton's effort, unfortunately. Though Collins was later to tell him that 'your paines hath encouraged Mr Smith much in the considerable Progresse he hath made in those tables of rootes he was about, for which I am much obliged to returne you my thankes' [1: 347], the tables of roots were never published.

Of his own mathematical studies Newton in these letters has nothing to say and it is tempting to accept the essential truth of Collins' remark to Gregory in October 1675 that 'both he and Dr Barrow [begin] to thinke Math^ll Specula-tions to grow at least nice and dry if not somewhat barren' [1: 356]. An earlier aside to the same correspondent the previous May reveals that Collins had tried

---

(22) Newton added a trivial restatement of this formula in a following letter to Smith on 27 August 1675 [1: 350–1], there observing in the case of the cube root $(\sqrt[3]{A} \approx \frac{1}{3}[2B + A/B^2])$ 'To finde the cube root of A to 11 decimal places: seek the Root by Logarithms to 5 decimal places, and suppose it B. Then square B, not by Logarithms, but by common Arithmetick, y^t you may have its exact square to 10 decimal places, and by this square Divide A to 11 decimal places, and to the Quotient add $2B$: The third part...shall be the root cubical of A to 11 decimal places. Your surest way will be to finde first the whole series of y^e Roots B by Loga-rithms, & try whether it be Regular by Differencing it: Then square those Roots by Nepeirs bones, and lastly Divide each number A by the correspondent square, and add $2B$ to each Quotient, and try the resulting series $[\approx 3\sqrt[3]{A}]$ againe by differencing it, whether it be Regular. If it be regular, I suppose you know the differences will at last come to be equal: what is said of Cubes is easily applyable to square-squares'.

(23) At this point Newton observed that Smith needed to extract only '76 [square roots] & 88 [Cube roots] & 94 [square square roots], whereof 10 are exact' [1: 349]. This relates to his 'former Direc[t]ion' in his letter to Smith on 8 May (see page 18 above) that 'the Square roots of all numbers between 2500 and 10000 and of all cube numbers between 1250 and 10000 and of all Squaresquare numbers between 625 and 10000 are only to be thus computed. For the...roots of numbers less...may be found by halfing'. Evidently, within these limits there will indeed be 76 square roots (of 2500, 2600, ..., 10000), 88 cube roots (of 1300, 1400, ..., 10000) and 94 fourth roots (of 700, 800, ..., 10000) to be extracted, ten of which (on including 1250 and 625) are exact.

to pump Newton regarding the current state of his researches while the latter was on a visit to London a little before Easter, but with little success for he 'did not seem to have any intent to publish anything as yet' [1: 341]. Newton's observation on that occasion that 'he had considered the finding out of what æquations are solved by Ordinates falling from the Intersections of any two Geometricall figures, ... in any Position at pleasure, on either of the Axes or any of the Diameters of either of the figures' was not seen by Collins to relate to the 'Problems for construing æquations' which he had been given three years before, but was circulated among his ring of correspondents as an important new insight.[25] A year later Newton was firmly to set the matter straight:

The other Problem I think I told you required no art but much calculation to resolve it, & therefore I have never thought of it since I saw you. There is nothing requisite to ye solution but this: To find two equations expressing ye nature of ye two curve lines, supposing their bases co-incident & their ordinates parallel; & putting ye same letter suppose x for ye bases in both æquations, & another letter suppose y for ye ordinates, to exterminate one of those letters. For y^e resulting equation will give you y^e several valors of y^e other letter, wch valors limit all y^e intersection points of y^e two curves.[26]

---

(24) This is, of course, the first stage in the 'Newton–Raphson' iterative resolution of the equation $x^n - A = 0$. More generally, if $f(x) = 0$ and $f(B) \approx 0$ 'very nearly', then

$$0 = f(B+b) \approx f(B) + bf'(B) \quad \text{and so} \quad x \text{ (or } B+b) \approx B - f(B)/f'(B),$$

iteration of which will in most cases generate ever closer approximations to the exact root x. As we shall see in the sixth volume, Newton implicitly invoked this procedure some ten years later in finding approximate solutions of Kepler's equation $f(x) \equiv x - e\sin x - N = 0$ (compare note (38) below), here iterating $x_{i+1} = x_i + (N - x_i + e\sin x_i)/(1 - e\cos x_i)$. (See his *Philosophiæ Naturalis Principia Mathematica* (London, ₁1687): Liber 1, Prop. XXXI, Scholium: 111–13.) With a debt to Newton explicitly acknowledged, the equivalent procedure for the general algebraic equation $f(x) \equiv x^n + ax^{n-1} + ... = 0$ was expounded at length by Joseph Raphson in his *Analysis Æquationum Universalis seu ad Æquationes resolvendas Methodus generalis ... ex nova infinitarum serierum methodo deducta et demonstrata* (London, ₁1690). On taking $n = 2$ and $A = B^2 + C$ Newton's present formula yields $\sqrt{[B^2 + C]} \approx B + \frac{1}{2}C/B$, an approximation known widely in classical times: the generalization to n-th roots was known at least as early as the fifteenth century, when it was developed by the Samarkand mathematician Jamšīd al-Kāšī. (See D. T. Whiteside, 'Patterns of Mathematical Thought in the later Seventeenth Century', *Archive for History of Exact Sciences*, **1**, 1961: 179–388, especially 207, note (38).)

(25) When, in preparation for his visit to London in October 1676, Leibniz collected together some points from his correspondence with Oldenburg on which he sought further enlightenment, he noted (perhaps from a lost postscript to Oldenburg's letter of 6 April 1673; see [2: 235, note (4)]) both Newton's 'exercitatio de constructionibus Conicis ... pro æquationibus' (II, 3, 2, §2) and the present 'Problema: datis duabus Sectionibus Conicis in quacunꝗ ad se invicem positione ductis; invenire quæ æquatio solvi possit per ordinatas, ab earum intersectione cadentes, in diametros vel axes figuræ [2: 236] (compare II: 177, note (15)). Later, having been introduced to Collins during his London visit, he was (in correction of II: 450, note (2)) to remark of the former that 'habet Collinius MSum ... Neutonianum' (see Hanover. Leibniz-Handschriften **35**, xv, 2: 12ᵛ; incorrectly transcribed on [2: 236]).

(26) Newton to Collins, 5 September 1676 [2: 95–6].

In other ways, too, Collins continued to impress upon his correspondents the extent and originality of Newton's mathematical researches—notably in a letter written to Oldenburg early in April 1675 for onward transmission to Leibniz,[27] in his stern (but none too cogent) critique[28] replying to Tschirnhaus' championing of Descartes' mathematical depth and originality, and in a (lost) letter in the spring of 1676 to the Danish geometer Georg Mohr.[29] Directly inspired by the last, Leibniz wrote to Oldenburg on the following 2 May asking if he might have Newton's demonstrations of two series expansions sent by Collins to Mohr, a request which was at once conveyed to Cambridge.[30] To ignore this plea for information from a foreigner whom he had never met and whose name could then have meant little to him would have been forgivable, but Newton instead—perhaps at Oldenburg's gentle insistence—revealed his generous side and began without delay to compose the first lengthy account for Leibniz of his methods for infinite series, his *epistola prior* of 13 June 1676 [**2**: 20–32].

Newton was no doubt aware in a general way that Leibniz, like so many others at this time, had found out a technique for reducing quantities to infinite series which was equivalent to one or other of his own.[31] What he now set out for Leibniz was a résumé of the opening pages of his 1671 tract, again presenting his

(27) Collins to Oldenburg, 10 April 1675 (Royal Society MS LXXXI, No. 36), sent on to Leibniz in Oldenburg's Latin version two days later (C. I. Gerhardt, *Der Briefwechsel von G. W. Leibniz mit Mathematikern*, **1** (Berlin, 1899): 113).

(28) Royal Society MS LXXX, No. 39. A selection from the Newtonian passages in this roughly ordered draft is reproduced in [**2**: 15–17].

(29) See Leibniz to Oldenburg, 2/12 May 1676 [**2**: 3–4], from which it appears that Collins had earlier—directly or by way of Oldenburg—sent to Mohr the Newtonian series expansions of $z = \sin^{-1} x$ and $x = \sin z$.

(30) Oldenburg's letter to Newton of 15 May 1676 is lost, but in a minute written on the verso of Newton's letter to him of 11 May he noted: 'Answ. by Dr Sidnam May 15. 76....In my letter...I imparted to Mr Newton y^e particulars contain'd in M. Leibnitz his letter to me of May 12. 1676' [**2**: 7]. Leibniz had also requested Oldenburg to ensure the preservation of the recently deceased James Gregory's researches in number theory (and by implication in all things mathematical). Collins at once began to draw up a detailed summary of Gregory's mathematical achievement, with long verbatim 'Extracts from Mr Gregories Letters, To be lent Monsr Leibnitz to peruse who is desired to return the same', the celebrated *Historiola*. (Apart from some stray sheets in private possession, Collins' much corrected draft—it was never more—is now Royal Society MS LXXXI, No. 46: extracts from the latter are reproduced on [**2**: 18–20].) Subsequently, presumably at Oldenburg's desire, Collins drastically curtailed it, composing in its stead the 'Abridgement' (now partly Royal Society MS LXXXI, No. 45—extracts from this are given on [**2**: 47–9]—with the rest in private possession) which, in Oldenburg's Latin version, was sent to Leibniz on 26 July (see Gerhardt's *Briefwechsel* (note (27) above): 167–79). When in London the following October, however, Leibniz was allowed by Collins to see the unabridged *Historiola* and to make extensive, still unpublished notes (now Hanover, L.-Hs. 35, VIII, 23: 1^r–2^v) upon it. Compare J. E. Hofmann, *Die Entwicklungsgeschichte der Leibnizschen Mathematik während des Aufenthaltes in Paris (1672–1676)* (Munich, 1949): 130–4, 183–7.

three basic methods of reduction: the 'Mercator' expansion of algebraic unit-fractions as infinite series 'per divisionem', his own expansion of radicals 'per extractionem in speciebus' in a fashion analogous to their numerical extraction in decimals, and his technique for extracting the roots of general affected equations in species. This approach he had already outlined seven years before in his *De Analysi*, which, as we have seen,[32] had achieved a restricted circulation at the time among the intimates of John Collins. The one novelty now introduced was his classical formulation of the binomial theorem for general index m/n, namely

$$(P+PQ)^{\frac{m}{n}} = P^{\frac{m}{n}}\left(1+\frac{m}{n}Q+\frac{m(m-n)}{2n^2}Q^2+\frac{m(m-n)(m-2n)}{6n^3}Q^3\ldots\right):\text{[33]}$$

under this the reduction 'per divisionem' is subsumed, in Example 6 following, as the particular case where $m=-1$, $n=1$.[34] His method for extracting the roots of a given equation as infinite series he illustrated by the now familiar examples $x^3-2x-5=0$ and $y^3+axy+a^2y-x^3-2a^3=0$.[35] In sequel, though he asserted that the detailed application of these techniques to such typical problems as squaring and rectifying curves or measuring the content and

(31) As he put it, speculation on infinite series was in the air: 'de [speculatione infinitarum serierum] jam cœpit esse rumor' [2, 20].

(32) See II: 166–8.

(33) In modern equivalent, of course. Newton himself wrote

$$\text{`}\overline{P+PQ}\text{\}}^{\frac{m}{n}} = P^{\frac{m}{n}}+\frac{m}{n}AQ+\frac{m-n}{2n}BQ+\frac{m-2n}{3n}CQ+\ldots \quad \&c.$$

Ubi $P+PQ$ significat quantitatem cujus radix vel etiam dimensio quævis vel radix dimensionis investiganda est, P primum terminum quantitatis ejus, Q reliquos terminos divisos per primum, & $\dfrac{m}{n}$ numeralem indicem dimensionis ipsius $P+PQ$ sive dimensio illa integra sit, sive (ut ita loquar) fracta, sive affirmativa, sive negativa.... Deniĝ pro terminis inter operandum inventis in Quoto, usurpo A, B, C, ... &c nempe A pro primo termino $P^{\frac{m}{n}}$, B pro secundo $\dfrac{m}{n}AQ$, & sic deinceps' [2: 21]. It is implicitly assumed that $|Q| < 1$.

(34) In Newton's other illustrative examples the index m/n is chosen to be $\frac{1}{2}$, $\frac{1}{3}$, $-\frac{1}{3}$, $\frac{4}{3}$, -3 and $-\frac{3}{5}$. His 'Exempl: 2', $(c^5+c^4x-x^5)^{\frac{1}{5}}$, is reduced to an infinite series both by putting $P = c$, $Q = x/c-x^5/c^5$ and by setting $P = -x^5$, $Q = -c^4/x^4-c^5/x^5$. (This was much mangled in the version sent on to Leibniz by Oldenburg on 26 July following; see Gerhardt's *Briefwechsel von G. W. Leibniz* (note (27) above): 181, note *. None the less, Leibniz was impressed enough by the binomial formula to note in its margin 'Hoc pulchrum, et hinc etiam elegantissimum compendium pro mea circuli dimensione ope transformationis facta. et pro aliis transformationibus' (*ibid.*: 183, note *).)

(35) See II: 218, 224 and III: 44, 54 respectively. To Leibniz, of course, who had (compare II: 248, note (1)) not yet seen the *De Analysi* or the revised 1671 fluxional tract, these examples were still wholly fresh and he was suitably appreciative of the power of the method they exemplified in his next letter to Oldenburg on 17/27 August following [2: 57–8].

surface-area of solids would be 'too tedious to describe' [2: 25], in default he listed several particular results which he had found by his method of series coupled, where appropriate, with his method of quadratures. In quick succession the series expansions of $z = r\sin^{-1}[x/r]$, its inverse $x = r\sin[z/r]$ and the versed sine $r(1 - \cos[z/r])$ are paraded, together with that of $x = e^{z/b} - 1$, inverse of the Mercator series development of $z = b\log(1 + x)$: all, as we have seen,[36] had long been known to Newton but were still unpublished. The series expanding the subtense $S(n\theta) = d\sin\tfrac{1}{2}n\theta$ of an arc of a circle of diameter $d = 2r$ in terms of $S(\theta) = x$, one which is $\dfrac{1}{n}$-th its size, here appears for the first time as the (unproved) generalization of the Viète sequence $S(\theta) = x$, $S(3\theta) = 3x - x^3/r^2$, ...

$$S(n\theta) = \sum_{0 \leqslant i \leqslant \frac{1}{2}(n-1)} (-1)^i \frac{n}{2i+1} \binom{\tfrac{1}{2}(n-1)+i}{2i} \frac{x^{2i+1}}{r^{2i}}, \quad n \text{ odd.}[37]$$ (To insert this expansion in his letter—unlike the surrounding series, it was derived neither by a binomial development nor by a root extraction—was not very logical of Newton and certainly inconsiderate.) Next he resolved the generalized Kepler problem for the semi-ellipse $y = \sqrt{[r^2 - (r/c)\,x^2]}$:[38] specifically, by a line through the given point $(0, q-t)$, $q = c^{\frac{1}{2}}r^{\frac{1}{2}}$, on the main axis to cut off an elliptic segment of given area. This yields the condition $z = qr\sin^{-1}[y/r] - y(q-t)$, where z is twice the area intersected: an equation simply inverted by Newton as

$$y = z/t - \tfrac{1}{6}qr^{-2}t^{-4}z^3 \ldots$$

after expanding $\sin^{-1}[y/r]$ as an infinite series. A related argument, borrowed

---

(36) Compare I: 125, note (17) and II: 236.

(37) See I: 478. Two numerical errors in Newton's expansion [2: 25]—namely '36' and '49' instead of '49' and '81' respectively—were repeated in the copy transmitted by Oldenburg to Leibniz in July (see Gerhardt, *Briefwechsel* (note (27) above): 186). In a 'scribble...not fit to be seen by any body' added to his letter to Oldenburg on 26 October 1676 (with some minor corrections to the *epistola posterior* sent two days before) Newton remarked that 'ye series of æquations for ye sections of an angle by whole numbers, wch M. Tschurnhause saith he can derive by an easy method one from an other [compare ULC. Add. 3971.1: 29^r–30^r, a copy (in Collins' hand) of 'D. Leibnitius de Circino Æquationum &c'], is contained in y^t one æquation wch I put...in my former letter' [2: 163]. He then gave the expansion of $S(n\theta)$, n odd, which runs in the contrary sense to the present one, viz:

$$S(n\theta) = \sum_{0 \leqslant i \leqslant \frac{1}{2}(n-1)} (-1)^{\frac{1}{2}(n-1)-i} \frac{n}{n-i} \binom{n-i}{i} \frac{x^{n-2i}}{r^{n-2i-1}}.$$

(38) The simple 'Astronomicum...Kepleri Problema' [2: 26] is the particular case of the following where the given point $(0, q-t)$ is a focus, that is, where $t = q - \sqrt{[1 - r^2/q^2]}$. This problem (which has no finite solution in terms of elementary functions) was first put out as a challenge by Johannes Kepler in his *Astronomia Nova* ΑΙΤΙΟΛΟΓΗΤΟΣ *seu Physica Cœlestis, tradita commentariis de Motibus Stellæ Martis* (Prague, 1609): Caput LX, 'Data anomalia media invenire anomaliam eccentri et sic coæquatam illi respondentem': 295–300 [= *Gesammelte Werke*, 3 (Munich, 1937): 376–8], but probably first became known to him in Wren's 1658 version which ordained that 'ex Medio motu Anomaliam coæquatam ut indigaret, secanda

no doubt from the 1671 tract,[39] rectifies the general arc $z = s(x)$ of this same ellipse by determining $s(x) = \int_0^x \sqrt{[(c^2r - (c-r)\,x^2)/c(cr - x^2)]} \cdot dx$, and then reducing this to $s(x) = x + \frac{1}{6}c^{-2}x^3 \ldots$ by developing the radical as an infinite series in x and integrating it term by term.[40] The corollary $x = s^{-1}(z) = z - \frac{1}{6}c^{-2}z^3 \ldots$ follows at once by inverting this. Also from his 1671 tract Newton repeated for Leibniz' benefit his series expansions for the subtangent, area and general arc of the quadratrix $y = x \cot [x/a]$.[41] From his letter to Collins on 20 August 1672 [**1**: 229], likewise, he borrowed his Gregorian series for the 'second segment' of the spheroid $x^2/a^2 + (y^2 + z^2)/c^2 = 1$, now adding new terms and elaborating a recursive scheme for generating all the coefficients to infinity in the expansion.[42] A final section in the *epistola prior* discusses certain approximate constructions which arise from rounding off series expansions of various geometrical arc-lengths. For instance, by considering only the first two terms in the series for the subtense $A = r \sin^{-1}[z/r]$ of an arc z of a circle of radius r, and in that for the subtense $B = r \sin^{-1}[\frac{1}{2}z/r]$ of the half arc, Newton was able to deduce—much as he had for Dary two years before[43]—the 'Theorema Hugenianum' that $z = \frac{1}{3}(8B - A)$ 'errore tantum existente $\dfrac{z^5}{7680r^4}$ — &c in excessu' [**2**: 30]. A following rectification of the general arc of the central conic $y^2 = rx \mp (r/q)\,x^2$ is again Huygenian: the arc-length $s = \int_0^x \frac{1}{2}\sqrt{[r/x(1 \mp x/q) + 4(1 \mp r/q)]} \cdot dx$ is con-

---

est semiellipsis per Focum in datâ ratione. . . . Mirum est quantum in hoc problemate sudaverit Keplerus, Orbitas suas—*volvens*[!] *nitendo neque proficit hilum.* Tandem anhelus, Geometrarum opem implorat; interim veritus ne propter arcuum & sinuum ἑτερογένειαν inveniatur Problema inexplicabile' (John Wallis, *Tractatus Duo* (note (6) above): 80).

(39) See III: 326, note (744); and compare note (17) above.

(40) At this point in his letter [**2**: 26–7] Newton took some trouble in recursively defining the coefficient $a_{m,n} = \dfrac{1}{2n-1}\dbinom{\frac{1}{2}}{m}\dbinom{n-2}{m-1}$ which he located at row m, column n in his triangular layout of $s(x)$, that is,

$$\int_0^x \left(1 - \frac{1 - r/c}{cr}x^2\right)^{\frac{1}{2}} \left(1 - \frac{1}{cr}x^2\right)^{-\frac{1}{2}} \cdot dx = \sum_{1 \leqslant n \leqslant \infty} \sum_{1 \leqslant m \leqslant n-1} a_{m,n} \frac{x^{2n-1}}{c^{2m}(cr)^{-m+n-1}}.$$

(41) See III: 326–8. These results were partially anticipated two years before in the *De Analysi* (II: 238–40).

(42) Compare III: 567–8. In modern equivalent the 'second segment' is

$$2\int_0^x \int_0^y z \cdot dy\,dx = 2cxy \sum_{1 \leqslant m,\, n \leqslant \infty} b_{m,n} \left(\frac{x}{a}\right)^{2m-2} \left(\frac{y}{c}\right)^{2n-2},$$

where $b_{m,n}$ (the coefficient of the term on row m, column n in Newton's array) is defined recursively by $b_{1,1} = 1$ and $b_{m,n} = b_{n,m} = \dfrac{(2m + 2n - 7)\,(2m - 3)}{(2m - 2)\,(2m - 1)} b_{m-1,n}.$

(43) See notes (16) and (18) above.

structed by supposing it to be set off in the tangent $x = 0$ at the conic's vertex and then determining the meet $(0, sx/(s-y))$ of the line joining $(0, s)$ and (x, y) with the conic's axis $y = 0$. Here Newton accurately rounded off $sx/(s-y)$ to $O(x^2)$ as $\frac{3}{2}r + (\frac{19}{10} \mp \frac{21}{10}r/q)\, x$.[44] A final, somewhat unwieldly 'mechanical' construction [2: 31] evaluates, to $O(x^{\frac{3}{2}})$, the conic's general area

$$2\int_0^x r^{\frac{1}{2}}x^{\frac{1}{2}}(1 \mp x/q)^{\frac{1}{2}}.dx \quad \text{as} \quad r^{\frac{1}{2}}x^{\frac{3}{2}}(\tfrac{4}{3} \mp \tfrac{2}{5}x/q - \tfrac{1}{14}x^2/q^2 \ldots).\text{[45]}$$

Oldenburg duly transmitted his 'apographum' of Newton's letter[46] to Leibniz on 26 July 1676, and was rewarded by a long reply from Paris on 17/27 August following [2: 57—64], itself passed on to Newton by way of Collins on 9 September.[47] We have outlined the content of Leibniz' letter elsewhere:[48] in brief, after a summary account of two types of integral transform,[49] Leibniz communicated the series for the inverse-tangent (newly discovered by him on applying the latter of his transforms to the circle) together with trivial variants on Newton's series for the exponential and versine functions. In conclusion he

(44) Eight years later Newton was implicitly to invoke the approximation $sx/(s-y) \approx \frac{3}{2}r$ for the particular case of the circle ($q = r$ and $y^2 = rx - x^2$) in the figure illustrating Theorem 2 of his *De Motu Corporum*. (See D. T. Whiteside, 'Newtonian Dynamics', *History of Science*, 5, 1966: 112; and compare Theorem XIII of Huygens' *De Circuli Magnitudine Inventa* (note (16)).)

(45) Compare III: 290, note (660).

(46) Reproduced by C. I. Gerhardt in his *Briefwechsel von G. W. Leibniz* (note (27)): 179–92 (compare [2: 54]). Newton's letter [2: 31–2] accompanying his *epistola prior* had been read to the Royal Society a month before, on 15 June, together with 'some communications of an algebraical nature for Monsieur LEIBNITZ, who by an express letter to Mr. Oldenburg had desired them' (Thomas Birch, *History of the Royal Society of London*, 3 (London, 1757): 319).

(47) Collins sent his own 'Duplicate' (now ULC. Add. 3971.1: 34^r–36^v) of Leibniz' original (British Museum. Add. 4294: 67–71) since 'Mr Oldenburgh is gone into the Country for 10 dayes' [2: 99] (compare [2: 89]). For his own part Newton professed he was less than keen to have a copy of the letter: 'I doubt I shal put you to too much trouble to transcribe Mr Leibnitz's whole letter if it be so long, & therefore I shall desire you to send me only a general account of it, wth such passages as you think may concern me' (Newton to Collins, 5 September 1676 [2: 96]).

(48) In note (13) on pages 530–2 above.

(49) Namely, the rationalizing transformation of $\int y.dx$ into its equal $\int v.dz$ by $z = ry/x$, and evaluation of the general sector $\frac{1}{2}\int[y.dx - x.dy]$ as $\frac{1}{2}[t.f(t) - \int f(t).dt]$ on taking

$$t = y - x(dy/dx) = f^{-1}(x).$$

(50) '...desideraverim ut Clarissimus Neutonus nonnulla quoque amplius explicet, ut originem theorematis quod initio ponit, item modum, quo quantitates *p. q. r.* in suis operationi-bus invenit, ac denique quomodo in Methodo regressuum se gerat...; neque enim explicat, quomodo id ex methodo sua derivetur. Nondum mihi licuit ejus literas qua merentur diligentia legere,...unde non satis...affirmare ausim, an nonnulla eorum quæ suppressit, ex sola earum lectione consequi possim; Sed optandum tamen foret, ipsum ea potius supplere Neutonum, quia credibile est non posse eum scribere quin aliquid semper præclari nos doceat Vir, ut apparet, egregiarum meditationum plenus' [2: 62].

added the request that Newton should amplify certain 'suppressed' aspects of his previous account of his researches into series, clarifying in particular the source of his ideas on the binomial theorem, the details of his resolution of generally affected algebraic equations and the precise nature of his method for inverting series.[50] To this flattering appeal Newton responded once more with the longest mathematical letter he ever wrote, the *epistola posterior*.

During the next six weeks Newton apparently broke off all correspondence[51] and we may picture him, with Wickins' secretarial assistance, drafting the long reply to Leibniz [2: 110–29] which was at last put in the post for Oldenburg on 24 October.[52] This letter, arguably the most important one he ever wrote and fairly certainly the most difficult to master in its mathematical subtleties, is a veritable treatise on the construction and application of infinite series: as such, Newton intended at least once (in 1684) to give it pride of place in an extended tract *De computo serierum*.[53] Those of his contemporaries who were privileged to read a copy of it—notably Collins and Wallis—were no less convinced that it merited immediate publication.[54] We can only agree with their judgement

(51) Collins' letter of 9 September remained unanswered till 8 November when (in his last extant letter to him) Newton made a rare apology: 'I doubt you think I have forgot to answer your last letter, & to return you thanks for ye pains you took in copying out for me ye large letters of...M. Leibnitz & M. Tschurnhause' [2: 179].

(52) Some idea of the effort Newton put into composing this letter may be had by comparing two rejected sheets from an early autograph draft no longer extant—one is now loose in ULC. Add. 3977.4 while the other (ULC. Add. 3964.5: 3^r–4^v) is reproduced as 1, 1, §2 above—with their widely variant, abridged equivalents in the letter as sent (British Museum. Add. 4294: 119 ff.). From this draft Newton began to write up a final autograph version which was temporarily abandoned after its second paragraph was begun so that Wickins could transcribe the penultimate copy of the letter now extant in his hand (ULC. Add. 3977.4). After making final adjustments to Wickins' script, Newton completed the definitive autograph version of his letter and sent it to Oldenburg on 24 October—just too late to be passed on to Leibniz before he left England *en route* to Hanover by way of Holland. In his covering letter Newton announced to Oldenburg that 'To Mr Leibnitz's ingenious letter I have returned an answer wch I doubt is too tedious. I could wish I had left out some things since to avoid greater tediousness I left out something else [compare 1, 1, §2 above] on wch they have some dependance. But I had rather you should have it any way, then write it over again being at present otherwise [!] incumbred' [1: 110].

(53) See 3, 2, §§1/2 above, especially pages 532–6.

(54) Little after Oldenburg had shown him Newton's first letter for Leibniz, Collins wrote to its author on 9 September 1676 that 'I thinke...you would doe well to publish [your method of infinite series] in Latin or permitt a Translation and the comming foorth thereof in English' [2: 99], but in his reply two months later Newton firmly quashed the suggestion 'being I beleive censured by divers for my letters in y^e *Transactions* about such things as no body els would have let come out wthout a substantial discours' [2: 179]. John Wallis, at this time busy preparing a first version of his *Algebra* for the press (see [2: 244, note 3]) had some time before autumn 1677 been sent a copy of the *epistola prior* by Collins, 'I seeing you therein[!] mentioned' [2: 242], and not long after—no doubt from Oldenburg—acquired a copy of the present letter also, apparently incorporating passages from both in his draft 'Account of the

though, in hindsight, our pleasure is restrained when we perceive how closely Newton reproduced whole pages of his 1671 fluxional tract for Leibniz' benefit. Since the abridged version of the letter which he intended to publish in 1684 has been given above[55] we need here only outline the main points in Newton's *epistola*.

The autobiographical sketch [**2**: 111–15] with which the letter opens, where Newton retraced his discovery (following in the path marked out by Wallis in his *Arithmetica Infinitorum*) of the general integral expansion

$$\int_0^x (1 \mp x^2)^m \, . \, dx = \sum_{0 \leqslant i \leqslant \infty} (\mp 1)^i \binom{m}{i} \frac{x^{2i+1}}{2i+1},$$

2*m* integral, was avowedly based not on memory but on his preserved manuscript record of his researches 'sub initio studiorum meorum Mathematicorum' a decade before.[56] This has long—and justly—been accepted as a trustworthy account of his invention of the binomial theorem but we may now test its essential truth by comparing it with the still extant autograph manuscripts which Newton himself had before him as he relived his discovery.[57] The accuracy of his following observations on the circumstances which led him to

learned paines of Englishmen' in algebra. (Collins himself had deliberately withheld the latter from Wallis 'in regard you lye under a censure from diverse for printing discourses that come to you in private Letters without permission or consent as is said of the parties concerned. Mr Newton last yeare sent up these Letters, you have seen with particular leave upon my importunity to print the same' [**2**: 242]. Writing back on 2 October 1677, however, Wallis was not chastened: 'I am stil of opinion y^t Mr Newton should perfect his notions, & print them suddenly. These letters, if printed, will need a little review by himself; for there be some slips in hasty writing them' [**2**: 238].) So matters remained during Collins' lifetime. When the revised *Treatise of Algebra Both Historical and Practical* finally appeared in 1685 it included (pages 318–20, 330–3 and 338–47) Englished excerpts from 'Two Letters of [Mr. *Isaac Newton*] (which I have seen) written to Mr. *Oldenburg*...(dated *June* 13 and *Octob.* 24. 1676,) full of very ingenious discoveries, and well deserving to be made more publick'. Newton's prior approval of such publication was never, it would appear, requested but ten years later, on 30 May 1695, Wallis sent Newton his 'fair copy of your two letters, which I wish were printed....I would have subjoined them (with your good leave) to the second volume of my *Opera Math*: [1693] if I had thought of it a little sooner' [**4**: 129]. Under pressure of a further letter on 3 July [**4**: 139] Newton drafted a reply [**4**: 140–1] giving Wallis permission to publish, though a year afterwards the latter admitted to Halley that 'he did not seem forward for [it]' [**4**: 186]. The full Latin versions of Newton's two letters finally appeared in print in Wallis' *Opera Mathematica*, **3** (Oxford, 1699): 622–9/634–45.

(55) See pages 618–33.

(56) 'Cùm primùm appuli ad studia Mathematica et cœperam ea mediocriter callere,' he wrote in a cancelled opening to the second paragraph of his letter (see the photocopy inserted at [**2**: 111]), 'incidi in opera Celeberrimi Wallisij nostri, & considerando series quarum intercalatione ipse exhibet aream Circuli et Hyperbolæ,...ad has interpolandas...inveni... regulam [quam] applicui ad series interserendas', subsequently adding that 'Hic fuit primus meus ingressus in has meditationes: qui e memoria sane exciderat nisi oculos in adversaria quædam ante paucas septimanas retulissem' [**2**: 112].

compose his *De Analysi* in 1669 and its augmented revise two years later we may likewise confirm by our more complete modern knowledge of contemporary events: this we have, indeed, amply established in our introductions to the reproduced text of those two basic Newtonian tracts on methods of infinite series and fluxions.[58] The remainder of the *epistola posterior* [**2**: 115–29] is effectively—often word for word[59]—a generous, well polished selection of the highlights of this latter (1671) treatise. Sadly, however, as subsequent history was to prove, the general fluxional principles expounded at length in Problems 1 and 2 of that tract were conveyed to Leibniz only in two anagrams [**2**: 115 and 127] which, even had he been able to unravel them, concealed bare enunciations of these two inverse problems which would have told him little.[60] What prevented Newton on this occasion from being more explicit on the subject of his fluxional insights was, almost certainly, lack of self-confidence and the memory of the hail of criticism he had had to endure when he made public his equally novel theory of light a few years before. In all other respects the tone of Newton's letter is one of friendly helpfulness, even in his criticism of certain 'oversights' in Leibniz' previous letter: indeed, he may have intended to give Leibniz a full account of his fluxional calculus at a subsequent date.[61]

(57) See I, **1**, 3, §3.2 and §4; also compare the logarithmic computations reproduced as I, **1**, 3, §3.3 and §5. In the extant manuscripts of these binomial researches there is, however, nothing to support Newton's present claim to have then observed that the binomial coefficients $\binom{m}{i}$, $m = 0, 1, 2, 3, 4, \ldots$ 'erant figuræ potestatum numeri 11, nempe harum $\overline{11}|^0$. $\overline{11}|^1$. $\overline{11}|^2$. $\overline{11}|^3$. $\overline{11}|^4 \ldots$'. Nor did he in 1676 hint at the effort it cost him to lay out the extended Wallisian interpolation schemes he had found necessary in the winter of 1664/5 to deduce the expansion of $\binom{m}{i}$ as '$\dfrac{m-0}{1} \times \dfrac{m-1}{2} \times \ldots \times \dfrac{m-4}{5}$ &c' [**2**: 111]. His further assertion that he thereafter proved the validity of his binomial expansions for unit-fractional powers both by deriving the original binomial after appropriately multiplying the binomial series into itself the necessary number of times and also by physically extracting n-th roots along traditional lines 'more Arithmetico' is also not confirmed by extant evidence: such constructions of series 'dividendo' and 'radicem extrahendo' first occur in the *De Analysi* (II: 212–14).

(58) See the introductions to II, **2**, 3 and III, **1**, 2.

(59) For instance, the integral comparisons [**2**: 119–20] and the calculations of conic area [**2**: 121–4] are repeated bodily from Problem 9 of the 1671 tract (see III: 244–54 and 222–30), while the cissoid rectification [**2**: 117] is, *verbatim*, Example 5 of the latter's twelfth problem (see III: 320). Likewise Newton's exposition in 1676 of the parallelogram rule [**2**: 126–7] is borrowed with but slight alteration from his equivalent account (III: 50 ff.) five years before.

(60) Newton, when he sent his *epistola posterior* to Oldenburg for copying and onward transmission to Leibniz, carefully recorded the solution of these anagrams in his Waste Book (ULC. Add. 4004: 81^v; see II: 190–1, note (25)).

(61) Newton to Oldenburg, 26 October 1676: 'I feare I have been something too severe in taking notice of some oversights in M. Leibnitz letter.... But yet they being I think real oversights I suppose he cannot be offended at it....I believe M. Leibnitz will not dislike ye Theorem towards ye beginning of my letter [see [**2**: 115–17] and III: 237, note (540)]

There was, however, to be no such future interchange of mathematical insights. To be sure, Oldenburg—as he noted in the margin of the transcript he later sent on to Leibniz—diligently copied Newton's *epistola posterior* on 'Nov. 4. 1676'[62] but hung on to it while awaiting Newton's answer to some queries he raised over 'things miswritten'[63] and till he could (or so he told Leibniz in his covering letter of 2 May 1677 [**2**: 208]) find a safe courier for this all-important letter. Leibniz did not, in fact, receive it till early in June 1677. Though he at once wrote a hasty acknowledgement to Oldenburg on 11 June [**2**: 212–19], attempting to keep the correspondence alive by communicating his own general method of tangents—notation apart, all but identical with Newton's—and some quick counter-criticisms of Newton's new *epistola*, and followed this up with a brief sequel on 12 July [**2**: 231–2], the half year's delay was too long. Oldenburg acknowledged their receipt in London on 9 August [**2**: 235] and then retired to his Kent farm to die within the month, leaving Collins—in almost the last act of their epistolary exchange—to post transcripts of Leibniz' communications on to Newton in Cambridge [compare **2**: 237].[64] Thereafter the tenuous link, by way of London, between Hanover and Cambridge was broken. Leibniz' criticisms were answered by Newton only seven years later, and then in the pages of his unpublished 'Matheseos Universalis Specimina'.

---

for squaring Curve lines Geometrically. Sometime when I have more leisure it's possible I may send him a fuller account of it: explaining how it is to be ordered for comparing curvilinear figures wth one another, & how ye simplest figure is to be found wth wch a propounded curve may be compared' [**2**: 162–3]. As we shall see in the sixth volume this 'fuller account' was composed only fifteen years afterwards (in late 1691) as the first version of his *De Quadratura Curvarum*.

(62) See Gerhardt's *Briefwechsel von G. W. Leibniz* (note (27) above): 202.

(63) Compare Newton's reply to Oldenburg on 14 November 1676 [**2**: 181].

(64) Collins to Newton, 30 August 1677: 'This letter from Libnitz [of 12 July] was transcribed from...the Originall borrowed from Mr Oldenburgh, and [I] send [it] you in his retirement into the Country.'

INDEX OF NAMES